Applet	Concept Illustrated	Description	Applet Activity
Standard deviation	Investigates how distribution shape and spread affect standard deviation.	Students visualize relationship between mean and standard deviation by adding and deleting data points; applet updates mean and standard deviation.	**2.4,** 96; **2.5,** 97; **2.6,** 97; **2.7,** 119
Confidence intervals for a proportion	Not all confidence intervals contain the population proportion. Investigates the meaning of 95% and 99% confidence.	Simulates selecting the population and confidence interval population proport plots confidence in and proportion cor	370
Confidence intervals for a mean (the impact of confidence level)	Not all confidence intervals contain the population mean. Investigates the meaning of 95% and 99% confidence.	Simulates selecting 100 random samples from population; finds 95% and 99% confidence intervals for each. Students specify sample size, distribution shape, and population mean and standard deviation; applet plots confidence intervals and reports number and proportion containing true mean.	**7.1,** 351; **7.2,** 351
Confidence intervals for a mean (not knowing standard deviation)	Confidence intervals obtained using the sample standard deviation are different from those obtained using the population standard deviation. Investigates effect of not knowing the population standard deviation.	Simulates selecting 100 random samples from the population and finds the 95% z-interval and 95% t-interval for each. Students specify sample size, distribution shape, and population mean and standard deviation; applet plots confidence intervals and reports number and proportion containing true mean.	**7.3,** 361; **7.4,** 361
Hypothesis tests for a proportion	Not all tests of hypotheses lead correctly to either rejecting or failing to reject the null hypothesis. Investigates the relationship between the level of confidence and the probabilities of making Type I and Type II errors.	Simulates selecting 100 random samples from population; calculates and plots z-statistic and P-value for each. Students specify population proportion, sample size, and null and alternative hypotheses; applet reports number and proportion of times null hypothesis is rejected at 0.05 and 0.01 levels.	**8.5,** 433; **8.6,** 434
Hypothesis tests for a mean	Not all tests of hypotheses lead correctly to either rejecting or failing to reject the null hypothesis. Investigates the relationship between the level of confidence and the probabilities of making Type I and Type II errors.	Simulates selecting 100 random samples from population; calculates and plots t statistic and P-value for each. Students specify population distribution shape, mean, and standard deviation; sample size, and null and alternative hypotheses; applet reports number and proportion of times null hypothesis is rejected at both 0.05 and 0.01 levels.	**8.1,** 407; **8.2,** 417; **8.3,** 417; **8.4,** 417
Correlation by eye	Correlation coefficient measures strength of linear relationship between two variables. Teaches user how to assess strength of a linear relationship from a scattergram.	Computes correlation coefficient r for a set of bivariate data plotted on a scattergram. Students add or delete points and guess value of r; applet compares guess to calculated value.	**11.2,** 652
Regression by eye	The least squares regression line has a smaller SSE than any other line that might approximate a set of bivariate data. Teaches students how to approximate the location of a regression line on a scattergram.	Computes least squares regression line for a set of bivariate data plotted on a scattergram. Students add or delete points and guess location of regression line by manipulating a line provided on the scattergram; applet plots least squares line and displays the equations and the SSEs for both lines.	**11.1,** 625

Get the Most Out of Pearson MyLab Statistics

Pearson MyLab Statistics, Pearson's online tutorial and assessment tool, creates personalized experiences for students and provides powerful tools for instructors. With a wealth of tested and proven resources, each course can be tailored to fit your specific needs. Talk to your Pearson Representative about ways to integrate Pearson MyLab Statistics into your course for the best results.

Data-Driven Reporting for Instructors

- Pearson MyLab Statistics' comprehensive on-line gradebook automatically tracks students' results to tests, quizzes, homework, and work in the study plan.

- The Reporting Dashboard makes it easier than ever to identify topics where students are struggling, or specific students who may need extra help.

Learning in Any Environment

- Because classroom formats and student needs continually change and evolve, Pearson MyLab Statistics has built-in flexibility to accommodate various course designs and formats.

- With a new, streamlined, mobile-friendly design, students and instructors can access courses from most mobile devices to work on exercises and review completed assignments.

Available in Pearson MyLab Statistics for Your Introductory Statistics Courses

Pearson MyLab Statistics is the market-leading online resource for learning and teaching statistics.

Leverage the Power of StatCrunch

Pearson MyLab Statistics leverages the power of StatCrunch—powerful, web-based statistics software. Integrated into Pearson MyLab Statistics, students can easily analyze data from their exercises and etext. In addition, access to the full online community allows users to take advantage of a wide variety of resources and applications at www.statcrunch.com.

Bring Statistics to Life

Virtually flip coins, roll dice, draw cards, and interact with animations on your mobile device with the extensive menu of experiments and applets in StatCrunch. Offering a number of ways to practice resampling procedures, such as permutation tests and bootstrap confidence intervals, StatCrunch is a complete and modern solution.

Real-World Statistics

Pearson MyLab Statistics video resources help foster conceptual understanding. StatTalk Videos, hosted by fun-loving statistician Andrew Vickers, demonstrate important statistical concepts through interesting stories and real-life events. This series of 24 videos includes assignable questions built in Pearson MyLab Statistics and an instructor's guide.

STATISTICS

THIRTEENTH EDITION

GLOBAL EDITION

James T. McClave

Info Tech, Inc.

University of Florida

Terry Sincich

University of South Florida

 Pearson

330 Hudson Street, NY NY 10013

Editorial Director: Chris Hoag
Editor in Chief: Deirdre Lynch
Acquisitions Editor: Patrick Barbera
Acquisitions Editor, Global Edition: Sourabh
    Maheshwari
Editorial Assistant: Justin Billing
Program Manager: Tatiana Anacki
Project Manager: Christine O'Brien
Assistant Project Editor, Global Edition:
    Aurko Mitra
Program Management Team Lead: Karen Wernholm
Project Management Team Lead: Peter Silvia
Senior Manufacturing Controller, Global Edition:
    Kay Holman
Media Producer: Jean Choe
Media Production Manager, Global Edition:
    Vikram Kumar

TestGen Content Manager: John Flanagan
MathXL Content Developer:
    Bob Carroll
Marketing Manager: Tiffany Bitzel
Marketing Assistant: Jennifer Myers
Senior Author Support/Technology
    Specialist: Joe Vetere
Rights and Permissions Project Manager:
    Gina Cheselka
Procurement Specialist: Carol Melville
Associate Director of Design: Andrea Nix
Program Design Lead:
    Barbara Atkinson
Text Design: Integra Software Services Pvt. Ltd.
Illustrations: Integra Software Services Pvt. Ltd.
Cover Design: Lumina Datamatics
Cover Image: vesna cvorovic/Shutterstock

Acknowledgments of third party content appear on page 884, which constitutes an extension of this copyright page.

Pearson Education Limited
KAO Two
KAO Park
Harlow
CM17 9NA
United Kingdom

and Associated Companies throughout the world

Visit us on the World Wide Web at: www.pearsonglobaleditions.com

© Pearson Education Limited 2018

ISBN 10: 1-292-16155-8
ISBN 13: 978-1-292-16155-6

British Library Cataloguing-in-Publication Data
A catalogue record for this book is available from the British Library

10 9 8 7 6 5 4 3 2 1

Typeset by Integra
Printed and bound by Vivar in Malaysia

Contents

Chapter 4   Discrete Random Variables   212

Chapter 5   Continuous Random Variables   260

Chapter 6   Sampling Distributions   310

Preface

This 13th edition of *Statistics* is an introductory text emphasizing inference, with extensive coverage of data collection and analysis as needed to evaluate the reported results of statistical studies and make good decisions. As in earlier editions, the text stresses the development of statistical thinking, the assessment of credibility, and the value of the inferences made from data, both by those who consume and those who produce them. It assumes a mathematical background of basic algebra.

The text incorporates the following features, developed from the American Statistical Association's (ASA) Guidelines for Assessment and Instruction in Statistics Education (GAISE) Project:

- Emphasize statistical literacy and develop statistical thinking
- Use real data in applications
- Use technology for developing conceptual understanding and analyzing data
- Foster active learning in the classroom
- Stress conceptual understanding rather than mere knowledge of procedures
- Emphasize intuitive concepts of probability

A briefer version of the book, *A First Course in Statistics*, is available for single semester courses that include minimal coverage of regression analysis, analysis of variance, and categorical data analysis.

New in the 13th Edition

- **Over 2,000 exercises, with revisions and updates to 25%.** Many new and updated exercises, based on contemporary studies and real data, have been added. Most of these exercises foster and promote critical thinking skills.
- **Updated technology.** All printouts from statistical software (SAS, SPSS, MINITAB, and the TI-83/TI-84 Plus Graphing Calculator) and corresponding instructions for use have been revised to reflect the latest versions of the software.
- **New Statistics in Action Cases.** Six of the 14 Statistics in Action cases are new or updated, each based on real data from a recent study.
- **Continued emphasis on Ethics.** Where appropriate, boxes have been added emphasizing the importance of ethical behavior when collecting, analyzing, and interpreting data with statistics.
- **Data Informed Development.** The authors analyzed aggregated student usage and performance data from Pearson MyLab Statistics for the previous edition of this text. The results of this analysis helped improve the quality and quantity of exercises that matter most to instructors and students.

Content-Specific Changes to This Edition

- **Chapter 1 (Statistics, Data, and Statistical Thinking).** Material on all basic sampling concepts (e.g., random sampling and sample survey designs) has been streamlined and moved to Section 1.5 (Collecting Data: Sampling and Related Issues).
- **Chapter 2 (Methods for Describing Sets of Data).** The section on summation notation has been moved to the appendix (Appendix A). Also, recent examples of misleading graphics have been added to Section 2.9 (Distorting the Truth with Descriptive Statistics).

- **Chapter 4 (Discrete Random Variables) and Chapter 5 (Continuous Random Variables).** Use of technology for computing probabilities of random variables with known probability distributions (e.g., binomial, Poisson, normal, and exponential distributions) has been incorporated into the relevant sections of these chapters. This reduces the use of tables of probabilities for these distributions.

- **Chapter 6 (Sampling Distributions).** In addition to the sampling distribution of the sample mean, we now cover (in new Section 6.4) the sampling distribution of a sample proportion.

- **Chapter 8 (Inferences Based on a Single Sample: Tests of Hypothesis).** The section on p-values in hypothesis testing (Section 8.3) has been moved up to emphasize the importance of their use in real-life studies. Throughout the remainder of the text, conclusions from a test of hypothesis are based on p-values.

Hallmark Strengths

We have maintained the pedagogical features of *Statistics* that we believe make it unique among introductory statistics texts. These features, which assist the student in achieving an overview of statistics and an understanding of its relevance in both the business world and everyday life, are as follows:

- **Use of Examples as a Teaching Device**—Almost all new ideas are introduced and illustrated by data-based applications and examples. We believe that students better understand definitions, generalizations, and theoretical concepts *after* seeing an application. All examples have three components: (1) "Problem," (2) "Solution," and (3) "Look Back" (or "Look Ahead"). This step-by-step process provides students with a defined structure by which to approach problems and enhances their problem-solving skills. The "Look Back" feature often gives helpful hints to solving the problem and/or provides a further reflection or insight into the concept or procedure that is covered.

- **Now Work**—A "Now Work" exercise suggestion follows each example. The Now Work exercise (marked with the icon [NW] in the exercise sets) is similar in style and concept to the text example. This provides the students with an opportunity to immediately test and confirm their understanding.

- **Statistics in Action**—Each chapter begins with a case study based on an actual contemporary, controversial, or high-profile issue. Relevant research questions and data from the study are presented and the proper analysis demonstrated in short "Statistics in Action Revisited" sections throughout the chapter. These motivate students to critically evaluate the findings and think through the statistical issues involved.

- **Applet Exercises**—The text is accompanied by applets (short computer programs) available at www.pearsonglobaleditions.com/mcclave and within Pearson MyLab Statistics. These point-and-click applets allow students to easily run simulations that visually demonstrate some of the more difficult statistical concepts (e.g., sampling distributions and confidence intervals). Each chapter contains several optional applet exercises in the exercise sets. They are denoted with the following icon: **Ⓓ**.

- **Real Data-Based Exercises**—The text includes more than 2,000 exercises based on applications in a variety of disciplines and research areas. All the applied exercises employ the use of current real data extracted from current publications (e.g., newspapers, magazines, current journals, and the Internet). Some students have difficulty learning the mechanics of statistical techniques when all problems are couched in terms of realistic applications. For this reason, all exercise sections are divided into four parts:

    Learning the Mechanics. Designed as straightforward applications of new concepts, these exercises allow students to test their abilities to comprehend a mathematical concept or a definition.

Applying the Concepts—Basic.   Based on applications taken from a wide variety of journals, newspapers, and other sources, these short exercises help students to begin developing the skills necessary to diagnose and analyze real-world problems.

Applying the Concepts—Intermediate.   Based on more detailed real-world applications, these exercises require students to apply their knowledge of the technique presented in the section.

Applying the Concepts—Advanced.   These more difficult real-data exercises require students to use their critical thinking skills.

- **Critical Thinking Challenges**—Placed at the end of the "Supplementary Exercises" sections only, students are asked to apply their critical thinking skills to solve one or two challenging real-life problems. These exercises expose students to real-world problems with solutions that are derived from careful, logical thought and selection of the appropriate statistical analysis tool.

- **Exploring Data with Statistical Computer Software and the Graphing Calculator**—Each statistical analysis method presented is demonstrated using output from three leading Windows-based statistical software packages: SAS, SPSS, and MINITAB. Students are exposed early and often to computer printouts they will encounter in today's high-tech world.

- **"Using Technology" Tutorials**—MINITAB software tutorials appear at the end of each chapter and include point-and-click instructions (with screen shots). These tutorials are easily located and show students how to best use and maximize MINITAB statistical software. In addition, output and keystroke instructions for the TI-83/TI-84 Plus Graphing Calculator are presented.

- **Profiles of Statisticians in History (Biography)**—Brief descriptions of famous statisticians and their achievements are presented in side boxes. With these profiles, students will develop an appreciation of the statistician's efforts and the discipline of statistics as a whole.

- **Data and Applets**—The Web site www.pearsonglobaleditions.com/mcclave has files for all the data sets marked with the data set icon **D** in the text. These include data sets for text examples, exercises, Statistics in Action, and Real-World cases. This site also contains the applets that are used to illustrate statistical concepts.

Flexibility in Coverage

The text is written to allow the instructor flexibility in coverage of topics. Suggestions for two topics, probability and regression, are given below.

- **Probability and Counting Rules**—One of the most troublesome aspects of an introductory statistics course is the study of probability. Probability poses a challenge for instructors because they must decide on the level of presentation, and students find it a difficult subject to comprehend. We believe that one cause for these problems is the mixture of probability and counting rules that occurs in most introductory texts. Consequently, we have included the counting rules (with examples) in an optional section (Section 3.7) of Chapter 3. Thus, the instructor can control the level of probability coverage.

- **Multiple Regression and Model Building**—This topic represents one of the most useful statistical tools for the solution of applied problems. Although an entire text could be devoted to regression modeling, we feel that we have presented coverage that is understandable, usable, and much more comprehensive than the presentations in other introductory statistics texts. We devote two full chapters to discussing the major types of inferences that can be derived from a regression analysis, showing how these results appear in the output from statistical software, and, most important, selecting multiple regression models to be used in an analysis. Thus, the instructor has the choice of one-chapter coverage of simple linear regression (Chapter 11), a two-chapter treatment of simple and multiple regression (excluding the sections on model building in Chapter 12), or complete coverage of regression analysis, including model building and regression diagnostics. This extensive coverage of such useful statistical tools will provide added evidence to the student of the relevance of statistics to real-world problems.

- **Role of Calculus in Footnotes**—Although the text is designed for students with a non-calculus background, footnotes explain the role of calculus in various derivations. Footnotes are also used to inform the student about some of the theory underlying certain methods of analysis. These footnotes allow additional flexibility in the mathematical and theoretical level at which the material is presented.

Get the most out of
Pearson MyLab
Statistics

Pearson MyLab Statistics is the world's leading online resource for teaching and learning statistics. Pearson MyLab Statistics helps students and instructors improve results, and provides engaging experiences and personalized learning for each student so learning can happen in any environment. Plus, it offers flexible and time-saving course management features to allow instructors to easily manage their classes while remaining in complete control, regardless of course format.

Personalized Support for Students

- Pearson MyLab Statistics comes with many learning resources—eText, animations, videos, and more—all designed to support your students as they progress through their course.

- The Adaptive Study Plan acts as a personal tutor, updating in real time based on student performance to provide personalized recommendations on what to work on next. With the new Companion Study Plan assignments, instructors can now assign the Study Plan as a prerequisite to a test or quiz, helping to guide students through concepts they need to master.

- Personalized Homework allows instructors to create homework assignments tailored to each student's specific needs, focused on just the topics they have not yet mastered.

Used by nearly 4 million students each year, the Pearson MyLab Statistics and Pearson MyLab Statistics family of products delivers consistent, measurable gains in student learning outcomes, retention, and subsequent course success.

Resources for Success

Student Resources

Excel® Manual (download only), by Mark Dummeldinger (University of South Florida). Available for download from www.pearsonglobaleditions.com/mcclave.

Study Cards for Statistics Software. This series of study cards, available for Excel®, MINITAB, JMP®, SPSS, R, StatCrunch®, and TI-83/84 Plus Graphing Calculators, provides students with easy step-by-step guides to the most common statistics software.

Instructor Resources

Instructor's Solutions Manual (download only), by Nancy Boudreau (Emeritus Associate Professor Bowling Green State University), includes complete worked-out solutions to all even-numbered text exercises. Careful attention has been paid to ensure that all methods of solution and notation are consistent with those used in the core text.

PowerPoint® Lecture Slides include figures and tables from the textbook. Available for download from Pearson's on-line catalog at www.pearsonglobaleditions.com/mcclave and in Pearson MyLab Statistics.

TestGen® (www.pearsoned.com/testgen) enables instructors to build, edit, print, and administer tests using a computerized bank of questions developed to cover all the objectives of the text. TestGen is algorithmically based, allowing instructors to create multiple but equivalent versions of the same question or test with the click of a button. Instructors can also modify test bank questions or add new questions. The software and test bank are available for download from Pearson Education's online catalog at www.pearsonglobaleditions.com/mcclave and in Pearson MyLab Statistics.

Online Test Bank, a test bank derived from TestGen®, is available for download from Pearson's online catalog at www.pearsonglobaleditions.com/mcclave and in Pearson MyLab Statistics.

Technology Resources

A companion website www.pearsonglobaleditions.com/mcclave holds a number of support materials, including:

- **Data sets** formatted as .csv, .txt, and TI files
- **Applets** (short computer programs) that allow students to run simulations that visually demonstrate statistical concepts
- **Chapter 14:** Nonparametric Statistics

Acknowledgments

This book reflects the efforts of a great many people over a number of years. First, we would like to thank the following professors, whose reviews and comments on this and prior editions have contributed to the 13th edition:

Reviewers Involved with the 13th Edition of *Statistics*

Sarol Aryal, *Montana State University—Billings*
Maggie McBride, *Montana State University—Billings*
Mehdi Razzaghi, *Bloomsburg University*
Kamel Rekab, *University of Missouri—Kansas City*
Jim Schott, *University of Central Florida*
Susan Schott, *University of Central Florida*
Dong Zhang, *Bloomsburg University*

Reviewers of Previous Editions

Bill Adamson, *South Dakota State*; Ibrahim Ahmad, *Northern Illinois University*; Roddy Akbari, *Guilford Technical Community College*; Ali Arab, *Georgetown University*; David Atkinson, *Olivet Nazarene University*; Mary Sue Beersman, *Northeast Missouri State University*; William H. Beyer, *University of Akron*; Marvin Bishop, *Manhattan College*; Patricia M. Buchanan, *Pennsylvania State University*; Dean S. Burbank, *Gulf Coast Community College*; Ann Cascarelle, *St. Petersburg College*; Jen Case, *Jacksonville State University*; Kathryn Chaloner, *University of Minnesota*; Hanfeng Chen, *Bowling Green State University*; Gerardo Chin-Leo, *The Evergreen State College*; Linda Brant Collins, *Iowa State University*; Brant Deppa, *Winona State University*; John Dirkse, *California State University—Bakersfield*; N. B. Ebrahimi, *Northern Illinois University*; John Egenolf, *University of Alaska—Anchorage*; Dale Everson, *University of Idaho*; Christine Franklin, *University of Georgia*; Khadiga Gamgoum, *Northern Virginia Community College*; Rudy Gideon, *University of Montana*; Victoria Marie Gribshaw, *Seton Hill College*; Larry Griffey, *Florida Community College*; David Groggel, *Miami University at Oxford*; John E. Groves, *California Polytechnic State University—San Luis Obispo*; Sneh Gulati, *Florida International University*; Dale K. Hathaway, *Olivet Nazarene University*; Shu-ping Hodgson, *Central Michigan University*; Jean L. Holton, *Virginia Commonwealth University*; Soon Hong, *Grand Valley State University*; Ina Parks S. Howell, *Florida International University*; Gary Itzkowitz, *Rowan College of New Jersey*; John H. Kellermeier, *State University College at Plattsburgh*; Golan Kibria, *Florida International University*; Timothy J. Killeen, *University of Connecticut*; William G. Koellner, *Montclair State University*; James R. Lackritz, *San Diego State University*; Diane Lambert, *AT&T/Bell Laboratories*; Edwin G. Landauer, *Clackamas Community College*; James Lang, *Valencia Junior College*; Glenn Larson, *University of Regina*; John J. Lefante, Jr., *University of South Alabama*; Pi-Erh Lin, *Florida State University*; R. Bruce Lind, *University of Puget Sound*; Rhonda Magel, *North Dakota State University*; Linda C. Malone, *University of Central Florida*; Allen E. Martin, *California State University—Los Angeles*; Rick Martinez, *Foothill College*; Brenda Masters, *Oklahoma State University*; Leslie Matekaitis, *Cal Genetics*; Maggie McBride, *Montana State University—Billings*; E. Donice McCune, *Stephen F. Austin State University*; Mark M. Meerschaert, *University of Nevada—Reno*; Greg Miller, *Stephen F. Austin State University*; Satya Narayan Mishra, *University of South Alabama*; Kazemi Mohammed, *University of North Carolina—Charlotte*; Christopher Morrell, *Loyola College in Maryland*; Mir Mortazavi, *Eastern New Mexico University*; A. Mukherjea, *University of South Florida*; Steve Nimmo, *Morningside College (Iowa)*; Susan Nolan, *Seton Hall University*; Thomas O'Gorman, *Northern Illinois University*; Bernard Ostle, *University of Central Florida*; William B. Owen, *Central Washington University*; Won J. Park, *Wright State University*; John J. Peterson, *Smith Kline & French Laboratories*; Ronald Pierce, *Eastern Kentucky University*;

Surajit Ray, *Boston University*; Betty Rehfuss, *North Dakota State University—Bottineau*; Andrew Rosalsky, *University of Florida*; C. Bradley Russell, *Clemson University*; Rita Schillaber, *University of Alberta*; Jim Schott, *University of Central Florida*; Susan C. Schott, *University of Central Florida*; George Schultz, *St. Petersburg Junior College*; Carl James Schwarz, *University of Manitoba*; Mike Seyfried, *Shippensburg University*; Arvind K. Shah, *University of South Alabama*; Lewis Shoemaker, *Millersville University*; Sean Simpson, *Westchester Community College*; Charles W. Sinclair, *Portland State University*; Robert K. Smidt, *California Polytechnic State University—San Luis Obispo*; Vasanth B. Solomon, *Drake University*; W. Robert Stephenson, *Iowa State University*; Engin Sungur, *University of Minnesota—Morris*; Thaddeus Tarpey, *Wright State University*; Kathy Taylor, *Clackamas Community College*; Sherwin Toribio, *University of Wisconsin—La Crosse*; Barbara Treadwell, *Western Michigan University*; Dan Voss, *Wright State University*; Augustin Vukov, *University of Toronto*; Dennis D. Wackerly, *University of Florida*; Barbara Wainwright, *Salisbury University*; Matthew Wood, *University of Missouri—Columbia*; Michael Zwilling, *Mt. Union College*

Other Contributors

Special thanks are due to our ancillary authors, Nancy Boudreau and Mark Dummeldinger, both of whom have worked with us for many years. Accuracy checkers Dave Bregenzer and Engin Sungur helped ensure a highly accurate, clean text. Finally, the Pearson Education staff of Deirdre Lynch, Patrick Barbera, Christine O'Brien, Justin Billing, Tatiana Anacki, Roxanne McCarley, Erin Kelly, Tiffany Bitzel, Jennie Myers Jean Choe, and Barbara Atkinson, as well as Integra-Chicago's Alverne Ball, helped greatly with all phases of the text development, production, and marketing effort.

Acknowledgments for the Global Edition

Pearson would like to thank and acknowledge the following people for their contributions to the Global Edition.

Contributors

Vikas Arora
Rahul Bhattacharya, *University of Calcutta*
Niladri Chatterjee, *Indian Institute of Technology Delhi*

Reviewers

Ruben Garcia, *Jakarta International College*
Mohammad Kacim, *Holy Spirit University of Kaslik*
Aneesh Kumar, *Mahatma Gandhi College—Iritty*
Santhosh Kumar, *Christ University—Bengaluru*
Abhishek Umrawal, *University of Delhi*

Applications Index

1 Statistics, Data, and Statistical Thinking

CONTENTS

Where We're Going

- Introduce the field of statistics (1.1)
- Demonstrate how statistics applies to real-world problems (1.2)
- Introduce the language of statistics and the key elements to any statistical problem (1.3)
- Differentiate between population and sample data (1.3)
- Differentiate between descriptive and inferential statistics (1.3)
- Identify the different types of data and data collection methods (1.4–1.5)
- Discover how critical thinking through statistics can help improve our quantitative literacy (1.6)

Statistics IN Action Social Media Network Usage—Are You Linked In?

The Pew Research Center, a nonpartisan organization funded by a Philadelphia-based charity, has conducted more than 100 surveys on Internet usage in the United States as part of the Pew Internet & American Life Project (PIALP). In a recent report titled *Social Media Update, 2013*, the PIALP examined adults' (ages 18 and up) attitudes and behavior toward online social media networks. Regarded merely as a fun, online activity for high school and college students just a few years ago, social media now exert tremendous influence over the way people around the world—of all ages—get and share information. The five social media sites investigated in this report include Facebook, Twitter, Instagram, Pinterest, and LinkedIn. The Pew Research Center contacted 1,445 Internet users via landline telephone or cell phone for the survey.

Several of the many survey questions asked are provided here as well as the survey results:

- *Social Networking:*

  When asked if they ever use an online social networking site, adults responded:

Yes	73%
No	27%

- *Facebook Usage:*

  When Facebook users were asked how often they visit the social media site, they responded:

Several times a day	40%
About once a day	24%
3–5 days a week	10%
1–2 days a week	13%
Every few weeks	6%
Less often	7%

- *Twitter Usage:*

  When asked if they ever use Twitter, adults responded:

Yes	18%
No	82%

- *Overall Social Media Usage:*

  When asked about how many of the five social networking sites they use, adults responded:

0	22%
1	36%
2	23%
3	12%
4	5%
5	2%

  (Average = 1.48 sites)

In the following "Statistics in Action Revisited" sections, we discuss several key statistical concepts covered in this chapter that are relevant to the Pew Internet & American Life Project survey.

Statistics IN Action Revisited

- Identifying the Population, Sample, and Inference (p. 37)
- Identifying the Data Collection Method and Data Type (p. 44)
- Critically Assessing the Ethics of a Statistical Study (p. 46)

1.1 The Science of Statistics

What does statistics mean to you? Does it bring to mind batting averages, Gallup polls, unemployment figures, or numerical distortions of facts (lying with statistics!)? Or is it simply a college requirement you have to complete? We hope to persuade you that statistics is a meaningful, useful science whose broad scope of applications to business, government, and the physical and social sciences is almost limitless. We also want to show that statistics can lie only when they are misapplied. Finally, we wish to demonstrate the key role statistics plays in critical thinking—whether in the classroom, on the job, or in everyday life. Our objective is to leave you with the impression that the time you spend studying this subject will repay you in many ways.

The *Random House College Dictionary* defines **statistics** as "the science that deals with the collection, classification, analysis, and interpretation of information or data."

Thus, a statistician isn't just someone who calculates batting averages at baseball games or tabulates the results of a Gallup poll. Professional statisticians are trained in *statistical science*. That is, they are trained in collecting information in the form of **data**, evaluating the information, and drawing conclusions from it. Furthermore, statisticians determine what information is relevant in a given problem and whether the conclusions drawn from a study are to be trusted.

> **Statistics** is the science of data. This involves collecting, classifying, summarizing, organizing, analyzing, presenting, and interpreting numerical and categorical information.

In the next section, you'll see several real-life examples of statistical applications that involve making decisions and drawing conclusions.

1.2 Types of Statistical Applications

"Statistics" means "numerical descriptions" to most people. Monthly housing starts, the failure rate of liver transplants, and the proportion of African-Americans who feel brutalized by local police all represent statistical descriptions of large sets of data collected on some phenomenon. (Later, in Section 1.4, we learn that not all data is numerical in nature.) Often the data are selected from some larger set of data whose characteristics we wish to estimate. We call this selection process *sampling*. For example, you might collect the ages of a sample of customers who shop for a particular product online to estimate the average age of *all* customers who shop online for the product. Then you could use your estimate to target the Web site's advertisements to the appropriate age group. Notice that statistics involves two different processes: (1) describing sets of data and (2) drawing conclusions (making estimates, decisions, predictions, etc.) about the sets of data on the basis of sampling. So, the applications of statistics can be divided into two broad areas: **descriptive statistics** and **inferential statistics**.

> **Descriptive statistics** utilizes numerical and graphical methods to look for patterns in a data set, to summarize the information revealed in a data set, and to present that information in a convenient form.

> **Inferential statistics** utilizes sample data to make estimates, decisions, predictions, or other generalizations about a larger set of data.

BIOGRAPHY FLORENCE NIGHTINGALE (1820–1910)

The Passionate Statistician

In Victorian England, the "Lady of the Lamp" had a mission to improve the squalid field hospital conditions of the British army during the Crimean War. Today, most historians consider Florence Nightingale to be the founder of the nursing profession. To convince members of the British Parliament of the need for supplying nursing and medical care to soldiers in the field, Nightingale compiled massive amounts of data from army files. Through a remarkable series of graphs (which included the first pie chart), she demonstrated that most of the deaths in the war either were due to illnesses contracted outside the battlefield or occurred long after battle action from wounds that went untreated. Florence Nightingale's compassion and self-sacrificing nature, coupled with her ability to collect, arrange, and present large amounts of data, led some to call her the Passionate Statistician. ■

Although we'll discuss both descriptive and inferential Statistics in the chapters that follow, the primary theme of the text is **inference**.

Let's begin by examining some studies that illustrate applications of statistics.

Study 1.1 "Best-Selling Girl Scout Cookies" (Source: *www.girlscouts.org*)

Since 1917, the Girl Scouts of America have been selling boxes of cookies. Currently, there are 12 varieties for sale: Thin Mints, Samoas, Lemonades, Tagalongs, Do-si-dos, Trefoils,

Savannah Smiles, Thanks-A-Lot, Dulce de Leche, Cranberry Citrus Crisps, Chocolate Chip, and Thank U Berry Much. Each of the approximately 150 million boxes of Girl Scout cookies sold each year is classified by variety. The results are summarized in Figure 1.1. From the graph, you can clearly see that the best-selling variety is Thin Mints (25%), followed by Samoas (19%) and Tagalongs (13%). Since the figure describes the various categories of boxes of Girl Scout cookies sold, the graphic is an example of descriptive statistics.

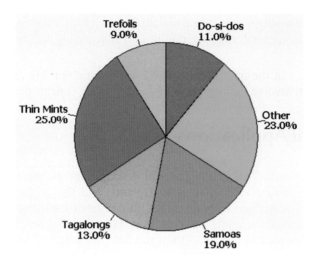

Figure 1.1

MINITAB graph of best-selling Girl Scout cookies (Based on www. girlscouts.org, 2011–12 sales.)

Study 1.2 "Are Action Video Game Players Better than Non-gamers at Complex, Divided Attention Tasks?" (Source: *Human Factors*, Vol. 56, No. 31, May 2014)

Researchers at the Universities of Illinois (Urbana-Champaign) and Central Florida conducted a study to determine whether video game players are better than non-video game players at crossing the street when presented with distractions. Each in a sample of 60 college students was classified as a video game player or a non-gamer. Participants entered a street crossing simulator and were asked to cross a busy street at an unsigned intersection. The simulator was designed to have cars traveling at various high rates of speed in both directions. During the crossing, the students also performed a memory task as a distraction. The researchers found no differences in either the street crossing performance or memory task score of video game players and non-gamers. "These results," say the researchers, "suggest that action video game players [and non-gamers] are equally susceptible to the costs of dividing attention in a complex task." Thus, inferential statistics was applied to arrive at this conclusion.

Study 1.3 "Does Rudeness Really Matter in the Workplace?" (Source: *Academy of Management Journal*, Oct. 2007)

Previous studies have established that rudeness in the workplace can lead to retaliatory and counterproductive behavior. However, there has been little research on how rude behaviors influence a victim's task performance. Consider a study where college students enrolled in a management course were randomly assigned to one of two experimental conditions: rudeness condition (45 students) and control group (53 students). Each student was asked to write down as many uses for a brick as possible in five minutes; this value (total number of uses) was used as a performance measure for each student. For those students in the rudeness condition, the facilitator displayed rudeness by berating the students in general for being irresponsible and unprofessional (due to a late-arriving confederate). No comments were made about the late-arriving confederate for students in the control group. As you might expect, the researchers discovered that the performance levels for students in the rudeness condition were generally lower than the performance levels for students in the control group; thus, they concluded that rudeness in the workplace negatively affects job performance. As in Study 1.2, this study is an example of the use of inferential statistics. The researchers used data collected on 98 college students in a simulated work environment to make an inference about the performance levels of all workers exposed to rudeness on the job.

These studies provide three real-life examples of the uses of statistics. Notice that each involves an analysis of data, either for the purpose of describing the data set (Study 1.1) or for making inferences about a data set (Studies 1.2 and 1.3).

1.3 Fundamental Elements of Statistics

Statistical methods are particularly useful for studying, analyzing, and learning about **populations** of **experimental units**.

> An **experimental** (or **observational**) **unit** is an object (e.g., person, thing, transaction, or event) about which we collect data.

> A **population** is a set of all units (usually people, objects, transactions, or events) that we are interested in studying.

For example, populations may include (1) *all* employed workers in the United States, (2) *all* registered voters in California, (3) *everyone* who is afflicted with AIDS, (4) *all* the cars produced last year by a particular assembly line, (5) the *entire* stock of spare parts available at Southwest Airlines' maintenance facility, (6) *all* sales made at the drive-in window of a McDonald's restaurant during a given year, or (7) the set of *all* accidents occurring on a particular stretch of interstate highway during a holiday period. Notice that the first three population examples (1–3) are sets (groups) of people, the next two (4–5) are sets of objects, the next (6) is a set of transactions, and the last (7) is a set of events. Notice also that *each set includes all the units in the population.*

In studying a population, we focus on one or more characteristics or properties of the units in the population. We call such characteristics **variables**. For example, we may be interested in the variables age, gender, and number of years of education of the people currently unemployed in the United States.

> A **variable** is a characteristic or property of an individual experimental (or observational) unit in the population.

The name *variable* is derived from the fact that any particular characteristic may vary among the units in a population.

In studying a particular variable, it is helpful to be able to obtain a numerical representation for it. Often, however, numerical representations are not readily available, so measurement plays an important supporting role in statistical studies. **Measurement** is the process we use to assign numbers to variables of individual population units. We might, for instance, measure the performance of the president by asking a registered voter to rate it on a scale from 1 to 10. Or we might measure the age of the U.S. workforce simply by asking each worker, "How old are you?" In other cases, measurement involves the use of instruments such as stopwatches, scales, and calipers.

If the population you wish to study is small, it is possible to measure a variable for every unit in the population. For example, if you are measuring the GPA for all incoming first-year students at your university, it is at least feasible to obtain every GPA. When we measure a variable for every unit of a population, it is called a **census** of the population. Typically, however, the populations of interest in most applications are much larger, involving perhaps many thousands, or even an infinite number, of units. Examples of large populations are those following the definition of population above, as well as all graduates of your university or college, all potential buyers of a new iPhone, and all pieces of first-class mail handled by the U.S. Post Office. For such populations, conducting a census would be prohibitively time consuming or costly. A reasonable alternative would be to select and study a *subset* (or portion) of the units in the population.

A **sample** is a subset of the units of a population.

For example, instead of polling all 145 million registered voters in the United States during a presidential election year, a pollster might select and question a sample of just 1,500 voters. (See Figure 1.2.) If he is interested in the variable "presidential preference," he would record (measure) the preference of each vote sampled.

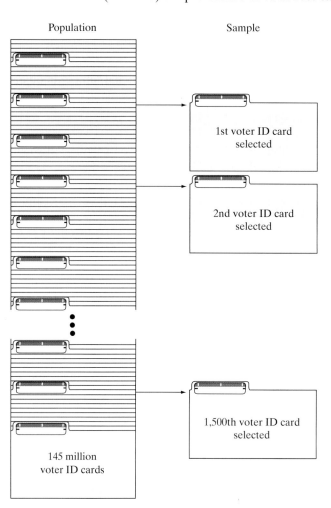

Figure 1.2

A sample of voter registration cards for all registered voters

After the variables of interest for every unit in the sample (or population) are measured, the data are analyzed, either by descriptive or inferential statistical methods. The pollster, for example, may be interested only in *describing* the voting patterns of the sample of 1,500 voters. More likely, however, he will want to use the information in the sample to make inferences about the population of all 145 million voters.

A **statistical inference** is an estimate, prediction, or some other generalization about a population based on information contained in a sample.

That is, *we use the information contained in the smaller sample to learn about the larger population.** Thus, from the sample of 1,500 voters, the pollster may estimate the percentage of all the voters who would vote for each presidential candidate if the election were held on the day the poll was conducted, or he might use the results to predict the outcome on election day.

*The terms *population* and *sample* are often used to refer to the sets of measurements themselves as well as to the units on which the measurements are made. When a single variable of interest is being measured, this usage causes little confusion. But when the terminology is ambiguous, we'll refer to the measurements as *population data sets* and *sample data sets*, respectively.

Example   1.1

Key Elements of a Statistical Problem—Ages of Broadway Ticketbuyers

Problem According to *Variety* (Jan. 10, 2014), the average age of Broadway ticketbuyers is 42.5 years. Suppose a Broadway theatre executive hypothesizes that the average age of ticketbuyers to her theatre's plays is less than 42.5 years. To test her hypothesis, she samples 200 ticketbuyers to her theatre's plays and determines the age of each.

a. Describe the population.

b. Describe the variable of interest.

c. Describe the sample.

d. Describe the inference.

Solution

a. The population is the set of all units of interest to the theatre executive, which is the set of all ticketbuyers to her theatre's plays.

b. The age (in years) of each ticketbuyer is the variable of interest.

c. The sample must be a subset of the population. In this case, it is the 200 ticketbuyers selected by the executive.

d. The inference of interest involves the *generalization* of the information contained in the sample of 200 ticketbuyers to the population of all her theatre's ticketbuyers. In particular, the executive wants to *estimate* the average age of the ticketbuyers to her theatre's plays in order to determine whether it is less than 42.5 years. She might accomplish this by calculating the average age of the sample and using that average to estimate the average age of the population.

Look Back A key to diagnosing a statistical problem is to identify the data set collected (in this example, the ages of the 200 ticketbuyers) as a population or a sample.

■ **Now Work Exercise 1.13**

Example   1.2

Key Elements of a Statistical Problem— Pepsi vs. Coca-Cola

Problem "Cola wars" is the popular term for the intense competition between Coca-Cola and Pepsi displayed in their marketing campaigns, which have featured movie and television stars, rock videos, athletic endorsements, and claims of consumer preference based on taste tests. Suppose, as part of a Pepsi marketing campaign, 1,000 cola consumers are given a blind taste test (i.e., a taste test in which the two brand names are disguised). Each consumer is asked to state a preference for brand A or brand B.

a. Describe the population.

b. Describe the variable of interest.

c. Describe the sample.

d. Describe the inference.

Solution

a. Since we are interested in the responses of cola consumers in a taste test, a cola consumer is the experimental unit. Thus, the population of interest is the collection or set of all cola consumers.

b. The characteristic that Pepsi wants to measure is the consumer's cola preference, as revealed under the conditions of a blind taste test, so *cola preference* is the variable of interest.

c. The sample is the 1,000 cola consumers selected from the population of all cola consumers.

d. The inference of interest is the *generalization* of the cola preferences of the 1,000 sampled consumers to the population of all cola consumers. In particular, the preferences of the consumers in the sample can be used to *estimate* the percentages of cola consumers who prefer each brand.

Look Back In determining whether the study is inferential or descriptive, we assess whether Pepsi is interested in the responses of only the 1,000 sampled customers (descriptive statistics) or in the responses of the entire population of consumers (inferential statistics).

■ **Now Work Exercise 1.16b**

The preceding definitions and examples identify four of the five elements of an inferential statistical problem: a population, one or more variables of interest, a sample, and an inference. But making the inference is only part of the story; we also need to know its **reliability**—that is, how good the inference is. The only way we can be certain that an inference about a population is correct is to include the entire population in our sample. However, because of *resource constraints* (i.e., insufficient time or money), we usually can't work with whole populations, so we base our inferences on just a portion of the population (a sample). Thus, we introduce an element of *uncertainty* into our inferences. Consequently, whenever possible, it is important to determine and report the reliability of each inference made. Reliability, then, is the fifth element of inferential statistical problems.

The **measure of reliability** that accompanies an inference separates the science of statistics from the art of fortune-telling. A palm reader, like a statistician, may examine a sample (your hand) and make inferences about the population (your life). However, unlike statistical inferences, the palm reader's inferences include no measure of reliability.

Suppose, like the theatre executive in Example 1.1, we are interested in the *error of estimation* (i.e., the difference between the average age of a population of ticketbuyers and the average age of a sample of ticketbuyers). Using statistical methods, we can determine a *bound on the estimation error*. This bound is simply a number that our estimation error (the difference between the average age of the sample and the average age of the population) is not likely to exceed. We'll see in later chapters that this bound is a measure of the uncertainty of our inference. The reliability of statistical inferences is discussed throughout this text. For now, we simply want you to realize that an inference is incomplete without a measure of its reliability.

> A **measure of reliability** is a statement (usually quantitative) about the degree of uncertainty associated with a statistical inference.

Let's conclude this section with a summary of the elements of descriptive and of inferential statistical problems and an example to illustrate a measure of reliability.

> **Four Elements of Descriptive Statistical Problems**
> 1. The population or sample of interest
> 2. One or more variables (characteristics of the population or sample units) that are to be investigated
> 3. Tables, graphs, or numerical summary tools
> 4. Identification of patterns in the data

> **Five Elements of Inferential Statistical Problems**
> 1. The population of interest
> 2. One or more variables (characteristics of the population units) that are to be investigated
> 3. The sample of population units
> 4. The inference about the population based on information contained in the sample
> 5. A measure of the reliability of the inference

Example   1.3

Reliability of an
Inference—Pepsi vs.
Coca-Cola

Problem Refer to Example 1.2, in which the preferences of 1,000 cola consumers were indicated in a taste test. Describe how the reliability of an inference concerning the preferences of all cola consumers in the Pepsi bottler's marketing region could be measured.

Solution When the preferences of 1,000 consumers are used to estimate those of all consumers in a region, the estimate will not exactly mirror the preferences of the population. For example, if the taste test shows that 56% of the 1,000 cola consumers preferred Pepsi, it does not follow (nor is it likely) that exactly 56% of all cola drinkers in the region prefer Pepsi. Nevertheless, we can use sound statistical reasoning (which we'll explore later in the text) to ensure that the sampling procedure will generate estimates that are almost certainly within a specified limit of the true percentage of all cola consumers who prefer Pepsi. For example, such reasoning might assure us that the estimate of the preference for Pepsi is almost certainly within 5% of the preference of the population. The implication is that the actual preference for Pepsi is between 51% [i.e., $(56 - 5)\%$] and 61% [i.e., $(56 + 5)\%$]—that is, $(56 \pm 5)\%$. This interval represents a measure of the reliability of the inference.

Look Ahead The interval 56 ± 5 is called a *confidence interval*, since we are confident that the true percentage of cola consumers who prefer Pepsi in a taste test falls into the range $(51, 61)$. In Chapter 7, we learn how to assess the degree of confidence (e.g., a 90% or 95% level of confidence) in the interval.

Statistics IN Action **Revisited** **Identifying the Population, Sample, and Inference**

Consider the 2013 Pew Internet & American Life Project survey on social networking. In particular, consider the survey results on the use of social networking sites like Facebook. The experimental unit for the study is an adult (the person answering the question), and the variable measured is the response ("yes" or "no") to the question.

The Pew Research Center reported that 1,445 adult Internet users participated in the study. Obviously, that number is not all of the adult Internet users in the United States. Consequently, the 1,445 responses represent a sample selected from the much larger population of all adult Internet users.

Earlier surveys found that 55% of adults used an online social networking site in 2006 and 65% in 2008. These are descriptive statistics that provide information on the popularity of social networking in past years. Since 73% of the surveyed adults in 2013 used an online social networking site, the Pew Research Center inferred that usage of social networking sites continues its upward trend, with more and more adults getting online each year. That is, the researchers used the descriptive statistics from the sample to make an inference about the current population of U.S. adults' use of social networking.

1.4 Types of Data

You have learned that statistics is the science of data and that data are obtained by measuring the values of one or more variables on the units in the sample (or population). All data (and hence the variables we measure) can be classified as one of two general types: **quantitative data** and **qualitative data**.

Quantitative data are data that are measured on a naturally occurring numerical scale.* The following are examples of quantitative data:

1. The temperature (in degrees Celsius) at which each piece in a sample of 20 pieces of heat-resistant plastic begins to melt

---

*Quantitative data can be subclassified as either *interval data* or *ratio data*. For ratio data, the origin (i.e., the value 0) is a meaningful number. But the origin has no meaning with interval data. Consequently, we can add and subtract interval data, but we can't multiply and divide them. Of the four quantitative data sets listed as examples, (1) and (3) are interval data while (2) and (4) are ratio data.

2. The current unemployment rate (measured as a percentage) in each of the 50 states

3. The scores of a sample of 150 law school applicants on the LSAT, a standardized law school entrance exam administered nationwide

4. The number of convicted murderers who receive the death penalty each year over a 10-year period

> **Quantitative data** are measurements that are recorded on a naturally occurring numerical scale.

In contrast, qualitative data cannot be measured on a natural numerical scale; they can only be classified into categories.* (For this reason, this type of data is also called *categorical data*.) Examples of qualitative data include the following:

1. The political party affiliation (Democrat, Republican, or Independent) in a sample of 50 voters

2. The defective status (defective or not) of each of 100 computer chips manufactured by Intel

3. The size of a car (subcompact, compact, midsize, or full size) rented by each of a sample of 30 business travelers

4. A taste tester's ranking (best, worst, etc.) of four brands of barbecue sauce for a panel of 10 testers

Often, we assign arbitrary numerical values to qualitative data for ease of computer entry and analysis. But these assigned numerical values are simply codes: They cannot be meaningfully added, subtracted, multiplied, or divided. For example, we might code Democrat = 1, Republican = 2, and Independent = 3. Similarly, a taste tester might rank the barbecue sauces from 1 (best) to 4 (worst). These are simply arbitrarily selected numerical codes for the categories and have no utility beyond that.

> **Qualitative** (or **categorical**) **data** are measurements that cannot be measured on a natural numerical scale; they can only be classified into one of a group of categories.

| Example 1.4

Data Types—Army Corps of Engineers Study of a Contaminated River

Problem Chemical and manufacturing plants often discharge toxic-waste materials such as DDT into nearby rivers and streams. These toxins can adversely affect the plants and animals inhabiting the river and the riverbank. The U.S. Army Corps of Engineers conducted a study of fish in the Tennessee River (in Alabama) and its three tributary creeks: Flint Creek, Limestone Creek, and Spring Creek. A total of 144 fish were captured, and the following variables were measured for each:

1. River/creek where each fish was captured

2. Species (channel catfish, largemouth bass, or smallmouth buffalo fish)

3. Length (centimeters)

4. Weight (grams)

5. DDT concentration (parts per million)

(For future analyses, these data are saved in the **FISHDDT** file.) Classify each of the five variables measured as quantitative or qualitative.

Solution The variables length, weight, and DDT concentration are quantitative because each is measured on a numerical scale: length in centimeters, weight in grams, and DDT in

---

*Qualitative data can be subclassified as either *nominal data* or *ordinal data*. The categories of an ordinal data set can be ranked or meaningfully ordered, but the categories of a nominal data set can't be ordered. Of the four qualitative data sets listed as examples, (1) and (2) are nominal and (3) and (4) are ordinal.

parts per million. In contrast, river/creek and species cannot be measured quantitatively: They can only be classified into categories (e.g., channel catfish, largemouth bass, and smallmouth buffalo fish for species). Consequently, data on river/creek and species are qualitative.

Look Ahead It is essential that you understand whether the data you are interested in are quantitative or qualitative, since the statistical method appropriate for describing, reporting, and analyzing the data depends on the data type (quantitative or qualitative).

■ **Now Work Exercise 1.12**

We demonstrate many useful methods for analyzing quantitative and qualitative data in the remaining chapters of the text. But first, we discuss some important ideas on data collection in the next section.

1.5 Collecting Data: Sampling and Related Issues

Once you decide on the type of data—quantitative or qualitative—appropriate for the problem at hand, you'll need to collect the data. Generally, you can obtain data in three different ways:

1. From a *published source*
2. From a *designed experiment*
3. From an *observational study* (e.g., a *survey*)

Sometimes, the data set of interest has already been collected for you and is available in a **published source**, such as a book, journal, or newspaper. For example, you may want to examine and summarize the divorce rates (i.e., number of divorces per 1,000 population) in the 50 states of the United States. You can find this data set (as well as numerous other data sets) at your library in the *Statistical Abstract of the United States*, published annually by the U.S. government. Similarly, someone who is interested in monthly mortgage applications for new home construction would find this data set in the *Survey of Current Business*, another government publication. Other examples of published data sources include *The Wall Street Journal* (financial data) and *Elias Sports Bureau* (sports information). The Internet (World Wide Web) now provides a medium by which data from published sources are readily obtained.*

A second method of collecting data involves conducting a **designed experiment**, in which the researcher exerts strict control over the units (people, objects, or things) in the study. For example, an often-cited medical study investigated the potential of aspirin in preventing heart attacks. Volunteer physicians were divided into two groups: the *treatment* group and the *control* group. Each physician in the treatment group took one aspirin tablet a day for one year, while each physician in the control group took an aspirin-free placebo made to look like an aspirin tablet. The researchers—not the physicians under study—controlled who received the aspirin (the treatment) and who received the placebo. As you'll learn in Chapter 10, a properly designed experiment allows you to extract more information from the data than is possible with an uncontrolled study.

Finally, observational studies can be employed to collect data. In an **observational study**, the researcher observes the experimental units in their natural setting and records the variable(s) of interest. For example, a child psychologist might observe and record the level of aggressive behavior of a sample of fifth graders playing on a school playground. Similarly, a zoologist may observe and measure the weights of newborn elephants born in captivity. Unlike a designed experiment, an observational study is a study in which the researcher makes no attempt to control any aspect of the experimental units.

*With published data, we often make a distinction between the *primary source* and a *secondary source*. If the publisher is the original collector of the data, the source is primary. Otherwise, the data is secondary-source data.

The most common type of observational study is a **survey,** where the researcher samples a group of people, asks one or more questions, and records the responses. Probably the most familiar type of survey is the political poll, conducted by any one of a number of organizations (e.g., Harris, Gallup, Roper, and CNN) and designed to predict the outcome of a political election. Another familiar survey is the Nielsen survey, which provides the major networks with information on the most-watched programs on television. Surveys can be conducted through the mail, with telephone interviews, or with in-person interviews. Although in-person surveys are more expensive than mail or telephone surveys, they may be necessary when complex information is to be collected.

A **designed experiment** is a data collection method where the researcher exerts full control over the characteristics of the experimental units sampled. These experiments typically involve a group of experimental units that are assigned the *treatment* and an untreated (or *control*) group.

An **observational study** is a data collection method where the experimental units sampled are observed in their natural setting. No attempt is made to control the characteristics of the experimental units sampled. (Examples include *opinion polls* and *surveys.*)

Regardless of which data collection method is employed, it is likely that the data will be a sample from some population. And if we wish to apply inferential statistics, we must obtain a **representative sample**.

A **representative sample** exhibits characteristics typical of those possessed by the target population.

For example, consider a political poll conducted during a presidential election year. Assume that the pollster wants to estimate the percentage of all 145 million registered voters in the United States who favor the incumbent president. The pollster would be unwise to base the estimate on survey data collected for a sample of voters from the incumbent's own state. Such an estimate would almost certainly be *biased* high; consequently, it would not be very reliable.

The most common way to satisfy the representative sample requirement is to select a random sample. A **simple random sample** ensures that every subset of fixed size in the population has the same chance of being included in the sample. If the pollster samples 1,500 of the 145 million voters in the population so that every subset of 1,500 voters has an equal chance of being selected, she has devised a random sample.

A **simple random sample** of n experimental units is a sample selected from the population in such a way that every different sample of size n has an equal chance of selection.

The procedure for selecting a simple random sample typically relies on a **random number generator**. Random number generators are available in table form online*, and they are built into most statistical software packages. The SAS, MINITAB, and SPSS statistical software packages all have easy-to-use random number generators for creating a random sample. The next example illustrates the procedure.

*One of many free online random number generators is available at *www.randomizer.org*.

Example	1.5

Generating a Simple Random Sample— Selecting Households for a Feasibility Study

↓	C1	C2
	HouseNum	Sample20
1	1	78
2	2	152
3	3	157
4	4	177
5	5	216
6	6	234
7	7	242
8	8	255
9	9	262
10	10	280
11	11	323
12	12	344
13	13	406
14	14	408
15	15	480
16	16	495
17	17	549
18	18	610
19	19	645
20	20	690
21	21	
22	22	
23	23	
24	24	
25	25	

Figure 1.3

Random Selection of 20 Households Using MINITAB

Problem Suppose you wish to assess the feasibility of building a new high school. As part of your study, you would like to gauge the opinions of people living close to the proposed building site. The neighborhood adjacent to the site has 711 homes. Use a random number generator to select a simple random sample of 20 households from the neighborhood to participate in the study.

Solution In this study, your population of interest consists of the 711 households in the adjacent neighborhood. To ensure that every possible sample of 20 households selected from the 711 has an equal chance of selection (i.e., to ensure a simple random sample), first assign a number from 1 to 711 to each of the households in the population. These numbers were entered into MINITAB. Now, apply the random number generator of MINITAB, requesting that 20 households be selected without replacement. Figure 1.3 shows the output from MINITAB. You can see that households numbered 78, 152, 157, 177, 216, ..., 690 are the households to be included in your sample.

Look Back It can be shown (proof omitted) that there are over 3×10^{38} possible samples of size 20 that can be selected from the 711 households. Random number generators guarantee (to a certain degree of approximation) that each possible sample has an equal chance of being selected.

■

In addition to simple random samples, there are more complex random sampling designs that can be employed. These include (but are not limited to) **stratified random sampling, cluster sampling, systematic sampling**, and **randomized response sampling**. Brief descriptions of each follow. (For more details on the use of these sampling methods, consult the references at the end of this chapter.)

Stratified random sampling is typically used when the experimental units associated with the population can be separated into two or more groups of units, called *strata*, where the characteristics of the experimental units are more similar within strata than across strata. Random samples of experimental units are obtained for each strata, then the units are combined to form the complete sample. For example, if you are gauging opinions of voters on a polarizing issue, like government-sponsored health care, you may want to stratify on political affiliation (Republicans and Democrats), making sure that representative samples of both Republicans and Democrats (in proportion to the number of Republicans and Democrats in the voting population) are included in your survey.

Sometimes it is more convenient and logical to sample natural groupings (*clusters*) of experimental units first, then collect data from all experimental units within each cluster. This involves the use of *cluster sampling*. For example, suppose a marketer for a large upscale restaurant chain wants to find out whether customers like the new menu. Rather than collect a simple random sample of all customers (which would be very difficult and costly to do), the marketer will randomly sample 10 of the 150 restaurant locations (clusters), then interview all customers eating at each of the 10 locations on a certain night.

Another popular sampling method is *systematic sampling*. This method involves systematically selecting every *k*th experimental unit from a list of all experimental units. For example, every fifth person who walks into a shopping mall could be asked whether he or she owns a smart phone. Or a quality control engineer at a manufacturing plant may select every 10th item produced on an assembly line for inspection.

A fourth alternative to simple random sampling is *randomized response sampling*. This design is particularly useful when the questions of the pollsters are likely to elicit false answers. For example, suppose each person in a sample of wage earners is asked whether he or she cheated on an income tax return. A cheater might lie, thus biasing an estimate of the true likelihood of someone cheating on his or her tax return.

To circumvent this problem, each person is presented with two questions, one being the object of the survey and the other an innocuous question, such as:

1. Did you ever cheat on your federal income tax return?

2. Did you drink coffee this morning?

One of the questions is chosen at random to answer by the wage earner by flipping a coin; however, which particular question is answered is unknown to the interviewer. In this way, the random response method attempts to elicit an honest response to a sensitive question. Sophisticated statistical methods are then employed to derive an estimate of percentage of "yes" responses to the sensitive question.

No matter what type of sampling design you employ to collect the data for your study, be careful to avoid **selection bias**. Selection bias occurs when some experimental units in the population have less chance of being included in the sample than others. This results in samples that are not representative of the population. Consider an opinion poll that employs either a telephone survey or mail survey. After collecting a random sample of phone numbers or mailing addresses, each person in the sample is contacted via telephone or the mail and a survey conducted. Unfortunately, these types of surveys often suffer from selection bias due to *nonresponse*. Some individuals may not be home when the phone rings, or others may refuse to answer the questions or mail back the questionnaire. As a consequence, no data is obtained for the nonrespondents in the sample. If the nonrespondents and respondents differ greatly on an issue, then **nonresponse bias** exits. For example, those who choose to answer a question on a school board issue may have a vested interest in the outcome of the survey—say, parents with children of school age, schoolteachers whose jobs may be in jeopardy, or citizens whose taxes might be substantially affected. Others with no vested interest may have an opinion on the issue but might not take the time to respond.

> **Selection bias** results when a subset of experimental units in the population has little or no chance of being selected for the sample.

> **Nonresponse bias** is a type of selection bias that results when data on all experimental units in a sample are not obtained.

Finally, even if your sample is representative of the population, the data collected may suffer from **measurement error**. That is, the values of the data (quantitative or qualitative) may be inaccurate. In sample surveys, opinion polls, and so on, measurement error often results from *ambiguous* or *leading questions*. Consider the survey question: "How often did you change the oil in your car last year?" It is not clear whether the researcher is wanting to know how often you personally changed the oil in your car or how often you took your car into a service station to get an oil change. The ambiguous question may lead to inaccurate responses. On the other hand, consider the question: "Does the new health plan offer more comprehensive medical services at less cost than the old one?" The way the question is phrased *leads* the reader to believe that the new plan is better and to a "yes" response—a response that is more desirable to the researcher. A better, more neutral way to phrase the question is: "Which health plan offers more comprehensive medical services at less cost, the old one or the new one?"

> **Measurement error** refers to inaccuracies in the values of the data collected. In surveys, the error may be due to ambiguous or leading questions and the interviewer's effect on the respondent.

We conclude this section with two examples involving actual sampling studies.

Example 1.6

Method of Data Collection—Internet Addiction Study

Problem What percentage of Web users are addicted to the Internet? To find out, a psychologist designed a series of 10 questions based on a widely used set of criteria for gambling addiction and distributed them through the Web site *ABCNews.com*. (A sample question: "Do you use the Internet to escape problems?") A total of 17,251 Web users responded to the questionnaire. If participants answered "yes" to at least half of the questions, they were viewed as addicted. The findings, released at an annual meeting of the American Psychological Association, revealed that 990 respondents, or 5.7%, are addicted to the Internet.

a. Identify the data collection method.

b. Identify the target population.

c. Are the sample data representative of the population?

Solution

a. The data collection method is a survey: 17,251 Internet users responded to the questions posed at the *ABCNews.com* Web site.

b. Since the Web site can be accessed by anyone surfing the Internet, presumably the target population is *all* Internet users.

c. Because the 17,251 respondents clearly make up a subset of the target population, they do form a sample. Whether or not the sample is representative is unclear, since we are given no information on the 17,251 respondents. However, a survey like this one in which the respondents are *self-selected* (i.e., each Internet user who saw the survey chose whether to respond to it) often suffers from *nonresponse bias*. It is possible that many Internet users who chose not to respond (or who never saw the survey) would have answered the questions differently, leading to a higher (or lower) percentage of affirmative answers.

Look Back Any inferences based on survey samples that employ self-selection are suspect due to potential nonresponse bias.

Example 1.7

Method of Data Collection—Study of Susceptibility to Hypnosis

Problem In a classic study, psychologists at the University of Tennessee investigated the susceptibility of people to hypnosis (*Psychological Assessment*, Mar. 1995). In a random sample of 130 undergraduate psychology students at the university, each experienced both traditional hypnosis and computer-assisted hypnosis. Approximately half were randomly assigned to undergo the traditional procedure first, followed by the computer-assisted procedure. The other half were randomly assigned to experience computer-assisted hypnosis first, then traditional hypnosis. Following the hypnosis episodes, all students filled out questionnaires designed to measure a student's susceptibility to hypnosis. The susceptibility scores of the two groups of students were compared.

a. Identify the data collection method.

b. Is the sample data representative of the target population?

Solution

a. Here, the experimental units are the psychology students. Since the researchers controlled which type of hypnosis—traditional or computer assisted—the students experienced first (through random assignment), a designed experiment was used to collect the data.

b. The sample of 130 psychology students was randomly selected from all psychology students at the University of Tennessee. If the target population is *all University of Tennessee psychology students*, it is likely that the sample is representative. However, the researchers warn that the sample data should not be used to make inferences about other, more general populations.

Look Ahead By using randomization in a designed experiment, the researcher is attempting to eliminate different types of bias, including self-selection bias. (We discuss this type of bias in the next section.)

■ **Now Work Exercise 1.27**

Statistics IN Action | **Revisited** | ## Identifying the Data Collection Method and Data Type

In the Pew Internet & American Life Project report, American adults were asked to respond to a variety of questions about social networking site usage. According to the report, the data were obtained through phone interviews in the United States of 1,445 adult Internet users. Consequently, the data collection method is a survey (observational study).

Both quantitative and qualitative data were collected in the survey. For example, the survey question asking adults if they ever use a social networking site is phrased to elicit a "yes" or "no" response. Since the responses produced for this question are categorical in nature, these data are qualitative. However, the question asking for the number of social networking sites used will give meaningful numerical responses, such as 0, 1, 2, and so on. Thus, these data are quantitative.

1.6 The Role of Statistics in Critical Thinking and Ethics

According to H. G. Wells, author of such science-fiction classics as *The War of the Worlds* and *The Time Machine,* "Statistical thinking will one day be as necessary for efficient citizenship as the ability to read and write." Written more than a hundred years ago, Wells's prediction is proving true today.

The growth in data collection associated with scientific phenomena, business operations, and government activities (quality control, statistical auditing, forecasting, etc.) has been remarkable in the past several decades. Every day the media present us with published results of political, economic, and social surveys. In increasing government emphasis on drug and product testing, for example, we see vivid evidence of the need for *quantitative literacy* (i.e., the ability to evaluate data intelligently). Consequently, each of us has to develop a discerning sense—an ability to use rational thought to interpret and understand the meaning of data. Quantitative literacy can help you make intelligent decisions, inferences, and generalizations; that is, it helps you *think critically* using statistics.

BIOGRAPHY H. G. WELLS (1866–1946)

Writer and Novelist

English-born Herbert George Wells published his first novel, *The Time Machine*, in 1895 as a parody of the English class division and as a satirical warning that human progress is inevitable. Although most famous as a science-fiction novelist, Wells was a prolific writer as a journalist, sociologist, historian, and philosopher. Wells's prediction about statistical thinking is just one of a plethora of observations he made about life on this world. Here are a few more of H. G. Wells's more famous quotes:

"Advertising is legalized lying."

"Crude classification and false generalizations are the curse of organized life."

"The crisis of today is the joke of tomorrow."

"Fools make researchers and wise men exploit them."

"The only true measure of success is the ratio between what we might have done and what we might have been on the one hand, and the thing we have made and the things we have made of ourselves on the other." ■

> **Statistical thinking** involves applying rational thought and the science of statistics to critically assess data and inferences. Fundamental to the thought process is that variation exists in populations of data.

To gain some insight into the role statistics plays in critical thinking, we present two examples of some misleading or faulty surveys.

Example **1.8**

Biased Sample— Motorcyclists and Helmets

Problem An article in the *New York Times* considered the question of whether motorcyclists should be required by law to wear helmets. In supporting his argument for no helmets, the editor of a magazine for Harley-Davidson bikers presented the results of one study that claimed "nine states without helmet laws had a lower fatality rate (3.05 deaths per 10,000 motorcycles) than those that mandated helmets (3.38)" and a survey that found "of 2,500 bikers at a rally, 98% of the respondents opposed such laws." Based on this information, do you think it is safer to ride a motorcycle without a helmet? What further statistical information would you like?

Solution You can use "statistical thinking" to help you critically evaluate the study. For example, before you can evaluate the validity of the 98% estimate, you would want to know how the data were collected. If a survey was, in fact, conducted, it's possible that the 2,500 bikers in the sample were not selected at random from the target population of all bikers, but rather were "self-selected." (Remember, they were all attending a rally—a rally, likely, for bikers who oppose the law.) If the respondents were likely to have strong opinions regarding the helmet law (e.g., to strongly oppose the law), the resulting estimate is probably biased high. Also, if the biased sample was intentional, with the sole purpose to mislead the public, the researchers would be guilty of **unethical statistical practice**.

You would also want more information about the study comparing the motorcycle fatality rate of the nine states without a helmet law to those states that mandate helmets. Were the data obtained from a published source? Were all 50 states included in the study, or were only certain states selected? That is, are you seeing sample data or population data? Furthermore, do the helmet laws vary among states? If so, can you really compare the fatality rates?

Look Back Questions such as these led a group of mathematics and statistics teachers attending an American Statistical Association course to discover a scientific and statistically sound study on helmets. The study reported a dramatic *decline* in motorcycle crash deaths after California passed its helmet law.

■ **Now Work Exercise 1.35c**

Example **1.9**

Manipulative or Ambiguous Survey Questions—Howard Stern on Sirius Radio

Problem A few years ago, talk-show host Howard Stern moved his controversial radio program from free, over-the-air (AM/FM) radio to Sirius satellite radio. At the time the move was perceived in the industry to boost satellite radio subscriptions. This led American Media Services, a developer of AM/FM radio properties, to solicit a nationwide random-digit dialing phone survey of 1,008 people. The purpose of the survey was to determine how much interest Americans really have in buying satellite radio service. After providing some background on Howard Stern's controversial radio program, one of the questions asked, "How likely are you to purchase a subscription to satellite radio after Howard Stern's move to Sirius?" The result: Eighty-six percent of the respondents stated that they weren't likely to buy satellite radio because of Stern's move. Consequently, American Media Services concluded that "the Howard Stern Factor is overrated" and that "few Americans expect to purchase satellite radio"—claims that made the headlines of news reports and Weblogs. Do you agree?

Solution First, we need to recognize that American Media Services has a vested interest in the outcome of the survey—the company makes its money from over-the-air broadcast radio stations. Second, although the phone survey was conducted using random-digit dialing, there is no information provided on the response rate. It's possible that nonrespondents (people who were not home or refused to answer the survey questions) tend to be people who use cell phones more than their landline phone, and, consequently, are more likely to use the latest in electronic technology, including satellite radio. Finally, the survey question itself is ambiguous. Do the respondents have negative feelings about satellite radio, Howard Stern, or both? If not for Howard Stern's program, would the respondents be more likely to buy satellite radio? To the critical thinker, it's unclear what the results of the survey imply.

Look Back Examining the survey results from the perspective of satellite radio providers, 14% of the respondents indicated that they would be likely to purchase satellite radio. Projecting the 14% back to the population of all American adults, this figure represents about 50 million people; what is interpreted as "few Americans" by American Media Services could be music to the ears of satellite radio providers.

> ■ **Now Work Exercise 1.33b**

Ethics IN Statistics

Intentionally selecting a biased sample in order to produce misleading statistics is considered *unethical statistical practice*.

As with many statistical studies, both the motorcycle helmet study and the satellite radio study are based on survey data. Most of the problems with these surveys result from the use of *nonrandom* samples. These samples are subject to errors such as *selection bias*, *nonresponse bias* (recall Example 1.6), and *measurement error*. Researchers who are aware of these problems yet continue to use the sample data to make inferences are practicing *unethical statistics*.

Statistics IN Action | **Revisited** | ## Critically Assessing the Ethics of a Statistical Study

The results from the Pew Internet & American Life Project (PIALP) survey on social networking use led the research center to make conclusions such as "Facebook is the dominant social networking platform." The survey included a sample of 1,445 adult Internet users. In order to assess the validity of this inference, a critical thinker would consider several issues.

First, is the sample representative of all U.S. adult Internet users? In the PIALP report, the claim is made that the survey results are based on telephone interviews with a nationally representative sample of Internet users. Two samples of adults were selected. The first sample consisted of landline telephone numbers randomly selected (using a random number generator) from all landline numbers listed in the United States. The second sample consisted of cellular telephone numbers randomly selected from all U.S. cell numbers that were not associated with directory-listed landline numbers. As many as seven attempts were made to contact and interview the adult at each number. These contact calls were staggered over times of day and days of the week in an attempt to reduce nonresponse bias. Clearly, there was no attempt to bias the results of the study by intentionally selecting a sample with a certain point of view. In fact, just the opposite occurred. Consequently, the sample of adults appears to be a truly random sample of all adult Internet users in the United States; therefore, the sample is very likely to be representative of the population of interest.

Second, is the sample of 1,445 adults large enough to draw a reliable inference? Although there are more than 250 million Internet users in the United States, we will learn in later chapters that a random sample of size 1,445 will allow the researchers to draw a reliable inference. In fact, the PIALP report states that the margin of sampling error for the study is about 3%. This means that, with a high level of confidence, anywhere from 70% to 76% (73% $\pm$ 3%) of all the adult Internet users in the United States have visited a social networking site.

Third, is the wording of the survey questions ambiguous or leading? Such questions are often crafted to elicit an answer that supports a specific point of view. For example, consider a hypothetical question about Twitter: "Twitter is becoming more and more popular with affluent people. Have you ever used or desired to use Twitter in order to keep up with your friends?" The question is meant to elicit a "yes" answer and bias the results of the study. If you examine the questions in the actual study (see p. 30), you will conclude that all are very straightforward and clear. Consequently, the survey results appear to be both ethical and valid.

In the remaining chapters of the text, you'll become familiar with the tools essential for building a firm foundation in statistics and statistical thinking.

CHAPTER NOTES

Key Terms

Census 33
Cluster sampling 41
Data 31
Descriptive statistics 31
Designed experiment 39
Experimental (or observational) unit 33
Inference 31
Inferential statistics 31
Measure of reliability 36
Measurement 33
Measurement error 42
Nonresponse bias 42
Observational study 39
Population 33
Published source 39
Qualitative (or categorical) data 38

Quantitative data 37
Random number generator 40
Randomized response sampling 41
Reliability (of an inference) 36
Representative sample 40
Sample 34
Selection bias 42
Simple random sample 40
Statistical inference 34
Statistical thinking 44
Statistics 30
Stratified random sample 41
Survey 40
Systematic sample 41
Unethical statistical practice 45
Variable 33

Key Ideas

Types of Statistical Applications

Descriptive
1. Identify **population** or **sample** (collection of **experimental units**)
2. Identify **variable(s)**
3. Collect **data**
4. **Describe** data

Inferential
1. Identify **population** (collection of *all* **experimental units**)
2. Identify **variable(s)**
3. Collect **sample** data (*subset* of population)
4. **Inference** about population based on sample
5. **Measure of reliability** of inference

Types of Data
1. **Quantitative** (numerical in nature)
2. **Qualitative** (categorical in nature)

Data Collection Methods
1. **Published source**
2. **Observational** (e.g., survey)
3. **Designed experiment**

Types of Random Samples
1. **Simple random sample**
2. **Stratified random sample**
3. **Cluster sample**
4. **Systematic sample**
5. **Random response sample**

Problems with Nonrandom Samples
1. **Selection bias**
2. **Nonresponse bias**
3. **Measurement error**

Exercises 1.1–1.36

Understanding the Principles

1.1 What is the primary job of a statistician?

1.2 What are the broad types of statistical applications?

1.3 What is sampling?

1.4 Distinguish between a population, an observational unit, and a sample?

1.5 Explain the role of variables in statistical studies of a population.

1.6 Explain the role of reliability in statistical inference.

1.7 List the four elements of descriptive statistical problems.

1.8 What are the different ways of collecting data for a statistical analysis?

1.9 What is the difference between stratified random sampling and cluster sampling?

1.10 What are the two types of bias that may be associated with a sampling design?

1.11 Suppose you are given a dataset, measured in inches, of the heights of students in a class. You decide to code the data as A, if height ≥ 70 inches; B, if 65 ≤ height < 70; C, if 60 ≤ height < 65; and D, if height < 60 inches. Explain whether the raw data is quantitative or qualitative. After the data is coded as A, B, C, or D are they qualitative or quantitative? Explain your answers.

D Applet Exercise 1.1

The *Random Numbers* applet generates a list of *n* random numbers from 1 to *N*, where *n* is the size of the sample and *N* is the size of the population. The list generated often contains repetitions of one or more numbers.

a. Using the applet *Random Numbers*, enter 1 for the minimum value, 10 for the maximum value, and 10 for the number of samples. Then click on *Sample*. Look at the results, and list any numbers that are repeated and the number of times each of these numbers occurs.

b. Repeat part **a**, changing the maximum value to 20 and keeping the size of the sample fixed at 10. If you still have repetitions, repeat the process, increasing the maximum value by 10 each time but keeping the size of the sample fixed. What is the smallest maximum value for which you had no repetitions?

c. Describe the relationship between the population size (maximum value) and the number of repetitions in the list of random numbers as the population size increases and the sample size remains the same. What can you conclude about using a random number generator to choose a relatively small sample from a large population?

Ⓓ Applet Exercise 1.2

The *Random Numbers* applet can be used to select a random sample from a population, but can it be used to simulate data? In parts **a** and **b**, you will use the applet to create data sets. Then you will explore whether those data sets are realistic.

    a. Consider the number of customers waiting in line to order at a fast-food outlet. Use the *Random Numbers* applet to simulate this data set by setting the minimum value equal to 0, the maximum value equal to 99, and the sample size equal to 30. Explain what the numbers in the list produced by the applet represent in the context of the problem. Do the numbers produced by the applet seem reasonable? Explain.

    b. Use the *Random Numbers* applet to simulate grades on a statistics test by setting the minimum value equal to 0, the maximum value equal to 100, and the sample size equal to 30. Explain what the numbers in the list produced by the applet represent in this context. Do the numbers produced by the applet seem reasonable? Explain.

    c. Referring to parts **a** and **b**, why do the randomly generated data seem more reasonable in one situation than in the other? Comment on the usefulness of using a random-number generator to produce data.

Applying the Concepts—Basic

1.12 **College application.** Colleges and universities are requiring an increasing amount of information about applicants before making acceptance and financial aid decisions. Classify each of the following types of data required on a college application as quantitative or qualitative.
    a. High school GPA
    b. High school class rank
    c. Applicant's score on the SAT or ACT
    d. Gender of applicant
    e. Parents' income
    f. Age of applicant

1.13 **STEM experiences for girls.** The National Science Foundation (NSF) promotes girls' participation in informal science, technology, engineering, and mathematics (STEM) programs. What has been the impact of these informal STEM experiences? This was the question of interest in the published study *Cascading Influences: Long-Term Impacts of Informal STEM Experiences for Girls* (Mar. 2013). A sample of 159 young women who recently participated in a STEM program were recruited to complete an online survey. Of these, only 27% felt that participation in the STEM program increased their interest in science.
    a. Identify the population of interest to the researchers.
    b. Identify the sample.
    c. Use the information in the study to make an inference about the relevant population.

1.14 **Drafting NFL quarterbacks.** The *Journal of Productivity Analysis* (Vol. 35, 2011) published a study of how successful National Football League (NFL) teams are in drafting productive quarterbacks. Data were collected for all 331 quarterbacks drafted over a 38-year period. Several variables were measured for each QB, including draft position (one of the top 10 players picked, selection between picks 11–50, or selected after pick 50), NFL winning ratio (percentage of games won), and QB production score (higher scores indicate more productive QBs). The researchers discovered that draft position was only weakly related to a quarterback's performance in the NFL. They concluded that "quarterbacks taken higher [in the draft] do not appear to perform any better."
    a. What is the experimental unit for this study?
    b. Identify the type (quantitative or qualitative) of each variable measured.
    c. Is the study an application of descriptive or inferential statistics? Explain.

1.15 **Ground motion of earthquakes.** In the *Journal of Earthquake Engineering* (Nov. 2004), a team of civil and environmental engineers studied the ground motion characteristics of 15 earthquakes that occurred around the world since 1940. Three (of many) variables measured on each earthquake were the type of ground motion (short, long, or forward directive), the magnitude of the earthquake (on the Richter scale), and peak ground acceleration (feet per second). One of the goals of the study was to estimate the inelastic spectra of any ground motion cycle.
    a. Identify the experimental units for this study.
    b. Do the data for the 15 earthquakes represent a population or a sample? Explain.
    c. Define the variables measured and classify them as quantitative or qualitative.

1.16 **Sprint speed training.** *The Sport Journal* (Winter 2004) reported on a study of a speed-training program for high school football players. Each participant was timed in a 40-yard sprint both before and after training. The researchers measured two variables: (1) the difference between the before and after sprint times (in seconds), and (2) the category of improvement ("improved," "no change," and "worse") for each player.
    a. Identify the type (quantitative or qualitative) of each variable measured.
    b. A total of 14 high school football players participated in the speed-training program. Does the data set collected represent a population or a sample? Explain.

1.17 **Corrosion prevention of buried steel structures.** Engineers have designed tests on underground steel structures that measure the potential for corrosion. In *Materials Performance* (Mar. 2013), two tests for steel corrosion—called "instant-off" and "instant-on" potential—were compared. The tests were applied to buried piping at a petrochemical plant. Both the "instant-off" and "instant-on" corrosion measurements were made at each of 19 different randomly selected pipe locations. One objective of the study is to determine if one test is more desirable (i.e., can more accurately predict the potential for corrosion) than the other when applied to buried steel piping.
    a. What are the experimental units for this study?
    b. Describe the sample.
    c. Describe the population.
    d. Is this an example of descriptive or inferential statistics?

1.18 **Age of Patients.** Consider the set of all patients enrolled in your local diabetes centre during the past 30 days. Suppose you are interested in learning about the average age of this group.
  a. Define the population and the variable of interest.
  b. Is the variable qualitative or quantitative?
  c. Suppose you obtain the age of every patient who visited the centre. Would this data represent a census or a sample?
  d. Suppose you obtain the age of the first patient visiting the centre on each day for the last 30 days. Would this data represent a census or a sample?
  e. If you obtain the ages of all the patients who visited the centre in the last 30 days, how much reliability does your findings have as an "estimate" of the average age?
  f. How can you select a random sample of size 30 from the population?

1.19 **Medicinal value of plants.** Sea buckthorn (*Hippophae*), a plant that typically grows at high altitudes in Europe and Asia, has been found to have medicinal value. The medicinal properties of berries collected from sea buckthorn were investigated in *Academia Journal of Medicinal Plants* (Aug. 2013). The following variables were measured for each plant sampled. Identify each as producing quantitative or qualitative data.
  a. Species of sea buckthorn (*H. rhamnoides, H. gyantsensis, H. neurocarpa, H. tibetana,* or *H. salicifolia*)
  b. Altitude of collection location (meters)
  c. Total flavonoid content in berries (milligrams per gram)

1.20 **Extinct birds.** Biologists at the University of California (Riverside) are studying the patterns of extinction in the New Zealand bird population. (*Evolutionary Ecology Research*, July 2003.) At the time of the Maori colonization of New Zealand (prior to European contact), the following variables were measured for each bird species:
  a. Flight capability (volant or flightless)
  b. Type of habitat (aquatic, ground terrestrial, or aerial terrestrial)
  c. Nesting site (ground, cavity within ground, tree, cavity above ground)
  d. Nest density (high or low)
  e. Diet (fish, vertebrates, vegetables, or invertebrates)
  f. Body mass (grams)
  g. Egg length (millimeters)
  h. Extinct status (extinct, absent from island, present)
  Identify each variable as quantitative or qualitative.

1.21 **Study of quality of drinking water.** *Disasters* (Vol. 28, 2004) published a study of the effects of a tropical cyclone on the quality of drinking water on a remote Pacific island. Water samples (size 500 milliliters) were collected approximately four weeks after Cyclone Ami hit the island. The following variables were recorded for each water sample:
  a. Town where sample was collected
  b. Type of water supply (river intake, stream, or borehole)
  c. Acidic level (pH scale, 1 to 14)
  d. Turbidity level (nephalometric turbidity units = NTUs)
  e. Temperature (degrees centigrade)
  f. Number of fecal coliforms per 100 milliliters
  g. Free-chlorine residual (milligrams per liter)
  h. Presence of hydrogen sulphide (yes or no)
  Identify each variable as quantitative or qualitative.

Applying the Concepts–Intermediate

1.22 **Annual survey of computer crimes.** The Computer Security Institute (CSI) conducts an annual survey of computer crime at U.S. businesses. CSI sends survey questionnaires to computer security personnel at all U.S. corporations and government agencies. The latest CSI survey was sent by post or email to 5,412 firms, and 351 organizations responded. Forty-one percent of the respondents admitted unauthorized use of computer systems at their firms during the year. (*CSI Computer Crime and Security Survey, 2010/2011.*)
  a. Identify the population of interest to CSI.
  b. Identify the data collection method used by CSI. Are there any potential biases in the method used?
  c. Describe the variable measured in the CSI survey. Is it quantitative or qualitative?
  d. What inference can be made from the study result?

1.23 **Treasury deficit prior to the Civil War.** In *Civil War History* (June 2009), historian Jane Flaherty researched the condition of the U.S. Treasury on the eve of the Civil War in 1861. Between 1854 and 1857 (under President Franklin Pierce), the annual surplus/deficit was +18.8, +6.7, +5.3, and +1.3 million dollars, respectively. In contrast, between 1858 and 1861 (under President James Buchanan), the annual surplus/deficit was −27.3, −16.2, −7.2, and −25.2 million dollars, respectively. Flaherty used these data to aid in portraying the exhausted condition of the U.S. Treasury when Abraham Lincoln took office in 1861. Does this study represent a descriptive or inferential statistical study? Explain.

1.24 **Library Database.** Your college library maintains the records of books borrowed. Some of the many variables stored in the database include the titles of the books, their subject categories (Mathematics, Statistics, Computer Science), type classification of each (textbook, reference, or manuals), number of times a book has been borrowed in the last 12 months, and the original prices of the books. Which of the variables measured are qualitative? Which are quantitative?

1.25 **National Airport Authority.** The National Airport Authority (NAA) of a country stores the relevant data for all the airports in a country to perform periodic checks on their maintenance. Data from each airport are compiled into the airport inventory. Some of the variables maintained by the NAA for each airport are listed below. Classify each variable as quantitative or qualitative.
  a. Category of airport (big, medium, small)
  b. Average daily traffic (number of flights)
  c. Number of terminal buildings
  d. Number of runways
  e. Length of a runway (meters)
  f. Closed at night (yes or no)
  g. Condition of access road (good, fair, poor)

1.26 **Corporate sustainability.** *Corporate sustainability* refers to business practices designed around social and environmental considerations (e.g., "going green"). *Business and Society* (Mar. 2011) published a paper on how firm size and firm type affect sustainability behaviors. A survey was sent to approximately 23,500 senior managers at CPA firms, of which 1,293 senior managers responded. (Note: It is not clear how the 23,500 senior managers were selected.) Due to missing data (incomplete survey answers), only 992 surveys were analyzed. These data were used to infer whether larger firms are more likely to report sustainability policies than smaller firms and whether public firms are more likely to report sustainability policies than private firms.

a. Identify the population of interest to the researchers.

b. What method was used to collect the sample data?

c. Comment on the representativeness of the sample.

d. How will your answer to part **c** affect the validity of the inferences drawn from the study?

1.27 **CT scanning for lung cancer.** A new type of screening for lung cancer, computed tomography (CT), has been developed. Medical researchers believe that CT scans are more sensitive than regular X-rays in pinpointing small tumors. The H. Lee Moffitt Cancer Center at the University of South Florida conducted a clinical trial of 50,000 smokers nationwide to compare the effectiveness of CT scans with X-rays for detecting lung cancer. (*Todays' Tomorrows*, Fall 2002.) Each participating smoker was randomly assigned to one of two screening methods, CT or chest X-ray, and his or her progress tracked over time. The age at which the scanning method first detects a tumor is the variable of interest.

a. Identify the data collection method used by the cancer researchers.

b. Identify the experimental units of the study.

c. Identify the type (quantitative or qualitative) of the variable measured.

d. Identify the population and sample.

e. What is the inference that will ultimately be drawn from the clinical trial?

1.28 **Zillow.com estimates of home values.** Zillow.com is a real estate Web site that provides free estimates of the market value of homes. A trio of University of Texas at San Antonio professors compared Zillow estimates to actual sale prices of homes and published their results in *The Appraisal Journal* (Winter 2010). The analysis was based on data collected for 2,045 single-family residential properties in Arlington, Texas, that sold during a recent year. Sale price and Zillow estimated value (in dollars) were measured for each property. Zillow claims that this market has one of its highest accuracy ratings. However, the research revealed that Zillow overestimated the market value by more than 10% for nearly half of the properties.

a. What is the experimental unit for this study?

b. Describe the variables measured in the study. Do these variables produce quantitative or qualitative data?

c. Give a scenario where the 2,045 properties represent a population.

d. If the 2,045 properties represent a representative sample from a population, describe the population.

e. Suppose the relevant population is all single-family residential properties in the United States. Do you believe the 2,045 properties are representative of this population? Explain.

1.29 **Massage vs. rest in boxing.** Does a massage enable the muscles of tired athletes to recover from exertion faster than usual? To answer this question, researchers recruited eight amateur boxers to participate in an experiment. (*British Journal of Sports Medicine*, Apr. 2000.) After a 10-minute workout in which each boxer threw 400 punches, half the boxers were given a 20-minute massage and half just rested for 20 minutes. Before they returned to the ring for a second workout, the heart rate (beats per minute) and blood lactate level (micromoles) were recorded for each boxer. The researchers found no difference in the means of the two groups of boxers for either variable.

a. Identify the data collection method used by the researchers.

b. Identify the experimental units of the study.

c. Identify the variables measured and their type (quantitative or qualitative).

d. What is the inference drawn from the analysis?

e. Comment on whether this inference can be made about all athletes.

1.30 **Insomnia and education.** Is insomnia related to education status? Researchers at the Universities of Memphis, Alabama at Birmingham, and Tennessee investigated this question in the *Journal of Abnormal Psychology* (Feb. 2005). Adults living in Tennessee were selected to participate in the study, which used a random-digit telephone dialing procedure. Two of the many variables measured for each of the 575 study participants were number of years of education and insomnia status (normal sleeper or chronic insomniac). The researchers discovered that the fewer the years of education, the more likely the person was to have chronic insomnia.

a. Identify the population and sample of interest to the researchers.

b. Identify the data collection method. Are there any potential biases in the method used?

c. Describe the variables measured in the study as quantitative or qualitative.

d. What inference did the researchers make?

1.31 **Guilt in decision making.** The effect of the emotion of guilt on how a decision maker focuses on a problem was investigated in the *Journal of Behavioral Decision Making* (January 2007). A total of 155 volunteer students participated in the experiement, where each was randomly assigned to one of three emotional states (guilt, anger, or neutral) through a reading/writing task. Immediately after the task, the students were presented with a decision problem (e.g., whether or not to spend money on repairing a very old car). The researchers found that a higher proportion of students in the guilty-state group chose not to repair the car than those in the neutral-state and anger-state groups.

a. Identify the population, sample, and variables measured for this study.

b. Identify the data collection method used.

c. What inference was made by the researcher?

d. In later chapters you will learn that the reliability of an inference is related to the size of the sample used. In addition to sample size, what factors might affect the reliability of the inference drawn in this study?

Applying the Concepts—Advanced

1.32 **Can money spent on gifts buy love?** Is the gift you purchased for that special someone really appreciated? This was the question investigated in the *Journal of Experimental Social Psychology* (Vol. 45, 2009). The researchers examined the link between engagement ring price (dollars) and level of appreciation of the recipient (measured on a 7-point scale where 1 = "not at all" and 7 = "to a great extent"). Participants for the study were those who used a popular Web site for engaged couples. The Web site's directory was searched for those with "average" American names (e.g., "John Smith," "Sara Jones"). These individuals were then invited to participate in an online survey in exchange for a $10 gift certificate. Of the respondents, those who paid really high or really low prices for the ring were excluded, leaving a sample size of 33 respondents.

a. Identify the experimental units for this study.

b. What are the variables of interest? Are they quantitative or qualitative in nature?

c. Describe the population of interest.

d. Do you believe the sample of 33 respondents is representative of the population? Explain.

e. In a second, designed study, the researchers investigated whether the link between gift price and level of appreciation was stronger for birthday gift givers than for birthday gift receivers. The participants were randomly assigned to play the role of gift-giver or gift-receiver. Assume that the sample consists of 50 individuals. Use a random number generator to randomly assign 25 individuals to play the gift-receiver role and 25 to play the gift-giver role.

1.33 **Dating and disclosure.** As an adolescent, did you voluntarily disclose information about dating and romantic relationships to your parents? This was the research question of interest in the *Journal of Adolescence* (Apr. 2010). A sample of 222 high school students was recruited to participate in the study. Some of the many variables measured on each student were age (years), gender, dating experience (number of dates), and the extent to which the student was willing to tell his/her parent (without being asked) about a dating issue (e.g., how late the daters stayed out). The responses for the last variable were categorized as "never tell," "rarely tell," "sometimes tell," "almost always tell," and "always tell."

a. Identify the data type for each variable.

NW **b.** The study was unclear on exactly how the sample of students was selected, stating only that "participants were recruited from health or government classes in a primarily European American middle-class school district." Based on this information, what are the potential caveats to using the sample to make inferences on dating and disclosure to parents for all high school students?

1.34 **Jersey City drug market study.** An article in the *Journal of Quantitative Criminology* (Mar. 2014) discussed several data collection methods for testing the efficacy of a new program designed to reduce drug-related crimes in Jersey City (N.J.). The sample consisted of 56 "hot spots"—areas of Jersey City with a high crime rate but not necessarily a high drug crime rate. In Method 1, the new program will be implemented in half of the 56 hot spots (selected at

random); the other half will not implement the program (the "no treatment" group). In Method 2, the 56 hot spots will be divided into four groups based on the level of drug crimes—very high drug crime rate (10 hot spots), high rate (8 hot spots), moderate rate (26 hot spots), and low rate (12 hot spots). In each group, half the hot spots will be randomly assigned the new crime prevention program and the other half no treatment. For both methods, after several weeks the crime rates in the hot spots where the program was implemented will be compared with those with no treatment.

a. Identify the data collection method for Method 1.

b. Identify the data collection method for Method 2.

c. Is this an application of descriptive or inferential statistics? Explain.

d. Given that there is a great deal of variation in the drug crime rates at the 56 hot spots, which data collection method do you recommend? Explain.

Critical Thinking Challenges

1.35 **Your choice for a mom.** After running stories on current First Lady Michelle Obama and former vice presidential candidate Sarah Palin on consecutive weeks, *USA Weekend* magazine asked its readers on Mother's Day, "Who would you rather have as your mom, Sarah Palin or Michelle Obama?" Readers were asked to vote online at www.usaweekend.com. Based on over 34,000 votes cast, the results were: Obama—71%, Palin—29%. (*USA Weekend Magazine* press release, May 11, 2010.)

a. What type of data collection method is used in this study?

b. Is the data collected quantitative or qualitative? Explain.

NW **c.** Discuss the validity of the study results. What are the potential problems with running a poll where voting is done online?

1.36 **Survey exposés.** Many surveys are conducted on different social, political, and lifestyle issues in order to predict events, identify choices, etc. However, the results obtained from these surveys are often incorrect. It may be that the surveys are conducted by special interest groups with specific objectives in mind, or it may be that the survey data were not collected properly. Here are a few such survey results. (Actual facts are given in parentheses.)

Delhi assembly poll 2015: An ABP Nielsen survey predicted about 39 seats for the Aam Aadmi Party, and 28 seats for the Bharatiya Janata Party. (*Fact:* The AAP got 67 seats and the BJP, 3 seats.)

Colgate Recommendations by Dentists: Colgate claimed that 80% of all dentists recommended their brand, making it seem that only 20% of dentists recommended some other brand. (*Fact:* The dentists were allowed to recommend more than one brand. Hence other brands were also equally recommended.)

New Seven Wonders of the World: The seven new wonders were announced on July 7, 2007 based on more than 100 million global votes that were submitted through the internet and over the telephone over several years. They were the Pyramids of Giza (honorary), Great Wall of China (China), Petra (Jordan), Colosseum (Italy),

Chichén Itzá (Mexico), Machu Picchu (Peru), Taj Mahal (India), and Christ the Redeemer (Brazil). (*Fact:* The survey did not include traditional wonders such as the Leaning Tower of Pisa.)

Global Dietary Survey: A survey commissioned by The Global Nutrition and Health Alliance suggests that 72% people in Germany and UK have good dietary habits. (*Fact:* There remains a huge amount of confusion about what our food actually contains, which foods are good sources of specific nutrients, and why we need specific nutrients.)

a. Refer to the Delhi Assembly Poll study for the prediction of election results in the 2015 elections. What data collection method would you recommend?

b. Refer to the Colgate Recommendations by Dentists claim. Discuss why it is unethical to report the results as stated.

c. Refer to New Seven Wonders of the World study. What do you think is wrong with the survey with such a huge set of respondents?

d. Refer to the Global Dietary Survey. What do you think is the cause behind such misinformation?

Activity **Data in the News**

Scan your daily newspaper or weekly news magazine, or search the Internet for articles that contain data. The data might be a summary of the results of a public opinion poll, the results of a vote by the U.S. Senate, or a list of crime rates, birth or death rates, etc. For each article you find, answer the following questions:

a. Do the data constitute a sample or an entire population? If a sample has been taken, clearly identify both the sample and the population; otherwise, identify the population.

b. What type of data (quantitative or qualitative) has been collected?

c. What is the source of the data?

d. If a sample has been observed, is it likely to be representative of the population?

e. If a sample has been observed, does the article present an explicit (or implied) inference about the population of interest? If so, state the inference made in the article.

f. If an inference has been made, has a measure of reliability been included? What is it?

g. Use your answers to questions d–f to critically evaluate the article. Comment on the ethics of the study.

References

Careers in Statistics, American Statistical Association, 2013. (*www.amstat.org*)

Cochran, W. G. *Sampling Techniques*, 3rd ed. New York: Wiley, 1977.

Deming, W. E. *Sample Design in Business Research.* New York: Wiley, 1963.

Ethical Guidelines for Statistical Practice. American Statistical Association, 1999. (*www.amstat.org*)

Hansen, M. H., Hurwitz, W. N., and Madow, W. G. *Sample Survey Methods and Theory*, Vol. 1. New York: Wiley, 1953.

Hoerl, R., and Snee, R. *Statistical Thinking: Improving Business Performance.* Boston: Duxbury, 2002.

Huff, D. *How to Lie with Statistics.* New York: Norton, 1982 (paperback 1993).

Kirk, R. E., ed. *Statistical Issues: A Reader for the Behavioral Sciences.* Monterey, CA: Brooks/Cole, 1972.

Kish, L. *Survey Sampling.* New York: Wiley, 1965 (paperback: 1995).

Peck, R., Casella, G., Cobb, G., Hoerl, R., Nolan, D., Starbuck, R., and Stern, H. *Statisics: A Guide to the Unknown*, 4th ed. Monterey, CA: Brooks/Cole, 2006.

Scheaffer, R., Mendenhall, W., and Ott, R. L. *Elementary Survey Sampling*, 6th ed. Boston: Duxbury, 2005.

What Is a Survey? American Statistical Association (F. Scheuren, editor), 2nd. ed. 2005. (*www.amstat.org*)

When you start a MINITAB session, you will see a screen similar to Figure 1.M.1. The bottom portion of the screen is an empty spreadsheet—called a MINITAB worksheet—with columns representing variables and rows representing observations (or cases). The very top of the screen is the MINITAB main menu bar, with buttons for the different functions and procedures available in MINITAB. Once you have entered data into the spreadsheet, you can analyze the data by clicking the appropriate menu buttons. The results will appear in the Session window.

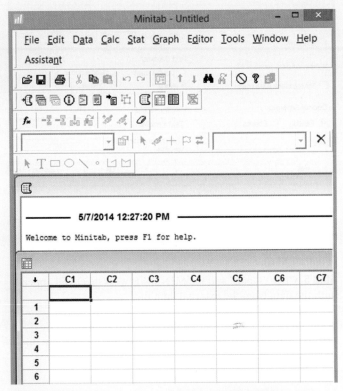

Figure 1.M.1 Initial screen viewed by the MINITAB user

Entering Data

Create a MINITAB data file by entering data directly into the worksheet. Figure 1.M.2 shows data entered for a variable called "RATIO." Name the variables (columns) by typing in the name of each variable in the box below the column number.

↓	C1	C2
	RATIO	
1	10.73	
2	8.89	
3	9.07	

Figure 1.M.2 Data entered into the MINITAB worksheet

Opening a MINITAB Data File

If the data have been previously saved as a MINITAB (.mtw or .mtp) file, access the data as follows.

Step 1   Click the "File" button on the menu bar, and then click "Open Worksheet," as shown in Figure 1.M.3. A dialog box similar to Figure 1.M.4 will appear.

Figure 1.M.3 Options for opening a MINITAB data file

Figure 1.M.4 Selecting the MINITAB data file to open

Step 2 Specify the location (folder) that contains the data, click on the MINITAB data file, then click "Open" (see Figure 1.M.4). The data will appear in the MINITAB worksheet, as shown in Figure 1.M.5.

↓	C1-T	C2	C3-T	C4	C5	C6
	RIVER	MILE	SPECIES	LENGTH	WEIGHT	DDT
1	FCM	5	CHANNELCATFISH	42.5	732	10.00
2	FCM	5	CHANNELCATFISH	44.0	795	16.00
3	FCM	5	CHANNELCATFISH	41.5	547	23.00
4	FCM	5	CHANNELCATFISH	39.0	465	21.00
5	FCM	5	CHANNELCATFISH	50.5	1252	50.00
6	FCM	5	CHANNELCATFISH	52.0	1255	150.00
7	LCM	3	CHANNELCATFISH	40.5	741	28.00
8	LCM	3	CHANNELCATFISH	48.0	1151	7.70
9	LCM	3	CHANNELCATFISH	48.0	1186	2.00
10	LCM	3	CHANNELCATFISH	43.5	754	19.00
11	LCM	3	CHANNELCATFISH	40.5	679	16.00
12	LCM	3	CHANNELCATFISH	47.5	985	5.40
13	SCM	1	CHANNELCATFISH	44.5	1133	2.60
14	SCM	1	CHANNELCATFISH	46.0	1139	3.10
15	SCM	1	CHANNELCATFISH	48.0	1186	3.50
16	SCM	1	CHANNELCATFISH	45.0	984	9.10
17	SCM	1	CHANNELCATFISH	43.0	965	7.80
18	SCM	1	CHANNELCATFISH	45.0	1084	4.10
19	TRM	275	CHANNELCATFISH	48.0	986	8.40
20	TRM	275	CHANNELCATFISH	45.0	1023	15.00

FISHDDT.MTP ***

Figure 1.M.5 The MINITAB worksheet showing the opened MINITAB data file

Accessing External Data from a File

Step 1 Click the "File" button on the menu bar, and then click "Open Worksheet" as shown in Figure 1.M.3. A dialog box similar to Figure 1.M.6 will appear.

Open Worksheet

Look in: Exercises

Name: AMMONIA, ASWELLS, ATLAS, ATLAS2, ATTIMES, BEANIE, Bible, BLACKBREAM, BODYCLUE, BOILERS, BONES

Recent Places, Desktop, Terry, Computer, Network

File name: BONES

Files of type: Data (*.dat)

Open / Cancel / Help

Look in Minitab Sample Data folder

Description... Options... Preview... ○ Merge ⊙ Open

Figure 1.M.6 Selecting the external data file in MINITAB

Step 2 Specify the disk drive and folder that contains the external data file and the file type (e.g., .dat), and then click on the file name, as shown in Figure 1.M.6.

Step 3 If the data set contains qualitative data or data with special characters, click on the "Options" button as shown in Figure 1.M.6. The Options dialog box, shown in Figure 1.M.7, will appear.

Open Worksheet - Options

Variable Names: ⊙ None ○ Automatic ○ Use row: ___

First Row of Data: ⊙ Automatic ○ Use row: ___

Number of data rows: 12

☐ Ignore blank data rows

Decimal Separator: ⊙ Period ○ Comma

Field Definition: ⊙ Free format ○ Single character separator: ⊙ Tab ○ Comma ○ Semicolon ○ Space ○ Period ○ None ○ Custom: ___

Text Delimiter: ○ Double quote ○ Single quote ⊙ None ○ Custom: ___

Convert Missing: From / To

Text Variable: ___ ___

Numeric or Date/Time Variable: ___ *

Help / OK / Cancel

Figure 1.M.7 Selecting the MINITAB data input options

Step 4 Specify the appropriate options for the data set, and then click "OK" to return to the "Open Worksheet" dialog box (Figure 1.M.6).

Step 5 Click "Open" and the MINITAB worksheet will appear with the data from the external data file.

Reminder: The variables (columns) can be named by typing in the name of each variable in the box under the column number.

Listing (Printing) Data

Step 1 Click on the "Data" button on the MINITAB main menu bar, and then click on "Display Data." (See Figure 1.M.8.) The resulting menu, or dialog box, appears as in Figure 1.M.9.

Step 2 Enter the names of the variables you want to print in the "Columns, constants, and matrices to display" box (you can do this by simply double clicking on the variables), and then click "OK." The printout will show up on your MINITAB session screen.

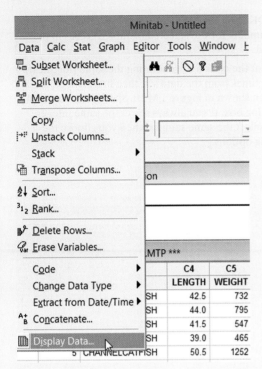

Figure 1.M.8 MINITAB options for displaying data

Figure 1.M.10 MINITAB menu options for sampling from a data set

Figure 1.M.9 MINITAB Display Data dialog box

Figure 1.M.11 MINITAB Menu Options for selecting a random sample from worksheet columns

MINITAB: Generating a Random Sample

Step 1 Click on the "Calc" button on the MINITAB menu bar and then click on "Random Data," and finally, click on "Sample From Columns," as shown in Figure 1.M.10. The resulting dialog box appears as shown in Figure 1.M.11.

Step 2 Specify the sample size (i.e., number of rows), the variable(s) to be sampled, and the column(s) where you want to save the sample.

Step 3 Click "OK" and the MINITAB worksheet will reappear with the values of the variable for the selected (sampled) cases in the column specified.

In MINITAB, you can also generate a sample of case numbers (e.g., a sample of cases from a population of cases numbered from 1 to 500).

Step 1 From the MINITAB menu, click on the "Calc" button and then click on "Random Data," and finally, click on the "Uniform" option (see Figure 1.M.10).

Step 2 In the resulting dialog box (shown in Figure 1.M.12), specify the number of cases (rows, i.e., the sample size), and the column where the case numbers selected will be stored.

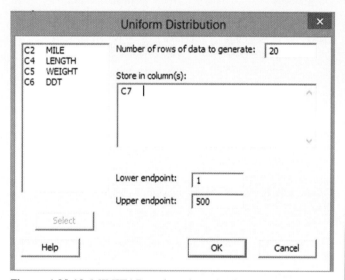

Figure 1.M.12 MINITAB options for selecting a random sample of case numbers

Step 3 Click "OK" and the MINITAB worksheet will reappear with the case numbers for the selected (sampled) cases in the column specified.

[*Note:* If you want the option of generating the same (identical) sample multiple times from the data set, then first click on the "Set Base" option shown in Figure 1.M.10. Specify an integer in the resulting dialog box. If you always select the same integer, MINITAB will select the same sample when you choose the random sampling options.]

2 Methods for Describing Sets of Data

CONTENTS

Where We've Been

- Examined the difference between inferential and descriptive statistics
- Described the key elements of a statistical problem
- Learned about the two types of data: quantitative and qualitative
- Discussed the role of statistical thinking in managerial decision making

Where We're Going

- Describe qualitative data with graphs (2.1)
- Describe quantitative data with graphs (2.2)
- Describe quantitative data with numerical measures (2.3–2.7)
- Describe the relationship between two quantitative variables with a graph (2.8)
- Detect descriptive methods that distort the truth (2.9)

Statistics IN Action   Body Image Dissatisfaction: Real or Imagined?

"Everything has beauty, but not everyone sees it" —ancient Chinese sage Confucius

"Body dissatisfaction can produce extreme body-shaping behaviors, such as eating disorders. Women and girls can't help being exposed to ultra-thin models in advertising whose body size is unrealistic and unhealthy. There is good evidence already that exposure to these unhealthy models leads a large proportion of women to feel dissatisfied with their own bodies." —Helga Dittmar, University of Sussex researcher

"Action figures (like G.I. Joe) present subtle messages of unrealistic role models of well-sculpted, heavily muscled, 'perfect' bodies that little boys see as their role models." —Sondra Kronberg, director of Eating Disorder Associates Treatment & Referral Centers

"By age 13, 53% of American girls are unhappy with their bodies; that figure grows to 78% by the time girls reach 17. In another study on fifth graders, 10-year-old girls and boys told researchers they were dissatisfied with their own bodies after watching a music video by Britney Spears or a clip from the TV show Friends. And adolescent girls who viewed commercials depicting unrealistically thin models felt 'less confident, more angry, and more dissatisfied with their weight and appearance.'" —statistics posted by the National Institute on Media and the Family

Are you dissatisfied with your physical appearance? Do you have a negative image of your own body? In today's media-driven society, many of us would answer *yes* to these questions (as the statistics in the previous quote show). Much research has been conducted on the body images of normal adolescents and adults. However, there is a lack of information on how patients with a body image disorder evaluate their own appearance, health, and fitness. To fill this gap, researchers from the Department of Psychiatry and Human Behavior at Brown University conducted and published a study in *Body Image: An International Journal of Research*, January 2010.

Data were collected on 92 patients diagnosed with *body dysmorphic disorder* (BDD). This disorder "is characterized by a distressing or impairing preoccupation with an imagined or slight defect in appearance that causes clinically significant distress or functional impairment." (Patients were also diagnosed with additional mental disorders, such as major depression or social phobia, called a comorbid disorder.) Each patient completed the Multidimensional Body-Self Relations Questionnaire (MBSRQ). The questionnaire elicits responses that assess how satisfied one is with his/her appearance, health, fitness, and weight. In this Statistics in Action, our focus is on the appearance evaluations (e.g., "How satisfied are you with your physical attractiveness and looks?") of the BDD patients. The scores for each of seven appearance items (questions) were recorded on 5-point Likert scales, where the possible responses are 1 = definitely dissatisfied, 2 = somewhat dissatisfied, 3 = neutral, 4 = somewhat satisfied, and 5 = definitely satisfied. These scores were summed and a total appearance score (ranging from 7 to 35 points) was recorded for each patient.

The data for the study (simulated on the basis of summary statistics presented in the journal article) are provided in the **BDD** file. For each of the 92 patients in the experiment, the following variables were measured:

1. *Total Appearance Score* (points)
2. *Gender* (M or F)
3. *Comorbid Disorder* (Major Depression, Social Phobia, Obsessive Compulsive, or Anorexia/Bulimia Nervosa)
4. *Dissatisfied with Looks* (Yes or No)

For this application, we are interested in evaluating the data collected on the BDD patients. Specifically, we want to know if (1) BDD females tend to be more dissatisfied with their looks than BDD males, (2) certain comorbid disorders lead to a higher level of dissatisfaction with body appearance, and (3) BDD patients have lower appearance scores than normal people. We apply the graphical and numerical descriptive techniques of this chapter to the BDD data to answer these questions in the following Statistics in Action Revisited sections:

Statistics IN Action   Revisited

- Interpreting Pie Charts for the Body Image Data (p. 63)
- Interpreting Histograms for the Body Image Data (p. 76)
- Interpreting Descriptive Statistics for the Body Image Data (p. 103)
- Detecting Outliers in the Body Image Data (p. 118)

 D *Data Set: BDD*

Suppose you wish to evaluate the mathematical capabilities of a class of 1,000 first-year college students, based on their quantitative Scholastic Aptitude Test (SAT) scores. How would you describe these 1,000 measurements? Characteristics of interest include the typical, or most frequent, SAT score; the variability in the scores; the highest and lowest scores; the "shape" of the data; and whether the data set contains any unusual scores. Extracting this information isn't easy. The 1,000 scores provide too many bits of information for our minds to comprehend. Clearly, we need some method for summarizing and characterizing the information in such a data set. Methods for describing data sets are also essential for statistical inference. Most populations make for large data sets. Consequently, we need methods for describing a data set that let us make inferences about a population on the basis of information contained in a sample.

Two methods for describing data are presented in this chapter, one *graphical* and the other *numerical*. Both play an important role in statistics. Section 2.1 presents both graphical and numerical methods for describing qualitative data. Graphical methods for describing quantitative data are illustrated in Sections 2.2, 2.8, and 2.9; numerical descriptive methods for quantitative data are presented in Sections 2.3–2.7. We end the chapter with a section on the *misuse* of descriptive techniques.

2.1 Describing Qualitative Data

Consider a study of aphasia published in the *Journal of Communication Disorders*. Aphasia is the "impairment or loss of the faculty of using or understanding spoken or written language." Three types of aphasia have been identified by researchers: Broca's, conduction, and anomic. The researchers wanted to determine whether one type of aphasia occurs more often than any other and, if so, how often. Consequently, they measured the type of aphasia for a sample of 22 adult aphasics. Table 2.1 gives the type of aphasia diagnosed for each aphasic in the sample.

For this study, the variable of interest, type of aphasia, is qualitative in nature. Qualitative data are nonnumerical in nature; thus, the value of a qualitative variable can only be classified into categories called *classes*. The possible types of aphasia—Broca's, conduction, and anomic—represent the classes for this qualitative variable. We can summarize such data numerically in two ways: (1) by computing the *class frequency*—the number of observations in the data set that fall into each class—or (2) by computing the *class relative frequency*—the proportion of the total number of observations falling into each class.

A **class** is one of the categories into which qualitative data can be classified.

The **class frequency** is the number of observations in the data set that fall into a particular class.

The **class relative frequency** is the class frequency divided by the total number of observations in the data set; that is,

$$\text{class relative frequency} = \frac{\text{class frequency}}{n}$$

Table 2.1	Data on 22 Adult Aphasics		
Subject	Type of Aphasia	Subject	Type of Aphasia
1	Broca's	12	Broca's
2	Anomic	13	Anomic
3	Anomic	14	Broca's
4	Conduction	15	Anomic
5	Broca's	16	Anomic
6	Conduction	17	Anomic
7	Conduction	18	Conduction
8	Anomic	19	Broca's
9	Conduction	20	Anomic
10	Anomic	21	Conduction
11	Conduction	22	Anomic

Based on Li, E. C., Williams, S. E., and Volpe, R. D. "The effects of topic and listener familiarity of discourse variables in procedural and narrative discourse tasks." *The Journal of Communication Disorders*, Vol. 28, No. 1, Mar. 1995, p. 44 (Table 1).

Ⓓ *Data Set:* APHASIA

The **class percentage** is the class relative frequency multiplied by 100; that is,

class percentage = (class relative frequency) × 100

Examining Table 2.1, we observe that 10 aphasics in the study were diagnosed as suffering from anomic aphasia, 5 from Broca's aphasia, and 7 from conduction aphasia. These numbers—10, 5, and 7—represent the class frequencies for the three classes and are shown in the summary table, Figure 2.1, produced with SPSS.

TYPE

		Frequency	Percent	Valid Percent	Cumulative Percent
Valid	Anomic	10	45.5	45.5	45.5
	Brocas	5	22.7	22.7	68.2
	Conduction	7	31.8	31.8	100.0
	Total	22	100.0	100.0	

Figure 2.1
SPSS summary table for types of aphasia

Figure 2.1 also gives the relative frequency of each of the three aphasia classes. From the class relative frequency definition, we calculate the relative frequency by dividing the class frequency by the total number of observations in the data set. Thus, the relative frequencies for the three types of aphasia are

$$\text{Anomic:} \frac{10}{22} = .455$$

$$\text{Broca's:} \frac{5}{22} = .227$$

$$\text{Conduction:} \frac{7}{22} = .318$$

These values, expressed as a percent, are shown in the SPSS summary table of Figure 2.1. From these relative frequencies, we observe that nearly half (45.5%) of the 22 subjects in the study are suffering from anomic aphasia.

Although the summary table of Figure 2.1 adequately describes the data of Table 2.1, we often want a graphical presentation as well. Figures 2.2 and 2.3 show two of the most widely used graphical methods for describing qualitative data: bar graphs and pie charts. Figure 2.2 shows the frequencies of the three types of aphasia in a **bar graph** produced with SAS. Note that the height of the rectangle, or "bar," over each class is equal to the class frequency. (Optionally, the bar heights can be proportional to class relative frequencies.)

Figure 2.2
SAS bar graph for type of aphasia

Figure 2.3
MINITAB pie chart for type of aphasia

In contrast, Figure 2.3 shows the relative frequencies of the three types of aphasia in a **pie chart** generated with MINITAB. Note that the pie is a circle (spanning 360°) and the size (angle) of the "pie slice" assigned to each class is proportional to the class relative frequency. For example, the slice assigned to anomic aphasia is 45.5% of 360°, or $(.455)(360°) = 163.8°$.

Before leaving the data set in Table 2.1, consider the bar graph shown in Figure 2.4, produced with SPSS. Note that the bars for the types of aphasia are arranged in descending order of height, from left to right across the horizontal axis. That is, the tallest bar (Anomic) is positioned at the far left and the shortest bar (Broca's) is at the far right. This rearrangement of the bars in a bar graph is called a **Pareto diagram**. One goal of a Pareto diagram (named for the Italian economist Vilfredo Pareto) is to make it easy to locate the "most important" categories—those with the largest frequencies.

Figure 2.4
SPSS Pareto diagram for type of aphasia

Summary of Graphical Descriptive Methods for Qualitative Data

Bar Graph: The categories (classes) of the qualitative variable are represented by bars, where the height of each bar is either the class frequency, class relative frequency, or class percentage.

Pie Chart: The categories (classes) of the qualitative variable are represented by slices of a pie (circle). The size of each slice is proportional to the class relative frequency.

Pareto Diagram: A bar graph with the categories (classes) of the qualitative variable (i.e., the bars) arranged by height in descending order from left to right.

Now Work Exercise 2.6

Let's look at a practical example that requires interpretation of the graphical results.

Example	**2.1** Ⓓ

Graphing and Summarizing Qualitative Data—Drug Designed to Reduce Blood Loss

Problem A group of cardiac physicians in southwest Florida has been studying a new drug designed to reduce blood loss in coronary bypass operations. Blood loss data for 114 coronary bypass patients (some who received a dosage of the drug and others who did not) are saved in the **BLOOD** file. Although the drug shows promise in reducing blood loss, the physicians are concerned about possible side effects and complications. So their data set includes not only the qualitative variable DRUG, which indicates whether or not the patient received the drug, but also the qualitative variable COMP, which specifies the type (if any) of complication experienced by the patient. The four values of COMP are (1) redo surgery, (2) post-op infection, (3) both, or (4) none.

a. Figure 2.5, generated by SAS, shows summary tables for the two qualitative variables, DRUG and COMP. Interpret the results.

b. Interpret the MINITAB and SPSS printouts shown in Figures 2.6 and 2.7, respectively.

The FREQ Procedure

DRUG	Frequency	Percent	Cumulative Frequency	Cumulative Percent
NO	57	50.00	57	50.00
YES	57	50.00	114	100.00

COMP	Frequency	Percent	Cumulative Frequency	Cumulative Percent
BOTH	6	5.26	6	5.26
INFECT	15	13.16	21	18.42
NONE	79	69.30	100	87.72
REDO	14	12.28	114	100.00

Figure 2.5

SAS summary tables for DRUG and COMP

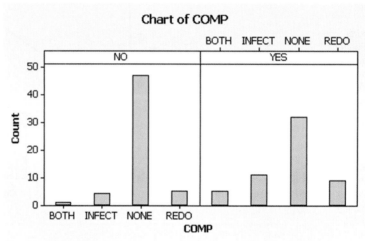

Figure 2.6

MINITAB side-by-side bar graphs for COMP by value of DRUG

COMP

DRUG			Frequency	Percent	Valid Percent	Cumulative Percent
NO	Valid	BOTH	1	1.8	1.8	1.8
		INFECT	4	7.0	7.0	8.8
		NONE	47	82.5	82.5	91.2
		REDO	5	8.8	8.8	100.0
		Total	57	100.0	100.0	
YES	Valid	BOTH	5	8.8	8.8	8.8
		INFECT	11	19.3	19.3	28.1
		NONE	32	56.1	56.1	84.2
		REDO	9	15.8	15.8	100.0
		Total	57	100.0	100.0	

Figure 2.7

SPSS summary tables for COMP by value of drug

Solution

a. The top table in Figure 2.5 is a summary frequency table for DRUG. Note that exactly half (57) of the 114 coronary bypass patients received the drug and half did not. The bottom table in Figure 2.5 is a summary frequency table for COMP. The class percentages are given in the Percent column. We see that about 69% of the 114 patients had no complications, leaving about 31% who experienced either a redo surgery, a post-op infection, or both.

b. Figure 2.6 is a MINITAB side-by-side bar graph of the data. The four bars in the left-side graph represent the frequencies of COMP for the 57 patients who did not receive the drug; the four bars in the right-side graph represent the frequencies of COMP for the 57 patients who did receive a dosage of the drug. The graph clearly shows that patients who did not receive the drug suffered fewer complications. The exact percentages are displayed in the SPSS summary tables of Figure 2.7. Over 56% of the patients who got the drug had no complications, compared with about 83% for the patients who got no drug.

Look Back Although these results show that the drug may be effective in reducing blood loss, Figures 2.6 and 2.7 imply that patients on the drug may have a higher risk of incurring complications. But before using this information to make a decision about the drug, the physicians will need to provide a measure of reliability for the inference. That is, the physicians will want to know whether the difference between the percentages of patients with complications observed in this sample of 114 patients is generalizable to the population of all coronary bypass patients.

■ **Now Work Exercise 2.17**

Statistics IN Action | **Revisited** | **Interpreting Pie Charts for the Body Image Data**

In the *Body Image: An International Journal of Research* (Jan. 2010) study, Brown University researchers measured several qualitative (categorical) variables for each of 92 body dysmorphic disorder (BDD) patients: *Gender* (M or F), *Comorbid Disorder* (Major Depression, Social Phobia, Obsessive Compulsive Disorder—OCD, or Anorexia/Bulimia Nervosa), and *Dissatisfied with Looks* (Yes or No). [*Note:* "Yes" values for Dissatisfied with Looks correspond to total appearance evaluation responses of 20 points (out of 35) or less. "No" values correspond to totals of 21 points or more.] Pie charts and bar graphs can be used to summarize and describe the responses for these variables. Recall that the data are saved in

the **BDD** file. We used MINITAB to create pie charts for these variables.

Figure SIA2.1 shows individual pie charts for each of the three qualitative variables, Gender, Disorder, and Dissatisfied with Looks. First, notice that of the 92 BDD patients, 65% are females and 35% are males. Of these patients, the most common comorbid disorder is major depression (42%), followed by social phobia (35%) and OCD (20%). The third pie chart shows that 77% of the BDD patients are dissatisfied in some way with their body.

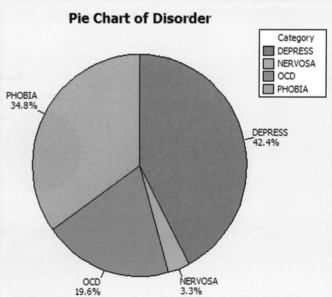

Figure SIA2.1
MINITAB pie charts for Gender, Disorder, and Dissatisfied with Looks

(Continued)

Figure SIA2.1
(*continued*)

Figure SIA2.2
MINITAB side-by-side pie
charts for Dissatisfied with
Looks by Gender

Figure SIA2.3
MINITAB side-by-side pie
charts for Dissatisfied with
Looks by Comorbid Disorder

Statistics IN Action
(continued)

Of interest in the study is whether BDD females tend to be more dissatisfied with their looks than BDD males. We can gain insight into this question by forming side-by-side pie charts of the Dissatisfied with Looks variable, one chart for females and one for males. These pie charts are shown in Figure SIA2.2. You can see that about 82% of the females are dissatisfied in some way with their body as compared to about 69% of the males. Thus, it does appear that BDD females tend to be more dissatisfied with their looks than males, at least for this sample of patients.

Also of interest is whether certain comorbid disorders lead to a higher level of dissatisfaction with body appearance. Figure SIA2.3 is a set of side-by-side pie charts for Dissatisfied with Looks, one chart for each of four comorbid disorders. The graphs show that the percentage of BDD patients who are dissatisfied in some way with their body image range from about 67% for those with a nervosa disorder to about 81% for those diagnosed with a social phobia.

 Data Set: BDD

⚠ **CAUTION** The information produced in these pie charts should be limited to describing the sample of 92 BDD patients. If one is interested in making inferences about the population of all BDD patients (as were the Brown University researchers), inferential statistical methods need to be applied to the data. These methods are the topics of later chapters. ▲

Exercises 2.1–2.24

Understanding the Principles

2.1 Explain the difference between class frequency, class relative frequency, and class percentage for a qualitative variable.

2.2 Explain the difference between a bar graph and a pie chart.

2.3 What is the advantage of a Pareto diagram over a bar graph?

Learning the Mechanics

2.4 Complete the following table:

Grades in Statistics	Frequency	Relative Frequency
A: 90–100		.2
B: 80–89	75	
C: 65–79	120	
D: 50–64		
F: Below 50	15	.05
Total		1.00

2.5 A qualitative variable with three classes (X, Y, and Z) is
Ⓓ   measured for each of 20 units randomly sampled from a
L02005 target population. The data (observed class for each unit) are as follows:

Y X X Z X Y Y Y X X Z X
Y Y X Z Y Y Y X

a. Compute the frequency for each of the three classes.
b. Compute the relative frequency for each of the three classes.
c. Display the results from part **a** in a frequency bar graph.
d. Display the results from part **b** in a pie chart.

Applying the Concepts—Basic

2.6 **Study of importance of libraries.** In *The Canadian Journal*
NW *of Information and Library Science* (Vol. 33, 2009), researchers from the University of Western Ontario reported on mall shoppers' opinions about libraries and their importance to today's society. Each in a sample of over 200 mall shoppers was asked the following question: "In today's world, with Internet access and online and large book sellers like Amazon, do you think libraries have become more, less, or the same in importance to their community?" The accompanying graphic summarizes the mall shoppers' responses.

a. What type of graph is shown?
b. Identify the qualitative variable described in the graph.
c. From the graph, identify the most common response.
d. Convert the graph into a Pareto diagram. Interpret the results.

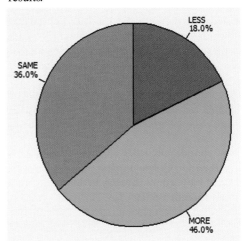

2.7 **Do social robots walk or roll?** According to the United Nations, social robots now outnumber industrial robots worldwide. A social (or service) robot is designed to entertain, educate, and care for human users. In a paper published by the *International Conference on Social Robotics* (Vol. 6414, 2010), design engineers investigated the trend in the design of social robots. Using a random sample of 106 social robots obtained through a Web search, the engineers found that 63 were built with legs only, 20 with wheels only, 8 with both legs and wheels, and 15 with neither legs nor wheels. This information is portrayed in the accompanying graphic.

a. What type of graph is used to describe the data?

b. Identify the variable measured for each of the 106 robot designs.

c. Use the graph to identify the social robot design that is currently used the most.

d. Compute class relative frequencies for the different categories shown in the graph.

e. Use the results from part **d** to construct a Pareto diagram for the data.

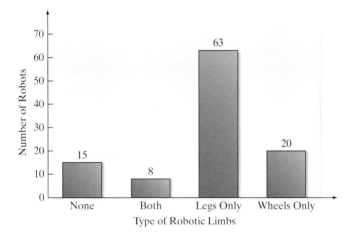

2.8 **Paying for music downloads.** If you use the Internet, have you ever paid to access or download music? This was one of the questions of interest in a recent *Pew Internet & American Life Project Survey* (October 2010). Telephone interviews were conducted on a representative sample of 1,003 adults living in the United States. For this sample, 248 adults stated that they do not use the Internet, 249 revealed that they use the Internet but have never paid to download music, and the remainder (506 adults) stated that they use the Internet and have paid to download music. The results are summarized in the MINITAB pie chart shown.

a. According to the pie chart, what proportion of the sample use the Internet and pay to download music? Verify the accuracy of this proportion using the survey results.

b. Now consider only the 755 adults in the sample that use the Internet. Create a graph that compares the proportion of these adults that pay to download music with the proportion that do not pay.

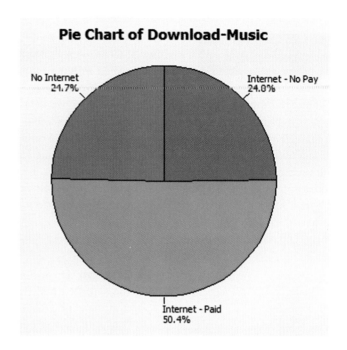

2.9 **Cheek teeth of extinct primates.** The characteristics of cheek teeth (e.g., molars) can provide anthropologists with information on the dietary habits of extinct mammals. The cheek teeth of an extinct primate species were the subject of research reported in the *American Journal of Physical Anthropology* (Vol. 142, 2010). A total of 18 cheek teeth extracted from skulls discovered in western Wyoming were analyzed. Each tooth was classified according to degree of wear (unworn, slight, light-moderate, moderate, moderate-heavy, or heavy). The 18 measurements are listed here.

Data on Degree of Wear

Unknown	Slight
Unknown	Slight
Unknown	Heavy
Moderate	Unworn
Slight	Light-moderate
Unknown	Light-moderate
Moderate-heavy	Moderate
Moderate	Unworn
Slight	Unknown

a. Identify the variable measured in the study and its type (quantitative or qualitative).

b. Count the number of cheek teeth in each wear category.

c. Calculate the relative frequency for each wear category.

d. Construct a relative frequency bar graph for the data.

e. Construct a Pareto diagram for the data.

f. Identify the degree of wear category that occurred most often in the sample of 18 cheek teeth.

2.10 **Humane education and classroom pets.** In grade school, did your teacher have a pet in the classroom? Many teachers do so in order to promote humane education.

The *Journal of Moral Education* (Mar. 2010) published a study designed to examine elementary school teachers' experiences with humane education and animals in the classroom. Based on a survey of 75 elementary school teachers, the following results were reported:

Survey Question	Response	Number of Teachers
Q1: Do you keep classroom pets?	Yes	61
	No	14
Q2: Do you allow visits by pets?	Yes	35
	No	40
Q3: What formal humane education program does your school employ?	Character counts	21
	Roots of empathy	9
	Teacher designed	3
	Other	14
	None	28

Based on *Journal of Moral Education*, Mar. 2010.

   a. Explain why the data collected are qualitative.

   b. For each survey question, summarize the results with a graph.

   c. For this sample of 75 teachers, write two or three sentences describing their experience with humane education and classroom pets.

2.11 **Estimating the rhino population.** The International Rhino Foundation estimates that there are 28,933 rhinoceroses living in the wild in Africa and Asia. A breakdown of the number of rhinos of each species is reported in the accompanying table:

Rhino Species	Population Estimate
African Black	5,055
African White	20,405
(Asian) Sumatran	100
(Asian) Javan	40
(Asian) Greater One-Horned	3,333
Total	28,933

Source: International Rhino Foundation, 2014.

   a. Construct a relative frequency table for the data.

   b. Display the relative frequencies in a bar graph.

   c. What proportion of the 28,933 rhinos are African rhinos? Asian?

2.12 **STEM experiences for girls.** The National Science Foundation (NSF) sponsored a study on girls' participation in informal science, technology, engineering, and mathematics (STEM) programs (see Exercise 1.13). The results of the study were published in *Cascading Influences: Long-Term Impacts of Informal STEM Experiences for Girls* (Mar. 2013). The researchers questioned 174 young women who recently participated in a STEM program. They used a pie chart to describe the geographic location (urban, suburban, or rural) of the STEM programs attended. Of the 174 STEM participants, 107 were in urban areas, 57 in suburban areas, and 10 in rural areas. Use this information to construct the pie chart. Interpret the results.

Applying the Concepts—Intermediate

2.13 **Microsoft program security issues.** The dominance of Microsoft in the computer software market has led to numerous malicious attacks (e.g., worms, viruses) on its programs. To help its users combat these problems, Microsoft periodically issues a security bulletin that reports the software affected by the vulnerability. In *Computers & Security* (July 2013), researchers focused on reported security issues with three Microsoft products: Office, Windows, and Explorer. In a sample of 50 security bulletins issued in 2012, 32 reported a security issue with Windows, 6 with Explorer, and 12 with Office. The researchers also categorized the security bulletins according to the expected repercussion of the vulnerability. Categories were Denial of service, Information disclosure, Remote code execution, Spoofing, and Privilege elevation. Suppose that of the 50 bulletins sampled, the following numbers of bulletins were classified into each respective category: 6, 8, 22, 3, 11.

   a. Construct a pie chart to describe the Microsoft products with security issues. Which product had the lowest proportion of security issues in 2012?

   b. Construct a Pareto diagram to describe the expected repercussions from security issues. Based on the graph, what repercussion would you advise Microsoft to focus on?

2.14 **Genealogy research.** The *Journal of Family History* (Jan. 2010) investigated the genealogy of a certain Korean clan. Of interest was whether or not a family name continued into the next generation in the clan genealogy (called a

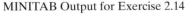

MINITAB Output for Exercise 2.14

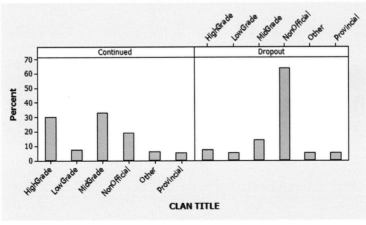

continued line) or dropped out (called a *dropout line*). Side-by-side pie charts were used to describe the rate at which certain occupational titles of clan individuals occurred within each line type. Similarly constructed MINITAB bar charts are shown on p. 67.

a. Identify the two qualitative variables graphed in the side-by-side pie charts.

b. Give a full interpretation of the charts. Identify the major differences (if any) between the two line groups.

Based on Sangkuk Lee, *Journal of Family History* (Jan 2010).

2.15 **Characteristics of ice melt ponds.** The National Snow and Ice Data Center (NSIDC) collects data on the albedo, depth, and physical characteristics of ice melt ponds in the Canadian Arctic. Environmental engineers at the University of Colorado are using these data to study how climate affects the sea ice. Data on 504 ice melt ponds located in the Barrow Strait in the Canadian Arctic are saved in the **PONDICE** file. One variable of interest is the type of ice observed for each pond, classified as first-year ice, multiyear ice, or landfast ice. An SAS summary table and a horizontal bar graph that describe the types of ice of the 504 melt ponds are shown below.

a. Of the 504 melt ponds, what proportion had landfast ice?

b. The University of Colorado researchers estimated that about 17% of melt ponds in the Canadian Arctic have first-year ice. Do you agree?

c. Convert the horizontal bar graph into a Pareto diagram. Interpret the graph.

2.16 **Railway track allocation.** One of the problems faced by transportation engineers is the assignment of tracks to trains at a busy railroad station. Overused and/or underused tracks cause increases in maintenance costs and inefficient allocation of resources. The *Journal of Transportation Engineering* (May 2013) investigated the optimization of track allocation at a Chinese railway station with 11 tracks. Using a new algorithm designed to minimize waiting time and bottlenecks, engineers assigned tracks to 53 trains in a single day as shown in the accompanying table. Construct a Pareto diagram for the data. Use the diagram to help the engineers determine

if the allocation of tracks to trains is evenly distributed and, if not, which tracks are underutilized and overutilized.

Track Assigned	Number of Trains
Track #1	3
Track #2	4
Track #3	4
Track #4	4
Track #5	7
Track #6	5
Track #7	5
Track #8	7
Track #9	4
Track #10	5
Track #11	5
Total	53

Source: Wu, J., et al. "Track allocation optimization at a railway station: Mean-variance model and case study." *Journal of Transportation Engineering*, Vol. 139, No. 5, May 2013 (extracted from Table 4).

2.17 **Satellites in orbit.** According to the Union of Concerned Scientists (www.ucsusa.org), there are 502 low earth orbit (LEO) and 432 geosynchronous orbit (GEO) satellites in space. Each satellite is owned by an entity in either the government, military, commercial, or civil sector. A breakdown of the number of satellites in orbit for each sector is displayed in the accompanying table. Use this information to construct a pair of graphs that compare the ownership sectors of LEO and GEO satellites in orbit. What observations do you have about the data?

	LEO Satellites	GEO Satellites
Government	229	59
Military	109	91
Commercial	118	281
Civil	46	1
Total	502	432

Source: Union of Concerned Scientists, Nov. 2012.

SAS output for Exercise 2.15

The FREQ Procedure

ICETYPE	Frequency	Percent	Cumulative Frequency	Cumulative Percent
First-year	88	17.46	88	17.46
Landfast	196	38.89	284	56.35
Multi-year	220	43.65	504	100.00

2.18 **Curbing street gang gun violence.** Operation Ceasefire is a program implemented to reduce street gang gun violence in the city of Boston. The effectiveness of the program was examined in the *Journal of Quantitative Criminology* (Mar. 2014). Over a 5-year period (2006–2010), there were 80 shootings involving a particular Boston street gang (called the Lucerne Street Doggz). The information in the table breaks down these shootings by year and by who was shot (gang members or non-gang members). Note: The Ceasefire program was implemented at the end of 2007.

Year	Total Shootings	Shootings of Gang Members	Shootings of Non-Gang Members
2006	37	30	7
2007	30	22	8
2008	4	3	1
2009	5	4	1
2010	4	3	1
Totals	80	62	18

Source: Braga, A. A, Hureau, D. M., and Papachristos, A. V. "Deterring gang-involved gun violence: Measuring the impact of Boston's Operation Ceasefire on street gang behavior." *Journal of Quantitative Criminology,* Vol. 30, No. 1, Mar. 2014 (adapted from Figure 3).

  a. Construct a Pareto diagram for the total annual shootings involving the Boston street gang.

  b. Construct a Pareto diagram for the annual shootings of the Boston street gang members.

  c. Based on the Pareto diagrams, comment on the effectiveness of Operation Ceasefire.

2.19 **Interactions in a children's museum.** The nature of child-led and adult-led interactions in a children's museum was investigated in *Early Childhood Education Journal* (Mar. 2014). Interactions by visitors to the museum were classified as (1) show-and-tell, (2) learning, (3) teaching, (4) refocusing, (5) participatory play, (6) advocating, or (7) disciplining. Over a 3-month period, the researchers observed 170 meaningful interactions, of which 81 were led by children and 89 were led by adult caregivers. The number of interactions observed in each category is provided in the accompanying table. Use side-by-side bar graphs to compare the interactions led by children and adults. Do you observe any trends?

Type of Interaction	Child-Led	Adult-Led
Show-and-tell	26	0
Learning	21	0
Teaching	0	64
Refocusing	21	10
Participatory Play	12	9
Advocating	0	4
Disciplining	1	2
Totals	81	89

Source: McMunn-Dooley, C., and Welch, M. M. "Nature of interactions among young children and adult caregivers in a children's museum." *Early Childhood Education Journal,* Vol. 42, No. 2, Mar. 2014 (adapted from Figure 2).

2.20 **Motivation and right-oriented bias.** Evolutionary theory suggests that motivated decision makers tend to exhibit a right-oriented bias. (For example, if presented with two equally valued brands of detergent on a supermarket shelf, consumers are more likely to choose the brand on the right.) In *Psychological Science* (November 2011), researchers tested this theory using data on all penalty shots attempted in World Cup soccer matches (a total of 204 penalty shots). The researchers believe that goalkeepers, motivated to make a penalty shot save but with little time to make a decision, will tend to dive to the right. The results of the study (percentages of dives to the left, middle, or right) are provided in the table. Note that the percentages in each row, corresponding to a certain match situation, add to 100%. Construct side-by-side bar graphs showing the distribution of dives for the three match situations. What inferences can you draw from the graphs?

Match Situation	Dive Left	Stay Middle	Dive Right
Team behind	29%	0%	71%
Tied	48%	3%	49%
Team ahead	51%	1%	48%

Source: Roskes, M., et al. "The right side? Under time pressure, approach motivation leads to right-oriented bias." *Psychological Science,* Vol. 22, No. 11, Nov. 2011 (adapted from Figure 2).

2.21 **Do you believe in the Bible?** Each year the National Opinion Research Center conducts the General Social Survey (GSS), eliciting opinions of Americans on a wide variety of social topics. One question in the survey asked about a person's belief in the Bible. A sample of 4,826 respondents selected from one of the following answers: (1) The Bible is the actual word of God, to be taken literally—1,527 respondents; (2) the Bible is the inspired word of God, but not everything is to be taken literally—2,231 respondents; (3) the Bible is an ancient book of fables legends, history, and moral precepts—996 respondents; and (4) the Bible has some other origin—72 respondents.

  a. Summarize the responses to the Bible question in the form of a relative frequency table.

  b. Summarize the responses to the Bible question in a pie chart.

  c. Write a few sentences that give a practical interpretation of the results shown in the summary table and graph.

Applying the Concepts—Advanced

2.22 **Museum management.** What criteria do museums use to evaluate their performance? In a worldwide survey reported in *Museum Management and Curatorship* (June 2010), managers of 30 leading museums of contemporary art were asked to provide the performance measure used most often. The data are provided in the table on p. 70. The researcher concluded that "there is a large amount of variation within the museum community with regard to... performance measurement and evaluation." Do you agree? Use a graph to support your conclusion.

Museum Number	Performance Measure	Museum Number	Performance Measure
1	Total visitors	16	Paying visitors
2	Paying visitors	17	Total visitors
3	Funds raised	18	Big shows
4	Big shows	19	Big shows
5	Funds raised	20	Total visitors
6	Total visitors	21	Funds raised
7	Total visitors	22	Total visitors
8	Big shows	23	Members
9	Members	24	Paying visitors
10	Funds raised	25	Big shows
11	Funds raised	26	Funds raised
12	Members	27	Members
13	Big shows	28	Total visitors
14	Paying visitors	29	Funds raised
15	Paying visitors	30	Total visitors

(11 selected observations from 223)

Well Class	Aquifer	Detect MTBE?
Private	Bedrock	Below Limit
Private	Bedrock	Below Limit
Public	Unconsolidated	Detect
Public	Unconsolidated	Below Limit
Public	Unconsolidated	Below Limit
Public	Unconsolidated	Below Limit
Public	Unconsolidated	Detect
Public	Unconsolidated	Below Limit
Public	Unconsolidated	Below Limit
Public	Bedrock	Detect
Public	Bedrock	Detect

Based on Ayotte, J. D., Argue, D. M., and McGarry, F. J. "Methyl *tert*-Butyl ether occurrence and related factors in public and private wells in southeast New Hampshire." *Environmental Science & Technology*, Vol. 39, No. 1, Jan. 2005, pp. 9–16.

2.23 **Groundwater contamination in wells.** In New Hampshire, about half the counties mandate the use of reformulated gasoline. This has led to an increase in the contamination of groundwater with methyl *tert*-butyl ether (MTBE). *Environmental Science & Technology* (Jan. 2005) reported on the factors related to MTBE contamination in public and private New Hamsphire wells. Data were collected on a sample of 223 wells. These data are saved in the **MTBE** file. Three of the variables are qualitative in nature: well class (public or private), aquifer (bedrock or unconsolidated), and detectable level of MTBE (below limit or detect). [*Note:* A detectable level of MTBE occurs if the MTBE value exceeds .2 microgram per liter.] The data on 11 selected wells are shown in the accompanying table.

a. Use graphical methods to describe each of the three qualitative variables for all 223 wells.

b. Use side-by-side bar charts to compare the proportions of contaminated wells for private and public well classes.

c. Use side-by-side bar charts to compare the proportions of contaminated wells for bedrock and unconsolidated aquifers.

d. What inferences can be made from the bar charts of parts **a–c**?

2.24 **Extinct New Zealand birds.** Refer to the *Evolutionary Ecology Research* (July 2003) study of the patterns of extinction in the New Zealand bird population, Exercise 1.20 (p. 49). Data on flight capability (volant or flightless), habitat (aquatic, ground terrestrial, or aerial terrestrial), nesting site (ground, cavity within ground, tree, or cavity above ground), nest density (high or low), diet (fish, vertebrates, vegetables, or invertebrates), body mass (grams), egg length (millimeters), and extinct status (extinct, absent from island, or present) for 132 bird species that existed at the time of the Maori colonization of New Zealand are saved in the **NZBIRDS** file. Use a graphical method to investigate the theory that extinct status is related to flight capability, habitat, and nest density.

2.2 Graphical Methods for Describing Quantitative Data

Recall from Section 1.4 that quantitative data sets consist of data that are recorded on a meaningful numerical scale. To describe, summarize, and detect patterns in such data, we can use three graphical methods: *dot plots, stem-and-leaf displays,* and *histograms.* Since most statistical software packages can be used to construct these displays, we'll focus here on their interpretation rather than their construction.

For example, the Environmental Protection Agency (EPA) performs extensive tests on all new car models to determine their mileage ratings. Suppose that the 100 measurements in Table 2.2 represent the results of such tests on a certain new car model. How can we summarize the information in this rather large sample?

A visual inspection of the data indicates some obvious facts. For example, most of the mileages are in the 30s, with a smaller fraction in the 40s. But it is difficult to provide much additional information on the 100 mileage ratings without resorting to some method of summarizing the data. One such method is a dot plot.

Dot Plots

A MINITAB **dot plot** for the 100 EPA mileage ratings is shown in Figure 2.8. The horizontal axis of the figure is a scale for the quantitative variable in miles per gallon (mpg).

Table 2.2		EPA Mileage Ratings on 100 Cars		
36.3	41.0	36.9	37.1	44.9
32.7	37.3	41.2	36.6	32.9
40.5	36.5	37.6	33.9	40.2
36.2	37.9	36.0	37.9	35.9
38.5	39.0	35.5	34.8	38.6
36.3	36.8	32.5	36.4	40.5
41.0	31.8	37.3	33.1	37.0
37.0	37.2	40.7	37.4	37.1
37.1	40.3	36.7	37.0	33.9
39.9	36.9	32.9	33.8	39.8
36.8	30.0	37.2	42.1	36.7
36.5	33.2	37.4	37.5	33.6
36.4	37.7	37.7	40.0	34.2
38.2	38.3	35.7	35.6	35.1
39.4	35.3	34.4	38.8	39.7
36.6	36.1	38.2	38.4	39.3
37.6	37.0	38.7	39.0	35.8
37.8	35.9	35.6	36.7	34.5
40.1	38.0	35.2	34.8	39.5
34.0	36.8	35.0	38.1	36.9

 Data Set: EPAGAS

Figure 2.8
MINITAB dot plot for 100 EPA mileage ratings

The rounded (to the nearest half gallon) numerical value of each measurement in the data set is located on the horizontal scale by a dot. When data values repeat, the dots are placed above one another, forming a pile at that particular numerical location. As you can see, this dot plot verifies that almost all of the mileage ratings are in the 30s, with most falling between 35 and 40 miles per gallon.

Stem-and-Leaf Display

Another graphical representation of these same data, a MINITAB **stem-and-leaf display**, is shown in Figure 2.9. In this display, the *stem* is the portion of the measurement (mpg) to the left of the decimal point, while the remaining portion, to the right of the decimal point, is the *leaf*.

In Figure 2.9, the stems for the data set are listed in the second column, from the smallest (30) to the largest (44). Then the leaf for each observation is listed to the right, in the row of the display corresponding to the observation's stem.* For example, the leaf 3 of the first observation (36.3) in Table 2.2 appears in the row corresponding to the stem 36. Similarly, the leaf 7 for the second observation (32.7) in Table 2.2 appears in the row corresponding to the stem 32, while the leaf 5 for the third observation (40.5) appears in the row corresponding to the stem 40. (The stems and leaves for these first three observations are highlighted in Figure 2.9.) Typically, the leaves in each row are ordered as shown in the MINITAB stem-and-leaf display.

*The first column of the MINITAB stem-and-leaf display represents the cumulative number of measurements from the class interval to the nearest extreme class interval.

Stem-and-Leaf Display: MPG

```
Stem-and-leaf of MPG   N  = 100
Leaf Unit = 0.10
```

```
   1     30   0
   2     31   8
   6     32   5799
  12     33   126899
  18     34   024588
  29     35   01235667899
  49     36   01233445566777888999
 (21)    37   000011122334456677899
  30     38   0122345678
  20     39   00345789
  12     40   0123557
   5     41   002
   2     42   1
   1     43
   1     44   9
```

Figure 2.9

MINITAB stem-and-leaf display for 100 mileage ratings

The stem-and-leaf display presents another compact picture of the data set. You can see at a glance that the 100 mileage readings were distributed between 30.0 and 44.9, with most of them falling in stem rows 35 to 39. The 6 leaves in stem row 34 indicate that 6 of the 100 readings were at least 34.0, but less than 35.0. Similarly, the 11 leaves in stem row 35 indicate that 11 of the 100 readings were at least 35.0, but less than 36.0. Only five cars had readings equal to 41 or larger, and only one was as low as 30.

The definitions of the stem and leaf for a data set can be modified to alter the graphical description. For example, suppose we had defined the stem as the tens digit for the gas mileage data, rather than the ones and tens digits. With this definition, the stems and leaves corresponding to the measurements 36.3 and 32.7 would be as follows:

Stem	Leaf		Stem	Leaf
3	6		3	2

Note that the decimal portion of the numbers has been dropped. Generally, only one digit is displayed in the leaf.

If you look at the data, you'll see why we didn't define the stem this way. All the mileage measurements fall into the 30s and 40s, so all the leaves would fall into just two stem rows in this display. The resulting picture would not be nearly as informative as Figure 2.9.

Now Work Exercise 2.31

Histograms

An SPSS **histogram** for the 100 EPA mileage readings given in Table 2.2 is shown in Figure 2.10. The horizontal axis of the figure, which gives the miles per gallon for a given automobile, is divided into **class intervals**, commencing with the interval from 30–31 and proceeding in intervals of equal size to 44–45 mpg. The vertical axis gives the number (or *frequency*) of the 100 readings that fall into each interval. It appears that about 21 of the 100 cars, or 21%, attained a mileage between 37 and 38 mpg. This class interval contains the highest frequency, and the intervals tend to contain a smaller number of the measurements as the mileages get smaller or larger.

Histograms can be used to display either the frequency or relative frequency of the measurements falling into the class intervals. The class intervals, frequencies, and relative frequencies for the EPA car mileage data are shown in the summary table, Table 2.3.*

---

*SPSS, like many software packages, will classify an observation that falls on the borderline of a class interval into the next-highest interval. For example, the gas mileage of 37.0, which falls on the border between the class intervals 36–37 and 37–38, is classified into the 37–38 class. The frequencies in Table 2.3 reflect this convention.

Figure 2.10

SPSS histogram for 100 EPA gas mileage ratings

By summing the relative frequencies in the intervals 35–36, 36–37, 37–38, and 38–39, you find that 65% of the mileages are between 35 and 39. Similarly, only 2% of the cars obtained a mileage rating over 42.0. Many other summary statements can be made by further examining the histogram and accompanying summary table. Note that the sum of all class frequencies will always equal the sample size n.

In interpreting a histogram, consider two important facts. First, the proportion of the total area under the histogram that falls above a particular interval on the x-axis is equal to the relative frequency of measurements falling into that interval. For example, the relative frequency for the class interval 37–38 in Figure 2.10 is .20. Consequently, the rectangle above the interval contains .20 of the total area under the histogram.

Second, imagine the appearance of the relative frequency histogram for a very large set of data (representing, say, a population). As the number of measurements in a data set is increased, you can obtain a better description of the data by decreasing the width of the

Table 2.3	Class Intervals, Frequencies, and Relative Frequencies for the Gas Mileage Data	
Class Interval	Frequency	Relative Frequency
30–31	1	0.01
31–32	1	0.01
32–33	4	0.04
33–34	6	0.06
34–35	6	0.06
35–36	11	0.11
36–37	20	0.20
37–38	21	0.21
38–39	10	0.10
39–40	8	0.08
40–41	7	0.07
41–42	3	0.03
42–43	1	0.01
43–44	0	0.00
44–45	1	0.01
Totals	100	1.00

class intervals. When the class intervals become small enough, a relative frequency histogram will (for all practical purposes) appear as a smooth curve. (See Figure 2.11.)

Figure 2.11
The effect of the size of a data set on the outline of a histogram

a. Small data set   b. Larger data set   c. Very large data set

Some recommendations for selecting the number of intervals in a histogram for smaller data sets are given in the following box:

Determining the Number of Classes in a Histogram

Number of Observations in Data Set	Number of Classes
Fewer than 25	5–6
25–50	7–14
More than 50	15–20

While histograms provide good visual descriptions of data sets—particularly very large ones—they do not let us identify individual measurements. In contrast, each of the original measurements is visible to some extent in a dot plot and is clearly visible in a stem-and-leaf display. The stem-and-leaf display arranges the data in ascending order, so it's easy to locate the individual measurements. For example, in Figure 2.9 we can easily see that two of the gas mileage measurements are equal to 36.3, but we can't see that fact by inspecting the histogram in Figure 2.10. However, stem-and-leaf displays can become unwieldy for very large data sets. A very large number of stems and leaves causes the vertical and horizontal dimensions of the display to become cumbersome, diminishing the usefulness of the visual display.

Example   2.2 Ⓓ

Graphing a Quantitative Variable—The "Water-Level Task"

Problem Over 60 years ago, famous child psychologist Jean Piaget devised a test of basic perceptual and conceptual skills dubbed the "water-level task." Subjects were shown a drawing of a glass being held at a 45° angle and asked to draw a line representing the true surface of the water. Today, research psychologists continue to use the task to test the perception of both adults and children. In one study, the water-level task was given to several groups that included 20 male bartenders and 20 female waitresses (*Psychological Science*, Mar. 1995). For each participant, the researchers measured the deviation (in angle degrees) of the judged line from the true line. These deviations (simulated on the basis of summary results presented in the journal article) are shown in Table 2.4. [*Note:* Deviations can be negative if the judged angle is smaller than the angle of the true line.]

Table 2.4   Water-Level Task Deviations (angle degrees)

Bartenders:	−9	6	10	6	10	−3	7	8	6	14	7	8	−5	2	−1	0	2	3	0	2
Waitresses:	7	10	25	8	10	8	12	9	35	10	12	11	7	10	21	−1	4	0	16	−1

Ⓓ *Data Set:* WLTASK

a. Use a statistical software package to create a frequency histogram for the combined data in Table 2.4. Then, shade the area under the histogram that corresponds to deviations recorded for waitresses. Interpret the result.

b. Use a statistical software package to create a stem-and-leaf display for these combined data. Again, shade each leaf of the display that corresponds to a deviation recorded for a waitress. Interpret the result.

Solution

a. We used SPSS to generate the frequency histogram, shown in Figure 2.12. Note that SPSS formed 20 classes, with class intervals -20 to $-15, -15$ to $-10, \ldots, 30$ to 35, and 35 to 40. This histogram clearly shows the clustering of the deviation angles between $0°$ and $15°$, with a few deviations in the upper end of the distribution (greater than $20°$). SPSS used green bars to shade the areas of the histogram that correspond to the measurements for waitresses. The graph clearly shows that waitresses tend to have greater (positive) deviations than do bartenders and fewer deviations near $0°$ relative to bartenders.

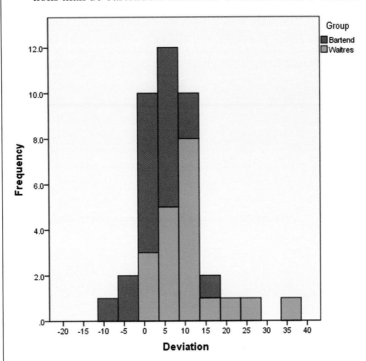

Figure 2.12

SPSS histogram for task deviations

b. We used MINITAB to produce the stem-and-leaf display in Figure 2.13. Note that the stem (the second column on the printout) represents the first digit (including 0) in the deviation angle measurement while the leaf (the third column on the printout) represents the second digit. Thus, the leaf 5 in the stem 2 row represents the deviation angle of $25°$. The shaded leaves represent deviations recorded for waitresses. As with the histogram, the stem-and-leaf display shows that deviations for waitresses tend to appear in the upper tail of the distribution. Together, the graphs imply that waitresses tend to overestimate the angle of the true line relative to bartenders.

Look Back As is usually the case with data sets that are not too large (say, fewer than 100 measurements), the stem-and-leaf display provides more detail than the histogram without being unwieldy. For instance, the stem-and-leaf display in Figure 2.13 clearly indicates the values of the individual measurements in the data set. For example, the largest deviation angle (representing the measurement $35°$) is shown in the last stem row. By contrast, histograms are most useful for displaying very large data sets when the overall shape of the distribution of measurements is more important than the identification of individual measurements.

Stem-and-Leaf Display: Deviation

```
Stem-and-leaf of Deviation   N  = 40
Leaf Unit = 1.0


   2    -0   95
   7    -0   31110
  14     0   0022234
 (12)    0   666777788889
  14     1   0000001224
   4     1   6
   3     2   1
   2     2   5
   1     3
   1     3   5
```

Figure 2.13

MINITAB stem-and-leaf display for task deviations

■ **Now Work Exercise 2.44**

Most statistical software packages can be used to generate histograms, stem-and-leaf displays, and dot plots. All three are useful tools for graphically describing data sets. We recommend that you generate and compare the displays whenever you can.

Summary of Graphical Descriptive Methods for Quantitative Data

Dot Plot: The numerical value of each quantitative measurement in the data set is represented by a dot on a horizontal scale. When data values repeat, the dots are placed above one another vertically.

Stem-and-Leaf Display: The numerical value of the quantitative variable is partitioned into a "stem" and a "leaf." The possible stems are listed in order in a column. The leaf for each quantitative measurement in the data set is placed in the corresponding stem row. Leaves for observations with the same stem value are listed in increasing order horizontally.

Histogram: The possible numerical values of the quantitative variable are partitioned into class intervals, each of which has the same width. These intervals form the scale of the horizontal axis. The frequency or relative frequency of observations in each class interval is determined. A vertical bar is placed over each class interval, with the height of the bar equal to either the class frequency or class relative frequency.

Statistics IN Action | Revisited

Interpreting Histograms for the Body Image Data

In the *Body Image: An International Journal of Research* (Jan. 2010) study of 92 BDD patients, the researchers asked each patient to respond to a series of questions on body image (e.g., "How satisfied are you with your physical attractiveness and looks?"). Recall that the scores were summed to yield an Appearance Evaluation score that ranged from 7 to 35 points. This score represents a quantitative variable. Consequently, to graphically investigate whether BDD females tend to be more dissatisfied with their looks than BDD males, we can form side-by-side histograms for the total score, one histogram for females and one for males. These histograms are shown in Figure SIA2.4.

Like the pie charts in the previous Statistics in Action Revisited section, the histograms tend to support the theory.

For females, the histogram for appearance evaluation score is centered at about 17 points, while for males the histogram is centered higher, at about 20 points. Also from the histograms you can see that about 55% of the female patients had a score of less than 20, compared to only about 45% of the males. Again, the histograms seem to indicate that BDD females tend to be more dissatisfied with their looks than males. In later chapters, we'll learn how to attach a measure of reliability to such an inference.

Ⓓ *Data Set:* BDD

Figure SIA2.4
MINITAB side-by-side histograms for Appearance Evaluation by Gender

Exercises 2.25–2.48

Understanding the Principles

2.25 Between a bar chart and a stem-and-leaf display, which one can you use for interval data?

2.26 Explain the difference between a bar graph and a histogram.

2.27 Explain the difference between the stem and the leaf in a stem-and-leaf display.

2.28 Explain the difference between a stem-and-leaf display and a histogram for representing numeric data.

2.29 How many classes are recommended in a histogram of a data set with less than 25 observations?

Learning the Mechanics

2.30 Consider the MINITAB histogram shown below.

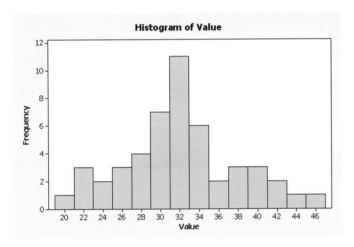

a. Is this a frequency histogram or a relative frequency histogram? Explain.

b. How many class intervals were used in the construction of this histogram?

c. How many measurements are there in the data set described by this histogram?

2.31 Consider the stem-and-leaf display shown here:
NW

Stem	Leaf
5	1
4	457
3	00036
2	1134599
1	2248
0	012

a. How many observations were in the original data set?

b. In the bottom row of the stem-and-leaf display, identify the stem, the leaves, and the numbers in the original data set represented by this stem and its leaves.

c. Re-create all the numbers in the data set, and construct a dot plot.

2.32 Graph the relative frequency histogram for the 500 measurements summarized in the accompanying relative frequency table.

Class Interval	Relative Frequency
.5–2.5	.10
2.5–4.5	.15
4.5–6.5	.25
6.5–8.5	.20
8.5–10.5	.05
10.5–12.5	.10
12.5–14.5	.10
14.5–16.5	.05

2.33 Refer to Exercise 2.32. Calculate the number of the 500 measurements falling into each of the measurement classes. Then graph a frequency histogram of these data.

Applying the Concepts—Basic

2.34 **Irrelevant speech effects.** In a psychological study of short-term memory, *irrelevant speech effects* refer to the degree to which the memorization process is impaired by irrelevant background speech (for example, trying to memorize a list of numbers while listening to a speech in an unfamiliar language). An analysis of irrelevant speech effects was carried out and published in *Acoustical Science & Technology* (Vol. 35, 2014). Subjects performed the memorization task under two conditions: (1) with irrelevant background speech and (2) in silence. The difference in the error rates for the two conditions—called the relative difference in error rate (RDER)—was computed for each subject. A MINITAB histogram summarizing the RDER values for 71 subjects is displayed here.

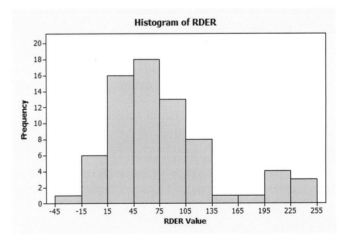

a. Convert the frequency histogram into a relative frequency histogram.

b. What proportion of the subjects had RDER values between 75 and 105?

c. What proportion of the subjects had RDER values below 15?

2.35 **Stability of compounds in new drugs.** Testing the metabolic stability of compounds used in drugs is the cornerstone of new drug discovery. Two important values computed from the testing phase are the fraction of compound unbound to plasma (*fup*) and the fraction of compound unbound to microsomes (*fumic*). A key formula for assessing stability assumes that the fup/fumic ratio is 1. Pharmacologists at Pfizer Global Research and Development investigated this phenomenon and reported the results in *ACS Medicinal Chemistry Letters* (Vol. 1, 2010). The fup/fumic ratio was determined for each of 416 drugs in the Pfizer database. A graph describing the fup/fumic ratios is shown below.

a. What type of graph is displayed?

b. What is the quantitative variable summarized in the graph?

c. Determine the proportion of fup/fumic ratios that fall above 1.

d. Determine the proportion of fup/fumic ratios that fall below .4.

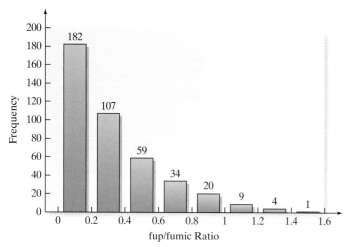

2.36 **Corporate sustainability of CPA firms.** Refer to the *Business and Society* (Mar. 2011) study on the sustainability behaviors of CPA corporations, Exercise 1.26 (p. 50). Recall that *corporate sustainability* refers to business practices designed around social and environmental considerations. Data on the level of support for corporate sustainability were obtained for 992 senior managers. Level of support was measured quantitatively. Simulation was used to convert the data from the study to a scale ranging from 0 to 160 points, where higher point values indicate a higher level of support for sustainability.

a. A histogram for level of support for sustainability is shown below. What type of histogram is produced, frequency or relative frequency?

b. Use the graph to estimate the percentage of the 992 senior managers who reported a high (100 points or greater) level of support for corporate sustainability.

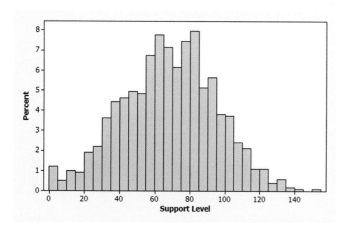

2.37 **Shaft graves in ancient Greece.** Archeologists have discovered a rise in *shaft graves* during the Middle Helladic period in ancient Greece (i.e., around 2000 BC). Shaft graves are named for the beautifully decorated sword shafts that are buried along with the bodies. An analysis of shaft graves was carried out and published in the *American Journal of Archaeology* (Jan. 2014). The table below gives the number of shafts buried at each of 13 recently discovered grave sites. Construct a dot plot for the data. What number of sword shafts was observed most often in the sample of 13 graves?

1	2	3	1	5	6	2	4	1	2	4	2	9

Source: Harrell, K. "The fallen and their swords: A new explanation for the rise of the shaft graves." *American Journal of Archaeology,* Vol. 118, No. 1, January 2014 (Figure 1).

2.38 **Cheek teeth of extinct primates.** Refer to the *American Journal of Physical Anthropology* (Vol. 142, 2010) study of the characteristics of cheek teeth (e.g., molars) in an extinct primate species, Exercise 2.9 (p. 66). In addition to degree of wear, the researchers recorded the dentary depth of molars (in millimeters) for 18 cheek teeth extracted from skulls. These depth measurements are listed in the accompanying table. Summarize the data graphically with a stem-and-leaf display. Is there a particular molar depth that occurs more frequently in the sample?

Data on Dentary Depth (mm) of Molars	
18.12	16.55
19.48	15.70
19.36	17.83
15.94	13.25
15.83	16.12
19.70	18.13
15.76	14.02
17.00	14.04
13.96	16.20

Based on Boyer, D. M., Evans, A. R., and Jernvall, J. "Evidence of dietary differentiation among late Paleocene–early Eocene Plesiadapids (Mammalia, primates)." *American Journal of Physical Anthropology,* Vol. 142, © 2010 (Table A3).

2.39 **Music performance anxiety.** The nature of performance anxiety by music students was investigated in the *British Journal of Music Education* (Mar. 2014). Symptoms of music performance anxiety include increased heart rate, shallow breathing, anxious thoughts, and the avoidance of practice. A Performance Anxiety Inventory (PAI)—measured on a scale from 20 to 80 points—was developed to measure music performance anxiety. The table below gives average PAI values for participants in eight different studies.

54	42	51	39	41	43	55	40

Source: Patston, T. "Teaching stage fright? Implications for music educators." *British Journal of Music Education,* Vol. 31, No. 1, Mar. 2014 (adapted from Figure 1).

a. Construct a stem-and-leaf plot for the data.

b. Locate the PAI value of 42 on the plot, part **a.**

c. Based on the graph, which of the following PAI score ranges is most likely to occur, 20–29, 30–39, 40–49, 50–59, 60–69, or 70–79?

Applying the Concepts—Intermediate

2.40 **Is honey a cough remedy?** Does a teaspoon of honey before bed really calm a child's cough? To test the folk remedy, pediatric researchers at Pennsylvania State University carried out a designed study involving a sample of 105 children who were ill with an upper respiratory tract infection (*Archives of Pediatrics and Adolescent Medicine*, Dec. 2007). On the first night, parents rated their children's cough symptoms on a scale from 0 (no problems at all) to 30 (extremely severe). On the second night, the parents were instructed to give their sick child a dosage of liquid "medicine" prior to bedtime. Unknown to the parents, some were given a dosage of dextromethorphan (DM)—an over-the-counter cough medicine—while others were given a similar dose of honey. Also, a third group of parents (the control group) gave their sick children no dosage at all. Again, the parents rated their children's cough symptoms, and the improvement in total cough symptoms score was determined for each child. The data (improvement scores) for the study are shown in the table below.

a. Construct a dot plot for the coughing improvement scores for the 35 children in the honey dosage group.

b. Refer to part **a.** What coughing improvement score occurred most often in the honey dosage group?

Honey Dosage:	12	11	15	11	10	13	10	4	15	16	9
	14	10	6	10	8	11	12	12	8	12	9
	11	15	10	15	9	13	8	12	10	8	9
	5	12									
DM Dosage:	4	6	9	4	7	7	7	9	12	10	11	6
	3	4	9	12	7	6	8	12	12	4	12	13
	7	10	13	9	4	4	10	15	9			
No Dosage (Control):	5	8	6	1	0	8	12	8	7	7	1	6	7	7
	12	7	9	7	9	5	11	9	5	6	8	8	6	7
	10	9	4	8	7	3	1	4	3					

Based on Paul, I. M., et al. "Effect of honey, dextromethorphan, and no treatment on nocturnal cough and sleep quality for coughing children and their parents." *Archives of Pediatrics and Adolescent Medicine*, Vol. 161, No. 12, Dec. 2007 (data simulated).

c. A MINITAB dot plot for the improvement scores of all three groups is shown below. Note that the green dots represent the children who received a dose of honey, the red dots represent those who got the DM dosage, and the black dots represent the children in the control group. What conclusions can pediatric researchers draw from the graph? Do you agree with the statement (extracted from the article), "Honey may be a preferable treatment for the cough and sleep difficulty associated with childhood upper respiratory tract infection"?

2.41 **Sanitation inspection of cruise ships.** To minimize the potential for gastrointestinal disease outbreaks, all passenger cruise ships arriving at U.S. ports are subject to unannounced sanitation inspections. Ships are rated on a 100-point scale by the Centers for Disease Control and Prevention. A score of 86 or higher indicates that the ship is providing an accepted standard of sanitation. The latest (as of Aug. 2013) sanitation scores for 186 cruise ships are saved in the **SANIT** file. The first five and last five observations in the data set are listed in the following table:

Ship Name	Sanitation Score
Adonia	96
Adventure of the Seas	93
AID Aaura	86
AID ABella	95
AID Aluna	93
⋮	⋮
Voyager of the Seas	96
VSP beta	100
Westerdam	98
Zaandam	100
Zuiderdam	96

Based on National Center for Environmental Health, Centers for Disease Control and Prevention, Aug. 5, 2013.

a. Generate a stem-and-leaf display of the data. Identify the stems and leaves of the graph.

b. Use the stem-and-leaf display to estimate the proportion of ships that have an accepted sanitation standard.

c. Locate the inspection score of 69 (*MS Columbus 2*) on the stem-and-leaf display.

2.42 **Crab spiders hiding on flowers.** Crab spiders use camouflage to hide on flowers while lying in wait to prey on other insects. Ecologists theorize that this natural camouflage also enables the spiders to hide from their own predators, such as birds and lizards. Researchers at the French Museum of Natural History conducted a field test of this theory and published the results in *Behavioral Ecology* (Jan. 2005). They collected a sample of 10 adult female crab spiders, each sitting on the yellow central part of a daisy. The chromatic contrast between each spider and the flower it was sitting on was measured numerically

MINITAB dot plot for Exercise 2.40

with a spectroradiometer, on which higher values indicate a greater contrast (and, presumably, easier detection by predators). The data for the 10 crab spiders are shown in the following table.

57	75	116	37	96	61	56	2	43	32

Based on Thery, M., et al. "Specific color sensitivities of prey and predator explain camouflage in different visual systems." *Behavioral Ecology*, Vol. 16, No. 1, Jan. 2005 (Table 1).

a. Summarize the chromatic contrast measurements for the 10 spiders with a stem-and-leaf display.

b. For birds, the detection threshold is 70. (A contrast of 70 or greater allows the bird to see the spider.) Locate the spiders that can be seen by bird predators by circling their respective contrast values on the stem-and-leaf display.

c. Use the result of part **b** to make an inference about the likelihood of a bird detecting a crab spider sitting on the yellow central part of a daisy.

2.43 **Sound waves from a basketball.** An experiment was conducted to characterize sound waves in a spherical cavity (*American Journal of Physics*, June 2010). A fully inflated basketball, hanging from rubber bands, was struck with a metal rod, producing a series of metallic-sounding pings. Of particular interest were the frequencies of sound waves resulting from the first 24 resonances (echoes). A mathematical formula, well known in physics, was used to compute the theoretical frequencies. These frequencies (measured in hertz) are listed in the table. Use a graphical method to describe the distribution of sound frequencies for the first 24 resonances.

Resonance	Frequency
1	979
2	1572
3	2113
4	2122
5	2659
6	2795
7	3181
8	3431
9	3638
10	3694
11	4038
12	4203
13	4334
14	4631
15	4711
16	4993
17	5130
18	5210
19	5214
20	5633
21	5779
22	5836
23	6259
24	6339

Based on Russell, D. A. "Basketballs as spherical acoustic cavities." *American Journal of Physics*, Vol. 78, No. 6, June 2010 (Table I).

2.44 **Research on eating disorders.** Data from a psychology experiment were reported and analyzed in *The American Statistician* (May 2001). Two samples of female students participated in the experiment. One sample consisted of 11 students known to suffer from the eating disorder bulimia; the other sample consisted of 14 students with normal eating habits. Each student completed a questionnaire from which a "fear of negative evaluation" (FNE) score was produced. (The higher score, the greater was the fear of negative evaluation.) The data are displayed in the following table:

Bulimic students:	21	13	10	20	25	19	16	21	24	13	14		
Normal students:	13	6	16	13	8	19	23	18	11	19	7	10	15	20

Based on Randles, R. H. "On neutral responses (zeros) in the sign test and ties in the Wilcoxon–Mann–Whitney test." *The American Statistician*, Vol. 55, No. 2, May 2001 (Figure 3).

a. Construct a dot plot or stem-and-leaf display for the FNE scores of all 25 female students.

b. Highlight the bulimic students on the graph you made in part **a**. Does it appear that bulimics tend to have a greater fear of negative evaluation? Explain.

c. Why is it important to attach a measure of reliability to the inference made in part **b**?

2.45 **Research on brain specimens.** The *postmortem interval* (PMI) is defined as the time elapsed (in days) between death and an autopsy. Knowledge of the PMI is considered essential to conducting medical research on human cadavers. The data in the accompanying table are the PMIs of 22 human brain specimens obtained at autopsy in a recent study (*Brain and Language*, June 1995). Describe the PMI data graphically with a dot plot. On the basis of the plot, make a summary statement about the PMIs of the 22 human brain specimens.

Postmortem Intervals for 22 Human Brain Specimens							
5.5	14.5	6.0	5.5	5.3	5.8	11.0	6.1
7.0	14.5	10.4	4.6	4.3	7.2	10.5	6.5
3.3	7.0	4.1	6.2	10.4	4.9		

Based on Hayes, T. L., and Lewis, D. A. "Anatomical specialization of the anterior motor speech area: Hemispheric differences in magnopyramidal neurons." *Brain and Language*, Vol. 49, No. 3, June 1995, p. 292 (Table 1).

Applying the Concepts—Advanced

2.46 **State SAT scores.** Educators are constantly evaluating the efficacy of public schools in the education and training of U.S. students. One quantitative assessment of change over time is the difference in scores on the SAT, which has been used for decades by colleges and universities as one criterion for admission. The **SAT** file contains average SAT scores for each of the 50 states and the District of Columbia for 2011 and 2014. Selected observations are shown in the following table:

State	2011	2014
Alabama	1623	1608
Alaska	1513	1495
Arizona	1539	1551
Arkansas	1692	1697
California	1513	1505
⋮	⋮	⋮
Wisconsin	1767	1771
Wyoming	1692	1757

Based on College Entrance Examination Board, 2014.

a. Use graphs to display the two SAT score distributions. How have the distributions of state scores changed from 2011 to 2014?

b. As another method of comparing the 2011 and 2014 average SAT scores, compute the **paired difference** by subtracting the 2011 score from the 2014 score for each state. Summarize these differences with a graph.

c. Interpret the graph you made in part **b**. How do your conclusions compare with those of part **a**?

d. Identify the state with the largest improvement in the SAT score between 2011 and 2014.

2.47 **Phishing attacks to e-mail accounts.** *Phishing* is the term used to describe an attempt to extract personal/financial information (e.g., PIN numbers, credit card information, bank account numbers) from unsuspecting people through fraudulent e-mail. An article in *Chance* (Summer 2007) demonstrates how statistics can help identify phishing attempts and make e-commerce safer. Data from an actual phishing attack against an organization were used to determine whether the attack may have been an "inside job" that originated within the company. The company set up a publicized e-mail account—called a "fraud box"—which enabled employees to notify it if they suspected an e-mail phishing attack. The interarrival times, i.e., the time differences (in seconds), for 267 fraud box e-mail notifications were recorded. Researchers showed that if there is

minimal or no collaboration or collusion from within the company, the interarrival times would have a frequency distribution similar to the one shown in the accompanying figure. The 267 interarrival times are saved in the **PHISH** file. Construct a frequency histogram for the interarrival times. Give your opinion on whether the phishing attack against the organization was an "inside job."

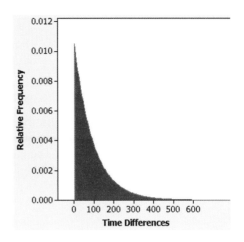

2.48 **Mineral flotation in water study.** A high concentration of calcium and gypsum in water can affect the water quality and limit mineral flotation. In *Minerals Engineering* (Vol. 46–47, 2013), chemical and materials engineers published a study of the impact of calcium and gypsum on the flotation properties of silica in water. Solutions of deionized water were prepared both with and without calcium/gypsum, and the level of flotation of silica in the solution was measured using a variable called *zeta potential* (measured in millivolts, mV). Assume that 50 specimens for each type of liquid solution were prepared and tested for zeta potential. The data (simulated, based on information provided in the journal article) are provided in the table. Create side-by-side graphs to compare the zeta potential distributions for the two types of solutions. How does the addition of calcium/gypsum to the solution affect water quality (measured by zeta potential of silica)?

Without calcium/gypsum									
−47.1	−53.0	−50.8	−54.4	−57.4	−49.2	−51.5	−50.2	−46.4	−49.7
−53.8	−53.8	−53.5	−52.2	−49.9	−51.8	−53.7	−54.8	−54.5	−53.3
−50.6	−52.9	−51.2	−54.5	−49.7	−50.2	−53.2	−52.9	−52.8	−52.1
−50.2	−50.8	−56.1	−51.0	−55.6	−50.3	−57.6	−50.1	−54.2	−50.7
−55.7	−55.0	−47.4	−47.5	−52.8	−50.6	−55.6	−53.2	−52.3	−45.7

With calcium/gypsum									
−9.2	−11.6	−10.6	−8.0	−10.9	−10.0	−11.0	−10.7	−13.1	−11.5
−11.3	−9.9	−11.8	−12.6	−8.9	−13.1	−10.7	−12.1	−11.2	−10.9
−9.1	−12.1	−6.8	−11.5	−10.4	−11.5	−12.1	−11.3	−10.7	−12.4
−11.5	−11.0	−7.1	−12.4	−11.4	−9.9	−8.6	−13.6	−10.1	−11.3
−13.0	−11.9	−8.6	−11.3	−13.0	−12.2	−11.3	−10.5	−8.8	−13.4

2.3 Numerical Measures of Central Tendency

Center
a.

|← Spread →|
b.

Figure 2.14
Numerical descriptive measures

When we speak of a data set, we refer to either a sample or a population. If statistical inference is our goal, we'll ultimately wish to use sample **numerical descriptive measures** to make inferences about the corresponding measures for a population.

As you'll see, a large number of numerical methods are available to describe quantitative data sets. Most of these methods measure one of two data characteristics:

1. The **central tendency** of the set of measurements—that is, the tendency of the data to cluster, or center, about certain numerical values. (See Figure 2.14a.)

2. The **variability** of the set of measurements—that is, the spread of the data. (See Figure 2.14b.)

In this section, we concentrate on **measures of central tendency**. In the next section, we discuss measures of variability.

The most popular and best understood measure of central tendency for a quantitative data set is the *arithmetic mean* (or simply the mean) of the data set.

The **mean** of a set of quantitative data is the sum of the measurements, divided by the number of measurements contained in the data set.

In everyday terms, the mean is the average value of the data set and is often used to represent a "typical" value. We denote the **mean** of a sample of measurements by $\bar{x}$ (read "x-bar") and represent the formula for its calculation as shown in the following box:

Formula for a Sample Mean

$$\bar{x} = \frac{\sum\limits_{i=1}^{n} x_i}{n}$$

[*Note:* $\sum\limits_{i=1}^{n} x_i = (x_1 + x_2 + \cdots + x_n)$. For more details and examples on this **summation notation**, see Appendix A.]

Example 2.3

Computing the Sample Mean

Problem Calculate the mean of the following five sample measurements: 5, 3, 8, 5, 6.

Solution Using the definition of sample mean and the summation notation, we find that

$$\bar{x} = \frac{\sum\limits_{i=1}^{5} x_i}{5} = \frac{5 + 3 + 8 + 5 + 6}{5} = \frac{27}{5} = 5.4$$

Thus, the mean of this sample is 5.4.

Look Back There is no specific rule for rounding when calculating $\bar{x}$ because $\bar{x}$ is specifically defined to be the sum of all measurements, divided by n; that is, it is a specific fraction. When $\bar{x}$ is used for descriptive purposes, it is often convenient to round the calculated value of $\bar{x}$ to the number of significant figures used for the original measurements. When $\bar{x}$ is to be used in other calculations, however, it may be necessary to retain more significant figures.

■ Now Work Exercise 2.57

Example   2.4   **EPAGAS**

Finding the Mean on a Printout—Mean Gas Mileage

Figure 2.15
SAS numerical descriptive measures for 100 EPA gas mileages

Problem Calculate the sample mean for the 100 EPA mileages given in Table 2.2.

Solution The mean gas mileage for the 100 cars is denoted

$$\bar{x} = \frac{\sum_{i=1}^{100} x_i}{100}$$

Rather than compute $\bar{x}$ by hand (or even with a calculator), we employed SAS to compute the mean. The SAS printout is shown in Figure 2.15. The sample mean, highlighted on the printout, is $\bar{x} = 36.9940$.

The MEANS Procedure						
Analysis Variable : MPG						
Mean	Std Dev	Variance	N	Minimum	Maximum	Median
36.9940000	2.4178971	5.8462263	100	30.0000000	44.9000000	37.0000000

Look Back Given this information, you can visualize a distribution of gas mileage readings centered in the vicinity of $\bar{x} \approx 37$. An examination of the relative frequency histogram (Figure 2.10) confirms that $\bar{x}$ does in fact fall near the center of the distribution.

The sample mean $\bar{x}$ will play an important role in accomplishing our objective of making inferences about populations on the basis of information about the sample. For this reason, we need to use a different symbol for the *mean of a population*—the mean of the set of measurements on every unit in the population. We use the Greek letter μ (mu) for the population mean.

> **Symbols for the Sample Mean and the Population Mean**
>
> In this text, we adopt a general policy of using Greek letters to represent numerical descriptive measures of the population and Roman letters to represent corresponding descriptive measures of the sample. The symbols for the mean are
>
> $$\bar{x} = \text{Sample mean} \qquad \mu = \text{Population mean}$$

We'll often use the sample mean $\bar{x}$ to estimate (make an inference about) the population mean μ. For example, the EPA mileages for the population consisting of *all* cars has a mean equal to some value μ. Our sample of 100 cars yielded mileages with a mean of $\bar{x} = 36.9940$. If, as is usually the case, we don't have access to the measurements for the entire population, we could use $\bar{x}$ as an estimator or approximator for μ. Then we'd need to know something about the reliability of our inference. That is, we'd need to know how accurately we might expect $\bar{x}$ to estimate μ. In Chapter 7, we'll find that this accuracy depends on two factors:

1. The *size of the sample*. The larger the sample, the more accurate the estimate will tend to be.

2. The *variability, or spread, of the data*. All other factors remaining constant, the more variable the data, the less accurate is the estimate.

Another important measure of central tendency is the **median**.

> The **median** of a quantitative data set is the middle number when the measurements are arranged in ascending (or descending) order.

Figure 2.16
Location of the median

The median is of most value in describing large data sets. If a data set is characterized by a relative frequency histogram (Figure 2.16), the median is the point on the x-axis such that half the area under the histogram lies above the median and half lies below. [*Note:* In Section 2.2, we observed that the relative frequency associated with a particular interval on the x-axis is proportional to the amount of area under the histogram that lies above the interval.] We denote the *median* of a *sample* by M. Like with the population mean, we use a Greek letter (η) to represent the population median.

Calculating a Sample Median M

Arrange the n measurements from the smallest to the largest.

1. If n is odd, M is the middle number.

2. If n is even, M is the mean of the middle two numbers.

Symbols for the Sample and Population Median

$$M = \text{Sample median}$$

$$\eta = \text{Population median}$$

Example	**2.5**

Computing the Median

Problem Consider the following sample of $n = 7$ measurements: 5, 7, 4, 5, 20, 6, 2.

a. Calculate the median M of this sample.

b. Eliminate the last measurement (the 2), and calculate the median of the remaining $n = 6$ measurements.

Solution

a. The seven measurements in the sample are ranked in ascending order: 2, 4, 5, 5, 6, 7, 20. Because the number of measurements is odd, the median is the middle measurement. Thus, the median of this sample is $M = 5$.

b. After removing the 2 from the set of measurements, we rank the sample measurements in ascending order as follows: 4, 5, 5, 6, 7, 20. Now the number of measurements is even, so we average the middle two measurements. The median is $M = (5 + 6)/2 = 5.5$.

Look Back When the sample size n is even (as in part **b**), exactly half of the measurements will fall below the calculated median M. However, when n is odd (as in part **a**), the percentage of measurements that fall below M is approximately 50%. The approximation improves as n increases.

■ **Now Work Exercise 2.55**

In certain situations, the median may be a better measure of central tendency than the mean. In particular, the median is less sensitive than the mean to extremely large or small measurements. Note, for instance, that all but one of the measurements in part **a** of Example 2.5 are close to $x = 5$. The single relatively large measurement, $x = 20$, does not affect the value of the median, 5, but it causes the mean, $\bar{x} = 7$, to lie to the right of most of the measurements.

As another example of data for which the central tendency is better described by the median than the mean, consider the household incomes of a community being studied by a sociologist. The presence of just a few households with very high incomes will affect the mean more than the median. Thus, the median will provide a more accurate picture of the typical income for the community. The mean could exceed the vast majority of the sample measurements (household incomes), making it a misleading measure of central tendency.

Example 2.6

EPAGAS

Finding the Median on a Printout—Median Gas Mileage

Problem Calculate the median for the 100 EPA mileages given in Table 2.2. Compare the median with the mean computed in Example 2.4.

Solution For this large data set, we again resort to a computer analysis. The median is highlighted on the SAS printout displayed in Figure 2.15 (p. 83). You can see that the median is 37.0. Thus, half of the 100 mileages in the data set fall below 37.0 and half lie above 37.0. Note that the median, 37.0, and the mean, 36.9940, are almost equal, a relationship that indicates a lack of **skewness** in the data. In other words, the data exhibit a tendency to have as many measurements in the left tail of the distribution as in the right tail. (Recall the histogram of Figure 2.10.)

Look Back In general, extreme values (large or small) affect the mean more than the median, since these values are used explicitly in the calculation of the mean. The median is not affected directly by extreme measurements, since only the middle measurement (or two middle measurements) is explicitly used to calculate the median. Consequently, if measurements are pulled toward one end of the distribution, the mean will shift toward that tail more than the median will.

■

A data set is said to be **skewed** if one tail of the distribution has more extreme observations than the other tail.

A comparison of the mean and the median gives us a general method for detecting skewness in data sets, as shown in the next box. With *rightward skewed* data, the right tail (high end) of the distribution has more extreme observations. These few, but large, measurements tend to pull the mean away from the median toward the right; that is, rightward skewness typically indicates that the mean is greater than the median. Conversely, with *leftward skewed* data, the left tail (low end) of the distribution has more extreme observations. These few, but small, measurements also tend to pull the mean away from the median, but toward the left; consequently, leftward skewness typically implies that the mean is smaller than the median.

Detecting Skewness by Comparing the Mean and the Median

If the data set is skewed to the right, then typically the median is less than the mean.

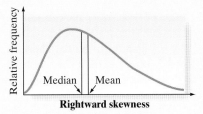

If the data set is symmetric, then the mean equals the median.

If the data set is skewed to the left, then typically the mean is less than the median.

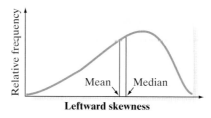

Leftward skewness

Now Work Exercise 2.54

A third measure of central tendency is the **mode** of a set of measurements.

The **mode** is the measurement that occurs most frequently in the data set.

Therefore, the mode shows where the data tend to concentrate.

Example	2.7

Finding the Mode

Problem Each of 10 taste testers rated a new brand of barbecue sauce on a 10-point scale, where 1 = awful and 10 = excellent. Find the mode for the following 10 ratings:

$$8 \quad 7 \quad 9 \quad 6 \quad 8 \quad 10 \quad 9 \quad 9 \quad 5 \quad 7$$

Solution Since 9 occurs most often (three times), the mode of the ten taste ratings is 9.

Look Back Note that the data are actually qualitative in nature (e.g., "awful," "excellent"). The mode is particularly useful for describing qualitative data. The modal category is simply the category (or class) that occurs most often.

■ **Now Work Exercise 2.56**

Because it emphasizes data concentration, the mode is also used with quantitative data sets to locate the region in which much of the data is concentrated. A retailer of men's clothing would be interested in the modal neck size and sleeve length of potential customers. The modal income class of the laborers in the United States is of interest to the U.S. Department of Labor.

For some quantitative data sets, the mode may not be very meaningful. For example, consider the EPA mileage ratings in Table 2.2. A reexamination of the data reveals that the gas mileage of 37.0 occurs most often (four times). However, the mode of 37.0 is not particularly useful as a measure of central tendency.

A more meaningful measure can be obtained from a relative frequency histogram for quantitative data. The measurement class containing the largest relative frequency is called the **modal class**. Several definitions exist for locating the position of the mode within a modal class, but the simplest is to define the mode as the midpoint of the modal class. For example, examine the frequency histogram for the EPA mileage ratings in Figure 2.10 (p. 73). You can see that the modal class is the interval 37–38. The mode (the midpoint) is 37.5 This modal class (and the mode itself) identifies the area in which the data are most concentrated and, in that sense, is a measure of central tendency. However, for most applications involving quantitative data, the mean and median provide more descriptive information than the mode.

Example	2.8	

Comparing the Mean, Median, and Mode—Earthquake Aftershocks

Problem Seismologists use the term "aftershock" to describe the smaller earthquakes that follow a main earthquake. Following the Northridge earthquake, the Los Angeles area experienced a record 2,929 aftershocks in a three-week period. The magnitudes (measured on the Richter scale) of these aftershocks as well as their interarrival times (in minutes) were recorded by the U.S. Geological Survey. (The data are saved in the **QUAKE** file.) Today seismologists continue to use these data to model future earthquake characteristics. Find and interpret the mean, median, and mode for both of these variables. Which measure of central tendency is better for describing the magnitude distribution? The distribution of interarrival times?

Solution Measures of central tendency for the two variables, magnitude and interarrival time, were produced using MINITAB. The means, medians, and modes are displayed in Figure 2.17.

Descriptive Statistics: MAGNITUDE, INT-TIME

Variable	N	Mean	Median	Mode	N for Mode
MAGNITUDE	2929	2.1197	2.0000	1.8	298
INT-TIME	2928	9.771	6.000	2	354

Figure 2.17
MINITAB descriptive statistics for earthquake data

For magnitude, the mean, median, and mode are 2.12, 2.00, and 1.8, respectively, on the Richter scale. The average magnitude is 2.12; half the magnitudes fall below 2.0; and the most commonly occurring magnitude is 1.8. These values are nearly identical, with the mean slightly larger than the median. This implies a slight rightward skewness in the data, which is shown graphically in the MINITAB histogram for magnitude displayed in Figure 2.18a. Because the distribution is nearly symmetric, any of the three measures would be adequate for describing the "center" of the earthquake aftershock magnitude distribution.

The mean, median, and mode of the interarrival times of the aftershocks are 9.77, 6.0, and 2.0 minutes, respectively. On average, the aftershocks arrive 9.77 minutes apart; half the aftershocks have interarrival times below 6.0 minutes; and the most commonly occurring interarrival time is 2.0 minutes. Note that the mean is much larger than the median, implying that the distribution of interarrival times is highly skewed to the right. This extreme rightward skewness is shown graphically in the histogram, in Figure 2.18b. The skewness is due to several exceptionally large interarrival times. Consequently, we would probably want to use the median of 6.0 minutes as the "typical" interarrival time for the aftershocks. You can see that the mode of 2.0 minutes is not very descriptive of the "center" of the interarrival time distribution.

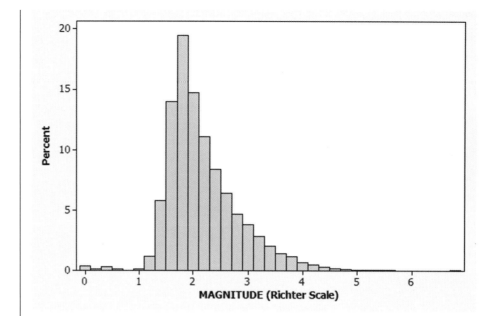

Figure 2.18a
MINITAB Histogram for
Magnitudes of Aftershocks

Figure 2.18b
MINITAB Histogram for
Inter-Arrival Times of
Aftershocks

Look Back The choice of which measure of central tendency to use will depend on the properties of the data set analyzed and the application of interest. Consequently, it is vital that you understand how the mean, median, and mode are computed.

■ **Now Work Exercise 2.68**

Exercises 2.49–2.72

Understanding the Principles

2.49 Give three different measures of central tendency.

2.50 Explain the difference between the mean and the median as measures of central tendency.

2.51 Which of the three measures of central tendency may not be unique for a given data set?

2.52 What are the typical relative positions of the mean and the median for a skewed distribution?

2.53 Between the mean and the median, which one is affected by the minimum and maximum values of the observed data?

2.54 Describe how the mean compares with the median for a distribution as follows:
[NW]
a. Skewed to the left
b. Skewed to the right
c. Symmetric

Learning the Mechanics

2.55 Calculate the mean and median of the following grade [NW] point averages:
[D]
L02055
3.2 2.5 2.1 3.7 2.8 2.0

2.56 Calculate the mode, mean, and median of the following [NW] data:
[D]
L02056
18 10 15 13 17 15 12 15 18 16 11

2.57 Calculate the mean for samples for which
[NW]

    a. $n = 10, \sum x = 85$

    b. $n = 16, \sum x = 400$

    c. $n = 45, \sum x = 35$

    d. $n = 18, \sum x = 242$

2.58 Construct one data set consisting of five measurements, and another consisting of six measurements, for which the medians are equal.

2.59 Calculate the mean, median, and mode for each of the following samples:
    a. $7, -2, 3, 3, 0, 4$
    b. $2, 3, 5, 3, 2, 3, 4, 3, 5, 1, 2, 3, 4$
    c. $51, 50, 47, 50, 48, 41, 59, 68, 45, 37$

ⓓ Applet Exercise 2.1

Use the applet entitled *Mean versus Median* to find the mean and median of each of the three data sets presented in Exercise 2.59. For each data set, set the lower limit to a number less than all of the data, set the upper limit to a number greater than all of the data, and then click on *Update*. Click on the approximate location of each data item on the number line. You can get rid of a point by dragging it to the trash can. To clear the graph between data sets, simply click on the trash can.

    a. Compare the means and medians generated by the applet with those you calculated by hand in Exercise 2.59. If there are differences, explain why the applet might give values slightly different from the hand-calculated values.

    b. Despite providing only approximate values of the mean and median of a data set, describe some advantages of using the applet to find those values.

ⓓ Applet Exercise 2.2

Use the applet *Mean versus Median* to illustrate your descriptions in Exercise 2.54. For each part **a**, **b**, and **c**, create a data set with 10 items that has the given property. Using the applet, verify that the mean and median have the relationship you described in Exercise 2.54.

ⓓ Applet Exercise 2.3

Use the applet *Mean versus Median* to study the effect that an extreme value has on the difference between the mean and median. Begin by setting appropriate limits and plotting the following data on the number line provided in the applet:

0 6 7 7 8 8 8 9 9 10

    a. Describe the shape of the distribution and record the value of the mean and median. On the basis of the shape of the distribution, do the mean and median have the relationship that you would expect?

    b. Replace the extreme value of 0 with 2, then 4, and then 6. Record the mean and median each time. Describe what is happening to the mean as 0 is replaced, in turn, by the higher numbers stated. What is happening to the median? How is the difference between the mean and the median changing?

    c. Now replace 0 with 8. What values does the applet give you for the mean and the median? Explain why the mean and the median should now be the same.

Applying the Concepts—Basic

2.60 **Shaft graves in ancient Greece.** Refer to the *American*
[D] *Journal of Archaeology* (Jan. 2014) study of sword shaft
SHAFTS graves in ancient Greece, Exercise 2.37 (p. 78). The number of sword shafts buried at each of 13 recently discovered grave sites is reproduced in the following table.

1	2	3	1	5	6	2	4	1	2	4	2	9

Source: Harrell, K. "The fallen and their swords: A new explanation for the rise of the shaft graves." *American Journal of Archaeology,* Vol. 118, No. 1, January 2014 (Figure 1).

    a. Calculate the mean of the data. Interpret the result.
    b. Calculate the median of the data. Interpret the result.
    c. Find the mode of the data. Interpret the result.

2.61 **Characteristics of a rock fall.** In *Environmental Geology*
[D] (Vol. 58, 2009) computer simulation was employed to
ROCKS estimate how far a block from a collapsing rock wall will bounce—called *rebound length*—down a soil slope. Based on the depth, location, and angle of block-soil impact marks left on the slope from an actual rockfall, the following 13 rebound lengths (in meters) were estimated. Compute the mean and median of the rebound lengths and interpret these values.

10.94	13.71	11.38	7.26	17.83	11.92	11.87	5.44	13.35	4.90	5.85	5.10	6.77

Based on Paronuzzi, P. "Rockfall-induced block propagation on a soil slope, northern Italy," *Environmental Geology*, Vol. 58, 2009 (Table 2).

2.62 **Music performance anxiety.** Refer to the *British Journal of*
[D] *Music Education* (Mar. 2014) study of music performance
PAI anxiety, Exercise 2.39 (p. 78). Scores on the Performance Anxiety Inventory (PAI) scale for participants in eight different studies are reproduced in the table.

54	42	51	39	41	43	55	40

Source: Patston, T. "Teaching stage fright? Implications for music educators." *British Journal of Music Education,* Vol. 31, No. 1, Mar. 2014 (adapted from Figure 1).

    a. Find and interpret the mean of the PAI scores.
    b. Find and interpret the median of the PAI scores.
    c. Suppose the PAI score of 39 results from a study involving development-delayed children; consequently, the researchers eliminated this data value from the analysis. What impact does this have on the value of the mean? The median?

2.63 **Corporate sustainability of CPA firms.** Refer to the
[D] *Business and Society* (Mar. 2011) study on the sustain-
SUSTAIN ability behaviors of CPA corporations, Exercise 2.36 (p. 78). Recall that the level of support for corporate sustainability (measured on a quantitative scale ranging

MINITAB output for Exercise 2.63

Descriptive Statistics: Support

```
                                                           N for
Variable    N    Mean   Minimum  Median  Maximum  Mode    Mode
Support    992  67.755    0.000  68.000  155.000    64      20
```

from 0 to 160 points) was obtained for each of 992 senior managers at CPA firms. Numerical measures of central tendency for level of support are shown in the accompanying MINITAB printout.

a. Locate the mean on the printout. Comment on the accuracy of the statement: "On average, the level of support for corporate sustainability for the 992 senior managers is 67.76 points."

b. Locate the median on the printout. Comment on the accuracy of the statement: "Half of the 992 senior managers reported a level of support for corporate sustainability below 68 points."

c. Locate the mode on the printout. Comment on the accuracy of the statement: "Most of the 992 senior managers reported a level of support for corporate sustainability below 64 points."

d. Based on the values of the measures of central tendency, make a statement about the type of skewness (if any) that exists in the distribution of 992 support levels. Check your answer by examining the histogram shown in Exercise 2.36.

2.64 **Is honey a cough remedy?** Refer to the *Archives of Pediatrics and Adolescent Medicine* (Dec. 2007) study of honey as a remedy for coughing, Exercise 2.40 (p. 79). Recall that the 105 ill children in the sample were randomly divided into three groups: those who received a dosage of an over-the-counter cough medicine (DM), those who received a dosage of honey (H), and those who received no dosage (control group). The coughing improvement scores for the patients are reproduced in the table below.

Honey Dosage:	12	11	15	11	10	13	10	4	15	16	9
	14	10	6	10	8	11	12	12	8	12	9
	11	15	10	15	9	13	8	12	10	8	9
	5	12									
DM Dosage:	4	6	9	4	7	7	7	9	12	10	11
	6	3	4	9	12	7	6	8	12	12	4
	12	13	7	10	13	9	4	4	10	15	9
No Dosage (Control):	5	8	6	1	0	8	12	8	7	7	1
	6	7	7	12	7	9	7	9	5	11	9
	5	6	8	8	6	7	10	9	4	8	7
	3	1	4	3							

Based on Paul, I. M., et al. "Effect of honey, dextromethorphan, and no treatment on nocturnal cough and sleep quality for coughing children and their parents." *Archives of Pediatrics and Adolescent Medicine*, Vol. 161, No. 12, Dec. 2007 (data simulated).

a. Find the median improvement score for the honey dosage group.

b. Find the median improvement score for the DM dosage group.

c. Find the median improvement score for the control group.

d. Based on the results, parts **a–c**, what conclusions can pediatric researchers draw? (We show how to support these conclusions with a measure of reliability in subsequent chapters.)

Applying the Concepts—Intermediate

2.65 **Cheek teeth of extinct primates.** Refer to the *American Journal of Physical Anthropology* (Vol. 142, 2010) study MOLARS of the characteristics of cheek teeth (e.g., molars) in an extinct primate species, Exercise 2.38 (p. 78). The data on dentary depth of molars (in millimeters) for 18 cheek teeth extracted from skulls are reproduced below.

Data on Dentary Depth (mm) of Molars	
18.12	16.55
19.48	15.70
19.36	17.83
15.94	13.25
15.83	16.12
19.70	18.13
15.76	14.02
17.00	14.04
13.96	16.20

Based on Boyer, D. M., Evans, A. R., and Jernvall, J. "Evidence of dietary differentiation among late Paleocene–early Eocene Plesiadapids (Mammalia, primates)." *American Journal of Physical Anthropology*, Vol. 142, © 2010 (Table A3).

a. Find and interpret the mean of the data set. If the largest depth measurement in the sample were doubled, how would the mean change? Would it increase or decrease?

b. Find and interpret the median of the data set. If the largest depth measurement in the sample were doubled, how would the median change? Would it increase or decrease?

c. Note that there is no single measurement that occurs more than once. How does this fact impact the mode?

2.66 **Ranking driving performance of professional golfers.** A group of Northeastern University researchers developed PGA a new method for ranking the total driving performance of golfers on the Professional Golf Association (PGA) tour (*The Sport Journal*, Winter 2007). The method requires knowing a golfer's average driving distance (yards) and driving accuracy (percent of drives that land in the fairway). The values of these two variables are used to compute a driving performance index. Data for the top 40 PGA golfers (ranked by the new method) are saved in the **PGA** file.

The first five and last five observations are listed in the accompanying table.

Rank	Player	Driving Distance (yards)	Driving Accuracy (%)	Driving Performance Index
1	Woods	316.1	54.6	3.58
2	Perry	304.7	63.4	3.48
3	Gutschewski	310.5	57.9	3.27
4	Wetterich	311.7	56.6	3.18
5	Hearn	295.2	68.5	2.82
⋮	⋮	⋮	⋮	⋮
36	Senden	291	66	1.31
37	Mickelson	300	58.7	1.30
38	Watney	298.9	59.4	1.26
39	Trahan	295.8	61.8	1.23
40	Pappas	309.4	50.6	1.17

Based on Wiseman, F., et. al. "A new method for ranking total driving performance on the PGA Tour," *Sports Journal*, Vol. 10, No. 1, Winter 2007 (Table 2).

a. Find the mean, median, and mode for the 40 driving performance index values.

b. Interpret each of the measures of central tendency calculated in part **a**.

c. Use the results from part **a** to make a statement about the type of skewness in the distribution of driving performance indexes. Support your statement with a graph.

2.67 **Symmetric or skewed?** Would you expect the data sets that follow to possess relative frequency distributions that are symmetric, skewed to the right, or skewed to the left? Explain.

a. The salaries of all persons employed by a large university

b. The grades on an easy test

c. The grades on a difficult test

d. The amounts of time students in your class studied last week

e. The ages of automobiles on a used-car lot

f. The amounts of time spent by students on a difficult examination (maximum time is 50 minutes)

2.68 NW ANTS **Mongolian desert ants.** The *Journal of Biogeography* (Dec. 2003) published an article on the first comprehensive study of ants in Mongolia (Central Asia). Botanists placed seed baits at 11 study sites and observed the ant species attracted to each site. Some of the data recorded at each study site are provided in the table below.

a. Find the mean, median, and mode for the number of ant species discovered at the 11 sites. Interpret each of these values.

b. Which measure of central tendency would you recommend to describe the center of the number-of-ant-species distribution? Explain.

c. Find the mean, median, and mode for the percentage of total plant cover at the five Dry Steppe sites only.

d. Find the mean, median, and mode for the percentage of total plant cover at the six Gobi Desert sites only.

e. On the basis of the results of parts **c** and **d**, does the center of the distribution for total plant cover percentage appear to be different at the two regions?

2.69 SAND **Permeability of sandstone during weathering.** Natural stone, such as sandstone, is a popular building construction material. An experiment was carried out in order to better understand the decay properties of sandstone when exposed to the weather (*Geographical Analysis*, Vol. 42, 2010). Blocks of sandstone were cut into 300 equal-sized slices and the slices randomly divided into three groups of 100 slices each. Slices in group A were not exposed to any type of weathering; slices in group B were repeatedly sprayed with a 10% salt solution (to simulate wetting by driven rain) under temperate conditions; and slices in group C were soaked in a 10% salt solution and then dried (to simulate blocks of sandstone exposed during a wet winter and dried during a hot summer). All sandstone slices were then tested for permeability, measured in milliDarcies (mD). These permeability values measure pressure decay as a function of time. The data for the study (simulated) are saved in the **SAND** file. Measures of central tendency for the permeability measurements of each sandstone group are displayed in the MINITAB printout on p. 92.

Data for Exercise 2.68

Site	Region	Annual Rainfall (mm)	Max. Daily Temp. (°C)	Total Plant Cover (%)	Number of Ant Species	Species Diversity Index
1	Dry Steppe	196	5.7	40	3	.89
2	Dry Steppe	196	5.7	52	3	.83
3	Dry Steppe	179	7.0	40	52	1.31
4	Dry Steppe	197	8.0	43	7	1.48
5	Dry Steppe	149	8.5	27	5	.97
6	Gobi Desert	112	10.7	30	49	.46
7	Gobi Desert	125	11.4	16	5	1.23
8	Gobi Desert	99	10.9	30	4	
9	Gobi Desert	125	11.4	56	4	.76
10	Gobi Desert	84	11.4	22	5	1.26
11	Gobi Desert	115	11.4	14	4	.69

Based on Pfeiffer, M., et al. "Community organization and species richness of ants in Mongolia along an ecological gradient from steppe to Gobi desert." *Journal of Biogeography*, Vol. 30, No. 12, Dec. 2003 (Table 1 and 2).

MINITAB output for Exercise 2.69

Descriptive Statistics: PermA, PermB, PermC

```
                                                    N for
Variable   N    Mean   Median                 Mode  Mode
PermA     100  73.62   70.45      59.9, 60, 60.1, 60.4    2
PermB     100 128.54  139.30  146.4, 146.6, 147.9, 148.3  3
PermC     100  83.07   78.65                  70.9   3

The data contain at least five mode values. Only the smallest four are shown.
```

a. Interpret the mean and median of the permeability measurements for group A sandstone slices.
b. Interpret the mean and median of the permeability measurements for group B sandstone slices.
c. Interpret the mean and median of the permeability measurements for group C sandstone slices.
d. Interpret the mode of the permeability measurements for group C sandstone slices.
e. The lower the permeability value, the slower the pressure decay in the sandstone over time. Which type of weathering (type B or type C) appears to result in faster decay?

2.70 **Mineral flotation in water study.** Refer to the *Minerals Engineering* (Vol. 46–47, 2013) study of the impact of calcium and gypsum on the flotation properties of silica in water, Exercise 2.48 (p. 81). The *zeta potential* (mV) was determined for each of 50 liquid solutions prepared without calcium/gypsum and for 50 liquid solutions prepared with calcium/gypsum.
a. Find the mean, median, and mode for the zeta potential measurements of the liquid solutions prepared without calcium/gypsum. Interpret these values.
b. Find the mean, median, and mode for the zeta potential measurements of the liquid solutions prepared with calcium/gypsum. Interpret these values.
c. In Exercise 2.48, you used graphs to compare the zeta potential distributions for the two types of solutions. Now use the measures of central tendency to make the comparison. How does the addition of calcium/gypsum to the solution impact water quality (measured by zeta potential of silica)?

Applying the Concepts—Advanced

2.71 **Contact lenses for myopia.** Myopia (i.e., nearsightedness) is a visual condition that affects over 100 million Americans. Two treatments that may slow myopia progression are the use of (1) corneal reshaping contact lenses and (2) bifocal soft contact lenses. In *Optometry and Vision Science* (Jan., 2013), university optometry professors compared the two methods for treating myopia. A sample of 14 myopia patients participated in the study. Each patient was fitted with a contact lens of each type for the right eye, and the peripheral refraction was measured for each type of lens. The differences (bifocal soft minus corneal reshaping) are shown in the following table. (These data are simulated, based on information provided in the journal article.)

Peripheral refraction differences

−0.15	−8.11	−0.79	−0.80	−0.81	−0.39	−0.68
−1.13	−0.32	0.01	−0.63	0.05	−0.41	−1.11

a. Find measures of central tendency for the difference measurements and interpret their values.
b. Note that the data contain one unusually large (negative) difference relative to the other difference measurements. Find this difference. (In Section 2.7, we call this value an outlier.)
c. The large negative difference of −8.11 is actually a typographical error. The actual difference for this patient is −0.11. Rerun the analysis in part **a** using the corrected difference. Which measure of central tendency is most affected by the correcting of the outlier?

2.72 **Active nuclear power plants.** The U.S. Energy Information Administration monitors all nuclear power plants operating in the United States. The table below lists the number of active nuclear power plants operating in each of a sample of 20 states.
a. Find the mean, median, and mode of this data set.

Data for Exercise 2.72

State	Number of Power Plants	State	Number of Power Plants
Alabama	5	New Hampshire	1
Arizona	3	New York	6
California	4	North Carolina	5
Florida	5	Ohio	3
Georgia	4	Pennsylvania	9
Illinois	11	South Carolina	7
Kansas	1	Tennessee	3
Louisiana	2	Texas	4
Massachusetts	1	Vermont	1
Mississippi	1	Wisconsin	3

Based on *Statistical Abstract of the United States*, 2012 (Table 942). U.S. Energy Information Administration, *Electric Power Annual.*

b. Eliminate the largest value from the data set and repeat part **a.** What effect does dropping this measurement have on the measures of central tendency found in part **a**?

c. Arrange the 20 values in the table from lowest to highest. Next, eliminate the lowest two values and the highest two values from the data set, and find the mean of the remaining data values. The result is called a *10% trimmed mean*, since it is calculated after removing the highest 10% and the lowest 10% of the data values. What advantages does a trimmed mean have over the regular arithmetic mean?

2.4 Numerical Measures of Variability

a. Drug A

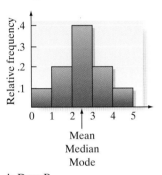

b. Drug B

Figure 2.19

Response time histograms for two drugs

Measures of central tendency provide only a partial description of a quantitative data set. The description is incomplete without a **measure of the variability**, or **spread**, of the data set. Knowledge of the data set's variability, along with knowledge of its center, can help us visualize the shape of the data set as well as its extreme values.

For example, suppose we want to compare response time to a stimulus for subjects treated with two different drugs, A and B. The histograms for the response times (in seconds) for each of the two drugs are shown in Figure 2.19. If you examine the two histograms, you'll notice that both data sets are symmetric, with equal modes, medians, and means. However, Drug A (Figure 2.19a) has response times spread with almost equal relative frequency over the measurement classes, while Drug B (Figure 2.19b) has most of its response times clustered about its center. Thus, the response times for Drug B are *less variable* than those for Drug A. Consequently, you can see that we need a measure of variability as well as a measure of central tendency to describe a data set.

Perhaps the simplest measure of the variability of a quantitative data set is its *range*.

> The **range** of a quantitative data set is equal to the largest measurement minus the smallest measurement.

The range is easy to compute and easy to understand, but it is a rather insensitive measure of data variation when the data sets are large. This is because two data sets can have the same range and be vastly different with respect to data variation. The phenomenon is demonstrated in Figure 2.19: Both distributions of data shown in the figure have the same range, but we already know that the response times for Drug B are much less variable than those for Drug A. Thus, you can see that the range does not always detect differences in data variation for large data sets.

Let's see if we can find a measure of data variation that is more sensitive than the range. Consider the two samples in Table 2.5: Each has five measurements. (We have ordered the numbers for convenience.) Note that both samples have a mean of 3 and that we have calculated the distance and direction, or *deviation*, between each measurement and the mean. What information do these deviations contain? If they tend to be large in magnitude, as in sample 1, the data are spread out, or highly variable. If the deviations are mostly small, as in sample 2, the data are clustered around the mean, $\bar{x}$, and therefore do not exhibit much variability. You can see that these deviations,

Table 2.5	Two Hypothetical Data Sets	
	Sample 1	Sample 2
Measurements	1, 2, 3, 4, 5	2, 3, 3, 3, 4
Mean	$\bar{x} = \dfrac{1 + 2 + 3 + 4 + 5}{5} = \dfrac{15}{5} = 3$	$\bar{x} = \dfrac{2 + 3 + 3 + 3 + 4}{5} = \dfrac{15}{5} = 3$
Deviations of measurement values from $\bar{x}$	$(1 - 3), (2 - 3), (3 - 3), (4 - 3),$ $(5 - 3)$ or $-2, -1, 0, 1, 2$	$(2 - 3), (3 - 3), (3 - 3), (3 - 3),$ $(4 - 3)$ or $-1, 0, 0, 0, 1$

a. Sample 1

b. Sample 2

Figure 2.20

Dot plots for deviations in Table 2.5

displayed graphically in Figure 2.20, provide information about the variability of the sample measurements.

The next step is to condense the information in these distances into a single numerical measure of variability. Averaging the deviations from $\bar{x}$ won't help because the negative and positive deviations cancel; that is, the sum of the deviations (and thus the average deviation) is always equal to zero.

Two methods come to mind for dealing with the fact that positive and negative deviations from the mean cancel. The first is to treat all the deviations as though they were positive, ignoring the sign of the negative values. We won't pursue this line of thought because the resulting measure of variability (the mean of the absolute values of the deviations) presents analytical difficulties beyond the scope of this text. A second method of eliminating the minus signs associated with the deviations is to square them. The quantity we can calculate from the squared deviations provides a meaningful description of the variability of a data set and presents fewer analytical difficulties in making inferences.

To use the squared deviations calculated from a data set, we first calculate the *sample variance*.

The **sample variance** for a sample of n measurements is equal to the sum of the squared deviations from the mean, divided by $(n-1)$. The symbol s^2 is used to represent the sample variance.

Formula for the Sample Variance:

$$s^2 = \frac{\sum_{i=1}^{n}(x_i - \bar{x})^2}{n-1}$$

Note: A shortcut formula for calculating s^2 is

$$s^2 = \frac{\sum_{i=1}^{n}x_i^2 - \frac{\left(\sum_{i=1}^{n}x_i\right)^2}{n}}{n-1}$$

Referring to the two samples in Table 2.5, you can calculate the variance for sample 1 as follows:

$$s^2 = \frac{(1-3)^2 + (2-3)^2 + (3-3)^2 + (4-3)^2 + (5-3)^2}{5-1}$$

$$= \frac{4+1+0+1+4}{4} = 2.5$$

The second step in finding a meaningful measure of data variability is to calculate the *standard deviation* of the data set.

The **sample standard deviation**, s, is defined as the positive square root of the sample variance, s^2, or, mathematically,

$$s = \sqrt{s^2}$$

The population variance, denoted by the symbol σ^2 (sigma squared), is the average of the squared deviations from the mean, μ, of the measurements on *all* units in the population, and σ (sigma) is the square root of this quantity.

Symbols for Variance and Standard Deviation

s^2 = Sample variance

s = Sample standard deviation

σ^2 = Population variance

σ = Population standard deviation

Notice that, unlike the variance, the standard deviation is expressed in the original units of measurement. For example, if the original measurements are in dollars, the variance is expressed in the peculiar units "dollars squared," but the standard deviation is expressed in dollars.

You may wonder why we use the divisor $(n - 1)$ instead of n when calculating the sample variance. Wouldn't using n seem more logical, so that the sample variance would be the average squared distance from the mean? The trouble is, using n tends to produce an underestimate of the population variance σ^2. So we use $(n - 1)$ in the denominator to provide the appropriate correction for this tendency.* Since sample statistics such as s^2 are used primarily to estimate population parameters such as σ^2, $(n - 1)$ is preferred to n in defining the sample variance.

Example   **2.9**

Computing Measures of Variation

Problem Calculate the variance and standard deviation of the following sample: 2, 3, 3, 3, 4.

Solution If you use the formula in the box to compute s^2 and s, you first need to find $\bar{x}$. From Table 2.6, we see that $\Sigma x = 15$. Thus, $\bar{x} = \dfrac{\Sigma x}{n} = \dfrac{15}{5} = 3$. Now, for each measurement, find $(x - \bar{x})$ and $(x - \bar{x})^2$, as shown.

Table 2.6	Calculating s^2	
x	$(x - \bar{x})$	$(x - \bar{x})^2$
2	−1	1
3	0	0
3	0	0
3	0	0
4	1	1
$\Sigma x = 15$		$\Sigma (x - \bar{x})^2 = 2$

Then we use[†]

$$s^2 = \frac{\Sigma (x - \bar{x})^2}{n - 1} = \frac{2}{5 - 1} = \frac{2}{4} = .5$$

$$s = \sqrt{.5} = .71$$

Look Back As the sample size n increases, these calculations can become very tedious. As the next example shows, we can use the computer to find s^2 and s.

■ **Now Work Exercise 2.80a**

---

*Appropriate here means that s^2, with a divisor of $(n - 1)$, is an *unbiased estimator* of σ^2. We define and discuss unbiasedness of estimators in Chapter 6.

[†]In calculating s^2, how many decimal places should you carry? Although there are no rules for the rounding procedure, it is reasonable to retain twice as many decimal places in s^2 as you ultimately wish to have in s. If, for example, you wish to calculate s to the nearest hundredth (two decimal places), you should calculate s^2 to the nearest ten-thousandth (four decimal places).

Example 2.10
EPAGAS

Finding Measures of Variation on a Printout

Problem Use the computer to find the sample variance s^2 and the sample standard deviation s for the 100 gas mileage readings given in Table 2.2.

Solution The SAS printout describing the gas mileage data is reproduced in Figure 2.21. The variance and standard deviation, highlighted on the printout, are $s^2 = 5.85$ and $s = 2.42$ (rounded to two decimal places). The value $s = 2.42$ represents a typical deviation of a gas mileage from the sample mean, $\bar{x} = 36.99$.

Figure 2.21
SAS numerical descriptive measures for 100 EPA mileages

The MEANS Procedure

Analysis Variable : MPG

Mean	Std Dev	Variance	N	Minimum	Maximum	Median
36.9940000	2.4178971	5.8462263	100	30.0000000	44.9000000	37.0000000

■

You now know that the standard deviation measures the variability of a set of data, and you know how to calculate the standard deviation. The larger the standard deviation, the more variable the data are. The smaller the standard deviation, the less variation there is in the data. But how can we practically interpret the standard deviation and use it to make inferences? This is the topic of Section 2.5.

Exercises 2.73–2.92

Understanding the Principles

2.73 What is the range of a data set?

2.74 Can there be two data sets that are different with respect to data variation but both have the same range?

2.75 What is advantage of the range as a measure of variability?

2.76 Why do we square the deviation from the mean before adding them to compute the variance?

2.77 If data set A has more variance than data set B, which one would have the lower standard deviation?

Learning the Mechanics

2.78 Calculate the variance and standard deviation for samples for which

a. $n = 10$, $\sum x^2 = 84$, $\sum x = 20$

b. $n = 40$, $\sum x^2 = 380$, $\sum x = 100$

c. $n = 20$, $\sum x^2 = 18$, $\sum x = 17$

2.79 Calculate the range, variance, and standard deviation for the following samples:
a. 39, 42, 40, 37, 41
b. 100, 4, 7, 96, 80, 3, 1, 10, 2
c. 100, 4, 7, 30, 80, 30, 42, 2

2.80 Calculate the range, variance, and standard deviation for
[NW] the following samples:
a. 4, 2, 1, 0, 1
b. 1, 6, 2, 2, 3, 0, 3
c. 8, −2, 1, 3, 5, 4, 4, 1, 3
d. 0, 2, 0, 0, −1, 1, −2, 1, 0, −1, 1, −1, 0, −3, −2, −1, 0, 1

2.81 Using only integers between −10 and 10, construct two data sets each with 10 observations such that the two sets have the same mean, but different variances. The two data sets should not have any common units.

2.82 Using only integers between −10 and 10, construct two data sets each with 10 observations such that the two sets have the same range, but different means. Moreover, the two data sets should not have any common units.

2.83 Consider the following sample of five measurements: 2, 1, 1, 0, 3.
a. Calculate the range, s^2, and s.
b. Add 3 to each measurement and repeat part **a**.
c. Subtract 4 from each measurement and repeat part **a**.
d. Considering your answers to parts **a**, **b**, and **c**, what seems to be the effect on the variability of a data set of adding the same number to or subtracting the same number from each measurement?

2.84 Compute s^2, and s for each of the data sets listed. Where appropriate, specify the units in which your answer is expressed.
a. 3, 1, 10, 10, 4
b. 8 feet, 10 feet, 32 feet, 5 feet
c. −1, −4, −3, 1, −4, −6
d. 1/5 ounce, 1/5 ounce, 1/5 ounce, 2/5 ounce, 1/5 ounce, 4/5 ounce

Ⓓ Applet Exercise 2.4

Use the applet entitled *Standard Deviation* to find the standard deviation of each of the four data sets listed in Exercise 2.80. For each data set, set the lower limit to a number less than all of the data, set the upper limit to a number greater than all of the data, and then click on *Update*. Click on the approximate location of each data item on the number line. You can get rid of a point by dragging it to the trash can. To clear the graph between data sets, simply click on the trash can.

a. Compare the standard deviations generated by the applet with those you calculated by hand in Exercise 2.80. If there are differences, explain why the applet might give values slightly different from the hand-calculated values.

b. Despite the fact that it provides a slightly different value of the standard deviation of a data set, describe some advantages of using the applet.

Ⓓ Applet Exercise 2.5

Use the applet *Standard Deviation* to study the effect that multiplying or dividing each number in a data set by the same number has on the standard deviation. Begin by setting appropriate limits and plotting the given data on the number line provided in the applet.

$$0 \quad 1 \quad 1 \quad 1 \quad 2 \quad 2 \quad 3 \quad 4$$

a. Record the standard deviation. Then multiply each data item by 2, plot the new data items, and record the standard deviation. Repeat the process, first multiplying each of the original data items by 3 and then by 4. Describe what happens to the standard deviation as the data items are multiplied by ever higher numbers. Divide each standard deviation by the standard deviation of the original data set. Do you see a pattern? Explain.

b. Divide each of the original data items by 2, plot the new data, and record the standard deviation. Repeat the process, first dividing each of the original data items by 3 and then by 4. Describe what happens to the standard deviation as the data items are divided by ever higher numbers. Divide each standard deviation by the standard deviation of the original data set. Do you see a pattern? Explain.

c. Using your results from parts **a** and **b**, describe what happens to the standard deviation of a data set when each of the data items in the set is multiplied or divided by a fixed number n. Experiment by repeating parts **a** and **b** for other data sets if you need to.

Ⓓ Applet Exercise 2.6

Use the applet *Standard Deviation* to study the effect that an extreme value has on the standard deviation. Begin by setting appropriate limits and plotting the following data on the number line provided in the applet:

$$0 \quad 6 \quad 7 \quad 7 \quad 8 \quad 8 \quad 8 \quad 9 \quad 9 \quad 10$$

a. Record the standard deviation. Replace the extreme value of 0 with 2, then 4, and then 6. Record the standard deviation each time. Describe what happens to the standard deviation as 0 is replaced by ever higher numbers.

b. How would the standard deviation of the data set compare with the original standard deviation if the 0 were replaced by 16? Explain.

Applying the Concepts—Basic

2.85 **Shell lengths of sea turtles.** *Aquatic Biology* (Vol. 9, 2010) Ⓓ reported on a study of green sea turtles inhabiting the Grand TURTLES Cayman South Sound lagoon. The data on curved carapace (shell) length (in centimeters) for 76 captured turtles are saved in the **TURTLES** file. Descriptive statistics for the data are shown on the accompanying MINITAB printout.

Descriptive Statistics: Length

Variable	N	Mean	StDev	Variance	Minimum	Maximum	Range
Length	76	55.47	11.34	128.57	30.37	81.63	51.26

a. Locate the range of the shell lengths on the printout.

b. Locate the variance of the shell lengths on the printout.

c. Locate the standard deviation of the shell lengths on the printout.

d. If the target of your interest is these specific 76 captured turtles, what symbols would you use to represent the variance and standard deviation?

2.86 **Shaft graves in ancient Greece.** Refer to the *American* Ⓓ *Journal of Archaeology* (Jan. 2014) study of sword shaft SHAFTS graves in ancient Greece, Exercise 2.60 (p. 89). The number of sword shafts buried at each of 13 recently discovered grave sites is reproduced in the following table.

1	2	3	1	5	6	2	4	1	2	4	2	9

Source: Harrell, K. "The fallen and their swords: A new explanation for the rise of the shaft graves." *American Journal of Archaeology,* Vol. 118, No. 1, January 2014 (Figure 1).

a. Calculate the range of the sample data.

b. Calculate the variance of the sample data.

c. Calculate the standard deviation of the sample data.

d. Which of the measures of variation computed in parts **a–c** have the same units of measure (number of sword shafts) as the original variable?

2.87 **Music performance anxiety.** Refer to the *British Journal of* Ⓓ *Music Education* (Mar. 2014) study of music performance PAI anxiety, Exercise 2.62 (p. 89). Scores (measured in points) on the Performance Anxiety Inventory (PAI) scale for participants in eight different studies are reproduced in the table.

54	42	51	39	41	43	55	40

Source: Patston, T. "Teaching stage fright? Implications for music educators." *British Journal of Music Education,* Vol. 31, No. 1, Mar. 2014 (adapted from Figure 1).

a. Find the variance of the sample PAI scores. Give the units of measurement for the variance.

b. Find the standard deviation of the sample PAI scores. Give the units of measurement for the standard deviation.

Applying the Concepts—Intermediate

2.88 **Characteristics of a rockfall.** Refer to the *Environmental* Ⓓ *Geology* (Vol. 58, 2009) study of how far a block from ROCKS a collapsing rock wall will bounce, Exercise 2.61 (p. 89).

The rebound lengths (meters) for a sample of 13 rock bounces are reproduced in the table below.

10.94	13.71	11.38	7.26	17.83	11.92	11.87
5.44	13.35	4.90	5.85	5.10	6.77	

Based on Paronuzzi, P. "Rockfall-induced block propagation on a soil slope, northern Italy." *Environmental Geology*, Vol. 58, 2009 (Table 2).

a. Compute the range of the 13 rebound lengths. Give the units of measurement of the range.

b. Compute the variance of the 13 rebound lengths. Give the units of measurement of the variance.

c. Compute the standard deviation of the 13 rebound lengths. Give the units of measurement of the standard deviation.

2.89 Is honey a cough remedy? Refer to the *Archives of Pediatrics and Adolescent Medicine* (Dec. 2007) study of honey as a remedy for coughing, Exercise 2.64 (p. 90). The coughing improvement scores for the patients in the over-the-counter cough medicine dosage (DM) group, honey dosage group, and control group are reproduced in the accompanying table.

Honey Dosage:	12	11	15	11	10	13	10	4	15	16	9
	14	10	6	10	8	11	12	12	8	12	9
	11	15	10	15	9	13	8	12	10	8	9
	5	12									
DM Dosage:	4	6	9	4	7	7	7	9	12	10	11
	6	3	4	9	12	7	6	8	12	12	4
	12	13	7	10	13	9	4	4	10	15	9
No Dosage (Control):	5	8	6	1	0	8	12	8	7	7	1
	6	7	7	12	7	9	7	9	5	11	9
	5	6	8	86	7	10	9	4	8	7	3
	1	4	3								

Based on Paul, I. M., et al. "Effect of honey, dextromethorphan, and no treatment on nocturnal cough and sleep quality for coughing children and their parents." *Archives of Pediatrics and Adolescent Medicine*, Vol. 161, No. 12, Dec. 2007 (data simulated).

a. Find the standard deviation of the improvement scores for the honey dosage group.

b. Find the standard deviation of the improvement scores for the DM dosage group.

c. Find the standard deviation of the improvement scores for the control group.

d. Based on the results, parts **a–c**, which group appears to have the most variability in coughing improvement scores? The least variability?

2.90 Permeability of sandstone during weathering. Refer to the *Geographical Analysis* (Vol. 42, 2010) study of the decay properties of sandstone when exposed to the weather, Exercise 2.69 (p. 91). Recall that slices of sandstone blocks were tested for permeability under three conditions: no exposure to any type of weathering (A), repeatedly sprayed with a 10% salt solution (B), and soaked in a 10% salt solution and dried (C). Measures of variation for the permeability measurements (mV) of each sandstone group are displayed in the accompanying MINITAB printout.

Descriptive Statistics: PermA, PermB, PermC

Variable	N	StDev	Variance	Minimum	Maximum	Range
PermA	100	14.48	209.53	55.20	122.40	67.20
PermB	100	21.97	482.75	50.40	150.00	99.60
PermC	100	20.05	401.94	52.20	129.00	76.80

a. Find the range of the permeability measurements for group A sandstone slices. Verify its value using the minimum and maximum values shown on the printout.

b. Find the standard deviation of the permeability measurements for group A sandstone slices. Verify its value using the variance shown on the printout.

c. Which condition (A, B, or C) has the more variable permeability data?

2.91 Cheek teeth of extinct primates. Refer to the *American Journal of Physical Anthropology* (Vol. 142, 2010) study of the characteristics of cheek teeth (e.g., molars) in an extinct primate species, Exercise 2.65 (p. 90). The data on dentary depth of molars (in millimeters) for 18 cheek teeth extracted from skulls are reproduced in the table.

Data on Dentary Depth (mm) of Molars

18.12	15.76	13.25
19.48	17.00	16.12
19.36	13.96	18.13
15.94	16.55	14.02
15.83	15.70	14.04
19.70	17.83	16.20

Based on Boyer, D. M., Evans, A. R., and Jernvall, J. "Evidence of dietary differentiation among late Paleocene–early Eocene Plesiadapids (Mammalia, primates)," *American Journal of Physical Anthropology*, Vol. 142, 2010 (Table A3).

a. Find the range of the data set. If the largest depth measurement in the sample were doubled, how would the range change? Would it increase or decrease?

b. Find the variance of the data set. If the largest depth measurement in the sample were doubled, how would the variance change? Would it increase or decrease?

c. Find the standard deviation of the data set. If the largest depth measurement in the sample were doubled, how would the standard deviation change? Would it increase or decrease?

2.92 Active nuclear power plants. Refer to Exercise 2.72 (p. 92) and the U.S. Energy Information Administration's data on the number of nuclear power plants operating in each of 20 states.

a. Find the range, variance, and standard deviation of this data set.

b. Eliminate the largest value from the data set and repeat part **a**. What effect does dropping this measurement have on the measures of variation found in part **a**?

c. Eliminate the smallest and largest value from the data set and repeat part **a**. What effect does dropping both of these measurements have on the measures of variation found in part **a**?

2.5 Using the Mean and Standard Deviation to Describe Data

We've seen that if we are comparing the variability of two samples selected from a population, the sample with the larger standard deviation is the more variable of the two. Thus, we know how to interpret the standard deviation on a relative or comparative basis, but we haven't explained how it provides a measure of variability for a single sample.

To understand how the standard deviation provides a measure of variability of a data set, consider the following questions: How many measurements are within one standard deviation of the mean? How many measurements are within two standard deviations? For any particular data set, we can answer these questions by counting the number of measurement in each of the intervals. However, finding an answer that applies to *any* set of data, whether a population or a sample, is more problematic.

Rules 2.1 and 2.2 give two sets of answers to the questions of how many measurements fall within one, two, and three standard deviations of the mean. The first, which applies to *any* set of data, is derived from a theorem proved by the Russian mathematician P. L. Chebyshev (1821–1894). The second, which applies to **mound-shaped, symmetric distributions** of data (for which the mean, median, and mode are all about the same), is based upon empirical evidence that has accumulated over the years. However, the percentages given for the intervals in Rule 2.2 are not only more precise when the data are mound-shaped but provide remarkably good approximations even when the distribution of the data is slightly skewed or asymmetric. Note that the rules apply to either population or sample data sets.

BIOGRAPHY PAFNUTY L. CHEBYSHEV (1821–1894)

The Splendid Russian Mathematician

P. L. Chebyshev was educated in mathematical science at Moscow University, eventually earning his master's degree. Following his graduation, Chebyshev joined St. Petersburg (Russia) University as a professor, becoming part of the well-known "Petersburg mathematical school." It was here that he proved his famous theorem about the probability of a measurement being within k standard deviations of the mean (Rule 2.1). His fluency in French allowed him to gain international recognition in probability theory. In fact, Chebyshev once objected to being described as a "splendid Russian mathematician," saying he surely was a "worldwide mathematician." One student remembered Chebyshev as "a wonderful lecturer" who "was always prompt for class. As soon as the bell sounded, he immediately dropped the chalk and, limping, left the auditorium." ■

Rule 2.1   **Using the Mean and Standard Deviation to Describe Data: Chebyshev's Rule**

Chebyshev's rule applies to any data set, regardless of the shape of the frequency distribution of the data.

a. It is possible that very few of the measurements will fall within one standard deviation of the mean [i.e., within the interval $(\overline{x} - s, \overline{x} + s)$ for samples and $(\mu - \sigma, \mu + \sigma)$ for populations].

b. At least $\frac{3}{4}$ of the measurements will fall within two standard deviations of the mean [i.e., within the interval $(\overline{x} - 2s, \overline{x} + 2s)$ for samples and $(\mu - 2\sigma, \mu + 2\sigma)$ for populations].

c. At least $\frac{8}{9}$ of the measurements will fall within three standard deviations of the mean [i.e., within the interval $(\overline{x} - 3s, \overline{x} + 3s)$ for samples and $(\mu - 3\sigma, \mu + 3\sigma)$ for populations].

d. Generally, for any number k greater than 1, at least $(1 - 1/k^2)$ of the measurements will fall within k standard deviations of the mean [i.e., within the interval $(\overline{x} - ks, \overline{x} + ks)$ for samples and $(\mu - k\sigma, \mu + k\sigma)$ for populations].

| Rule 2.2 | Using the Mean and Standard Deviation to Describe Data: The Empirical Rule |

The **empirical rule** is a rule of thumb that applies to data sets with frequency distributions that are mound shaped and symmetric, as follows:

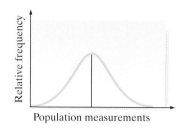

a. Approximately 68% of the measurements will fall within one standard deviation of the mean [i.e., within the interval $(\bar{x} - s, \bar{x} + s)$ for samples and $(\mu - \sigma, \mu + \sigma)$ for populations].

b. Approximately 95% of the measurements will fall within two standard deviations of the mean [i.e., within the interval $(\bar{x} - 2s, \bar{x} + 2s)$ for samples and $(\mu - 2\sigma, \mu + 2\sigma)$ for populations].

c. Approximately 99.7% (essentially all) of the measurements will fall within three standard deviations of the mean [i.e., within the interval $(\bar{x} - 3s, \bar{x} + 3s)$ for samples and $(\mu - 3\sigma, \mu + 3\sigma)$ for populations].

Example 2.11 Ⓓ

Interpreting the Standard Deviation—Rat-in-Maze Experiment

Problem Thirty students in an experimental psychology class use various techniques to train a rat to move through a maze. At the end of the course, each student's rat is timed as it negotiates the maze. The results (in minutes) are listed in Table 2.7. Determine the fraction of the 30 measurements in the intervals $\bar{x} \pm s$, $\bar{x} \pm 2s$, and $\bar{x} \pm 3s$, and compare the results with those predicted using Rules 2.1 and 2.2.

Table 2.7	Times (in Minutes) of 30 Rats Running through a Maze								
1.97	.60	4.02	3.20	1.15	6.06	4.44	2.02	3.37	3.65
1.74	2.75	3.81	9.70	8.29	5.63	5.21	4.55	7.60	3.16
3.77	5.36	1.06	1.71	2.47	4.25	1.93	5.15	2.06	1.65

Ⓓ *Data Set:* RATMAZE

Solution First, we entered the data into the computer and used MINITAB to produce summary statistics. The mean and standard deviation of the sample data, highlighted on the printout shown in Figure 2.22, are

$$\bar{x} = 3.74 \text{ minutes} \qquad s = 2.20 \text{ minutes}$$

rounded to two decimal places.

Figure 2.22
MINITAB descriptive statistics for rat maze times

Descriptive Statistics: RUNTIME

Variable	N	Mean	StDev	Minimum	Median	Maximum
RUNTIME	30	3.744	2.198	0.600	3.510	9.700

Now we form the interval

$$(\bar{x} - s, \bar{x} + s) = (3.74 - 2.20, 3.74 + 2.20) = (1.54, 5.94)$$

A check of the measurements shows that 23 of the times are within this one standard-deviation interval around the mean. This number represents 23/30, or ≈ 77%, of the sample measurements.

The next interval of interest is

$$(\bar{x} - 2s, \bar{x} + 2s) = (3.74 - 4.40, 3.74 + 4.40) = (-.66, 8.14)$$

All but two of the times are within this interval, so 28/30, or approximately 93%, are within two standard deviations of $\bar{x}$.

Finally, the three-standard-deviation interval around $\bar{x}$ is

$$(\bar{x} - 3s, \bar{x} + 3s) = (3.74 - 6.60, 3.74 + 6.60) = (-2.86, 10.34)$$

All of the times fall within three standard deviations of the mean.

These one-, two-, and three-standard-deviation percentages (77%, 93%, and 100%) agree fairly well with the approximations of 68%, 95%, and 100% given by the empirical rule (Rule 2.2).

Look Back If you look at the MINITAB frequency histogram for this data set (Figure 2.23), you'll note that the distribution is not really mound shaped, nor is it extremely skewed. Thus, we get reasonably good results from the mound-shaped approximations. Of course, we know from Chebyshev's rule (Rule 2.1) that no matter what the shape of the distribution, we would expect at least 75% and at least 89% of the measurements to lie within two and three standard deviations of $\bar{x}$, respectively.

Figure 2.23

MINITAB histogram of rat maze times

■ **Now Work Exercise 2.98**

Example 2.12 ⒟

EPAGAS

Checking the Calculation of the Sample Standard Deviation

Problem Chebyshev's rule and the empirical rule are useful as a check on the calculation of the standard deviation. For example, suppose we calculated the standard deviation for the gas mileage data (Table 2.2) to be 5.85. Are there any "clues" in the data that enable us to judge whether this number is reasonable?

Solution The range of the mileage data in Table 2.2 is $44.9 - 30.0 = 14.9$. From Chebyshev's rule and the empirical rule, we know that most of the measurements (approximately 95% if the distribution is mound shaped) will be within two standard deviations of the mean. And regardless of the shape of the distribution and the number of measurements, almost all of them will fall within three standard deviations of the mean. Consequently, we would expect the range of the measurements to be between 4 (i.e., $\pm 2s$) and 6 (i.e., $\pm 3s$) standard deviations in length. (See Figure 2.24.) For the car mileage data, this means that s should fall between

$$\frac{\text{Range}}{6} = \frac{14.9}{6} = 2.48 \quad \text{and} \quad \frac{\text{Range}}{4} = \frac{14.9}{4} = 3.73$$

Figure 2.24
The relation between the range and the standard deviation

Hence, the standard deviation should not be much larger than 1/4 of the range, particularly for the data set with 100 measurements. Thus, we have reason to believe that the calculation of 5.85 is too large. A check of our work reveals that 5.85 is the variance s^2, not the standard deviation s. (See Example 2.10.) We "forgot" to take the square root (a common error); the correct value is $s - 2.42$. Note that this value is slightly smaller than the range divided by 6 (2.48). The larger the data set, the greater is the tendency for very large or very small measurements (extreme values) to appear, and when they do, the range may exceed six standard deviations.

Look Back   In examples and exercises, we'll sometimes use $s \approx \text{range}/4$ to obtain a crude, and usually conservatively large, approximation for s. However, we stress that this is no substitute for calculating the exact value of s when possible.

■ **Now Work Exercise 2.99**

In the next example, we use the concepts in Chebyshev's rule and the empirical rule to build the foundation for making statistical inferences.

Example   2.13

Making a Statistical Inference — Car Battery Guarantee

Problem A manufacturer of automobile batteries claims that the average length of life for its grade A battery is 60 months. However, the guarantee on this brand is for just 36 months. Suppose the standard deviation of the life length is known to be 10 months and the frequency distribution of the life-length data is known to be mound shaped.

a. Approximately what percentage of the manufacturer's grade A batteries will last more than 50 months, assuming that the manufacturer's claim is true?

b. Approximately what percentage of the manufacturer's batteries will last less than 40 months, assuming that the manufacturer's claim is true?

c. Suppose your battery lasts 37 months. What could you infer about the manufacturer's claim?

Solution If the distribution of life length is assumed to be mound shaped with a mean of 60 months and a standard deviation of 10 months, it would appear as shown in Figure 2.25. Note that we can take advantage of the fact that mound-shaped distributions are (approximately) symmetric about the mean, so that the percentages given by the empirical rule can be split equally between the halves of the distribution on each side of the mean.

For example, since approximately 68% of the measurements will fall within one standard deviation of the mean, the distribution's symmetry implies that approximately $(1/2)(68\%) = 34\%$ of the measurements will fall between the mean and one standard deviation on each side. This concept is illustrated in Figure 2.25. The figure also shows that 2.5% of the measurements lie beyond two standard deviations in each

Figure 2.25
Battery life-length distribution: manufacturer's claim assumed true

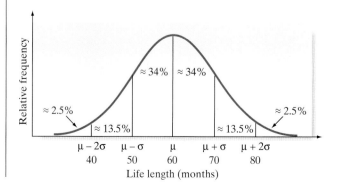

direction from the mean. This result follows from the fact that if approximately 95% of the measurements fall within two standard deviations of the mean, then about 5% fall outside two standard deviations; if the distribution is approximately symmetric, then about 2.5% of the measurements fall beyond two standard deviations on each side of the mean.

a. It is easy to see in Figure 2.25 that the percentage of batteries lasting more than 50 months is approximately 34% (between 50 and 60 months) plus 50% (greater than 60 months). Thus, approximately 84% of the batteries should have a life exceeding 50 months.

b. The percentage of batteries that last less than 40 months can also be easily determined from Figure 2.25: Approximately 2.5% of the batteries should fail prior to 40 months, assuming that the manufacturer's claim is true.

c. If you are so unfortunate that your grade A battery fails at 37 months, you can make one of two inferences: Either your battery was one of the approximately 2.5% that fail prior to 40 months, or something about the manufacturer's claim is not true. Because the chances are so small that a battery fails before 40 months, you would have good reason to have serious doubts about the manufacturer's claim. A mean smaller than 60 months or a standard deviation longer than 10 months would each increase the likelihood of failure prior to 40 months.*

Look Back The approximations given in Figure 2.25 are more dependent on the assumption of a **mound-shaped distribution** than are the assumptions stated in the empirical rule (Rule 2.2) because the approximations in Figure 2.25 depend on the (approximate) symmetry of the mound-shaped distribution. We saw in Example 2.11 that the empirical rule can yield good approximations even for skewed distributions. This will *not* be true of the approximations in Figure 2.25; the distribution *must* be mound shaped and (approximately) symmetric.

■

Example 2.13 is our initial demonstration of the statistical inference-making process. At this point, you should realize that we'll use sample information (in Example 2.13, your battery's failure at 37 months) to make inferences about the population (in Example 2.13, the manufacturer's claim about the length of life for the population of all batteries). We'll build on this foundation as we proceed.

Statistics IN Action | Revisited   **Interpreting Descriptive Statistics for the Body Image Data**

We return to the analysis of the data from the *Body Image: An International Journal of Research* (Jan. 2010) study of 92 BDD patients. Recall that the quantitative variable of interest in the study is Total Appearance Evaluation score (ranging from 7 to 35 points). One of the questions of interest to the Brown University researchers was whether female BDD patients were less satisfied in their body image than normal females (i.e., females with no disorders), and whether male BDD patients were less satisfied in their body image than normal males. The analysis involved comparing the mean score for the BDD patients to the mean score for a normal group (called a "norm"). The appearance evaluation "norms"

for females and males were 23.5 and 24.4, respectively.

MINITAB descriptive statistics for the data in the **BDD** file are displayed in Figure SIA2.5, with means and standard deviations highlighted. The sample mean score for females is 17.05 while the sample mean for males is 19.0. Note that both of these values fall well below the "norm" for each respective gender. Consequently, *for this sample of BDD patients*, both females and males have a lower average opinion about their body image than do normal females and males.

(*Continued*)

*The assumption that the distribution is mound shaped and symmetric may also be incorrect. However, if the distribution were skewed to the right, as life-length distributions often tend to be, the percentage of measurements more than two standard deviations below the mean would be even less than 2.5%.

Statistics IN Action *(continued)*	Descriptive Statistics: AEScore

Descriptive Statistics: AEScore

Variable	Gender	N	Mean	StDev	Variance	Minimum	Maximum
AEScore	F	60	17.050	4.757	22.625	7.000	30.000
	M	32	19.000	5.424	29.419	7.000	32.000

Figure SIA2.5

MINITAB descriptive statistics for appearance evaluation by gender

To interpret the standard deviations, we substitute into the formula, $\bar{x} \pm 2s$, to obtain the intervals:

Females:	$\bar{x} \pm 2s = 17.05 \pm 2(4.76) = 17.05 \pm 9.52$
	$= (7.53, 26.57)$
Males:	$\bar{x} \pm 2s = 19.00 \pm 2(5.42) = 19.00 \pm 10.84$
	$= (8.16, 29.84)$

From Chebyshev's rule (Rule 2.1), we know that at least 75% of the females in the sample of BDD patients will have an Appearance Evaluation score anywhere between 7.53 and 26.57 points. Similarly, we know that at least 75% of the males in the sample of BDD patients will have an Appearance Evaluation score anywhere between 8.16 and 29.84 points. You can see that the interval for the females lies slightly below the interval for males. How does this information help in determining whether the female BDD patients and male BDD patients have mean scores below the "norm" for their gender? In Chapters 7 and 8 we will learn about an inferential method that uses both the sample mean and sample standard deviation to determine whether the population mean of the BDD patients is really smaller than the "norm" value.

 Data Set: BDD

Exercises 2.93–2.113

Understanding the Principles

2.93 What do you mean by a mound-shaped, symmetric distribution?

2.94 As per Chebyshev's rule how many measurements fall within $(\mu - 2\sigma, \mu + 2\sigma)$?

2.95 The output from a statistical computer program indicates that the mean and standard deviation of a data set consisting of 200 measurements are $1,500 and $300, respectively.
  a. What are the units of measurement of the variable of interest? On the basis of the units, what type of data is this, quantitative or qualitative?
  b. What can be said about the number of measurements between $900 and $2,100? between $600 and $2,400? between $1,200 and $1,800? between $1,500 and $2,100?

2.96 For any set of data, what can be said about the percentage of the measurements contained in each of the following intervals?
  a. $\bar{x} - s$ to $\bar{x} + s$
  b. $\bar{x} - 2s$ to $\bar{x} + 2s$
  c. $\bar{x} - 3s$ to $\bar{x} + 3s$

2.97 For a set of data with a mound-shaped relative frequency distribution, what can be said about the percentage of the measurements contained in each of the intervals specified in Exercise 2.96?

Learning the Mechanics

2.98 The following is a sample of 25 measurements.

L02098

| 7 | 6 | 6 | 11 | 8 | 9 | 11 | 9 | 10 | 8 | 7 | 7 | 5 |
| 9 | 10 | 7 | 7 | 7 | 7 | 9 | 12 | 10 | 10 | 8 | 6 |

NW **a.** Compute $\bar{x}, s^2$, and s for this sample.
  b. Count the number of measurements in the intervals $\bar{x} \pm s, \bar{x} \pm 2s$, and $\bar{x} \pm 3s$. Express each count as a percentage of the total number of measurements.
  c. Compare the percentages found in part **b** with the percentages given by the empirical rule and Chebyshev's rule.
  d. Calculate the range and use it to obtain a rough approximation for s. Does the result compare favorably with the actual value for s found in part **a**?

2.99 Given a data set with a largest value of 760 and a smallest
NW value of 135, what would you estimate the standard deviation to be? Explain the logic behind the procedure you used to estimate the standard deviation. Suppose the standard deviation is reported to be 25. Is this number reasonable? Explain.

Applying the Concepts—Basic

2.100 Do social robots walk or roll? Refer to the *International Conference on Social Robotics* (Vol. 6414, 2010) study on ROBOTS the current trend in the design of social robots, Exercise 2.7 (p. 66). Recall that in a random sample of social robots obtained through a Web search, 28 were built with wheels. The numbers of wheels on each of the 28 robots are listed in the table on p. 105.
  a. Generate a histogram for the sample data set. Is the distribution of number of wheels mound shaped and symmetric?
  b. Find the mean and standard deviation for the sample data set.
  c. Form the interval $\bar{x} \pm 2s$.

d. According to Chebychev's rule, what proportion of sample observations will fall within the interval in part **c**?

e. According to the empirical rule, what proportion of sample observations will fall within the interval in part **c**?

f. Determine the actual proportion of sample observations that fall within the interval in part **c**. Even though the histogram in part **a** is not perfectly symmetric, does the empirical rule provide a good estimate of the proportion?

| 4 | 4 | 3 | 3 | 3 | 6 | 4 | 2 | 2 | 2 | 1 | 3 | 3 | 3 |
| 3 | 4 | 4 | 3 | 2 | 8 | 2 | 2 | 3 | 4 | 3 | 3 | 4 | 2 |

Source: Chew, S., et al. "Do social robots walk or roll?" *International Conference on Social Robotics,* Vol. 6414, 2010 (adapted from Figure 2).

2.101 Sanitation inspection of cruise ships. Refer to Exercise 2.41 (p. 79) and the Centers for Disease Control and Prevention listing of the sanitation scores for 186 cruise ships. The data are saved in the **SANIT** file.

a. Find the mean and standard deviation of the sanitation scores.

b. Calculate the intervals $\bar{x} \pm s, \bar{x} \pm 2s$, and $\bar{x} \pm 3s$.

c. Find the percentage of measurements in the data set that fall within each of the intervals in part **b**. Do these percentages agree with either Chebyshev's rule or the empirical rule?

2.102 Motivation of drug dealers. Researchers at Georgia State University investigated the personality characteristics of drug dealers in order to shed light on their motivation for participating in the illegal drug market (*Applied Psychology in Criminal Justice*, Sept. 2009). The sample consisted of 100 convicted drug dealers who attended a court-mandated counseling program. Each dealer was scored on the Wanting Recognition (WR) Scale, which provides a quantitative measure of a person's level of need for approval and sensitivity to social situations. (Higher scores indicate a greater need for approval.) The sample of drug dealers had a mean WR score of 39, with a standard deviation of 6. Assume the distribution of WR scores for drug dealers is mound shaped and symmetric.

a. Give a range of WR scores that will contain about 95% of the scores in the drug dealer sample.

b. What proportion of the drug dealers will have WR scores above 51?

c. Give a range of WR sores that will contain nearly all the scores in the drug dealer sample.

2.103 Dentists' use of anesthetics. A study published in *Current Allergy & Clinical Immunology* (Mar. 2004) investigated allergic reactions of dental patients to local anesthetics. Based on a survey of dental practitioners, the study reported that the mean number of units (ampoules) of local anesthetics used per week by dentists was 79, with a standard deviation of 23. Suppose we want to determine the percentage of dentists who use less than 102 units of local anesthetics per week.

a. Assuming that nothing is known about the shape of the distribution for the data, what percentage of dentists use less than 102 units of local anesthetics per week?

b. Assuming that the data have a mound-shaped distribution, what percentage of dentists use less than 102 units of local anesthetics per week?

2.104 Irrelevant speech effects. Refer to the *Acoustical Science & Technology* (Vol. 35, 2014) study of irrelevant speech effects, Exercise 2.34 (p. 77). Recall that subjects performed a memorization task under two conditions: (1) with irrelevant background speech and (2) in silence. The difference in the error rates for the two conditions—called the relative difference in error rate (RDER)—was computed for each subject. Descriptive statistics for the RDER values are shown in the SAS printout.

The MEANS Procedure

	Analysis Variable : RDER			
N	Mean	Std Dev	Minimum	Maximum
71	78.1885070	63.2429128	-19.2710000	254.1090000

a. Based on the histogram shown in Exercise 2.34, which rule is more appropriate for describing the distribution of the RDER values? Why?

b. Apply the rule you selected in part **a** to the data. Specifically, estimate the proportion of RDER values that fall in the interval $\bar{x} \pm 2s$.

2.105 Corporate sustainability of CPA firms. Refer to the *Business and Society* (Mar. 2011) study on the sustainability behaviors of CPA corporations, Exercise 2.36 (pp. 78). Recall that the level of support for corporate sustainability (measured on a quantitative scale ranging from 0 to 160 points) was obtained for each 992 senior managers at CPA firms. Numerical measures of both central tendency and variation for level of support are shown in the MINITAB printout below. Give an estimate of the number of the 992 senior managers who have support levels between 41 and 95 points. (Use the histogram in Exercise 2.36 to help you decide which rule to apply.)

Applying the Concepts—Intermediate

2.106 Characteristics of antiwar demonstrators. The characteristics of antiwar demonstrators in the United States were examined in the *American Journal of Sociology* (Jan. 2014). Using data collected for over 5,000 antiwar demonstrators over a recent 3-year period, the researchers found that the mean number of protest organizations joined by the demonstrators was .90 with a standard deviation of 1.10.

MINITAB Output for Exercise 2.105

Descriptive Statistics: Support

Variable	N	Mean	StDev	Minimum	Median	Maximum
Support	992	67.755	26.871	0.000	68.000	155.000

a. Give an interval that captures the number of protest organizations joined for at least 75% of the demonstrators.

b. One of the antiwar demonstrators had joined 7 protest organizations. Is this typical of the sample of antiwar demonstrators? Explain.

2.107 Hand washing versus hand rubbing. As an alternative to hand washing, some hospitals allow health workers to rub their hands with an alcohol-based antiseptic. The *British Medical Journal* (Aug. 17, 2002) reported on a study to compare the effectiveness of washing the hands with soap and rubbing the hands with alcohol. One group of health care workers used hand rubbing, while a second group used hand washing to clean their hands. The bacterial count (number of colony-forming units) on the hand of each worker was recorded. The table gives descriptive statistics on bacteria counts for the two groups of health care workers.

	Mean	Standard Deviation
Hand rubbing	35	59
Hand washing	69	106

a. For hand rubbers, form an interval that contains at least 75% of the bacterial counts. (*Note*: The bacterial count cannot be less than 0.)

b. Repeat part **a** for hand washers.

c. On the basis of your results in parts **a** and **b**, make an inference about the effectiveness of the two hand-cleaning methods.

2.108 Extinct New Zealand birds. Refer to the *Evolutionary Ecology Research* (July 2003) study of the patterns of extinction in the New Zealand bird population, presented in Exercise 2.24 (p. 70). Consider the data on the egg length (measured in millimeters) for the 132 bird species saved in the **NZBIRDS** file.

a. Find the mean and standard deviation of the egg lengths.

b. Form an interval that can be used to predict the egg length of a bird species found in New Zealand.

2.109 Velocity of Winchester bullets. The *American Rifleman* reported on the velocity of ammunition fired from the FEG P9R pistol, a 9-mm gun manufactured in Hungary. Field tests revealed that Winchester bullets fired from the pistol had a mean velocity (at 15 feet) of 936 feet per second and a standard deviation of 10 feet per second. Tests were also conducted with Uzi and Black Hills ammunition.

a. Describe the velocity distribution of Winchester bullets fired from the FEG P9R pistol.

b. A bullet whose brand is unknown is fired from the FEG P9R pistol. Suppose the velocity (at 15 feet) of the bullet is 1,000 feet per second. Is the bullet likely to be manufactured by Winchester? Explain.

2.110 Permeability of sandstone during weathering. Refer to the *Geographical Analysis* (Vol. 42, 2010) study of the decay properties of sandstone when exposed to the weather, Exercises 2.69 and 2.90 (pp. 91 and 98). Recall that slices of sandstone blocks were measured for permeability under three conditions: no exposure to any type of weathering (A),

repeatedly sprayed with a 10% salt solution (B), and soaked in a 10% salt solution and dried (C).

a. Combine the mean (from Exercise 2.69) and standard deviation (from Exercise 2.90) to make a statement about where most of the permeability measurements for group A sandstone slices will fall. Which rule (and why) did you use to make this inference?

b. Repeat part **a** for group B sandstone slices.

c. Repeat part **a** for group C sandstone slices.

d. Based on all your analyses, which type of weathering (type A, B, or C) appears to result in faster decay (i.e., higher permeability measurements)?

Applying the Concepts—Advanced

2.111 Shopping vehicle and judgment. While shopping at the grocery store, are you more likely to buy a vice product (e.g., a candy bar) when pushing a shopping cart or carrying a shopping basket? This was the question of interest in a study published in the *Journal of Marketing Research* (Dec. 2011). The researchers believe that when your arm is flexed (as when carrying a basket) you are more likely to choose a vice product than when your arm is extended (as when pushing a cart). To test this theory in a laboratory setting, the researchers recruited 22 consumers and had each push their hand against a table while they were asked a series of shopping questions. Half of the consumers were told to put their arm in a flex position (similar to a shopping basket), and the other half were told to put their arm in an extended position (similar to a shopping cart). Participants were offered several choices between a vice and a virtue (e.g., a movie ticket vs. a shopping coupon, pay later with a larger amount vs. pay now), and a choice score (on a scale of 0 to 100) was determined for each. (Higher scores indicate a greater preference for vice options.) The average choice score for consumers with a flexed arm was 59, while the average for consumers with an extended arm was 43.

a. Suppose the standard deviations of the choice scores for the flexed arm and extended arm conditions are 4 and 2, respectively. Does this information support the researchers' theory? Explain.

b. Suppose the standard deviations of the choice scores for the flexed arm and extended arm conditions are 10 and 15, respectively. Does this information support the researchers' theory? Explain.

2.112 Animal-assisted therapy for heart patients. A study was presented at the *American Heart Association Conference* (Nov. 2005) to gauge whether animal-assisted therapy can improve the physiological responses of heart failure patients. A team of nurses from the UCLA Medical Center randomly divided 76 heart patients into three groups. Each patient in group T was visited by a human volunteer accompanied by a trained dog, each patient in group V was visited by a volunteer only, and the patients in group C were not visited at all. The anxiety level of each patient was measured (in points) both before and after the visits. The table on p. 107 gives summary statistics for the drop in anxiety level for patients in the three groups. Suppose the anxiety level of a patient selected from the study had a drop of 22.5 points. From which group is the patient more likely to have come? Explain.

	Sample Size	Mean Drop	Std. Dev.
Group T: Volunteer + trained dog	26	10.5	7.6
Group V: Volunteer only	25	3.9	7.5
Group C: Control group (no visit)	25	1.4	7.5

Based on Cole, K., et al. "Animal assisted therapy decreases hemodynamics, plasma epinephrine and state anxiety in hospitalized heart failure patients." *American Journal of Critical Care*, 2007, 16: 575–585.

2.113 Land purchase decision. A buyer for a lumber company must decide whether to buy a piece of land containing 5,000 pine trees. If 1,000 of the trees are at least 40 feet tall, the buyer will purchase the land; otherwise, he won't. The owner of the land reports that the height of the trees has a mean of 30 feet and a standard deviation of 3 feet. On the basis of this information, what is the buyer's decision?

2.6 Numerical Measures of Relative Standing

We've seen that numerical measures of central tendency and variability describe the general nature of a quantitative data set (either a sample or a population). In addition, we may be interested in describing the *relative* quantitative location of a particular measurement within a data set. Descriptive measures of the relationship of a measurement to the rest of the data are called **measures of relative standing**.

One measure of the relative standing of a measurement is its **percentile ranking**, or **percentile score**. For example, suppose you scored an 80 on a test and you want to know how you fared in comparison with others in your class. If the instructor tells you that you scored at the 90th percentile, it means that 90% of the grades were lower than yours and 10% were higher. Thus, if the scores were described by the relative frequency histogram in Figure 2.26, the 90th percentile would be located at a point such that 90% of the total area under the relative frequency histogram lies below the 90th percentile and 10% lies above. If the instructor tells you that you scored in the 50th percentile (the median of the data set), 50% of the test grades would be lower than yours and 50% would be higher.

Percentile rankings are of practical value only with large data sets. Finding them involves a process similar to the one used in finding a median. The measurements are ranked in order, and a rule is selected to define the location of each percentile. Since we are interested primarily in interpreting the percentile rankings of measurements (rather than in finding particular percentiles for a data set), we define the *p*th *percentile* of a data set as follows.

Figure 2.26
Location of 90th percentile for test grades

> For any set of *n* measurements (arranged in ascending or descending order), the **pth percentile** is a number such that $p\%$ of the measurements fall below that number and $(100 - p)\%$ fall above it.

Example 2.14 Ⓓ WLTASK

Finding and Interpreting Percentiles—The "Water-Level Task"

Problem Refer to the water-level task deviations for the 40 subjects in Table 2.4. An SPSS printout describing the data is shown in Figure 2.27. Locate the 25th percentile and 95th percentile on the printout, and interpret the associated values.

Solution Both the 25th percentile and 95th percentile are highlighted on the SPSS printout. The values in question are 2.0 and 24.8, respectively. Our interpretations are as follows: 25% of the 40 deviations fall below 2.0 and 95% of the deviations fall below 24.8.

Case Processing Summary

	Cases					
	Valid		Missing		Total	
	N	Percent	N	Percent	N	Percent
Deviation	40	100.0%	0	.0%	40	100.0%

Figure 2.27
SPSS percentile statistics for water-level task deviations

Percentiles

		Percentiles						
		5	10	25	50	75	90	95
Weighted Average (Definition 1)	Deviation	-4.90	-1.00	2.00	7.00	10.00	15.80	24.80
Tukey's Hinges	Deviation			2.00	7.00	10.00		

Look Back The method for computing percentiles with small data sets varies according to the software used. As the sample size increases, the percentiles from the different software packages will converge to a single number.

■ Now Work Exercise 2.115

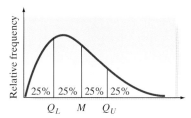

Figure 2.28
The quartiles for a data set

Percentiles that partition a data set into four categories, each category containing exactly 25% of the measurements, are called **quartiles**. The *lower quartile (Q_L)* is the 25th percentile, the *middle quartile (M)* is the median or 50th percentile, and the *upper quartile (Q_U)* is the 75th percentile, as shown in Figure 2.28. Therefore, in Example 2.14, we have (from the SPSS printout, Figure 2.27) $Q_L = 2.0$, $M = 7.0$, and $Q_U = 10.0$. Quartiles will prove useful in finding unusual observations in a data set (Section 2.7).

> The **lower quartile (Q_L)** is the 25th percentile of a data set. The **middle quartile (M)** is the median or 50th percentile. The **upper quartile (Q_U)** is the 75th percentile.

Another measure of relative standing in popular use is the *z***-score**. As you can see in the following definition, the *z*-score makes use of the mean and standard deviation of a data set in order to specify the relative location of the measurement.

> The **sample *z*-score** for a measurement x is
> $$z = \frac{x - \bar{x}}{s}$$
> The **population *z*-score** for a measurement x is
> $$z = \frac{x - \mu}{\sigma}$$

Note that the *z*-score is calculated by subtracting $\bar{x}$ (or μ) from the measurement x and then dividing the result by s (or σ). The final result, the *z*-**score**, represents the distance between a given measurement x and the mean, expressed in standard deviations.

Example   2.15

Finding a *z*-Score

Problem Suppose a sample of 2,000 high school seniors' verbal SAT scores is selected. The mean and standard deviation are

$$\bar{x} = 550 \quad s = 75$$

Suppose Joe Smith's score is 475. What is his sample *z*-score?

Solution Joe Smith's verbal SAT score lies below the mean score of the 2,000 seniors, as shown in Figure 2.29.

Figure 2.29
Verbal SAT scores of high school seniors

325	475	550		775
$\bar{x} - 3s$	Joe Smith's score	$\bar{x}$		$\bar{x} + 3s$

We compute

$$z = \frac{x - \bar{x}}{s} = \frac{475 - 550}{75} = -1.0$$

which tells us that Joe Smith's score is 1.0 standard deviation *below* the sample mean; in short, his sample *z*-score is -1.0.

Look Back The numerical value of the z-score reflects the relative standing of the measurement. A large positive z-score implies that the measurement is larger than almost all other measurements, whereas a large negative z-score indicates that the measurement is smaller than almost every other measurement. If a z-score is 0 or near 0, the measurement is located at or near the mean of the sample or population.

■ **Now Work Exercise 2.117**

We can be more specific if we know that the frequency distribution of the measurements is mound shaped. In this case, the following interpretation of the z-score can be given:

Interpretation of z-Scores for Mound-Shaped Distributions of Data

1. Approximately 68% of the measurements will have a z-score between −1 and 1.
2. Approximately 95% of the measurements will have a z-score between −2 and 2.
3. Approximately 99.7% (almost all) of the measurements will have a z-score between −3 and 3.

Note that this interpretation of z-scores is identical to that given by the empirical rule for mound-shaped distributions (Rule 2.2). The statement that a measurement falls into the interval from $(\mu - \sigma)$ to $(\mu + \sigma)$ is equivalent to the statement that a measurement has a population z-score between −1 and 1, since all measurements between $(\mu - \sigma)$ and $(\mu + \sigma)$ are within one standard deviation of μ. These z-scores are displayed in Figure 2.30.

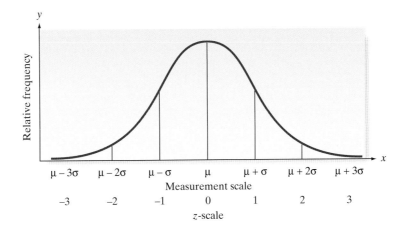

Figure 2.30
Population z-scores for a mound-shaped distribution

Exercises 2.114–2.131

Understanding the Principles

2.114 For a quantitative data set, explain the
  a. 20th percentile.
  b. middle quartile.
  c. population z-score for a measurement.

2.115 Give the percentage of measurements in a data set that are
[NW] above and below each of the following percentiles:
  a. 75th percentile
  b. 50th percentile
  c. 20th percentile
  d. 84th percentile

2.116 For a mound-shaped data, what percentage of the measurements will have negative z-score?

Learning the Mechanics

2.117 Compute the z-score corresponding to each of the follow-
[NW] ing values of x:
  a. $x = 40, s = 5, \bar{x} = 30$
  b. $x = 90, \mu = 89, \sigma = 2$
  c. $\mu = 50, \sigma = 5, x = 50$
  d. $s = 4, x = 20, \bar{x} = 30$

e. In parts **a–d**, state whether the z-score locates x within a sample or within a population.

f. In parts **a–d**, state whether each value of x lies above or below the mean and by how many standard deviations.

2.118 What will be the z-score of an observation $x = 10$ if it is picked from a distribution with the following means (μ) and standard deviations (σ)?

a. $\mu = 30$ and $\sigma = 10$

b. $\mu = -10$ and $\sigma = 10$

c. $\mu = 30$ and $\sigma = 20$

d. $\mu = 20$ and $\sigma = 30$

2.119 Suppose that 40 and 90 are two elements of a population data set and that their z-scores are -2 and 3, respectively. Using only this information, is it possible to determine the population's mean and standard deviation? If so, find them. If not, explain why it's not possible.

Applying the Concepts—Basic

2.120 Analysis of students' performance. 250 students appeared for a test of 500 marks. The scores they obtained have a mean of 383, a first quartile of 355, a third quartile of 402, and a 90th percentile of 437. Interpret each of these numerical descriptive measures.

2.121 Age distribution. According to a demographic survey of India, 69% of the estimated population of 1.25 billion are aged above 14 years. Give a percentile ranking for the age of 14 years in the distribution of all ages across the overall population of India.

2.122 Stability of compounds in new drugs. Refer to the Pfizer Global Research and Development study (reported in *ACS Medicinal Chemistry Letters*, Vol. 1, 2010) of the metabolic stability of drugs, Exercise 2.35 (p. 77). Recall that the stability of each of 416 drugs was measured by the fup/fumic ratio.

a. In Exercise 2.35, you determined the proportion of fup/fumic ratios that fall above 1. Use this proportion to determine the percentile rank of 1.

b. In Exercise 2.35, you determined the proportion of fup/fumic ratios that fall below .4. Use this proportion to determine the percentile rank of .4.

2.123 Sanitation inspection of cruise ships. Refer to the sanitation levels of cruise ships presented in Exercise 2.41 (p. 79).

a. Give a measure of relative standing for the *Nautilus Explorer's* score of 74. Interpret the result.

b. Give a measure of relative standing for the *Star Princess'* score of 92. Interpret the result.

2.124 Motivation of drug dealers. Refer to the *Applied Psychology in Criminal Justice* (Sept. 2009) study of convicted drug dealers' motivations, Exercise 2.102 (p. 105). Recall that the sample of drug dealers had a mean Wanting Recognition (WR) score of 39 points, with a standard deviation of 6 points.

a. Find and interpret the z-score for a drug dealer with a WR score of 30 points.

b. What proportion of the sampled drug dealers had WR scores below 39 points? (Assume the distribution of WR scores is mound shaped and symmetric.)

Applying the Concepts—Intermediate

2.125 Extinct New Zealand birds. Refer to the *Evolutionary Ecology Research* (July 2003) study of the patterns of extinction in the New Zealand bird population, first presented in Exercise 2.24 (p. 70). Again, consider the data on the egg length (measured in millimeters) for the 132 bird species.

a. Find the 10th percentile for the egg length distribution and interpret its value.

b. The *moas (P. australis)* is a bird species with an egg length of 205 millimeters. Find the z-score for this species of bird and interpret its value.

2.126 Lead in drinking water. The U.S. Environmental Protection Agency (EPA) sets a limit on the amount of lead permitted in drinking water. The EPA *Action Level* for lead is .015 milligram per liter (mg/L) of water. Under EPA guidelines, if 90% of a water system's study samples have a lead concentration less than .015 mg/L, the water is considered safe for drinking. I (coauthor Sincich) received a report on a study of lead levels in the drinking water of homes in my subdivision. The 90th percentile of the study sample had a lead concentration of .00372 mg/L. Are water customers in my subdivision at risk of drinking water with unhealthy lead levels? Explain.

2.127 Voltage sags and swells. The power quality of a transformer is measured by the quality of the voltage. Two causes of poor power quality are "sags" and "swells." A sag is an unusual dip, and a swell is an unusual increase in the voltage level of a transformer. The power quality of transformers built in Turkey was investigated in *Electrical Engineering* (Vol. 95, 2013). For a sample of 103 transformers built for heavy industry, the mean number of sags per week was 353 and the mean number of swells per week was 184. Assume the standard deviation of the sag distribution is 30 sags per week and the standard deviation of the swell distribution is 25 swells per week. Suppose one of the transformers is randomly selected and found to have 400 sags and 100 swells in a week.

a. Find the z-score for the number of sags for this transformer. Interpret this value.

b. Find the z-score for the number of swells for this transformer. Interpret this value.

2.128 Blue versus red exam study. In a study of how external clues influence performance, psychology professors at the University of Alberta and Pennsylvania State University gave two different forms of a midterm examination to a large group of introductory psychology students. The questions on the exam were identical and in the same order, but one exam was printed on blue paper and the other on red paper (*Teaching Psychology*, May 1998). Grading only the difficult questions on the exam, the researchers found that scores on the blue exam had a distribution with a mean of 53% and a standard deviation of 15%, while scores on the red exam had a distribution with a mean of 39% and a standard deviation of 12%. (Assume that both distributions are approximately mound shaped and symmetric.)

a. Give an interpretation of the standard deviation for the students who took the blue exam.

b. Give an interpretation of the standard deviation for the students who took the red exam.

c. Suppose a student is selected at random from the group of students who participated in the study and the student's score on the difficult questions is 20%. Which exam form is the student more likely to have taken, the blue or the red exam? Explain.

2.129 Ranking Ph.D. programs in economics. The *Southern Economic Journal* (Apr. 2008) published a guide to graduate study in economics by ranking the Ph.D. programs at 129 colleges and universities. Each program was evaluated according to the number of publications published by faculty teaching in the Ph.D. program and by the quality of the publications. Data obtained from the Social Science Citation Index (SSCI) were used to calculate an overall productivity score for each Ph.D. program. The mean and standard deviation of these 129 productivity scores were then used to compute a *z*-score for each economics program. Harvard University had the highest *z*-score ($z = 5.08$) and, hence, was the top-ranked school; Howard University was ranked last since it had the lowest *z*-score ($z = -0.81$). The data (*z*-scores) for all 129 economic programs are saved in the **ECOPHD** file.

ECOPHD

a. Interpret the *z*-score for Harvard University.
b. Interpret the *z*-score for Howard University.
c. The authors of the *Southern Economic Journal* article note that "only 44 of the 129 schools have positive *z*-scores, indicating that the distribution of overall productivity is skewed to the right." Do you agree? (Check your answer by constructing a histogram for the *z*-scores in the **ECOPHD** file.)

Applying the Concepts—Advanced

2.130 Ranking Ph.D. programs in economics (cont'd). Refer to the *Southern Economic Journal* (Apr. 2008) study of Ph.D. programs in economics, Exercise 2.129. The authors also made the following observation:

"A noticeable feature of this skewness is that distinction between schools diminishes as the rank declines. For example, the top-ranked school, Harvard, has a *z*-score of 5.08, and the fifth-ranked school, Yale, has a *z*-score of 2.18, a substantial difference. However, ... the 70th-ranked school, the University of Massachusetts, has a *z*-score of -0.43, and the 80th-ranked school, the University of Delaware, has a *z*-score of -0.50, a very small difference. [Consequently] the ordinal rankings presented in much of the literature that ranks economics departments miss the fact that below a relatively small group of top programs, the differences in [overall] productivity become fairly small." Do you agree?

2.131 GPAs of students. At one university, the students are given *z*-scores at the end of each semester, rather than the traditional GPAs. The mean and standard deviation of all students' cumulative GPAs, on which the *z*-scores are based, are 2.7 and .5, respectively.

a. Translate each of the following *z*-scores to corresponding GPA scores: $z = 2.0, z = -1.0, z = .5, z = -2.5$.
b. Students with *z*-scores below -1.6 are put on probation. What is the corresponding probationary GPA?
c. The president of the university wishes to graduate the top 16% of the students with *cum laude* honors and the top 2.5% with *summa cum laude* honors. Where (approximately) should the limits be set in terms of *z*-scores? In terms of GPAs? What assumption, if any, did you make about the distribution of the GPAs at the university?

2.7 Methods for Detecting Outliers: Box Plots and *z*-Scores

Sometimes it is important to identify inconsistent or unusual measurements in a data set. An observation that is unusually large or small relative to the data values we want to describe is called an **outlier**.

Outliers are often attributable to one of several causes. First, the measurement associated with the outlier may be invalid. For example, the experimental procedure used to generate the measurement may have malfunctioned, the experimenter may have misrecorded the measurement, or the data might have been coded incorrectly in the computer. Second, the outlier may be the result of a misclassified measurement. That is, the measurement belongs to a population different from that from which the rest of the sample was drawn. Finally, the measurement associated with the outlier may be recorded correctly and from the same population as the rest of the sample but represent a rare (chance) event. Such outliers occur most often when the relative frequency distribution of the sample data is extremely skewed because a skewed distribution has a tendency to include extremely large or small observations relative to the others in the data set.

> An observation (or measurement) that is unusually large or small relative to the other values in a data set is called an **outlier**. Outliers typically are attributable to one of the following causes:
>
> 1. The measurement is observed, recorded, or entered into the computer incorrectly.
> 2. The measurement comes from a different population.
> 3. The measurement is correct but represents a rare (chance) event.

Two useful methods for detecting outliers, one graphical and one numerical, are **box plots** and z-scores. The box plot is based on the quartiles (defined in Section 2.6) of a data set. Specifically, a box plot is based on the *interquartile range (IQR)*—the distance between the lower and upper quartiles:

$$IQR = Q_U - Q_L$$

The **interquartile range (IQR)** is the distance between the lower and upper quartiles:

$$IQR = Q_U - Q_L$$

An annotated MINITAB box plot for the gas mileage data (see Table 2.2) is shown in Figure 2.31.* Note that a rectangle (the *box*) is drawn, with the bottom and top of the rectangle (the **hinges**) drawn at the quartiles Q_L and Q_U, respectively. Recall that Q_L represents the 25th percentile and Q_U represents the 75th percentile. By definition, then, the "middle" 50% of the observations—those between Q_L and Q_U—fall inside the box. For the gas mileage data, these quartiles are at 35.625 and 38.375. Thus,

$$IQR = 38.375 - 35.625 = 2.75$$

The median is shown at 37 by a horizontal line within the box.

To guide the construction of the "tails" of the box plot, two sets of limits, called **inner fences** and **outer fences**, are used. Neither set of fences actually appears on the plot. Inner fences are located at a distance of 1.5(IQR) from the hinges. Emanating from the hinges of the box are vertical lines called the **whiskers**. The two whiskers extend to the most extreme observation inside the inner fences. For example, the inner fence on the lower side of the gas mileage box plot is

$$\text{Lower inner fence} = \text{Lower hinge} - 1.5(IQR)$$
$$= 35.625 - 1.5(2.75)$$
$$= 35.625 - 4.125 = 31.5$$

Figure 2.31

Annotated MINITAB box plot for EPA gas mileages

The smallest measurement *inside* this fence is the second-smallest measurement, 31.8. Thus, the lower whisker extends to 31.8. Similarly, the upper whisker extends to 42.1, the largest measurement inside the upper inner fence:

$$\text{Upper inner fence} = \text{Upper hinge} + 1.5(IQR)$$
$$= 38.375 + 1.5(2.75)$$
$$= 38.375 + 4.125 = 42.5$$

*Although box plots can be generated by hand, the amount of detail required makes them particularly well suited for computer generation. We use computer software to generate the box plots in this section.

Values that are beyond the inner fences are deemed *potential outliers* because they are extreme values that represent relatively rare occurrences. In fact, for mound-shaped distributions, less than 1% of the observations are expected to fall outside the inner fences. Two of the 100 gas mileage measurements, 30.0 and 44.9, fall beyond the inner fences, one on each end of the distribution. Each of these potential outliers is represented by a common symbol (an asterisk in MINITAB).

The other two imaginary fences, the outer fences, are defined at a distance 3(IQR) from each end of the box. Measurements that fall beyond the outer fences (also represented by an asterisk in MINITAB) are very extreme measurements that require special analysis. Since less than one-hundredth of 1% (.01% or .0001) of the measurements from mound-shaped distributions are expected to fall beyond the outer fences, these measurements are considered to be *outliers*. Because there are no measurements of gas mileage beyond the outer fences, there are no outliers.

Recall that outliers may be incorrectly recorded observations, members of a population different from the rest of the sample, or, at the least, very unusual measurements from the same population. The box plot of Figure 2.31 detected two potential outliers: the two gas mileage measurements beyond the inner fences. When we analyze these measurements, we find that they are correctly recorded. Perhaps they represent mileages that correspond to exceptional models of the car being tested or to unusual gas mixtures. Outlier analysis often reveals useful information of this kind and therefore plays an important role in the statistical inference-making process.

In addition to detecting outliers, box plots provide useful information on the variation in a data set. The elements (and nomenclature) of box plots are summarized in the box. Some aids to the interpretation of box plots are also given.

Elements of a Box Plot[†]

1. A rectangle (the **box**) is drawn with the ends (the **hinges**) drawn at the lower and upper quartiles (Q_L and Q_U). The median M of the data is shown in the box, usually by a line or a symbol (such as "+").

2. The points at distances 1.5(IQR) from each hinge mark the **inner fences** of the data set. Lines (the **whiskers**) are drawn from each hinge to the most extreme measurement inside the inner fence. Thus,

$$\text{Lower inner fence} = Q_L - 1.5(\text{IQR})$$
$$\text{Upper inner fence} = Q_U + 1.5(\text{IQR})$$

3. A second pair of fences, the **outer fences**, appears at a distance of 3(IQR) from the hinges. One symbol (e.g., "*") is used to represent measurements falling between the inner and outer fences, and another (e.g., "0") is used to represent measurements that lie beyond the outer fences. Thus, outer fences are not shown unless one or more measurements lie beyond them. We have

$$\text{Lower outer fence} = Q_L - 3(\text{IQR})$$
$$\text{Upper outer fence} = Q_U + 3(\text{IQR})$$

4. The symbols used to represent the median and the extreme data points (those beyond the fences) will vary with the software you use to construct the box plot. (You may use your own symbols if you are constructing a box plot by hand.) You should consult the program's documentation to determine exactly which symbols are used.

[†]The originator of the box plot, John Tukey, initially used the constants 1 and 2 in the equations for the inner and outer fences, respectively. Through repetition and experience, Tukey modified the formulas using the constants 1.5 and 3.

Aids to the Interpretation of Box Plots

1. The line (median) inside the box represents the "center" of the distribution of data.

2. Examine the length of the box. The IQR is a measure of the sample's variability and is especially useful for the comparison of two samples. (See Example 2.16.)

3. Visually compare the lengths of the whiskers. If one is clearly longer, the distribution of the data is probably skewed in the direction of the longer whisker.

4. Analyze any measurements that lie beyond the fences. Less than 5% should fall beyond the inner fences, even for very skewed distributions. Measurements beyond the outer fences are probably outliers, with one of the following explanations:

   a. The measurement is incorrect. It may have been observed, recorded, or entered into the computer incorrectly.

   b. The measurement belongs to a population different from the population that the rest of the sample was drawn from. (See Example 2.17.)

   c. The measurement is correct *and* from the same population as the rest of the sample. Generally, we accept this explanation only after carefully ruling out all others.

Example   2.16   Ⓓ
WLTASK
Computer-Generated Box Plot—The "Water-Level Task"

Problem Refer to the *Psychological Science* experiment, called the "water-level task," Examples 2.2 and 2.14. Use statistical software to produce a box plot for the 40 water-level task deviation angles shown in Table 2.4. Identify any outliers in the data set.

Solution The 40 deviations were entered into MINITAB, and a box plot produced in Figure 2.32. Recall (from Example 2.14) that $Q_L = 2, M = 7$, and $Q_U = 10$. The middle line in the box represents the median deviation of $M = 7$. The lower hinge of the box represents $Q_L = 2$, while the upper hinge represents $Q_U = 10$. Note that there are two extreme observations (represented by asterisks) that are beyond the upper inner fence, but inside the upper outer fence. These two values are outliers. Examination of the data reveals that these measurements correspond to deviation angles of 25° and 35°.

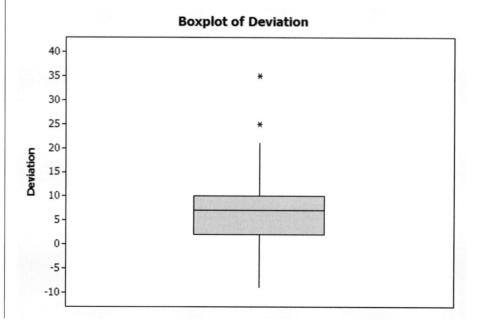

Figure 2.32
MINITAB box plot for water-level task deviations

Look Ahead Before removing the outliers from the data set, a good analyst will make a concerted effort to find the cause of the outliers. For example, both the outliers are deviation measurements for waitresses. An investigation may discover that both waitresses were new on the job the day the experiment was conducted. If the study target population is experienced bartenders and waitresses, these two observations would be dropped from the analysis.

─────■ **Now Work Exercise 2.137**

| Example | 2.17 | D |

Comparing Box Plots—Stimulus Reaction Study

Problem A Ph.D. student in psychology conducted a stimulus reaction experiment as a part of her dissertation research. She subjected 50 subjects to a threatening stimulus and 50 to a nonthreatening stimulus. The reaction times of all 100 students, recorded to the nearest tenth of a second, are listed in Table 2.8. Box plots of the two resulting samples of reaction times, generated with SAS, are shown in Figure 2.33. Interpret the box plots.

Table 2.8   Reaction Times of Students

Nonthreatening Stimulus

2.0	1.8	2.3	2.1	2.0	2.2	2.1	2.2	2.1	2.1
2.0	2.0	1.8	1.9	2.2	2.0	2.2	2.4	2.1	2.0
2.2	2.1	2.2	1.9	1.7	2.0	2.0	2.3	2.1	1.9
2.0	2.2	1.6	2.1	2.3	2.0	2.0	2.0	2.2	2.6
2.0	2.0	1.9	1.9	2.2	2.3	1.8	1.7	1.7	1.8

Threatening Stimulus

1.8	1.7	1.4	2.1	1.3	1.5	1.6	1.8	1.5	1.4
1.4	2.0	1.5	1.8	1.4	1.7	1.7	1.7	1.4	1.9
1.9	1.7	1.6	2.5	1.6	1.6	1.8	1.7	1.9	1.9
1.5	1.8	1.6	1.9	1.3	1.5	1.6	1.5	1.6	1.5
1.3	1.7	1.3	1.7	1.7	1.8	1.6	1.7	1.7	1.7

D *Data Set:* REACTION

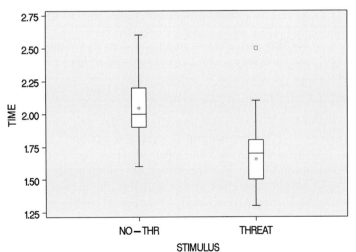

Box Plots of Time by Stimulus

Figure 2.33

SAS box plots for reaction-time data

Solution In SAS, the median is represented by the horizontal line through the box, while the asterisk (*) represents the mean. Analysis of the box plots on the same numerical scale reveals that the distribution of times corresponding to the threatening stimulus lies below that of the nonthreatening stimulus. The implication is that the reaction times tend to be faster to the threatening stimulus. Note, too, that the upper whiskers of both samples are longer than the lower whiskers, indicating that the reaction times are positively skewed.

No observations in the two samples fall between the inner and outer fences. However, there is one outlier: the observation of 2.5 seconds corresponding to the threatening stimulus that is beyond the outer fence (denoted by the square symbol in SAS). When the researcher examined her notes from the experiments, she found that the subject whose time was beyond the outer fence had mistakenly been given the nonthreatening stimulus. You can see in Figure 2.33 that his time would have been within the upper whisker if moved to the box plot corresponding to the non-threatening stimulus. The box plots should be reconstructed, since they will both change slightly when this misclassified reaction time is moved from one sample to the other.

Look Ahead The researcher concluded that the reactions to the threatening stimulus were faster than those to the nonthreatening stimulus. However, she was asked by her Ph.D. committee whether the results were *statistically significant*. Their question addresses the issue of whether the observed difference between the samples might be attributable to chance or sampling variation rather than to real differences between the populations. To answer this question, the researcher must use inferential statistics rather than graphical descriptions. We discuss how to compare two samples by means of inferential statistics in Chapter 9.

The next example illustrates how z-scores can be used to detect outliers and make inferences.

Example   2.18

Inference Using *z*-Scores

Problem Suppose a female bank employee believes that her salary is low as a result of sex discrimination. To substantiate her belief, she collects information on the salaries of her male counterparts in the banking business. She finds that their salaries have a mean of $64,000 and a standard deviation of $2,000. Her salary is $57,000. Does this information support her claim of sex discrimination?

Solution The analysis might proceed as follows: First, we calculate the z-score for the woman's salary with respect to those of her male counterparts. Thus,

$$z = \frac{\$57,000 - \$64,000}{\$2,000} = -3.5$$

The implication is that the woman's salary is 3.5 standard deviations *below* the mean of the male salary distribution. Furthermore, if a check of the male salary data shows that the frequency distribution is mound shaped, we can infer that very few salaries in this distribution should have a z-score less than -3, as shown in Figure 2.34. Clearly, a z-score of -3.5 represents an outlier. Either her salary is from a distribution different from the male salary distribution, or it is a very unusual (highly improbable) measurement from a salary distribution no different from the male distribution.

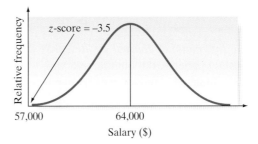

Figure 2.34
Male salary distribution

Look Back Which of the two situations do you think prevails? Statistical thinking would lead us to conclude that her salary does not come from the male salary distribution, lending support to the female bank employee's claim of sex discrimination. However, the careful investigator should require more information before inferring that sex discrimination is the cause. We would want to know more about the data collection technique the woman used and more about her competence at her job. Also, perhaps other factors, such as length of employment, should be considered in the analysis.

■ **Now Work Exercise 2.138**

Examples 2.17 and 2.18 exemplify an approach to statistical inference that might be called the **rare-event approach**. An experimenter hypothesizes a specific frequency distribution to describe a population of measurements. Then a sample of measurements is drawn from the population. If the experimenter finds it unlikely that the sample came from the hypothesized distribution, the hypothesis is judged to be false. Thus, in Example 2.18, the woman believes that her salary reflects discrimination. She hypothesizes that her salary should be just another measurement in the distribution of her male counterparts' salaries if no discrimination exists. However, it is so unlikely that the sample (in this case, her salary) came from the male frequency distribution that she rejects that hypothesis, concluding that the distribution from which her salary was drawn is different from the distribution for the men.

This rare-event approach to inference making is discussed further in later chapters. Proper application of the approach requires a knowledge of probability, the subject of our next chapter.

We conclude this section with some rules of thumb for detecting outliers.

Rules of Thumb for Detecting Outliers*

Box Plots: Observations falling between the inner and outer fences are deemed *suspect outliers*. Observations falling beyond the outer fence are deemed *highly suspect outliers*.

Suspect outliers	*Highly suspect outliers*
Between $Q_L - 1.5(\text{IQR})$ and $Q_L - 3(\text{IQR})$ or Between $Q_U + 1.5(\text{IQR})$ and $Q_U + 3(\text{IQR})$	Below $Q_L - 3(\text{IQR})$ or Above $Q_U + 3(\text{IQR})$

z-Scores: Observations with z-scores greater than 3 in absolute value are considered *outliers*. For some highly skewed data sets, observations with z-scores greater than 2 in absolute value *may be outliers*.

Possible Outliers	*Outliers*
$\lvert z \rvert > 2$	$\lvert z \rvert > 3$

---

*The z-score and box plot methods both establish rule-of-thumb limits outside of which a measurement is deemed to be an outlier. Usually, the two methods produce similar results. However, the presence of one or more outliers in a data set can inflate the computed value of *s*. Consequently, it will be less likely that an errant observation would have a z-score larger than $\lvert 3 \rvert$. In contrast, the values of the quartiles used to calculate the intervals for a box plot are not affected by the presence of outliers.

Detecting Outliers in the Body Image Data

In the *Body Image: An International Journal of Research* (Jan. 2010) study of 92 BDD patients, the quantitative variable of interest is Total Appearance Evaluation score (ranging from 7 to 35 points). Are there any unusual scores in the BDD data set? We will apply both the box plot and *z*-score method to aid in identifying any outliers in the data. Since from previous analyses, there appears to be a difference in the distribution of appearance evaluation scores for males and females, we will analyze the data by gender.

To employ the *z*-score method, we require the mean and standard deviation of the data for each gender. These values were already computed in the previous Statistics in Action Revisited section. For females, $\bar{x} = 17.05$ and $s = 4.76$; for males, $\bar{x} = 19.0$ and $s = 5.42$ (see Figure SIA2.5). Then, the 3-standard-deviation interval for each gender is:

Females:	$\bar{x} \pm 3s = 17.05 \pm 3(4.76) = 17.05 \pm 14.28 = (2.77, 31.33)$
Males:	$\bar{x} \pm 3s = 19.00 \pm 3(5.42) = 19.00 \pm 16.26 = (2.74, 35.26)$

If you examine the appearance evaluation scores in the **BDD** file, you will find that none of the scores fall beyond the 3-standard-deviation interval for each group. Consequently,

if we use the *z*-score approach, there are no highly suspect outliers in the data.

Box plots for the data are shown in Figure SIA2.6. Although several suspect outliers (asterisks) are shown on the box plot for each gender, there are no highly suspect outliers (zeros) shown. That is, no data points fall beyond the outer fences of the box plots. As with the *z*-score approach, the box plot method does not detect any highly suspect outliers.

[*Note*: If we were to detect one or more highly suspect outliers, we should investigate whether or not to include the observation in any analysis that leads to an inference about the population of BDD patients. Is the outlier a legitimate value (in which case it will remain in the data set for analysis) or is the outlier associated with a subject that is not a member of the population of interest—say, a person who is misdiagnosed with BDD (in which case it will be removed from the data set prior to analysis)?]

D *Data Set: BDD*

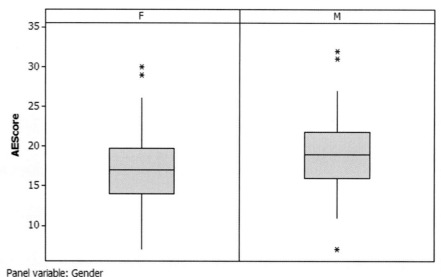

Boxplot of AEScore

Panel variable: Gender

Figure SIA2.6
MINITAB box plots for Appearance Evaluation score by Gender

Exercises 2.132–2.153

Understanding the Principles

2.132 What is the percentage of a population that lies outside the interquartile range?

2.133 What are the hinges of a box plot?

2.134 What does the central line parallel to the hinges of a box plot indicate?

2.135 Define an outlier.

Learning the Mechanics

2.136 Suppose a data set consisting of exam scores has a lower quartile $Q_L = 60$, a median $M = 75$, and an upper quartile $Q_U = 85$. The scores on the exam range from 18 to 100. Without having the actual scores available to you, construct as much of the box plot as possible.

2.137 Consider the following horizontal box plot:

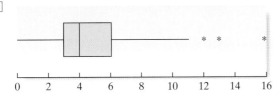

a. What is the median of the data set (approximately)?
b. What are the upper and lower quartiles of the data set (approximately)?
c. What is the interquartile range of the data set (approximately)?
d. Is the data set skewed to the left, skewed to the right, or symmetric?
e. What percentage of the measurements in the data set lie to the right of the median? To the left of the upper quartile?
f. Identify any outliers in the data.

2.138 A sample data set has a mean of 57 and a standard deviation of 11. Determine whether each of the following sample measurements is an outlier.
a. 65
b. 21
c. 72
d. 98

2.139 Consider the following sample data set. Construct a box plot for the data and use it to identify any outliers.

L02139

121	171	158
173	184	163
157	85	145
165	172	196
170	159	172
161	187	100
142	166	171

2.140 Consider the following sample data set.

L02140

171	152	170
168	169	171
190	183	185
140	173	206
172	174	169
199	151	180
167	170	188

a. Construct a box plot.
b. Identify any outliers that may exist in the data.

Applet Exercise 2.7

Use the applet *Standard Deviation* to determine whether an item in a data set may be an outlier. Begin by setting appropriate limits and plotting the given data on the number line provided in the applet. Here is the data set:

10 80 80 85 85 85 85 90 90 90 90 90 95 95 95 95 100 100

a. The green arrow shows the approximate location of the mean. Multiply the standard deviation given by the applet by 3. Is the data item 10 more than three standard deviations away from the green arrow (the mean)? Can you conclude that the 10 is an outlier?
b. Using the mean and standard deviation from part **a**, move the point at 10 on your plot to a point that appears

to be about three standard deviations from the mean. Repeat the process in part **a** for the new plot and the new suspected outlier.
c. When you replaced the extreme value in part **a** with a number that appeared to be within three standard deviations of the mean, the standard deviation got smaller and the mean moved to the right, yielding a new data set whose extreme value was *not* within three standard deviations of the mean. Continue to replace the extreme value with higher numbers until the new value is within three standard deviations of the mean in the new data set. Use trial and error to estimate the smallest number that can replace the 10 in the original data set so that the replacement is not considered to be an outlier.

Applying the Concepts—Basic

2.141 Blond hair types in the Southwest Pacific. A mutation of a pigmentation gene is theorized to be the cause of blond-hair genotypes observed in several Southwest Pacific islands. The distribution of this mutated gene was investigated in the *American Journal of Physical Anthropology* (Apr. 2014). For each in a sample of 550 Southwest Pacific islanders, the effect of the mutation on hair pigmentation was measured with the melanin (M) index, where M ranges between −4 and 4. The box plots below show the distribution of M for three different genotypes—CC, CT, and TT. Identify any outliers in the data. For which genotypes do these outliers occur?

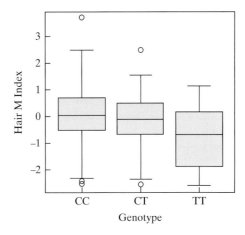

2.142 Dentists' use of anesthetics. Refer to the *Current Allergy & Clinical Immunology* study of the use of local anesthetics in dentistry, presented in Exercise 2.103 (p. 105). Recall that the mean number of units (ampoules) of local anesthetics used per week by dentists was 79, with a standard deviation of 23. Consider a dentist who used 175 units of local anesthetics in a week.
a. Find the z-score for this measurement.
b. Would you consider the measurement to be an outlier? Explain.
c. Give several reasons the outlier may have occurred.

2.143 Research on brain specimens. Refer to the *Brain and Language* data on postmortem intervals (PMIs) of 22 human brain specimens, presented in Exercise 2.45 (p. 80). The mean and standard deviation of the PMI values are 7.3 and 3.18, respectively.
a. Find the z-score for the PMI value of 3.3.
b. Is the PMI value of 3.3 considered an outlier? Explain.

MINITAB Output for Exercise 2.144

Descriptive Statistics: Support

Variable	N	Mean	StDev	Variance	Minimum	Median	Maximum
Support	992	67.755	26.871	722.036	0.000	68.000	155.000

2.144 Corporate sustainability of CPA firms. Refer to the *Business and Society* (Mar. 2011) study on the sustainability behaviors of CPA corporations, Exercise 2.105 (p. 105). Numerical descriptive measures for level of support for corporate sustainability for the 992 senior managers are repeated in the above MINITAB printout. One of the managers reported a support level of 155 points. Would you consider this support level to be typical of the study sample? Explain.

2.145 Voltage sags and swells. Refer to the *Electrical Engineering* (Vol. 95, 2013) study of power quality (measured by "sags" and "swells") in Turkish transformers, Exercise 2.127 (p. 110). For a sample of 103 transformers built for heavy industry, the mean and standard deviation of the number of sags per week were 353 and 30, respectively; also, the mean and standard deviation of the number of swells per week were 184 and 25, respectively. Consider a transformer that has 400 sags and 100 swells in a week.
a. Would you consider 400 sags per week unusual statistically? Explain.
b. Would you consider 100 swells per week unusual statistically? Explain.

2.146 Treating psoriasis with the "Doctorfish of Kangal." Psoriasis is a skin disorder with no known cure. An alternative treatment for psoriasis is ichthyotherapy, also known as therapy with the "Doctorfish of Kangal." Fish from the hot pools of Kangal, Turkey, feed on the skin scales of bathers, reportedly reducing the symptoms of psoriasis. In one study, 67 patients diagnosed with psoriasis underwent three weeks of ichthyotherapy. (*Evidence-Based Research in Complementary and Alternative Medicine*, Dec. 2006). The Psoriasis Area Severity Index (PASI) of each patient was measured both before and after treatment. (The lower the PASI score, the better is the skin condition.) Box plots of the PASI scores, both before (baseline) and after three weeks of ichthyotherapy treatment, are shown in the accompanying diagram.
a. Find the approximate 25th percentile, the median, and the 75th percentile for the PASI scores before treatment.

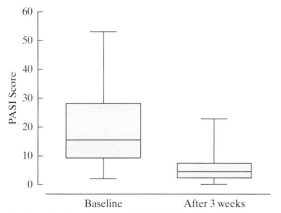

Based on Grassberger, M., and Hoch, W. "Ichthyotherapy as alternative treatment for patients with psoriasis: A pilot study." *Evidence-Based Research in Complementary and Alternative Medicine*, Vol. 3, No. 4, Dec. 2004 (Figure 3). National Center for Complementary and Alternative Medicine.

b. Find the approximate 25th percentile, the median, and the 75th percentile for the PASI scores after treatment.
c. Comment on the effectiveness of ichthyotherapy in treating psoriasis.

Applying the Concepts—Intermediate

2.147 Sanitation inspection of cruise ships. Refer to the data on sanitation levels of cruise ships, presented in Exercise 2.41 (p. 79).
a. Use the box plot method to detect any outliers in the data.
b. Use the z-score method to detect any outliers in the data.
c. Do the two methods agree? If not, explain why.

2.148 Characteristics of a rockfall. Refer to the *Environmental Geology* (Vol. 58, 2009) study of how far a block from a collapsing rockwall will bounce, Exercise 2.61 (p. 89). The computer-simulated rebound lengths (in meters) for 13 block-soil impact marks left on a slope from an actual rockfall are reproduced in the table. Do you detect any outliers in the data? Explain.

10.94	13.71	11.38	7.26	17.83	11.92	11.87	5.44	13.35	4.90	5.85	5.10	6.77

Based on Paronuzzi, P. "Rockfall-induced block propagation on a soil slope, northern Italy." *Environmental Geology*, Vol. 58, 2009 (Table 2).

2.149 Comparing SAT Scores. Refer to Exercise 2.46 (p. 80), in which we compared state average SAT scores in 2011 and 2014.
a. Construct side-by-side box plots of the SAT scores for the two years.
b. Compare the variability of the SAT scores for the two years.
c. Are any state SAT scores outliers in either year? If so, identify them.

2.150 Is honey a cough remedy? Refer to the *Archives of Pediatrics and Adolescent Medicine* (Dec. 2007) study of honey as a remedy for coughing, Exercise 2.64 (p. 90). Recall that coughing improvement scores were recorded for children in three groups: the over-the-counter cough medicine dosage (DM) group, the honey dosage group, and the control (no medicine) group.
a. For each group, construct a box plot of the improvement scores.
b. How do the median improvement scores compare for the three groups?
c. How does the variability in improvement scores compare for the three groups?
d. Do you detect any outliers in any of the three coughing improvement score distributions?

2.151 Permeability of sandstone during weathering. Refer to the *Geographical Analysis* (Vol. 42, 2010) study of the decay properties of sandstone when exposed to the weather, Exercise 2.110 (p. 106). Recall that slices of sandstone blocks were tested for permeability under three conditions: no exposure to any type of weathering (A), repeatedly sprayed with a 10% salt solution (B), and soaked in a 10% salt solution and dried (C).

a. Identify any outliers in the permeability measurements for group A sandstone slices.

b. Identify any outliers in the permeability measurements for group B sandstone slices.

c. Identify any outliers in the permeability measurements for group C sandstone slices.

d. If you remove the outliers detected in parts **a–c**, how will descriptive statistics like the mean, median, and standard deviation be affected? If you are unsure of your answer, carry out the analysis.

Applying the Concepts—Advanced

2.152 Defective items in a batch. A manufacturer sells his products in boxes containing 50 items. He claims that on an average, a box will contain no more than four defective items. In order to verify his claim, a box is selected randomly and its items checked. It has been found that the box contains eight defective items. Assume that the standard deviation of the number of defective items per box is 1.5.

a. If the mean number of defectives per box is really 4 what is the z-score corresponding to eight?

b. Considering your answer to part **a**, do you have any reason to believe that the manufacturer's claim is correct?

c. What would be your answer if the standard deviation is assumed to be 1.25 instead of 1.5?

2.153 Untutored second language acquisition. A student from Turkey named Alex is attending a university in the United States. While in this country, he has never been tutored in English. After his second year at university, Alex was interviewed numerous times over a 1-year period. His language was analyzed for syntactic complexity (measured as number of clauses per analysis-of-speech unit) and lexical diversity (a complex measure, called D, on a 100-point scale). *Applied Linguistics* (May 2014) reported Alex's syntactic complexity and lexical diversity values from his first interview as 1.73 and 55.88, respectively. For comparison purposes, the researchers recorded similar language scores for 3 native English speakers, all who attended the same university and had the same major as Alex. The syntactic complexity scores of the 3 English speakers were 1.88, 1.82, and 2.22. The lexical diversity scores of the 3 English speakers were 64.62, 58.03, and 56.97. According to the researchers, "Alex is within native speaker range, even allowing for the variation that naturally takes place within conversation." Do you agree? Explain.

2.8 Graphing Bivariate Relationships (Optional)

The claim is often made that the crime rate and the unemployment rate are "highly correlated." Another popular belief is that smoking and lung cancer are "related." Some people even believe that the Dow Jones Industrial Average and the lengths of fashionable skirts are "associated." The words *correlated*, *related*, and *associated* imply a relationship between two variables—in the examples just mentioned, two *quantitative* variables.

One way to describe the relationship between two quantitative variables—called a **bivariate relationship**—is to construct a **scatterplot**. A scatterplot is a two-dimensional plot, with one variable's values plotted along the vertical axis and the others along the horizontal axis. For example, Figure 2.35 is a scatterplot relating (1) the cost of mechanical work (heating, ventilating, and plumbing) to (2) the floor area of the building for a sample of 26 factory and warehouse buildings. Note that the scatterplot suggests a general tendency for mechanical cost to increase as building floor area increases.

When an increase in one variable is generally associated with an increase in the second variable, we say that the two variables are "positively related" or

Figure 2.35

Scatterplot of cost vs. floor area

"positively correlated."* Figure 2.35 implies that mechanical cost and floor area are positively correlated. Alternatively, if one variable has a tendency to decrease as the other increases, we say the variables are "negatively correlated." Figure 2.36 shows several hypothetical scatterplots that portray a positive bivariate relationship (Figure 2.36a), a negative bivariate relationship (Figure 2.36b), and a situation in which the two variables are unrelated (Figure 2.36c).

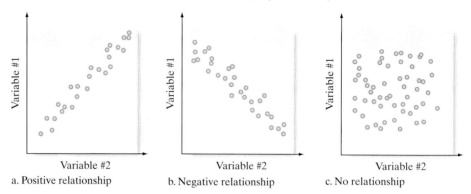

Figure 2.36
Hypothetical bivariate relationships

a. Positive relationship   b. Negative relationship   c. No relationship

| Example | 2.19 | D |

Graphing Bivariate Data—Hospital Application

Problem A medical item used to administer to a hospital patient is called a *factor*. For example, factors can be intravenous (IV) tubing, IV fluid, needles, shave kits, bedpans, diapers, dressings, medications, and even code carts. The coronary care unit at Bayonet Point Hospital (St. Petersburg, Florida) recently investigated the relationship between the number of factors administered per patient and the patient's length of stay (in days). Data on these two variables for a sample of 50 coronary care patients are given in Table 2.9. Use a scatterplot to describe the relationship between the two variables of interest: number of factors and length of stay.

Table 2.9   Medfactors Data on Patients' Factors and Length of Stay

Number of Factors	Length of Stay (days)	Number of Factors	Length of Stay (days)
231	9	354	11
323	7	142	7
113	8	286	9
208	5	341	10
162	4	201	5
117	4	158	11
159	6	243	6
169	9	156	6
55	6	184	7
77	3	115	4
103	4	202	6
147	6	206	5
230	6	360	6
78	3	84	3
525	9	331	9
121	7	302	7
248	5	60	2
233	8	110	2
260	4	131	5
224	7	364	4
472	12	180	7
220	8	134	6
383	6	401	15
301	9	155	4
262	7	338	8

Based on Bayonet Point Hospital, Coronary Care Unit.

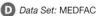 *Data Set:* MEDFAC

*A formal definition of correlation is given in Chapter 11. There we will learn that correlation measures the strength of the linear (or straight-line) relationship between two quantitative variables.

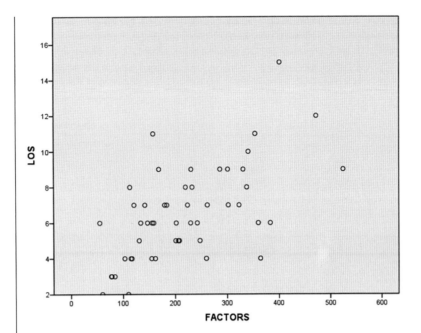

Figure 2.37
SPSS scatterplot for medical
factors data in Table 2.9

Solution Rather than construct the plot by hand, we resort to a statistical software package. The SPSS plot of the data in Table 2.9, with length of stay (LOS) on the vertical axis and number of factors (FACTORS) on the horizontal axis, is shown in Figure 2.37.

Although the plotted points exhibit a fair amount of variation, the scatterplot clearly shows an increasing trend. It appears that a patient's length of stay is positively correlated with the number of factors administered to the patient.

Look Back If hospital administrators can be confident that the sample trend shown in Figure 2.37 accurately describes the trend in the population, then they may use this information to improve their forecasts of lengths of stay for future patients.

■ **Now Work Exercise 2.157**

The scatterplot is a simple, but powerful, tool for describing a bivariate relationship. However, keep in mind that it is only a graph. No measure of reliability can be attached to inferences made about bivariate populations based on scatterplots of sample data. The statistical tools that enable us to make inferences about bivariate relationships are presented in Chapter 11.

Exercises 2.154–2.169

Understanding the Principles

2.154 What is a bivariate scatterplot?

2.155 What is the use of a bivariate scatterplot?

2.156 How can a bivariate scatterplot be used to measure the relationship between the two variables?

Learning the Mechanics

2.157 Construct a scatterplot for the data in the table that follows.
Ⓓ Do you detect a trend?
L02157
NW

Variable #1:	5	3	−1	2	7	6	4	0	8
Variable #2:	14	3	10	1	8	5	3	2	12

2.158 Construct a scatterplot for the data in the table that follows.
Ⓓ Do you detect a trend?
L02158

Variable #1:	.5	1	1.5	2	2.5	3	3.5	4	4.5	5
Variable #2:	2	1	3	4	6	10	9	12	17	17

Applying the Concepts—Basic

2.159 Comparing SAT scores. Refer to Exercise 2.46 (p. 80) and the data on state SAT scores saved in the **SAT** file. Consider a scatterplot for the data, with 2014 SAT score on the vertical axis and 2011 SAT score on the horizontal axis. What type of trend would you expect to observe? Why? Create the scatterplot and check your answer.
Ⓓ
SAT

2.160 Does elevation impact hitting performance in baseball?
(D) PARKS The Colorado Rockies play their major league home baseball games in Coors Field, Denver. Each year, the Rockies are among the leaders in team batting statistics (e.g., home runs, batting average, and slugging percentage). Many baseball experts attribute this phenomenon to the "thin air" of Denver—called the "mile-high city" due to its elevation. *Chance* (Winter 2006) investigated the effects of elevation on slugging percentage in Major League Baseball. Data were compiled on players' composite slugging percentage at each of 29 cities for a recent season as well as each city's elevation (feet above sea level). (Selected observations are shown in the next table.) Construct a scatterplot for the data. Do you detect a trend?

City	Slug Pct	Elevation
Anaheim	.480	160
Arlington	.605	616
Atlanta	.530	1050
Baltimore	.505	130
Boston	.505	20
⋮	⋮	⋮
Denver	.625	5277
⋮	⋮	⋮
Seattle	.550	350
San Francisco	.510	63
St. Louis	.570	465
Tampa	.500	10
Toronto	.535	566

Based on Schaffer, J., and Heiny, E. L. "The effects of elevation on slugging percentage in Major League Baseball." *Chance*, Vol. 19, No. 1, Winter 2006 (Figure 2).

2.161 Lobster trap placement.
(D) TRAPS Strategic placement of lobster traps is one of the keys for a successful lobster fisherman. An observational study of teams fishing for the red spiny lobster in Baja California Sur, Mexico, was conducted and the results published in *Bulletin of Marine Science* (Apr. 2010). Two variables measured for each of 8 teams from the Punta Abreojos (PA) fishing cooperative were total catch of lobsters (in kilograms) during the season and average percentage of traps allocated per day to exploring areas of unknown catch (called *search frequency*). These data are listed in the accompanying table. Graph the data in a scatterplot. What type of trend, if any, do you observe?

Lobster fishing study data

Total Catch	Search Frequency
2,785	35
6,535	21
6,695	26
4,891	29
4,937	23
5,727	17
7,019	21
5,735	20

Source: Shester, G. G. "Explaining catch variation among Baja California lobster fishers through spatial analysis of trap-placement decisions." *Bulletin of Marine Science*, Vol. 86, No. 2, Apr. 2010 (Table 1).

2.162 Feeding behavior of fish.
(D) BREAM The feeding behavior of black bream (a type of fish) spawned in aquariums was studied in *Brain, Behavior and Evolution* (Apr. 2000). Zoologists recorded the number of aggressive strikes of two black bream feeding at the bottom of the aquarium in the 10-minute period following the addition of food. The number of strikes and age of the fish (in days) were recorded approximately each week for nine weeks, as shown in the table below.

Week	Number of Strikes	Age of Fish (days)
1	85	120
2	63	136
3	34	150
4	39	155
5	58	162
6	35	169
7	57	178
8	12	184
9	15	190

Based on Shand, J., et al. "Variability in the location of the retinal ganglion cell area centralis is correlated with ontogenetic changes in feeding behavior in the Blackbream, *Acanthopagrus 'butcher'.*" *Brain and Behavior*, Vol. 55, No. 4, Apr. 2000 (Figure 10H).

a. Construct a scatterplot of the data, with number of strikes on the y-axis and age of the fish on the x-axis.
b. Examine the scatterplot of part **a.** Do you detect a trend?

2.163 New method for blood typing.
(D) BTYPE In *Analytical Chemistry* (May 2010), medical researchers tested a new method of typing blood using low cost paper. Blood drops were applied to the paper and the rate of absorption (called *blood wicking*) was measured. The table gives the wicking lengths (in millimeters) for six blood drops, each at a different antibody concentration. Construct a plot to investigate the relationship between wicking length and antibody concentration. What do you observe?

Droplet	Length (mm)	Concentration
1	22.50	0.0
2	16.00	0.2
3	13.50	0.4
4	14.00	0.6
5	13.75	0.8
6	12.50	1.0

Based on Khan, M. S. "Paper diagnostic for instant blood typing." *Analytical Chemistry*, Vol. 82, No. 10, May 2010 (Figure 4b).

Applying the Concepts—Intermediate

2.164 Sound waves from a basketball.
(D) BBALL Refer to the *American Journal of Physics* (June 2010) study of sound waves in a spherical cavity, Exercise 2.43 (p. 80). The frequencies of sound waves (estimated using a mathematical formula) resulting from the first 24 resonances (echoes) after striking a basketball with a metal rod are reproduced in the table on p. 125. Graph the data in a scatterplot, with frequency on the vertical axis and resonance number on the horizontal axis. Since a mathematical formula was used to estimate frequency, the researcher expects an increasing trend with very little variation. Does the graph support the researcher's theory?

Based on Russell, D. A. "Basketballs as spherical acoustic cavities." *American Journal of Physics*, Vol. 78, No. 6, June 2010 (Table I).

Data for Exercise 2.164

Resonance	Frequency
1	979
2	1572
3	2113
4	2122
5	2659
6	2795
7	3181
8	3431
9	3638
10	3694
11	4038
12	4203
13	4334
14	4631
15	4711
16	4993
17	5130
18	5210
19	5214
20	5633
21	5779
22	5836
23	6259
24	6339

Source: Russell, D. A. "Basketballs as spherical acoustic cavities." *American Journal of Physics,* Vol. 48, No. 6, June 2010 (Table I).

2.165 Plants that grow on Swiss cliffs. A rare plant that grows on the limestone cliffs of the Northern Swiss Jura mountains was studied in *Alpine Botany* (Nov. 2012). The researchers collected data from a sample of 12 limestone cliffs. Several of the variables measured for each cliff included the altitude above sea level (meters), plant population size (number of plants growing), and molecular variance (i.e., the variance in molecular weight of the plants). These data are provided in the accompanying table. The researchers are interested in whether either altitude or population size is related to molecular variance.

Cliff Number	Altitude	Population Size	Molecular Variance
1	468	147	59.8
2	589	209	24.4
3	700	28	42.2
4	664	177	59.5
5	876	248	65.8
6	909	53	17.7
7	1032	33	12.5
8	952	114	27.6
9	832	217	35.9
10	1099	10	13.3
11	982	8	3.6
12	1053	15	3.2

Source: Rusterholz, H., Aydin, D., and Baur, B. "Population structure and genetic diversity of relict populations of *Alyssum montanum* on limestone cliffs in the Northern Swiss Jura mountains." *Alpine Botany,* Vol. 122, No. 2, Nov. 2012 (Tables 1 and 2).

a. Use a scatterplot to investigate the relationship between molecular variance and altitude. Do you detect a trend?

b. Use a scatterplot to investigate the relationship between molecular variance and population size. Do you detect a trend?

2.166 Forest fragmentation study. Ecologists classify the cause of forest fragmentation as either anthropogenic (i.e., due to human development activities, such as road construction or logging) or natural in origin (e.g., due to wetlands or wildfire). *Conservation Ecology* (Dec. 2003) published an article on the causes of fragmentation for 54 South American forests. Using advanced high-resolution satellite imagery, the researchers developed two fragmentation indexes for each forest—one for anthropogenic fragmentation and one for fragmentation from natural causes. The values of these two indexes (where higher values indicate more fragmentation) for five of the forests in the sample are shown in the table below. The data for all 54 forests are saved in the **FRAG** file.

Ecoregion (forest)	Anthropogenic Index	Natural Origin Index
Araucaria moist forests	34.09	30.08
Atlantic Coast *restingas*	40.87	27.60
Bahia coastal forests	44.75	28.16
Bahia interior forests	37.58	27.44
Bolivian *Yungas*	12.40	16.75

Based on Wade, T. G., et al. "Distribution and causes of global forest fragmentation." *Conservation Ecology,* Vol. 72, No. 2, Dec. 2003 (Table 6).

a. Ecologists theorize that an approximately linear (straight-line) relationship exists between the two fragmentation indexes. Graph the data for all 54 forests. Does the graph support the theory?

b. Delete the data for the three forests with the largest anthropogenic indexes, and reconstruct the graph of part **a.** Comment on the ecologists' theory.

2.167 Mongolian desert ants. Refer to the *Journal of Biogeography* (Dec. 2003) study of ants in Mongolia, presented in Exercise 2.68 (p. 91). Data on annual rainfall, maximum daily temperature, percentage of plant cover, number of ant species, and species diversity index recorded at each of 11 study sites are saved in the **ANTS** file.

a. Construct a scatterplot to investigate the relationship between annual rainfall and maximum daily temperature. What type of trend (if any) do you detect?

b. Use scatterplots to investigate the relationship that annual rainfall has with each of the other four variables in the data set. Are the other variables positively or negatively related to rainfall?

Applying the Concepts—Advanced

2.168 Spreading rate of spilled liquid. A contract engineer at DuPont Corp. studied the rate at which a spilled volatile liquid will spread across a surface (*Chemical Engineering Progress,* Jan. 2005). Suppose that 50 gallons of methanol spills onto a level surface outdoors. The engineer uses derived empirical formulas (assuming a state of turbulence-free convection) to calculate the mass (in pounds) of the spill after a period ranging from 0 to 60 minutes. The calculated mass values are given in the table on the next page. Is there evidence to indicate that the mass of the spill tends to diminish as time increases?

Data for Exercise 2.168

Time (minutes)	Mass (pounds)	Time (minutes)	Mass (pounds)
0	6.64	22	1.86
1	6.34	24	1.60
2	6.04	26	1.37
4	5.47	28	1.17
6	4.94	30	0.98
8	4.44	35	0.60
10	3.98	40	0.34
12	3.55	45	0.17
14	3.15	50	0.06
16	2.79	55	0.02
18	2.45	60	0.00
20	2.14		

Based on Barry, J. "Estimating rates of spreading and evaporation of volatile liquids." *Chemical Engineering Progress*, Vol. 101, No. 1. Jan. 2005.

2.169 Ranking driving performance of professional golfers. Refer to *The Sport Journal* (Winter 2007) analysis of a new method for ranking the total driving performance of golfers on the PGA tour, Exercise 2.66 (p. 90). Recall that the method uses both the average driving distance (in yards) and the driving accuracy (percent of drives that land in the fairway). Data on these two variables for the top 40 PGA golfers are saved in the **PGA** file. A professional golfer is practicing a new swing to increase his average driving distance. However, he is concerned that his driving accuracy will be lower. Is his concern reasonable? Explain.

2.9 Distorting the Truth with Descriptive Statistics

A picture may be "worth a thousand words," but pictures can also color messages or distort them. In fact, the pictures displayed in statistics—histograms, bar charts, and other graphical images—are susceptible to distortion, so we have to examine each of them with care. Accordingly, we begin this section by mentioning a few of the pitfalls to watch for in interpreting a chart or a graph. Then we discuss how numerical descriptive statistics can be used to distort the truth.

Graphical Distortions

One common way to change the impression conveyed by a graph is to alter the scale on the vertical axis, the horizontal axis, or both. For example, consider the data on collisions of large marine vessels operating in European waters over a certain five-year period, summarized in Table 2.10. Figure 2.38 is a MINITAB bar graph showing the frequency of collisions for each of the three locations listed in the table. The graph shows that in-port collisions occur more often than collisions at sea or collisions in restricted waters.

Table 2.10 Collisions of Marine Vessels by Location	
Location	Number of Ships
At Sea	376
In Restricted Waters	273
In Port	478
Total	1,127

Based on *The Dock and Harbour Authority*. Data Set: MARINE

Now, suppose you want to use the same data to exaggerate the difference between the number of in-port collisions and the number of collisions in restricted waters. One way to do this is to increase the distance between successive units on the vertical axis—that is, *stretch* the vertical axis by graphing only a few units per inch. A telltale sign of stretching is a long vertical axis, but this indication is often hidden by starting the vertical axis at some point above the origin, 0. Such a graph is shown in the SPSS printout in Figure 2.39. By starting the bar chart at 250 collisions (instead of 0), it appears that the frequency of in-port collisions is many times greater than the frequency of collisions in restricted waters.

Ethics IN Statistics
Intentionally distorting a graph to portray a particular viewpoint is considered unethical statistical practice.

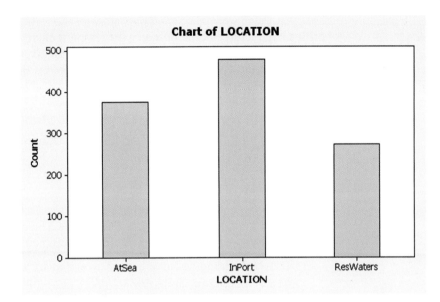

Figure 2.38

MINITAB bar graph of vessel collisions by location

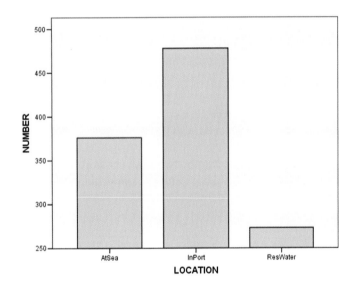

Figure 2.39

SPSS bar graph of vessel collisions by location, with adjusted vertical axis

Another method of achieving visual distortion with bar graphs is by making the width of the bars proportional to their height. For example, look at the bar chart in Figure 2.40a, which depicts the percentage of the total number of motor vehicle deaths in a year that occurred on each of four major highways. Now suppose we make both the width and the height grow as the percentage of fatal accidents grows. This change is shown in Figure 2.40b. The distortion is that the reader may tend to equate the *area* of the bars with the percentage of deaths occurring at each highway when, in fact, the true relative frequency of fatal accidents is proportional only to the *height* of the bars.

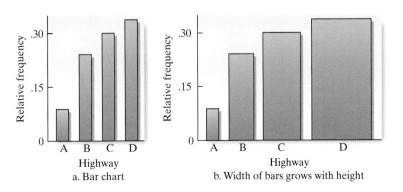

Figure 2.40

Relative frequency of fatal motor vehicle accidents on each of four major highways

Although we've discussed only a few of the ways that graphs can be used to convey misleading pictures of phenomena, the lesson is clear: Look at all graphical descriptions of data with a critical eye. In particular, check the axes and the size of the units on each axis. Ignore the visual changes, and concentrate on the actual numerical changes indicated by the graph or chart

Misleading Numerical Descriptive Statistics

The information in a data set can also be distorted by using numerical descriptive measures, as Example 2.20 shows.

Example 2.20

Misleading Descriptive Statistics

Problem Suppose you're considering working for a small law firm—one that currently has a senior member and three junior members. You inquire about the salary you could expect to earn if you join the firm. Unfortunately, you receive two answers:

Answer A: The senior member tells you that an "average employee" earns $107,500.

Answer B: One of the junior members later tells you that an "average employee" earns $95,000.

Which answer can you believe?

Solution The confusion exists because the phrase "average employee" has not been clearly defined. Suppose the four salaries paid are $95,000 for each of the three junior members and $145,000 for the senior member. Then,

$$\text{Mean} = \frac{3(\$95,000) + \$145,000}{4} = \frac{\$430,000}{4} = \$107,500$$

$$\text{Median} = \$95,000$$

You can now see how the two answers were obtained: The senior member reported the mean of the four salaries, and the junior member reported the median. The information you received was distorted because neither person stated which measure of central tendency was being used.

Look Back On the basis of our earlier discussion of the mean and median, we would probably prefer the median as the number that best describes the salary of the "average employee."

Another distortion of information in a sample occurs when *only* a measure of central tendency is reported. Both a measure of central tendency and a measure of variability are needed to obtain an accurate mental image of a data set.

Suppose, for instance, that you want to buy a new car and are trying to decide which of two models to purchase. Since energy and economy are both important issues, you decide to purchase model A because its EPA mileage rating is 32 miles per gallon in the city, whereas the mileage rating for model B is only 30 miles per gallon in the city.

However, you may have acted too quickly. How much variability is associated with the ratings? As an extreme example, suppose that further investigation reveals that the standard deviation for model A mileages is 5 miles per gallon, whereas that for model B is only 1 mile per gallon. If the mileages form a mound-shaped distribution, they might appear as shown in Figure 2.41. Note that the larger amount of variability associated with model A implies that more risk is involved in purchasing that model. That is, the particular car you purchase is more likely to have a mileage rating that will differ greatly from the EPA rating of 32 miles per gallon if you purchase model A, while a model B car is not likely to vary from the 30-miles-per-gallon rating by more than 2 miles per gallon.

We conclude this section with another example on distorting the truth with numerical descriptive measures.

Figure 2.41

Mileage distributions for two car models

<table>
<tr><td>

Example   **2.21**

More Misleading Descriptive Statistics—Delinquent Children

</td>
<td>

Problem *Children Out of School in America* is a report on the delinquency of school-age children written by the Children's Defense Fund (CDF). Consider the following three reported results of the CDF survey.

• Reported result: Twenty-five percent of the 16- and 17-year-olds in the Portland, Maine, Bayside East Housing Project were out of school. Actual data: *Only eight children were surveyed; two were found to be out of school.*

• Reported result: Of all the secondary school students who had been suspended more than once in census tract 22 in Columbia, South Carolina, 33% had been suspended two times and 67% had been suspended three or more times. Actual data: *CDF found only three children in that entire census tract who had been suspended; one child was suspended twice and the other two children three or more times.*

• Reported result: In the Portland Bayside East Housing Project, 50% of all the secondary school children who had been suspended more than once had been suspended three or more times. Actual data: *The survey found just two secondary school children who had been suspended in that area; one of them had been suspended three or more times.*

Identify the potential distortions in the results reported by the CDF.

</td></tr>
</table>

Solution In each of these examples, the reporting of percentages (i.e., relative frequencies) instead of the numbers themselves is misleading. No inference we might draw from the examples cited would be reliable. (We'll see how to measure the reliability of estimated percentages in Chapter 7.) In short, either the report should state the numbers alone instead of percentages, or, better yet, it should state that the numbers were too small to report by region.

Ethics IN Statistics
Purposeful reporting of numerical descriptive statistics in order to mislead the target audience is considered unethical statistical practice.

Look Back If several regions were combined, the numbers (and percentages) would be more meaningful.

Exercises 2.170–2.173

Applying the Concepts—Intermediate

2.170 Museum management. Refer to the *Museum Management and Curatorship* (June 2010) study of how museums evaluate their performance, Exercise 2.22 (p. 69). Recall that managers of 30 museums of contemporary art identified the performance measure used most often. A summary of the results is reproduced in the table. Consider the bar graph shown. Identify two ways in which the bar graph might mislead the viewer by overemphasizing the importance of one of the performance measures.

Performance Measure	Number of Museums	Proportion of Museums
Total visitors	8	.267
Paying visitors	5	.167
Big shows	6	.200
Funds raised	7	.233
Members	4	.133

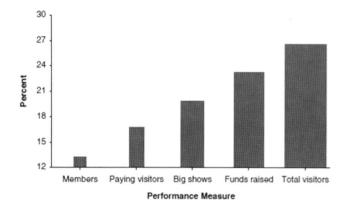

2.171 Trend in Iraq War casualties. While the United States was still actively fighting in the Iraq War, a news media outlet produced a graphic showing a dramatic decline in the annual number of American casualties. The number of deaths for the years 2003, 2004, 2005, and 2006 were (approximately) 475, 850, 820, and 130, respectively.

a. Create a scatterplot showing the dramatic decline in the number of American deaths per year.

b. The graphic was based on data collected through February 2006. Knowing this fact, why is the time series plot misleading?

c. What information would you like to have in order to construct a graph that accurately reflects the trend in American casualties from the Iraq War?

2.172 BP oil leak. In the summer of 2010, an explosion on the Deepwater Horizon oil drilling rig caused a leak in one of British Petroleum (BP) Oil Company's wells in the Gulf of Mexico. Crude oil rushed unabated for 3 straight months into the Gulf until BP could fix the leak. During the disaster, BP used suction tubes to capture some of the gushing oil. In May 2011, in an effort to demonstrate the daily improvement in the process, a BP representative presented a graphic on the daily number of 42-gallon barrels (bbl) of oil collected by the suctioning process. A graphic similar to the one used by BP is shown below,

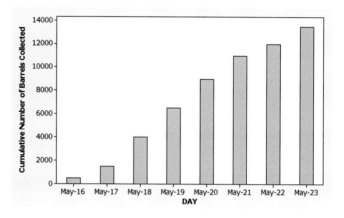

a. Note that the vertical axis represents the "cumulative" number of barrels collected per day. This is calculated by adding the amounts of the previous days' oil collection to the current day's oil collection. Explain why this graph is misleading.

b. Estimates of the actual number of barrels of oil collected per day for each of the 8 days are listed in the accompanying table. Construct a graph for this data that accurately depicts BP's progress in its daily collection of oil. What conclusions can you draw from the graph?

Estimates of Daily Collection of Oil

Day	Number of Barrels (bbl)
May 16	500
May 17	1,000
May 18	3,000
May 19	2,500
May 20	2,500
May 21	2,000
May 22	1,000
May 23	1,500

2.173 Irrelevant speech effects. Refer to the *Acoustical Science & Technology* (Vol. 35, 2014) study of irrelevant speech effects, Exercise 2.104 (p. 105). Recall that subjects performed a memorization task under two conditions: (1) with irrelevant background speech and (2) in silence. The difference in the error rates for the two conditions—called the relative difference in error rate (RDER)—was computed for each subject. Descriptive statistics for the RDER values are shown in the MINITAB printout on p. 131. The media has requested that the researchers provide a single statistic that best represents the center of the distribution of RDER values. This statistic will be used to publicize the study findings.

MINITAB Output for Exercise 2.173

Descriptive Statistics: RDER

Variable	N	Mean	StDev	Minimum	Q1	Median	Q3	Maximum
RDER	71	78.19	63.24	-19.27	37.81	60.31	102.80	254.11

a. Suppose you are in support of the irrelevant speech effect theory. Consequently, you want to magnify the difference in error rates between the two conditions. Which statistic would you select, and why?

b. Suppose you do not believe in the irrelevant speech effect theory. Consequently, you want to diminish the difference in error rates between the two conditions. Which statistic would you select, and why?

c. If you truly are neutral in regard to the irrelevant speech effect theory, explain why it is better to obtain descriptive statistics for the absolute difference between error rates (i.e., the absolute value of RDER).

CHAPTER NOTES

Key Terms

Note: Starred (*) terms are from the optional sections in this chapter.

Bar graph 60
Bivariate relationship* 121
Box plot* 112
Central tendency 82
Chebyshev's rule 99
Class 59
Class percentage 60
Class interval 72
Class frequency 59
Class relative frequency 59
Dot plot 70
Empirical rule 100
Hinge* 112
Histogram 72
Inner fence* 112
Interquartile range (IQR)* 112
Lower quartile (Q_L)* 108
Mean 82
Measures of central tendency 82
Measures of relative standing 107
Measures of variability (or spread) 93
Median 83
Middle quartile (M)* 108
Modal class 86

Mode 86
Mound-shaped distribution 103
Numerical descriptive measures 82
Outer fence* 112
Outlier* 111
Pareto diagram 61
pth percentile 107
Percentile ranking (or percentile score) 107
Pie chart 61
Quartiles* 108
Range 93
Rare-event approach* 117
Scatterplot* 121
Skewness/skewed 85
Spread 93
Standard deviation 94
Stem-and-leaf display 71
Symmetric distribution 99
Summation notation 82
Upper quartile (Q_U)* 108
Variability 82
Variance 94
Whiskers* 112
z-score 108

Key Symbols

	Sample	*Population*
Mean:	$\bar{x}$	μ
Variance:	s^2	σ^2
Std. Dev.	s	σ
Median:	M	η
Lower Quartile:	Q_L	
Upper Quartile:	Q_U	
Interquartile Range:	IQR	

Key Ideas

Describing Qualitative Data

1. Identify **category** classes
2. Determine **class frequencies**
3. **Class relative frequency** = (class frequency)$/n$
4. Graph relative frequencies

Pie Chart:

Bar Graph:

Pareto Diagram:

Graphing Quantitative Data

One Variable:
1. Identify class intervals
2. Determine class interval frequencies
3. Class interval relative frequency = (class interval frequency)$/n$
4. Graph class interval relative frequencies

Dot Plot:

Stem-and-Leaf Display:

Box plot:

Histogram:

Two Variables:

Scatterplot:

Numerical Description of Quantitative Data

Central Tendency

Mean: $\bar{x} = (\Sigma x_i)/n$
Median: Middle value (or mean of two
   middle values) when data
   ranked in order
Mode: Value that occurs most often

Variation

Range: Difference between largest and
   smallest value
Variance:

$$s^2 = \frac{\Sigma (x_i - \bar{x})^2}{n - 1} = \frac{\Sigma x_i^2 - \dfrac{(\Sigma x_i)^2}{n}}{n - 1}$$

Std Dev.: $s = \sqrt{s^2}$
Interquartile Range: IQR $= Q_U - Q_L$

Relative Standing

Percentile Score: Percentage of values
   that fall below *x*-score
z-score: $z = (x - \bar{x})/s$

Rules for Describing Quantitative Data

Interval	Chebyshev's Rule	Empirical Rule
$\bar{x} \pm s$	At least 0%	$\approx 68\%$
$\bar{x} \pm 2s$	At least 75%	$\approx 95\%$
$\bar{x} \pm 3s$	At least 89%	$\approx$ All

Rules for Detecting Quantitative Outliers

Method	Suspect	Highly Suspect
Box plot:	Values between inner and outer fences	Values beyond outer fences
z-score:	$2 < \lvert z \rvert < 3$	$\lvert z \rvert > 3$

Guide to Selecting the Data Description Method

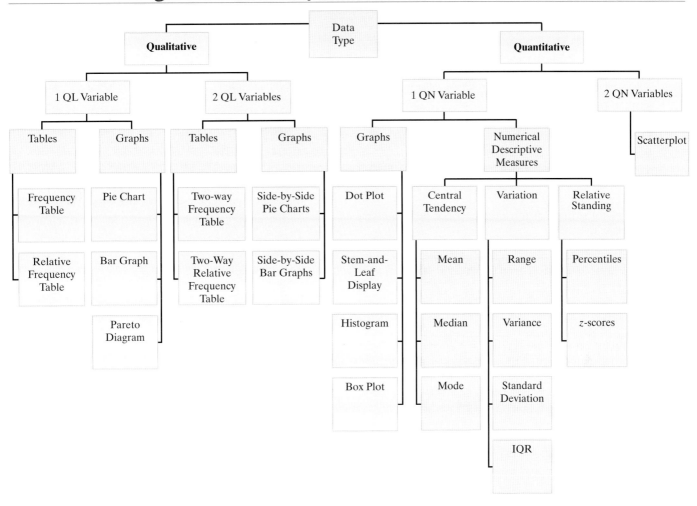

Supplementary Exercises 2.174–2.213

Note: Starred (*) exercises refer to the optional section in this chapter.

Understanding the Principles

2.174 Explain why we generally prefer the standard deviation to the range as a measure of variation.

2.175 Discuss conditions under which the median is preferred to the mean as a measure of central tendency.

2.176 Give a situation in which we would prefer using a box plot over z-scores to detect an outlier.

2.177 Give a situation in which we would prefer using a stem-and-leaf display over a histogram in describing quantitative data graphically.

2.178 Give a technique that is used to distort information shown on a graph.

Learning the Mechanics

2.179 Consider the following three measurements: 50, 70, 80. Find the z-score for each measurement if they are from a population with a mean and standard deviation equal to
  a. $\mu = 60, \sigma = 10$
  b. $\mu = 60, \sigma = 5$
  c. $\mu = 40, \sigma = 10$
  d. $\mu = 40, \sigma = 100$

2.180 Refer to Exercise 2.179. For parts **a–d**, determine whether the three measurements 50, 70, and 80 are outliers.

2.181 Construct a relative frequency histogram for the data summarized in the following table:

Measurement Class	Relative Frequency	Measurement Class	Relative Frequency
.00–.75	.02	5.25–6.00	.15
.75–1.50	.01	6.00–6.75	.12
1.50–2.25	.03	6.75–7.50	.09
2.25–3.00	.05	7.50–8.25	.05
3.00–3.75	.10	8.25–9.00	.04
3.75–4.50	.14	9.00–9.75	.01
4.50–5.25	.19		

2.182 If the range of a set of data is 20, find a rough approximation to the standard deviation of the data set.

2.183 For each of the data sets in parts **a–c,** compute $\bar{x}$, s^2, and s. If appropriate, specify the units in which your answers are expressed.
   a. 4, 6, 6, 5, 6, 7
   b. −$1, $4, −$3, $0, −$3, −$6
   c. 3/5%, 4/5%, 2/5%, 1/5%, 1/16%
   d. Calculate the range of each data set in parts **a–c.**

2.184 For each of the data sets in parts **a–d,** compute $\bar{x}$, s^2, and s.
   a. 13, 1, 10, 3, 3
   b. 13, 6, 6, 0
   c. 1, 0, 1, 10, 11, 11, 15
   d. 3, 3, 3, 3
   e. For each of the data sets in parts **a–d**, form the interval $\bar{x} \pm 2s$ and calculate the percentage of the measurements that fall into that interval.

*__2.185__ Construct a scatterplot for the data listed here. Do you
   D detect any trends?
   L02185

Variable #1:	174	268	345	119	400	520	190	448	307	252
Variable #2:	8	10	15	7	22	31	15	20	11	9

Applying the Concepts—Basic

2.186 **Excavating ancient pottery.** Archaeologists excavating the ancient Greek settlement at Phylakopi classified the pottery found in trenches. (*Chance*, Fall 2000.) The accompanying table describes the collection of 837 pottery pieces uncovered in a particular layer at the excavation site. Construct and interpret a graph that will aid the archaeologists in understanding the distribution of the types of pottery found at the site.

Pot Category	Number Found
Burnished	133
Monochrome	460
Slipped	55
Painted in curvilinear decoration	14
Painted in geometric decoration	165
Painted in naturalistic decoration	4
Cycladic white clay	4
Conical cup clay	2
Total	837

Based on Berg, I., and Bliedon, S. "The pots of Phylakopi: Applying statistical techniques to archaeology." *Chance*, Vol. 13, No. 4, Fall 2000.

2.187 **Japanese reading levels.** University of Hawaii language professors C. Hitosugi and R. Day incorporated a 10-week extensive reading program into a second-semester Japanese language course in an effort to improve students' Japanese reading comprehension. (*Reading in a Foreign Language*, Apr. 2004.) The professors collected 266 books originally written for Japanese children and required their students to read at least 40 of them as part of the grade in the course. The books were categorized into reading levels (color coded for easy selection) according to length and complexity. The reading levels for the 266 books are summarized in the following table:

Reading Level	Number
Level 1 (Red)	39
Level 2 (Blue)	76
Level 3 (Yellow)	50
Level 4 (Pink)	87
Level 5 (Orange)	11
Level 6 (Green)	3
Total	266

Source: Hitosugi, C. I., and Day, R. R. "Extensive reading in Japanese," *Reading in a Foreign Language*, Vol. 16, No. 1. Apr. 2004 (Table 2). Reprinted with permission from the National Foreign Language Resource Center, University of Hawaii.

   a. Calculate the proportion of books at reading level 1 (red).
   b. Repeat part **a** for each of the remaining reading levels.
   c. Verify that the proportions in parts **a** and **b** sum to 1.
   d. Use the previous results to form a bar graph for the reading levels.
   e. Construct a Pareto diagram for the data. Use the diagram to identify the reading level that occurs most often.

2.188 **Reading Japanese books.** Refer to Exercise 2.187.
   D Fourteen students participated in a 10-week extensive
   JREAD reading program in a second-semester Japanese language course. The number of books read by each student and the student's grade in the course are listed in the next table.
   a. Construct a stem-and-leaf display for the number of books read by the students.
   b. Highlight (or circle) the leaves in the display that correspond to students who earned an A grade in the course. What inference can you make about these students?

Number of Books	Course Grade
53	A
42	A
40	A
40	B
39	A
34	A
34	A
30	A
28	B
24	A
22	C
21	B
20	B
16	B

Source: Hitosugi, C. I., and Day, R. R. "Extensive reading in Japanese." *Reading in a Foreign Language*, Vol. 16, No. 1, Apr. 2004 (Table 4). Reprinted with permission from the National Foreign Language Resource Center, University of Hawaii.

c. Find the mean, median, and mode of the number of books read. Interpret these values.

d. What do the mean and median indicate about the skewness of the distribution of the data?

e. Find the mean and standard deviation of the number of books read by students who earned an A grade.

f. Find the mean and standard deviation of the number of books read by students who earned either a B or C grade.

g. Refer to parts **a** and **b**. Which of the two groups of students has a more variable distribution for number of books read?

h. Find the *z*-score for an A student who read 40 books. Interpret the result.

i. Find the *z*-score for a B or C student who read 40 books. Interpret the result.

j. Which of the two groups of students is more likely to have read 40 books? Explain.

2.189 Body length of armadillos. A group of environmentalists reported the results of a study of giant armadillos inhabiting the southeastern region of Venezuela. A sample of 80 armadillos was captured and the body length (not including the tail) of each was measured (in centimeters). A MINITAB graph summarizing the data is shown below.

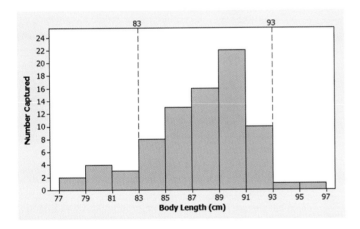

a. What type of graph is employed?

b. How many armadillos have body lengths between 87 and 91 centimeters?

c. What proportion of the armadillos have body lengths between 87 and 91 centimeters?

d. The dotted vertical lines on the graph show the minimum and maximum sizes of giant armadillos that can be legally captured for commercial purposes. What proportion of the captured armadillos are illegal?

2.190 Crash tests on new cars. Each year, the National Highway Traffic Safety Administration (NHTSA) crash tests new car models to determine how well they protect the driver and front-seat passenger in a head-on collision. The NHTSA has developed a "star" scoring system for the frontal crash test, with results ranging from one star (*) to five stars (*****). The more stars in the rating, the better the level of crash protection in a head-on collision. The NHTSA crash test results for 98 cars in a recent model year are stored in the data file named **CRASH**. The driver-side star ratings for the 98 cars are summarized in the MINITAB printout below. Use the information in the printout to form a pie chart. Interpret the graph.

Tally for Discrete Variables: DRIVSTAR

DRIVSTAR	Count	Percent
2	4	4.08
3	17	17.35
4	59	60.20
5	18	18.37
N=	98	

2.191 Crash tests on new cars. Refer to Exercise 2.190 and the NHTSA crash test data. One quantitative variable recorded by the NHTSA is driver's severity of head injury (measured on a scale from 0 to 1,500). Numerical descriptive statistics for the 98 driver head-injury ratings are displayed in the MINITAB printout at the bottom of the page.

a. Interpret each of the statistics shown on the printout.

b. Find the *z*-score for a driver head-injury rating of 408. Interpret the result.

2.192 Estimating the age of glacial drifts. Tills are glacial drifts consisting of a mixture of clay, sand, gravel, and boulders. Engineers from the University of Washington's Department of Earth and Space Sciences studied the chemical makeup of buried tills in order to estimate the age of the glacial drifts in Wisconsin (*American Journal of Science*, Jan. 2005). The ratio of the elements aluminum (Al) and beryllium (Be) in sediment is related to the duration of burial. The Al–Be ratios for a sample of 26 buried till specimens are given in the accompanying table.

3.75	4.05	3.81	3.23	3.13	3.30	3.21	3.32	4.09	3.90	5.06	3.85	3.88
4.06	4.56	3.60	3.27	4.09	3.38	3.37	2.73	2.95	2.25	2.73	2.55	3.06

Based on *American Journal of Science*, Vol. 305, No. 1, Jan. 2005, p. 16 (Table 2).

MINITAB Output for Exercise 2.191

Descriptive Statistics: DRIVHEAD

Variable	N	Mean	StDev	Minimum	Q1	Median	Q3	Maximum
DRIVHEAD	98	603.7	185.4	216.0	475.0	605.0	724.3	1240.0

a. Find and interpret the z-score associated with the largest ratio, the smallest ratio, and the mean ratio.

b. Would you consider the largest ratio to be unusually large? Why or why not?

c. Construct a box plot for the data and identify any outliers.

2.193 Training zoo animals. "The Training Game" is an activity used in psychology in which one person shapes an arbitrary behavior by selectively reinforcing the movements of another person. A group of 15 psychology students at the Georgia Institute of Technology played the game at Zoo Atlanta while participating in an experimental psychology laboratory in which they assisted in the training of animals (*Teaching of Psychology*, May 1998). At the end of the session, each student was asked to rate the statement: "'The Training Game is a great way for students to understand the animal's perspective during training." Responses were recorded on a 7-point scale ranging from 1 (strongly disagree) to 7 (strongly agree). The 15 responses were summarized as follows:

$$\text{mean} = 5.87, \text{mode} = 6, \text{median} = 6$$

a. Interpret the measures of central tendency in the words of the problem.

b. What type of skewness (if any) is likely to be present in the distribution of student responses? Explain.

2.194 Groundwater contamination in wells. Refer to the *Environmental Science & Technology* (Jan. 2005) study of the factors related to MTBE contamination in 223 New Hampshire wells, presented in Exercise 2.23 (p. 70). The data are saved in the **MTBE** file. Two of the many quantitative variables measured for each well are the pH level (standard units) and the MTBE level (micrograms per liter).

a. Construct a histogram for the pH levels of the sampled wells. From the histogram, estimate the proportion of wells with pH values less than 7.0.

b. For those wells with detectible levels of MTBE, construct a histogram for the MTBE values. From the histogram, estimate the proportion of contaminated wells with MTBE values that exceed 5 micrograms per liter.

c. Find the mean and standard deviation for the pH levels of the sampled wells, and construct the interval $\bar{x} \pm 2s$. Estimate the percentage of wells with pH levels that fall within the interval. What rule did you apply to obtain the estimate? Explain.

d. Find the mean and standard deviation for the MTBE levels of the sampled wells, and construct the interval $\bar{x} \pm 2s$. Estimate the percentage of wells with MTBE levels that fall within the interval. What rule did you apply to obtain the estimate? Explain.

2.195 Sociology fieldwork methods. University of New Mexico professor Jane Hood investigated the fieldwork methods used by qualitative sociologists. (*Teaching Sociology*, July 2006.) Searching for all published journal articles, dissertations, and conference proceedings over the previous seven years in the *Sociological Abstracts* database, she discovered that fieldwork methods could be categorized as follows: Interview, Observation plus Participation, Observation Only, and Grounded Theory. The accompanying table shows the number of papers in each category. Use an appropriate graph to portray the results of the study. Interpret the graph.

Fieldwork Method	Number of Papers
Interview	5,079
Observation + Participation	1,042
Observation Only	848
Grounded Theory	537

Based on Hood, J. C. "Teaching against the text: The case of qualitative methods." *Teaching Sociology*, Vol. 34, Issue 3, p. 207 (Exhibit 2).

Applying the Concepts—Intermediate

2.196 Radioactive lichen. Lichen has a high absorbance capacity for radiation fallout from nuclear accidents. Since lichen is a major food source for Alaskan caribou, and caribou are, in turn, a major food source for many Alaskan villagers, it is important to monitor the level of radioactivity in lichen. Researchers at the University of Alaska, Fairbanks, collected data on nine lichen specimens at various locations for this purpose. The amount of the radioactive element cesium-137 was measured (in microcuries per milliliter) for each specimen. The data values, converted to logarithms, are given in the following table (note that the closer the value is to zero, the greater is the amount of cesium in the specimen).

Location			
Bethel	−5.50	−5.00	
Eagle Summit	−4.15	−4.85	
Moose Pass	−6.05		
Turnagain Pass	−5.00		
Wickersham Dome	−4.10	−4.50	−4.60

Based on Lichen Radionuclide Baseline Research Project, 2003, p. 25. *Orion*, University of Alaska–Fairbanks.

a. Construct a dot plot for the nine measurements.

b. Construct a stem-and-leaf display for the nine measurements.

c. Construct a histogram plot of the nine measurements.

d. Which of the three graphs in parts **a–c**, respectively, is most informative?

e. What proportion of the measurements has a radioactivity level of −5.00 or lower?

2.197 Improving SAT scores. The National Education Longitudinal Survey (NELS) tracks a nationally representative sample of U.S. students from eighth grade through high school and college. Research published in *Chance* (Winter 2001) examined the SAT scores of 265 NELS students who paid a private tutor to help them improve their scores. The table summarizes the changes in both the SAT-Mathematics and SAT-Verbal scores for these students.

	SAT-Math	SAT-Verbal
Mean change in score	19	7
Standard deviation of score changes	65	49

a. Suppose one of the 265 students who paid a private tutor is selected at random. Give an interval that is likely to contain the change in this student's SAT-Math score.

b. Repeat part **a** for the SAT-Verbal score.

c. Suppose the selected student's score increased on one of the SAT tests by 140 points. Which test, the SAT-Math or SAT-Verbal, is the one most likely to have had the 140-point increase? Explain.

2.198 Ammonia in car exhaust. Three-way catalytic converters have been installed in new vehicles in order to reduce pollutants from motor vehicle exhaust emissions. However, these converters unintentionally increase the level of ammonia in the air. *Environmental Science & Technology* (Sept. 1, 2000) published a study on the ammonia levels near the exit ramp of a San Francisco highway tunnel. The data in the table represent daily ammonia concentrations (parts per million) on eight randomly selected days during the afternoon drive time in the summer of a recent year. Find an interval that is likely to contain the ammonia level for a randamly selected day during afternoon drive time.

1.53	1.50	1.37	1.51	1.55	1.42	1.41	1.48

2.199 College protests of labor exploitation. The United Students Against Sweatshops (USAS) was formed by students on U.S. and Canadian college campuses in 1999 to protest labor exploitation in the apparel industry. Clark University sociologist Robert Ross analyzed the USAS movement in the *Journal of World-Systems Research* (Winter 2004). Between 1999 and 2000, there were 18 student "sit-ins" for a "sweat-free campus" organized at several universities. The following table gives the duration (in days) of each sit-in, as well as the number of student arrests.

a. Do the data support the theory that sit-ins of longer duratian are more likely to lead to arrests? Support your answer with a graph.

Sit-in	Year	University	Duration (days)	Number of Arrests	Tier Ranking
1	1999	Duke	1	0	1st
2	1999	Georgetown	4	0	1st
3	1999	Wisconsin	1	0	1st
4	1999	Michigan	1	0	1st
5	1999	Fairfield	1	0	1st
6	1999	North Carolina	1	0	1st
7	1999	Arizona	10	0	1st
8	2000	Toronto	11	0	1st
9	2000	Pennsylvania	9	0	1st
10	2000	Macalester	2	0	1st
11	2000	Michigan	3	0	1st
12	2000	Wisconsin	4	54	1st
13	2000	Tulane	12	0	1st
14	2000	SUNY Albany	1	11	2nd
15	2000	Oregon	3	14	2nd
16	2000	Purdue	12	0	2nd
17	2000	Iowa	4	16	2nd
18	2000	Kentucky	1	12	2nd

Based on Ross, R. J. S. "From antisweatshop to global justice to antiwar: How the new new left is the same and different from the old new left." *Journal of World-Systems Research*, Vol. X, No. 1, Winter 2004 (Table 1 and 3).

b. Repeat part **a**, but graph only the data for sit-ins in which there was at least one arrest. Do you detect a trend?

c. Comment on the reliability of the trend you detected in part **b**.

2.200 Speed of light from galaxies. Astronomers theorize that cold dark matter caused the formation of galaxies and clusters of galaxies in the universe. The theoretical model for cold dark matter requires an estimate of the velocities of light emitted from galaxy clusters. *The Astronomical Journal* (July 1995) published a study of observed velocities of galaxies in four different clusters. Galaxy velocity was measured in kilometers per second (km/s), using a spectrograph and high-power telescope.

a. The observed velocities of 103 galaxies located in the cluster named A2142 are summarized in the accompanying histogram. Comment on whether the empirical rule is applicable to describing the velocity distribution for this cluster.

Source: Oegerle, W. R., Hill, J. M., and Fitchett, M. J. "Observations of high dispersion clusters of galaxies: Constraints on cold dark matter." *The Astronomical Journal*, Vol. 110, No. 1, July 1995, p. 37. Reproduced by permission of the AAS.

b. The mean and standard deviation of the 103 velocities observed in galaxy cluster A2142 were reported as $\bar{x} = 27{,}117$ km/s and $s = 1{,}280$ km/s, respectively. Use this information to construct an interval that captures approximately 95% of the velocities of the galaxies in the cluster.

c. Recommend a single velocity value to be used in the CDM model for galaxy cluster A2142. Explain your reasoning.

2.201 Whistling dolphins. Marine scientists who study dolphin communication have discovered that bottlenose dolphins exhibit an individualized whistle contour known as their *signature whistle*. A study was conducted to categorize the signature whistles of 10 captive adult bottlenose dolphins in socially interactive contexts (*Ethology*, July 1995). A total of 185 whistles were recorded during the study period; each whistle contour was analyzed and assigned to a category by a contour similarity (CS) technique. The results are reported in the table on p. 138. Use a graphical method to summarize the results. Do you detect any patterns in the data that might be helpful to marine scientists?

Data for Exercise 2.201

Whistle Category	Number of Whistles
Type a	97
Type b	15
Type c	9
Type d	7
Type e	7
Type f	2
Type g	2
Type h	2
Type i	2
Type j	4
Type k	13
Other types	25

Based on McCowan, B., and Reiss, D. "Quantitative comparison of whistle repertoires from captive adult bottlenose dolphins (Delphiniae, *Tursiops truncates*): A re-evaluation of the signature whistle hypothesis." *Ethology*, Vol. 100, No. 3, July 1995, p. 200 (Table 2).

2.202 Sentence complexity study. A study published in *Applied Psycholinguistics* (June 1998) compared the language skills of young children (16–30 months old) from low-income and middle-income families. A total of 260 children—65 in the low-income and 195 in the middle-income group—completed the Communicative Development Inventory (CDI) exam. One of the variables measured on each child was sentence complexity score. Summary statistics for the scores of the two groups are reproduced in the accompanying table. Use this information to sketch a graph of the sentence complexity score distribution for each income group. (Assume that the distributions are mound shaped and symmetric.) Compare the distributions. What can you infer?

	Low Income	Middle Income
Sample Size	65	195
Mean	7.62	15.55
Standard Deviation	8.91	12.24
Minimum	0	0
Maximum	36	37

Based on Arriaga, R. I., et al. "Scores on the MacArthur Communicative Development Inventory of children from low-income and middle-income families." *Applied Psycholinguistics*, Vol. 19, No. 2, June 1998, p. 217 (Table 7).

2.203 Freckling of superalloy ingots. Freckles are defects that sometimes form during the solidification of alloy ingots. A freckle index has been developed to measure the level of freckling on the ingot. A team of engineers conducted several experiments to measure the freckle index of a certain type of superalloy (*Journal of Metallurgy*, Sept. 2004). The data for $n = 18$ alloy tests are shown in the table.

12.6	22.0	4.1	16.4	1.4	2.4
16.8	10.0	3.2	30.1	6.8	14.6
2.5	12.0	33.4	22.2	8.1	15.1

Based on Yang, W. H. "A freckle criterion for the solidification of superalloys with a tilted solidification front." *JOM: Journal of the Minerals, Metals and Materials Society*, Vol. 56, No. 9, Sept. 2004.

a. Construct a box plot for the data and use it to find any outliers.

b. Find and interpret the z-scores associated with the alloys you identified in part **a.**

2.204 Archaeologists' study of ring diagrams. Archaeologists gain insight into the social life of ancient tribes by measuring the distance (in centimeters) of each artifact found at a site from the "central hearth" (i.e., the middle of an artifact scatter). A graphical summary of these distances, in the form of a histogram, is called a *ring diagram*. (*World Archaeology*, Oct. 1997). Ring diagrams for two archaeological sites (A and G) in Europe are shown below.

a. Identify the type of skewness (if any) present in the data collected at the two sites.

b. Archaeologists have associated unimodal ring diagrams with open-air hearths and multimodal ring diagrams with hearths inside dwellings. Can you identify the type of hearth (*open air or inside dwelling*) that was most likely present at each site?

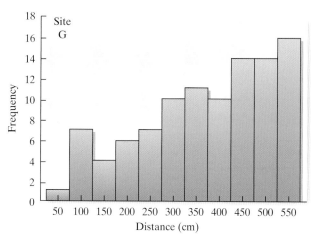

2.205 Oil spill impact on seabirds. The *Journal of Agricultural, Biological, and Environmental Statistics* (Sept. 2000) published a study on the impact of the *Exxon Valdez* tanker oil spill on the seabird population in Prince William Sound, Alaska. Data were collected on 96 shoreline locations (called *transects*) of constant width, but variable length. For each transect, the number of seabirds found is recorded, as are the length (in kilometers) of the transect and whether

MINITAB Output for Exercise 2.205

Descriptive Statistics: Density

Variable	Oil	N	Mean	StDev	Minimum	Q1	Median	Q3	Maximum
Density	no	36	3.27	6.70	0.000	0.000	0.890	3.87	36.23
	yes	60	3.495	5.968	0.0000	0.000	0.700	5.233	32.836

or not the transect was in an oiled area. (The first five and last five observations in the **EVOS** file are listed in the accompanying table.)

Transect	Seabirds	Length	Oil
1	0	4.06	No
2	0	6.51	No
3	54	6.76	No
4	0	4.26	No
5	14	3.59	No
⋮	⋮	⋮	⋮
92	7	3.40	Yes
93	4	6.67	Yes
94	0	3.29	Yes
95	0	6.22	Yes
96	27	8.94	Yes

Based on McDonald, T. L., Erickson, W. P., and McDonald, L. L. "Analysis of count data from before–after control-impact studies." *Journal of Agricultural, Biological, and Environmental Statistics,* Vol. 5, No. 3, Sept. 2000, pp. 277–8 (Table A.1).

a. Identify the variables measured as quantitative or qualitative.
b. Identify the experimental unit.
c. Use a pie chart to describe the percentage of transects in oiled and unoiled areas.
*d. Use a graphical method to examine the relationship between observed number of seabirds and transect length.
e. Observed seabird density is defined as observed count divided by length of transect. MINITAB descriptive statistics for seabird densities in unoiled and oiled transects are displayed in the printout at the top of the page. Assess whether the distribution of seabird densities differs for transects in oiled and unoiled areas.
f. For unoiled transects, give an interval of values that is likely to contain at least 75% of the seabird densities.
g. For oiled transects, give an interval of values that is likely to contain at least 75% of the seabird densities.
h. Which type of transect, an oiled or unoiled one, is more likely to have a seabird density of 16? Explain.

2.206 Benford's Law of Numbers. According to *Benford's law,* certain digits (1, 2, 3,..., 9) are more likely to occur as the first significant digit in a randomly selected number than are other digits. For example, the law predicts that the number "1" is the most likely to occur (30% of the time) as the first digit. In a study reported in the *American Scientist* (July–Aug. 1998) to test Benford's law, 743 first-year college students were asked to write down a six-digit number at random. The first significant digit of each number was

recorded and its distribution summarized in the following table. Describe the first digit of the "random guess" data with an appropriate graph. Does the graph support Benford's law? Explain.

First Digit	Number of Occurrences
1	109
2	75
3	77
4	99
5	72
6	117
7	89
8	62
9	43
Total	743

Based on Hill, T. P. "The first digit phenomenon." *American Scientist,* Vol. 86, No. 4, July–Aug. 1998, p. 363 (Figure 5).

Applying the Concepts—Advanced

2.207 Speed of light from galaxies. Refer to *The Astronomical Journal* study of galaxy velocities, presented in Exercise 2.200 (p. 137). A second cluster of galaxies, named A1775, is thought to be a *double cluster*—that is, two clusters of galaxies in close proximity. Fifty-one velocity observations (in kilometers per second, km/s) from cluster A1775 are listed in the table.

22,922	20,210	21,911	19,225	18,792	21,993	23,059
20,785	22,781	23,303	22,192	19,462	19,057	23,017
20,186	23,292	19,408	24,909	19,866	22,891	23,121
19,673	23,261	22,796	22,355	19,807	23,432	22,625
22,744	22,426	19,111	18,933	22,417	19,595	23,408
22,809	19,619	22,738	18,499	19,130	23,220	22,647
22,718	22,779	19,026	22,513	19,740	22,682	19,179
19,404	22,193					

Based on Oegerle, W. R., Hill, J. M., and Fitchett, M. J. "Observations of high dispersion clusters of galaxies: Constraints on cold dark matter." *The Astronomical Journal,* Vol. 110, No. 1, July 1995 (Table 1 and Figure 1).

a. Use a graphical method to describe the velocity distribution of galaxy cluster A1775.
b. Examine the graph you created in part **a.** Is there evidence to support the double-cluster theory? Explain.

c. Calculate numerical descriptive measures (e.g., mean and standard deviation) for galaxy velocities in cluster A1775. Depending on your answer to part **b,** you may need to calculate two sets of numerical descriptive measures, one for each of the clusters (say, A1775A and A1775B) within the double cluster.

d. Suppose you observe a galaxy velocity of 20,000 km/s. Is this galaxy likely to belong to cluster A1775A or A1775B? Explain.

2.208 Standardized test "average." *U.S. News & World Report* reported on many factors contributing to the breakdown of public education. One study mentioned in the article found that over 90% of the nation's school districts reported that their students were scoring "above the national average" on standardized tests. Using your knowledge of measures of central tendency, explain why the schools' reports are incorrect. Does your analysis change if the term "average" refers to the mean? To the median? Explain what effect this misinformation might have on the perception of the nation's schools.

2.209 Zinc phosphide in pest control. A chemical company produces a substance composed of 98% cracked corn particles and 2% zinc phosphide for use in controlling rat populations in sugarcane fields. Production must be carefully controlled to maintain the 2% zinc phosphide because too much zinc phosphide will cause damage to the sugarcane and too little will be ineffective in controlling the rat population. Records from past production indicate that the distribution of the actual percentage of zinc phosphide present in the substance is approximately mound shaped, with a mean of 2.0% and a standard deviation of .08%. Suppose one batch chosen randomly actually contains 1.80% zinc phosphide. Does this indicate that there is too little zinc phosphide in this production? Explain your reasoning.

2.210 Grades in statistics. The final grades given by two professors in introductory statistics courses have been carefully examined. The students in the first professor's class had a grade point average of 3.0 and a standard deviation of .2. Those in the second professor's class had grade points with an average of 3.0 and a standard deviation of 1.0. If you had a choice, which professor would you take for this course?

2.211 Salaries of professional athletes. The salaries of superstar professional athletes receive much attention in the media. The multimillion-dollar long-term contract is now commonplace among this elite group. Nevertheless, rarely does a season pass without negotiations between one or more of the players' associations and team owners for additional salary and fringe benefits for *all* players in their particular sports.

a. If a players' association wanted to support its argument for higher "average" salaries, which measure of central tendency do you think it should use? Why?

b. To refute the argument, which measure of central tendency should the owners apply to the players' salaries? Why?

Critical Thinking Challenges

2.212 No Child Left Behind Act. According to the government, federal spending on K–12 education has increased dramatically over the past 20 years, but student performance has stayed essentially the same. Hence, former President George Bush signed into law the No Child Left Behind Act, a bill that promised improved student achievement for all U.S. children. *Chance* (Fall 2003) reported on a graphic that was designed to support the legislation. The graphic, obtained from the U.S. Department of Education Web site (www.ed.gov), is reproduced here. The bars in the graph represent annual federal spending on education, in billions of dollars (left-side vertical axis). The horizontal line represents the annual average 4th-grade children's reading ability score (right-side vertical axis). Critically assess the information portrayed in the graph. Does it, in fact, support the government's position that our children are not making classroom improvements despite federal spending on education? Use the following facts (divulged in the *Chance* article) to help you frame your answer: (1) The U.S. student population has also increased dramatically over the past 20 years, (2) 4th-grade reading test scores are designed to have an average of 250 with a standard deviation of 50, and (3) the reading test scores of 7th and 12th graders and the mathematics scores of 4th graders did improve substantially over the past 20 years.

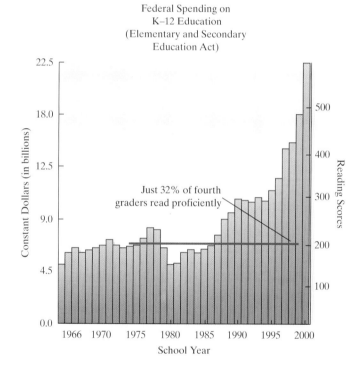

2.213 The Hite Report. Researcher Shere Hite shocked conservative America with her famous Hite Report on the permissive sexual attitudes of American men and women. In her book *Women and Love: A Cultural Revolution in Progress* (Knopf Press, 1988), Hite reveals some startling statistics describing how women feel about contemporary relationships:

- Eighty-four percent are not emotionally satisfied with their relationship.
- Ninety-five percent report "emotional and psychological harassment" from their men.
- Seventy percent of those married five years or more are having extramarital affairs.
- Only 13% of those married more than two years are "in love."

Hite conducted the survey by mailing out 100,000 questionnaires to women across the country over a seven-year period. The questionnaires were mailed to a wide variety of organizations, including church groups, women's voting and political groups, women's rights organizations, and counseling and walk-in centers for women. Organizational leaders were asked to circulate the questionnaires to their members. Hite also relied on volunteer respondents who wrote in for copies of the questionnaire. Each questionnaire consisted of 127 open-ended questions, many with numerous subquestions and follow-ups. Hite's instruc-

tions read, "It is not necessary to answer every question! Feel free to skip around and answer those questions you choose." Approximately 4,500 completed questionnaires were returned, a response rate of 4.5%. These questionnaires form the data set from which the preceding percentages were determined. Hite claims that the 4,500 women respondents are a representative sample of all women in the United States and, therefore, that the survey results imply that vast numbers of women are "suffering a lot of pain in their love relationships with men." Critically assess the survey results. Do you believe they are reliable?

Activity   Describing Data from Popular Sources

We list here several sources of real-life data sets. (Many in the list have been obtained from Wasserman and Bernero's *Statistics Sources,* and many can be accessed via the Internet.) This index of data sources is far from complete, but is a useful reference for anyone interested in finding almost any type of data.

First we list some almanacs:
Baseball Almanac (www.baseball-almanac.com)
Information Please Almanac (www.infoplease.com)
World Almanac and Book of Facts (www.worldalmanac.com)

U.S. Government publications are also rich sources of data:
Agricultural Statistics (www.nass.usda.gov)
Digest of Educational Statistics (nces.ed.gov)
Handbook of Labor Statistics (www.bls.gov)
Housing and Urban Development Statistics (www.huduser.org)
Social Indicators (www.unstats.un.org)
Uniform Crime Reports (www.fbi.gov)
Vital Statistics of the United States (www.cdc.gov)
Economic Indicators (www.economicindicators.gov)
Monthly Labor Review (www.bls.gov)
Survey of Current Business (www.bea.gov)
Statistical Abstract of the United States (www.census.gov)

Many data sources are published on an annual basis:
Commodity Yearbook (www.crbtrader.com)
The World Bank Data Catalog (data.worldbank.org)
Yearbook of American and Canadian Churches (www.thearda.com)
Standard and Poor's Corporation, Trade and Securities: Statistics (published monthly)

Some sources contain data that are international in scope:
Compendium of Social Statistics (www.esewa.un.org)
Stokholm International Peace Research Institute Yearbook (www.sipri.org)
United Nations Statistical Yearbook (www.unstats.un.org)
World Handbook of Political and Social Indicators (www.lepsr.umich.edu)

Here are a few miscellaneous data sources:
Federal Drug Data Sources (www.whitehousedrugpolicy.gov)
Internet Data Sources for Social Scientists (www.ciser.cornell.edu)
Climate Data Sources (www.realclimate.org)
World Health Organization Statistics (www.who.int)

Utilizing the data sources listed, sources suggested by your instructor, or your own resourcefulness, find one real-life quantitative data set that stems from an area of particular interest to you.

a. Describe the data set by using a relative frequency histogram, stem-and-leaf display, or dot plot.

b. Find the mean, median, variance, standard deviation, and range of the data set.

c. Use Chebyshev's rule and the empirical rule to describe the distribution of this data set. Count the actual number of observations that fall within one, two, and three standard deviations of the mean of the data set, and compare these counts with the description of the data set you developed in part **b**.

References

Huff, D. *How to Lie with Statistics*. New York: Norton, 1954.

Tufte, E. R. *Beautiful Evidence.* Cheshire, CT: Graphics Press, 2006.

Tufte, E. R. *Envisioning Information*. Cheshire, CT: Graphics Press, 1990.

Tufte, E. R. *Visual Display of Quantitative Information*. Cheshire, CT: Graphics Press, 1983.

Tufte, E. R. *Visual Explanations*. Cheshire, CT: Graphics Press, 1997.

Tukey, J. *Exploratory Data Analysis*. Reading, MA: Addison-Wesley, 1977.

USING TECHNOLOGY   **MINITAB: Describing Data**

Graphing Data

Step 1 Click on the "Graph" button on the MINITAB menu bar.

Step 2 Click on the graph of your choice (bar, pie, scatterplot, histogram, dot plot, or stem-and-leaf) to view the appropriate dialog box. The dialog box for a histogram is shown in Figure 2.M.1.

Step 3 Make the appropriate variable selections and click "OK" to view the graph.

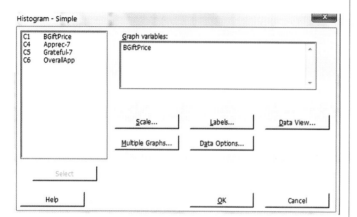

Figure 2.M.1 MINITAB histogram dialog box

Numerical Descriptive Statistics

Step 1 Click on the "Stat" button on the main menu bar, click on "Basic Statistics," and then click on "Display Descriptive Statistics." The resulting dialog box appears in Figure 2.M.2.

Step 2 Select the quantitative variables you want to analyze and place them in the "Variables" box. You can control which descriptive statistics appear by clicking the "Statistics" button on the dialog box and making your selections. (As an option, you

Figure 2.M.2 MINITAB descriptive statistics dialog box

can create histograms and dot plots for the data by clicking the "Graphs" button and making the appropriate selections.)

Step 3 Click "OK" to view the descriptive statistics printout.

TI-83/TI–84 Plus Graphing Calculator: Describing Data

Histogram from Raw Data

Step 1 *Enter the data*

- Press **STAT** and select **1:Edit**

Note: If the list already contains data, clear the old data. Use the up arrow to highlight "L1." Press **CLEAR ENTER**. Use the arrow and **ENTER** keys to enter the data set into **L1.**

Step 2 *Set up the histogram plot*

- Press **2nd** and press **Y =** for **STAT PLOT**
- Press **1** for **Plot 1**
- Set the cursor so that **ON** is flashing
- For **Type**, use the arrow and Enter keys to highlight and select the histogram
- For **Xlist**, choose the column containing the data (in most cases, **L1**)

Note: Press **2nd 1** for **L1** Freq should be set to 1.

Step 3 *Select your window settings*

- Press **WINDOW** and adjust the settings as follows:

$$\textbf{X min} = \text{lowest class boundary}$$
$$\textbf{X max} = \text{highest class boundary}$$
$$\textbf{X sel} = \text{class width}$$
$$\textbf{Y min} = 0$$
$$\textbf{Y max} \geq \text{greatest class frequency}$$
$$\textbf{Y scl} = 1$$
$$\textbf{X res} = 1$$

Step 4 *View the graph*

- Press **GRAPH**

Optional *Read class frequencies and class boundaries*

Step 5 You can press **TRACE** to read the class frequencies and class boundaries. Use the arrow keys to move between bars.

Example The following figures show TI-83/TI-84 window settings and histogram for the following sample data:

86, 70, 62, 98, 73, 56, 53, 92, 86, 37, 62, 83, 78, 49, 78, 37, 67, 79, 57

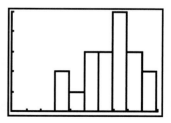

Histogram from a Frequency Table

Step 1 *Enter the data*

• Press **STAT** and select **1:Edit**

Note: If a list already contains data, clear the old data. Use the up arrow to highlight the list name, "L1" or "L2."

• Press **CLEAR ENTER**

• Enter the midpoint of each class into **L1**

• Enter the class frequencies or relative frequencies into **L2**

Step 2 *Set up the histogram plot*

• Press **2nd** and **Y =** for **STATPLOT**

• Press **1** for **Plot 1**

• Set the cursor so that **ON** is flashing

• For **Type,** use the arrow and Enter keys to highlight and select the histogram

• For **Xlist,** choose the column containing the midpoints

• For **Freq,** choose the column containing the frequencies or relative frequencies

Step 3–4 *Follow steps 3–4 given above*

Note: To set up the Window for relative frequencies, be sure to set **Y max** to a value that is greater than or equal to the largest relative frequency.

One-Variable Descriptive Statistics

Step 1 *Enter the data*

• Press **STAT** and select **1:Edit**

Note: If the list already contains data, clear the old data. Use the up arrow to highlight "L1." Press **CLEAR ENTER.**

• Use the arrow and ENTER keys to enter the data set into L1

Step 2 *Calculate descriptive statistics*

• Press **STAT**

• Press the right arrow key to highlight **CALC**

• Press **ENTER** for **1-Var Stats**

• Enter the name of the list containing your data

• Press **2nd 1** for **L1** (or **2nd 2** for **L2**, etc.)

• Press **ENTER**

You should see the statistics on your screen. Some of the statistics are off the bottom of the screen. Use the down arrow to scroll through to see the remaining statistics. Use the up arrow to scroll back up.

Example Consider finding descriptive statistics for the following sample data set:

86, 70, 62, 98, 73, 56, 53, 92, 86, 37, 62, 83, 78, 49, 78, 37, 67, 79, 57

The output screens for this example are shown below:

Sorting Data (to Find the Mode)

The descriptive statistics do not include the mode. To find the mode, sort your data as follows:

• Press **STAT**

• Press **2** for **SORTA(**

• Enter the name of the list your data are in. If your data are in **L1**, press **2nd 1**

• Press **ENTER**

• The screen will say: **DONE**

• To see the sorted data, press **STAT** and select **1:Edit**

• Scroll down through the list and locate the data value that occurs most frequently

Box Plot

Step 1 *Enter the data*

• Press **STAT** and select **1:Edit**

Note: If the list already contains data, clear the old data. Use the up arrow to highlight "L1."

• Press **CLEAR ENTER**

• Use the arrow and **ENTER** keys to enter the data set into **L1**

Step 2 *Set up the box plot*

• Press **2nd Y =** for **STAT PLOT**

• Press **1** for **Plot 1**

• Set the cursor so that "**ON**" is flashing

• For **TYPE,** use the right arrow to scroll through the plot icons and select the box plot in the middle of the second row

• For **XLIST,** choose **L1**

• Set **FREQ** to **1**

Step 3 *View the graph*

• Press **ZOOM** and select **9:ZoomStat**

Optional *Read the five number summary*

• Press **TRACE**

• Use the left and right arrow keys to move between **minX, Q1, Med, Q3,** and **maxX**

Example Make a box plot for the given data:

86, 70, 62, 98, 73, 56, 53, 92, 86, 37, 62, 83, 78, 49, 78, 37, 67, 79, 57

The output screen for this example is shown on p. 144.

Scatterplots

Step 1 *Enter the data*

- Press **STAT** and select **1:Edit**

Note: If a list already contains data, clear the old data. Use the up arrow to highlight the list name, "**L1**" or "**L2**."

- Press **CLEAR ENTER**
- Enter your *x*-data in **L1** and your *y*-data in **L2**

Step 2 *Set up the scatterplot*

- Press **2nd Y =** for **STAT PLOT**
- Press **1** for **Plot 1**
- Set the cursor so that **ON** is flashing
- For **Type,** use the arrow and Enter keys to highlight and select the scatterplot (first icon in the first row)

- For **Xlist,** choose the column containing the *x*-data
- For **Ylist,** choose the column containing the *y*-data

Step 3 *View the scatterplot*

- Press **ZOOM 9** for **ZoomStat**

Example The figures below show a table of data entered on the T1-84 and the scatterplot of the data obtained using the steps given above and in the first column.

3 Probability

CONTENTS

Where We've Been

- Identified the objective of inferential statistics: to make inferences about a population on the basis of information in a sample
- Introduced graphical and numerical descriptive measures for both quantitative and qualitative data

Where We're Going

- Develop probability as a measure of uncertainty (3.1)
- Introduce basic rules for finding probabilities (3.2–3.6)
- Use a probability as a measure of reliability for an inference (3.2–3.6)
- Provide more advanced rules for finding probabilities (3.7–3.8)

Statistics in Action   Lotto Buster! Can You Improve Your Chance of Winning?

"Welcome to the Wonderful World of Lottery Busters." So began the premier issue of Lottery Buster, *a monthly publication for players of the state lottery games.* Lottery Buster *provides interesting facts and figures on the 42 state lotteries and 2 multistate lotteries currently operating in the United States and, more importantly, tips on how to increase a player's odds of winning the lottery.*

In 1963, New Hampshire became the first state in modern times to authorize a state lottery as an alternative to increasing taxes. (Prior to that time, beginning in 1895, lotteries were banned in America because they were thought to be corrupt.) Since then, lotteries have become immensely popular, for two reasons. First, they lure you with the opportunity to win millions of dollars with a $1 investment, and second, when you lose, at least you believe that your money is going to a good cause. Many state lotteries, like Florida's, earmark a high percentage of lottery revenues to fund state education.

The popularity of the state lottery has brought with it an avalanche of "experts" and "mathematical wizards" (such as the editors of *Lottery Buster*) who provide advice on how to win the lottery—for a fee, of course! Many offer guaranteed "systems" of winning through computer software products with catchy names such as Lotto Wizard, Lottorobics, Win4D, and Lotto-Luck.

For example, most knowledgeable lottery players would agree that the "golden rule" or "first rule" in winning lotteries is *game selection*. State lotteries generally offer three types of games: Instant (scratch-off tickets or online) games, Daily Numbers (Pick-3 or Pick-4), and the weekly Pick-6 Lotto game.

One version of the Instant game involves scratching off the thin opaque covering on a ticket with the edge of a coin to determine whether you have won or lost. The cost of the ticket ranges from 50¢ to $1, and the amount won ranges from $1 to $100,000 in most states to as much as $1 million in others. *Lottery Buster* advises against playing the Instant game because it is "a pure chance play, and you can win only by dumb luck. No skill can be applied to this game."

The Daily Numbers game permits you to choose either a three-digit (Pick-3) or four-digit (Pick-4) number at a cost of $1 per ticket. Each night, the winning number is drawn. If your number matches the winning number, you win a large sum of money, usually $100,000. You do have some control over the Daily Numbers game (since you pick the numbers that you play); consequently, there are strategies available to increase your chances of winning. However, the Daily Numbers game, like the Instant game, is not available for out-of-state play.

To play Pick-6 Lotto, you select six numbers of your choice from a field of numbers ranging from 1 to N, where N depends on which state's game you are playing. For example, Florida's current Lotto game involves picking six numbers ranging from 1 to 53. The cost of a ticket is $1, and the payoff, if your six numbers match the winning numbers drawn, is $7 million or more, depending on the number of tickets purchased. (To date, Florida has had the largest state weekly payoff, over $200 million.) In addition to capturing the grand prize, you can win second-, third-, and fourth-prize payoffs by matching five, four, and three of the six numbers drawn, respectively. And you don't have to be a resident of the state to play the state's Lotto game.

In this chapter, several Statistics in Action Revisited examples demonstrate how to use the basic concepts of probability to compute the odds of winning a state lottery game and to assess the validity of the strategies suggested by lottery "experts."

Statistics in Action   Revisited

- The Probability of Winning Lotto (p. 156)
- The Probability of Winning Lotto with a Wheeling System (p. 167)
- The Probability of Winning Cash 3 or Play 4 (p. 181)

Recall that one branch of statistics is concerned with decisions made about a population on the basis of information learned about a sample. You can see how this is accomplished more easily if you understand the relationship between population and sample—a relationship that becomes clearer if we reverse the statistical procedure of making inferences from sample to population. In this chapter, we assume that the population is *known* and calculate the chances of obtaining various samples from the population. Thus, we show that probability is the reverse of statistics: *In probability, we use information about the population to infer the probable nature of the sample.*

Probability plays an important role in making inferences. Suppose, for example, you have an opportunity to invest in an oil exploration company. Past records show that, out of 10 previous oil drillings (a sample of the company's experiences), all 10 came up dry.

What do you conclude? Do you think the chances are better than 50:50 that the company will hit a gusher? Should you invest in this company? Chances are, your answer to these questions will be an emphatic "No!" However, if the company's exploratory prowess is sufficient to hit a producing well 50% of the time, a record of 10 dry wells out of 10 drilled is an event that is just too improbable.

Or suppose you're playing poker with what your opponents assure you is a well-shuffled deck of cards. In three consecutive five-card hands, the person on your right is dealt four aces. On the basis of this sample of three deals, do you think the cards are being adequately shuffled? Again, your answer is likely to be "No," because dealing three hands of four aces is just too improbable if the cards were properly shuffled.

Note that the decisions concerning the potential success of the oil-drilling company and the adequacy of card shuffling both involve knowing the chance—or probability—of a certain sample result. Both situations were contrived so that you could easily conclude that the probabilities of the sample results were small. Unfortunately, the probabilities of many observed sample results aren't so easy to evaluate intuitively. In these cases, we need the assistance of a theory of probability.

3.1 Events, Sample Spaces, and Probability

Let's begin our treatment of probability with straightforward examples that are easily described. With the aid of these simple examples, we can introduce important definitions that will help us develop the notion of probability more easily.

Suppose a coin is tossed once and the up face is recorded. The result we see is called an *observation*, or *measurement*, and the process of making an observation is called an *experiment*. Notice that our definition of experiment is broader than the one used in the physical sciences, which brings to mind test tubes, microscopes, and other laboratory equipment. Statistical experiments may include, in addition to these things, recording an Internet user's preference for a Web browser, recording a voter's opinion on an important political issue, measuring the amount of dissolved oxygen in a polluted river, observing the level of anxiety of a test taker, counting the number of errors in an inventory, and observing the fraction of insects killed by a new insecticide. The point is that a statistical experiment can be almost any act of observation, as long as the outcome is uncertain.

> An **experiment** is an act or process of observation that leads to a single outcome that cannot be predicted with certainty.

Consider another simple experiment consisting of tossing a die and observing the number on the up face. The six possible outcomes of this experiment are as follows:

1. Observe a 1.

2. Observe a 2.

3. Observe a 3.

4. Observe a 4.

5. Observe a 5.

6. Observe a 6.

Note that if this experiment is conducted once, *you can observe one and only one of these six basic outcomes, and the outcome cannot be predicted with certainty*. Also, these outcomes cannot be decomposed into more basic ones. Because observing the outcome of an experiment is similar to selecting a sample from a population, the basic possible outcomes of an experiment are called **sample points**.*

> A **sample point** is the most basic outcome of an experiment.

*Alternatively, the term *simple event* can be used.

Example	3.1

Listing Sample Points for a Coin-Tossing Experiment

Problem Two coins are tossed, and their up faces are recorded. List all the sample points for this experiment.

Solution Even for a seemingly trivial experiment, we must be careful when listing the sample points. At first glance, we might expect one of three basic outcomes: Observe two heads; Observe two tails; or Observe one head and one tail. However, further reflection reveals that the last of these, Observe one head and one tail, can be decomposed into two outcomes: Head on coin 1, Tail on coin 2; and Tail on coin 1, Head on coin 2.

A useful tool for illustrating this notion is a **tree diagram**. Figure 3.1 shows a tree diagram for this experiment. At the top of the "tree" there are two branches, representing

Figure 3.1
Tree diagram for the coin-tossing experiment

```
                        1st Coin
                           ●
                       H /     \ T
                        /       \
                2nd Coin
              ●              ●
           H /  \ T       H /  \ T
            /    \         /    \
         H H     H T     T H     T T
```

the two outcomes (H or T) for the first tossed coin. Each of these outcomes results in two more branches, representing the two outcomes (H or T) for the second tossed coin. Consequently, after tossing both coins, you can see that we have four sample points:

1. Observe HH
2. Observe HT
3. Observe TH
4. Observe TT

where H in the first position means "Head on coin 1," H in the second position means "Head on coin 2," and so on.

Look Back Even if the coins are identical in appearance, they are, in fact, two distinct coins. Thus, the sample points must account for this distinction.

■ **Now Work Exercise 3.15a**

```
  ○ H        ○ T

                   S
```
a. Experiment: Observe the up face on a coin

```
  ○ 1   ○ 2   ○ 3
  ○ 4   ○ 5   ○ 6

                   S
```
b. Experiment: Observe the up face on a die

```
  ○ HH       ○ HT
  ○ TH       ○ TT

                   S
```
c. Experiment: Observe the up faces on two coins

Figure 3.2
Venn diagrams for the three experiments from Table 3.1

We often wish to refer to the collection of all the sample points of an experiment. This collection is called the *sample space* of the experiment. For example, there are six sample points in the sample space associated with the die-toss experiment. The sample spaces for the experiments discussed thus far are shown in Table 3.1.

The **sample space** of an experiment is the collection of all its sample points.

Just as graphs are useful in describing sets of data, a pictorial method for presenting the sample space will often be useful. Figure 3.2 shows such a representation for each of the experiments in Table 3.1. In each case, the sample space is shown as a closed figure, labeled S, containing all possible sample points. Each sample point is represented by a solid dot (i.e., a "point") and labeled accordingly. Such graphical representations are called **Venn diagrams**.

Table 3.1	Experiments and Their Sample Spaces

Experiment:	Observe the up face on a coin.
Sample Space:	1. Observe a head.
	2. Observe a tail.

This sample space can be represented in set notation as a set containing two sample points:

$$S: \{H, T\}$$

Here, H represents the sample point Observe a head and T represents the sample point Observe a tail.

Experiment:	Observe the up face on a die.
Sample Space:	1. Observe a 1.
	2. Observe a 2.
	3. Observe a 3.
	4. Observe a 4.
	5. Observe a 5.
	6. Observe a 6.

This sample space can be represented in set notation as a set of six sample points:

$$S: \{1, 2, 3, 4, 5, 6\}$$

Experiment:	Observe the up faces on two coins.
Sample Space:	1. Observe HH.
	2. Observe HT.
	3. Observe TH.
	4. Observe TT.

This sample space can be represented in set notation as a set of four sample points:

$$S: \{HH, HT, TH, TT\}$$

BIOGRAPHY

JOHN VENN (1834–1923)

The English Logician

Born in Hull, England, John Venn is probably best known for his pictorial representation of unions and intersections (the Venn diagram). While lecturing in moral science at Cambridge University, Venn wrote *The Logic of Chance* and two other treatises on logic. In these works, Venn probably had his greatest contribution to the field of probability and statistics: the notion that the probability of an event is simply the long-run proportion of times the event occurs. Besides being a well-known mathematician, Venn was a historian, philosopher, priest, and skilled machine builder. (His machine for bowling cricket balls once beat a top star on the Australian cricket team.) ■

Now that we know that an experiment will result in *only one* basic outcome—called a sample point—and that the sample space is the collection of all possible sample points, we're ready to discuss the probabilities of the sample points. You have undoubtedly used the term *probability* and have some intuitive idea about its meaning. Probability is generally used synonymously with "chance," "odds," and similar concepts. For example, if a fair coin is tossed, we might reason that both the sample points Observe a head and Observe a tail have the same *chance* of occurring. Thus, we might state, "The probability of observing a head is 50%" or "The odds of seeing a head are 50:50." Both of these statements are based on an informal knowledge of probability. We'll begin our treatment of probability by using such informal concepts and then solidify what we mean later.

The probability of a sample point is a number between 0 and 1 that measures the likelihood that the outcome will occur when the experiment is performed. This number is usually taken to be the relative frequency of the occurrence of a sample point in a very long series of repetitions of an experiment.* For example, if we are assigning probabilities to the two sample points in the coin-toss experiment (Observe a head and Observe a tail), we might reason that if we toss a balanced coin a very large number of times, the sample points Observe a head and Observe a tail will occur with the same relative frequency of .5.

Our reasoning is supported by Figure 3.3. The figure plots the relative frequency of the number of times that a head occurs in simulations (by computer) of the toss of a coin N times, where N ranges from as few as 25 tosses to as many as 1,500 tosses. You can see that when N is large (e.g., $N = 1,500$), the relative frequency is converging to .5. Thus, the probability of each sample point in the coin-tossing experiment is .5.

For some experiments, we may have little or no information on the relative frequency of occurrence of the sample points; consequently, we must assign probabilities to the sample points on the basis of general information about the experiment.

* The result derives from an axiom in probability theory called the **law of large numbers**. Phrased informally, the law states that the relative frequency of the number of times that an outcome occurs when an experiment is replicated over and over again (i.e., a large number of times) approaches the true (or theoretical) probability of the outcome.

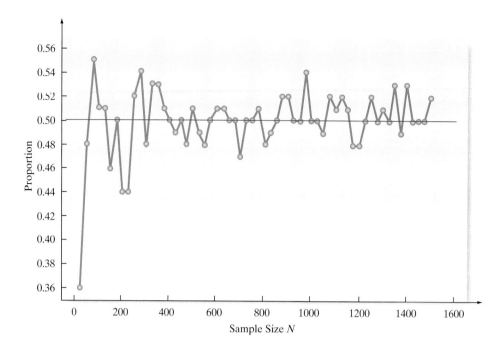

Figure 3.3

Proportion of heads in N tosses of a coin

Figure 3.4

Experiment: invest in a business venture and observe whether it succeeds (S) or fails (F)

For example, if the experiment is about investing in a business venture and observing whether it succeeds or fails, the sample space would appear as in Figure 3.4.

We are unlikely to be able to assign probabilities to the sample points of this experiment based on a long series of repetitions, since unique factors govern each performance of this kind of experiment. Instead, we may consider factors such as the personnel managing the venture, the general state of the economy at the time, the rate of success of similar ventures, and any other information deemed pertinent. If we finally decide that the venture has an 80% chance of succeeding, we assign a probability of .8 to the sample point Success. This probability can be interpreted as a measure of our degree of belief in the outcome of the business venture; that is, it is a subjective probability. Notice, however, that such probabilities should be based on expert information that is carefully assessed. If not, we may be misled on any decisions based on these probabilities or based on any calculations in which they appear. [*Note:* For a text that deals in detail with the subjective evaluation of probabilities, see Winkler (1972) or Lindley (1985).]

No matter how you assign the probabilities to sample points, the probabilities assigned must obey two rules:

Probability Rules for Sample Points

Let p_i represent the probability of sample point i. Then

1. All sample point probabilities *must* lie between 0 and 1 (i.e., $0 \le p_i \le 1$).

2. The probabilities of all the sample points within a sample space *must* sum to 1 (i.e., $\Sigma p_i = 1$).

Assigning probabilities to sample points is easy for some experiments. For example, if the experiment is to toss a fair coin and observe the face, we would probably all agree to assign a probability of $^1/_2$ to the two sample points, Observe a head and Observe a tail. However, many experiments have sample points whose probabilities are more difficult to assign.

| **Example** | **3.2** |

Sample Point Probabilities—Hotel Water Conservation

Problem *Going Green* is the term used to describe water conservation programs at hotels and motels. Many hotels now offer their guests the option of participating in these Going Green programs by reusing towels and bed linens. Suppose you randomly select one hotel from a registry of all hotels in Orange County, California, and check whether the hotel participates in a water conservation program. Show how this problem might be formulated in the framework of an experiment with sample points and a sample space. Indicate how probabilities might be assigned to the sample points.

Solution The experiment can be defined as the selection of an Orange County hotel and the observation of whether a water conservation program is offered to guests at the hotel. There are two sample points in the sample space for this experiment:

C: {The hotel offers a water conservation program.}

N: {The hotel does not offer a water conservation program.}

The difference between this and the coin-toss experiment becomes apparent when we attempt to assign probabilities to the two sample points. What probability should we assign to sample point C? If you answer .5, you are assuming that the events C and N occur with equal likelihood, just like the sample points Observe a head and Observe a tail in the coin-toss experiment. But assigning sample point probabilities for the hotel water conservation experiment is not so easy. In fact, a recent report by the Orange County Water District stated that 70% of the hotels in the county now participate in a water conservation program of some type. In that case, it might be reasonable to approximate the probability of the sample point C as .7 and that of the sample point N as .3.

Look Back Here we see that the sample points are not always equally likely, so assigning probabilities to them can be complicated, particularly for experiments that represent real applications (as opposed to coin- and die-toss experiments).

◾ **Now Work Exercise 3.22**

Although the probabilities of sample points are often of interest in their own right, it is usually probabilities of *collections* of sample points that are important. Example 3.3 demonstrates this point.

| **Example** | **3.3** |

Probability of a Collection of Sample Points— Die-Tossing Experiment

Problem A fair die is tossed, and the up face is observed. If the face is even, you win $1. Otherwise, you lose $1. What is the probability that you win?

Solution Recall that the sample space for this experiment contains six sample points:

$$S: \{1,2,3,4,5,6\}$$

Since the die is balanced, we assign a probability of $1/6$ to each of the sample points in this sample space. An even number will occur if one of the sample points Observe a 2, Observe a 4, or Observe a 6 occurs. A collection of sample points such as this is called an *event*, which we denote by the letter A. Since the event A contains three sample points—all with probability $1/6$—and since no sample points can occur simultaneously, we reason that the probability of A is the sum of the probabilities of the sample points in A. Thus, the probability of A is $1/6 + 1/6 + 1/6 = 1/2$.

Look Back On the basis of this notion of probability, *in the long run*, you will win $1 half the time and lose $1 half the time.

◾ **Now Work Exercise 3.14**

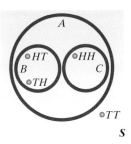

Figure 3.13
Venn diagram for coin-toss experiment

and that $B \cap C$ contains no sample points. (See Figure 3.13.) Thus, B and C are mutually exclusive, and it follows that

$$P(A) = P(B \cup C) = P(B) + P(C) = \frac{1}{2} + \frac{1}{4} = \frac{3}{4}$$

Look Back Although this example is quite simple, it shows us that writing events with verbal descriptions that include the phrases "at least" or "at most" as unions of mutually exclusive events is very useful. This practice enables us to find the probability of the event by adding the probabilities of the mutually exclusive events.

Statistics in Action | **Revisited** | **The Probability of Winning Lotto with a Wheeling System**

Refer to Florida's Pick-6 Lotto game, in which you select six numbers of your choice from a field of numbers ranging from 1 to 53. In Section 3.1, we learned that the probability of winning Lotto on a single ticket is only 1 in approximately 23 million. The "experts" at *Lottery Buster* recommend many strategies for increasing the odds of winning the lottery. One strategy is to employ a *wheeling system*. In a complete wheeling system, you select more than six numbers—say, seven—and play every combination of six of those seven numbers.

For example, suppose you choose to "wheel" the following seven numbers: 2, 7, 18, 23, 30, 32, and 51. Every combination of six of these seven numbers is listed in Table SIA3.1. You can see that there are seven different possibilities. (Use the combinations rule with $N = 7$ and $n = 6$ to verify this.) Thus, we would purchase seven tickets (at a cost of $7) corresponding to these different combinations in a complete wheeling system.

Ticket #5 or Ticket #6 or Ticket #7 is the winning combination. Note that this probability is stated with the use of the word *or*, implying a *union* of seven events. Letting T_1 represent the event that Ticket #1 wins, and defining $T_2, T_3, \dots, T_7$ in a similar fashion, we want to find

$$P(T_1 \text{ or } T_2 \text{ or } T_3 \text{ or } T_4 \text{ or } T_5 \text{ or } T_6 \text{ or } T_7)$$

Recall (Section 3.1) that the 22,957,480 possible combinations in Pick-6 Lotto are mutually exclusive and equally likely to occur. Consequently, the probability of the union of the seven events is simply the sum of the probabilities of the individual events, where each event has probability $1/(22,957,480)$:

$$P(\text{win Lotto with 7 Wheeled Numbers})$$
$$= P(T_1 \text{ or } T_2 \text{ or } T_3 \text{ or } T_4 \text{ or } T_5 \text{ or } T_6 \text{ or } T_7)$$
$$= 7/(22,957,480) = .0000003$$

In terms of odds, we now have 3 chances in 10 million of winning the Lotto with the complete wheeling system. Technically, the "experts" are correct: Our odds of winning Lotto have increased (from 1 in 23 million). However, the probability of winning is so close to 0 that we question whether the $7 spent on lottery tickets is worth the negligible increase in odds. In fact, it can be shown that to increase your chance of winning the 6/53 Lotto to 1 chance in 100 (i.e., .01) by means of a complete wheeling system, you would have to wheel 26 of your favorite numbers—a total of 230,230 combinations at a cost of $230,230!

Table SIA3.1	Wheeling the Seven Numbers 2, 7, 18, 23, 30, 32, and 51					
Ticket #1	2	7	18	23	30	32
Ticket #2	2	7	18	23	30	51
Ticket #3	2	7	18	23	32	51
Ticket #4	2	7	18	30	32	51
Ticket #5	2	7	23	30	32	51
Ticket #6	2	18	23	30	32	51
Ticket #7	7	18	23	30	32	51

To determine whether this strategy does, in fact, increase our odds of winning, we need to find the probability that one of these seven combinations occurs during the 6/53 Lotto draw. That is, we need to find the probability that either Ticket #1 or Ticket #2 or Ticket #3 or Ticket #4 or

---

Exercises 3.38–3.67

Understanding the Principles

3.38 Define in words mutually exclusive events.

3.39 Define in words the union of two events.

3.40 Define in words the intersection of two events.

3.41 Define in words the complement of an event.

3.42 State the rule of complements.

3.43 State the additive rule of probability for mutually exclusive events.

3.44 State the additive rule of probability for any two events.

Learning the Mechanics

3.45 A fair coin is tossed three times, and the events A and B
[NW] are defined as follows:

> A: {At least one head is observed.}
>
> B: {The number of heads observed is odd.}

a. Identify the sample points in the events A, B, $A \cup B$, A^c, and $A \cap B$.

b. Find $P(A), P(B), P(A \cup B), P(A^c)$, and $P(A \cap B)$ by summing the probabilities of the appropriate sample points.

c. Use the additive rule to find $P(A \cup B)$. Compare your answer with the one you obtained in part **b**.

d. Are the events A and B mutually exclusive? Why?

3.46 Suppose $P(B) = .7, P(D) = .5,$ and $P(B \cap D) = .3$. Find the probabilities below.
a. $P(D^c)$
b. $P(B^c)$
c. $P(B \cup D)$

3.47 Consider the following Venn diagram, where
[NW] $P(E_2) = P(E_3) = \frac{1}{5}, P(E_4) = P(E_5) = \frac{1}{20},$
$P(E_6) = \frac{1}{10}, P(E_7) = \frac{1}{5}$:

Find each of the following probabilities:
a. $P(A)$ **e.** $P(A^c)$
b. $P(B)$ **f.** $P(B^c)$
c. $P(A \cup B)$ **g.** $P(A \cup A^c)$
d. $P(A \cap B)$ **h.** $P(A^c \cap B)$

3.48 Consider the following Venn diagram, where

> $P(E_1) = .2, P(E_2) = P(E_3) = .04, P(E_4) = .06,$
>
> $P(E_5) = .08, P(E_6) = .15, P(E_7) = .25,$ and
>
> $P(E_8) = .18$

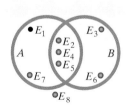

Find the following probabilities:
a. $P(A^c)$
b. $P(B^c)$
c. $P(A \cap B^c)$
d. $P(A \cap B)$
e. $P(A \cup B)$
f. $P(A^c \cup B^c)$
g. Are events A and B mutually exclusive? Why?

3.49 A pair of fair dice is tossed. Define the events A and B as follows. Define the following events:

> A: {An 8 is rolled} (The sum of the numbers of dots on the upper faces of the two dice is equal to 8.)
>
> B: {At least one of the two dice is showing a 6}

a. Identify the sample points in the event $A, B, A \cap B$, $A \cup B$, and A^c.

b. Find $P(A), P(B), P(A \cap B), P(A \cup B)$, and $P(A^c)$ by summing the probabilities of the appropriate sample points.

c. Use the additive rule to find $P(A \cup B)$. Compare your answer with that for the same event in part **b**.

d. Are A and B mutually exclusive? Why?

3.50 Three fair coins are tossed. We wish to find the probability of the event A: {Observe at most two tails.}

a. Express A as the union of three mutually exclusive events. Using the expression you wrote, find the probability of A.

b. Express A as the complement of an event. Using the expression you wrote, find the probability of A.

3.51 The outcomes of two variables are (Low, Medium, High)
[NW] and (On, Off), respectively. An experiment is conducted in which the outcomes of each of the two variables are observed. The probabilities associated with each of the six possible outcome pairs are given in the following table:

	Low	Medium	High
On	.50	.10	.05
Off	.25	.07	.03

Consider the following events:

> A: {On}
> B: {Medium or On}
> C: {Off and Low}
> D: {High}

a. Find $P(A)$.
b. Find $P(B)$.
c. Find $P(C)$.
d. Find $P(D)$.
e. Find $P(A^c)$.
f. Find $P(A \cup B)$.
g. Find $P(A \cap B)$.
h. Consider each possible pair of events taken from the events A, B, C, and D. List the pairs of events that are mutually exclusive. Justify your choices.

ⓓ Applet Exercise 3.3

Use the applets entitled *Simulating the Probability of Rolling a 6* and *Simulating the Probability of Rolling a 3 or 4* to explore the additive rule of probability.

a. Explain why the applet *Simulating the Probability of Rolling a 6* can also be used to simulate the probability of rolling a 3. Then use the applet with $n = 1000$ to simulate the probability of rolling a 3. Record the cumulative proportion. Repeat the process to simulate the probability of rolling a 4.

b. Use the applet *Simulating the Probability of Rolling a 3 or 4* with $n = 1000$ to simulate the probability of rolling a 3 or 4. Record the cumulative proportion.

c. Add the two cumulative proportions from part **a**. How does this sum compare with the cumulative proportion in part **b**? How does your answer illustrate the additive rule for probability?

🅓 Applet Exercise 3.4

Use the applets entitled *Simulating the Probability of Rolling a 6* and *Simulating the Probability of Rolling a 3 or 4* to simulate the probability of the complement of an event.

   a. Explain how the applet *Simulating the Probability of Rolling a 6* can also be used to simulate the probability of the event *rolling a 1, 2, 3, 4, or 5*. Then use the applet with $n = 1000$ to simulate this probability.
   b. Explain how the applet *Simulating the Probability of Rolling a 3 or 4* can also be used to simulate the probability of the event *rolling a 1, 2, 5, or 6*. Then use the applet with $n = 1000$ to simulate this probability.
   c. Which applet could be used to simulate the probability of the event *rolling a 1, 2, 3, or 4*? Explain.

Applying the Concepts—Basic

3.52 Social networking. According to the *Pew Research Internet Project* (Dec. 2013) survey, the two most popular social networking sites in the United States are Facebook and LinkedIn. Of all adult Internet users, 71% have a Facebook account, 22% have a LinkedIn account, and 18% visit both Facebook and LinkedIn.
   a. Draw a Venn diagram to illustrate the use of social networking sites in the United States.
   b. Find the probability that an adult Internet user visits either Facebook or LinkedIn.
   c. Use your answer to part **b** to find the probability that an adult Internet user does not have an account with either social networking site.

3.53 Do social robots walk or roll? Refer to the *International Conference on Social Robotics* (Vol. 6414, 2010) study of the trend in the design of social robots, Exercise 3.17 (p. 157). Recall that in a random sample of 106 social robots, 63 were built with legs only, 20 with wheels only, 8 with both legs and wheels, and 15 with neither legs nor wheels. Use the rule of complements to find the probability that a randomly selected social robot is designed with either legs or wheels.

3.54 Study of analysts' forecasts. The *Journal of Accounting Research* (Mar. 2008) published a study on relationship incentives and degree of optimism among analysts' forecasts. Participants were analysts at either a large or small brokerage firm who made their forecasts either early or late in the quarter. Also, some analysts were only concerned with making an accurate forecast, while others were also interested in their relationship with management. Suppose one of these analysts is randomly selected. Consider the following events:

$A = \{$The analyst is only concerned with making an accurate forecast.$\}$
$B = \{$The analyst makes the forecast early in the quarter.$\}$
$C = \{$The analyst is from a small brokerage firm.$\}$

Describe each of the following events in terms of unions, intersections, and complements (e.g., $A \cup B$, $A \cap B$, A^C, etc.).

   a. The analyst makes an early forecast and is only concerned with accuracy.
   b. The analyst is not only concerned with accuracy.
   c. The analyst is from a small brokerage firm or makes an early forecast.
   d. The analyst makes a late forecast and is not only concerned with accuracy.

3.55 Gene expression profiling. Gene expression profiling is a state-of-the-art method for determining the biology of cells. In *Briefings in Functional Genomics and Proteomics* (Dec. 2006), biologists reviewed several gene expression profiling methods. The biologists applied two of the methods (A and B) to data collected on proteins in human mammary cells. The probability that the protein is cross-referenced (i.e., identified) by method A is .41, the probability that the protein is cross-referenced by method B is .42, and the probability that the protein is cross-referenced by both methods is .40.
   a. Draw a Venn diagram to illustrate the results of the gene-profiling analysis.
   b. Find the probability that the protein is cross-referenced by either method A or method B.
   c. On the basis of your answer to part **b**, find the probability that the protein is not cross-referenced by either method.

3.56 Scanning errors at Wal-Mart. The National Institute of Standards and Technology (NIST) mandates that for every 100 items scanned through the electronic checkout scanner at a retail store, no more than 2 should have an inaccurate price. A study of the accuracy of checkout scanners at Wal-Mart stores in California (*Tampa Tribune*, Nov. 22, 2005) showed that, of the 60 Wal-Mart stores investigated, 52 violated the NIST scanner accuracy standard. If 1 of the 60 Wal-Mart stores is randomly selected, what is the probability that the store does not violate the NIST scanner accuracy standard?

3.57 Sleep apnea and sleep stage transitioning. Sleep apnea is a common sleep disorder characterized by collapses of the upper airway during sleep. *Chance* (Winter 2009) investigated the role of sleep apnea in how people transition from one sleep stage to another. The various stages of sleep for a large group of sleep apnea patients were monitored in 30-second intervals, or "epochs." For each epoch, sleep stage was categorized as Wake, REM, or Non-REM. The table below provides a summary of the results. Each cell of the table gives the number of epochs that occurred when transitioning from the previous sleep stage to the current sleep stage. Consider the previous and current sleep stage of a randomly selected epoch from the study.

| | Previous Sleep Stage | | |
Current Stage	Non-REM	REM	Wake
Non-REM	31,880	160	1,733
REM	346	7,609	175
Wake	1,588	358	6,079
Totals	33,814	8,127	7,987

Based on Caffo, B. S., et al. "An overview of observational sleep research with application to sleep stage transitioning." *Chance*, Vol. 22, No. 1, Winter 2009 (Table 2).

a. List the sample points for this experiment.
b. Assign reasonable probabilities to the sample points.
c. What is the probability that the current sleep stage of the epoch is REM?
d. What is the probability that the previous sleep stage of the epoch is Wake?
e. What is the probability that the epoch transitioned from the REM sleep stage to the Non-REM sleep stage?

Applying the Concepts—Intermediate

3.58 **Attempted suicide methods.** A study of attempted suicide methods was published in the journal *Death Studies* (Vol. 38, 2014). Data for 59 suicide victims who had previously attempted suicide were collected. The table gives a breakdown of the number of suicide victims according to method used and gender. Suppose one of the suicide victims is selected at random.

Method	Males	Females	Totals
Self-poisoning	22	13	35
Hanging	4	0	4
Jumping from a height	7	4	11
Cutting	6	3	9
Totals	39	20	59

a. Find $P(A)$, where $A = \{$male$\}$.
b. Find $P(B)$, where $B = \{$jumping from a height$\}$.
c. Are A and B mutually exclusive events?
d. Find $P(A^C)$.
e. Find $P(A \cup B)$.
f. Find $P(A \cap B)$.

3.59 **Guilt in decision making.** The effect of guilt emotion on how a decision maker focuses on the problem was investigated in the Jan. 2007 issue of the *Journal of Behavioral Decision Making* (see Exercise 1.31, p. 50). A total of 171 volunteer students participated in the experiment, where each was randomly assigned to one of three emotional states (guilt, anger, or neutral) through a reading/writing task. Immediately after the task, the students were presented with a decision problem where the stated option has predominantly negative features (e.g., spending money on repairing a very old car). The results (number responding in each category) are summarized in the accompanying table. Suppose one of the 171 participants is selected at random.

Emotional State	Choose Stated Option	Do Not Choose Stated Option	Totals
Guilt	45	12	57
Anger	8	50	58
Neutral	7	49	56
Totals	60	111	171

Based on Gangemi, A., and Mancini, F. "Guilt and focusing in decision-making." *Journal of Behavioral Decision Making*, Vol. 20, Jan. 2007 (Table 2).

a. Find the probability that the respondent is assigned to the guilty state.

b. Find the probability that the respondent chooses the stated option (repair the car).
c. Find the probability that the respondent is assigned to the guilty state and chooses the stated option.
d. Find the probability that the respondent is assigned to the guilty state or chooses the stated option.

3.60 **Abortion provider survey.** The Alan Guttmacher Institute Abortion Provider Survey is a survey of all 358 known nonhospital abortion providers in the United States (*Perspectives on Sexual and Reproductive Health*, Jan./Feb. 2003). For one part of the survey, the 358 providers were classified according to case load (number of abortions performed per year) and whether they permit their patients to take the abortion drug misoprostol at home or require the patients to return to the abortion facility to receive the drug. The responses are summarized in the accompanying table. Suppose we select, at random, one of the 358 providers and observe the provider's case load (fewer than 50, or 50 or more) and home use of the drug (yes or no).

Permit Drug at Home	Number of Abortions		
	Fewer than 50	50 or More	Totals
Yes	170	130	300
No	48	10	58
Totals	218	140	358

Based on Henshaw, S. K., and Finer, L. B. "The accessibility of abortion services in the United States, 2001." *Perspectives on Sexual and Reproductive Health*, Vol. 35, No. 1, Jan./Feb. 2003 (Table 4).

a. Find the probability that the provider permits home use of the abortion drug.
b. Find the probability that the provider permits home use of the drug or has a case load of fewer than 50 abortions.
c. Find the probability that the provider permits home use of the drug and has a case load of fewer than 50 abortions.

3.61 **Fighting probability of fallow deer bucks.** In *Aggressive Behavior* (Jan./Feb. 2007), zoologists investigated the likelihood of fallow deer bucks fighting during the mating season. During the observation period, the researchers recorded 205 encounters between two bucks. Of these, 167 involved one buck clearly initiating the encounter with the other. In these 167 initiated encounters, the zoologists kept track of whether or not a physical contact fight occurred and whether the initiator ultimately won or lost the encounter. (The buck that is driven away by the other is considered the loser.) A summary of the 167 initiated encounters is provided in the table on p. 171. Suppose we select one of these 167 encounters and note the outcome (fight status and winner).

a. What is the probability that a fight occurs and the initiator wins?
b. What is the probability that no fight occurs?
c. What is the probability that there is no clear winner?
d. What is the probability that a fight occurs or the initiator loses?

e. Are the events "No clear winner" and "Initiator loses" mutually exclusive?

	Initiator Wins	No Clear Winner	Initiator Loses	Totals
Fight	26	23	15	64
No Fight	80	12	11	103
Totals	106	35	26	167

Based on Bartos, L. et al. "Estimation of the probability of fighting in fallow deer (*Dama dama*) during the rut." *Aggressive Behavior*, Vol. 33, Jan./ Feb. 2007, pp. 7–13.

3.62 **Cell phone handoff behavior.** A "handoff" is a term used in wireless communications to describe the process of a cell phone moving from a coverage area of one base station to another. Each base station has multiple channels (called color codes) that allow it to communicate with the cell phone. The *Journal of Engineering, Computing and Architecture* (Vol. 3., 2009) published a study of cell phone handoff behavior. During a sample driving trip that involved crossing from one base station to another, the different color codes accessed by the cell phone were monitored and recorded. The table below shows the number of times each color code was accessed for two identical driving trips, each using a different cell phone model. (*Note:* The table is similar to the one published in the article.) Suppose you randomly select one point during the combined driving trips.

	Color Code				
	0	5	b	c	Totals
Model 1	20	35	40	0	85
Model 2	15	50	6	4	75
Totals	35	85	46	4	160

a. What is the probability that the cell phone is using color code 5?
b. What is the probability that the cell phone is using color code 5 or color code 0?
c. What is the probability that the cell phone used is Model 2 and the color code is 0?

3.63 **Chemical signals of mice.** The ability of a mouse to recognize the odor of a potential predator (e.g., a cat) is essential to the mouse's survival. The chemical makeup of these odors—called kairomones—was the subject of a study published in *Cell* (May 14, 2010). Typically, the sources of these odors are major urinary proteins (Mups). Cells collected from lab mice were exposed to Mups from rodent species A, Mups from rodent species B, and kairomones (from a cat). The accompanying Venn diagram shows the proportion of cells that chemically responded to each of the three odors. (*Note*: A cell may respond to more than a single odor.)
a. What is the probability that a lab mouse responds to all three source odors?
b. What is the probability that a lab mouse responds to the kairomone?

c. What is the probability that a lab mouse responds to Mups A and Mups B, but not the kairomone?

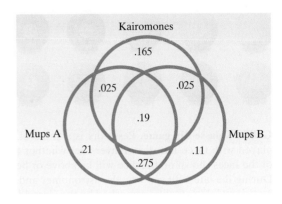

3.64 **Employee behavior problems.** The *Organizational Development Journal* (Summer 2006) reported on the results of a survey of human resource officers (HROs) at major firms. The focus of the study was employee behavior, namely absenteeism, promptness to work, and turnover. The study found that 55% of the HROs had problems with absenteeism. Also, 41% of the HROs had problems with turnover. Suppose that 22% of the HROs had problems with both absenteeism and turnover.
a. Find the probability that a human resource officer selected from the group surveyed had problems with employee absenteeism or employee turnover.
b. Find the probability that a human resource officer selected from the group surveyed did not have problems with employee absenteeism.
c. Find the probability that a human resource officer selected from the group surveyed did not have problems with employee absenteeism nor with employee turnover.

Applying the Concepts—Advanced

3.65 **Cloning credit or debit cards.** Wireless identity theft is a technique of stealing an individual's personal information from radio-frequency-enabled cards (e.g., credit or debit cards). Upon capturing this data, thieves are able to program their own cards to respond in an identical fashion via *cloning*. A method for detecting cloning attacks in radio-frequency identification (RFID) applications was explored in *IEEE Transactions on Information Forensics and Security* (Mar. 2013). The method was illustrated using a simple ball drawing game. Consider a group of 10 balls, 5 representing genuine RFID cards and 5 representing clones of one or more of these cards. A coloring system was used to distinguish among the different genuine cards. Because there were 5 genuine cards, 5 colors—yellow, blue, red, purple, and orange—were used. Balls of the same color represent either the genuine card or a clone of the card. Suppose the 10 balls are colored as follows: 3 yellow, 2 blue, 1 red, 3 purple, 1 orange. (See figure on p. 172.) Note that the singleton red and orange balls must represent the genuine cards (i.e., there are no clones of these cards). If two balls of the same color are drawn (without replacement) from the 10 balls, then a cloning attack is detected. For this example, find the probability of detecting a cloning attack.

3.66 **Galileo's passe-dix game.** Passe-dix is a game of chance played with three fair dice. Players bet whether the sum of the faces shown on the dice will be above or below 10. During the late 16th century, the astronomer and mathematician Galileo Galilei was asked by the Grand Duke of Tuscany to explain why "the chance of throwing a total of 9 with three fair dice was less than that of throwing a total of 10." (*Interstat*, Jan. 2004). The Grand Duke believed that the chance should be the same, since "there are an equal number of partitions of the numbers 9 and 10."

Find the flaw in the Grand Duke's reasoning and answer the question posed to Galileo.

3.67 **Encoding variability in software.** At the 2012 *Gulf Petrochemicals and Chemicals Association (GPCA) Forum*, Oregon State University software engineers presented a paper on modeling and implementing variation in computer software. The researchers employed the compositional choice calculus (CCC)—a formal language for representing, generating, and organizing variation in tree-structured artifacts. The CCC language was compared to two other coding languages—the annotative choice calculus (ACC) and the computational feature algebra (CFA). Their research revealed the following: any type of expression (e.g., plain expressions, dimension declarations, or lambda abstractions) found in either ACC or CFA can be found in CCC; plain expressions exist in both ACC and CFA; dimension declarations exist in ACC but not CFA; lambda abstractions exist in CFA but not ACC. Based on this information, draw a Venn diagram that illustrates the relationships among the three languages. (*Hint:* An expression represents a sample point in the Venn diagram.)

3.5 Conditional Probability

Figure 3.14

Reduced sample space for the die-toss experiment: given that event B has occurred

The event probabilities we've been discussing give the relative frequencies of the occurrences of the events when the experiment is repeated a very large number of times. Such probabilities are often called **unconditional probabilities** because no special conditions are assumed other than those that define the experiment.

Often, however, we have additional knowledge that might affect the outcome of an experiment, so we may need to alter the probability of an event of interest. A probability that reflects such additional knowledge is called the **conditional probability** of the event. For example, we've seen that the probability of observing an even number (event A) on a toss of a fair die is $^1/_2$. But suppose we're given the information that on a particular throw of the die, the result was a number less than or equal to 3 (event B). Would the probability of observing an even number on that throw of the die still be equal to $^1/_2$? It can't be because making the assumption that B has occurred reduces the sample space from six sample points to three sample points (namely, those contained in event B). This reduced sample space is as shown in Figure 3.14.

Because the sample points for the die-toss experiment are equally likely, each of the three sample points in the reduced sample space is assigned an equal *conditional probability* of $^1/_3$. Since the only even number of the three in the reduced sample space B is the number 2 and the die is fair, we conclude that the probability that A occurs *given that B occurs* is $^1/_3$. We use the symbol $P(A|B)$ to represent the probability of event A given that event B occurs. For the die-toss example,

$$P(A|B) = \frac{1}{3}$$

To get the probability of event A given that event B occurs, we proceed as follows: We divide the probability of the part of A that falls within the reduced sample space B, namely, $P(A \cap B)$, by the total probability of the reduced sample space, namely, $P(B)$. Thus, for the die-toss example with event A: {Observe an even number.} and event B: {Observe a number less than or equal to 3.}, we find that

$$P(A|B) = \frac{P(A \cap B)}{P(B)} = \frac{P(2)}{P(1) + P(2) + P(3)} = \frac{^1/_6}{^3/_6} = \frac{1}{3}$$

The formula for $P(A|B)$ is true in general:

Conditional Probability Formula

To find the *conditional probability that event A occurs given that event B occurs,* divide the probability that *both* A and B occur by the probability that B occurs; that is,

$$P(A|B) = \frac{P(A \cap B)}{P(B)}$$

[We assume that $P(B) \neq 0.$]

This formula adjusts the probability of $A \cap B$ from its original value in the complete sample space S to a conditional probability in the reduced sample space B. On the one hand, if the sample points in the complete sample space are equally likely, then the formula will assign equal probabilities to the sample points in the reduced sample space, as in the die-toss experiment. On the other hand, if the sample points have unequal probabilities, the formula will assign conditional probabilities proportional to the probabilities in the complete sample space. The latter situation is illustrated in the next three examples.

Example   3.13

The Conditional Probability Formula— Executives Who Cheat at Golf

Problem To develop programs for business travelers staying at convention hotels, Hyatt Hotels Corp. commissioned a study of executives who play golf. The study revealed that 55% of the executives admitted that they had cheated at golf. Also, 20% of the executives admitted that they had cheated at golf and had lied in business. Given that an executive had cheated at golf, what is the probability that the executive also had lied in business?

Solution Let's define events A and B as follows:

$$A = \{\text{Executive had cheated at golf.}\}$$
$$B = \{\text{Executive had lied in business.}\}$$

From the study, we know that 55% of executives had cheated at golf, so $P(A) = .55$. Now, executives who *both* cheat at golf (event A) *and* lie in business (event B) represent the compound event $A \cap B$. From the study, $P(A \cap B) = .20$. We want to know the probability that an executive lied in business (event B) given that he or she cheated at golf (event A); that is, we want to know the conditional probability $P(B|A)$. Applying the conditional probability formula, we have

$$P(B|A) = \frac{P(A \cap B)}{P(A)} = \frac{.20}{.55} = .364$$

Thus, given that a certain executive had cheated at golf, the probability that the executive also had lied in business is .364.

Look Back One of the keys to applying the formula correctly is to write the information in the study in the form of probability statements involving the events of interest. The word "and" in the clause "cheat at golf *and* lie in business" implies an intersection of the two events A and B. The word "given" in the phrase "*given* that an executive cheats at golf" implies that event A is the given event.

■ **Now Work Exercise 3.87**

Example   3.14

Applying the Conditional Probability Formula in a Two-Way Table—Smoking and Cancer

Problem Many medical researchers have conducted experiments to examine the relationship between cigarette smoking and cancer. Consider an individual randomly selected from the adult male population. Let A represent the event that the individual smokes, and let A^c denote the complement of A (the event that the individual does not smoke). Similarly, let B represent the event that the individual develops cancer, and let B^c be the complement of that event. Then the four sample points associated with the experiment are shown in Figure 3.15, and their probabilities for a certain section of the United States are given in Table 3.5. Use these sample point probabilities to examine the relationship between smoking and cancer.

Figure 3.15
Sample space for Example 3.15

Table 3.5	Probabilities of Smoking and Developing Cancer	
	Develops Cancer	
Smoker	Yes, B	No, B^c
Yes, A	.05	.20
No, A^c	.03	.72

Solution One method of determining whether the given probabilities indicate that smoking and cancer are related is to compare the *conditional probability* that an adult male acquires cancer given that he smokes with the conditional probability that an adult male acquires cancer given that he does not smoke [i.e., compare $P(B|A)$ with $P(B|A^c)$].

First, we consider the reduced sample space A corresponding to adult male smokers. This reduced sample space is shaded in Figure 3.15. The two sample points $A \cap B$ and $A \cap B^c$ are contained in this reduced sample space, and the adjusted probabilities of these two sample points are the two conditional probabilities

$$P(B|A) = \frac{P(A \cap B)}{P(A)} \quad \text{and} \quad P(B^c|A) = \frac{P(A \cap B^c)}{P(A)}$$

The probability of event A is the sum of the probabilities of the sample points in A:

$$P(A) = P(A \cap B) + P(A \cap B^c) = .05 + .20 = .25$$

Then the values of the two conditional probabilities in the reduced sample space A are

$$P(B|A) = \frac{.05}{.25} = .20 \quad \text{and} \quad P(B^c|A) = \frac{.20}{.25} = .80$$

These two numbers represent the probabilities that an adult male smoker develops cancer and does not develop cancer, respectively.

In a like manner, the conditional probabilities of an adult male nonsmoker developing cancer and not developing cancer are

$$P(B|A^c) = \frac{P(A^c \cap B)}{P(A^c)} = \frac{.03}{.75} = .04$$

$$P(B^c|A^c) = \frac{P(A^c \cap B^c)}{P(A^c)} = \frac{.72}{.75} = .96$$

Two of the conditional probabilities give some insight into the relationship between cancer and smoking: the probability of developing cancer given that the adult male is a smoker, and the probability of developing cancer given that the adult male is not a smoker. The conditional probability that an adult male smoker develops cancer (.20) is five times the probability that a nonsmoker develops cancer (.04). This relationship does not imply that smoking *causes* cancer, but it does suggest a pronounced link between smoking and cancer.

Look Back Notice that the conditional probabilities $P(B^c|A) = .80$ and $P(B|A) = .20$ are in the same 4-to-1 ratio as the original unconditional probabilities, .20 and .05. The conditional probability formula simply adjusts the unconditional probabilities so that they add to 1 in the reduced sample space A of adult male smokers. Similarly, the conditional probabilities $P(B^c|A^c) = .96$ and $P(B|A^c) = .04$ are in the same 24-to-1 ratio as the unconditional probabilities .72 and .03.

■ **Now Work Exercise 3.73a–b**

Example   3.15

Conditional Probability in a Two-Way Table— Consumer Complaints

Problem The Federal Trade Commission's investigation of consumer product complaints has generated much interest on the part of manufacturers in the quality of their products. A manufacturer of an electromechanical kitchen utensil conducted an analysis of a large number of consumer complaints and found that they fell into the six categories shown in Table 3.6. If a consumer complaint is received, what is the probability that the cause of the complaint was the appearance of the product, given that the complaint originated during the guarantee period?

Table 3.6	Distribution of Product Complaints			
	Reason for Complaint			
	Electrical	Mechanical	Appearance	Totals
During Guarantee Period	18%	13%	32%	63%
After Guarantee Period	12%	22%	3%	37%
Totals	30%	35%	35%	100%

Solution Let A represent the event that the cause of a particular complaint is the appearance of the product, and let B represent the event that the complaint occurred during the guarantee period. Checking Table 3.6, you can see that $(18 + 13 + 32)\% = 63\%$ of the complaints occur during the guarantee period. Hence, $P(B) = .63$. The percentage of complaints that were caused by appearance and occurred during the guarantee period (the event $A \cap B$) is 32%. Therefore, $P(A \cap B) = .32$.

Using these probability values, we can calculate the conditional probability $P(A|B)$ that the cause of a complaint is appearance given that the complaint occurred during the guarantee time:

$$P(A|B) = \frac{P(A \cap B)}{P(B)} = \frac{.32}{.63} = .51$$

Thus, slightly more than half of the complaints that occurred during the guarantee period were due to scratches, dents, or other imperfections in the surface of the kitchen devices.

Look Back Note that the answer $\frac{.32}{.63}$ is the proportion for the event of interest A (.32), divided by the row total proportion for the given event B (.63). That is, it is the proportion of the time that A occurs together with the given event B.

■ **Now Work Exercise 3.86a**

3.6   The Multiplicative Rule and Independent Events

The probability of an intersection of two events can be calculated with the *multiplicative rule*, which employs the conditional probabilities we defined in the previous section. Actually, we've already developed the formula in another context. Recall that the conditional probability of B given A is

$$P(B|A) = \frac{P(A \cap B)}{P(A)}$$

Multiplying both sides of this equation by $P(A)$, we obtain a formula for the probability of the intersection of events A and B. This formula is often called the **multiplicative rule of probability**.

> **Multiplicative Rule of Probability**
> $P(A \cap B) = P(A)P(B|A)$ or, equivalently, $P(A \cap B) = P(B)P(A|B)$

Example 3.16 **The Multiplicative Rule—a Famous Psychological Experiment**	**Problem** In a classic psychology study conducted in the early 1960s, Stanley Milgram performed a series of experiments in which a teacher is asked to shock a learner who is attempting to memorize word pairs whenever the learner gives the wrong answer. The shock levels increase with each successive wrong answer. (Unknown to the teacher, the shocks are not real.) Two events of interest are

$$A: \{\text{The teacher "applies" a severe shock (450 volts).}\}$$
$$B: \{\text{The learner protests verbally prior to receiving the shock.}\}$$

A recent application of Milgram's shock study revealed that $P(B) = .5$ and $P(A|B) = .7$. On the basis of this information, what is the probability that a learner will protest verbally and a teacher will apply a severe shock? That is, find $P(A \cap B)$.

Solution We want to calculate $P(A \cap B)$. Using the formula for the multiplicative rule, we obtain

$$P(A \cap B) = P(B) \, P(A|B) = (.5)(.7) = .35$$

Thus, about 35% of the time, the learner will give verbal protest and the teacher will apply a severe shock.

Look Back The multiplicative rule can be expressed in two ways: $P(A \cap B) = P(A) \, P(B|A)$, or $P(A \cap B) = P(B) \, P(A|B)$. Select the formula that involves a given event for which you know the probability (e.g., event B in the example).

■ **Now Work Exercise 3.81**

Intersections often contain only a few sample points. In this case, the probability of an intersection is easy to calculate by summing the appropriate sample point probabilities. However, the formula for calculating intersection probabilities is invaluable when the intersection contains numerous sample points, as the next example illustrates.

Example 3.17 **Applying The Multiplicative Rule—Study of Welfare Workers**	**Problem** A county welfare agency employs 10 welfare workers who interview prospective food stamp recipients. Periodically, the supervisor selects, at random, the forms completed by two workers and subsequently audits them for illegal deductions. Unknown to the supervisor, three of the workers have regularly been giving illegal deductions to applicants. What is the probability that both of the workers chosen have been giving illegal deductions?

Solution Define the following two events:

$$A: \{\text{First worker selected gives illegal deductions.}\}$$
$$B: \{\text{Second worker selected gives illegal deductions.}\}$$

We want to find the probability that both workers selected have been giving illegal deductions. This event can be restated as {First worker gives illegal deductions *and* second worker gives illegal deductions.}. Thus, we want to find the probability of the intersection $A \cap B$. Applying the multiplicative rule, we have

$$P(A \cap B) = P(A)P(B|A)$$

To find $P(A)$, it is helpful to consider the experiment as selecting 1 worker from the 10. Then the sample space for the experiment contains 10 sample points (representing the 10 welfare workers), in which the 3 workers giving illegal deductions are denoted by the symbol I (I_1, I_2, I_3) and the 7 workers not giving illegal deductions are denoted by the symbol N ($N_1, \ldots, N_7$). The resulting Venn diagram is illustrated in Figure 3.16.

Since the first worker is selected at random from the 10, it is reasonable to assign equal probabilities to the 10 sample points. Thus, each sample point has a probability of $^1/_{10}$. The sample points in event A are $\{I_1, I_2, I_3\}$—the three workers who are giving illegal deductions. Thus,

$$P(A) = P(I_1) + P(I_2) + P(I_3) = \frac{1}{10} + \frac{1}{10} + \frac{1}{10} = \frac{3}{10}$$

To find the conditional probability $P(B|A)$, we need to alter the sample space S. Since we know that A has occurred (i.e., the first worker selected is giving illegal deductions), only 2 of the 9 remaining workers in the sample space are giving illegal deductions. The Venn diagram for this new sample space is shown in Figure 3.17.

Each of these nine sample points is equally likely, so each is assigned a probability of $^1/_9$. Since the event $B|A$ contains the sample points $\{I_1, I_2\}$, we have

$$P(B|A) = P(I_1) + P(I_2) = \frac{1}{9} + \frac{1}{9} = \frac{2}{9}$$

Substituting $P(A) = {}^3/_{10}$ and $P(B|A) = {}^2/_9$ into the formula for the multiplicative rule, we find that

$$P(A \cap B) = P(A)P(B|A) = \left(\frac{3}{10}\right)\left(\frac{2}{9}\right) = \frac{6}{90} = \frac{1}{15}$$

Thus, there is a 1-in-15 chance that both workers chosen by the supervisor have been giving illegal deductions to food stamp recipients.

Look Back The key words *both* and *and* in the statement "both A and B occur" imply an intersection of two events. This, in turn, implies that we should *multiply* probabilities to obtain the answer.

■ **Now Work Exercise 3.77**

Figure 3.16
Venn diagram for finding $P(A)$

Figure 3.17
Venn diagram for finding $P(B|A)$

The sample-space approach is only one way to solve the problem posed in Example 3.17. An alternative method employs the **tree diagram** (introduced in Example 3.1). Tree diagrams are helpful for calculating the probability of an intersection.

To illustrate, a tree diagram for Example 3.17 is displayed in Figure 3.18. The tree begins at the far left with two branches. These branches represent the two possible outcomes N (no illegal deductions) and I (illegal deductions) for the first worker

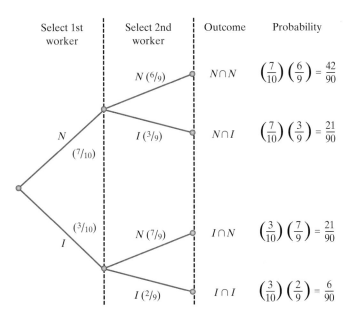

Figure 3.18
Tree diagram for Example 3.17

selected. The unconditional probability of each outcome is given (in parentheses) on the appropriate branch. That is, for the first worker selected, $P(N) = \frac{7}{10}$ and $P(I) = \frac{3}{10}$. (These unconditional probabilities can be obtained by summing sample point probabilities as in Example 3.17.)

The next level of the tree diagram (moving to the right) represents the outcomes for the second worker selected. The probabilities shown here are conditional probabilities, since the outcome for the first worker is assumed to be known. For example, if the first worker is giving illegal deductions (I), the probability that the second worker is also giving illegal deductions (I) is $\frac{2}{9}$, because, of the 9 workers left to be selected, only 2 remain who are giving illegal deductions. This conditional probability, $\frac{2}{9}$, is shown in parentheses on the bottom branch of Figure 3.18.

Finally, the four possible outcomes of the experiment are shown at the end of each of the four tree branches. These events are intersections of two events (outcome of first worker *and* outcome of second worker). Consequently, the multiplicative rule is applied to calculate each probability, as shown in Figure 3.18. You can see that the intersection $\{I \cap I\}$—the event that both workers selected are giving illegal deductions—has probability $\frac{6}{90} = \frac{1}{15}$, which is the same as the value obtained in Example 3.17.

In Section 3.5, we showed that the probability of event A may be substantially altered by the knowledge that an event B has occurred. However, this will not always be the case; in some instances, the assumption that event B has occurred will *not* alter the probability of event A at all. When this occurs, we say that the two events A and B are *independent events*.

Events A and B are **independent events** if the occurrence of B does not alter the probability that A has occurred; that is, events A and B are independent if

$$P(A \mid B) = P(A)$$

When events A and B are independent, it is also true that

$$P(B \mid A) = P(B)$$

Events that are not independent are said to be **dependent**.

Example 3.18 **Checking for Independence— Die-Tossing Experiment** **Figure 3.19** Venn diagram for die-toss experiment	**Problem** Consider the experiment of tossing a fair die, and let $A = \{$Observe an even number.$\}$ $B = \{$Observe a number less than or equal to 4.$\}$ Are A and B independent events?

Solution The Venn diagram for this experiment is shown in Figure 3.19. We first calculate

$$P(A) = P(2) + P(4) + P(6) = \frac{1}{2}$$

$$P(B) = P(1) + P(2) + P(3) + P(4) = \frac{4}{6} = \frac{2}{3}$$

$$P(A \cap B) = P(2) + P(4) = \frac{2}{6} = \frac{1}{3}$$

Now, assuming that B has occurred, we see that the conditional probability of A given B is

$$P(A \mid B) = \frac{P(A \cap B)}{P(B)} = \frac{\frac{1}{3}}{\frac{2}{3}} = \frac{1}{2} = P(A)$$

Thus, assuming that the occurrence of event B does not alter the probability of observing an even number, that probability remains $\frac{1}{2}$. Therefore, the events A and B are independent.

Look Back Note that if we calculate the conditional probability of B given A, our conclusion is the same:

$$P(B\,|\,A) = \frac{P(A \cap B)}{P(A)} = \frac{{}^{1}/_{3}}{{}^{1}/_{2}} = \frac{2}{3} = P(B)$$

■

Example	**3.19**

Checking for Independence— Consumer Complaint Study

Problem Refer to the consumer product complaint study in Example 3.15. The percentages of complaints of various types during and after the guarantee period are shown in Table 3.6 (p. 175). Define the following events:

A: {The cause of the complaint is the appearance of the product.}

B: {The complaint occurred during the guarantee term.}

Are A and B independent events?

Solution Events A and B are independent if $P(A\,|\,B) = P(A)$. In Example 3.15, we calculated $P(A\,|\,B)$ to be .51, and from Table 3.6, we see that

$$P(A) = .32 + .03 = .35$$

Therefore, $P(A\,|\,B)$ is not equal to $P(A)$, and A and B are dependent events.

■ **Now Work Exercise 3.73c**

To gain an intuitive understanding of independence, think of situations in which the occurrence of one event does not alter the probability that a second event will occur. For example, new medical procedures are often tested on laboratory animals. The scientists conducting the tests generally try to perform the procedures on the animals so that the results for one animal do not affect the results for the others. That is, the event that the procedure is successful on one animal is *independent* of the result for another. In this way, the scientists can get a more accurate idea of the efficacy of the procedure than if the results were dependent, with the success or failure for one animal affecting the results for other animals.

As a second example, consider an election poll in which 1,000 registered voters are asked their preference between two candidates. Pollsters try to use procedures for selecting a sample of voters so that the responses are independent. That is, the objective of the pollster is to select the sample so that one polled voter's preference for candidate A does not alter the probability that a second polled voter prefers candidate A.

Now consider the world of sports. Do you think the results of a batter's successive trips to the plate in baseball, or of a basketball player's successive shots at the basket, are independent? For example, if a basketball player makes two successive shots, is the probability of making the next shot altered from its value if the result of the first shot is not known? If a player makes two shots in a row, the probability of a third successful shot is likely to be different from what we would assign if we knew nothing about the first two shots. Why should this be so? Research has shown that many such results in sports tend to be *dependent* because players (and even teams) tend to get on "hot" and "cold" streaks, during which their probabilities of success may increase or decrease significantly.

We will make three final points about independence.

Point 1: **The property of independence**, unlike the property of mutual exclusivity, generally **cannot be shown on, or gleaned from, a Venn diagram**. This means that *you can't trust your intuition*. In general, the only way to check for independence is by performing the calculations of the probabilities in the definition.

Figure 3.20
Mutually exclusive events are dependent events

Point 2: Suppose that events A and B are mutually exclusive, as shown in Figure 3.20, and that both events have nonzero probabilities. Are these events independent or dependent? That is, does the assumption that B occurs alter the probability of the occurrence of A? It certainly does, because if we assume that B has occurred, it is impossible for A to have occurred simultaneously. That is, $P(A|B) = 0$. Thus, **mutually exclusive events are dependent events**, since $P(A) \neq P(A|B)$.

Point 3: The probability of the intersection of independent events is very easy to calculate. Referring to the formula for calculating the probability of an intersection, we find that

$$P(A \cap B) = P(A)P(B|A)$$

Thus, since $P(B|A) = P(B)$ when A and B are independent, we have the following useful rule:

Probability of Intersection of Two Independent Events

If events A and B are independent, then the probability of the intersection of A and B equals the product of the probabilities of A and B; that is,

$$P(A \cap B) = P(A)P(B)$$

The converse is also true: If $P(A \cap B) = P(A)P(B)$, then events A and B are independent.

In the die-toss experiment, we showed in Example 3.18 that the two events
A: {Observe an even number.}   and B: {Observe a number less than or equal to 4.}
are independent if the die is fair. Thus,

$$P(A \cap B) = P(A)P(B) = \left(\frac{1}{2}\right)\left(\frac{2}{3}\right) = \frac{1}{3}$$

This result agrees with the one we obtained in the example:

$$P(A \cap B) = P(2) + P(4) = \frac{2}{6} = \frac{1}{3}$$

Example   **3.20**

Probability of Independent Events Occurring Simultaneously—Divorced Couples Study

Problem Recall from Example 3.5 (p. 153) that the American Association for Marriage and Family Therapy (AAMFT) found that 25% of divorced couples are classified as "fiery foes" (i.e., they communicate through their children and are hostile toward each other).

a. What is the probability that in a sample of 2 divorced couples, both are classified as "fiery foes"?

b. What is the probability that in a sample of 10 divorced couples, all 10 are classified as "fiery foes"?

Solution

a. Let F_1 represent the event that divorced couple 1 is classified as a pair of "fiery foes" and F_2 represent the event that divorced couple 2 is also classified as a pair of "fiery foes." Then the event that *both* couples are "fiery foes" is the intersection of the two events, $F_1 \cap F_2$. On the basis of the AAMFT survey, which found that 25% of divorced couples are "fiery foes," we could reasonably conclude that $P(F_1) = .25$ and $P(F_2) = .25$. However, in order to compute the probability of $F_1 \cap F_2$ from the multiplicative rule, we must make the assumption that the two events are independent. Since the classification of any divorced couple is not likely to affect the classification of another divorced couple, this assumption is reasonable. Assuming independence, we have

$$P(F_1 \cap F_2) = P(F_1)P(F_2) = (.25)(.25) = .0625$$

b. To see how to compute the probability that 10 of 10 divorced couples will all be classified as "fiery foes," first consider the event that 3 of 3 couples are "fiery foes." If F_3 represents the event that the third divorced couple are "fiery foes," then we want to compute the probability of the intersection $F_1 \cap F_2 \cap F_3$. Again assuming independence of the classifications, we have

$$P(F_1 \cap F_2 \cap F_3) = P(F_1)P(F_2)P(F_3) = (.25)(.25)(.25) = .015625$$

Similar reasoning leads us to the conclusion that the intersection of 10 such events can be calculated as follows:

$$P(F_1 \cap F_2 \cap F_3 \cap \ldots \cap F_{10}) = P(F_1)P(F_2)\ldots P(F_{10}) = (.25)^{10} = .000001$$

Thus, the probability that 10 of 10 divorced couples sampled are all classified as "fiery foes" is 1 in 1 million, assuming that the probability of each couple being classified as "fiery foes" is .25 and that the classification decisions are independent.

Look Back The very small probability in part **b** makes it extremely unlikely that 10 of 10 divorced couples are "fiery foes." If this event should actually occur, we should question the probability of .25 provided by the AAMFT and used in the calculation—it is likely to be much higher. (This conclusion is another application of the rare-event approach to statistical inference.)

■ **Now Work Exercise 3.89**

Statistics IN Action | **Revisited** | **The Probability of Winning Cash 3 or Play 4**

In addition to the biweekly Lotto 6/53, the Florida Lottery runs several other games. Two popular daily games are Cash 3 and Play 4. In Cash 3, players pay $1 to select three numbers in order, where each number ranges from 0 to 9. If the three numbers selected (e.g., 2–8–4) match exactly the order of the three numbers drawn, the player wins $500. Play 4 is similar to Cash 3, but players must match four numbers (each number ranging from 0 to 9). For a $1 Play 4 ticket (e.g., 3–8–3–0), the player will win $5,000 if the numbers match the order of the four numbers drawn.

During the official drawing for Cash 3, 10 table tennis balls numbered 0, 1, 2, 3, 4, 5, 6, 7, 8, and 9 are placed into each of three chambers. The balls in the first chamber are colored pink, the balls in the second chamber are blue, and the balls in the third chamber are yellow. One ball of each color is randomly drawn, with the official order as pink–blue–yellow. In Play 4, a fourth chamber with orange balls is added, and the official order is pink–blue–yellow–orange. Since the draws of the colored balls are random and independent, we can apply an extension of the probability rule for the intersection of two independent events to

find the odds of winning Cash 3 and Play 4. The probability of matching a numbered ball being drawn from a chamber is 1/10; therefore,

$$P(\text{Win Cash 3}) = P(\text{match pink } and \text{ match blue} \\ and \text{ match yellow})$$
$$= P(\text{match pink}) \times P(\text{match blue}) \times P(\text{match yellow})$$
$$= (1/10)(1/10)(1/10) = 1/1000 = .001$$

$$P(\text{Win Play 4}) = P(\text{match pink } and \text{ match blue} \\ and \text{ match yellow } and \text{ match orange})$$
$$= P(\text{match pink}) \times P(\text{match blue}) \times P(\text{match yellow}) \times P(\text{match orange})$$
$$= (1/10)(1/10)(1/10)$$
$$= 1/10,000 = .0001$$

(Continued)

Exercises 3.68–3.103

Understanding the Principles

3.68 Explain the difference between an unconditional probability and a conditional probability.

3.69 Give the multiplicative rule of probability for
　a. two independent events.
　b. any two events.

3.70 Give the formula for finding $P(B|A)$.

3.71 Defend or refute each of the following statements:
　a. Dependent events are always mutually exclusive.
　b. Mutually exclusive events are always dependent.
　c. Independent events are always mutually exclusive.

Learning the Mechanics

3.72 For two events, A and B, $P(A) = .4, P(B) = .4$, and $P(A|B) = .75$.
　a. Find $P(A \cap B)$.
　b. Find $P(B|A)$.

3.73 For two events, A and B, $P(A) = .4, P(B) = .2$, and
[NW] $P(A \cap B) = .1$.
　a. Find $P(A|B)$.
　b. Find $P(B|A)$.
　c. Are A and B independent events?

3.74 An experiment results in one of three mutually exclusive events A, B, and C. It is known that $P(A) = .50$, $P(B) = .30$, and $P(C) = .20$, Find each of the following probabilities:
　a. $P(A \cup B)$
　b. $P(A \cap B)$
　c. $P(A|B)$
　d. $P(B \cup C)$
　e. Are B and C independent events? Explain.

3.75 For two independent events, A and B, $P(A) = .2$ and $P(B) = .6$.
　a. Find $P(A \cap B)$.
　b. Find $P(A|B)$.
　c. Find $P(A \cup B)$.

3.76 An experiment results in one of five sample points with probabilities $P(E_1) = .22, P(E_2) = .21, P(E_3) = .21$, $P(E_4) = .24$, and $P(E_5) = .12$. The following events have been defined:

$$A: \{E_1, E_3\}$$
$$B: \{E_2, E_3, E_4\}$$
$$C: \{E_1, E_5\}$$

　a. Find $P(A)$.
　b. Find $P(B)$.
　c. Find $P(A \cap B)$.
　d. Find $P(A|B)$.
　e. Find $P(B \cap C)$.

　f. Find $P(C|B)$.
　g. Consider each pair of events A and B, A and C, and B and C. Are any of the pairs of events independent? Why?

3.77 Consider the experiment defined by the accompany-
[NW] ing Venn diagram, with the sample space S containing five sample points. The sample points are assigned the following probabilities: $P(E_1) = .1, P(E_2) = .1$, $P(E_3) = .2, P(E_4) = .5, P(E_5) = .1$.

　a. Calculate $P(A), P(B)$, and $P(A \cap B)$.
　b. Suppose we know that event A has occurred, so the reduced sample space consists of the three sample points in A: E_1, E_2, and E_3. Use the formula for conditional probability to determine the probabilities of these three sample points given that A has occurred. Verify that the conditional probabilities are in the same ratio to one another as the original sample point probabilities and that they sum to 1.
　c. Calculate the conditional probability $P(B|A)$ in two ways: First, sum $P(E_2|A)$ and $P(E_3|A)$, since these sample points represent the event that B occurs given that A has occurred. Second, use the formula for conditional probability:

$$P(B|A) = \frac{P(A \cap B)}{P(A)}$$

　Verify that the two methods yield the same result.

3.78 A sample space contains six sample points and events A, B, and C, as shown in the Venn diagram below. The probabilities of the sample points are $P(1) = .20, P(2) = .05$, $P(3) = .30, P(4) = .10, P(5) = .10$, and $P(6) = .25$.

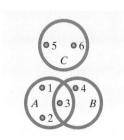

a. Which pairs of events, if any, are mutually exclusive? Why?

b. Which pairs of events, if any, are independent? Why?

c. Find $P(A \cup B)$ by adding the probabilities of the sample points and then by using the additive rule. Verify that the answers agree. Repeat for $P(A \cup C)$.

3.79 Two fair dice are tossed, and the following events are defined:

A: {The sum of the numbers showing is odd.}
B: {The sum of the numbers showing is 9, 11, or 12.}

Are events A and B independent? Why?

3.80 A box contains two white, two red, and two blue poker chips. Two chips are randomly chosen without replacement, and their colors are noted. Define the following events:

A: {Both chips are of the same color.}
B: {Both chips are red.}
C: {At least one chip is red or white.}

Find $P(B|A)$, $P(B|A^c)$, $P(B|C)$, $P(A|C)$, and $P(C|A^c)$.

Ⓓ Applet Exercise 3.5

Use the applet entitled *Simulating the Probability of Rolling a 6* to simulate conditional probabilities. Begin by running the applet twice with $n = 10$, without resetting between runs. The data on your screen represent 20 rolls of a die. The diagram above the *Roll* button shows the frequency of each of the six possible outcomes. Use this information to find each of the following probabilities:

a. The probability of 6 given that the outcome is 5 or 6
b. The probability of 6 given that the outcome is even
c. The probability of 4 or 6 given that the outcome is even
d. The probability of 4 or 6 given that the outcome is odd

Applying the Concepts—Basic

3.81 **Blood diamonds.** According to *Global Research News* [NW] (Mar. 4, 2014), one-fourth of all rough diamonds produced in the world are *blood diamonds*. (Any diamond that is mined in a war zone—often by children—in order to finance a warlord's activity, an insurgency, or an invading army's effort is considered a blood diamond.) Also, 90% of the world's rough diamonds are processed in Surat, India, and of these diamonds, one-third are blood diamonds.

a. Find the probability that a rough diamond is not a blood diamond.
b. Find the probability that a rough diamond is processed in Surat and is a blood diamond.

3.82 **Do social robots walk or roll?** Refer to the *International Conference on Social Robotics* (Vol. 6414, 2010) study of the trend in the design of social robots, Exercise 3.17 (p. 157). Recall that in a random sample of 106 social robots, 63 were built with legs only, 20 with wheels only, 8 with both legs and wheels, and 15 with neither legs nor wheels. If a social robot is designed with wheels, what is the probability that the robot also has legs?

3.83 **Crop damage by wild boars.** Refer to the *Current Zoology* (Apr. 2014) study of crop damage by wild boars, Exercise 3.18 (p. 158). Recall that researchers identified 157 incidents of crop damage in the study area caused by wild boars over a five-year period. The table giving the types of crops destroyed and corresponding percentage of incidents is reproduced

below. The researchers also determined that of the cereal crops damaged, 71% involved wheat, barley, or oats. For one randomly selected incident, what is the likelihood that the wild boars damaged a wheat, barley, or oats cereal crop?

Type	Percentage
Cereals	45%
Orchards	5
Legumes	20
Vineyards	15
Other crops	15
Total	100%

3.84 **Cardiac stress testing.** In addition to standard exercise electrocardiography (SEE), two widely applied methods of stress testing of cardiac patients are stress echocardiography (SE) and single-photon emission computed tomography (SPECT). These methods were evaluated in the *American Journal of Medical Quality* (Mar./Apr. 2014). A sample of 866 heart catheterization patients in the Penn State Hershey Medical Center all received some form of stress testing prior to the catheterization. Of these, SE was performed on 272 patients, SPECT on 560 patients, and SEE on 38 patients. Four patients received both SE and SPECT but not SEE. Consider a patient randomly selected from those in the study.

a. What is the probability that the patient received SE prior to the catheterization?
b. What is the probability that the patient received both SE and SPECT prior to the catheterization?
c. Given that the patient received SPECT, what is the probability the patient also received SE prior to the catheterization?

3.85 **National firearms survey.** The Harvard School of Public Health conducted a study of privately held firearm stock in the United States. In a representative household telephone survey of 2,770 adults, 26% reported that they own at least one gun (*Injury Prevention*, Jan. 2007). Of those that own a gun, 5% own a handgun. Suppose one of the 2,770 adults surveyed is randomly selected.

a. What is the probability that the adult owns at least one gun?
b. What is the probability that the adult owns at least one gun and the gun is a handgun?

3.86 **Guilt in decision making.** Refer to the *Journal of Behavioral Decision Making* (Jan. 2007) study of the effect of guilt emotion on how a decision maker focuses on the problem, Exercise 3.59 (p. 170). The results (number responding in each category) for the 171 study participants are reproduced in the accompanying table. Suppose one of the 171 participants is selected at random.

Emotional State	Choose Stated Option	Do Not Choose Stated Option	Totals
Guilt	45	12	57
Anger	8	50	58
Neutral	7	49	56
Totals	60	111	171

Based on Gangemi, A., and Mancini, F. "Guilt and focusing in decision-making." *Journal of Behavioral Decision Making*, Vol. 20, Jan. 2007 (Table 2).

a. Given the respondent is assigned to the guilty state, what is the probability that the respondent chooses the stated option?

b. If the respondent does not choose to repair the car, what is the probability of the respondent being in the anger state?

c. Are the events {repair the car} and {guilty state} independent?

3.87 **Speeding linked to fatal car crashes.** According to the National Highway Traffic Safety Administration's National Center for Statistics and Analysis (NCSA), "Speeding is one of the most prevalent factors contributing to fatal traffic crashes" (*NHTSA Technical Report*, Aug. 2005). The probability that speeding is a cause of a fatal crash is .3. Furthermore, the probability that speeding and missing a curve are causes of a fatal crash is .12. Given that speeding is a cause of a fatal crash, what is the probability that the crash occurred on a curve?

3.88 **Appraisals and negative emotions.** According to psychological theory, each negative emotion one has is linked to some thought (appraisal) about the environment. This theory was tested in the journal *Cognition and Emotion* (Vol. 24, 2010). Undergraduate students were asked to respond to events that occurred within the past 15 minutes. Each student rated the negative emotions he or she felt (e.g., anger, sadness, fear, or guilt) and gave an appraisal of the event (e.g., unpleasant or unfair). A total of 10,797 responses were evaluated in the study, of which 594 indicated a perceived unfairness appraisal. Of these unfairness appraisals, the students reported feeling angry in 127 of them. Consider a response randomly selected from the study.

a. What is the probability that the response indicates a perceived unfairness appraisal?

b. Given a perceived unfairness appraisal, what is the probability of an angry emotion?

Applying the Concepts—Intermediate

3.89 **Health risks to beachgoers.** Refer to the University of Florida study of health risks to beachgoers, Exercise 3.24 (p. 158). According to the study, 6 out of 1,000 people exposed to wet sand for a 10-minute period will acquire gastroenteritis and 7 out of 1,000 people exposed to ocean water for a 10-minute period will acquire gastroenteritis. Suppose you and a friend go to the beach. Over the next 10 minutes, you play in the wet sand while your friend swims in the ocean. What is the probability that at least one of you acquires gastroenteritis? (What assumption did you need to make in order to find the probability?)

3.90 **Sleep apnea and sleep stage transitioning.** Refer to the sleep apnea study published in *Chance* (Winter 2009), Exercise 3.57 (p. 169). Recall that the various stages of sleep for a large group of sleep apnea patients were monitored in 30-second "epochs," and the sleep stage for the previous and current epoch was determined. A summary table is reproduced in the next column. One of the epochs in the study is selected.

a. Given that the previous sleep stage for the epoch was the Wake state, what is the probability that the current sleep stage is REM?

b. Given that the current sleep stage for the epoch is REM, what is the probability that the previous sleep stage was not the Wake state?

Current Stage	Previous Sleep Stage		
	Non-REM	REM	Wake
Non-REM	31,880	160	1,733
REM	346	7,609	175
Wake	1,588	358	6,079
Totals	33,814	8,127	7,987

Based on Caffo, B. S., et al. "An overview of observational sleep research with application to sleep stage transitioning." *Chance*, Vol. 22, No. 1, Winter 2009 (Table 2).

c. Are the events {previous stage is REM} and {current stage is REM} mutually exclusive?

d. Are the events {previous stage is REM} and {current stage is REM} independent?

e. Are the events {previous stage is Wake} and {current stage is Wake} independent?

3.91 **Firefighters' use of gas detection devices.** Two deadly gases that can be present in fire smoke are hydrogen cyanide and carbon monoxide. *Fire Engineering* (Mar. 2013) reported the results of a survey of 244 firefighters conducted by the Fire Smoke Coalition. The purpose of the survey was to assess the base level of knowledge of firefighters regarding the use of gas detection devices at the scene of a fire. The survey revealed the following: Eighty percent of firefighters had no standard operating procedures (SOP) for detecting/monitoring hydrogen cyanide in fire smoke; 49% had no SOP for detecting/ monitoring carbon monoxide in fire smoke. Assume that 94% of firefighters had no SOP for detecting either hydrogen cyanide or carbon monoxide in fire smoke. What is the probability that a firefighter has no SOP for detecting hydrogen cyanide and no SOP for detecting carbon monoxide in fire smoke?

3.92 **Compensatory advantage in education.** According to *Sociology of Education* (Apr. 2014), compensatory advantage occurs when individuals from privileged backgrounds are less dependent on prior negative outcomes than individuals from disadvantaged families. The researcher applied this theory to educational success. Let S_t represent the event that a student is successfully promoted to the next grade level in year t, and let F_t represent the event that a student fails to be promoted to the next grade level in year t. Now consider two probabilities: P_1 = the probability that a student is successfully promoted to the next grade level in year $t + 1$ given that the student was successfully promoted the previous year (i.e., in year t) and P_2 = the probability that a student is successfully promoted to the next grade level in year $t + 1$ given that the student failed in the previous year. Compensatory advantage theory states that the difference between the two probabilities, $P_1 - P_2$, is greater for disadvantaged families (say, the lower-class L) than for privileged families (say, the upper-class U).

a. Write P_1 as a conditional probability using symbols.

b. Write P_2 as a conditional probability using symbols.

c. The tables on p. 185 give the grade promotion results (number of students) for both upper- and lower-class students at a particular school. Does the compensatory advantage theory hold at this school?

Results for Exercise 3.92

Upper Class	Successfully promoted (S_t)	Failure (F_t)
Successfully promoted (S_{t+1})	45	8
Failure (F_{t+1})	5	2
Totals	50	10

Lower Class	Successfully promoted (S_t)	Failure (F_t)
Successfully promoted (S_{t+1})	90	15
Failure (F_{t+1})	10	5
Totals	100	20

3.93 **Are you really being served red snapper?** Red snapper is a rare and expensive reef fish served at upscale restaurants. Federal law prohibits restaurants from serving a cheaper, look-alike variety of fish (e.g., vermillion snapper or lane snapper) to customers who order red snapper. Researchers at the University of North Carolina used DNA analysis to examine fish specimens labeled "red snapper" that were purchased from vendors across the country (*Nature*, July 15, 2004). The DNA tests revealed that 77% of the specimens were not red snapper, but the cheaper, look-alike variety of fish.

    a. Assuming that the results of the DNA analysis are valid, what is the probability that you are actually served red snapper the next time you order it at a restaurant?

    b. If there are five customers at a restaurant, all who have ordered red snapper, what is the probability that at least one customer is actually served red snapper?

3.94 **Fighting probability of fallow deer bucks.** Refer to the *Aggressive Behavior* (Jan./Feb. 2007) study of fallow deer bucks fighting during the mating season, presented in Exercise 3.61 (p. 170). Recall that researchers recorded 167 encounters between two bucks, one of which clearly initiated the encounter with the other. A summary of the fight status of the initiated encounters is provided in the accompanying table. Suppose we select 1 of these 167 encounters and note the outcome (fight status and winner).

    a. Given that a fight occurs, what is the probability that the initiator wins?

    b. Given no fight, what is the probability that the initiator wins?

    c. Are the events "no fight" and "initiator wins" independent?

	Initiator Wins	No Clear Winner	Initiator Loses	Totals
Fight	26	23	15	64
No Fight	80	12	11	103
Totals	106	35	26	167

Based on Bartos, L. et al. "Estimation of the probability of fighting in fallow deer (*Dama dama*) during the rut." *Aggressive Behavior*, Vol. 33, Jan./Feb. 2007, pp. 7–13.

3.95 **Extinct New Zealand birds.** Refer to the *Evolutionary Ecology Research* (July 2003) study of the patterns of extinction in the New Zealand bird population, presented in Exercise 2.24 (p. 70). Consider the data on extinction status (extinct, absent from island, present) for the 132 bird species. The data are summarized in the accompanying MINITAB printout. Suppose you randomly select 10 of the 132 bird species (without replacement) and record the extinction status of each.

    a. What is the probability that the first species you select is extinct? (*Note*: Extinct = Yes on the MINITAB printout.)

    b. Suppose the first 9 species you select are all extinct. What is the probability that the 10th species you select is extinct?

Tally for Discrete Variables: Extinct

```
Extinct   Count   Percent
 Absent      16     12.12
    No       78     59.09
   Yes       38     28.79
    N=      132
```

3.96 **Muscle, fat, and bone issues while aging.** In *Ageing Research Reviews* (May 2014), nutritionists published a study of the health issues faced by aging adults. Three of the most common health conditions are obesity (high body fat mass), osteoporosis (low bone mineral density in neck and spine), and sarcopenia (low skeletal muscle mass index). The researchers provided names for aging adults who suffer from one or more of these conditions. These are shown in the accompanying Venn diagram. Statistics from the CDC are used to estimate the following probabilities:

$P(\text{Obesity}) = .35$
$P(\text{Osteoporosis}) = .30$
$P(\text{Sarcopenia}) = .15$
$P(\text{Osteopenic Obesity}) = .01$
$P(\text{Osteopenic Sarcopenia}) = .03$
$P(\text{Sarcopenic Obesity}) = .05$
$P(\text{Osteo-sarcopenic Obesity}) = .02$

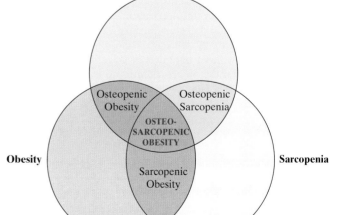

a. Find the probability that an adult suffers from both obesity and sarcopenia.

b. Find the probability that an adult suffers from either obesity or sarcopenia.

3.97 Ambulance response time. *Geographical Analysis* (Jan. 2010) presented a study of Emergency Medical Services (EMS) ability to meet the demand for an ambulance. In one example, the researchers presented the following scenario. An ambulance station has one vehicle and two demand locations, A and B. The probability that the ambulance can travel to a location in under eight minutes is .58 for location A and .42 for location B. The probability that the ambulance is busy at any point in time is .3.

a. Find the probability that EMS can meet demand for an ambulance at location A.

b. Find the probability that EMS can meet demand for an ambulance at location B.

3.98 Intrusion detection systems. A computer intrusion detection system (IDS) is designed to provide an alarm whenever there is unauthorized access into a computer system. A probabilistic evaluation of a system with two independently operating intrusion detection systems (a double IDS) was published in the *Journal of Research of the National Institute of Standards and Technology* (Nov.–Dec. 2003). Consider a double IDS with system A and system B. If there is an intruder, system A sounds an alarm with probability .9 and system B sounds an alarm with probability .95. If there is no intruder, the probability that system A sounds an alarm (i.e., a false alarm) is .2 and the probability that system B sounds an alarm is .1.

a. Use symbols to express the four probabilities just given.

b. If there is an intruder, what is the probability that both systems sound an alarm?

c. If there is no intruder, what is the probability that both systems sound an alarm?

d. Given that there is an intruder, what is the probability that at least one of the systems sounds an alarm?

3.99 Detecting traces of TNT. University of Florida researchers in the Department of Materials Science and Engineering have invented a technique that rapidly detects traces of TNT (*Today*, Spring 2005). The method, which involves shining a laser on a potentially contaminated object, provides instantaneous results and gives no false positives. In this application, a false positive would occur if the laser detected traces of TNT when, in fact, no TNT were actually present on the object. Let *A* be the event that the laser light detects traces of TNT. Let *B* be the event that the object contains no traces of TNT. The probability of a false positive is 0. Write this probability in terms of *A* and *B*, using symbols such as " ∪ ", "∩", and " | ".

Applying the Concepts—Advanced

3.100 Random mutation of cells. *Chance* (Spring 2010) presented an article on the random mutation hypothesis developed by microbiologists. Under this hypothesis, when a wild-type organic cell (e.g., a bacteria cell) divides, there is a chance that at least one of the two "daughter" cells is a mutant. When a mutant cell divides, both offspring will be mutant. The schematic in the next column shows a possible

pedigree from a single cell that has divided. Note that one "daughter" cell is mutant (◉) and one is a normal cell (○).

a. Consider a single, normal cell that divides into two offspring. List the different possible pedigrees.

b. Assume that a "daughter" cell is equally likely to be mutant or normal. What is the probability that a single, normal cell that divides into two offspring will result in at least one mutant cell?

c. Now assume that the probability of a mutant "daughter" cell is .2. What is the probability that a single, normal cell that divides into two offspring will result in at least one mutant cell?

d. The schematic below shows a possible second-generation pedigree from a single cell that has divided. Note that the first-generation mutant cell automatically produces two mutant cells in the second generation. List the different possible second-generation pedigrees. (*Hint:* Use your answer to part **a.**)

e. Assume that a "daughter" cell is equally likely to be mutant or normal. What is the probability that a single, normal cell that divides into two offspring will result in at least one mutant cell after the second generation?

3.101 Testing a psychic's ability. Consider an experiment in which 10 identical small boxes are placed side by side on a table. A crystal is placed at random inside one of the boxes. A self-professed "psychic" is asked to pick the box that contains the crystal.

a. If the "psychic" simply guesses, what is the probability that she picks the box with the crystal?

b. If the experiment is repeated seven times, what is the probability that the "psychic" guesses correctly at least once?

c. A group called the Tampa Bay Skeptics tested a self-proclaimed "psychic" by administering the preceding experiment seven times. The "psychic" failed to pick the correct box all seven times (*Tampa Tribune*, Sept. 20, 1998). What would you infer about this person's psychic ability?

3.102 Risk of a natural-gas pipeline accident. *Process Safety Progress* (Dec. 2004) published a risk analysis for a natural-gas pipeline between Bolivia and Brazil. The most likely scenario for an accident would be natural-gas leakage from a hole in the pipeline. The probability that the leak ignites immediately (causing a jet fire) is .01. If the leak does not immediately ignite, it may result in the delayed ignition of a gas cloud. Given no immediate ignition, the probability of delayed ignition (causing a flash fire) is .01. If there is no delayed ignition, the gas cloud will disperse harmlessly. Suppose a leak occurs in the natural-gas pipeline. Find the probability that either a jet fire or a flash fire will occur. Illustrate with a tree diagram.

3.103 Encryption systems with erroneous ciphertexts. In cryptography, ciphertext is encrypted or encoded text that is unreadable by a human or computer without the proper algorithm to decrypt it into plaintext. The impact of erroneous ciphertexts on the performance of an encryption system was investigated in *IEEE Transactions on Information Forensics and Security* (Apr. 2013). For one data encryption system, the probability of receiving an erroneous ciphertext is assumed to be β, where $0 < \beta < 1$. The researchers showed that if an erroneous ciphertext occurs, the probability of an error in restoring plaintext using the decryption system is .5. When no error occurs in the received ciphertext, the probability of an error in restoring plaintext using the decryption system is $\alpha\beta$, where $0 < \alpha < 1$. Use this information to give an expression for the probability of an error in restoring plaintext using the decryption system.

3.7 Some Additional Counting Rules (Optional)

In Section 3.1, we pointed out that experiments sometimes have so many sample points that it is impractical to list them all. However, many of these experiments possess sample points with identical characteristics. If you can develop a **counting rule** to count the number of sample points for such an experiment, it can be used to aid in the solution of the problems. The combinations rule for selecting n elements from N elements without regard to order was presented in Section 3.1 (p. 155). In this optional section, we give three additional counting rules: the *multiplicative rule*, the *permutations rule*, and the *partitions rule*.

| **Example** | **3.21** |

Applying the Multiplicative Rule— Airline Shipping

Problem A product can be shipped by four airlines, and each airline can ship via three different routes.

a. How many distinct ways exist to ship the product?

b. Assuming the airline and route are selected at random, what is the probability that Route 2 is used?

Solution

a. A pictorial representation of the different ways to ship the product will aid in counting them. The tree diagram (see Section 3.6) is useful for this purpose. Consider the tree diagram shown in Figure 3.21. At the starting point (stage 1), there are four choices— the different airlines—to begin the journey. Once we have chosen an airline (stage 2), there are three choices—the different routes—to complete the shipment and reach the final destination. Thus, the tree diagram clearly shows that there are $(4)(3) = 12$ distinct ways to ship the product. These are the sample points for the experiment.

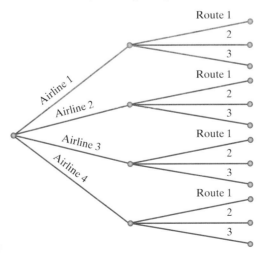

Figure 3.21

Tree diagram for shipping problem

b. Since the airline and route are randomly selected, it is reasonable to assume that the sample points are equiprobable. Thus, each has probability $1/12$. Route 2 is selected on 4 of the 12 shipping options, hence

$$P(\text{Route 2}) = 4/12 = 1/3.$$

Look Back The method of solving this example can be generalized to any number of stages with sets of different elements. This method is given by the *multiplicative rule*.

▪ **Now Work Exercise 3.114c**

> **The Multiplicative Rule**
>
> You have k sets of elements, n_1 in the first set, n_2 in the second set, ..., and n_k in the kth set. Suppose you wish to form a sample of k elements by *taking one element from each* of the k sets. Then the number of different samples that can be formed is the product
>
> $$n_1 n_2 n_3 \ldots n_k$$

Example **3.22**

Applying the Multiplicative Rule—Hiring Executives

Problem There are 20 candidates for three different executive positions: E_1, E_2, and E_3. How many different ways could you fill the positions?

Solution For this example, there are $k = 3$ sets of elements, corresponding to

Set 1: Candidates available to fill position E_1
Set 2: Candidates remaining (after filling E_1) that are available to fill E_2
Set 3: Candidates remaining (after filling E_1 and E_2) that are available to fill E_3

The numbers of elements in the sets are $n_1 = 20$, $n_2 = 19$, and $n_3 = 18$. Therefore, the number of different ways of filling the three positions is given by the multiplicative rule as $n_1 n_2 n_3 = (20)(19)(18) = 6{,}840$.

Example **3.23**

Applying the Multiplicative Rule—10 Coin Tosses

Problem Consider the experiment discussed in Example 3.10 of tossing a coin 10 times. Show how we found that there were $2^{10} = 1{,}024$ sample points for this experiment.

Solution There are $k = 10$ sets of elements for this experiment. Each set contains two elements: a head and a tail. Thus, there are

$$(2)(2)(2)(2)(2)(2)(2)(2)(2)(2) = 2^{10} = 1{,}024$$

different outcomes (sample points) of this experiment.

Example **3.24**

Applying the Permutations Rule—Selecting Soldiers for a Mission

Problem Suppose there are five dangerous military missions, each requiring one soldier. In how many different ways can 5 soldiers from a squadron of 100 be assigned to these five missions?

Solution We can solve this problem by using the multiplicative rule. The entire set of 100 soldiers is available for the first mission, and after the selection of 1 soldier for that mission, 99 are available for the second mission, etc. Thus, the total number of different ways of choosing 5 soldiers for the five missions is

$$n_1 n_2 n_3 n_4 n_5 = (100)(99)(98)(97)(96) = 9{,}034{,}502{,}400$$

Look Back The arrangement of elements in a distinct order is called a **permutation**. There are more than 9 billion different *permutations* of 5 elements (soldiers) drawn from a set of 100 elements!

> **Permutations Rule**
>
> Given a *single set* of N different elements, you wish to select n elements from the N and *arrange* them within n positions. The number of different **permutations** of the N elements taken n at a time is denoted by P_n^N and is equal to
>
> $$P_n^N = N(N-1)(N-2) \cdot \cdots \cdot (N-n+1) = \frac{N!}{(N-n)!}$$
>
> where $n! = n(n-1)(n-2) \ldots (3)(2)(1)$ and is called n *factorial*. (For example, $5! = 5 \cdot 4 \cdot 3 \cdot 2 \cdot 1 = 120$.) The quantity $0!$ is defined to be 1.

Example **3.25** **Applying the Permutations Rule—Driving Routes**	**Problem** Suppose you wish to drive, in sequence, from a starting point to each of five cities, and you wish to compare the distances—and, ultimately, the costs—of the different routes. Denote the five cities as C_1, C_2, C_3, C_4, and C_5.

a. How many different routes would you have to compare?

b. Suppose the driving distance from C_2 to C_1 is the most costly. How many of the routes include this distance?

c. If the routes are selected at random, what is the probability that the route includes driving from C_2 to C_1?

Solution

a. Represent a route moving from the starting point to C_2 to C_1 to C_3 to C_4 to C_5 as $\{C_2C_1C_3C_4C_5\}$. Then the total number of routes would equal the number of ways you could arrange $N = 5$ cities into $n = 5$ positions, where position order matters. This number is

$$P_n^N = P_5^5 = \frac{5!}{(5-5)!} = \frac{5!}{0!} = \frac{(5 \cdot 4 \cdot 3 \cdot 2 \cdot 1)}{1} = 120$$

(Recall that $0! = 1$.) This value represents the number of sample points in the experiment.

b. To determine the number of routes going directly from C_2 to C_1, first consider the different routing configurations that would include this distance. These would include all routes with C_2 in the position next to, and to the left of, C_1 as shown here:

$$\{C_2C_1\_\_\_\_\_\}, \quad \{\_\_C_2C_1\_\_\_\}, \quad \{\_\_\_\_C_2C_1\_\}, \quad \{_____C_2C_1\}$$

For each of these configurations, the remaining 3 cities can be arranged in the 3 open positions in 6 different ways, since

$$P_n^N = P_3^3 = \frac{3!}{(3-3)!} = \frac{3!}{0!} = \frac{(3 \cdot 2 \cdot 1)}{1} = 6$$

Since there are 4 configurations of this form, each with 6 routing possibilities, there are a total of $4 \cdot 6 = 24$ different routes going directly from C_2 to C_1.

c. Due to the random selection of routes, the 120 sample points (part **a**) are equiprobable, i.e., each sample point has probability 1/120. Then the probability that the route selected includes driving from C_2 to C_1 is the number of sample points that include this event (i.e., 24, as shown in part **b**) divided by the total number of sample points, 120. That is,

$$P(\text{route includes driving from } C_2 \text{ to } C_1) = \frac{24}{120} = \frac{1}{5} = .2$$

■ **Now Work Exercise 3.115c**

Example **3.26**	**Problem** Suppose you are supervising four construction workers. You must assign three to job 1 and one to job 2. In how many different ways can you make this assignment?
Applying the Partitions Rule—Assigning Workers	**Solution** To begin, suppose each worker is to be assigned to a distinct job. Then, using the multiplicative rule, you obtain $(4)(3)(2)(1) = 24$ ways of assigning the workers to four distinct jobs. The 24 ways are listed in four groups in Table 3.7 (where *ABCD* signifies that worker *A* was assigned the first distinct job, worker *B* the second, etc.).

Table 3.7   Workers Assigned to Four Jobs, Example 3.26

Group 1	Group 2	Group 3	Group 4
ABCD	*ABDC*	*ACDB*	*BCDA*
ACBD	*ADBC*	*ADCB*	*BDCA*
BACD	*BADC*	*CADB*	*CBDA*
BCAD	*BDAC*	*CDAB*	*CDBA*
CABD	*DABC*	*DACB*	*DBCA*
CBAD	*DBAC*	*DCAB*	*DCBA*

Now suppose the first three positions represent job 1 and the last position represents job 2. You can see that all the listings in group 1 represent the same outcome of the experiment of interest. That is, workers *A*, *B*, and *C* are assigned to job 1 and worker *D* to job 2. Similarly, group 2 listings are equivalent, as are group 3 and group 4 listings. Thus, there are only four different assignments of four workers to the two jobs. These assignments are shown in Table 3.8.

Table 3.8   Workers Assigned to Two Jobs, Example 3.26

Job 1	Job 2
ABC	*D*
ABD	*C*
ACD	*B*
BCD	*A*

Look Back To generalize this result, we point out that the final result can be found by

$$\frac{(4)(3)(2)(1)}{(3)(2)(1)(1)} = 4$$

The numerator, $(4)(3)(2)(1)$, is the number of different ways (*permutations*) the workers could be assigned four distinct jobs. In the denominator, the division by $(3)(2)(1)$ is to remove the duplicated permutations resulting from the fact that three workers are assigned the same job, and the division by 1 is associated with the worker assigned to job 2.

■ **Now Work Exercise 3.116**

Partitions Rule

Suppose you wish to partition a *single* set of *N* different elements into *k* sets, with the first set containing n_1 elements, the second containing n_2 elements, …, and the *k*th set containing n_k elements. Then the number of different partitions is

$$\frac{N!}{n_1! n_2! \dots n_k!}, \text{ where } n_1 + n_2 + n_3 + \cdots + n_k = N$$

Example **3.27**	**Problem** Suppose you have 12 construction workers and you wish to assign 3 to jobsite 1, 4 to jobsite 2, and 5 to jobsite 3. In how many different ways can you make this assignment?
Applying the Partitions Rule—Assigning Workers	**Solution** For this example, $k = 3$ (corresponding to the $k = 3$ different jobsites), $N = 12, n_1 = 3, n_2 = 4$, and $n_3 = 5$. Then the number of different ways to assign the workers to the jobsites is

$$\frac{N!}{n_1! n_2! n_3!} = \frac{12!}{3!4!5!} = \frac{12 \cdot 11 \cdot 10 \cdot \cdots \cdot 3 \cdot 2 \cdot 1}{(3 \cdot 2 \cdot 1)(4 \cdot 3 \cdot 2 \cdot 1)(5 \cdot 4 \cdot 3 \cdot 2 \cdot 1)} = 27,720$$

Example   3.28

Applying the Combinations Rule— Selecting Soldiers for a Mission

Problem Five soldiers from a squadron of 100 are to be chosen for a dangerous mission. In how many ways can groups of 5 be formed?

Solution This problem is equivalent to sampling $n = 5$ elements from a set of $N = 100$ elements. Here, we apply the combinations rule from Section 3.1. The number of ways is the number of possible combinations of 5 soldiers selected from 100, or

$$\binom{100}{5} = \frac{100!}{(5!)(95!)} = \frac{100 \cdot 99 \cdot 98 \cdot 97 \cdot 96 \cdot 95 \cdot 94 \cdots 2 \cdot 1}{(5 \cdot 4 \cdot 3 \cdot 2 \cdot 1)(95 \cdot 94 \cdots 2 \cdot 1)}$$

$$= \frac{100 \cdot 99 \cdot 98 \cdot 97 \cdot 96}{5 \cdot 4 \cdot 3 \cdot 2 \cdot 1} = 75{,}287{,}520$$

Look Back Compare this result with that of Example 3.24, where we found that the number of permutations of 5 elements drawn from 100 was more than 9 billion. Because the order of the elements does not affect combinations, there are *fewer* combinations than permutations.

■ **Now Work Exercise 3.115a**

When working a probability problem, you should carefully examine the experiment to determine whether you can use one or more of the rules we have discussed in this section. Let's see how these rules (summarized in the next box) can help solve a probability problem.

Summary of Counting Rules

1. *Multiplicative rule.* If you are drawing *one element from each of k sets of elements*, where the sizes of the sets are $n_1, n_2, \cdots, n_k$, then the number of different results is

$$n_1 n_2 n_3 \ldots n_k$$

2. *Permutations rule.* If you are drawing *n elements from a set of N elements and arranging the n elements in a distinct order*, then the number of different results is

$$P_n^N = \frac{N!}{(N-n)!}$$

3. *Partitions rule.* If you are partitioning the *elements of a set of N elements into k groups consisting* of $n_1, n_2, \ldots, n_k$ elements $(n_1 + \cdots + n_k = N)$, then the number of different results is

$$\frac{N!}{n_1! n_2! \cdots n_k!}$$

4. *Combinations rule.* If you are drawing *n elements from a set of N elements without regard to the order of the n elements*, then the number of different results is

$$\binom{N}{n} = \frac{N!}{n!(N-n)!}$$

[*Note:* The combinations rule is a special case of the partitions rule when $k = 2$.]

Example   **3.29**

**Applying Several
Counting Rules—Ranking
Detergents**

Problem   A consumer testing service is commissioned to rank the top 3 brands of laundry detergent. Ten brands are to be included in the study.

a. In how many different ways can the consumer testing service arrive at the final ranking?

b. If the testing service can distinguish no difference among the brands and therefore arrives at the final ranking by random selection, what is the probability that company Z's brand is ranked first? in the top three?

Solution

a. Since the testing service is drawing 3 elements (brands) from a set of 10 elements and arranging the 3 elements into a distinct order, we use the permutations rule to find the number of different results:

$$P_3^{10} = \frac{10!}{(10-3)!} = 10 \cdot 9 \cdot 8 = 720$$

b. The steps for calculating the probability of interest are as follows:

Step 1   The experiment is to select and rank 3 brands of detergent from 10 brands.

Step 2   There are too many sample points to list. However, we know from part **a** that there are 720 different outcomes (i.e., sample points) of this experiment.

Step 3   If we assume that the testing service determines the rankings at random, each of the 720 sample points should have an equal probability of occurrence. Thus,

$$P(\text{Each sample point}) = \frac{1}{720}$$

Step 4   One event of interest to company Z is that its brand receives top ranking. We will call this event A. The list of sample points that result in the occurrence of event A is long, but the *number* of sample points contained in event A is determined by breaking event A into two parts:

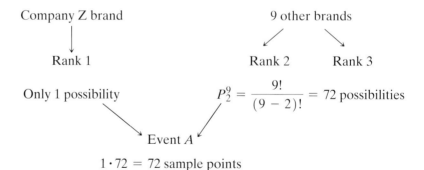

Thus, event A can occur in 72 different ways.

Now define B as the event that company Z's brand is ranked in the top three. Since event B specifies only that brand Z appear in the top three, we repeat the previous calculations, fixing brand Z in position 2 and then in position 3. We conclude that the number of sample points contained in event B is $3(72) = 216$.

Step 5   The final step is to calculate the probabilities of events A and B. Since the 720 sample points are equally likely to occur, we find that

$$P(A) = \frac{\text{Number of sample points in } A}{\text{Total number of sample points}} = \frac{72}{720} = \frac{1}{10}$$

Similarly,

$$P(B) = \frac{216}{720} = \frac{3}{10}$$

| Example | 3.30 |

Applying Several Counting Rules—Top 3 Detergents

Problem Refer to Example 3.29. Suppose the consumer testing service is to choose the top 3 laundry detergents from the group of 10, *but is not to rank them.*

a. In how many different ways can the testing service choose the 3 brands to be designated as top detergents?

b. Assuming that the testing service makes its choice by a random selection and that company X has 2 brands in the group of 10, what is the probability that exactly 1 of company X's brands is selected in the top 3? At least 1?

Solution

a. The testing service is selecting 3 elements (brands) from a set of 10 elements *without regard to order*, so we can apply the combinations rule to determine the number of different results:

$$\binom{10}{3} = \frac{10!}{3!(10-3)!} = \frac{10 \cdot 9 \cdot 8}{3 \cdot 2 \cdot 1} = 120$$

b. The steps for calculating the probabilities of interest are as follows:

Step 1 The experiment is to select (*but not rank*) three brands from 10.

Step 2 There are 120 sample points for this experiment.

Step 3 Since the selection is made at random,

$$P(\text{Each sample point}) = \frac{1}{120}$$

Step 4 Now, define events A and B as follows:

$$A: \{\text{Exactly one company X brand is selected.}\}$$

$$B: \{\text{At least one company X brand is selected.}\}$$

Since each of the sample points is equally likely to occur, we need only know the number of sample points in A and B to determine their probabilities.

For event A to occur, exactly 1 brand from company X must be selected, along with 2 of the remaining 8 brands from other companies. We thus break A into two parts:

2 company X brands   8 other brands
↓   ↓
Select 1   Select 2

$$\binom{2}{1} = 2 \text{ different} \qquad\qquad \binom{8}{2} = \frac{8!}{2!(8-2)!} = 28 \text{ different}$$
$$\text{possibilities} \qquad\qquad\qquad\qquad\qquad\qquad \text{possibilities}$$

Event A

$$2 \cdot 28 = 56 \text{ sample points}$$

Note that the 1 company X brand can be selected in 2 ways, whereas the 2 other brands can be selected in 28 ways. (We use the combinations rule because the order of selection is not important.) Now we use the multiplicative rule to combine 1 of the 2 ways to select a company X brand with 1 of the 28 ways to select two other brands, yielding a total of 56 sample points for event A.

To count the sample points contained in B, we first note that

$$B = A \cup C$$

where A is defined as before and we have

$$C: \{\text{Both company X brands are selected.}\}$$

We have determined that A contains 56 sample points. Using the same method for event C, we find that

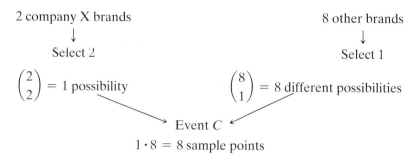

2 company X brands → Select 2

$\binom{2}{2} = 1$ possibility

8 other brands → Select 1

$\binom{8}{1} = 8$ different possibilities

Event C

$1 \cdot 8 = 8$ sample points

Thus, event C contains eight sample points, and $P(C) = \dfrac{8}{120}$.

Step 5 Since all the sample points are equally likely,

$$P(A) = \frac{\text{Number of sample ponts in } A}{\text{Total number of sample points}} = \frac{56}{120} = \frac{7}{15}$$

$$P(B) = P(A \cup C) = P(A) + P(C) - P(A \cap C)$$

where $P(A \cap C) = 0$ because the testing service cannot select exactly one *and* exactly two of company X's brands; that is, A and C are mutually exclusive events. Thus,

$$P(B) = \frac{56}{120} + \frac{8}{120} = \frac{64}{120} = \frac{8}{15}$$

■ **Now Work Exercise 3.128**

Exercises 3.104–3.131

Understanding the Principles

3.104 Give a scenario where the multiplicative rule applies.

3.105 Give a scenario where the permutations rule applies.

3.106 Give a scenario where the partitions rule applies.

3.107 What is the difference between the permutations rule and the combinations rule?

Learning the Mechanics

3.108 Find the numerical values of

a. $\binom{6}{3}$ **b.** P_2^5 **c.** P_2^4 **d.** $\binom{100}{98}$

e. $\binom{50}{50}$ **f.** $\binom{50}{0}$ **g.** P_3^5 **h.** P_0^{10}

3.109 Use the multiplicative rule to determine the number of sample points in the sample space corresponding to the experiment of tossing the following.

a. a 4-sided die 4 times
b. a 4-sided die 5 times
c. a 4-sided die 6 times
d. a 4-sided die n times

3.110 Determine the number of sample points contained in the sample space when you toss the following:

a. 1 die
b. 2 dice
c. 4 dice
d. n dice

3.111 An experiment consists of choosing objects without regard to order. How many sample points are there if you choose the following?

a. 4 objects from 8   **b.** 2 objects from 5
c. 2 objects from 34   **d.** 7 objects from 9
e. n objects from t

Applying the Concepts—Basic

3.112 Cheek teeth of extinct primates. Refer to the *American Journal of Physical Anthropology* (Vol. 142, 2010) study MOLARS of the dietary habits of extinct mammals, Exercise 3.25 (p. 158). Recall that 18 cheek teeth extracted from skulls of an extinct primate species discovered in western Wyoming were analyzed. Each tooth was classified according to degree of wear (unworn, slight, light-moderate, moderate, moderate-heavy, or heavy), with the 18 measurements shown in the accompanying table.

Data on Degree of Wear	
Unknown	Slight
Unknown	Slight
Unknown	Heavy
Moderate	Unworn
Slight	Light-moderate
Unknown	Light-moderate
Moderate-heavy	Moderate
Moderate	Unworn
Slight	Unknown

a. Suppose the researcher will randomly select one tooth from each wear category for a more detailed analysis. How many different samples are possible?

b. Repeat part **a**, but do not include any teeth classified as "unknown" in the sample.

c. If the tooth is selected at random, what is the probability it is not classified as "unknown"?

3.113 **Choosing portable grill displays.** Refer to the *Journal of Consumer Research* (Mar. 2003) study of how people attempt to influence the choices of others, Exercise 3.29 (p. 159). Recall that students selected three portable grill displays to be compared from an offering of five different grill displays.

a. Use a counting rule to count the number of ways the three displays can be selected from the five available displays to form a three-grill-display combination.

b. The researchers informed students to select the three displays in order to convince people to choose Grill #2. Consequently, Grill #2 was a required selection. Use a counting rule to count the number of different ways the three grill displays can be selected from the five displays if Grill #2 must be selected. (Your answer should agree with the answer in 3.29a.)

c. Now suppose the three selected grills will be set up in a specific order for viewing by a customer. (The customer views one grill first, then the second, and finally the third grill.) Again, Grill #2 must be one of the three selected. How many different ways can the three grill displays be selected if customers view the grills in order?

3.114 **Monitoring impedance to leg movements.** In an experiment designed to monitor impedance to leg movement, Korean engineers attached electrodes to the ankles and knees of volunteers. Of interest were the voltage readings between pairs of electrodes (*IEICE Transactions on Information & Systems*, Jan. 2005). These readings were used to determine the signal-to-noise ratio (SNR) of impedance changes such as knee flexes and hip extensions.

a. Six voltage electrodes were attached to key parts of the ankle. How many electrode pairs on the ankle are possible?

b. Ten voltage electrodes were attached to key parts of the knee. How many electrode pairs on the knee are possible?

[NW] **c.** Determine the number of possible electrode pairs such that one electrode is attached to the knee and one is attached to the ankle.

3.115 **Picking a basketball team.** Suppose you are to choose a
[NW] basketball team (five players) from eight available athletes.

a. How many ways can you choose a team (ignoring positions)?

b. How many ways can you choose a team composed of two guards, two forwards, and a center?

c. How many ways can you choose a team composed of one each of a point guard, shooting guard, power forward, small forward, and center?

3.116 **Selecting project teams.** Suppose you are managing 10 em-
[NW] ployees and you need to form three teams to work on different projects. Assume that each employee may serve on any team. In how many different ways can the teams be formed if the number of members on each project team are as follows:

a. 3, 3, 4   **b.** 2, 3, 5
c. 1, 4, 5   **d.** 2, 4, 4

3.117 **Using game simulation to teach a course.** Refer to the *Engineering Management Research* (May 2012) study of using a simulation game to teach a production course, Exercise 3.32 (p. 159). Recall that a purchase order card in the simulated

game consists of a choice of one of two color television brands (*A* or *B*), one of two colors (red or black), and the quantity ordered (1, 2, or 3 TVs). Use a counting rule to determine the number of different purchase order cards that are possible. Does your answer agree with the list you produced in Exercise 3.32?

3.118 **ZIP codes.** An eight-digit code is used to distinguish between different regions. The code uses numbers between 1 and 9, inclusive.

a. How many different codes are available for use? (Assume that any eight-digit number can be used as a code.)

b. The first three digits of a code for a particular region are 333. If no other region has these first three digits as part of its code, how many codes can exist in this region?

3.119 **Traveling between cities.** A salesperson living in city *A* wishes to visit five cities *B*, *C*, *D*, *E*, and *F*.

a. If the cities are all connected by airlines, how many different travel plans could be constructed to visit each city exactly once and then return home?

b. Suppose all cities are connected, except that *D* and *E* are not directly connected. How many different flight plans would be available to the salesperson?

Applying the Concepts—Intermediate

3.120 **Football uniform combinations.** Nike manufactures uniforms for both major college and professional (NFL) football teams. The company achieved notoriety for reinventing the look of the Oregon Ducks football team. Oregon now has five different helmets, seven different jerseys, and six different pants to choose from each game. How many different uniform combinations are available for Oregon football players?

3.121 **Kiwifruit as an iron supplement.** An article published in *BMC Public Health* (Vol. 10, 2010) studied the effectiveness of gold kiwifruit as an iron supplement in women with an iron deficiency. Fifty women were assigned to receive an iron-fortified breakfast cereal with a banana, and 50 women were assigned to receive an iron-fortified breakfast cereal with gold kiwifruit. At the conclusion of the 16-week study, the medical researchers selected two women from each group and performed a full physical examination on each. How many possible ways could the selections be made?

3.122 **Randomization in a study of TV commercials.** Gonzaga University professors conducted a study of over 1,500 television commercials and published their results in the *Journal of Sociology, Social Work and Social Welfare* (Vol. 2, 2008). Commercials from eight networks—ABC, FAM, FOX, MTV, ESPN, CBS, CNN, and NBC—were sampled during an eight-day period, with one network randomly selected each day. The table below shows the order determined by random draw:

ABC—July 6 (Wed)
FAM—July 7 (Thr)
FOX—July 9 (Sat)
MTV—July 10 (Sun)
ESPN—July 11 (Mon)
CBS—July 12 (Tue)
CNN—July 16 (Sat)
NBC—July 17 (Sun)

a. Determine the number of possible orderings of the networks over the eight days.

b. What is the probability that ESPN is selected on Monday, July 11th?

c. What is the probability that MTV is selected on a Sunday?

3.123 Multilevel marketing schemes. Successful companies employ multilevel marketing (MLM) schemes to distribute their products. In one company, each distributor recruits several additional distributors to work at the first level. Each of these first-level distributors, in turn, recruits distributors to work at the second level. Distributors at any level continue to recruit distributors, forming additional levels. A distributor at any particular level makes a 5% commission on the sales of all distributors at lower levels in his or her "group."

a. Elaine is a distributor for an MLM company. Elaine recruits seven distributors to work for her at the first level. Each of these distributors recruits an additional six distributors to work at the second level. How many second-level distributors are under Elaine?

b. Refer to part **a.** Suppose each distributor at Elaine's second level recruits six third-level distributors, and each of these third-level distributors recruits four level-four distributors. How many fourth-level distributors are under Elaine?

3.124 Mathematical theory of partitions. Mathematicians at the University of Florida solved a 30-year-old mathematics problem with the use of the theory of partitions (*Explore,* Fall 2000). In mathematical terminology, a partition is a representation of an integer as a sum of positive integers. (For example, the number 3 has three possible partitions: $3, 2 + 1,$ and $1 + 1 + 1.$) The researchers solved the problem by using "colored partitions" of a number, where the colors correspond to the four suits—red hearts, red diamonds, black spades, and black clubs—in a standard 52-card bridge deck. Consider forming colored partitions of an integer.

a. How many colored partitions of the number 3 are possible? [*Hint:* One partition is 3♥; another is 2♦ + 1♣.]

b. How many colored partitions of the number 5 are possible?

3.125 The "marriage" problem. A mathematics assignment problem in which each element from one set is matched with one and only one element from another set is often referred to as a "marriage" problem. *Mathematical Social Sciences* (May 2014) considered the following marriage problem. After passing their preliminary exams, doctoral students in a Ph.D. program are each assigned to a professor for research assistance. Each professor can accept only one additional student per year. Suppose there are three students, *a, b,* and *c,* who have passed their preliminary exams and three professors, *A, B,* and *C.* Also suppose that professor *A* prefers student *a,* professor *B* prefers student *b,* and professor *C* prefers student *c.*

a. If the assignment is made at random, what are the chances that the preferred matches occur?

b. If the assignment is made at random, what are the chances that professor *A* is matched with student *a*?

c. If the assignment is made at random, what are the chances that none of the professors are matched with their preferred student?

3.126 Florida license plates. In the mid-1980s, the state of Florida ran out of combinations of letters and numbers for its license plates. Then, each license plate contained three letters of the alphabet, followed by three digits selected from the 10 digits 0, 1, 2,..., 9.

a. How many different license plates did this system allow?

b. New Florida tags were obtained by reversing the procedure, starting with three digits followed by three letters. How many new tags did the new system provide?

c. Since the new tag numbers were added to the old numbers, what is the total number of licenses available for registration in Florida?

3.127 Selecting a maintenance support system. In the *Journal of Quality in Maintenance Engineering* (Vol. 9, 2003), researchers used an analytic hierarchy process to help build a preferred maintenance organization for ARTHUR, the Norwegian Army's high-tech radar system. The process requires the builder to select alternatives in three different stages (called echelons). In the first echelon, the builder must choose one of two mobile units (regular soldiers or soldiers with engineering training). In the second echelon, the builder chooses one of three heavy mobile units (units in the Norwegian Army, units from Supplier 2, or shared units). Finally, in the third echelon, the builder chooses one of three maintenance workshops (Norwegian Army, Supplier 1, or Supplier 2).

a. How many maintenance organization alternatives exist when choices are made in the three echelons?

b. The researchers determined that only four of the alternatives in part **a** are feasible alternatives for ARTHUR. If one of the alternatives is randomly selected, what is the probability that it is a feasible alternative?

3.128 Volleyball positions. Intercollegiate volleyball rules require that after the opposing team has lost its serve, each of the six members of the serving team must rotate into new positions on the court. Hence, each player must be able to play all six different positions. How many different team combinations, by player and position, are possible during a volleyball game? If players are initially assigned to the positions in a random manner, find the probability that the best server on the team is in the serving position.

3.129 Studying exam questions. A college professor hands out a list of 10 questions, 5 of which will appear on the final examination for the course. One of the students taking the course is pressed for time and can prepare for only 6 of the 10 questions on the list. Suppose the professor chooses the 5 questions at random from the 10.

a. What is the probability that the student will be prepared for all 5 questions that appear on the final examination?

b. What is the probability that the student will be prepared for fewer than 2 questions?

c. What is the probability that the student will be prepared for exactly 4 questions?

Applying the Concepts—Advanced

3.130 Modeling the behavior of granular media. *Granular media* are substances made up of many distinct grains—including sand, rice, ball bearings, and flour. The properties of these materials were theoretically modeled in *Engineering Computations: International Journal for Computer-Aided*

Engineering and Software (Vol. 30, No. 2, 2013). The model assumes there is a system of N non-interacting granular particles. The particles are grouped according to energy level. Assume there are r energy levels, with N_i particles at energy level i, $i = 1, 2, 3, \ldots, r$. Consequently, $N = N_1 + N_2 + \cdots + N_r$. A *microset* is defined as a possible grouping of the particles among the energy levels. For example, suppose $N = 7$ and $r = 3$. Then one possible microset is $N_1 = 1, N_2 = 2$, and $N_3 = 4$. That is, there is one particle at energy level 1, two particles at energy level 2, and four particles at energy level 3. Determine the number of different microsets possible when $N = 7$ and $r = 3$.

3.131 A straight flush in poker. Consider 5-card poker hands dealt from a standard 52-card bridge deck. Two important events are

A: {You draw a flush.}

B: {You draw a straight.}

[*Note:* A *flush* consists of any 5 cards of the same suit. A *straight* consists of any 5 cards with values in sequence. In a straight, the cards may be of any suit, and an ace may be considered as having a value of 1 or, alternatively, a value higher than a king.]

a. How many different 5-card hands can be dealt from a 52-card bridge deck?

b. Find $P(A)$.

c. Find $P(B)$.

d. The event that both A and B occur—that is, $A \cap B$—is called a *straight flush*. Find $P(A \cap B)$.

3.8 Bayes's Rule (Optional)

An early attempt to employ probability in making inferences is the basis for a branch of statistical methodology known as **Bayesian statistical methods**. The logic employed by the English philosopher Thomas Bayes in the mid-1700s involves converting an unknown conditional probability, say, $P(B|A)$, to one involving a known conditional probability, say, $P(A|B)$. The method is illustrated in the next example.

Example 3.31

Applying Bayes's Logic—Intruder Detection System

Problem An unmanned monitoring system uses high-tech video equipment and microprocessors to detect intruders. A prototype system has been developed and is in use outdoors at a weapons munitions plant. The system is designed to detect intruders with a probability of .90. However, the design engineers expect this probability to vary with the weather. The system automatically records the weather condition each time an intruder is detected. On the basis of a series of controlled tests in which an intruder was released at the plant under various weather conditions, the following information is available: Given that the intruder was, in fact, detected by the system, the weather was clear 75% of the time, cloudy 20% of the time, and raining 5% of the time. When the system failed to detect the intruder, 60% of the days were clear, 30% cloudy, and 10% rainy. Use this information to find the probability of detecting an intruder, given rainy weather. (Assume that an intruder has been released at the plant.)

Solution Define D to be the event that the intruder is detected by the system. Then D^c is the event that the system failed to detect the intruder. Our goal is to calculate the conditional probability $P(D|\text{Rainy})$. From the statement of the problem, the following information is available:

$$P(D) = .90 \qquad\qquad PD^c = .10$$
$$P(\text{Clear}|D) = .75 \qquad P(\text{Clear}|D^c) = .60$$
$$P(\text{Cloudy}|D) = .20 \qquad P(\text{Cloudy}|D^c) = .30$$
$$P(\text{Rainy}|D) = .05 \qquad P(\text{Rainy}|D^c) = .10$$

Note that $P(D|\text{Rainy})$ is not one of the conditional probabilities that is known. However, we can find

$$P(\text{Rainy} \cap D) = P(D)P(\text{Rainy}|D) = (.90)(.05) = .045$$

and

$$P(\text{Rainy} \cap D^c) = P(D^c)P(\text{Rainy}|D^c) = (.10)(.10) = .01$$

by the multiplicative probability rule. These two probabilities are highlighted on the tree diagram for the problem in Figure 3.22.

Now, the event Rainy is the union of two mutually exclusive events: $(\text{Rainy} \cap D)$ and $(\text{Rainy} \cap D^c)$. Thus, applying the additive probability rule, we have

$$P(\text{Rainy}) = P(\text{Rainy} \cap D) + P(\text{Rainy} \cap D^c) = .045 + .01 = .055$$

We now apply the formula for conditional probability to obtain

$$P(D|\text{Rainy}) = \frac{P(\text{Rainy} \cap D)}{P(\text{Rainy})} = \frac{P(\text{Rainy} \cap D)}{P(\text{Rainy} \cap D) + P(\text{Rainy} \cap D^c)} = \frac{.045}{.055} = .818$$

Therefore, under rainy weather conditions, the prototype system can detect the intruder with a probability of .818, a value lower than the designed probability of .90.

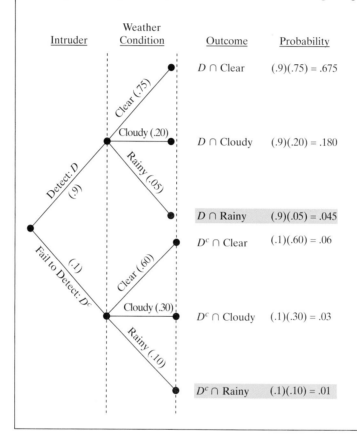

Figure 3.22

Tree Diagram for Example 3.31

The technique utilized in Example 3.31, called **Bayes's rule**, can be applied when an observed event A occurs with any one of several mutually exclusive and exhaustive events, $B_1, B_2, \ldots, B_k$. The formula for finding the appropriate conditional probabilities is given in the next box.

Bayes's Rule

Given k mutually exclusive and exhaustive events, $B_1, B_2, \ldots, B_k$ such that $P(B_1) + P(B_2) + \cdots + P(B_k) = 1$, and given an observed event A, it follows that
$$P(B_i|A) = P(B_i \cap A)/P(A)$$

$$= \frac{P(B_i)P(A|B_i)}{P(B_1)P(A|B_1) + P(B_2)P(A|B_2) + \cdots + P(B_k)P(A|B_k)}$$

In applying Bayes's rule to Example 3.31, the observed event $A = \{\text{Rainy}\}$ and the $k = 2$ mutually exclusive and exhaustive events are the complementary events

$D = \{\text{intruder detected}\}$ and $D^c = \{\text{intruder not detected}\}$. Hence, we obtain the formula

$$P(D|\text{Rainy}) = \frac{P(D)P(\text{Rainy}|D)}{P(D)P(\text{Rainy}|D) + P(D^c)P(\text{Rainy}|D^c)}$$

$$= \frac{(.90)(.05)}{(.90)(.05) + (.10)(.10)} = .818$$

Example **3.32**

Bayes's Rule Application— Wheelchair Control

Problem Electric wheelchairs are difficult to maneuver for many disabled people. In a paper presented at the *1st International Workshop on Advances in Service Robotics* (Mar. 2003), researchers applied Bayes's rule to evaluate an "intelligent" robotic controller that aims to capture the intent of a wheelchair user and aid in navigation. Consider the following scenario. From a certain location in a room, a wheelchair user will either (1) turn sharply to the left and navigate through a door, (2) proceed straight to the other side of the room, or (3) turn slightly right and stop at a table. Denote these three events as D (for door), S (Straight), and T (for table). Based on previous trips, $P(D) = .5$, $P(S) = .2$, and $P(T) = .3$. The wheelchair is installed with a robot-controlled joystick. When the user intends to go through the door, she points the joystick straight 30% of the time; when the user intends to go straight, she points the joystick straight 40% of the time; and, when the user intends to go to the table, she points the joystick straight 5% of the time. If the wheelchair user points the joystick straight, what is her most likely destination?

Solution Let $J = \{\text{joystick is pointed straight}\}$. The user intention percentages can be restated as the following conditional probabilities: $P(J|D) = .3$, $P(J|S) = .4$, and $P(J|T) = .05$. Since the user has pointed the joystick straight, we want to find the following probabilities: $P(D|J)$, $P(S|J)$, and $P(T|J)$. Now, the three events, D, S, and T, represent mutually exclusive and exhaustive events, where $P(D) = .5$, $P(S) = .2$, and $P(T) = .3$. Consequently, we can apply Bayes's rule as follows:

$$P(D|J) = P(J|D) \cdot P(D)/[P(J|D) \cdot P(D) + P(J|S) \cdot P(S) + P(J|T) \cdot P(T)]$$
$$= (.3)(.5)/[(.3)(.5) + (.4)(.2) + (.05)(.3)] = .15/.245 = .612$$
$$P(S|J) = P(J|S) \cdot P(S)/[P(J|D) \cdot P(D) + P(J|S) \cdot P(S) + P(J|T) \cdot P(T)]$$
$$= (.4)(.2)/[(.3)(.5) + (.4)(.2) + (.05)(.3)] = .08/.245 = .327$$
$$P(T|J) = P(J|T) \cdot P(T)/[P(J|D) \cdot P(D) + P(J|S) \cdot P(S) + P(J|T) \cdot P(T)]$$
$$= (.05)(.3)/[(.3)(.5) + (.4)(.2) + (.05)(.3)] = .015/.245 = .061$$

Note that the largest conditional probability is $P(D|J) = .612$. Thus, if the joystick is pointed straight, the wheelchair user is most likely headed through the door.

■ **Now Work Exercise 3.136**

Exercises 3.132–3.147

Understanding the Principles

3.132 Explain the difference between the two probabilities $P(A|B)$ and $P(B|A)$.

3.133 Why is Bayes's rule unnecessary for finding $P(B|A)$ if events A and B are independent?

3.134 Why is Bayes's rule unnecessary for finding $P(B|A)$ if events A and B are mutually exclusive?

Learning the Mechanics

3.135 Suppose the events B_1 and B_2 are mutually exclusive and complementary events, such that $P(B_1) = .31$ and $P(B_2) = .69$. Consider another event A such that $P(A|B_1) = .1$ and $P(A|B_2) = .5$.

Complete parts **a** through **e** below.
  a. Find $P(B_1 \cap A)$.
  b. Find $P(B_2 \cap A)$.
  c. Find $P(A)$ using the results in parts **a** and **b**.
  d. Find $P(B_1|A)$.
  e. Find $P(B_2|A)$.

3.136 Suppose the events B_1, B_2, and B_3 are mutually exclusive and complementary events such that $P(B_1) = .2$, $P(B_2) = .15$, and $P(B_3) = .65$. Consider another event A such that $P(A|B_1) = .4$, $P(A|B_2) = .25$, and $P(A|B_3) = .6$. Use Bayes's rule to find
  a. $P(B_1|A)$
  b. $P(B_2|A)$
  c. $P(B_3|A)$

3.137 Suppose the events $B_1, B_2,$ and B_3 are mutually exclusive and complementary events such that $P(B_1) = .55, P(B_2) = .35,$ and $P(B_3) = .1.$ Consider another event A such that $P(A) = .2.$ If A is independent of $B_1, B_2,$ and $B_3,$ use Bayes's rule to show that $P(B_1|A) = P(B_1) = .55.$

Applying the Concepts—Basic

3.138 **Reverse-engineering gene identification.** In *Molecular Systems Biology* (Vol. 3, 2007), geneticists at the University of Naples (Italy) used reverse engineering to identify genes. They calculated $P(G|D),$ where D is a gene expression data set of interest and G is a graphical identifier. Several graphical identifiers were investigated. Suppose that, for two different graphical identifiers G_1 and $G_2, P(D|G_1) = .5$ and $P(D|G_2) = .3.$ Also, $P(D) = .34, P(G_1) = .2,$ and $P(G_2) = .8.$
a. Use Bayes's rule to find $P(G_1|D).$
b. Use Bayes's rule to find $P(G_2|D).$

3.139 **Drug testing in athletes.** When Olympic athletes are tested for illegal drug use (i.e., doping), the results of a single test are used to ban the athlete from competition. In *Chance* (Spring 2004), University of Texas biostatisticians demonstrated the application of Bayes's rule to making inferences about testosterone abuse among Olympic athletes. They used the following example: In a population of 1,000 athletes, suppose 100 are illegally using testosterone. Of the users, suppose 50 would test positive for testosterone. Of the nonusers, suppose 9 would test positive.
a. Given that the athlete is a user, find the probability that a drug test for testosterone will yield a positive result. (This probability represents the *sensitivity* of the drug test.)
b. Given that the athlete is a nonuser, find the probability that a drug test for testosterone will yield a negative result. (This probability represents the *specificity* of the drug test.)
c. If an athlete tests positive for testosterone, use Bayes's rule to find the probability that the athlete is really doping. (This probability represents the *positive predictive value* of the drug test.)

3.140 **Fingerprint expertise.** A study published in *Psychological Science* (Aug. 2011) tested the accuracy of experts and novices in identifying fingerprints. Participants were presented pairs of fingerprints and asked to judge whether the prints in each pair matched. The pairs were presented under three different conditions: prints from the same individual (*match condition*), non-matching but similar prints (*similar distracter condition*), and non-matching and very dissimilar prints (*non-similar distracter condition*). The percentages of correct decisions made by the two groups under each of the three conditions are listed in the table.

Condition	Fingerprint experts	Novices
Match	92.12%	74.55%
Similar Distracter	99.32%	44.82%
Non-similar Distracter	100%	77.03%

Source: Tangen, J. M., et al. "Identifying fingerprint expertise." *Psychological Science*, Vol. 22, No. 8, Aug. 2011 (Figure 1).

a. Given a pair of matched prints, what is the probability that an expert will fail to identify the match?

b. Given a pair of matched prints, what is the probability that a novice will fail to identify the match?
c. Assume the study included 10 participants—5 experts, and 5 novices. Suppose that a pair of matched prints are presented to a randomly selected study participant and the participant fails to identify the match. Is the participant more likely to be an expert or a novice?

3.141 **Tests for Down syndrome.** Currently, there are three diagnostic tests available for chromosome abnormalities in a developing fetus: triple serum marker screening, ultrasound, and amniocentesis. The safest (to both the mother and fetus) and least expensive of the three is the ultrasound test. Two San Diego State University statisticians investigated the accuracy of using ultrasound to test for Down syndrome (*Chance*, Summer 2007). Let D denote that the fetus has a genetic marker for Down syndrome and N denote that the ultrasound test is normal (i.e., no indication of chromosome abnormalities). Then, the statisticians desire the probability $P(D|N).$ Use Bayes's rule and the following probabilities (provided in the article) to find the desired probability: $P(D) = 1/80, P(D^C) = 79/80, P(N|D) = 1/2, P(N^C|D) = 1/2, P(N|D^C) = 1,$ and $P(N^C|D^C) = 0.$

Applying the Concepts—Intermediate

3.142 **HIV testing and false positives.** Bayes's rule was applied to the problem of HIV testing in *The American Statistician* (Aug. 2008). In North America, the probability of a person having HIV is .008. A test for HIV yields either a positive or negative result. Given that a person has HIV, the probability of a positive test result is .99. (This probability is called the *sensitivity* of the test.) Given that a person does not have HIV, the probability of a negative test result is also .99. (This probability is called the *specificity* of the test.) The authors of the article are interested in the probability that a person actually has HIV given that the test is positive.
a. Find the probability of interest for a North American by using Bayes's rule.
b. In East Asia, the probability of a person having HIV is only .001. Find the probability of interest for an East Asian by using Bayes's rule. (Assume that both the sensitivity and specificity of the test are .99.)
c. Typically, if one tests positive for HIV, a follow-up test is administered. What is the probability that a North American has HIV given that both tests are positive? (Assume that the tests are independent.)
d. Repeat part **c** for an East Asian.

3.143 **Mining for dolomite.** Dolomite is a valuable mineral that is found in sedimentary rock. During mining operations, dolomite is often confused with shale. The radioactivity features of rock can aid miners in distinguishing between dolomite and shale rock zones. For example, if the gamma ray reading of a rock zone exceeds 60 API units, the area is considered to be mostly shale (and is not mined); if the gamma ray reading of a rock zone is less than 60 API units, the area is considered to be abundant in dolomite (and is mined). Data on 771 core samples in a rock quarry collected by the Kansas Geological Survey revealed that 476 of the samples are dolomite and 295 of the samples are shale. Of the 476 dolomite core samples, 34 had a gamma ray reading greater than 60. Of the 295 shale core samples, 280 had a gamma ray reading greater than 60. Suppose you obtain a gamma ray reading greater than 60 at a certain depth of the rock quarry. Should this area be mined?

3.144 Nondestructive evaluation. Nondestructive evaluation (NDE) describes methods that quantitatively characterize materials, tissues, and structures by noninvasive means, such as X-ray computed tomography, ultrasonics, and acoustic emission. Recently, NDE was used to detect defects in steel castings (*JOM*, May 2005). Assume that the probability that NDE detects a "hit" (i.e., predicts a defect in a steel casting) when, in fact, a defect exists is .97. (This is often called the probability of detection.) Assume also that the probability that NDE detects a "hit" when, in fact, no defect exists is .005. (This is called the probability of a false call.) Past experience has shown that a defect occurs once in every 100 steel castings. If NDE detects a "hit" for a particular steel casting, what is the probability that an actual defect exists?

3.145 Intrusion detection systems. Refer to the *Journal of Research of the National Institute of Standards and Technology* (Nov.–Dec. 2003) study of a double intrusion detection system with independent systems, presented in Exercise 3.98 (p. 186). Recall that if there is an intruder, system A sounds an alarm with probability .9 and system B sounds an alarm with probability .95. If there is no intruder, system A sounds an alarm with probability .2 and system B sounds an alarm with probability .1. Now, assume that the probability of an intruder is .4. If both systems sound an alarm, what is the probability that there is an intruder?

3.146 Confidence of feedback information for improving quality. In the semiconductor manufacturing industry, companies strive to improve product quality. One key to improved quality is having confidence in the feedback generated by production equipment. A study of the confidence level of feedback information was published in *Engineering Applications of Artificial Intelligence* (Vol. 26, 2013). At any point in time during the production process, a report can be generated. The report is classified as either "OK" or "not OK." As an example, consider the following probabilities: The probability an "OK" report is generated in any time period (t) is .8. The probability of an "OK" report at one time period ($t + 1$), given an "OK" report in the previous time period (t), is .9. Also, the probability of an "OK" report at one time period ($t + 1$), given a "not OK" report in the previous time period (t), is .5. If an "OK" report is generated in one time period, what is the probability that an "OK" report was generated in the previous time period?

Applying the Concepts—Advanced

3.147 Forensic analysis of JFK assassination bullets. Following the assassination of President John F. Kennedy (JFK) in 1963, the House Select Committee on Assassinations (HSCA) conducted an official government investigation. The HSCA concluded that although there was a probable conspiracy, involving at least one additional shooter other than Lee Harvey Oswald, the additional shooter missed all limousine occupants. A recent analysis of assassination bullet fragments, reported in the *Annals of Applied Statistics* (Vol. 1, 2007), contradicted these findings, concluding that evidence used to rule out a second assassin by the HSCA is fundamentally flawed. It is well documented that at least two different bullets were the source of bullet fragments used in the assassination. Let E = {bullet evidence used by the HSCA}, T = {two bullets used in the assassination}, and T^C = {more than two bullets used in the assassination}. Given the evidence (E), which is more likely to have occurred—two bullets used (T) or more than two bullets used (T^C)?

a. The researchers demonstrated that the ratio, $P(T|E)/P(T^C|E)$, is less than 1. Explain why this result supports the theory of more than two bullets used in the assassination of JFK.

b. To obtain the result in part **a**, the researchers first showed that

$$P(T|E)/P(T^C|E) = [P(E|T) \cdot P(T)]/[P(E|T^C) \cdot P(T^C)]$$

Demonstrate this equality using Bayes's theorem.

CHAPTER NOTES

Key Terms

Note: Items marked with an asterisk (*) are from the optional sections in this chapter.

Key Symbols

S	**Sample space** (collection of all sample points)
A: $\{1,2\}$	Set of **sample points in event** A
$P(A)$	**Probability** of event A
$A \cup B$	**Union** of events A and B (either A or B occurs)
$A \cap B$	**Intersection** of events A and B (both A and B occur)
A^c	**Complement** of A (event A does not occur)
$A \mid B$	Event A occurs, **given that** event B occurs
$\binom{N}{n}$	Number of **combinations** of N elements taken n at a time
$N!$	N **factorial** $= N(N-1)(N-2) \ldots (2)(1)$

Key Ideas

Probability Rules for k Sample Points, $S_1, S_2, S_3, \ldots, S_k$

1. $0 \leq P(E_i) \leq 1$
2. $\sum P(E_i) = 1$

Combinations Rule

Counting number of samples of n elements selected from N elements

$$\binom{N}{n} = \frac{N!}{n!(N-n)!}$$
$$= \frac{N(N-1)(N-2) \ldots (N-n+1)}{n(n-1)(n-2) \ldots (2)(1)}$$

Bayes's Rule

For mutually exclusive events, $B_1, B_2, B_3, \ldots, B_k$, such that $P(B_1) + P(B_2) + \cdots + P(B_k) = 1$,

$$P(B_i \mid A) = \frac{P(B_i)P(A \mid B_i)}{P(B_1)P(A \mid B_1) + P(B_2)P(A \mid B_2) + \cdots + P(B_k)P(A \mid B_k)}$$

Guide to Selecting Probability Rules

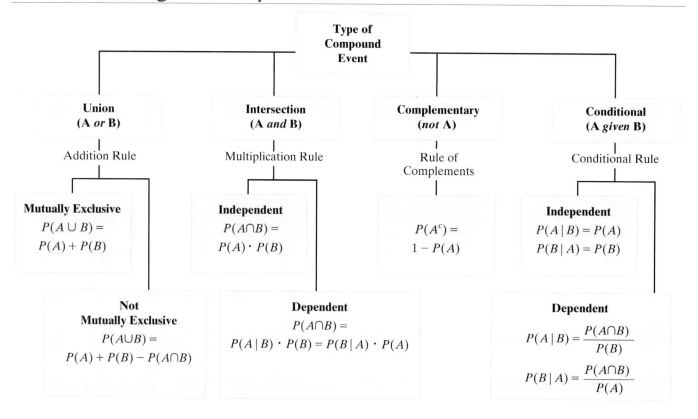

Supplementary Exercises 3.148–3.196

Note: Starred (*) exercises refer to the optional sections in this chapter.

Understanding the Principles

3.148 Which of the following pairs of events are mutually exclusive?
  a. $A = \{$The Tampa Bay Rays win the World Series next year.$\}$
  $B = \{$Evan Longoria, Rays infielder, hits 75 home runs runs next year.$\}$
  b. $A = \{$Psychiatric patient Tony responds to a stimulus within 5 seconds.$\}$
  $B = \{$Psychiatric patient Tony has the fastest stimulus response time of 2.3 seconds.$\}$
  c. $A = \{$High school graduate Cindy enrolls at the University of South Florida next year.$\}$
  $B = \{$High school graduate Cindy does not enroll in college next year.$\}$

3.149 Use the symbols $\cap$, $\cup$, $|$, and ^{c} to convert the following statements into compound events involving events A and B, where $A = \{$You purchase a notebook computer$\}$ and $B = \{$You vacation in Europe$\}$:
  a. You purchase a notebook computer or vacation in Europe.
  b. You will not vacation in Europe.
  c. You purchase a notebook computer and vacation in Europe.
  d. Given that you vacation in Europe, you will not purchase a notebook computer.

Learning the Mechanics

3.150 A sample space consists of four sample points $S_1, S_2, S_3,$ and S_4, where $P(S_1) = .3, P(S_2) = .3, P(S_3) = .2,$ and $P(S_4) = .2.$
  a. Show that the sample points obey the two probability rules for a sample space.
  b. If an event $A = \{S_1, S_4\}$, find $P(A)$.

3.151 A and B are mutually exclusive events, with $P(A) = .3$ and $P(B) = .4.$
  a. Find $P(A|B)$.
  b. Are A and B independent events?

3.152 For two events A and B, suppose $P(A) = .5, P(B) = .8$ and $P(A \cap B) = .4$. Find $P(A \cup B)$.

3.153 Given that $P(A \cap B) = .3$ and $P(B|A) = .9$, find $P(A)$.

3.154 The Venn diagram in the next column illustrates a sample space containing six sample points and three events A, B, and C. The probabilities of the sample points are:
$$P(1) = .2, P(2) = .1, P(3) = .2, P(4) = .2, P(5) = .1,$$
and $P(6) = .2$
  a. Find $P(A \cap B)$, $P(B \cap C)$, $P(A \cup C)$, $P(A \cup B \cup C)$, $P(B^c)$, $P(A^c \cap B)$, $P(B|C)$, and $P(B|A)$.
  b. Are A and B independent? Mutually exclusive? Why?
  c. Are B and C independent? Mutually exclusive? Why?

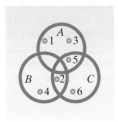

S

3.155 A fair die is tossed, and the up face is noted. If the number is less than 3, the die is tossed again; otherwise, a fair coin is tossed. Consider the following events:
$$A: \{\text{A tail appears on the coin}\}$$
$$B: \{\text{The die is tossed only one time}\}$$
  a. List the sample points in the sample space.
  b. Give the probability for each of the sample points.
  c. Find $P(A)$ and $P(B)$.
  d. Identify the sample points in $A^c, B^c, A \cap B,$ and $A \cup B$.
  e. Find $P(A^c), P(B^c), P(A \cap B), P(A \cup B), P(A|B),$ and $P(B|A)$.
  f. Are A and B mutually exclusive events? Independent events? Why?

3.156 A balanced die is thrown once. If a 4 appears, a ball is drawn from urn 1; otherwise, a ball is drawn from urn 2. urn 1 contains two red, four white and six black balls. Urn 2 contains one red and five black balls.
  a. Find the probability that a red ball is drawn.
  b. Find the probability that urn 1 was used given that a red ball was drawn.

3.157 Two events, A and B, are independent, with $P(A) = .5$ and $P(B) = .3.$
  a. Are A and B mutually exclusive? Why?
  b. Find $P(A|B)$ and $P(B|A)$.
  c. Find $P(A \cup B)$.

3.158 Find the numerical value of
  a. $7!$
  b. $\binom{8}{7}$
  c. $\binom{9}{1}$
  d. P_3^5
  e. $\binom{7}{4}$
  f. $0!$
  g. P_3^9
  h. P_3^{80}

Ⓓ Applet Exercise 3.6

Use the applet entitled *Random Numbers* to generate a list of 50 numbers between 1 and 100, inclusive. Use the list to find each of the following probabilities:
  a. The probability that a number chosen from the list is less than or equal to 50

b. The probability that a number chosen from the list is even

c. The probability that a number chosen from the list is both less than or equal to 50 and even

d. The probability that a number chosen from the list is less than or equal to 50, given that the number is even

e. Do your results from parts **a–d** support the conclusion that the events *less than or equal to 50* and *even* are independent? Explain.

Applying the Concepts—Basic

3.159 **Going online for health information.** A *cyberchondriac* is defined as a person who regularly searches the Web for health care information. A Harris Poll surveyed 1,010 U.S. adults by telephone and asked each respondent how often (in the past month) he or she looked for health care information online. The results are summarized in the following table. Consider the response category of a randomly selected person who participated in the Harris poll.

Response (# per month)	Percentage of Respondents
None	25
1 or 2	31
3–5	25
6–9	5
10 or more	14
Total	100%

Based on "Cyberchondriac" poll. *The Harris Poll*, July 29, 2008 (Table 3).

a. List the sample points for the experiment.

b. Assign reasonable probabilities to the sample points.

c. Find the probability that the respondent looks for health care information online more than two times per month.

3.160 **Study of ancient pottery.** Refer to the *Chance* (Fall 2000) study of ancient pottery found at the Greek settlement of Phylakopi, presented in Exercise 2.186 (p. 134). Of the 837 pottery pieces uncovered at the excavation site, 183 were painted. These painted pieces included 14 painted in a curvilinear decoration, 165 painted in a geometric decoration, and 4 painted in a naturalistic decoration. Suppose 1 of the 837 pottery pieces is selected and examined.

a. What is the probability that the pottery piece is painted?

b. Given that the pottery piece is painted, what is the probability that it is painted in a curvilinear decoration?

3.161 **Post office violence.** *The Wall Street Journal* (Sept. 1, 2000) reported on an independent study of postal workers and violence at post offices. In a sample of 12,000 postal workers, 600 were physically assaulted on the job in a recent year. Use this information to estimate the probability that a randomly selected postal worker will be physically assaulted on the job during the year.

3.162 **Sterile couples in Jordan.** A sterile family is a couple that has no children by their deliberate choice or because they are biologically infertile. Researchers at Yarmouk University (in Jordan) estimated the proportion of sterile couples in that country to be .06 (*Journal of Data Science*,

July 2003). Also, 64% of the sterile couples in Jordan are infertile. Find the probability that a Jordanian couple is both sterile and infertile.

3.163 **NHTSA new car crash testing.** Refer to the National Highway Traffic Safety Administration (NHTSA) crash tests of new car models, presented in Exercise 2.190 (p. 135). Recall that the NHTSA has developed a "star" scoring system, with results ranging from one star (*) to five stars (*****). The more stars in the rating, the better the level of crash protection in a head-on collision. A summary of the driver-side star ratings for 98 cars is reproduced in the accompanying MINITAB printout. Assume that one of the 98 cars is selected at random. State whether each of the following is true or false:

a. The probability that the car has a rating of two stars is 4.

b. The probability that the car has a rating of four or five stars is .7857.

c. The probability that the car has a rating of one star is 0.

d. The car has a better chance of having a two-star rating than of having a five-star rating.

Tally for Discrete Variables: DRIVSTAR

DRIVSTAR	Count	Percent
2	4	4.08
3	17	17.35
4	59	60.20
5	18	18.37
N=	98	

3.164 **Selecting a sample.** A random sample of eight students is to be selected from 40 sociology majors for participation in a special program. In how many different ways can the sample be drawn?

3.165 **Fungi in beech forest trees.** The current status of the beech tree species in East Central Europe was evaluated by Hungarian university professors in *Applied Ecology and Environmental Research* (Vol. 1, 2003). Of 188 beech trees surveyed, 49 had been damaged by fungi. Depending on the species of fungus, damage will occur on either the trunk, branches, or leaves of the tree. In the damaged trees, the trunk was affected 85% of the time, the leaves 10% of the time, and the branches 5% of the time.

a. Give a reasonable estimate of the probability of a beech tree in East Central Europe being damaged by fungi.

b. A fungus-damaged beech tree is selected at random, and the area (trunk, leaf, or branch) affected is observed. List the sample points for this experiment, and assign a reasonable probability to each one.

3.166 **Do you have a library card?** According to a Harris poll, 68% of all American adults have a public library card. The percentages do differ by gender—62% of males have library cards compared to 73% of females.

a. Consider the three probabilities: $P(A) = .68$, $P(A|B) = .62$, and $P(A|C) = .73$. Define events $A, B,$ and C.

b. Assuming that two-thirds of all American adults are males and one-thirds are females, what is the probability that an American adult is a male who owns a library card?

3.167 Beach erosional hot spots. Beaches that exhibit high erosion rates relative to the surrounding beach are defined as *erosional hot spots*. The U.S. Army Corps of Engineers is conducting a study of beach hot spots. Through an online questionnaire, data are collected on six beach hot spots. The data are listed in the next table.

Beach Hot Spot	Beach Condition	Nearshore Bar Condition	Long-Term Erosion Rate (miles/year)
Miami Beach, FL	No dunes/flat	Single, shore parallel	4
Coney Island, NY	No dunes/flat	Other	13
Surfside, CA	Bluff/scarp	Single, shore parallel	35
Monmouth Beach, NJ	Single dune	Planar	Not estimated
Ocean City, NJ	Single dune	Other	Not estimated
Spring Lake, NJ	Not observed	Planar	14

Based on *Identification and Characterization of Erosional Hotspots*, William & Mary Virginia Institute of Marine Science, U.S. Army Corps of Engineers Project Report, Mar. 18, 2002.

a. Suppose you record the nearshore bar condition of each beach hot spot. Give the sample space for this experiment.

b. Find the probabilities of the sample points in the sample space you defined in part **a**.

c. What is the probability that a beach hot spot has either a planar or single, shore-parallel nearshore bar condition?

d. Now suppose you record the beach condition of each beach hot spot. Give the sample space for this experiment.

e. Find the probabilities of the sample points in the sample space you defined in part **d**.

f. What is the probability that the condition of the beach at a particular beach hot spot is not flat?

3.168 Chemical insect attractant. An entomologist is studying the effect of a chemical sex attractant (pheromone) on insects. Several insects are released at a site equidistant from the pheromone under study and a control substance. If the pheromone has an effect, more insects will travel toward it rather than toward the control. Otherwise, the insects are equally likely to travel in either direction. Suppose the pheromone under study has no effect, so that it is equally likely that an insect will move toward the pheromone or toward the control. Suppose four insects are released.

a. List or count the number of different ways the insects can travel.

b. What is the chance that all four travel toward the pheromone?

c. What is the chance that exactly three travel toward the pheromone?

d. What inference would you make if the event in part **c** actually occurs? Explain.

3.169 Toxic chemical incidents. *Process Safety Progress* (Sept. 2004) reported on an emergency response system for incidents involving toxic chemicals in Taiwan. The system has logged over 250 incidents since being implemented. The next table gives a breakdown of the locations where these incidents occurred. Consider the location of a toxic chemical incident in Taiwan.

a. List the sample points for this experiment.

b. Assign reasonable probabilities to the sample points.

c. What is the probability that the incident occurs in a school laboratory?

d. What is the probability that the incident occurs in either a chemical or a nonchemical plant?

e. What is the probability that the incident does not occur in transit?

f. What is the probability that in a sample of 3 independently selected incidents, all 3 occur in a school laboratory?

g. Would you expect the event in part **f** to occur? Explain.

Location	Percent of Incidents
School laboratory	6%
In transit	26%
Chemical plant	21%
Nonchemical plant	35%
Other	12%
Total	100%

Based on Chen, J. R., et al. "Emergency response of toxic chemicals in Taiwan: The system and case studies." *Process Safety Progress*, Vol. 23, No. 3, Sept. 2004 (Figure 5a).

Applying the Concepts—Intermediate

3.170 Federal civil trial appeals. The *Journal of the American Law and Economics Association* (Vol. 3, 2001) published the results of a study of appeals of federal civil trials. The accompanying table, extracted from the article, gives a breakdown of 2,143 civil cases that were appealed by either the plaintiff or the defendant. The outcome of the appeal, as well as the type of trial (judge or jury), was determined for each civil case. Suppose one of the 2,143 cases is selected at random and both the outcome of the appeal and the type of trial are observed.

	Jury	Judge	Totals
Plaintiff trial win—reversed	194	71	265
Plaintiff trial win—affirmed/dismissed	429	240	669
Defendant trial win—reversed	111	68	179
Defendant trial win—affirmed/dismissed	731	299	1,030
Totals	1,465	678	2,143

a. Find $P(A)$, where $A = \{$jury trial$\}$.

b. Find $P(B)$, where $B = \{$plaintiff trial win is reversed$\}$.

c. Are A and B mutually exclusive events?

d. Find $P(A^c)$.

e. Find $P(A \cup B)$.

f. Find $P(A \cap B)$.

3.171 Winning at roulette. *Roulette* is a very popular game in many American casinos. In Roulette, a ball spins on a circular wheel that is divided into 38 arcs of equal length, bearing the numbers 00, 0, 1, 2, . . . , 35, 36. The number of the arc on which the ball stops is the outcome of one play of the game. The numbers are also colored in the manner shown in the following table.

| Red: 1, 3, 5, 7, 9, 12, 14, 16, 18, 19, 21, 23, 25, 27, 30, 32, 34, 36 |
| Black: 2, 4, 6, 8, 10, 11, 13, 15, 17, 20, 22, 24, 26, 28, 29, 31, 33, 35 |
| Green: 00, 0 |

Players may place bets on the table in a variety of ways, including bets on odd, even, red, black, high, low, etc. Consider the following events:

A: {The outcome is an odd number (00 and 0 are considered neither odd nor even.)}
B: {The outcome is a black number.}
C: {The outcome is a low number (1–18).}

a. Define the event $A \cap B$ as a specific set of sample points.
b. Define the event $A \cup B$ as a specific set of sample points.
c. Find $P(A), P(B), P(A \cap B), P(A \cup B)$, and $P(C)$ by summing the probabilities of the appropriate sample points.
d. Define the event $A \cap B \cap C$ as a specific set of sample points.
e. Use the additive rule to find $P(A \cup B)$. Are events A and B mutually exclusive? Why?
f. Find $P(A \cap B \cap C)$ by summing the probabilities of the sample points given in part **d.**
g. Define the event $(A \cup B \cup C)$ as a specific set of sample points.
h. Find $P(A \cup B \cup C)$ by summing the probabilities of the sample points given in part **g.**

3.172 Cigar smoking and cancer. The *Journal of the National Cancer Institute* (Feb. 16, 2000) published the results of a study that investigated the association between cigar smoking and death from tobacco-related cancers. Data were obtained for a national sample of 137,243 American men. The results are summarized in the table below. Each male in the study was classified according to his cigar-smoking status and whether or not he died from a tobacco-related cancer.

a. Find the probability that a randomly selected man never smoked cigars and died from cancer.
b. Find the probability that a randomly selected man was a former cigar smoker and died from cancer.
c. Find the probability that a randomly selected man was a current cigar smoker and died from cancer.
d. Given that a male was a current cigar smoker, find the probability that he died from cancer.
e. Given that a male never smoked cigars, find the probability that he died from cancer.

| | Died from Cancer | | |
Cigars	Yes	No	Totals
Never Smoked	782	120,747	121,529
Former Smoker	91	7,757	7,848
Current Smoker	141	7,725	7,866
Totals	1,014	136,229	137,243

Based on Shapiro, J. A., Jacobs, E. J., and Thun, M. J. "Cigar smoking in men and risk of death from tobacco-related cancers." *Journal of the National Cancer Institute,* Vol. 92, No. 4, Feb. 16, 2000 (Table 2).

*3.173 Errors in estimating job costs. A construction company employs three sales engineers. Engineers 1, 2, and 3 estimate the costs of 30%, 20%, and 50%, respectively, of all jobs bid on by the company. For $i = 1, 2, 3$, define E_i to be the event that a job is estimated by engineer i. The following probabilities describe the rates at which the engineers make serious errors in estimating costs:

$P(\text{error}|E_1) = .01, \ P(\text{error}|E_2) = .03,$ and
$P(\text{error}|E_3) = .02$

a. If a particular bid results in a serious error in estimating job cost, what is the probability that the error was made by engineer 1?
b. If a particular bid results in a serious error in estimating job cost, what is the probability that the error was made by engineer 2?
c. If a particular bid results in a serious error in estimating job cost, what is the probability that the error was made by engineer 3?
d. Based on the probabilities given in parts **a–c**, which engineer is most likely responsible for making the serious error?

3.174 Elderly wheelchair user study. The *American Journal of Public Health* (Jan. 2002) reported on a study of elderly wheelchair users who live at home. A sample of 306 wheelchair users, age 65 or older, were surveyed about whether they had an injurious fall during the year and whether their home featured any one of five structural modifications: bathroom modifications, widened doorways/hallways, kitchen modifications, installed railings, and easy-open doors. The responses are summarized in the accompanying table. Suppose we select, at random, one of the 306 wheelchair users surveyed.

Home Features	Injurious Fall(s)	No Falls	Totals
All 5	2	7	9
At least 1, but not all	26	162	188
None	20	89	109
Totals	48	258	306

Based on Berg, K., Hines, M., and Allen, S. "Wheelchair users at home: Few home modifications and many injurious falls." *American Journal of Public Health,* Vol. 92, No. 1, Jan. 2002 (Table 1).

a. Find the probability that the wheelchair user had an injurious fall.
b. Find the probability that the wheelchair user had all five features installed in the home.
c. Find the probability that the wheelchair user had no falls and none of the features installed in the home.
d. Given no features installed in the home, find the probability of an injurious fall.

*3.175 Matching medical students with residencies. The National Resident Matching Program (NRMP) is a service provided to match graduating medical students with residency appointments at hospitals. After students and hospital officials have evaluated each other, they submit rank-order lists of their preferences to the NRMP. Using a matching algorithm, the NRMP then generates final, nonnegotiable assignments of students to the residency programs of hospitals (www.nrmp.org). Assume that three graduating medical students (#1, #2, and #3) have applied for positions at three different hospitals (*A*, *B*, and *C*), each of which has one and only one resident opening.

a. How many different assignments of medical students to hospitals are possible? List them.

b. Suppose student #1 prefers hospital B. If the NRMP algorithm is entirely random, what is the probability that the student is assigned to hospital B?

3.176 Selecting new-car options. A company sells midlevel models of automobiles in eight different styles. A buyer can get an automobile in one of four colors and with either standard or automatic transmission. Would it be reasonable to expect a dealer to stock at least one automobile in every combination of style, color, and transmission? At a minimum, how many automobiles would the dealer have to stock?

3.177 Shooting free throws. In college basketball games, a player may be afforded the opportunity to shoot two consecutive foul shots (free throws).

a. Suppose a player who makes (i.e., scores on) 75% of his foul shots has been awarded two free throws. If the two throws are considered independent, what is the probability that the player makes both shots? Exactly one? Neither shot?

b. Suppose a player who makes 75% of his first attempted foul shots has been awarded two free throws and the outcome on the second shot is dependent on the outcome of the first shot. In fact, if this player makes the first shot, he makes 80% of the second shots; and if he misses the first shot, he makes 75% of the second shots. In this case, what is the probability that the player makes both shots? Exactly one? Neither shot?

c. In parts **a** and **b,** we considered two ways of modeling the probability that a basketball player makes two consecutive foul shots. Which model do you think gives a more realistic explanation of the outcome of shooting foul shots; that is, do you think two consecutive foul shots are independent or dependent? Explain.

3.178 Lie detector test. Consider a lie detector called the Computerized Voice Stress Analyzer (CVSA). The manufacturer claims that the CVSA is 98% accurate and, unlike a polygraph machine, will not be thrown off by drugs and medical factors. However, laboratory studies by the U.S. Defense Department found that the CVSA had an accuracy rate of 49.8%, slightly less than pure chance. Suppose the CVSA is used to test the veracity of four suspects. Assume that the suspects' responses are independent.

a. If the manufacturer's claim is true, what is the probability that the CVSA will correctly determine the veracity of all four suspects?

b. If the manufacturer's claim is true, what is the probability that the CVSA will yield an incorrect result for at least one of the four suspects?

c. Suppose that in a laboratory experiment conducted by the U.S. Defense Department on four suspects, the CVSA yielded incorrect results for two of the suspects. Make an inference about the true accuracy rate of the new lie detector.

3.179 Maize seeds. The genetic origin and properties of maize (modern-day corn) were investigated in *Economic Botany*. Seeds from maize ears carry either single spikelets or paired spikelets but not both. Progeny tests on approximately 600 maize ears revealed the following information: Forty percent of all seeds carry single spikelets, while 60% carry paired spikelets. A seed with single spikelets will produce maize ears with single spikelets 29% of the time and paired spikelets 71% of the time. A seed with paired spikelets will produce maize ears with single spikelets 26% of the time and paired spikelets 74% of the time.

a. Find the probability that a randomly selected maize ear seed carries a single spikelet and produces ears with single spikelets.

b. Find the probability that a randomly selected maize ear seed produces ears with paired spikelets.

*3.180 Selecting committee members. The Republican governor of a state with no income tax is appointing a committee of five members to consider changes in the income tax law. There are 15 state representatives—seven Democrats and eight Republicans—available for appointment to the committee. Assume that the governor selects the committee of five members randomly from the 15 representatives.

a. In how many different ways can the committee members be selected?

b. What is the probability that no Democrat is appointed to the committee? If this were to occur, what would you conclude about the assumption that the appointments were made at random? Why?

c. What is the probability that the majority of the committee members are Republican? If this were to occur, would you have reason to doubt that the governor made the selections randomly? Why?

3.181 Series and parallel systems. Consider the two systems shown in the schematic below. System A operates properly only if all three components operate properly. (The three components are said to operate *in series*.) The probability of failure for system A components 1, 2, and 3 is .1, .08, and .13, respectively. Assume that the components operate independently of each other.

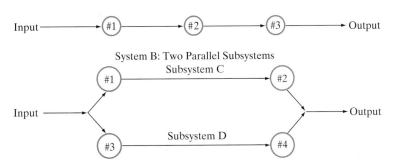

System A: Three Components in Series

Input → #1 → #2 → #3 → Output

System B: Two Parallel Subsystems

Subsystem C
#1 → #2

Input

Subsystem D
#3 → #4

Output

System B comprises two subsystems said to operate *in parallel*. Each subsystem has two components that operate in series. System B will operate properly as long as at least one of the subsystems functions properly. The probability of failure for each component in the system is .2. Assume that the components operate independently of each other.

a. Find the probability that System A operates properly.

b. What is the probability that at least one of the components in System A will fail and therefore that the system will fail?

c. Find the probability that System B operates properly.

d. Find the probability that exactly one subsystem in System B fails.

e. Find the probability that System B fails to operate properly.

f. How many parallel subsystems like the two shown would be required to guarantee that the system would operate properly at least 99% of the time?

3.182 Monitoring the quality of power equipment. *Mechanical Engineering* (Feb. 2005) reported on the need for wireless networks to monitor the quality of industrial equipment. For example, consider Eaton Corp., a company that develops distribution products. Eaton estimates that 90% of the electrical switching devices it sells can monitor the quality of the power running through the device. Eaton further estimates that, of the buyers of electrical switching devices capable of monitoring quality, 90% do not wire the equipment up for that purpose. Use this information to estimate the probability that an Eaton electrical switching device is capable of monitoring power quality and is wired up for that purpose.

***3.183 Repairing a computer system.** The local area network (LAN) for the computing system at a large university is temporarily shut down for repairs. Previous shutdowns have been due to hardware failure, software failure, or power failure. Maintenance engineers have determined that the probabilities of hardware, software, and power problems are .01, .05, and .02, respectively. They have also determined that if the system experiences hardware problems, it shuts down 73% of the time. Similarly, if software problems occur, the system shuts down 12% of the time; and, if power failure occurs, the system shuts down 88% of the time. What is the probability that the current shutdown of the LAN is due to hardware failure? software failure? power failure?

***3.184 Dream experiment in psychology.** A clinical psychologist is given tapes of 20 subjects discussing their recent dreams. The psychologist is told that 10 are high-anxiety individuals and 10 are low-anxiety individuals. The psychologist's task is to select the 10 high-anxiety subjects.

a. Count the number of sample points for this experiment.

b. Assuming that the psychologist is guessing, assign probabilities to each of the sample points.

c. Find the probability that the psychologist guesses all classifications correctly.

d. Find the probability that the psychologist guesses at least 9 of the 10 high-anxiety subjects correctly.

3.185 Sociology fieldwork methods. Refer to University of New Mexico professor Jane Hood's study of the fieldwork methods used by qualitative sociologists, presented in Exercise 2.195 (p. 136). Recall that she discovered that fieldwork methods could be classified into four distinct categories: Interview, Observation plus Participation, Observation Only, and Grounded Theory. The table that follows, reproduced from *Teaching Sociology* (July 2006), gives the number of sociology field research papers in each category. Suppose we randomly select one of these research papers and determine the method used. Find the probability that the method used is either Interview or Grounded Theory.

Fieldwork Method	Number of Papers
Interview	5,079
Observation plus Participation	1,042
Observation Only	848
Grounded Theory	537

Based on Hood, J. C. "Teaching against the text: The case of qualitative methods." *Teaching Sociology*, Vol. 34, Issue 3 (Exhibit 2).

3.186 Forest fragmentation study. Refer to the *Conservation Ecology* (Dec. 2003) study of the causes of forest fragmentation, presented in Exercise 2.166 (p. 125). Recall that the researchers used advanced high-resolution satellite imagery to develop fragmentation indexes for each forest. A 3 × 3 grid was superimposed over an aerial photo of the forest, and each square (pixel) of the grid was classified as forest (F), as earmarked for anthropogenic land use (A), or as natural land cover (N). An example of one such grid is shown here. The edges of the grid (where an "edge" is an imaginary line that separates any two adjacent pixels) are classified as F–A, F–N, A–A, A–N, N–N, or F–F edges.

A	A	N
N	F	F
N	F	F

a. Note that there are 12 edges inside the grid. Classify each edge as F–A, F–N, A–A, A–N, N–N, or F–F.

b. The researchers calculated the fragmentation index by considering only the F-edges in the grid. Count the number of F-edges. (These edges represent the sample space for the experiment.)

c. If an F-edge is selected at random, find the probability that it is an F–A edge. (This probability is proportional to the anthropogenic fragmentation index calculated by the researchers.)

d. If an F-edge is selected at random, find the probability that it is an F–N edge. (This probability is proportional to the natural fragmentation index calculated by the researchers.)

Applying the Concepts—Advanced

3.187 Odds of winning a horse race. Handicappers for horse races express their beliefs about the probability of each horse winning a race in terms of **odds.** If the probability of event E is $P(E)$, then the *odds in favor of E* are $P(E)$ to $1 - P(E)$. Thus, if a handicapper assesses a probability of .25 that Smarty Jones will win the Belmont Stakes, the odds in favor of Smarty Jones are $^{25}/_{100}$ to $^{75}/_{100}$, or 1 to 3. It follows that the *odds against E* are $1 - P(E)$ to $P(E)$, or 3 to 1 against a win by Smarty Jones. In general, if the odds in favor of event E are a to b, then $P(E) = a/(a + b)$.

a. A second handicapper assesses the probability of a win by Smarty Jones to be $^1/_5$. According to the second handicapper, what are the odds in favor of a Smarty Jones win?

b. A third handicapper assesses the odds in favor of Smarty Jones to be 2 to 3. According to the third handicapper, what is the probability of a Smarty Jones win?

c. A fourth handicapper assesses the odds against Smarty Jones winning to be 5 to 3. Find this handicapper's assessment of the probability that Smarty Jones will win.

3.188 Sex composition patterns of children in families. In having children, is there a genetic factor that causes some families to favor one sex over the other? That is, does having boys or girls "run in the family"? This was the question of interest in *Chance* (Fall 2001). Using data collected on children's sex for over 4,000 American families that had at least two children, the researchers compiled the accompanying table. Make an inference about whether having boys or girls "runs in the family."

Sex Composition of First Two Children	Frequency
Boy–Boy	1,085
Boy–Girl	1,086
Girl–Boy	1,111
Girl–Girl	926
Total	4,208

Based on Rodgers, J. L., and Doughty, D. "Does having boys or girls run in the family?" *Chance*, Vol. 14, No. 4, Fall 2001 (Table 3).

3.189 Finding an organ transplant match. One of the problems encountered with organ transplants is the body's rejection of the transplanted tissue. If the antigens attached to the tissue cells of the donor and receiver match, the body will accept the transplanted tissue. Although the antigens in identical twins always match, the probability of a match in other siblings is .2, and that of a match in two people from the population at large is .001. Suppose you need a kidney and you have a brother and three sisters.

a. If one of your four siblings offers a kidney, what is the probability that the antigens will match?

b. If all four siblings offer a kidney, what is the probability that all four antigens will match?

c. If all four siblings offer a kidney, what is the probability that none of the antigens will match?

d. Repeat parts **b** and **c**, this time assuming that the four donors were obtained from the population at large.

3.190 Chance of winning at blackjack. Blackjack, a favorite game of gamblers, is played by a dealer and at least one opponent. At the outset of the game, 2 cards of a 52-card bridge deck are dealt to the player and 2 cards to the dealer.

Drawing an ace and a face card is called *blackjack*. If the dealer does not draw a blackjack and the player does, the player wins. If both the dealer and player draw blackjack, a "push" (i.e., a tie) occurs.

a. What is the probability that the dealer will draw a blackjack?

b. What is the probability that the player wins with a blackjack?

3.191 Accuracy of pregnancy tests. Sixty percent of all women who submit to pregnancy tests are really pregnant. A certain pregnancy test gives a *false positive* result with probability .03 and a *valid positive result* with probability .98. If a particular woman's test is positive, what is the probability that she really is pregnant? [*Hint:* If A is the event that a woman is pregnant and B is the event that the pregnancy test is positive, then B is the union of the two mutually exclusive events $A \cap B$ and $A^c \cap B$. Also, the probability of a false positive result may be written as $P(B|A^c) = .03$.]

3.192 Chance of winning at "craps." A version of the dice game "craps" is played in the following manner. A player starts by rolling two balanced dice. If the roll (the sum of the two numbers showing on the dice) results in a 7 or 11, the player wins. If the roll results in a 2 or a 3 (called *craps*), the player loses. For any other roll outcome, the player continues to throw the dice until the original roll outcome recurs (in which case the player wins) or until a 7 occurs (in which case the player loses).

a. What is the probability that a player wins the game on the first roll of the dice?

b. What is the probability that a player loses the game on the first roll of the dice?

c. If the player throws a total of 4 on the first roll, what is the probability that the game ends (win or lose) on the next roll?

*3.193 The perfect bridge hand. According to a morning news program, a very rare event recently occurred in Dubuque, Iowa. Each of four women playing bridge was astounded to note that she had been dealt a perfect bridge hand. That is, one woman was dealt all 13 spades, another all 13 hearts, another all the diamonds, and another all the clubs. What is the probability of this rare event?

3.194 Odd Man Out. Three people play a game called "Odd Man Out." In this game, each player flips a fair coin until the outcome (heads or tails) for one of the players is not the same as that for the other two players. This player is then "the odd man out" and loses the game. Find the probability that the game ends (i.e., either exactly one of the coins will fall heads or exactly one of the coins will fall tails) after only one toss by each player. Suppose one of the players, hoping to reduce his chances of being the odd man out, uses a two-headed coin. Will this ploy be successful? Solve by listing the sample points in the sample space.

Critical Thinking Challenges

3.195 "Let's Make a Deal." Marilyn vos Savant, who is listed in the *Guinness Book of World Records Hall of Fame* as having the "Highest IQ," writes a weekly column in the Sunday newspaper supplement *Parade Magazine*.

Her column, "Ask Marilyn," is devoted to games of skill, puzzles, and mind-bending riddles. In one issue (*Parade Magazine*, Feb. 24, 1991), vos Savant posed the following question:

Suppose you're on a game show and you're given a choice of three doors. Behind one door is a car; behind the others, goats. You pick a door—say, #1—and the host, who knows what's behind the doors, opens another door—say #3—which has a goat. He then says to you, "Do you want to pick door #2?" Is it to your advantage to switch your choice?

Marilyn's answer: "Yes, you should switch. The first door has a $1/3$ chance of winning [the car], but the second has a $2/3$ chance [of winning the car]." Predictably, vos Savant's surprising answer elicited thousands of critical letters, many of them from Ph.D. mathematicians, that disagreed with her. Who is correct, the Ph.D.s or Marilyn?

3.196 Most likely coin-toss sequence. In *Parade Magazine*'s (Nov. 26, 2000) column "Ask Marilyn," the following question was posed: "I have just tossed a [balanced] coin 10 times, and I ask you to guess which of the following three sequences was the result. One (and only one) of the sequences is genuine."

1. H H H H H H H H H H
2. H H T T H T T H H H
3. T T T T T T T T T T

Marilyn's answer to the question posed was "Though the chances of the three specific sequences occurring randomly are equal ... it's reasonable for us to choose sequence (2) as the most likely genuine result." Do you agree?

Activity | **Exit Polls**

Exit polls are conducted in selected locations as voters leave their polling places after voting. In addition to being used to predict the outcome of elections before the votes are counted, these polls seek to gauge tendencies among voters. The results are usually stated in terms of conditional probabilities.

The accompanying table shows the results of exit polling that suggest that men were move likely to vote for Mitt Romney, while women were more likely to vote for Barack Obama in the 2012 presidential election. In addition, the table suggests that more women than men voted in the election. The six percentages in the last three columns represent conditional probabilities, where the given event is gender.

2012 Presidential Election, Vote by Gender

	Obama	Romney	Other
Male (47%)	45%	52%	3%
Female (53%)	55%	44%	1%

Based on CNN.com

1. Find similar exit poll results in which voters are categorized by race, income, education, or some other criterion for a recent national, state, or local election. Choose two different examples, and interpret the percentages given as probabilities, or conditional probabilities where appropriate.

2. Use the multiplicative rule of probability to find the probabilities related to the percentages given. [For example, in the accompanying table, find P(Obama and Male).] Then interpret each of these probabilities and use them to determine the total percentage of the electorate that voted for each candidate.

3. Describe a situation in which a political group might use a form of exit polling to gauge voters' opinions on a "hot-button" topic (e.g., global warming). Identify the political group, the "hot-button" topic, the criterion used to categorize voters, and how the voters' opinions will be determined. Then describe how the results will be summarized as conditional probabilities. How might the results of the poll be used to support a particular agenda?

References

Bennett, D. J. *Randomness.* Cambridge, MA: Harvard University Press, 1998.

Epstein, R. A. *The Theory of Gambling and Statistical Logic*, rev. ed. New York: Academic Press, 1977.

Feller, W. *An Introduction to Probability Theory and Its Applications*, 3rd ed., Vol. 1. New York: Wiley, 1968.

Lindley, D. V. *Making Decisions*, 2nd ed. London: Wiley, 1985.

Parzen, E. *Modern Probability Theory and Its Applications.* New York: Wiley, 1960.

Wackerly, D., Mendenhall, W., and Scheaffer, R. L. *Mathematical Statistics with Applications*, 7th ed. Boston: Duxbury, 2008.

Williams, B. *A Sampler on Sampling.* New York: Wiley, 1978.

Winkler, R. L. *An Introduction to Bayesian Inference and Decision.* New York: Holt, Rinehart and Winston, 1972.

Wright, G., and Ayton, P., eds. *Subjective Probability.* New York: Wiley, 1994.

TI-83/TI-84 Plus Graphing Calculator:
Combinations and Permutations

Combinations—Choosing *n* Elements from *N* Elements

Step 1 Enter the value of *N*.

Step 2 Press **MATH**, then select **PRB**.

Step 3 Press **nCr** for combinations.

Step 4 Enter the value of *n*, then press **ENTER**.

Permutations—Choosing *n* Elements from *N* Elements

Step 1 Enter the value of *N*.

Step 2 Press **MATH**. Then select **PRB**.

Step 3 Press **nPr** for permutations.

Step 4 Enter the value of *n*, then press **ENTER**.

4 Discrete Random Variables

CONTENTS

Where We've Been

- Used probability to make an inference about a population from data in an observed sample
- Used probability to measure the reliability of the inference

Where We're Going

- Develop the notion of a random variable (4.1)
- Learn that many types of numerical data are observed values of discrete random variables (4.1)
- Learn the probabilistic properties of discrete random variables (4.2–4.3)
- Study several important types of discrete random variables and their probability models (4.4–4.6)

Statistics IN Action Probability in a Reverse Cocaine Sting: Was Cocaine Really Sold?

The American Statistician (May 1991) described an interesting application of a discrete probability distribution in a case involving illegal drugs. It all started with a "bust" in a midsized Florida city. During the bust, police seized approximately 500 foil packets of a white, powdery substance, presumably cocaine. Since it is not a crime to buy or sell nonnarcotic cocaine look-alikes (e.g., inert powders), detectives had to prove that the packets actually contained cocaine in order to convict their suspects of drug trafficking. When the police laboratory randomly selected and chemically tested 4 of the packets, all 4 tested positive for cocaine. This finding led to the conviction of the traffickers.

After the conviction, the police decided to use the remaining foil packets (i.e., those not tested) in reverse sting operations. Two of these packets were randomly selected and sold by undercover officers to a buyer. Between the sale and the arrest, however, the buyer disposed of the evidence. The key question is, Beyond a reasonable doubt, did the defendant really purchase cocaine?

In court, the defendant's attorney argued that his client should not be convicted because the police could not prove

that the missing foil packets contained cocaine. The police contended, however, that since 4 of the original packets tested positive for cocaine, the 2 packets sold in the reverse sting were also highly likely

to contain cocaine. In this chapter, two Statistics in Action Revisited examples demonstrate how to use probability models to solve the dilemma posed by the police's reverse sting. (The case represented Florida's first cocaine-possession conviction without the actual physical evidence.)

Statistics IN Action   Revisited

- Using the Binomial Model to Solve the Cocaine Sting Case (p. 238)
- Using the Hypergeometric Model to Solve the Cocaine Sting Case (p. 249)

HH
○
(2)

HT
○
(1)

TT
○
(0)

TH
○
(1)

S

Figure 4.1

Venn diagram for coin-tossing experiment

You may have noticed that many of the examples of experiments in Chapter 3 generated quantitative (numerical) observations. The unemployment rate, the percentage of voters favoring a particular candidate, the cost of textbooks for a school term, and the amount of pesticide in the discharge waters of a chemical plant are all examples of numerical measurements of some phenomenon. Thus, most experiments have sample points that correspond to values of some numerical variable.

To illustrate, consider the coin-tossing experiment of Chapter 3. Figure 4.1 is a Venn diagram showing the sample points when two coins are tossed and the up faces (heads or tails) of the coins are observed. One possible numerical outcome is the total number of heads observed. This value (0, 1, or 2) is shown in parentheses on the Venn diagram, with one numerical value associated with each sample point. In the jargon of probability, the variable "total number of heads observed in two tosses of a coin" is called a *random variable*.

> A **random variable** is a variable that assumes numerical values associated with the random outcomes of an experiment, where one (and only one) numerical value is assigned to each sample point.

The term *random variable* is more meaningful than the term *variable* alone because the adjective *random* indicates that the coin-tossing experiment may result in one of the several possible values of the variable—0, 1, and 2—according to the *random* outcomes of the experiment: *HH, HT, TH,* and *TT.* Similarly, if the experiment is to count the number of customers who use the drive-up window of a bank each day, the random variable (the number of customers) will vary from day to day, partly because of the random phenomena that influence whether customers use the window. Thus, the possible values of this random variable range from 0 to the maximum number of customers the window can serve in a day.

We define two different types of random variables, *discrete* and *continuous*, in Section 4.1. Then we spend the remainder of the chapter discussing specific types of discrete random variables and the aspects that make them important to the statistician. We discuss continuous random variables in Chapter 5.

4.1 Two Types of Random Variables

Recall that the sample-point probabilities corresponding to an experiment must sum to 1. Dividing one unit of probability among the sample points in a sample space and, consequently, assigning probabilities to the values of a random variable are not always as easy as the examples in Chapter 3 might lead you to believe. If the number of sample points can be completely listed, the job is straightforward. But if the experiment results in an infinite number of sample points, which is impossible to list, the task of assigning probabilities to the sample points is impossible without the aid of a probability model. The next three examples demonstrate the need for different probability models, depending on the number of values that a random variable can assume.

Example 4.1

Values of a Discrete Random Variable—Wine Ratings

Problem A panel of 10 experts for the *Wine Spectator* (a national publication) is asked to taste a new white wine and assign it a rating of 0, 1, 2, or 3. A score is then obtained by adding together the ratings of the 10 experts. How many values can this random variable assume?

Solution A sample point is a sequence of 10 numbers associated with the rating of each expert. For example, one sample point is

$$\{1, 0, 0, 1, 2, 0, 0, 3, 1, 0\}$$

[*Note:* Using one of the counting rules from Section 3.7, there are a total of $4^{10} = 1,048,576$ sample points for this experiment.] Now, the random variable assigns a score to each one of these sample points by adding the 10 numbers together. Thus, the smallest score is 0 (if all 10 ratings are 0), and the largest score is 30 (if all 10 ratings are 3). Since every integer between 0 and 30 is a possible score, the random variable denoted by the symbol x can assume 31 values. Note that the value of the random variable for the sample point shown above is $x = 8$.*

Look Back Use of the random variable reduces the amount of information required in the experiment—from over 1 million sample points to 31 values of the random variable. Also, this is an example of a *discrete random variable*, since there is a finite number of distinct possible values. Whenever all the possible values that a random variable can assume can be listed (or *counted*), the random variable is *discrete*.

Example 4.2

Values of a Discrete Random Variable—EPA Application

Problem Suppose the Environmental Protection Agency (EPA) takes readings once a month on the amount of pesticide in the discharge water of a chemical company. If the amount of pesticide exceeds the maximum level set by the EPA, the company is forced to take corrective action and may be subject to penalty. Consider the random variable x to be the number of months before the company's discharge exceeds the EPA's maximum level. What values can x assume?

Solution The company's discharge of pesticide may exceed the maximum allowable level on the first month of testing, the second month of testing, etc. It is possible that the company's discharge will *never* exceed the maximum level. Thus, the set of possible

---

*The standard mathematical convention is to use a capital letter (e.g., X) to denote the theoretical random variable. The possible values (or realizations) of the random variable are typically denoted with a lowercase letter (e.g., x). Thus, in Example 4.1, the random variable X can take on the values $x = 0, 1, 2, \ldots, 30$. Since this notation can be confusing for introductory statistics students, we simplify the notation by using the lowercase x to represent the random variable throughout.

values for the number of months until the level is first exceeded is the set of all positive integers 1, 2, 3, 4, ...

Look Back If we can list the values of a random variable x, even though the list is never ending, we call the list **countable** and the corresponding random variable **discrete**. Thus, the number of months until the company's discharge first exceeds the limit is a **discrete random variable**.

■ Now Work Exercise 4.7

Example **4.3**	**Problem** Refer to Example 4.2. A second random variable of interest is the amount x of pesticide (in milligrams per liter) found in the monthly sample of discharge waters from the same chemical company. What values can this random variable assume?
Values of a Continuous Random Variable—Another EPA Application	**Solution** Some possible values of x are 1.7, 28.42, and 100.987 milligrams per liter. Unlike the *number* of months before the company's discharge exceeds the EPA's maximum level, the set of all possible values for the *amount* of discharge *cannot* be listed (i.e., is not countable). The possible values for the amount x of pesticide would correspond to the points on the interval between 0 and the largest possible value the amount of the discharge could attain, the maximum number of milligrams that could occupy 1 liter of volume. (Practically, the interval would be much smaller, say, between 0 and 500 milligrams per liter.) **Look Ahead** When the values of a random variable are not countable but instead correspond to the points on some interval, we call the variable a *continuous random variable*. Thus, the *amount* of pesticide in the chemical plant's discharge waters is a *continuous random variable*.

■ Now Work Exercise 4.9

> Random variables that can assume a *countable* number of values are called **discrete**.

> Random variables that can assume values corresponding to any of the points contained in an interval are called **continuous**.

The following are examples of discrete random variables:

1. The number of seizures an epileptic patient has in a given week: $x = 0, 1, 2, \ldots$

2. The number of voters in a sample of 500 who favor impeachment of the president: $x = 0, 1, 2, \ldots, 500$

3. The shoe size of a tennis player: $x = \ldots 5, 5\frac{1}{2}, 6, 6\frac{1}{2}, 7, 7\frac{1}{2} \ldots$

4. The change received for paying a bill: $x = 1¢, 2¢, 3¢, \ldots, \$1, \$1.01, \$1.02, \ldots$

5. The number of customers waiting to be served in a restaurant at a particular time: $x = 0, 1, 2, \ldots$

Note that several of the examples of discrete random variables begin with the words *The number of.* ... This wording is very common, since the discrete random variables most frequently observed are counts. The following are examples of continuous random variables:

1. The length of time (in seconds) between arrivals at a hospital clinic: $0 \le x \le \infty$ (infinity)

2. The length of time (in minutes) it takes a student to complete a one-hour exam: $0 \le x \le 60$

3. The amount (in ounces) of carbonated beverage loaded into a 12-ounce can in a can-filling operation: $0 \le x \le 12$

4. The depth (in feet) at which a successful oil-drilling venture first strikes oil: $0 \le x \le c$, where c is the maximum depth obtainable

5. The weight (in pounds) of a food item bought in a supermarket: $0 \leq x \leq 500$ [*Note:* Theoretically, there is no upper limit on x, but it is unlikely that it would exceed 500 pounds.]

Discrete random variables and their probability distributions are discussed in this chapter. Continuous random variables and their probability distributions are the topic of Chapter 5.

Exercises 4.1–4.16

Understanding the Principles

4.1 Define a discrete random variable.

4.2 Define a continuous random variable.

Applying the Concepts—Basic

4.3 **Type of Random Variable.** Classify the following random variables according to whether they are discrete or continuous:
a. The number of students in a class
b. The height of students in a college
c. The proportion of girls in a class of 100 students
d. The monthly income of employees in a company
e. The water flowing through the Hirakud Dam
f. The amount of rainfall in a particular city.

4.4 **Type of Random Variable.** Identify the following random variables as discrete or continuous:
a. The life span of an electric bulb
b. The heart rate (number of beats per minute) of an adult male
c. The number of defective push pins produced by a machine in a company
d. The temperature in a particular city
e. The population of fishes in a pond.
f. The number of leaks in a 100 kilometer oil pipeline

4.5 **Type of Random Variable.** Identify the following variables as discrete or continuous:
a. The time span of failure of an electronic component
b. The number of accidents committed per month in your community
c. The number of machine breakdowns during a given day
d. The number of matches won by a player
e. The number of faulty blades in a pack of 100
f. The amount of water in a bottle

4.6 **NHTSA crash tests.** The National Highway Traffic Safety Administration (NHTSA) has developed a driver-side "star" scoring system for crash-testing new cars. Each crash-tested car is given a rating ranging from one star (*) to five stars (*****): the more stars in the rating, the better is the level of crash protection in a head-on collision. Suppose that a car is selected and its driver-side star rating is determined. Let x equal the number of stars in the rating. Is x a discrete or continuous random variable?

4.7 **Customers in line at a Subway shop.** The number of customers, x, waiting in line to order sandwiches at a Subway shop at noon is of interest to the store manager. What values can x assume? Is x a discrete or continuous random variable?

4.8 **Sound waves from a basketball.** Refer to the *American Journal of Physics* (June 2010) experiment on sound waves produced from striking a basketball, Exercise 2.43 (p. 80). Recall that the frequencies of sound wave echoes resulting from striking a hanging basketball with a metal rod were recorded. Classify the random variable, frequency (measured in hertz) of an echo, as discrete or continuous.

4.9 **Mongolian desert ants.** Refer to the *Journal of Biogeography* (Dec. 2003) study of ants in Mongolia, presented in Exercise 2.68 (p. 91). Two of the several variables recorded at each of 11 study sites were annual rainfall (in millimeters) and number of ant species. Identify these variables as discrete or continuous.

4.10 **Motivation of drug dealers.** Refer to the *Applied Psychology in Criminal Justice* (Sept. 2009) study of the personality characteristics of drug dealers, Exercise 2.102 (p. 105). For each of 100 convicted drug dealers, the researchers measured several variables, including the number of prior felony arrests x. Is x a discrete or continuous random variable? Explain.

Applying the Concepts—Intermediate

4.11 **Nutrition.** Give an example of a discrete random variable of interest to a nutritionist.

4.12 **Physics.** Give an example of a discrete random variable of interest to a physicist.

4.13 **Medical.** Give an example of a discrete random variable of interest to a doctor in a hospital.

4.14 **Artist.** Give an example of a discrete random variable of interest to an artist.

4.15 **Irrelevant speech effects.** Refer to the *Acoustical Science & Technology* (Vol. 35, 2014) study of the degree to which the memorization process is impaired by irrelevant background speech (called irrelevant speech effects), Exercise 2.34 (p. 77). Recall that subjects performed a memorization task under two conditions: (1) with irrelevant background speech and (2) in silence. Let x represent the difference in the error rates for the two conditions—called the relative difference in error rate (RDER). Explain why x is a continuous random variable.

4.16 **Shaft graves in ancient Greece.** Refer to the *American Journal of Archaeology* (Jan. 2014) study of shaft graves in ancient Greece, Exercise 2.37 (p. 78). Let x represent the number of decorated sword shafts buried at a discovered grave site. Explain why x is a discrete random variable.

4.2 Probability Distributions for Discrete Random Variables

A complete description of a discrete random variable requires that we *specify all the values the random variable can assume* and *the probability associated with each value.* To illustrate, consider Example 4.4.

Example	4.4

Finding a Probability Distribution—Coin-Tossing Experiment

HH	HT
○	○
x = 2	x = 1

TH	TT
○	○
x = 1	x = 0

S

Figure 4.2
Venn diagram for the two-coin-toss experiment

Problem Recall the experiment of tossing two coins (p. 213), and let x be the number of heads observed. Find the probability associated with each value of the random variable x, assuming that the two coins are fair.

Solution The sample space and sample points for this experiment are reproduced in Figure 4.2. Note that the random variable x can assume values 0, 1, 2. Recall (from Chapter 3) that the probability associated with each of the four sample points is $1/4$. Then, identifying the probabilities of the sample points associated with each of these values of x, we have

$$P(x = 0) = P(TT) = \frac{1}{4}$$

$$P(x = 1) = P(TH) + P(HT) = \frac{1}{4} + \frac{1}{4} = \frac{1}{2}$$

$$P(x = 2) = P(HH) = \frac{1}{4}$$

Thus, we now know the values the random variable can assume (0, 1, 2) and how the probability is *distributed over* those values ($1/4$, $1/2$, $1/4$). This dual specification completely describes the random variable and is referred to as the *probability distribution*, denoted by the symbol $p(x)$.* The probability distribution for the coin-toss example is shown in tabular form in Table 4.1 and in graphic form in Figure 4.3. Since the probability distribution for a discrete random variable is concentrated at specific points (values of x), the graph in Figure 4.3a represents the probabilities as the heights of vertical lines over the corresponding values of x. Although the representation of the probability distribution as a histogram, as in Figure 4.3b, is less precise (since the probability is spread over a unit interval), the histogram representation will prove useful when we approximate probabilities of certain discrete random variables in Section 4.4.

Table 4.1
Probability Distribution for Coin-Toss Experiment: Tabular Form

x	$p(x)$
0	$1/4$
1	$1/2$
2	$1/4$

Figure 4.3
Probability distribution for coin-toss experiment: graphical form

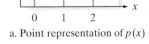
a. Point representation of $p(x)$

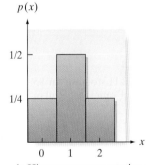
b. Histogram representation of $p(x)$

Look Ahead We could also present the probability distribution for x as a formula, but this would unnecessarily complicate a very simple example. We give the formulas for the probability distributions of some common discrete random variables later in the chapter.

■ **Now Work Exercise 4.23**

*In standard mathematical notation, the probability that a random variable X takes on a value x is denoted $P(X = x) = p(x)$. Thus, $P(X = 0) = p(0)$, $P(X = 1) = p(1)$, etc. In this text, we adopt the simpler $p(x)$ notation.

> The **probability distribution of a discrete random variable** is a graph, table, or formula that specifies the probability associated with each possible value that the random variable can assume.

Two requirements must be satisfied by all probability distributions for discrete random variables:

> **Requirements for the Probability Distribution of a Discrete Random Variable x**
>
> **1.** $p(x) \geq 0$ for all values of x.
>
> **2.** $\sum p(x) = 1$
>
> where the summation of $p(x)$ is over all possible values of x.*

Example	4.5

Probability Distribution from a Graph—Playing Craps

Figure 4.4

MINITAB Histogram for $5 Wager on Come-Out Roll in Craps

Problem Craps is a popular casino game in which a player throws two dice and bets on the outcome (the sum total of the dots showing on the upper faces of the two dice). Consider a $5 wager. On the first toss (called the *come-out* roll), if the total is 7 or 11 the roller wins $5. If the outcome is a 2, 3, or 12, the roller loses $5 (i.e., the roller wins −$5). For any other outcome (4, 5, 6, 8, 9, or 10), a *point* is established and no money is lost or won on that roll (i.e., the roller wins $0). In a computer simulation of repeated tosses of two dice, the outcome x of the come-out roll wager (−$5, $0, or +$5) was recorded. A relative frequency histogram summarizing the results is shown in Figure 4.4. Use the histogram to find the approximate probability distribution of x.

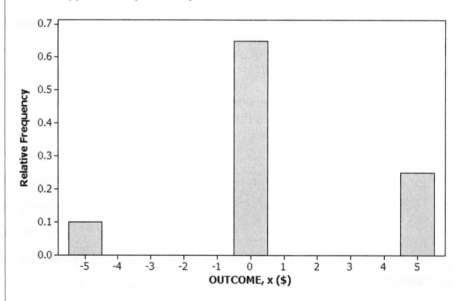

Solution The histogram shows that the relative frequencies of the outcomes $x = -\$5, x = \$0$, and $x = \$5$ are .1, .65, and .25, respectively. For example, in repeated tosses of two dice, 25% of the outcomes resulted in a sum of 7 or 11 (a $5 win for the roller). Based on our long-run definition of probability given in Chapter 3, these relative frequencies estimate the probabilities of the three outcomes. Consequently, the approximate probability distribution of x, the outcome of the come-out wager in craps, is $p(-\$5) = .1, p(\$0) = .65$, and $p(\$5) = .25$. Note that these probabilities sum to 1.

*Unless otherwise indicated, summations will always be over all possible values of x.

Look Back When two dice are tossed, there is a total of 36 possible outcomes. (Can you list these outcomes, or sample points?) Of these, 4 result in a sum of 2, 3, or 12; 24 result in a sum of 4, 5, 6, 8, 9, or 10; and 8 result in a sum of 7 or 11. Using the rules of probability established in Chapter 3, you can show that the actual probability distribution for x is $p(-\$5) = 4/36 = .1111, p(\$0) = 24/36 = .6667$, and $p(\$5) = 8/36 = .2222$.

■ **Now Work Exercise 4.19**

Examples 4.4 and 4.5 illustrate how the probability distribution for a discrete random variable can be derived, but for many practical situations the task is much more difficult. Fortunately, numerous experiments and associated discrete random variables observed in nature possess identical characteristics. Thus, you might observe a random variable in a psychology experiment that would possess the same probability distribution as a random variable observed in an engineering experiment or a social sample survey. We classify random variables according to type of experiment, derive the probability distribution for each of the different types, and then use the appropriate probability distribution when a particular type of random variable is observed in a practical situation. The probability distributions for most commonly occurring discrete random variables have already been derived. This fact simplifies the problem of finding the probability distributions for random variables, as the next example illustrates.

Example 4.6 **Probability Distribution Using a Formula—Texas Droughts**	**Problem** A drought is a period of abnormal dry weather that causes serious problems in the farming industry of the region. University of Arizona researchers used historical annual data to study the severity of droughts in Texas (*Journal of Hydrologic Engineering*, Sept./Oct. 2003). The researchers showed that the distribution of x, the number of consecutive years that must be sampled until a dry (drought) year is observed, can be modeled using the formula $$p(x) = (.3)(.7)^{x-1}, x = 1, 2, 3, \ldots$$ Find the probability that exactly 3 years must be sampled before a drought year occurs. **Solution** We want to find the probability that $x = 3$. Using the formula, we have $$p(3) = (.3)(.7)^{3-1} = (.3)(.7)^2 = (.3)(.49) = .147$$ Thus, there is about a 15% chance that exactly 3 years must be sampled before a drought year occurs in Texas. **Look Back** The probability of interest can also be derived using the principles of probability developed in Chapter 3. The event of interest is $N_1 N_2 D_3$, where N_1 represents no drought occurs in the first sampled year, N_2 represents no drought occurs in the second sampled year, and D_3 represents a drought occurs in the third sampled year. The researchers discovered that the probability of a drought occurring in any sampled year is .3 (and, consequently, the probability of no drought occurring in any sampled year is .7). Using the multiplicative rule of probability for independent events, the probability of interest is $(.7)(.7)(.3) = .147$.

In Sections 4.4–4.6, we describe three important types of discrete random variables, give their probability distributions, and explain where and how they can be applied in practice. (Mathematical derivations of the probability distributions will be omitted, but these details can be found in the chapter references.)

But first, in Section 4.3, we discuss some descriptive measures of these sometimes complex probability distributions. Since probability distributions are analogous to the relative frequency distributions of Chapter 2, it should be no surprise that the mean and standard deviation are useful descriptive measures.

Exercises 4.17–4.39

Understanding the Principles

4.17 How do you represent the probability distribution of a random variable denoting the number of heads when an unbiased coin tossed once?

Learning the Mechanics

4.18 Consider the following probability distribution:

x:	−5	−2	0	1
p(x)	.1	.2	.3	.4

a. List the values that x may assume.
b. What is the probability that x is greater than 0?
c. What is the probability that x = −2?

4.19 A discrete random variable x can assume five possible values: 20, 21, 22, 23, and 24. The MINITAB histogram at the bottom of the page shows the likelihood of each value.
a. What is p(22)?
b. What is the probability that x equals 20 or 24?
c. What is $P(x \le 23)$?

4.20 Explain why each of the following is or is not a valid probability distribution for a discrete random variable x:

a.

x	0	1	2	3
p(x)	.2	.3	.3	.2

b.

x	−2	−1	0
p(x)	.25	.50	.20

c.

x	4	9	20
p(x)	−.3	1.0	.3

d.

x	2	3	5	6
p(x)	.15	.20	.40	.35

4.21 The random variable x has the following discrete probability distribution:

x:	5	6	7	8	9
p(x:)	.2	.3	.2	.1	.2

Since the values that x can assume are mutually exclusive events, the event {x ≤ 7} is the union of three mutually exclusive events:

$$\{x = 5\} \ \cup \ \{x = 6\} \ \cup \ \{x = 7\}$$

a. Find $P(x \le 7)$.
b. Find $P(x > 7)$.
c. Find $P(x \le 9)$.
d. Find $P(x = 8)$.
e. Find $P(x \le 6 \text{ or } x > 7)$.

4.22 The random variable x has the discrete probability distribution shown here:

x:	−4	−3	−2	−1	0
p(x:)	.10	.15	.40	.30	.05

a. Find $P(x \le -1)$.
b. Find $P(x > -1)$.
c. Find $P(x < 0)$.
d. Find $P(x < -2)$.
e. Find $P(-4 < x < -1)$.
f. Find $P(x < -3)$.

4.23 Toss three fair coins, and let x equal the number of heads observed.
a. Identify the sample points associated with this experiment, and assign a value of x to each sample point.
b. Calculate p(x) for each value of x.
c. Construct a probability histogram for p(x).
d. What is $P(x = 2 \text{ or } x = 3)$?

Histogram for Exercise 4.19

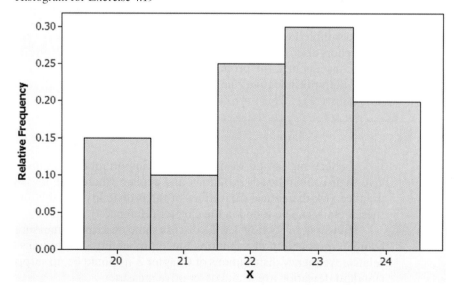

Applet Exercise 4.1

Use the applet entitled *Random Numbers* to generate a list of 25 numbers between 1 and 3, inclusive. Let x represent a number chosen from this list.

    a. What are the possible values of x?

    b. Give the probability distribution for x in table form.

    c. Let y be a number randomly chosen from the set $\{1, 2, 3\}$. Give the probability distribution for y in table form.

    d. Compare the probability distributions of x and y in parts **b** and **c**. Why should these distributions be approximately the same?

Applet Exercise 4.2

Run the applet entitled *Simulating the Probability of a Head with a Fair Coin* 10 times with $n = 2$, resetting between runs, to simulate flipping two coins 10 times. Count and record the number of heads each time. Let x represent the number of heads on a single flip of the coins.

    a. What are the possible values of x?

    b. Use the results of the simulation to find the probability distribution for x in table form.

    c. Explain why the probability of exactly two heads should be close to .25.

Applying the Concepts—Basic

4.24 **Size of TV households.** According to a region based survey, 26% of all TV households have a size of 1 person, 36% have a household size of 2, 17% have a household size of 3, and 21% have a household size of 4 or more. Let x represent the size (number of people) in a randomly selected TV household.

    a. List the possible values of the discrete random variable x.

    b. What is the probability that $x = 2$?

    c. What is the probability that $x \geq 3$?

4.25 **Do social robots walk or roll?** Refer to the *International Conference on Social Robotics* (Vol. 6414, 2010) study of the trend in the design of social robots, Exercise 2.7 (p. 66). Recall that in a random sample of 106 social (or service) robots designed to entertain, educate, and care for human users, 63 were built with legs only, 20 with wheels only, 8 with both legs and wheels, and 15 with neither legs nor wheels. Assume the following: Of the 63 robots with legs only, 50 have two legs, 5 have three legs, and 8 have four legs; of the 8 robots with both legs and wheels, all 8 have two legs. Suppose one of the 106 social robots is randomly selected. Let x equal the number of legs on the robot.

    a. List the possible values of x.

    b. Find the probability distribution of x.

    c. Find $E(x)$ and give a practical interpretation of its value.

4.26 **NHTSA crash tests.** Refer to the National Highway Traffic Safety Administration (NHTSA) crash tests of new car models, presented in Exercise 4.6 (p. 216). A summary of the driver-side star ratings for the 69 cars is reproduced in the MINITAB printout in the next column.

Assume that 1 of the 69 cars is selected at random, and let x equal the number of stars in the car's driver-side star rating.

    a. Use the information in the printout to find the probability distribution for x.

    b. Find $P(x = 3)$.

    c. Find $P(x < 4)$.

MINITAB Output for Exercise 4.26

Tally for Discrete Variables: DRIVSTAR

DRIVSTAR	Count	Percent
2	10	14.49
3	15	21.74
4	26	37.68
5	18	26.09
N=	69	

4.27 **Downloading apps to your cell phone.** According to a study, nearly 40% of adult cell phone owners have downloaded an application ("app") to their cell phone. The accompanying table gives the probability distribution for x, the number of apps used at least once a week by cell phone owners who have downloaded an app to their phone.

Number of Apps Used, x	$p(x)$
0	.30
1	.24
2	.06
3	.06
4	.05
5	.10
6	.07
7	.05
8	.03
9	.02
10	.02

    a. Show that the properties of a probability distribution for a discrete random variable are satisfied.

    b. Find $P(x > 3)$.

4.28 **Controlling the water hyacinth.** An insect that naturally feeds on the water hyacinth, one of the world's worst aquatic weeds, is the delphacid. Female delphacids lay anywhere from one to four eggs onto a water hyacinth blade. The *Annals of the Entomological Society of America* (Jan. 2005) published a study of the life cycle of a South American delphacid species. The following table gives the percentages of water hyacinth blades that have one, two, three, and four delphacid eggs:

	One Egg	Two Eggs	Three Eggs	Four Eggs
Percentage of Blades	40	54	2	4

Source: Sosa, A. J., et al. "Life history of *Megamelus scutellaris* with description of immature stages," *Annals of the Entomological Society of America*, Vol. 98, No. 1, Jan. 2005 (adapted from Table 1).

a. One of the water hyacinth blades in the study is randomly selected, and x, the number of delphacid eggs on the blade, is observed. Give the probability distribution of x.

b. What is the probability that the blade has at least three delphacid eggs?

4.29 **Gender in two-child families.** *Human Biology* (Feb. 2009) published a study on the gender of children in two-child families. In populations where it is just as likely to have a boy as a girl, the probabilities of having two girls, two boys, or a boy and a girl are well known. Let x represent the number of boys in a two-child family.

a. List the possible ways (sample points) in which a two-child family can be gender-configured. (For example, BG represents the event that the first child is a boy and the second is a girl.)

b. Assuming boys are just as likely as girls to be born, assign probabilities to the sample points in part **a**.

c. Use the probabilities, part **a**, to find the probability distribution for x.

d. The article reported on the results of the National Health Interview Survey (NHIS) of almost 43,000 two-child families. The table gives the proportion of families with each gender configuration. Use this information to revise the probability distribution for x.

Gender Configuration	Proportion
Girl-girl (GG)	.222
Boy-girl (BG)	.259
Girl-boy (GB)	.254
Boy-boy (BB)	.265

4.30 **Environmental vulnerability of amphibians.** Many species of amphibians are declining due to dramatic changes in the environment. The environmental vulnerability of amphibian species native to Mexico was examined in *Amphibian & Reptile Conservation* (Aug. 2013). The researchers quantified this decline using a modified Environmental Vulnerability Score (EVS) algorithm. Scores ranged from 3 to 20 points, with higher scores indicating a greater level of environmental vulnerability. EVS scores for 382 snake species were determined and are summarized in the table below. Let x represent the EVS score for a randomly selected snake species in Mexico.

a. Specify the probability distribution for EVS score x.

b. EVS scores greater than 14 are indicative of species that are endangered due to high environmental vulnerability. Find $P(x > 14)$.

Applying the Concepts—Intermediate

4.31 **The "last name" effect in purchasing.** The *Journal of Consumer Research* (Aug. 2011) published a study demonstrating the "last name" effect—i.e., the tendency for consumers with last names that begin with a later letter of the alphabet to purchase an item before consumers with last names that begin with earlier letters. To facilitate the analysis, the researchers assigned a number, x, to each consumer based on the first letter of the consumer's last name. For example, last names beginning with "A" were assigned $x = 1$, last names beginning with "B" were assigned $x = 2$, and last names beginning with "Z" were assigned $x = 26$.

a. If the first letters of consumers' last names are equally likely, find the probability distribution for x.

b. Do you believe the probability distribution in part **a** is realistic? Explain. How might you go about estimating the true probability distribution for x?

4.32 **Solar energy cells.** According to the Earth Policy Institute (July 2013), 60% of the world's solar energy cells are manufactured in China. Consider a random sample of 5 solar energy cells, and let x represent the number in the sample that are manufactured in China. In Section 4.4, we show that the probability distribution for x is given by the formula

$$p(x) = \frac{(5!)(.6)^x(.4)^{5-x}}{(x!)(5-x)!},$$

where $n! = (n)(n-1)(n-2) \ldots (2)(1)$

a. Explain why x is a discrete random variable.

b. Find $p(x)$ for $x = 0, 1, 2, 3, 4$, and 5.

c. Show that the properties for a discrete probability distribution are satisfied.

d. Find the probability that at least 4 of the 5 solar energy cells in the sample are manufactured in China.

4.33 **Contaminated gun cartridges.** A weapons manufacturer uses a liquid propellant to produce gun cartridges. During the manufacturing process, the propellant can get mixed with another liquid to produce a contaminated cartridge. A statistician hired by the company to investigate the level of contamination in the stored cartridges found that 25% of the cartridges in particular lot were contaminated. Suppose you randomly sample (without replacement) gun cartridges from this lot until you find a contaminated one. Let x be the number of cartridges sampled until a contaminated one is found. It is known that the probability distribution for x is given by the formula

$$p(x) = (.25)(.75)^{x-1}, x = 1, 2, 3, \ldots$$

a. Find $p(2)$. Interpret this result.

b. Find $p(4)$. Interpret this result.

c. Find $P(x \geq 2)$. Interpret this result.

4.34 **Variable speed limit control for freeways.** A common transportation problem in large cities is congestion on the freeways. In the *Canadian Journal of Civil Engineering* (Jan. 2013), civil engineers investigated the use of variable

Table for Exercise 4.30

EVS Score	3	4	5	6	7	8	9	10	11	12	13	14	15	16	17	18	19	20	Total
Number of Snake Species	1	1	7	10	9	19	17	30	25	31	46	52	50	44	24	9	7	0	382

Source: Wilson, L. D., et al. "A conservation reassessment of the amphibians of Mexico based on the EVS measure." *Amphibian & Reptile Conservation*, Vol. 7, No. 1, Aug. 2013 (Table 4).

speed limits (VSL) to control the congestion problem. The study site was an urban freeway in Edmonton, Canada. A portion of the freeway was divided equally into three sections, and variable speed limits were posted (independently) in each section. Simulation was used to find the optimal speed limits based on various traffic patterns and weather conditions. Probability distributions of the speed limits for the three sections were determined. For example, one possible set of distributions is as follows (probabilities in parentheses). *Section 1*: 30 mph (.05), 40 mph (.25), 50 mph (.25), 60 mph (.45); *Section 2*: 30 mph (.10), 40 mph (.25), 50 mph (.35), 60 mph (.30); *Section 3*: 30 mph (.15), 40 mph (.20), 50 mph (.30), 60 mph (.35).

a. Verify that the properties of discrete probability distributions are satisfied for each individual section of the freeway.

b. Consider a vehicle that will travel through the three sections of the freeway at a steady (fixed) speed. Let *x* represent this speed. Find the probability distribution for *x*.

c. Refer to part **b**. What is the probability that the vehicle can travel at least 50 mph through the three sections of the freeway?

4.35 **Choosing portable grill displays.** Refer to the *Journal of Consumer Research* (Mar. 2003) marketing study of influencing consumer choices by offering undesirable alternatives, presented in Exercise 3.29 (p. 159). Recall that each of 124 college students selected showroom displays for portable grills. Five different displays (representing five different-sized grills) were available, but the students were instructed to select only three displays in order to maximize purchases of Grill #2 (a smaller grill). The table that follows shows the grill display combinations and number of each selected by the 124 students. Suppose 1 of the 124 students is selected at random. Let *x* represent the sum of the grill numbers selected by that student. (This sum is an indicator of the size of the grills selected.)

a. Find the probability distribution for *x*.

b. What is the probability that *x* exceeds 10?

Grill Display Combination	Number of Students
1–2–3	35
1–2–4	8
1–2–5	42
2–3–4	4
2–3–5	1
2–4–5	34

Based on Hamilton, R. W. "Why do people suggest what they do not want? Using context effects to influence others' choices." *Journal of Consumer Research*, Vol. 29, Mar. 2003 (Table 1).

4.36 **Reliability of a manufacturing network.** A team of industrial management university professors investigated the reliability of a manufacturing system that involves multiple production lines (*Journal of Systems Sciences & Systems Engineering*, Mar. 2013). An example of such a network is a system for producing integrated circuit (IC) cards with two production lines set up in sequence. Items (IC cards) first pass through Line 1 and then are processed by Line 2. The probability distribution of the maximum capacity level of each line is shown in the next column. Assume the lines operate independently.

Line	Maximum Capacity, x	p(x)
1	0	.01
	12	.02
	24	.02
	36	.95
2	0	.002
	35	.002
	70	.996

a. Verify that the properties of discrete probability distributions are satisfied for each line in the system.

b. Find the probability that the maximum capacity level for Line 1 will exceed 30 items.

c. Repeat part **b** for Line 2.

d. Now consider the network of two production lines. What is the probability that a maximum capacity level of 30 items is maintained throughout the network?

Applying the Concepts—Advanced

4.37 **Voter preferences for a committee.** A "Condorcet" committee of, say, 3 members is a committee that is preferred by voters over any other committee of 3 members. A scoring function was used to determine "Condorcet" committees in *Mathematical Social Sciences* (Nov. 2013). Consider a committee with members A, B, and C. Suppose there are 10 voters who each have a preference for a 3-member committee. For example, one voter may prefer a committee made up of members A, C, and G. Then this voter's preference score for the {A, B, C} committee is 2, because 2 of the members (A and B) are on this voter's preferred list. For a 3-member committee, voter preference scores range from 0 (no members on the preferred list) to 3 (all members on the preferred list). The table below shows the preferred lists of 10 voters for a 3-member committee selected from potential members A, B, C, D, E, F, and G.

Voter:	1	2	3	4	5	6	7	8	9	10
	A	B	A	B	D	C	A	B	A	A
	C	C	B	C	E	E	C	C	B	C
	D	E	D	F	F	G	G	D	C	G

a. Find the preference score for committee {A, B, C} for each voter.

b. For a randomly selected voter, let *x* represent the preference score for committee {A, B, C}. Determine the probability distribution for *x*.

c. What is the probability that the preference score, *x*, exceeds 2?

d. Is {A, B, C} a "Condorcet" committee?

4.38 **Punnett square for earlobes.** Geneticists use a grid—called a *Punnett square*—to display all possible gene combinations in genetic crosses. (The grid is named for Reginald Punnett, a British geneticist who developed the method in the early 1900s.) The figure on p. 224 is a Punnett square for a cross involving human earlobes. In humans, free earlobes (E) are dominant over attached earlobes (e). Consequently, the gene pairs EE and Ee will result in free earlobes, while the gene pair ee results in attached earlobes. Consider a couple with genes as shown in the

Punnett square below. Suppose the couple has seven children. Let x represent the number of children with attached earlobes (i.e., with the gene pair ee). Find the probability distribution of x.

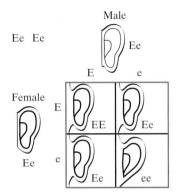

4.39 **Robot-sensor system configuration.** Engineers at Broadcom Corp. and Simon Fraser University collaborated on research involving a robot-sensor system in an unknown environment (*The International Journal of Robotics Research*, Dec. 2004). As an example, the engineers presented the three-point, single-link robotic system shown in the accompanying figure.

Each point (A, B, or C) in the physical space of the system has either an "obstacle" status or a "free" status. There are two single links in the system: $A \leftrightarrow B$ and $B \leftrightarrow C$. A link has a "free" status if and only if both points in the link are "free"; otherwise the link has an "obstacle" status. Of interest is the random variable x: the total number of links in the system that are "free."

a. List the possible values of x for the system.

b. The researchers stated that the probability of any point in the system having a "free" status is .5. Assuming that the three points in the system operate independently, find the probability distribution for x.

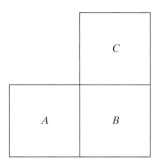

4.3 Expected Values of Discrete Random Variables

If a discrete random variable x were observed a very large number of times and the data generated were arranged in a relative frequency distribution, the relative frequency distribution would be indistinguishable from the probability distribution for the random variable. Thus, the probability distribution for a random variable is a theoretical model for the relative frequency distribution of a population. To the extent that the two distributions are equivalent (and we will assume that they are), the probability distribution for x possesses a mean μ and a variance σ^2 that are identical to the corresponding descriptive measures for the population. This section explains how you can find the mean value for a random variable. We illustrate the procedure with an example.

Examine the probability distribution for x (the number of heads observed in the toss of two fair coins) in Figure 4.5. Try to locate the mean of the distribution intuitively. We may reason that the mean μ of this distribution is equal to 1 as follows: In a large number of experiments—say, 100,000—$^1/_4$ (or 25,000) should result in $x = 0$ heads, $^1/_2$ (or 50,000) in $x = 1$ head, and $^1/_4$ (or 25,000) in $x = 2$ heads. Therefore, the average number of heads is

$$\mu = \frac{0(25,000) + 1(50,000) + 2(25,000)}{100,000} = 0(^1/_4) + 1(^1/_2) + 2(^1/_4)$$

$$= 0 + ^1/_2 + ^1/_2 = 1$$

Note that to get the population mean of the random variable x, we multiply each possible value of x by its probability $p(x)$, and then we sum this product over all possible values of x. The *mean of x* is also referred to as the *expected value of x*, denoted $E(x)$.

The **mean**, or **expected value**, of a discrete random variable x is

$$\mu = E(x) = \Sigma x p(x)$$

Figure 4.5
Probability distribution for a two-coin toss

Expected is a mathematical term and should not be interpreted as it is typically used. Specifically, *a random variable might never be equal to its "expected value."* Rather, the expected value is the mean of the probability distribution, or a measure of its central tendency. You can think of μ as the mean value of x in a *very large* (actually, *infinite*) number of repetitions of the experiment in which the values of x occur in proportions equivalent to the probabilities of x.

Example	4.7

Finding an Expected Value—An Insurance Application

Problem Suppose you work for an insurance company and you sell a $10,000 one-year term insurance policy at an annual premium of $290. Actuarial tables show that the probability of death during the next year for a person of your customer's age, sex, health, etc., is .001. What is the expected gain (amount of money made by the company) for a policy of this type?

Solution The experiment is to observe whether the customer survives the upcoming year. The probabilities associated with the two sample points, Live and Die, are .999 and .001, respectively. The random variable you are interested in is the gain x, which can assume the values shown in the following table:

Gain x	Sample Point	Probability
$290	Customer lives	.999
−$9,710	Customer dies	.001

If the customer lives, the company gains the $290 premium as profit. If the customer dies, the gain is negative because the company must pay $10,000, for a net "gain" of $$(290 - 10,000) = -\$9,710$. The expected gain is therefore

$$\mu = E(x) = \Sigma xp(x) = (290)(.999) + (-9,710)(.001) = \$280$$

In other words, if the company were to sell a very large number of $10,000 one-year policies to customers possessing the characteristics described, it would (on the average) net $280 per sale in the next year.

Look Back Note that $E(x)$ need not equal a possible value of x. That is, the expected value is $280, but x will equal either $290 or −$9,710 each time the experiment is performed (a policy is sold and a year elapses). The expected value is a measure of central tendency—and in this case represents the average over a very large number of one-year policies—but is not a possible value of x.

■ **Now Work Exercise 4.47**

We learned in Chapter 2 that the mean and other measures of central tendency tell only part of the story about a set of data. The same is true about probability distributions: We need to measure variability as well. Since a probability distribution can be viewed as a representation of a population, we will use the population variance to measure its variability.

The *population variance* σ^2 is defined as the average of the squared distance of x from the population mean μ. Since x is a random variable, the squared distance, $(x - \mu)^2$, is also a random variable. Using the same logic we employed to find the mean value of x, we calculate the mean value of $(x - \mu)^2$ by multiplying all possible values of $(x - \mu)^2$ by $p(x)$ and then summing over all possible x values.* This quantity,

$$E[(x - \mu)^2] = \sum_{\text{all } x} (x - \mu)^2 p(x)$$

is also called the *expected value of the squared distance from the mean;* that is, $\sigma^2 = E[(x - \mu)^2]$. The standard deviation of x is defined as the square root of the variance σ^2.

> The **variance** of a random variable x is
>
> $$\sigma^2 = E[(x - \mu)^2] = \Sigma(x - \mu)^2 p(x) = \Sigma x^2 p(x) - \mu^2$$

*It can be shown that $E[(x - \mu)^2] = E(x^2) - \mu^2$, where $E(x^2) = \Sigma x^2 p(x)$. Note the similarity between this expression and the shortcut formula $\Sigma(x - \bar{x})^2 = \Sigma x^2 - (\Sigma x)^2/n$ given in Chapter 2.

> The **standard deviation** of a discrete random variable is equal to the square root of the variance, or $\sigma = \sqrt{\sigma^2}$.

Knowing the mean μ and standard deviation σ of the probability distribution of x, in conjunction with Chebyshev's rule (Rule 2.1) and the empirical rule (Rule 2.2), we can make statements about the likelihood of values of x falling within the intervals $\mu \pm \sigma, \mu \pm 2\sigma$, and $\mu \pm 3\sigma$. These probabilities are given in the following box:

Probability Rules for a Discrete Random Variable

Let x be a discrete random variable with probability distribution $p(x)$, mean μ, and standard deviation σ. Then, depending on the shape of $p(x)$, the following probability statements can be made:

	Chebyshev's Rule	Empirical Rule
	Applies to any probability distribution (see Figure 4.6a)	Applies to probability distributions that are mound shaped and symmetric (see Figure 4.6b)
$P(\mu - \sigma < x < \mu + \sigma)$	≥ 0	$\approx .68$
$P(\mu - 2\sigma < x < \mu + 2\sigma)$	$\geq 3/4$	$\approx .95$
$P(\mu - 3\sigma < x < \mu + 3\sigma)$	$\geq 8/9$	≈ 1.00

Figure 4.6

Shapes of two probability distributions for a discrete random variable x

a. Skewed distribution

b. Mound shaped, symmetric

Example 4.8

Finding μ and σ—Skin Cancer Treatment

Problem Medical research has shown that a certain type of chemotherapy is successful 70% of the time when used to treat skin cancer. Suppose five skin cancer patients are treated with this type of chemotherapy, and let x equal the number of successful cures out of the five. The probability distribution for the number x of successful cures out of five is given in the following table:

x	0	1	2	3	4	5
$p(x)$	.002	.029	.132	.309	.360	.168

a. Find $\mu = E(x)$. Interpret the result.

b. Find $\sigma = \sqrt{E[(x - \mu)^2]}$. Interpret the result.

c. Graph $p(x)$. Locate μ and the interval $\mu \pm 2\sigma$ on the graph. Use either Chebyshev's rule or the empirical rule to approximate the probability that x falls into this interval. Compare your result with the actual probability.

d. Would you expect to observe fewer than two successful cures out of five?

Solution

a. Applying the formula for μ, we obtain

$$\mu = E(x) = \Sigma x p(x)$$
$$= 0(.002) + 1(.029) + 2(.132) + 3(.309) + 4(.360) + 5(.168) = 3.50$$

On average, the number of successful cures out of five skin cancer patients treated with chemotherapy will equal 3.5. Remember that this expected value has meaning only when the experiment—treating five skin cancer patients with chemotherapy—is repeated a large number of times.

b. Now we calculate the variance of x:

$$\sigma^2 = E[(x - \mu)^2] = \Sigma(x - \mu)^2 p(x)$$
$$= (0 - 3.5)^2(.002) + (1 - 3.5)^2(.029) + (2 - 3.5)^2(.132)$$
$$+ (3 - 3.5)^2(.309) + (4 - 3.5)^2(.360) + (5 - 3.5)^2(.168)$$
$$= 1.05$$

Thus, the standard deviation is

$$\sigma = \sqrt{\sigma^2} = \sqrt{1.05} = 1.02$$

This value measures the spread of the probability distribution of x, the number of successful cures out of five. A more useful interpretation is obtained by answering parts **c** and **d**.

c. The graph of $p(x)$ is shown in Figure 4.7, with the mean μ and the interval $\mu \pm 2\sigma = 3.50 \pm 2(1.02) = 3.50 \pm 2.04 = (1.46, 5.54)$ also indicated. Note particularly that $\mu = 3.5$ locates the center of the probability distribution. Since this distribution is a theoretical relative frequency distribution that is moderately mound shaped (see Figure 4.7), we expect (from Chebyshev's rule) at least 75% and, more likely (from the empirical rule), approximately 95% of observed x values to fall between 1.46 and 5.54. You can see from the figure that the actual probability that x falls in the interval $\mu \pm 2\sigma$ includes the sum of $p(x)$ for the values $x = 2, x = 3, x = 4$, and $x = 5$. This probability is $p(2) + p(3) + p(4) + p(5) = .132 + .309 + .360 + .168 = .969$. Therefore, 96.9% of the probability distribution lies within two standard deviations of the mean. This percentage is consistent with both Chebyshev's rule and the empirical rule.

d. Fewer than two successful cures out of five implies that $x = 0$ or $x = 1$. Both of these values of x lie outside the interval $\mu \pm 2\sigma$, and the empirical rule tells us that such a result is unlikely (approximate probability of .05). The exact probability, $P(x \leq 1)$, is $p(0) + p(1) = .002 + .029 = .031$. Consequently, in a single experiment in which five skin cancer patients are treated with chemotherapy, we would not expect to observe fewer than two successful cures.

■ **Now Work Exercise 4.46**

$p(x)$

Figure 4.7
Graph of $p(x)$ for Example 4.8

Exercises 4.40–4.56

Understanding the Principles

4.40 What do you mean by expectation of a discrete random variable.

4.41 Will the expectation of a negative random variable be always negative?

4.42 For a mound-shaped, symmetric distribution, what is the probability that x falls into the interval $\mu \pm 3\sigma$?

Learning the Mechanics

4.43 Consider the probability distribution for the random variable x shown here:

x	−1	1	2	5
y	.2	.4	.2	.2

a. Find $\mu = E(x)$.
b. Find $\sigma^2 = E[(x - \mu)^2]$.
c. Find σ.
d. Interpret the value you obtained for μ.
e. In this case, can the random variable x ever assume the value μ? Explain.
f. In general, can a random variable ever assume a value equal to its expected value? Explain.

4.44 Consider the probability distribution for the random variable x shown here:

x	10	15	20	25	30	35
$p(x)$	.05	.20	.30	.25	.10	.10

a. Find μ, σ^2, and σ.
b. Graph $p(x)$.
c. Locate μ and the interval $\mu \pm 2\sigma$ on your graph. What is the probability that x will fall within the interval $\mu \pm 2\sigma$?

4.45 Consider the probability distributions shown here:

x	3	4	5
$p(x)$	.3	.4	.3

y	3	4	5
$p(y)$	.1	.8	.1

a. Use your intuition to find the mean for each distribution. How did you arrive at your choice?
b. Which distribution appears to be more variable? Why?
c. Calculate μ and σ^2 for each distribution. Compare these answers with your answers in parts **a** and **b**.

4.46 Consider the probability distribution shown here:

NW

x	−4	−3	−2	−1	0	1	2	3	4
$p(x)$	.02	.07	.10	.15	.30	.18	.10	.06	.02

a. Calculate μ, σ^2, and σ.
b. Graph $p(x)$. Locate μ, $\mu - 2\sigma$, and $\mu + 2\sigma$ on the graph.
c. What is the probability that x will fall into the interval $\mu \pm 2\sigma$?

Applying the Concepts—Basic

4.47 **NHTSA car crash tests.** Refer to Exercise 4.26 (p. 221), in which you found the probability distribution for x, the number of stars in a randomly selected car's driver-side crash rating. Find $\mu = E(x)$ for this distribution and interpret the result practically.

4.48 **Downloading apps to your cell phone.** Refer to the survey of cell phone owners, Exercise 4.27 (p. 221). There you were given the probability distribution for x, the number of apps used at least once a week by cell phone owners who have downloaded an app to their phone.
a. Find the mean and variance of x.
b. Give an interval that will contain the value of x with a probability of at least .75.

4.49 **Controlling the water hyacinth.** Refer to the *Annals of the Entomological Society of America* (Jan. 2005) study of the number of insect eggs, x, on a blade of water hyacinth, presented in Exercise 4.28 (p. 221). The probability distribution for x is reproduced in the accompanying table. Find the mean of the probability distribution for x, and interpret its value.

Number of eggs, x	1	2	3	4
$p(x)$	.40	.54	.02	.04

Source: Sosa, A. J., et al. "Life history of *Megamelus scutellaris* with description of immature stages," *Annals of the Entomological Society of America*, Vol. 98, No. 1, Jan. 2005 (adapted from Table 1).

4.50 **Gender in two-child families.** Refer to the *Human Biology* (Feb. 2009) study on the gender of children in two-child families, Exercise 4.29 (p. 222). Recall that the National Health Interview Survey (NHIS) of almost 43,000 two-child families yielded the following table. You used this information in Exercise 4.29d to find the probability distribution for x, the number of boys in a two-child family. Find $E(x)$ and give a practical interpretation of its value.

Gender Configuration	Proportion
Girl-girl (GG)	.222
Boy-girl (BG)	.259
Girl-boy (GB)	.254
Boy-boy (BB)	.265

4.51 **Environmental vulnerability of amphibians.** Refer to the *Amphibian & Reptile Conservation* (Aug. 2013) study of Environmental Vulnerability Scores (EVS) for amphibian species native to Mexico, Exercise 4.30 (p. 222). The EVS values for 382 snake species are summarized in the table below. Let x represent the EVS score for a randomly selected snake species in Mexico. Find the mean of x, and interpret the result.

Applying the Concepts—Intermediate

4.52 **Reliability of a manufacturing network.** Refer to the *Journal of Systems Sciences & Systems Engineering* (Mar. 2013) study of the reliability of a manufacturing system that involves multiple production lines, Exercise 4.36 (p. 223). Consider, again, the network for producing integrated circuit (IC) cards with two production lines set up in sequence. The probability distribution of the maximum capacity level of each line is given in Exercise 4.36.
a. Find the mean maximum capacity for each line. Interpret the results practically.
b. Find the standard deviation of the maximum capacity for each line. Interpret the results practically.

4.53 **Beach erosional hot spots.** Refer to the U.S. Army Corps of Engineers' study of beach erosional hot spots, presented in Exercise 3.167 (p. 205). The data on the nearshore bar condition for six beach hot spots are reproduced in the

Table for Exercise 4.51

EVS Score	3	4	5	6	7	8	9	10	11	12	13	14	15	16	17	18	19	20	Total
Number of Snake Species	1	1	7	10	9	19	17	30	25	31	46	52	50	44	24	9	7	0	382

accompanying table. Suppose you randomly select two of these six beaches and count x, the total number in the sample with a planar nearshore bar condition.

Beach Hot Spot	Nearshore Bar Condition
Miami Beach, FL	Single, shore parallel
Coney Island, NY	Other
Surfside, CA	Single, shore parallel
Monmouth Beach, NJ	Planar
Ocean City, NJ	Other
Spring Lake, NJ	Planar

Based on *Identification and Characterization of Erosional Hotspots*, William & Mary Virginia Institute of Marine Science, U.S. Army Corps of Engineers Project Report, Mar. 18, 2002.

a. List all possible pairs of beach hot spots that can be selected from the six.

b. Assign probabilities to the outcomes in part **a**.

c. For each outcome in part **a**, determine the value of x.

d. Form a probability distribution table for x.

e. Find the expected value of x. Interpret the result.

4.54 **Expected Lotto winnings.** The chance of winning a famous lottery game is 1 in approximately 50 million. Suppose you buy a $1 lottery ticket in anticipation of winning the $7 million grand prize. Calculate your expected net winnings for this single ticket. Interpret the result.

4.55 **Expected winnings in roulette.** In the popular casino game of roulette, you can bet on whether the ball will fall in an arc on the wheel colored red, black, or green. You showed (Exercise 3.171, p. 205) that the probability of a red outcome is 18/38, that of a black outcome is 18/38, and that of a green outcome is 2/38. Suppose you make a $5 bet on red. Find your expected net winnings for this single bet. Interpret the result.

Applying the Concepts—Advanced

4.56 **Parlay card betting.** Odds makers try to predict which professional and college football teams will win and by how much (the *spread*). If the odds makers do this accurately, adding the spread to the underdog's score should make the final score a tie. Suppose a bookie will give you $700000 for every $100000 you risk if you pick the winners in three different football games (adjusted by the spread) on a "parlay" card. What is the bookie's expected earnings per dollar wagered? Interpret this value.

The Binomial Random Variable

Many experiments result in *dichotomous* responses (i.e., responses for which there exist two possible alternatives, such as Yes–No, Pass–Fail, Defective–Nondefective, or Male–Female). A simple example of such an experiment is the coin-toss experiment. A coin is tossed a number of times, say, 10. Each toss results in one of two outcomes, Head or Tail, and the probability of observing each of these two outcomes remains the same for each of the 10 tosses. Ultimately, we are interested in the probability distribution of x, the number of heads observed. Many other experiments are equivalent to tossing a coin (either balanced or unbalanced) a fixed number n of times and observing the number x of times that one of the two possible outcomes occurs. Random variables that possess these characteristics are called **binomial random variables**.

Public opinion and consumer preference polls (e.g., the CNN, Gallup, and Harris polls) frequently yield observations on binomial random variables. For example, suppose a sample of 100 students is selected from a large student body and each person is asked whether he or she favors (a Head) or opposes (a Tail) a certain campus issue. Suppose we are interested in x, the number of students in the sample who favor the issue. Sampling 100 students is analogous to tossing the coin 100 times. Thus, you can see that opinion polls that record the number of people who favor a certain issue are real-life equivalents of coin-toss experiments. We have been describing a **binomial experiment**, identified by the following characteristics:

Characteristics of a Binomial Random Variable

1. The experiment consists of n identical trials.

2. There are only two possible outcomes on each trial. We will denote one outcome by S (for Success) and the other by F (for Failure).

3. The probability of S remains the same from trial to trial. This probability is denoted by p, and the probability of F is denoted by $q = 1 - p$.

4. The trials are independent.

5. The binomial random variable x is the number of S's in n trials.

The Bernoulli Distribution

Son of a magistrate and spice maker in Basel, Switzerland, Jacob Bernoulli completed a degree in theology at the University of Basel. While at the university, however, he studied mathematics secretly and against the will of his father. Jacob taught mathematics to his younger brother Johan, and they both went on to become distinguished European mathematicians. At first the brothers collaborated on the problems of the time (e.g., calculus); unfortunately, they later became bitter mathematical rivals. Jacob applied his philosophical training and mathematical intuition to probability and the theory of games of chance, where he developed the law of large numbers. In his book *Ars Conjectandi*, published in 1713 (eight years after his death), the binomial distribution was first proposed. Jacob showed that the binomial distribution is a sum of independent 0–1 variables, now known as Bernoulli random variables. ■

Example 4.9	**Problem** For the following examples, decide whether x is a binomial random variable:

Assessing Whether
x Is a Binomial

a. A university scholarship committee must select two students to receive a scholarship for the next academic year. The committee receives 10 applications for the scholarships—6 from male students and 4 from female students. Suppose the applicants are all equally qualified, so that the selections are randomly made. Let x be the number of female students who receive a scholarship.

b. Before marketing a new product on a large scale, many companies conduct a consumer-preference survey to determine whether the product is likely to be successful. Suppose a company develops a new diet soda and then conducts a taste-preference survey in which 100 randomly chosen consumers state their preferences from among the new soda and the two leading sellers. Let x be the number of the 100 who choose the new brand over the two others.

c. Some surveys are conducted by using a method of sampling other than simple random sampling. For example, suppose a television cable company plans to conduct a survey to determine the fraction of households in a certain city that would use its new fiber-optic service (FiOS). The sampling method is to choose a city block at random and then survey every household on that block. This sampling technique is called *cluster sampling*. Suppose 10 blocks are so sampled, producing a total of 124 household responses. Let x be the number of the 124 households that would use FiOS.

Solution

a. In checking the binomial characteristics, a problem arises with both characteristic 3 (probabilities the same across trials) and characteristic 4 (independence). On the one hand, given that the first student selected is female, the probability that the second chosen is female is $^3/_9$. On the other hand, given that the first selection is a male student, the probability that the second is female is $^4/_9$. Thus, the conditional probability of a Success (choosing a female student to receive a scholarship) on the second trial (selection) depends on the outcome of the first trial, and the trials are therefore dependent. Since the trials are *not independent*, this variable is not a binomial random variable. (This variable is actually a *hypergeometric* random variable, the topic of optional Section 4.6.)

b. Surveys that produce dichotomous responses and use random-sampling techniques are classic examples of binomial experiments. In this example, each randomly selected consumer either states a preference for the new diet soda or does not. The sample of 100 consumers is a very small proportion of the totality of potential consumers, so the response of one would be, for all practical purposes, independent of another.* Thus, x is a binomial random variable.

c. This example is a survey with dichotomous responses (Yes or No to using FiOS), but the sampling method is not simple random sampling. Again, the binomial characteristic of independent trials would probably not be satisfied. The responses of

---

*In most real-life applications of the binomial distribution, the population of interest has a finite number of elements (trials), denoted N. When N is large and the sample size n is small relative to N, say, $n/N \leq .05$, the sampling procedure, for all practical purposes, satisfies the conditions of a binomial experiment.

households within a particular block would be dependent, since households within a block tend to be similar with respect to income, level of education, and general interests. Thus, the binomial model would not be satisfactory for *x* if the cluster sampling technique were employed.

Look Back Nonbinomial variables with two outcomes on every trial typically occur because they do not satisfy characteristic 3 or characteristic 4 of a binomial distribution listed in the box on p. 229.

■ **Now Work Exercise 4.72a**

Example   4.10

Deriving the Binomial Probability Distribution— Passing a Physical Fitness Exam

Problem The Heart Association claims that only 10% of U.S. adults over 30 years of age meet the minimum requirements established by the President's Council on Fitness, Sports, and Nutrition. Suppose four adults are randomly selected and each is given the fitness test.

a. Use the steps given in Chapter 3 (box on p. 153) to find the probability that none of the four adults passes the test.

b. Find the probability that three of the four adults pass the test.

c. Let *x* represent the number of the four adults who pass the fitness test. Explain why *x* is a binomial random variable.

d. Use the answers to parts **a** and **b** to derive a formula for $p(x)$, the probability distribution of the binomial random variable *x*.

Solution

a. 1. The first step is to define the experiment. Here we are interested in observing the fitness test results of each of the four adults: pass (*S*) or fail (*F*).

   2. Next, we list the sample points associated with the experiment. Each sample point consists of the test results of the four adults. For example, *SSSS* represents the sample point denoting that all four adults pass, while *FSSS* represents the sample point denoting that adult 1 fails, while adults 2, 3, and 4 pass the test. The tree diagram, Figure 4.8, shows that there are 16 sample points. These 16 sample points are listed in Table 4.2.

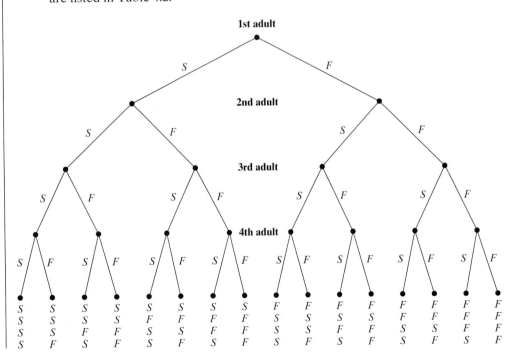

Figure 4.8

Tree Diagram Showing Outcomes for Fitness Tests

Table 4.2	Sample Points for Fitness Test of Example 4.10			
SSSS	*FSSS*	*FFSS*	*SFFF*	*FFFF*
	SFSS	*FSFS*	*FSFF*	
	SSFS	*FSSF*	*FFSF*	
	SSSF	*SFFS*	*FFFS*	
		SFSF		
		SSFF		

3. We now assign probabilities to the sample points. Note that each sample point can be viewed as the intersection of four adults' test results, and assuming that the results are independent, the probability of each sample point can be obtained by the multiplicative rule as follows:

$$P(SSSS) = P[\,(\text{adult 1 passes}) \cap (\text{adult 2 passes})$$
$$\cap (\text{adult 3 passes}) \cap (\text{adult 4 passes})\,]$$
$$= P(\text{adult 1 passes}) \times P(\text{adult 2 passes})$$
$$\times P(\text{adult 3 passes}) \times P(\text{adult 4 passes})$$
$$= (.1)(.1)(.1)(.1) = (.1)^4 = .0001$$

All other sample-point probabilities are calculated by similar reasoning. For example,

$$P(FSSS) = (.9)(.1)(.1)(.1) = .0009$$

You can check that this reasoning results in sample-point probabilities that add to 1 over the 16 points in the sample space.

4. Finally, we add the appropriate sample-point probabilities to obtain the desired event probability. The event of interest is that all four adults fail the fitness test. In Table 4.2, we find only one sample point, *FFFF*, contained in this event. All other sample points imply that at least one adult passes. Thus,

$$P(\text{All four adults fail}) = P(FFFF) = (.9)^4 = .6561$$

b. The event that three of the four adults pass the fitness test consists of the four sample points in the second column of Table 4.2: *FSSS, SFSS, SSFS,* and *SSSF.* To obtain the event probability, we add the sample-point probabilities:

$$P(3 \text{ of } 4 \text{ adults pass}) = P(FSSS) + P(SFSS) + P(SSFS) + P(SSSF)$$
$$= (.1)^3(.9) + (.1)^3(.9) + (.1)^3(.9) + (.1)^3(.9)$$
$$= 4(.1)^3(.9) = .0036$$

Note that each of the four sample-point probabilities is the same because each sample point consists of three *S*'s and one *F*; the order does not affect the probability because the adults' test results are (assumed) independent.

c. We can characterize this experiment as consisting of four identical trials: the four test results. There are two possible outcomes to each trial, *S* or *F*, and the probability of passing, $p = .1$, is the same for each trial. Finally, we are assuming that each adult's test result is independent of all the others', so that the four trials are independent. Then it follows that x, the number of the four adults who pass the fitness test, is a binomial random variable.

d. The event probabilities in parts **a** and **b** provide insight into the formula for the probability distribution $p(x)$. First, consider the event that three adults pass (part **b**). We found that

$$P(x = 3) =$$

$$(\text{Number of sample points for which } x = 3) \times (.1)^{\text{Number of successes}} \times (.9)^{\text{Number of failures}}$$
$$= 4(.1)^3(.9)^1$$

In general, we can use combinatorial mathematics to count the number of sample points. For example,

Number of sample points for which $x = 3$

$\quad$ = Number of different ways of selecting 3 successes in the 4 trials

$$= \binom{4}{3} = \frac{4!}{3!(4-3)!} = \frac{4 \cdot 3 \cdot 2 \cdot 1}{(3 \cdot 2 \cdot 1) \cdot 1} = 4$$

The formula that works for any value of x can be deduced as follows:

$$P(x = 3) = \binom{4}{3}(.1)^3(.9)^1 = \binom{4}{x}(.1)^x(.9)^{4-x}$$

The component $\binom{4}{x}$ counts the number of sample points with x successes, and the component $(.1)^x(.9)^{4-x}$ is the probability associated with each sample point having x successes.

$\quad$ For the general binomial experiment, with n trials and probability p of Success on each trial, the probability of x successes is

$$p(x) = \binom{n}{x} \underbrace{p^x(1-p)^{n-x}}$$

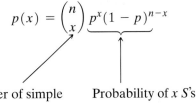

Number of simple
events with
x S's

Probability of x S's
and $(n - x)$ F's in
any simple event

Look Ahead In theory, you could always resort to the principles developed in this example to calculate binomial probabilities; just list the sample points and sum their probabilities. However, as the number of trials (n) increases, the number of sample points grows very rapidly. (The number of sample points is 2^n.) Thus, we prefer the formula for calculating binomial probabilities, since its use avoids listing sample points. ∎

$\quad$ The **binomial distribution*** is summarized in the following box:

The Binomial Probability Distribution

$$p(x) = \binom{n}{x}p^x q^{n-x} \quad (x = 0, 1, 2, \ldots, n)$$

where
$\quad p$ = Probability of a success on a single trial
$\quad q = 1 - p$
$\quad n$ = Number of trials
$\quad x$ = Number of successes in n trials
$n - x$ = Number of failures in n trials
$$\binom{n}{x} = \frac{n!}{x!(n-x)!}$$

$\quad$ As noted in Chapter 3, the symbol 5! means $5 \cdot 4 \cdot 3 \cdot 2 \cdot 1 = 120$. Similarly, $n! = n(n-1)(n-2) \ldots 3 \cdot 2 \cdot 1$. (Remember, $0! = 1$.)

---

*The binomial distribution is so named because the probabilities $p(x), x = 0, 1, \ldots, n$, are terms of the binomial expansion, $(q + p)^n$.

Example 4.11

Applying the Binomial Distribution—Physical Fitness Problem

Problem Refer to Example 4.10. Use the formula for a binomial random variable to find the probability distribution of x, where x is the number of adults who pass the fitness test. Graph the distribution.

Solution For this application, we have $n = 4$ trials. Since a success S is defined as an adult who passes the test, $p = P(S) = .1$ and $q = 1 - p = .9$. Substituting $n = 4, p = .1$, and $q = .9$ into the formula for $p(x)$, we obtain

$$p(0) = \frac{4!}{0!(4-0)!}(.1)^0(.9)^{4-0} = \frac{4 \cdot 3 \cdot 2 \cdot 1}{(1)(4 \cdot 3 \cdot 2 \cdot 1)}(.1)^0(.9)^4 = 1(.1)^0(.9)^4 = .6561$$

$$p(1) = \frac{4!}{1!(4-1)!}(.1)^1(.9)^{4-1} = \frac{4 \cdot 3 \cdot 2 \cdot 1}{(1)(3 \cdot 2 \cdot 1)}(.1)^1(.9)^3 = 4(.1)(.9)^3 = .2916$$

$$p(2) = \frac{4!}{2!(4-2)!}(.1)^2(.9)^{4-2} = \frac{4 \cdot 3 \cdot 2 \cdot 1}{(2 \cdot 1)(2 \cdot 1)}(.1)^2(.9)^2 = 6(.1)^2(.9)^2 = .0486$$

$$p(3) = \frac{4!}{3!(4-3)!}(.1)^3(.9)^{4-3} = \frac{4 \cdot 3 \cdot 2 \cdot 1}{(3 \cdot 2 \cdot 1)(1)}(.1)^3(.9)^1 = 4(.1)^3(.9) = .0036$$

$$p(4) = \frac{4!}{4!(4-4)!}(.1)^4(.9)^{4-4} = \frac{4 \cdot 3 \cdot 2 \cdot 1}{(4 \cdot 3 \cdot 2 \cdot 1)(1)}(.1)^4(.9)^0 = 1(.1)^4(.9) = .0001$$

Look Back Note that these probabilities, listed in Table 4.3, sum to 1. A graph of this probability distribution is shown in Figure 4.9.

Figure 4.9
Probability distribution for physical fitness example: Graphical form

Table 4.3	Probability Distribution for Physical Fitness Example: Tabular Form
x	*p(x)*
0	.6561
1	.2916
2	.0486
3	.0036
4	.0001

■ **Now Work Exercise 4.63**

Example 4.12

Finding μ and σ—Physical Fitness Problem

Problem Refer to Examples 4.10 and 4.11. Calculate μ and σ, the mean and standard deviation, respectively, of the number of the four adults who pass the test. Interpret the results.

Solution From Section 4.3, we know that the mean of a discrete probability distribution is

$$\mu = \sum xp(x)$$

Referring to Table 4.3, the probability distribution for the number x who pass the fitness test, we find that

$$\mu = 0(.6561) + 1(.2916) + 2(.0486) + 3(.0036) + 4(.0001) = .4$$
$$= 4(.1) = np$$

Thus, in the long run, the average number of adults (out of four) who pass the test is only .4. [*Note: The relationship $\mu = np$ holds in general for a binomial random variable.*]

The variance is

$$\sigma^2 = \Sigma(x - \mu)^2 p(x) = \Sigma(x - .4)^2 p(x)$$
$$= (0 - .4)^2(.6561) + (1 - .4)^2(.2916) + (2 - .4)^2(.0486)$$
$$+ (3 - .4)^2(.0036) + (4 - .4)^2(.0001)$$
$$= .104976 + .104976 + .124416 + .024336 + .001296$$
$$= .36 = 4(.1)(.9) = npq$$

[*Note: The relationship $\sigma^2 = npq$ holds in general for a binomial random variable.*]
Finally, the standard deviation of the number who pass the fitness test is

$$\sigma = \sqrt{\sigma^2} = \sqrt{.36} = .6$$

Since the distribution shown in Figure 4.9 is skewed right, we should apply Chebyshev's rule to describe where most of the x-values fall. According to the rule, at least 75% of the x values will fall into the interval $\mu \pm 2\sigma = .4 \pm 2(.6) = (-.8, 1.6)$. Since x cannot be negative, we expect (i.e., in the long run) the number of adults out of four who pass the fitness test to be less than 1.6.

Look Back Examining Figure 4.9, you can see that all observations equal to 0 or 1 will fall within the interval $(-.8, 1.6)$. The probabilities corresponding to these values (from Table 4.3) are .6561 and .2916, respectively. Summing them, we obtain $.6561 + .2916 = .9477 \approx .95$. This result is closer to the value stated by the empirical rule. In practice, researchers have found that the proportion of observations that fall within two standard deviations of the mean for many skewed distributions will be close to .95.

We emphasize that you need not use the expectation summation rules to calculate μ and σ^2 for a binomial random variable. You can find them easily from the formulas $\mu = np$ and $\sigma^2 = npq$.

Mean, Variance, and Standard Deviation for a Binomial Random Variable

Mean: $\mu = np$
Variance: $\sigma^2 = npq$
Standard deviation: $\sigma = \sqrt{npq}$

Using Tables and Technology for Binomial Probabilities

Calculating binomial probabilities becomes tedious when n is large. For some values of n and p, the binomial probabilities have been tabulated in Table I of Appendix B. Part of that table is shown in Table 4.4; a graph of the binomial probability distribution for $n = 10$ and $p = .10$ is shown in Figure 4.10.

Table 4.4   Reproduction of Part of Table I of Appendix B: Binomial Probabilities for $n = 10$

k \ p	.01	.05	.10	.20	.30	.40	.50	.60	.70	.80	.90	.95	.99
0	.904	.599	.349	.107	.028	.006	.001	.000	.000	.000	.000	.000	.000
1	.996	.914	.736	.376	.149	.046	.011	.002	.000	.000	.000	.000	.000
2	1.000	.988	.930	.678	.383	.167	.055	.012	.002	.000	.000	.000	.000
3	1.000	.999	.987	.879	.650	.382	.172	.055	.011	.001	.000	.000	.000
4	1.000	1.000	.998	.967	.850	.633	.377	.166	.047	.006	.000	.000	.000
5	1.000	1.000	1.000	.994	.953	.834	.623	.367	.150	.033	.002	.000	.000
6	1.000	1.000	1.000	.999	.989	.945	.828	.618	.350	.121	.013	.001	.000
7	1.000	1.000	1.000	1.000	.998	.988	.945	.833	.617	.322	.070	.012	.000
8	1.000	1.000	1.000	1.000	1.000	.988	.989	.954	.851	.624	.264	.086	.004
9	1.000	1.000	1.000	1.000	1.000	1.000	.999	.994	.972	.893	.651	.401	.096

Table I actually contains a total of nine tables, labeled (**a**) through (**i**), one each corresponding to $n = 5, 6, 7, 8, 9, 10, 15, 20$, and 25, respectively. In each of these tables, the columns correspond to values of p and the rows correspond to values of the random variable x. The entries in the table represent **cumulative binomial probabilities**. For example, the entry in the column corresponding to $p = .10$ and the row corresponding to $x = 2$ is .930 (highlighted), and its interpretation is

$$P(x \le 2) = P(x = 0) + P(x = 1) + P(x = 2) = .930$$

This probability is also highlighted in the graphical representation of the binomial distribution with $n = 10$ and $p = .10$ in Figure 4.10

You can also use Table I to find the probability that x equals a specific value. For example, suppose you want to find the probability that $x = 2$ in the binomial distribution with $n = 10$ and $p = .10$. This probability is found by subtraction as follows:

$$P(x = 2) = [P(x = 0) + P(x = 1) + P(x = 2)] - [P(x = 0) + P(x = 1)]$$
$$= P(x \le 2) - P(x \le 1) = .930 - .736 = .194$$

The probability that a binomial random variable exceeds a specified value can be found from Table I together with the notion of complementary events. For example, to find the probability that x exceeds 2 when $n = 10$ and $p = .10$, we use

$$P(x > 2) = 1 - P(x \le 2) = 1 - .930 = .070$$

Note that this probability is represented by the *un*highlighted portion of the graph in Figure 4.10.

All probabilities in Table I are rounded to three decimal places. Thus, although none of the binomial probabilities in the table is exactly zero, some are small enough (less than .0005) to round to .000. For example, using the formula to find $P(x = 0)$ when $n = 10$ and $p = .6$, we obtain

$$P(x = 0) = \binom{10}{0}(.6)^0(.4)^{10-0} = .4^{10} = .00010486$$

but this is rounded to .000 in Table I of Appendix A. (See Table 4.4.)

Similarly, none of the table entries is exactly 1.0, but when the cumulative probabilities exceed .9995, they are rounded to 1.000. The row corresponding to the largest possible value for $x, x = n$, is omitted because all the cumulative probabilities in that row are equal to 1.0 (exactly). For example, in Table 4.4 with $n = 10$, $P(x \le 10) = 1.0$, no matter what the value of p.

Of course, you can also use technology such as statistical software or a graphing calculator to find binomial probabilities. The following example further illustrates the use of Table I and statistical software.

Figure 4.10
Binomial probability distribution for $n = 10$ and $p = .10$, with $P(x \le 2)$ highlighted

---

Example **4.13**	

Using the Binomial Table and Software—Voting for Mayor

Problem Suppose a poll of 20 voters is taken in a large city. The purpose is to determine x, the number who favor a certain candidate for mayor. Suppose that 60% of all the city's voters favor the candidate.

a. Find the mean and standard deviation of x.

b. Use Table I of Appendix B to find the probability that $x \le 10$. Verify this probability using MINITAB.

c. Use Table I to find the probability that $x > 12$.

d. Use Table I to find the probability that $x = 11$. Verify this probability using MINITAB.

e. Graph the probability distribution of x, and locate the interval $\mu \pm 2\sigma$ on the graph.

Solution

a. The number of voters polled is presumably small compared with the total number of eligible voters in the city. Thus, we may treat x, the number of the 20 who favor the mayoral candidate, as a binomial random variable. The value of p is the fraction of the total number of voters who favor the candidate (i.e., $p = .6$). Therefore, we calculate the mean and variance:

$$\mu = np = 20(.6) = 12$$
$$\sigma^2 = npq = 20(.6)(.4) = 4.8$$
$$\sigma = \sqrt{4.8} = 2.19$$

b. Looking in the row for $k = 10$ and the column for $p = .6$ of Table I (Appendix B) for $n = 20$, we find the value .245. Thus,

$$P(x \leq 10) = .245$$

This value agrees with the cumulative probability shaded at the top of the MINITAB printout shown in Figure 4.11.

Cumulative Distribution Function

Binomial with n = 20 and p = 0.6

```
  x    P( X <= x )
 10       0.244663
```

Probability Density Function

Binomial with n = 20 and p = 0.6

```
  x    P( X = x )
 11       0.159738
```

Figure 4.11

MINITAB Output for Example 4.13

c. To find the probability

$$P(x > 12) = \sum_{x=13}^{20} p(x)$$

we use the fact that for all probability distributions,

$$\sum_{\text{all } x} p(x) = 1.$$

Therefore,

$$P(x > 12) = 1 - P(x \leq 12) = 1 - \sum_{x=0}^{12} p(x)$$

Consulting Table I of Appendix B, we find the entry in row $k = 12$, column $p = .6$ to be .584. Thus,

$$P(x > 12) = 1 - .584 = .416$$

d. To find the probability that exactly 11 voters favor the candidate, recall that the entries in Table I are cumulative probabilities and use the relationship

$$P(x = 11) = [p(0) + p(1) + \cdots + p(11)] - [p(0) + p(1) + \cdots + p(10)]$$
$$= P(x \leq 11) - P(x \leq 10)$$

Then

$$P(x = 11) = .404 - .245 = .159$$

Again, this value agrees with the probability shaded at the bottom of the MINITAB printout, Figure 4.11.

e. The probability distribution for x is shown in Figure 4.12. Note that

$$\mu - 2\sigma = 12 - 2(2.2) = 7.6 \quad \mu + 2\sigma = 12 + 2(2.2) = 16.4$$

Figure 4.12
The binomial probability distribution for x in Example 4.13; $n = 20$ and $p = .6$

The interval $\mu - 2\sigma$ to $\mu + 2\sigma$ also is shown in Figure 4.12. The probability that x falls into the interval $\mu + 2\sigma$ is $P(x = 8, 9, 10, \ldots, 16) = P(x \le 16) - P(x \le 7) = .984 - .021 = .963$. Note that this probability is very close to the .95 given by the empirical rule. Thus, we expect the number of voters in the sample of 20 who favor the mayoral candidate to be between 8 and 16.

■ **Now Work Exercise 4.64**

Using the Binomial Model to Solve the Cocaine Sting Case

Refer to the reverse cocaine sting case described on p. 213. During a drug bust, police seized 496 foil packets of a white, powdery substance that appeared to be cocaine. The police laboratory randomly selected 4 packets and found that all 4 tested positive for cocaine. This finding led to the conviction of the drug traffickers. Following the conviction, the police used 2 of the remaining 492 foil packets (i.e., those not tested) in a reverse sting operation. The 2 randomly selected packets were sold by undercover officers to a buyer who disposed of the evidence before being arrested. Is there evidence beyond a reasonable doubt that the 2 packets contained cocaine?

To solve the dilemma, we will assume that, of the 496 original packets confiscated, 331 contained cocaine and 165 contained an inert (legal) powder. (A statistician hired as an expert witness on the case showed that the chance of the defendant being found not guilty is maximized when 331 packets contain cocaine and 165 do not.) First, we'll find the probability that 4 packets randomly selected from the original 496 will test positive for cocaine. Then we'll find the probability that the 2 packets sold in the reverse sting did not contain cocaine. Finally, we'll find the probability that both events occur (i.e., that the first 4 packets selected test positive for cocaine, but that the next 2 packets selected do not). In each of these probability calculations, we will apply the binomial probability distribution to approximate the probabilities.

Let x be the number of packets that contain cocaine in a sample of n selected from the 496 packets. Here, we are defining

a success as a packet that contains cocaine. If n is small, say, $n = 2$ or $n = 4$, then x has an approximate binomial distribution with probability of success $p = 331/496 \approx .67$.

The probability that the first 4 packets selected contain cocaine [i.e., $P(x = 4)$] is obtained from the binomial formula with $n = 4$ and $p = .67$:

$$P(x = 4) = p(4) = \binom{4}{4}p^4(1 - p)^0 = \frac{4!(.67)^4(.33)^0}{4!0!}$$
$$= (.67)^4 = .202$$

Thus, there is about a 20% chance that all 4 of the randomly selected packets contain cocaine.

Given that 4 of the original packets tested positive for cocaine, the probability that the 2 packets randomly selected and sold in the reverse sting *do not* contain cocaine is approximated by a binomial distribution with $n = 2$ and $p = .67$. Since a success is a packet with cocaine, we find $P(x = 0)$:

$$P(x = 0) = p(0) = \binom{2}{0}p^0(1 - p)^2 = \frac{2!(.67)^0(.33)^2}{0!2!}$$
$$= (.33)^2 = .109$$

Statistics IN Action
(continued)

Finally, to compute the probability that, of the original 496 foil packets, the first 4 selected (at random) test positive for cocaine and the next two selected (at random) test negative, we employ the multiplicative law of probability. Let A be the event that the first 4 packets test positive. Let B be the event that the next 2 packets test negative. We want to find the probability of both events occurring [i.e., $P(A \text{ and } B) = P(A \cap B) = P(B|A)P(A)$]. Note that this probability is the product of the two previously calculated probabilities:

$$P(A \text{ and } B) = (.109)(.201) = .022$$

Consequently, there is only a .022 probability (i.e., about 2 chances in a hundred) that the first 4 packets will test positive for cocaine and the next 2 packets will test negative for cocaine. A reasonable jury would likely believe that an event with such a small probability is unlikely to occur and conclude that the 2 "lost" packets contained cocaine. In other words, most of us would infer that the defendant in the reverse cocaine sting was guilty of drug trafficking.

[*Epilogue*: Several of the defendant's lawyers believed that the .022 probability was too high for jurors to conclude guilt "beyond a reasonable doubt." The argument was made moot, however, when, to the surprise of the defense, the prosecution revealed that the remaining 490 packets had not been used in any other reverse sting operations and offered to test a sample of them. On the advice of the statistician, the defense requested that an additional 20 packets be tested. All 20 tested positive for cocaine! As a consequence of this new evidence, the defendant was convicted by the jury.]

Exercises 4.57–4.84

Understanding the Principles

4.57 Name two parameters of a binomial random variable.

4.58 Give the formula for $p(x)$ for a binomial random variable with $n = 10$ and $p = .3$.

4.59 Consider the following binomial probability distribution:

$$p(x) = \binom{7}{x}(.4)^x(.6)^{7-x}(x = 0, 1, 2, \ldots, 7)$$

a. How many trials (n) are in the experiment?
b. What is the value of p, the probability of success?

Learning the Mechanics

4.60 Refer to Exercise 4.59.
a. Graph the probability distribution.
b. Find the mean and standard deviation of x.
c. Show the mean and the two-standard-deviation interval on each side of the mean on the graph you drew in part **a**.

4.61 Compute the following:
a. $\dfrac{5!}{2!(5-2)!}$ **b.** $\binom{6}{2}$ **c.** $\binom{5}{0}$

d. $\binom{7}{7}$ **e.** $\binom{5}{3}$

4.62 Suppose x is a binomial random variable with $n = 4$ and $p = .4$.
a. Calculate the value of $p(x), x = 0, 1, 2, 3, 4$, using the formula for a binomial probability distribution.
b. Using your answers to part **a**, give the probability distribution for x in tabular form.

4.63 If x is a binomial random variable, compute $p(x)$ for each of the following cases:
[NW]
a. $n = 5, x = 1, p = .2$
b. $n = 4, x = 2, q = .4$
c. $n = 3, x = 0, p = .7$
d. $n = 5, x = 3, p = .1$
e. $n = 4, x = 2, q = .6$
f. $n = 3, x = 1, p = .9$

4.64 If x is a binomial random variable, use Table I in Appendix B
[NW] or technology to find the following probabilities:
a. $P(x = 2)$ for $n = 10, p = .4$
b. $P(x \le 5)$ for $n = 15, p = .6$
c. $P(x > 1)$ for $n = 5, p = .1$

4.65 If x is a binomial random variable, use Table I in Appendix B or technology to find the following probabilities:
a. $P(x = 9)$ for $n = 24, p = .7$
b. $P(x \le 5)$ for $n = 15, p = .5$
c. $P(x = 2)$ for $n = 18, p = .5$

4.66 If x is a binomial random variable, calculate μ, σ^2, and σ for each of the following:
a. $n = 8, p = .3$
b. $n = 100, p = .2$
c. $n = 90, p = .4$
d. $n = 60, p = .9$
e. $n = 50, p = .7$
f. $n = 1,000, p = .05$

4.67 The binomial probability distribution is a family of probability distributions with each single distribution depending on the values of n and p. Assume that x is a binomial random variable with $n = 10$.
a. Determine a value of p such that the probability distribution of x is symmetric.
b. Determine a value of p such that the probability distribution of x is skewed to the right.
c. Determine a value of p such that the probability distribution of x is skewed to the left.
d. Graph each of the binomial distributions you obtained in parts **a**, **b**, and **c**. Locate the mean for each distribution on its graph.
e. In general, for what values of p will a binomial distribution be symmetric? skewed to the right? skewed to the left?

Ⓓ Applet Exercise 4.3

Use the applets entitled *Simulating the Probability of a Head with an Unfair Coin (P(H) = 0.2)* and *Simulating the Probability of a*

Head with an Unfair Coin (P(H) = 0.8) to study the mean μ of a binomial distribution.

a. Run each applet once with $n = 1,000$ and record the cumulative proportions. How does the cumulative proportion for each applet compare with the value of $P(H)$ given for the applet?

b. Using the cumulative proportion from each applet as p, compute $\mu = np$ for each applet, where $n = 1,000$. What does the value of μ represent in terms of the results obtained from running each applet in part **a**?

c. In your own words, describe what the mean μ of a binomial distribution represents.

Ⓓ Applet Exercise 4.4

Open the applet entitled *Sample from a Population*. On the pull-down menu to the right of the top graph, select *Binary*. Set $n = 10$ as the sample size and repeatedly choose samples from the population. For each sample, record the number of 1's in the sample. Let x be the number of 1's in a sample of size 10. Explain why x is a binomial random variable.

Ⓓ Applet Exercise 4.5

Use the applet entitled *Simulating the Stock Market* to estimate the probability that the stock market will go up each of the next two days. Repeatedly run the applet for $n = 2$, recording the number of ups each time. Use the proportion of 2's among your results as the estimate of the probability. Compare your answer with the binomial probability where $x = 2, n = 2$, and $p = 0.5$.

Applying the Concepts—Basic

4.68 **Working on summer vacation.** Suppose in a school it is found that 30% of adults do not work at all while on summer vacation. In a random sample of 8 adults, let x represent the number who do not work during summer vacation.

a. For this experiment, define the event that represents a "success."

b. Explain why x is (approximately) a binomial random variable.

c. Give the value of p for this binomial experiment.

d. Find $P(x = 3)$.

e. Find the probability that 2 or fewer of the 8 adults do not work during summer vacation.

4.69 **Superstitions survey.** Are Americans superstitious? A *Harris* (Feb. 2013) poll of over 2,000 adult Americans was designed to answer this question. One survey item concerned the phrase "see a penny, pick it up, all day long you'll have good luck." The poll found that just one-third of Americans (33%) believe finding and picking up a penny is good luck. Consider a random sample of 20 U.S. adults and let x represent the number who believe finding and picking up a penny is good luck.

a. For this experiment, define the event that represents a "success."

b. Explain why x is (approximately) a binomial random variable.

c. Give the value of p for this binomial experiment.

d. Find $P(x < 10)$.

e. Find $P(x = 17)$.

f. Find $P(x > 5)$.

4.70 **Where will you get your next pet?** According to an Associated Press/Petside.com poll, half of all pet owners would get their next dog or cat from a shelter (*USA Today*, May 12, 2010). Consider a random sample of 20 pet owners and define x as the number of pet owners who would acquire their next dog or cat from a shelter. Assume that x is a binomial random variable.

a. For this binomial experiment, what is n?

b. Find $P(x = 7)$.

c. Find $P(x \geq 3)$.

d. Find $P(x > 5)$.

4.71 **Chemical signals of mice.** Refer to the *Cell* (May 14, 2010) study of the ability of a mouse to recognize the odor of a potential predator, Exercise 3.63 (p. 171). Recall that the sources of these odors are typically major urinary proteins (Mups). In an experiment, 40% of lab mice cells exposed to chemically produced cat Mups responded positively (i.e., recognized the danger of the lurking predator). Consider a sample of 100 lab mice cells, each exposed to chemically produced cat Mups. Let x represent the number of cells that respond positively.

a. Explain why the probability distribution of x can be approximated by the binomial distribution.

b. Find $E(x)$ and interpret its value, practically.

c. Find the variance of x.

d. Give an interval that is likely to contain the value of x.

4.72 **Analysis of bottled water.** Is the bottled water you're drinking really purified water? A study of various brands of bottled water conducted by the Natural Resources Defense Council found that 25% of bottled water is just tap water packaged in a bottle (www.nrdc.org). Consider a sample of five bottled-water brands, and let x equal the number of these brands that use tap water.

⟦NW⟧ **a.** Explain why x is (approximately) a binomial random variable.

b. Give the probability distribution for x as a formula.

c. Find $P(x = 2)$.

d. Find $P(x \leq 1)$.

4.73 **Caesarian births.** A medical research reports that 75% of all births in a certain region take place by Caesarian section each year.

a. In a random sample of 1,000 births, how many, on average, will take place by Caesarian section?

b. What is the standard deviation of the number of Caesarian section births in a sample of 1,000 births?

c. Use your answers to parts **a** and **b** to form an interval that is likely to contain the number of Caesarian section births in a sample of 1,000 births.

Apply the Concepts—Intermediate

4.74 **Fingerprint expertise.** Refer to the *Psychological Science* (Aug. 2011) study of fingerprint identification, Exercise 3.140 (p. 200). The study found that when presented with prints from the same individual, a fingerprint expert will correctly identify the match 92% of the time. In contrast, a novice will correctly identify the match 75% of the time. Consider a sample of five different pairs of fingerprints where each pair is a match.

a. What is the probability that an expert will correctly identify the match in all five pairs of fingerprints?

b. What is the probability that a novice will correctly identify the match in all five pairs of fingerprints?

4.75 Hotel guest satisfaction. J. D. Power and Associates' North American Hotel Guest Satisfaction Index Study revealed that 66% of hotel guests are aware of the hotel's "green" conservation program. Among these guests, 72% actually participate in the program by reusing towels and bed linens. In a random sample of 15 hotel guests, consider the number (x) of guests who are aware and participate in the hotel's conservation efforts.

a. Explain why x is (approximately) a binomial random variable.

b. Use the rules of probability to determine the value of p for this binomial experiment.

c. Assume p = .45. Find the probability that at least 10 of the 15 hotel guests are aware of and participate in the hotel's conservation efforts.

4.76 Immediate feedback to incorrect exam answers. Researchers from the Educational Testing Service (ETS) found that providing immediate feedback to students answering open-ended questions can dramatically improve students' future performance on exams (*Educational and Psychological Measurement*, Feb. 2010). The ETS researchers used questions from the Graduate Record Examination (GRE) in the experiment. After obtaining feedback, students could revise their answers. Consider one of these questions. Initially, 50% of the students answered the question correctly. After providing immediate feedback to students who answered incorrectly, 70% answered correctly. Consider a bank of 100 open-ended questions similar to those on the GRE.

a. In a random sample of 20 students, what is the probability that more than half initially answer the question correctly?

b. Refer to part **a**. After providing immediate feedback, what is the probability that more than half of the students answer the question correctly?

4.77 Making your vote count. Democratic and Republican presidential state primary elections differ in the way winning candidates are awarded delegates. In Republican states, the winner is awarded all the state's delegates; conversely, the Democratic state winner is awarded delegates in proportion to the percentage of votes. This difference led to a *Chance* (Fall 2007) article on making your vote count. Consider a scenario where you are one of five county commissioners voting on an issue, where each commissioner is equally likely to vote for or against.

a. Your vote counts (i.e., is the decisive vote) only if the other four voters split, 2 in favor and 2 against. Use the binomial distribution to find the probability that your vote counts.

b. If you convince two other commissioners to "vote in bloc" (i.e., you all agree to vote among yourselves first, and whatever the majority decides is the way all three will vote, guaranteeing that the issue is decided by the bloc), your vote counts only if these 2 commissioners split their bloc votes, 1 in favor and 1 against. Again, use the binomial distribution to find the probability that your vote counts.

4.78 Marital name changing. As a female, how likely are you to follow tradition and change your last name to your husband's when you get married? This was one of the questions investigated in research published in *Advances in Applied Sociology* (Nov. 2013). According to the study, the probability that an American female will change her last name upon marriage is .9. Consider a sample of 25 American females. How many of these females would you expect to change their last name upon marriage? How likely is it that 20 or more of these females will change their last name upon marriage?

4.79 Victims of domestic abuse. According to researchers at Dan Jones & Associates, 1 in every 3 women has been a victim of domestic abuse (Domestic Violence: Incidence and Prevalence Study, Sept.–Dec., 2005). This probability was obtained from a survey of 1,000 adult women residing in Utah. Suppose we randomly sample 10 women and find that 6 have been abused.

a. What is the probability of observing 5 or more abused women in a sample of 10 if the proportion p of women who are victims of domestic abuse is really $p = \frac{1}{3}$?

b. Many experts on domestic violence believe that the proportion of women who are domestically abused is closer to $p = .10$. Calculate the probability of observing 5 or more abused women in a sample of 10 if $p = .10$.

c. Why might your answers to parts **a** and **b** lead you to believe that $p = \frac{1}{3}$?

4.80 Underwater acoustic communication. A subcarrier is one telecommunication signal carrier that is piggybacked on top of another carrier so that effectively two signals are carried at the same time. Subcarriers are classified as data subcarriers (used for data transmissions), pilot subcarriers (used for channel synchronization), and null subcarriers (used for direct current). In the *IEEE Journal of Oceanic Engineering* (Apr. 2013), researchers studied the characteristics of subcarriers for underwater acoustic communications. Based on an experiment conducted off the coast of Martha's Vineyard (MA), they estimated that 25% of subcarriers are pilot subcarriers, 10% are null subcarriers, and 65% are data subcarriers. Consider a sample of 50 subcarriers transmitted for underwater acoustic communications.

a. How many of the 50 subcarriers do you expect to be pilot subcarriers? Null subcarriers? Data subcarriers?

b. How likely is it to observe 10 or more pilot subcarriers?

c. If you observe more than 25 pilot subcarriers, what would you conclude? Explain.

4.81 Testing a psychic's ESP. Refer to Exercise 3.101 (p. 186) and the experiment conducted by the Tampa Bay Skeptics to see whether an acclaimed psychic has extrasensory perception (ESP). Recall that a crystal was placed, at random, inside 1 of 10 identical boxes lying side by side on a table. The experiment was repeated seven times, and x, the number of correct decisions, was recorded. (Assume that the seven trials are independent.)

a. If the psychic is guessing (i.e., if the psychic does *not* possess ESP), what is the value of p, the probability of a correct decision on each trial?

b. If the psychic is guessing, what is the expected number of correct decisions in seven trials?

c. If the psychic is guessing, what is the probability of no correct decisions in seven trials?

d. Now suppose the psychic has ESP and $p = .5$. What is the probability that the psychic guesses incorrectly in all seven trials?

e. Refer to part **d**. Recall that the psychic failed to select the box with the crystal on all seven trials. Is this evidence against the psychic having ESP? Explain.

Applying the Concepts—Advanced

4.82 **Assigning a passing grade.** A literature professor decides to give a 25-question true–false quiz to determine who has read an assigned novel. She wants to choose the passing grade such that the probability of passing a student who guesses on every question is less than .1. What score should she set as the lowest passing grade?

4.83 **GA golf ball specifications.** According to the U.S. Golf Association (USGA), "The weight of the [golf] ball shall not be greater than 1.620 ounces avoirdupois (45.93 grams).… The diameter of the ball shall not be less than 1.680 inches.… The velocity of the ball shall not be greater than 250 feet per second" (USGA, 2014). The USGA periodically checks the specifications of golf balls sold in the United States by randomly sampling balls from pro shops around the country. Two dozen of each kind are sampled, and if more than three do not meet size or velocity requirements, that kind of ball is removed from the USGA's approved-ball list.

a. What assumptions must be made and what information must be known in order to use the binomial probability distribution to calculate the probability that the USGA will remove a particular kind of golf ball from its approved-ball list?

b. Suppose 10% of all balls produced by a particular manufacturer are less than 1.680 inches in diameter, and assume that the number of such balls, x, in a sample of two dozen balls can be adequately characterized by a binomial probability distribution. Find the mean and standard deviation of the binomial distribution.

c. Refer to part **b**. If x has a binomial distribution, then so does the number, y, of balls in the sample that meet the USGA's minimum diameter. [*Note:* $x + y = 24$.] Describe the distribution of y. In particular, what are p, q, and n? Also, find $E(y)$ and the standard deviation of y.

4.84 **Does having boys run in the family?** *Chance* (Fall 2001) reported that the eight men in the Rodgers family produced 24 biological children over four generations. Of these 24 children, 21 were boys and 3 were girls. How likely is it for a family of 24 children to have 21 boys? Use the binomial distribution and the fact that 50% of the babies born in the United States are male to answer the question. Do you agree with the statement, "Rodgers men produce boys"?

4.5 The Poisson Random Variable (Optional)

A type of probability distribution that is often useful in describing the number of rare events that will occur during a specific period or in a specific area or volume is the **Poisson distribution** (named after the 18th-century physicist and mathematician Siméon Poisson). Typical examples of random variables for which the Poisson probability distribution provides a good model are

1. The number of traffic accidents per month at a busy intersection

2. The number of noticeable surface defects (scratches, dents, etc.) found by quality inspectors on a new automobile

3. The number of parts per million (ppm) of some toxin found in the water or air emissions from a manufacturing plant

4. The number of diseased trees per acre of a certain woodland

5. The number of death claims received per day by an insurance company

6. The number of unscheduled admissions per day to a hospital

Characteristics of a Poisson Random Variable

1. The experiment consists of counting the number of times a certain event occurs during a given unit of time or in a given area or volume (or weight, distance, or any other unit of measurement).

2. The probability that an event occurs in a given unit of time, area, or volume is the same for all the units.

3. The number of events that occur in one unit of time, area, or volume is independent of the number that occur in other units.

4. The mean (or expected) number of events in each unit is denoted by the Greek letter lambda (λ).

The characteristics of the **Poisson random variable** are usually difficult to verify for practical examples. The examples given here satisfy them well enough that the Poisson distribution provides a good model in many instances.* As with all probability models, the real test of the adequacy of the Poisson model is whether it provides a reasonable approximation to reality—that is, whether empirical data support it.

The probability distribution, mean, and variance for a Poisson random variable are shown in the next box.

Probability Distribution, Mean, and Variance for a Poisson Random Variable

$$p(x) = \frac{\lambda^x e^{-\lambda}}{x!} \quad (x = 0, 1, 2, \dots) \quad \mu = \lambda \quad \sigma^2 = \lambda$$

where

λ = Mean number of events during a given unit of time, area, volume, etc.

$e = 2.71828\dots$

BIOGRAPHY   SIMÉON D. POISSON (1781–1840)

A Lifetime Mathematician

Growing up in France during the French Revolution, Siméon-Denis Poisson was sent away by his father to become an apprentice surgeon, but he lacked the manual dexterity to perform the delicate procedures required and returned home. He eventually enrolled in the École Polytechnique to study mathematics. In his final year of study, Poisson wrote a paper on the theory of equations that was of such quality that he was allowed to graduate without taking the final examination. Two years later, he was named a professor at the university. During his illustrious career, Poisson published between 300 and 400 mathematics papers. He is most known for his 1837 paper in which he first set forth the distribution of a rare event—the Poisson distribution (although the distribution was actually described years earlier by one of the Bernoulli brothers). Poisson dedicated his life to mathematics, once stating that "Life is good for only two things: to study mathematics and to teach it." ■

The calculation of Poisson probabilities is made easier by the use of technology such as statistical software, as illustrated in Example 4.14.

Example   4.14

Finding Poisson Probabilities—Whale Sightings

Problem Ecologists often use the number of reported sightings of a rare species of animal to estimate the remaining population size. For example, suppose the number x of reported sightings per week of blue whales is recorded. Assume that x has (approximately) a Poisson probability distribution. Furthermore, assume that the average number of weekly sightings is 2.6.

a. Find the mean and standard deviation of x, the number of blue-whale sightings per week. Interpret the results.

b. Use MINITAB to find the probability that exactly five sightings are made during a given week.

c. Use MINITAB to find the probability that fewer than two sightings are made during a given week.

d. Use MINITAB to find the probability that more than five sightings are made during a given week.

Solution

a. The mean and variance of a Poisson random variable are both equal to λ. Thus, for this example,

$$\mu = \lambda = 2.6$$
$$\sigma^2 = \lambda = 2.6$$

---

*The Poisson probability distribution also provides a good approximation to a binomial probability distribution with mean $\lambda = np$ when n is large and p is small (say, $np \le 7$).

Then the standard deviation of x is

$$\sigma = \sqrt{2.6} = 1.61$$

Remember that the mean measures the central tendency of the distribution and does not necessarily equal a possible value of x. In this example, the mean is 2.6 sightings, and although there cannot be 2.6 sightings during a given week, the average number of weekly sightings is 2.6. Similarly, the standard deviation of 1.61 measures the variability of the number of sightings per week. Perhaps a more helpful measure is the interval $\mu \pm 2\sigma$, which in this case stretches from $-.62$ to 5.82. We expect the number of sightings to fall into this interval most of the time—with at least 75% relative frequency (according to Chebyshev's rule) and probably closer to 95% relative frequency (the empirical rule). The mean and the two-standard-deviation interval around it are shown in Figure 4.13.

Figure 4.13

Probability distribution for number of blue-whale sightings

b. Since x represents the number of blue-whale sightings per week, then we want to find $P(x = 5)$. Like for binomial probabilities, tables and computer software give Poisson probabilities of the form $P(x = k)$ or cumulative probabilities of the form $P(x \le k)$. Setting $\lambda = 2.6$ as an option in MINITAB and specifying the value "5" for a Poisson probability, we obtain the value (shaded) shown at the top of the MINITAB printout in Figure 4.14:

$$P(x = 5) = .0735$$

Probability Density Function

```
Poisson with mean = 2.6

x   P( X = x )
5   0.0735394
```

Cumulative Distribution Function

```
Poisson with mean = 2.6

x   P( X <= x )
1    0.267385
```

Cumulative Distribution Function

```
Poisson with mean = 2.6

x   P( X <= x )
5    0.950963
```

Figure 4.14

MINITAB Output for Example 4.14

c. Here, we want to find $P(x < 2)$. Since x is a discrete whole number, we know that

$$P(x < 2) = P(x \le 1)$$

This probability is now in the form of a cumulative probability and can be found using statistical software. Setting $\lambda = 2.6$ as an option in MINITAB and specifying the value "1" for a cumulative Poisson probability, we obtain the value (shaded) shown in the middle of the MINITAB printout in Figure 4.14:

$$P(x < 2) = P(x \le 1) = .2674$$

d. To find the probability that more than five sightings are observed in a given week, we consider the complementary event

$$P(x > 5) = 1 - P(x \le 5)$$

Again, we select $\lambda = 2.6$ as an option in MINITAB and now specify the value "5" for a cumulative Poisson probability. This yields the probability .951, shaded at the bottom of the MINITAB printout in Figure 4.14. Now, we compute:

$$P(x > 5) = 1 - P(x \le 5) = 1 - .951 = .049$$

Note from Figure 4.13 that this probability is the area in the interval $\mu \pm 2\sigma$, or $2.6 \pm 2(1.61) = (-.62, 5.82)$. Then the number of sightings should exceed 5—or, equivalently, should be more than 2 standard deviations from the mean—during only about 4.9% of all weeks. This percentage agrees remarkably well with that given by the empirical rule for mound-shaped distributions, which informs us to expect about 5% of the measurements (values of the random variable x) to lie further than 2 standard deviations from the mean.

■ **Now Work Exercise 4.99**

Exercises 4.85–4.106

Understanding the Principles

4.85 What is the relation between the mean and the variance of a Poisson random variable?

4.86 Consider a Poisson random variable with probability distribution

$$p(x) = \frac{20^x e^{-20}}{x!} \quad (x = 0, 1, 2, \ldots)$$

What is the value of λ?

4.87 Consider the Poisson probability distribution shown here:

$$p(x) = \frac{5^x e^{-5}}{x!} \quad (x = 0, 1, 2, \ldots)$$

What is the value of λ?

Learning the Mechanics

4.88 Refer to Exercise 4.86.
   a. Graph the probability distribution.
   b. Find the mean and standard deviation of x.

4.89 Refer to Exercise 4.87.
   a. Graph the probability distribution.
   b. Find the mean and standard deviation of x.

4.90 Given that x is a random variable for which a Poisson probability distribution provides a good approximation, compute the following:
   a. $P(x \le 1)$ when $\lambda = 1$
   b. $P(x \le 1)$ when $\lambda = 2$
   c. $P(x \le 1)$ when $\lambda = 3$
   d. What happens to the probability of the event $\{x \le 1\}$ as λ increases from 1 to 3? Is this intuitively reasonable?

4.91 Assume that x is a random variable having a Poisson probability distribution with a mean of 2. Find the following probabilities:
   a. $P(x \le 2)$
   b. $P(x \ge 2)$
   c. $P(x = 2)$
   d. $P(x = 0)$
   e. $P(x < 1)$
   f. $P(x < 4)$

4.92 Suppose x is a random variable for which a Poisson probability distribution with $\lambda = 2$ provides a good characterization.
   a. Graph $p(x)$ for $x = 0, 1, 2, \ldots, 7$.
   b. Find μ and σ for x, and locate μ and the interval $\mu \pm 2\sigma$ on the graph.
   c. What is the probability that x will fall within the interval $\mu \pm 2\sigma$?

4.93 Suppose x is a random variable for which a Poisson probability distribution with $\lambda = .5$ provides a good characterization.
   a. Graph $p(x)$ for $x = 0, 1, 2, 3$.
   b. Find μ and σ for x, and locate μ and the interval $\mu \pm 2\sigma$ on the graph.
   c. What is the probability that x will fall within the interval $\mu \pm 2\sigma$?

4.94 As mentioned in this section, when n is large, p is small, and $np \le 7$, the Poisson probability distribution provides a good approximation to the binomial probability distribution. Since we provide exact binomial probabilities (Table I in Appendix B) for relatively small values of n, you can investigate the adequacy of the approximation for $n = 30$. Use Table I to find $p(0), p(1)$, and $p(2)$ for $n = 30$ and $p = .05$. Calculate the corresponding Poisson approximations, using $\lambda = \mu = np$. [*Note:* These approximations are reasonably good for n as small as 30, but to use such an approximation in a practical situation we would prefer to have $n \geq 100$.]

Applying the Concepts—Basic

4.95 **Eye fixation experiment.** Cognitive scientists at the University of Massachusetts designed an experiment to measure x, the number of times a reader's eye fixated on a single word before moving past that word (*Memory and Cognition*, Sept. 1997). For this experiment, x was found to have a mean of 1. Suppose one of the readers in the experiment is randomly selected, and assume that x has a Poisson distribution.
  a. Find $P(x = 0)$.
  b. Find $P(x > 1)$.
  c. Find $P(x \leq 2)$.

4.96 **Noise in laser imaging.** Penumbrol imaging is a technique used by nuclear engineers for imaging objects (e.g., X-rays and lasers) that emit high-energy photons. In *IEICE Transactions on Information & Systems* (Apr. 2005), researchers demonstrated that penumbrol images are always degraded by noise, where the number x of noise events occurring in a unit of time follows a Poisson process with mean λ. Suppose that $\lambda = 9$ for a particular image.
  a. Find and interpret the mean of x.
  b. Find the standard deviation of x.
  c. The signal-to-noise ratio (SNR) for a penumbrol image is defined as SNR $= \mu/\sigma$, where μ and σ are the mean and standard deviation, respectively, of the noise process. Find the SNR for x.

4.97 **Spare line replacement units (LRU).** The Department of U.S. Defense Reliability Analysis Center publishes Selected Topics in Assurance Related Technologies (START) sheets to help improve the quality of manufactured components and systems. One START sheet, titled "Application of the Poisson Distribution" (Vol. 9, No. 1, 2002), focuses on a spare line replacement unit (LRU). The number of LRUs that fail in any 10,000-hour period is assumed to follow a Poisson distribution with a mean of 2.1.
  a. Find the probability that there are no LRU failures during the next 10,000 hours of operation.
  b. Find the probability that there are at least one LRU failures during the next 10,000 hours of operation.

4.98 **Rare planet transits.** A "planet transit" is a rare celestial event in which a planet appears to cross in front of its star as seen from Earth. The planet transit causes a noticeable dip in the star's brightness, allowing scientists to detect a new planet even though it is not directly visible. After one year of a mission in which 3,000 stars were monitored, NASA announced that 5 planet transits were detected (NASA, American Astronomical Society, Jan. 4, 2010). Assume that the number of planet transits discovered for every 3,000 stars follows a Poisson distribution with $\lambda = 5$. What is the probability that, in the next 3,000 stars monitored by the mission, more than 8 planet transits will be seen?

4.99 **Airline fatalities.** U.S. airlines average about 4.5 fatalities per month. (*Statistical Abstract of the United States: 2012*). Assume that the probability distribution for x, the number of fatalities per month, can be approximated by a Poisson probability distribution.

a. What is the probability that no fatalities will occur during any given month?
b. What is the probability that one fatality will occur during any given month?
c. Find $E(x)$ and the standard deviation of x.

Applying the Concepts—Intermediate

4.100 **Traffic fatalities and sporting events.** The relationship between close sporting events and game-day traffic fatalities was investigated in the *Journal of Consumer Research* (Dec. 2011). The researchers found that closer football and basketball games are associated with more traffic fatalities. The methodology used by the researchers involved modeling the traffic fatality count for a particular game as a Poisson random variable. For games played at the winner's location (home court or home field), the mean number of traffic fatalities was .5. Use this information to find the probability that at least 3 game-day traffic fatalities will occur at the winning team's location.

4.101 **LAN videoconferencing.** A network administrator is installing a videoconferencing module in a local area network (LAN) computer system. Of interest is the capacity of the LAN to handle users who attempt to call in for videoconferencing during peak hours. Calls are blocked if the user finds that all LAN lines are "busy." The capacity is directly related to the rate at which calls are blocked ("Traffic Engineering Model for LAN Video Conferencing," Intel, 2005). Let x equal the number of calls blocked during the peak hour (busy) videoconferencing call time. The network administrator believes that x has a Poisson distribution with mean $\lambda = 6$.
  a. Find the probability that fewer than 2 calls are blocked during the peak hour.
  b. Find $E(x)$ and interpret its value.
  c. Is it likely that no calls will be blocked during the peak hour?
  d. If, in fact, $x = 0$ during a randomly selected peak hour, what would you infer about the value of λ? Explain.

4.102 **Making high-stakes insurance decisions.** The *Journal of Economic Psychology* (Sept. 2008) published the results of a high-stakes experiment where subjects were asked how much they would pay for insuring a valuable painting. The painting was threatened by fire and theft, hence the need for insurance. To make the risk realistic, the subjects were informed that if it rained on exactly 24 days in July, the painting was considered to be stolen; and, if it rained on exactly 23 days in August, the painting was considered to be destroyed by fire. Although the probability of these two events, "fire" and "theft," was ambiguous for the subjects, the researchers estimated their probabilities of occurrence at .0001. Rain frequencies for the months of July and August were shown to follow a Poisson distribution with a mean of 10 days per month.
  a. Find the probability that it will rain on exactly 24 days in July.
  b. Find the probability that it will rain on exactly 23 days in August.
  c. Are the probabilities, parts **a** and **b**, good approximations to the probabilities of "fire" and "theft"?

4.103 **Davy Crockett's use of words.** Davy Crockett, a U.S. congressman during the 1830s, published *A Narrative of the Life of David Crockett, Written by Himself*. In a *Chance*

(Spring 1999) article, researchers determined that x, the number of times (per 1,000 words) that the word *though* appears in the *Life* narrative, has a mean of .25. Assume that x has an approximate Poisson distribution.

a. Find the variance of x.

b. Calculate the interval $\mu \pm 2\sigma$. What proportion of the time will x fall into this interval?

c. In 1836, Davy Crockett died at the Alamo defending Texans from an attack by the Mexican general Santa Anna. The following year, the narrative *Colonel Crockett's Exploits and Adventures in Texas* was published. Some historians have doubts whether Crockett was the actual author of this narrative. In the *Texas* narrative, "though" appears twice in the first 1,000 words. Use this information and your answer to part **b** to make an inference about whether Crockett actually wrote the *Texas* narrative. (Assume that the *Life* narrative was really written by Davy Crockett.)

4.104 Flaws in plastic-coated wire. The British Columbia Institute of Technology provides on its Web site (www.math.bcit.ca) practical applications of statistics to mechanical engineering. The following is a Poisson application. A roll of plastic-coated wire has an average of .8 flaws per 4-meter length of wire. Suppose a quality control engineer will sample a 4-meter length of wire from a roll of wire 220 meters in length. If no flaws are found in the sample, the engineer will accept the entire roll of wire. What is the probability that the roll will be rejected? What assumption did you make to find this probability?

Applying the Concepts—Advanced

4.105 Waiting for a car wash. A certain automatic car wash takes exactly 5 minutes to wash a car. On the average, 10 cars per hour arrive at the car wash. Suppose that, 30 minutes before closing time, five cars are in line. If the car wash is in continuous use until closing time, is it likely anyone will be in line at closing time?

4.106 Elevator passenger arrivals. A study of the arrival process of people using elevators at a multilevel office building was conducted and the results reported in *Building Services Engineering Research and Technology* (Oct. 2012). Suppose that at one particular time of day, elevator passengers arrive in batches of size 1 or 2 (i.e., either 1 or 2 people arriving at the same time to use the elevator). The researchers assumed that the number of batches, n, arriving over a specific time period follows a Poisson process with mean $\lambda = 1.1$. Now let x_n represent the number of passengers (either 1 or 2) in batch n and assume the batch size has probabilities $p = P(x_n = 1) = .4$ and $q = P(x_n = 2) = .6$. Then the total number of passengers arriving over a specific time period is $y = \Sigma_{i=1}^{n} x_i$. The researchers showed that if $x_1, x_2, \ldots, x_n$ are independent and identically distributed random variables and also independent of n, then y follows a *compound Poisson* distribution.

a. Find $P(y = 0)$, i.e., the probability of no arrivals during the time period. [*Hint:* $y = 0$ only when $n = 0$.]

b. Find $P(y = 1)$, i.e., the probability of only 1 arrival during the time period. [*Hint:* $y = 1$ only when $n = 1$ and $x_1 = 1$.]

4.6 The Hypergeometric Random Variable (Optional)

The **hypergeometric probability distribution** provides a realistic model for some types of enumerative (countable) data. The characteristics of the hypergeometric distribution are listed in the following box:

Characteristics of a Hypergeometric Random Variable

1. The experiment consists of randomly drawing n elements without replacement from a set of N elements, r of which are S's (for Success) and $(N - r)$ of which are F's (for Failure).

2. The hypergeometric random variable x is the number of S's in the draw of n elements.

Note that both the hypergeometric and binomial characteristics stipulate that each draw, or trial, results in one of two outcomes. The basic difference between these random variables is that the hypergeometric trials are dependent, while the binomial trials are independent. The draws are dependent because the probability of drawing an S (or an F) is dependent on what occurred on preceding draws.

To illustrate the dependence between trials, we note that the probability of drawing an S on the first draw is r/N. Then the probability of drawing an S on the second draw depends on the outcome of the first. It will be either $(r - 1)/(N - 1)$ or $r/(N - 1)$, depending on whether the first draw was an S or an F. Consequently, the results of the draws represent dependent events.

For example, suppose we define x as the number of women hired in a random selection of three applicants from a total of six men and four women. This random variable satisfies the characteristics of a **hypergeometric random variable** with $N = 10$ and $n = 3$. The possible outcomes on each trial are either the selection of a female (S) or the selection of a male (F). Another example of a hypergeometric random variable is the number x of defective large-screen plasma televisions in a random selection of $n = 4$ from a shipment of $N = 8$ TVs. Finally, as a third example, suppose $n = 5$ stocks are randomly selected from a list of $N = 15$ stocks. Then the number x of the five companies selected that pay regular dividends to stockholders is a hypergeometric random variable.

The hypergeometric probability distribution is summarized in the following box:

Probability Distribution, Mean, and Variance of the Hypergeometric Random Variable

$$p(x) = \frac{\binom{r}{x}\binom{N-r}{n-x}}{\binom{N}{n}} \quad [x = \text{Maximum}\,[0, n - (N - r)], \ldots, \text{Minimum}(r, n)]$$

$$\mu = \frac{nr}{N} \quad \sigma^2 = \frac{r(N-r)n(N-n)}{N^2(N-1)}$$

where

N = Total number of elements

r = Number of S's in the N elements

n = Number of elements drawn

x = Number of S's drawn in the n elements

Example **4.15** **Applying the Hypergeometric Distribution—Selecting Teaching Assistants**	**Problem** Suppose a professor randomly selects three new teaching assistants from a total of 10 applicants—six male and four female students. Let x be the number of females who are hired. **a.** Find the mean and standard deviation of x. **b.** Find the probability that no females are hired. **Solution** **a.** Since x is a hypergeometric random variable with $N = 10, n = 3$, and $r = 4$, the mean and variance are $$\mu = \frac{nr}{N} = \frac{(3)(4)}{10} = 1.2$$ $$\sigma^2 = \frac{r(N-r)n(N-n)}{N^2(N-1)} = \frac{4(10-4)3(10-3)}{(10)^2(10-1)}$$ $$= \frac{(4)(6)(3)(7)}{(100)(9)} = .56$$ The standard deviation is $$\sigma = \sqrt{.56} = .75$$ **b.** The probability that no female students are hired by the professor, assuming that the selection is truly random, is $$P(x = 0) = p(0) = \frac{\binom{4}{0}\binom{10-4}{3-0}}{\binom{10}{3}}$$ $$= \frac{\dfrac{4!}{0!(4-0)!}\dfrac{6!}{3!(6-3)!}}{\dfrac{10!}{3!(10-3)!}} = \frac{(1)(20)}{120} = \frac{1}{6} = .1667$$

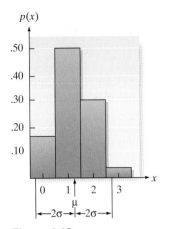

$p(x)$

.50
.40
.30
.20
.10

0 1 2 3 → x

μ

←—2σ—→←—2σ—→

Figure 4.16
Probability distribution for x in
Example 4.15

We verified this calculation using statistical software. The probability .166667, shaded on the MINITAB printout, Figure 4.15, agrees with our answer.

Probability Density Function

Hypergeometric with N = 10, M = 4, and n = 3

Figure 4.15
MINITAB Output
for Example 4.15

x   P(X = x)
0   0.166667

Look Back The entire probability distribution for x is shown in Figure 4.16. The mean $\mu = 1.2$ and the interval $\mu \pm 2\sigma = (-.3, 2.7)$ are indicated. You can see that if this random variable were to be observed over and over again a large number of times, most of the values of x would fall within the interval $\mu \pm 2\sigma$.

■ **Now Work Exercise 4.116d**

Using the Hypergeometric Model to Solve the Cocaine Sting Case

The reverse cocaine sting case described on p. 213 and solved in Section 4.4 can also be solved by applying the hypergeometric distribution. In fact, the probabilities obtained with the hypergeometric distribution are *exact* probabilities, compared with the approximate probabilities obtained with the binomial distribution.

Our objective, you will recall, is to find the probability that 4 packets (randomly selected from 496 packets confiscated in a drug bust) will contain cocaine and 2 packets (randomly selected from the remaining 492) will not contain cocaine. We assumed that, of the 496 original packets, 331 contained cocaine and 165 contained an inert (legal) powder. Since we are sampling *without replacement* from the 496 packets, the probability of a success (i.e., the probability of a packet containing cocaine) does not remain *exactly* the same from trial to trial. For example, for the first randomly selected packet, the probability of a success is $331/496 = .66734$. If we find that the first 3 packets selected contain cocaine, the probability of success for the 4th packet selected is now $328/493 = .66531$. You can see that these probabilities are not exactly the same. Hence, the binomial distribution will only approximate the distribution of x, the number of packets that contain cocaine in a sample of size n.

To find the probability that 4 packets randomly selected from the original 496 will test positive for cocaine under the hypergeometric distribution, we first identify the parameters of the distribution:

$N = 496$ is the total number of packets in the population
$S = 331$ is the number of successes (cocaine packets) in the population
$n = 4$ is the sample size

$x = 4$ is the number of successes (cocaine packets) in the sample

Substituting into the formula for $p(x)$ (on p. 248), we obtain

$$P(x = 4) = p(4) = \frac{\binom{331}{4}\binom{165}{0}}{\binom{496}{4}} = \frac{\left(\frac{331!}{4!327!}\right)\left(\frac{165!}{0!165!}\right)}{\left(\frac{496!}{4!492!}\right)}$$

$$= .197$$

To find the probability that 2 packets randomly selected from the remaining 492 will test negative for cocaine, *assuming that the first 4 packets tested positive*, we identify the parameters of the relevant hypergeometric distribution:

$N = 492$ is the total number of packets in the population
$S = 327$ is the number of successes (cocaine packets) in the population
$n = 2$ is the sample size
$x = 0$ is the number of successes (cocaine packets) in the sample

Again, we substitute into the formula for $p(x)$ (p. 248) to obtain

$$P(x = 0) = p(0) = \frac{\binom{327}{0}\binom{165}{2}}{\binom{492}{2}} = \frac{\left(\frac{327!}{0!327!}\right)\left(\frac{165!}{2!163!}\right)}{\left(\frac{492!}{2!490!}\right)}$$

$$= .112$$

(Continued)

By the multiplicative law of probability, the probability that the first 4 packets test positive for cocaine and the next 2 packets test negative is the product of the two probabilities just given:

P (first 4 positive and next 2 negative) $= (.197)(.112)$
$= .0221$

Note that this exact probability is almost identical to the approximate probability computed with the binomial distribution in Section 4.4.

Exercises 4.107–4.127

Understanding the Principles

4.107 Explain the difference between sampling with replacement and sampling without replacement.

4.108 Give the characteristics of a hypergeometric distribution.

4.109 How do binomial and hypergeometric random variables differ? In what respects are they similar?

Learning the Mechanics

4.110 Given that x is a hypergeometric random variable with $N = 7, n = 3$, and $r = 4$, compute the following:
 a. $P(x = 1)$
 b. $P(x = 0)$
 c. $P(x = 3)$
 d. $P(x = 4)$

4.111 Given that x is a hypergeometric random variable, compute $p(x)$ for each of the following cases:
 a. $N = 6, n = 4, r = 4, x = 2$
 b. $N = 10, n = 6, r = 4, x = 4$
 c. $N = 3, n = 3, r = 3, x = 3$
 d. $N = 5, n = 3, r = 3, x = 1$

4.112 Given that x is a hypergeometric random variable with $N = 5, n = 3$, and $r = 2$:
 a. Display the probability distribution for x in tabular form.
 b. Compute μ and σ for x.
 c. Graph $p(x)$, and locate μ and the interval $\mu \pm 2\sigma$ on the graph.
 d. What is the probability that x will fall within the interval $\mu \pm 2\sigma$?

4.113 Given that x is a hyper-geometric random variable with $N = 10, n = 6$, and $r = 4$, find the following probabilities:
 a. $P(x = 1)$
 b. $P(x = 3)$
 c. $P(x \leq 3)$
 d. $P(x \geq 3)$
 e. $P(x < 2)$
 f. $P(x \geq 5)$

4.114 Given that x is a hypergeometric random variable with $N = 6, n = 3$, and $r = 2$:
 a. Display the probability distribution for x in tabular form.
 b. Compute the mean and variance of x.
 c. Graph $p(x)$, and locate μ and the interval $\mu \pm 2\sigma$ on the graph.
 d. What is the probability that x will fall within the interval $\mu \pm 2\sigma$?

4.115 Suppose you plan to sample 10 items from a population of 120 items and would like to determine the probability of observing 4 defective items in the sample. Which probability distribution should you use to compute this probability under the conditions listed here? Justify your answers.
 a. The sample is drawn without replacement.
 b. The sample is drawn with replacement.

Applying the Concepts — Basic

4.116 **Do social robots walk or roll?** Refer to the *International Conference on Social Robotics* (Vol. 6414, 2010) study of the trend in the design of social robots, Exercise 4.25 (p. 221). The study found that of 106 social robots, 63 were built with legs only, 20 with wheels only, 8 with both legs and wheels, and 15 with neither legs nor wheels. Suppose you randomly select 10 of the 106 social robots and count the number, x, with neither legs nor wheels.
 a. Demonstrate why the probability distribution for x should not be approximated by the binomial distribution.
 b. Show that the properties of the hypergeometric probability distribution are satisfied for this experiment.
 c. Find μ and σ for the probability distribution for x.
 d. Calculate the probability that $x = 2$. ⬛NW

4.117 **Mail rooms contaminated with anthrax.** In *Chance* (Spring 2002), research statisticians discussed the problem of sampling mail rooms for the presence of anthrax spores. Let x equal the number of mail rooms contaminated with anthrax spores in a random sample of 3 mail rooms selected from a population of 100 mail rooms. If 20 of the 100 mail rooms are contaminated with anthrax, researchers showed that the probability distribution for x is given by the formula

$$p(x) = \frac{\binom{20}{x}\binom{80}{3-x}}{\binom{100}{3}}$$

 a. What is the name given to this probability distribution?
 b. Specify r, N, and n for this distribution.
 c. Find $p(0)$.
 d. Find $p(1)$.
 e. Find $p(2)$.
 f. Find $p(3)$.

4.118 **ESL students and plagiarism.** Refer to the *Journal of Education and Human Development* (Vol. 3, 2009) investigation of plagiarism among English-as-a-second-language (ESL) students, Exercise 3.30 (p. 159). Recall that of six ESL students taking a master's course in linguistics, three admitted to plagiarizing on an essay. Two of these six students will be selected to write an original article for the college newspaper. If the two students are selected at random, what is the probability that at least one plagiarized on the essay?

4.119 On-site treatment of hazardous waste. The Resource Conservation and Recovery Act mandates the tracking and disposal of hazardous waste produced at U.S. facilities. *Professional Geographer* (Feb. 2000) reported the hazardous-waste generation and disposal characteristics of 209 facilities. Only 8 of these facilities treated hazardous waste on-site.
a. In a random sample of 10 of the 209 facilities, what is the expected number in the sample that treat hazardous waste on-site? Interpret this result.
b. Find the probability that 4 of the 10 selected facilities treat hazardous waste on-site.

4.120 Guilt in decision making. Refer to the *Journal of Behavioral Decision Making* (Jan. 2007) study of how guilty feelings impact decisions, Exercise 3.59 (p. 170). Recall that 57 students were assigned to a guilty state through a reading/writing task. Immediately after the task, the students were presented with a decision problem where the stated option has predominantly negative features (e.g., spending money on repairing a very old car). Of these 57 students, 45 chose the stated option. Suppose 10 of the 57 guilty-state students are selected at random. Define x as the number in the sample of 10 who chose the stated option.
a. Find $P(x = 5)$.
b. Find $P(x = 8)$.
c. What is the expected value (mean) of x?

4.121 Contaminated gun cartridges. Refer to the investigation of contaminated gun cartridges at a weapons manufacturer presented in Exercise 4.33 (p. 222). In a sample of 160 cartridges from a certain lot, 36 were found to be contaminated and 124 were "clean." If you randomly select 7 of these 160 cartridges, what is the probability that all 7 will be "clean"?

Applying the Concepts—Intermediate

4.122 Lot inspection sampling. Imagine that you are purchasing small lots of a manufactured product. If it is very costly to test a single item, it may be desirable to test a sample of items from the lot instead of testing every item in the lot. Suppose each lot contains 15 items. You decide to sample 8 items per lot and reject the lot if you observe 1 or more defectives.
a. If the lot contains 1 defective item, what is the probability that you will accept the lot?
b. What is the probability that you will accept the lot if it contains 2 defective items?

4.123 Museum management. Refer to the *Museum Management and Curatorship* (June 2010) study of the criteria used to evaluate museum performance, Exercise 3.28 (p. 159). Recall that the managers of 30 leading museums of contemporary art were asked to provide the performance measure used most often. Of these 30 museums, 8 specified "total visitors" as the performance measure. Consider a random sample of 5 museums selected from the 30. How likely is it that none of the museums in the sample will specify "total visitors" as the performance measure?

4.124 Cell phone handoff behavior. Refer to the *Journal of Engineering, Computing and Architecture* (Vol. 3., 2009) study of cell phone handoff behavior, Exercise 3.62 (p. 171). Recall that a "handoff" describes the process of a cell phone moving from one base channel (identified by a color code) to another. During a particular driving trip a cell phone changed channels (color codes) 85 times. Color code "b" was accessed 40 times on the trip. You randomly select 7 of the 85 handoffs. How likely is it that the cell phone accesses color code "b" only twice for these 7 handoffs?

4.125 Establishing boundaries in academic engineering. How academic engineers establish boundaries (e.g., differentiating between engineers and other scientists, defining the different disciplines within engineering, determining the quality of journal publications) was investigated in *Engineering Studies* (Aug. 2012). Participants were 10 tenured or tenure-earning engineering faculty members in a School of Engineering at a large research-oriented university. The table gives a breakdown of the department affiliation of the engineers. Each participant was interviewed at length, and responses were used to establish boundaries. Suppose we randomly select 3 participants from the original 10 to form a university committee charged with developing boundary guidelines. Let x represent the number of committee members who are from the Department of Engineering Physics. Identify the probability distribution for x, and give a formula for the distribution.

Department	Number of Participants
Chemical Engineering	1
Civil Engineering	2
Engineering Physics	4
Mechanical Engineering	2
Industrial Engineering	1

Applying the Concepts—Advanced

4.126 Gender discrimination suit. The *Journal of Business & Economic Statistics* (July 2000) presented a case in which a charge of gender discrimination was filed against the U.S. Postal Service. At the time, there were 302 U.S. Postal Service employees (229 men and 73 women) who applied for promotion. Of the 72 employees who were awarded promotion, 5 were female. Make an inference about whether or not females at the U.S. Postal Service were promoted fairly.

4.127 Awarding of home improvement grants. A community development agency makes home improvement grants each year to homeowners in depressed city neighborhoods. One year, grants were awarded by the city council via a "random selection" of 142 homeowners' applications from among a total of 750 applications: 603 from the north side, and 147 from the south side, of the city. Oddly, all 142 grants awarded were from the north side—clearly a highly improbable outcome if, in fact, the 142 winners were randomly selected from among the 750 applicants.
a. Suppose the 142 winning applications were randomly selected from among the total of 750, and let x equal the number in the sample from the north side. Find the mean and standard deviation of x.
b. Use the results of part **a** to support a contention that the grant winners were not randomly selected.

CHAPTER NOTES

Key Terms

Note: Starred (*) terms are from the optional sections in this chapter.

Binomial distribution 233
Binomial experiment 229
Binomial random variable 229
Continuous random
   variable 215
Countable 215
Cumulative binomial
   probabilities 236
Discrete random variable 215
Expected value 224
*Hypergeometric probability
   distribution 247

*Hypergeometric random
   variable 247
Mean value of a discrete
   random variable 224
*Poisson distribution 242
*Poisson random variable 243
Probability distribution of a
   discrete random variable 218
Random variable 213
Standard deviation of a discrete
   random variable 226
Variance of a random variable 225

S Outcome of binomial trial denoted "success"
F Outcome of binomial trial denoted "failure"
p $P(S)$ in binomial trial
q $P(F)$ in binomial trial $= 1 - p$
e Constant used in Poisson probability distribution;
         $e = 2.71828 \ldots$

Key Ideas

Properties of Discrete Probability Distributions
   1. $p(x) \geq 0$
   2. $\sum_{\text{all } x} p(x) = 1$

Key Symbols

$p(x)$ Probability distribution for discrete random variable x
$f(x)$ Probability distribution for continuous random variable x

Guide to Selecting a Discrete Probability Distribution

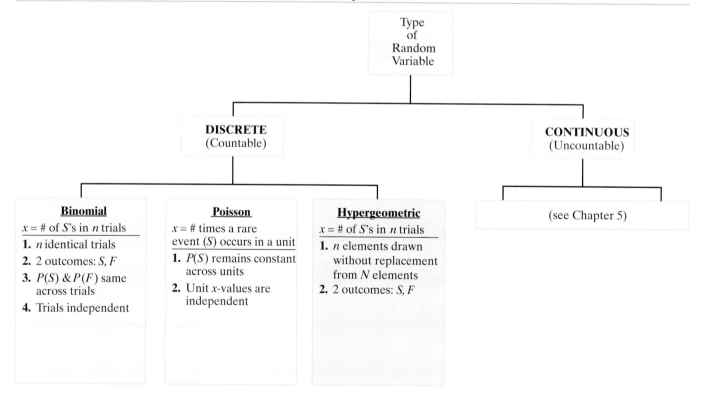

Key Formulas

Random Variable	Probability Distribution	Mean	Variance
General Discrete:	Table, formula, or graph for $p(x)$	$\displaystyle\sum_{\text{all }x} x \cdot p(x)$	$\displaystyle\sum_{\text{all }x}(x-\mu)^2 \cdot p(x)$
Binomial:	$p(x) = \binom{n}{x} p^x q^{n-x}$	np	npq
Poisson:	$p(x) = \dfrac{\lambda^x e^{-\lambda}}{x!}$	λ	λ
Hypergeometric:	$p(x) = \dfrac{\binom{r}{x}\binom{N-r}{n-x}}{\binom{N}{x}}$	$\dfrac{nr}{N}$	$\dfrac{r(N-r)\,n(N-n)}{N^2\,(N-1)}$

Supplementary Exercises 4.128–4.160

Note: Starred (*) exercises refer to the optional sections in this chapter.

Understanding the Principles

4.128 Which of the following describe discrete random variables, and which describe continuous random variables?
 a. The heights of rubber plants
 b. The number of second-hand cars on sale
 c. The number of flats decorated by an interior designer
 d. The number of non-vegetarian guests in a party

4.129 Identify the type of random variable—binomial, Poisson, or hypergeometric—described by each of the following probability distributions:
 *a. $p(x) = \dfrac{.5^x e^{-.5}}{x!} \quad (x = 0, 1, 2, ...)$
 b. $p(x) = \binom{6}{x}(.2)^x(.8)^{6-x} \quad (x = 0, 1, 2, ..., 6)$
 c. $p(x) = \dfrac{10!}{x!(10-x)!}(.9)^x(.1)^{10-x}$
 $(x = 0, 1, 2, ..., 10)$

4.130 For each of the following examples, decide whether x is a binomial random variable and explain your decision:
 a. A manufacturer of computer chips randomly selects 200 chips from each hour's production in order to estimate the proportion of defectives. Let x represent the number of defectives in the 200 chips sampled.
 b. Of six applicants for a job, two will be selected. Although all applicants appear to be equally qualified, only four have the ability to fulfill the expectations of the company. Suppose that the two selections are made at random from the six applicants, and let x be the number of qualified applicants selected.
 c. The number of calls received at a fire service station is denoted by x. Let x represent the number of calls received on the hot line during a specified workday.
 d. In a state there are two types of voters, one who favors the state income tax and the other who does not

in light of the state's current fiscal condition. Suppose a poll of 2,000 registered voters is conducted to determine how many would favor a state income tax in light of the state's current fiscal condition. Let x be the number in the sample who would favor the tax.

Learning the Mechanics

4.131 Suppose x is a binomial random variable with $n = 22$ and $p = .8$.
 a. Find $P(x = 16)$.
 b. Find $P(x \le 18)$.
 c. Find $P(x > 18)$.
 d. Find $P(12 \le x \le 16)$.
 e. Find $P(15 < x < 20)$.
 f. Find μ, σ^2, and σ.
 g. What is the probability that x is in the interval $\mu \pm 2\sigma$?

*4.132 Given that x is a hypergeometric random variable, compute $p(x)$ for each of the following cases:
 a. $N = 9, n = 5, r = 4, x = 0$
 b. $N = 10, n = 6, r = 4, x = 3$
 c. $N = 6, n = 3, r = 4, x = 3$

4.133 Suppose x is a binomial random variable. Find $p(x)$ for each of the following combinations of x, n, and p:
 a. $x = 1, n = 4, p = .2$
 b. $x = 4, n = 21, p = .4$
 c. $x = 0, n = 4, p = .5$
 d. $x = 4, n = 6, p = .6$
 e. $n = 16, x = 12, p = .8$
 f. $n = 12, x = 8, p = .7$

4.134 Consider the discrete probability distribution shown here:

x	5	10	12	14
$p(x)$	.2	.3	.1	.4

 a. Calculate μ, σ^2, and σ.
 b. What is $P(x < 12)$?

c. Calculate $\mu \pm 2\sigma$.

d. What is the probability that x is in the interval $\mu \pm 2\sigma$?

*4.135 Suppose x is a Poisson random variable. Compute $p(x)$ for each of the following cases:

a. $\lambda = 3, x = 4$

b. $\lambda = 2, x = 3$

c. $\lambda = 1.5, x = 2$

Applying the Concepts—Basic

4.136 **Use of laughing gas.** According to the American Dental Association, 60% of all dentists use nitrous oxide ("laughing gas") in their practice. If x equals the number of dentists in a random sample of five dentists who use laughing gas in practice, then the probability distribution of x is

x	0	1	2	3	4	5
$p(x)$	.0102	.0768	.2304	.3456	.2592	.0778

a. Find the probability that the number of dentists using laughing gas in the sample of five is less than unity.

b. Find $E(x)$ and interpret the result.

c. Show that the distribution of x is binomial with $n = 5$ and $p = .6$.

4.137 **Benford's Law of Numbers.** Refer to the *American Scientist* (July–Aug. 1998) study of which integer is most likely to occur as the first significant digit in a randomly selected number (Benford's law), presented in Exercise 2.206 (p. 139). The table giving the frequency of each integer selected as the first digit in a six-digit random number is reproduced here:

First Digit	Frequency of Occurrence
1	109
2	75
3	77
4	99
5	72
6	117
7	89
8	62
9	43
Total	743

a. Construct a probability distribution for the first significant digit x.

b. Find $E(x)$.

c. If possible, give a practical interpretation of $E(x)$.

4.138 **Parents who condone spanking.** According to a nationwide survey, 60% of parents with young children condone spanking their child as a regular form of punishment (*Tampa Tribune*, Oct. 5, 2000). Consider a random sample of four people, each of whom is a parent with young children. Assume that x, the number in the sample who condone spanking, is a binomial random variable.

a. What is the probability that none of the four parents condones spanking as a regular form of punishment for their children?

b. What is the probability that at least one condones spanking as a regular form of punishment?

c. Give the mean and standard deviation of x. Interpret the results.

4.139 **Tracking missiles with satellite imagery.** The Space-Based Infrared System (SBIRS) uses satellite imagery to detect and track missiles (*Chance*, Summer 2005). The probability that an intruding object (e.g., a missile) will be detected on a flight track by SBIRS is .8. Consider a sample of 20 simulated tracks, each with an intruding object. Let x equal the number of these tracks on which SBIRS detects the object.

a. Demonstrate that x is (approximately) a binomial random variable.

b. Give the values of p and n for the binomial distribution.

c. Find $P(x = 15)$, the probability that SBIRS will detect the object on exactly 15 tracks.

d. Find $P(x \geq 15)$, the probability that SBIRS will detect the object on at least 15 tracks.

e. Find $E(x)$ and interpret the result.

*4.140 **Emergency rescue vehicle use.** An emergency rescue vehicle is used an average of 1.4 times daily. Use the Poisson distribution to find

a. The probability that the vehicle will be used exactly twice tomorrow

b. The probability that the vehicle will be used more than twice

c. The probability that the vehicle will be used exactly three times

4.141 **Married-women study.** According to *Women's* Day magazine (Jan. 2007), 50% of married women would marry their current husbands again if given the chance. Consider the number x in a random sample of 12 married women who would marry their husbands again. Identify the discrete probability distribution that best models the distribution of x. Explain.

*4.142 **Dutch elm disease.** A nursery advertises that it has 12 elm trees for sale. Unknown to the nursery, 4 of the trees have already been infected with Dutch elm disease and will die within a year.

a. If a buyer purchases two trees, what is the probability that both trees will be healthy?

b. Refer to part **a.** What is the probability that at least 1 of the trees is infected?

4.143 **Dust mite allergies.** A dust mite allergen level that exceeds 2 micrograms per gram (μg/g) of dust has been associated with the development of allergies. Consider a random sample of four homes, and let x be the number of homes with a dust mite level that exceeds 2 μg/g. The probability distribution for x, based on a study, is shown in the following table:

x	0	1	2	3	4
$p(x)$	.07	.31	.38	.17	.07

a. Verify that the probabilities for x in the table sum to 1.

b. Find the probability that three or four of the homes in the sample have a dust mite level that exceeds 2 μg/g.

c. Find the probability that fewer than two homes in the sample have a dust mite level that exceeds 2 μg/g.

d. Find $E(x)$. Give a meaningful interpretation of the result.

e. Find σ.

f. Find the exact probability that x is in the interval $\mu \pm 2\sigma$. Compare your answer with Chebyshev's rule and the empirical rule.

*4.144 Deep-draft vessel casualties. Economists at the University of New Mexico modeled the number of casualties (deaths or missing persons) experienced by a deep-draft U.S. flag vessel over a three-year period as a Poisson random variable x. The researchers estimated $E(x)$ to be .03 (*Management Science*, Jan. 1999).
  a. Find the variance of x.
  b. Discuss the conditions that would make the researchers' Poisson assumption plausible.
  c. What is the probability that a deep-draft U.S. flag vessel will have no casualties over a three-year period?

4.145 Belief in an afterlife. A national poll conducted by *The New York Times* (May 7, 2000) revealed that 80% of Americans believe that after you die, some part of you lives on, either in a next life on earth or in heaven. Consider a random sample of 12 Americans and count x, the number who believe in life after death.
  a. Find $P(x = 3)$.
  b. Find $P(x \leq 7)$.
  c. Find $P(x > 4)$.

Applying the Concepts—Intermediate

4.146 When to replace a maintenance system. An article in the *Journal of Quality of Maintenance Engineering* (Vol. 19, 2013) studied the problem of finding the optimal replacement policy for a maintenance system. Consider a system that is tested every 12 hours. The test will determine whether there are any flaws in the system. Assume the probability of no flaw being detected is .85. If a flaw (failure) is detected, the system is repaired. Following the 5th failed test, the system is completely replaced. Now let x represent the number of tests until the system needs to be replaced.
  a. Give the probability distribution for x as a formula.
  b. Find the probability that the system needs to be replaced after 8 total tests.

*4.147 Extinct New Zealand birds. Refer to the *Evolutionary Ecology Research* (July 2003) study of the patterns of extinction in the New Zealand bird population, presented in Exercise 3.95 (p. 185). Of the 132 bird species in the population, 38 are extinct. Suppose you randomly select 10 of the 132 bird species (without replacement) and record the extinction status of each.
  a. What is the probability that exactly 5 of the 10 species you select are extinct?
  b. What is the probability that at most 1 species is extinct?

4.148 Student gambling on sports. A study of gambling activity at the University of West Georgia (UWG) discovered that 60% of the male students wagered on sports the previous year (*The Sport Journal*, Fall 2006). Consider a random sample of 50 UWG male students. How many of these students would you expect to have gambled on sports the previous year? Give a range that is likely to include the number of male students who have gambled on sports.

4.149 Parents' behavior at a gym meet. *Pediatric Exercise Science* (Feb. 1995) published an article on the behavior of parents at competitive youth gymnastic meets. On the basis of a survey of the gymnasts, the researchers estimated the probability of a parent "yelling" at his or her child before, during, or after the meet as .05. In a random sample of 20 parents attending a gymnastic meet, find the probability that at least 1 parent yells at his or her child before, during, or after the meet.

4.150 Testing for spoiled wine. Suppose that you are purchasing cases of wine (10 bottles per case) and that, periodically, you select a test case to determine the adequacy of the bottles' seals. To do this, you randomly select and test 4 bottles in the case. If a case contains 1 spoiled bottle of wine, what is the probability that this bottle will turn up in your sample?

*4.151 Vinyl chloride emissions. The Environmental Protection Agency (EPA) limits the amount of vinyl chloride in plant air emissions to no more than 10 parts per million. Suppose the mean emission of vinyl chloride for a particular plant is 5 parts per million. Assume that the number x of parts per million of vinyl chloride in air samples follows a Poisson probability distribution.
  a. What is the standard deviation of x for the plant?
  b. Is it likely that a sample of air from the plant would yield a value of x that would exceed the EPA limit? Explain.

4.152 Countries that allow a free press. The degree to which democratic and nondemocratic countries attempt to control the news media was examined in the *Journal of Peace Research* (Nov. 1997). The article reported that 80% of all democratic regimes allow a free press. In contrast, 10% of all nondemocratic regimes allow a free press.
  a. In a random sample of 50 democratic regimes, how many would you expect to allow a free press? Give a range that is highly likely to include the number of democratic regimes with a free press.
  b. In a random sample of 50 nondemocratic regimes, how many would you expect to allow a free press? Give a range that is highly likely to include the number of nondemocratic regimes with a free press.

4.153 Chickens with fecal contamination. The United States Department of Agriculture (USDA) reports that, under its standard inspection system, one in every 100 slaughtered chickens passes inspection for fecal contamination. In Exercise 3.21 (p. 158), you found the probability that a randomly selected slaughtered chicken passes inspection for fecal contamination. Now find the probability that, in a random sample of 6 slaughtered chickens, at least one passes inspection for fecal contamination.

*4.154 Crime Watch neighborhood. In many cities, neighborhood Crime Watch groups are formed in an attempt to reduce the amount of criminal activity. Suppose one neighborhood that has experienced an average of 8 crimes per year organizes such a group. During the first year following the creation of the group, 2 crimes are committed in the neighborhood.
  a. Use the Poisson distribution to calculate the probability that 2 or fewer crimes are committed in a year, assuming that the average number is still 8 crimes per year.
  b. Do you think this event provides some evidence that the Crime Watch group has been effective in this neighborhood?

4.155 The Showcase Showdown. On the popular television game show *The Price Is Right*, contestants can play "The Showcase Showdown." The game involves a large wheel with 20 nickel values, 5, 10, 15, 20, . . . , 95, 100, marked on it. Contestants spin the wheel once or twice, with the objective of obtaining the highest total score *without going over a dollar (100)*. [According to the *American Statistician* (Aug. 1995), the optimal strategy for the first spinner in a three-player game is to spin a second time only if the value of the initial spin is 65 or less.] Let *x* represent the score of a single contestant playing "The Showcase Showdown." Assume a "fair" wheel (i.e., a wheel with equally likely outcomes). If the total of the player's spins exceeds 100, the total score is set to 0.

a. If the player is permitted only one spin of the wheel, find the probability distribution for *x*.

b. Refer to part **a**. Find $E(x)$ and interpret this value.

c. Refer to part **a**. Give a range of values within which *x* is likely to fall.

d. Suppose the player will spin the wheel twice, no matter what the outcome of the first spin. Find the probability distribution for *x*.

e. What assumption did you make to obtain the probability distribution in part **d**? Is it a reasonable assumption?

f. Find μ and σ for the probability distribution of part **d**, and interpret the results.

g. Refer to part **d**. What is the probability that in two spins the player's total score exceeds a dollar (i.e., is set to 0)?

h. Suppose the player obtains a 20 on the first spin and decides to spin again. Find the probability distribution for *x*.

i. Refer to part **h**. What is the probability that the player's total score exceeds a dollar?

j. Given that the player obtains a 65 on the first spin and decides to spin again, find the probability that the player's total score exceeds a dollar.

k. Repeat part **j** for different first-spin outcomes. Use this information to suggest a strategy for the one-player game.

Applying the Concepts—Advanced

4.156 **Network forensic analysis.** A network forensic analyst is responsible for identifying worms, viruses, and infected nodes in the computer network. A new methodology for finding patterns in data that signify infections was investigated in *IEEE Transactions on Information Forensics and Security* (May 2013). The method uses multiple filters to check strings of information. For this exercise, consider a data string of length 4 bytes (positions), where each byte is either a 0 or a 1 (e.g., 0010). Also, consider two possible strings, named S_1 and S_2. In a simple single filter system, the probability that S_1 and S_2 differ in any one of the bytes is .5. Derive a formula for the probability that the two strings differ on exactly *x* of the 4 bytes. Do you recognize this probability distribution?

4.157 **How many questionnaires to mail?** The probability that a person responds to a mailed questionnaire is .4. How many questionnaires should be mailed if you want to be reasonably certain that at least 100 will be returned?

4.158 **Reliability of a "one-shot" device.** A "one-shot" device can be used only once; after use, the device (e.g., a nuclear weapon, space shuttle, automobile air bag) either is destroyed or must be rebuilt. The destructive nature of a one-shot device makes repeated testing either impractical or too costly. Hence, the reliability of such a device must

be determined with minimal testing. Consider a one-shot device that has some probability *p* of failure. Of course, the true value of p is unknown, so designers will specify a value of *p* that is the largest defective rate that they are willing to accept. Designers will conduct *n* tests of the device and determine the success or failure of each test. If the number of observed failures, *x*, is less than or equal to some specified value *k*, then the device is considered to have the desired failure rate. Consequently, the designers want to know the minimum sample size *n* needed so that observing *K* or fewer defectives in the sample will demonstrate that the true probability of failure for the one-shot device is no greater than *p*.

a. Suppose the desired failure rate for a one-shot device is $p = .12$. Suppose also that designers will conduct $n = 30$ tests of the device and conclude that the device is performing to specifications if $K = 1$ (i.e., if 1 or no failure is observed in the sample). Find $P(x \le pbc)$.

b. In reliability analysis, $1 - P(x \le K)$ is often called the level of confidence for concluding that the true failure rate is less than or equal to *p*. Find the level of confidence for the one-shot device described in part **a**. In your opinion, is this an acceptable level? Explain.

c. Demonstrate that the confidence level can be increased by either (1) increasing the sample size *n* or (2) decreasing the number *K* of failures allowed in the sample.

d. Typically, designers want a confidence level of .90, .95, or .99. Find the values of *n* and *K* to use so that the designers can conclude with at least 95% confidence that the failure rate for the one-shot device of part **a** is no greater than $p = .12$.

Note: The U.S. Department of Defense Reliability Analysis Center (DoD RAC) provides designers with free access to tables and toolboxes that give the minimum sample size n required to obtain a desired confidence level for a specified number of observed failures in the sample.

*★**4.159** **Emergency room bed availability.** The mean number of patients admitted per day to the emergency room of a small hospital is 2.7. If, on a given day, there are only four beds available for new patients, what is the probability that the hospital will not have enough beds to accommodate its newly admitted patients?

Critical Thinking Challenge

*★**4.160** **Space shuttle disaster.** On January 28, 1986, the space shuttle *Challenger* exploded, killing all seven astronauts aboard. An investigation concluded that the explosion was caused by the failure of the O ring seal in the joint between the two lower segments of the right solid rocket booster. In a report made one year prior to the catastrophe, the National Aeronautics and Space Administration (NASA) claimed that the probability of such a failure was about $^1/_{60,000}$, or about once in every 60,000 flights. But a risk-assessment study conducted for the Air Force at about the same time assessed the probability to be $^1/_{35}$, or about once in every 35 missions. (*Note*: The shuttle had flown 24 successful missions prior to the disaster.) Given the events of January 28, 1986, which risk assessment—NASA's or the Air Force's—appears to be more appropriate?

Activity   Simulating Binomial Probabilities Using Poker Chips

Consider the following random variables:

1. The number x of people who recover from a certain disease out of a sample of five patients

2. The number x of voters in a sample of five who favor a method of tax reform

3. The number x of hits a baseball player gets in five official times at bat

In each case, x is a binomial random variable (or approximately so) with $n = 5$ trials. Assume that in each case the probability of success is .3. (What is a success in each of the examples?) If this were true, the probability distribution for x would be the same for each of the three examples. To obtain a relative frequency histogram for x, conduct the following experiment: Place 10 poker chips (pennies, marbles, or any 10 *identical* items) in a bowl, and mark 3 of the 10 "Success." The remaining 7

will be marked "Failure." Randomly select a chip from the 10, observing whether it is a success or failure. Then return the chip, randomly select a second chip from the 10 available chips, and record this outcome. Repeat the process until a total of five trials has been conducted. Count the number x of successes observed in the five trials. Repeat the entire process 100 times to obtain 100 observed values of x.

a. Use the 100 values of x obtained from the simulation to construct a relative frequency histogram for x. Note that this histogram is an approximation to $p(x)$.

b. Calculate the exact values of $p(x)$ for $n = 5$ and $p = .3$, and compare these values with the approximations found in part **a**.

c. If you were to repeat the simulation an extremely large number of times (say, 100,000), how do you think the relative frequency histogram and true probability distribution would compare?

References

Hogg, R. V., McKean, J. W., and Craig, A. *Introduction to Mathematical Statistics*, 6th ed. Upper Saddle River, NJ: Prentice Hall, 2005.

Larsen, R. J., and Marx, M. L. *An Introduction to Mathematical Statistics and Its Applications,* 4th ed. Upper Saddle River, N.J.: Prentice Hall, 2005.

Parzen, E. *Modern Probability Theory and Its Applications*. New York: Wiley, 1960.

Wackerly, D., Mendenhall, W., and Scheaffer, R. L. *Mathematical Statistics with Applications*, 7th ed. North Scituate, MA: Duxbury, 2008.

USING TECHNOLOGY   MINITAB: Discrete Probabilities

Step 1 Select the "Calc" button on the MINITAB menu bar, click on "Probability Distributions," and then finally select the discrete distribution of your choice (e.g., "Binomial"), as shown in Figure 4.M.1.

Step 2 Select either "Probability" or "Cumulative probability" on the resulting dialog box.

Step 3 Specify the parameters of the distribution (e.g., sample size n, probability of success p).

[*Note*: For the Poisson option, enter the mean λ. For the hypergeometric option, enter N, r, and n.]

Step 4 Specify the value x of interest in the "Input constant" box.

Step 5 Click "OK." The probability for the value of x will appear on the MINITAB session window.

[*Note*: Figure 4.M.2 gives the specifications for finding $p(x = 2)$ in a binomial distribution with $n = 5$ and $p = .3$.]

TI-83/TI-84 Plus Graphing Calculator: Discrete Random Variables and Probabilities

Calculating the Mean and Standard Deviation of a Discrete Random Variable

Step 1 *Enter the data*

• Press **STAT** and select **1:Edit**

Figure 4.M.1 MINITAB menu options for discrete probabilities

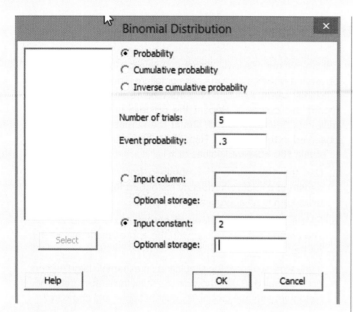

Figure 4.M.2 MINITAB binomial distribution dialog box

(*Note:* If the lists already contain data, clear the old data. Use the up **ARROW** to highlight 'L1.')

- Press **CLEAR ENTER**.
- Use the up **ARROW** to highlight 'L2'
- Press **CLEAR ENTER**
- Use the **ARROW** and **ENTER** keys to enter the x-values of the variable into **L1**
- Use the **ARROW** and **ENTER** keys to enter the probabilities, $P(x)$, into **L2**

Step 2 *Access the Calc Menu*

- Press **STAT**
- Arrow right to **CALC**
- Select **1-Var Stats**
- Press **ENTER**
- Press **2nd 1** for **L1**
- Press **COMMA**
- Select **2nd 2** for **L2**
- Press **ENTER**

The mean and standard deviation will be displayed on the screen, as will the quartiles, min, and max.

Calculating Binomial Probabilities

I. $P(x = k)$

To compute the probability of k successes in n trials where p is the probability of success for each trial, use the **binompdf(** command. *Binompdf* stands for "binomial probability density function." This command is under the **DISTR**ibution menu and has the format **binompdf(*n, p, k*)**.

Example Compute the probability of 5 successes in 8 trials where the probability of success for a single trial is 40%. In this example, $n = 8$, $p = .4$, and $k = 5$.

Step 1 *Enter the binomial parameter*

- Press **2nd VARS** for **DISTR**
- Press the down **ARROW** key until **binompdf** is highlighted
- Press **ENTER**
- After **binompdf(**, type **8, .4, 5**) (*Note*: Be sure to use the **COMMA** key between each parameter)
- Press **ENTER**
- You should see

Thus, $P(x = k)$ is about 12.4%.

II. $P(x \leq k)$

To compute the probability of k or fewer successes in n trials where the p is probability of success for each trial, use the **binomcdf(** command. *Binomcdf* stands for "binomial *cumulative* probability density function." This command is under the **DISTR**ibution menu and has the format **binomcdf(*n, p, k*)**.

Example Compute the probability of 5 or fewer successes in 8 trials where the probability of success for a single trial is 40%. In this example, $n = 8$, $p = .4$, and $k = 5$.

Step 2 *Enter the binomial parameters*

- Press **2nd VARS** for **DISTR**
- Press down the **ARROW** key until **binomcdf** is highlighted
- Press **ENTER**
- After **binomcdf(**, type **8, .4, 5**)
- Press **ENTER**
- You should see

Thus, $P(x \leq 5)$ is about 95%.

III. $P(x < k), P(x > k), P(x \geq k)$

To find the probability of less than k successes $P(x < k)$, more than k successes $P(x > k)$, or at least k successes $P(x \geq k)$, variations of the **binomcdf(** command must be used as shown below.

- $P(x < k)$ use **binomcdf(*n, p, k* − 1)**
- $P(x > k)$ use **1 − binomcdf(*n, p, k*)**
- $P(x \geq k)$ use **1 − binomcdf(*n, p, k* − 1)**

Calculating Poisson Probabilities

I. $P(x \leq k)$

To compute $P(x = k)$, the probability of exactly k successes in a specified interval where λ is the mean number of successes in the interval, use the **poissonpdf(** command. *Poissonpdf* stands for "Poisson probability density function." This command is under the **DISTR**ibution menu and has the format **poissonpdf()**.

Example Suppose that the number, x, of reported sightings per week of blue whales is recorded. Assume that x has approximately a Poisson probability distribution and that the average number of weekly sightings is 2.6. Compute the probability that exactly five sightings are made during a given week. In this example, $\lambda = 2.6$ and $k = 5$.

Step 1   *Enter the Poisson parameters*

- Press **2nd VARS** for **DISTR**
- Press the down **ARROW** key until **poissonpdf** is highlighted
- Press **ENTER**
- After **poissonpdf(** , type **2.6,5** (*Note*: Be sure to use the **COMMA** key between each parameter.)
- Press **ENTER**
- You should see

```
poissonpdf(2.6,5
)
        .0735393591
```

Thus, the $P(x = 5)$ is about 7.4%.

II. $P(x \leq k)$

To compute the probability of k or fewer successes in a specified interval where λ is the mean number of successes in the interval, use the **poissoncdf(** command. *Poissoncdf* stands for "Poisson *cumulative* probability density function." This command is under the **DISTR**ibution menu and has the format **poissoncdf(**.

Example In the preceding example, compute the probability that five or fewer sightings are made during a given week. In this example, $\lambda = 2.6$ and $k = 5$.

Step 1   *Enter the Poisson parameters*

- Press **2nd VARS** for **DISTR**
- Press the down **ARROW** key until **poissoncdf** is highlighted
- Press **ENTER**
- After **poissoncdf(**, type **2.6, 5**)

```
poissoncdf(2.6,5
)
        .9509628481
```

- Press **ENTER**
- You should see the screen at the bottom of the first column

Thus, the $P(x = 5)$ is about 95.1%.

III. $P(x < k), P(x > k), P(x \geq k)$

To find the probability of less than k successes, more than k successes, or at least k successes, variations of the **poissoncdf(** command must be used as shown below.

- $P(x < k)$ use **poissoncdf(λ, $k - 1$)**
- $P(x > k)$ use **1 − poissoncdf(λ, k)**
- $P(x \geq k)$ use **1 − poissoncdf(λ, $k - 1$)**

Calculating Hypergeometric Probabilities

Step 1   Compute $\binom{r}{x}$ using the keystrokes for combinations (see Using Technology, Chapter 3).

Step 2   Press **STO>, ALPHA, 4,** then press **ENTER** to save the result of $\binom{r}{x}$ as the variable **T.**

Step 3   Compute $\binom{N-r}{n-x}$ using the keystrokes for combinations (see Using Technology, Chapter 3).

Step 4   Press **STO>, ALPHA, 5,** then press **ENTER** to save the result of $\binom{N-r}{n-x}$ as the variable **U.**

Step 5   Compute $\binom{N}{x}$ using the keystrokes for combinations (see Using Technology, Chapter 3).

Step 6   Press **STO>, ALPHA, 6,** then press **ENTER** to save the result of $\binom{N}{x}$ as the variable **V.**

Step 7   Press **ALPHA, 4, ×, ALPHA, 5, ÷, ALPHA, 6,** then press **ENTER**.

The probability will be displayed on the window.

5 Continuous Random Variables

CONTENTS

Where We've Been

- Learned how to find probabilities of discrete events by using the probability rules of Chapter 3
- Discussed probability models (distributions) for discrete random variables in Chapter 4

Where We're Going

- Develop the notion of a probability distribution for a continuous random variable. (5.1)
- Study several important types of continuous random variables and their probability models. (5.2–5.3, 5.6)
- Introduce the normal probability distribution as one of the most useful distributions in statistics. (5.3–5.5)

Statistics in Action — Super Weapons Development—Is the Hit Ratio Optimized?

The U.S. Army is working with a major defense contractor to develop a "super" weapon. The weapon is designed to fire a large number of sharp tungsten bullets—called flechettes—with a single shot that will destroy a large number of enemy soldiers. Flechettes are about the size of an average nail, with small fins at one end to stabilize them in flight. Since World War I, when France dropped them in large quantities from aircraft on to masses of ground troops, munitions experts have experimented with using flechettes in a variety of guns. The problem with using flechettes as ammunition is accuracy: Current weapons that fire large quantities of flechettes have unsatisfactory hit ratios when fired at long distances.

The defense contractor (not named here for both confidentiality and security reasons) has developed a prototype gun that fires 1,100 flechettes with a single round. In range tests, three 2-feet-wide targets were set up a distance of 500 meters (approximately 1,500 feet) from the weapon. With a number line used as a reference, the centers of the three targets were at 0, 5, and 10 feet, respectively, as shown in Figure SIA5.1. The prototype gun was aimed at the middle target (center at 5 feet) and fired once. The point where each of the 1,100 flechettes landed at the 500-meter distance was measured with the use of a horizontal and vertical grid. For the purposes of this application, only the horizontal measurements are considered. These 1,100 measurements are saved in the **MOAGUN** file. (The data are simulated for confidentiality reasons.) For example, a flechette with a value of $x = 5.5$ hit the middle target, but a flechette with a value of $x = 2.0$ did not hit any of the three targets. (See Figure SIA5.1.)

The defense contractor is interested in the likelihood of any one of the targets being hit by a flechette and, in particular, wants to set the gun specifications to maximize the number of hits. The weapon is designed to have a mean horizontal value equal to the aim point (e.g., $\mu = 5$ feet when aimed at the center target). By changing specifications, the contractor can vary the standard deviation σ. The **MOAGUN** file contains flechette measurements for three different range tests: one with a standard deviation of $\sigma = 1$ foot, one with $\sigma = 2$ feet, and one with $\sigma = 4$ feet.

In this chapter, two Statistics in Action Revisited examples demonstrate how we can use one of the probability models discussed in the chapter—the normal probability distribution—to aid the defense contractor in developing its "super" weapon.

Figure SIA5.1
Target placement on gun range

Statistics in Action — Revisited

- Using the Normal Model to Maximize the Probability of a Hit with the Super Weapon (p. 276)
- Assessing whether the Normal Distribution Is Appropriate for Modeling the Super Weapon Hit Data (p. 284)

D *Data Set:* MOAGUN

In this chapter, we'll consider some continuous random variables that are commonly encountered. Recall that a **continuous random variable** is a random variable that can assume any value within some interval or intervals. For example, the length of time between a person's visits to a doctor, the thickness of sheets of steel produced in a rolling mill, and the yield of wheat per acre of farmland are all continuous random variables. The methodology we employ to describe continuous random variables will necessarily be somewhat different from that used to describe discrete random variables. First we discuss the general form of **continuous probability distributions,** and then we explore three specific types that are used in making statistical decisions. The normal probability distribution, which plays a basic and important role in both the theory and application of statistics, is essential to the study of most of the subsequent chapters in this book. The other types have practical applications, but a study of these topics is optional.

5.1 Continuous Probability Distributions

The graphical form of the probability distribution for a continuous random variable x is a smooth curve that might appear as shown in Figure 5.1. This curve, a function of x, is denoted by the symbol $f(x)$ and is variously called a **probability density function,** a **frequency function,** or a **probability distribution**.

The areas under a probability distribution correspond to probabilities for x. For example, the area A beneath the curve between the two points a and b, as shown in Figure 5.1, is the probability that x assumes a value between a and $b (a < x < b)$. Because there is no area over a single point, say, $x = a$, it follows that (according to our model) the probability associated with a particular value of x is equal to 0; that is, $P(x = a) = 0$, and hence $P(a < x < b) = P(a \le x \le b)$. In other words, the probability is the same regardless of whether you include the endpoints of the interval. Also, because areas over intervals represent probabilities, it follows that the total area under a probability distribution—the total probability assigned to the set of all values of x—should equal 1. Note that probability distributions for continuous random variables possess different shapes, depending on the relative frequency distributions of real data that the probability distributions are supposed to model.

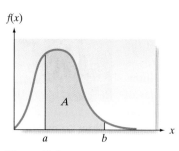

Figure 5.1

A probability distribution $f(x)$ for a continuous random variable x

> The **probability distribution for a continuous random variable, x,** can be represented by a smooth curve—a function of x, denoted $f(x)$. The curve is called a **density function** or **frequency function.** The probability that x falls between two values, a and b, i.e., $P(a < x < b)$, is the area under the curve between a and b.

Example 5.1

Finding a Continuous Probability—Paper Friction in a Photocopier

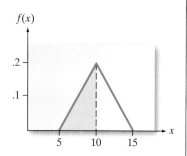

Figure 5.2

Density Function for Friction Coefficient, Example 5.1

Problem Researchers at the University of Rochester studied the friction that occurs in the paper-feeding process of a photocopier and published their results in the *Journal of Engineering for Industry*. The friction coefficient, x, is a continuous random variable that measures the degree of friction between two adjacent sheets of paper in the feeder stack. The random variable can be modeled using the smooth curve (called a *triangular density function*) shown in Figure 5.2. Note that the friction coefficient, x, ranges between the values 5 and 15. Find the probability that the friction coefficient is less than 10.

Solution To find the probability that the friction coefficient is less than 10, we need to compute $P(5 < x < 10)$. This is the area between $x = 5$ and $x = 10$ shaded on the graph in Figure 5.2. Note that this area is represented by a right triangle with a base of length 5 and a height of .2. Since the area of a triangle is $(1/2)(\text{base}) \times (\text{height})$, then

$$P(5 < x < 10) = (1/2)(\text{base}) \times (\text{height}) = (.5)(5)(.2) = .5$$

Look Ahead The areas under most probability distributions are obtained by means of calculus or numerical methods.* Because these methods often involve difficult procedures, we give the areas for some of the most common probability distributions in tabular form in Appendix B. Then, to find the area between two values of x, say, $x = a$ and $x = b$, you simply have to consult the appropriate table, or use technology such as statistical software or a graphing calculator.

■

---

*Students with a knowledge of calculus should note that the probability that x assumes a value in the interval $a < x < b$ is $P(a < x < b) = \int_a^b f(x)\, dx$, assuming that the integral exists. As with the requirements for a discrete probability distribution, we require that $f(x) \ge 0$ and $\int_{-\infty}^{\infty} f(x)\, dx = 1$.

For each of the continuous random variables presented in this chapter, we will give the formula for the probability distribution, along with its mean μ and standard deviation σ. These two numbers will enable you to make some approximate probability statements about a random variable even when you do not have access to a table of areas under the probability distribution.

5.2 The Uniform Distribution

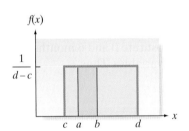

Figure 5.3
The uniform probability distribution

All the probability problems discussed in Chapter 3 had sample spaces that contained a finite number of sample points. In many of these problems—for example, in those involving a die toss or coin toss—the sample points were assigned equal probabilities. For continuous random variables, there is an infinite number of values in the sample space, but in some cases the values may appear to be equally likely. For example, if a short exists in a 5-meter stretch of electrical wire, it may have an equal probability of being in any particular 1-centimeter segment along the line. Or if a safety inspector plans to choose a time at random during the 4 afternoon work hours to pay a surprise visit to a certain area of a plant, then each 1-minute time interval in this 4-hour work period will have an equally likely chance of being selected for the visit.

Continuous random variables that appear to have *equally likely outcomes over their range of possible values* possess a **uniform probability distribution,** perhaps the simplest of all continuous probability distributions. Suppose the random variable x can assume values only in an interval $c \le x \le d$. Then the uniform frequency function has a rectangular shape, as shown in Figure 5.3. Note that the possible values of x consist of all points in the interval between point c and point d, inclusive. The height of $f(x)$ is constant in that interval and equals $1/(d - c)$. Therefore, the total area under $f(x)$ is given by

$$\text{Total area of rectangle} = (\text{Base})(\text{Height}) = (d - c)\left(\frac{1}{d - c}\right) = 1$$

The uniform probability distribution provides a model for continuous random variables that are *evenly distributed* over a certain interval. That is, a **uniform random variable** is just as likely to assume a value in one interval as it is to assume a value in any other interval of equal size. There is no clustering of values around any one value; instead, there is an even spread over the entire region of possible values.

The uniform distribution is sometimes referred to as the **randomness distribution,** since one way of generating a uniform random variable is to perform an experiment in which a point is *randomly selected* on the horizontal axis between the points c and d. If we were to repeat this experiment infinitely often, we would create a uniform probability distribution like that shown in Figure 5.3. The random selection of points in an interval can also be used to generate random numbers such as those in Table I of Appendix A. Recall that random numbers are selected in such a way that every number has an equal probability of selection. Therefore, random numbers are realizations of a uniform random variable. (Random numbers were used to draw random samples in Section 3.7.) The formulas for the uniform probability distribution, its mean, and its standard deviation are shown in the next box.

Probability Distribution for a Uniform Random Variable x

Probability density function: $f(x) = \dfrac{1}{d - c}$ $(c \le x \le d)$

Mean: $\mu = \dfrac{c + d}{2}$ Standard deviation: $\sigma = \dfrac{d - c}{\sqrt{12}}$

$P(a < x < b) = (b - a)/(d - c), c \le a < b \le d$

Suppose the interval $a < x < b$ lies within the domain of x; that is, it falls within the larger interval $c \le x \le d$. Then the probability that x assumes a value within the interval $a < x < b$ is equal to the area of the rectangle over the interval, namely, $(b - a)/(d - c)$.* (See the shaded area in Figure 5.3.)

Example   5.2

Applying the Uniform Distribution— Used-Car Warranties

Problem An unprincipled used-car dealer sells a car to an unsuspecting buyer, even though the dealer knows that the car will have a major breakdown within the next 6 months. The dealer provides a warranty of 45 days on all cars sold. Let x represent the length of time until the breakdown occurs. Assume that x is a uniform random variable with values between 0 and 6 months.

a. Calculate and interpret the mean and standard deviation of x.

b. Graph the probability distribution of x, and show the mean on the horizontal axis. Also show one- and two-standard-deviation intervals around the mean.

c. Calculate the probability that the breakdown occurs while the car is still under warranty.

Solution

a. To calculate the mean and standard deviation for x, we substitute 0 and 6 months for c and d, respectively, in the formulas for uniform random variables. Thus,

$$\mu = \frac{c + d}{2} = \frac{0 + 6}{2} = 3 \text{ months}$$

$$\sigma = \frac{d - c}{\sqrt{12}} = \frac{6 - 0}{\sqrt{12}} = \frac{6}{3.464} = 1.73 \text{ months}$$

Our interpretations of μ and σ follow:
The average length of time x until breakdown for all similar used cars is $\mu = 3$ months. From Chebyshev's rule (Rule 2.1, p. 99), we know that at least 75% of the values of x in the distribution will fall into the interval

$$\mu \pm 2\sigma = 3 \pm 2(1.73)$$
$$= 3 \pm 3.46$$

or between $-.46$ and 6.46 months. Consequently, we expect the length of time until breakdown to be less than 6.46 months at least 75% of the time.

b. The uniform probability distribution is

$$f(x) = \frac{1}{d - c} = \frac{1}{6 - 0} = \frac{1}{6} \quad (0 \leq x \leq 6)$$

The graph of this function is shown in Figure 5.4. The mean and the one- and two-standard-deviation intervals around the mean are shown on the horizontal axis. Note that the entire distribution of x lies within the $\mu \pm 2\sigma$ interval. (This result demonstrates, once again, the conservativeness of Chebyshev's rule.)

Figure 5.4
Distribution for x in Example 5.2

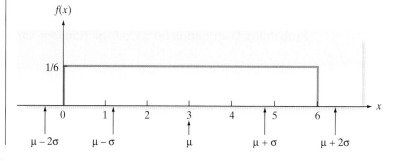

*The student with a knowledge of calculus should note that

$$P(a < x < b) = \int_a^b f(x)\, d(x) = \int_a^b [1/(d - c)]\, dx = (b - a)/(d - c)$$

c. To find the probability that the car is still under warranty when it breaks down, we must find the probability that x is less than 45 days, or (about) 1.5 months. As indicated in Figure 5.5, we need to calculate the area under the frequency function $f(x)$ between the points $x = 0$ and $x = 1.5$. Therefore, in this case $a = 0$ and $b = 1.5$. Applying the formula in the box, we have

$$P(x < 1.5) = (1.5 - 0)/(6 - 0) = 1.5/6 = .25$$

Figure 5.5
Probability that car breaks down within 1.5 months of purchase

That is, there is a 25% chance that the car will break down while under warranty.

Look Back The calculated probability in part **c** is the area of a rectangle with base $1.5 - 0 = 1.5$ and height $= 1/6$. Alternatively, the probability that the unsuspecting car buyer will be able to have the car repaired at the dealer's expense is $P(0 < x < 1.5) = (\text{Base})(\text{Height}) = (1.5)(1/6) = .25$.

■ **Now Work Exercise 5.9**

Exercises 5.1–5.20

Understanding the Principles

5.1 *True or False:* The probability that the uniform random variable will assume a value in any interval of equal length is different for each interval.

5.2 The uniform distribution is a _____ distribution. (Fill in the blank.)

Learning the Mechanics

5.3 Suppose x is a random variable best described by a uniform probability distribution with $c = 20$ and $d = 40$.
  a. Find $f(x)$.
  b. Find the mean and standard deviation of x.
  c. Graph $f(x)$, and locate μ and the interval $\mu \pm 2\sigma$ on the graph. Note that the probability that x assumes a value within the interval $\mu \pm 2\sigma$ is equal to 1.

5.4 Refer to Exercise 5.3. Find the following probabilities:
  a. $P(15 \leq x \leq 30)$ **b.** $P(25 < x < 35)$
  c. $P(x \geq 30)$ **d.** $P(x \leq 20)$
  e. $P(x \leq 30)$ **f.** $P(25.5 \leq x \leq 30.5)$

5.5 Suppose x is a random variable best described by a uniform probability distribution with $c = 4$ and $d = 6$.
  a. Find $f(x)$.
  b. Find the mean and standard deviation of x.
  c. Find $P(\mu - \sigma \leq x \leq \mu + \sigma)$.
  d. Find $P(x > 4.78)$.
  e. Find $P(4.4 \leq x \leq 5.7)$.
  f. Find $P(x < 4)$.

5.6 Refer to Exercise 5.5. Find the value of a makes each of the following probability statements true:
  a. $P(x \geq a) = .5$
  b. $P(x \leq a) = .2$

  c. $P(x \leq a) = 0$
  d. $P(4.5 \leq x \leq a) = .5$

5.7 The random variable x is best described by a uniform probability distribution with $c = 100$ and $d = 200$. Find the probability that x assumes a value
  a. More than two standard deviations from μ
  b. Less than three standard deviations from μ
  c. Within two standard deviations of μ

5.8 The random variable x is best described by a uniform probability distribution with mean 100 and standard deviation 10. Find c, d, and $f(x)$. Graph the probability distribution.

Ⓓ Applet Exercise 5.1

Open the applet entitled *Sample from a Population*. On the pull-down menu to the right of the top graph, select *Uniform*. The box to the left of the top graph displays the population mean, median, and standard deviation.
  a. Run the applet for each available value of n on the pull-down menu for the sample size. Go from the smallest to the largest value of n. For each value of n, observe the shape of the graph of the sample data and record the mean, median, and standard deviation of the sample.
  b. Describe what happens to the shape of the graph and the mean, median, and standard deviation of the sample as the sample size increases.

Ⓓ Applet Exercise 5.2

Suppose we set the *Random Numbers* applet to generate one number between 1 and 100, inclusive. We let the value of the random variable x be the number generated when the *Sample* button is clicked. Explain why the distribution of x is approximately uniform even though x is a discrete rather than continuous random variable.

Applying the Concepts—Basic

5.9 **Uranium in the Earth's crust.** The *American Mineralogist* (Oct. 2009) published a study of the evolution of uranium minerals in the Earth's crust. Researchers estimate that the trace amount of uranium x in reservoirs follows a uniform distribution ranging between 1 and 3 parts per million.
   a. Find $E(x)$ and interpret its value.
   b. Compute $P(2 < x < 2.5)$.
   c. Compute $P(x < 1.75)$.

5.10 **Requests to a Web server.** According to Brighton Webs LTD, a British company that specializes in data analysis, the arrival time of requests to a Web server within each hour can be modeled by a uniform distribution (http://www.brighton-webs.co.uk). Specifically, the number of seconds x from the start of the hour that the request is made is uniformly distributed between 0 and 3,600 seconds. Find the probability that a request is made to a Web server sometime during the last 20 minutes of the hour.

5.11 **New method for detecting anthrax.** Researchers at the University of South Florida Center for Biological Defense have developed a safe method for rapidly detecting anthrax spores in powders and on surfaces (*USF Magazine*, Summer 2002). Consider a powder specimen that has exactly 10 anthrax spores. Suppose that the number of anthrax spores in the sample detected by the new method follows an approximate uniform distribution between 10 and 20.
   a. Find the probability that 16 or fewer anthrax spores are detected in the powder specimen
   b. Find the probability that between 12 and 15 anthrax spores are detected in the powder specimen.

5.12 **School attendance.** A "discrete" uniform distribution was the subject of an article published in the *Journal of Probability* (Vol. 2014). The random variable of interest was the number of days x in a month that a secondary school student attends a language class. Assume that x can be approximated by a uniform probability distribution with range 0 to 20 days. Is it likely that a randomly selected student will attend 2 or fewer days in the month?

5.13 **Maintaining pipe wall temperature.** Maintaining a constant temperature in a pipe wall in some hot process applications is critical. A new technique that utilizes bolt-on trace elements to maintain the temperature was presented in the *Journal of Heat Transfer* (Nov. 2000). Without bolt-on trace elements, the pipe wall temperature of a switch condenser used to produce plastic has a uniform distribution ranging from 260° to 290°F. When several bolt-on trace elements are attached to the piping, the wall temperature is uniform from 278° to 285°F.
   a. Ideally, the pipe wall temperature should range between 280° and 284°F. What is the probability that the temperature will fall into this ideal range when no bolt-on trace elements are used? when bolt-on trace elements are attached to the pipe?
   b. When the temperature is 268°F or lower, the hot liquid plastic hardens (or "plates"), causing a buildup in the piping. What is the probability of the plastic plating when no bolt-on trace elements are used? when bolt-on trace elements are attached to the pipe?

Applying the Concepts—Intermediate

5.14 **Social network densities.** Social networks involve interactions (connections) between members of the network. Sociologists define network density as the ratio of actual network connections to the number of possible one to one connections. For example, a network with 10 members has $\binom{10}{2} = 45$ total possible connections. If that network has only 5 connections, the network density is 5/45 = .111. Sociologists at the University of Michigan assumed that the density x of a social network would follow a uniform distribution between 0 and 1 (*Social Networks*, 2010).
   a. On average, what is the density of a randomly selected social network?
   b. What is the probability that the randomly selected network has a density higher than .7?
   c. Consider a social network with only 2 members. Explain why the uniform model would not be a good approximation for the distribution of network density.

5.15 **Load on timber beams.** Timber beams are widely used in home construction. When the load (measured in pounds) per unit length has a constant value over part of a beam, the load is said to be uniformly distributed over that part of the beam. Uniformly distributed beam loads were used to derive the stiffness distribution of the beam in the *American Institute of Aeronautics and Astronautics Journal* (May 2013). Consider a cantilever beam with a uniformly distributed load between 100 and 115 pounds per linear foot.
   a. What is the probability that a beam load exceeds 110 pounds per linear foot?
   b. What is the probability that a beam load is less than 102 pounds per linear foot?
   c. Find a value L such that the probability that the beam load exceeds L is only .1.

5.16 **Cycle availability of a system.** In the jargon of system maintenance, "cycle availability" is defined as the probability that a system is functioning at any point in time. The U.S. Department of Defense developed a series of performance measures for assessing system cycle availability (*START*, Vol. 11, 2004). Under certain assumptions about the failure time and maintenance time of a system, cycle availability is shown to be uniformly distributed between 0 and 10. Find the following parameters for cycle availability: mean, standard deviation, 20th percentile, lower quartile, and upper quartile. Interpret the results.

5.17 **Soft-drink dispenser.** The manager of a local soft-drink bottling company believes that when a new beverage-dispensing machine is set to dispense 8 ounces, it in fact dispenses an amount x at random anywhere between 7.5 and 8.5 ounces, inclusive. Suppose x has a uniform probability distribution.
   a. Is the amount dispensed by the beverage machine a discrete or a continuous random variable? Explain.
   b. Graph the frequency function for x, the amount of beverage the manager believes is dispensed by the new machine when it is set to dispense 8 ounces.
   c. Find the mean and standard deviation for the distribution graphed in part **b**, and locate the mean and the interval $\mu \pm 2\sigma$ on the graph.
   d. Find $P(x \geq 8)$.

e. Find $P(x < 7)$

f. Find $P(7.75 < x < 8.25)$.

g. What is the probability that each of the next two bottles filled by the new machine will contain more than 8.25 ounces of beverage? Assume that the amount of beverage dispensed in one bottle is independent of the amount dispensed in another bottle.

5.18 **Time delays at a bus stop.** A bus is scheduled to stop at a certain bus stop every half hour on the hour and the half hour. At the end of the day, buses still stop after every 30 minutes, but because delays often occur earlier in the day, the bus is never early and is likely to be late. The director of the bus line claims that the length of time a bus is late is uniformly distributed and the maximum time that a bus is late is 20 minutes.

a. If the director's claim is true, what is the expected number of minutes a bus will be late?

b. If the director's claim is true, what is the probability that the last bus on a given day will be more than 19 minutes late?

c. If you arrive at the bus stop at the end of a day at exactly half-past the hour and must wait more than 19 minutes for the bus, what would you conclude about the director's claim? Why?

Applying the Concepts—Advanced

5.19 **Gouges on a spindle.** A tool-and-die machine shop produces extremely high-tolerance spindles. The spindles are 18-inch slender rods used in a variety of military equipment. A piece of equipment used in the manufacture of the spindles malfunctions on occasion and places a single gouge somewhere on the spindle. However, if the spindle can be cut so that it has 14 consecutive inches without a gouge, then the spindle can be salvaged for other purposes. Assuming that the location of the gouge along the spindle is best described by a uniform distribution, what is the probability that a defective spindle can be salvaged?

5.20 **Reliability of a robotic device.** The *reliability* of a piece of equipment is frequently defined to be the probability p that the equipment performs its intended function successfully for a given period under specific conditions (Render and Heizer, *Principles of Operations Management*, 1995). Because p varies from one point in time to another, some reliability analysts treat p as if it were a random variable. Suppose an analyst characterizes the uncertainty about the reliability of a particular robotic device used in an automobile assembly line by means of the following distribution:

$$f(p) = \begin{cases} 1 & 0 \le p \le 1 \\ 0 & \text{otherwise} \end{cases}$$

a. Graph the analyst's probability distribution for p.

b. Find the mean and variance of p.

c. According to the analyst's probability distribution for p, what is the probability that p is greater than .95? Less than .95?

d. Suppose the analyst receives the additional information that p is definitely between .90 and .95, but that there is complete uncertainty about where it lies between these values. Describe the probability distribution the analyst should now use to describe p.

The Normal Distribution

One of the most commonly observed continuous random variables has a **bell-shaped probability distribution** (or **bell curve**), as shown in Figure 5.6. It is known as a **normal random variable** and its probability distribution is called a **normal distribution**.

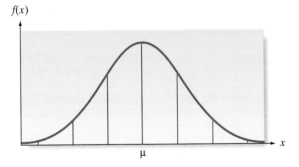

Figure 5.6

A normal probability distribution

The normal distribution plays a very important role in the science of statistical inference. Moreover, many phenomena generate random variables with probability distributions that are very well approximated by a normal distribution. For example, the error made in measuring a person's blood pressure may be a normal random variable, and the probability distribution for the yearly rainfall in a certain region might be approximated by a normal probability distribution. You can determine the adequacy of the normal approximation to an existing population of data by comparing the relative frequency distribution of a large sample of the data with the normal probability distribution. Methods for detecting disagreement between a set of data and the assumption of normality are presented in Section 5.4.

CARL F. GAUSS (1777–1855)

The Gaussian Distribution

The normal distribution began in the 18th century as a theoretical distribution for errors in disciplines in which fluctuations in nature were believed to behave randomly. Although he may not have been the first to discover the formula, the normal distribution was named the Gaussian distribution after Carl Friedrich Gauss. A well-known and respected German mathematician, physicist, and astronomer, Gauss applied the normal distribution while studying the motion of planets and stars. Gauss's prowess as a mathematician was exemplified by one of his most important discoveries: At the young age of 22, Gauss constructed a regular 17-gon by ruler and compasses—a feat that was the most major advance in mathematics since the time of the ancient Greeks. In addition to publishing close to 200 scientific papers, Gauss invented the heliograph as well as a primitive telegraph. ■

The normal distribution is perfectly symmetric about its mean μ, as can be seen in the examples in Figure 5.7. Its spread is determined by the value of its standard deviation σ. The formula for the normal probability distribution is shown in the next box. When plotted, this formula yields a curve like that shown in Figure 5.6.

Figure 5.7
Several normal distributions with different means and standard deviations

---

Probability Distribution for a Normal Random Variable x

Probability density function: $f(x) = \dfrac{1}{\sigma\sqrt{2\pi}}e^{-(1/2)[(x-\mu)/\sigma]^2}$

where

μ = Mean of the normal random variable x

σ = Standard deviation

π = 3.1416...

e = 2.71828...

$P(x < a)$ is obtained from a table of normal probabilities.

Note that the mean μ and standard deviation σ appear in this formula, so that no separate formulas for μ and σ are necessary. To graph the normal curve, we have to know the numerical values of μ and σ. Computing the area over intervals under the normal probability distribution is a difficult task.* Consequently, we will use either technology or the computed areas listed in Table II of Appendix B. Although there are an infinitely large number of normal curves—one for each pair of values of μ and σ—we have formed a single table that will apply to any normal curve.

Table II is based on a normal distribution with mean $\mu = 0$ and standard deviation $\sigma = 1$, called a *standard normal distribution*. A random variable with a standard normal distribution is typically denoted by the symbol z. The formula for the probability distribution of z is

$$f(z) = \frac{1}{\sqrt{2\pi}}e^{-(1/2)z^2}$$

Figure 5.8 shows the graph of a standard normal distribution.

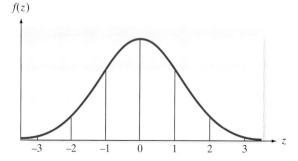

Figure 5.8
Standard normal distribution: $\mu = 0, \sigma = 1$

*The student with a knowledge of calculus should note that there is no closed-form expression for $P(a < x < b) = \int_a^b f(x)\,dx$ for the normal probability distribution. The value of this definite integral can be obtained to any desired degree of accuracy by numerical approximation. For this reason, it is tabulated for the user.

> The **standard normal distribution** is a normal distribution with $\mu = 0$ and $\sigma = 1$. A random variable with a standard normal distribution, denoted by the symbol z, is called a **standard normal random variable**.

Since we will ultimately convert all normal random variables to standard normal variables in order to use Table II to find probabilities, it is important that you learn to use Table II well. A partial reproduction of that table is shown in Table 5.1. Note that the values of the standard normal random variable z are listed in the left-hand column. The entries in the body of the table give the area (probability) between 0 and z. Examples 5.3–5.6 illustrate the use of the table.

Table 5.1   Reproduction of Part of Table II in Appendix B

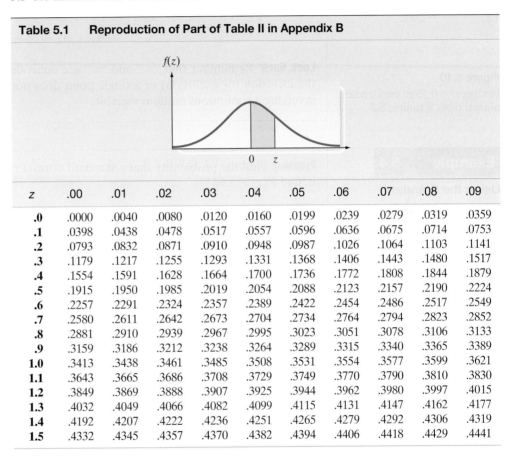

z	.00	.01	.02	.03	.04	.05	.06	.07	.08	.09
.0	.0000	.0040	.0080	.0120	.0160	.0199	.0239	.0279	.0319	.0359
.1	.0398	.0438	.0478	.0517	.0557	.0596	.0636	.0675	.0714	.0753
.2	.0793	.0832	.0871	.0910	.0948	.0987	.1026	.1064	.1103	.1141
.3	.1179	.1217	.1255	.1293	.1331	.1368	.1406	.1443	.1480	.1517
.4	.1554	.1591	.1628	.1664	.1700	.1736	.1772	.1808	.1844	.1879
.5	.1915	.1950	.1985	.2019	.2054	.2088	.2123	.2157	.2190	.2224
.6	.2257	.2291	.2324	.2357	.2389	.2422	.2454	.2486	.2517	.2549
.7	.2580	.2611	.2642	.2673	.2704	.2734	.2764	.2794	.2823	.2852
.8	.2881	.2910	.2939	.2967	.2995	.3023	.3051	.3078	.3106	.3133
.9	.3159	.3186	.3212	.3238	.3264	.3289	.3315	.3340	.3365	.3389
1.0	.3413	.3438	.3461	.3485	.3508	.3531	.3554	.3577	.3599	.3621
1.1	.3643	.3665	.3686	.3708	.3729	.3749	.3770	.3790	.3810	.3830
1.2	.3849	.3869	.3888	.3907	.3925	.3944	.3962	.3980	.3997	.4015
1.3	.4032	.4049	.4066	.4082	.4099	.4115	.4131	.4147	.4162	.4177
1.4	.4192	.4207	.4222	.4236	.4251	.4265	.4279	.4292	.4306	.4319
1.5	.4332	.4345	.4357	.4370	.4382	.4394	.4406	.4418	.4429	.4441

Example 5.3

Using the Standard Normal Table to Find $P(-z_0 < z < z_0)$

Problem Find the probability that the standard normal random variable z falls between -1.33 and $+1.33$.

Solution The standard normal distribution is shown again in Figure 5.9. Since all probabilities associated with standard normal random variables can be depicted as areas under the standard normal curve, you should always draw the curve and then equate the desired probability to an area.

In this example, we want to find the probability that z falls between -1.33 and $+1.33$, which is equivalent to the area between -1.33 and $+1.33$, shown highlighted in Figure 5.9.

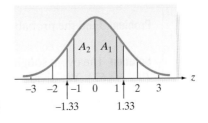

Figure 5.9

Areas under the standard normal curve for Example 5.3

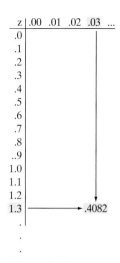

Figure 5.10
Finding $z = 1.33$ in the standard normal table, Example 5.3

Table II gives the area between $z = 0$ and any positive value of z, so that if we look up $z = 1.33$ (the value in the 1.3 row and .03 column, as shown in Figure 5.10), we find that the area between $z = 0$ and $z = 1.33$ is .4082. This is the area labeled A_1 in Figure 5.9. To find the area A_2 located between $z = 0$ and $z = -1.33$, we note that the symmetry of the normal distribution implies that the area between $z - 0$ and any point to the left is equal to the area between $z = 0$ and the point equidistant to the right. Thus, in this example the area between $z = 0$ and $z = -1.33$ is equal to the area between $z = 0$ and $z = +1.33$. That is,

$$A_1 = A_2 = .4082$$

The probability that z falls between -1.33 and $+1.33$ is the sum of the areas of A_1 and A_2. We summarize in probabilistic notation:

$$P(-1.33 < z < -1.33) = P(-1.33 < z < 0) + P(0 < z \le 1.33)$$
$$= A_1 + A_2 = .4082 + .4082 = .8164$$

Look Back Remember that "<" and "≤" are equivalent in events involving z because the inclusion (or exclusion) of a single point does not alter the probability of an event involving a continuous random variable.

■ **Now Work Exercise 5.25e–f**

| **Example** | **5.4** |

Using the Standard Normal Table to Find $P(z > z_0)$

Problem Find the probability that a standard normal random variable exceeds 1.64; that is, find $P(z > 1.64)$.

Solution The area under the standard normal distribution to the right of 1.64 is the highlighted area labeled A_1 in Figure 5.11. This area represents the probability that z exceeds 1.64. However, when we look up $z = 1.64$ in Table II, we must remember that the probability given in the table corresponds to the area between $z = 0$ and $z = 1.64$ (the area labeled A_2 in Figure 5.11). From Table II, we find that $A_2 = .4495$. To find the area A_1 to the right of 1.64, we make use of two facts:

1. The standard normal distribution is symmetric about its mean, $z = 0$.

2. The total area under the standard normal probability distribution equals 1.

Taken together, these two facts imply that the areas on either side of the mean, $z = 0$, equal .5; thus, the area to the right of $z = 0$ in Figure 5.11 is $A_1 + A_2 = .5$. Then

$$P(z > 1.64) = A_1 = .5 - A_2 = .5 - .4495 = .0505$$

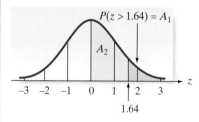

Figure 5.11
Areas under the standard normal curve for Example 5.4

Look Back To attach some practical significance to this probability, note that the implication is that the chance of a standard normal random variable exceeding 1.64 is only about .05.

■ **Now Work Exercise 5.26a**

| **Example** | **5.5** |

Using the Standard Normal Table to Find $P(z < z_0)$

Problem Find the probability that a standard normal random variable lies to the left of .67.

Solution The event sought is shown as the highlighted area in Figure 5.12. We want to find $P(z < .67)$. We divide the highlighted area into two parts: the area A_1 between $z = 0$ and $z = .67$, and the area A_2 to the left of $z = 0$. We must always make such a division when the desired area lies on both sides of the mean ($z = 0$) because Table II contains areas between $z = 0$ and the point you look up. We look up $z = .67$ in Table II to find that $A_1 = .2486$. The symmetry of the standard normal distribution also implies that half the distribution lies on each side of the mean, so the area A_2 to the left of $z = 0$ is .5. Then

$$P(z < .67) = A_1 + A_2 = .2486 + .5 = .7486$$

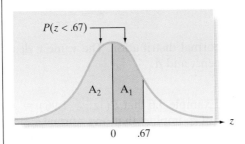

Figure 5.12

Areas under the standard normal curve for Example 5.5

Look Back Note that this probability is approximately .75. Thus, about 75% of the time, the standard normal random variable z will fall below .67. This statement implies that $z = .67$ represents the approximate 75th percentile (or upper quartile) of the standard normal distribution.

■ **Now Work Exercise 5.26h**

Example   5.6

Using the Standard Normal Table to Find $P(|z| > z_0)$

Problem Find the probability that a standard normal random variable exceeds 1.96 in absolute value.

Solution The event sought is shown highlighted in Figure 5.13. We want to find

$$P(|z| > 1.96) = P(z < -1.96 \text{ or } z > 1.96)$$

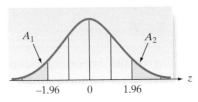

Figure 5.13

Areas under the standard normal curve for Example 5.6

Note that the total highlighted area is the sum of the two areas A_1 and A_2—areas that are equal because of the symmetry of the normal distribution.

We look up $z = 1.96$ and find the area between $z = 0$ and $z = 1.96$ to be .4750. Then A_2, the area to the right of 1.96, is $.5 - .4750 = .0250$, so that

$$P(|z| > 1.96) = A_1 + A_2 = .0250 + .0250 = .05$$

Look Back We emphasize, again, the importance of sketching the standard normal curve in finding normal probabilities.

■

To apply Table II to a normal random variable x with any mean μ and any standard deviation σ, we must first convert the value of x to a z-score. The population z-score for a measurement was defined in Section 2.6 as the *distance* between the measurement and the population mean, divided by the population standard deviation. Thus, the z-score gives the distance between a measurement and the mean in units equal to the standard deviation. In symbolic form, the z-score for the measurement x is

$$z = \frac{x - \mu}{\sigma}$$

Note that when $x = \mu$, we obtain $z = 0$.

An important property of the normal distribution is that if x is normally distributed with any mean and any standard deviation, z is *always* normally distributed with mean 0 and standard deviation 1. That is, z is a standard normal random variable.

> **Converting a Normal Distribution to a Standard Normal Distribution**
>
> If x is a normal random variable with mean μ and standard deviation σ, then the random variable z defined by the formula
>
> $$z = \frac{x - \mu}{\sigma}$$
>
> has a standard normal distribution. The value z describes the number of standard deviations between x and μ.

Recall from Example 5.6 that $P(|z| > 1.96) = .05$. This probability, coupled with our interpretation of z, implies that any normal random variable lies more than 1.96 standard deviations from its mean only 5% of the time. Compare this statement to the empirical rule (Chapter 2), which tells us that about 5% of the measurements in mound-shaped distributions will lie beyond two standard deviations from the mean. The normal distribution actually provides the model on which the empirical rule is based, along with much "empirical" experience with real data that often approximately obey the rule, whether drawn from a normal distribution or not.

Example **5.7**	**Problem** Assume that the length of time, x, between charges of a cellular phone is normally distributed with a mean of 10 hours and a standard deviation of 1.5 hours. Find the probability that the cell phone will last between 8 and 12 hours between charges.

Finding a Normal Probability—Cell Phone Application

Solution The normal distribution with mean $\mu = 10$ and $\sigma = 1.5$ is shown in Figure 5.14. The desired probability that the cell phone lasts between 8 and 12 hours is highlighted. In order to find that probability, we must first convert the distribution to a standard normal distribution, which we do by calculating the z-score:

$$z = \frac{x - \mu}{\sigma}$$

Figure 5.14
Areas under the normal curve for Example 5.7

The z-scores corresponding to the important values of x are shown beneath the x values on the horizontal axis in Figure 5.14. Note that $z = 0$ corresponds to the mean of $\mu = 10$ hours, whereas the x values 8 and 12 yield z-scores of -1.33 and $+1.33$, respectively. Thus, the event that the cell phone lasts between 8 and 12 hours is equivalent to the event that a standard normal random variable lies between -1.33 and $+1.33$. We found this probability in Example 5.3 (see Figure 5.9) by doubling the area corresponding to $z = 1.33$ in Table II. That is,

$$P(8 \le x \le 12) = P(-1.33 \le z \le 1.33) = 2(.4082) = .8164$$

■ **Now Work Exercise 5.31a**

Table II in Appendix B provides good approximations to probabilities under the normal curve. However, if you do not have access to a normal table, you can always rely on statistical software or a graphing calculator to compute the desired probability. With most statistical software, you will need to specify the mean and standard deviation of the normal distribution, as well as the key values of the variable for which you desire probabilities. In Example 5.7, we desire $P(8 \le x \le 12)$, where $\mu = 10$ and $\sigma = 1.5$. To find this probability using MINITAB's normal probability function, we enter 10 for the mean and 1.5 for the standard deviation, and then find two cumulative probabilities: $P(x \le 12)$ and $P(x < 8)$. These two probabilities are shown (shaded) on the MINITAB printout in Figure 5.15. The difference between the two probabilities yields the desired result:

$$P(8 \le x \le 12) = P(x \le 12) - P(x < 8) = .908789 - .0912112 = .8175778$$

Note that this probability agrees with the value computed using Table II to two decimal places. The difference is due to rounding of the probabilities given in Table II.

Cumulative Distribution Function

Normal with mean = 10 and standard deviation = 1.5

```
x   P( X <= x )
12     0.908789
```

Cumulative Distribution Function

Normal with mean = 10 and standard deviation = 1.5

```
x   P( X <= x )
8     0.0912112
```

Figure 5.15

MINITAB output with cumulative normal probabilities

The steps to follow in calculating a probability corresponding to a normal random variable are shown in the following box:

Steps for Finding a Probability Corresponding to a Normal Random Variable

1. Sketch the normal distribution and indicate the mean of the random variable x. Then shade the area corresponding to the probability you want to find.

2. Convert the boundaries of the shaded area from x values to standard normal random variable z values by using the formula

$$z = \frac{x - \mu}{\sigma}$$

Show the z values under the corresponding x values on your sketch.

3. Use technology or Table II in Appendix B to find the areas corresponding to the z values. If necessary, use the symmetry of the normal distribution to find areas corresponding to negative z values and the fact that the total area on each side of the mean equals .5 to convert the areas from Table II to the probabilities of the event you have shaded.

Example   5.8

Using Normal Probabilities to Make an Inference—Advertised Gas Mileage

Problem Suppose an automobile manufacturer introduces a new model that has an advertised mean in-city mileage of 27 miles per gallon. Although such advertisements seldom report any measure of variability, suppose you write the manufacturer for the details of the tests and you find that the standard deviation is 3 miles per gallon. This information leads you to formulate a probability model for the random variable x, the in-city mileage for this car model. You believe that the probability distribution of x can be approximated by a normal distribution with a mean of 27 and a standard deviation of 3.

a. If you were to buy this model of automobile, what is the probability that you would purchase one that averages less than 20 miles per gallon for in-city driving? In other words, find $P(x < 20)$.

b. Suppose you purchase one of these new models and it does get less than 20 miles per gallon for in-city driving. Should you conclude that your probability model is incorrect?

Solution

a. The probability model proposed for x, the in-city mileage, is shown in Figure 5.16. We are interested in finding the area A to the left of 20, since that area corresponds to the probability that a measurement chosen from this distribution falls below 20. In other words, if this model is correct, the area A represents the fraction of cars that can be

Figure 5.16
Area under the normal curve for Example 5.8

expected to get less than 20 miles per gallon for in-city driving. To find A, we first calculate the z value corresponding to $x = 20$. That is,

$$z = \frac{x - \mu}{\sigma} = \frac{20 - 27}{3} = -\frac{7}{3} = -2.33$$

Then

$$P(x < 20) = P(z < -2.33)$$

as indicated by the highlighted area in Figure 5.16. Since Table II gives only areas to the right of the mean (and because the normal distribution is symmetric about its mean), we look up 2.33 in Table II and find that the corresponding area is .4901. This is equal to the area between $z = 0$ and $z = -2.33$, so we find that

$$P(x < 20) = A = .5 - .4901 = .0099 \approx .01$$

According to this probability model, you should have only about a 1% chance of purchasing a car of this make with an in-city mileage under 20 miles per gallon.

b. Now you are asked to make an inference based on a sample: the car you purchased. You are getting less than 20 miles per gallon for in-city driving. What do you infer? We think you will agree that one of two possibilities exists:

1. The probability model is correct. You simply were unfortunate to have purchased one of the cars in the 1% that get less than 20 miles per gallon in the city.

2. The probability model is incorrect. Perhaps the assumption of a normal distribution is unwarranted, or the mean of 27 is an overestimate, or the standard deviation of 3 is an underestimate, or some combination of these errors occurred. At any rate, the form of the actual probability model certainly merits further investigation.

   You have no way of knowing with certainty which possibility is correct, but the evidence points to the second one. We are again relying on the rare-event approach to statistical inference that we introduced earlier. The sample (one measurement in this case) was so unlikely to have been drawn from the proposed probability model that it casts serious doubt on the model. We would be inclined to believe that the model is somehow in error.

Look Back In applying the rare-event approach, the calculated probability must be small (say, less than or equal to .05) in order to infer that the observed event is, indeed, unlikely.

■ **Now Work Exercise 5.40**

Occasionally you will be given a probability and will want to find the values of the normal random variable that correspond to that probability. For example, suppose the scores on a college entrance examination are known to be normally distributed and a certain prestigious university will consider for admission only those applicants whose scores exceed the 90th percentile of the test score distribution. To determine the minimum score for consideration for admission, you will need to be able to use Table II or statistical software in reverse, as demonstrated in the next example.

Example	**5.9**

Finding a z-Value Associated with a 1-Tail Normal Probability

Problem Find the value of z — call it z_0 — in the standard normal distribution that will be exceeded only 10% of the time. That is, find z_0 such that $P(z \geq z_0) = .10$.

Solution In this case, we are given a probability, or an area, and are asked to find the value of the standard normal random variable that corresponds to the area. Specifically, we want to find the value z_0 such that only 10% of the standard normal distribution exceeds z_0. (See Figure 5.17.)

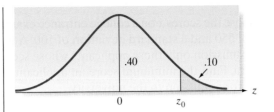

Figure 5.17
Areas under the standard
normal curve for Example 5.9

We know that the total area to the right of the mean, $z = 0$, is .5, which implies that z_0 must lie to the right of 0 ($z_0 > 0$). To pinpoint the value, we use the fact that the area to the right of z_0 is .10, which implies that the area between $z = 0$ and z_0 is $.5 - .1 = .4$. But areas between $z = 0$ and some other z value are exactly the types given in Table II. Therefore, we look up the area .4000 in the body of Table II and find that the corresponding z value is (to the closest approximation) $z_0 = 1.28$. The implication is that the point 1.28 standard deviations above the mean is the 90th percentile of a normal distribution.

We can also arrive at this answer using statistical software. In MINITAB, we use the inverse cumulative distribution function for a normal random variable and specify the cumulative probability,

$$P(z \le z_0) = .9$$

The value of z_0 is shown (shaded) on the MINITAB printout in Figure 5.18. You can see that this value agrees with our solution using the normal table.

Inverse Cumulative Distribution Function

```
Normal with mean = 0 and standard deviation = 1

P( X <= x )        x
      0.9    1.28155
```

Figure 5.18
MINITAB Output for
Example 5.9

Look Back As with earlier problems, it is critical to correctly draw the normal probability sought on the normal curve. The placement of z_0 to the left or right of 0 is the key. Be sure to shade the probability (area) involving z_0. If it does not agree with the probability sought (i.e., if the shaded area is greater than .5 and the probability sought is smaller than .5), then you need to place z_0 on the opposite side of 0.

■

Example   5.10

**Finding a z-Value
Associated with a 2-Tail
Normal Probability**

Problem Find the value of z_0 such that 95% of the standard normal z values lie between $-z_0$ and $+z_0$; that is, find $P(-z_0 \le z \le z_0) = .95$.

Solution Here we wish to move an equal distance z_0 in the positive and negative directions from the mean $z = 0$ until 95% of the standard normal distribution is enclosed. This means that the area on each side of the mean will be equal to $\frac{1}{2}(.95) = .475$, as shown in Figure 5.19. Since the area between $z = 0$ and z_0 is .475, we look up .475 in the body of Table II to find the value $z_0 = 1.96$. Thus, as we found in reverse order in Example 5.6, 95% of a normal distribution lies between $+1.96$ and -1.96 standard deviations of the mean.

Figure 5.19
Areas under the standard normal
curve for Example 5.10

■ **Now Work Exercise 5.28a**

Now that you have learned to find a standard normal z value that corresponds to a specified probability, we demonstrate a practical application in Example 5.11.

Example	5.11

Finding the Value of a Normal Random Variable—College Entrance Exam Application

Problem Suppose the scores x on a college entrance examination are normally distributed with a mean of 550 and a standard deviation of 100. A certain prestigious university will consider for admission only those applicants whose scores exceed the 90th percentile of the distribution. Find the minimum score an applicant must achieve in order to receive consideration for admission to the university.

Solution In this example, we want to find a score x_0 such that 90% of the scores (x values) in the distribution fall below x_0 and only 10% fall above x_0. That is,

$$P(x \leq x_0) = .90$$

Converting x to a standard normal random variable where $\mu = 550$ and $\sigma = 100$, we have

$$P(x \leq x_0) = P\left(z \leq \frac{x_0 - \mu}{\sigma}\right)$$

$$= P\left(z \leq \frac{x_0 - 550}{100}\right) = .90$$

In Example 5.9 (see Figure 5.17), we found the 90th percentile of the standard normal distribution to be $z_0 = 1.28$. That is, we found that $P(z \leq 1.28) = .90$. Consequently, we know that the minimum test score x_0 corresponds to a z-score of 1.28; in other words,

$$\frac{x_0 - 550}{100} = 1.28$$

If we solve this equation for x_0, we find that

$$x_0 = 550 + 1.28(100) = 550 + 128 = 678$$

This x value is shown in Figure 5.20. Thus, the 90th percentile of the test score distribution is 678. That is to say, an applicant must score at least 678 on the entrance exam to receive consideration for admission by the university.

Look Back As the example shows, in practical applications of the normal table in reverse, first find the value of z_0 and then use the z-score formula in reverse to convert the value to the units of x.

Figure 5.20
Area under the normal curve for Example 5.11

■ **Now Work Exercise 5.42b**

Using the Normal Model to Maximize the Probability of a Hit with the Super Weapon

Recall that a defense contractor has developed a prototype gun for the U.S. Army that fires 1,100 flechettes with a single round. The specifications of the weapon are set so that when the gun is aimed at a target 500 meters away, the mean horizontal grid value of the flechettes is equal to the aim point. In the range test, the weapon was aimed at the center target in Figure SIA5.1; thus, $\mu = 5$ feet. For three different tests, the standard deviation was set at $\sigma = 1$ foot, $\sigma = 2$ feet, and $\sigma = 4$ feet. From past experience, the defense contractor has found that the distribution of the horizontal flechette measurements is closely approximated by a normal distribution. Therefore, we can use

the normal distribution to find the probability that a single flechette shot from the weapon will hit any one of the three targets. Recall from Figure SIA5.1 that the three targets range from -1 to 1, 4 to 6, and 9 to 11 feet on the horizontal grid.

Consider first the middle target. Letting x represent the horizontal measurement for a flechette shot from the gun, we see that the flechette will hit the target if $4 \leq x \leq 6$. Using the normal probability table (Table II, Appendix B), we then find that the probability that this flechette will hit the target when $\mu = 5$ and $\sigma = 1$ is

Middle: $P(4 \le x \le 6) = P\left(\dfrac{4-5}{1} < z < \dfrac{6-5}{1}\right)$

$\sigma = 1$

$= P(-1 < z < 1)$
$= 2(.3413) = .6826$

Similarly, we find the probabilities that the flechette hits the left and right targets shown in Figure SIA5.1:

Left: $P(-1 \le x \le 1) = P\left(\dfrac{-1-5}{1} < z < \dfrac{1-5}{1}\right)$

$\sigma = 1$

$= P(-6 < z < -4) \approx 0$

Right: $P(9 \le x \le 11) = P\left(\dfrac{9-5}{1} < z < \dfrac{11-5}{1}\right)$

$\sigma = 1$

$= P(4 < z < 6) \approx 0$

You can see that there is about a 68% chance that a flechette will hit the middle target, but virtually no chance that one will hit the left or right target when the standard deviation is set at 1 foot.

To find these three probabilities for $\sigma = 2$ and $\sigma = 4$, we use the normal probability function in MINITAB. Figure SIA5.2 is a MINITAB worksheet giving the cumulative probabilities of a normal random variable falling below the x values in the first column. The cumulative probabilities for $\sigma = 2$ and $\sigma = 4$ are given in the columns named "sigma2" and "sigma4," respectively.

Using the cumulative probabilities in the figure to find the three probabilities when $\sigma = 2$, we have

Middle: $P(4 \le x \le 6) = P(x \le 6) - P(x \le 4)$

$\sigma = 2$

$= .6915 - .3085 = .3830$

Left: $P(-1 \le x \le 1) = P(x \le 1) - P(x \le -1)$

$\sigma = 2$

$= .0227 - .0013 = .0214$

Right: $P(9 \le x \le 11) = P(x \le 11) - P(x \le 9)$

$\sigma = 2$

$= .9987 - .9773 = .0214$

Thus, when $\sigma = 2$, there is about a 38% chance that a flechette will hit the middle target, a 2% chance that one will hit the left target, and a 2% chance that one will hit the right target. The probability that a flechette will hit either the middle or the left or the right target is simply the sum of these three probabilities (an application of the additive rule of probability). This sum is $.3830 + .0214 + .0214 = .4258$; consequently, there is about a 42% chance of hitting any one of the three targets when specifications are set so that $\sigma = 2$.

Now we use the cumulative probabilities in Figure SIA5.2 to find the three hit probabilities when $\sigma = 4$:

Middle: $P(4 \le x \le 6) = P(x \le 6) - P(x \le 4)$

$\sigma = 4$

$= .5987 - .4013 = .1974$

Left: $P(-1 \le x \le 1) = P(x \le 1) - P(x \le -1)$

$\sigma = 4$

$= .1587 - .0668 = .0919$

Right: $P(9 \le x \le 11) = P(x \le 11) - P(x \le 9)$

$\sigma = 4$

$= .9332 - .8413 = .0919$

Thus, when $\sigma = 4$, there is about a 20% chance that a flechette will hit the middle target, a 9% chance that one will hit the left target, and a 9% chance that one will hit the right target. The probability that a flechette will hit any one of the three targets is $.1974 + .0919 + .0919 = .3812$.

These probability calculations reveal a few patterns. First, the probability of hitting the middle target (the target at which the gun is aimed) is reduced as the standard deviation is increased. Obviously, then, if the U.S. Army wants to maximize the chance of hitting the target that the prototype gun is aimed at, it will want specifications set with a small value of σ. But if the Army wants to hit multiple targets with a single shot of the weapon, σ should be increased. With a larger σ, not as many of the flechettes will hit the target aimed at, but more will hit peripheral targets. Whether σ should be set at 4 or 6 (or some other value) depends on how high of a hit rate is required for the peripheral targets.

↓	C1	C2	C3	C4	C5
	x	sigma1	sigma2	sigma4	
1	-1	0.00000	0.001350	0.066807	
2	1	0.00003	0.022750	0.158655	
3	4	0.15866	0.308538	0.401294	
4	6	0.84134	0.691462	0.598706	
5	9	0.99997	0.977250	0.841345	
6	11	1.00000	0.998650	0.933193	
7					

Figure SIA5.2

MINITAB worksheet with cumulative normal probabilities

Exercises 5.21–5.54

Understanding the Principles

5.21 *True or False:* The normal distribution is a symmetric probability distribution.

5.22 If x has a normal distribution with mean μ and standard deviation σ, find the mean of the random variable $z = (x - \mu)/\sigma$.

5.23 How do you transform a normal random variable to a standard normal random variable?

Learning the Mechanics

5.24 Find the area under the standard normal probability distribution between the following pairs of z-scores:
    a. $z = 0$ and $z = 1.90$
    b. $z = 0$ and $z = .90$
    c. $z = 0$ and $z = 2.90$
    d. $z = 0$ and $z = 0$
    e. $z = -1.90$ and $z = 0$
    f. $z = -.90$ and $z = 0$
    g. $z = -1.59$ and $z = 0$
    h. $z = -.48$ and $z = 0$

5.25 Find each of the following probabilities for a standard normal random variable z:
    a. $P(z = 1)$
    b. $P(z \leq 1)$
    c. $P(z < 1)$
    d. $P(z > 1)$
[NW]   **e.** $P(-1 \leq z \leq 1)$
[NW]   **f.** $P(-2 \leq z \leq 2)$
    g. $P(-2.16 \leq z \leq .55)$
[NW]   **h.** $P(-.42 < z < 1.96)$

5.26 Find the following probabilities for the standard normal random variable z:
[NW]   **a.** $P(z > 1.46)$
    b. $P(z < -1.56)$
    c. $P(.67 \leq z \leq 2.41)$
    d. $P(-1.96 \leq z < -.33)$
    e. $P(z \geq 0)$
    f. $P(-2.33 < z < 1.50)$
    g. $P(z \geq -2.33)$
[NW]   **h.** $P(z < 2.33)$

5.27 Give the z-score for a measurement from a normal distribution for the following:
    a. 2 standard deviation above the mean
    b. 2 standard deviation below the mean
    c. Equal to the mean
    d. 2.5 standard deviations above the mean
    e. 3 standard deviations above the mean

5.28 Find a value z_0 of the standard normal random variable z such that
[NW]   **a.** $P(z \leq z_0) = .0401$
    b. $P(-z_0 \leq z \leq z_0) = .95$
    c. $P(-z_0 \leq z \leq z_0) = .90$
    d. $P(-z_0 \leq z \leq z_0) = .8740$
    e. $P(-z_0 \leq z \leq 0) = .2967$
    f. $P(-2 < z < z_0) = .9710$
    g. $P(z \geq z_0) = .5$
    h. $P(z \geq z_0) = .0057$

5.29 Find a value z_0 of the standard normal random variable z such that
    a. $P(z \geq z_0) = .10$
    b. $P(z \geq z_0) = .003$
    c. $P(z \leq z_0) = .01$
    d. $P(z \geq z_0) = .20$
    e. $P(z > z_0) = .02$

5.30 Suppose the random variable x is best described by a normal distribution with $\mu = 36$ and $\sigma = 6$. Find the z-score that corresponds to each of the following x values:
    a. $x = 36$
    b. $x = 42$
    c. $x = 47.5$
    d. $x = 18$
    e. $x = 66$
    f. $x = 50$

5.31 Suppose x is a normally distributed random variable with $\mu = 11$ and $\sigma = 2$. Find each of the following:
[NW]   **a.** $P(10 \leq x \leq 12)$
    b. $P(6 \leq x \leq 10)$
    c. $P(13 \leq x \leq 16)$
    d. $P(7.8 \leq x \leq 12.6)$
    e. $P(x \geq 13.24)$
    f. $P(x \geq 7.62)$

5.32 Suppose x is a normally distributed random variable with $\mu = 40$ and $\sigma = 10$. Find a value x_0 of the random variable x such that
    a. $P(x \geq x_0) = .5$
    b. $P(x < x_0) = .025$
    c. $P(x > x_0) = .10$
    d. $P(x > x_0) = .95$

5.33 Refer to Exercise 5.32. Find x_0 such that
    a. 10% of the values of x are less than x_0.
    b. 80% of the values of x are less than x_0.
    c. 1% of the values of x are greater than x_0.

5.34 Suppose x is a normally distributed random variable with mean 100 and standard deviation 8. Draw a rough graph of the distribution of x. Locate μ and the interval $\mu \pm 2\sigma$ on the graph. Find the following probabilities:
    a. $P(\mu - 3\sigma \leq x \leq \mu + 3\sigma)$
    b. $P(x \geq \mu + 3\sigma)$
    c. $P(x \leq 84)$
    d. $P(84 \leq x \leq 124)$
    e. $P(84 \leq x \leq 96)$

5.35 The random variable x has a normal distribution with standard deviation 20. It is known that the probability that x exceeds 150 is .90. Find the mean μ of the probability distribution.

5.36 The random variable x has a normal distribution with $\mu = 200$ and $\sigma = 20$.
    a. Find the probability that x assumes a value more than two standard deviations from its mean. More than three standard deviations from μ.
    b. Find the probability that x assumes a value within one standard deviation of its mean. Within two standard deviations of μ.
    c. Find the value of x that represents the 80th percentile of this distribution. The 10th percentile.

Ⓓ Applet Exercise 5.3

Open the applet entitled *Sample from a Population*. On the pull-down menu to the right of the top graph, select *Bell shaped*. The box to the left of the top graph displays the population mean, median, and standard deviation.

a. Run the applet for each available value of n on the pull-down menu for the sample size. Go from the smallest to the largest value of n. For each value of n, observe the shape of the graph of the sample data and record the mean, median, and standard deviation of the sample.

b. Describe what happens to the shape of the graph and the mean, median, and standard deviation of the sample as the sample size increases.

Applying the Concepts—Basic

5.37 **Dental anxiety study.** To gauge their fear of going to a dentist, a random sample of adults completed the Modified Dental Anxiety Scale questionnaire (*BMC Oral Health*, Vol. 9, 2009). Scores on the scale range from zero (no anxiety) to 25 (extreme anxiety). The mean score was 11 and the standard deviation was 4. Assume that the distribution of all scores on the Modified Dental Anxiety Scale is approximately normal with $\mu = 11$ and $\sigma = 4$.

a. Suppose you score a 10 on the Modified Dental Anxiety Scale. Find the z-value for your score.

b. Find the probability that someone scores between 10 and 15 on the Modified Dental Anxiety Scale.

c. Find the probability that someone scores above 20 on the Modified Dental Anxiety Scale.

5.38 **Tomato as a taste modifier.** Miraculin—a protein naturally produced in a rare tropical fruit—can convert a sour taste into a sweet taste; thus, it has the potential to be an alternative low-calorie sweetener. In *Plant Science* (May 2010), a group of Japanese environmental scientists investigated the ability of a hybrid tomato plant to produce miraculin. For a particular generation of the tomato plant, the amount x of miraculin produced (measured in micrograms per gram of fresh weight) had a mean of 105.3 and a standard deviation of 8.0. Assume that x is normally distributed.

a. Find $P(x > 120)$.

b. Find $P(100 < x < 110)$.

c. Find the value a for which $P(x < a) = .25$.

5.39 **Deep mixing of soil.** Deep mixing is a ground improvement method developed for soft soils like clay, silt, and peat. Swedish civil engineers investigated the properties of soil improved by deep mixing with lime-cement columns in the journal *Giorisk* (Vol. 7, 2013). The mixed soil was tested by advancing a cylindrical rod with a cone tip down into the soil. During penetration, the cone penetrometer measures the cone tip resistance (megapascals, MPa). The researchers established that tip resistance for the deep mixed soil followed a normal distribution with $\mu = 2.2$ MPa and $\sigma = .9$ MPa.

a. Find the probability that the tip resistance will fall between 1.3 and 4.0 MPa.

b. Find the probability that the tip resistance will exceed 1.0 MPa.

c. Find a value of tip resistance, T, such that 35% of all soil samples have tip resistance values that exceed T.

5.40 **Casino gaming.** In *Chance* (Spring 2005), University [NW] of Denver statistician R. C. Hannum discussed casino gaming and the laws of probability. Casino games of pure chance (e.g., craps, roulette, baccarat, and keno) always yield a "house advantage." For example, in the game of double-zero roulette, the expected casino win percentage is 5.26% on bets made on whether the outcome will be either black or red. (This percentage implies that for every $5 bet on black or red, the casino will earn a net of about 25 cents.) It can be shown that in 100 roulette plays on black/red, the average casino win percentage is normally distributed with mean 5.26% and standard deviation 10%. Let x represent the average casino win percentage after 100 bets on black/red in double-zero roulette.

a. Find $P(x > 0)$. (This is the probability that the casino wins money.)

b. Find $P(5 < x < 15)$.

c. Find $P(x < 1)$.

d. If you observed an average casino win percentage of -25% after 100 roulette bets on black/red, what would you conclude?

5.41 **Shopping vehicle and judgment.** Refer to the *Journal of Marketing Research* (Dec. 2011) study of whether you are more likely to choose a vice product (e.g., a candy bar) when your arm is flexed (as when carrying a shopping basket) than when your arm is extended (as when pushing a shopping cart), Exercise 2.111 (p. 106). The study measured choice scores (on a scale of 0 to 100, where higher scores indicate a greater preference for vice options) for consumers shopping under each of the two conditions. Recall that the average choice score for consumers with a flexed arm was 59, while the average for consumers with an extended arm was 43. For both conditions, assume that the standard deviation of the choice scores is 5. Also assume that both distributions are approximately normal.

a. In the flexed arm condition, what is the probability that a consumer has a choice score of 60 or greater?

b. In the extended arm condition, what is the probability that a consumer has a choice score of 60 or greater?

5.42 **Shell lengths of sea turtles.** Refer to the *Aquatic Biology* (Vol. 9, 2010) study of green sea turtles inhabiting the Grand Cayman South Sound lagoon, Exercise 2.85 (p. 97). Researchers discovered that the curved carapace (shell) length of these turtles is approximately normally distributed with mean 55.7 centimeters and standard deviation 11.5 centimeters.

a. The minimum and maximum size limits for captured turtles in the legal marine turtle fishery are 40 cm and 60 cm, respectively. How likely are you to capture a green sea turtle that is considered illegal?

[NW] **b.** What maximum limit, L, should be set so that only 10% of the turtles captured have shell lengths greater than L?

5.43 **Flicker in an electrical power system.** An assessment of the quality of the electrical power system in Turkey was the topic of an article published in *Electrical Engineering* (Mar. 2013). One measure of quality is the degree to which voltage fluctuations cause light flicker in the system. The perception of light flicker x when the system is set at 380 kV was measured periodically (over 10-minute intervals). For transformers supplying heavy industry plants, the light flicker distribution was found to follow (approximately) a normal distribution with $\mu = 2.2\%$ and $\sigma = .5\%$. If the perception of light flicker exceeds 3%, the transformer is shut down and the system is reset. How likely is it for a transformer supplying a heavy industry plant to be shut down due to light flicker?

Applying the Concepts—Intermediate

5.44 **Optimal goal target in soccer.** When attempting to score a goal in soccer, where should you aim your shot? Should you aim for a goalpost (as some soccer coaches teach), the middle of the goal, or some other target? To answer these questions, *Chance* (Fall 2009) utilized the normal probability distribution. Suppose the accuracy *x* of a professional soccer player's shots follows a normal distribution with a mean of zero feet and a standard deviation of 3 feet. (For example, if the player hits his target, $x = 0$; if he misses his target by 2 feet to the right, $x = 2$; and if he misses 1 foot to the left, $x = -1$.) Now, a regulation soccer goal is 24 feet wide. Assume that a goalkeeper will stop (save) all shots within 9 feet of where he is standing; all other shots on goal will score. Consider a goalkeeper who stands in the middle of the goal.

a. If the player aims for the right goalpost, what is the probability that he will score?

b. If the player aims for the center of the goal, what is the probability that he will score?

c. If the player aims for halfway between the right goalpost and the outer limit of the goalkeeper's reach, what is the probability that he will score?

5.45 **Voltage sags and swells.** Refer to the *Electrical Engineering* (Vol. 95, 2013) study of the power quality of a transformer, Exercise 2.127 (p. 110). Recall that two causes of poor power quality are "sags" and "swells." (A sag is an unusual dip and a swell is an unusual increase in the voltage level of a transformer.) For Turkish transformers built for heavy industry, the mean number of sags per week was 353 and the mean number of swells per week was 184. As in Exercise 2.127, assume the standard deviation of the sag distribution is 30 sags per week and the standard deviation of the swell distribution is 25 swells per week. Also, assume that the number of sags and number of swells are both normally distributed. Suppose one of the transformers is randomly selected and found to have 400 sags and 100 swells in a week.

a. What is the probability that the number of sags per week is less than 400?

b. What is the probability that the number of swells per week is greater than 100?

5.46 **Safety of underground tunnels.** Research published in the journal *Tunnelling and Underground Space Technology* (July 2014) evaluated the safety of underground tunnels built in rigid soils. A Factor of Safety (FS), measured as the ratio of capacity over demand, was determined for three different areas of tunnels made from shotcrete: tunnel face, tunnel walls, and tunnel crown. FS was determined to be normally distributed in each area, with means and standard deviations shown in the table. Tunnel failure is considered to occur when FS is lower than or equal to 1. Which tunnel area is more likely to result in failure? Why?

	Mean (μ)	Standard Deviation (σ)
Tunnel Face	1.2	.16
Tunnel Walls	1.4	.20
Tunnel Crown	2.1	.70

5.47 **Ambulance response time.** Ambulance response time is measured as the time (in minutes) between the initial call to emergency medical services (EMS) and when the patient is reached by ambulance. *Geographical Analysis* (Vol. 41, 2009) investigated the characteristics of ambulance response time for EMS calls in Edmonton, Alberta. For a particular EMS station (call it Station A), ambulance response time is known to be normally distributed with $\mu = 7.5$ minutes and $\sigma = 2.5$ minutes.

a. Regulations require that 90% of all emergency calls should be reached in 9 minutes or less. Are the regulations met at EMS Station A? Explain.

b. A randomly selected EMS call in Edmonton has an ambulance response time of 2 minutes. Is it likely that this call was serviced by Station A? Explain.

5.48 **Rating employee performance.** Almost all companies utilize some type of year-end performance review for their employees. Human Resources (HR) at the University of Texas Health Science Center provides guidelines for supervisors rating their subordinates. For example, raters are advised to examine their ratings for a tendency to be either too lenient or too harsh. According to HR, "if you have this tendency, consider using a normal distribution—10% of employees (rated) exemplary, 20% distinguished, 40% competent, 20% marginal, and 10% unacceptable." Suppose you are rating an employee's performance on a scale of 1 (lowest) to 100 (highest). Also, assume the ratings follow a normal distribution with a mean of 100 and a standard deviation of 30.

a. What is the lowest rating you should give to an "exemplary" employee if you follow the University of Texas HR guidelines?

b. What is the lowest rating you should give to a "competent" employee if you follow the University of Texas HR guidelines?

5.49 **California's electoral college votes.** During a presidential election, each state is allotted a different number of votes to the electoral college depending on population. For example, California is allotted 55 votes (the most) while several states (including the District of Columbia) are allotted 3 votes each (the least). When a presidential candidate wins the popular vote in a state, the candidate wins all the electoral college votes in that state. To become president, a candidate must win 270 of the total of 538 votes in the electoral college. *Chance* (Winter 2010) demonstrated the impact of winning California on the presidential election. Assuming a candidate wins California's 55 votes, the number of additional electoral college votes the candidate will win can be approximated by a normal distribution with $\mu = 241.5$ votes and $\sigma = 49.8$ votes. If a presidential candidate wins the popular vote in California, what are the chances that he or she becomes the next U.S. president?

5.50 **Alcohol, threats, and electric shocks.** A group of Florida State University psychologists examined the effects of alcohol on the reactions of people to a threat (*Journal of Abnormal Psychology*, Vol. 107, 1998). After obtaining a specified blood alcohol level, the psychologists placed experimental subjects in a room and threatened them with electric shocks. Using sophisticated equipment to monitor the subjects' eye movements, the psychologists recorded the startle response (measured in milliseconds) of each subject. The mean and standard deviation of the startle responses were 37.9 and 12.4, respectively. Assume that the startle response x for a person with the specified blood alcohol level is approximately normally distributed.

a. Find the probability that x is between 40 and 50 milliseconds.

b. Find the probability that x is less than 30 milliseconds.

c. Give an interval for x centered around 37.9 milliseconds so that the probability that x falls in the interval is .95.

d. Ten percent of the experimental subjects have startle responses above what value?

Applying the Concepts—Advanced

5.51 **Box plots and the standard normal distribution.** What relationship exists between the standard normal distribution and the box-plot methodology (Section 2.7) for describing distributions of data by means of quartiles? The answer depends on the true underlying probability distribution of the data. Assume for the remainder of this exercise that the distribution is normal.

a. Calculate the values z_L and z_U of the standard normal random variable z that correspond, respectively, to the hinges of the box plot (i.e., the lower and upper quartiles Q_L and Q_U) of the probability distribution.

b. Calculate the z values that correspond to the inner fences of the box plot for a normal probability distribution.

c. Calculate the z values that correspond to the outer fences of the box plot for a normal probability distribution.

d. What is the probability that an observation lies beyond the inner fences of a normal probability distribution? The outer fences?

e. Can you now better understand why the inner and outer fences of a box plot are used to detect outliers in a distribution? Explain.

5.52 **Load on frame structures.** In the *Journal of the International Association for Shell and Spatial Structures* (Apr. 2004), Japanese environmental researchers studied the performance of truss-and-frame structures subjected to uncertain loads. The load was assumed to have a normal distribution with a mean of 20 thousand pounds. Also, the probability that the load is between 10 and 30 thousand pounds is .95. On the basis of this information, find the standard deviation of the load distribution.

5.53 **Range of women's heights.** In *Chance* (Winter 2007), Yale Law School professor Ian Ayres published the results of a study he conducted with his son and daughter on

whether college students could estimate a range for women's heights. The students were shown a graph of a normal distribution of heights and were asked, "The average height of women over 20 years old in the United States is 64 inches. Using your intuition, please give your best estimate of the range of heights that would include $C\%$ of women over 20 years old. Please make sure that the center of the range is the average height of 64 inches." The value of C was randomly selected as 50%, 75%, 90%, 95%, or 99% for each student surveyed.

a. Give your estimate of the range for $C = 50\%$ of women's heights.

b. Give your estimate of the range for $C = 75\%$ of women's heights.

c. Give your estimate of the range for $C = 90\%$ of women's heights.

d. Give your estimate of the range for $C = 95\%$ of women's heights.

e. Give your estimate of the range for $C = 99\%$ of women's heights.

f. The standard deviation of heights for women over 20 years old is known to be 2.6 inches. Use this information to revise your answers to parts **a–e**.

g. Which value of C has the most accurate estimated range? (*Note:* The researchers found that college students were most accurate for $C = 90\%$ and $C = 95\%$.)

5.54 **Executive coaching and meeting effectiveness.** Poor executive leadership during business meetings can result in counterproductive outcomes (e.g., negative employee attitudes, ambiguous objectives). Can executive coaching help improve business meeting effectiveness? This was the question of interest in an article published in *Consulting Psychology Journal: Practice and Research* (Vol. 61, 2009). Certain behaviors by leaders during meetings were categorized as content behaviors (e.g., seeking information, disagreeing/attacking) or process behaviors (e.g., asking clarifying questions, summarizing). The goal of executive coaching was to reduce content behaviors in favor of process behaviors. The study reported that prior to receiving executive coaching, the percentage of observed content behaviors of leaders has a mean of 75% with a standard deviation of 8.5%. In contrast, after receiving executive coaching, the percentage of observed content behaviors of leaders has a mean of 52% with a standard deviation of 7.5%. Assume that the percentage of observed content behaviors is approximately normally distributed for both leaders with and without executive coaching. Suppose you observe 70% content behaviors by the leader of a business meeting. Give your opinion on whether the leader has received executive coaching.

5.4 Descriptive Methods for Assessing Normality

In the chapters that follow, we learn how to make inferences about the population on the basis of information contained in the sample. Several of these techniques are based on the assumption that the population is approximately normally distributed. Consequently, it will be important to determine whether the sample data come from a normal population before we can apply these techniques properly.

A number of descriptive methods can be used to check for normality. In this section, we consider the four methods summarized in the following box:

Determining whether the Data Are from an Approximately Normal Distribution

1. Construct either a histogram or stem-and-leaf display for the data, and note the shape of the graph. If the data are approximately normal, the shape of the histogram or stem-and-leaf display will be similar to the normal curve shown in Figure 5.6 (i.e., the display will be mound shaped and symmetric about the mean).

2. Compute the intervals $\bar{x} \pm s$, $\bar{x} \pm 2s$, and $\bar{x} \pm 3s$, and determine the percentage of measurements falling into each. If the data are approximately normal, the percentages will be approximately equal to 68%, 95%, and 100%, respectively.

3. Find the interquartile range IQR and standard deviation s for the sample, and then calculate the ratio IQR/s. If the data are approximately normal, then IQR/$s \approx 1.3$.

4. Construct a *normal probability plot* for the data. If the data are approximately normal, the points will fall (approximately) on a straight line.

The first two methods come directly from the properties of a normal distribution established in Section 5.3. Method 3 is based on the fact that for normal distributions, the z values corresponding to the 25th and 75th percentiles are $-.67$ and $.67$, respectively. (See Example 5.5.) Since $\sigma = 1$ for a standard normal distribution,

$$\frac{\text{IQR}}{\sigma} = \frac{Q_U - Q_L}{\sigma} = \frac{.67 - (-.67)}{1} = 1.34$$

The final descriptive method for checking normality is based on a *normal probability plot*. In such a plot, the observations in a data set are ordered from smallest to largest and are then plotted against the expected z-scores of observations calculated under the assumption that the data come from a normal distribution. When the data are, in fact, normally distributed, a linear (straight-line) trend will result. A nonlinear trend in the plot suggests that the data are nonnormal.

A **normal probability plot** for a data set is a scatterplot with the ranked data values on one axis and their corresponding expected z-scores from a standard normal distribution on the other axis. [*Note:* Computation of the expected standard normal z-scores is beyond the scope of this text. Therefore, we will rely on available statistical software packages to generate a normal probability plot.]

Example 5.12
EPAGAS

Checking for Normal Data—EPA Estimated Gas Mileages

Problem The EPA mileage ratings on 100 cars, first presented in Chapter 2 (p. 71), are reproduced in Table 5.2. Recall that these data are saved in the **EPAGAS** file. Numerical and graphical descriptive measures for the data are shown on the MINITAB and SPSS printouts presented in Figure 5.21a–c. Determine whether the EPA mileage ratings are from an approximate normal distribution.

Solution As a first check, we examine the MINITAB histogram of the data shown in Figure 5.21a. Clearly, the mileages fall into an approximately mound shaped, symmetric distribution centered around the mean of about 37 mpg. Note that a normal curve is superimposed on the figure. Therefore, check #1 in the box indicates that the data are approximately normal.

Table 5.2		EPA Gas Mileage Ratings for 100 Cars (miles per gallon)							
36.3	41.0	36.9	37.1	44.9	36.8	30.0	37.2	42.1	36.7
32.7	37.3	41.2	36.6	32.9	36.5	33.2	37.4	37.5	33.6
40.5	36.5	37.6	33.9	40.2	36.4	37.7	37.7	40.0	34.2
36.2	37.9	36.0	37.9	35.9	38.2	38.3	35.7	35.6	35.1
38.5	39.0	35.5	34.8	38.6	39.4	35.3	34.4	38.8	39.7
36.3	36.8	32.5	36.4	40.5	36.6	36.1	38.2	38.4	39.3
41.0	31.8	37.3	33.1	37.0	37.6	37.0	38.7	39.0	35.8
37.0	37.2	40.7	37.4	37.1	37.8	35.9	35.6	36.7	34.5
37.1	40.3	36.7	37.0	33.9	40.1	38.0	35.2	34.8	39.5
39.9	36.9	32.9	33.8	39.8	34.0	36.8	35.0	38.1	36.9

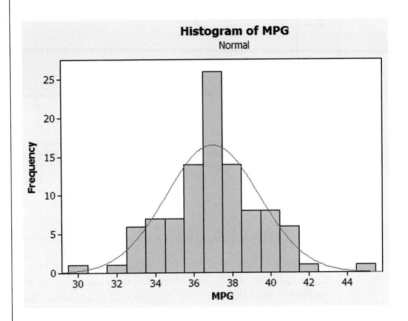

Figure 5.21a

MINITAB histogram for gas mileage data

Figure 5.21b

MINITAB descriptive statistics for gas mileage data

Descriptive Statistics: MPG

Variable	N	Mean	StDev	Minimum	Q1	Median	Q3	Maximum
MPG	100	36.994	2.418	30.000	35.625	37.000	38.375	44.900

To apply check #2, we obtain $\bar{x} = 37$ and $s = 2.4$ from the MINITAB printout of Figure 5.21b. The intervals $\bar{x} \pm s, \bar{x} \pm 2s,$ and $\bar{x} \pm 3s$ are shown in Table 5.3, as is the percentage of mileage ratings that fall into each interval. These percentages agree almost exactly with those from a normal distribution.

Check #3 in the box requires that we find the ratio IQR/s. From Figure 5.21b, the 25th percentile (labeled Q_1 by MINITAB) is $Q_L = 35.625$ and the 75th percentile (labeled Q_3 by MINITAB) is $Q_U = 38.375$. Then IQR $= Q_U - Q_L = 2.75$, and the ratio is

$$\frac{\text{IQR}}{s} = \frac{2.75}{2.4} = 1.15$$

Since this value is approximately equal to 1.3, we have further confirmation that the data are approximately normal.

A fourth descriptive method is to interpret a normal probability plot. An SPSS normal probability plot of the mileage data is shown in Figure 5.21c. Notice that the ordered mileage values (shown on the horizontal axis) fall reasonably close to a straight line when plotted against the expected values from a normal distribution. Thus, check #4 also suggests that the EPA mileage data are approximately normally distributed.

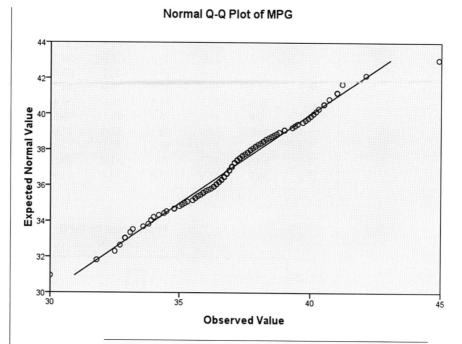

Figure 5.21c
SPSS normal probability plot for gas mileage data

Table 5.3   Describing the 100 EPA Mileage Ratings

Interval	Percentage in Interval
$\bar{x} \pm s = (34.6, 39.4)$	68
$\bar{x} \pm 2s = (32.2, 41.8)$	96
$\bar{x} \pm 3s = (29.8, 44.2)$	99

Look Back The checks for normality given in the box on p. 282 are simple yet powerful techniques to apply, but they are only descriptive in nature. It is possible (although unlikely) that the data are nonnormal even when the checks are reasonably satisfied. Thus, we should be careful not to claim that the 100 EPA mileage ratings are, in fact, normally distributed. We can only state that it is reasonable to believe that the data are from a normal distribution.*

■ **Now Work Exercise 5.62**

As we will learn in the next chapter, several inferential methods of analysis require the data to be approximately normal. If the data are clearly nonnormal, inferences derived from the method may be invalid. Therefore, it is advisable to check the normality of the data prior to conducting any analysis.

Statistics IN Action **Revisited** ## Assessing whether the Normal Distribution Is Appropriate for Modeling the Super Weapon Hit Data

In *Statistics in Action Revisited* in Section 5.3, we used the normal distribution to find the probability that a single flechette from a super weapon that shoots 1,100 flechettes at once hits one of three targets at 500 meters. Recall that for three range tests, the weapon was always aimed at the center target (i.e., the specification mean was set at $\mu = 5$ feet), but the specification standard deviation was varied at $\sigma = 1$ foot, $\sigma = 2$ feet, and $\sigma = 4$ feet. Table SIA5.1 shows the calculated normal probabilities of hitting the three targets for the different values of σ, as well as the actual results of the three range tests. (Recall that the actual data are saved in the **MOAGUN** file.) You can see that the proportion of the 1,100 flechettes that actually hit each target—called the hit ratio—agrees very well with the estimated probability of a hit derived from the normal distribution.

---

*Statistical tests of normality that provide a measure of reliability for the inference are available. However, these tests tend to be very sensitive to slight departures from normality (i.e., they tend to reject the hypothesis of normality for any distribution that is not perfectly symmetrical and mound shaped). Consult the references (see especially Ramsey & Ramsey, 1990) if you want to learn more about these tests.

Table SIA5.1 Summary of Normal Probability Calculations and Actual Range Test Results

Target	Specification	Normal Probability	Actual Number of Hits	Hit Ratio (Hits/1,100)
LEFT (−1 to 1)	$\sigma = 1$	.0000	0	.000
	$\sigma = 2$	.0214	30	.027
	$\sigma = 4$	.0919	73	.066
MIDDLE (4 to 6)	$\sigma = 1$	.6826	764	.695
	$\sigma = 2$	.3820	409	.372
	$\sigma = 4$	.1974	242	.220
RIGHT (9 to 11)	$\sigma = 1$	.0000	0	.000
	$\sigma = 2$	.0214	23	.021
	$\sigma = 4$	.0919	93	.085

Consequently, it appears that our assumption that the horizontal hit measurements are approximately normally distributed is reasonably satisfied. Further evidence that this assumption is satisfied is provided by the MINITAB histograms of the horizontal hit measurements shown in Figures SIA5.3a–c. The normal curves superimposed on the histograms fit the data very well.

 Data Set: MOAGUN

Figure SIA5.3a
MINITAB histogram for the horizontal hit measurements when $\sigma = 1$

Figure SIA5.3b
MINITAB histogram for the horizontal hit measurements when $\sigma = 2$

Figure SIA5.3c
MINITAB histogram for the horizontal hit measurements when $\sigma = 4$

Exercises 5.55–5.76

Understanding the Principles

5.55 How do you construct a normal probability plot of a data set from a normal population?

5.56 Give four methods for determining whether the sample data come from a normal population.

5.57 If a population data set is normally distributed, what is the proportion of measurements you would expect to fall within the following intervals?
  a. $\mu \pm .5\sigma$
  b. $\mu \pm 2.5\sigma$
  c. $\mu \pm 1.25\sigma$

5.58 How do you use IQR to check that the data came from a standard normal distribution?

Learning the Mechanics

5.59 Normal probability plots for three data sets are shown at the bottom of the page. Which plot indicates that the data are approximately normally distributed?

5.60 Consider a sample data set with the following summary statistics: $s = 190$, $Q_L = 144$, and $Q_U = 390$.
  a. Calculate IQR.
  b. Calculate IQR/s.
  c. Is the value of IQR/s approximately equal to 1.3? What does this imply?

5.61 Examine the following sample data.

L05061

32	48	25	135	53	37	5	39	213	165
109	40	1	146	39	25	21	66	64	57
197	60	112	10	155	134	301	304	107	82
35	81	60	95	401	308	180	3	200	59

  a. Construct a stem-and-leaf plot to assess whether the data are from an approximately normal distribution.
  b. Find the values of Q_L, Q_U, and s for the sample data.
  c. Use the results from part **b** to assess the normality of the data.
  d. Generate a normal probability plot for the data, and use it to assess whether the data are approximately normal.

5.62 Examine the sample data in the next column.
  [NW] **a.** Construct a stem-and-leaf plot to assess whether the data are from an approximately normal distribution.
   L05062 **b.** Compute s for the sample data.

5.9	5.3	1.6	7.4	8.6	3.2	2.1
4.0	7.3	8.4	5.9	6.7	4.5	6.3
6.0	9.7	3.5	3.1	4.3	3.3	8.4
4.6	8.2	6.5	1.1	5.0	9.4	6.4

  c. Find the values of Q_L and Q_U and the value of s from part **b** to assess whether the data come from an approximately normal distribution.
  d. Generate a normal probability plot for the data, and use it to assess whether the data are approximately normal.

Applying the Concepts—Basic

5.63 **Irrelevant speech effects.** Refer to the analysis of irrelevant speech effects in Exercise 2.34 (p. 77). Recall that irrelevant speech effects refer to the degree to which the memorization process is impaired by irrelevant background speech. In a study published in *Acoustical Science & Technology* (Vol. 35, 2014), subjects performed a memorization task under two conditions: (1) with irrelevant background speech and (2) in silence. The difference in the error rates for the two conditions—called the relative difference in error rate (RDER)—was computed for each subject. A MINITAB histogram and descriptive statistics pintout summarizing the RDER values for 71 subjects are displayed below and on the next page. Would you recommend using a normal distribution to model the distribution of RDER values? Explain.

Plots for Exercise 5.59

MINITAB output for Exercise 5.63

Descriptive Statistics: RDER

Variable	N	Mean	StDev	Minimum	Q1	Median	Q3	Maximum	IQR
RDER	71	78.19	63.24	-19.27	37.81	60.31	102.80	254.11	64.99

5.64 **Characteristics of antiwar demonstrators.** Refer to the *American Journal of Sociology* (Jan. 2014) study of the characteristics of antiwar demonstrators in the United States, Exercise 2.106 (p. 105). Based on data collected for over 5,000 antiwar demonstrators over a recent 3-year period, the researchers found that the mean number of protest organizations joined by the demonstrators was .90 with a standard deviation of 1.10 and a median of 1. Explain why x, the number of protest organizations joined by a randomly selected demonstrator, cannot be exactly normally distributed.

5.65 **Software file updates.** Software configuration management was used to monitor a software engineering team's performance at Motorola, Inc. (*Software Quality Professional*, Nov. 2004). One of the variables of interest was the number of updates to a file changed because of a problem report. Summary statistics for $n = 421$ files yielded the following results: $\bar{x} = 4.71, s = 6.09, Q_L = 1$, and $Q_U = 6$. Are these data approximately normally distributed? Explain.

5.66 **Shear strength of rock fractures.** Understanding the characteristics of rock masses, especially the nature of the fractures, are essential when building dams and power plants. The shear strength of rock fractures was investigated in *Engineering Geology* (May 12, 2010). The Joint Roughness Coefficient (JRC) was used to measure shear strength. Civil engineers collected JRC data for over 750 rock fractures. The results (simulated from information provided in the article) are summarized in the SPSS histogram shown below. Should the engineers use the normal probability distribution to model the behavior of shear strength for rock fractures? Explain.

Based on Pooyan Asadollahi and Fulvio Tonon, "Constitutive model for rock fractures: Revisiting Barton's empirical model." *Engineering Geology*, Vol. 113, no. 1, pp. 11–32.

5.67 **Estimating glacier elevations.** Digital elevation models (DEMs) are now used to estimate elevations and slopes of remote regions. In *Arctic, Antarctic, and Alpine Research* (May 2004), geographers analyzed reading errors from maps produced by DEMs. Two readers of a DEM map of White Glacier (in Canada) estimated elevations at 400 points in the area. The difference between the elevation estimates of the two readers had a mean of $\mu = .28$ meter and a standard deviation of $\sigma = 1.6$ meters. A histogram of the difference (with a normal histogram superimposed on the graph) is shown below.

a. On the basis of the histogram, the researchers concluded that the difference between elevation estimates is not normally distributed. Why?

b. Will the interval $\mu \pm 2\sigma$ contain more than 95%, exactly 95%, or less than 95% of the 400 elevation differences? Explain.

"Uncertainty in digital elevation models of Axel Heiberg Island. Arctic Canada," *Arctic, Antarctic, and Alpine Research*, Vol. 36, No. 2, May 2004 (Figure 3). © Regents of the University of Colorado. Reprinted with permission.

5.68 **Drug content assessment.** Scientists at GlaxoSmithKline Medicines Research Center used high-performance liquid chromatography (HPLC) to determine the amount of drug in a tablet produced by the company (*Analytical Chemistry*, Dec. 15, 2009). Drug concentrations (measured as a percentage) for 50 randomly selected tablets are listed in the accompanying table.

91.28	92.83	89.35	91.90	82.85	94.83	89.83	89.00	84.62
86.96	88.32	91.17	83.86	89.74	92.24	92.59	84.21	89.36
90.96	92.85	89.39	89.82	89.91	92.16	88.67	89.35	86.51
89.04	91.82	93.02	88.32	88.76	89.26	90.36	87.16	91.74
86.12	92.10	83.33	87.61	88.20	92.78	86.35	93.84	91.20
93.44	86.77	83.77	93.19	81.79				

Based on Borman, P. J., et al., "Design and analysis of method equivalence studies." *Analytical Chemistry*, Vol. 81, No. 24, Dec. 15, 2009 (Table 3).

a. Descriptive statistics for the drug concentrations are shown at the top of the following SPSS printout on p. 288.

SPSS Output for Exercise 5.68

Descriptives

			Statistic	Std. Error
Content	Mean		89.2906	.45021
	95% Confidence Interval for Mean	Lower Bound	88.3859	
		Upper Bound	90.1953	
	5% Trimmed Mean		89.3963	
	Median		89.3750	
	Variance		10.134	
	Std. Deviation		3.18344	
	Minimum		81.79	
	Maximum		94.83	
	Range		13.04	
	Interquartile Range		4.84	
	Skewness		-.544	.337
	Kurtosis		-.389	.662

Use this information to assess whether the data are approximately normal.

b. An SPSS normal probability plot is shown above. Use this information to assess whether the data are approximately normal.

Applying the Concepts—Intermediate

5.69 **Mineral flotation in water study.** Refer to the *Minerals Engineering* (Vol. 46–47, 2013) study of the impact of calcium and gypsum on the flotation properties of silica in water, Exercise 2.48 (p. 81). Recall that 50 solutions of deionized water were prepared both with and without calcium/gypsum, and the level of flotation of silica in the solution was measured using a variable called *zeta potential* (measured in millivolts, mV). The data (simulated, based on information provided in the journal article) are reproduced in the tables at the bottom of the page. Which of the two zeta potential distributions, without calcium/gypsum or with calcium/gypsum, can be approximated by a normal distribution?

5.70 **Habitats of endangered species.** An evaluation of the habitats of endangered salmon species was performed in *Conservation Ecology* (Dec. 2003). The researchers identified 734 sites (habitats) for Chinook, coho, or steelhead salmon species in Oregon and assigned a habitat quality score to each site. (Scores range from 0 to 36 points, with lower scores indicating poorly maintained or degraded habitats.) The data are saved in the **HABITAT** file. Give your opinion on whether the habitat quality score is normally distributed.

5.71 **Baseball batting averages.** Major League Baseball (MLB) has two leagues: the American League (AL), which utilizes the designated hitter (DH) to bat for the pitcher, and the National League (NL), which does not allow the DH. A player's batting average is computed by dividing the player's total number of hits by his official number of at bats. The batting averages for all AL and NL players with at least 100 official at bats during the 2013 season are stored in the **MLBAL** and **MLBNL** files, respectively. Determine whether each batting average distribution is approximately normal.

5.72 **Ranking the driving performance of professional golfers.** Refer to *The Sport Journal* (Winter 2007) article on a new method for ranking the driving performance of PGA golfers, presented in Exercise 2.66 (p. 90). Recall that the method incorporates a golfer's average driving distance

Data for Exercise 5.69

Without calcium/gypsum

−47.1	−53.0	−50.8	−54.4	−57.4	−49.2	−51.5	−50.2	−46.4	−49.7
−53.8	−53.8	−53.5	−52.2	−49.9	−51.8	−53.7	−54.8	−54.5	−53.3
−50.6	−52.9	−51.2	−54.5	−49.7	−50.2	−53.2	−52.9	−52.8	−52.1
−50.2	−50.8	−56.1	−51.0	−55.6	−50.3	−57.6	−50.1	−54.2	−50.7
−55.7	−55.0	−47.4	−47.5	−52.8	−50.6	−55.6	−53.2	−52.3	−45.7

With calcium/gypsum

−9.2	−11.6	−10.6	−8.0	−10.9	−10.0	−11.0	−10.7	−13.1	−11.5
−11.3	−9.9	−11.8	−12.6	−8.9	−13.1	−10.7	−12.1	−11.2	−10.9
−9.1	−12.1	−6.8	−11.5	−10.4	−11.5	−12.1	−11.3	−10.7	−12.4
−11.5	−11.0	−7.1	−12.4	−11.4	−9.9	−8.6	−13.6	−10.1	−11.3
−13.0	−11.9	−8.6	−11.3	−13.0	−12.2	−11.3	−10.5	−8.8	−13.4

(yards) and driving accuracy (percentage of drives that land in the fairway) into a driving performance index. Data on these three variables for the top 40 PGA golfers are saved in the **PGA** file. Determine which of the variables—driving distance, driving accuracy, and driving performance index—are approximately normally distributed.

5.73 Permeability of sandstone during weathering. Refer to the *Geographical Analysis* (Vol. 42, 2010) study of the decay properties of sandstone when exposed to the weather, Exercise 2.69 (p. 91). Recall that blocks of sandstone were cut into 300 equal-sized slices and the slices randomly divided into three groups of 100 slices each. Slices in group A were not exposed to any type of weathering; slices in group B were repeatedly sprayed with a 10% salt solution (to simulate wetting by driven rain) under temperate conditions; and slices in group C were soaked in a 10% salt solution and then dried (to simulate blocks of sandstone exposed during a wet winter and dried during a hot summer). All sandstone slices were then tested for permeability, measured in milliDarcies (mD). The data for the study (simulated) are saved in the **SAND** file. Is it plausible to assume that the permeability measurements in any of the three experimental groups are approximately normally distributed?

5.74 Cruise ship sanitation scores. Refer to the data on the Aug. 2013 sanitation scores for 186 cruise ships, presented in Exercise 2.41 (p. 79). The data are saved in the **SANIT** file. Assess whether the sanitation scores are approximately normally distributed.

Applying the Concepts—Advanced

5.75 Blond hair types in the Southwest Pacific. Refer to the *American Journal of Physical Anthropology* (Apr. 2014) study of a mutation of blond-hair genotypes, Exercise 2.141

(p. 119). Recall that for each of 550 Southwest Pacific islanders, the effect of the mutation on hair pigmentation was measured with the melanin (M) index, where M ranges between −4 and 4. Box plots showing the distribution of M for three different genotypes—CC, CT, and TT—are reproduced below (left column). Use the guidelines of this section and Exercise 5.51 (p. 281) to determine whether any of these distributions are normally distributed.

5.76 Ranking Ph.D. programs in economics. Refer to the *Southern Economic Journal* (Apr. 2008) rankings of Ph.D. programs in economics at 129 colleges and universities, Exercise 2.129 (p. 111). Recall that the number of publications published by faculty teaching in the Ph.D. program and the quality of the publications were used to calculate an overall productivity score for each program. The mean and standard deviation of these 129 productivity scores were then used to compute a z-score for each economics program. The data (z-scores) for all 129 economic programs are saved in the **ECOPHD** file. A MINITAB normal probability plot for the z-scores is shown in the accompanying printout.

a. Use the graph to assess whether the data are approximately normal.

b. Based on the graph, determine the nature of the skewness of the data.

Hair M by genotype

Probability Plot of Z-Score
Normal

5.5 Approximating a Binomial Distribution with a Normal Distribution (Optional)

When n is large, a normal probability distribution may be used to provide a good approximation to the probability distribution of a binomial random variable. To show how this approximation works, we refer to Example 4.13, in which we used the binomial distribution to model the number x of 20 eligible voters who favor a candidate. We assumed that 60% of all the eligible voters favored the candidate. The mean and standard deviation of x were found to be $\mu = 12$ and $\sigma = 2.2$, respectively. The binomial distribution for $n = 20$ and $p = .6$ is shown in Figure 5.22, and the approximating normal distribution with mean $\mu = 12$ and standard deviation $\sigma = 2.2$ is superimposed.

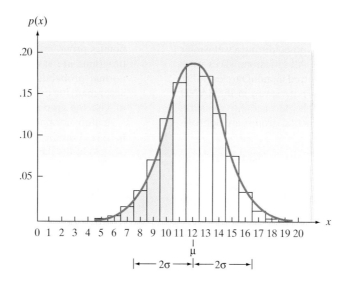

Figure 5.22

Binomial distribution for $n = 20, p = .6$ and normal distribution with $\mu = 12, \sigma = 2.2$

As part of Example 4.13, we used Table I to find the probability that $x \leq 10$. This probability, which is equal to the sum of the areas contained in the rectangles (shown in Figure 5.22) that correspond to $p(0), p(1), p(2), \ldots, p(10)$, was found to equal .245. The portion of the normal curve that would be used to approximate the area $p(0) + p(1) + p(2) + \cdots + p(10)$ is highlighted in Figure 5.22. *Note that this highlighted area lies to the left of 10.5 (not 10), so we may include all of the probability in the rectangle corresponding to p(10).* Because we are approximating a discrete distribution (the binomial) with a continuous distribution (the normal), we call the use of 10.5 (instead of 10 or 11) a **correction for continuity**. That is, we are correcting the discrete distribution so that it can be approximated by the continuous one. The use of the correction for continuity leads to the calculation of the following standard normal z-value:

$$z = \frac{x - \mu}{\sigma} = \frac{10.5 - 12}{2.2} = -.68$$

Using Table II, we find the area between $z = 0$ and $z = .68$ to be .2517. Then the probability that x is less than or equal to 10 is approximated by the area under the normal distribution to the left of 10.5, shown highlighted in Figure 5.22. That is,

$$P(x \leq 10) \approx P(z \leq -.68) = .5 - P(-.68 < z \leq 0) = .5 - .2517 = .2483$$

The approximation differs only slightly from the exact binomial probability, .245. Of course, when tables of exact binomial probabilities are available, we will use the exact value rather than a normal approximation.

The normal distribution will not always provide a good approximation to binomial probabilities. The following is a useful rule of thumb to determine when n is large enough for the approximation to be effective: *The interval $\mu \pm 3\sigma$ should lie within* *the range of the binomial random variable x (i.e., from 0 to n) in order for the normal approximation to be adequate.* The rule works well because almost all of the normal distribution falls within three standard deviations of the mean, so if this interval is contained within the range of x values, there is "room" for the normal approximation to work.

As shown in Figure 5.23a for the preceding example with $n = 20$ and $p = .6$, the interval $\mu \pm 3\sigma = 12 + 3(2.2) = (5.4, 18.6)$ lies within the range from 0 to 20. However, if we were to try to use the normal approximation with $n = 10$ and $p = .1$, the interval $\mu \pm 3\sigma$ becomes $1 \pm 3(.95)$, or $(-1.85, 3.85)$. As shown in Figure 5.23b, this interval is not contained within the range of x, since $x = 0$ is the lower bound for a binomial random variable. Note in Figure 5.23b that the normal distribution will not "fit" in the range of x; therefore, it will not provide a good approximation to the binomial probabilities.

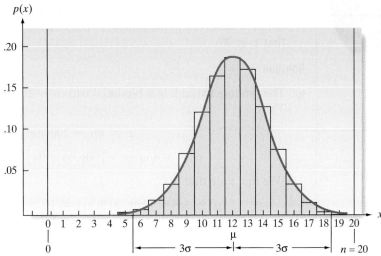

a. $n = 20, p = .6$: Normal approximation is good

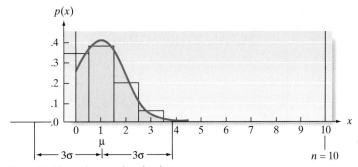

Figure 5.23

Rule of thumb for normal approximation to binomial probabilities

b. $n = 10, p = .1$: Normal approximation is poor

BIOGRAPHY ABRAHAM DE MOIVRE (1667–1754)

Advisor to Gamblers

French-born mathematician Abraham de Moivre moved to London when he was 21 years old to escape religious persecution. In England, he earned a living first as a traveling teacher of mathematics and then as an advisor to gamblers, underwriters, and annuity brokers. De Moivre's major contributions to probability theory are contained in two of his books: *The Doctrine of Chances* (1718) and *Miscellanea Analytica* (1730). In these works, he defines statistical independence, develops the formula for the normal probability distribution, and derives the normal curve as an approximation to the binomial distribution. Despite his eminence as a mathematician, de Moivre died in poverty. He is famous for using an arithmetic progression to predict the day of his death. ■

5.141 Galaxy velocity study. Recall *The Astronomical Journal* (July 1995) study of galaxy velocity, Exercise 5.125 (p. 302). The observed velocity of a galaxy located in galaxy cluster A2142 was found to have a normal distribution with a mean of 27,117 kilometers per second (km/s) and a standard deviation of 1,280 km/s. A galaxy with a velocity of 24,350 km/s is observed. Comment on the likelihood of this galaxy being located in cluster A2142.

5.142 **Parents who condone spanking. In Exercise 4.138 (p. 254), you learned that 60% of parents with young children condone spanking their child as a regular form of punishment (*Tampa Tribune*, Oct. 5, 2000). A child psychologist with 200 parent clients claims that no more than 40 of the parents condone spanking. Do you believe this claim? Explain.

5.143 **Victims of domestic abuse. In Exercise 4.79 (p. 241), you learned that some researchers believe that one in every three women has been a victim of domestic abuse (Domestic Violence: Incidence and Prevalence study, Sept.–Dec., 2005).

a. For a random sample of 240 women, what is the approximate probability that more than half are victims of domestic abuse?

b. For a random sample of 240 women, what is the approximate probability that fewer than 78 are victims of domestic abuse?

c. Would you expect to observe fewer than 50 domestically abused women in a sample of 240? Explain.

5.144 Mile run times. A physical-fitness association is including the mile run in its secondary-school fitness test for students. The time for this event is approximately normally distributed with a mean of 470 seconds and a standard deviation of 45 seconds. If the association wants to designate the fastest 10% of secondary-school students as "excellent," what time should the association set for this criterion?

5.145 Wear-out of used display panels. Wear-out failure time of electronic components is often assumed to have a normal distribution. Can the normal distribution be applied to the wear-out of used manufactured products, such as colored display panels? A lot of 50 used display panels was purchased by an outlet store. Each panel displays 12 to 18 color characters. Prior to acquisition, the panels had been used for about one-third of their expected lifetimes. The data in the table below give the failure times (in years) of the 50 used panels. Use the techniques of this chapter to determine whether the used panel wear-out times are approximately normally distributed.

5.146 Fitness of cardiac patients. The physical fitness of a patient is often measured by the patient's maximum oxygen uptake (recorded in milliliters per kilogram, ml/kg). The mean maximum oxygen uptake for cardiac patients who regularly participate in sports or exercise programs was found to be 24.1, with a standard deviation of 6.30 (*Adapted Physical*

Activity Quarterly, Oct. 1997). Assume that this distribution is approximately normal.

a. What is the probability that a cardiac patient who regularly participates in sports has a maximum oxygen uptake of at least 20 ml/kg?

b. What is the probability that a cardiac patient who regularly exercises has a maximum oxygen uptake of 10.5 ml/kg or lower?

c. Consider a cardiac patient with a maximum oxygen uptake of 10.5. Is it likely that this patient participates regularly in sports or exercise programs? Explain.

5.147 Forest development following wildfires. *Ecological Applications* (May 1995) published a study on the development of forests following wildfires in the Pacific Northwest. One variable of interest to the researcher was tree diameter at breast height 110 years after the fire.

a. The population of Douglas fir trees was shown to have an approximately normal diameter distribution with $\mu = 50$ centimeters (cm) and $\sigma = 12$ cm. Find the diameter d such that 30% of the Douglas fir trees in the population have diameters that exceed d.

b. Another species of tree, western hemlock, was found to have a breast height diameter distribution that resembled an exponential distribution with $\theta = 30$ centimeters. Find the probability that a western hemlock tree growing in the forest damaged by wildfire 110 years ago has a diameter that exceeds 25 centimeters.

5.148 Sedimentary deposits in reservoirs. Geologists have successfully used statistical models to evaluate the nature of sedimentary deposits (called *facies*) in reservoirs. One of the models' key parameters is the proportion P of facies bodies in a reservoir. An article in *Mathematical Geology* (Apr. 1995) demonstrated that the number of facies bodies that must be sampled to estimate P satisfactorily is approximately normally distributed with $\mu = 99$ and $\sigma = 4.3$. How many facies bodies are required to estimate P satisfactorily for 99% of the reservoirs evaluated?

5.149 Normal curve approximation. A. K. Shah published a simple approximation for areas under the normal curve in *The American Statistician* (Feb. 1985). Shah showed that the area A under the standard normal curve between 0 and z is

$$A \approx \begin{cases} z(4.4 - z)/10 & \text{for } 0 \leq z \leq 2.2 \\ .49 & \text{for } 2.2 < z < 2.6 \\ .50 & \text{for } z \geq 2.6 \end{cases}$$

a. Use Shah's approximation to find
i. $P(0 < z < 1.2)$
ii. $P(0 < z < 2.5)$
iii. $P(z > .8)$
iv. $P(z < 1.0)$

b. Find the exact probabilities in part **a**.

Data for Exercise 5.145

0.01	1.21	1.71	2.30	2.96	0.19	1.22	1.75	2.30	2.98	0.51
1.24	1.77	2.41	3.19	0.57	1.48	1.79	2.44	3.25	0.70	1.54
1.88	2.57	3.31	0.73	1.59	1.90	2.61	1.19	0.75	1.61	1.93
2.62	3.50	0.75	1.61	2.01	2.72	3.50	1.11	1.62	2.16	2.76
3.50	1.16	1.62	2.18	2.84	3.50					

Based on Irony, T. Z., Lauretto, M., Pereira, C., and Stern, J. M. "A Weibull wearout test: Full Bayesian approach," paper presented at *Mathematical Sciences Colloquium*, Binghamton University, Binghamton, UK, Dec. 2001.

c. Shah showed that the approximation has a maximum absolute error of .0052. Verify this for the approximations in part **a**.

*5.150 Spruce budworm infestation. An infestation of a certain species of caterpillar called the spruce budworm can cause extensive damage to the timberlands of the northern United States. It is known that an outbreak of this type of infestation occurs, on the average, every 40 years. Assuming that this phenomenon obeys an exponential probability law, what is the probability that catastrophic outbreaks of spruce budworm infestation will occur within 8 years of each other?

5.151 Maximum time to take a test. A professor believes that if a class is allowed to work on an examination as long as desired, the times spent by the students would be approximately normal with mean 45 minutes and standard deviation 6.5 minutes. About how long should be allotted for the examination if the professor wants almost all (say, 97.5%) of the class to finish?

*5.152 Reliability of CD-ROMs. In *Reliability Ques* (Mar. 2004), the exponential distribution was used to model the lengths of life of CD-ROM drives in a two-drive system. The two CD-ROM drives operate independently, and at least one drive must be operating for the system to operate successfully. Both drives have a mean length of life of 25,000 hours.

a. The reliability $R(t)$ of a single CD-ROM drive is the probability that the life of the drive exceeds t hours. Give a formula for $R(t)$.

b. Use the result from part **a** to find the probability that the life of the single CD-ROM drive exceeds 8,760 hours (the number of hours of operation in a year).

c. The reliability $S(t)$ of the two-CD-ROM-drive system is the probability that the life of at least one drive exceeds t hours. Give a formula for $S(t)$. [*Hint:* Use the rule of complements and the fact that the two drives operate independently.]

d. Use the result from part **c** to find the probability that the two-drive CD-ROM system has a life whose length exceeds 8,760 hours.

e. Compare the probabilities you found in parts **b** and **d**.

Applying the Concepts—Advanced

5.153 Dye discharged in paint. A machine used to regulate the amount of dye dispensed for mixing shades of paint can be set so that it discharges an average of μ milliliters (mL) of dye per can of paint. The amount of dye discharged is known to have a normal distribution with a standard deviation of .4 mL. If more than 6 mL of dye are discharged when making a certain shade of blue paint, the shade is unacceptable. Determine the setting for μ so that only 1% of the cans of paint will be unacceptable.

5.154 Accidents at a plant. The number of serious accidents in a manufacturing plant has (approximately) a Poisson probability distribution with a mean of three serious accidents per month. It can be shown that if x, the number of events per unit time, has a Poisson distribution with mean λ, then the time between two successive events has an exponential probability distribution with mean $\theta = 1/\lambda$.

a. If an accident occurs today, what is the probability that the next serious accident will not occur within the next month?

b. What is the probability that more than one serious accident will occur within the next month?

5.155 Water retention of soil cores. A team of soil scientists investigated the water retention properties of soil cores sampled from an uncropped field consisting of silt loam (*Soil Science*, Jan. 1995). At a pressure of .1 megapascal (MPa), the water content of the soil (measured in cubic meters of water per cubic meter of soil) was determined to be approximately normally distributed with $\mu = .27$ and $\sigma = .04$. In addition to water content readings at a pressure of .1 MPa, measurements were obtained at pressures 0, .005, .01, .03, and 1.5 MPa. Consider a soil core with a water content reading of .14. Is it likely that this reading was obtained at a pressure of .1 MPa? Explain.

Critical Thinking Challenges

*5.156 Weights of corn chip bags. The net weight per bag of a certain brand of corn chips is listed as 12 ounces. The weight of chips actually dispensed in each bag by an automated machine, when operating to specifications, is a normal random variable with mean 11.5 ounces and standard deviation .27 ounce. A quality control inspector will randomly select 1,500 bags from the automated process and measure the weight of each bag precisely. Ideally, all of the bags should contain at least 10 ounces of corn chips. However, if at least 97% of the bags contain 12 ounces or more, the process is considered "under control" and no adjustments to the machine are deemed necessary. Assess the likelihood that the quality control inspector will find fewer than 97% of the bags with 12 ounces or more even if the process is "under control."

5.157 IQs and *The Bell Curve*. The controversial book, *The Bell Curve* (Free Press, 1994), employs statistical analyses heavily in an attempt to support the authors' position on "intelligence and class structure in American life." Since the book's publication, many expert statisticians have raised doubts about the authors' statistical methods and the inferences drawn from them. (See, for example, "Wringing *The Bell Curve*: A cautionary tale about the relationships among race, genes, and IQ," *Chance*, Summer 1995.) One of the many controversies sparked by the book is the authors' tenet that level of intelligence (or lack thereof) is a cause of a wide range of intractable social problems, including constrained economic mobility. The measure of intelligence chosen by the authors is the well-known intelligence quotient (IQ). Psychologists traditionally treat IQ as a random variable having a normal distribution with mean $\mu = 100$ and standard deviation $\sigma = 15$.

The Bell Curve refers to five cognitive classes of people defined by percentiles of the normal distribution. Class I ("very bright") consists of those with IQs above the 95th percentile; Class II ("bright") are those with IQs between the 75th and 95th percentiles; Class III ("normal") includes IQs between the 25th and 75th percentiles; Class IV ("dull") are those with IQs between the 5th and 25th percentiles; and Class V ("very dull") are IQs below the 5th percentile.

a. Assuming that the distribution of IQ is accurately represented by the normal curve, determine the proportion of people with IQs in each of the five cognitive classes.

b. Although the authors define the cognitive classes in terms of percentiles, they stress that IQ scores should be compared with *z*-scores, not percentiles. In other words, it is more informative to give the difference in *z*-scores for two IQ scores than it is to give the difference in percentiles. Do you agree?

c. Researchers have found that scores on many intelligence tests are decidedly nonnormal. Some distributions are skewed toward higher scores, others toward lower scores. How would the proportions in the five cognitive classes defined in part **a** differ for an IQ distribution that is skewed right? Skewed left?

Activity **The Normal Approximation to the Binomial**

For large values of *n*, the computational effort involved in working with the binomial probability distribution is considerable. Fortunately, in many instances the normal distribution provides a good approximation to the binomial distribution. This activity was designed to enable you to demonstrate to yourself how well the normal distribution approximates the binomial distribution.

a. Let the random variable *x* have a binomial probability with $n = 10$ and $p = .5$. Using the binomial distribution, find the probability that *x* takes on a value in each of the following intervals: $\mu \pm \sigma$, $\mu \pm 2\sigma$, and $\mu \pm 3\sigma$.

b. Find the probabilities requested in part **a** by using a normal approximation to the given binomial distribution.

c. Find the magnitude of the difference between each of the three probabilities as determined by the binomial distribution and by the normal approximation.

d. Letting *x* have a binomial distribution with $n = 20$ and $p = .5$, repeat parts **a**, **b**, and **c**. Notice that the probability estimates provided by the normal distribution are more accurate for $n = 20$ than for $n = 10$.

e. Letting *x* have a binomial distribution with $n = 20$ and $p = .01$, repeat parts **a**, **b**, and **c**. Notice that the probability estimates provided by the normal distribution are poor in this case. Explain why this occurs.

References

Hogg, R. V., McKean, J. W., and Craig, A. T. *Introduction to Mathematical Statistics*, 6th ed. Upper Saddle River, NJ: Prentice Hall, 2005.

Lindgren, B. W. *Statistical Theory*, 4th ed. New York: Chapman & Hall, 1993.

Mood, A. M., Graybill, F. A., and Boes, D. C. *Introduction to the Theory of Statistics*, 3rd ed. New York: McGraw-Hill, 1974.

Ramsey, P. P., and Ramsey, P. H. "Simple tests of normality in small samples." *Journal of Quality Technology*, Vol. 22, 1990.

Ross, S. M. *Stochastic Processes*, 2nd ed. New York: Wiley, 1996.

Wackerly, D., Mendenhall, W., and Scheaffer, R. L. *Mathematical Statistics with Applications*, 7th ed. North Scituate, MA: Duxbury, 2008.

Winkler, R. L., and Hays, W. *Statistics: Probability, Inference, and Decision*, 2nd ed. New York: Holt, Rinehart and Winston, 1975.

USING TECHNOLOGY **MINITAB: Continuous Random Variable Probabilities and Normal Probability Plots**

Continuous Probabilities

Step 1 Select the "Calc" button on the MINITAB menu bar, click on "Probability Distributions," and then finally select the Continuous distribution of your choice (e.g., "Normal"), as shown in Figure 5.M.1.

Step 2 Select either "Probability" or "Cumulative probability" on the resulting dialog box.

Step 3 Specify the parameters of the distribution (e.g., the range for a uniform distribution, μ and σ for a normal distribution, and the mean θ for an exponential distribution).

Step 4 Specify the value *x* of interest in the "Input constant" box.

Step 5 Click "OK". The probability for the value of *x* will appear on the MINITAB session window.

Figure 5.M.1 MINITAB menu options for continuous probabilities

Figure 5.M.2 MINITAB normal distribution dialog box

[*Note:* Figure 5.M.2 gives the specifications for finding $P(x \le 23)$ in a normal distribution with $\mu = 24.5$ and $\sigma = 1$.]

Normal Probability Plot

Step 1 Select "Graph" on the MINITAB menu bar, and then click on "Probability Plot," as shown in Figure 5.M.3.

Figure 5.M.3 MINITAB options for a normal probability plot

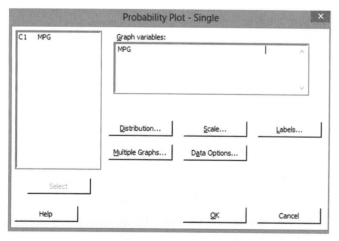

Figure 5.M.4 MINITAB normal probability plot dialog box

Step 2 Select "Single" (for one variable) on the next box, and the dialog box will appear as shown in Figure 5.M.4.

Step 3 Specify the variable of interest in the "Graph variables" box, and then click the "Distribution" button and select the "Normal" option. Click "OK" to return to the Probability Plot dialog box.

Step 4 Click "OK" to generate the normal probability plot.

TI–83/TI–84 Plus Graphing Calculator: Normal Random Variable and Normal Probability Plots

Graphing the Area under the Standard Normal Curve

Step 1 *Turn off all plots*

- Press **2nd PRGM** and select **1:ClrDraw**
- Press **ENTER,** and 'Done' will appear on the screen
- Press **2nd Y**= and select **4:PlotsOff**
- Press **ENTER**, and 'Done' will appear on the screen

Step 2 *Set the viewing window (Recall that almost all of the area under the standard normal curve falls between −5 and 5. A height of 0.5 is a good choice for Ymax.)*

Note: When entering a negative number, be sure to use the negative sign (**−**), and not the minus sign.

- Set **Xmin** = −5
- **Xmax** = 5
- **Xscl** = 1
- **Ymin** = 0
- **Ymax** = .5
- **Yscl** = 0
- **Xres** = 1

Step 3 *View graph*

- Press **2nd VARS**
- Arrow right to **DRAW**
- Press **ENTER** to select **1:ShadeNorm(**
- Enter your lower limit (e.g., −5)
- Press **COMMA**
- Enter your upper limit (e.g., 1.5)
- Press **)**
- Press **ENTER**

    The graph will be displayed along with the area, lower limit, and upper limit.

    Thus, $P(Z < 1.5) = .9332$

Finding Normal Probabilities without a Graph

To compute probabilities for a normal distribution, use the **normalcdf(** command. *Normalcdf* stands for "normal cumulative density function." This command is under the **DISTR**ibution menu and has the format **normalcdf(***lower limit, upper limit, mean, standard deviation***).**

Step 1 *Find the probability*

- Press **2nd VARS** for **DISTR** and select **Normalcdf(**
- After **Normalcdf(**, type in the lower limit
- Press **COMMA**
- Enter the upper limit
- Press **COMMA**
- Enter the mean
- Press **COMMA**
- Enter the standard deviation
- Press **)**
- Press **ENTER**

  The probability will be displayed on the screen.

  Example What is $P(x < 115)$ for a normal distribution with $\mu = 100$ and $\sigma = 10$? In this example, the lower limit is $-\infty$, the upper limit is 115, the mean is 100, and the standard deviation is 10. To represent $-\infty$ on the calculator, enter **(−) 1**, press **2nd** and press the **COMMA** key for **EE**, and then press **99**. The screen appears as follows:

  Thus, $P(x < 115) = .9332$

Finding Normal Probabilities with a Graph

Step 1 *Turn off all plots*

- Press **Y =** and **CLEAR** all functions from the Y registers
- Press **2nd Y =** and select **4:PlotsOff**
- Press **ENTER ENTER**, and 'Done' will appear on the screen

Step 2 *Set the viewing window (These values depend on the mean and standard deviation of the data.)* Note: When entering a negative number, be sure to use the negative sign **(−),** not the minus sign.

- Press **WINDOW**
- Set **Xmin** $= \mu - 5\sigma$
- **Xmax** $= \mu + 5\sigma$
- **Xscl** $= \sigma$
- **Ymin** $= -.125/\sigma$
- **Ymax** $= .5/\sigma$
- **Yscl** $= 1$
- **Xres** $= 1$

Step 3 *View graph*

- Press **2nd VARS**
- **ARROW** right to **DRAW**
- Press **ENTER** to select **1:ShadeNorm(**
- Enter the lower limit
- Press **COMMA**
- Enter the upper limit
- Press **COMMA**
- Enter the mean

- Press **COMMA**
- Enter the standard deviation
- Press **)**
- Press **ENTER**

  The graph will be displayed along with the area, lower limit, and upper limit.

  Example What is $P(x < 115)$ for a normal distribution with $\mu = 100$ and $\sigma = 10$? In this example, the lower limit is $-\infty$, the upper limit is 115, the mean is 100, and the standard deviation is 10. To represent $-\infty$ on the calculator, enter **(−) 1**, press **2nd** and press the **comma** key for **EE**, and then press **99**. The screens appear as follows:

Graphing a Normal Probability Plot

Step 1 *Enter the data*

- Press **STAT** and select **1:Edit**

  Note: If the list already contains data, clear the old data. Use the up arrow to highlight **L1**

- Press **CLEAR ENTER**
- Use the **ARROW** and **ENTER** keys to enter the data set into **L1**

Step 2 *Set up the normal probability plot*

- Press **Y =** and **CLEAR** all functions from the Y registers
- Press **2nd** and press **Y =** for **STAT PLOT**
- Press **1** for **Plot 1**
- Set the cursor so that **ON** is flashing
- For **Type,** use the **ARROW** and **ENTER** keys to highlight and select the last graph in the bottom row
- For **Data List,** choose the column containing the data (in most cases, L1)

  (Note: Press **2nd 1** for **L1)**
- For **Data Axis,** choose **X** and press **ENTER**

Step 3 *View plot*

- Press **ZOOM 9**

  Your data will be displayed against the expected z-scores from a normal distribution. If you see a "generally" linear relationship, your data set is approximately normal.

6 Sampling Distributions

CONTENTS

Where We've Been

- Learned that the objective of most statistical investigations is inference—that is, making decisions about a population on the basis of information about a sample

- Discovered that sample statistics such as the sample mean and sample variance can be used to make decisions

- Found that probability distributions for random variables are used to construct theoretical models of populations

Where We're Going

- Establish that a sample statistic is a random variable with a probability distribution (6.1)

- Define a *sampling distribution* as the probability distribution of a sample statistic (6.1)

- Give two important properties of sampling distributions (6.2)

- Learn that the sampling distribution of both the sample mean and the sample proportion tends to be approximately normal (6.3, 6.4)

Statistics IN Action   The Insomnia Pill: Is It Effective?

More than 20 years ago, neuroscientists at the Massachusetts Institute of Technology (MIT) began experimenting with melatonin—a hormone secreted by the pineal gland in the brain—as a sleep-inducing hormone. Their original study, published in the *Proceedings of the National Academy of Sciences* (Mar. 1994), brought encouraging news to insomniacs and travelers who suffer from jet lag: Melatonin was discovered to be effective in reducing sleep onset latency (the amount of time between a person lying down to sleep and the actual onset of stage one sleep). Furthermore, since the hormone is naturally produced, it is nonaddictive.

Since then, pharmaceutical companies that produce melatonin pills for treating insomnia have received negative reports on the efficacy of the drug. Many people don't think the melatonin pills work. Consequently, the MIT researchers undertook a follow-up study and published their results in *Sleep Medicine Reviews* (Feb. 2005). They reported that commercially available melatonin pills contain too high a dosage of the drug. When the melatonin receptors in the brain are exposed to too much of the hormone, they become unresponsive. The researchers' analysis of all previous studies on melatonin as a sleep inducer confirmed that, taken in small doses, melatonin is effective in reducing sleep onset latency.

In this Statistics in Action, our focus is on the original MIT study. Young male volunteers were given various doses of melatonin or a placebo (a dummy medication containing no melatonin). Then they were placed in a dark room at midday and told to close their eyes for 30 minutes. The variable of interest was sleep onset latency (in minutes).

According to the lead investigator, Professor Richard Wurtman, "Our volunteers fall asleep in 5 or 6 minutes on melatonin, while those on placebo take about 15 minutes." Wurtman warned, however, that uncontrolled doses of melatonin could cause mood-altering side effects.

Now, consider a random sample of 40 young males, each of whom is given a dosage of the sleep-inducing hormone melatonin. The times taken (in minutes) to fall asleep for these 40 males are listed in Table SIA6.1 and saved in the **INSOMNIA** file. The researchers know that with the placebo (i.e., no hormone), the mean sleep onset latency is $\mu = 15$ minutes and the standard deviation is $\sigma = 10$ minutes. They want to use the data to make an inference about the true value of μ for those taking the melatonin. Specifically, the researchers want to know whether melatonin is an effective drug against insomnia.

In this chapter, a Statistics in Action Revisited example demonstrates how we can use one of the topics discussed in the chapter—the Central Limit Theorem—to make an inference about the effectiveness of melatonin as a sleep-inducing hormone.

Statistics IN Action   Revisited

- Making an Inference about the Mean Sleep Onset Latency for Insomnia Pill Takers (p. 328)

Table SIA6.1   Times Taken (in Minutes) for 40 Male Volunteers to Fall Asleep

7.6	2.1	1.4	1.5	3.6	17.0	14.9	4.4	4.7	20.1
7.7	2.4	8.1	1.5	10.3	1.3	3.7	2.5	3.4	10.7
2.9	1.5	15.9	3.0	1.9	8.5	6.1	4.5	2.2	2.6
7.0	6.4	2.8	2.8	22.8	1.5	4.6	2.0	6.3	3.2

[*Note: These data are simulated sleep times based on summary information provided in the MIT study.*]   Ⓓ *Data Set: INSOMNIA*

In Chapters 4 and 5, we assumed that we knew the probability distribution of a random variable, and using this knowledge, we were able to compute the mean, variance, and probabilities associated with the random variable. However, in most practical applications, this information is not available. To illustrate, in Example 4.13 (p. 236), we calculated the probability that the binomial random variable x, the number of 20 polled voters who favored a certain mayoral candidate, assumed specific values. To do this, it was necessary to assume some value for p, the proportion of all voters who favored the candidate. Thus, for the purposes of illustration, we assumed that $p = .6$ when, in all likelihood, the exact value of p would be unknown. In fact, the probable purpose of taking the poll is to estimate p. Similarly, when we modeled the in-city gas mileage of a certain automobile model, we used the normal probability distribution with an *assumed* mean and standard deviation of 27 and 3 miles per gallon, respectively. In most situations, the true mean and standard deviation are unknown quantities that have to be estimated. Numerical quantities that describe probability distributions are called *parameters*. Thus, p, the probability of a success in a binomial experiment, and μ and σ, the mean and standard deviation, respectively, of a normal distribution, are examples of parameters.

> A **parameter** is a numerical descriptive measure of a population. Because it is based on the observations in the population, its value is almost always unknown.

We have also discussed the sample mean $\bar{x}$, sample variance s^2, sample standard deviation s, and the like, which are numerical descriptive measures calculated from the sample. (See Table 6.1 for a list of the statistics covered so far in this text.) We will often use the information contained in these *sample statistics* to make inferences about the parameters of a population.

> A **sample statistic** is a numerical descriptive measure of a sample. It is calculated from the observations in the sample.

Table 6.1   List of Population Parameters and Corresponding Sample Statistics

	Population Parameter	Sample Statistic
Mean:	μ	$\bar{x}$
Median:	η	M
Variance:	σ^2	s^2
Standard deviation:	σ	s
Binomial proportion:	p	$\hat{p}$

Note that the term *statistic* refers to a *sample* quantity and the term *parameter* refers to a *population* quantity.

Before we can show you how to use sample statistics to make inferences about population parameters, we need to be able to evaluate their properties. Does one sample statistic contain more information than another about a population parameter? On what basis should we choose the "best" statistic for making inferences about a parameter? The purpose of this chapter is to answer these questions.

6.1 The Concept of a Sampling Distribution

If we want to estimate a parameter of a population—say, the population mean μ—we can use a number of sample statistics for our estimate. Two possibilities are the sample mean $\bar{x}$ and the sample median M. Which of these do you think will provide a better estimate of μ?

Before answering this question, consider the following example: Toss a fair die, and let x equal the number of dots showing on the up face. Suppose the die is tossed three times, producing the sample measurements 2, 2, 6. The sample mean is then $\bar{x} = 3.33$, and the sample median is $M = 2$. Since the population mean of x is $\mu = 3.5$, you can see that, for this sample of three measurements, the sample mean $\bar{x}$ provides an estimate that falls closer to μ than does the sample median (see Figure 6.1a). Now suppose we toss the die three more times and obtain the sample measurements 3, 4, 6. Then the mean and median of this sample are $\bar{x} = 4.33$ and $M = 4$, respectively. This time, M is closer to μ. (See Figure 6.1b.)

This simple example illustrates an important point: Neither the sample mean nor the sample median will *always* fall closer to the population mean. Consequently, we cannot compare these two sample statistics or, in general, any two sample statistics on

Figure 6.1

Comparing the sample mean ($\bar{x}$) and sample median (M) as estimators of the population mean (μ)

a. Sample 1: $\bar{x}$ is closer than M to μ b. Sample 2: M is closer than $\bar{x}$ to μ

the basis of their performance with a single sample. Instead, we need to recognize that sample statistics are themselves random variables because different samples can lead to different values for the sample statistics. As random variables, sample statistics must be judged and compared on the basis of their probability distributions (i.e., the *collection* of values and associated probabilities of each statistic that would be obtained if the sampling experiment were repeated a *very large number of times*). We will illustrate this concept with another example.

Suppose it is known that in a certain part of Canada the daily high temperature recorded for all past months of January has a mean $\mu = 10°F$ and a standard deviation $\sigma = 5°F$. Consider an experiment consisting of randomly selecting 25 daily high temperatures from the records of past months of January and calculating the sample mean $\bar{x}$. If this experiment were repeated a very large number of times, the value of $\bar{x}$ would vary from sample to sample. For example, the first sample of 25 temperature measurements might have a mean $\bar{x} = 9.8$, the second sample a mean $\bar{x} = 11.4$, the third sample a mean $\bar{x} = 10.5$, etc. If the sampling experiment were repeated a very large number of times, the resulting histogram of sample means would be approximately the probability distribution of $\bar{x}$. If $\bar{x}$ is a good estimator of μ, we would expect the values of $\bar{x}$ to cluster around μ as shown in Figure 6.2. This probability distribution is called a *sampling distribution* because it is generated by repeating a sampling experiment a very large number of times.

> The **sampling distribution** of a sample statistic calculated from a sample of n temperature measurements is the probability distribution of the statistic.

In actual practice, the sampling distribution of a statistic is obtained mathematically or (at least approximately) by simulating the sample on a computer, using a procedure similar to that just described.

If $\bar{x}$ has been calculated from a sample of $n = 25$ measurements selected from a population with mean $\mu = 10$ and standard deviation $\sigma = 5$, the sampling distribution (Figure 6.2) provides information about the behavior of $\bar{x}$ in repeated sampling.

Figure 6.2

Sampling distribution for $\bar{x}$ based on a sample of $n = 25$ temperature measurements

For example, the probability that you will draw a sample of 25 measurements and obtain a value of $\bar{x}$ in the interval $9 \leq \bar{x} \leq 10$ will be the area under the sampling distribution over that interval.

Since the properties of a statistic are typified by its sampling distribution, it follows that, to compare two sample statistics, you compare their sampling distributions. For example, if you have two statistics, A and B, for estimating the same parameter (for purposes of illustration, suppose the parameter is the population variance σ^2), and if their sampling distributions are as shown in Figure 6.3, you would prefer statistic A over statistic B. You would do so because the sampling distribution for statistic A centers over σ^2 and has less spread (variation) than the sampling distribution for statistic B. Then, when you draw a single sample in a practical sampling situation, the probability is higher that statistic A will fall nearer σ^2.

Figure 6.3

Two sampling distributions for estimating the population variance σ^2

Remember that, in practice, we will not know the numerical value of the unknown parameter σ^2, so we will not know whether statistic A or statistic B is closer to σ^2 for a particular sample. We have to rely on our knowledge of the theoretical sampling distributions to choose the best sample statistic and then use it sample after sample. The procedure for finding the sampling distribution for a statistic is demonstrated in Examples 6.1 and 6.2.

Example 6.1

Finding a Sampling Distribution—Coin Tossing Experiment

Problem In a single toss of a fair coin, let x equal the number of heads observed. Now consider a sample of $n = 2$ tosses. Find the sampling distribution of $\bar{x}$, the sample mean.

Solution For one coin toss, the result is either a head (H) or tail (T). Thus, the value of x is either $x = 1$ or $x = 0$. The four possible outcomes (sample points) for two coin tosses and corresponding probabilities are listed Table 6.2. Since a fair coin is tossed, these four possible outcomes are equally likely.

Table 6.2   Outcomes for $n = 2$ Coin Tosses

Outcome (Toss 1, Toss 2)	Probability	$\bar{x}$
H H $(x = 1, x = 1)$	$^1/_4$	1
H T $(x = 1, x = 0)$	$^1/_4$	.5
T H $(x = 0, x = 1)$	$^1/_4$	.5
T T $(x = 0, x = 0)$	$^1/_4$	0

The value of $\bar{x}$ for each outcome is also listed in Table 6.2. You can see that the values are either 0, .5, or 1. Now, $P(\bar{x} = 1) = P(\text{H H}) = {}^1/_4$. Similarly, $P(\bar{x} = 0) = P(\text{T T}) = {}^1/_4$. Finally, using the probability law for unions of mutually exclusive events, $P(\bar{x} = .5) = P(\text{H T or T H}) = P(\text{H T}) + P(\text{T H}) = {}^1/_4 + {}^1/_4 = {}^1/_2$. Thus, the sampling distribution of $\bar{x}$ is:

$\bar{x}$	0	.5	1
$p(\bar{x})$	$^1/_4$	$^1/_2$	$^1/_4$

Example 6.2

Finding a Sampling Distribution—Come-Out Roll in Craps

Problem Consider the popular casino game of craps, in which a player throws two dice and bets on the outcome (the sum total of the dots showing on the upper faces of the two dice). In Example 4.5 (p. 218), we looked at the possible outcomes of a $5 wager on the first toss (called the *come-out* roll). Recall that if the sum total of the dice is 7 or 11, the roller wins $5; if the total is a 2, 3, or 12, the roller loses $5 (i.e., the roller "wins" −$5);

and, for any other total (4, 5, 6, 8, 9, or 10), no money is lost or won on that roll (i.e., the roller wins $0). Let x represent the result of the come-out roll wager (−$5, $0, or +$5). We showed in Example 4.5 that the actual probability distribution of x is:

Outcome of wager, x	−5	0	5
$p(x)$	1/9	6/9	2/9

Now, consider a random sample of $n = 3$ come-out rolls.

a. Find the sampling distribution of the sample mean, $\bar{x}$.

b. Find the sampling distribution of the sample median, M.

Solution The outcomes for every possible sample of $n = 3$ come-out rolls are listed in Table 6.3, along with the sample mean and median. The probability of each sample is obtained using the Multiplicative Rule. For example, the probability of the sample $(0, 0, 5)$ is $p(0) \cdot p(0) \cdot p(5) = (6/9)(6/9)(2/9) = 72/729 = .099$. The probability for each sample is also listed in Table 6.3. Note that the sum of these probabilities is equal to 1.

Table 6.3   All Possible Samples of $n = 3$ Come-Out Rolls in Craps

Possible Samples	$\bar{x}$	M	Probability
−5, −5, −5	−5	−5	$(1/9)(1/9)(1/9) = 1/729$
−5, −5, 0	−3.33	−5	$(1/9)(1/9)(6/9) = 6/729$
−5, −5, 5	−1.67	−5	$(1/9)(1/9)(2/9) = 2/729$
−5, 0, −5	−3.33	−5	$(1/9)(6/9)(1/9) = 6/729$
−5, 0, 0	−1.67	0	$(1/9)(6/9)(6/9) = 36/729$
−5, 0, 5	0	0	$(1/9)(6/9)(2/9) = 12/729$
−5, 5, −5	−1.67	−5	$(1/9)(2/9)(1/9) = 2/729$
−5, 5, 0	0	0	$(1/9)(2/9)(6/9) = 12/729$
−5, 5, 5	1.67	5	$(1/9)(2/9)(2/9) = 4/729$
0, −5, −5	−3.33	−5	$(6/9)(1/9)(1/9) = 6/729$
0, −5, 0	−1.67	0	$(6/9)(1/9)(6/9) = 36/729$
0, −5, 5	0	0	$(6/9)(1/9)(2/9) = 12/729$
0, 0, −5	−1.67	0	$(6/9)(6/9)(1/9) = 36/729$
0, 0, 0	0	0	$(6/9)(6/9)(6/9) = 216/729$
0, 0, 5	1.67	0	$(6/9)(6/9)(2/9) = 72/729$
0, 5, −5	0	0	$(6/9)(2/9)(1/9) = 12/729$
0, 5, 0	1.67	0	$(6/9)(2/9)(6/9) = 72/729$
0, 5, 5	3.33	5	$(6/9)(2/9)(2/9) = 24/729$
5, −5, −5	−1.67	−5	$(2/9)(1/9)(1/9) = 2/729$
5, −5, 0	0	0	$(2/9)(1/9)(6/9) = 12/729$
5, −5, 5	1.67	5	$(2/9)(1/9)(2/9) = 4/729$
5, 0, −5	0	0	$(2/9)(6/9)(1/9) = 12/729$
5, 0, 0	1.67	0	$(2/9)(6/9)(6/9) = 72/729$
5, 0, 5	3.33	5	$(2/9)(6/9)(2/9) = 24/729$
5, 5, −5	1.67	5	$(2/9)(2/9)(1/9) = 4/729$
5, 5, 0	3.33	5	$(2/9)(2/9)(6/9) = 24/729$
5, 5, 5	5	5	$(2/9)(2/9)(2/9) = 8/729$

$$Sum = 729/729 = 1$$

a. From Table 6.3, you can see that $\bar{x}$ can assume the values −5, −3.33, −1.67, 0, 1.67, 3.33, and 5. Because $\bar{x} = -5$ occurs only in one sample, $P(\bar{x} = -5) = 1/729$. Similarly, $\bar{x} = -3.33$ occurs in three samples, $(-5, -5, 0), (-5, 0, -5)$, and $(0, -5, -5)$. Therefore, $P(\bar{x} = -3.33) = 6/729 + 6/729 + 6/729 = 18/729$. Calculating the probabilities of the remaining values of $\bar{x}$ and arranging them in a table, we obtain the following probability distribution:

$\bar{x}$	-5	-3.33	-1.67	0	1.67	3.33	5
$p(\bar{x})$	$1/729 =$ .0014	$18/729 =$ .0247	$114/729 =$ .1564	$288/729 =$ .3951	$228/729 =$ .3127	$72/729 =$ .0988	$8/729 =$ .0110

This is the sampling distribution for $\bar{x}$ because it specifies the probability associated with each possible value of $\bar{x}$. You can see that the most likely mean outcome after 3 randomly selected come-out rolls is $\bar{x} = \$0$; this result occurs with probability $288/729 = .3951$.

b. In Table 6.3, you can see that the median M can assume one of three values: $-5, 0$, and 5. The value $M = -5$ occurs in 7 different samples. Therefore, $P(M = -5)$ is the sum of the probabilities associated with these 7 samples; that is, $P(M = -5) = 1/729 + 6/729 + 2/729 + 6/729 + 2/729 + 6/729 + 2/729 = 25/729$. Similarly, $M = 0$ occurs in 13 samples and $M = 5$ occurs in 7 samples. These probabilities are obtained by summing the probabilities of their respective sample points. After performing these calculations, we obtain the following probability distribution for the median M:

M	-5	0	5
$p(M)$	$25/729 = .0343$	$612/729 = .8395$	$92/729 = .1262$

Once again, the most likely median outcome after 3 randomly selected come-out rolls is $\bar{x} = \$0$—a result that occurs with probability $612/729 = .8395$.

Look Back The sampling distributions of parts **a** and **b** are found by first listing all possible distinct values of the statistic and then calculating the probability of each value. Note that if the values of x were equally likely, the 27 sample points in Table 6.3 would all have the same probability of occurring, namely, $1/27$.

■ **Now Work Exercise 6.3**

Examples 6.1 and 6.2 demonstrate the procedure for finding the exact sampling distribution of a statistic when the number of different samples that could be selected from the population is relatively small. In the real world, populations often consist of a large number of different values, making samples difficult (or impossible) to enumerate. When this situation occurs, we may choose to obtain the approximate sampling distribution for a statistic by simulating the sampling over and over again and recording the proportion of times different values of the statistic occur. Example 6.3 illustrates this procedure.

Example **6.3**

Simulating a Sampling Distribution—Thickness of Steel Sheets

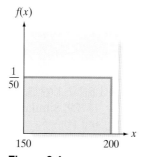

$f(x)$

$\frac{1}{50}$

150   200

Figure 6.4

Uniform distribution for thickness of steel sheets

Problem The rolling machine of a steel manufacturer produces sheets of steel of varying thickness. The thickness of a steel sheet follows a uniform distribution with values between 150 and 200 millimeters. Suppose we perform the following experiment over and over again: Randomly sample 11 steel sheets from the production line and record the thickness x of each. Calculate the two sample statistics

$$\bar{x} = \text{Sample mean} = \frac{\Sigma x}{11}$$

$$M = \text{Median} = \text{Sixth sample measurement when the 11 thicknesses are arranged in ascending order}$$

Obtain approximations to the sampling distributions of $\bar{x}$ and M.

Solution The population of thicknesses follows the uniform distribution shown in Figure 6.4. We used MINITAB to generate 1,000 samples from this population, each with $n = 11$ observations. Then we computed $\bar{x}$ and M for each sample. Our goal is to obtain approximations to the sampling distributions of $\bar{x}$ and M in order to find out which sample statistic ($\bar{x}$ or M) contains more information about μ. [*Note:* In this particular example, we *know* that the population mean is $\mu = 175$ mm. (See Section 5.2.)] The first 10 of the 1,000 samples generated are presented in Table 6.4. For instance, the first computer-generated sample from the uniform distribution contained the following measurements (arranged in ascending order): 151, 157, 162, 169, 171, 173, 181, 182, 187, 188, and 193 millimeters. The sample mean $\bar{x}$ and median M computed for this sample are

$$\bar{x} = \frac{151 + 157 + \cdots + 193}{11} = 174.0$$

$$M = \text{Sixth ordered measurement} = 173$$

The MINITAB relative frequency histograms for $\bar{x}$ and M for the 1,000 samples of size $n = 11$ are shown in Figure 6.5. These histograms represent approximations to the true sampling distributions of $\bar{x}$ and M.

Table 6.4   **First 10 Samples of $n = 11$ Thickness Measurements from Uniform Distribution**

Sample	Thickness Measurements											Mean	Median
1	173	171	187	151	188	181	182	157	162	169	193	174.00	173
2	181	190	182	171	187	177	162	172	188	200	193	182.09	182
3	192	195	187	187	172	164	164	189	179	182	173	180.36	182
4	173	157	150	154	168	174	171	182	200	181	187	172.45	173
5	169	160	167	170	197	159	174	174	161	173	160	169.46	169
6	179	170	167	174	173	178	173	170	173	198	187	176.55	173
7	166	177	162	171	154	177	154	179	175	185	193	172.09	175
8	164	199	152	153	163	156	184	151	198	167	180	169.73	164
9	181	193	151	166	180	199	180	184	182	181	175	179.27	181
10	155	199	199	171	172	157	173	187	190	185	150	176.18	173

D *Data Set:* SIMUNI

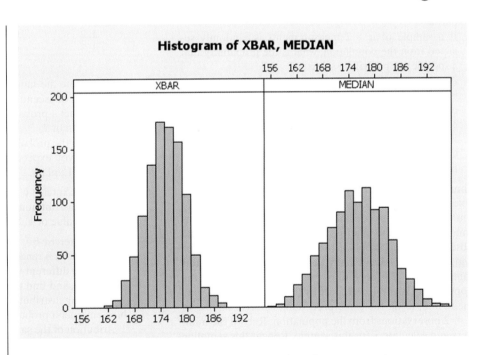

Figure 6.5
MINITAB histograms for sample mean and sample median, Example 6.3

Look Back You can see that the values of $\bar{x}$ tend to cluster around μ to a greater extent than do the values of M. Thus, on the basis of the observed sampling distributions, we conclude that $\bar{x}$ contains more information about μ than M does—at least for samples of $n = 11$ measurements from the uniform distribution.

■ **Now Work Exercise 6.8**

As noted earlier, many sampling distributions can be derived mathematically, but the theory necessary to do so is beyond the scope of this text. Consequently, when we need to know the properties of a statistic, we will present its sampling distribution and simply describe its properties. Several of the important properties we look for in sampling distributions are discussed in the next section

Exercises 6.1–6.9

Understanding the Principles

6.1 What is the difference between a population parameter and a sample statistic?

6.2 What is a sampling distribution of a sample statistic?

Learning the Mechanics

6.3 The probability distribution shown here describes a population of measurements that can assume values of 0, 2, 4, and 6, each of which occurs with the same relative frequency:

x	0	2	4	6
$p(x)$	$1/4$	$1/4$	$1/4$	$1/4$

a. List all the different samples of $n = 2$ measurements that can be selected from this population.
b. Calculate the mean of each different sample listed in part **a**.
c. If a sample of $n = 2$ measurements is randomly selected from the population, what is the probability that a specific sample will be selected?
d. Assume that a random sample of $n = 2$ measurements is selected from the population. List the different values of $\bar{x}$ found in part **b**, and find the probability of each. Then give the sampling distribution of the sample mean $\bar{x}$ in tabular form.
e. Construct a probability histogram for the sampling distribution of $\bar{x}$.

6.4 Simulate sampling from the population described in Exercise 6.3 by marking the values of x, one on each of four identical coins (or poker chips, etc.). Place the coins (marked 0, 2, 4, and 6) into a bag, randomly select one, and observe its value. Replace this coin, draw a second coin, and observe its value. Finally, calculate the mean $\bar{x}$ for this sample of $n = 2$ observations randomly selected from the population (Exercise 6.3, part **b**). Replace the coins, mix them, and, using the same procedure, select a sample of $n = 2$ observations from the population. Record the numbers and calculate $\bar{x}$ for this sample. Repeat this sampling process until you acquire 100 values of $\bar{x}$. Construct a relative frequency distribution for these 100 sample means. Compare this distribution with the exact sampling distribution of $\bar{x}$ found in part **e** of Exercise 6.3. [*Note:* The distribution obtained in this exercise is an approximation to the exact sampling distribution. However, if you were to repeat the sampling procedure, drawing two coins not 100 times, but 10,000 times, then the relative frequency

distribution for the 10,000 sample means would be almost identical to the sampling distribution of $\bar{x}$ found in Exercise 6.3, part **e**.]

6.5 Consider the population described by the probability distribution shown below. The random variable x is observed twice with independent observations.

x	0	1	2	3
$p(x)$	.3	.2	.3	.2

Sample	Probability	Sample	Probability
0, 0	.09	2, 0	.09
0, 1	.06	2, 1	.06
0, 2	.09	2, 2	.09
0, 3	.06	2, 3	.06
1, 0	.06	3, 0	.06
1, 1	.04	3, 1	.04
1, 2	.06	3, 2	.06
1, 3	.04	3, 3	.04

a. Complete the sampling distribution table. Type the values of x in ascending order.
b. Construct a probability histogram for the sampling distribution of $\bar{x}$.
c. What is the probability that $\bar{x}$ is 3 or larger?
d. Would you expect to observe a value of $\bar{x}$ equal to 3 or larger? Explain.

6.6 Refer to Exercise 6.5 and find $E(x) = \mu$. Then use the sampling distribution of $\bar{x}$ found in Exercise 6.5 to find the expected value of $\bar{x}$. Note that $E(\bar{x}) = \mu$.

6.7 Refer to Exercise 6.5. Assume that a random sample of $n = 2$ measurements is randomly selected from the population.
a. List the different values that the sample median M may assume, and find the probability of each. Then give the sampling distribution of the sample median.
b. Construct a probability histogram for the sampling distribution of the sample median, and compare it with the probability histogram for the sample mean (Exercise 6.5, part **b**).

6.8 In Example 6.3, we used the computer to generate 1,000 samples, each containing $n = 11$ observations, from a uniform distribution over the interval from 150 to 200. Now use the computer to generate 500 samples, each containing $n = 15$ observations, from that same population.

a. Calculate the sample mean for each sample. To approximate the sampling distribution of $\bar{x}$, construct a relative frequency histogram for the 500 values of $\bar{x}$.

b. Repeat part **a** for the sample median. Compare this approximate sampling distribution with the approximate sampling distribution of $\bar{x}$ found in part **a**.

6.9 Consider a population that contains values of x equal to 00, 01, 02, 03, ..., 96, 97, 98, 99. Assume that these values

occur with equal probability. Use the computer to generate 500 samples, each containing $n = 20$ measurements, from this population. Calculate the sample mean $\bar{x}$ and sample variance s^2 for each of the 500 samples.

a. To approximate the sampling distribution of $\bar{x}$, construct a relative frequency histogram for the 500 values of $\bar{x}$.

b. Repeat part **a** for the 500 values of s^2.

6.2 Properties of Sampling Distributions: Unbiasedness and Minimum Variance

The simplest type of statistic used to make inferences about a population parameter is a *point estimator*—a rule or formula that tells us how to use the sample data to calculate a single number that is intended to estimate the value of some population parameter. For example, the sample mean $\bar{x}$ is a point estimator of the population mean μ. Similarly, the sample variance s^2 is a point estimator of the population variance σ^2.

> A **point estimator** of a population parameter is a rule or formula that tells us how to use the sample data to calculate a single number that can be used as an *estimate* of the population parameter.

Often, many different point estimators can be found to estimate the same parameter. Each will have a sampling distribution that provides information about the point estimator. By examining the sampling distribution, we can determine how large the difference between an estimate and the true value of the parameter (called the **error of estimation**) is likely to be. We can also tell whether an estimator is more likely to overestimate or to underestimate a parameter.

Example   6.4

Comparing Two Statistics

Problem Suppose two statistics, A and B, exist to estimate the same population parameter θ (theta). (Note that θ could be any parameter: μ, σ^2, σ, etc.) Suppose the two statistics have sampling distributions as shown in Figure 6.6. On the basis of these sampling distributions, which statistic is more attractive as an estimator of θ?

Solution As a first consideration, we would like the sampling distribution to center over the value of the parameter we wish to estimate. One way to characterize this property is in terms of the mean of the sampling distribution. Consequently, we say that a statistic is *unbiased* if the mean of the sampling distribution is equal to the parameter it is intended to estimate. This situation is shown in Figure 6.6a, where the mean μ_A of statistic A is equal to θ. If the mean of a sampling distribution is *not* equal to the parameter it is intended to estimate, the statistic is said to be *biased*. The sampling distribution for a

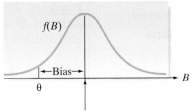

Figure 6.6

Sampling distributions of unbiased and biased estimators

a. Unbiased sample statistic for the parameter θ

b. Biased sample statistic for the parameter θ

biased statistic is shown in Figure 6.6b. The mean μ_B of the sampling distribution for statistic B is not equal to θ; in fact, it is shifted to the right of θ.

Look Back You can see that biased statistics tend either to overestimate or to underestimate a parameter. Consequently, when other properties of statistics tend to be equivalent, we will choose an unbiased statistic to estimate a parameter of interest.*

■ **Now Work Exercise 6.14**

Unbiased and Biased Estimators

If the sampling distribution of a sample statistic $(\hat{\theta})$ has a mean equal to the population parameter (θ) the statistic is intended to estimate, the statistic is said to be an **unbiased estimate** of the parameter. If the mean of the sampling distribution is not equal to the parameter, the statistic is said to be a **biased estimate** of the parameter.

$$\text{Unbiased: } \quad E(\hat{\theta}) = \theta$$
$$\text{Biased: } \quad E(\hat{\theta}) \neq \theta$$

The standard deviation of a sampling distribution measures another important property of statistics: the spread of these estimates generated by repeated sampling. Suppose two statistics, A and B, are both unbiased estimators of the population parameter. Since the means of the two sampling distributions are the same, we turn to their standard deviations in order to decide which will provide estimates that fall closer to the unknown population parameter we are estimating. Naturally, we will choose the sample statistic that has the smaller standard deviation. Figure 6.7 depicts sampling distributions for A and B. Note

Figure 6.7
Sampling distributions for two unbiased estimators

that the standard deviation of the distribution for A is smaller than the standard deviation of the distribution for B, indicating that, over a large number of samples, the values of A cluster more closely around the unknown population parameter than do the values of B. Stated differently, the probability that A is close to the parameter value is higher than the probability that B is close to the parameter value.

In sum, to make an inference about a population parameter, we use the sample statistic with a sampling distribution that is unbiased and has a small standard deviation (usually smaller than the standard deviation of other unbiased sample statistics). The derivation of this sample statistic will not concern us, because the "best" statistic for estimating specific parameters is a matter of record. We will simply present an unbiased estimator with its standard deviation for each population parameter we consider. [*Note:* The standard deviation of the sampling distribution of a statistic is also called the **standard error of the statistic**.]

Example	6.5

Biased and Unbiased Estimators—Craps Application

Problem Refer to Example 6.2 and the outcome x of a \$5 wager in craps. We found the sampling distributions of the sample mean $\bar{x}$ and the sample median M for random samples of $n = 3$ dice rolls from a population defined by the following probability distribution:

x	-5	0	5
$p(x)$	$1/9$	$6/9$	$2/9$

---

*Unbiased statistics do not exist for all parameters of interest, but they do exist for all the parameters considered in this text.

The sampling distributions of $\bar{x}$ and M were found to be as follows:

$\bar{x}$	-5	-3.33	-1.67	0	1.67	3.33	5
$p(\bar{x})$	$1/729$	$18/729$	$114/729$	$288/729$	$228/729$	$72/729$	$8/729$

M	-5	0	5
$p(M)$	$25/729$	$612/729$	$92/729$

a. Show that $\bar{x}$ is an unbiased estimator of μ in this situation.

b. Show that M is a biased estimator of μ in this situation.

Solution

a. The expected value of a discrete random variable x (see Section 4.3) is defined as $E(x) = \Sigma x p(x)$, where the summation is over all values of x. Then

$$E(x) = \mu = \Sigma x p(x) = (-5)\left(\frac{1}{9}\right) + (0)\left(\frac{6}{9}\right) + (5)\left(\frac{2}{9}\right) = \frac{5}{9} = .556$$

The expected value of the discrete random variable $\bar{x}$ is

$$E(\bar{x}) = \Sigma(\bar{x})p(\bar{x})$$

summed over all values of $\bar{x}$, or

$$E(\bar{x}) = (-5)\left(\frac{1}{729}\right) + (-3.33)\left(\frac{18}{729}\right) + (-1.67)\left(\frac{114}{729}\right) + \cdots + (5)\left(\frac{8}{729}\right) = .556$$

Since $E(\bar{x}) = \mu$, $\bar{x}$ is an unbiased estimator of μ.

b. The expected value of the sample median M is

$$E(M) = \Sigma M p(M) = (-5)\left(\frac{25}{729}\right) + (0)\left(\frac{612}{729}\right) + (5)\left(\frac{92}{729}\right) = \frac{335}{729} = .460$$

Since the expected value of M is not equal to μ ($\mu = .556$), the sample median M is a biased estimator of μ.

─■ **Now Work Exercise 6.16a–d**

Example 6.6

Variance of Estimators—Craps Application

Problem Refer to Example 6.5 and find the standard deviations of the sampling distributions of $\bar{x}$ and M. Which statistic would appear to be a better estimator of μ?

Solution The variance of the sampling distribution of $\bar{x}$ (we denote it by the symbol $\sigma_{\bar{x}}^2$) is found to be

$$E\{[\bar{x} - E(\bar{x})]^2\} = \Sigma(\bar{x} - \mu)^2 p(\bar{x})$$

where, from Example 6.5,

$$E(\bar{x}) = \mu = .556$$

Then

$$\sigma_{\bar{x}}^2 = (-5-.556)^2\left(\frac{1}{729}\right) + (-3.33-.556)^2\left(\frac{18}{729}\right) + \cdots + (5-.556)^2\left(\frac{8}{729}\right)$$

$$= 2.675$$

and

$$\sigma_{\bar{x}} = \sqrt{2.675} = 1.64$$

Similarly, the variance of the sampling distribution of M (we denote it by σ_M^2) is

$$\sigma_M^2 = E\{[M - E(M)]^2\}$$

where, from Example 6.5, the expected value of M is $E(M) = .460$. Then

$$\sigma_M^2 = E\{[M - E(M)]^2\} = \Sigma[M - E(M)]^2 p(M)$$
$$= (-5-.460)(^{25}/_{729}) + (0-.460)(^{612}/_{729}) + (5-.460)(^{92}/_{729}) = 3.801$$

and

$$\sigma_M = \sqrt{3.801} = 1.95$$

Which statistic appears to be the better estimator for the population mean μ, the sample mean $\bar{x}$ or the median M? To answer this question, we compare the sampling distributions of the two statistics. The sampling distribution of the sample median M is biased (i.e., it is located to the left of the mean μ), and its standard deviation $\sigma_M = 1.95$ is larger than the standard deviation of the sampling distribution of $\bar{x}$, $\sigma_{\bar{x}} = 1.64$. Consequently, for the population in question, the sample mean $\bar{x}$ would be a better estimator of the population mean μ than the sample median M would be.

Look Back Ideally, we desire an estimator that is unbiased *and* has the smallest variance among all unbiased estimators. We call this statistic the **minimum-variance unbiased estimator (MVUE)**.

■ **Now Work Exercise 6.16e, f**

Exercises 6.10–6.20

Understanding the Principles

6.10 What is a point estimator of a population parameter?

6.11 What is the difference between a biased and unbiased estimator?

6.12 What is the MVUE for a parameter?

6.13 What are the properties of an ideal estimator?

Learning the Mechanics

6.14 Consider the following probability distribution:
NW

x	0	1	4
$p(x)$	$^1/_3$	$^1/_3$	$^1/_3$

a. Find μ and σ^2.
b. Find the sampling distribution of the sample mean $\bar{x}$ for a random sample of $n = 2$ measurements from this distribution.
c. Show that $\bar{x}$ is an unbiased estimator of μ. [*Hint:* Show that $E(\bar{x}) = \Sigma\bar{x}p(\bar{x}) = \mu$.]
d. Find the sampling distribution of the sample variance s^2 for a random sample of $n = 2$ measurements from this distribution.
e. Show that s^2 is an unbiased estimator for σ^2.

6.15 Consider the following probability distribution:

x	2	4	9
$p(x)$	$^1/_3$	$^1/_3$	$^1/_3$

a. Calculate μ for this distribution.
b. Find the sampling distribution of the sample mean $\bar{x}$ for a random sample of $n = 3$ measurements from this distribution, and show that $\bar{x}$ is an unbiased estimator of μ.

c. Find the sampling distribution of the sample median M for a random sample of $n = 3$ measurements from this distribution, and show that the median is a biased estimator of μ.
d. If you wanted to use a sample of three measurements from this population to estimate μ, which estimator would you use? Why?

6.16 Consider the following probability distribution:
NW

x	0	1	2
$p(x)$	$^1/_3$	$^1/_3$	$^1/_3$

a. Find μ.
b. For a random sample of $n = 3$ observations from this distribution, find the sampling distribution of the sample mean.
c. Find the sampling distribution of the median of a sample of $n = 3$ observations from this population.
d. Refer to parts **b** and **c**, and show that both the mean and median are unbiased estimators of μ for this population.
e. Find the variances of the sampling distributions of the sample mean and the sample median.
f. Which estimator would you use to estimate μ? Why?

6.17 Use the computer to generate 750 samples, each containing $n = 20$ measurements, from a population that contains values of x equal to 1, 2, ..., 48, 49, 50. Assume that these values of x are equally likely. Calculate the sample mean $\bar{x}$ and median M for each sample. Construct relative frequency histograms for the 750 values of $\bar{x}$ and the 750 values of M. Use these approximations to the sampling distributions of $\bar{x}$ and M to answer the following questions:
a. Does it appear that $\bar{x}$ and M are unbiased estimators of the population mean? [*Note:* $\mu = 25.5$.]
b. Which sampling distribution displays greater variation?

6.18   Refer to Exercise 6.5.
  a. Show that $\bar{x}$ is an unbiased estimator of μ.
  b. Find $\sigma_{\bar{x}}^2$.
  c. Find the probability that the $\bar{x}$ will fall within $2\sigma_{\bar{x}}\mu$

6.19   Refer to Exercise 6.5.
  a. Find the sampling distribution of s^2.
  b. Find the population variance σ^2.
  c. Show that s^2 is an unbiased estimator of σ^2.

  d. Find the sampling distribution of the sample standard deviation s.
  e. Show that s is a biased estimator of σ.

6.20   Refer to Exercise 6.7, in which we found the sampling distribution of the sample median. Is the median an unbiased estimator of the population mean μ?

6.3 The Sampling Distribution of x̄ and the Central Limit Theorem

Estimating the mean useful life of automobiles, the mean number of crimes per month in a large city, and the mean yield per acre of a new soybean hybrid are practical problems with something in common. In each case, we are interested in making an inference about the mean μ of some population. As we mentioned in Chapter 2, the sample mean $\bar{x}$ is, in general, a good estimator of μ. We now develop pertinent information about the sampling distribution for this useful statistic. We will show that $\bar{x}$ is the minimum-variance unbiased estimator (MVUE) of μ.

Example   6.7

Describing the Sampling Distribution of x̄

Problem Suppose a population has the uniform probability distribution given in Figure 6.8. Then the mean and standard deviation of this probability distribution are, respectively, $\mu = 175$ and $\sigma = 14.43$. (See Section 5.2 for the formulas for μ and σ.) Now suppose a sample of 11 measurements is selected from this population. Describe the sampling distribution of the sample mean $\bar{x}$ based on the 1,000 sampling experiments discussed in Example 6.3.

Figure 6.8
Sampled uniform population

Solution Recall that in Example 6.3 we generated 1,000 samples of $n = 11$ measurements each. The MINITAB histogram for the 1,000 sample means is shown in Figure 6.9, with a normal probability distribution superimposed. You can see that this normal probability distribution approximates the computer-generated sampling distribution very well.

  To fully describe a normal probability distribution, it is necessary to know its mean and standard deviation. MINITAB gives these statistics for the 1,000 $\bar{x}$'s in the upper right corner of the histogram of Figure 6.9. You can see that the mean is 175.2 and the standard deviation is 4.383.

  To summarize our findings based on 1,000 samples, each consisting of 11 measurements from a uniform population, the sampling distribution of $\bar{x}$ appears to be approximately normal with a mean of about 175 and a standard deviation of about 4.38.

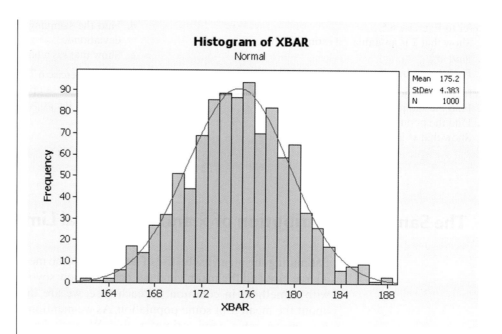

Figure 6.9
MINITAB histogram for
sample mean in 1,000 samples

Look Back Note that the simulated value $\mu_{\bar{x}} = 175.2$ is very close to $\mu = 175$ for the uniform distribution; that is, the simulated sampling distribution of $\bar{x}$ appears to provide an unbiased estimate of μ.

The true sampling distribution of $\bar{x}$ has the properties given in the next box, assuming only that a random sample of n observations has been selected from *any* population.

Properties of the Sampling Distribution of $\bar{x}$

1. The mean of the sampling distribution of $\bar{x}$ equals the mean of the sampled population. That is, $\mu_{\bar{x}} = E(\bar{x}) = \mu$.

2. The standard deviation of the sampling distribution of $\bar{x}$ equals

$$\frac{\text{Standard deviation of sampled population}}{\text{Square root of sample size}}$$

That is, $\sigma_{\bar{x}} = \sigma/\sqrt{n}$*

The standard deviation $\sigma_{\bar{x}}$ is often referred to as the **standard error of the mean**.

You can see that our approximation to $\mu_{\bar{x}}$ in Example 6.7 was precise, since property 1 assures us that the mean is the same as that of the sampled population: 175. Property 2 tells us how to calculate the standard deviation of the sampling distribution of $\bar{x}$. Substituting $\sigma = 14.43$ (the standard deviation of the sampled uniform distribution) and the sample size $n = 11$ into the formula for $\sigma_{\bar{x}}$, we find that

$$\sigma_{\bar{x}} = \frac{\sigma}{\sqrt{n}} = \frac{14.43}{\sqrt{11}} = 4.35$$

Thus, the approximation we obtained in Example 6.7, $\sigma_{\bar{x}} \approx 4.38$, is very close to the exact value, $\sigma_{\bar{x}} = 4.35$. It can be shown (proof omitted) that the value of $\sigma_{\bar{x}}^2$ is the smallest variance among all unbiased estimators of μ; thus, $\bar{x}$ is the MVUE for μ.

What about the shape of the sampling distribution? Two theorems provide this information. One is applicable whenever the original population data are normally

---

*If the sample size n is large relative to the number N of elements in the population (e.g., 5% or more), $\sigma/\sqrt{n}$ must be multiplied by the finite population correction factor $\sqrt{(N-n)/(N-1)}$. In most sampling situations, this correction factor will be close to 1 and can be ignored.

distributed. The other, applicable when the sample size n is large, represents one of the most important theoretical results in statistics: the **Central Limit Theorem**.

Theorem 6.1

If a random sample of n observations is selected from a population with a normal distribution, the sampling distribution of $\bar{x}$ will be a normal distribution.

Theorem 6.2: Central Limit Theorem

Consider a random sample of n observations selected from a population (*any* population) with mean μ and standard deviation σ. Then, when n is sufficiently large, the sampling distribution of $\bar{x}$ will be approximately a normal distribution with mean $\mu_{\bar{x}} = \mu$ and standard deviation $\sigma_{\bar{x}} = \sigma/\sqrt{n}$. The larger the sample size, the better will be the normal approximation to the sampling distribution of $\bar{x}$.*

Thus, for sufficiently large samples, the sampling distribution of $\bar{x}$ is approximately normal. How large must the sample size n be so that the normal distribution provides a good approximation to the sampling distribution of $\bar{x}$? The answer depends on the shape of the distribution of the sampled population, as shown by Figure 6.10. Generally speaking, the greater the skewness of the sampled population distribution, the larger the sample size must be before the normal distribution is an adequate approximation to the sampling distribution of $\bar{x}$. For most sampled populations, sample sizes of $n \geq 30$ will suffice for the normal approximation to be reasonable.

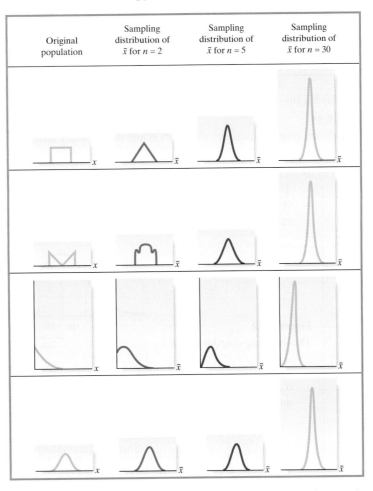

Figure 6.10

Sampling distributions of $\bar{x}$ for different populations and different sample sizes

*Moreover, because of the Central Limit Theorem, the sum of a random sample of n observations, Σx, will possess a sampling distribution that is approximately normal for large samples. This distribution will have a mean equal to $n\mu$ and a variance equal to $n\sigma^2$. Proof of the Central Limit Theorem is beyond the scope of this book, but it can be found in many mathematical statistics texts.

Example	6.8

Using the Central Limit Theorem to Find a Probability

Problem Suppose we have selected a random sample of $n = 36$ observations from a population with mean equal to 80 and standard deviation equal to 6. It is known that the population is not extremely skewed.

a. Sketch the relative frequency distributions for the population and for the sampling distribution of the sample mean $\bar{x}$.

b. Find the probability that $\bar{x}$ will be larger than 82.

Solution

a. We do not know the exact shape of the population relative frequency distribution, but we do know that it should be centered about $\mu = 80$, its spread should be measured by $\sigma = 6$, and it is not highly skewed. One possibility is shown in Figure 6.11a. From the Central Limit Theorem, we know that the sampling distribution of $\bar{x}$ will be approximately normal, since the sampled population distribution is not extremely skewed. We also know that the sampling distribution will have mean

$$\mu_{\bar{x}} = \mu = 80$$

and standard deviation $\qquad \sigma_{\bar{x}} = \dfrac{\sigma}{\sqrt{n}} = \dfrac{6}{\sqrt{36}} = 1$

The sampling distribution of $\bar{x}$ is shown in Figure 6.11b.

a. Population relative frequency distribution

b. Sampling distribution of $\bar{x}$

Figure 6.11
A population relative frequency distribution and the sampling distribution for $\bar{x}$

b. The probability that $\bar{x}$ will exceed 82 is equal to the highlighted area in Figure 6.12. To find this area, we need to find the z value corresponding to $\bar{x} = 82$. Recall that the standard normal random variable z is the difference of any normally distributed random variable and its mean, expressed in units of its standard deviation. Since $\bar{x}$ is

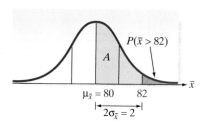

Figure 6.12
The sampling distribution of $\bar{x}$

a normally distributed random variable with mean $\mu_{\bar{x}} = \mu$ and standard deviation $\sigma_{\bar{x}} = \sigma/\sqrt{n}$, it follows that the standard normal z value corresponding to the sample mean $\bar{x}$ is

$$z = \frac{(\text{Normal random variable}) - (\text{Mean})}{\text{Standard deviation}} = \frac{\bar{x} - \mu_{\bar{x}}}{\sigma_{\bar{x}}}$$

Therefore, for $\bar{x} = 82$, we have

$$z = \frac{\bar{x} - \mu_{\bar{x}}}{\sigma_{\bar{x}}} = \frac{82 - 80}{1} = 2$$

The area A in Figure 6.12 corresponding to $z = 2$ is given in the table of areas under the normal curve (see Table II of Appendix B) as .4772. Therefore, the tail area corresponding to the probability that $\bar{x}$ exceeds 82 is

$$P(\bar{x} > 82) = P(z > 2) = .5 - .4772 = .0228$$

Look Back The key to finding the probability in part **b** is to recognize that the distribution of $\bar{x}$ is normal with $\mu_{\bar{x}} = \mu$ and $\sigma_{\bar{x}} = \sigma/\sqrt{n}$.

■ **Now Work Exercise 6.30**

Example   6.9

Application of the Central Limit Theorem—Testing a Manufacturer's Claim

Problem A manufacturer of automobile batteries claims that the distribution of the lengths of life of its best battery has a mean of 54 months and a standard deviation of 6 months. Suppose a consumer group decides to check the claim by purchasing a sample of 50 of the batteries and subjecting them to tests that estimate the battery's life.

a.   Assuming that the manufacturer's claim is true, describe the sampling distribution of the mean lifetime of a sample of 50 batteries.

b.   Assuming that the manufacturer's claim is true, what is the probability that the consumer group's sample has a mean life of 52 or fewer months?

Solution

a.   Even though we have no information about the shape of the probability distribution of the lives of the batteries, we can use the Central Limit Theorem to deduce that the sampling distribution for a sample mean lifetime of 50 batteries is approximately normally distributed. Furthermore, the mean of this sampling distribution is the same as the mean of the sampled population, which is $\mu = 54$ months according to the manufacturer's claim. Finally, the standard deviation of the sampling distribution is given by

$$\sigma_{\bar{x}} = \frac{\sigma}{\sqrt{n}} = \frac{6}{\sqrt{50}} = .85 \text{ month}$$

Note that we used the claimed standard deviation of the sampled population, $\sigma = 6$ months. Thus, if we assume that the claim is true, then the sampling distribution for the mean life of the 50 batteries sampled must be as shown in Figure 6.13.

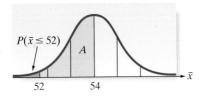

Figure 6.13
Sampling distribution of $\bar{x}$ in Example 6.9 for $n = 50$

b.   If the manufacturer's claim is true, the probability that the consumer group observes a mean battery life of 52 or fewer months for its sample of 50 batteries, $P(\bar{x} \leq 52)$, is equivalent to the highlighted area in Figure 6.13. Since the sampling distribution is approximately normal, we can find this area by computing the standard normal z value:

$$z = \frac{\bar{x} - \mu_{\bar{x}}}{\sigma_{\bar{x}}} = \frac{\bar{x} - \mu}{\sigma_{\bar{x}}} = \frac{52 - 54}{.85} = -2.35$$

Here, $\mu_{\bar{x}}$, the mean of the sampling distribution of $\bar{x}$, is equal to μ, the mean of the lifetimes of the sampled population, and $\sigma_{\bar{x}}$ is the standard deviation of the sampling distribution of $\bar{x}$. Note that z is the familiar standardized distance (z-score) of Section 2.6, and since $\bar{x}$ is approximately normally distributed, it will possess the standard normal distribution of Section 5.3.

The area A shown in Figure 6.13 between $\bar{x} = 52$ and $\bar{x} = 54$ (corresponding to $z = -2.35$) is found in Table II of Appendix B to be .4906. Therefore, the area to the left of $\bar{x} = 52$ is

$$P(\bar{x} \leq 52) = .5 - A = .5 - .4906 = .0094$$

Thus, the probability that the consumer group will observe a sample mean of 52 or less is only .0094 if the manufacturer's claim is true.

> **Look Back** If the 50 batteries tested do exhibit a mean of 52 or fewer months, the consumer group will have strong evidence that the manufacturer's claim is untrue because such an event is very unlikely to occur if the claim is true. (This is still another application of the *rare-event approach* to statistical inference.)

■ **Now Work Exercise 6.34**

We conclude this section with three comments on the sampling distribution of $\bar{x}$. First, from the formula $\sigma_{\bar{x}} = \sigma/\sqrt{n}$, we see that the standard deviation of the sampling distribution of $\bar{x}$ gets smaller as the sample size n gets larger. For example, we computed $\sigma_{\bar{x}} = .85$ when $n = 50$ in Example 6.9. However, for $n = 100$, we obtain $\sigma_{\bar{x}} = \sigma/\sqrt{n} = 6/\sqrt{100} = .60$. This relationship will hold true for most of the sample statistics encountered in this text. That is, *the standard deviation of the sampling distribution decreases as the sample size increases.* Consequently, the larger the sample size, the more accurate the sample statistic (e.g., $\bar{x}$) is in estimating a population parameter (e.g., μ). We will use this result in Chapter 7 to help us determine the sample size needed to obtain a specified accuracy of estimation.

Our second comment concerns the Central Limit Theorem. In addition to providing a very useful approximation for the sampling distribution of a sample mean, the Central Limit Theorem offers an explanation for the fact that many relative frequency distributions of data possess mound-shaped distributions. Many of the measurements we take in various areas of research are really means or sums of a large number of small phenomena. For example, a year's growth of a pine seedling is the total of the numerous individual components that affect the plant's growth. Similarly, we can view the length of time a construction company takes to build a house as the total of the times taken to complete a multitude of distinct jobs, and we can regard the monthly demand for blood at a hospital as the total of the many individual patients' needs. Whether or not the observations entering into these sums satisfy the assumptions basic to the Central Limit Theorem is open to question; however, it is a fact that many distributions of data in nature are mound shaped and possess the appearance of normal distributions.

Finally, it is important to understand when to use σ or $\sigma/\sqrt{n}$ in your statistical analysis. If the statistical inference you want to make concerns a single value of a random variable—for example, *the life length x of one randomly selected battery*—then use σ, the standard deviation of the probability distribution for x in your calculations. Alternatively, if the statistical inference you want to make concerns the sample mean—for example, *the mean life length $\bar{x}$ for a random sample of n batteries*—then use $\sigma/\sqrt{n}$, the standard deviation of the sampling distribution of $\bar{x}$.

| Statistics IN Action | Revisited | ## Making an Inference about the Mean Sleep Onset Latency for Insomnia Pill Takers |

In a Massachusetts Institute of Technology (MIT) study, each member of a sample of 40 young male volunteers was given a dosage of the sleep-inducing hormone melatonin, was placed in a dark room at midday, and was told to close his eyes for 30 minutes. The researchers measured the time (in minutes) elapsed before each volunteer fell asleep—called sleep onset latency. Recall that the data (shown in Table SIA6.1) are saved in the **INSOMNIA** file.

Previous research established that with the placebo (i.e., no hormone), the mean sleep onset latency is $\mu = 15$ minutes and the standard deviation is $\sigma = 10$ minutes. If the true value of μ for those taking the melatonin is $\mu < 15$

(i.e., if, on average, the volunteers fall asleep faster with the drug than with the placebo), then the researchers can infer that melatonin is an effective drug against insomnia.

Descriptive statistics for the 40 sleep times are displayed in the SPSS printout shown in Figure SIA6.1. You can see that the mean for the sample is $\bar{x} = 5.935$ minutes. If the drug is not effective in reducing sleep times, then the distribution of sleep times will be no different from the distribution with the placebo. That is, if the drug is not effective, the mean and standard deviation of the population of sleep times are $\mu = 15$ and $\sigma = 10$. If this is true, how likely is it to observe a sample mean that is below 6 minutes?

Descriptive Statistics

	N	Minimum	Maximum	Mean	Std. Deviation
SLEEPTIME	40	1.3	22.8	5.935	5.3917
Valid N (listwise)	40				

Figure SIA6.1
SPSS descriptive statistics for sleep time data

To answer this question, we desire the probability $P(\bar{x} < 6)$. To find this probability, we invoke the Central Limit Theorem. According to the theorem, the sampling distribution of $\bar{x}$ has the following mean and standard deviation:

$$\mu_{\bar{x}} = \mu = 15$$
$$\sigma_{\bar{x}} = \sigma/\sqrt{n} = 10/\sqrt{40} = 1.58$$

The theorem also states that $\bar{x}$ is approximately normally distributed. Therefore, we find the desired probability (using the standard normal table) as follows:

$$P(\bar{x} < 6) = P\left(z < \frac{6 - \mu_x}{\sigma_{\bar{x}}}\right)$$

$$= P\left(z < \frac{6 - 15}{1.58}\right) = P(z < -5.70) \approx 0$$

In other words, the probability that we observe a sample mean below $\bar{x} = 6$ minutes *if the mean and standard deviation of the sleep times are*, respectively, $\mu = 15$ and $\sigma = 10$ *(i.e., if the drug is not effective)*, is almost 0. Therefore, either the drug is not effective and the researchers have observed an extremely rare event (one with almost no chance of happening), or the true value of μ for those taking the melatonin pill is much less than 15 minutes. The rare-event approach to making statistical inferences, of course, would favor the second conclusion. Melatonin appears to be an effective insomnia pill, one that lowers the average time it takes the volunteers to fall asleep.

 Data Set: INSOMNIA

Exercises 6.21–6.48

Understanding the Principles

6.21 What do the symbols $\mu_{\bar{x}}$ and $\sigma_{\bar{x}}$ represent?

6.22 How does the mean of the sampling distribution of $\bar{x}$ relate to the mean of the population from which the sample is selected?

6.23 How does the standard deviation of the sampling distribution of $\bar{x}$ relate to the standard deviation of the population from which the sample is selected?

6.24 Another name given to the standard deviation of $\bar{x}$ is the _____. (Fill in the blank.)

6.25 State the Central Limit Theorem.

6.26 Will the sampling distribution of $\bar{x}$ always be approximately normally distributed? Explain.

Learning the Mechanics

6.27 Suppose a random sample of n measurements is selected from a population with mean $\mu = 225$ and variance $\sigma^2 = 225$. For each of the following values of n, give the mean and standard deviation of the sampling distribution of the sample mean $\bar{x}$.
  a. $n = 9$ **b.** $n = 49$
  c. $n = 225$ **d.** $n = 500$
  e. $n = 1,000$ **f.** $n = 2,500$

6.28 Suppose a random sample of $n = 16$ measurements is selected from a population with mean μ and standard deviation σ. For each of the following values of μ and σ, give the values of $\mu_{\bar{x}}$ and $\sigma_{\bar{x}}$.
  a. $\mu = 12, \sigma = 3$
  b. $\mu = 144, \sigma = 16$
  c. $\mu = 24, \sigma = 28$
  d. $\mu = 12, \sigma = 96$

6.29 Consider the following probability distribution:

x	3	4	6	9
$p(x)$	.3	.1	.5	.1

  a. Find μ, σ^2, and σ.
  b. Find the sampling distribution of $\bar{x}$ for random samples of $n = 2$ measurements from this distribution by listing all possible values of $\bar{x}$, and find the probability associated with each.
  c. Use the results of part **b** to calculate $\mu_{\bar{x}}$ and $\sigma_{\bar{x}}$. Confirm that $\mu_{\bar{x}} = \mu$ and that $\sigma_{\bar{x}} = \sigma/\sqrt{n} = \sigma/\sqrt{2}$.

6.30 A random sample of $n = 64$ observations is drawn from a population with a mean equal to 20 and standard deviation equal to 16.
  a. Give the mean and standard deviation of the (repeated) sampling distribution of $\bar{x}$.
  b. Describe the shape of the sampling distribution of $\bar{x}$. Does your answer depend on the sample size?
  c. Calculate the standard normal z-score corresponding to a value of $\bar{x} = 16$.
  d. Calculate the standard normal z-score corresponding to $\bar{x} = 23$.
  e. Find $P(\bar{x} < 16)$.
  f. Find $P(\bar{x} > 23)$.
  g. Find $P(16 < \bar{x} < 23)$.

6.31 A random sample of $n = 100$ observations is selected from a population with $\mu = 31$ and $\sigma = 25$.
  a. Find $\mu_{\bar{x}}$ and $\sigma_{\bar{x}}$.
  b. Describe the shape of the sampling distribution of $\bar{x}$.
  c. Find $P(\bar{x} \geq 28)$.

d. Find $P(22.1 \le \bar{x} \le 26.8)$.

e. Find $P(\bar{x} \le 28.2)$.

f. Find $P(\bar{x} \ge 27.0)$.

6.32 A random sample of $n = 2{,}500$ observations is selected from a population with $\mu = 500$ and $\sigma = 150$.

a. What are the largest and smallest values of $\bar{x}$ that you would expect to see?

b. How far, at the most, would you expect $\bar{x}$ to deviate from μ?

c. Did you have to know μ to answer part **b**? Explain.

6.33 Consider a population that contains values of x equal to $0, 1, 2, \ldots, 97, 98, 99$. Assume that the values of x are equally likely. For each of the following values of n, use the computer to generate 500 random samples and calculate $\bar{x}$ for each sample. For each sample size, construct a relative frequency histogram of the 500 values of $\bar{x}$. What changes occur in the histograms as the value of n increases? What similarities exist? Use $n = 2, n = 10, n = 25, n = 50$, and $n = 100$.

Ⓓ Applet Exercise 6.1

Open the applet entitled *Sampling Distribution*. On the pull-down menu to the right of the top graph, select *Binary*.

a. Run the applet for the sample size $n = 10$ and the number of samples $N = 1000$. Observe the shape of the graph of the sample proportions, and record the mean, median, and standard deviation of the sample proportions.

b. How does the mean of the sample proportions compare with the mean $\mu = 0.5$ of the original distribution?

c. Use the formula $\sigma = \sqrt{np(1-p)}$, where $n = 1$ and $p = 0.5$, to compute the standard deviation of the original distribution. Divide the result by $\sqrt{10}$, the square root of the sample size used in the sampling distribution. How does this result compare with the standard deviation of the sample proportions?

d. Explain how the graph of the distribution of sample proportions suggests that the distribution may be approximately normal.

e. Explain how the results of parts **b–d** illustrate the Central Limit Theorem.

Ⓓ Applet Exercise 6.2

Open the applet entitled *Sampling Distributions*. On the pull-down menu to the right of the top graph, select *Uniform*. The box to the left of the top graph displays the population mean, median, and standard deviation of the original distribution.

a. Run the applet for the sample size $n = 30$ and the number of samples $N = 1000$. Observe the shape of the graph of the sample means, and record the mean, median, and standard deviation of the sample means.

b. How does the mean of the sample means compare with the mean of the original distribution?

c. Divide the standard deviation of the original distribution by $\sqrt{30}$, the square root of the sample size used in the sampling distribution. How does this result compare with the standard deviation of the sample proportions?

d. Explain how the graph of the distribution of sample means suggests that the distribution may be approximately normal.

e. Explain how the results of parts **b–d** illustrate the Central Limit Theorem.

Applying the Concepts—Basic

6.34 **Corporate sustainability of CPA firms.** Refer to the *Business and Society* (Mar. 2011) study on the sustainability behaviors of CPA corporations, Exercise 1.26 (p. 50). Recall that *corporate sustainability* refers to business practices designed around social and environmental considerations. The level of support senior managers have for corporate sustainability was measured quantitatively on a scale ranging from 0 to 160 points, where higher point values indicate a higher level of support for sustainability. The study provided the following information on the distribution of levels of support for sustainability: $\mu = 68, \sigma = 27$. Now consider a random sample of 30 senior managers and let $\bar{x}$ represent the sample mean level of support.

a. Give the value of $\mu_{\bar{x}}$, the mean of the sampling distribution of $\bar{x}$, and interpret the result.

b. Give the value of $\sigma_{\bar{x}}$, the standard deviation of the sampling distribution of $\bar{x}$, and interpret the result.

c. What does the Central Limit Theorem say about the shape of the sampling distribution of $\bar{x}$?

d. Find $P(\bar{x} > 65)$.

6.35 **Voltage sags and swells.** Refer to the *Electrical Engineering* (Vol. 95, 2013) study of the power quality (sags and swells) of a transformer, Exercise 2.127 (p. 110). For transformers built for heavy industry, the distribution of the number of sags per week has a mean of 353 with a standard deviation of 30. Of interest is $\bar{x}$, the sample mean number of sags per week for a random sample of 45 transformers.

a. Find $E(\bar{x})$ and interpret its value.

b. Find $\text{Var}(\bar{x})$.

c. Describe the shape of the sampling distribution of $\bar{x}$.

d. How likely is it to observe a sample mean number of sags per week that exceeds 400?

6.36 **Phishing attacks to e-mail accounts.** In Exercise 2.47 (p. 81), you learned that *phishing* describes an attempt to extract personal/financial information from unsuspecting people through fraudulent e-mail. Data from an actual phishing attack against an organization were presented in *Chance* (Summer 2007). The interarrival times, i.e., the time differences (in seconds), for 267 fraud box e-mail notifications were recorded and are saved in the **PHISH** file. For this exercise, consider these interarrival times to represent the population of interest.

a. In Exercise 2.47 you constructed a histogram for the interarrival times. Describe the shape of the population of interarrival times.

b. Find the mean and standard deviation of the population of interarrival times.

c. Now consider a random sample of $n = 40$ interarrival times selected from the population. Describe the shape of the sampling distribution of $\bar{x}$, the sample mean. Theoretically, what are $\mu_{\bar{x}}$ and $\sigma_{\bar{x}}$?

d. Find $P(\bar{x} < 90)$.

e. Use a random number generator to select a random sample of $n = 40$ interarrival times from the population, and calculate the value of $\bar{x}$. (Every student in the class should do this.)

f. Refer to part **e**. Obtain the values of $\bar{x}$ computed by the students and combine them into a single data set. Form a histogram for these values of $\bar{x}$. Is the shape approximately normal?

g. Refer to part **f.** Find the mean and standard deviation of the $\bar{x}$-values. Do these values approximate $\mu_{\bar{x}}$ and $\sigma_{\bar{x}}$, respectively?

6.37 **Physical activity of obese young adults.** In a study on the physical activity of young adults, pediatric researchers measured overall physical activity as the total number of registered movements (counts) over a period of time and then computed the number of counts per minute (cpm) for each subject (*International Journal of Obesity*, Jan. 2007). The study revealed that the overall physical activity of obese young adults has a mean of $\mu = 320$ cpm and a standard deviation of $\sigma = 100$ cpm. (In comparison, the mean for young adults of normal weight is 540 cpm.) In a random sample of $n = 100$ obese young adults, consider the sample mean counts per minute, $\bar{x}$.

a. Describe the sampling distribution of $\bar{x}$.
b. What is the probability that the mean overall physical activity level of the sample is between 300 and 310 cpm?
c. What is the probability that the mean overall physical activity level of the sample is greater than 360 cpm?

6.38 **Cost of unleaded fuel.** According to the American Automobile Association (AAA), the average cost of a gallon of regular unleaded fuel at gas stations in May 2014 was $3.65 (*AAA Fuel Gauge Report*). Assume that the standard deviation of such costs is $.15. Suppose that a random sample of $n = 100$ gas stations is selected from the population and the cost per gallon of regular unleaded fuel is determined for each. Consider $\bar{x}$, the sample mean cost per gallon.

a. Calculate $\mu_{\bar{x}}$ and $\sigma_{\bar{x}}$.
b. What is the approximate probability that the sample has a mean fuel cost between $3.65 and $3.67?
c. What is the approximate probability that the sample has a mean fuel cost that exceeds $3.67?
d. How would the sampling distribution of $\bar{x}$ change if the sample size n were doubled from 100 to 200? How do your answers to parts **b** and **c** change when the sample size is doubled?

6.39 **Requests to a Web server.** In Exercise 5.10 (p. 266) you learned that Brighton Webs LTD modeled the arrival time of requests to a Web server within each hour using a uniform distribution. Specifically, the number of seconds x from the start of the hour that the request is made is uniformly distributed between 0 and 3,600 seconds. In a random sample of $n = 75$ Web server requests, let $\bar{x}$ represent the sample mean number of seconds from the start of the hour that the request is made.

a. Find $E(\bar{x})$ and interpret its value.
b. Find $\text{Var}(\bar{x})$.
c. Describe the shape of the sampling distribution of $\bar{x}$.
d. Find the probability that $\bar{x}$ is between 1,900 and 1,950 seconds.
e. Find the probability that $\bar{x}$ exceeds 2,150 seconds..

Applying the Concepts—Intermediate

6.40 **Shell lengths of sea turtles.** Refer to the *Aquatic Biology* (Vol. 9, 2010) study of green sea turtles inhabiting the Grand Cayman South Sound lagoon, Exercise 2.85 (p. 97). Research shows that the curved carapace (shell) lengths of these turtles has a distribution with mean $\mu = 50$ cm and standard deviation $\sigma = 10$ cm. In the study, $n = 76$ green sea turtles were captured from the lagoon; the mean shell length for the sample was $\bar{x} = 55.5$ cm. How likely is it to observe a sample mean of 55.5 cm or larger?

6.41 **Tomato as a taste modifier.** Miraculin is a protein naturally produced in a rare tropical fruit that can convert a sour taste into a sweet taste. Refer to the *Plant Science* (May 2010) investigation of the ability of a hybrid tomato plant to produce miraculin, Exercise 5.38 (p. 279). Recall that the amount x of miraculin produced in the plant had a mean of 105.3 micrograms per gram of fresh weight with a standard deviation of 8.0. Consider a random sample of $n = 64$ hybrid tomato plants and let $\bar{x}$ represent the sample mean amount of miraculin produced. Would you expect to observe a value of $\bar{x}$ less than 103 micrograms per gram of fresh weight? Explain.

6.42 **Uranium in the Earth's crust.** Refer to the *American Mineralogist* (Oct. 2009) study of the evolution of uranium minerals in the Earth's crust, Exercise 5.9 (p. 266). Recall that researchers estimate that the trace amount of uranium x in reservoirs follows a uniform distribution ranging between 1 and 3 parts per million. In a random sample of $n = 60$ reservoirs, let $\bar{x}$ represent the sample mean amount of uranium.

a. Find $E(\bar{x})$ and interpret its value.
b. Find $\text{Var}(\bar{x})$.
c. Describe the shape of the sampling distribution of $\bar{x}$.
d. Find the probability that $\bar{x}$ is between 1.5 ppm and 2.5 ppm.
e. Find the probability that $\bar{x}$ exceeds 2.2 ppm.

6.43 **Critical part failures in NASCAR vehicles.** Refer to *The Sport Journal* (Winter 2007) analysis of critical part failures at NASCAR races, Exercise 5.106 (p. 299). Recall that researchers found that the time x (in hours) until the first critical part failure is exponentially distributed with $\mu = .10$ and $\sigma = .10$. Now, consider a random sample of $n = 50$ NASCAR races and let $\bar{x}$ represent the sample mean time until the first critical part failure.

a. Find $E(\bar{x})$ and $\text{Var}(\bar{x})$.
b. Although x has an exponential distribution, the sampling distribution of $\bar{x}$ is approximately normal. Why?
c. Find the probability that the sample mean time until the first critical part failure exceeds .13 hour.

6.44 **Motivation of drug dealers.** Refer to the *Applied Psychology in Criminal Justice* (Sept. 2009) investigation of the personality characteristics of drug dealers, Exercise 2.102 (p. 105). Convicted drug dealers were scored on the Wanting Recognition (WR) Scale—a scale that provides a quantitative measure of a person's level of need for approval and sensitivity to social situations. (Higher scores indicate a greater need for approval.) Based on the study results, we can assume that the WR scores for the population of convicted drug dealers have a mean of 40 and a standard deviation of 5. Suppose that in a sample of $n = 100$ people, the mean WR scale score is $\bar{x} = 42$. Is this sample likely to have been selected from the population of convicted drug dealers? Explain.

6.45 **Is exposure to a chemical in Teflon-coated cookware hazardous?** Perfluorooctanoic acid (PFOA) is a chemical used in Teflon®-coated cookware to prevent food from sticking. The Environmental Protection Agency is investigating the potential risk of PFOA as a cancer-causing agent (*Science News Online*, Aug. 27, 2005). It is known that the blood concentration of PFOA in the general population has a mean of $\mu = 6$ parts per billion (ppb) and a standard deviation of $\sigma = 10$ ppb. *Science News Online* reported on tests for PFOA exposure conducted on a sample of 326 people who live near DuPont's Teflon-making Washington (West Virginia) Works facility.
   a. What is the probability that the average blood concentration of PFOA in the sample is greater than 7.5 ppb?
   b. The actual study resulted in $\bar{x} = 300$ ppb. Use this information to make an inference about the true mean (μ) PFOA concentration for the population that lives near DuPont's Teflon facility.

6.46 **Characteristics of antiwar demonstrators.** Refer to the *American Journal of Sociology* (Jan. 2014) study of the characteristics of antiwar demonstrators in the United States, Exercise 2.106 (p. 105). The researchers found that the mean number of protest organizations joined by antiwar demonstrators over a recent 3-year period was .90 with a standard deviation of 1.10. Assume that these values represent the true mean μ and true standard deviation σ of the population of all antiwar demonstrators over the 3 years.
   a. In a sample of 50 antiwar demonstrators selected from the population, what is the expected value of $\bar{x}$, the sample mean number of protest organizations joined by the demonstrators?
   b. Find values of the sample mean, L and U, such that $P(L < \bar{x} < U) = .95$.

Applying the Concepts—Advanced

6.47 **Hand washing versus hand rubbing.** Refer to the *British Medical Journal* (Aug. 17, 2002) study comparing the effectiveness of hand washing with soap and hand rubbing with alcohol, presented in Exercise 2.107 (p. 106). Health care workers who used hand rubbing had a mean bacterial count of 35 per hand with a standard deviation of 59. Health care workers who used hand washing had a mean bacterial count of 69 per hand with a standard deviation of 106. In a random sample of 50 health care workers, all using the same method of cleaning their hands, the mean bacterial count per hand ($\bar{x}$) is less than 30. Give your opinion on whether this sample of workers used hand rubbing with alcohol or hand washing with soap.

6.48 **Video game players and divided attention tasks.** *Human Factors* (May 2014) published the results of a study designed to determine whether video game players are better than non-video game players at crossing the street when presented with distractions. Participants (college students) entered a street crossing simulator. The simulator was designed to have cars traveling at various high rates of speed in both directions. During the crossing, the students also performed a memory task as a distraction. The researchers found that students who are video game players took an average of 5.1 seconds to cross the street, with a standard deviation of .8 seconds. Assume that the time, x, to cross the street for the population of video game players has $\mu = 5.1$ and $\sigma = .8$. Now consider a sample of 30 students and let $\bar{x}$ represent the sample mean time (in seconds) to cross the street in the simulator.
   a. Find $P(\bar{x} > 5.5)$.
   b. The 30 students in the sample are all non-video game players. What inference can you make about μ and/or σ for the population of non-video game players? Explain.

6.4 The Sampling Distribution of the Sample Proportion

Suppose you want to estimate the proportion of voters in favor of a bill to legalize gambling in your state, or the percentage of store customers who use a store credit card to make a purchase, or the fraction of giant Pandas born in captivity. In each case, the data of interest are categorical in nature with two outcomes (e.g., favor or do not favor); consequently, we want to make an inference about a binomial proportion, p.

Just as the sample mean is a good estimator of the population mean, the sample proportion—denoted $\hat{p}$—is a good estimator of the population proportion p. How good the estimator $\hat{p}$ is will depend on the sampling distribution of the statistic. This sampling distribution has properties similar to those of the sampling distribution of $\bar{x}$, as shown in the following example.

Example **6.10**

Simulating the Sampling Distribution of $\hat{p}$—Voting for Mayor

Problem Refer to Example 4.13 (p. 236) and the problem of sampling city voters to determine whether they are in favor of a certain candidate for mayor. We assumed that 60% of all the city's voters favor the candidate. Suppose we will randomly sample 100 voters and ask each if he or she is or is not in favor of the candidate. Use a computer and statistical software to simulate the sampling distribution of $\hat{p}$, the sample proportion of voters in favor of the candidate. What properties do you observe?

Solution Consider a single random sample of 100 voters. If we define a "success" as a voter in favor of the candidate, and if 58 of the 100 are in favor of the candidate, then our estimate of the binomial proportion p is $\hat{p} = 58/100 = .58$. By repeatedly taking samples in this manner and calculating $\hat{p}$ each time, we can simulate the sampling distribution.

We used SPSS to generate 1,000 samples of size $n = 100$ from a binomial population with probability of success $p = .6$. For each sample, we calculated $\hat{p} = x/n$, where x is the number of successes in the sample. The estimated proportions for the first 20 samples are shown in Table 6.5. An SPSS histogram for all 1,000 estimated proportions is shown in Figure 6.14. This histogram approximates the sampling distribution of $\hat{p}$.

Table 6.5	Results for First 20 Samples of n = 100 Voters	
Sample	Number of Success, x	Sample Proportion, $\hat{p}$
1	60	.60
2	58	.58
3	56	.56
4	61	.61
5	54	.54
6	67	.67
7	54	.54
8	55	.55
9	56	.56
10	69	.69
11	59	.59
12	64	.64
13	53	.53
14	65	.65
15	66	.66
16	61	.61
17	60	.60
18	55	.55
19	57	.57
20	61	.61

Descriptive Statistics

	N	Minimum	Maximum	Mean	Std. Deviation
phat	1000	.44	.74	.5988	.04823
Valid N (listwise)	1000				

Figure 6.14

SPSS histogram of 1,000 values of $\hat{p}$, Example 6.10

First, note that the sampling distribution appears to be normally distributed. Thus, the sampling distribution of $\hat{p}$ has the same property as the sampling distribution of $\bar{x}$—normally distributed for large n. Second, you can see that the sampling distribution is centered around the value $p = .6$. That is, it appears that the mean of the sampling distribution is equal to p, the proportion being estimated.

Look Back The normal distribution property identified with computer simulation becomes obvious once you recognize that the sample proportion $\hat{p}$ can be written as an average of 0s and 1s, where 1 represents a "success" (a voter in favor of the candidate) and 0 represents a "failure" (a voter not in favor of the candidate). In the first sample shown in Table 6.5, there were 60 successes and 40 failures; thus

$$\hat{p} = \frac{(60)(1) + (40)(0)}{100} = 60/100 = .60$$

Since $\hat{p}$ is a mean, the Central Limit Theorem can be applied when the sample size n is large.

The true sampling distribution of $\hat{p}$ has the properties listed in the next box.

Sampling Distribution of $\hat{p}$

1. Mean of the sampling distribution is equal to the true binomial proportion, p; that is, $E(\hat{p}) = p$. Consequently, $\hat{p}$ is an unbiased estimator of p.

2. Standard deviation of the sampling distribution is equal to $\sqrt{p(1 - p)/n}$; that is, $\sigma_{\hat{p}} = \sqrt{p(1 - p)/n}$.

3. For large samples, the sampling distribution is approximately normal. (A sample is considered large if $n\hat{p} \geq 15$ and $n(1 - \hat{p}) \geq 15$.)

In Example 6.10, $p = .6$. Thus, the true mean of the sampling distribution of $\hat{p}$ is equal to .6. Also, the true standard deviation of the sampling distribution of $\hat{p}$ is

$$\sigma_{\hat{p}} = \sqrt{p(1 - p)/n} = \sqrt{(.6)(.4)/100} = .049$$

Note that our estimates of the mean and standard deviation using simulation —.599 and .048, respectively—closely approximate the true values.

Example 6.11

Finding a Probability Involving $\hat{p}$—Voters in Favor of a Candidate

Problem Refer to Example 6.10 (p. 332). Again, assume that 60% of all city voters are in favor of the candidate. In a random sample of 100 voters, what is the probability that fewer than half are in favor of the candidate?

Solution We want to find the probability that the sample proportion is less than .5, i.e., $P(\hat{p} < .5)$. From previous work, we know that the sampling distribution of $\hat{p}$ is normally distributed (since n is large), with mean $p = .6$ and standard deviation $\sigma_{\hat{p}} = .049$. Consequently,

$$P(\hat{p} < .5) = P\{(\hat{p} - p)/\sigma_{\hat{p}} < (.5 - p)/\sigma_{\hat{p}}\} = P\{z < (.5 - .6)/.049\} = P(z < -2.04)$$

Using MINITAB software (see Figure 6.15), we find $P(\hat{p} < .5) = P(z < -2.04) = 0.21$.

Cumulative Distribution Function

```
Normal with mean = 0 and standard deviation = 1

   x    P( X <= x )
-2.04    0.0206752
```

Figure 6.15
MINITAB output for a normal probability, Example 6.11

■ **Now Work Exercise 6.52**

In the next two chapters, we utilize the sampling distributions of both $\bar{x}$ and $\hat{p}$ to make inferences about population means and proportions.

Exercises 6.49–6.62

Learning the Mechanics

6.49 Consider a sampling distribution with $p = .14$. and samples of size n each. For each value of n, give the mean and standard deviation of the sampling distribution of the sample proportion, $\hat{p}$.
  a. For a random sample of size $n = 4,000$.
  b. For a random sample of size $n = 1,000$.
  c. For a random sample of size $n = 500$.

6.50 Suppose a random sample of $n = 591$ measurements is selected from a binomial population with probability of success p. For each of the given values of p, give the mean and standard deviation of the sampling distribution of the sample proportion, $\hat{p}$.
  a. $p = .2$
  b. $p = .5$
  c. $p = .8$

6.51 A random sample of $n = 70$ measurements is drawn from a binomial population with probability of success .3.
  a. Give the mean and standard deviation of the sampling distribution of the sample proportion, $\hat{p}$.
  b. Describe the shape of the sampling distribution of $\hat{p}$.
  c. Calculate the standard normal z-score corresponding to a value of $\hat{p} = .33$.
  d. Find $P(\hat{p} > .33)$.

6.52 A random sample of $n = 250$ measurements is drawn from a binomial population with probability of success .85.
  [NW]
  a. Find $E(\hat{p})$ and $\sigma_{\hat{p}}$.
  b. Describe the shape of the sampling distribution of $\hat{p}$.
  c. Find $P(\hat{p} < .9)$.

6.53 A random sample of $n = 2,000$ measurements is drawn from a binomial population with probability of success .25. What are the smallest and largest values of $\hat{p}$ you would expect to see?

6.54 Consider a population with values of x equal to 0 or 1. Assume that the values of x are equally likely. For each of the following values of n, use a computer to generate 750 random samples and calculate the sample proportion of 1s observed for each sample. Then construct a histogram of the 750 sample proportions for each sample.
  a. $n = 5$
  b. $n = 50$
  c. $n = 250$
  d. Refer to histograms, part **a–c**. What changes occur as the value of n increases? What similarities exist?

Applying the Concepts—Basic

6.55 **Do social robots walk or roll?** Refer to the *International Conference on Social Robotics* (Vol. 6414, 2010) study of the trend in the design of social robots, Exercise 2.7 (p. 66). The researchers obtained a random sample of 106 social robots through a Web search and determined the number that were designed with legs, but no wheels. Let $\hat{p}$ represent the sample proportion of social robots designed with legs, but no wheels. Assume that in the population of all social robots, 40% are designed with legs, but no wheels.
  a. Give the mean and standard deviation of the sampling distribution of $\hat{p}$.
  b. Describe the shape of the sampling distribution of $\hat{p}$.
  c. Find $P(\hat{p} > .59)$.
  d. Recall that the researchers found that 63 of the 106 robots were built with legs only. Does this result cast doubt on the assumption that 40% of all social robots are designed with legs, but no wheels? Explain.

6.56 **Paying for music downloads.** According to a recent *Pew Internet & American Life Project Survey* (Oct. 2010), 67% of adults who use the Internet have paid to download music. In a random sample of $n = 1,000$ adults who use the Internet, let $\hat{p}$ represent the proportion who have paid to download music.
  a. Find the mean of the sampling distribution of $\hat{p}$.
  b. Find the standard deviation of the sampling distribution of $\hat{p}$.
  c. What does the Central Limit Theorem say about the shape of the sampling distribution of $\hat{p}$?
  d. Compute the probability that $\hat{p}$ is less than .75.
  e. Compute the probability that $\hat{p}$ is greater than .50.

6.57 **Working on summer vacation.** According to an *Adweek/ Harris* (July 2011) poll of U.S. adults, 35% do not work during their summer vacation. Assume that the true proportion of all U.S. adults that do not work during summer vacation is $p = .35$. Now consider a random sample of 500 U.S. adults.
  a. What is the probability that between 50% and 60% of the sampled adults do not work during summer vacation?
  b. What is the probability that over 40% of the sampled adults do not work during summer vacation?

6.58 **Superstitions survey.** A *Harris* (Feb. 2013) poll found that one-third of Americans (33%) believe finding and picking up a penny is good luck. Assume that the true proportion of all Americans who believe finding and picking up a penny is good luck is $p = .33$. In a random sample of 1,000 Americans, let $\hat{p}$ represent the proportion who believe finding and picking up a penny is good luck.
  a. Describe the properties of the sampling distribution of $\hat{p}$.
  b. Find $P(\hat{p} < .40)$. Interpret this result.
  c. Find $P(\hat{p} > .30)$. Interpret this result.

Applying the Concepts—Intermediate

6.59 **Crop damage by wild boars.** Refer to the *Current Zoology* (Apr. 2014) study of crop damage incurred by wild boars in southern Italy, Exercise 3.18 (p. 158). If crop damage by wild boars occurs, the probability that the type of crop damaged is cereals is .45, legumes .20, and orchards .05. In a random sample of 100 crop damage incidents caused by wild boars, what is the probability that at least 50 incidents involve
  a. cereal crops?
  b. legumes?
  c. orchards?

6.60 **Downloading "apps" to your cell phone.** Refer to Exercise 4.27 (p. 221). The study found that 40% of adult cell phone owners have downloaded an application ("app") to their cell phone. Assume this percentage applies to the population of all adult cell phone owners.

a. In a random sample of 50 adult cell phone owners, how likely is it to find that more than 60% have downloaded an "app" to their cell phone?

b. Refer to part **a.** Suppose you observe a sample proportion of .62. What inference can you make about the true proportion of adult cell phone owners who have downloaded an "app"?

c. Suppose the sample of 50 cell phone owners is obtained at a convention for the International Association for the Wireless Telecommunications Industry. How will your answer to part **b** change, if at all?

6.61 **Hotel guest satisfaction.** Refer to the results of the North American Hotel Guest Satisfaction Index Study referenced in Exercise 4.75 (p. 241). Recall that 66% of hotel guests were aware of the hotel's "green" conservation program; of these guests, 72% actually participated in the program by reusing towels and bed linens. In a random sample of 100 hotel guests, find the probability that fewer than 42 were aware of and participated in the hotel's conservation efforts.

6.62 **Fingerprint expertise.** Refer to the *Psychological Science* (Aug. 2011) study of fingerprint identification, Exercise 4.74 (p. 240). Recall that when presented with prints from the same individual, a fingerprint expert will correctly identify the match 92% of the time. Consider a forensic database of 1,000 different pairs of fingerprints, where each pair is a match.

a. What proportion of the 1,000 pairs would you expect an expert to correctly identify as a match?

b. What is the probability that an expert will correctly identify fewer than 900 of the fingerprint matches?

CHAPTER NOTES

Key Terms

Biased estimate 320
Central Limit Theorem 325
Error of estimation 319
Minimum-variance unbiased estimator (MVUE) 322
Parameter 312
Point estimator 319

Sample statistic 312
Sampling distribution 313
Standard error of the mean 324
Standard error of the statistic 320
Unbiased estimate 320

Key Ideas

Sampling distribution of a statistic — the theoretical probability distribution of the statistic in repeated sampling
Unbiased estimator — a statistic with a sampling distribution mean equal to the population parameter being estimated
Central Limit Theorem — the sampling distribution of the sample mean, $\bar{x}$, or the sample proportion, $\hat{p}$, is approximately normal for large n
$\bar{x}$ is the **minimum-variance unbiased estimator (MVUE)** of μ
$\hat{p}$ is the **MVUE** of p

Key Formulas

	Mean	Standard Deviation	z-Score
Sampling distribution of $\bar{x}$	$\mu_{\bar{x}} = \mu$	$\sigma_{\bar{x}} = \dfrac{\sigma}{\sqrt{n}}$	$z = \dfrac{\bar{x} - \mu_{\bar{x}}}{\sigma_{\bar{x}}} = \dfrac{\bar{x} - \mu}{\sigma/\sqrt{n}}$
Sampling distribution of $\hat{p}$	p	$\sigma_p = \sqrt{p(1-p)/n}$	$z = \dfrac{\hat{p} - p}{\sqrt{p(1-p)/n}}$

Key Symbols

θ	Population parameter (general)
$\mu_{\bar{x}}$	True mean of sampling distribution of $\bar{x}$
$\sigma_{\bar{x}}$	True standard deviation of sampling distribution of $\bar{x}$
p	True mean of the sampling distribution of $\hat{p}$
$\sigma_{\hat{p}}$	True standard deviation of the sampling distribution of $\hat{p}$

Generating the Sampling Distribution of $\bar{x}$

Select sample size n (large) from target population

Calculate $\bar{x}$

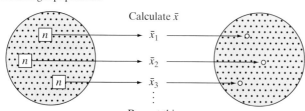

$\bar{x}_1$
$\bar{x}_2$
$\bar{x}_3$
⋮

Population:
Mean $= \mu$
Std. Dev. $= \sigma$
Unknown shape

Repeat this process an infinite number of times

Sampling distribution of $\bar{x}$
(i.e., theoretical population of x's):
Mean $= \mu_{\bar{x}} = \mu$
Std. Dev. $= \sigma_{\bar{x}} = \sigma/\sqrt{n}$
Normal distribution (Central Limit Theorem)

Supplementary Exercises 6.63–6.93

Understanding the Principles

6.63 **True or False.** The sample mean, $\bar{x}$, will always be equal to $\mu_{\bar{x}}$.

6.64 **True or False.** The sampling distribution of $\bar{x}$ is normally distributed, regardless of the size of the sample n.

6.65 **True or False.** The standard error of $\bar{x}$ will always be smaller than σ.

6.66 Describe how you could obtain the simulated sampling distribution of a sample statistic.

6.67 **True or False.** The sampling distribution of $\hat{p}$ is normally distributed for large n

Learning the Mechanics

6.68 The standard deviation (or, as it is usually called, the *standard error*) of the sampling distribution for the sample mean, $\bar{x}$, is equal to the standard deviation of the population from which the sample was selected, divided by the square root of the sample size. That is,

$$\sigma_{\bar{x}} = \frac{\sigma}{\sqrt{n}}$$

  a. As the sample size is increased, what happens to the standard error of $\bar{x}$? Why is this property considered important?

  b. Suppose a sample statistic has a standard error that is not a function of the sample size. In other words, the standard error remains constant as n changes. What would this imply about the statistic as an estimator of a population parameter?

  c. Suppose another unbiased estimator (call it A) of the population mean is a sample statistic with a standard error equal to

$$\sigma_A = \frac{\sigma}{\sqrt[3]{n}}$$

  Which of the sample statistics, $\bar{x}$ or A, is preferable as an estimator of the population mean? Why?

  d. Suppose that the population standard deviation σ is equal to 25 and that the sample size is 64. Calculate the standard errors of $\bar{x}$ and A. Assuming that the sampling distribution of A is approximately normal, interpret the standard errors. Why is the assumption of (approximate) normality unnecessary for the sampling distribution of $\bar{x}$?

6.69 Consider a sample statistic A. As with all sample statistics, A is computed by utilizing a specified function (formula) of the sample measurements. (For example, if A were the sample mean, the specified formula would sum the measurements and divide by the number of measurements.)

  a. Describe what we mean by the phrase "the sampling distribution of the sample statistic A."

  b. Suppose A is to be used to estimate a population parameter α. What is meant by the assertion that A is an unbiased estimator of α?

  c. Consider another sample statistic, B. Assume that B is also an unbiased estimator of the population parameter α. How can we use the sampling distributions of A and B to decide which is the better estimator of α?

  d. If the sample sizes on which A and B are based are large, can we apply the Central Limit Theorem and assert that the sampling distributions of A and B are approximately normal? Why or why not?

6.70 A random sample of 50 observations is to be drawn from a large population of measurements. It is known that 40% of the measurements in the population are 1's, 20% are 2's, 30% are 3's, and 10% are 4's.

  a. Give the mean and standard deviation of the (repeated) sampling distribution of $\bar{x}$, the sample mean of the 50 observations.

  b. Describe the shape of the sampling distribution of $\bar{x}$. Does your answer depend on the sample size?

6.71 A random sample of $n = 64$ observations is selected from a population with $\mu = 22.5$ and $\sigma = 2.8$. Approximate each of the following probabilities:

  a. $P(\bar{x} \le 22.5)$
  b. $P(\bar{x} \le 22)$
  c. $P(\bar{x} \ge 23.5)$
  d. $P(21.9 \le \bar{x} \le 22.8)$

6.72 A random sample of $n = 750$ observations is selected from a binomial population with $p = .45$.

  a. Give the mean and standard deviation of the (repeated) sampling distribution of $\hat{p}$, the sample proportion of successes for the 750 observations.

  b. Describe the shape of the sampling distribution of $\hat{p}$. Does your answer depend on the sample size?

6.73 A random sample of $n = 300$ observations is selected from a binomial population with $p = .8$. Approximate each of the following probabilities:

  a. $P(\hat{p} < .66)$
  b. $P(\hat{p} > .55)$
  c. $P(.58 < \hat{p} < .62)$

6.74 Use a statistical software package to generate 150 random samples of size $n = 5$ from a population characterized by a normal probability distribution with a mean of 75 and a standard deviation of 8. Compute $\bar{x}$ for each sample, and plot a frequency distribution for the 150 values of $\bar{x}$. Repeat this process for $n = 10, 25, 50$, and 75. How does the fact that the sampled population is normal affect the sampling distribution of $\bar{x}$?

6.75 Use a statistical software package to generate 150 random samples of size $n = 5$ from a population characterized by a uniform probability distribution with $c = 5$ and $d = 20$. Compute $\bar{x}$ for each sample, and plot a frequency distribution for the 150 $\bar{x}$ values. Repeat this process for $n = 10, 25, 50$, and 75. Explain how your plots illustrate the Central Limit Theorem.

6.76 Suppose x equals the number of heads observed when a single coin is tossed; that is, $x = 0$ or $x = 1$. The population corresponding to x is the set of 0's and 1's generated when the coin is tossed repeatedly a large number of times. Suppose we select $n = 4$ observations from this population. (That is, we toss the coin four times and observe 4 values of x.)

  a. List the different samples (combinations of 0's and 1's) that could be obtained.

  b. Calculate the value of $\bar{x}$ for each of the samples.

c. List the values that $\bar{x}$ can assume, and find the probabilities of observing these values.

d. Construct a graph of the sampling distribution of $\bar{x}$.

6.77 A random sample of size n is to be drawn from a large population with mean 150 and standard deviation 20, and the sample mean $\bar{x}$ is to be calculated. To see the effect of different sample sizes on the standard deviation of the sampling distribution of $\bar{x}$, plot $\sigma/\sqrt{n}$ against n for $n = 1, 2, 5, 10, 25, 50,$ and 75.

Applying the Concepts—Basic

6.78 Salaries of travel managers. According to a 2012 Business Travel News survey, the average salary of a travel manager is $110,550. Assume that the standard deviation of such salaries is $25,000. Consider a random sample of 40 travel managers, and let $\bar{x}$ represent the mean salary for the sample.

a. What is $\mu_{\bar{x}}$?

b. What is $\sigma_{\bar{x}}$?

c. Describe the shape of the sampling distribution of $\bar{x}$.

d. Find the z-score for the value $\bar{x} = \$100,000$.

e. Find $P(\bar{x} > 100,000)$.

6.79 Children's attitudes toward reading. In the journal *Knowledge Quest* (Jan./Feb. 2002), education professors investigated children's attitudes toward reading. One study measured third through sixth graders' attitudes toward recreational reading on a 140-point scale (where higher scores indicate a more positive attitude). The mean score for this population of children was 106 points, with a standard deviation of 16.4 points. Consider a random sample of 36 children from this population, and let $\bar{x}$ represent the mean recreational reading attitude score for the sample.

a. What is $\mu_{\bar{x}}$?

b. What is $\sigma_{\bar{x}}$?

c. Describe the shape of the sampling distribution of $\bar{x}$.

d. Find the z-score for the value $\bar{x} = 100$ points.

e. Find $P(\bar{x} < 100)$.

6.80 Dentists' use of laughing gas. According to the American Dental Association, 60% of all dentists use nitrous oxide in their practice. In a random sample of 100 dentists, let $\hat{p}$ represent the proportion who use laughing gas in their practice.

a. Find $E(\hat{p})$.

b. Find $\sigma_{\hat{p}}$.

c. Describe the shape of the sampling distribution of $\hat{p}$.

d. Find $P(\hat{p} > .65)$.

6.81 Marital name changing. Refer to the *Advances in Applied Sociology* (Nov. 2013) study of marital name changing, Exercise 4.78 (p. 241). According to the study, the probability that an American female will change her last name upon marriage is .9. Consider a sample of 250 American females. Let $\hat{p}$ represent the proportion of those females who change their last name upon marriage.

a. Find the mean and standard deviation of $\hat{p}$.

b. Calculate the interval, $E(\hat{p}) \pm 2\sigma_{\hat{p}}$.

c. If samples of 250 American females are drawn repeatedly a large number of times and the sample proportion, $\hat{p}$, determined for each sample, what proportion of the $\hat{p}$ values will fall within the interval, part **b**?

6.82 Violence and stress. Interpersonal violence (e.g., rape) generally leads to psychological stress for the victim. *Clinical Psychology Review* (Vol. 15, 1995) reported on the results of all recently published studies of the relationship between interpersonal violence and psychological stress. The distribution of the time elapsed between the violent incident and the initial sign of stress has a mean of 5.1 years and a standard deviation of 6.1 years. Consider a random sample of $n = 150$ victims of interpersonal violence. Let $\bar{x}$ represent the mean time elapsed between the violent act and the first sign of stress for the sampled victims.

a. Give the mean and standard deviation of the sampling distribution of $\bar{x}$.

b. Will the sampling distribution of $\bar{x}$ be approximately normal? Explain.

c. Find $P(\bar{x} > 5.5)$.

d. Find $P(4 < \bar{x} < 5)$.

6.83 Research on eating disorders. Refer to *The American Statistician* (May 2001) study of female students who suffer from bulimia, presented in Exercise 2.44 (p. 80). Recall that each student completed a questionnaire from which a "fear of negative evaluation" (FNE) score was produced. (The higher the score, the greater is the fear of negative evaluation.) Suppose the FNE scores of bulimic students have a distribution with mean $\mu = 18$ and standard deviation $\sigma = 5$. Now, consider a random sample of 45 female students with bulimia.

a. What is the probability that the sample mean FNE score is greater than 17.5?

b. What is the probability that the sample mean FNE score is between 18 and 18.5?

c. What is the probability that the sample mean FNE score is less than 18.5?

Applying the Concepts—Intermediate

6.84 Producing machine bearings. To determine whether a metal lathe that produces machine bearings is properly adjusted, a random sample of 36 bearings is collected and the diameter of each is measured.

a. If the standard deviation of the diameters of the bearings measured over a long period of time is .005 inch, what is the approximate probability that the mean diameter $\bar{x}$ of the sample of 36 bearings will lie within .0002 inch of the population mean diameter of the bearings?

b. If the population of diameters has an extremely skewed distribution, how will your approximation in part **a** be affected?

6.85 Quality control. Refer to Exercise 6.84. The mean diameter of the bearings produced by the machine is supposed to be .5 inch. The company decides to use the sample mean to decide whether the process is in control (i.e., whether it is producing bearings with a mean diameter of .5 inch). The machine will be considered out of control if the mean of the sample of $n = 36$.

Diameters is less than .5025 inch or larger than .5045 inch. If the true mean diameter of the bearings produced by the machine is .505 inch, what is the approximate probability that the test will imply that the process is out of control?

6.86 Errors in filling prescriptions A large number of preventable errors (e.g., overdoses, botched operations, misdiagnoses) are being made by doctors and nurses in U.S. hospitals. A study of a major metropolitan hospital revealed that of every 100 medications prescribed or dispensed, 1 was in error, but only 1 in 500 resulted in an error that caused significant problems for the patient. It is known that the hospital prescribes and dispenses 40,000 medications per year.

a. What is the expected proportion of errors per year at this hospital? The expected proportion of significant errors per year?

b. Within what limits would you expect the proportion of significant errors per year to fall?

6.87 Susceptibility to hypnosis. The Computer-Assisted Hypnosis Scale (CAHS) is designed to measure a person's susceptibility to hypnosis. CAHS scores range from 0 (no susceptibility) to 12 (extremely high susceptibility). A study in *Psychological Assessment* (Mar. 1995) reported a mean CAHS score of 4.59 and a standard deviation of 2.95 for University of Tennessee undergraduates. Assume that $\mu = 4.59$ and $\sigma = 2.95$ for this population. Suppose a psychologist uses the CAHS to test a random sample of 50 subjects.

a. Would you expect to observe a sample mean CAHS score of $\bar{x} = 6$ or higher? Explain.

b. Suppose the psychologist actually observes $\bar{x} = 6.2$. On the basis of your answer to part **a,** make an inference about the population from which the sample was selected.

6.88 Supercooling temperature of frogs. Many species of terrestrial tree frogs can survive prolonged exposure to low winter temperatures during hibernation. In freezing conditions, the frog's body temperature is called its *supercooling temperature*. A study in *Science* revealed that the supercooling temperature of terrestrial frogs frozen at $-6°C$ has a relative frequency distribution with a mean of $-2°C$ and a standard deviation of $.3°C$. Consider the mean supercooling temperature $\bar{x}$ of a random sample of $n = 42$ terrestrial frogs frozen at $-6°C$.

a. Find the probability that $\bar{x}$ exceeds $-2.05°C$.

b. Find the probability that $\bar{x}$ falls between $-2.20°C$ and $-2.10°C$.

6.89 Levelness of concrete slabs. Geotechnical engineers use water-level "manometer" surveys to assess the levelness of newly constructed concrete slabs. Elevations are typically measured at eight points on the slab; of interest is the maximum differential between elevations. The *Journal of Performance of Constructed Facilities* (Feb. 2005) published an article on the levelness of slabs in California residential developments. Elevation data collected on over 1,300 concrete slabs *before tensioning* revealed that the maximum differential x has a mean of $\mu = .53$ inch and a standard deviation of $\sigma = .193$ inch. Consider a sample of $n = 50$ slabs selected from those surveyed, and let $\bar{x}$ represent the mean of the sample.

a. Describe fully the sampling distribution of $\bar{x}$.

b. Find $P(\bar{x} > .58)$.

c. The study also revealed that the mean maximum differential of concrete slabs measured *after tensioning and loading* is $\mu = .58$ inch. Suppose the sample data yield $\bar{x} = .59$ inch. Comment on whether the sample measurements were obtained before tensioning or after tensioning and loading.

Applying the Concepts—Advanced

6.90 Machine repair time. [*Note:* This exercise refers to an optional section in Chapter 5.] An article in *Industrial Engineering* (Aug. 1990) discussed the importance of modeling machine downtime correctly in simulation studies. As an illustration, the researcher considered a single-machine-tool system with repair times (in minutes) that can be modeled by an exponential distribution with $\theta = 60$. (See optional Section 5.6.) Of interest is the mean repair time $\bar{x}$ of a sample of 100 machine breakdowns.

a. Find $E(\bar{x})$ and the variance of $\bar{x}$.

b. What probability distribution provides the best model of the sampling distribution of $\bar{x}$? Why?

c. Calculate the probability that the mean repair time $\bar{x}$ is no longer than 30 minutes.

6.91 Flaws in aluminum siding. [*Note:* This exercise refers to an optional section in Chapter 4.] A building contractor has decided to purchase a load of factory-reject aluminum siding as long as the average number of flaws per piece of siding in a sample of size 45 from the factory's reject pile is 1.5 or less. If it is known that the number of flaws per piece of siding in the factory's reject pile has a Poisson probability distribution with a mean of 1.75, find the approximate probability that the contractor will not purchase a load of siding. [*Hint:* If x is a Poisson random variable with mean λ, then $\sigma_{\bar{x}}^2$ equals λ/n.]

6.92 Soft-drink bottles. A soft-drink bottler purchases glass bottles from a vendor. The bottles are required to have an internal pressure of at least 150 pounds per square inch (psi). A prospective bottle vendor claims that its production process yields bottles with a mean internal pressure of 155 psi and a standard deviation of 2 psi. The bottler strikes an agreement with the vendor that permits the bottler to sample from the vendor's production process to verify the vendor's claim. The bottler randomly selects 50 bottles from the last 10,000 produced, measures the internal pressure of each, and finds the mean pressure for the sample to be .75 psi below the process mean cited by the vendor.

a. Assuming the vendor's claim to be true, what is the probability of obtaining a sample mean this far or farther below the process mean? What does your answer suggest about the validity of the vendor's claim?

b. If the process standard deviation were 2 psi as claimed by the vendor, but the mean were 154 psi, would the observed sample result be more or less likely than the result in part **a**? What if the mean were 156 psi?

c. If the process standard deviation were 1.5 psi as claimed by the vendor, but the mean were 155 psi, would the observed sample result be more or less likely than the result in part **a**? What if the mean were 3 psi?

Critical Thinking Challenge

6.93 Fecal pollution at Huntington Beach. The state of California mandates fecal indicator bacteria monitoring at all public beaches. When the concentration of fecal bacteria in the water exceeds a certain limit (400 colony-forming units of fecal coliform per 100 milliliters), local health officials must post a sign (called surf zone posting) warning beachgoers of potential health risks upon entering the water. For fecal bacteria, the state uses a single-sample standard; that is, if the fecal limit is exceeded in a single sample of water, surf zone posting is mandatory. This single-sample standard policy has led to a recent rash of beach closures in California.

Joon Ha Kim and Stanley B. Grant, engineers at the University of California at Irvine, conducted a study of the surf water quality at Huntington Beach in California and reported the results in *Environmental Science & Technology* (Sept. 2004). The researchers found that beach closings were occurring despite low pollution levels in some instances, while in others, signs were not posted when the fecal limit was exceeded. They attributed these "surf zone posting errors" to the variable nature of water quality in the surf zone (for example, fecal bacteria concentration

tends to be higher during ebb tide and at night) and the inherent time delay between when a water sample is collected and when a sign is posted or removed. In order to prevent posting errors, the researchers recommend using an averaging method, rather than a single sample, to determine unsafe water quality. (For example, one simple averaging method is to take a random sample of multiple water specimens and compare the average fecal bacteria level of the sample with the limit of 400 cfu/100 mL in order to determine whether the water is safe.)

Discuss the pros and cons of using the single-sample standard versus the averaging method. Part of your discussion should address the probability of posting a sign when in fact the water is safe and the probability of posting a sign when in fact the water is unsafe. (Assume that the fecal bacteria concentrations of water specimens at Huntington Beach follow an approximately normal distribution.)

Activity Simulating a Sampling Distribution—Cell Phone Usage

[*Note:* This activity is designed for small groups or the entire class.] Consider the length of time a student spends making a cell phone call, sending/retrieving a text message, or accessing e-mail on his/her cell phone. Let x represent the length of time, in seconds, for a single cell phone activity (call, text, or e-mail). Here, we are interested in the sampling distribution of $\bar{x}$, the mean length of time for a sample of size n cell phone activities.

1. Keep track of the time lengths for all cell phone activities you engage in over the next week.

2. Pool your time length data with data from other class members or the entire class so that the pooled data set has at least 100 observations. Designate someone in the group to calculate the mean and standard deviation of the pooled data set.

3. Devise a convenient way to choose random samples from the pooled data set. (For example, you could assign each observation a number beginning with "1" and use a random number generator to select a sample.)

4. Choose a random sample of size $n = 30$ from the pooled data, and find the mean of the sample. Group members should repeat the process of choosing a sample of size $n = 30$ from the pooled data and finding the sample mean until the group has accumulated at least 25 sample means. (Call this data set *Sample Means*.)

5. Find the mean and standard deviation of the *Sample Means* data set. Also, form a histogram for the *Sample Means* data set. Explain how the Central Limit Theorem is illustrated in this activity.

References

Hogg, R. V., McKean, J. W., and Craig, A. T. *Introduction to Mathematical Statistics*, 6th ed. Upper Saddle River, NJ: Prentice Hall, 2005.

Larsen, R. J., and Marx, M. L. *An Introduction to Mathematical Statistics and Its Applications*, 4th ed. Upper Saddle River, NJ: Prentice Hall, 2005.

Lindgren, B. W. *Statistical Theory*, 3rd ed. New York: Macmillan, 1976.

Wackerly, D., Mendenhall, W., and Scheaffer, R. L. *Mathematical Statistics with Applications*, 7th ed. North Scituate, MA: Duxbury, 2008.

MINITAB: Simulating a Sampling Distribution

Step 1 Select "Calc" on the MINITAB menu bar, and then click on "Random Data" (see Figure 6.M.1).

Figure 6.M.1 MINITAB options for generating random data

Step 2 On the resulting menu list, click on the distribution of your choice (e.g., "Uniform"). A dialog box similar to the one (the Uniform Distribution) shown in Figure 6.M.2 will appear.

Uniform Distribution

Number of rows of data to generate: 1000

Store in column(s):

C1-C40

Lower endpoint: 0.0

Upper endpoint: 1.0

Select

Help   OK   Cancel

Figure 6.M.2 MINITAB dialog box for simulating the uniform distribution

Step 3 Specify the number of samples (e.g., 1,000) to generate in the "Number of rows to generate" box and the columns where the data will be stored in the "Store in columns" box. (The number of columns will be equal to the sample size, e.g., 40.) Finally, specify the parameters of the distribution (e.g., the lower and upper range of the uniform distribution). When you click "OK," the simulated data will appear on the MINITAB worksheet.

Step 4 Calculate the value of the sample statistic of interest for each sample. To do this, click on the "Calc" button on the MINITAB menu bar, and then click on "Row Statistics," as shown in Figure 6.M.3. The resulting dialog box appears in Figure 6.M.4.

Figure 6.M.3 MINITAB selections for generating sample statistics for the simulated data

Row Statistics

Statistic

○ Sum   ○ Median
● Mean   ○ Sum of squares
○ Standard deviation   ○ N total
○ Minimum   ○ N nonmissing
○ Maximum   ○ N missing
○ Range

Input variables:

C1-C40

Select

Store result in: C41

Help   OK   Cancel

Figure 6.M.4 MINITAB row statistics dialog box

Step 5 Check the sample statistic (e.g., the mean) you want to calculate, specify the "Input variables" (or columns), and specify the column where you want the value of the sample statistic to be saved. Click "OK" and the value of the statistic for each sample will appear on the MINITAB worksheet.

[*Note:* Use the MINITAB menu choices provided in the Using Technology section in Chapter 2, pp. 114-115, to generate a histogram of the sampling distribution of the statistic or to find the mean and variance of the sampling distribution.]

7 Inferences Based on a Single Sample *Estimation with Confidence Intervals*

CONTENTS

Where We've Been

- Learned that populations are characterized by numerical descriptive measures—called *parameters*
- Found that decisions about population parameters are based on *statistics* computed from the sample
- Discovered that *inferences* about parameters are subject to uncertainty and that this uncertainty is reflected in the *sampling distribution* of a statistic

Where We're Going

- Estimate a population parameter (mean, proportion, or variance) on the basis of a sample selected from the population (7.1)
- Use the sampling distribution of a statistic to form a confidence interval for the population parameter (7.2–7.4, 7.6)
- Show how to select the proper sample size for estimating a population parameter (7.5)

Statistics in Action Medicare Fraud Investigations

U.S. Department of Justice (USDOJ) press release (May 8, 2008): *Eleven people have been indicted in . . . a targeted criminal, civil and administrative effort against individuals and health care companies that fraudulently bill the Medicare program. The indictments in the Central District of California resulted from the creation of a multi-agency team of federal, state and local investigators designed specifically to combat Medicare fraud through the use of real-time analysis of Medicare billing data.*

USDOJ press release (June 27, 2008): *The owners of four Miami-based health care corporations were sentenced and remanded to prison yesterday for their roles in schemes to defraud the Medicare program. Collectively, the three defendants through their companies collected more than $14 million from the Medicare program for unnecessary medicine, durable medical equipment (DME) and home health care services.*

USDOJ press release (Oct. 7, 2008): *Eight Miami–Dade County residents have been charged in a 16-count indictment for their alleged roles in a Medicare fraud scheme involving fake HIV infusion treatments.*

As the above press releases imply, the U.S. Department of Justice (USDOJ) and its Medicare Fraud Strike Force conduct investigations into suspected fraud and abuse of the Medicare system by health care providers. According to published reports, the Strike Force was responsible for almost 25% of the Medicare fraud charges brought nationwide in 2007.

One way in which Medicare fraud occurs is through the use of "upcoding," which refers to the practice of providers coding Medicare claims at a higher level of care than was actually provided to the patient. For example, suppose a particular kind of claim can be coded at three levels, where Level 1 is a routine office visit, Level 2 is a thorough examination involving advanced diagnostic tests, and Level 3 involves performing minor surgery. The amount of Medicare payment is higher for each increased level of claim. Thus, upcoding would occur if Level 1 services were billed at Level 2 or Level 3 payments or if Level 2 services were billed at Level 3 payment.

The USDOJ relies on sound statistical methods to help identify Medicare fraud. Once the USDOJ has determined

that possible upcoding has occurred, it next seeks to further investigate whether it is the result of legitimate practice (perhaps the provider is a specialist giving higher levels of care) or the result of fraudulent action on the part of the provider. To further its investigation, the USDOJ will next ask a statistician to select a sample of the provider's claims. For example, the statistician might determine that a random sample of 52 claims from the 1,000 claims in question will provide a sufficient sample to estimate the overcharge reliably. The USDOJ then asks a health care expert to audit each of the medical files corresponding to the sampled claims and determine whether the level of care matches the level billed by the provider and, if not, to determine what level should have been billed. Once the audit has been completed, the USDOJ will calculate the overcharge.

In this chapter, we present a recent Medicare fraud case investigated by the USDOJ. Results for the audit of 52 sampled claims, with the amount paid for each claim, the amount disallowed by the auditor, and the amount that should have been paid for each claim, are saved in the **MCFRAUD** file.* Knowing that a total of $103,500 was paid for the 1,000 claims, the USDOJ wants to use the sample results to extrapolate the overpayment amount to the entire population of 1,000 claims.

Statistics in Action Revisited

- Estimating the Mean Overpayment (p. 360)
- Estimating the Coding Error Rate (p. 369)
- Determining Sample Size (p. 376)

 Data Set: MCFRAUD

7.1 Identifying and Estimating the Target Parameter

In this chapter, our goal is to estimate the value of an unknown population parameter, such as a population mean or a proportion from a binomial population. For example, we might want to know the mean gas mileage for a new car model, the average expected life of a flat-screen computer monitor, or the proportion of Iraq War veterans with post-traumatic stress syndrome.

You'll see that different techniques are used for estimating a mean or proportion, depending on whether a sample contains a large or small number of measurements.

*Data provided (with permission) from Info Tech, Inc., Gainesville, Florida.

Nevertheless, our objectives remain the same: We want to use the sample information to estimate the population parameter of interest (called the *target parameter*) and to assess the reliability of the estimate.

> The unknown population parameter (e.g., mean or proportion) that we are interested in estimating is called the **target parameter**.

Often, there are one or more key words in the statement of the problem that indicate the appropriate target parameter. Some key words associated with the parameters covered in this chapter are listed in the following box:

> **Determining the Target Parameter**
>
> | Parameter | Key Words or Phrases | Type of Data |
> |---|---|---|
> | μ | Mean; average | Quantitative |
> | p | Proportion; percentage; fraction; rate | Qualitative |
> | σ^2 (optional) | Variance; variability; spread | Quantitative |

For the examples given in the first paragraph of this section, the words *mean* in "mean gas mileage" and *average* in "average life expectancy" imply that the target parameter is the population mean μ. The word *proportion* in "proportion of Iraq War veterans with post-traumatic stress syndrome" indicates that the target parameter is the binomial proportion p.

In addition to key words and phrases, the type of data (quantitative or qualitative) collected is indicative of the target parameter. With quantitative data, you are likely to be estimating the mean or variance of the data. With qualitative data with two outcomes (success or failure), the binomial proportion of successes is likely to be the parameter of interest.

A single number calculated from the sample that estimates a target population parameter is called a **point estimator**. For example, we'll use the sample mean, $\bar{x}$, to estimate the population mean μ. Consequently, $\bar{x}$ is a point estimator. Similarly, we'll learn that the sample proportion of successes, denoted $\hat{p}$, is a point estimator for the binomial proportion p and that the sample variance s^2 is a point estimator for the population variance σ^2. Also, we will attach a measure of reliability to our estimate by obtaining an **interval estimator**—a range of numbers that contains the target parameter with a high degree of confidence. For this reason the interval estimate is also called a **confidence interval**.

> A **point estimator** of a population parameter is a rule or formula that tells us how to use the sample data to calculate a *single* number that can be used as an *estimate* of the target parameter.

> An **interval estimator (or confidence interval)** is a formula that tells us how to use the sample data to calculate an *interval* that *estimates* the target parameter.

We consider methods for estimating a population mean in Sections 7.2 and 7.3. Estimating a population proportion is presented in Section 7.4. We show how to determine the sample sizes necessary for reliable estimates of the target parameters in Section 7.5. Finally, in optional Section 7.6 we discuss estimation of a population variance.

7.2   Confidence Interval for a Population Mean: Normal (*z*) Statistic

Suppose a large hospital wants to estimate the average length of time patients remain in the hospital. Hence, the hospital's target parameter is the population mean μ. To accomplish this objective, the hospital administrators plan to randomly sample 100 of all previous patients' records and to use the sample mean $\bar{x}$ of the lengths of stay to estimate μ, the mean of *all* patients' visits. The sample mean $\bar{x}$ represents a *point estimator* of the population mean μ. How can we assess the accuracy of this large-sample point estimator?

According to the Central Limit Theorem, the sampling distribution of the sample mean is approximately normal for large samples, as shown in Figure 7.1. Let us calculate the interval estimator:

$$\bar{x} \pm 1.96\sigma_{\bar{x}} = \bar{x} \pm \frac{1.96\sigma}{\sqrt{n}}$$

Probability = .95

$f(\bar{x})$

μ

$\leftarrow 1.96\sigma_{\bar{x}} \rightarrow \leftarrow 1.96\sigma_{\bar{x}} \rightarrow$

$\bar{x}$

Figure 7.1

Sampling distribution of $\bar{x}$

That is, we form an interval from 1.96 standard deviations below the sample mean to 1.96 standard deviations above the mean. *Prior to drawing the sample*, what are the chances that this interval will enclose μ, the population mean?

To answer this question, refer to Figure 7.1. If the 100 measurements yield a value of $\bar{x}$ that falls between the two lines on either side of μ (i.e., within 1.96 standard deviations of μ), then the interval $\bar{x} \pm 1.96\sigma_{\bar{x}}$ will contain μ; if $\bar{x}$ falls outside these boundaries, the interval $\bar{x} \pm 1.96\sigma_{\bar{x}}$ will not contain μ. From Section 6.3, we know that the area under the normal curve (the sampling distribution of $\bar{x}$) between these boundaries is exactly .95. Thus, the interval $\bar{x} \pm 1.96\sigma_{\bar{x}}$ will contain μ with a probability equal to .95.

Example   7.1

Estimating the Mean Hospital Length of Stay, σ Known

Problem Consider the large hospital that wants to estimate the average length of stay of its patients, μ. The hospital randomly samples $n = 100$ of its patients and finds that the sample mean length of stay is $\bar{x} = 4.5$ days. Also, suppose it is known that the standard deviation of the length of stay for all hospital patients is $\sigma = 4$ days. Use the interval estimator $\bar{x} \pm 1.96\sigma_{\bar{x}}$ to calculate a confidence interval for the target parameter, μ.

Solution Substituting $\bar{x} = 4.5$ and $\sigma = 4$ into the interval estimator formula, we obtain:

$$\bar{x} \pm 1.96\sigma_{\bar{x}} = \bar{x} \pm (1.96)\sigma/\sqrt{n} = 4.5 \pm (1.96)(4/\sqrt{100}) = 4.5 \pm .78$$

Or, (3.72, 5.28). We can also obtain this confidence interval using statistical software, as shown (highlighted) on the MINITAB printout, Figure 7.2.

One-Sample Z

The assumed standard deviation = 4

Figure 7.2

MINITAB Output Showing 95% Confidence Interval for μ, σ Known

N	Mean	SE Mean	95% CI
100	4.500	0.400	(3.716, 5.284)

Look Back Since we know the probability that the interval $\bar{x} \pm 1.96\sigma_{\bar{x}}$ will contain μ is .95, we call the interval estimator a *95% confidence interval* for μ.

■ **Now Work Exercise 7.9a**

The interval $\bar{x} \pm 1.96\sigma_{\bar{x}}$ in Example 7.1 is called a *large-sample* 95% confidence interval for the population mean μ. The term *large-sample* refers to the sample being of sufficiently large size that we can apply the Central Limit Theorem and the **normal (*z*) statistic** to determine the form of the sampling distribution of $\bar{x}$. Empirical research suggests that a sample size n exceeding a value between 20 and 30 will usually yield a sampling distribution of $\bar{x}$ that is approximately normal. This result led many practitioners to adopt the rule of thumb that a sample size of $n \geq 30$ is required to use large-sample confidence interval procedures. Keep in mind, though, that 30 is not a magical number and, in fact, is quite arbitrary.

Also, note that the large-sample interval estimator requires knowing the value of the population standard deviation, σ. In most (if not nearly all) practical applications, however, the value of σ will be unknown. For large samples, the fact that σ is unknown poses only a minor problem since the sample standard deviation s provides a very good approximation to σ.* The next example illustrates the more realistic large-sample confidence interval procedure.

| **Example** | **7.2** | |

HOSPLOS

Estimating the Mean Hospital Length of Stay, σ Unknown

Problem Refer to Example 7.1 and the problem of estimating μ, the average length of stay of a hospital's patients. The lengths of stay (in days) for the $n = 100$ sampled patients are shown in Table 7.1. Use the data to find a 95% confidence interval for μ and interpret the result.

Table 7.1   Lengths of Stay (in Days) for 100 Patients

2	3	8	6	4	4	6	4	2	5
8	10	4	4	4	2	1	3	2	10
1	3	2	3	4	3	5	2	4	1
2	9	1	7	17	9	9	9	4	4
1	1	1	3	1	6	3	3	2	5
1	3	3	14	2	3	9	6	6	3
5	1	4	6	11	22	1	9	6	5
2	2	5	4	3	6	1	5	1	6
17	1	2	4	5	4	4	3	2	3
3	5	2	3	3	2	10	2	4	2

D *Data Set:* HOSPLOS

Solution The hospital almost surely does not know the true standard deviation, σ, of the population of lengths of stay. However, since the sample size is large, we will use the sample standard deviation, s, as an estimate for σ in the confidence interval formula. An SAS printout of summary statistics for the sample of 100 lengths of stay is shown at the top of Figure 7.3. From the shaded portion of the printout, we find $\bar{x} = 4.53$ days and $s = 3.68$ days. Substituting these values into the interval estimator formula, we obtain:

$$\bar{x} \pm (1.96)\sigma/\sqrt{n} \approx \bar{x} \pm (1.96)s/\sqrt{n} = 4.53 \pm (1.96)(3.68)/\sqrt{100} = 4.53 \pm .72$$

Or, (3.81, 5.25). That is, we estimate the mean length of stay for all hospital patients to fall within the interval 3.81 to 5.25 days.

```
Sample Statistics for LOS

     N          Mean        Std. Dev.        Std. Error
---------------------------------------------------------
    100         4.53          3.68             0.37

Hypothesis Test

    Null hypothesis:    Mean of LOS  =  0
    Alternative:        Mean of LOS ^= 0

            t Statistic       Df        Prob > t
        ---------------------------------------------
              12.318          99         <.0001


95 % Confidence Interval for the Mean

        Lower Limit:              3.80
        Upper Limit:              5.26
```

Figure 7.3

SAS printout with summary statistics and 95% confidence interval for data on 100 hospital stays

*In the terminology of Chapter 6, it can be shown that s is an *unbiased* estimator of σ.

Look Back The confidence interval is also shown at the bottom of the SAS printout, Figure 7.3. Note that the endpoints of the interval vary slightly from those computed in the example. This is due to the fact that when σ is unknown and n is large, the sampling distribution of $\bar{x}$ will deviate slightly from the normal (z) distribution. In practice, these differences can be ignored.

■ Now Work Exercise 7.15

Can we be sure that μ, the true mean, is in the interval from 3.81 to 5.25 in Example 7.2? We cannot be certain, but we can be reasonably confident that it is. This confidence is derived from the knowledge that if we were to draw repeated random samples of 100 measurements from this population and form the interval $\bar{x} \pm 1.96\sigma_{\bar{x}}$ each time, 95% of the intervals would contain μ. We have no way of knowing (without looking at all the patients' records) whether our sample interval is one of the 95% that contain μ or one of the 5% that do not, but the odds certainly favor its containing μ. The probability, .95, that measures the confidence we can place in the interval estimate is called a *confidence coefficient*. The percentage, 95%, is called the *confidence level* for the interval estimate.

> The **confidence coefficient** is the probability that an interval estimator encloses the population parameter—that is, the relative frequency with which the interval estimator encloses the population parameter when the estimator is used repeatedly a very large number of times. The **confidence level** is the confidence coefficient expressed as a percentage.

Now we have seen how an interval can be used to estimate a population mean. When we use an interval estimator, we can usually calculate the probability that the estimation *process* will result in an interval that contains the true value of the population mean. That is, the probability that the interval contains the parameter in repeated usage is usually known. Figure 7.4 shows what happens when 10 different samples are drawn from a population and a confidence interval for μ is calculated from each. The location of μ is indicated by the vertical line in the figure. Ten confidence intervals, each based on one of 10 samples, are shown as horizontal line segments. Note that the confidence intervals move from sample to sample, sometimes containing μ and other times missing μ. *If our confidence level is 95%, then in the long run, 95% of our sample confidence intervals will contain μ.*

Suppose you wish to choose a confidence coefficient other than .95. Notice in Figure 7.1 that the confidence coefficient .95 is equal to the total area under the sampling distribution, less .05 of the area, which is divided equally between the two tails. Using this idea, we can construct a confidence interval with any desired confidence coefficient by increasing or decreasing the area (call it α) assigned to the tails of the sampling distribution. (See Figure 7.5.) For example, if we place the area $\alpha/2$ in each tail and if $z_{\alpha/2}$ is the z value such that $\alpha/2$ will lie to its right, then the confidence interval with confidence coefficient $(1 - \alpha)$ is

$$\bar{x} \pm z_{\alpha/2}\sigma_{\bar{x}}$$

Figure 7.4

Confidence intervals for μ: 10 samples

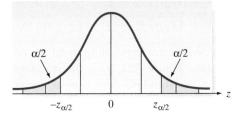

Figure 7.5

Locating $z_{\alpha/2}$ on the standard normal curve

for each. These 15 measurements (in millions of characters) are listed in Table 7.5, followed by a MINITAB summary statistics printout in Figure 7.12.

a. Form a 99% confidence interval for the mean number of characters printed before the printhead fails. Interpret the result.

b. What assumption is required for the interval you found in part **a** to be valid? Is that assumption reasonably satisfied?

Figure 7.12
MINITAB printout with descriptive statistics and 99% confidence interval for Example 7.5

Table 7.5	Number of Characters (in Millions) for $n = 15$ Printhead Tests

| 1.13 | 1.55 | 1.43 | .92 | 1.25 | 1.36 | 1.32 | .85 | 1.07 | 1.48 | 1.20 | 1.33 | 1.18 | 1.22 | 1.29 |

D *Data Set: PRNTHD*

```
Variable   N     Mean    StDev   SE Mean        99% CI
NUMCHAR    15   1.23867  0.19316  0.04987   (1.09020, 1.38714)
```

Solution

a. For this small sample ($n = 15$), we use the *t*-statistic to form the confidence interval. We use a confidence coefficient of .99 and $n - 1 = 14$ degrees of freedom to find $t_{\alpha/2}$ in Table III:

$$t_{\alpha/2} = t_{.005} = 2.977$$

[*Note:* The small sample forces us to extend the interval almost three standard deviations (of $\bar{x}$) on each side of the sample mean in order to form the 99% confidence interval.] From the MINITAB printout shown in Figure 7.12, we find that $\bar{x} = 1.24$ and $s = .19$. Substituting these (rounded) values into the confidence interval formula, we obtain

$$\bar{x} \pm t_{.005}\left(\frac{s}{\sqrt{n}}\right) = 1.24 \pm 2.977\left(\frac{.19}{\sqrt{15}}\right)$$
$$= 1.24 \pm .15 \text{ or } (1.09, 1.39)$$

This interval is highlighted in Figure 7.12.

Our interpretation is as follows: The manufacturer can be 99% confident that the printhead has a mean life of between 1.09 and 1.39 million characters. If the manufacturer were to advertise that the mean life of its printheads is (at least) 1 million characters, the interval would support such a claim. Our confidence is derived from the fact that 99% of the intervals formed in repeated applications of this procedure will contain μ.

b. Since *n* is small, we must assume that the number of characters printed before the printhead fails is a random variable from a normal distribution. That is, we assume that the population from which the sample of 15 measurements is selected is distributed normally. One way to check this assumption is to graph the distribution of data in Table 7.5. If the sample data are approximately normally distributed, then the population from which the sample is selected is very likely to be normal. A MINITAB stem-and-leaf plot for the sample data is displayed in Figure 7.13.

Figure 7.13
MINITAB stem-and-leaf display of data in Table 7.5

```
Stem-and-Leaf Display: NUMBER

Stem-and-leaf of NUMBER  N  = 15
Leaf Unit = 0.010

    1    8    5
    2    9    2
    3   10    7
    5   11    38
  (4)   12    0259
    6   13    236
    3   14    38
    1   15    5
```

The distribution is mound shaped and nearly symmetric. Therefore, the assumption of normality appears to be reasonably satisfied.

Look Back Other checks for normality, such as a normal probability plot and the ratio IQR/s, may also be used to verify the normality condition.

■ **Now Work Exercise 7.37**

We have emphasized throughout this section that an assumption that the population is approximately normally distributed is necessary for making small-sample inferences about μ when the t-statistic is used. Although many phenomena do have approximately normal distributions, it is also true that many random phenomena have distributions that are not normal or even mound shaped. Empirical evidence acquired over the years has shown that the t-distribution is rather insensitive to moderate departures from normality. That is, the use of the t-statistic when sampling from slightly skewed mound-shaped populations generally produces credible results; however, for cases in which the distribution is distinctly nonnormal, we must either take a large sample or use a *nonparametric method* (the topic of Chapter 14).

> **What Do You Do When the Population Relative Frequency Distribution Departs Greatly from Normality?**
>
> *Answer:* Use the nonparametric statistical methods of Chapter 14 (available online).

Statistics IN Action Revisited — Estimating the Mean Overpayment

Refer to the Medicare fraud investigation described in the Statistics in Action (p. 343). Recall that the United States Department of Justice (USDOJ) obtained a random sample of 52 claims from a population of 1,000 Medicare claims. For each claim, the amount paid, the amount disallowed (denied) by the auditor, and the amount that should have been paid (allowed) were recorded and saved in the **MCFRAUD** file. The USDOJ wants to use these data to calculate an estimate of the overpayment for all 1,000 claims in the population.

One way to do this is to first use the sample data to estimate the mean overpayment per claim for the population, then use the estimated mean to extrapolate the overpayment amount to the population of all 1,000 claims. The difference between the amount paid and the amount allowed by the auditor

represents the overpayment for each claim. This value is recorded as the amount denied in the **MCFRAUD** file. These overpayment amounts are listed in Table SIA7.1.

MINITAB software is used to find a 95% confidence interval for μ, the mean overpayment amount. The MINITAB printout is displayed in Figure SIA7.1. The 95% confidence interval for μ, (highlighted on the printout) is (16.51, 27.13). Thus, the USDOJ can be 95% confident that the mean overpayment amount for the population of 1,000 claims is between $16.51 and $27.13.

Table SIA7.1   Overpayment Amounts for Sample of 52 Claims

$0.00	$31.00	$0.00	$37.20	$37.20	$0.00	$43.40	$0.00
$37.20	$43.40	$0.00	$37.20	$0.00	$24.80	$0.00	$0.00
$37.20	$0.00	$37.20	$0.00	$37.20	$37.20	$37.20	$0.00
$37.20	$37.20	$0.00	$0.00	$37.20	$0.00	$43.40	$37.20
$0.00	$37.20	$0.00	$37.20	$37.20	$0.00	$37.20	$37.20
$0.00	$37.20	$43.40	$0.00	$37.20	$37.20	$37.20	$0.00
$43.40	$0.00	$43.40	$0.00				

Ⓓ *Data Set:* MCFRAUD

Figure SIA7.1
MINITAB confidence interval for mean overpayment

```
Variable         N   Mean  StDev  SE Mean      95% CI
Denied_Amount   52  21.82  19.08     2.65  (16.51, 27.13)
```

Statistics IN Action
(continued)

Now, let x_i represent the overpayment amount for the i claim. If the true mean μ were known, then the total overpayment amount for all 1,000 claims would be equal to

$$\sum_{i=1}^{1,000} x_i = (1,000)\left[\sum_{i=1}^{1,000} x_i/(1,000)\right] = (1,000)\mu$$

Consequently, to estimate the total overpayment amount for the 1,000 claims, the USDOJ will simply multiply the end points of the interval by 1,000. This yields the 95% confidence interval ($16,510, $27,130).* Typically, the USDOJ is willing to give the Medicare provider in question the benefit of the doubt by demanding a repayment equal to the lower 95% confidence bound—in this case, $16,510.

Exercises 7.29–7.51

Understanding the Principles

7.29 State the two problems (and corresponding solutions) that arise with using a small sample to estimate μ.

7.30 Compare the shapes of the z- and t-distributions.

7.31 Explain the differences in the sampling distributions [NW] of $\bar{x}$ for large and small samples under the following assumptions:
  a. The variable of interest, x, is normally distributed.
  b. Nothing is known about the distribution of the variable x.

Ⓓ Applet Exercise 7.3

Use the applet entitled *Confidence Intervals for a Mean (the impact of not knowing the standard deviation)* to compare proportions of z-intervals and t-intervals that contain the mean for a population that is normally distributed.
  a. Using $n = 5$ and the normal distribution with mean 50 and standard deviation 10, run the applet several times. How do the proportions of z-intervals and t-intervals that contain the mean compare?
  b. Repeart part **a** first for $n = 10$ and then for $n = 20$. Compare your results with those you obtained in part **a**.
  c. Describe any patterns you observe between the proportion of z-intervals that contain the mean and the proportion of t-intervals that contain the mean as the sample size increases.

Ⓓ Applet Exercise 7.4

Use the applet entitled *Confidence Intervals for a Mean (the impact of not knowing the standard deviation)* to compare proportions of z-intervals and t-intervals that contain the mean for a population with a skewed distribution.
  a. Using $n = 5$ and the right-skewed distribution with mean 50 and standard deviation 10, run the applet several times. How do the proportions of z-intervals and t-intervals that contain the mean compare?
  b. Repeat part **a** first for $n = 10$ and then for $n = 20$. Compare your results with those you obtained in part **a**.

  c. Describe any patterns you observe between the proportion of z-intervals that contain the mean and the proportion of t-intervals that contain the mean as the sample size increases.
  d. How does the skewness of the underlying distribution affect the proportions of z-intervals and t-intervals that contain the mean?

Learning the Mechanics

7.32 Suppose you have selected a random sample of $n = 7$ measurements from a normal distribution. Compare the standard normal z-values with the corresponding t-values if you were forming the following confidence intervals:
  a. 80% confidence interval
  b. 90% confidence interval
  c. 95% confidence interval
  d. 98% confidence interval
  e. 99% confidence interval
  f. Use the table values you obtained in parts **a–e** to sketch the z- and t-distributions. What are the similarities and differences?

7.33 Let t_0 be a specific value of t. Use technology or Table III in Appendix B to find t_0 values such that following statements are true.
  a. $P(t \geq t_0) = .025$, where df $= 10$
  b. $P(t \geq t_0) = .01$, where df $= 18$
  c. $P(t \leq t_0) = .005$, where df $= 7$
  d. $P(t \leq t_0) = .05$, where df $= 14$

7.34 Let t_0 be a specific value of t. Use technology or Table III of Appendix B to find t_0 values such that the following statements are true:
  a. $P(-t_0 < t < t_0) = .95$, where df $= 19$
  b. $P(t \leq t_0) = .01$, where df $= 6$
  c. $P(t \leq -t_0 \text{ or } t \geq t_0) = .10$, where df $= 19$
  d. $P(t \leq -t_0 \text{ or } t \geq t_0) = .01$, where df $= 16$

7.35 The following random sample was selected from a normal distribution: 4, 6, 3, 5, 9, 3.
  a. Construct a 90% confidence interval for the population mean μ.

*This interval represents an approximation to the true 95% confidence interval for the total amount of overpayment. The precise interval involves use of a continuity correction for population size. Consult the references for the use of a finite population correction factor in interval estimation.

b. Construct a 95% confidence interval for the population mean μ.

c. Construct a 99% confidence interval for the population mean μ.

d. Assume that the sample mean $\bar{x}$ and sample standard deviation s remain exactly the same as those you just calculated, but that they are based on a sample of $n = 25$ observations rather than $n = 6$ observations. Repeat parts **a–c**. What is the effect of increasing the sample size on the width of the confidence intervals?

7.36 The following sample of 16 measurements was selected from a population that is approximately normally distributed:

91	80	99	110	95	106	78	121	106	100
97	82	100	83	115	104				

a. Construct an 80% confidence interval for the population mean.

b. Construct a 95% confidence interval for the population mean, and compare the width of this interval with that of part **a**.

c. Carefully interpret each of the confidence intervals, and explain why the 80% confidence interval is narrower.

Applying the Concepts—Basic

7.37 **Music performance anxiety.** Refer to the *British Journal of Music Education* (Mar. 2014) study of performance anxiety by music students, Exercise 2.39 (p. 78). Recall that the Performance Anxiety Inventory (PAI) was used to measure music performance anxiety on a scale from 20 to 80 points. The table below gives PAI values for participants in eight different studies.

54	42	51	39	41	43	55	40

Source: Patston, T. "Teaching stage fright? — Implications for music educators." *British Journal of Music Education*, Vol. 31, No. 1, Mar. 2014 (adapted from Figure 1).

a. Compute the mean PAI value, $\bar{x}$, for the sample of 8 studies. (See your answer to Exercise 2.62a.)

b. Compute the standard deviation of the PAI values, s, for the sample of 8 studies. (See your answer to Exercise 2.87b.)

c. Use the results, parts **a** and **b**, to form a 95% confidence interval for μ, the true mean PAI value for the population of all similar music performance anxiety studies.

d. For the interval, part **c**, to be valid, how should the population of PAI values for all music performance anxiety studies be distributed?

e. If you were to repeatedly sample eight music performance anxiety studies and form a 95% confidence interval for μ for each sample, what proportion of the intervals will actually contain the true value of μ?

7.38 **Giraffes have excellent vision.** Due to habitat, giraffes travel in small groups. Hence, they require excellent vision in order to detect predators. The eyesight of giraffes was studied in *African Zoology* (Oct. 2013). The researchers measured a variety of eye characteristics for 27 giraffes native to Zimbabwe, Africa. One variable measured was eye mass (in grams). The study reported $\bar{x} = 53.4$ g and $s = 8.6$ g.

a. Use this information to find a 99% confidence interval for the true mean eye mass of giraffes native to Zimbabwe.

b. Suppose it is known that the mean eye mass of an African water buffalo is $\mu = 31$ grams. Is it likely that the true mean eye mass of a giraffe is larger or smaller than this mean? Explain.

c. Suppose it is known that the mean eye mass of an African elephant is $\mu = 58$ grams. Is it likely that the true mean eye mass of a giraffe is larger or smaller than this mean? Explain.

7.39 **Radon exposure in Egyptian tombs.** Many ancient Egyptian tombs were cut from limestone rock that contained uranium. Since most tombs are not well-ventilated, guards, tour guides, and visitors may be exposed to deadly radon gas. In *Radiation Protection Dosimetry* (Dec. 2010), a study of radon exposure in tombs in the Valley of Kings, Luxor, Egypt (recently opened for public tours), was conducted. The radon levels—measured in becquerels per cubic meter (Bq/m^3)—in the inner chambers of a sample of 12 tombs were determined. For this data, assume that $\bar{x} = 3{,}643 \, Bq/m^3$ and $s = 1{,}187 \, Bq/m^3$. Use this information to estimate, with 95% confidence, the true mean level of radon exposure in tombs in the Valley of Kings. Interpret the resulting interval.

7.40 **Rainfall and desert ants.** Refer to the *Journal of Biogeography* (Dec. 2003) study of ants and their habitat in the desert of Central Asia, presented in Exercise 2.68 (p. 91). Recall that botanists randomly selected five sites in the Dry Steppe region and six sites in the Gobi Desert where ants were observed. One of the variables of interest is the annual rainfall (in millimeters) at each site. Summary statistics for the annual rainfall at each site are provided in the SAS printout below.

a. Give a point estimate for the average annual rainfall amount at ant sites in the Dry Steppe region of Central Asia.

b. Give the *t*-value used in a small-sample 90% confidence interval for the true average annual rainfall amount at ant sites in the Dry Steppe region.

c. Use the result you obtained in part **b** and the values of $\bar{x}$ and s shown on the SAS printout to form a 90% confidence interval for the target parameter.

d. Give a practical interpretation for the interval you found in part **c**.

e. Use the data in the **ANTS** file to check the validity of the confidence interval you found in part **c**.

f. Repeat parts **a–e** for the Gobi Desert region of Central Asia.

		The MEANS Procedure	
		Analysis Variable : RAIN	
REGION	N Obs	Mean	Std Dev
DryStepp	5	183.4000000	20.6470337
Gobi	6	110.0000000	15.9749804

7.41 Lobster trap placement. An observational study of teams fishing for the red spiny lobster in Baja California Sur, Mexico, was conducted and the results published in *Bulletin of Marine Science* (Apr. 2010). One of the variables of interest was the average distance separating traps—called *trap spacing*—deployed by the same team of fishermen. Trap spacing measurements (in meters) for a sample of seven teams of red spiny lobster fishermen are shown in the accompanying table. Of interest is the mean trap spacing for the population of red spiny lobster fishermen fishing in Baja California Sur, Mexico.

93	99	105	94	82	70	86

Based on Shester, G. G. "Explaining catch variation among Baja California lobster fishers through spatial analysis of trap-placement decisions." *Bulletin of Marine Science*, Vol. 86, No. 2, Apr. 2010 (Table 1), pp. 479–498.

a. Identify the target parameter for this study.
b. Compute a point estimate of the target parameter.
c. What is the problem with using the normal (*z*) statistic to find a confidence interval for the target parameter?
d. Find a 95% confidence interval for the target parameter.
e. Give a practical interpretation of the interval, part **d**.
f. What conditions must be satisfied for the interval, part **d**, to be valid?

7.42 Shell lengths of sea turtles. Refer to the *Aquatic Biology* (Vol. 9, 2010) study of green sea turtles inhabiting the Grand Cayman South Sound lagoon, Exercise 7.24 (p. 354). Time-depth recorders were deployed on 6 of the 76 captured turtles. The time-depth recorders allowed the environmentalists to track the movement of the sea turtles in the lagoon. These 6 turtles had a mean shell length of 52.9 cm with a standard deviation of 6.8 cm.

a. Use the information on the 6 tracked turtles to estimate, with 99% confidence, the true mean shell length of all green sea turtles in the lagoon. Interpret the result.
b. What assumption about the distribution of shell lengths must be true in order for the confidence interval, part **a**, to be valid? Is this assumption reasonably satisfied? (Use the data saved in the **TURTLES** file to help you answer this question.)

7.43 Duration of daylight in western Pennsylvania. What area of the United States has the least amount of daylight, on average? Having grown up in western Pennsylvania, co-author Sincich wonders if it is his hometown of Sharon, PA. Data on the number of minutes of daylight per day in Sharon, PA, for 12 randomly selected days (one each month) in a recent year were obtained from the *Naval Oceanography Portal* Web site (*aa.usno.navy.mil/USNO/astronomical-applications/data-services*). The data are listed in the table. Descriptive statistics and a 95% confidence interval for the mean are produced in the SPSS printout in the next column.

a. Locate the confidence interval on the printout and give the value of the confidence coefficient.
b. Use the descriptive statistics on the printout to calculate the 95% confidence interval. Be sure your answer agrees with the interval shown on the printout.

591	618	704	831	896	909	865	839	748	672	583	565

Based on Astronomical Applications Dept., U.S. Naval Observatory, Washington, DC.

c. Practically interpret the confidence interval.
d. Comment on the method of sampling. Do you think the sample is representative of the target population?

SPSS Output for Exercise 7.43

Descriptives

			Statistic	Std. Error
DAYLIGHT	Mean		735.08	37.402
	95% Confidence Interval for Mean	Lower Bound	652.76	
		Upper Bound	817.40	
	5% Trimmed Mean		734.87	
	Median		726.00	
	Variance		16786.992	
	Std. Deviation		129.565	
	Minimum		565	
	Maximum		909	
	Range		344	
	Interquartile Range		261	
	Skewness		.016	.637
	Kurtosis		-1.752	1.232

Applying the Concepts—Intermediate

7.44 Do social robots walk or roll? Refer to the *International Conference on Social Robotics* (Vol. 6414, 2010) study on the current trend in the design of social robots, Exercise 2.7 (p. 66). Recall that in a random sample of social robots obtained through a Web search, 28 were built with wheels. The number of wheels on each of the 28 robots is reproduced in the accompanying table.

a. Estimate μ, the average number of wheels used on all social robots built with wheels, with 99% confidence.
b. Practically interpret the interval, part **a**.
c. Refer to part **a**. In repeated sampling, what proportion of all similarly constructed confidence intervals will contain the true mean, μ?

4	4	3	3	3	6	4	2	2	2	1	3	3	3
3	4	4	3	2	8	2	2	3	4	3	3	4	2

Source: Chew, S., et al. "Do social robots walk or roll?" *International Conference on Social Robotics*, Vol. 6414, 2010 (adapted from Figure 2).

7.45 Pitch memory of amusiacs. A team of psychologists and neuroscientists tested the pitch memory of individuals diagnosed with amusia (a disorder that impacts one's perception of music) and reported their results in *Advances in Cognitive Psychology* (Vol. 6, 2010). Each in a sample of 17 amusiacs listened to a series of tone pairs and then were asked to determine if the tones were the same or different. In one trial, the tones were separated by 1 second. In a second trial, the tones were separated by 5 seconds. Scores in the two trials were compared for each amusiac. The mean score difference was .11 with a standard deviation of .19. Use this information to form a 90% confidence interval for the true mean score difference for all amusiacs. Interpret the result. What assumption about the population of score differences must hold true for the interval to be valid?

7.46 Shaft graves in ancient Greece. Refer to the *American Journal of Archaeology* (Jan. 2014) study of shaft graves in ancient Greece, Exercise 2.37 (p. 78). Recall that shaft graves are named for the beautifully decorated sword shafts that are buried along with the bodies. The table on p. 364

gives the number of shafts buried at each of 13 recently discovered grave sites.

1	2	3	1	5	6	2	4	1	2	4	2	9

Source: Harrell, K. "The fallen and their swords: A new explanation for the rise of the shaft graves." *American Journal of Archaeology*, Vol. 118, No. 1, Jan. 2014 (Figure 1).

a. Estimate the average number of shafts buried in ancient Greece graves using a 90% confidence interval. Give a practical interpretation of the interval.

b. What assumption about the data on shaft graves is required for the inference, part **a**, to be valid?

7.47 **Oxygen bubbles in molten salt.** Molten salt is used in an electro-refiner to treat nuclear fuel waste. Eventually, the salt needs to be purified (for reuse) or disposed of. A promising method of purification involves oxidation. Such a method was investigated in *Chemical Engineering Research and Design* (Mar. 2013). An important aspect of the purification process is the rising velocity of oxygen bubbles in the molten salt. An experiment was conducted in which oxygen was inserted (at a designated sparging rate) into molten salt and photographic images of the bubbles taken. A random sample of 25 images yielded the data on bubble velocity (measured in meters per second) shown in the table. [*Note:* These data are simulated based on information provided in the article.]

0.275	0.261	0.209	0.266	0.265	0.312	0.285	0.317	0.229
0.251	0.256	0.339	0.213	0.178	0.217	0.307	0.264	0.319
0.298	0.169	0.342	0.270	0.262	0.228	0.220		

a. Use statistical software to find a 95% confidence interval for the mean bubble rising velocity of the population. Interpret the result.

b. The researchers discovered that the mean bubble rising velocity is $\mu = .338$ when the sparging rate of oxygen is 3.33×10^{-6}. Do you believe that the data in the table were generated at this sparging rate? Explain.

7.48 **Minimizing tractor skidding distance.** In planning for a new forest road to be used for tree harvesting, planners must select the location that will minimize tractor skidding distance. In the *Journal of Forest Engineering* (July 1999), researchers wanted to estimate the true mean skidding distance along a new road in a European forest. The skidding distances (in meters) were measured at 20 randomly selected road sites. These values are given in the accompanying table.

| 488 | 350 | 457 | 199 | 285 | 409 | 435 | 574 | 439 | 546 |
| 385 | 295 | 184 | 261 | 273 | 400 | 311 | 312 | 141 | 425 |

Based on Tujek, J., & Pacola, E. "Algorithms for skidding distance modeling on a raster Digital Terrain Model." *Journal of Forest Engineering*, Vol. 10, No. 1, July 1999 (Table 1).

a. Estimate, with a 95% confidence interval, the true mean skidding distance of the road.

b. Give a practical interpretation of the interval you found in part **a**.

c. What conditions are required for the inference you made in part **b** to be valid? Are these conditions reasonably satisfied?

d. A logger working on the road claims that the mean skidding distance is at least 425 meters. Do you agree?

7.49 **Reproduction of bacteria-infected spider mites.** Zoologists in Japan investigated the reproductive traits of spider mites with a bacterial infection (*Heredity*, Jan. 2007). Male and female pairs of infected spider mites were mated in a laboratory and the number of eggs produced by each female recorded. Summary statistics for several samples are provided in the accompanying table. Note that, in some samples, one or both infected spider mites were treated with antibiotic prior to mating.

a. For each type of female–male pair, construct and interpret a 90% confidence interval for the population mean number of eggs produced by the female spider mite.

b. Identify the type of female–male pair that appears to produce the highest mean number of eggs.

Female–Male Pairs	Sample Size	Mean # of Eggs	Standard Deviation
Both untreated	29	20.9	3.34
Male treated	23	20.3	3.50
Female treated	18	22.9	4.37
Both treated	21	18.6	2.11

Based on Gotoh, T., Noda, H., & Ito, S. "Cardinium symbionts cause cytoplasmic incompatibility in spider mites." *Heredity*, Vol. 98, No. 1, Jan. 2007 (Table 2).

7.50 **Antigens for a parasitic roundworm in birds.** *Ascaridia galli* is a parasitic roundworm that attacks the intestines of birds, especially chickens and turkeys. Scientists are working on a synthetic vaccine (antigen) for the parasite. The success of the vaccine hinges on the characteristics of DNA in peptide (protein) produced by the antigen. In the journal *Gene Therapy and Molecular Biology* (June 2009), scientists tested alleles of antigen-produced protein for level of peptide. For a sample of 4 alleles, the mean peptide score was 1.43 and the standard deviation was .13.

a. Use this information to construct a 90% confidence interval for the true mean peptide score in alleles of the antigen-produced protein.

b. Interpret the interval for the scientists.

c. What is meant by the phrase "90% confidence"?

Applying the Concepts—Advanced

7.51 **Study on waking sleepers early.** Scientists have discovered increased levels of the hormone adrenocorticotropin in people just before they awake from sleeping (*Nature*, Jan. 7, 1999). In the study described, 15 subjects were monitored during their sleep after being told that they would be woken at a particular time. One hour prior to the designated wake-up time, the adrenocorticotropin level (pg/mL) was measured in each, with the following results:

$$\bar{x} = 37.3 \qquad s = 13.9$$

a. Use a 95% confidence interval to estimate the true mean adrenocorticotropin level of sleepers one hour prior to waking.

b. Interpret the interval you found in part **a** in the words of the problem.

c. The researchers also found that if the subjects were woken three hours earlier than they anticipated, the average adrenocorticotropin level was 25.5 pg/mL. Assume that $\mu = 25.5$ for all sleepers who are woken three hours earlier than expected. Use the interval from part **a** to make an inference about the mean adrenocorticotropin level of sleepers under two conditions: one hour before the anticipated wake-up time and three hours before the anticipated wake-up time.

7.4 Large-Sample Confidence Interval for a Population Proportion

The number of public-opinion polls has grown at an astounding rate in recent years. Almost daily, the news media report the results of some poll. Pollsters regularly determine the percentage of people in favor of the president's welfare-reform program, the fraction of voters in favor of a certain candidate, the fraction of customers who favor a particular brand of wine, and the proportion of people who smoke cigarettes. In each case, we are interested in estimating the percentage (or proportion) of some group with a certain characteristic. In this section, we consider methods for making inferences about population proportions when the sample is large.

Example 7.6

Estimating a Population Proportion—Fraction Who Trust the President

Problem Public-opinion polls are conducted regularly to estimate the fraction of U.S. citizens who trust the president. Suppose 1,000 people are randomly chosen and 637 answer that they trust the president. How would you estimate the true fraction of *all* U.S. citizens who trust the president?

Solution What we have really asked is how you would estimate the probability p of success in a binomial experiment in which p is the probability that a person chosen trusts the president. One logical method of estimating p for the population is to use the proportion of successes in the sample. That is, we can estimate p by calculating

$$\hat{p} = \frac{\text{Number of people sampled who trust the president}}{\text{Number of people sampled}}$$

where $\hat{p}$ is read "p hat." Thus, in this case,

$$\hat{p} = \frac{637}{1,000} = .637$$

Look Back To determine the reliability of the estimator $\hat{p}$, we need to know its sampling distribution. That is, if we were to draw samples of 1,000 people over and over again, each time calculating a new estimate $\hat{p}$, what would be the frequency distribution of all the $\hat{p}$-values? The answer lies in viewing $\hat{p}$ as the average, or mean, number of successes per trial over the n trials. If each success is assigned a value equal to 1 and each failure is assigned a value of 0, then the sum of all n sample observations is x, the total number of successes, and $\hat{p} = x/n$ is the average, or mean, number of successes per trial in the n trials. The Central Limit Theorem tells us that the *relative frequency distribution of the sample mean for any population is approximately normal for sufficiently large samples.*

■ **Now Work Exercise 7.59a**

The repeated sampling distribution of $\hat{p}$ has the characteristics listed in the next box and shown in Figure 7.14.

Figure 7.14
Sampling distribution of $\hat{p}$

> **Sampling Distribution of $\hat{p}$**
> 1. The mean of the sampling distribution of $\hat{p}$ is p; that is, $\hat{p}$ is an unbiased estimator of p.
> 2. The standard deviation of the sampling distribution of $\hat{p}$ is $\sqrt{pq/n}$, that is, $\sigma_p = \sqrt{pq/n}$, where $q = 1 - p$.
> 3. For large samples, the sampling distribution of $\hat{p}$ is approximately normal. A sample size is considered large if both $n\hat{p} \geq 15$ and $n\hat{q} \geq 15$.

The fact that $\hat{p}$ is a "sample mean fraction of successes" allows us to form confidence intervals about p in a manner that is completely analogous to that used for large-sample estimation of μ.

> **Large-Sample Confidence Interval for p**
>
> $$\hat{p} \pm z_{\alpha/2}\sigma_{\hat{p}} = \hat{p} \pm z_{\alpha/2}\sqrt{\frac{pq}{n}} \approx \hat{p} \pm z_{\alpha/2}\sqrt{\frac{\hat{p}\hat{q}}{n}}$$
>
> where $\hat{p} = \dfrac{x}{n}$ and $\hat{q} = 1 - \hat{p}$
>
> *Note:* When n is large, $\hat{p}$ can approximate the value of p in the formula for $\sigma_{\hat{p}}$.

> **Conditions Required for a Valid Large-Sample Confidence Interval for p**
> 1. A random sample is selected from the target population.
> 2. The sample size n is large. (This condition will be satisfied if both $n\hat{p} \geq 15$ and $n\hat{q} \geq 15$. Note that $n\hat{p}$ and $n\hat{q}$ are simply the number of successes and number of failures, respectively, in the sample.

Thus, if 637 of 1,000 U.S. citizens say they trust the president, a 95% confidence interval for the proportion of *all* U.S. citizens who trust the president is

$$\hat{p} \pm z_{\alpha/2}\sigma_{\hat{p}} = .637 \pm 1.96\sqrt{\frac{pq}{1,000}}$$

where $q = 1 - p$. Just as we needed an approximation for σ in calculating a large-sample confidence interval for μ, we now need an approximation for p. As Table 7.6 shows, the approximation for p does not have to be especially accurate because the value of $\sqrt{pq}$ needed for the confidence interval is relatively insensitive to changes in p. Therefore, we can use $\hat{p}$ to approximate p. Keeping in mind that $\hat{q} = 1 - \hat{p}$, we substitute these values into the formula for the confidence interval:

$$\hat{p} \pm 1.96\sqrt{pq/1,000} \approx \hat{p} \pm 1.96\sqrt{\hat{p}\hat{q}/1,000}$$
$$= .637 \pm 1.96\sqrt{(.637)(.363)/1,000} = .637 \pm .030$$
$$= (.607, .667)$$

Then we can be 95% confident that the interval from 60.7% to 66.7% contains the true percentage of *all* U.S. citizens who trust the president. That is, in repeated constructions of confidence intervals, approximately 95% of all samples would produce confidence

Table 7.6   Values of pq for Several Different Values of p

p	pq	$\sqrt{pq}$
.5	.25	.50
.6 or .4	.24	.49
.7 or .3	.21	.46
.8 or .2	.16	.40
.9 or .1	.09	.30

intervals that enclose p. Note that the guidelines for interpreting a confidence interval about μ also apply to interpreting a confidence interval for p because p is the "population mean fraction of successes" in a binomial experiment.

Example **7.7**	**Problem** Many public polling agencies conduct surveys to determine the current consumer sentiment concerning the state of the economy. For example, the Bureau of Economic and Business Research (BEBR) at the University of Florida conducts quarterly surveys to gauge consumer sentiment in the Sunshine State. Suppose that BEBR randomly samples 484 consumers and finds that only 157 are optimistic about the state of the economy. Use a 90% confidence interval to estimate the proportion of all consumers in Florida who are optimistic about the state of the economy. Based on the confidence interval, can BEBR infer that a minority of Florida consumers is optimistic about the economy?

A Large-Sample Confidence Interval for p—Proportion Optimistic about the Economy

Problem Many public polling agencies conduct surveys to determine the current consumer sentiment concerning the state of the economy. For example, the Bureau of Economic and Business Research (BEBR) at the University of Florida conducts quarterly surveys to gauge consumer sentiment in the Sunshine State. Suppose that BEBR randomly samples 484 consumers and finds that only 157 are optimistic about the state of the economy. Use a 90% confidence interval to estimate the proportion of all consumers in Florida who are optimistic about the state of the economy. Based on the confidence interval, can BEBR infer that a minority of Florida consumers is optimistic about the economy?

Solution The number x of the 484 sampled consumers who are optimistic about the Florida economy is a binomial random variable if we can assume that the sample was randomly selected from the population of Florida consumers and that the poll was conducted identically for each consumer sampled.

The point estimate of the proportion of Florida consumers who are optimistic about the economy is

$$\hat{p} = \frac{x}{n} = \frac{157}{484} = .324$$

We first check to be sure that the sample size is sufficiently large that the normal distribution provides a reasonable approximation to the sampling distribution of $\hat{p}$. We require the number of successes in the sample, $n\hat{p}$, and the number of failures, $n\hat{q}$, both to be at least 15. Since the number of successes is $n\hat{p} = 157$ and the number of failures is $n\hat{q} = 327$, we may conclude that the normal approximation is reasonable.

We now proceed to form the 90% confidence interval for p, the true proportion of Florida consumers who are optimistic about the state of the economy:

$$\hat{p} \pm z_{\alpha/2}\sigma_{\hat{p}} = \hat{p} \pm z_{\alpha/2}\sqrt{\frac{pq}{n}} \approx \hat{p} \pm z_{\alpha/2}\sqrt{\frac{\hat{p}\hat{q}}{n}}$$

$$= .324 \pm 1.645\sqrt{\frac{(.324)(.676)}{484}} = .324 \pm .035 = (.289, .359)$$

(This interval is also shown on the MINITAB printout of Figure 7.15.) Thus, we can be 90% confident that the proportion of all Florida consumers who are confident about the economy is between .289 and .359. As always, our confidence stems from the fact that 90% of all similarly formed intervals will contain the true proportion p and not from any knowledge about whether this particular interval does.

Can we conclude on the basis of this interval that a minority of Florida consumers is optimistic about the economy? If we wished to use this interval to infer that a minority is optimistic, the interval would have to support the inference that p is less than .5—that is, that less than 50% of Florida consumers are optimistic about the economy. Note that the interval contains only values below .5. Therefore, we can conclude on the basis of this 90% confidence interval that the true value of p is less than .5.

Test and CI for One Proportion

```
Sample   X    N   Sample p        90% CI
1       157  484  0.324380  (0.289379, 0.359381)

Using the normal approximation.
```

Figure 7.15

MINITAB printout with 90% confidence interval for p

Look Ahead If the confidence interval includes .5 (e.g., an interval from .42 to .54), then we could not conclude that the true proportion of consumers who are optimistic is less than .5. (This is because it is possible that p is as high as .54.)

■ **Now Work Exercise 7.65**

We conclude this section with a warning and an illustrative example.

⚠ **CAUTION** Unless n is extremely large, the large-sample procedure presented in this section performs poorly when p is near 0 or near 1. ▲

The problem stated in the above warning can be illustrated as follows. Suppose you want to estimate the proportion of people who die from a bee sting using a sample size of $n = 100$. This proportion is likely to be near 0, say $p \approx .001$. If so, then $np \approx 100(.001) = .1$ is less than the recommended value of 15 (see conditions in the box on p. 366). Consequently, a confidence interval for p based on a sample of $n = 100$ will probably be misleading.

To overcome this potential problem, an *extremely* large sample size is required. Since the value of n required to satisfy "extremely large" is difficult to determine, statisticians (see Agresti & Coull, 1998) have proposed an alternative method, based on the Wilson (1927) point estimator of p. **Wilson's adjustment for estimating p** is outlined in the next box. Researchers have shown that this confidence interval works well for any p, even when the sample size n is very small.

Adjusted $(1 - \alpha)$ 100% Confidence Interval for a Population Proportion p

$$\widetilde{p} \pm z_{\alpha/2}\sqrt{\frac{\widetilde{p}(1 - \widetilde{p})}{n + 4}}$$

where $\widetilde{p} = \dfrac{x + 2}{n + 4}$ is the adjusted sample proportion of observations with the characteristic of interest, x is the number of successes in the sample, and n is the sample size.

Example   7.8
Using the Adjusted Confidence Interval Procedure—Proportion Who Are Victims of a Violent Crime

Problem According to *True Odds: How Risk Affects Your Everyday Life* (Walsh, 1997), the probability of being the victim of a violent crime is less than .01. Suppose that, in a random sample of 200 Americans, 3 were victims of a violent crime. Use a 95% confidence interval to estimate the true proportion of Americans who were victims of a violent crime.

Solution Let p represent the true proportion of Americans who were victims of a violent crime. Since p is near 0, an "extremely large" sample is required to estimate its value by the usual large-sample method. Here, the number of "successes," 3, is less than 15. Thus, we doubt whether the sample size of 200 is large enough to apply the large-sample method. Alternatively, we will apply Wilson's adjustment outlined in the box.

Because the number of "successes" (i.e., number of victims of a violent crime) in the sample is $x = 3$, the adjusted sample proportion is

$$\widetilde{p} = \frac{x + 2}{n + 4} = \frac{3 + 2}{200 + 4} = \frac{5}{204} = .025$$

Note that this adjusted sample proportion is obtained by adding a total of four observations—two "successes" and two "failures"—to the sample data. Substituting $\widetilde{p} = .025$ into the equation for a 95% confidence interval, we obtain

$$\widetilde{p} \pm 1.96\sqrt{\frac{\widetilde{p}(1 - \widetilde{p})}{n + 4}} = .025 \pm 1.96\sqrt{\frac{(.025)(.975)}{204}}$$

$$= .025 \pm .021$$

or (.004, .046). Consequently, we are 95% confident that the true proportion of Americans who are victims of a violent crime falls between .004 and .046.

Look Back: If we apply the standard large-sample confidence interval formula, where $\hat{p} = 3/200 = .015$, we obtain

$$\hat{p} \pm 1.96\sqrt{\frac{\hat{p}\hat{q}}{200}} = .015 \pm 1.96\sqrt{\frac{(.015)(.985)}{200}}$$

$$= .015 \pm .017$$

or, $(-.002, .032)$. Note that the interval estimate contains negative (nonsensical) values for the true proportion. Such a result is typical when the large-sample method is misapplied.

■ **Now Work Exercise 7.71b**

Statistics IN Action | **Revisited** | **Estimating the Coding Error Rate**

In the previous Statistics in Action Revisited (p. 360), we showed how to estimate the mean overpayment amount for claims in a Medicare fraud study. In addition to estimating overcharges, the USDOJ also is interested in estimating the *coding error rate* of a Medicare provider. The coding error rate is defined as the proportion of Medicare claims that are coded incorrectly. Thus, for this inference, the USDOJ is interested in estimating a population proportion, p. Typically, the USDOJ finds that about 50% of the claims in a Medicare fraud case are incorrectly coded.

If you examine the sample data in Table SIA7.1, you can verify that of the 52 audited claims, 31 were determined to be coded incorrectly, resulting in an overcharge. These are the claims with a disallowed amount greater than $0. Therefore, an estimate of the coding error rate, p, for this Medicare provider is

$$\hat{p} = 31/52 = .596$$

A 95% confidence interval for p can be obtained with the use of a confidence interval formula or with statistical software.

The **MCFRAUD** file includes a qualitative variable (called "Coding Error") at two levels, where "Yes" represents that the claim is coded incorrectly and "No" represents that the claim is coded correctly. Thus, the USDOJ desires an estimate of the proportion of "Yes" values for the "Coding Error" variable. A confidence interval for the proportion of incorrectly coded claims is highlighted on the accompanying MINITAB printout, Figure SIA7.2. The interval, (.45, .73), implies that the true coding error rate for the 1,000 claims in the population falls between .45 and .73, with 95% confidence. Note that the value .5—the proportion of incorrectly coded claims expected by the USDOJ—falls within the 95% confidence interval.

D *Data Set:* MCFRAUD

Figure SIA7.2
MINITAB confidence interval for coding error rate

```
Event = Yes

Variable       X   N  Sample p       95% CI
Coding_Error  31  52  0.596154  (0.451016, 0.729940)
```

Exercises 7.52–7.73

Understanding the Principles

7.52 Describe the sampling distribution of $\hat{p}$ on the basis of large samples of size n. That is, give the mean, the standard deviation, and the (approximate) shape of the distribution of $\hat{p}$ when large samples of size n are (repeatedly) selected from the binomial distribution with probability p of success.

7.53 Explain the meaning of the phrase "$\hat{p}$ is an unbiased estimator of p."

7.54 If p is near 0 or 1, how large a sample is needed to employ the large-sample confidence interval procedure?

D Applet Exercise 7.5

Use the applet entitled *Confidence Intervals for a Proportion* to investigate the effect of the value of p on the number of confidence intervals that contain the population proportion p for a fixed sample size. For this exercise, use sample size $n = 10$.

    a. Run the applet several times without clearing for $p = .1$. What proportion of the 95% confidence intervals contain p? What proportion of the 99% confidence intervals contain p? Do the results surprise you? Explain.

    b. Repeat part **a** for each value of p: $p = .2, p = .3, p = .4. p = .5, p = .6, p = .7, p = .8,$ and $p = .9$.

c. Which value of p yields the greatest proportion of each type of interval that contains p?

d. Based on your results, what values of p will yield more reliable confidence intervals for a fixed sample size n? Explain.

Ⓓ Applet Exercise 7.6

Use the applet entitled *Confidence Intervals for a Proportion* to investigate the effect of the sample size on the number of confidence intervals that contain the population proportion p for a value of p close to 0 or 1.

a. Run the applet several times without clearing for $p = .5$ and $n = 50$. Record the proportion of the 99% confidence intervals containing p.

b. Now set $p = .1$ and run the applet several times without clearing for $n = 50$. How does the proportion of the 99% confidence intervals containing p compare with that in part **a**?

c. Repeat part **b**, keeping $p = .1$ and increasing the sample size by 50 until you find a sample size that yields a similar proportion of the 99% confidence intervals containing p as that in part **a**.

d. Based on your results, describe how the value of p affects the sample size needed to guarantee a certain level of confidence.

Learning the Mechanics

7.55 A random sample of size $n = 250$ yielded $p = .80$.

a. Is the sample size large enough to use the large sample approximation to construct a confidence interval for p? Explain.

b. Construct a 95% confidence interval for p.

c. Interpret the 95% confidence interval.

d. Explain what is meant by the phrase "95% confidence interval."

7.56 A random sample of size $n = 169$ yielded $\hat{p} = .65$.

a. Is the sample size large enough to use the methods of this section to construct a confidence interval for p? Explain.

b. Construct a 90% confidence interval for p.

c. What assumption is necessary to ensure the validity of this confidence interval?

7.57 For the binomial sample information summarized in each part, indicate whether the sample size is large enough to use the methods of this chapter to construct a confidence interval for p.

a. $n = 500, \hat{p} = .05$

b. $n = 100, \hat{p} = .05$

c. $n = 10, \hat{p} = .5$

d. $n = 10, \hat{p} = .3$

7.58 A random sample of 50 consumers taste-tested a new snack food. Their responses were coded (0: do not like; 1: like; 2: indifferent) and recorded as follows:

1	0	0	1	2	0	1	1	0	0
0	1	0	2	0	2	2	0	0	1
1	0	0	0	0	1	0	2	0	0
0	1	0	0	1	0	0	1	0	1
0	2	0	0	1	1	0	0	0	1

a. Use an 80% confidence interval to estimate the proportion of consumers who like the snack food.

b. Provide a statistical interpretation for the confidence interval you constructed in part **a**.

Applying the Concepts—Basic

7.59 **Paying for music downloads.** If you use the Internet, have you ever paid to access or download music? This was one of the questions of interest in a recent *Pew Internet and American Life Project Survey* (Oct. 2010). Telephone interviews were conducted on a representative sample of 1,003 adults living in the United States. For this sample, 506 adults stated that they have paid to download music.

NW **a.** Use the survey information to find a point estimate for the true proportion of U.S. adults who have paid to download music.

b. Find an interval estimate for the proportion, part **a**. Use a 90% confidence interval.

c. Give a practical interpretation of the interval, part **b**. Your answer should begin with "We are 90% confident ..."

d. Explain the meaning of the phrase "90% confident."

7.60 **Interactions in a children's museum.** Refer to the *Early Childhood Education Journal* (Mar. 2014) study of interactions in a children's museum, Exercise 2.19 (p. 69). (Interactions by visitors to the museum included show-and-tell, learning, teaching, refocusing, participatory play, advocating, and disciplining interactions.) Recall that the researchers observed a sample of 170 meaningful interactions, of which 81 were led by children and 89 were led by adult caregivers.

a. Give a point estimate of the true proportion of all meaningful interactions in a children's museum that are led by children.

b. Form a 90% confidence interval around the proportion, part **a**.

c. Interpret the interval, part **b**, in the words of the study.

d. Suppose an educator claims that the true proportion of all meaningful interactions in a children's museum that are led by children is 35%. Make an inference about this claim.

7.61 **Is Starbucks coffee overpriced?** The *Minneapolis Star Tribune* (Aug. 12, 2008) reported that 73% of Americans say that Starbucks coffee is overpriced. The source of this information was a national telephone survey of 1,000 American adults conducted by Rasmussen Reports.

a. Identify the population of interest in this study.

b. Identify the sample for the study.

c. Identify the parameter of interest in the study.

d. Find and interpret a 95% confidence interval for the parameter of interest.

7.62 **Nannies who work for celebrities.** The International Nanny Association reports that in a sample of 528 in-home child care providers (nannies), 20 work for either a nationally known, a locally known, or an internationally known celebrity (*2011 International Nanny Association Salary and Benefits Survey*). Use Wilson's adjustment to find a 95% confidence interval for the true proportion of all nannies who work for a celebrity. Interpret the resulting interval.

7.63 **National Firearms Survey.** Refer to the Harvard School of Public Health survey to determine the size and composition of privately held firearm stock in the United States, presented in Exercise 3.85 (p. 183). Recall that, in a representative household telephone survey of 2,770 adults,

26% reported that they own at least one gun (*Injury Prevention*, Jan. 2007). The researchers want to estimate the true percentage of adults in the United States that own at least one gun.

a. Identify the population of interest to the researchers.
b. Identify the parameter of interest to the researchers.
c. Compute an estimate of the population parameter.
d. Form a 99% confidence interval around the estimate.
e. Interpret the confidence interval practically.
f. Explain the meaning of the phrase "99% confident."

7.64 **Are you really being served red snapper?** Refer to the *Nature* (July 15, 2004) study of fish specimens labeled "red snapper," presented in Exercise 3.93 (p. 185). Recall that federal law prohibits restaurants from serving a cheaper look-alike variety of fish to customers who order red snapper. In an effort to estimate the true proportion of fillets that are really red snapper, researchers analyzed the meat from each in a sample of 22 "red snapper" fish fillets purchased from vendors across the United States. DNA tests revealed that 17 of the 22 fillets (or 77%) were not red snapper but the cheaper look-alike variety of fish.

a. Identify the parameter of interest to the researchers.
b. Explain why a large-sample confidence interval is inappropriate to apply in this study.
c. Use Wilson's adjustment to construct a 95% confidence interval for the parameter of interest.
d. Give a practical interpretation of the confidence interval.

7.65 **What we do when we are sick at home.** *USA Today* (Feb. 15, 2007) reported on the results of an opinion poll in which adults were asked what one thing they are most likely to do when they are home sick with a cold or the flu. In the survey, 63% said that they are most likely to sleep and 18% said that they would watch television. Although the sample size was not reported, typically opinion polls include approximately 1,000 randomly selected respondents.

a. Assuming a sample size of 1,000 for this poll, construct a 95% confidence interval for the true percentage of all adults who would choose to sleep when they are at home sick.
b. If the true percentage of adults who would choose to sleep when they are at home sick is 70%, would you be surprised? Explain.

Applying the Concepts—Intermediate

7.66 **Curbing street gang gun violence.** Refer to the *Journal of Quantitative Criminology* (Mar. 2014) study of a program designed to curb street gang gun violence, Exercise 2.18 (p. 69). After implementation of Boston's Operation Ceasefire program, the researchers examined 80 shootings by members of a particular Boston street gang over the next five years. Of these, only 18 involved shootings of non-gang members. Use a 99% confidence interval to estimate the true proportion of shootings by members of this Boston street gang that involved non-gang members. Interpret the result.

7.67 **Do social robots walk or roll?** Refer to the *International Conference on Social Robotics* (Vol. 6414, 2010) study of the trend in the design of social robots, Exercise 2.7 (p. 66). The researchers obtained a random sample of 106 social robots through a Web search and determined that 63 were designed with legs but no wheels.

a. Find a 99% confidence interval for the proportion of all social robots designed with legs but no wheels. Interpret the result.
b. In Exercise 6.55, you assumed that 40% of all social robots are designed with legs but no wheels. Comment on the validity of this assumption.

7.68 **Study of aircraft bird strikes.** As worldwide air traffic volume has grown over the years, the problem of airplanes striking birds and other flying wildlife has increased dramatically. The *International Journal for Traffic and Transport Engineering* (Vol. 3, 2013) reported on a study of aircraft bird strikes at Aminu Kano International Airport in Nigeria. During the survey period, a sample of 44 aircraft bird strikes were analyzed. The researchers found that 36 of the 44 bird strikes at the airport occurred above 100 feet. Suppose an airport air traffic controller estimates that less than 70% of aircraft bird strikes occur above 100 feet. Comment on the accuracy of this estimate. Use a 95% confidence interval to support your inference.

7.69 **Teenagers' use of emoticons in school writing.** Due to the popularity of instant messaging and social networking, informal elements such as emoticons (e.g., the symbol ":)" to represent a smile) and abbreviations (e.g., "LOL" for "laughing out loud") have worked their way into teenagers' school writing assignments. A *Pew Internet and American Life Project* (Apr. 2008) survey interviewed 700 randomly selected U.S. teenagers by telephone on their writing habits. Overall, 448 of the teenagers admitted using at least one informal element in school writing assignments. Based on the survey results, construct a 99% confidence interval for the proportion of all U.S. teenagers who have used at least one informal element in school writing assignments. Give a practical interpretation of the interval.

7.70 **Characteristics of ice-melt ponds.** Refer to the University of Colorado study of ice-melt ponds in the Canadian Arctic, presented in Exercise 2.15 (p. 68). Environmental engineers are using data collected by the National Snow and Ice Data Center to learn how climate affects the sea ice. Data on 504 ice melt ponds revealed 88 as having "first-year ice." Recall that the researchers estimated that about 17% of melt ponds in the Canadian Arctic have first-year ice. Use the methodology of this chapter to estimate, with 90% confidence, the percentage of all ice-melt ponds in the Canadian Arctic that have first-year ice. Give a practical interpretation of the results.

7.71 **Splinting in mountain-climbing accidents.** The most common injury that occurs among mountain climbers is trauma to the lower extremity (leg). Consequently, rescuers must be proficient in immobilizing and splinting of fractures. In *High Altitude Medicine & Biology* (Vol. 10, 2009), researchers provided official recommendations for mountain emergency medicine. As part of the document, the researchers examined the likelihood of needing certain types of splints. A Scottish Mountain Rescue study reported that there was 1 femoral shaft splint needed among 333 live casualties. The researchers will use this study to estimate the proportion of all mountain casualties that require a femoral shaft splint.

a. Is the sample large enough to apply the large-sample estimation method of this section? Show why or why not.

NW **b.** Use Wilson's adjustment to find a 95% confidence interval for the true proportion of all mountain casualties that require a femoral shaft splint. Interpret the result.

7.72 Do you think you smell? If you falsely believe you emit an unpleasant, foul, or offensive body odor, you may suffer from olfactory reference syndrome (ORS). The disorder disables patients, who often isolate themselves and consider suicide. Psychiatrists disagree over how prevalent ORS is in the human population. *Depression and Anxiety* (June 2010) discussed one self-reported survey of 2,481 university students in Japan. The study reported that 52 of the students were "concerned with emitting a strange bodily odor."

a. Use the survey results to estimate the true proportion of all people in the world who suffer from ORS. Place 95% confidence bounds around the estimate and interpret the result.

b. Give several reasons why the inference, part **a**, may be invalid. Explain.

Applying the Concepts—Advanced

7.73 Eye shadow, mascara, and nickel allergies. Pigmented makeup products like mascara and eye shadow may contain metal (e.g., nickel) allergens. Is a nickel allergy more likely to occur in women who report cosmetic dermatitis from using eye shadow or mascara? This was the question of interest in a paper published in the *Journal of the European Academy of Dermatology and Venereology* (June 2010). In a sample of 131 women with cosmetic dermatitis from using eye shadow, 12 were diagnosed with a nickel allergy. In a sample of 250 women with cosmetic dermatitis from using mascara, 25 were diagnosed with a nickel allergy. Suppose you are informed that the true proportion with a nickel allergy for one of the two groups (eye shadow or mascara) is .12. Can you determine which group is referenced? Explain.

7.5 Determining the Sample Size

Recall from Section 1.5 that one way to collect the relevant data for a study used to make inferences about a population is to implement a designed (planned) experiment. Perhaps the most important design decision faced by the analyst is to determine the size of the sample. We show in this section that the appropriate sample size for making an inference about a population mean or proportion depends on the desired reliability.

Estimating a Population Mean

Consider Example 7.1 (p. 345), in which we estimated the mean length of stay for patients in a large hospital. A sample of 100 patients' records produced the 95% confidence interval $\bar{x} \pm 1.96\sigma_{\bar{x}} = 4.5 \pm .78$. Consequently, our estimate $\bar{x}$ was within .78 day of the true mean length of stay, μ, for all the hospital's patients at the 95% confidence level. That is, the 95% confidence interval for μ was $2(.78) = 1.56$ days wide when 100 accounts were sampled. This is illustrated in Figure 7.16a.

Now suppose we want to estimate μ to within .25 day with 95% confidence. That is, we want to narrow the width of the confidence interval from 1.56 days to .50 day, as

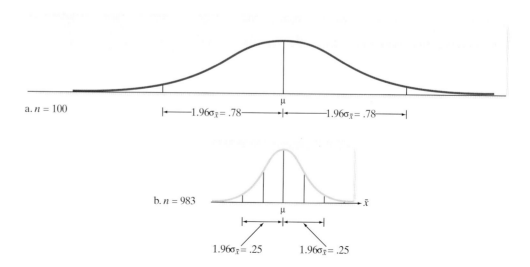

Figure 7.16

Relationship between sample size and width of confidence interval: hospital-stay example shown in Figure 7.16b. How much will the sample size have to be increased to accomplish this? If we want the estimator $\bar{x}$ to be within .25 day of μ, then we must have

$$1.96\sigma_{\bar{x}} = .25 \text{ or, equivalently, } 1.96\left(\frac{\sigma}{\sqrt{n}}\right) = .25$$

Note that we are using $\sigma_{\bar{x}} = \dfrac{\sigma}{\sqrt{n}}$ in the formula, since we are dealing with the sampling distribution of $\bar{x}$ (the estimator of μ).

The necessary sample size is obtained by solving this equation for n. First, we need the value of σ. In Example 7.1, we assumed that the standard deviation of length of stay was $\sigma = 4$ days. Thus,

$$1.96\left(\frac{\sigma}{\sqrt{n}}\right) = 1.96\left(\frac{4}{\sqrt{n}}\right) = .25$$

$$\sqrt{n} = \frac{1.96(4)}{.25} = 31.36$$

$$n = (31.36)^2 = 983.45$$

Consequently, over 983 patients' records will have to be sampled to estimate the mean length of stay, μ, to within .25 day with (approximately) 95% confidence. The confidence interval resulting from a sample of this size will be approximately .50 day wide. (See Figure 7.16b.)

In general, we express the reliability associated with a confidence interval for the population mean μ by specifying the **sampling error** within which we want to estimate μ with $100(1 - \alpha)\%$ confidence. The sampling error (denoted SE) is then equal to the half-width of the confidence interval, as shown in Figure 7.17.

The procedure for finding the sample size necessary to estimate μ with a specific sampling error is given in the following box. Note that if σ is unknown (as is usually the case, in practice), you will need to estimate the value of σ.

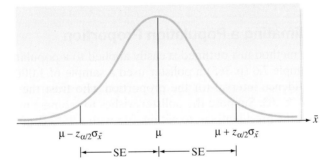

Figure 7.17

Specifying the sampling error SE as the half-width of a confidence interval

---

Determination of Sample Size for $100(1 - \alpha)\%$ Confidence Intervals for μ

In order to estimate μ with a sampling error SE and with $100(1 - \alpha)\%$ confidence, the required sample size is found as follows:

$$z_{\alpha/2}\left(\frac{\sigma}{\sqrt{n}}\right) = SE$$

The solution for n is given by the equation

$$n = \frac{(z_{\alpha/2})^2\sigma^2}{(SE)^2}$$

Note: The value of σ is usually unknown. It can be estimated by the standard deviation s from a previous sample. Alternatively, we may approximate the range R of observations in the population and (conservatively) estimate $\sigma \approx R/4$. In any case, you should round the value of n obtained *upward* to ensure that the sample size will be sufficient to achieve the specified reliability.

Example **7.9**

Sample Size for Estimating μ— Mean Inflation Pressure ot Footballs

Problem Suppose the manufacturer of official NFL footballs uses a machine to inflate the new balls to a pressure of 13.5 pounds. When the machine is properly calibrated, the mean inflation pressure is 13.5 pounds, but uncontrollable factors cause the pressures of individual footballs to vary randomly from about 13.3 to 13.7 pounds. For quality control purposes, the manufacturer wishes to estimate the mean inflation pressure to within .025 pound of its true value with a 99% confidence interval. What sample size should be specified for the experiment?

Solution We desire a 99% confidence interval that estimates μ with a sampling error of SE = .025 pound. For a 99% confidence interval, we have $z_{\alpha/2} = z_{.005} = 2.575$. To estimate σ, we note that the range of observations is $R = 13.7 - 13.3 = .4$ and we use $\sigma \approx R/4 = .1$. Next, we employ the formula derived in the box to find the sample size n:

$$n = \frac{(z_{\alpha/2})^2 \sigma^2}{(\text{SE})^2} \approx \frac{(2.575)^2(.1)^2}{(.025)^2} = 106.09$$

We round this up to $n = 107$. Realizing that σ was approximated by $R/4$, we might even advise that the sample size be specified as $n = 110$ to be more certain of attaining the objective of a 99% confidence interval with a sampling error of .025 pound or less.

Look Back To determine the value of the sampling error SE, look for the value that follows the key words "estimate μ to within . . . ."

■ **Now Work Exercise 7.84**

Sometimes the formula will lead to a solution that indicates a small sample size is sufficient to achieve the confidence interval goal. Unfortunately, the procedures and assumptions for small samples differ from those for large samples, as we discovered in Section 7.3. Therefore, if the formulas yield a small sample size, one simple strategy is to select a sample size $n = 30$.

Estimating a Population Proportion

The method just outlined is easily applied to a population proportion p. To illustrate, in Example 7.6 (p. 365) a pollster used a sample of 1,000 U.S. citizens to calculate a 95% confidence interval for the proportion who trust the president, obtaining the interval $.637 \pm .03$. Suppose the pollster wishes to estimate more precisely the proportion who trust the president, say, to within .015 with a 95% confidence interval.

The pollster wants a confidence interval for p with a sampling error SE = .015. The sample size required to generate such an interval is found by solving the following equation for n:

$$z_{\alpha/2}\sigma_{\hat{p}} = \text{SE} \qquad \text{or} \qquad z_{\alpha/2}\sqrt{\frac{pq}{n}} = .015 \qquad \text{(see Figure 7.18)}$$

Since a 95% confidence interval is desired, the appropriate z value is $z_{\alpha/2} = z_{.025} = 1.96$. We must approximate the value of the product pq before we can solve the equation for n. As shown in Table 7.6 (p. 366), the closer the values of p and q to .5, the larger is the

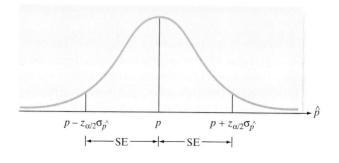

Figure 7.18
Specifying the sampling error SE of a confidence interval for a population proportion p

product pq. Thus, to find a conservatively large sample size that will generate a confidence interval with the specified reliability, we generally choose an approximation of p close to .5. In the case of the proportion of U.S. citizens who trust the president, however, we have an initial sample estimate of $\hat{p} = .637$. A conservatively large estimate of pq can therefore be obtained by using, say, $p = .60$. We now substitute into the equation and solve for n:

$$1.96 \sqrt{\frac{(.60)(.40)}{n}} = .015$$

$$n = \frac{(1.96)^2(.60)(.40)}{(.015)^2} = 4{,}097.7 \approx 4{,}098$$

The pollster must sample about 4,098 U.S. citizens to estimate the percentage who trust the president with a confidence interval of width .03.

The procedure for finding the sample size necessary to estimate a population proportion p with a specified sampling error SE is given in the following box:

Determination of Sample Size for $100(1 - \alpha)\%$ Confidence Interval for p

In order to estimate a binomial probability p with sampling error SE and with $100(1 - \alpha)\%$ confidence, the required sample size is found by solving the following equation for n:

$$z_{\alpha/2} \sqrt{\frac{pq}{n}} = \text{SE}$$

The solution for n can be written as follows:

$$n = \frac{(z_{\alpha/2})^2(pq)}{(\text{SE})^2}$$

Note: Because the value of the product pq is unknown, it can be estimated by the sample fraction of successes, $\hat{p}$, from a previous sample. Remember (Table 7.6) that the value of pq is at its maximum when p equals .5, so you can obtain conservatively large values of n by approximating p by .5 or values close to .5. In any case, you should round the value of n obtained *upward* to ensure that the sample size will be sufficient to achieve the specified reliability.

Example 7.10	**Problem** A cellular telephone manufacturer that entered the post-regulation market quickly has an initial problem with excessive customer complaints and consequent returns of cell phones for repair or replacement. The manufacturer wants to estimate the magnitude of the problem in order to design a quality control program. How many cellular telephones should be sampled and checked in order to estimate the fraction defective, p, to within .01 with 90% confidence?

**Sample Size
for Estimating
p — Fraction of Defective
Cell Phones**

Solution In order to estimate p to within .01 of its true value, we set the half-width of the confidence interval equal to SE $= .01$, as shown in Figure 7.19.

The equation for the sample size n requires an estimate of the product pq. We could most conservatively estimate $pq = .25$ (i.e., use $p = .5$), but this estimate may be too conservative. By contrast, a value of .1, corresponding to 10% defective, will probably be conservatively large for this application. The solution is therefore

$$n = \frac{(z_{\alpha/2})^2(pq)}{(\text{SE})^2} = \frac{(1.645)^2(.1)(.9)}{(.01)^2} = 2{,}435.4 \approx 2{,}436$$

Thus, the manufacturer should sample 2,436 telephones in order to estimate the fraction defective, p, to within .01 with 90% confidence.

Figure 7.19
Specified reliability for
estimate of fraction defective
in Example 7.10

Look Back Remember that this answer depends on our approximation of pq, for which we used .09. If the fraction defective is closer to .05 than .10, we can use a sample of 1,286 telephones (check this) to estimate p to within .01 with 90% confidence.

─── ■ **Now Work Exercise 7.88**

Ethics IN Statistics

In sampling, intentional omission of experimental units (e.g., respondents in a survey) in order to bias the results toward a particular view or outcome is considered *unethical statistical practice.*

The cost of sampling will also play an important role in the final determination of the sample size to be selected to estimate either μ or p. Although more complex formulas can be derived to balance the reliability and cost considerations, we will solve for the necessary sample size and note that the sampling budget may be a limiting factor. (Consult the references for a more complete treatment of this problem.) Once the sample size n is determined, be sure to devise a sampling plan that will ensure that a representative sample is selected from the target population.

Statistics IN Action   **Revisited**   ## Determining Sample Size

In the previous Statistics in Action applications in this chapter, we used confidence intervals (1) to estimate μ, the mean overpayment amount for claims in a Medicare fraud study, and (2) to estimate p, the coding error rate (i.e., proportion of claims that are incorrectly coded) of a Medicare provider. Both of these confidence intervals were based on selecting a random sample of 52 claims from the population of claims handled by the Medicare provider. How does the USDOJ determine how many claims to sample for auditing?

Consider the problem of estimating the coding error rate, p. As stated in a previous Statistics in Action Revisited, the USDOJ typically finds that about 50% of the claims in a Medicare fraud case are incorrectly coded. Suppose the USDOJ wants to estimate the true coding error rate of a Medicare provider to within .1 with 95% confidence. How many claims should be randomly sampled for audit in order to attain the desired estimate?

Here, the USDOJ desires a sampling error of $SE = .1$, a confidence level of $1 - \alpha = .95$ (for which $z_{\alpha/2} = 1.96$),

and uses an estimate $p \approx .50$. Substituting these values into the sample size formula (p. 373), we obtain.

$$n = (z_{\alpha/2})^2(pq)/(SE)^2$$
$$= (1.96)^2(.5)(.5)/(.1)^2$$
$$= 96.04$$

Consequently, the USDOJ should audit about 97 randomly selected claims to attain a 95% confidence interval for p with a sampling error of .10.

[*Note:* You may wonder why the sample actually used in the fraud analysis included only 52 claims. The sampling strategy employed involved more than selecting a simple random sample; rather, it used a more sophisticated sampling scheme, called stratified random sampling. The 52 claims represented the sample for just one of the strata.]

Exercises 7.74–7.98

Understanding the Principles

7.74 How does the sampling error SE compare with the width of a confidence interval?

7.75 **True or false.** For a specified sampling error SE, increasing the confidence level $(1 - \alpha)$ will lead to a larger n in determining the sample size.

7.76 **True or false.** For a fixed confidence level $(1 - \alpha)$, increasing the sampling error SE will lead to a smaller n in determining the sample size.

Learning the Mechanics

7.77 If you wish to estimate a population mean with a sampling distribution error $SE = .29$ using a 95% confidence interval and you know from prior sampling that σ^2 is approximately equal to 3.8, how many observations would have to be included in your sample?

7.78 If nothing is known about p, .5 can be substituted for p in the sample-size formula for a population proportion. But when this is done, the resulting sample size may be larger than needed. Under what circumstances will using $p = .5$ in the sample-size formula yield a sample size larger than is needed to construct a confidence interval for p with a specified bound and a specified confidence level?

7.79 Suppose you wish to estimate a population mean correct to within .12 with a confidence level of .90. You do not know σ^2, but you know that the observations will range in value between 36 and 44.
   a. Find the approximate sample size that will produce the desired accuracy of the estimate. You wish to be conservative to ensure that the sample size will be ample for achieving the desired accuracy of the estimate. [*Hint:* assume that the range of the observations will equal 4σ.]
   b. Calculate the approximate sample size, making the less conservative assumption that the range of the observations is equal to 6σ.

7.80 In each case, find the approximate sample size required to construct a 99% confidence interval for p that has sampling error SE = .07.
   a. Assume that p is near .4.
   b. Assume that you have no prior knowledge about p, but you wish to be certain that your sample is large enough to achieve the specified accuracy for the estimate.

7.81 A 90% confidence interval for p is given as (.48, .72). How large was the sample used to construct this interval?

7.82 It costs you $10 to draw a sample of size $n = 1$ and measure the attribute of interest. You have a budget of $1,500.
   a. Do you have sufficient funds to estimate the population mean for the attribute of interest with a 95% confidence interval 6 units in width? Assume $\sigma = 15$.
   b. If you used a 90% confidence level, would your answer to part **a** change? Explain.

7.83 Suppose you wish to estimate the mean of a normal population with a 95% confidence interval and you know from prior information that $\sigma^2 \approx 1$.
   a. To see the effect of the sample size on the width of the confidence interval, calculate the width of the confidence interval for $n = 36, 64, 81, 225$, and 900.
   b. Plot the width as a function of sample size n on graph paper. Connect the points by a smooth curve, and note how the width decreases as n increases.

Applying the Concepts—Basic

7.84 **Giraffes have excellent vision.** Refer to the *African Zoology* [NW] (Oct. 2013) study of a giraffe's eyesight, Exercise 7.38 (p. 362). Recall that the researchers measured the eye mass for a sample of 27 giraffes native to Zimbabwe, Africa, and found $\bar{x} = 53.4$ grams and $s = 8.6$ grams. Suppose the objective is to sample enough giraffes in order to obtain an estimate of the mean eye mass to within 3 grams of its true value with a 99% confidence interval.

   a. Identify the confidence coefficient for this study.
   b. Identify the desired sampling error for this study.
   c. Find the sample size required to obtain the desired estimate of the true mean.

7.85 **Shaft graves in ancient Greece.** Refer to the *American* [D] *Journal of Archaeology* (Jan. 2014) study of shaft graves SHAFTS in ancient Greece, Exercise 7.46 (p. 363). Recall that you estimated μ, the average number of shafts buried in ancient Greece graves, using data collected for 13 recently discovered grave sites and a 90% confidence interval. However, you would like to reduce the width of the interval for μ.
   a. Will increasing the confidence level to .95 reduce the width of the interval?
   b. Will increasing the sample size reduce the width of the interval?
   c. Determine the sample size required to estimate μ to within .5 shaft with 90% confidence.

7.86 **Risk of home burglary in cul-de-sacs.** Research published in the *Journal of Quantitative Criminology* (Mar. 2010) revealed that the risk of burglaries in homes located on cul-de-sacs is lower than for homes on major roads. Suppose you want to estimate the true percentage of cul-de-sac homes in your home city that were burglarized in the past year. Devise a sampling plan so that your estimate will be accurate to within 2% of the true value using a confidence coefficient of 95%. How many cul-de-sac homes need to be sampled and what information do you need to collect for each sampled home?

7.87 **Lobster trap placement.** Refer to the *Bulletin of Marine Science* (Apr. 2010) study of lobster trap placement, Exercise 7.41 (p. 363). Recall that you used a 95% confidence interval to estimate the mean trap spacing (in meters) for the population of red spiny lobster fishermen fishing in Baja California Sur, Mexico. How many teams of fishermen would need to be sampled in order to reduce the width of the confidence interval to 5 meters? Use the sample standard deviation from Exercise 7.41 in your calculation.

7.88 **Aluminum cans contaminated by fire.** A gigantic [NW] warehouse located in Tampa, Florida, stores approximately 60 million empty aluminum beer and soda cans. Recently, a fire occurred at the warehouse. The smoke from the fire contaminated many of the cans with blackspot, rendering them unusable. A University of South Florida statistician was hired by the insurance company to estimate p, the true proportion of cans in the warehouse that were contaminated by the fire. How many aluminum cans should be randomly sampled to estimate the true proportion to within .02 with 90% confidence?

7.89 **Do social robots walk or roll?** Refer to the *International Conference on Social Robotics* (Vol. 6414, 2010) study of the trend in the design of social robots, Exercise 7.67 (p. 371). Recall that you used a 99% confidence interval to estimate the proportion of all social robots designed with legs but no wheels. How many social robots would need to be sampled in order to estimate the proportion to within .075 of its true value?

Applying the Concepts—Intermediate

7.90 **Duration of daylight in western Pennsylvania.** Refer to the *Naval Oceanography Portal* data on number of minutes of daylight per day in Sharon, PA, Exercise 7.43 (p. 363). An estimate of the mean number of minutes of daylight per day was obtained using data collected for 12 randomly selected days (one each month) in a recent year.
   a. Determine the number of days that need to be sampled in order to estimate the desired mean to within 45 minutes of its true value with 95% confidence.
   b. Based on your answer, part **a**, develop a sampling plan that will likely result in a random sample that is representative of the population.
   c. Go to the Web site, http://aa.usno.navy.mil/USNO/astronomical-applications/data-services, and collect the data for Sharon, PA using your sampling plan.
   d. Use the data, part **c**, to construct a 95% confidence interval for the desired mean. Does your interval have the desired width?

7.91 **Pitch memory of amusiacs.** Refer to the *Advances in Cognitive Psychology* (Vol. 6, 2010) study of pitch memory of amusiacs, Exercise 7.45 (p. 363). Recall that diagnosed amusiacs listened to a series of tone pairs and were asked to determine if the tones were the same or different. In the first trial, the tones were separated by 1 second; in the second trial, the tones were separated by 5 seconds. The variable of interest was the difference between scores on the two trials. How many amusiacs would need to participate in the study in order to estimate the true mean score difference for all amusiacs to within .05 with 90% confidence?

7.92 **Shopping on Black Friday.** Refer to the *International Journal of Retail and Distribution Management* (Vol. 39, 2011) survey of Black Friday shoppers, Exercise 7.22 (p. 353). One question was "How many hours do you usually spend shopping on Black Friday?"
   a. How many Black Friday shoppers should be included in a sample designed to estimate (with 95% confidence) the average number of hours spent shopping on Black Friday if you want the estimate to deviate no more than .5 hour from the true mean?
   b. Devise a sampling plan for collecting the data that will likely result in a representative sample.

7.93 **Study of aircraft bird strikes.** Refer to the *International Journal for Traffic and Transport Engineering* (Vol. 3, 2013) study of aircraft bird strikes at a Nigerian airport, Exercise 7.68 (p. 371). Recall that an air traffic controller wants to estimate the true percentage of aircraft bird strikes that occur above 100 feet. Determine how many aircraft bird strikes need to be analyzed in order to estimate the true percentage to within 5% if you use a 95% confidence interval.

7.94 **Bacteria in bottled water.** Is the bottled water you drink safe? The Natural Resources Defense Council warns that the bottled water you are drinking may contain more bacteria and other potentially carcinogenic chemicals than are allowed by state and federal regulations. Of the more than 1,000 bottles studied, nearly one-fourth exceeded government levels. Suppose that the Natural Resources Defense Council wants an updated estimate of the population proportion of bottled water that violates at least one government standard. Determine the sample size (number of bottles) needed to estimate this proportion to within ± .05 with 99% confidence.

7.95 **Do you think you smell?** Refer to the *Depression and Anxiety* (June 2010) study of patients who suffer from olfactory reference syndrome (ORS), Exercise 7.72 (p. 372). Recall that psychiatrists disagree over how prevalent ORS is in the human population. Suppose you want to estimate the true proportion of U.S. adults who suffer from ORS using a 99% confidence interval. Determine the size of the sample necessary to attain a sampling error no larger than .04.

7.96 **Caffeine content of coffee.** According to a Food and Drug Administration (FDA) study, a cup of coffee contains an average of 115 milligrams (mg) of caffeine, with the amount per cup ranging from 60 to 180 mg. Suppose you want to repeat the FDA experiment in order to obtain an estimate of the mean caffeine content in a cup of coffee correct to within 10 mg with 99% confidence. How many cups of coffee would have to be included in your sample?

7.97 **Eye shadow, mascara, and nickel allergies.** Refer to the *Journal of the European Academy of Dermatology and Venereology* (June 2010) study of the link between nickel allergies and use of mascara or eye shadow, Exercise 7.73 (p. 372). Recall that two groups of women were sampled: one group with cosmetic dermatitis from using eye shadow and another group with cosmetic dermatitis from using mascara. In either group, how many women would need to be sampled in order to yield an estimate (with 95% confidence) of the population percentage with a nickel allergy that falls no more than 3% from the true value?

Applying the Concepts—Advanced

7.98 **Preventing production of defective items.** It costs more to produce defective items—since they must be scrapped or reworked—than it does to produce non-defective items. This simple fact suggests that manufacturers should ensure the quality of their products by perfecting their production processes instead of depending on inspection of finished products (Deming, 1986). In order to better understand a particular metal stamping process, a manufacturer wishes to estimate the mean length of items produced by the process during the past 24 hours.
   a. How many parts should be sampled in order to estimate the population mean to within .2 millimeter (mm) with 95% confidence? Previous studies of this machine have indicated that the standard deviation of lengths produced by the stamping operation is about 2 mm.
   b. Time permits the use of a sample size no larger than 225. If a 95% confidence interval for μ is constructed with $n = 225$, will it be wider or narrower than would have been obtained using the sample size determined in part **a**? Explain.
   c. If management requires that μ be estimated to within .2 mm and that a sample size of no more than 225 be used, what is (approximately) the maximum confidence level that could be attained for a confidence interval that meets management's specifications?

7.6   Confidence Interval for a Population Variance (Optional)

In the previous sections, we considered interval estimation for population means or proportions. In this optional section, we discuss a confidence interval for a population variance, σ^2.

Recall Example 1.4 (p. 38) and the U.S. Army Corps of Engineers study of contaminated fish in the Tennessee River, Alabama. It is important for the Corps of Engineers to know how stable the weights of the contaminated fish are. That is, how large is the variation in the fish weights? The keyword "variation" indicates that the target population parameter is σ^2, the variance of the weights of all contaminated fish inhabiting the Tennessee River. Of course, the exact value of σ^2 will be unknown. Consequently, the Corps of Engineers wants to estimate its value with a high level of confidence.

Intuitively, it seems reasonable to use the sample variance, s^2, to estimate σ^2. However, unlike with sample means and proportions, the sampling distribution of s^2 does not follow a normal (z) distribution or a Student's t distribution. Rather, when certain assumptions are satisfied (we discuss these later), the sampling distribution of s^2 possesses approximately a **chi-square (χ^2) distribution**. The chi-square probability distribution, like the t distribution, is characterized by a quantity called the *degrees of freedom* (df) associated with the distribution. Several chi-square distributions with different df values are shown in Figure 7.20. You can see that unlike z and t distributions, the chi-square distribution is not symmetric about 0.

The upper-tail areas for this distribution have been tabulated and are given in Table IV in Appendix B, a portion of which is reproduced in Table 7.7. The table gives the values of χ^2, denoted as χ^2_α, that locate an area of α in the upper tail of the

Figure 7.20
Several χ^2 Probability Distributions

Table 7.7   Reproduction of Part of Table IV in Appendix B: Critical Values of Chi Square

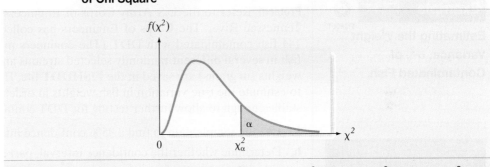

Degrees of Freedom	$\chi^2_{.100}$	$\chi^2_{.050}$	$\chi^2_{.025}$	$\chi^2_{.010}$	$\chi^2_{.005}$
1	2.70554	3.84146	5.02389	6.63490	7.87944
2	4.60517	5.99147	7.37776	9.21034	10.5966
3	6.25139	7.81473	9.34840	11.3449	12.8381
4	7.77944	9.48773	11.1433	13.2767	14.8602
5	9.23635	11.0705	12.8325	15.0863	16.7496
6	10.6446	12.5916	14.4494	16.8119	18.5476
7	12.0170	14.0671	16.0128	18.4753	20.2777
8	13.3616	15.5073	17.5346	20.0902	21.9550
9	14.6837	16.9190	19.0228	21.6660	23.5893
10	15.9871	18.3070	20.4831	23.2093	25.1882
11	17.2750	19.6751	21.9200	24.7250	25.7569
12	18.5494	21.0261	23.3367	26.2170	28.2995
13	19.8119	22.3621	24.7356	27.6883	29.8194
14	21.0642	23.6848	26.1190	29.1413	31.3193
15	22.3072	24.9958	27.4884	30.5779	32.8013
16	23.5418	26.2862	28.8454	31.9999	34.2672
17	24.7690	27.5871	30.1910	33.4087	35.7185
18	25.9894	28.8693	31.5264	34.8053	37.1564
19	27.2036	30.1435	32.8523	36.1908	38.5822

chi-square distribution; that is, $P(\chi^2 > \chi_\alpha^2) = \alpha$. As with the t-statistic, the degrees of freedom associated with s^2 are $(n-1)$. Thus, for $n = 10$ and an upper-tail value $\alpha = .05$, you will have $n-1 = 9$ df and $\chi_{.05}^2 = 16.9190$ (highlighted in Table 7.7). [*Note:* Values of χ_α^2 can also be obtained using the inverse chi-square option of statistical software.]

The chi-square distribution is used to find a confidence interval for σ^2, as shown in the box. An illustrative example follows.

A $100(1-\alpha)$ Confidence Interval for σ^2

$$\frac{(n-1)s^2}{\chi_{\alpha/2}^2} \le \sigma^2 \le \frac{(n-1)s^2}{\chi_{(1-\alpha/2)}^2}$$

where $\chi_{\alpha/2}^2$ and $\chi_{(1-\alpha/2)}^2$ are values corresponding to an area of $\alpha/2$ in the right (upper) and left (lower) tails, respectively, of the chi-square distribution based on $(n-1)$ degrees of freedom.

Conditions Required for a Valid Confidence Interval for σ^2

1. A random sample is selected from the target population.
2. The population of interest has a relative frequency distribution that is approximately normal.

Example **7.11** **D**

FISHDDT

Estimating the Weight Variance, σ^2, of Contaminated Fish

Problem Refer to the U.S. Army Corps of Engineers study of contaminated fish in the Tennessee River. The Corps of Engineers has collected data for a random sample of 144 fish contaminated with DDT. (The engineers made sure to capture contaminated fish in several different randomly selected streams and tributaries of the river.) The fish weights (in grams) are saved in the **FISHDDT** file. The Army Corps of Engineers wants to estimate the true variation in fish weights in order to determine whether the fish are stable enough to allow further testing for DDT contamination.

a. Use the sample data to find a 95% confidence interval for the parameter of interest.

b. Determine whether the confidence interval, part **a**, is valid.

Solution

a. Here the target parameter is σ^2, the variance of the population of weights of contaminated fish. First, we need to find the sample variance, s^2, to compute the interval estimate. The MINITAB printout, Figure 7.21 gives descriptive statistics for the sample weights. You can see that the variance (highlighted) is $s^2 = 141,787$.

For $\alpha = .05$ (a 95% confidence interval), we require the critical values $\chi_{\alpha/2}^2 = \chi_{.025}^2$ and $\chi_{(1-\alpha/2)}^2 = \chi_{.975}^2$ for a chi-square distribution with $(n - 1) = 143$ degrees of freedom. Examining Table IV in Appendix B, we see that these values are given for df = 100 and df = 150, but not for df = 143. We could approximate these critical chi-square values using the entries for df = 150 (the row closest to

---

Figure 7.21
MINITAB Descriptive Statistics for Fish Weights, Example 7.11

Descriptive Statistics: WEIGHT

Variable	N	Mean	StDev	Variance	Minimum	Maximum
WEIGHT	144	1049.7	376.5	141787.0	173.0	2302.0

df = 143). Or we could use statistical software to obtain the exact values. The exact values, obtained using MINITAB (and shown on Figure 7.22), are $\chi^2_{.025} = 111.79$ and $\chi^2_{.975} = 178.0$.

Inverse Cumulative Distribution Function

```
Chi-Square with 143 DF

P( X <= x )        x
       0.025   111.787
```

Inverse Cumulative Distribution Function

```
Chi-Square with 143 DF

P( X <= x )        x
       0.975   177.998
```

Figure 7.22

MINITAB Output with critical χ^2 values, Example 7.11

Substituting the appropriate values into the formula given in the box, we obtain:

$$\frac{(143)(141{,}787)}{178} \le \sigma^2 \le \frac{(143)(141{,}787)}{111.79}$$

Or

$$113{,}907 \le \sigma^2 \le 181{,}371$$

Thus, the Army Corps of Engineers can be 95% confident that the variance in weights of the population of contaminated fish ranges between 113,907 and 181,371. [*Note:* You can obtain this interval directly using statistical software as well. This interval is shown (highlighted) on the MINITAB printout, Figure 7.23. Our calculated values agree, except for rounding.]

Test and CI for One Variance: WEIGHT

```
Method

The chi-square method is only for the normal distribution.
The Bonett method is for any continuous distribution.


Statistics

Variable    N  StDev  Variance
WEIGHT    144    377    141787


95% Confidence Intervals

                          CI for        CI for
Variable  Method          StDev        Variance
WEIGHT    Chi-Square   (338, 426)  (113909, 181377)
          Bonett       (333, 431)  (111188, 185831)
```

Figure 7.23

MINITAB Output with 95% Confidence Interval for σ^2, Example 7.11

b. According to the box, two conditions are required for the confidence interval to be valid. First, the sample must be randomly selected from the population. The Army Corps of Engineers did, indeed, collect a random sample of contaminated fish, making sure to sample fish from different locations in the Tennessee River. Second, the population data (the fish weights) must be approximately normally distributed. A MINITAB histogram for the sampled fish weights (with a normal curve superimposed) is displayed in Figure 7.24. Clearly, the data appear to be approximately normally distributed. Thus, the confidence interval is valid.

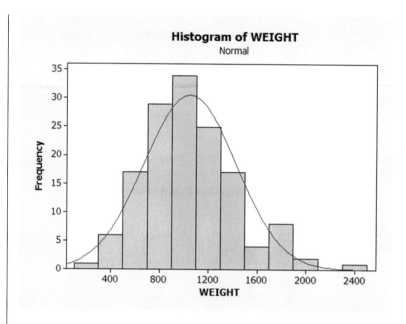

Figure 7.24
MINITAB histogram of fish weights, Example 7.11

Look Ahead Will this confidence interval be practically useful in helping the Corps of Engineers decide whether the weights of the fish are stable? Only if it is clear what a weight variance of, say, 150,000 grams^2 implies. Most likely, the Corps of Engineers will want the interval in the same units as the weight measurement—grams. Consequently, a confidence interval for σ, the standard deviation of the population of fish weights, is desired. We demonstrate how to obtain this interval estimate in the next example.

■ **Now Work Exercise 7.107a**

Example   7.12   Ⓓ

FISHDDT

Estimating the Weight Standard Deviation, σ, of Contaminated Fish

Problem Refer to Example 7.11. Find a 95% confidence interval for σ, the true standard deviation of the contaminated fish weights.

Solution A confidence interval for σ is obtained by taking the square roots of the lower and upper endpoints of a confidence interval for σ^2. Consequently, the 95% confidence interval for σ is:

$$\sqrt{113,907} \le \sigma \le \sqrt{181,371}$$

Or,

$$337.5 \le \sigma \le 425.9$$

[*Note:* This interval is also shown on the MINITAB printout, Figure 7.23.]
Thus, the engineers can be 95% confident that the true standard deviation of fish weights is between 337.5 grams and 425.9 grams.

Look Back Suppose the Corps of Engineers' threshold is $\sigma = 500$ grams. That is, if the standard deviation in fish weights is 500 grams or higher, further DDT contamination tests will be suspended due to the unstableness of the fish weights. Since the 95% confidence interval for σ lies below 500 grams, the engineers will continue the DDT contamination tests on the fish.

■ **Now Work Exercise 7.107b**

⚠ **CAUTION** The procedure for estimating either σ^2 or σ requires an assumption regardless of whether the sample size n is large or small (see the conditions in the box). The sampled data must come from a population that has an approximate normal distribution. Unlike small sample confidence intervals for μ based on the t-distribution, *slight to moderate departures from normality will render the chi-square confidence interval for σ^2 invalid.* ▲

Exercises 7.99–7.117

Understanding the Principles

7.99 What sampling distribution is used to find an interval estimate for σ^2?

7.100 What conditions are required for a valid confidence interval for σ^2?

7.101 How many degrees of freedom are associated with a chi-square sampling distribution for a sample of size n?

Learning the Mechanics

7.102 For the combinations of α and degrees of freedom (df) in parts **a** through **d** below, use either Table IV in Appendix B or statistical software to find the values of $\chi^2_{\alpha/2}$ and $\chi^2_{(1-\alpha/2)}$ that would be used to form a confidence interval for σ^2.
  a. $\alpha = .05$, df = 6
  b. $\alpha = .10$, df = 14
  c. $\alpha = .01$, df = 22
  d. $\alpha = .05$, df = 22

7.103 Given the following values of $\bar{x}$, s, and n, form a 90% confidence interval for σ^2.
  a. $\bar{x} = 21$, $s = 2.7$, $n = 50$
  b. $\bar{x} = 1.7$, $s = .04$, $n = 14$
  c. $\bar{x} = 160$, $s = 31.7$, $n = 24$
  d. $\bar{x} = 9.8$, $s = 1.6$, $n = 5$

7.104 Refer to Exercise 7.103. For each part, **a–d**, form a 90% confidence interval for σ.

7.105 A random sample of $n = 6$ observations from a normal distribution resulted in the data shown in the table. Compute a 95% confidence interval for σ^2.

8	2	3	7	11	6

Applying the Concepts—Basics

7.106 Corporate sustainability of CPA firms. Refer to the *Business and Society* (Mar. 2011) study on the sustainability behaviors of CPA corporations, Exercise 7.18 (p. 352). Recall that the level of support for corporate sustainability (measured on a quantitative scale ranging from 0 to 160 points) was obtained for each in a sample of 992 senior managers at CPA firms. The accompanying MINITAB printout gives 90% confidence intervals for both the variance and standard deviation of level of support for all senior managers at CPA firms.

```
Statistics

Variable    N    StDev    Variance
Support    992    26.9         722


90% Confidence Intervals

                              CI for       CI for
Variable   Method            StDev        Variance
Support    Chi-Square     (25.9, 27.9)   (672, 779)
```

  a. Locate the 90% confidence interval for σ^2 on the printout.
  b. Use the sample variance on the printout to calculate the 90% confidence interval for σ^2. Does your result agree with the interval shown on the printout?

  c. Locate the 90% confidence interval for σ on the printout.
  d. Use the result, part **a**, to calculate the 90% confidence interval for σ. Does your result agree with the interval shown on the printout?
  e. Give a practical interpretation of the 90% confidence interval for σ.
  f. What assumption about the distribution of level of support is required for the inference, part **e**, to be valid? Is this assumption reasonably satisfied?

7.107 Characteristics of a rockfall. Refer to the *Environmental Geology* (Vol. 58, 2009) simulation study of how far a block from a collapsing rock wall will bounce down a soil slope, Exercise 2.61 (p. 89). Rebound lengths (in meters) were estimated for 13 rock bounces. The data are repeated in the table. A MINITAB analysis of the data is shown in the printout below.

10.94	13.71	11.38	7.26	17.83	11.92	11.87
5.44	13.35	4.90	5.85	5.10	6.77	

Based on Paronuzzi, P. "Rockfall-induced block propagation on a soil slope, northern Italy." *Environmental Geology*, Vol. 58, 2009 (Table 2).

  a. Locate a 95% confidence interval for σ^2 on the printout. Interpret the result.
  b. Locate a 95% confidence interval for σ on the printout. Interpret the result.
  c. What conditions are required for the intervals, parts **a** and **b**, to be valid?

```
Test and CI for One Variance: REB-LENGTH

Method

The chi-square method is only for the normal distribution.
The Bonett method is for any continuous distribution.


Statistics

Variable      N   StDev   Variance
REB-LENGTH   13    4.09       16.8


95% Confidence Intervals

                            CI for        CI for
Variable      Method        StDev        Variance
REB-LENGTH    Chi-Square   (2.94, 6.76)  (8.6, 45.7)
              Bonett       (2.97, 6.64)  (8.8, 44.1)
```

7.108 Motivation of drug dealers. Refer to the *Applied Psychology in Criminal Justice* (Sept. 2009) study of the personality characteristics of convicted drug dealers, Exercise 7.17 (p. 352). A random sample of 100 drug dealers had a mean Wanting Recognition (WR) score of 39 points, with a standard deviation of 6 points. The researchers also are interested in σ^2, the variation in WR scores for all convicted drug dealers.
  a. Identify the target parameter, in symbols and words.
  b. Compute a 99% confidence interval for σ^2.
  c. What does it mean to say that the target parameter lies within the interval with "99% confidence"?
  d. What assumption about the data must be satisfied in order for the confidence interval to be valid?

e. To obtain a practical interpretation of the interval, part **b**, explain why a confidence interval for the standard deviation, σ, is desired.

f. Use the results, part **b**, to compute a 99% confidence interval for σ. Give a practical interpretation of the interval.

7.109 Facial structure of CEOs. Refer to the *Psychological Science* (Vol. 22, 2011) study of a chief executive officer's facial structure, Exercise 7.21 (p. 353). Recall that the facial width-to-height ratio (WHR) was determined by computer analysis for each in a sample of 55 CEOs at publicly traded *Fortune* 500 firms, with the following results: $\bar{x} = 1.96, s = .15$.

a. Find and interpret a 95% confidence interval for the standard deviation, σ, of the facial WHR values for all CEOs at publicly traded *Fortune* 500 firms. Interpret the result.

b. For the interval, part **a**, to be valid, the population of WHR values should be distributed how? Draw a sketch of the required distribution to support your answer.

7.110 Antigens for a parasitic roundworm in birds. Refer to the *Gene Therapy and Molecular Biology* (June 2009) study of DNA in peptide (protein) produced by antigens for a parasitic roundworm in birds, Exercise 7.50 (p. 364). Recall that scientists tested each in a sample of 4 alleles of antigen-produced protein for level of peptide. The results were: $\bar{x} = 1.43$ and $s = .13$. Use this information to construct a 90% confidence interval for the true variation in peptide scores for alleles of the antigen-produced protein. Interpret the interval for the scientists.

7.111 Oil content of fried sweet potato chips. The characteristics of sweet potato chips fried at different temperatures were investigated in the *Journal of Food Engineering* (Sept. 2013). A sample of 6 sweet potato slices were fried at 130° using a vacuum fryer. One characteristic of interest to the researchers was internal oil content (measured in gigagrams). The results were: $\bar{x} = .178 \, \text{g/g}$ and $s = .011 \, \text{g/g}$. Use this information to construct a 95% confidence interval for the true standard deviation of the internal oil content distribution for the sweet potato chips. Interpret the result practically.

Applying the Concepts—Intermediate

7.112 Radon exposure in Egyptian tombs. Refer to the *Radiation Protection Dosimetry* (Dec. 2010) study of radon exposure in tombs carved from limestone in the Egyptian Valley of Kings, Exercise 7.39 (p. 362). The radon levels in the inner chambers of a sample of 12 tombs were determined, yielding the following summary statistics: $\bar{x} = 3{,}643 \, \text{Bq/m}^3$ and $s = 4{,}487 \, \text{Bq/m}^3$. Use this information to estimate, with 95% confidence, the true standard deviation of radon levels in tombs in the Valley of Kings. Interpret the resulting interval. Be sure to give the units of measurement in your interpretation.

7.113 Cheek teeth of extinct primates. Refer to the *American Journal of Physical Anthropology* (Vol. 142, 2010) study of the characteristics of cheek teeth (e.g., molars) in an extinct primate species, Exercise 2.38 (p. 78). Recall that the researchers recorded the dentary depth of molars (in millimeters) for a sample of 18 cheek teeth extracted from skulls. The data are repeated in the table. Estimate the

true standard deviation in molar depths for the population of cheek teeth in extinct primates using a 95% confidence interval. Give a practical interpretation of the result. Are the conditions required for a valid confidence interval reasonably satisfied?

18.12	16.55
19.48	15.70
19.36	17.83
15.94	13.25
15.83	16.12
19.70	18.13
15.76	14.02
17.00	14.04
13.96	16.20

Based on Boyer, D. M., Evans, A. R., and Jernvall, J. "Evidence of dietary differentiation among Late Paleocene–Early Eocene Plesiadapids (Mammalia, Primates)." *American Journal of Physical Anthropology*, Vol. 142, 2010 (Table A3).

7.114 Shell lengths of sea turtles. Refer to the *Aquatic Biology* (Vol. 9, 2010) study of green sea turtles inhabiting the Grand Cayman South Sound lagoon, Exercise 7.24 (p. 354). Recall that the data on shell length, measured in centimeters, for 76 captured turtles are saved in the **TURTLES** file. Use the sample data to estimate the true variance in shell lengths of all green sea turtles in the lagoon with 90% confidence. Interpret the result.

7.115 Lobster trap placement. Refer to the *Bulletin of Marine Science* (Apr. 2010) study of red spiny lobster trap placement, Exercise 7.41 (p. 363). Trap spacing measurements (in meters) for a sample of seven teams of red spiny lobster fishermen are repeated in the table below. The researchers want to know how variable the trap spacing measurements are for the population of red spiny lobster fishermen fishing in Baja California Sur, Mexico. Provide the researchers with an estimate of the target parameter using a 99% confidence interval.

93	99	105	94	82	70	86

Based on Shester, G. G. "Explaining catch variation among Baja California lobster fishers through spatial analysis of trap-placement decisions." *Bulletin of Marine Science*, Vol. 86, No. 2, Apr. 2010 (Table 1), pp. 479–498.

7.116 Is honey a cough remedy? Refer to the *Archives of Pediatrics and Adolescent Medicine* (Dec. 2007) study of honey as a remedy for coughing, Exercise 2.40 (p. 79). Recall that the 105 ill children in the sample were randomly divided into groups. One group received a dosage of an over-the-counter cough medicine (DM); another group received a dosage of honey (H). The coughing improvement scores (as determined by the children's parents) for the patients in the two groups are reproduced in the table on p. 385. The pediatric researchers desire information on the variation in coughing improvement scores for each of the two groups.

a. Find a 90% confidence interval for the standard deviation in improvement scores for the honey dosage group.

b. Repeat part **a** for the DM dosage group.

c. Based on the results, parts **a** and **b**, what conclusions can the pediatric researchers draw about which group has the smaller variation in improvement scores? (We demonstrate a more statistically valid method for comparing variances in Chapter 9.)

Honey Dosage:	12 11 15 11 10 13 10 4 15 16 9 14 10 6 10 8 11 12 12 8 12 9 11 15 10 15 9 13 8 12 10 8 9 5 12
DM Dosage:	4 6 9 4 7 7 7 9 12 10 11 6 3 4 9 12 7 6 8 12 12 4 12 13 7 10 13 9 4 4 10 15 9

Based on Paul, I. M., et al. "Effect of honey, dextromethorphan, and no treatment on nocturnal cough and sleep quality for coughing children and their parents." *Archives of Pediatrics and Adolescent Medicine*, Vol. 161, No. 12, Dec. 2007 (data simulated).

7.117 Phishing attacks to e-mail accounts. Refer to the *Chance* (Summer 2007) study of an actual phishing attack against an organization, Exercise 6.36 (p. 330). Recall that *phishing* describes an attempt to extract personal/financial information from unsuspecting people through fraudulent e-mail. The interarrival times (in seconds) for 267 fraud box e-mail notifications are saved in the accompanying file. Like with Exercise 6.36, consider these interarrival times to represent the population of interest.
a. Obtain a random sample of $n = 10$ interarrival times from the population.
b. Use the sample, part **a**, to obtain an interval estimate of the population variance of the interarrival times. What is the measure of reliability for your estimate?
c. Find the true population variance for the data. Does the interval, part **b**, contain the true variance? Give one reason why it may not.

CHAPTER NOTES

Key Terms

Note: Asterisks (*) denote terms from the optional section in this chapter.

*Chi-square (χ^2) distribution 379
Confidence coefficient 347
Confidence interval 344
Confidence level 344
Degrees of freedom (df) 355
Interval estimator 344

Point estimator 344
Sampling error 373
Target parameter 344
t-statistic 355
Wilson's adjustment for estimating *p* 368
z-statistic 345

Key Symbols

θ	General population parameter (theta)
μ	Population mean
σ^2	Population variance
σ	Population standard deviation
p	Population proportion; P (Success) in binomial trial
q	$1 - p$
$\bar{x}$	Sample mean (estimator of μ)
s^2	Sample variance (estimator of σ^2)
$\hat{p}$	Sample proportion (estimator of p)
$\mu_{\bar{x}}$	Mean of the population sampling distribution of $\bar{x}$
$\sigma_{\bar{x}}$	Standard deviation of the sampling distribution of $\bar{x}$
$\sigma_{\hat{p}}$	Standard deviation of the sampling distribution of $\hat{p}$
SE	Sampling error in estimation
α	$(1 - \alpha)$ represents the confidence coefficient
$z_{\alpha/2}$	z-value used in a $100(1 - \alpha)\%$ large-sample confidence interval for μ or p
$t_{\alpha/2}$	Student's t-value used in a $100(1 - \alpha)\%$ small-sample confidence interval for μ
$\chi^2_{\alpha/2}$	*Chi-Square value used in a $100(1 - \alpha)\%$ confidence interval for σ^2

Key Ideas

Confidence Interval: An interval that encloses an unknown population parameter with a certain level of confidence, $(1 - \alpha)$
Confidence Coefficient: The probability $(1 - \alpha)$ that a randomly selected confidence interval encloses the true value of the population parameter.

Population Parameters, Estimators, & Standard Errors

Parameter (θ)	Estimator ($\hat{\theta}$)	Standard Error of Estimator ($\sigma_{\hat{\theta}}$)	Estimated Std. Error ($\hat{\sigma}_{\hat{\theta}}$)
Mean, μ	$\bar{x}$	$\sigma/\sqrt{n}$	$s/\sqrt{n}$
Proportion, p	$\hat{p}$	$\sqrt{pq}/n$	$\sqrt{\hat{p}\hat{q}}/n$
*Variance, σ^2	s^2	—	—
Standard Deviation, r	S	—	—

Determining the Sample Size n:

Estimating μ: $\quad n = (z_{\alpha/2})^2(\sigma^2)/(\text{SE})^2$
Estimating p: $\quad n = (z_{\alpha/2})^2(pq)/(\text{SE})^2$

Key Words for Identifying the Target Parameter:

μ — Mean, Average
p — Proportion, Fraction, Percentage, Rate, Probability
*σ^2 — Variance, Variation, Spread

Commonly Used z-values for a Large-Sample Confidence Interval for μ or p:

90% CI:	$(1 - \alpha) = .10$	$z_{.05} = 1.645$
95% CI:	$(1 - \alpha) = .05$	$z_{.025} = 1.96$
98% CI:	$(1 - \alpha) = .02$	$z_{.01} = 2.326$
99% CI:	$(1 - \alpha) = .01$	$z_{.005} = 2.575$

Illustrating the Notion of "95% Confidence"

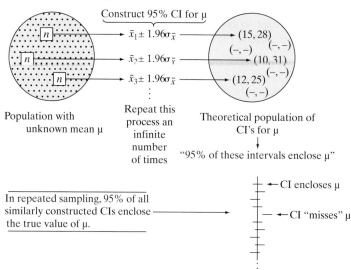

Guide to Forming a Confidence Interval

Supplementary Exercises 7.118–7.152

Note: List the assumptions necessary for the valid implementation of the statistical procedures you use in solving all these exercises. Starred (*) exercises are from the optional section in this chapter.

Understanding the Principles

7.118 Interpret the phrase "95% confident" in the following statement: "We are 95% confident that the proportion of all PCs with a computer virus falls between .12 and .18."

7.119 For each of the following, identify the target parameter as μ, p, or σ^2.
   a. Average score on the SAT
   b. Mean time waiting at a supermarket checkout lane
   c. Proportion of voters in favor of legalizing marijuana
   d. Percentage of NFL players who have ever made the Pro Bowl
   e. Dropout rate of American college students
   *f. Variation in IQ scores of sociopaths

7.120 In each of the following instances, determine whether you would use a z- or t-statistic (or neither) to form a 99% confidence interval for μ; then look up the appropriate z or t value.

a. Random sample of size $n = 17$ from a normal distribution with unknown mean μ and standard deviation σ

b. Random sample of size $n = 192$ from a normal distribution with unknown mean μ and standard deviation σ

c. Random sample of size $n = 16$ from a normal distribution with unknown mean and standard deviation $\sigma = 4$

d. Random sample of size $n = 80$ from a distribution about which nothing is known

e. Random sample of size $n = 12$ from a distribution about which nothing is known

Learning the Mechanics

*7.121 Use Table IV, Appendix B, or statistical software to find $\chi^2_{\alpha/2}$ and $\chi^2_{1-\alpha/2}$ for each of the following:

a. $n = 10, \alpha = .05$

b. $n = 20, \alpha = .05$

c. $n = 50, \alpha = .01$

7.122 Let t_0 represent a particular value of t from Table III of Appendix B. Find the table values such that the following statements are true:

a. $P(t \le t_0) = .05$, where df $= 14$

b. $P(t \ge t_0) = .005$, where df $= 19$

c. $P(t \le -t_0 \text{ or } t \ge t_0) = .10$, where df $= 9$

d. $P(t \le -t_0 \text{ or } t \ge t_0) = .01$, where df $= 28$

7.123 In a random sample of 225 measurements, 125 possess the characteristic of interest, A.

a. Use a 95% confidence interval to estimate the true proportion p of measurements in the population with characteristic A.

b. How large a sample would be needed to estimate p to within .02 with 95% confidence?

7.124 A random sample of 400 measurements is selected from a population, and the sample mean and standard deviation are $\bar{x} = 40.5$ and $s = 26.0$, respectively.

a. Use a 99% confidence interval to estimate the mean of the population, μ.

b. How large a sample would be needed to estimate μ to within .5 with 99% confidence?

c. What is meant by the phrase "99% confidence" as it is used in this exercise?

*d. Find a 99% confidence interval for σ^2.

Applying the Concepts—Basic

7.125 CDC health survey. The Centers for Disease Control and Prevention (CDCP) in Atlanta, Georgia, conduct an annual survey of the general health of the U.S. population as part of their Behavioral Risk Factor Surveillance System. Using random-digit dialing, the CDCP telephones U.S. citizens over 18 years of age and asks them the following four questions:

1. Is your health generally excellent, very good, good, fair, or poor?

2. How many days during the previous 30 days was your physical health not good because of injury or illness?

3. How many days during the previous 30 days was your mental health not good because of stress, depression, or emotional problems?

4. How many days during the previous 30 days did your physical or mental health prevent you from performing your usual activities?

Identify the parameter of interest for each question.

7.126 Personal networks of older adults. In sociology, a personal network is defined as the people with whom you make frequent contact. A stratified random sample of men and women born between 1908 and 1937 was used to gauge the size of the personal network of older adults. Each adult in the sample was asked to "please name the people (e.g., in your neighborhood) you have frequent contact with and who are also important to you." Based on the number of people named, the personal network size for each adult was determined. The responses of 2,819 adults in the sample yielded the following statistics on network size: $\bar{x} = 14.6; s = 9.8$ (*Sociological Methods & Research*, Aug. 2001).

a. Give a point estimate for the mean personal network size of all older adults.

b. Form a 95% confidence interval for the mean personal network size of all older adults.

c. Give a practical interpretation of the interval you found in part **b**.

d. Give the conditions required for the interval in part **b** to be valid.

7.127 Ancient Greek pottery. Refer to the *Chance* (Fall 2000) study of 837 pieces of pottery found at the ancient Greek settlement at Phylakopi, presented in Exercise 2.186 (p. 134). Of the 837 pieces, 183 were painted with either a curvilinear, geometric, or naturalistic decoration. Find a 90% confidence interval for the population proportion of all pottery artifacts at Phylakopi that are painted. Interpret the resulting interval.

7.128 Extinct New Zealand birds. Refer to the *Evolutionary Ecology Research* (July 2003) study of the patterns of extinction in the New Zealand bird population, presented in Exercise 2.108 (p. 106). Suppose you are interested in estimating the mean egg length (in millimeters) for the New Zealand bird population.

a. What is the target parameter?

b. Recall that the egg lengths for 132 bird species are saved in the **NZBIRDS** file. Obtain a random sample of 50 egg lengths from the data set.

c. Find the mean and standard deviation of the 50 egg lengths you obtained in part **b**.

d. Use the information from part **c** to form a 99% confidence interval for the true mean egg length of a bird species found in New Zealand.

e. Give a practical interpretation of the interval you found in part **d**.

7.129 Scanning errors at Wal-Mart. Refer to the National Institute for Standards and Technology (NIST) study of the accuracy of checkout scanners at Wal-Mart stores in California, presented in Exercise 3.56 (p. 169). NIST sets standards so that no more than 2 of every 100 items scanned through an electronic checkout scanner can have an inaccurate price. Recall that in a sample of 60 Wal-Mart stores, 52 violated the NIST scanner accuracy standard. Suppose you want to estimate the true proportion of Wal-Mart stores in California that violate the NIST standard.

a. Explain why the large-sample methodology of Section 7.4 is inappropriate for this study.

b. Determine the number of Wal-Mart stores that must be sampled in order to estimate the true proportion to within .08 with 95% confidence, using the large-sample method.

7.130 Speed training in football. Refer to *The Sport Journal* (Winter 2004) study on the effectiveness of a speed-training program for football players, Exercise 7.26 (p. 354). The decreases in 40-yard sprint times (time after training minus time before training) for 38 players are saved in the **SPRINT** file. An athlete's sprint performance will be classified as "Improved" if the "after" time is less than the "before" time, and classified as "Not Improved" if otherwise.
a. Find an estimate for the true proportion of all high school athletes who attain improved sprint times after participating in the speed-training program.
b. Convert the estimate, part **a**, into a 95% confidence interval. Give a practical interpretation of the result.
c. How many high school athletes should be sampled to estimate the true proportion to within .03 with 95% confidence?

7.131 Assessing the bending strength of a wooden roof. The white wood material used for the roof of an ancient Japanese temple is imported from Northern Europe. The wooden roof must withstand as much as 100 centimeters of snow in the winter. Architects at Tohoku University (in Japan) conducted a study to estimate the mean bending strength of the white wood roof (*Journal of the International Association for Shell and Spatial Structures*, Aug. 2004). A sample of 25 pieces of the imported wood was tested and yielded the following statistics on breaking strength (in MPa): $\bar{x} = 75.4, s = 10.9$.
a. Estimate the true mean breaking strength of the white wood with a 90% confidence interval. Interpret the result.
b. Suppose you want to estimate the true mean breaking strength of the white wood to within 4 MPa, using a 90% confidence interval. How many pieces of the imported wood need to be tested?

7.132 Water pollution testing. The EPA wants to test a randomly selected sample of n water specimens and estimate μ, the mean daily rate of pollution produced by a mining operation. If the EPA wants a 95% confidence interval estimate with a sampling error of 1 milligram per liter (mg/L), how many water specimens are required in the sample? Assume that prior knowledge indicates that pollution readings in water samples taken during a day are approximately normally distributed with a standard deviation equal to 5 (mg/L).

7.133 Radioactive lichen. Refer to the Lichen Radionuclide Baseline Research project at the University of Alaska, presented in Exercise 2.196 (p. 136). Recall that the researchers collected 9 lichen specimens and measured the amount (in microcuries per milliliter) of the radioactive element cesium-137 for each. (The natural logarithms of the data values are saved in the **LICHEN** file.) A MINITAB printout with summary statistics for the actual data is shown below.

Variable	N	Mean	StDev	SE Mean	95% CI
CESIUM	9	0.009027	0.004854	0.001618	(0.005296, 0.012759)

a. Give a point estimate for the mean amount of cesium in lichen specimens collected in Alaska.

b. Give the *t*-value used in a small-sample 95% confidence interval for the true mean amount of cesium in Alaskan lichen specimens.
c. Use the result you obtained in part **b** and the values of $\bar{x}$ and s shown on the MINITAB printout to form a 95% confidence interval for the true mean amount of cesium in Alaskan lichen specimens.
d. Check the interval you found in part **c** with the 95% confidence interval shown on the MINITAB printout.
e. Give a practical interpretation for the interval you obtained in part **c**.
f. Suppose the researchers want to increase the sample size in order to estimate the mean μ to within .001 microcurie per milliliter of its true value, using a 95% confidence interval. Compute the sample size necessary to obtain the desired estimate.

7.134 Attention time given to twins. Psychologists have found that twins, in their early years, tend to have lower IQs and pick up language more slowly than nontwins (*Wisconsin Twin Research Newsletter*, Winter 2004). The slower intellectual growth of most twins may be caused by benign parental neglect. Suppose it is desired to estimate the mean attention time given to twins per week by their parents. A sample of 50 sets of $2\frac{1}{2}$-year-old twin boys is taken, and at the end of 1 week, the attention time given to each pair is recorded. The data (in hours) are listed in the following table. Find a 90% confidence interval for the mean attention time given to all twin boys by their parents. Interpret the confidence interval.

20.7	16.7	22.5	12.1	2.9
23.5	6.4	1.3	39.6	35.6
10.9	7.1	46.0	23.4	29.4
44.1	13.8	24.3	9.3	3.4
15.7	46.6	10.6	6.7	5.4
14.0	20.7	48.2	7.7	22.2
20.3	34.0	44.5	23.8	20.0
43.1	14.3	21.9	17.5	9.6
36.4	0.8	1.1	19.3	14.6
32.5	19.1	36.9	27.9	14.0

7.135 Oven cooking study. A group of Harvard University School of Public Health researchers studied the impact of cooking on the size of indoor air particles (*Environmental Science & Technology*, Sept. 1, 2000). The decay rate (measured in μm/hour) for fine particles produced from oven cooking or toasting was recorded on six randomly selected days. The data is provided in the table.

.95	.83	1.20	.89	1.45	1.12

Based on Abt, E., et al. "Relative contribution of outdoor and indoor particle sources to indoor concentrations." *Environmental Science & Technology*, Vol. 34, No. 17, Sept. 1, 2000 (Table 3).

a. Find and interpret a 95% confidence interval for the true average decay rate of fine particles produced from oven cooking or toasting.
b. Explain what the phrase "95% confident" implies in the interpretation of part **a**.
*c. Estimate the true standard deviation of decay rate with a 95% confidence interval. Interpret the result.

d. What must be true about the distribution of the population of decay rates for the inferences you made in parts **a** and **c** to be valid?

e. Suppose that we want to estimate the average decay rate of fine particles produced from oven cooking or toasting to within .04 with 95% confidence. How large a sample should be selected?

7.136 Scary-movie study. According to a University of Michigan study, many adults have experienced lingering "fright" effects from a scary movie or TV show they saw as a teenager. In a survey of 150 college students, 39 said they still experience "residual anxiety" from a scary TV show or movie.

a. Give a point estimate $\hat{p}$ for the true proportion of college students who experience "residual anxiety" from a scary TV show or movie.

b. Find a 90% confidence interval for p.

c. Interpret the interval you found in part **b**.

Applying the Concepts—Intermediate

7.137 Brown-bag lunches at work. In a study reported in *The Wall Street Journal*, the Tupperware Corporation surveyed 1,007 U.S. workers. Of the people surveyed, 665 indicated that they take their lunch to work with them. Of these 665 taking their lunch, 200 reported that they take it in brown bags.

a. Find a 95% confidence interval estimate of the population proportion of U.S. workers who take their lunch to work with them. Interpret the interval.

b. Consider the population of U.S. workers who take their lunch to work with them. Find a 95% confidence interval estimate of the population proportion who take brown-bag lunches. Interpret the interval.

7.138 Research on brain specimens. In Exercise 2.45 (p. 80), you learned that the postmortem interval (PMI) is the elapsed time between death and the performance of an autopsy on the cadaver. *Brain and Language* (June 1995) reported on the PMIs of 22 randomly selected human brain specimens obtained at autopsy. The data are reproduced in the following table. A coroner claims that the true mean PMI of human brain specimens obtained at autopsy is 10 days. Do you agree? Use a 95% confidence interval to make an inference.

5.5	14.5	6.0	5.5	5.3	5.8	11.0	6.4
7.0	14.5	10.4	4.6	4.3	7.2	10.5	6.5
3.3	7.0	4.1	6.2	10.4	4.9		

Based on Hayes, T. L., and Lewis, D. A. "Anatomical specialization of the anterior motor speech area: Hemispheric differences in magnopyramidal neurons," *Brain and Language*, Vol. 49, No. 3, p. 292 (Table 1).

7.139 Inbreeding of tropical wasps. Tropical swarm-founding wasps rely on female workers to raise their offspring. One possible explanation for this strange behavior is inbreeding, which increases relatedness among the wasps, presumably making it easier for the workers to pick out their closest relatives as propagators of their own genetic material. To test this theory, 197 swarm-founding wasps were captured in Venezuela, frozen at −70°C, and then subjected to a series of genetic tests. The data were used to generate an inbreeding coefficient x for each wasp specimen, with the following results: $\bar{x} = .025$ and $s = .662$.

a. Construct a 99% confidence interval for the mean inbreeding coefficient of this species of wasp.

b. A coefficient of 0 implies that the wasp has no tendency to inbreed. Use the confidence interval you constructed in part **a** to make an inference about the tendency for this species of wasp to inbreed.

7.140 Exercise workout dropouts. Researchers at the University of Florida's Department of Exercise and Sport Sciences conducted a study of variety in exercise workouts (*Journal of Sport Behavior*, 2001). A sample of 120 men and women were randomly divided into three groups, with 40 people per group. Group 1 members varied their exercise routine in workouts, group 2 members performed the same exercise at each workout, and group 3 members had no set schedule or regulations for their workouts.

a. By the end of the study, 15 people had dropped out of the first exercise group. Estimate the dropout rate for exercisers who vary their routine in workouts. Use a 90% confidence interval and interpret the result.

b. By the end of the study, 23 people had dropped out of the third exercise group. Estimate the dropout rate for exercisers who have no set schedule for their workouts. Use a 90% confidence interval and interpret the result.

7.141 USGA golf ball tests. The United States Golf Association (USGA) tests all new brands of golf balls to ensure that they meet USGA specifications. One test conducted is intended to measure the average distance traveled when the ball is hit by a machine called "Iron Byron." Suppose the USGA wishes to estimate the mean distance for a new brand to within 1.5 yard with 95% confidence. Assume that past tests have indicated that the standard deviation of the distances Iron Byron hits golf balls is approximately 10 yards. How many golf balls should be hit by Iron Byron to achieve the desired accuracy in estimating the mean?

7.142 Studies on treating Alzheimer's disease. Alzheimer's disease is a progressive disease of the brain. The journal *eCAM* (Nov. 2006) published an article that critiqued the quality of the methodology used in studies on Alzheimer's treatment. For each in a sample of 13 studies, the quality of the methodology was measured on the Wong scale, with scores ranging from 9 (low quality) to 27 (high quality). The data are shown in the table below.

a. Estimate, with a 99% confidence interval, the mean quality μ of all studies on the treatment of Alzheimer's disease. Interpret the result.

b. According to the researchers, a study with a Wong score below 18 used a methodology that "fails to support the author's conclusions" about the treatment of Alzheimer's. Use Wilson's adjustment to estimate the proportion of all studies on the treatment of Alzheimer's disease with a Wong score below 18. Construct a 99% confidence interval around the estimate and interpret the result.

22	21	18	19	20	15	19	20	15	20	17	20	21

Based on Chiappelli, F., et al. "Evidence-based research in complementary and alternative medicine III: Treatment of patients with Alzheimer's disease." *eCAM*, Vol. 3, No. 4, Nov. 2006 (Table 1).

7.143 Asthma drug study. The chemical benzalkonium chloride (BAC) is an antibacterial agent that is added to some asthma medications to prevent contamination. However, adding BAC to asthma drugs can cause airway constriction in patients. In a sample of 18 asthmatic patients, each of whom received a heavy dose of BAC, 10 experienced a significant drop in breathing capacity (*Journal of Allergy and Clinical Immunology*, Jan. 2001). Based on this information, a 95% confidence interval for the true percentage of asthmatic patients who experience breathing difficulties after taking BAC is (.326, .785).

a. Why might the confidence interval lead to an erroneous inference?

b. How many asthma patients must be included in the study in order to estimate the true percentage who experience a significant drop in breathing capacity to within 4% with a 95% confidence interval?

7.144 Training zoo animals. Refer to Exercise 2.193 (p. 136) and the *Teaching of Psychology* (May 1998) study in which students assisted in the training of zoo animals. A sample of 15 psychology students rated "The Training Game" as a "great" method of understanding the animal's perspective during training on a 7–1 point scale (where 1 = strongly disagree and 7 = strongly agree). The mean response was 5.87, with a standard deviation of 1.51.

a. Construct a 95% confidence interval for the true mean response of the students.

b. Suppose you want to reduce the width of the 95% confidence interval to half the size obtained in part **a**. How many students are required in the sample in order to obtain the desired confidence interval width?

7.145 Salmonella in ice cream bars. Recently, a case of salmonella (bacterial) poisoning was traced to a particular brand of ice cream bar, and the manufacturer removed the bars from the market. Despite this response, many consumers refused to purchase *any* brand of ice cream bars for some time after the event (McClave, personal correspondence). One manufacturer conducted a survey of consumers 6 months after the poisoning. A sample of 244 ice cream bar consumers was contacted, and 23 indicated that they would not purchase ice cream bars because of the potential for food poisoning.

a. What is the point estimate of the true fraction of the entire market who refuse to purchase bars 6 months after the poisoning?

b. Is the sample size large enough to use the normal approximation for the sampling distribution of the estimator of the binomial probability? Justify your response.

c. Construct a 95% confidence interval for the true proportion of the market who still refuse to purchase ice cream bars 6 months after the event.

d. Interpret both the point estimate and confidence interval in terms of this application.

7.146 Salmonella in ice cream bars (cont'd). Refer to Exercise 7.145. Suppose it is now 1 year after the poisoning was traced to ice cream bars. The manufacturer wishes to estimate the proportion who still will not purchase bars to within .04, using a 90% confidence interval. How many consumers should be sampled?

7.147 Latex allergy in health care workers. Refer to the *Current Allergy & Clinical Immunology* (Mar. 2004) study of health care workers who use latex gloves, presented in Exercise 7.15 (p. 352). In addition to the 46 hospital employees who were diagnosed with a latex allergy on the basis of a skin-prick test, another 37 health care workers were diagnosed with the allergy by means of a latex-specific serum test. Of these 83 workers with a confirmed latex allergy, only 36 suspected that they had the allergy when they were asked about it on a questionnaire. Make a statement about the likelihood that a health care worker with a latex allergy suspects that he or she actually has the allergy. Attach a measure of reliability to your inference.

7.148 Jitter in a water power system. *Jitter* is a term used to describe the variation in conduction time of a water power system. Low throughput jitter is critical to successful waterline technology. An investigation of throughput jitter in the opening switch of a prototype system (*Journal of Applied Physics*) yielded the following descriptive statistics on conduction time for $n = 18$ trials: $\bar{x} = 334.8$ nanoseconds, $s = 6.3$ nanoseconds. (Conduction time is defined as the length of time required for the downstream current to equal 10% of the upstream current.)

a. Construct a 95% confidence interval for the true standard deviation of conduction times of the prototype system.

b. Practically interpret the confidence interval, part **a**.

c. A system is considered to have low throughput jitter if the true conduction time standard deviation is less than 7 nanoseconds. Does the prototype system satisfy this requirement? Explain.

Applying the Concepts—Advanced

7.149 IMA salary survey. Each year, *Management Accounting* reports the results of a salary survey of the members of the Institute of Management Accountants (IMA). One year, the 2,112 members responding had a salary distribution with a 20th percentile of $35,100, a median of $50,000, and an 80th percentile of $73,000.

a. Use this information to determine the minimum sample size that could be used in next year's survey to estimate the mean salary of IMA members to within $2,000 with 98% confidence. [*Hint:* To estimate *s*, first apply Chebyshev's theorem to find *k* such that at least 60% of the data fall within *k* standard deviations of μ. Then find $s \approx$ (80th percentile −20th percentile)/2*k*.]

b. Explain how you estimated the standard deviation required for the calculation of the sample size.

c. List any assumptions you make.

7.150 Air bags pose danger for children. By law, all new cars must be equipped with both driver-side and passenger-side safety air bags. There is concern, however, over whether air bags pose a danger for children sitting on the passenger side. In a National Highway Traffic Safety Administration (NHTSA) study of 55 people killed by the explosive force of air bags, 35 were children seated on the front-passenger side. This study led some car owners with the information about children to disconnect the passenger-side air bag.

a. Use the study to estimate the risk of an air bag fatality on a child seated on the front passenger seat.

b. NHTSA investigators determined that 24 of the 35 children killed by the air bags were not wearing seat belts or were improperly restrained. How does this information affect your assessment of the risk of an air bag fatality?

.93	.88	.85	.91	.91	.84	.90	.98	.88
.89	.98	.87	.91	.92	.99	1.14	1.06	.93

Based on Barnett, A. "Misapplications review: Jail terms." *Interfaces,* Vol. 25, No. 2, Mar.–Apr. 1995, p. 20.

Critical Thinking Challenge

7.151 Scallops, sampling, and the law. *Interfaces* (Mar.–Apr. 1995) discussed the case of a ship that fishes for scallops off the coast of New England. In order to protect baby scallops from being harvested, the U.S. Fisheries and Wildlife Service requires that "the average meat per scallop weigh at least $\frac{1}{36}$ of a pound." The ship was accused of violating this weight standard. Author Arnold Barnett lays out the scenario:

The vessel arrived at a Massachusetts port with 11,000 bags of scallops, from which the harbormaster randomly selected 18 bags for weighing. From each such bag, his agents took a large scoopful of scallops; then, to estimate the bag's average meat per scallop, they divided the total weight of meat in the scoopful by the number of scallops it contained. Based on the 18 [numbers] thus generated, the harbormaster estimated that each of the ship's scallops possessed an average $\frac{1}{39}$ of a pound of meat (that is, they were about seven percent lighter than the minimum requirement). Viewing this outcome as conclusive evidence that the weight standard had been violated, federal authorities at once confiscated 95 percent of the catch (which they then sold at auction). The fishing voyage was thus transformed into a financial catastrophe for its participants.

The actual scallop weight measurements for each of the 18 sampled bags are listed in the next table and saved in the **SCALLOPS** file. For ease of exposition, Bennett expressed each number as a multiple of $\frac{1}{36}$ of a pound, the minimum permissible average weight per scallop. Consequently, numbers below 1 indicate individual bags that do not meet the standard.

The ship's owner filed a lawsuit against the federal government, declaring that his vessel had fully complied with the weight standard. A Boston law firm was hired to represent the owner in legal proceedings, and Bennett was retained by the firm to provide statistical litigation support and, if necessary, expert witness testimony.

a. Recall that the harbormaster sampled only 18 of the ship's 11,000 bags of scallops. One of the questions the lawyers asked Bennett was "Can a reliable estimate of the mean weight of all the scallops be obtained from a sample of size 18?" Give your opinion on this issue.

b. As stated in the article, the government's decision rule is to confiscate a catch if the sample mean weight of the scallops is less than $\frac{1}{36}$ of a pound. Do you see any flaws in this rule?

c. Develop your own procedure for determining whether a ship is in violation of the minimum-weight restriction. Apply your rule to the data. Draw a conclusion about the ship in question.

7.152 A sampling dispute goes to court. Sampling of Medicare and Medicaid claims by the federal and state agencies who administer those programs has become common practice to determine whether providers of those services are submitting valid claims. (See the *Statistics in Action* for this chapter.) The reliability of inferences based on those samples depends on the methodology used to collect the sample of claims. Consider estimating the true proportion, p, of the population of claims that are invalid. (Invalid claims should not have been reimbursed by the agency.) Of course, to estimate a binomial parameter, p, within a given level of precision, we use the formula provided in Section 7.5 to determine the necessary sample size. In a recent actual case, the statistician determined a sample size large enough to ensure that the bound on the error of the estimate would not exceed .05, using a 95% confidence interval. He did so by assuming that the true error rate was $p = .5$, which, as discussed in Section 7.5, provides the maximum sample size needed to achieve the desired bound on the error.

a. Determine the sample size necessary to estimate p to within .05 of the true value using a 95% confidence interval.

b. After the sample was selected and the sampled claims were audited, it was determined that the estimated error rate was $\hat{p} = .20$ and a 95% confidence interval for p was (.15, .25). Was the desired bound on the error of the estimate met?

c. An economist hired by the Medicare provider noted that, since the desired bound on the error of .05 is equal to 25% of the estimated $\hat{p} = .20$ invalid claim rate, the "true" bound on the error was .25, not .05. He argued that a significantly larger sample would be necessary to meet the "relative error" (the bound on the error divided by the error rate) goal of .05, that the statistician's use of the "absolute error" of .05 was inappropriate, and that more sampling was required. The statistician argued that the relative error is a moving target because it depends on the sample estimate of the invalid claim rate, which cannot be known prior to selecting the sample. He noted that if the estimated invalid claim rate turned out to be larger than .5, the relative error would then be lower than the absolute error bound. As a consequence, the case went to trial over the relative versus absolute error dispute. Give your opinion on the matter.

[*Note:* The court concluded that "absolute error was the fair and accurate measure of the margin of error." As a result, a specified absolute bound on the error continues to be the accepted method for determining the sample size necessary to provide a reliable estimate of Medicare and Medicaid providers' claim submission error rates.]

Activity   Conducting a Pilot Study

Choose a population pertinent to your major area of interest—a population that has an unknown mean or, if the population is binomial, that has an unknown probability of success. For example, a marketing major may be interested in the proportion of consumers who prefer a certain product. A sociology major may be interested in estimating the proportion of people in a certain socioeconomic group or the mean income of people living in a particular part of a city. A political science major may wish to estimate the proportion of an electorate in favor of a certain candidate, a certain amendment, or a certain presidential policy. A pre-med student might want to find the average length of time patients stay in the hospital or the average number of people treated daily in the emergency room. We could

continue with examples, but the point should be clear: Choose something of interest to you.

Define the parameter you want to estimate and conduct a *pilot study* to obtain an initial estimate of the parameter of interest and, more importantly, an estimate of the variability associated with the estimator. A pilot study is a small experiment (perhaps 20 to 30 observations) used to gain some information about the population of interest. The purpose of the study is to help plan more elaborate, future experiments. Using the results of your pilot study, determine the sample size necessary to estimate the parameter to within a reasonable bound (of your choice) with a 95% confidence interval.

References

Agresti, A., and Coull, B. A. "Approximate is better than 'exact' for interval estimation of binomial proportions." *The American Statistician*, Vol. 52, No. 2, May 1998, pp. 119–126.

Cochran, W. G. *Sampling Techniques*, 3rd ed. New York: Wiley, 1977.

Freedman, D., Pisani, R., and Purves, R. *Statistics*. New York: Norton, 1978.

Kish, L. *Survey Sampling*. New York: Wiley, 1965.

Mendenhall, W., Beaver, R. J., and Beaver, B. *Introduction to Probability and Statistics*, 13th ed. Belmont, CA: Brooks/Cole, 2009.

Wilson, E. G. "Probable inference, the law of succession, and statistical inference." *Journal of the American Statistical Association*, Vol. 22, 1927, pp. 209–212.

USING TECHNOLOGY   MINITAB: Confidence Intervals

MINITAB can be used to obtain one-sample confidence intervals for a population mean, a population proportion, and a population variance.

Confidence Interval for a Mean

Step 1 Access the MINITAB data worksheet that contains the quantitative variable of interest.

Step 2 Click on the "Stat" button on the MINITAB menu bar and then click on "Basic Statistics" and "1-Sample t," as shown in Figure 7.M.1.

Figure 7.M.1 MINITAB menu options—confidence interval for the mean

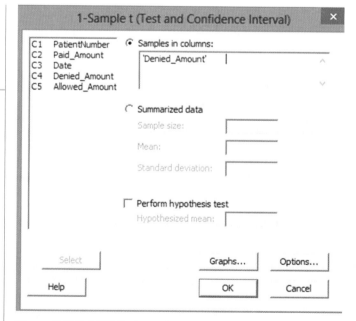

Figure 7.M.2 MINITAB 1-sample t dialog box

Step 3 On the resulting dialog box (shown in Figure 7.M.2), click on "Samples in columns," and then specify the quantitative variable of interest in the open box.

Step 4 Click on the "Options" button at the bottom of the dialog box and specify the confidence level in the resulting dialog box, as shown in Figure 7.M.3.

Figure 7.M.3 MINITAB 1-sample t options

Step 5 Click "OK" to return to the "1-Sample t" dialog box, and then click "OK" again to produce the confidence interval.

Note: If you want to produce a confidence interval for the mean from summary information (e.g., the sample mean, sample standard deviation, and sample size), click on "Summarized data" in the "1-Sample t" dialog box, as shown in Figure 7.M.4. Enter the values of the summary statistics and then click "OK."

Figure 7.M.4 MINITAB 1-sample t using summary statistics

Important: The MINITAB 1-sample *t* procedure uses the *t*-statistic to generate the confidence interval. When the sample size *n* is small, this is the appropriate method. When the sample size *n* is large, the *t*-value will be approximately equal to the large-sample *z*-value and the resulting interval will still be valid. If you have a large sample and you know the value of the population standard deviation σ (which is rarely the case), select "1-Sample Z" from the "Basic Statistics" menu options (see Figure 7.M.1) and make the appropriate selections.

Confidence Interval for a Proportion

Step 1 Access the MINITAB data worksheet that contains the qualitative variable of interest.

Step 2 Click on the "Stat" button on the MINITAB menu bar and then click on "Basic Statistics" and "1-Proportion."

Step 3 On the resulting dialog box (shown in Figure 7.M.5), click on "Samples in Columns" and then specify the qualitative variable of interest in the open box.

Figure 7.M.5 MINITAB 1 proportion dialog box

Step 4 Click on the "Options" button at the bottom of the dialog box and specify the confidence level in the resulting dialog box, as shown in Figure 7.M.6. Also, check the "Use test and interval based on normal distribution" box at the bottom.

Figure 7.M.6 MINITAB 1 proportion options

Step 5 Click "OK" to return to the "1 Proportion" dialog box and then click "OK" again to produce the confidence interval.

Note: If you want to produce a confidence interval for a proportion from summary information (e.g., the number of successes and the sample size), click on "Summarized data" in the "1 Proportion" dialog box (see Figure 7.M.5). Enter the value for the number of trials (i.e., the sample size) and the number of events (i.e., the number of successes) and then click "OK."

Confidence Interval for a Variance

Step 1 Access the MINITAB data worksheet that contains the quantitative variable of interest.

Step 2 Click on the "Stat" button on the main MINITAB menu bar, then click on "Basic Statistics" and "1 Variance."

Step 3 On the resulting dialog box (shown in Figure 7.M.7), select "Samples in columns" in the "Data" box, then specify the quantitative variable of interest in the "Columns" box.

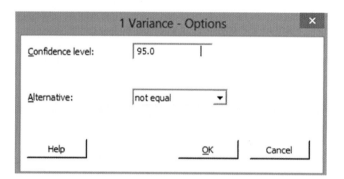

Figure 7.M.7 MINITAB 1 variance dialog box

Step 4 Click the "Options" button at the bottom of the dialog box and specify the confidence level in the resulting menu, as shown in Figure 7.M.8.

Figure 7.M.8 MINITAB 1 variance options

Step 5 Click "OK" to return to the "1 Variance" dialog box and then click "OK" again to produce the confidence interval.

Note: If you want to produce a confidence interval for a variance from summary information (e.g., the sample variance), select "Sample variance" in the "Data" box of the "1 Variance" dialog box (see Figure 7.M.9). Enter the values of the sample size and sample variance in the appropriate boxes, then click "OK."

Figure 7.M.9 MINITAB 1 variance dialog box with summary statistics option

TI-83/TI-84 Plus Graphing Calculator: Confidence Intervals

Creating a Confidence Interval for a Population Mean (Known σ or $n \geq 30$)

Step 1 *Enter the data (Skip to Step 2 if you have summary statistics, not raw data.)*

- Press **STAT** and select **1:Edit**

Note: If the list already contains data, clear the old data. Use the up **ARROW** to highlight "**L1**."

- Press **CLEAR ENTER**

- Use the **ARROW** and **ENTER** keys to enter the data set into **L1**

Step 2 *Access the statistical tests menu*

- Press **STAT**

- Arrow right to **TESTS**

- Arrow down to **ZInterval**

- Press **ENTER**

Step 3 *Choose "**Data**," or "**Stats**" ("Data" is selected when you have entered the raw data into a List. "Stats" is selected when you are given only the mean, standard deviation, and sample size.)*

- Press **ENTER**

  If you selected "Data," enter a value for σ. (The best approximation is s, the sample standard deviation.)

- Set **List** to **L1**

- Set **Freq** to **1**

- Set **C-Level** to the confidence level

- Arrow down to "**Calculate**"

- Press **ENTER**

If you selected "Stats," enter a value for σ. (The best approximation is s, the sample standard deviation.)

- Enter the sample mean and sample size

- Set **C-Level** to the confidence level

- Arrow down to "**Calculate**"

- Press **ENTER**

(The screen below at the right is set up for an example with a standard deviation of 20, a mean of 200, and a sample size of 40.) The confidence interval will be displayed along with the sample mean and the sample size.

Creating a Confidence Interval for a Population Mean ($n < 30$)

Step 1 *Enter the data (Skip to Step 2 if you have summary statistics, not raw data.)*

- Press **STAT** and select **1:Edit**

Note: If the list already contains data, clear the old data. Use the up **ARROW** to highlight "**L1**."

- Press **CLEAR ENTER**
- Use the **ARROW** and **ENTER** keys to enter the data set into **L1**

Step 2 *Access the statistical tests menu*

- Press **STAT**
- Arrow right to **TESTS**
- Arrow down to **TInterval**
- Press **ENTER**

```
EDIT CALC TESTS
2↑T-Test…
3:2-SampZTest…
4:2-SampTTest…
5:1-PropZTest…
6:2-PropZTest…
7:ZInterval…
8↓TInterval…
```

Step 3 *Choose "Data" or "Stats" ("Data" is selected when you have entered the raw data into a List. "Stats" is selected when you are given only the mean, standard deviation, and sample size.)*

- Press **ENTER**
- If you selected "Data," set **List** to **L1**
- Set **Freq** to **1**
- Set **C-Level** to the confidence level
- Arrow down to "**Calculate**"

```
TInterval
Inpt:DATA Stats
List:L₁
Freq:1
C-Level:.95
Calculate
```

- Press **ENTER**

  If you selected "Stats" enter the mean, standard deviation, and sample size.

- Set **C-Level** to the confidence level
- Arrow down to "**Calculate**"

```
TInterval
Inpt:Data Stats
x̄:100
Sx:10
n:19
C-Level:.95
Calculate
```

- Press **ENTER** (The screen at the bottom of the first column is set up for an example with a mean of 100 and a standard deviation of 10.)

The confidence interval will be displayed with the mean, standard deviation, and sample size.

Creating a Confidence Interval for a Population Proportion (Large Samples)

Step 1 *Access the statistical tests menu*

- Press **STAT**
- Arrow right to **TESTS**
- Arrow down to **1-PropZInt**
- Press **ENTER**

```
1-PropZInt
x:532
n:1100
C-Level:.95
Calculate
```

Step 2 *Enter the values for* **x, n**, *and* **C-Level**

- where **x** = number of successes
- **n** = sample size
- **C-Level** = level of confidence
- Arrow down to "**Calculate**"
- Press **ENTER**

(The screens at the right are set up for an example with $x = 532$, $n = 1,100$, and confidence level of .95.)

```
1-PropZInt
(.4541,.51317)
p̂=.4836363636
n=1100
```

Note: A confidence interval for a population variance is not available in the TI-83/TI-84 Plus graphing calculator.

8 Inferences Based on a Single Sample *Tests of Hypothesis*

CONTENTS

Where We've Been

- Used sample information to provide a *point estimate* of a population parameter
- Used the sampling distribution of a statistic to assess the reliability of an estimate through a *confidence interval*

Where We're Going

- Introduce the concepts of a *test of hypothesis* (8.1–8.2)
- Provide a measure of reliability for the hypothesis test—called the *significance level* of the test (8.2, 8.3)
- Test a specific value of a population parameter (mean, proportion, or variance) (8.4–8.6, 8.8)

Statistics IN Action — Diary of a KLEENEX® User—How Many Tissues in a Box?

In 1924, Kimberly-Clark Corporation invented a facial tissue for removing cold cream and began marketing it as KLEENEX® brand tissues. Today, KLEENEX® is recognized as the top-selling brand of tissue in the world. A wide variety of KLEENEX® products is available, ranging from extra-large tissues to tissues with lotion. Over the years, Kimberly-Clark Corporation has packaged the tissues in boxes of different sizes and shapes and varied the number of tissues packaged in each box. For example, currently an upright "Everyday" box contains 80 two-ply tissues, an "anti-viral" box contains 68 tissues (with a moisture-activated middle layer "that kills 99.9% of cold and flu viruses"), and a convenient "On the Go" pack contains 10 miniature tissues.

How does Kimberly-Clark Corp. decide how many tissues to put in each box? According to the *Wall Street Journal*, marketing experts at the company use the results of a survey of KLEENEX® customers to help determine how many tissues are packed in a box. In the mid-1980s, when Kimberly-Clark Corp. developed a cold-care box, designed especially for people who have a cold, the company conducted its initial survey of customers for this purpose. Hundreds of customers were asked to keep count of their KLEENEX® use in diaries. According to the *Wall Street Journal* report, the survey results left "little doubt that the company should put 60 tissues in each box." The cold-care box is now called the "anti-viral" box. The number 60 was "the average number of times people blow their nose during a cold." The company increased the number of tissues packaged in this box to 68 based on the results of a more recent survey.

From summary information provided in the *Wall Street Journal* article, we constructed a data set that represents the results of a survey similar to the one just described. In the data file named **TISSUES**, we recorded the number of tissues used by each of 250 consumers during a period when they had a cold. We apply the hypothesis-testing methodology presented in this chapter to that data set in several Statistics in Action Revisited examples.

Statistics IN Action — Revisited

- Identifying the Key Elements of a Hypothesis Test Relevant to the KLEENEX® Survey (p. 406)
- Testing a Population Mean in the KLEENEX® Survey (p. 417)
- Testing a Population Proportion in the KLEENEX® Survey (p. 432)

Ⓓ *Data Set:* TISSUES

Suppose you wanted to determine whether the mean level of a driver's blood alcohol exceeds the legal limit after two drinks, or whether the majority of registered voters approve of the president's performance. In both cases, you are interested in making an inference about how the value of a parameter relates to a specific numerical value. Is it less than, equal to, or greater than the specified number? This type of inference, called a **test of hypothesis**, is the subject of this chapter.

We introduce the elements of a test of hypothesis in Sections 8.1–8.3. We then show how to conduct tests of hypothesis about a population mean in Sections 8.4–8.5. Large-sample tests about binomial probabilities are the subject of Section 8.6, and some advanced methods for determining the reliability of a test are covered in optional Section 8.7. Finally, we show how to conduct a test about a population variance in optional Section 8.8.

8.1 The Elements of a Test of Hypothesis

Suppose building specifications in a certain city require that the average breaking strength of residential sewer pipe be more than 2,400 pounds per foot of length (i.e., per linear foot). Each manufacturer who wants to sell pipe in that city must demonstrate that its product meets the specification. Note that we are interested in making an inference about the mean μ of a population. However, in this example we are less interested in estimating the value of μ than we are in testing a **hypothesis** about its value—that is, *we want to decide whether the mean breaking strength of the pipe exceeds 2,400 pounds per linear foot.*

> A statistical **hypothesis** is a statement about the numerical value of a population parameter.

The method used to reach a decision is based on the rare-event concept explained in earlier chapters. We define two hypotheses: (1) The **null hypothesis** represents the status quo to the party performing the sampling experiment—the hypothesis that will be assumed to be true unless the data provide convincing evidence that it is false. (2) The **alternative**, or **research, hypothesis** is that which will be accepted only if the data provide convincing evidence of its truth. From the point of view of the city conducting the tests, the null hypothesis is that the manufacturer's pipe does *not* meet specifications unless the tests provide convincing evidence otherwise. The null and alternative hypotheses are therefore

> *Null hypothesis* (H_0): $\mu \leq 2{,}400$
>
>   (i.e., the manufacturer's pipe does not meet specifications)
>
> *Alternative (research) hypothesis* (H_a): $\mu > 2{,}400$
>
>   (i.e., the manufacturer's pipe meets specifications)

> The **null hypothesis**, denoted H_0, represents the hypothesis that will be assumed to be true unless the data provide convincing evidence that it is false. This usually represents the "status quo" or some statement about the population parameter that the researcher wants to test.

> The **alternative (research) hypothesis**, denoted H_a, represents the hypothesis that will be accepted only if the data provide convincing evidence of its truth. This usually represents the values of a population parameter for which the researcher wants to gather evidence to support.

How can the city decide when enough evidence exists to conclude that the manufacturer's pipe meets specifications? Because the hypotheses concern the value of the population mean μ, it is reasonable to use the sample mean $\bar{x}$ to make the inference, just as we did when we formed confidence intervals for μ in Sections 7.2 and 7.3. The city will conclude that the pipe meets specifications only when the sample mean $\bar{x}$ convincingly indicates that the population mean exceeds 2,400 pounds per linear foot.

"Convincing" evidence in favor of the alternative hypothesis will exist when the value of $\bar{x}$ exceeds 2,400 by an amount that cannot be readily attributed to sampling variability. To decide, we compute a **test statistic**, i.e., a numerical value computed from the sample. Here, the test statistic is the z-value that measures the distance between the value of $\bar{x}$ and the value of μ specified in the alternative hypothesis. When the null hypothesis contains more than one value of μ, as in this case (H_0: $\mu \leq 2{,}400$), we use the value of μ closest to the values specified in the alternative hypothesis. The idea is that if the hypothesis that μ *equals* 2,400 can be rejected in favor of $\mu > 2{,}400$, then μ *less than or equal to* 2,400 can certainly be rejected. Thus, the test statistic is

$$z = \frac{\bar{x} - 2{,}400}{\sigma_{\bar{x}}} = \frac{\bar{x} - 2{,}400}{\sigma / \sqrt{n}}$$

Note that a value of $z = 1$ means that $\bar{x}$ is 1 standard deviation above $\mu = 2{,}400$, a value of $z = 1.5$ means that $\bar{x}$ is 1.5 standard deviations above $\mu = 2{,}400$, and so on. How large must z be before the city can be convinced that the null hypothesis can be rejected in favor of the alternative and conclude that the pipe meets specifications?

> The **test statistic** is a sample statistic, computed from information provided in the sample, that the researcher uses to decide between the null and alternative hypotheses.

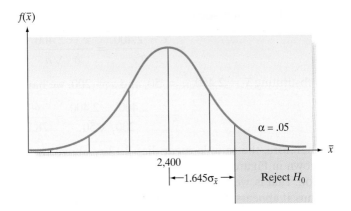

Figure 8.1
The sampling distribution of $\bar{x}$, assuming $\mu = 2{,}400$

If you examine Figure 8.1, you will note that the chance of observing $\bar{x}$ more than 1.645 standard deviations above 2,400 is only .05—*if in fact the true mean μ is 2,400.* Thus, if the sample mean is more than 1.645 standard deviations above 2,400, either H_0 is true and a relatively rare event has occurred (.05 probability), or H_a is true and the population mean exceeds 2,400. Because we would most likely reject the notion that a rare event has occurred, we would reject the null hypothesis ($\mu \leq 2{,}400$) and conclude that the alternative hypothesis ($\mu > 2{,}400$) is true. What is the probability that this procedure will lead us to an incorrect decision?

Such an incorrect decision—deciding that the null hypothesis is false when in fact it is true—is called a **Type I error.** As indicated in Figure 8.1, the risk of making a Type I error is denoted by the symbol α—that is,

$$\alpha = P\,(\text{Type I error})$$
$$= P\,(\text{Rejecting the null hypothesis when in fact the null hypothesis is true})$$

> A **Type I error** occurs if the researcher rejects the null hypothesis in favor of the alternative hypothesis when, in fact, H_0 is true. The probability of committing a Type I error is denoted by α.

In our example,

$$\alpha = P\,(z > 1.645 \text{ when in fact } \mu = 2{,}400) = .05$$

We now summarize the elements of the test:

$H_0\colon \mu \leq 2{,}400$ (Pipe does not meet specifications.)
$H_a\colon \mu > 2{,}400$ (Pipe meets specifications.)

Test statistic: $z = \dfrac{\bar{x} - 2{,}400}{\sigma_{\bar{x}}}$

Rejection region: $z > 1.645$, which corresponds to $\alpha = .05$

Note that the **rejection region** refers to the values of the test statistic for which we will *reject the null hypothesis.*

> The **rejection region** of a statistical test is the set of possible values of the test statistic for which the researcher will reject H_0 in favor of H_a.

To illustrate the use of the test, suppose we test 50 sections of sewer pipe and find the mean and standard deviation for these 50 measurements to be

$$\bar{x} = 2{,}460 \text{ pounds per linear foot}$$
$$s = 200 \text{ pounds per linear foot}$$

As in the case of estimation, we can use s to approximate σ when s is calculated from a large set of sample measurements.

The test statistic is

$$z = \frac{\bar{x} - 2{,}400}{\sigma_{\bar{x}}} = \frac{\bar{x} - 2{,}400}{\sigma/\sqrt{n}} \approx \frac{\bar{x} - 2{,}400}{s/\sqrt{n}}$$

Substituting $\bar{x} = 2{,}460$, $n = 50$, and $s = 200$, we have

$$z \approx \frac{2{,}460 - 2{,}400}{200/\sqrt{50}} = \frac{60}{28.28} = 2.12$$

Therefore, the sample mean lies $2.12\sigma_{\bar{x}}$ above the hypothesized value of μ, 2,400, as shown in Figure 8.2. Because this value of z exceeds 1.645, it falls into the rejection region. That is, we reject the null hypothesis that $\mu = 2{,}400$ and conclude that $\mu > 2{,}400$. Thus, it appears that the company's pipe has a mean strength that exceeds 2,400 pounds per linear foot.

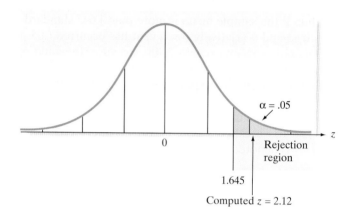

Figure 8.2

Location of the test statistic for a test of the hypothesis $H_0\colon \mu = 2{,}400$

How much faith can be placed in this conclusion? What is the probability that our statistical test could lead us to reject the null hypothesis (and conclude that the company's pipe meets the city's specifications) when in fact the null hypothesis is true? The answer is $\alpha = .05$—that is, we selected the level of risk, α, of making a Type I error when we constructed the test. Thus, the chance is only 1 in 20 that our test would lead us to conclude the manufacturer's pipe satisfies the city's specifications when in fact the pipe does *not* meet specifications.

Now, suppose the sample mean breaking strength for the 50 sections of sewer pipe turned out to be $\bar{x} = 2{,}430$ pounds per linear foot. Assuming that the sample standard deviation is still $s = 200$, the test statistic is

$$z = \frac{2{,}430 - 2{,}400}{200/\sqrt{50}} = \frac{30}{28.28} = 1.06$$

Therefore, the sample mean $\bar{x} = 2{,}430$ is only 1.06 standard deviations above the null hypothesized value of $\mu = 2{,}400$. As shown in Figure 8.3, this value does not fall into the rejection region ($z > 1.645$). Therefore, we know that we cannot reject H_0 using $\alpha = .05$. Even though the sample mean exceeds the city's specification of 2,400 by 30 pounds per linear foot, it does not exceed the specification by enough to provide *convincing* evidence that the *population mean* exceeds 2,400.

A **Type II error** occurs if the researcher accepts the null hypothesis when, in fact, H_0 is false. The probability of committing a Type II error is denoted by β.

Should we accept the null hypothesis $H_0\colon \mu \leq 2{,}400$ and conclude that the manufacturer's pipe does not meet specifications? To do so would be to risk a **Type II error**—that of concluding that the null hypothesis is true (the pipe does not meet specifications) when in fact it is false (the pipe does meet specifications). We denote the probability of

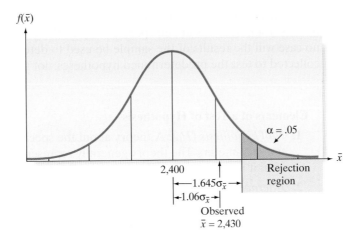

Figure 8.3
Location of test statistic
when $\bar{x} = 2,430$

committing a Type II error by β, and we show in Section 8.7 that β is often difficult to determine precisely. Rather than make a decision (accept H_0) for which the probability of error (β) is unknown, we avoid the potential Type II error by avoiding the conclusion that the null hypothesis is true. Instead, we will simply state that *the sample evidence is insufficient to reject H_0 at* $\alpha = .05$. Because the null hypothesis is the "status-quo" hypothesis, the effect of not rejecting H_0 is to maintain the status quo. In our pipe-testing example, the effect of having insufficient evidence to reject the null hypothesis that the pipe does not meet specifications is probably to prohibit the use of the manufacturer's pipe unless and until there is sufficient evidence that the pipe does meet specifications—that is, until the data indicate convincingly that the null hypothesis is false, we usually maintain the status quo implied by its truth.

Table 8.1 summarizes the four possible outcomes (i.e., **conclusions**) of a test of hypothesis. The "true state of nature" columns in Table 8.1 refer to the fact that either the null hypothesis H_0 is true or the alternative hypothesis H_a is true. Note that the true state of nature is unknown to the researcher conducting the test. The "decision" rows in Table 8.1 refer to the action of the researcher, assuming that he or she will conclude either that H_0 is true or that H_a is true, based on the results of the sampling experiment. Note that a Type I error can be made *only* when the null hypothesis is rejected in favor of the alternative hypothesis, and a Type II error can be made *only* when the null hypothesis is accepted. Our policy will be to make a decision only when we know the probability of making the error that corresponds to that decision. Because α is usually specified by the analyst, we will generally be able to reject H_0 (accept H_a) when the sample evidence supports that decision. However, because β is usually not specified, *we will generally avoid the decision to accept* H_0, *preferring instead to state that the sample evidence is insufficient to reject* H_0 *when the test statistic is not in the rejection region.*

Table 8.1   Conclusions and Consequences for a Test of Hypothesis

	True State of Nature	
Conclusion	H_0 True	H_a True
Accept H_0 (Assume H_0 True)	Correct decision	Type II error (probability β)
Reject H_0 (Assume H_a True)	Type I error (probability α)	Correct decision

⚠ **CAUTION** Be careful not to "accept H_0" when conducting a test of hypothesis because the measure of reliability, $\beta = P$ (Type II error), is almost always unknown. If the test statistic does not fall into the rejection region, it is better to state the conclusion as "insufficient evidence to reject H_0."* ▲

*In many practical business applications of hypothesis testing, nonrejection leads management to behave as if the null hypothesis were accepted. Accordingly, the distinction between acceptance and nonrejection is frequently blurred in practice. We discuss the issues connected with the acceptance of the null hypothesis and the calculation of β in more detail in Section 8.7.

The elements of a test of hypothesis are summarized in the following box. Note that the first four elements are all specified *before* the sampling experiment is performed. In no case will the results of the sample be used to determine the hypotheses; the data are collected to test the predetermined hypotheses, not to formulate them.

Elements of a Test of Hypothesis

1. *Null hypothesis* (H_0): A theory about the specific values of one or more population parameters. The theory generally represents the status quo, which we adopt until it is proven false. The theory is always stated as H_0: parameter = value.

2. *Alternative (research) hypothesis* (H_a): A theory that contradicts the null hypothesis. The theory generally represents that which we will adopt only when sufficient evidence exists to establish its truth.

3. *Test statistic:* A sample statistic used to decide whether to reject the null hypothesis.

4. *Rejection region:* The numerical values of the test statistic for which the null hypothesis will be rejected. The rejection region is chosen so that the probability is α that it will contain the test statistic when the null hypothesis is true, thereby leading to a Type I error. The value of α is usually chosen to be small (e.g., .01, .05, or .10) and is referred to as the **level of significance** of the test.

5. *Assumptions:* Clear statement(s) of any assumptions made about the population(s) being sampled.

6. *Experiment and calculation of test statistic:* Performance of the sampling experiment and determination of the numerical value of the test statistic.

7. *Conclusion:*

   a. If the numerical value of the test statistic falls in the rejection region, we reject the null hypothesis and conclude that the alternative hypothesis is true. We know that the hypothesis-testing process will lead to this conclusion incorrectly (Type I error) only $100\alpha\%$ of the time when H_0 is true.

   b. If the test statistic does not fall in the rejection region, we do not reject H_0. Thus, we reserve judgment about which hypothesis is true. We do not conclude that the null hypothesis is true because we do not (in general) know the probability β that our test procedure will lead to an incorrect acceptance of H_0 (Type II error).

As with confidence intervals, the methodology for testing hypotheses varies depending on the target population parameter. In this chapter, we develop methods for testing a population mean, a population proportion, and (optionally) a population variance. As a reminder, the key words and the type of data associated with these target parameters are again listed in the accompanying box.

Determining the Target Parameter

Parameter	Key Words or Phrases	Type of Data
μ	Mean; average	Quantitative
p	Proportion; percentage; fraction; rate	Qualitative
σ^2	Variance; variability; spread	Quantitative

8.2 Formulating Hypotheses and Setting Up the Rejection Region

In Section 8.1 we learned that the null and alternative hypotheses form the basis for inference using a test of hypothesis. The null and alternative hypotheses may take one of several forms. In the sewer pipe example, we tested the null hypothesis that the population mean strength of the pipe is less than or equal to 2,400 pounds per linear foot against the alternative hypothesis that the mean strength exceeds 2,400—that is, we tested

$$H_0: \mu \leq 2{,}400 \quad \text{(Pipe does not meet specifications.)}$$
$$H_a: \mu > 2{,}400 \quad \text{(Pipe meets specifications.)}$$

This is a **one-tailed** (or **one-sided**) **statistical test** because the alternative hypothesis specifies that the population parameter (the population mean μ in this example) is strictly greater than a specified value (2,400 in this example). If the null hypothesis had been $H_0: \mu \geq 2{,}400$ and the alternative hypothesis had been $H_a: \mu < 2{,}400$, the test would still be one-sided because the parameter is still specified to be on "one side" of the null hypothesis value. Some statistical investigations seek to show that the population parameter is *either larger or smaller* than some specified value. Such an alternative hypothesis is called a **two-tailed** (or **two-sided**) **hypothesis**.

While alternative hypotheses are always specified as strict inequalities, such as $\mu < 2{,}400$, $\mu > 2{,}400$, or $\mu \neq 2{,}400$, *null hypotheses are usually specified as equalities*, such as $\mu = 2{,}400$. Even when the null hypothesis is an inequality, such as $\mu \leq 2{,}400$, we specify $H_0: \mu = 2{,}400$, reasoning that if sufficient evidence exists to show that $H_a: \mu > 2{,}400$ is true when tested against $H_0: \mu = 2{,}400$, then surely sufficient evidence exists to reject $\mu < 2{,}400$ as well. Therefore, the null hypothesis is specified as the value of μ closest to a one-sided alternative hypothesis and as the only value *not* specified in a two-tailed alternative hypothesis. The steps for selecting the null and alternative hypotheses are summarized in the following box.

Steps for Selecting the Null and Alternative Hypotheses

1. Select the *alternative hypothesis* as that which the sampling experiment is intended to establish. The alternative hypothesis will assume one of three forms:

   a. One-tailed, **upper-tailed**   (e.g., $H_a: \mu > 2{,}400$)
   b. One-tailed, **lower-tailed**   (e.g., $H_a: \mu < 2{,}400$)
   c. Two-tailed   (e.g., $H_a: \mu \neq 2{,}400$)

2. Select the *null hypothesis* as the status quo, that which will be presumed true unless the sampling experiment conclusively establishes the alternative hypothesis. The null hypothesis will be specified as that parameter value closest to the alternative in one-tailed tests and as the complementary (or only unspecified) value in two-tailed tests.

$$(\text{e.g., } H_0: \mu = 2{,}400)$$

A **one-tailed test** of hypothesis is one in which the alternative hypothesis is directional and includes the symbol "<" or ">."

A **two-tailed test** of hypothesis is one in which the alternative hypothesis does not specify departure from H_0 in a particular direction and is written with the symbol "$\neq$."

Example 8.1

Formulating H_0 and H_a for a Test of a Population Mean—Quality Control

Problem A metal lathe is checked periodically by quality control inspectors to determine whether it is producing machine bearings with a mean diameter of .5 inch. If the mean diameter of the bearings is larger or smaller than .5 inch, then the process is out of control and must be adjusted. Formulate the null and alternative hypotheses for a test to determine whether the bearing production process is out of control.

Solution The hypotheses must be stated in terms of a population parameter. Here, we define μ as the true mean diameter (in inches) of all bearings produced by the metal lathe. If either $\mu > .5$ or $\mu < .5$, then the lathe's production process is out of control. Because the inspectors want to be able to detect either possibility (indicating that the process is in need of adjustment), these values of μ represent the alternative (or research) hypothesis. Alternatively, because $\mu = .5$ represents an in-control process (the status quo), this represents the null hypothesis. Therefore, we want to conduct the two-tailed test:

$$H_0\colon \mu = .5 \text{ (i.e., the process is in control)}$$
$$H_a\colon \mu \neq .5 \text{ (i.e., the process is out of control)}$$

Look Back Here, the alternative hypothesis is not necessarily the hypothesis that the quality control inspectors desire to support. However, they will make adjustments to the metal lathe settings only if there is strong evidence to indicate that the process is out of control. Consequently, $\mu \neq .5$ must be stated as the alternative hypothesis.

■ **Now Work Exercise 8.11a**

Example 8.2

Formulating H_0 and H_a for a Test of a Population Proportion—Cigarette Advertisements

Problem Cigarette advertisements are required by federal law to carry the following statement: "Warning: The surgeon general has determined that cigarette smoking is dangerous to your health." However, this warning is often located in inconspicuous corners of the advertisements and printed in small type. Suppose the Federal Trade Commission (FTC) claims that 80% of cigarette consumers fail to see the warning. A marketer for a large tobacco firm wants to gather evidence to show that the FTC's claim is too high, i.e., that fewer than 80% of cigarette consumers fail to see the warning. Specify the null and alternative hypotheses for a test of the FTC's claim.

Solution The marketer wants to make an inference about p, the true proportion of all cigarette consumers who fail to see the surgeon general's warning. In particular, the marketer wants to collect data to show that fewer than 80% of cigarette consumers fail to see the warning, i.e., $p < .80$. Consequently, $p < .80$ represents the alternative hypothesis and $p = .80$ (the claim made by the FTC) represents the null hypothesis. That is, the marketer desires the one-tailed (lower-tailed) test:

$$H_0\colon p = .80 \text{ (i.e., the FTC's claim is true)}$$
$$H_a\colon p < .80 \text{ (i.e., the FTC's claim is false)}$$

Look Back Whenever a claim is made about the value of a particular population parameter and the researcher wants to test the claim, believing that it is false, the claimed value will represent the null hypothesis.

■ **Now Work Exercise 8.15**

The rejection region for a two-tailed test differs from that for a one-tailed test. When we are trying to detect departure from the null hypothesis in *either* direction, we must establish a rejection region in both tails of the sampling distribution of the test statistic. Figures 8.4a and 8.4b show the one-tailed rejection regions for lower- and upper-tailed tests, respectively. The two-tailed rejection region is illustrated in Figure 8.4c. Note that a rejection region is established in each tail of the sampling distribution for a two-tailed test.

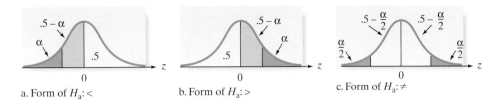

Figure 8.4
Rejection regions corresponding to one- and two-tailed tests

a. Form of H_a: < b. Form of H_a: > c. Form of H_a: ≠

The rejection regions corresponding to typical values selected for α are shown in Table 8.2 for one- and two-tailed tests based on large samples. Note that the smaller α you select, the more evidence (the larger z) you will need before you can reject H_0.

Table 8.2   Rejection Regions for Common Values of α, Large n

	Alternative Hypotheses		
	Lower-Tailed	Upper-Tailed	Two-Tailed
$\alpha = .10$	$z < -1.28$	$z > 1.28$	$z < -1.645$ or $z > 1.645$
$\alpha = .05$	$z < -1.645$	$z > 1.645$	$z < -1.96$ or $z > 1.96$
$\alpha = .01$	$z < -2.33$	$z > 2.33$	$z < -2.575$ or $z > 2.575$

Example   8.3

Setting Up a Hypothesis Test for μ — Mean Drug Response Time

Problem The effect of drugs and alcohol on the nervous system has been the subject of considerable research. Suppose a research neurologist is testing the effect of a drug on response time by injecting 100 rats with a unit dose of the drug, subjecting each rat to a neurological stimulus, and recording its response time. The neurologist knows that the mean response time for rats not injected with the drug (the "control" mean) is 1.2 seconds. She wishes to test whether the mean response time for drug-injected rats differs from 1.2 seconds. Set up the test of hypothesis for this experiment, using $\alpha = .01$.

Solution The key word *mean* in the statement of the problem implies that the target parameter is μ, the mean response time for all drug-injected rats. Since the neurologist wishes to detect whether μ differs from the control mean of 1.2 seconds in *either* direction—that is, $\mu < 1.2$ or $\mu > 1.2$—we conduct a two-tailed statistical test. Following the procedure for selecting the null and alternative hypotheses, we specify as the alternative hypothesis that the mean differs from 1.2 seconds, since determining whether the drug-injected mean differs from the control mean is the purpose of the experiment. The null hypothesis is the presumption that drug-injected rats have the same mean response time as control rats unless the research indicates otherwise. Thus,

H_0: $\mu = 1.2$ (Mean response time is 1.2 seconds)

H_a: $\mu \neq 1.2$ (Mean response time is less than 1.2 or greater than 1.2 seconds)

The test statistic measures the number of standard deviations between the observed value of $\bar{x}$ and the null-hypothesized value $\mu = 1.2$:

$$\text{Test statistic:} \quad z = \frac{\bar{x} - 1.2}{\sigma_{\bar{x}}}$$

The rejection region must be designated to detect a departure from $\mu = 1.2$ in *either* direction, so we will reject H_0 for values of z that are either too small (negative) or too large (positive). To determine the precise values of z that constitute the rejection region, we first select α, the probability that the test will lead to incorrect rejection of the null hypothesis. Then we divide α equally between the lower and upper tail of the distribution of z, as shown in Figure 8.5. In this example, $\alpha = .01$, so $\alpha/2 = .005$ is placed in each tail. The areas in the tails correspond to $z = -2.575$ and $z = 2.575$, respectively (from Table 8.2), so

Rejection region: $z < -2.575$ or $z > 2.575$ (see Figure 8.5)

Figure 8.5
Two-tailed rejection region:
$\alpha = .01$

Assumptions: Since the sample size of the experiment is large enough $(n > 30)$, the Central Limit Theorem will apply, and no assumptions need be made about the population of response time measurements. The sampling distribution of the sample mean response of 100 rats will be approximately normal, regardless of the distribution of the individual rats' response times.

Look Back Note that the test is set up *before* the sampling experiment is conducted. The data are not used to develop the test. Evidently, the neurologist wants to conclude that the mean response time for the drug-injected rats differs from the control mean only when the evidence is very convincing because the value of α has been set quite low at .01. If the experiment results in the rejection of H_0, she can confidently conclude that the mean response time of the drug-injected rats differs from the control mean because there is only a .01 probability of a Type I error.

■ **Now Work Exercise 8.13b, c**

Once the test is set up, the neurologist of Example 8.3 is ready to perform the sampling experiment and conduct the test. The hypothesis test is performed in Section 8.4.

Statistics IN Action Revisited

Identifying the Key Elements of a Hypothesis Test Relevant to the KLEENEX® Survey

In Kimberly-Clark Corporation's survey of people with colds, each of 250 customers was asked to keep count of his or her use of KLEENEX® tissues in diaries. One goal of the company was to determine how many tissues to package in a cold-care (now "anti-viral") box of KLEENEX®; consequently, the total number of tissues used was recorded for each person surveyed. Since number of tissues is a quantitative variable, the parameter of interest is either μ, the mean number of tissues used by all customers with colds, or σ^2, the variance of the number of tissues used.

Recall that recently, the company increased from 60 to 68 the number of tissues it packages in a cold-care box of

KLEENEX® tissues. This decision was based on a claim made by marketing experts that the average number of times a person will blow his or her nose during a cold exceeds the previous mean of 60. The key word *average* implies that the target parameter is μ, and the marketers are claiming that $\mu > 60$. In order to test the claim, we set up the following null and alternative hypotheses:

$$H_0: \mu = 60 \qquad H_a: \mu > 60$$

We'll conduct this test in the next Statistics in Action Revisited on p. 417.

Exercises 8.1–8.21

Understanding the Principles

8.1 What is the difference between the null and the alternative hypothesis?

8.2 Which element of a test of hypothesis is used to decide whether to reject the null hypothesis in favor of the alternative hypothesis?

8.3 What is the size of a test of hypothesis?

8.4 What are the different errors associated with a hypothesis testing problem?

8.5 Give a real-life example of committing a Type I error and Type II error.

8.6 If a right-tailed test is rejected at the 1% level of significance, do we need to test at the 5% level of significance? Why?

8.7 If the null hypothesis is not rejected, does that imply that it is true? Explain.

Learning the Mechanics

8.8 Consider a test of $H_0: \mu = 5$. In each of the following cases, give the rejection region for the test in terms of the z-statistic:
  a. $H_a: \mu > 5, \alpha = .10$
  b. $H_a: \mu > 5, \alpha = .05$
  c. $H_a: \mu < 5, \alpha = .01$
  d. $H_a: \mu \neq 5, \alpha = .10$

8.9 For each of the following rejection regions, sketch the sampling distribution for z and indicate the location of the rejection region.
  a. $z > 1.96$
  b. $z > 1.645$
  c. $z > 2.575$
  d. $z < -1.28$
  e. $z < -1.645$ or $z > 1.645$
  f. $z < -2.575$ or $z > 2.575$
  g. For each of the rejection regions specified in parts **a–f**, what is the probability that a Type I error will be made?

ⓓ Applet Exercise 8.1

Use the applet entitled *Hypotheses Test for a Mean* to investigate the frequency of Type I and Type II errors. For this exercise, use $n = 100$ and the normal distribution with mean 50 and standard deviation 10.
  a. Set the null mean equal to 50 and the alternative to *not equal*. Run the applet one time. How many times was the null hypothesis rejected at level .05? In this case, the null hypothesis is true. Which type of error occurred each time the true null hypothesis was rejected? What is the probability of rejecting a true null hypothesis at level .05? How does the proportion of times the null hypothesis was rejected compare with this probability?
  b. Clear the applet, then set the null mean equal to 47, and keep the alternative at *not equal*. Run the applet one time. How many times was the null hypothesis *not* rejected at level .05? In this case, the null hypothesis is false. Which type of error occurred each time the null hypothesis was *not* rejected? Run the applet several more times without clearing. Based on your results, what can you conclude about the probability of failing to reject the null hypothesis for the given conditions?

Applying the Concepts—Basic

8.10 **Walking to improve health.** In a study investigating a link between walking and improved health (*Social Science & Medicine*, Apr. 2014), researchers reported that adults walked an average of 5.5 days in the past month for the purpose of health or recreation. Specify the null and alternative hypotheses for testing whether the true average number of days in the past month that adults walked for the purpose of health or recreation is lower than 5.5 days.

8.11 **Americans' favorite sport.** *The Harris Poll* (Dec. 2013) conducted an online survey of American adults to determine their favorite sport. Your friend believes professional (NFL) football is the favorite sport for 45% of American adults. Specify the null and alternative hypotheses for testing this belief. Be sure to identify the parameter of interest.

8.12 **Infants' listening time.** Researchers writing in *Analysis of Verbal Behavior* (Dec. 2007) reported that the mean listening time of 16-month-old infants exposed to nonmeaningful monosyllabic words (e.g., "giff," "cham," "gack") is 8 seconds. Set up the null and alternative hypotheses for testing the claim.

8.13 **Play Golf America program.** In the Play Golf America program, teaching professionals at participating golf clubs provide a free 10-minute lesson to new customers. According to the Professional Golf Association (PGA), golf facilities that participate in the program gain, on average, $2,400 in green fees, lessons, or equipment expenditures. A teaching professional at a golf club believes that the average gain in green fees, lessons, or equipment expenditures for participating golf facilities exceeds $2,400.
  a. In order to support the claim made by the teaching professional, what null and alternative hypotheses should you test?
  b. Suppose you select $\alpha = .05$. Interpret this value in the words of the problem.
  c. For $\alpha = .05$, specify the rejection region of a large-sample test.

8.14 **Effectiveness of online courses.** The Sloan Survey of Online Learning, "Going the Distance: Online Education in the United States, 2011," reported that 68% of college presidents believe that their online education courses are as good as or superior to courses that utilize traditional face-to-face instruction.
  a. Give the null hypothesis for testing the claim made by the Sloan Survey.
  b. Give the rejection region for a two-tailed test using $\alpha = .01$.

8.15 **DNA-reading tool for quick identification of species.** A biologist and a zoologist at the University of Florida were the first scientists to test the effectiveness of a high-tech handheld device designed to instantly identify the DNA of an animal species (*PLOS Biology*, Dec. 2005). They used the DNA-reading device on tissue samples collected from mollusks with brightly colored shells. The scientists discovered that the error rate of the device is less than 5 percent. Set up the null and alternative hypotheses as if you want to support the findings.

Applying the Concepts—Intermediate

8.16 **Calories in school lunches.** India's Mid-Day Meal scheme mandates that high schools that are part of this scheme must serve lunches that contain at least 700 calories and 20 grams of protein. Suppose a nutritionist believes that the true mean number of calories served at lunch at all high schools that are part of this scheme is less than 700 calories.
  a. Identify the parameter of interest.
  b. Specify the null and alternative hypotheses for testing this claim.
  c. Describe a Type I error in the words of the problem.
  d. Describe a Type II error in the words of the problem.

8.17 **Automated checking.** A software company develops a software that can scan passports, check finger prints, and read eye pupils. The company claimed that the software can detect deceitful activity correctly 70% of the time.

a. Identify the parameter of interest.

b. Give the null and alternative hypotheses for testing the claim made by the software developer company.

c. Describe a Type I error in the words of the problem.

d. Describe a Type II error in the words of the problem.

8.18 **Virtual reality hypnosis for pain.** The *International Journal of Clinical and Experimental Hypnosis* (Vol. 58, 2010) investigated using virtual reality hypnosis (VRH) to reduce the pain of trauma patients. Patients reported their pain intensity on the Graphic Rating Scale (GRS) both prior to and one hour after a VRH session. The researchers reported that hypnosis reduced the pain intensity of trauma patients by an average of $\mu = 10$ points on the GRS. Suppose you want to test whether trauma patients who receive normal analgesic care (i.e., medication, but no hypnosis) will have a smaller reduction in average pain intensity than the VRH trauma patients.

a. Set up H_0 and H_a for the test.

b. Describe a Type I error for this test.

c. Describe a Type II error for this test.

8.19 **Melting point of a mercury compound.** In order to determine the melting point (in degrees Celsius) of a mercury compound, an analyst recorded the readings obtained during his experiment. The analyst claimed that the average melting point of the compound was 150°C.

a. Give the null and alternative hypotheses for testing whether the average melting point is less than 150°C.

b. Describe a Type I error for this test.

c. Describe a Type II error for this test.

Applying the Concepts—Advanced

8.20 **Jury trial outcomes.** Sometimes, the outcome of a jury trial defies the "commonsense" expectations of the general public (e.g., the 1995 O. J. Simpson verdict and the 2011 Casey Anthony verdict). Such a verdict is more acceptable if we understand that the jury trial of an accused murderer is analogous to the statistical hypothesis-testing process. The null hypothesis in a jury trial is that the accused is innocent. (The status quo hypothesis in the U.S. system of justice is innocence, which is assumed to be true until proven *beyond a reasonable doubt*.) The alternative hypothesis is guilt, which is accepted only when

sufficient evidence exists to establish its truth. If the vote of the jury is unanimous in favor of guilt, the null hypothesis of innocence is rejected and the court concludes that the accused murderer is guilty. Any vote other than a unanimous one for guilt results in a "not guilty" verdict. The court never accepts the null hypothesis; that is, the court never declares the accused "innocent." A "not guilty" verdict (as in the O. J. Simpson case) implies that the court could not find the defendant guilty *beyond a reasonable doubt*.

a. Define Type I and Type II errors in a murder trial.

b. Which of the two errors is the more serious? Explain.

c. The court does not, in general, know the values of α and β, but ideally, both should be small. One of these probabilities is assumed to be smaller than the other in a jury trial. Which one, and why?

d. The court system relies on the belief that the value of α is made very small by requiring a unanimous vote before guilt is concluded. Explain why this is so.

e. For a jury prejudiced against a guilty verdict as the trial begins, will the value of α increase or decrease? Explain.

f. For a jury prejudiced against a guilty verdict as the trial begins, will the value of β increase or decrease? Explain.

8.21 **Intrusion detection systems.** Refer to the *Journal of Research of the National Institute of Standards and Technology* (Nov.–Dec. 2003) study of a computer intrusion detection system (IDS), presented in Exercise 3.98 (p. 186). Recall that an IDS is designed to provide an alarm whenever unauthorized access (e.g., an intrusion) to a computer system occurs. The probability of the system giving a false alarm (i.e., providing a warning when, in fact, no intrusion occurs) is defined by the symbol α, while the probability of a missed detection (i.e., no warning given when, in fact, an intrusion occurs) is defined by the symbol β. These symbols are used to represent Type I and Type II error rates, respectively, in a hypothesis-testing scenario.

a. What is the null hypothesis H_0?

b. What is the alternative hypothesis H_a?

c. According to actual data on the EMERALD system collected by the Massachusetts Institute of Technology Lincoln Laboratory, only 1 in 1,000 computer sessions with no intrusions resulted in a false alarm. For the same system, the laboratory found that only 500 of 1,000 intrusions were actually detected. Use this information to estimate the values of α and β.

8.3 Observed Significance Levels: *p*-Values

According to the statistical test procedure described in Section 8.2, the rejection region and, correspondingly, the value of α are selected prior to conducting the test, and the conclusions are stated in terms of rejecting or not rejecting the null hypothesis. A second method of presenting the results of a statistical test is one that reports the extent to which the test statistic disagrees with the null hypothesis and leaves to the reader the task of deciding whether to reject the null hypothesis. This measure of disagreement is called the *observed significance level* (or *p-value*) for the test.

> The **observed significance level**, or p-value, for a specific statistical test is the probability (assuming H_0 is true) of observing a value of the test statistic that is at least as contradictory to the null hypothesis, and supportive of the alternative hypothesis, as the actual one computed from the sample data.

Recall testing $H_0: \mu = 2{,}400$ versus $H_a: \mu = 2{,}400$, where μ is the mean breaking strength of sewer pipe (Section 8.1). The value of the test statistic computed for the sample of $n = 50$ sections of sewer pipe was $z = 2.12$. Because the test is one-tailed—that is, the alternative (research) hypothesis of interest is $H_a: \mu > 2{,}400$—values of the test statistic even more contradictory to H_0 than the one observed would be values larger than $z = 2.12$. Therefore, the observed significance level (p-value) for this test is

$$p\text{-value} = P(z > 2.12)$$

or, equivalently, the area under the standard normal curve to the right of $z = 2.12$ (see Figure 8.6).

The area A in Figure 8.6 is given in Table II in Appendix B as .4830. Therefore, the upper-tail area corresponding to $z = 2.12$ is

$$p\text{-value} = .5 - .4830 = .0170$$

Figure 8.6

Finding the p-value for an upper-tailed test when $z = 2.12$

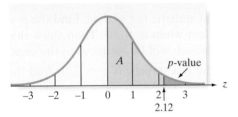

Consequently, we say that these test results are "very significant" (i.e., they disagree rather strongly with the null hypothesis, $H_0: \mu = 2{,}400$, and favor $H_a: \mu > 2{,}400$). The probability of observing a z-value as large as 2.12 is only .0170, if in fact the true value of μ is 2,400.

If you are inclined to select $\alpha = .05$ for this test, then you would reject the null hypothesis because the p-value for the test, .0170, is less than .05. In contrast, if you choose $\alpha = .01$, you would not reject the null hypothesis because the p-value for the test is larger than .01. Thus, the use of the observed significance level is identical to the test procedure described in the preceding sections except that the choice of α is left to you.

The steps for calculating the p-value corresponding to a test statistic for a population mean are given in the next box.

> **Steps for Calculating the p-Value for a Test of Hypothesis**
>
> **1.** Determine the value of the test statistic z corresponding to the result of the sampling experiment.
>
> **2. a.** If the test is one-tailed, the p-value is equal to the tail area beyond z in the same direction as the alternative hypothesis. Thus, if the alternative hypothesis is of the form $>$, the p-value is the area to the right of, or above, the observed z-value. Conversely, if the alternative is of the form $<$, the p-value is the area to the left of, or below, the observed z-value. (See Figure 8.7.)
>
> **b.** If the test is two-tailed, the p-value is equal to twice the tail area beyond the observed z-value in the direction of the sign of z—that is, if z is positive, the p-value is twice the area to the right of, or above, the observed z-value. Conversely, if z is negative, the p-value is twice the area to the left of, or below, the observed z-value. (See Figure 8.8.)

Figure 8.7
Finding the p-value for a one-tailed test

a. Lower-tailed test, H_a: $\mu < \mu_0$

b. Upper-tailed test, H_a: $\mu > \mu_0$

Figure 8.8
Finding the p-value for a two-tailed test: p-value $= 2\left(\frac{p}{2}\right)$

a. Test statistic z negative

b. Test statistic z positive

Example **8.4**

Comparing Rejection Regions to p-Values

Problem Consider the one-tailed test of hypothesis, H_0: $\mu = 100$ versus H_a: $\mu > 100$.

a. Suppose the test statistic is $z = 1.44$. Find the p-value of the test and the rejection region for the test when $\alpha = .05$. Then show that the conclusion using the rejection region approach will be identical to the conclusion based on the p-value.

b. Now suppose the test statistic is $z = 3.01$; find the p-value and rejection region for the test when $\alpha = .05$. Again, show that the conclusion using the rejection region approach will be identical to the conclusion based on the p-value.

Solution

a. The p-value for the test is the probability of observing a test statistic more contradictory to the null hypothesis (i.e., more supportive of the alternative hypothesis) than the value $z = 1.44$. Since we are conducting an upper-tailed test (H_a: $\mu > 100$), the probability we seek is:

$$p\text{-value} = P(z > 1.44) = 1 - P(z < 1.44)$$

The latter probability in the expression is a cumulative probability that can be obtained from the standard normal table (Table II, Appendix B) or by using statistical software. The MINITAB printout giving this probability is shown in Figure 8.9.

Cumulative Distribution Function

```
Normal with mean = 0 and standard deviation = 1

   x   P( X <= x )
1.44     0.925066
```

Figure 8.9
MINITAB normal probability for Example 8.4a

Therefore, we compute

$$p\text{-value} = P(z > 1.44) = 1 - P(z < 1.44) = 1 - .925 = .075$$

This p-value is shown on Figure 8.10.

Since $\alpha = .05$ and the test is upper-tailed, the rejection region for the test is $z > 1.645$ (see Table 8.2). This rejection region is also shown in Figure 8.10. Observe that the test statistic ($z = 1.44$) falls outside the rejection region, implying that we fail to reject H_0. Also, $\alpha = .05$ is less than p-value $= .075$. This result also implies that we should fail to reject H_0. Consequently, both decision rules agree—there is insufficient evidence to reject H_0.

Figure 8.10

Test for Example 8.4a—Fail to reject H_0

b. For $z = 3.01$, the observed significance level of the test is:

$$p\text{-value} = P(z > 3.01) = 1 - P(z < 3.01)$$

A MINITAB printout giving the cumulative probability, $P(z < 3.01)$, is shown in Figure 8.11.

Cumulative Distribution Function

Normal with mean = 0 and standard deviation = 1

```
    x    P( X <= x )
  3.01     0.998694
```

Figure 8.11

MINITAB normal probability for Example 8.4b

Thus, we have

$$p\text{-value} = P(z > 3.01) = 1 - P(z < 3.01) = 1 - .9987 = .0013.$$

This p-value is shown on Figure 8.12.

Again, for $\alpha = .05$ and an upper-tailed test, the rejection region is $z > 1.645$. This rejection region is also shown in Figure 8.12. Now the test statistic ($z = 3.01$) falls within the rejection region, leading us to reject H_0. And, $\alpha = .05$ now exceeds the p-value (.0013), which also implies that we should reject H_0. Once again, both decision rules agree—and they always will if the same value of α is used to make the decision.

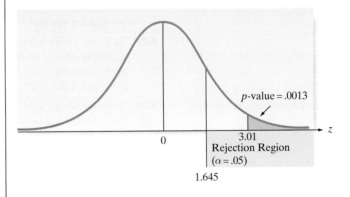

Figure 8.12

Test for Example 8.4b— Reject H_0

Look Ahead Since p-values are easily computed using statistical software, most analysts and researchers utilize the p-value approach to hypothesis testing. In future examples, we will adopt this approach generally, although we'll continue to show the appropriate rejection regions where necessary for illustrative purposes.

■ **Now Work Exercise 8.25**

When publishing the results of a statistical test of hypothesis in journals, case studies, reports, and so on, many researchers make use of p-values. Instead of selecting α beforehand and then conducting a test, as outlined in this chapter, the researcher computes (usually with the aid of a statistical software package) and reports the value of the appropriate test statistic and its associated p value. It is left to the reader of the report to judge the significance of the result (i.e., the reader must determine whether to reject the null hypothesis in favor of the alternative hypothesis, based on the reported p-value). Usually, the null hypothesis is rejected if the observed significance level is *less than* the fixed significance level, α, chosen by the reader. The inherent advantage of reporting test results in this manner is twofold: (1) Readers are permitted to select the maximum value of α that they would be willing to tolerate if they actually carried out a standard test of hypothesis in the manner outlined in this chapter, and (2) a measure of the degree of significance of the result (i.e., the p-value) is provided.

Ethics IN Statistics

Selecting the value of α *after* computing the observed significance level (p-value) in order to guarantee a preferred conclusion is considered *unethical statistical practice.*

Reporting Test Results as p-Values: How to Decide Whether to Reject H_0

1. Choose the maximum value of α that you are willing to tolerate.

2. If the observed significance level (p-value) of the test is less than the chosen value of α, reject the null hypothesis. Otherwise, do not reject the null hypothesis.

Note: Some statistical software packages (e.g., SPSS) will conduct only two-tailed tests of hypothesis. For these packages, you obtain the p-value for a one-tailed test as shown in the box:

Converting a Two-Tailed p-Value from a Printout to a One-Tailed p-Value

$$p = \frac{\text{Reported } p\text{-value}}{2} \qquad \text{if} \begin{cases} H_a \text{ is of form } > \text{ and } z \text{ is positive} \\ H_a \text{ is of form } < \text{ and } z \text{ is negative} \end{cases}$$

$$p = 1 - \left(\frac{\text{Reported } p\text{-value}}{2} \right) \qquad \text{if} \begin{cases} H_a \text{ is of form } > \text{ and } z \text{ is negative} \\ H_a \text{ is of form } < \text{ and } z \text{ is positive} \end{cases}$$

Exercises 8.22–8.30

Learning the Mechanics

8.22 Consider the test of H_0: $\mu = 5$. For each of the following, find the p-value of the test:
   a. H_a: $\mu > 5$, $z = 3.16$
   b. H_a: $\mu < 5$, $z = -3.16$
   c. H_a: $\mu \neq 5$, $z = 3.16$

8.23 If a hypothesis test were conducted using $\alpha = .05$, for which of the following p-values would the null hypothesis be rejected?
   a. .07   **b.** .20
   c. .04   **d.** .001
   e. .002   **f.** .032

8.24 For each α and observed significance level (p-value) pair, indicate whether the null hypothesis would be rejected.
   a. $\alpha = .05$, p-value $= .005$
   b. $\alpha = .10$, p-value $= .03$
   c. $\alpha = .05$, p-value $= .002$
   d. $\alpha = .01$, p-value $= .028$
   e. $\alpha = .025$, p-value $= .50$

8.25 In a test of the hypothesis H_0: $\mu = 30$ versus H_a: $\mu > 30$, a sample of $n = 100$ observations possessed mean $\bar{x} = 29.3$ and standard deviation $s = 4$. Find and interpret the p-value for this test.

8.26 In a test of H_0: $\mu = 50$ against H_a: $\mu > 50$, the sample data yielded the test statistic $z = 2.24$. Find and interpret the p-value for the test.

8.27 In a test of the hypothesis H_0: $\mu = 20$ versus H_a: $\mu \neq 20$, a sample of $n = 50$ observations possessed mean $\bar{x} = 20.7$ and standard deviation $s = 3$. Find and interpret the p-value for this test.

8.28 In a test of H_0: $\mu = 100$ against H_a: $\mu \neq 100$, the sample data yielded the test statistic $z = 2.17$. Find the p-value for the test.

8.29 In a test of H_0: $\mu = 75$ performed using the computer, SPSS reports a two-tailed p-value of .0456. Make the appropriate conclusion for each of the following situations:
   a. H_a: $\mu < 75$, $z = -2$, $\alpha = .05$
   b. H_a: $\mu < 75$, $z = -2$, $\alpha = .10$
   c. H_a: $\mu > 75$, $z = 2$, $\alpha = .10$
   d. H_a: $\mu \neq 75$, $z = 2$, $\alpha = .01$

8.30 An analyst tested the null hypothesis $\mu \geq 30$ against the alternative hypothesis that $\mu < 30$. The analyst reported a p-value of .07. What is the smallest value of α for which the null hypothesis would be rejected?

8.4 Test of Hypothesis about a Population Mean: Normal (z) Statistic

When testing a hypothesis about a population mean μ, the test statistic we use will depend on whether the sample size n is large (say, $n \geq 30$) or small and whether we know the value of the population standard deviation, σ. In this section, we consider the large-sample case.

Because the sample size is large, the Central Limit Theorem guarantees that the sampling distribution of $\bar{x}$ is approximately normal. Consequently, the test statistic for a test based on large samples will be based on the normal z-statistic. Although the z-statistic requires that we know the true population standard deviation σ, this is rarely, if ever, the case. However, we established in Chapter 7 that when n is large, the sample standard deviation s provides a good approximation to σ, and the z-statistic can be approximated as follows:

$$z = \frac{\bar{x} - \mu_0}{\sigma_{\bar{x}}} = \frac{\bar{x} - \mu_0}{\sigma/\sqrt{n}} \approx \frac{\bar{x} - \mu_0}{s/\sqrt{n}}$$

where μ_0 represents the value of μ specified in the null hypothesis.

The setup of a large-sample test of hypothesis about a population mean is summarized in the following boxes. Both the one- and two-tailed tests are shown.

Large-Sample Test of Hypothesis about μ Based on a Normal (z) Statistic

	σ known	σ unknown
Test statistic:	$z_c = \dfrac{(\bar{x} - \mu_0)}{(\sigma/\sqrt{n})}$	$z_c = \dfrac{(\bar{x} - \mu_0)}{(s/\sqrt{n})}$

	One-Tailed Tests		Two-Tailed Test
	H_0: $\mu = \mu_0$	H_0: $\mu = \mu_0$	H_0: $\mu = \mu_0$
	H_a: $\mu < \mu_0$	H_a: $\mu > \mu_0$	H_a: $\mu \neq \mu_0$
Rejection region:	$z_c < -z_\alpha$	$z_c > z_\alpha$	$\lvert z_c \rvert > z_{\alpha/2}$
p-value:	$P(z < z_c)$	$P(z > z_c)$	$2P(z > z_c)$ if z_c is positive
			$2P(z < z_c)$ if z_c is negative

Decision: Reject H_0 if $\alpha > p$-value or if test statistic (z_c) falls in rejection region where $P(z > z_\alpha) = \alpha$, $P(z > z_{\alpha/2}) = \alpha/2$, and $\alpha = P(\text{Type I error}) = P(\text{Reject } H_0 \mid H_0 \text{ true})$.

[*Note:* The symbol for the numerical value assigned to μ under the null hypothesis is μ_0.]

Conditions Required for a Valid Large-Sample Hypothesis Test for μ

1. A random sample is selected from the target population.
2. The sample size n is large (e.g., $n \geq 30$). (Due to the Central Limit Theorem, this condition guarantees that the test statistic will be approximately normal regardless of the shape of the underlying probability distribution of the population.)

Once the test has been set up, the sampling experiment is performed and the test statistic calculated. The next box contains possible conclusions for a test of hypothesis, depending on the result of the sampling experiment.

Possible Conclusions for a Test of Hypothesis

1. If the calculated test statistic falls in the rejection region, reject H_0 and conclude that the alternative hypothesis H_a is true. State that you are rejecting H_0 at the α level of significance. Remember that the confidence is in the testing *process*, not the particular result of a single test.
2. If the test statistic does not fall in the rejection region, conclude that the sampling experiment does not provide sufficient evidence to reject H_0 at the α level of significance. [Generally, we will not "accept" the null hypothesis unless the probability β of a Type II error has been calculated (see Section 8.7).]

Example   8.5 Ⓓ

DRUGRAT

Carrying Out a Hypothesis Test for μ—Mean Drug Response Time

Problem Refer to the neurological response-time test set up in Example 8.3 (p. 405). The sample of 100 drug-injected rats yielded the results (in seconds) shown in Table 8.3. At $\alpha = .01$, use these data to conduct the test of hypothesis,

$$H_0: \mu = 1.2$$
$$H_a: \mu \neq 1.2$$

Table 8.3   Drug Response Times for 100 Rats, Example 8.5

1.90	2.17	0.61	1.17	0.66	1.86	1.41	1.30	0.70	0.56
2.00	1.27	0.98	1.55	0.64	0.60	1.55	0.93	0.48	0.39
0.86	1.19	0.79	1.37	1.31	0.85	0.71	1.21	1.23	0.89
1.84	0.80	0.64	1.08	0.74	0.93	1.71	1.05	1.44	0.42
0.70	0.54	1.40	1.06	0.54	0.17	0.98	0.89	1.28	0.68
0.98	1.14	1.16	1.64	1.16	1.01	1.09	0.77	1.58	0.99
0.57	0.27	0.51	1.27	1.81	0.88	0.31	0.92	0.93	1.66
0.21	0.79	0.94	0.45	1.19	1.60	0.14	0.99	1.08	1.57
0.55	1.65	0.81	1.00	2.55	1.96	1.31	1.88	1.51	1.48
0.61	0.05	1.21	0.48	1.63	1.45	0.22	0.49	1.29	1.40

Ⓓ *Data Set: DRUGRAT*

Solution To carry out the test, we need to find the values of $\bar{x}$ and s. (In this study, σ is obviously unknown. So we will use s to estimate σ.) These values, $\bar{x} = 1.0517$ and $s = .4982$, are shown (highlighted) on the MINITAB printout of Figure 8.13. Now we substitute these sample statistics into the test statistic and obtain

$$z = \frac{\bar{x} - 1.2}{\sigma_{\bar{x}}} = \frac{\bar{x} - 1.2}{\sigma/\sqrt{n}} \approx \frac{1.0517 - 1.2}{.4982/\sqrt{100}} = -2.98$$

The implication is that the sample mean, 1.0517, is (approximately) three standard deviations below the null-hypothesized value of 1.2 in the sampling distribution of $\bar{x}$.

Recall from Example 8.3 that the rejection region for the test at $\alpha = .01$ is

$$Rejection\ region: |z| > 2.575$$

From Figure 8.5 (p. 406), you can see that $z = -2.98$ falls in the lower-tail rejection region, which consists of all values of $z < -2.575$. Therefore, this sampling experiment provides sufficient evidence to reject H_0 and conclude, at the $\alpha = .01$ level of significance, that the mean response time for drug-injected rats differs from the control mean of 1.2 seconds. It appears that the rats receiving an injection of the drug have a mean response time that is less than 1.2 seconds.

One-Sample Z: TIME   I

Test of mu = 1.2 vs not = 1.2
The assumed standard deviation = 0.4982

Variable	N	Mean	StDev	SE Mean	99% CI	Z	P
TIME	100	1.0517	0.4982	0.0498	(0.9234, 1.1800)	-2.98	0.003

Figure 8.13
MINITAB Analysis of Drug Response Times, Example 8.5

Look Back Four points about the test of hypothesis in this example apply to all statistical tests:

1. Since z is less than -2.575, it is tempting to state our conclusion at a significance level lower than $\alpha = .01$. We resist this temptation because the level of α is determined *before* the sampling experiment is performed. If we decide that we are willing to tolerate a 1% Type I error rate, the result of the sampling experiment should have no effect on that decision. *In general, the same data should not be used both to set up and to conduct the test.*

2. When we state our conclusion at the .01 level of significance, we are referring to the failure rate of the *procedure*, not the result of this particular test. We know that the test procedure will lead to the rejection of the null hypothesis only 1% of the time when in fact $\mu = 1.2$. *Therefore, when the test statistic falls into the rejection region, we infer that the alternative $\mu \neq 1.2$ is true and express our confidence in the procedure by quoting either the α level of significance or the $100(1 - \alpha)\%$ confidence level.*

3. Although a test may lead to a "statistically significant" result (i.e., rejecting H_0 at significance level α, as in the preceding test), it may not be "practically significant." For example, suppose the neurologist tested $n = 100,000$ drug-injected rats, resulting in $\bar{x} = 1.1995$ and $s = .05$. Now a two-tailed hypothesis test of H_0: $\mu = 1.2$ results in a test statistic of

$$z = \frac{(1.1995 - 1.2)}{.05/\sqrt{100,000}} = -3.16$$

This result at $\alpha = .01$ leads us to "reject H_0" and conclude that the mean μ is "statistically different" from 1.2. However, for all practical purposes, the sample mean $\bar{x} = 1.1995$ and the hypothesized mean $\mu = 1.2$ are the same. Because the result is not "practically significant," the neurologist is not likely to consider a unit dose of the drug as an inhibitor to response time in rats. Consequently, *not all "statistically significant" results are "practically significant."*

4. Finally, the inference derived from a two-tailed test of hypothesis will match the inference obtained from a confidence interval whenever the same value of α is used for both. To see this, a 99% confidence interval for μ is shown on the MINITAB printout, Figure 8.13. Note that the interval, (.92, 1.18), does not include the hypothesized value, $\mu = 1.2$. Consequently, both the confidence interval (using $\alpha = .01$) and the two-tailed hypothesis test (using $\alpha = .01$) lead to the same conclusion—namely, that the true mean drug response time is not equal to 1.2 seconds.

■ **Now Work Exercise 8.40**

Example **8.6**

Using p-Values—Test ^{DRUGRAT}
of Mean Drug
Response Time

Problem Find the observed significance level (*p*-value) for the test of the mean drug response time in Examples 8.3 and 8.5. Interpret the result.

Solution Again, we are testing H_0: $\mu = 1.2$ seconds versus H_a: $\mu \neq 1.2$ seconds. The observed value of the test statistic in Example 8.5 was $z = -2.98$, and any value of z less than -2.98 or greater than 2.98 (because this is a two-tailed test) would be even more contradictory to H_0. Therefore, the observed significance level for the test is

$$p\text{-value} = P(z < -2.98 \text{ or } z > 2.98) = P(|z| > 2.98)$$

Thus, we calculate the area below the observed z-value, $z = -2.98$, and double it. Consulting Table II in Appendix B, we find that $P(z < -2.98) = .5 - .4986 = .0014$. Therefore, the *p*-value for this two-tailed test is

$$2P(z < -2.98) = 2(.0014) = .0028$$

This *p*-value can also be obtained using statistical software. The rounded *p*-value is shown (highlighted) on the MINITAB printout, Figure 8.13. Since $\alpha = .01$ is greater than *p*-value $= .0028$, our conclusion is identical to that in Example 8.5—reject H_0.

Look Back We can interpret this *p*-value as a strong indication that the drug has an effect on mean rat response because we would observe a test statistic this extreme or more extreme only 28 in 10,000 times if the drug had no effect ($\mu = 1.2$). The extent to which the mean differs from 1.2 could be better determined by calculating a confidence interval for μ. As we saw in Example 8.5, the 99% confidence interval for μ is (.92, 1.18).

■ **Now Work Exercise 8.36**

Example 8.7 D

LOS

Using *p*-Values—Test of Mean Hospital Length of Stay

Problem Knowledge of the amount of time a patient occupies a hospital bed—called *length of stay* (LOS)—is important for allocating resources. At one hospital, the mean LOS was determined to be 5 days. A hospital administrator believes the mean LOS may now be less than 5 days due to a newly adopted managed health care system. To check this, the LOSs (in days) for 100 randomly selected hospital patients were recorded; these data are listed in Table 8.4. Suppose we want to test the hypothesis that the true mean length of stay at the hospital is less than 5 days; that is,

$$H_0: \mu = 5 \,(\text{Mean LOS after adoption is 5 days})$$
$$H_a: \mu < 5 \,(\text{Mean LOS after adoption is less than 5 days})$$

Assuming that $\sigma = 3.68$, use the data in the table to conduct the test at $\alpha = .05$.

Table 8.4   Lengths of Stay for 100 Hospital Patients

2	3	8	6	4	4	6	4	2	5
8	10	4	4	4	2	1	3	2	10
1	3	2	3	4	3	5	2	4	1
2	9	1	7	17	9	9	9	4	4
1	1	1	3	1	6	3	3	2	5
1	3	3	14	2	3	9	6	6	3
5	1	4	6	11	22	1	9	6	5
2	2	5	4	3	6	1	5	1	6
17	1	2	4	5	4	4	3	2	3
3	5	2	3	3	2	10	2	4	2

D *Data Set:* LOS

Solution The data were entered into a computer, and MINITAB was used to conduct the analysis. The MINITAB printout for the lower-tailed test is displayed in Figure 8.14. Both the test statistic, $z = -1.28$, and the *p*-value of the test, $p = .101$, are highlighted on the MINITAB printout. Since the *p*-value exceeds our selected α value, $\alpha = .05$, we cannot reject the null hypothesis. Hence, there is insufficient evidence (at $\alpha = .05$) to conclude that the true mean LOS at the hospital is less than 5 days.

One-Sample Z: LOS

```
Test of mu = 5 vs < 5
The assumed standard deviation = 3.68


                                    95% Upper
Variable    N    Mean   StDev  SE Mean    Bound      Z      P
LOS        100  4.530   3.678    0.368    5.135  -1.28  0.101
```

Figure 8.14
MINITAB lower-tailed test of mean LOS, Example 8.7

Look Back A hospital administrator, desirous of a mean length of stay less than 5 days, may be tempted to select an α level that leads to a rejection of the null hypothesis *after* determining *p*-value $= .101$. There are two reasons one should resist this temptation. First, the administrator would need to select $\alpha > .101$ (say, $\alpha = .15$) in order to conclude that $H_a: \mu < 5$ is true. A Type I error rate of 15% is considered too large by most researchers. Second, and more importantly, such a strategy is considered unethical statistical practice. (See marginal note, p. 412.)

■ Now Work Exercise 8.45

In the next section, we demonstrate how to conduct a test for μ when the sample size is small.

| Statistics IN Action | Revisited | **Testing a Population Mean in the KLEENEX® Survey** |

Refer to Kimberly-Clark Corporation's survey of 250 people who kept a count of their use of KLEENEX® tissues in diaries (p. 406). We want to test the claim made by marketing experts that μ, the average number of tissues used by people with colds, is greater than 60 tissues. That is, we want to test

$$H_0: \mu = 60 \qquad H_a: \mu > 60$$

We will select $\alpha = .05$ as the level of significance for the test.

The survey results for the 250 sampled KLEENEX® users are stored in the **TISSUES** data file. A MINITAB analysis of the data yielded the printout displayed in Figure SIA8.1.

The observed significance level of the test, highlighted on the printout, is p-value = 000. Since this p-value is less than $\alpha = .05$, we have sufficient evidence to reject H_0; therefore, we conclude that the mean number of tissues used by a person with a cold is greater than 60 tissues. This result supports the company's decision to put more than 60 (in fact, 68) tissues in a cold-care (anti-viral) box of KLEENEX.

D *Data Set:* TISSUES

Figure SIA8.1
MINITAB test of $\mu = 60$ for KLEENEX survey

```
Test of mu = 60 vs > 60

                                    95% Lower
Variable    N   Mean  StDev  SE Mean   Bound     T      P
NUMUSED   250  68.68  25.03    1.58    66.06   5.48  0.000
```

Exercises 8.31–8.50

Understanding the Principles

8.31 Which condition ensures normality of the sample mean in large samples?

8.32 For what values of the test statistic do you reject H_0? fail to reject H_0?

Learning the Mechanics

8.33 A random sample of 121 observations from a population with standard deviation 50 yielded a sample mean of 120.
   a. Test the null hypothesis that $\mu = 110$ against the alternative hypothesis that $\mu > 110$, using $\alpha = .05$. Interpret the results of the test.
   b. Test the null hypothesis that $\mu = 110$ against the alternative hypothesis that $\mu \neq 110$, using $\alpha = .05$. Interpret the results of the test.
   c. Compare the p-values of the two tests you conducted. Explain why the results differ.

8.34 A random sample of 64 observations produced the following summary statistics: $\bar{x} = .36$ and $s^2 = .034$.
   a. Test the null hypothesis that $\mu = .397$ against the alternative hypothesis that $\mu < .397$, using $\alpha = .10$.
   b. Test the null hypothesis that $\mu = .397$ against the alternative hypothesis that $\mu \neq .397$, using $\alpha = .10$. Interpret the result.

8.35 Suppose you are interested in conducting the statistical test of $H_0: \mu = 100$ against $H_a: \mu > 100$, and you have decided to use the following decision rule: Reject H_0 if the sample mean of a random sample of 50 items is more than 115. Assume that the standard deviation of the population is 40.
   a. Express the decision rule in terms of z.
   b. Find α, the probability of making a Type I error, by using this decision rule.

D Applet Exercise 8.2

Use the applet entitled *Hypotheses Test for a Mean* to investigate the effect of the underlying distribution on the proportion of Type I errors. For this exercise, take $n = 100$, mean = 50, standard deviation = 10, null mean = 50, and alternative $<$.
   a. Select the normal distribution and run the applet several times without clearing. What happens to the proportion of times the null hypothesis is rejected at the .05 level as the applet is run more and more times?
   b. Clear the applet and then repeat part **a**, using the right-skewed distribution. Do you get similar results? Explain.
   c. Describe the effect that the underlying distribution has on the probability of making a Type I error.

D Applet Exercise 8.3

Use the applet entitled *Hypotheses Test for a Mean* to investigate the effect of the underlying distribution on the proportion of Type II errors. For this exercise, take $n = 100$, mean = 50, standard deviation = 10, null mean = 52, and alternative $<$.
   a. Select the normal distribution and run the applet several times without clearing. What happens to the proportion of times the null hypothesis is rejected at the .01 level as the applet is run more and more times? Is this what you would expect? Explain.
   b. Clear the applet and then repeat part **a**, using the right-skewed distribution. Do you get similar results? Explain.
   c. Describe the effect that the underlying distribution has on the probability of making a Type II error.

D Applet Exercise 8.4

Use the applet entitled *Hypotheses Test for a Mean* to investigate the effect of the null mean on the probability of making a Type II error. For this exercise, take $n = 100$, mean = 50, standard deviation = 10, and alternative < with the normal distribution. Set the null mean to 55 and run the applet several times without clearing. Record the proportion of Type II errors that occurred at the .01 level. Clear the applet and repeat for null means of 54, 53, 52, and 51. What can you conclude about the probability of a Type II error as the null mean gets closer to the actual mean? Can you offer a reasonable explanation for this behavior?

Applying the Concepts—Basic

8.36 **Stability of compounds in new drugs.** Refer to the *ACS Medicinal Chemistry Letters* (Vol. 1, 2010) study of the metabolic stability of drugs, Exercise 2.35 (p. 77). Recall that two important values computed from the testing phase are the fraction of compound unbound to plasma (*fup*) and the fraction of compound unbound to microsomes (*fumic*). A key formula for assessing stability assumes that the fup/fumic ratio is 1. Pharmacologists at Pfizer Global Research and Development tested 416 drugs and reported the fup/fumic ratio for each. These data are saved in the **FUP** file, and summary statistics are provided in the accompanying MINITAB printout. Suppose the pharmacologists want to determine if the true mean ratio, μ, differs from 1.

a. Specify the null and alternative hypothesis for this test.

b. Descriptive statistics for the sample ratios are provided in the accompanying MINITAB printout. Note that the sample mean, $\bar{x} = .327$, is less than 1. Consequently, a pharmacologist wants to reject the null hypothesis. What are the problems with using such a decision rule?

c. Locate values of the test statistic and corresponding *p*-value on the printout.

d. Select a value of α, the probability of a Type I error. Interpret this value in the words of the problem.

e. Give the appropriate conclusion, based on the results of parts **c** and **d**.

f. What conditions must be satisfied for the test results to be valid?

One-Sample T: RATIO

Test of mu = 1 vs not = 1

Variable	N	Mean	StDev	SE Mean	95% CI	T	P
RATIO	416	0.3269	0.2915	0.0143	(0.2988, 0.3550)	-47.09	0.000

8.37 **Corporate sustainability of CPA firms.** Refer to the *Business and Society* (Mar. 2011) study on the sustainability behaviors of CPA corporations, Exercise 7.18 (p. 352). Recall that the level of support for corporate sustainability

(measured on a quantitative scale ranging from 0 to 160 points) was obtained for each in a sample of 992 senior managers at CPA firms. The data (where higher point values indicate a higher level of support for sustainability) are saved in the referenced file. The CEO of a CPA firm claims that the true mean level of support for sustainability is 75 points.

a. Specify the null and alternative hypotheses for testing this claim.

b. For this problem, what is a Type I error? A Type II error?

c. The SPSS printout at the bottom of this page gives the results of the test. Locate the *p*-value on the printout.

d. At $\alpha = .05$, give the appropriate conclusion.

e. What assumptions, if any, about the distribution of support levels must hold true in order for the inference derived from the test to be valid? Explain.

8.38 **Speeding and young drivers.** Psychologists conducted a survey of 258 student drivers and their attitudes toward speeding and reported the results in the *British Journal of Educational Psychology* (Vol. 80, 2010). One of the variables of interest was the response to the question, "Are you confident that you can resist your friends' persuasion to drive faster?" Each response was measured on a 7-point scale, from 1 = "definitely no" to 7 = "definitely yes." The data were collected 5 months after the students had attended a safe-driver presentation. The psychologists reported a sample mean response of 4.98 and a sample standard deviation of 1.62. Suppose it is known that the true mean response of students who *do not* attend a safe-driver presentation is $\mu = 4.7$.

a. Set up the null and alternative hypotheses for testing whether the true mean student-driver response 5 months after a safe-driver presentation is larger than 4.7.

b. Calculate the test statistic for the hypothesis test.

c. Find the rejection region for the hypothesis test, using $\alpha = .05$.

d. State the appropriate conclusion, in the words of the problem.

e. Do the test results indicate that the safe-driver presentation was effective in helping students feel more confident that they can resist their friends' persuasion to drive faster? Explain.

f. The distribution of response scores (on a 7-point scale) for all student drivers is unlikely to be normal. Does this impact the validity of the hypothesis test? Why or why not?

SPSS Output for Exercise 8.37

One-Sample Statistics

	N	Mean	Std. Deviation	Std. Error Mean
SUPPORT	992	67.76	26.871	.853

One-Sample Test

	Test Value = 75					
					95% Confidence Interval of the Difference	
	t	df	Sig. (2-tailed)	Mean Difference	Lower	Upper
SUPPORT	-8.492	991	.000	-7.245	-8.92	-5.57

8.39 **Emotional empathy in young adults.** According to a theory in psychology, young female adults show more emotional empathy toward others than do males. The *Journal of Moral Education* (June 2010) tested this theory by examining the attitudes of a sample of 30 female college students. Each student completed the Ethic of Care Interview, which consisted of a series of statements on empathy attitudes. For the statement on emotional empathy (e.g., "I often have tender, concerned feelings for people less fortunate than me"), the sample mean response was 3.28. Assume the population standard deviation for females is .5. [*Note*: Empathy scores ranged from 0 to 4, where 0 = "never" and 4 = "always".] Suppose it is known that male college students have an average emotional empathy score of $\mu = 3$.

   a. Specify the null and alternative hypotheses for testing whether female college students score higher than 3.0 on the emotional empathy scale.
   b. Compute the test statistic.
   c. Find the observed significance level (*p*-value) of the test.
   d. At $\alpha = .01$, what is the appropriate conclusion?
   e. How small of an α-value can you choose and still have sufficient evidence to reject the null hypothesis?

8.40 **Heart rate during laughter.** Laughter is often called "the best medicine," since studies have shown that laughter can reduce muscle tension and increase oxygenation of the blood. In the *International Journal of Obesity* (Jan. 2007), researchers at Vanderbilt University investigated the physiological changes that accompany laughter. Ninety subjects (18–34 years old) watched film clips designed to evoke laughter. During the laughing period, the researchers measured the heart rate (beats per minute) of each subject, with the following summary results: $\bar{x} = 73.5, s = 6$. It is well known that the mean resting heart rate of adults is 71 beats per minute.

   a. Set up H_0 and H_a for testing whether the true mean heart rate during laughter exceeds 71 beats per minute.
   b. If $\alpha = 05$, find the rejection region for the test.
   c. Calculate the value of the test statistic.
   d. Make the appropriate conclusion.

8.41 **Facial structure of CEOs.** Refer to the *Psychological Science* (Vol. 22, 2011) study on using a chief executive officer's facial structure to predict a firm's financial performance, Exercise 7.21 (p. 353). Recall that the facial width-to-height ratio (WHR) for each in a sample of 55 CEOs at publicly traded *Fortune* 500 firms was determined. The sample resulted in $\bar{x} = 1.96$ and $s = .15$. An analyst wants to predict the financial performance of a *Fortune* 500 firm based on the value of the true mean facial WHR of CEOs. The analyst wants to use the value of $\mu = 2.2$. Do you recommend he use this value? Conduct a test of hypothesis for μ to help you answer the question. Specify all the elements of the test: H_0, H_a, test statistic, *p*-value, α, and your conclusion.

Applying the Concepts—Intermediate

8.42 **Packaging of a children's health food.** Junk foods (e.g., potato chips) are typically packaged to appeal to children. Can similar packaging of a healthy food product influence children's desire to consume the product? This was the question of interest in an article published in the *Journal of Consumer Behaviour* (Vol. 10, 2011). A fictitious brand of a healthy food product—sliced apples—was packaged to appeal to children (a smiling cartoon apple on the front of the package). The researchers showed the packaging to a sample of 408 schoolchildren and asked each whether he or she was willing to eat the product. Willingness to eat was measured on a 5-point scale, with 1 = "not willing at all" and 5 = "very willing." The data are summarized as follows: $\bar{x} = 3.69, s = 2.44$. Suppose the researchers know that the mean willingness to eat an actual brand of sliced apples (which is not packaged for children) is $\mu = 3$.

   a. Conduct a test to determine whether the true mean willingness to eat the brand of sliced apples packaged for children exceeds 3. Use $\alpha = .05$ to make your conclusion.
   b. The data (willingness to eat values) are not normally distributed. How does this affect (if at all) the validity of your conclusion in part **a**? Explain.

8.43 **Dating and disclosure.** Refer to the *Journal of Adolescence* (Apr. 2010) study of adolescents' disclosure of their dating and romantic relationships, Exercise 1.33 (p. 51). Recall that a sample of 222 high school students was recruited to participate in the study. One of the variables of interest was the level of disclosure to an adolescent's mother (measured on a 5-point scale, where 1 = "never tell," 2 = "rarely tell," 3 = "sometimes tell," 4 = "almost always tell," and 5 = "always tell"). The sampled high school students had a mean disclosure score of 3.26 and a standard deviation of .93. The researchers hypothesize that the true mean disclosure score of all adolescents will exceed 3. Do you believe the researchers? Conduct a formal test of hypothesis using $\alpha = .01$.

8.44 **Birth order and IQ.** An international team of economists investigated the possible link between IQ and birth order in *CESifo Economic Studies* (Vol. 57, 2011). The data source for the research was the Medical Birth Registry of Norway. It is known that the mean IQ (measured in stanines) for all Norway residents is 5.2 points. In the study, a sample of 581 Norway residents who were the 6th-born or later in their families had a mean IQ score of 4.7 points with a standard deviation of 1.8 points. Is this sufficient evidence to conclude that the mean IQ score of all Norway residents who were the 6th-born or later in their families is lower than the country mean of 5.2 points? Use $\alpha = .01$ as a measure of reliability for your inference.

8.45 **Bone fossil study.** Archeologists have found that humerus bones from the same species of animal tend to have approximately the same length-to-width ratios. It is known **BONES** that species A exhibits a mean ratio of 8.5. Suppose 41 fossils of humerus bones were unearthed at an archeological site in East Africa, where species A is believed to have lived. (Assume that the unearthed bones were all from the same unknown species.) The length-to-width ratios of the bones are listed in the following table.

10.73	8.89	9.07	9.20	10.33	9.98	9.84	9.59
8.48	8.71	9.57	9.29	9.94	8.07	8.37	6.85
8.52	8.87	6.23	9.41	6.66	9.35	8.86	9.93
8.91	11.77	10.48	10.39	9.39	9.17	9.89	8.17
8.93	8.80	10.02	8.38	11.67	8.30	9.17	12.00
9.38							

a. Test whether the population mean ratio of all bones of this particular species differs from 8.5. Use $\alpha = .01$.

b. What are the practical implications of the test you conducted in part **a**?

8.46 **Cooling method for gas turbines.** A popular cooling method for a gas turbine engine uses high-pressure inlet fogging. The performance of a sample of 67 gas turbines augmented with high-pressure inlet fogging was investigated in the *Journal of Engineering for Gas Turbines and Power* (Jan. 2005). One measure of performance is heat rate (kilojoules per kilowatt per hour). Heat rates for the 67 gas turbines are listed in the next table. Suppose that a standard gas turbine has, on average, a heat rate of 10,000 kJ/kWh.

a. Conduct a test to determine whether the mean heat rate of gas turbines augmented with high-pressure inlet fogging exceeds 10,000 kJ/kWh. Use $\alpha = .05$.

b. Identify a Type I error for this study. Identify a Type II error.

14622	13196	11948	11289	11964	10526	10387	10592	10460	10086
14628	13396	11726	11252	12449	11030	10787	10603	10144	11674
11510	10946	10508	10604	10270	10529	10360	14796	12913	12270
11842	10656	11360	11136	10814	13523	11289	11183	10951	9722
10481	9812	9669	9643	9115	9115	11588	10888	9738	9295
9421	9105	10233	10186	9918	9209	9532	9933	9152	9295
16243	14628	12766	8714	9469	11948	12414			

8.47 **Irrelevant speech effects.** Refer to the *Acoustical Science & Technology* (Vol. 35, 2014) study of irrelevant speech effects, Exercise 7.14 (p. 351). Recall that subjects performed a memorization task under two conditions: (1) with irrelevant background speech and (2) in silence. The difference in the error rates for the two conditions—called the relative difference in error rate (RDER)—was computed for each subject. Descriptive statistics for the RDER values are reproduced in the following SAS printout. Conduct a test to determine if the average difference in error rates for all subjects who perform the memorization tasks exceeds 75 percent. Use $\alpha = .01$ and interpret the results practically.

The MEANS Procedure

Analysis Variable : RDER

N	Mean	Std Dev	Minimum	Maximum
71	78.1885070	63.2429128	-19.2710000	254.1090000

8.48 **Time required to complete a task.** When asked, "How much time will you require to complete this task?", cognitive theory posits that people will typically underestimate the time required. Would the opposite theory hold if the question was phrased in terms of how much work could be completed in a given amount of time? This was the question of interest to researchers writing in *Applied Cognitive Psychology* (Vol. 25, 2011). For one study conducted by the researchers, each in a sample of 40 University of Oslo students was asked how many minutes it would take to read a 32-page report. In a second study, 42 students were asked how many pages of a lengthy report they could read in 48 minutes. (The students in either study did not actually read the report.) Numerical descriptive statistics (based

on summary information published in the article) for both studies are provided in the accompanying table.

	Estimated Time (minutes)	Estimated Number of Pages
Sample size, n	40	42
Sample mean, $\bar{x}$	60	28
Sample standard deviation, s	41	14

a. The researchers determined that the actual mean time it takes to read the report is $\mu = 48$ minutes. Is there evidence to support the theory that the students, on average, will overestimate the time it takes to read the report? Test using $\alpha = .10$.

b. The researchers also determined that the actual mean number of pages of the report that is read within the allotted time is $\mu = 32$ pages. Is there evidence to support the theory that the students, on average, will underestimate the number of report pages that can be read? Test using $\alpha = .10$.

c. The researchers noted that the distributions of both estimated time and estimated number of pages are highly skewed (i.e., not normally distributed). Does this fact affect the inferences derived in parts **a** and **b**? Explain.

Applying the Concepts—Advanced

8.49 **Social interaction of mental patients.** The *Community Mental Health Journal* (Aug. 2000) presented the results of a survey of over 6,000 clients of the Department of Mental Health and Addiction Services (DMHAS) in Connecticut. One of the many variables measured for each mental health patient was frequency of social interaction (on a 5-point scale, where $1 =$ very infrequently, $3 =$ occasionally, and $5 =$ very frequently). The 6,681 clients who were evaluated had a mean social interaction score of 2.95 with a standard deviation of 1.10.

a. Conduct a hypothesis test (at $\alpha = .01$) to determine whether the true mean social interaction score of all Connecticut mental health patients differs from 3.

b. Examine the results of the study from a practical view, and then discuss why "statistical significance" does not always imply "practical significance."

c. Because the variable of interest is measured on a 5-point scale, it is unlikely that the population of ratings will be normally distributed. Consequently, some analysts may perceive the test from part **a** to be invalid and search for alternative methods of analysis. Defend or refute this position.

8.50 **Instructing English-as-a-first-language learners.** A study published in the journal *Applied Linguistics* (Feb. 2014) investigated the effects of implicit and explicit classroom interventions during the instruction of Japanese to English-as-a-first-language (EFL) students. EFL students at a Japanese university were divided into implicit ($n = 44$) and explicit ($n = 37$) groups. [*Note:* Implicit instruction emphasized reading comprehension only, while explicit instruction included both deductive and inductive activities designed to bring learners' attention to rules and patterns.] Following an instruction period, all students were tested on their ability to use epistemic terms such as *think*, *maybe*, *seem*, and *may* in an essay. A summary of the epistemic type scores for the two groups of EFL students are provided in the table on p. 421.

Table for Exercise 8.50

	Mean Score (points)	Standard Deviation
Implicit ($n = 44$)	5.9	2.4
Explicit ($n = 37$)	9.7	3.1

Source: Fordyce, K. "The differential effects of explicit and implicit instruction on EFL learners' use of epistemic stance." *Applied Linguistics,* Vol. 35, No. 1, Feb. 2014 (Table 6).

a. A 95% confidence interval for μ_{Imp}, the true mean epistemic type score of EFL students taught with implicit instruction, is (5.17, 6.63). Based on this interval, is there evidence that μ_{Imp} differs from 6?

b. Use the summary information in the table to test the hypothesis that μ_{Imp} differs from 6 at $\alpha = .05$.

c. Explain why the inferences in parts **a** and **b** agree.

d. A 95% confidence interval for μ_{Exp}, the true mean epistemic type score of EFL students taught with explicit instruction, is (8.67, 10.73). Based on this interval, is there evidence that μ_{Exp} differs from 6?

e. Use the summary information in the table to test the hypothesis that μ_{Exp} differs from 6 at $\alpha = .05$.

f. Explain why the inferences in parts **d** and **e** agree.

8.5 Test of Hypothesis about a Population Mean: Student's *t*-Statistic

Most water-treatment facilities monitor the quality of their drinking water on an hourly basis. One variable monitored is pH, which measures the degree of alkalinity or acidity in the water. A pH below 7.0 is acidic, one above 7.0 is alkaline, and a pH of 7.0 is neutral. One water-treatment plant has a target pH of 8.5. (Most try to maintain a slightly alkaline level.) The mean and standard deviation of 1 hour's test results, based on 17 water samples at this plant, are

$$\bar{x} = 8.42 \qquad s = .16$$

Does this sample provide sufficient evidence that the mean pH level in the water differs from 8.5?

This inference can be placed in a test-of-hypothesis framework. We establish the target pH as the null hypothesized value and then utilize a two-tailed alternative that the true mean pH differs from the target:

$$H_0: \mu = 8.5 \ (\text{Mean pH level is 8.5.})$$
$$H_a: \mu \neq 8.5 \ (\text{Mean pH level differs from 8.5.})$$

Recall from Section 7.3 that when we are faced with making inferences about a population mean from the information in a small sample, two problems emerge:

1. The normality of the sampling distribution for $\bar{x}$ does not follow from the Central Limit Theorem when the sample size is small. We must assume that the distribution of measurements from which the sample was selected is approximately normally distributed in order to ensure the approximate normality of the sampling distribution of $\bar{x}$.

2. If the population standard deviation σ is unknown, as is usually the case, then we cannot assume that s will provide a good approximation for σ when the sample size is small. Instead, we must use the *t*-distribution rather than the standard normal *z*-distribution to make inferences about the population mean μ.

Therefore, as the test statistic of a small-sample test of a population mean, we use the *t*-statistic:

$$\text{\textit{Test statistic}: } t = \frac{\bar{x} - \mu_0}{s/\sqrt{n}} = \frac{\bar{x} - 8.5}{s/\sqrt{n}}$$

In this equation, μ_0 is the null-hypothesized value of the population mean μ. In the example here, $\mu_0 = 8.5$.

To find the rejection region, we must specify the value of α, the probability that the test will lead to rejection of the null hypothesis when it is true, and then consult the *t* table (Table III of Appendix B). With $\alpha = .05$, the two-tailed rejection region is

Rejection region: $t_{\alpha/2} = t_{.025} = 2.120$ with $n - 1 = 16$ degrees of freedom

Reject H_0 if $t < -2.120$ or $t > 2.120$

$\frac{\alpha}{2} = .025$ $\frac{\alpha}{2} = .025$

Rejection region   0   Rejection region

−2.120   2.120

Observed
$t = -2.05$

Figure 8.15
Two-tailed rejection region for small-sample t-test

The rejection region is shown in Figure 8.15.

We are now prepared to calculate the test statistic and reach a conclusion:

$$t = \frac{\bar{x} - \mu_0}{s/\sqrt{n}} = \frac{8.42 - 8.50}{.16/\sqrt{17}} = \frac{-.08}{.039} = -2.05$$

Since the calculated value of t does not fall into the rejection region (Figure 8.15), we cannot reject H_0 at the $\alpha = .05$ level of significance. Thus, the water-treatment plant should not conclude that the mean pH differs from the 8.5 target on the basis of the sample evidence.

It is interesting to note that the calculated t value, -2.05, is *less than* the .05-level z-value, -1.96. The implication is that if we had *incorrectly* used a z statistic for this test, we would have rejected the null hypothesis at the .05 level and concluded that the mean pH level differs from 8.5. The important point is that the statistical procedure to be used must always be closely scrutinized and all the assumptions understood. Many statistical lies are the result of misapplications of otherwise valid procedures.

The technique for conducting a small-sample test of hypothesis about a population mean is summarized in the following boxes:

Small-Sample Test of Hypothesis about μ Based on a Student's t-Statistic

Test statistic: $t_c = \dfrac{(\bar{x} - \mu_0)}{(s/\sqrt{n})}$

	One-Tailed Tests		Two-Tailed Test
	$H_0: \mu = \mu_0$	$H_0: \mu = \mu_0$	$H_0: \mu = \mu_0$
	$H_a: \mu < \mu_0$	$H_a: \mu > \mu_0$	$H_a: \mu \neq \mu_0$
Rejection region:	$t_c < -t_\alpha$	$t_c > t_\alpha$	$\lvert t_c \rvert > t_{\alpha/2}$
p-value:	$P(t < t_c)$	$P(t > t_c)$	$2P(t > t_c)$ if t_c is positive
			$2P(t < t_c)$ if t_c is negative

Decision: Reject H_0 if $\alpha > p$-value or if test statistic (t_c) falls in rejection region where $P(t > t_\alpha) = \alpha, P(t > t_{\alpha/2}) = \alpha/2, \alpha = P(\text{Type I error}) = P(\text{Reject } H_0 \mid H_0 \text{ true})$, and t is based on $(n-1)$ degrees of freedom.

[*Note*: The symbol for the numerical value assigned to μ under the null hypothesis is μ_0.]

Conditions Required for a Valid Small-Sample Hypothesis Test for μ

1. A random sample is selected from the target population.
2. The population from which the sample is selected has a distribution that is approximately normal.

Example	8.8	D

A Small-Sample Test for μ—Does a New Engine Meet Air-Pollution Standards?

ENGINE

Problem A major car manufacturer wants to test a new engine to determine whether it meets new air-pollution standards. The mean emission μ of all engines of this type must be less than 20 parts per million of carbon. Ten engines are manufactured for testing purposes, and the emission level of each is determined. The data (in parts per million) are listed in Table 8.5. Do the data supply sufficient evidence to allow the manufacturer to conclude that this type of engine meets the pollution standard? Assume that the manufacturer is willing to risk a Type I error with probability $\alpha = .01$.

Table 8.5   **Emission Levels for Ten Engines**

15.6	16.2	22.5	20.5	16.4	19.4	19.6	17.9	12.7	14.9

D *Data Set:* ENGINE

Solution The manufacturer wants to support the research hypothesis that the mean emission level μ for all engines of this type is less than 20 parts per million. The elements of this small-sample one-tailed test are as follows:

$$H_0: \mu = 20 \ (\text{Mean emission level is 20 ppm.})$$
$$H_a: \mu < 20 \ (\text{Mean emission level is less than 20 ppm.})$$

$$\textit{Test statistic: } t = \frac{\bar{x} - 20}{s/\sqrt{n}}$$

Rejection region: For $\alpha = .01$ and df $= n - 1 = 9$, the one-tailed rejection region (see Figure 8.16) is $t < -t_{.01} = -2.821$.

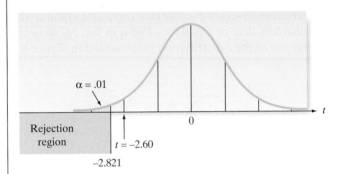

Figure 8.16

A *t*-distribution with 9 df and the rejection region for Example 8.8

Assumption: The relative frequency distribution of the population of emission levels for all engines of this type is approximately normal. Based on the shape of the MINITAB stem-and-leaf display of the data shown in Figure 8.17, this assumption appears to be reasonably satisfied.

To calculate the test statistic, we analyzed the **ENGINE** data with MINITAB. The MINITAB printout is shown at the bottom of Figure 8.17. From the printout, we obtain $\bar{x} = 17.57$ and $s = 2.95$. Substituting these values into the test statistic formula and rounding, we get

$$t = \frac{\bar{x} - 20}{s/\sqrt{n}} = \frac{17.57 - 20}{2.95/\sqrt{10}} = -2.60$$

Since the calculated *t* falls outside the rejection region (see Figure 8.16), the manufacturer cannot reject H_0. There is insufficient evidence to conclude that $\mu < 20$ parts per million and that the new type of engine meets the pollution standard.

Stem-and-Leaf Display: E-LEVEL

```
Stem-and-leaf of E-LEVEL   N  = 10
Leaf Unit = 1.0


  1    1  2
  3    1  45
 (3)   1  667
  4    1  99
  2    2  0
  1    2  2
```

One-Sample T: E-LEVEL

```
Test of mu = 20 vs < 20
```

Figure 8.17

MINITAB analysis of 10 emission levels, Example 8.8

Variable	N	Mean	StDev	SE Mean	95% Upper Bound	T	P
E-LEVEL	10	17.5700	2.9522	0.9336	19.2814	-2.60	0.014

Look Back Are you satisfied with the reliability associated with this inference? The probability is only $\alpha = .01$ that the test would support the research hypothesis if in fact it were false.

■ Now Work Exercise 8.57a–b

Example	8.9

The p-Value for a Small-Sample Test of μ

Problem Find the observed significance level for the test described in Example 8.8. Interpret the result.

Solution The test of Example 8.8 was a lower-tailed test: H_0: $\mu = 20$ versus H_a: $\mu < 20$. Since the value of t computed from the sample data was $t = -2.60$, the observed significance level (or p-value) for the test is equal to the probability that t would assume a value less than or equal to -2.60 if in fact H_0 were true. This is equal to the area in the lower tail of the t-distribution (highlighted in Figure 8.18).

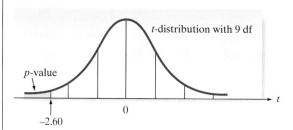

Figure 8.18
The observed significance level for the test of Example 8.8

One way to find this area (i.e., the p-value for the test) is to consult the t-table (Table III in Appendix B). Unlike the table of areas under the normal curve, Table III gives only the t values corresponding to the areas .100, .050, .025, .010, .005, .001, and .0005. Therefore, we can only approximate the p-value for the test. Since the observed t value was based on nine degrees of freedom, we use the df = 9 row in Table III and move across the row until we reach the t values that are closest to the observed $t = -2.60$. [*Note:* We ignore the minus sign.] The t values corresponding to p-values of .010 and .025 are 2.821 and 2.262, respectively. Since the observed t value falls between $t_{.010}$ and $t_{.025}$, the p-value for the test lies between .010 and .025. In other words, $.010 < p\text{-value} < .025$. Thus, we would reject the null hypothesis H_0: $\mu = 20$ parts per million for any value of α larger than .025 (the upper bound of the p-value).

A second, more accurate, way to obtain the p-value is to use a statistical software package to conduct the test of hypothesis. Both the test statistic (-2.60) and the p-value (.014) are highlighted on the MINITAB printout of Figure 8.17.

You can see that the actual p-value of the test falls within the bounds obtained from Table III. Thus, the two methods agree and we will reject H_0: $\mu = 20$ in favor of H_a: $\mu < 20$ for any α level larger than .025.

■ Now Work Exercise 8.57c

Small-sample inferences typically require more assumptions and provide less information about the population parameter than do large-sample inferences. Nevertheless, the t-test is a method of testing a hypothesis about a population mean of a normal distribution when only a small number of observations is available.

> **What Can Be Done If the Population Relative Frequency Distribution Departs Greatly from Normal?**
>
> *Answer:* Use one of the nonparametric statistical methods of Chapter 14.

Exercises 8.51–8.72

Understanding the Principles

8.51 When do we use the *t*-test instead of the *z*-test in small samples?

8.52 What are the assumptions behind a *t*-test in testing a hypothesis about a population mean?

Learning the Mechanics

8.53 For each of the following rejection regions, sketch the sampling distribution of *t* and indicate the location of the rejection region on your sketch:
  a. $t > 1.64$, where df = 6
  b. $t < -1.872$, where df = 12
  c. $t < -2.161$ or $t > 2.161$, where df = 25

8.54 For each of the rejection regions defined in Exercise 8.53, what is the probability that a Type I error will be made?

8.55 A random sample of *n* observations is selected from a normal population to test the null hypothesis that $\mu = 15$. Specify the rejection region for each of the following combinations of H_a, α, and *n*:
  a. $H_a: \mu \neq 15; \alpha = .05; n = 16$
  b. $H_a: \mu > 15; \alpha = .01; n = 26$
  c. $H_a: \mu > 15; \alpha = .10; n = 12$
  d. $H_a: \mu < 15; \alpha = .01; n = 15$
  e. $H_a: \mu \neq 15; \alpha = .10; n = 22$
  f. $H_a: \mu < 15; \alpha = .05; n = 6$

8.56 The following sample of six measurements was randomly selected from a normally distributed population: $2, 5, -2, 7, 2, 4$.
  a. Test the null hypothesis that the mean of the population is 2 against the alternative hypothesis, $\mu < 2$. Use $\alpha = .05$.
  b. Test the null hypothesis that the mean of the population is 2 against the alternative hypothesis, $\mu \neq 2$. Use $\alpha = .05$.
  c. Find the observed significance level for each test.

8.57 A sample of five measurements, randomly selected from a [NW] normally distributed population, resulted in the following summary statistics: $\bar{x} = 4.8, s = 1.3$.
  a. Test the null hypothesis that the mean of the population is 6 against the alternative hypothesis, $\mu < 6$. Use $\alpha = .05$.
  b. Test the null hypothesis that the mean of the population is 6 against the alternative hypothesis, $\mu \neq 6$. Use $\alpha = .05$.
  c. Find the observed significance level for each test.

8.58 Suppose you conduct a *t*-test for the null hypothesis $H_0: \mu = 1,500$ versus the alternative hypothesis $H_a: \mu > 1,500$, based on a sample of 17 observations. The test results are $t = 1.89$, *p*-value = .038.
  a. What assumptions are necessary for the validity of this procedure?
  b. Interpret the results of the test.
  c. Suppose the alternative hypothesis had been the two-tailed $H_a: \mu \neq 1,500$. If the *t*-statistic were unchanged, what would the *p*-value be for this test? Interpret the *p*-value for the two-tailed test.

Applying the Concepts—Basic

8.59 **Lobster trap placement.** Refer to the *Bulletin of Marine Science* (Apr. 2010) observational study of lobster trap placement by teams fishing for the red spiny lobster in Baja California Sur, Mexico, Exercise 7.41 (p. 363). Trap spacing measurements (in meters) for a sample of seven teams of red spiny lobster fishermen are reproduced in the accompanying table. Let μ represent the average of the trap spacing measurements for the population of red spiny lobster fishermen fishing in Baja California Sur, Mexico. In Exercise 7.41 you computed the mean and standard deviation of the sample measurements to be $\bar{x} = 89.9$ meters and $s = 11.6$ meters, respectively. Suppose you want to determine if the true value of μ differs from 95 meters.

93	99	105	94	82	70	86

Based on Shester, G. G. "Explaining catch variation among Baja California lobster fishers through spatial analysis of trap-placement decisions." *Bulletin of Marine Science*, Vol. 86, No. 2, April 2010 (Table 1), pp. 479–498.

  a. Specify the null and alternative hypotheses for this test.
  b. Since $\bar{x} = 89.9$ is less than 95, a fisherman wants to reject the null hypothesis. What are the problems with using such a decision rule?
  c. Compute the value of the test statistic.
  d. Find the approximate *p*-value of the test.
  e. Select a value of α, the probability of a Type I error. Interpret this value in the words of the problem.
  f. Give the appropriate conclusion, based on the results of parts **d** and **e**.
  g. What conditions must be satisfied for the test results to be valid?
  h. In Exercise 7.41 you found a 95% confidence interval for μ. Does the interval support your conclusion in part **f**?

8.60 **Music performance anxiety.** Refer to the *British Journal of Music Education* (Mar. 2014) study of performance anxiety by music students, Exercise 7.37 (p. 362). Recall that the Performance Anxiety Inventory (PAI) was used to measure music performance anxiety on a scale from 20 to 80 points. The table below gives PAI values for participants in eight different studies. A MINITAB printout of the data analysis is shown.

54	42	51	39	41	43	55	40

Source: Patston, T. "Teaching stage fright?—Implications for music educators." *British Journal of Music Education*, Vol. 31, No. 1, Mar. 2014 (adapted from Figure 1).

One-Sample T: PAI

Test of mu = 40 vs > 40

Variable	N	Mean	StDev	SE Mean	95% Lower Bound	T	P
PAI	8	45.63	6.59	2.33	41.21	2.41	0.023

a. Set up the null and alternative hypotheses for determining whether the mean PAI value, μ, for all similar studies of music performance anxiety exceeds 40.

b. Find the rejection region for the test, part **a**, using $\alpha = .05$.

c. Compute the test statistic.

d. State the appropriate conclusion for the test.

e. What conditions are required for the test results to be valid?

f. Locate the *p*-value for the test on the MINITAB printout and use it to make a conclusion. (Your conclusion should agree with your answer in part **d**.)

g. How would your conclusion change if you used $\alpha = .01$?

8.61 **Dental anxiety study.** Refer to the *BMC Oral Health* (Vol. 9, 2009) study of adults who completed the Dental Anxiety Scale, presented in Exercise 5.37 (p. 279). Recall that scores range from 0 (no anxiety) to 25 (extreme anxiety). Summary statistics for the scores of 15 adults who completed the questionnaire are $\bar{x} = 10.7$ and $s = 3.6$. Conduct a test of hypothesis to determine whether the mean Dental Anxiety Scale score for the population of college students differs from $\mu = 11$. Use $\alpha = .05$.

8.62 **Crab spiders hiding on flowers.** Refer to the *Behavioral Ecology* (Jan. 2005) experiment on crab spiders' use of camouflage to hide from predators (e.g., birds) on flowers, presented in Exercise 2.42 (p. 79). Researchers at the French Museum of Natural History collected a sample of 10 adult female crab spiders, each sitting on the yellow central part of a daisy, and measured the chromatic contrast between each spider and the flower. The data (for which higher values indicate a greater contrast, and, presumably, an easier detection by predators) are shown in the accompanying table. The researchers discovered that a contrast of 70 or greater allows birds to see the spider. Of interest is whether the true mean chromatic contrast of crab spiders on daisies is less than 70.

57	75	116	37	96	61	56	2	43	32

Based on Thery, M., et al. "Specific color sensitivities of prey and predator explain camouflage in different visual systems." *Behavioral Ecology*, Vol. 16, No. 1, Jan. 2005 (Table 1).

a. Define the parameter of interest, μ.

b. Set up the null and alternative hypotheses of interest.

c. Find $\bar{x}$ and s for the sample data, and then use these values to compute the test statistic.

d. Give the rejection region for $\alpha = .10$.

e. State the appropriate conclusion in the words of the problem.

8.63 **Cheek teeth of extinct primates.** Refer to the *American Journal of Physical Anthropology* (Vol. 142, 2010) study of the characteristics of cheek teeth (e.g., molars) in an extinct primate species, Exercise 2.38 (p. 78). Recall that the researchers recorded the dentary depth of molars (in millimeters) for a sample of 18 cheek teeth extracted from skulls. These depth measurements are listed in the accompanying table. Anthropologists know that the mean dentary depth of molars in an extinct primate species—called Species A—is 15 millimeters. Is there evidence to indicate that the sample of 18 cheek teeth come from some other extinct primate species (i.e., some species other than Species A)? Use the SPSS printout at the bottom of the page to answer the question.

18.12	16.55
19.48	15.70
19.36	17.83
15.94	13.25
15.83	16.12
19.70	18.13
15.76	14.02
17.00	14.04
13.96	16.20

Based on Boyer, D. M., Evans, A. R., and Jernvall, J. "Evidence of dietary differentiation among Late Paleocene–Early Eocene Plesiadapids (Mammalia, Primates)." *American Journal of Physical Anthropology*, Vol. 142, © 2010.

8.64 **Radon exposure in Egyptian tombs.** Refer to the *Radiation Protection Dosimetry* (Dec. 2010) study of radon exposure in Egyptian tombs, Exercise 7.39 (p. 362). The radon levels—measured in becquerels per cubic meter (Bq/m^3)—in the inner chambers of a sample of 12 tombs are listed in the table. For the safety of the guards and visitors, the Egypt Tourism Authority (ETA) will temporarily close the tombs if the true mean level of radon exposure in the tombs rises to 6,000 Bq/m^3. Consequently, the ETA wants to conduct a test to determine if the true mean level of radon exposure in the tombs is less than 6,000 Bq/m^3, using a Type I error probability of .10. An SAS analysis of the data is shown on p. 427. Specify all the elements of the test: H_0, H_a, test statistic, *p*-value, α, and your conclusion.

50	910	180	580	7800	4000
390	12100	3400	1300	11900	1100

SPSS Output for Exercise 8.63

One-Sample Statistics

	N	Mean	Std. Deviation	Std. Error Mean
M2Depth	18	16.4994	1.97042	.46443

One-Sample Test

	Test Value = 15					
					95% Confidence Interval of the Difference	
	t	df	Sig. (2-tailed)	Mean Difference	Lower	Upper
M2Depth	3.229	17	.005	1.49944	.5196	2.4793

SAS Output for Exercise 8.64

The TTEST Procedure

Variable: RADON

N	Mean	Std Dev	Std Err	Minimum	Maximum
12	3642.5	4486.9	1295.3	50.0000	12100.0

Mean	95% CL Mean		Std Dev	95% CL Std Dev	
3642.5	-Infty	5968.6	4486.9	3178.5	7618.3

DF	t Value	Pr < t
11	-1.82	0.0480

Applying the Concepts—Intermediate

8.65 Yield strength of steel connecting bars. To protect against earthquake damage, steel beams are typically fitted and connected with plastic hinges. However, these plastic hinges are prone to deformations and are difficult to inspect and repair. An alternative method of connecting steel beams—one that uses high-strength steel bars with clamps—was investigated in *Engineering Structures* (July 2013). Mathematical models for predicting the performance of these steel connecting bars assume the bars have a mean yield strength of 300 megapascals (MPa). To verify this assumption, the researchers conducted material property tests on the steel connecting bars. In a sample of three tests, the yield strengths were 354, 370, and 359 MPa. Do the data indicate that the true mean yield strength of the steel bars exceeds 300 MPa? Test using $\alpha = .01$.

8.66 Pitch memory of amusiacs. Amusia is a congenital disorder that adversely impacts one's perception of music. Refer to the *Advances in Cognitive Psychology* (Vol. 6, 2010) study of the pitch memory of individuals diagnosed with amusia, Exercise 7.45 (p. 363). Recall that each in a sample of 17 amusiacs listened to a series of tone pairs and was asked to determine if the tones were the same or different. In the first trial, the tones were separated by 1 second; in a second trial, the tones were separated by 5 seconds. The difference in accuracy scores for the two trials was determined for each amusiac (where the difference is the score on the first trial minus the score on the second trial). The mean score difference was .11 with a standard deviation of .19.

a. In theory, the longer the delay between tones, the less likely one is to detect a difference between the tones. Consequently, the true mean score difference should exceed 0. Set up the null and alternative hypotheses for testing the theory.

b. Carry out the test, part **a**, using $\alpha = .05$. Is there evidence to support the theory?

8.67 Free recall memory strategy. Psychologists who study memory often use a measure of "free recall" (e.g., the number of correctly recalled items in a list of to-be-remembered items). The strategy used to memorize the list—for example, category clustering—is often just as important. Researchers at Central Michigan University developed an algorithm for computing measures of category clustering in *Advances in Cognitive Psychology* (Oct. 2012). One measure, called ratio of repetition, was recorded for a sample of 8 participants in a memory study. These ratios are listed in the table. Test the theory that the average ratio of repetition for all participants in a similar memory study differs from .5. Select an appropriate Type I error rate for your test.

.25	.43	.57	.38	.38	.60	.47	.30

Source: Senkova, O., & Otani, H. "Category clustering calculator for free recall." *Advances in Cognitive Psychology,* Vol. 8, No. 4, Oct. 2012 (Table 3).

8.68 Increasing hardness of polyester composites. Polyester resins reinforced with fiberglass are used to fabricate wall panels of restaurants. It is theorized that adding cement kiln dust (CKD) to the polyester composite will increase wall panel hardness. In a study published in *Advances in Applied Physics* (Vol. 2, 2014), hardness (joules per squared centimeters) was determined for three polyester composite mixtures that used a 40% CKD weight ratio. The hardness values were reported as 83, 84, and 79 j/cm^2. Research has shown that the mean hardness value of polyester composite mixtures that use a 20% CKD weight ratio is $\mu = 76$ j/cm^2. In your opinion, does using a 40% CKD weight ratio increase the mean hardness value of polyester composite mixtures? Support your answer statistically.

8.69 Minimizing tractor skidding distance. Refer to the *Journal of Forest Engineering* (July 1999) study of minimizing tractor skidding distances along a new road in a European forest, presented in Exercise 7.48 (p. 364). The skidding distances (in meters) were measured at 20 randomly selected road sites. The data are repeated in the accompanying table. Recall that a logger working on the road claims that the mean skidding distance is at least 425 meters. Is there sufficient evidence to refute this claim? Use $\alpha = .10$.

488	350	457	199	285	409	435	574	439	546
385	295	184	261	273	400	311	312	141	425

Based on Tujek, J., and Pacola, E. "Algorithms for skidding distance modeling on a raster Digital Terrain Model." *Journal of Forest Engineering,* Vol. 10, No. 1, July 1999 (Table 1).

8.70 Dissolved organic compound in lakes. The level of dissolved oxygen in the surface water of a lake is vital to maintaining the lake's ecosystem. Environmentalists from the University of Wisconsin monitored the dissolved oxygen levels over time for a sample of 25 lakes in the state (*Aquatic Biology*, May 2010). To ensure a representative sample, the environmentalists focused on several lake characteristics, including dissolved organic compound (DOC). The DOC data (measured in grams per cubic meters) for the 25 lakes are listed in the table on p. 428. The population of Wisconsin lakes has a mean DOC value of 15 grams/m^3. Use a hypothesis test (at $\alpha = .10$) to make an inference about whether the sample is representative of all Wisconsin lakes for the characteristic dissolved organic compound.

Data for Exercise 8.70

Lake	DOC
Allequash	9.6
Big Muskellunge	4.5
Brown	13.2
Crampton	4.1
Cranberry Bog	22.6
Crystal	2.7
EastLong	14.7
Helmet	3.5
Hiawatha	13.6
Hummingbird	19.8
Kickapoo	14.3
Little Arbor Vitae	56.9
Mary	25.1
Muskellunge	18.4
Northgate Bog	2.7
Paul	4.2
Peter	30.2
Plum	10.3
Reddington Bog	17.6
Sparkling	2.4
Tenderfoot	17.3
Trout Bog	38.8
Trout Lake	3.0
Ward	5.8
West Long	7.6

Based on Langman, O. C., et al. "Control of dissolved oxygen in northern temperate lakes over scales ranging from minutes to days." *Aquatic Biology*, Vol. 9, May 2010 (Table 1).

8.71 Walking straight into circles. When people get lost in unfamiliar terrain, do they really walk in circles, as is commonly believed? To answer this question, researchers conducted a field experiment and reported the results in *Current Biology* (Sept. 29, 2009). Fifteen volunteers were blindfolded and asked to walk as straight as possible in a certain direction in a large field. Walking trajectories were monitored every second for 50 minutes using GPS and the average directional bias (degrees per second) recorded for each walker. The data are shown in the table below. A strong tendency to veer consistently in the same direction will cause walking in circles. A mean directional bias of 0 indicates that walking trajectories were random. Consequently, the researchers tested whether the true mean bias differed significantly from 0. A MINITAB printout of the analysis is shown below.

a. Interpret the results of the hypothesis test for the researchers. Use $\alpha = .10$.

b. Although most volunteers showed little overall bias, the researchers produced maps of the walking paths showing that each occasionally made several small circles during the walk. Ultimately, the researchers supported the "walking in circles" theory. Explain why the data in the table are insufficient for testing whether an individual walks in circles.

−4.50	−1.00	−0.50	−0.15	0.00	0.01	0.02	0.05	0.15
0.20	0.50	0.50	1.00	2.00	3.00			

Based on Souman, J. L., Frissen, I., Sreenivasa, M. N., and Ernst, M. O. "Walking straight into circles." *Current Biology*, Vol. 19, No. 18, Sept. 29, 2009 (Figure 2).

Applying the Concepts—Advanced

8.72 Lengths of great white sharks. One of the most feared predators in the ocean is the great white shark. It is known that the white shark grows to a mean length of 21 feet; however, one marine biologist believes that great white sharks off the Bermuda coast grow much longer owing to unusual feeding habits. To test this claim, some full-grown great white sharks were captured off the Bermuda coast, measured, and then set free. However, because the capture of sharks is difficult, costly, and very dangerous, only three specimens were sampled. Their lengths were 24, 20, and 22 feet. Do these data support the marine biologist's claim at $\alpha = .10$?

MINITAB Output for Exercise 8.71

One-Sample T: BIAS

```
Test of mu = 0 vs not = 0


Variable   N   Mean   StDev   SE Mean      95% CI        T     P
BIAS       15  0.085  1.603   0.414    (-0.802, 0.973)  0.21  0.840
```

8.6 Large-Sample Test of Hypothesis about a Population Proportion

Inferences about population proportions (or percentages) are often made in the context of the probability p of "success" for a binomial distribution. We saw how to use large samples from binomial distributions to form confidence intervals for p in Section 7.4. We now consider tests of hypotheses about p.

Consider, for example, a method currently used by doctors to screen women for breast cancer. The method fails to detect cancer in 20% of the women who actually have the disease. Suppose a new method has been developed that researchers hope will detect cancer more accurately. This new method was used to screen a random sample of 140 women known to have breast cancer. Of these, the new method failed to detect cancer in 12 women. Does this sample provide evidence that the failure rate of the new method differs from the one currently in use?

We first view this experiment as a binomial one with 140 screened women as the trials and failure to detect breast cancer as "Success" (in binomial terminology). Let p represent the probability that the new method fails to detect the breast cancer; on the one hand, if the new method is no better than the current one, then the failure rate is $p = .2$. On the other hand, if the new method is either better or worse than the current method, then the failure rate is either smaller or larger than 20%; that is, $p \neq .2$.

We can now place the problem in the context of a test of hypothesis:

$$H_0: p = .2 \; (\text{Cancer detection rate} = .2.)$$
$$H_a: p \neq .2 \; (\text{Cancer detection rate differs from .2.})$$

Recall that the sample proportion $\hat{p}$ is really just the sample mean of the outcomes of the individual binomial trials and, as such, is approximately normally distributed (for large samples) according to the Central Limit Theorem. Thus, for large samples we can use the standard normal z as the test statistic:

$$\textit{Test statistic:} \quad z = \frac{\text{Sample proportion} - \text{Null hypothesized proportion}}{\text{Standard deviation of sample proportion}}$$

$$= \frac{\hat{p} - p_0}{\sigma_{\hat{p}}}$$

Note that we used the symbol p_0 to represent the null hypothesized value of p.

Rejection region: We use the standard normal distribution to find the appropriate rejection region for the specified value of α. Using $\alpha = .05$, the two-tailed rejection region is

$$z < -z_{\alpha/2} = -z_{.025} = -1.96 \quad \text{or} \quad z > z_{\alpha/2} = z_{.025} = 1.96$$

(See Figure 8.19.)

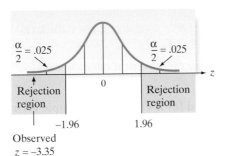

Figure 8.19

Rejection region for breast cancer example

We are now prepared to calculate the value of the test statistic. Before doing so, however, we want to be sure that the sample size is large enough to ensure that the normal approximation for the sampling distribution of $\hat{p}$ is reasonable. Recall from Section 7.4 that we require both np and nq to be at least 15. Because the null hypothesized value, p_0, is assumed to be the true value of p until our test procedure indicates otherwise, we require $np_0 \geq 15$ and $nq_0 \geq 15$ (where $q_0 = 1 - p_0$). Since $p_0 = .2$ for our test, we have

$$np_0 = (140)(.2) = 28$$

and

$$nq_0 = (140)(.8) = 112$$

Therefore, the normal distribution will provide a reasonable approximation for the sampling distribution of $\hat{p}$.

Returning to the hypothesis test at hand, the proportion of the screenings that failed to detect breast cancer is

$$\hat{p} = \frac{12}{140} = .086$$

Finally, we calculate the number of standard deviations (the z value) between the sampled and hypothesized values of the binomial proportion:

$$z = \frac{\hat{p} - p_0}{\sigma_{\hat{p}}} = \frac{\hat{p} - p_0}{\sqrt{p_0 q_0/n}} = \frac{.086 - .2}{\sqrt{(.2)(.8)/140}} = \frac{-.114}{.034} = -3.35$$

The implication is that the observed sample proportion is (approximately) 3.35 standard deviations below the null-hypothesized proportion, .2 (Figure 8.19). Therefore, we reject the null hypothesis, concluding at the .05 level of significance that the true failure rate of the new method for detecting breast cancer differs from .20. Since $\hat{p} = .086$, it appears that the new method is better (i.e., has a smaller failure rate) than the method currently in use. (To estimate the magnitude of the failure rate for the new method, a confidence interval can be constructed.)

The test of hypothesis about a population proportion p is summarized in the next box. Note that the procedure is entirely analogous to that used for conducting large-sample tests about a population mean.

Large-Sample Test of Hypothesis about p: Normal (z) Statistic

Test statistic: $\quad z_c = \dfrac{(\hat{p} - p_0)}{\sigma_{\hat{p}}} = \dfrac{(\hat{p} - p_0)}{\sqrt{(p_0 q_0/n)}}$

	One-Tailed Tests		Two-Tailed Test
	$H_0: p = p_0$	$H_0: p = p_0$	$H_0: p = p_0$
	$H_a: p < p_0$	$H_a: p > p_0$	$H_a: p \neq p_0$
Rejection region:	$z_c < -z_\alpha$	$z_c > z_\alpha$	$|z_c| > z_{\alpha/2}$
p-value:	$P(z < z_c)$	$P(z > z_c)$	$2P(z > z_c)$ if z_c is positive
			$2P(t < z_c)$ if z_c is negative

Decision: Reject H_0 if $\alpha > p$-value or if test statistic (z_c) falls in rejection region where $P(z > z_\alpha) = \alpha, P(z > z_{\alpha/2}) = \alpha/2$, and $\alpha = P(\text{Type I error}) = P(\text{Reject } H_0 | H_0 \text{ true})$.

[*Note:* The symbol for the numerical value assigned to p under the null hypothesis is p_0.]

Conditions Required for a Valid Large-Sample Hypothesis Test for p

1. A random sample is selected from a binomial population.
2. The sample size n is large. (This condition will be satisfied if both $np_0 \geq 15$ and $nq_0 \geq 15$.)

Example 8.10

A Hypothesis Test for p—Proportion of Defective Batteries

Problem The reputations (and hence sales) of many businesses can be severely damaged by shipments of manufactured items that contain a large percentage of defectives. For example, a manufacturer of alkaline batteries may want to be reasonably certain that less than 5% of its batteries are defective. Suppose 300 batteries are randomly selected from a very large shipment; each is tested and 10 defective batteries are found. Does this outcome provide sufficient evidence for the manufacturer to conclude that the fraction defective in the entire shipment is less than .05? Use $\alpha = .01$.

Solution The objective of the sampling is to determine whether there is sufficient evidence to indicate that the fraction defective, p, is less than .05. Consequently, we will test the null hypothesis that $p = .05$ against the alternative hypothesis that $p < .05$. The elements of the test are

$\qquad H_0: p = .05$ (Fraction of defective batteries equals .05.)

$\qquad H_a: p < .05$ (Fraction of defective batteries is less than .05.)

$\qquad$ *Test statistic:* $z = \dfrac{\hat{p} - p_0}{\sigma_{\hat{p}}}$

$\qquad$ *Rejection region:* $z < -z_{.01} = -2.33 \qquad$ (see Figure 8.20)

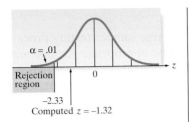

Figure 8.20

Rejection region for
Example 8.10

Before conducting the test, we check to determine whether the sample size is large enough to use the normal approximation to the sampling distribution of $\hat{p}$. Since $np_0 = (300)(.05) = 15$ and $nq_0 = (300)(.95) = 285$ are both at least 15, the normal approximation will be adequate.

We now calculate the test statistic:

$$z = \frac{\hat{p} - .05}{\sigma_{\hat{p}}} = \frac{(10/300) - .05}{\sqrt{p_0 q_0 / n}} = \frac{.03333 - .05}{\sqrt{p_0 q_0 / 300}}$$

Notice that we use p_0 to calculate $\sigma_{\hat{p}}$ because, in contrast to calculating $\sigma_{\hat{p}}$ for a confidence interval, the test statistic is computed on the assumption that the null hypothesis is true—that is, $p = p_0$. Therefore, substituting the values for $\hat{p}$ and p_0 into the z statistic, we obtain

$$z \approx \frac{-.01667}{\sqrt{(.05)(.95)/300}} = \frac{-.01667}{.0126} = -1.32$$

As shown in Figure 8.20, the calculated z-value does not fall into the rejection region. Therefore, there is insufficient evidence at the .01 level of significance to indicate that the shipment contains less than 5% defective batteries.

■ **Now Work Exercise 8.76a–b**

Example 8.11

Finding the *p*-Value for a Test about a Population Proportion *p*

Figure 8.21

The observed significance level for Example 8.11

Problem In Example 8.10, we found that we did not have sufficient evidence at the $\alpha = .01$ level of significance to indicate that the fraction defective, p, of alkaline batteries was less than $p = .05$. How strong was the weight of evidence favoring the alternative hypothesis ($H_a: p < .05$)? Find the observed significance level (*p*-value) for the test.

Solution The computed value of the test statistic z was $z = -1.32$. Therefore, for this lower-tailed test, the observed significance level is

$$p\text{-value} = P(z \leq -1.32)$$

This lower-tail area is shown in Figure 8.21. The area between $z = 0$ and $z = 1.32$ is given in Table II in Appendix B as .4066. Therefore, the observed significance level is $.5 - .4066 = .0934$.

Note: The *p*-value can also be obtained with statistical software. The MINITAB printout shown in Figure 8.22 gives the *p*-value (highlighted).

Test and CI for One Proportion

Test of p = 0.05 vs p < 0.05

Sample	X	N	Sample p	95% Upper Bound	Z-Value	P-Value
1	10	300	0.033333	0.050380	-1.32	0.093

Using the normal approximation.

Figure 8.22

MINITAB lower-tailed test of p, Example 8.11

Look Back Although we did not reject $H_0: p = .05$ at $\alpha = .01$, the probability of observing a z-value as small as or smaller than -1.32 is only .093 if in fact H_0 is true. Therefore, we would reject H_0 if we had selected $\alpha = .10$ prior to sampling (since the observed significance level is less than .10). This point reminds us of the importance of our choice of α prior to collecting the sample data.

■ **Now Work Exercise 8.76c**

Small samples

Since most surveys and studies employ large samples, the large-sample testing procedure based on the normal (z) statistic presented here will be appropriate for making inferences about a population proportion. *However, in the case of small samples (where either np_0 or nq_0 is less than 15), tests for a population proportion based on the z-statistic may not be valid—especially when conducting one-tailed tests.* A test of proportions that can be applied to small samples utilizes the binomial, rather than the normal, distribution. These are called *exact binomial tests* due to the fact that the exact (rather than approximate) *p*-value for the test is computed based on the binomial distribution.

For example, consider testing H_0: $p = .6$ against H_a: $p < .6$, where p is the true proportion of all voters in a large city that favor a certain mayoral candidate. Suppose that in a poll of $n = 20$ voters (randomly selected from all registered voters in the city), $x = 10$ favor the mayoral candidate. Is this sufficient evidence at $\alpha = .05$ to reject the null hypothesis?

First, note that $p_0 = .6$ and $np_0 = 20(.6) = .12$ is less than 15; consequently, we require a small-sample test of binomial proportions. Now, we observed $x = 10$ voters in favor of the mayoral candidate. Then, by definition (see p. 409), the observed significance level (*p*-value) of this lower-tailed test will be equal to the probability of observing 10 or fewer voters in favor of the candidate, given that the true population proportion in favor of the candidate is $p = .6$. That is,

$$p\text{-value} = P(x \leq 10 \mid p = .6)$$

In Example 4.13 (p. 236), we found this binomial probability using MINITAB. The MINITAB printout is reproduced in Figure 8.23. You can see that *p*-value = .245. Since $\alpha = .05 < p$-value, we fail to reject the null hypothesis.

Figure 8.23

MINITAB Output with *p*-Value for Exact Binomial Test

Cumulative Distribution Function

Binomial with n = 20 and p = 0.6

x	P(X <= x)
10	0.244663

All three statistical software packages presented in this text—MINITAB, SAS, and SPSS—have procedures for conducting exact binomial tests in the case of small samples.

Statistics in Action | **Revisited** | **Testing a Population Proportion in the KLEENEX® Survey**

In the previous "Statistics in Action Revisited" (p. 417), we investigated Kimberly-Clark Corporation's assertion that the company should put more than 60 tissues in an anti-viral box of KLEENEX® tissues. We did this by testing the claim that the mean number of tissues used by a person with a cold is $\mu = 60$, using data collected from a survey of 250 KLEENEX® users. Another approach to the problem is to consider the proportion of KLEENEX® users who use more than 60 tissues when they have a cold. Now the population parameter of interest is p, the proportion of all KLEENEX® users who use more than 60 tissues when they have a cold.

Kimberly-Clark Corporation's belief that the company should put more than 60 tissues in an anti-viral box will be supported if over half of the KLEENEX® users surveyed use more than 60 tissues (i.e., if $p > .5$). Is there evidence to indicate that the population proportion exceeds .5? To answer this question, we set up the following null and alternative hypotheses:

$$H_0: p = .5 \quad H_a: p > .5$$

Recall that the survey results for the 250 sampled KLEENEX® users are stored in the **TISSUES** data file. In addition to the number of tissues used by each person, the file contains a qualitative variable—called USED60—representing whether the person used fewer or more than 60 tissues. (The values of USED60 in the data set are "LESS" and "MORE.") A MINITAB analysis of this variable yielded the printout displayed in Figure SIA8.2 on p. 433.

Statistics IN Action
(continued)

On the MINITAB printout, x represents the number of the 250 people with colds that used more than 60 tissues. Note that $x = 154$. This value is used to compute the test statistic $z = 3.67$, highlighted on the printout. The p-value of the test, also highlighted on the printout, is p-value $= .000$. Since this value is less than $\alpha = .05$, there is sufficient evidence (at $\alpha = .05$) to reject H_0; we conclude that the proportion of all KLEENEX® users who use more than 60 tissues when they have a cold exceeds .5. This conclusion again supports, the company's decision to put more than 60 tissues in an anti-viral box of KLEENEX.

D *Data Set:* TISSUES

Test and CI for One Proportion: USED60

```
Test of p = 0.5 vs p > 0.5

Event = MORE

                                    95% Lower
Variable    X    N   Sample p        Bound   Z-Value  P-Value
USED60     154  250  0.616000     0.565404      3.67    0.000

Using the normal approximation.
```

Figure SIA8.2
MINITAB test of $p = .5$ for KLEENEX survey

Exercises 8.73–8.93

Understanding the Principles

8.73 Which statistical distribution is typically associated with making inferences about a population proportion p?

8.74 What are the assumptions for a valid large-sample test for p?

Learning the Mechanics

8.75 For the binomial sample sizes and null-hypothesized values of p in each part, determine whether the sample size is large enough to use the normal approximation methodology presented in this section to conduct a test of the null hypothesis $H_0: p = p_0$.
  a. $n = 600, p_0 = .06$
  b. $n = 200, p_0 = .99$
  c. $n = 40, p_0 = .3$
  d. $n = 30, p_0 = .2$
  e. $n = 20, p_0 = .4$

8.76 Suppose a random sample of 100 observations from a binomial population gives a value of $\hat{p} = .69$ and you wish to test the null hypothesis that the population parameter p is equal to .75 against the alternative hypothesis that p is less than .75.
  a. Noting that $\hat{p} = .69$, what does your intuition tell you? Does the value of $\hat{p}$ appear to contradict the null hypothesis?
  b. Use the large-sample z-test to test $H_0: p = .75$ against the alternative hypothesis $H_a: p < .75$. Use $\alpha = .05$. How do the test results compare with your intuitive decision from part **a**?
  c. Find and interpret the observed significance level of the test you conducted in part **b**.

8.77 Suppose the sample in Exercise 8.76 has produced $\hat{p} = .74$ and we wish to test $H_0: p = .8$ against the alternative $H_a: p < .8$.

a. Calculate the value of the z statistic for this test.
b. Note that the numerator of the z statistic $(\hat{p} - p_0 = .74 - .80 = -.06)$ is the same as for Exercise 8.76. Considering this, why is the absolute value of z for this exercise larger than that calculated in Exercise 8.76?
c. Complete the test, using $\alpha = .05$, and interpret the result.
d. Find the observed significance level for the test and interpret its value.

8.78 A random sample of 100 observations is selected from a binomial population with unknown probability of success p. The computed value of $\hat{p}$ is equal to .70.
  a. Test $H_0: p = .61$ against $H_a: p > .61$. Use $\alpha = .01$.
  b. Test $H_0: p = .61$ against $H_a: p > .61$. Use $\alpha = .10$.
  c. Test $H_0: p = .86$ against $H_a: p \neq .86$. Use $\alpha = .05$.
  d. Form a 95% confidence interval for p.
  e. Form a 99% confidence interval for p.

8.79 Refer to Exercise 7.58 (p. 370), in which 50 consumers taste-tested a new snack food.
  D
  SNACK a. Test $H_0: p = .5$ against $H_a: p > .5$, where p is the proportion of customers who do not like the snack food. Use $\alpha = .10$.
  b. Report the observed significance level of your test.

D Applet Exercise 8.5

Use the applet entitled *Hypotheses Test for a Proportion* to investigate the relationships between the probabilities of Type I and Type II errors occurring at levels .05 and .01. For this exercise, use $n = 100$, true $p = 0.5$, and alternative *not equal*.
  a. Set null $p = .5$. What happens to the proportion of times the null hypothesis is rejected at the .05 level and at the .01 level as the applet is run more and more times? What type of error has occurred when the null hypothesis is rejected in this situation? Based on your results, is this type of error more likely to occur at level .05 or at level .01? Explain.

b. Set null $p = .6$. What happens to the proportion of times the null hypothesis is *not* rejected at the .05 level and at the .01 level as the applet is run more and more times? What type of error has occurred when the null hypothesis is *not* rejected in this situation? Based on your results, is this type of error more likely to occur at level .05 or at level .01? Explain.

c. Use your results from parts **a** and **b** to make a general statement about the probabilities of Type I and Type II errors at levels .05 and .01.

ⓓ Applet Exercise 8.6

Use the applet entitled *Hypotheses Test for a Proportion* to investigate the effect of the true population proportion p on the probability of a Type I error occurring. For this exercise, use $n = 100$ and alternative *not equal*.

a. Set true $p = .5$ and null $p = .5$. Run the applet several times, and record the proportion of times the null hypothesis is rejected at the .01 level.

b. Clear the applet and repeat part **a** for true $p = .1$ and null $p = .1$. Then repeat one more time for true $p = .01$ and null $p = .01$.

c. Based on your results from parts **a** and **b**, what can you conclude about the probability of a Type I error occurring as the true population proportion gets closer to 0?

Applying the Concepts—Basic

8.80 **Paying for music downloads.** If you use the Internet, have you ever paid to access or download music? This was one of the questions of interest in a recent *Pew Internet and American Life Project Survey* (Oct. 2010). In a representative sample of 755 adults who use the Internet, 506 stated that they have paid to download music. Let p represent the true proportion of all Internet-using adults who have paid to download music.

a. Compute a point estimate of p.

b. Set up the null and alternative hypotheses for testing whether the true proportion of all Internet-using adults who have paid to download music exceeds .7.

c. Compute the test statistic for part **b**.

d. Find the rejection region for the test if $\alpha = .01$.

e. Find the p-value for the test.

f. Make the appropriate conclusion using the rejection region.

g. Make the appropriate conclusion using the p-value.

8.81 **Underwater sound-locating ability of alligators.** Alligators have shown the ability to determine the direction of an airborne sound. But can they locate underwater sounds? This was the subject of research published in the *Journal of Herpetology* (Dec. 2014). Alligators inhabiting the flood control canals in the Florida Everglades were monitored for movement toward a sound produced from a submerged diving bell. Movements within a 180° arc of the direction toward the sound were scored as movements toward the sound; all movements in other directions were scored as movements away from the sound. Consequently, the researchers assumed that the proportion of movements toward the sound expected by chance is $180°/360° = .5$. In a sample of $n = 50$ alligators, 42 moved toward the underwater sound.

a. Give the null and alternative hypotheses for testing whether the true proportion of alligators that move toward the underwater sound is higher than expected by chance.

b. In a sample of $n = 50$ alligators, assume that 42 moved toward the underwater sound. Use this information to compute an estimate of the true proportion of alligators that move toward the underwater sound.

c. Compute the test statistic for this study.

d. Compute the observed significance level (p-value) of the test.

e. Make the appropriate conclusion in the words of the problem.

8.82 **Gummy bears: red or yellow?** *Chance* (Winter 2010) presented a lesson in hypothesis testing carried out by medical students in a biostatistics class. Students were blind-folded and then given a red-colored or yellow-colored gummy bear to chew. (Half the students were randomly assigned to receive the red gummy bear and half to receive the yellow bear. The students could not see what color gummy bear they were given.) After chewing, the students were asked to guess the color of the candy based on the flavor. Of the 121 students who participated in the study, 97 correctly identified the color of the gummy bear.

a. If there is no relationship between color and gummy bear flavor, what proportion of the population of students will correctly identify the color?

b. Specify the null and alternative hypotheses for testing whether color and flavor are related.

c. Carry out the test and give the appropriate conclusion at $\alpha = .01$. Use the p-value of the test to make your decision.

8.83 **Dehorning of dairy calves.** For safety reasons, calf dehorning has become a routine practice at dairy farms. A report by Europe's Standing Committee on the Food Chain and Animal Health (SANKO) stated that 80% of European dairy farms carry out calf dehorning. A later study, published in the *Journal of Dairy Science* (Vol. 94, 2011), found that in a sample of 639 Italian dairy farms, 515 dehorn calves. Does the *Journal of Dairy Science* study support or refute the figure reported by SANKO? Explain.

8.84 **Teenagers' use of emoticons in school writing.** Refer to the *Pew Internet and American Life Project* (Apr. 2008) survey of the writing habits of U.S. teenagers, Exercise 7.69 (p. 371). Recall that in a random sample of 700 teenagers, 448 admitted to using at least one informal element in school writing assignments. [*Note*: Emoticons, such as using the symbol ":)" to represent a smile, and abbreviations, such as writing "LOL" for "laughing out loud," are considered informal elements.] Is there evidence to indicate that less than 65% of all U.S. teenagers have used at least one informal element in school writing assignments? Test the relevant hypotheses using $\alpha = .05$.

8.85 **TV subscription streaming.** "Streaming" of television programs is trending upward. According to *The Harris Poll* (Aug. 26, 2013), over one-third of Americans qualify as "subscription streamers," i.e., those who watch streamed TV programs through a subscription service such as Netflix, Hulu Plus, or Amazon Prime. The poll included 2,242 adult TV viewers, of which 785 are subscription streamers. Based on this result, can you conclude that the true fraction of adult TV viewers who are subscription streamers differs from one-third? Carry out the test using a Type I error rate of .10.

Applying the Concepts—Intermediate

8.86 Underwater acoustic communication. Refer to the *IEEE Journal of Oceanic Engineering* (Apr. 2013) study of the characteristics of subcarriers—telecommunication signals carried on top of another—for underwater acoustic communications, Exercise 4.80 (p. 241). Recall that a subcarrier can be classified as either a data subcarrier (used for data transmissions), a pilot subcarrier (used for channel estimation and synchronization), or a null subcarrier (used for direct current and guard banks transmitting no signal). In a sample of 1,024 subcarrier signals transmitted off the coast of Martha's Vineyard, 672 were determined to be data subcarriers, 256 pilot subcarriers, and 96 null subcarriers. Suppose a communications engineer who works near Martha's Vineyard believes that fewer than 70% of all subcarrier signals transmitted in the area are data subcarriers. Is there evidence to support this belief? Test using $\alpha = .05$.

8.87 Identifying organisms using a computer. *National Science Education Standards* recommend that all life science students be exposed to methods of identifying unknown biological specimens. Due to certain limitations of traditional identification methods, biology professors at Slippery Rock University (SRU) developed a computer-aided system for identifying common conifers (deciduous trees) called Confir ID (*The American Biology Teacher*, May 2010). A sample of 171 life science students were exposed to both a traditional method of identifying conifers and Confir ID and then asked which method they preferred. As a result, 138 students indicated their preference for Confir ID. In order to change the life sciences curriculum at SRU to include Confir ID, the biology department requires that more than 70% of the students prefer the new, computerized method. Should Confir ID be added to the curriculum at SRU? Explain your reasoning.

8.88 Organic certified coffee. Coffee markets that conform to organic standards focus on the environmental aspect of coffee growing, such as the use of shade trees and a reduced reliance on chemical pesticides. A study of organic coffee growers was published in *Food Policy* (Vol. 36, 2010). In a representative sample of 845 coffee growers from southern Mexico, 417 growers were certified to sell to organic coffee markets while 77 growers were transitioning to become organic certified. In the United States, 60% of the coffee growers are organic certified. Is there evidence to indicate that fewer than 60% of the coffee growers in southern Mexico are either organic certified or transitioning to become certified? State your conclusion so that there is only a 5% chance of making a Type I error.

8.89 Detection of motorcycles while driving. The factors that affect the visibility of powered two-wheelers (PTWs)—motorcycles—on the road were investigated in *Accident Analysis and Prevention* (Vol. 44, 2012). A visual search study was conducted in which viewers were presented with pictures of driving scenarios and asked to identify the presence or absence of a PTW. Of interest to the researchers is the detection rate, i.e., the proportion of pictures showing a PTW in which the viewer actually detected the presence of the PTW. Suppose that, in theory, the true detection rate for pictures of PTWs is .70. The study revealed that in a sample of 2,376 pictures that included a PTW, only 1,554 were detected by the viewers. Use this result to test the theory at $\alpha = .10$.

8.90 Defensibility of a landscape. Settlers often chose sites to live and build communities in based on the landscape's defensibility. The defensibility of archaeological sites in the Northwest Pacific was investigated in the *Journal of Anthropological Archaeology* (Vol. 35, 2014). A defensibility index was calculated for each in a sample of 23 sites that included a trench embankment. These index values (on a scale of 1 to 100) are listed in the accompanying table. The researcher determined that the mean defensibility index of the entire Northwest Pacific landscape is 47. Of interest is whether the defensibility index of a sampled site exceeds this mean.

70	70	68	67	67	66	65	63	63	59	57	57
57	55	55	54	52	52	51	46	46	44	37	

Source: Bocinsky, R. K. "Extrinsic site defensibility and landscape-based archaeological inference: An example from the Northwest Coast." *Journal of Anthropological Archaeology*, Vol. 35, 2014 (Table 3).

a. Estimate p, the true proportion of all trench embankment sites that have a defensibility index that exceeds the mean of the entire Northwest Pacific landscape.

SAS Output for Exercise 8.90

The FREQ Procedure

x	Frequency	Percent	Cumulative Frequency	Cumulative Percent
ABOVE	19	82.61	19	82.61
BELOW	4	17.39	23	100.00

Binomial Proportion	
x = ABOVE	
Proportion (P)	0.8261
ASE	0.0790
95% Lower Conf Limit	0.6712
95% Upper Conf Limit	0.9810
Exact Conf Limits	
95% Lower Conf Limit	0.6122
95% Upper Conf Limit	0.9505

Test of H0: Proportion = 0.6	
ASE under H0	0.1022
Z	2.2133
One-sided Pr > Z	0.0134
Two-sided Pr > |Z|	0.0269
Exact Test	
One-sided Pr >= P	0.0190
Two-sided = 2 * One-sided	0.0379

Sample Size = 23

b. An archaeologist wants to conduct a statistical test of hypothesis to determine if p is larger than .60. Give the null and alternative hypotheses for this test.

c. Demonstrate that the sample is not large enough to apply the large-sample test of a binomial proportion.

d. The SAS printout shown on p. 435 gives the results of an exact binomial proportion test of the hypotheses, part **b**. Locate the p-value of the test (bottom of the printout) and interpret the results.

8.91 **Astronomy students and the Big Bang Theory.** Indiana University professors investigated first-year college students' knowledge of astronomy (*Astronomy Education Review*, Vol. 2, 2003). One concept of interest was the Big Bang theory of the creation of the universe. In a sample of 148 freshman students, 37 believed that the Big Bang theory accurately described the creation of planetary systems. Based on this information, would you be willing to state that more than 20% of all freshman college students believe in the Big Bang theory? How confident are you of your decision?

Applying the Concepts—Advanced

8.92 **Killing insects with low oxygen.** A group of Australian entomological toxicologists investigated the impact of exposure to low oxygen on the mortality of insects (*Journal of Agricultural, Biological, and Environmental Statistics*, Sept. 2000). Thousands of adult rice weevils were placed in a chamber filled with wheat grain, and the chamber was exposed to nitrogen gas for 4 days. Insects were assessed as dead or alive 24 hours after exposure. At the conclusion of the experiment, 31,386 weevils were dead and 35 weevils were found alive. Previous studies had shown a 99% mortality rate in adult rice weevils exposed to carbon dioxide for 4 days. What advice would you give to the entomologists regarding the use of nitrogen or carbon dioxide for killing rice weevils?

8.93 **Male fetal deaths following 9/11/2001.** According to research published in *BMC Public Health* (May 25, 2010), the terrorist attacks of September 11, 2001, may have led to a spike in miscarriages of male babies. Researchers from the University of California, using data from the National Vital Statistics System, reported that the male fetal death rate in September of that year was significantly higher than expected. Suppose it is known that the miscarriage rate for pregnant women expecting a boy in the United States is 3 in 1,000. In a random sample of 2,000 pregnant women expecting a baby boy during September of 2001, how many would need to have experienced a fetal death (miscarriage) in order to support the claim (at $\alpha = .05$) that the male fetal death rate in September 2001 was higher than expected?

8.7 Calculating Type II Error Probabilities: More about β (Optional)

In our introduction to hypothesis testing in Section 8.1, we showed that the probability of committing a Type I error, α, can be controlled by the selection of the rejection region for the test. Thus, when the test statistic falls into the rejection region and we make the decision to reject the null hypothesis, we do so knowing the error rate for incorrect rejections of H_0. The situation corresponding to accepting the null hypothesis, and thereby risking a Type II error, is not generally as controllable. For that reason, we adopted a policy of nonrejection of H_0 when the test statistic does not fall into the rejection region, rather than risking an error of unknown magnitude.

To see how β, the probability of a Type II error, can be calculated for a test of hypothesis, recall the example in Section 8.1 in which a city tests a manufacturer's pipe to see whether it meets the requirement that the mean strength exceed 2,400 pounds per linear foot. The setup for the test is as follows:

$H_0: \mu = 2,400$

$H_a: \mu > 2,400$

Test statistic: $z = \dfrac{\bar{x} - 2,400}{\sigma/\sqrt{n}}$

Rejection region: $z > 1.645$ for $\alpha = .05$

Figure 8.24a shows the rejection region for the **null distribution**—that is, the distribution of the test statistic, assuming that the null hypothesis is true. The area in the rejection region is .05, and this area represents α, the probability that the test statistic leads to rejection of H_0 when in fact H_0 is true.

The Type II error probability β is calculated under the assumption that the null hypothesis is false because it is defined as the *probability of accepting H_0 when it is false*. Since H_0 is false for any value of μ exceeding 2,400, one value of β exists for each possible value of μ greater than 2,400 (an infinite number of possibilities). Figures 8.24b, 8.24c, and 8.24d show three of the possibilities, corresponding to alternative hypothesis values of μ equal to 2,425, 2,450, and 2,475, respectively. Note that β is the area in the

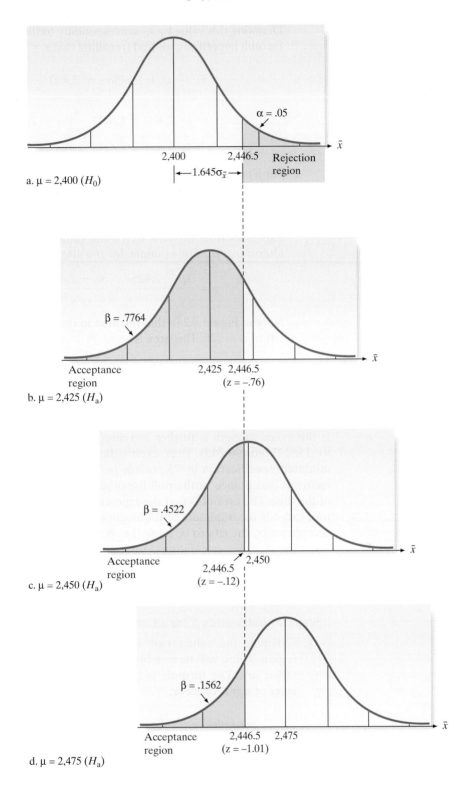

Figure 8.24

Values of α and β for various values of μ

nonrejection (or *acceptance*) *region* in each of these distributions and that β decreases as the true value of μ moves farther from the null-hypothesized value of $\mu = 2,400$. This behavior is sensible because the probability of incorrectly accepting the null hypothesis should decrease as the distance between the null and alternative values of μ increases.

In order to calculate the value of β for a specific value of μ in H_a, we proceed as follows:

1. We calculate the value of $\bar{x}$ that corresponds to the border between the acceptance and rejection regions. For the sewer pipe example, this is the value of $\bar{x}$ that lies 1.645 standard deviations above $\mu = 2,400$ in the sampling distribution of $\bar{x}$.

Denoting this value by $\bar{x}_0$, corresponding to the largest value of $\bar{x}$ that supports the null hypothesis, we find (recalling that $s = 200$ and $n = 50$) that

$$\bar{x}_0 = \mu_0 + 1.645\sigma_{\bar{x}} = 2{,}400 + 1.645\left(\frac{\sigma}{\sqrt{n}}\right)$$

$$\approx 2{,}400 + 1.645\left(\frac{s}{\sqrt{n}}\right) = 2{,}400 + 1.645\left(\frac{200}{\sqrt{50}}\right)$$

$$= 2{,}400 + 1.645(28.28) = 2{,}446.5$$

2. For a particular alternative distribution corresponding to a value of μ denoted by μ_a, we calculate the z-value corresponding to $\bar{x}_0$, the border between the rejection and acceptance regions. We then use this z-value and Table II of Appendix B to determine the area in the *acceptance region* under the alternative distribution. This area is the value of β corresponding to the particular alternative μ_a. For example, for the alternative $\mu_a = 2{,}425$, we calculate

$$z = \frac{\bar{x}_0 - 2{,}425}{\sigma_{\bar{x}}} \approx \frac{\bar{x}_0 - 2{,}425}{s/\sqrt{n}} = \frac{2{,}446.5 - 2{,}425}{28.28} = .76$$

Note in Figure 8.24b that the area in the acceptance region is the area to the left of $z = .76$. This area is

$$\beta = .5 + .2764 = .7764$$

Thus, the probability that the test procedure will lead to an incorrect acceptance of the null hypothesis $\mu = 2{,}400$ when in fact $\mu = 2{,}425$ is about .78. As the average strength of the pipe increases to 2,450, the value of β decreases to .4522 (Figure 8.24c). If the mean strength is further increased to 2,475, the value of β is further decreased to .1562 (Figure 8.24d). Thus, even if the true mean strength of the pipe exceeds the minimum specification by 75 pounds per linear foot, the test procedure will lead to an incorrect acceptance of the null hypothesis (rejection of the pipe) approximately 16% of the time. The upshot is that the pipe must be manufactured so that the mean strength well exceeds the minimum requirement if the manufacturer wants the probability of its acceptance by the city to be large (i.e., β to be small).

The steps for calculating β for a large-sample test about a population mean and a population proportion are summarized in the following boxes:

Steps for Calculating β for a Large-Sample Test about μ

1. Calculate the value(s) of $\bar{x}$ corresponding to the border(s) of the rejection region. There will be one border value for a one-tailed test and two for a two-tailed test. The formula is one of the following, corresponding to a test with level of significance α:

Upper-tailed test: $\bar{x}_0 = \mu_0 + z_\alpha\,\sigma_{\bar{x}} \approx \mu_0 + z_\alpha\left(\dfrac{s}{\sqrt{n}}\right)$

($H_a\!: \mu > \mu_0$)

Lower-tailed test: $\bar{x}_0 = \mu_0 - z_\alpha\,\sigma_{\bar{x}} \approx \mu_0 - z_\alpha\left(\dfrac{s}{\sqrt{n}}\right)$

($H_a\!: \mu < \mu_0$)

Two-tailed test: $\bar{x}_{0,\,L} = \mu_0 - z_{\alpha/2}\,\sigma_{\bar{x}} \approx \mu_0 - z_{\alpha/2}\left(\dfrac{s}{\sqrt{n}}\right)$

($H_a\!: \mu \neq \mu_0$)

$\bar{x}_{0,\,U} = \mu_0 + z_{\alpha/2}\,\sigma_{\bar{x}} \approx \mu_0 + z_{\alpha/2}\left(\dfrac{s}{\sqrt{n}}\right)$

2. Specify the value of μ_a in the alternative hypothesis for which the value of β is to be calculated. Then convert the border value(s) of $\bar{x}_0$ to z-value(s), using the alternative distribution with mean μ_a. The general formula for the z value is

$$z = \frac{\bar{x}_0 - \mu_a}{\sigma_{\bar{x}}}$$

3. Sketch the alternative distribution (centered at μ_a), and shade the area in the acceptance (nonrejection) region. Use the z-statistic(s) and Table II of Appendix A to find the shaded area, which is β.

Lower-tailed test: $\quad \beta = P\left(z > \dfrac{\bar{x}_0 - \mu_a}{\sigma_{\bar{x}}}\right)$

$(H_a: \mu < \mu_0)$

Upper-tailed test: $\quad \beta = P\left(z < \dfrac{\bar{x}_0 - \mu_a}{\sigma_{\bar{x}}}\right)$

$(H_a: \mu > \mu_0)$

Two-tailed test: $\quad \beta = P\left(\dfrac{\bar{x}_{0,L} - \mu_a}{\sigma_{\bar{x}}} < z < \dfrac{\bar{x}_{0,U} - \mu_a}{\sigma_{\bar{x}}}\right)$

$(H_a: \mu \neq \mu_0)$

Steps for Calculating β for a Large-Sample Test about p

1. Calculate the value(s) of $\hat{p}$ corresponding to the border(s) of the rejection region. There will be one border for a one-tailed test and two for a two-tailed test. The formula is one of the following, corresponding to a test with level of significance α.

Upper-tailed test: $\quad \hat{p}_0 = p_0 + z_\alpha \sigma_p = p_0 + z_\alpha \sqrt{\dfrac{p_0 q_0}{n}}$

$(H_a: p > p_0)$

Lower-tailed test: $\quad \hat{p}_0 = p_0 - z_\alpha \sigma_p = p_0 - z_\alpha \sqrt{\dfrac{p_0 q_0}{n}}$

$(H_a: p < p_0)$

Two-tailed test: $\quad \hat{p}_{0, L} = p_0 - z_{\alpha/2} \sigma_p = p_0 - z_{\alpha/2} \sqrt{\dfrac{p_0 q_0}{n}}$

$(H_a: p \neq p_0)$

$$\hat{p}_{0, U} = p_0 + z_{\alpha/2} \sigma_p = p_0 + z_{\alpha/2} \sqrt{\dfrac{p_0 q_0}{n}}$$

2. Specify the value of p_a in the alternative hypothesis for which the value of β is to be calculated. Then convert the border values of $\hat{p}_0$ to z-value(s), using the alternative distribution with mean p_a. The general formula for the z value is:

$$z = (\hat{p}_0 - p_a)/\sigma_p = \frac{\hat{p}_0 - p_a}{\sqrt{\dfrac{p_a q_a}{n}}}$$

3. Sketch the alternative distribution (centered at p_a), and shade the area in the acceptance (nonrejection) region. Use the z-statistic(s) and Table II of Appendix B to find the shaded area, which is β.

(continued)

$$\text{Upper-tailed test:} \quad \beta = P\left(z < \frac{\hat{p}_0 - p_a}{\sqrt{\dfrac{p_a q_a}{n}}}\right)$$

$$(H_a: p > p_0)$$

$$\text{Lower-tailed test:} \quad \beta = P\left(z > \frac{\hat{p}_0 - p_a}{\sqrt{\dfrac{p_a q_a}{n}}}\right)$$

$$(H_a: p < p_0)$$

$$\text{Two-tailed test:} \quad \beta = P\left(\frac{\hat{p}_{0,L} - p_a}{\sqrt{\dfrac{p_a q_a}{n}}} < z < \frac{p_{0,U} - p_a}{\sqrt{\dfrac{p_a q_a}{n}}}\right)$$

$$(H_a: p \neq p_0)$$

Following the calculation of β for a particular value of the parameter in H_a, you should interpret the value in the context of the hypothesis-testing application. It is often useful to interpret the value of $1 - \beta$, which is known as the **power of the test** corresponding to a particular alternative, say, μ_a for a population mean. Since β is the probability of accepting the null hypothesis when the alternative hypothesis is true with $\mu = \mu_a$, $1 - \beta$ is the probability of the complementary event, or the probability of rejecting the null hypothesis when the alternative $H_a: \mu = \mu_a$ is true. That is, the power $(1 - \beta)$ measures the likelihood that the test procedure will lead to the correct decision (reject H_0) for a particular value of the mean (or proportion) in the alternative hypothesis.

> The **power of the test** is the probability that the test will correctly lead to the rejection of the null hypothesis for a particular value of μ or p in the alternative hypothesis. The power is equal to $(1 - \beta)$ for the particular alternative considered.

For example, in the sewer pipe example, we found that $\beta = .7764$ when $\mu = 2{,}425$. This is the probability that the test leads to the (incorrect) acceptance of the null hypothesis when $\mu = 2{,}425$. Or, equivalently, the power of the test is $1 - .7764 = .2236$, which means that the test will lead to the (correct) rejection of the null hypothesis only 22% of the time when the pipe exceeds specifications by 25 pounds per linear foot. When the manufacturer's pipe has a mean strength of 2,475 (that is, 75 pounds per linear foot in excess of specifications), the power of the test increases to $1 - .1562 = .8438$. That is, the test will lead to the acceptance of the manufacturer's pipe 84% of the time if $\mu = 2{,}475$.

Example 8.12 **Finding the Power of the Test**	**Problem** Recall the drug experiment in Examples 8.3 and 8.5, in which we tested to determine whether the mean response time for rats injected with a drug differs from the control mean response time of $\mu = 1.2$ seconds. The test setup is repeated here:

$$H_0: \mu = 1.2$$

$$H_a: \mu \neq 1.2 \ (\text{i.e., } \mu < 1.2 \text{ or } \mu > 1.2)$$

$$\text{Test statistic: } z = \frac{\bar{x} - 1.2}{\sigma_{\bar{x}}}$$

$$\text{Rejection region: } z < -1.96 \text{ or } z > 1.96 \text{ for } \alpha = .05$$

$$z < -2.575 \text{ or } z > 2.575 \quad \text{for } \alpha = .01$$

Note that two rejection regions have been specified, corresponding to values of $\alpha = .05$ and $\alpha = .01$, respectively. Assume that $n = 100$ and $s = .5$.

a. Suppose drug-injected rats have a mean response time of 1.1 seconds; that is, $\mu = 1.1$. Calculate the values of β corresponding to the two rejection regions. Discuss the relationship between the values of α and β.

b. Calculate the power of the test for each of the rejection regions when $\mu = 1.1$.

Solution

a. We first consider the rejection region corresponding to $\alpha = .05$. The first step is to calculate the border values of $\bar{x}$ corresponding to the two-tailed rejection region, $z < -1.96$ or $z > 1.96$:

$$\bar{x}_{0,L} = \mu_0 - 1.96\sigma_{\bar{x}} \approx \mu_0 - 1.96\left(\frac{s}{\sqrt{n}}\right) = 1.2 - 1.96\left(\frac{.5}{10}\right) = 1.102$$

$$\bar{x}_{0,U} = \mu_0 + 1.96\sigma_{\bar{x}} \approx \mu_0 + 1.96\left(\frac{s}{\sqrt{n}}\right) = 1.2 + 1.96\left(\frac{.5}{10}\right) = 1.298$$

These border values are shown in Figure 8.25.

Next, we convert these values to z-values in the alternative distribution with $\mu_a = 1.1$:

$$z_L = \frac{\bar{x}_{0,L} - \mu_a}{\sigma_{\bar{x}}} \approx \frac{1.102 - 1.1}{.05} = .04$$

$$z_U = \frac{\bar{x}_{0,U} - \mu_a}{\sigma_{\bar{x}}} \approx \frac{1.298 - 1.1}{.05} = 3.96$$

a. $\mu = 1.2$ (H_0)
Two rejection regions
$\alpha = .05$ and $\alpha = .01$

b. $\mu = 1.1$ (H_a)
β for $\alpha = .05$ rejection region

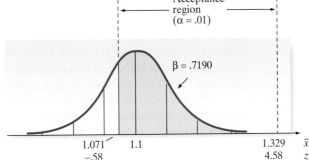

c. $\mu = 1.1$ (H_a)
β for $\alpha = .01$ rejection region

Figure 8.25

Calculation of β for drug-injected rate (Example 8.12)

These z-values are shown in Figure 8.25b. You can see that the acceptance (or non-rejection) region is the area between them. Using Table II of Appendix B, we find that the area between $z = 0$ and $z = .04$ is .0160 and the area between $z = 0$ and $z = 3.96$ is (approximately) .5 (since $z = 3.96$ is off the scale of Table II). Then the area between $z = .04$ and $z = 3.96$ is approximately

$$\beta = .5 - .0160 = .4840$$

Thus, the test with $\alpha = .05$ will lead to a Type II error about 48% of the time when the mean reaction time for drug-injected rats is .1 second less than the control mean response time.

For the rejection region corresponding to $\alpha = .01$, $z < -2.575$ or $z > 2.575$, we find that

$$\bar{x}_{0, L} = 1.2 - 2.575\left(\frac{.5}{10}\right) = 1.0712$$

$$\bar{x}_{0, U} = 1.2 + 2.575\left(\frac{.5}{10}\right) = 1.3288$$

These border values of the rejection region are shown in Figure 8.25c.

Converting these two border values to z-values in the alternative distribution with $\mu_a = 1.1$, we find that $z_L = -.58$ and $z_U = 4.58$. The area between these values is approximately

$$\beta = P(-.58 < z < 4.58) = .2190 + .5 = .7190$$

Thus, the chance that the test procedure with $\alpha = .01$ will lead to an incorrect acceptance of H_0 is about 72%.

Note that the value of β increases from .4840 to .7190 when we decrease the value of α from .05 to .01. This is a general property of the relationship between α and β: *As α is decreased (increased), β is increased (decreased).*

b. The power is defined to be the probability of (correctly) rejecting the null hypothesis when the alternative is true. When $\mu = 1.1$ and $\alpha = .05$, we find that

$$\text{Power} = 1 - \beta = 1 - .4840 = .5160$$

When $\mu = 1.1$ and $\alpha = .01$, we get

$$\text{Power} = 1 - \beta = 1 - .7190 = .2810$$

You can see that the power of the test is decreased as the level of α is decreased. This means that as the probability of incorrectly rejecting the null hypothesis is decreased, the probability of correctly accepting the null hypothesis for a given alternative is also decreased.

Look Back A key point in this example is that the value of α must be selected carefully, with the realization that a test is made less capable of detecting departures from the null hypothesis when the value of α is decreased.

◾ **Now Work Exercise 8.102a**

Note: Most statistical software packages now have options for computing the power of standard tests of hypothesis. Usually, you will need to specify the type of test (z-test or t-test), the form of $H_a (<, >,$ or $\neq)$, the standard deviation, the sample size, and the value of the parameter in H_a (or the difference between the value in H_0 and the value in H_a). The MINITAB power analysis for Example 8.12 when $\alpha = .05$ is displayed in Figure 8.26. The power of the test (.516) is highlighted on the printout.

We have shown that the probability of committing a Type II error, β, is inversely related to α (Example 8.12), and that the value of β decreases as the value of μ_a (or p_a) moves farther from the null-hypothesis value. Consider again the sewer pipe example. The sample size n also affects β. Remember that the standard deviation of the sampling distribution of $\bar{x}$ is inversely proportional to the square root of the sample size ($\sigma_{\bar{x}} = \sigma/\sqrt{n}$).

Power and Sample Size

```
1-Sample Z Test

Testing mean = null (versus not = null)
Calculating power for mean = null + difference
Alpha = 0.05  Assumed standard deviation = 0.5
```

	Sample	
Difference	Size	**Power**
0.1	100	0.516005

Figure 8.26
MINITAB power analysis
for Example 8.12

Thus, as illustrated in Figure 8.27, the variability of both the null and alternative sampling distributions is decreased as n is increased. If the value of α is specified and remains fixed, then the value of β decreases as n increases, as illustrated in Figure 8.27. Conversely, the power of the test for a given alternative hypothesis is increased as the sample size is increased.

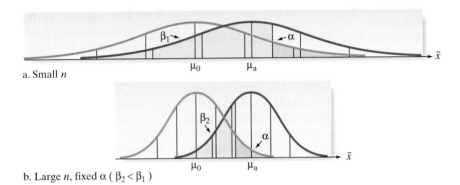

Figure 8.27
Relationship between α, β,
and n

a. Small n

b. Large n, fixed α ($\beta_2 < \beta_1$)

The properties of β and power are summarized in the following box:

Properties of β and Power

1. For fixed n and α, the value of β decreases and the power increases as the distance between the specified null value μ_0 (or p_0) and the specified alternative value μ_a (or p_a) increases. (See Figure 8.24.)
2. For fixed n, μ_0, and μ_a (or p_0 and p_a) the value of β increases and the power decreases as the value of α is decreased. (See Figure 8.25.)
3. For fixed α, μ_0, and μ_a, (or p_0 and p_a) the value of β decreases and the power increases as the sample size n is increased. (See Figure 8.27.)

Exercises 8.94–8.107

Understanding the Principles

8.94 How is the power of the test related to the acceptance region?

8.95 If the power of the test is .85, what can be said about the probability of committing a Type II error?

8.96 If the power of the test tends to unity, then what will be the conclusion for the given testing problem?

Learning the Mechanics

8.97 Suppose you want to test $H_0: \mu = 2{,}000$ against $H_a: \mu > 2{,}000$, using $\alpha = .05$. The population in question

is normally distributed with standard deviation 140. A random sample of size $n = 49$ will be used.
  a. Specify the role of $\bar{x}$ in this problem.
  b. Find the value of $\bar{x}_0$, that value of $\bar{x}$ above which the null hypothesis will be rejected.
  c. What will be the value of z when $\mu = 2{,}020$?
  d. Find β.
  e. Compute the power of this test for detecting the alternative $H_a: \mu = 2{,}020$.

8.98 Refer to Exercise 8.97.

    a. If $\mu = 2{,}040$ instead of $2{,}020$, what is the probability that the hypothesis test will incorrectly fail to reject H_0? That is, what is β?

    b. If $\mu = 2{,}040$, what is the probability that the test will correctly reject the null hypothesis? That is, what is the power of the test?

    c. Compare β and the power of the test when $\mu = 2{,}040$ to the values you obtained in Exercise 8.97 for $\mu = 2{,}020$. Explain the differences.

8.99 It is desired to test $H_0: \mu = 60$ against $H_a: \mu < 60$, using $\alpha = .10$. The population in question is uniformly distributed with standard deviation 22.5. A random sample of size 81 will be drawn from the population.

    a. Describe the (approximate) sampling distribution of $\bar{x}$ under the assumption that H_0 is true.

    b. Describe the (approximate) sampling distribution of $\bar{x}$ under the assumption that the population mean is 55.

    c. If μ were really equal to 55, what is the probability that the hypothesis test would lead the investigator to commit a Type II error?

    d. What is the power of this test for detecting the alternative $H_a: \mu = 55$?

8.100 Refer to Exercise 8.99.

    a. Find β for each of the following values of the population mean: 65.4, 63.4, 61.4, 59.4, and 57.4.

    b. Plot each value of β you obtained in part **a** against its associated population mean. Show β on the vertical axis and μ on the horizontal axis. Draw a curve through the five points on your graph.

    c. Use your graph from part **b** to find the approximate probability that the hypothesis test will lead to a Type II error when $\mu = 63.4$.

    d. Convert each of the β values you calculated in part **a** to the power of the test at the specified value of μ. Plot the power on the vertical axis against μ on the horizontal axis. Compare the graph of part **b** with the *power curve* you plotted here.

    e. Examine the graphs of parts **b** and **d**. Explain what they reveal about the relationships among the distance between the true mean μ and the null-hypothesized mean μ_0, the value of β, and the power.

8.101 Suppose you want to conduct the two-tailed test of $H_0: p = .8$ against $H_a: p \neq .8$ using $\alpha = .05$. A random sample of size 100 will be drawn from the population in question.

    a. Describe the sampling distribution of $\hat{p}$ under the assumption that H_0 is true.

    b. Describe the sampling distribution of $\hat{p}$ under the assumption that $p = .75$.

    c. If p were really equal to .75, find the value of β associated with the test.

    d. Find the value of β for the alternative $H_a: p = .69$.

Applying the Concepts—Intermediate

8.102 **Latex allergy in health care workers.** Refer to the *Current Allergy & Clinical Immunology* (March 2004) study of health care workers diagnosed with a latex allergy, Exercise 7.15 (p. 352). Consider testing the null hypothesis of $\mu = 20$ against the alternative hypothesis of $\mu < 20$ using $\alpha = .01$, where μ is the mean number of latex gloves used per week. The sample of $n = 46$ workers had a standard deviation of $s = 11.9$. Assume s is a good estimate of the true standard deviation σ.

    NW **a.** Find the power of the test if the true mean is $\mu = 19$.

    b. Repeat part **a** for the true mean values 18, 16, 14, and 12.

    c. Plot the power of the test on the vertical axis against the alternative mean on the horizontal axis. Draw a curve through the points. What pattern do you observe?

    d. Use the power curve, part **c**, to estimate the power for the mean value $\mu = 15$. Calculate the power for this value of μ, and compare it with your approximation.

    e. If the true value of the mean number of latex gloves used per week is really 15, what is the probability that the test will fail to reject $H_0: \mu = 20$?

8.103 **Increasing the sample size.** Refer to Exercise 8.102. Show what happens to the power curve when the sample size is increased from $n = 46$ to $n = 100$. Assume that the standard deviation is $\sigma = 11.9$.

8.104 **Gas mileage of the Honda Civic.** According to the Environmental Protection Agency (EPA) *Fuel Economy Guide*, the 2014 Honda Civic automobile obtains a mean of 36 miles per gallon (mpg) on the highway. Suppose Honda claims that the EPA has underestimated the Civic's mileage. To support its assertion, the company selects $n = 50$ model 2014 Civic cars and records the mileage obtained for each car over a driving course similar to the one used by the EPA. The following data result: $\bar{x} = 38.3$ mpg, $s = 6.4$ mpg.

    a. If Honda wishes to show that the mean mpg for 2014 Civic autos is greater than 36 mpg, what should the alternative hypothesis be? the null hypothesis?

    b. Do the data provide sufficient evidence to support the auto manufacturer's claim? Test, using $\alpha = .05$. List any assumptions you make in conducting the test.

    c. Calculate the power of the test for the mean values of 36.5, 37.0, 37.5, 38.0, and 38.5, assuming that $s = 6.4$ is a good estimate of σ.

    d. Plot the power of the test on the vertical axis against the mean on the horizontal axis. Draw a curve through the points.

    e. Use the power curve of part **d** to estimate the power for the mean value $\mu = 37.75$. Calculate the power for this value of μ, and compare it with your approximation.

    f. Use the power curve to approximate the power of the test when $\mu = 41$. If the true value of the mean mpg for this model is really 41, what (approximately) are the chances that the test will fail to reject the null hypothesis that the mean is 36?

8.105 **Cooling method for gas turbines.** Refer to the *Journal of Engineering for Gas Turbines and Power* (Jan. 2005) study of a new cooling method for gas turbine engines, presented in Exercise 8.46 (p. 420). The heat rates (kilojoules per kilowatt per hour) for 67 gas turbines cooled with the new method are saved in the **TURBINE** file. Assume that the standard deviation of the population of heat rates is $\sigma = 1{,}600$. In Exercise 8.46, you tested whether the true mean heat rate μ exceeded 10,000 kJ/kWh at $\alpha = .05$. If the true mean is really $\mu = 10{,}500$, what is the chance that you will make a Type II error?

8.106 Underwater sound-locating ability of alligators. Refer to the *Journal of Herpetology* (Dec. 2014) study of whether alligators can locate underwater sounds, Exercise 8.81 (p. 434). Recall that the researchers assumed that the proportion of alligators moving toward the underwater sound expected by chance is .5. Also, in a sample of $n = 50$ alligators, 42 moved toward the underwater sound. In Exercise 8.81, you used this information to test whether p, the true proportion of alligators that move toward the underwater sound, is higher than expected by chance. Calculate the power of the test if $p = .6$. Interpret this probability practically.

8.107 Gummy bears: red or yellow? Refer to the *Chance* (Winter 2010) experiment to determine if color of a gummy bear is related to its flavor, Exercise 8.82 (p. 434). You tested the null hypothesis of $p = .5$ against the two-tailed alternative hypothesis of $p \neq .5$ using $\alpha = .01$, where p represents the true proportion of blindfolded students who correctly identify the color of the gummy bear. Recall that of the 121 students who participated in the study, 97 correctly identified the color. Find the power of the test if the true proportion is $p = .65$.

8.8 Test of Hypothesis about a Population Variance (Optional)

Although many practical problems involve inferences about a population mean (or proportion), it is sometimes of interest to make an inference about a population variance σ^2. To illustrate, a quality control supervisor in a cannery knows that the exact amount each can contains will vary, since there are certain uncontrollable factors that affect the amount of fill. The mean fill per can is important, but equally important is the variance of the fill. If σ^2 is large, some cans will contain too little and others too much. Suppose regulatory agencies specify that the standard deviation of the amount of fill in 16-ounce cans should be less than .1 ounce. To determine whether the process is meeting this specification, the supervisor randomly selects 10 cans and weighs the contents of each. The results are given in Table 8.6.

Table 8.6	Fill Weights (ounces) of 10 Cans			
16.00	15.95	16.10	16.02	15.99
16.06	16.04	16.05	16.03	16.02

Ⓓ *Data Set:* FILLWT

Do these data provide sufficient evidence to indicate that the variability is as small as desired? To answer this question, we need a procedure for testing a hypothesis about σ^2.

Intuitively, it seems that we should compare the sample variance σ^2 with the hypothesized value of σ^2 (or s with σ) in order to make a decision about the population's variability. The quantity

$$\frac{(n-1)s^2}{\sigma^2}$$

is known to have a **chi-square** (χ^2) **distribution** when the population from which the sample is taken is *normally distributed*. (Chi-square distributions were introduced in optional Section 7.6.)

Since the distribution of $\frac{(n-1)s^2}{\sigma^2}$ is known, we can use this quantity as a test statistic in a test of hypothesis for a population variance, as illustrated in the next example.

Example 8.13

A One-Tailed Test for σ^2—Fill Weight Variance

FILLWT

Problem Refer to the fill weights for the sample of ten 16-ounce cans in Table 8.6. Is there sufficient evidence to conclude that the true standard deviation σ of the fill measurements of 16-ounce cans is less than .1 ounce?

Solution Here, we want to test whether $\sigma < .1$. Since the null and alternative hypotheses must be stated in terms of σ^2 (rather than σ), we want to test the null hypothesis that $\sigma^2 = (.1)^2 = .01$ against the alternative that $\sigma^2 < .01$. Therefore, the elements of the test are

$H_0: \sigma^2 = .01$ (Fill variance equals .01—i.e., process specifications are not met.)

$H_a: \sigma^2 < .01$ (Fill variance is less than .01—i.e., process specifications are met.)

Test statistic: $\chi^2 = \dfrac{(n-1)s^2}{\sigma^2}$

Assumption: The distribution of the amounts of fill is approximately normal.

Rejection region: The smaller the value of s^2 we observe, the stronger is the evidence in favor of H_a. Thus, we reject H_0 for "small values" of the test statistic. Recall that the chi-square distribution depends on $(n-1)$ degrees of freedom. With $\alpha = .05$ and $(n-1) = 9$ df, the critical χ^2 value for rejection is found in Table IV of Appendix B. A partial reproduction of this table is shown in Table 8.7.

Table 8.7   A Partial Reproduction of Table IV, Appendix B—Critical Values of χ^2

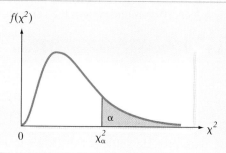

Degrees of Freedom	$\chi^2_{.995}$	$\chi^2_{.990}$	$\chi^2_{.975}$	$\chi^2_{.950}$	$\chi^2_{.900}$	$\chi^2_{.100}$	$\chi^2_{.050}$	$\chi^2_{.025}$	$\chi^2_{.010}$	$\chi^2_{.005}$
1	.0000393	.0001571	.0009821	.0039321	.0157908	2.70554	3.84146	5.02389	6.63490	7.87944
2	.0100251	.0201007	.0506356	.102587	.210720	4.60517	5.99147	7.37776	9.21034	10.5966
3	.0717212	.114832	.215795	.351846	.584375	6.25139	7.81473	9.34840	11.3449	12.8381
4	.206990	.297110	.484419	.710721	1.063623	7.77944	9.48773	11.1433	13.2767	14.8602
5	.411740	.554300	.831211	1.145476	1.61031	9.23635	11.0705	12.8325	15.0863	16.7496
6	.675727	.872085	1.237347	1.63539	2.20413	10.6446	12.5916	14.4494	16.8119	18.5476
7	.989265	1.239043	1.68987	2.16735	2.83311	12.0170	14.0671	16.0128	18.4753	20.2777
8	1.344419	1.646482	2.17973	2.73264	3.48954	13.3616	15.5073	17.5346	20.0902	21.9550
9	1.734926	2.087912	2.70039	3.32511	4.16816	14.6837	16.9190	19.0228	21.6660	23.5893
10	2.15585	2.55821	3.24697	3.94030	4.86518	15.9871	18.3070	20.4831	23.2093	25.1882

The area given in this table is the area to the *right* of the numerical χ^2 value. That is, the table gives the value χ^2_α that cuts off a tail area to the right equal to α. For this lower-tailed test, we need $\alpha = .05$ as the area to the *left* of the critical value, as shown in Figure 8.28; consequently, the area to the right is $(1 - \alpha) = .95$ and we will reject H_0 if $\chi^2 < \chi^2_{.95}$. Looking in the .95 column and 9 df row of Table 8.7, we find the critical value 3.32511. Therefore,

Reject H_0 if $\chi^2 < 3.32511$ (see Figure 8.28)

A SAS printout of the analysis is displayed in Figure 8.29. The value of s (highlighted on the printout) is $s = .0412$. Substituting into the formula for the test statistic, we have

$$\chi^2 = \frac{(n-1)s^2}{\sigma^2} = \frac{9(.0412)^2}{.01} = 1.53$$

Figure 8.28

Rejection region for
Example 8.13

```
Sample Statistics for FILLAMT

     N       Mean      Std. Dev.      Variance
    ------------------------------------------------
    10      16.026       0.0412         0.0017

Hypothesis Test

    Null hypothesis:          Variance of FILLAMT => 0.01
    Alternative:              Variance of FILLAMT <  0.01

         Chi-square       Df         Prob
    ------------------------------------------------
           1.524           9        0.0030
```

Figure 8.29

SAS test of fill amount variance,
Example 8.13

Conclusion: Since the value of the test statistic is less than 3.32511, the supervisor can conclude that the variance σ^2 of the population of all amounts of fill is less than .01 (i.e., $\sigma < .1$), with probability of a Type I error equal to $\alpha = .05$. If this procedure is repeatedly used, it will incorrectly reject H_0 only 5% of the time. Thus, the quality control supervisor is confident in the decision that the cannery is operating within the desired limits of variability.

Look Back Note that both the test statistic and the one-tailed p-value of the test are shown on the SAS printout. Note also that the p-value (.003) is less than $\alpha = .05$, thus confirming our conclusion to reject H_0.

■ **Now Work Exercise 8.121**

One-tailed and two-tailed tests of hypothesis for σ^2 are given in the following box.

Large-Sample Test of Hypothesis about σ^2: Chi-Square Statistic

Test statistic: $\quad \chi_c^2 = \dfrac{(n-1)s^2}{\sigma_0^2}$

	One-Tailed Tests		Two-Tailed Test
	$H_0: \sigma^2 = \sigma_0^2$	$H_0: \sigma^2 = \sigma_0^2$	$H_0: \sigma^2 = \sigma_0^2$
	$H_a: \sigma^2 < \sigma_0^2$	$H_a: \sigma^2 > \sigma_0^2$	$H_a: \sigma^2 \neq \sigma_0^2$
Rejection region:	$\chi_c^2 < \chi_{(1-\alpha)}^2$	$\chi_c^2 > \chi_\alpha^2$	$\chi_c^2 < \chi_{(1-\alpha/2)}^2$ or $\chi_c^2 > \chi_{\alpha/2}^2$
p-value:	$P(\chi^2 < \chi_c^2)$	$P(\chi^2 > \chi_c^2)$	$2 \min\{P(\chi^2 < \chi_c^2), P(\chi^2 > \chi_c^2)\}$

Decision: Reject H_0 if $\alpha > p$-value or if test statistic (χ_c^2) falls in rejection region where $P(\chi^2 > \chi_\alpha^2) = \alpha$, $P(\chi^2 < \chi_{(1-\alpha)}^2) = \alpha$, $\alpha = P(\text{Type I error}) = P(\text{Reject } H_0 \mid H_0 \text{ true})$ and the chi-square distribution is based on $(n-1)$ degrees of freedom.

[*Note:* The symbol for the numerical value assigned to σ^2 under the null hypothesis is σ_0^2.]

> **Conditions Required for a Valid Large-Sample Hypothesis Test for σ^2**
>
> 1. A random sample is selected from the target population.
> 2. The population from which the sample is selected has a distribution that is approximately normal.

Example 8.14 Ⓓ

FILLWT

A Two-Tailed Test for σ^2—Fill Weight Variance

Problem Refer to Example 8.13. Now conduct the two-tailed test of hypothesis, H_0: $\sigma^2 = .01$ against H_a: $\sigma^2 \neq .01$, using $\alpha = .05$.

Solution Since $\alpha = .05$, $\alpha/2 = .025$ and $(1 - \alpha/2) = .975$. The rejection region for this two-tailed test, as outlined in the box, requires that we find the critical chi-square values $\chi^2_{(1-\alpha/2)} = \chi^2_{.975}$ and $\chi^2_{\alpha/2} = \chi^2_{.025}$. Looking in the 9 df row of Table 8.7, we find $\chi^2_{.975} = 2.70039$ and $\chi^2_{.025} = 19.0228$. Therefore, the rejection region is:

$$\text{Reject } H_0 \text{ if } \chi^2 < 2.70039 \text{ or } \chi^2 > 19.0228$$

From Example 8.13, the test statistic is $\chi^2 = 1.53$. Since this value falls into the rejection region, there is sufficient evidence to reject H_0 in favor of H_a. That is, there is evidence to indicate that the variance in the fill weights, σ^2, differs from the hypothesized value of .01.

Look Back Remember to divide the value of α in half to find the critical chi-square values in the rejection region for a two-tailed test of a population variance. Of course, you can always resort to statistical software to find the p-value of the test and derive your conclusion by simply comparing α to the p-value.

⬛

⚠ **CAUTION** The procedure for conducting a hypothesis test for σ^2 in the preceding examples requires an assumption regardless of whether the sample size n is large or small. We must assume that the population from which the sample is selected has an approximate normal distribution. Unlike small-sample tests for μ based on the t-statistic, *slight to moderate departures from normality will render the χ^2 test invalid.* ▲

Exercises 8.108–8.128

Understanding the Principles

8.108 What test statistic is used to make inferences about σ^2?

8.109 What assumption is required for a valid test for σ^2?

8.110 *True or False.* The null hypotheses H_0: $\sigma^2 = .36$ and H_0: $\sigma = .6$ are equivalent.

8.111 *True or False.* Even when the sample size n is large, assumptions about the population are necessary to test the population variance σ^2.

Learning the Mechanics

8.112 Let χ^2_0 be a particular value of χ^2. Find the value of χ^2_0 such that
  a. $P(\chi^2 > \chi^2_0) = .05$ for $n = 12$
  b. $P(\chi^2 > \chi^2_0) = .10$ for $n = 9$
  c. $P(\chi^2 > \chi^2_0) = .05$ for $n = 5$
  d. $P(\chi^2 < \chi^2_0) = .025$ for $n = 7$
  e. $P(\chi^2 < \chi^2_0) = .10$ for $n = 10$

8.113 A random sample of n observations is selected from a normal population to test the null hypothesis that $\sigma^2 = 35$. Specify the rejection region for each of the following combinations of H_a, α, and n:
  a. H_a: $\sigma^2 \neq 35$; $\alpha = .01$; $n = 16$
  b. H_a: $\sigma^2 > 35$; $\alpha = .01$; $n = 24$
  c. H_a: $\sigma^2 > 35$; $\alpha = .05$; $n = 15$
  d. H_a: $\sigma^2 < 35$; $\alpha = .05$; $n = 13$
  e. H_a: $\sigma^2 > 35$; $\alpha = .01$; $n = 10$
  f. H_a: $\sigma^2 < 25$; $\alpha = .05$; $n = 25$

8.114 A random sample of six measurements gave $\bar{x} = 19.4$ and $s^2 = 2.208$.
  a. What assumptions must you make concerning the population in order to test a hypothesis about σ^2?
  b. Suppose the assumptions in part **a** are satisfied. Test the null hypothesis $\sigma^2 = 1$ against the alternative hypothesis $\sigma^2 > 1$. Use $\alpha = .01$.
  c. Refer to part **b**. Suppose the test statistic is $\chi^2 = 12.83$. Use Table IV of Appendix B or statistical software to find the p-value of the test.

d. Test the null hypothesis $\sigma^2 = 1$ against the alternative hypothesis $\sigma^2 \neq 1$. Use $\alpha = .01$.

8.115 Refer to Exercise 8.114. Suppose we had $n = 80$, $\bar{x} = 19.4$, and $s^2 = 2.208$.
   a. Test the null hypothesis $H_0: \sigma^2 = 1$ against the alternative hypothesis $H_a: \sigma^2 > 1$.
   b. Compare your test result with that of Exercise 8.114.

8.116 A random sample of $n = 7$ observations from a normal population produced the following measurements: 6, 2, 8, 5, 5, 7, and 11. Do the data provide sufficient evidence to indicate that $\sigma^2 < 2$? Test, using $\alpha = .05$.

Applying the Concepts—Basic

8.117 Trading skills of institutional investors. *The Journal of Finance* (Apr. 2011) published the results of an analysis of trading skills of institutional investors. The study focused on "round-trip" trades, i.e., trades in which the same stock was both bought and sold in the same quarter. In a random sample of 200 round-trip trades made by institutional investors, the sample standard deviation of the rates of return was 8.82%. One property of a consistent performance of institutional investors is a small variance in the rates of return of round-trip trades, say, a standard deviation of less than 10%.
   a. Specify the null and alternative hypotheses for determining whether the population of institutional investors performs consistently.
   b. Find the rejection region for the test using $\alpha = .05$.
   c. Interpret the value of α in the words of the problem.
   d. A MINITAB printout of the analysis is shown below. Locate the test statistic and p-value on the printout.

   e. Give the appropriate conclusion in the words of the problem.

Test and CI for One Variance

```
Method

Null hypothesis          Sigma = 10
Alternative hypothesis   Sigma < 10

The chi-square method is only for the normal distribution.
The Bonett method cannot be calculated with summarized data.

Statistics

  N   StDev  Variance
200   8.82      77.8

Tests

                  Test
Method        Statistic   DF  P-Value
Chi-Square       154.81   199    0.009
```

---

f. What assumptions about the data are required for the inference to be valid?

8.118 Irrelevant speech effects. Refer to the *Acoustical Science & Technology* (Vol. 35, 2014) study of irrelevant speech effects, Exercise 8.47 (p. 420). Recall that 71 subjects performed a memorization task under two conditions: (1) with irrelevant background speech and (2) in silence. Descriptive statistics on the relative difference in the error rates (RDER) for the two conditions yielded $\bar{x} = 78.2$ percent and $s = 63.2$ percent. Suppose the researchers theorize that the true standard deviation of the RDER values differs from 60 percent.
   a. Set up the null and alternative hypotheses for testing the researchers' theory.
   b. Compute the test statistic.
   c. Find the rejection region of the test using $\alpha = .01$.
   d. Use statistical software to find the p-value of the test.
   e. Make the appropriate practical conclusion. State your answer in the words of the problem.
   f. What assumption about the population of RDER values must hold true for the inference, part **e**, to be valid?

8.119 Characteristics of a rockfall. Refer to the *Environmental Geology* (Vol. 58, 2009) simulation study of how far a block from a collapsing rockwall will bounce down a soil slope, Exercise 2.61 (p. 89). Rebound lengths (in meters) were estimated for 13 rock bounces. The data are repeated in the table. Descriptive statistics for the rebound lengths are shown on the SAS printout at the bottom of the page. Consider a test of hypothesis for the variation in rebound lengths for the theoretical population of rock bounces from the collapsing rockwall. In particular, a geologist wants to determine if the variance differs from 10 m^2.

10.94	13.71	11.38	7.26	17.83	11.92	11.87	5.44	13.35
4.90	5.85	5.10	6.77					

Based on Paronuzzi, P. "Rockfall-induced block propagation on a soil slope, northern Italy." *Environmental Geology*, Vol. 58, 2009 (Table 2).

---

   a. Define the parameter of interest.
   b. Specify the null and alternative hypotheses.
   c. Compute the value of the test statistic.
   d. Determine the rejection region for the test using $\alpha = .10$.
   e. Find the approximate p-value of the test.
   f. Make the appropriate conclusion.
   g. What condition must be satisfied in order for the inference, part **f**, to be valid?

SAS Output for Exercise 8.119

The MEANS Procedure

Analysis Variable : LENGTH

N	Mean	Std Dev	Minimum	Maximum
13	9.7169231	4.0947291	4.9000000	17.8300000

8.120 Shaft graves in ancient Greece. Refer to the *American Journal of Archaeology* (Jan. 2014) study of sword shaft graves in ancient Greece, Exercise 2.37 (p. 78). The number of sword shafts buried at each of 13 recently discovered grave sites is reproduced in the following table. Let σ represent the standard deviation of the number of sword shafts for the population of all shaft graves in ancient Greece. Consider the test H_0: $\sigma = 3$ versus H_a: $\sigma < 3$, using $\alpha = .05$.

1	2	3	1	5	6	2	4	1	2	4	2	9

Source: Harrell, K. "The fallen and their swords: A new explanation for the rise of the shaft graves." *American Journal of Archaeology,* Vol. 118, No. 1, Jan. 2014 (Figure 1).

a. Use statistical software to find the sample standard deviation, *s*.
b. Calculate the test statistic.
c. Use statistical software to find the *p*-value of the test. Interpret the result.
d. Assume the sample grave with 9 sword shafts was discovered to be an outlying observation by the researchers, i.e., one from a different population. Remove the data point from the sample and repeat parts **a–c**. What conclusions can you now draw?

8.121 Latex allergy in health care workers. Refer to the *Current Allergy & Clinical Immunology* (Mar. 2004) study of $n = 46$ hospital employees who were diagnosed with a latex allergy from exposure to the powder on latex gloves, presented in Exercise 7.15 (p. 352). Recall that the number of latex gloves used per week by the sampled workers is summarized as follows: $\bar{x} = 19.3$ and $s = 11.9$. Let σ^2 represent the variance in the number of latex gloves used per week by all hospital employees. Consider testing H_0: $\sigma^2 = 100$ against H_a: $\sigma^2 \neq 100$.
a. Give the rejection region for the test at a significance level of $\alpha = .01$.
b. Calculate the value of the test statistic.
c. Use the results from parts **a** and **b** to draw the appropriate conclusion.

Applying the Concepts—Intermediate

8.122 Oil content of fried sweet potato chips. Refer to the *Journal of Food Engineering* (Sept. 2013) study of the characteristics of sweet potato chips fried at different temperatures, Exercise 7.111 (p. 384). Recall that a sample of 6 sweet potato slices were fried at 130° using a vacuum fryer and the internal oil content (gigagrams) was measured for each slice. The results were: $\bar{x} = .178\,g/g$ and $s = .011\,g/g$.
a. Conduct a test of hypothesis to determine if the standard deviation, σ, of the population of internal oil contents for sweet potato slices fried at 130° differs from .1. Use $\alpha = .05$.
b. In Exercise 7.111, you formed a 95% confidence interval for the true standard deviation of the internal oil content distribution for the sweet potato chips. Use this interval to make an inference about whether $\sigma = .1$. Does the result agree with the test, part **a**?

8.123 Strand bond performance of pre-stressed concrete. An experiment was carried out to investigate the strength of pre-stressed, bonded concrete after anchorage failure has occurred and the results published in *Engineering Structures* (June 2013). The maximum strand force, measured in kiloNewtons (kN), achieved after anchorage failure for 12 pre-stressed concrete strands is given in the accompanying table. Conduct a test of hypothesis to determine if the true standard deviation of the population of maximum strand forces is less than 5 kN. Test using $\alpha = .10$.

158.2	161.5	166.5	158.4	159.9	161.9	162.8	161.2	160.1	175.6	168.8	163.7

8.124 Drug content assessment. *Analytical Chemistry* (Dec. 15, 2009) presented research on a new method used by GlaxoSmithKline Medicines Research Center to determine the amount of drug in a tablet. Drug concentrations (measured as a percentage) for 50 randomly selected tablets are provided in the accompanying table. The standard method of assessing drug content yields a concentration variance of 9. Can the scientists at GlaxoSmithKline conclude that the new method of determining drug concentration is less variable than the standard method? Test using $\alpha = .01$.

91.28	92.83	89.35	91.90	82.85	94.83	89.83	89.00	84.62
86.96	88.32	91.17	83.86	89.74	92.24	92.59	84.21	89.36
90.96	92.85	89.39	89.82	89.91	92.16	88.67	89.35	86.51
89.04	91.82	93.02	88.32	88.76	89.26	90.36	87.16	91.74
86.12	92.10	83.33	87.61	88.20	92.78	86.35	93.84	91.20
93.44	86.77	83.77	93.19	81.79				

Source: Borman, P. J., Marion, J. C., Damjanov, I., & Jackson, P. "Design and analysis of method equivalence studies." *Analytical Chemistry,* Vol. 81, No. 24, Dec. 15, 2009 (Table 3).

8.125 Point spreads of football games. During a football league season, oddsmakers establish a point spread on each game for betting purposes. For example, the defending champion were established as a 1-point underdog against another team in a season. The final scores of the games were compared against the final point spreads established by the betting agency. The difference between the outcome of the game and the point spread (called a point-spread error) was calculated for 12 games. The mean and standard deviation of the point-spread errors are $\bar{x} = -1.6$ and $s = 13.3$. Suppose a researcher wants to know whether the true standard deviation of the point-spread errors exceeds 15. Conduct the analysis using $\alpha = .05$.

8.126 Cooling method for gas turbines. Refer to the *Journal of Engineering for Gas Turbines and Power* (Jan. 2005) study of the performance of augmented gas turbine engines, presented in Exercise 8.46 (p. 420). Recall that the performance of each in a sample of 67 gas turbines was measured by heat rate (kilojoules per kilowatt per hour). Suppose that standard gas turbines have heat rates with a standard deviation of 1,500 kJ/kWh. Is there sufficient evidence to indicate that the heat rates of the augmented gas turbine engine are more variable than the heat rates of the standard gas turbine engine? Test, using $\alpha = 0.5$.

8.127 Birth weights of cocaine babies. A group of researchers at the University of Texas-Houston conducted a comprehensive study of pregnant cocaine-dependent women (*Journal of Drug Issues*, Summer 1997). All the women in the study used cocaine on a regular basis (at least three times a week) for more than a year. One of the many variables measured was birth weight (in grams) of the baby delivered. For a sample of 16 cocaine-dependent women, the mean birth weight was 2,971 grams and the standard deviation was 410 grams. Test (at $\alpha = .01$) to determine whether the variance in birth weights of babies delivered by cocaine-dependent women is less than 200,000 grams^2.

Applying the Concepts—Advanced

8.128 Motivation of drug dealers. Refer to the *Applied Psychology in Criminal Justice* (Sept. 2009) study of the personality characteristics of convicted drug dealers, Exercise 7.17 (p. 352). A random sample of 100 drug dealers had a mean Wanting Recognition (WR) score of 39 points, with a standard deviation of 6 points. Recall that the WR score is a quantitative measure of a person's level of need for approval and sensitivity to social situations. (Higher scores indicate a greater need for approval.) A criminal psychologist claims that the range of WR scores for the population of convicted drug dealers is 42 points. Do you believe the psychologist's claim? (*Hint*: Assume the population of WR scores is normally distributed.)

CHAPTER NOTES

Key Terms

Note: Starred (*) terms are from the optional sections in this chapter.

Alternative (research)
  hypothesis 398
*Chi-square (χ^2)
  distribution 445
Conclusion 401
Hypothesis 397
Level of significance 402
Lower-tailed test 403
*Null distribution 436
Null hypothesis 398
Observed significance level
  (*p*-value) 409

One-tailed (one-sided)
  statistical test 403
*Power of the test 440
Rejection region 399
Test of hypothesis 397
Test statistic 398
Two-tailed (two-sided)
  hypothesis 403
Two-tailed test 403
Type I error 399
Type II error 400
Upper-tailed test 403

Key Symbols

μ	Population mean
p	Population proportion, $P(\text{Success})$, in binomial trial
σ^2	Population variance
$\bar{x}$	Sample mean (estimator of μ)
$\hat{p}$	Sample proportion (estimator of p)
s^2	Sample variance (estimator of σ^2)
H_0	Null hypothesis
H_a	Alternative hypothesis
α	Probability of a Type I error
β	Probability of a Type II error
χ^2	Chi-square (sampling distribution of s^2 for normal data)

Key Ideas

Key Words for Identifying the Target Parameter

μ — Mean, Average
p — Proportion, Fraction, Percentage, Rate, Probability
σ^2 — Variance, Variability, Spread

Elements of a Hypothesis Test

1. *Null hypothesis* (H_0)
2. *Alternative hypothesis* (H_a)
3. *Test statistic* (z, t, or X^2)
4. *Significance level* (α)
5. *p-value*
6. *Conclusion*

Forms of Alternative Hypothesis

Lower tailed: H_a: $\mu_0 < 50$
Upper tailed: H_a: $\mu_0 > 50$
Two tailed: H_a: $\mu_0 \neq 50$

Type I Error = Reject H_0 when H_0 is true (occurs with probability α)

Type II Error = Accept H_0 when H_0 is false (occurs with probability β)

Power of the Test = $P(\text{Reject } H_0 \text{ when } H_0 \text{ is false})$
= $1 - \beta$

Using p-Values to Decide

1. Choose significance level (α)
2. Obtain *p*-value of the test
3. If $\alpha > $ *p*-value, Reject H_0

Guide to Selecting a One-Sample Hypothesis Test

Supplementary Exercises 8.129–8.170

Note: List the assumptions necessary for the valid implementation of the statistical procedures you use in solving all these exercises. Starred (*) exercises refer to the optional sections in this chapter.

Understanding the Principles

8.129 What is the *p*-value associated with a test of hypothesis in statistics?

8.130 What is the criterion of the parent population for a large-sample and small-sample test of hypothesis about a population variance s^2?

8.131 What is the critical value of the test?

8.132 If the acceptance of the alternative hypothesis of a particular test would cause your firm to go out of business, would you want α to be small or large? Explain.

8.133 *Complete the following statement:* The larger the *p*-value associated with a test of hypothesis, the weaker is the support for the _____ hypothesis. Explain your answer.

Learning the Mechanics

8.134 A random sample of 20 observations selected from a normal population produced $\bar{x} = 82.6$ and $s^2 = 29.4$.
  a. Form a 95% confidence interval for μ.

b. Test $H_0: \mu = 90$ against $H_a: \mu < 90$. Use $\alpha = .05$.
  c. Test $H_0: \mu = 90$ against $H_a: \mu \neq 90$. Use $\alpha = .01$.
  d. Form a 90% confidence interval for μ.
  e. How large a sample would be required to estimate μ to within 1 unit with 99% confidence?

8.135 A random sample of $n = 300$ observations from a binomial population yields $\hat{p} = .39$.
  a. Test $H_0: p = .45$ against $H_a: p < .45$. Use $\alpha = .05$.
  b. Test $H_0: p = .45$ against $H_a: p \neq .45$. Use $\alpha = .05$.
  c. Form a 99% confidence interval for p.
  d. Form a 95% confidence interval for p.
  e. How large a sample would be required to estimate p to within .05 with 99% confidence?

8.136 A random sample of 185 measurements possessed a mean of $\bar{x} = 18.2$ and a standard deviation of $s = .89$.
  a. Form a 95% confidence interval for μ.
  b. Test $H_0: \mu = 18.3$ against $H_a: \mu \pm 18.3$. Use $\alpha = .05$.
  c. Test $H_0: \mu = 18.4$ against $H_a: \mu \pm 18.4$. Use $\alpha = .05$.

8.137 A random sample of 51 observations from a normal population possessed a mean of $\bar{x} = 88$ and a standard deviation of $s = 6.9$.
  a. Test $H_0: \sigma^2 = 20$ against $H_a: \sigma^2 > 20$. Use $\alpha = .05$.
  b. Test $H_0: \sigma^2 = 20$ against $H_a: \sigma^2 \neq 20$. Use $\alpha = .05$.

8.138 A t-test is conducted for the null hypothesis $H_0: \mu = 20$ versus the alternative hypothesis $H_a: \mu > 20$ for a random sample of $n = 25$ observations. The test results are $t = 1.165$ and p-value $= .1452$.
   a. Interpret the p-value.
   b. What assumptions are necessary for the validity of this test?
   c. Calculate and interpret the p-value, assuming that the alternative hypothesis was $H_a: \mu \neq 20$ instead.

Applying the Concepts—Basic

8.139 **Use of herbal therapy.** According to the *Journal of Advanced Nursing* (Jan. 2001), 45% of senior women (i.e., women over the age of 65) use herbal therapies to prevent or treat health problems. Also, senior women who use herbal therapies use an average of 2.5 herbal products in a year.
   a. Give the null hypothesis for testing the first claim by the journal.
   b. Give the null hypothesis for testing the second claim by the journal.

8.140 **FDA mandatory new-drug testing.** When a new drug is formulated, the pharmaceutical company must subject it to lengthy and involved testing before receiving the necessary permission from the Food and Drug Administration (FDA) to market the drug. The FDA requires the pharmaceutical company to provide substantial evidence that the new drug is safe for potential consumers.
   a. If the new-drug testing were to be placed in a test-of-hypothesis framework, would the null hypothesis be that the drug is safe or unsafe? the alternative hypothesis?
   b. Given the choice of null and alternative hypotheses in part **a**, describe Type I and Type II errors in terms of this application. Define α and β in terms of this application.
   c. If the FDA wants to be very confident that the drug is safe before permitting it to be marketed, is it more important that α or β be small? Explain.

8.141 **Sleep deprivation study.** In a certain study, 22 healthy college students deprived of one night's sleep received an array of tests intended to measure their thinking time, fluency, flexibility, and originality of thought. The overall test scores of the sleep-deprived students were compared with the average score expected from students who received their accustomed sleep. Suppose the overall scores of the 22 sleep-deprived students had a mean of $\bar{x} = 73$ and a standard deviation of 17. (Lower scores are associated with a decreased ability to think creatively.)
   a. Test the hypothesis that the true mean score of sleep-deprived subjects is less than 80, the mean score of subjects who received sleep prior to taking the test. Use $\alpha = .05$.
   b. Write the 95% confidence interval for the testing problem.

8.142 **Accuracy of price scanners at Wal-Mart.** Refer to Exercise 7.129 (p. 387) and the study of the accuracy of checkout scanners at Wal-Mart stores in California. Recall that the National Institute for Standards and Technology (NIST) mandates that, for every 100 items scanned through the electronic checkout scanner at a retail store, no more than 2 should have an inaccurate price. A study of random items purchased at California Wal-Mart stores found that 8.3% had the wrong price (*Tampa Tribune*, Nov. 22, 2005). Assume that the study included 1,000 randomly selected items.

   a. Identify the population parameter of interest in the study.
   b. Set up H_0 and H_a for a test to determine whether the true proportion of items scanned at California Wal-Mart stores exceeds the 2% NIST standard.
   c. Find the test statistic and rejection region (at $\alpha = .05$) for the test.
   d. Give a practical interpretation of the test.
   e. What conditions are required for the inference made in part **d** to be valid? Are these conditions met?

8.143 **Accounting and Machiavellianism.** *Behavioral Research in Accounting* (Jan. 2008) published a study of Machiavellian traits in accountants. (*Machiavellian* describes negative character traits that include manipulation, cunning, duplicity, deception, and bad faith.) A Machiavellian ("Mach") rating score was determined for each in a sample of 122 purchasing managers with the following results: $\bar{x} = 99.6$, $s = 12.6$. (*Note:* Scores range from a low of 40 to a high of 160, with the theoretical neutral Mach rating score of 100.) A director of purchasing at a major firm claims that the true mean Mach rating score of all purchasing managers is 85.
   a. Specify the null and alternative hypotheses for a test of the director's claim.
   b. Define a Type I error for this test.
   c. Interpret the value, $\alpha = .10$.
   d. Give the rejection region for the test using $\alpha = .10$.
   e. Find the value of the test statistic.
   f. Use the result, part **e**, to make the appropriate conclusion.
   g. Do you need to make any assumptions about the distribution of Mach rating scores for the population of all purchasing managers? Explain.

8.144 **The "Pepsi challenge."** "Take the Pepsi Challenge" was a marketing campaign used by the Pepsi-Cola Company. Coca-Cola drinkers participated in a blind taste test in which they tasted unmarked cups of Pepsi and Coke and were asked to select their favorite. Pepsi claimed that "in recent blind taste tests, more than half the Diet Coke drinkers surveyed said they preferred the taste of Diet Pepsi." Suppose 100 Diet Coke drinkers took the Pepsi Challenge and 65 preferred the taste of Diet Pepsi. Test the hypothesis that more than one third of all Diet Coke drinkers will select Diet Pepsi in a blind taste test. Use $\alpha = .01$.

8.145 **Masculinizing human faces.** *Nature* (Aug. 27, 1998) published a study of facial characteristics that are deemed attractive. In one experiment, 67 human subjects viewed side by side an image of a Caucasian male face and the same image 50% masculinized using special computer graphics. Each subject was asked to select the facial image they deemed more attractive. Fifty-eight of the 67 subjects felt that masculinization of face shape decreased attractiveness of the male face. The researchers used this sample information to test whether the subjects showed a preference for either the unaltered or the morphed male face.
   a. Set up the null and alternative hypotheses for this test.
   b. Compute the test statistic.
   c. The researchers reported a p-value ≈ 0 for the test. Do you agree?
   d. Make the appropriate conclusion in the words of the problem. Use $\alpha = .01$.

*8.146 **Identifying type of urban land cover.** *Geographical Analysis* (Oct. 2006) published a study of a new method for analyzing remote-sensing data from satellite pixels in order to identify urban land cover. The method uses a numerical measure of the distribution of gaps, or the sizes of holes, in the pixel, called *lacunarity*. Summary statistics for the lacunarity measurements in a sample of 100 grassland pixels are $\bar{x} = 225$ and $s = 20$. It is known that the mean lacunarity measurement for all grassland pixels is 220. The method will be effective in identifying land cover if the standard deviation of the measurements is 10% (or less) of the true mean (i.e., if the standard deviation is less than 22).

a. Give the null and alternative hypotheses for a test to determine whether, in fact, the standard deviation of all grassland pixels is less than 22.

b. A MINITAB analysis of the data is provided below. Locate and interpret the *p*-value of the test. Use $\alpha = .10$.

Test for One Standard Deviation

```
Method
Null hypothesis        Sigma = 22
Alternative hypothesis Sigma = < 22

The standard method is only for the normal distribution.

Statistics
N   StDev  Variance
100  20.0      400


Tests
Method       Chi-Square  DF  P-Value
Standard          81.82  99    0.105
```

8.147 **Alkalinity of river water.** In Exercise 5.129 (p. 303), you learned that the mean alkalinity level of water specimens collected from the Han River in Seoul, South Korea, is 50 milligrams per liter (*Environmental Science & Engineering*, Sept. 1, 2000). Consider a random sample of 100 water specimens collected from a tributary of the Han River. Suppose the mean and standard deviation of the alkalinity levels for the sample are, respectively, $\bar{x} = 67.8$ mpl and $s = 14.4$ mpl. Is there sufficient evidence (at $\alpha = .01$) to indicate that the population mean alkalinity level of water in the tributary exceeds 50 mpl?

8.148 **Reading Japanese books.** Refer to the *Reading in a Foreign Language* (Apr. 2004) experiment to improve the Japanese reading comprehension levels of University of Hawaii students presented in Exercise 2.188 (p. 134). Recall that 14 students participated in a 10-week extensive reading program in a second-semester Japanese course. The data on number of books read by each student are saved in the **JREAD** file. A MINITAB printout of the data analysis is shown below.

One-Sample T: BOOKS

Test of mu = 25 vs > 25

```
                                    95% Lower
Variable   N   Mean   StDev  SE Mean  Bound   T      P
BOOKS     14  31.64  10.49    2.80   26.68  2.37  0.017
```

---

a. State the null and alternative hypotheses for determining whether the average number of books read by all students who participated in the extensive reading program exceeds 25.

b. Find the rejection region for the test, using $\alpha = .05$.

c. Compute the test statistic.

d. State the appropriate conclusion for the test.

e. What conditions are required for the test results to be valid?

f. Locate the *p*-value on the MINITAB printout and use it to test the hypothesis. Your conclusion should agree with your answer in part **d**.

8.149 **Teacher perceptions of child behavior.** *Developmental Psychology* (Mar. 2003) published a study on teacher perceptions of the behavior of elementary school children. Teachers rated the aggressive behavior of a sample of 11,160 New York City public school children by responding to the statement "This child threatens or bullies others in order to get his/her own way." Responses were measured on a scale ranging from 1 (*never*) to 5 (*always*). Summary statistics for the sample of 11,160 children were reported as $\bar{x} = 2.15$ and $s = 1.05$. Let μ represent the mean response for the population of all New York City public school children. Suppose you want to test $H_0: \mu = 3$ against $H_a: \mu \neq 3$.

a. In the words of the problem, define a Type I error and a Type II error.

b. Use the sample information to conduct the test at a significance level of $\alpha = .05$.

c. Conduct the test from part **b** at a significance level of $\alpha = .10$.

Applying the Concepts—Intermediate

8.150 **Errors in medical tests.** Medical tests have been developed to detect many serious diseases. A medical test is designed to minimize the probability that it will produce a "false positive" or a "false negative." A false positive is a positive test result for an individual who does not have the disease, whereas a false negative is a negative test result for an individual who does have the disease.

a. If we treat a medical test for a disease as a statistical test of hypothesis, what are the null and alternative hypotheses for the medical test?

b. What are the Type I and Type II errors for the test? Relate each to false positives and false negatives.

c. Which of these errors has graver consequences? Considering this error, is it more important to minimize α or β? Explain.

8.151 **Post-traumatic stress of POWs.** *Psychological Assessment* (Mar. 1995) published the results of a study of World War II aviators captured by German forces after having been shot down. Having located a total of 239 World War II aviator POW survivors, the researchers asked each veteran to participate in the study; 33 responded to the letter of invitation. Each of the 33 POW survivors was administered the Minnesota Multiphasic Personality Inventory, one component of which measures level of post-traumatic stress disorder (PTSD). [*Note:* The higher the score, the higher is the level of PTSD.] The aviators produced a mean PTSD score of $\bar{x} = 9.00$ and a standard deviation of $s = 9.32$. Conduct a test to determine if the true mean PTSD score for all World War II aviator POWS is less than 16. [*Note*: The value 16 represents the mean PTSD score established for Vietnam POWS.] Use $\alpha = .10$.

8.152 Colored string preferred by chickens. Refer to the *Applied Animal Behaviour Science* (Oct. 2000) study of domestic chickens exposed to a pecking stimulus, presented in Exercise 7.25 (p. 354). Recall that the average number of pecks a chicken takes at a white string over a specified time interval is known to be $\mu = 7.5$ pecks. In an experiment in which 72 chickens were exposed to blue string, the average number of pecks was $\bar{x} = 1.13$ pecks, with a standard deviation of $s = 2.21$ pecks.
   a. On average, are chickens more apt to peck at white string than at blue string? Conduct the appropriate test of hypothesis, using $\alpha = .05$.
   b. Compare your answer to part **a** with your answer to Exercise 7.25**b**.
   c. Find the *p*-value for the test and interpret it.

8.153 Inbreeding of tropical wasps. Refer to the *Science* study of inbreeding in tropical swarm-founding wasps, presented in Exercise 7.139 (p. 389). A sample of 197 wasps, captured, frozen, and subjected to a series of genetic tests, yielded a sample mean inbreeding coefficient of $\bar{x} = .044$ with a standard deviation of $s = .884$. Recall that if the wasp has no tendency to inbreed, the true mean inbreeding coefficient μ for the species will equal 0.
   a. Test the hypothesis that the true mean inbreeding coefficient μ for this species of wasp exceeds 0. Use $\alpha = .05$.
   b. Compare the inference you made in part **a** with the inference you obtained in Exercise 7.139, using a confidence interval. Do the inferences agree? Explain.

8.154 A dental bonding agent. When bonding teeth, orthodontists must maintain a dry field. A bonding adhesive (called "Smartbond") has been developed to eliminate the necessity of a dry field. However, there is concern that the new bonding adhesive is not as strong as the current standard, a composite adhesive (*Trends in Biomaterials & Artificial Organs*, Jan. 2003). Tests on a sample of 10 extracted teeth bonded with the new adhesive resulted in a mean breaking strength (after 24 hours) of $\bar{x} = 5.07$ Mpa and a standard deviation of $s = .46$ Mpa, where Mpa = megapascal, a measure of force per unit area.
   a. Orthodontists want to know if the true mean breaking strength of the new bonding adhesive is less than 5.70 Mpa, the mean breaking strength of the composite adhesive. Conduct the appropriate analysis for the orthodontists. Use $\alpha = .01$.
   *b. In addition to requiring a good mean breaking strength, orthodontists are concerned about the variability in breaking strength of the new bonding adhesive. Conduct a test to determine whether the breaking strength variance differs from .5 Mpa.

8.155 Single-parent families. Examining data collected on 835 males from the National Youth Survey (a longitudinal survey of a random sample of U.S. households), researchers at Carnegie Mellon University found that 401 of the male youths were raised in a single-parent family (*Sociological Methods & Research*, Feb. 2001). Does this information allow you to conclude that more than 45% of male youths are raised in a single-parent family? Test at $\alpha = .05$.

*8.156 Weights of parrot fish. A marine biologist wishes to use parrot fish for experimental purposes due to the belief that their weight is fairly stable (i.e., the variability in weights among parrot fish is small). The biologist randomly samples 10 parrot fish and finds that their mean weight is 6.3 pounds and the standard deviation is 2.4 pounds. The biologist will use the parrot fish only if there is evidence that the variance of their weights is less than 6.
   a. Is there sufficient evidence for the biologist to claim that the variability in weights among parrot fish is small enough to justify their use in the experiment? Test at $\alpha = .01$.
   b. State any assumptions that are needed for the test mentioned in part **a** to be valid.

8.157 PCB in plant discharge. The EPA sets a limit of 5 parts per million (ppm) on PCB (polychlorinated biphenyl, a dangerous substance) in water. A major manufacturing firm producing PCB for electrical insulation discharges small amounts from the plant. The company management, attempting to control the PCB in its discharge, has given instructions to halt production if the mean amount of PCB in the effluent exceeds 4 ppm. A random sample of 50 water specimens produced the following statistics: $\bar{x} = 4.1$ ppm and $s = .5$ ppm.
   a. Do these statistics provide sufficient evidence to halt the production process? Use $\alpha = .01$.
   b. If you were the plant manager, would you want to use a large or a small value for α for the test in part **a**?

*8.158 PCB in plant discharge (cont'd). Refer to Exercise 8.157.
   a. In the context of the problem, define a Type II error.
   b. Calculate β for the test described in part **a** of Exercise 8.157, assuming that the true mean is $\mu = 4.1$ ppm.
   c. What is the power of the test to detect the effluent's departure from the standard of 4.0 ppm when the mean is 4.1 ppm?
   d. Repeat parts **b** and **c**, assuming that the true mean is 4.2 ppm. What happens to the power of the test as the plant's mean PCB departs farther from the standard?

8.159 PCB in plant discharge (cont'd). Refer to Exercises 8.157 and 8.158.
   a. Suppose a value of .10 is used to conduct the test. Does this change favor the manufacturer? Explain.
   b. Determine the value of β and the power of the test when $\alpha = .10$ and $\mu = 4.1$.
   c. What happens to the power of the test when α is increased?

8.160 Federal civil trial appeals. Refer to the *Journal of the American Law and Economics Association* (Vol. 3, 2001) study of appeals of federal civil trials, presented in Exercise 3.170 (p. 205). A breakdown of 678 civil cases that were originally tried in front of a judge and appealed by either the plaintiff or the defendant is reproduced in the table on p. 456. Do the data provide sufficient evidence to indicate that the percentage of civil cases appealed that are actually reversed is less than 25%? Test, using $\alpha = .01$.

Table for Exercise 8.160

Outcome of Appeal	Number of Cases
Plaintiff trial win—reversed	71
Plaintiff trial win—affirmed/dismissed	240
Defendant trial win—reversed	68
Defendant trial win—affirmed/dismissed	299
Total	678

8.161 Choosing portable grill displays. Refer to the *Journal of Consumer Research* (Mar. 2003) experiment on influencing the choices of others by offering undesirable alternatives, presented in Exercise 3.29 (p. 159). Recall that each of 124 college students selected three portable grills out of five to display on the showroom floor. The students were instructed to include Grill #2 (a smaller-sized grill) and select the remaining two grills in the display to maximize purchases of Grill #2. If the six possible grill display combinations (1–2–3, 1–2–4, 1–2–5, 2–3–4, 2–3–5, and 2–4–5) are selected at random, then the proportion of students selecting any display will be $1/6 = .167$. One theory tested by the researcher is that the students will tend to choose the three-grill display so that Grill #2 is a compromise between a more desirable and a less desirable grill. Of the 124 students, 85 students selected a three-grill display that was consistent with that theory. Use this information to test the theory proposed by the researcher at $\alpha = .05$.

8.162 **Interocular eye pressure. Ophthalmologists require an instrument that can rapidly measure interocular pressure for glaucoma patients. The device now in general use is known to yield readings of this pressure with a variance of 10.3. The variance of five pressure readings on the same eye by a newly developed instrument is equal to 9.8. Does this sample variance provide sufficient evidence to indicate that the new instrument is more reliable than the instrument currently in use? (Use $\alpha = .05$.)

8.163 Radioactive lichen. Refer to the Lichen Radionuclide Baseline Research project to monitor the level of radioactivity in lichen, presented in Exercise 7.133 (p. 388). Recall that University of Alaska researchers collected nine lichen specimens and measured the amount of the radioactive element cesium-137 (in microcuries per milliliter) in each specimen. Assume that in previous years the mean cesium amount in lichen was $\mu = .003$ microcurie per milliliter. Is there sufficient evidence to indicate that the mean amount of cesium in lichen specimens differs from this value? Use the SAS printout below to conduct a complete test of hypothesis at $\alpha = .10$.

```
Sample Statistics for CESIUM

    N       Mean       Std. Dev.      Std. Error
--------------------------------------------------------
    9      0.0090       0.0049          0.0016

Hypothesis Test

Null hypothesis:    Mean of CESIUM =  0.003
Alternative:        Mean of CESIUM ^= 0.003

        t Statistic     Df       Prob > t
    --------------------------------------------
           3.725         8        0.0058
```

8.164 Study of lunar soil. *Meteoritics* (Mar. 1995) reported the results of a study of lunar soil evolution. Data were obtained from the *Apollo 16* mission to the moon, during which a 62-cm core was extracted from the soil near the landing site. Monomineralic grains of lunar soil were separated out and examined for coating with dust and glass fragments. Each grain was then classified as coated or uncoated. Of interest is the "coat index"—that is, the proportion of grains that are coated. According to soil evolution theory, the coat index will exceed .5 at the top of the core, equal .5 in the middle of the core, and fall below .5 at the bottom of the core. Use the summary data in the accompanying table to test each part of the three-part theory. Use $\alpha = .05$ for each test.

	Location (depth)		
	Top (4.25 cm)	Middle (28.1 cm)	Bottom (54.5 cm)
Number of grains sampled	84	73	81
Number coated	64	35	29

Based on Basu, A., and McKay, D. S. "Lunar soil evolution processes and Apollo 16 core 60013/60014." *Meteoritics*, Vol. 30, No. 2, Mar. 1995, p. 166 (Table 2).

Applying the Concepts—Advanced

8.165 Polygraph test error rates. In a classic study reported in *Discover* magazine, a group of physicians subjected the *polygraph* (or *lie detector*) to the same careful testing given to medical diagnostic tests. They found that if 1,000 people were subjected to the polygraph and 500 told the truth and 500 lied, the polygraph would indicate that approximately 185 of the truth tellers were liars and that approximately 120 of the liars were truth tellers.

a. In the application of a polygraph test, an individual is presumed to be a truth teller (H_0) until "proven" a liar (H_a). In this context, what is a Type I error? A Type II error?

b. According to the study, what is the probability (approximately) that a polygraph test will result in a Type I error? A Type II error?

8.166 Parents who condone spanking. In Exercise 4.138 (p. 254) you read about a nationwide survey that claimed that 60% of parents with young children condone spanking their child as a regular form of punishment (*Tampa Tribune*, Oct. 5, 2000). In a random sample of 100 parents with young children, how many parents would need to say that they condone spanking as a form of punishment in order to refute the claim?

8.167 Solar joint inspections. X-rays and lasers are used to inspect solder-joint defects on printed circuit boards (PCBs). A particular manufacturer of laser-based inspection equipment claims that its product can inspect at least 10 solder joints per second, on average, when the joints are spaced .1 inch apart. The equipment was tested by a potential buyer on 48 different PCBs. In each case, the equipment was operated for exactly 1 second. The numbers of solder joints inspected on each run are listed in the table on the next page.

Data for exercise 8.167

10	9	10	10	11	9	12	8	8	9	6	10
7	10	11	9	9	13	9	10	11	10	12	8
9	9	9	7	12	6	9	10	10	8	7	9
11	12	10	0	10	11	12	9	7	9	9	10

a. The potential buyer doubts the manufacturer's claim. Do you agree?

*b. Assume that the standard deviation of the number of solder joints inspected on each run is 1.2, and the true mean number of solder joints that can be inspected is really equal to 9.5. How likely is the buyer to correctly conclude that the claim is false?

8.168 Effectiveness of skin cream. Pond's has discontinued the production of Age-Defying Complex, a cream with alpha-hydroxy acid, with Age-Defying Towlettes. Pond's advertised that the product could reduce wrinkles and improve the skin. In a study published in *Archives of Dermatology* (June 1996), 33 middle-aged women used a product with alpha-hydroxy acid for 22 weeks. At the end of the study period, a dermatologist judged whether each woman exhibited any improvement in the condition of her skin. The results for the 33 women (where I = improved skin and N = no improvement) are listed in the accompanying table. Can you conclude that the cream will improve the skin of more than 60% of middle-aged women?

I	I	N	I	N	N	I	I	I	I	I	I
N	I	I	I	N	I	I	N	I	N	I	I
I	I	I	I	I	N	I	I	N			

Critical Thinking Challenges

8.169 Verifying voter petitions. To get their names on the ballot of a local election, political candidates often must obtain petitions bearing the signatures of a minimum number of registered voters. According to the *St. Petersburg Times,* in Pinellas County, Florida, a certain political candidate obtained petitions with 18,200 signatures. To verify that the names on the petitions were signed by actual registered voters, election officials randomly sampled 100 of the signatures and checked each for authenticity. Only 2 were invalid signatures.

a. Is 98 out of 100 verified signatures sufficient to believe that more than 17,000 of the total 18,200 signatures are valid? Use $\alpha = .01$.

b. Repeat part a if only 16,000 valid signatures are required.

8.170 The Hot Tamale caper. "Hot Tamales" are chewy, cinnamon-flavored candies. A bulk vending machine is known to dispense, on average, 15 Hot Tamales per bag. *Chance* (Fall 2000) published an article on a classroom project in which students were required to purchase bags of Hot Tamales from the machine and count the number of candies per bag. One student group claimed it purchased five bags that had the following candy counts: 25, 23, 21, 21, and 20. There was some question as to whether the students had fabricated the data. Use a hypothesis test to gain insight into whether or not the data collected by the students were fabricated. Use a level of significance that gives the benefit of the doubt to the students.

Activity	**Challenging a Claim: Tests of Hypotheses**

Use the Internet or a newspaper or magazine to find an example of a claim made by a political or special-interest group about some characteristic (e.g., favor gun control) of the U.S. population. In this activity, you represent a rival group that believes the claim may be false.

1. In your example, what kinds of evidence might exist that would cause one to suspect that the claim might be false and therefore worthy of a statistical study? Be specific. If the claim were false, how would consumers be hurt?

2. Describe what data are relevant and how that data might be collected.

3. Explain the steps necessary to reject the group's claim at level α. State the null and alternative hypotheses. If you reject the claim, does it mean that the claim is false?

4. If you reject the claim when the claim is actually true, what type of error has occurred? What is the probability of this error occurring?

5. If you were to file a lawsuit against the group based on your rejection of its claim, how might the group use your results to defend itself?

References

Snedecor, G. W., and Cochran, W. G. *Statistical Methods,* 7th ed. Ames, IA: Iowa State University Press, 1980.

Wackerly, D., Mendenhall, W., and Scheaffer, R. *Mathematical Statistics with Applications,* 7th ed. Belmont, CA: Thomson, Brooks/Cole, 2008.

MINITAB: Tests of Hypotheses

Testing μ

Step 1 Access the MINITAB data worksheet that contains the sample data.

Step 2 Click on the "Stat" button on the MINITAB menu bar and then click on "Basic Statistics" and "1-Sample t," as shown in Figure 8.M.1.

Figure 8.M.1 MINITAB menu options for testing a mean

Step 3 On the resulting dialog box (shown in Figure 8.M.2), click on "Samples in columns" and then specify the quantitative variable of interest in the open box.

Figure 8.M.2 MINITAB 1-sample t test dialog box

Step 4 Check "Perform hypothesis test" and then specify the value of μ_0 for the null hypothesis in the "Hypothesized mean" box.

Step 5 Click on the "Options" button at the bottom of the dialog box and specify the form of the alternative hypothesis, as shown in Figure 8.M.3.

Figure 8.M.3 MINITAB 1-sample t test options

Step 6 Click "OK" to return to the "1-Sample t" dialog box and then click "OK" again to produce the hypothesis test.

Note: If you want to produce a test for the mean from summary information (e.g., the sample mean, sample standard deviation, and sample size), click on "Summarized data" in the "1-Sample t" dialog box, enter the values of the summary statistics and μ_0, and then click "OK."

Important: The MINITAB one-sample t-procedure uses the t-statistic to generate the hypothesis test. When the sample size n is small, this is the appropriate method. When the sample size n is large, the t-value will be approximately equal to the large-sample z-value, and the resulting test will still be valid. If you have a large sample and you know the value of the population standard deviation σ (which is rarely the case), select "1-Sample Z" from the "Basic Statistics" menu options (see Figure 8.M.1) and make the appropriate selections.

Testing p

Step 1 Access the MINITAB data worksheet that contains the sample data.

Step 2 Click on the "Stat" button on the MINITAB menu bar and then click on "Basic Statistics" and "1 Proportion" (see Figure 8.M.1).

Step 3 On the resulting dialog box (shown in Figure 8.M.4), click on "Samples in columns," and then specify the qualitative variable of interest in the open box.

Figure 8.M.4 MINITAB 1-proportion test dialog box

Step 4 Check "Perform hypothesis test" and then specify the null hypothesis value p_0 in the "Hypothesized proportion" box.

Step 5 Click "Options," then specify the form of the alternative hypothesis in the resulting dialog box, as shown in Figure 8.M.5. Also, check the "Use test and interval based on normal distribution" box at the bottom.

Figure 8.M.5 MINITAB 1-proportion test options

Step 6 Click "OK" to return to the "1 Proportion" dialog box and then click "OK" again to produce the test results.

Note: If you want to produce a confidence interval for a proportion from summary information (e.g., the number of successes and the sample size), click on "Summarized data" in the "1 Proportion" dialog box (see Figure 8.M.4). Enter the value for the number of trials (i.e., the sample size) and the number of events (i.e., the number of successes), and then click "OK."

Testing σ^2

Step 1 Access the MINITAB data worksheet that contains the sample data set.

Step 2 Click on the "Stat" button on the MINITAB menu bar and then click on "Basic Statistics" and "1 Variance" (see Figure 8.M.1).

Step 3 Once the resulting dialog box appears (see Figure 8.M.6), click on "Samples in columns" and then specify the quantitative variable of interest in the open box.

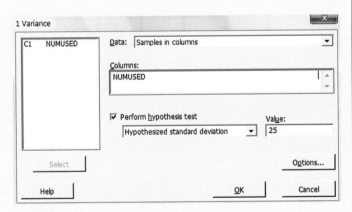

Figure 8.M.6 MINITAB 1-variance test dialog box

Step 4 Check "Perform hypothesis test" and specify the null hypothesis value of the standard deviation σ_0 in the open box.

Step 5 Click on the "Options" button at the bottom of the dialog box and specify the form of the alternative hypothesis (similar to Figure 8.M.3).

Step 6 Click "OK" twice to produce the hypothesis test.

Note: If you want to produce a test for the variance from summary information (e.g., the sample standard deviation and sample size), click on "Summarized data" in the "1 Variance" dialog box (Figure 8.M.6) and enter the values of the summary statistics.

TI-83/TI-84 Plus Graphing Calculator: Tests of Hypotheses

Note: The TI-83/TI-84 Plus graphing calculator cannot currently conduct a test for a population variance.

Hypothesis Test for a Population Mean (Large Sample Case)

Step 1 *Enter the data (Skip to Step 2 if you have summary statistics, not raw data.)*

- Press **STAT** and select **1:Edit**

Note: If the list already contains data, clear the old data. Use the up **ARROW** to highlight "**L1.**"

- Press **CLEAR ENTER**
- Use the **ARROW** and **ENTER** keys to enter the data set into **L1**

Step 2 *Access the statistical tests menu*

- Press **STAT**
- Arrow right to **TESTS**
- Press **ENTER** to select either **Z-Test** (if large sample and known) or **T-Test** (if unknown)

Step 3 *Choose "**Data**" or "**Stats**" ("Data" is selected when you have entered the raw data into a List. "Stats" is selected when you are given only the mean, standard deviation, and sample size.)*

- Press **ENTER**

If you selected "Data," enter the values for the hypothesis test where μ_0 = the value for μ in the null hypothesis, σ = assumed value of the population standard deviation.

- Set **List** to **L1**
- Set **Freq** to **1**
- Use the **ARROW** to highlight the appropriate alternative hypothesis
- Press **ENTER**
- Arrow down to "**Calculate**"
- Press **ENTER**

```
Z-Test
 Inpt:Data Stats
 μ0:10000
 σ:1000
 x̄:9755
 n:40
 μ:≠μ0 <μ0 >μ0
 Calculate Draw
```

If you selected "Stats," enter the values for the hypothesis test where μ_0 = the value for μ in the null hypothesis, σ = assumed value of the population standard deviation.

- Enter the sample mean, sample standard deviation, and sample size
- Use the **ARROW** to highlight the appropriate alternative hypothesis
- Press **ENTER**
- Arrow down to "**Calculate**"
- Press **ENTER**

```
Z-Test
μ<10000
z=-1.549516053
P=.0606288707
x̄=9755
n=40
```

The chosen test will be displayed as well as the z (or t) test statistic, the p-value, the sample mean, and the sample size.

Testing *p*

Step 1 *Enter the data (Skip to Step 2 if you have summary statistics, not raw data.)*

- Press **STAT** and select **1:Edit**

Note: If the list already contains data, clear the old data. Use the up **ARROW** to highlight "**L1.**"

- Press **CLEAR ENTER**
- Use the **ARROW** and **ENTER** keys to enter the data set into **L1**

Step 2 *Access the statistical tests menu*

- Press **STAT**
- Arrow right to **TESTS**
- Press **ENTER** after selecting **1-Prop Z Test**

Step 3 *Enter the hypothesized proportion p_0, the number of success x, and the sample size n*

- Use the **ARROW** to highlight the appropriate alternative hypothesis
- Press **ENTER**
- Arrow down to "**Calculate**"
- Press **ENTER**

The chosen test will be displayed as well as the z-test statistic and the p-value.

9 Inferences Based on Two Samples
Confidence Intervals and Tests of Hypotheses

CONTENTS

Where We've Been

- Explored two methods for making statistical inferences: *confidence intervals* and *tests of hypotheses*
- Studied confidence intervals and tests for a single population mean μ, a single population proportion p, and a single population variance σ^2
- Learned how to select the sample size necessary to estimate a population parameter with a specified margin of error

Where We're Going

- Learn how to identify the target parameter for comparing two populations (9.1)
- Learn how to compare two means by using confidence intervals and tests of hypotheses (9.2–9.3)
- Apply these inferential methods to problems in which we want to compare two population proportions, or two population variances (9.4, 9.6)
- Determine the sizes of the samples necessary to estimate the difference between two population parameters with a specified margin of error (9.5)

Statistics IN Action   ZixIt Corp. v. Visa USA Inc.—A Libel Case

The National Law Journal (Aug. 26–Sept. 2, 2002) reported on an interesting court case in volving ZixIt Corp., a start-up Internet credit card clearing center. ZixIt claimed that its new online credit card processing system would allow Internet shoppers to make purchases without revealing their credit card numbers. This claim violated the established protocols of most major credit card companies, including Visa. Without the company's knowledge, a Visa vice president for technology research and development began writing e-mails and Web site postings on a Yahoo! message board for ZixIt investors, challenging ZixIt's claim and urging investors to sell their ZixIt stock. The Visa executive posted over 400 e-mail and notes before he was caught. Once it was discovered that a Visa executive was responsible for the postings, ZixIt filed a lawsuit against Visa Corp., alleging that Visa—using the executive as its agent—had engaged in a "malicious two-part scheme to disparage and interfere with ZixIt" and its efforts to market the new online credit card processing system. In the libel case ZixIt asked for $699 million in damages.

Dallas lawyers Jeff Tillotson and Mike Lynn, of the law firm Lynn Tillotson & Pinker, were hired to defend Visa in the lawsuit. The lawyers, in turn, hired Dr. James McClave (co-author of this text) as their expert statistician. McClave testified in court on an "event study" he did matching the Visa executive's e-mail postings with movement of ZixIt's stock price the next business day. McClave's testimony,

showing that there was an equal number of days when the stock went up as went down after a posting, helped the lawyers representing Visa to prevail in the case. *The National Law Journal* reported that, after two and a half days of deliberation, "the jurors found [the Visa executive] was not acting in the scope of his employment and that Visa had not defamed ZixIt or interfered with its business."

In this chapter, we demonstrate several of the statistical analyses McClave used to infer that the Visa executive's postings had no effect on ZixIt's stock price. The daily ZixIt stock prices as well as the timing of the Visa executive's postings are saved in the **ZIXITVISA** file.* We apply the statistical methodology presented in this chapter to this data set in two Statistics in Action Revisited examples.

Statistics IN Action   Revisited

- Comparing Mean Price Changes (p. 473)
- Comparing Proportions (p. 497)

D *Data Set:* ZIXITVISA

9.1   Identifying the Target Parameter

Many experiments involve a comparison of two populations. For instance, a sociologist may want to estimate the difference in mean life expectancy between inner-city and suburban residents. Or a consumer group may want to test whether two major brands of food freezers differ in the average amount of electricity they use. Or a political candidate might want to estimate the difference in the proportions of voters in two districts who favor her candidacy. Or a professional golfer might be interested in comparing the variability in the distance that two competing brands of golf balls travel when struck with the same club. In this chapter, we consider techniques for using two samples to compare the populations from which they were selected.

The same procedures that are used to estimate and test hypotheses about a single population can be modified to make inferences about two populations. As in Chapters 7 and 8, the methodology used will depend on the sizes of the samples and the parameter of interest (i.e., the *target parameter*). Some key words and the type of data associated with the parameters covered in this chapter are listed in the following box.

---

*Data provided (with permission) from Info Tech, Inc., Gainesville, Florida.

Determining the Target Parameter

Parameter	Key Words or Phrases	Type of Data
$\mu_1 - \mu_2$	Mean difference; difference in averages	Quantitative
$p_1 - p_2$	Difference between proportions, percentages, fractions, or rates; compare proportions	Qualitative
$(\sigma_1)^2/(\sigma_2)^2$	Ratio of variances; difference in variability or spread; compare variation	Quantitative

You can see that the key words *difference* and *compare* help identify the fact that two populations are to be compared. In the previous examples, the words *mean* in *mean life expectancy* and *average* in *average amount of electricity* imply that the target parameter is the difference in population means, $\mu_1 - \mu_2$. The word *proportions* in *proportions of voters in two districts* indicates that the target parameter is the difference in proportions, $p_1 - p_2$. Finally, the key word *variability* in *variability in the distance* identifies the ratio of population variances, $(\sigma_1)^2/(\sigma_2)^2$, as the target parameter.

As with inferences about a single population, the type of data (quantitative or qualitative) collected on the two samples is also indicative of the target parameter. With quantitative data, you are likely to be interested in comparing the means or variances of the data. With qualitative data with two outcomes (success or failure), a comparison of the proportions of successes is likely to be of interest.

We consider methods for comparing two population means in Sections 9.2 and 9.3. A comparison of population proportions is presented in Section 9.4 and population variances in optional Section 9.6. We show how to determine the sample sizes necessary for reliable estimates of the target parameters in Section 9.5.

9.2 Comparing Two Population Means: Independent Sampling

In this section, we develop both large-sample and small-sample methodologies for comparing two population means. In the large-sample case, we use the *z*-statistic; in the small-sample case, we use the *t*-statistic.

Large Samples

Example   9.1 **D**
DIETS

A Large-Sample Confidence Interval for $(\mu_1 - \mu_2)$—Comparing Mean Weight Loss for Two Diets

Problem A dietitian has developed a diet that is low in fats, carbohydrates, and cholesterol. Although the diet was initially intended to be used by people with heart disease, the dietitian wishes to examine the effect this diet has on the weights of obese people. Two random samples of 100 obese people each are selected, and one group of 100 is placed on the low-fat diet. The other 100 are placed on a diet that contains approximately the same quantity of food, but is not as low in fats, carbohydrates, and cholesterol. For each person, the amount of weight lost (or gained) in a three-week period is recorded. The data, saved in the **DIETS** file, are listed in Table 9.1. Form a 95% confidence interval for the difference between the population mean weight losses for the two diets. Interpret the result.

Solution Recall that the general form of a large-sample confidence interval for a single mean μ is $\bar{x} \pm z_{\alpha/2}\sigma_{\bar{x}}$. That is, we add and subtract $z_{\alpha/2}$ standard deviations of the sample estimate $\bar{x}$ to and from the value of the estimate. We employ a similar procedure to form the confidence interval for the difference between two population means.

Let μ_1 represent the mean of the conceptual population of weight losses for all obese people who could be placed on the low-fat diet. Let μ_2 be similarly defined for the other diet. We wish to form a confidence interval for $(\mu_1 - \mu_2)$. An intuitively appealing estimator for $(\mu_1 - \mu_2)$ is the difference between the sample means, $(\bar{x}_1 - \bar{x}_2)$. Thus, we will form the confidence interval of interest with

$$(\bar{x}_1 - \bar{x}_2) \pm z_{\alpha/2}\sigma_{(\bar{x}_1 - \bar{x}_2)}$$

Table 9.1   Diet Study Data, Example 9.1

Weight Losses for Low-Fat Diet

8	10	10	12	9	3	11	7	9	2
21	8	9	2	2	20	14	11	15	6
13	8	10	12	1	7	10	13	14	4
8	12	8	10	11	19	0	9	10	4
11	7	14	12	11	12	4	12	9	2
4	3	3	5	9	9	4	3	5	12
3	12	7	13	11	11	13	12	18	9
6	14	14	18	10	11	7	9	7	2
16	16	11	11	3	15	9	5	2	6
5	11	14	11	6	9	4	17	20	10

Weight Losses for Regular Diet

6	6	5	5	2	6	10	3	9	11
14	4	10	13	3	8	8	13	9	3
4	12	6	11	12	9	8	5	8	7
6	2	6	8	5	7	16	18	6	8
13	1	9	8	12	10	6	1	0	13
11	2	8	16	14	4	6	5	12	9
11	6	3	9	9	14	2	10	4	13
8	1	1	4	9	4	1	1	5	6
14	0	7	12	9	5	9	12	7	9
8	9	8	10	5	8	0	3	4	8

Ⓓ *Data Set:* DIETS

Assuming that the two samples are independent, we write the standard deviation of the difference between the sample means (i.e., the *standard error* of $\bar{x}_1 - \bar{x}_2$) as

$$\sigma_{(\bar{x}_1 - \bar{x}_2)} = \sqrt{\frac{\sigma_1^2}{n_1} + \frac{\sigma_2^2}{n_2}}$$

Typically (as in this example), the population variances σ_1^2 and σ_2^2 are unknown. Since the samples are both large ($n_1 = n_2 = 100$), the sample variances s_1^2 and s_2^2 will be good estimators of their respective population variances. Thus, the estimated standard error is

$$\sigma_{(\bar{x}_1 - \bar{x}_2)} \approx \sqrt{\frac{s_1^2}{n_1} + \frac{s_2^2}{n_2}}$$

Summary statistics for the diet data are displayed at the top of the SPSS printout shown in Figure 9.1. Note that $\bar{x}_1 = 9.31$, $\bar{x}_2 = 7.40$, $s_1 = 4.67$, and $s_2 = 4.04$. Using these values and observing that $\alpha = .05$ and $z_{.025} = 1.96$, we find that the 95% confidence interval is, approximately,

$$(9.31 - 7.40) \pm 1.96\sqrt{\frac{(4.67)^2}{100} + \frac{(4.04)^2}{100}} = 1.91 \pm (1.96)(.62) = 1.91 \pm 1.22$$

or (.69, 3.13). This interval (rounded) is highlighted in Figure 9.1.

Group Statistics

DIET		N	Mean	Std. Deviation	Std. Error Mean
WTLOSS	LOWFAT	100	9.31	4.668	.467
	REGULAR	100	7.40	4.035	.404

Independent Samples Test

		Levene's Test for Equality of Variances		t-test for Equality of Means					95% Confidence Interval of the Difference	
		F	Sig.	t	df	Sig. (2-tailed)	Mean Difference	Std. Error Difference	Lower	Upper
WTLOSS	Equal variances assumed	1.367	.244	3.095	198	.002	1.910	.617	.693	3.127
	Equal variances not assumed			3.095	193.940	.002	1.910	.617	.693	3.127

Figure 9.1

SPSS analysis of diet study data

Using this estimation procedure over and over again for different samples, we know that approximately 95% of the confidence intervals formed in this manner will enclose the difference in population means $(\mu_1 - \mu_2)$. Therefore, we are highly confident that the mean weight loss for the low-fat diet is between .69 and 3.13 pounds more than the mean weight loss for the other diet. With this information, the dietitian better understands the potential of the low-fat diet as a weight-reduction diet.

Look Back If the confidence interval for $(\mu_1 - \mu_2)$ contains 0 [e.g., $(-2.5, 1.3)$], then it is possible for the difference between the population means to be 0 (i.e., $\mu_1 - \mu_2 = 0$). In this case, we could not conclude that a significant difference exists between the mean weight losses for the two diets.

■ **Now Work Exercise 9.6a**

The justification for the procedure used in Example 9.1 to estimate $(\mu_1 - \mu_2)$ relies on the properties of the sampling distribution of $(\bar{x}_1 - \bar{x}_2)$. The performance of the estimator in repeated sampling is pictured in Figure 9.2, and its properties are summarized in the following box:

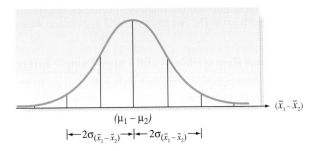

Figure 9.2

Sampling distribution of $(\bar{x}_1 - \bar{x}_2)$

Properties of the Sampling Distribution of $(\bar{x}_1 - \bar{x}_2)$

1. The mean of the sampling distribution of $(\bar{x}_1 - \bar{x}_2)$ is $(\mu_1 - \mu_2)$.

2. If the two samples are independent, the standard deviation of the sampling distribution is

$$\sigma_{(\bar{x}_1 - \bar{x}_2)} = \sqrt{\frac{\sigma_1^2}{n_1} + \frac{\sigma_2^2}{n_2}}$$

where σ_1^2 and σ_2^2 are the variances of the two populations being sampled and n_1 and n_2 are the respective sample sizes. We also refer to $\sigma_{(\bar{x}_1 - \bar{x}_2)}$ as the **standard error of the statistic** $(\bar{x}_1 - \bar{x}_2)$.

3. By the Central Limit Theorem, the sampling distribution of $(\bar{x}_1 - \bar{x}_2)$ is approximately normal *for large samples*.

In Example 9.1, we noted the similarity in the procedures for forming a large-sample confidence interval for one population mean and a large-sample confidence interval for the difference between two population means. When we are testing hypotheses, the procedures are again similar. The general large-sample procedures for forming confidence intervals and testing hypotheses about $(\mu_1 - \mu_2)$ are summarized in the following boxes:

Large, Independent Samples Confidence Interval for $(\mu_1 - \mu_2)$: Normal (z) Statistic

σ_1^2 and σ_2^2 known: $(\bar{x}_1 - \bar{x}_2) \pm z_{\alpha/2}\, \sigma_{(\bar{x}_1 - \bar{x}_2)} = (\bar{x}_1 - \bar{x}_2) \pm z_{\alpha/2}\sqrt{\dfrac{\sigma_1^2}{n_1} + \dfrac{\sigma_2^2}{n_2}}$

σ_1^2 and σ_2^2 unknown: $(\bar{x}_1 - \bar{x}_2) \pm z_{\alpha/2}\, \sigma_{(\bar{x}_1 - \bar{x}_2)} \approx (\bar{x}_1 - \bar{x}_2) \pm z_{\alpha/2}\sqrt{\dfrac{s_1^2}{n_1} + \dfrac{s_2^2}{n_2}}$

Applying the Concepts—Advanced

9.29 **Ethnicity and pain perception.** An investigation of ethnic differences in reports of pain perception was presented at the annual meeting of the American Psychosomatic Society (Mar. 2001). A sample of 55 blacks and 159 whites participated in the study. Subjects rated (on a 13-point scale) the intensity and unpleasantness of pain felt when a bag of ice was placed on their foreheads for two minutes. (Higher ratings correspond to higher pain intensity.) A summary of the results is provided in the following table.

	Blacks	Whites
Sample size	55	159
Mean pain intensity	8.2	6.9

a. Why is it dangerous to draw a statistical inference from the summarized data? Explain.

b. Give values of the missing sample standard deviations that would lead you to conclude (at $\alpha = .05$) that blacks, on average, have a higher pain intensity rating than whites.

c. Give values of the missing sample standard deviations that would lead you to an inconclusive decision (at $\alpha = .05$) regarding whether blacks or whites have a higher mean intensity rating.

9.30 **Mineral flotation in water study.** Refer to the *Minerals Engineering* (Vol. 46–47, 2013) study of the impact of calcium and gypsum on the flotation properties of silica in water, Exercise 2.48 (p. 81). Fifty solutions of deionized water were prepared both with and without calcium/gypsum, and the level of flotation of silica in the solution was measured using a variable called *zeta potential* (measured in millivolts, mV). The data (simulated, based on information provided in the journal article) are reproduced in the table. Can you conclude that the addition of calcium/gypsum to the solution affects silica flotation level?

Data for Exercise 9.30

Without calcium/gypsum

−47.1	−53.0	−50.8	−54.4	−57.4	−49.2	−51.5	−50.2	−46.4	−49.7
−53.8	−53.8	−53.5	−52.2	−49.9	−51.8	−53.7	−54.8	−54.5	−53.3
−50.6	−52.9	−51.2	−54.5	−49.7	−50.2	−53.2	−52.9	−52.8	−52.1
−50.2	−50.8	−56.1	−51.0	−55.6	−50.3	−57.6	−50.1	−54.2	−50.7
−55.7	−55.0	−47.4	−47.5	−52.8	−50.6	−55.6	−53.2	−52.3	−45.7

With calcium/gypsum

−9.2	−11.6	−10.6	−8.0	−10.9	−10.0	−11.0	−10.7	−13.1	−11.5
−11.3	−9.9	−11.8	−12.6	−8.9	−13.1	−10.7	−12.1	−11.2	−10.9
−9.1	−12.1	−6.8	−11.5	−10.4	−11.5	−12.1	−11.3	−10.7	−12.4
−11.5	−11.0	−7.1	−12.4	−11.4	−9.9	−8.6	−13.6	−10.1	−11.3
−13.0	−11.9	−8.6	−11.3	−13.0	−12.2	−11.3	−10.5	−8.8	−13.4

9.3 Comparing Two Population Means: Paired Difference Experiments

In Example 9.4, we compared two methods of teaching reading to "slow learners" by means of a 95% confidence interval. Suppose it is possible to measure the "reading IQs" of the "slow learners" *before* they are subjected to a teaching method. Eight pairs of "slow learners" with similar reading IQs are found, and one member of each pair is randomly assigned to the standard teaching method while the other is assigned to the new method. The data are given in Table 9.3. Do the data support the hypothesis that the population mean reading test score for "slow learners" taught by the new method is greater than the mean reading test score for those taught by the standard method?

Table 9.3	**Reading Test Scores for Eight Pairs of "Slow Learners"**	
Pair	New Method (1)	Standard Method (2)
1	77	72
2	74	68
3	82	76
4	73	68
5	87	84
6	69	68
7	66	61
8	80	76

Ⓓ *Data Set:* PAIREDSCORES

We want to test

$$H_0: (\mu_1 - \mu_2) = 0$$
$$H_a: (\mu_1 - \mu_2) > 0$$

Many researchers mistakenly use the t statistic for two independent samples (Section 9.2) to conduct this test. This invalid analysis is shown on the MINITAB printout of Figure 9.10. The test statistic, $t = 1.26$, and the p-value of the test, $p = .115$., are highlighted on the printout. At $\alpha = .10$, the p-value exceeds α. Thus, from *this* analysis, we might conclude that we do not have sufficient evidence to infer a difference in the mean test scores for the two methods.

```
Two-sample T for NEW vs STANDARD

            N    Mean   StDev  SE Mean
NEW         8   76.00    6.93      2.4
STANDARD    8   71.63    7.01      2.5


Difference = mu (NEW) - mu (STANDARD)
Estimate for difference:   4.38
95% lower bound for difference:   -1.76
T-Test of difference = 0 (vs >): T-Value = 1.26  P-Value = 0.115  DF = 14
Both use Pooled StDev = 6.9687
```

Figure 9.10

MINITAB printout of an invalid analysis of reading test scores in Table 9.3

If you examine the data in Table 9.3 carefully, however, you will find this result difficult to accept. The test score of the new method is larger than the corresponding test score for the standard method *for every one of the eight pairs of "slow learners."* This, in itself, seems to provide strong evidence to indicate that μ_1 exceeds μ_2. Why, then, did the t-test fail to detect the difference? The answer is, *the independent samples t-test is not a valid procedure to use with this set of data.*

The t-test is inappropriate because the assumption of independent samples is invalid. We have randomly chosen *pairs of test scores;* thus, once we have chosen the sample for the new method, we have *not* independently chosen the sample for the standard method. The dependence between observations within pairs can be seen by examining the pairs of test scores, which tend to rise and fall together as we go from pair to pair. This pattern provides strong visual evidence of a violation of the assumption of independence required for the two-sample t-test of Section 9.2. Note also that

$$s_p^2 = \frac{(n_1 - 1)s_1^2 + (n_2 - 1)s_2^2}{n_1 + n_2 - 2} = \frac{(8 - 1)(6.93)^2 + (8 - 1)(7.01)^2}{8 + 8 - 2} = 48.58$$

Hence, there is a *large variation within samples* (reflected by the large value of s_p^2) in comparison to the relatively *small difference between the sample means*. Because s_p^2 is so large, the t-test of Section 9.2 is unable to detect a difference between μ_1 and μ_2.

We now consider a valid method of analyzing the data of Table 9.3. In Table 9.4, we add the column of differences between the test scores of the pairs of "slow learners."

Table 9.4   Differences in Reading Test Scores

Pair	New Method	Standard Method	Difference (New Method − Standard Method)
1	77	72	5
2	74	68	6
3	82	76	6
4	73	68	5
5	87	84	3
6	69	68	1
7	66	61	5
8	80	76	4

We can regard these differences in test scores as a random sample of differences for all pairs (matched on reading IQ) of "slow learners," past and present. Then we can use this sample to make inferences about the mean of the population of differences, μ_d, which is equal to the difference $(\mu_1 - \mu_2)$. That is, the mean of the population (and sample) of differences equals the difference between the population (and sample) means. Thus, our test becomes

$$H_0: \mu_d = 0 \quad (\mu_1 - \mu_2 = 0)$$
$$H_a: \mu_d > 0 \quad (\mu_1 - \mu_2 > 0)$$

The test statistic is a one-sample t (Section 8.4), since we are now analyzing a single sample of differences for small n. Thus,

$$\text{Test statistic: } t = \frac{\bar{x}_d - 0}{s_d/\sqrt{n_d}}$$

where

$\bar{x}_d$ = Sample mean difference

s_d = Sample standard deviation of differences

n_d = Number of differences = Number of pairs

Assumptions: The population of differences in test scores is approximately normally distributed. The sample differences are randomly selected from the population differences. [*Note:* We do not need to make the assumption that $\sigma_1^2 = \sigma_2^2$.]

Rejection region: At significance level $\alpha = .05$, we will reject H_0 if $t > t_{.05}$, where $t_{.05}$ is based on $(n_d - 1)$ degrees of freedom.

t-distribution with 7 df

$\alpha = .05$

0

Rejection region

$t = 1.895$

Figure 9.11
Rejection region for Example 9.4

Referring to Table II in Appendix B, we find the t-value corresponding to $\alpha = .05$ and $n_d - 1 = 8 - 1 = 7$ df to be $t_{.05} = 1.895$. Then we will reject the null hypothesis if $t > 1.895$. (See Figure 9.11.) Note that the number of degrees of freedom decreases from $n_1 + n_2 - 2 = 14$ to 7 when we use the paired difference experiment rather than the two independent random samples design.

Summary statistics for the $n_d = 8$ differences are shown in the MINITAB printout of Figure 9.12. Note that $\bar{x}_d = 4.375$ and $s_d = 1.685$. Substituting these values into the formula for the test statistic, we have

$$t = \frac{\bar{x}_d - 0}{s_d/\sqrt{n_d}} = \frac{4.375}{1.685/\sqrt{8}} = 7.34$$

Because this value of t falls into the rejection region, we conclude (at $\alpha = .05$) that the population mean test score for "slow learners" taught by the new method exceeds the population mean score for those taught by the standard method. We can reach the same conclusion by noting that the p-value of the test, highlighted in Figure 9.12, is much smaller than $\alpha = .05$.

Figure 9.12
MINITAB paired difference analysis of reading test scores

```
Paired T for NEW - STANDARD

              N    Mean   StDev   SE Mean
NEW           8   76.00    6.93      2.45
STANDARD      8   71.63    7.01      2.48
Difference    8    4.375   1.685     0.596


95% lower bound for mean difference: 3.246
T-Test of mean difference = 0 (vs > 0): T-Value = 7.34  P-Value = 0.000
```

Now Work Exercises 9.37a and b

This kind of experiment, in which observations are paired and the differences are analyzed, is called a **paired difference experiment**. In many cases, a paired difference experiment can provide more information about the difference between population means than an independent samples experiment can. The idea is to compare population means by comparing the differences between pairs of experimental units (objects, people, etc.) that were similar prior to the experiment. The differencing removes sources of variation that tend to inflate σ^2. For example, when two children are taught to read by two different methods, the observed difference in achievement may be due to a difference in the effectiveness of the two teaching methods, *or* it may be due to differences in the initial reading levels and IQs of the two children (random error). To reduce the effect of differences in the children on the observed differences in reading achievement, the two methods of reading are imposed on two children who are more likely to possess similar intellectual capacity, namely, children with nearly equal IQs. The effect of this pairing is to remove the larger source of variation that would be present if children with different abilities were randomly assigned to the two samples. Making comparisons within groups of similar experimental units is called **blocking**, and the paired difference experiment is a simple example of a **randomized block experiment**. In our example, pairs of children with matching IQ scores represent the blocks.

Some other examples for which the paired difference experiment might be appropriate are the following:

1. Suppose you want to estimate the difference $(\mu_1 - \mu_2)$ in mean price per gallon between two major brands of premium gasoline. If you choose two independent random samples of stations for each brand, the variability in price due to geographic location may be large. To eliminate this source of variability, you could choose pairs of stations of similar size, one station for each brand, in close geographic proximity and use the sample of differences between the prices of the brands to make an inference about $(\mu_1 - \mu_2)$.

2. Suppose a college placement center wants to estimate the difference $(\mu_1 - \mu_2)$ in mean starting salaries for men and women graduates who seek jobs through the center. If it independently samples men and women, the starting salaries may vary because of their different college majors and differences in grade point averages. To eliminate these sources of variability, the placement center could match male and female job seekers according to their majors and grade point averages. Then the differences between the starting salaries of each pair in the sample could be used to make an inference about $(\mu_1 - \mu_2)$.

3. Suppose you wish to estimate the difference $(\mu_1 - \mu_2)$ in mean absorption rate into the bloodstream for two drugs that relieve pain. If you independently sample people, the absorption rates might vary because of age, weight, sex, blood pressure, etc. In fact, there are many possible sources of nuisance variability, and pairing individuals who are similar in all the possible sources would be quite difficult. However, it may be possible to obtain two measurements *on the same person*. First, we administer one of the two drugs and record the time until absorption. After a sufficient amount of time, the other drug is administered and a second measurement on absorption time is obtained. The differences between the measurements for each person in the sample could then be used to estimate $(\mu_1 - \mu_2)$. This procedure would be advisable only if the amount of time allotted between drugs is sufficient to guarantee little or no carry-over effect. Otherwise, it would be better to use different people matched as closely as possible on the factors thought to be most important.

Now Work Exercise 9.35

The hypothesis-testing procedures and the method of forming confidence intervals for the difference between two means in a paired difference experiment are summarized in the following boxes for both large and small n:

Paired Difference Confidence Interval for $\mu_d = \mu_1 - \mu_2$

Large Sample, Normal (z) Statistic

$$\bar{x}_d \pm z_{\alpha/2}\frac{\sigma_d}{\sqrt{n_d}} \approx \bar{x}_d \pm z_{\alpha/2}\frac{s_d}{\sqrt{n_d}}$$

Small Sample, Student's t-Statistic

$$\bar{x}_d \pm t_{\alpha/2}\frac{s_d}{\sqrt{n_d}}$$

where $t_{\alpha/2}$ is based on $(n_d - 1)$ degrees of freedom

---

Paired Difference Test of Hypothesis for $\mu_d = (\mu_1 - \mu_2)$

	One-Tailed Tests		Two-Tailed Test
	$H_0: \mu_d = D_0$	$H_0: \mu_d = D_0$	$H_0: \mu_d = D_0$
	$H_a: \mu_d < D_0$	$H_a: \mu_d > D_0$	$H_a: \mu_d \neq D_0$

Large Sample, Normal (z) Test Statistic: $z_c = \dfrac{(\bar{x}_d - D_0)}{(\sigma_d/\sqrt{n_d})} \approx \dfrac{(\bar{x}_d - D_0)}{(s_d/\sqrt{n_d})}$

Rejection region:	$z_c < -z_\alpha$	$z_c > z_\alpha$	$|z_c| > z_{\alpha/2}$
p-value:	$P(z < z_c)$	$P(z > z_c)$	$2P(z > z_c)$ if z_c is positive
			$2P(z < z_c)$ if z_c is negative

Small Sample, Student's t-Test Statistic: $t_c = \dfrac{(\bar{x}_d - D_0)}{(s_d/\sqrt{n_d})}$

Rejection region:	$t_c < -t_\alpha$	$t_c > t_\alpha$	$|t_c| > t_{\alpha/2}$
p-value:	$P(t < t_c)$	$P(t > t_c)$	$2P(t > t_c)$ if t_c is positive
			$2P(t < t_c)$ if t_c is negative

Decision: Reject H_0 if $\alpha > p$-value or if test statistic falls in rejection region where $P(z > z_\alpha) = \alpha, P(z > z_{\alpha/2}) = \alpha/2, P(t > t_\alpha) = \alpha, P(t > t_{\alpha/2}) = \alpha/2$, the distribution of t is based on $(n_d - 1)$df, and $\alpha = P(\text{Type I error}) = P(\text{Reject } H_0 | H_0 \text{ true})$.

[*Note:* The symbol for the numerical value assigned to the difference μ_d under the null hypothesis is D_0. For testing equal population means, $D_0 = 0$.]

Conditions Required for Valid Large-Sample Inferences about μ_d

1. A random sample of differences is selected from the target population of differences.
2. The sample size n_d is large (i.e., $n_d \geq 30$). (By the Central Limit Theorem, this condition guarantees that the test statistic will be approximately normal, regardless of the shape of the underlying probability distribution of the population.)

Conditions Required for Valid Small-Sample Inferences about μ_d

1. A random sample of differences is selected from the target population of differences.
2. The population of differences has a distribution that is approximately normal.

Example 9.5

Confidence Interval For μ_d—Comparing Mean Salaries of Males and Females

Problem An experiment is conducted to compare the starting salaries of male and female college graduates who find jobs. Pairs are formed by choosing a male and a female with the same major and similar grade point averages (GPAs). Suppose a random sample of 10 pairs is formed in this manner and the starting annual salary of each person is recorded. The results are shown in Table 9.5. Compare the mean starting salary μ_1 for males with the mean starting salary μ_2 for females, using a 95% confidence interval. Interpret the results.

Table 9.5	Data on Annual Salaries for Matched Pairs of College Graduates		
Pair	Male	Female	Difference Male − Female
1	$29,300	$28,800	$ 500
2	41,500	41,600	−100
3	40,400	39,800	600
4	38,500	38,500	0
5	43,500	42,600	900
6	37,800	38,000	−200
7	69,500	69,200	300
8	41,200	40,100	1,100
9	38,400	38,200	200
10	59,200	58,500	700

D *Data Set:* GRADS

Solution Since the data on annual salary are collected in pairs of males and females matched on GPA and major, a paired difference experiment is performed. To conduct the analysis, we first compute the differences between the salaries, as shown in Table 9.5. Summary statistics for these $n = 10$ differences are displayed at the top of the SAS printout shown in Figure 9.13.

```
                        The MEANS Procedure

                    Analysis Variable : DIFF

      Mean          Std Dev     N       Minimum       Maximum

    400.00         434.61      10     -200.00       1100.00




          Two Sample Paired t-test for the Means of MALE and FEMALE

Sample Statistics

    Group          N      Mean     Std. Dev.   Std. Error
    -----------------------------------------------------
    MALE          10     43930       11665        3688.8
    FEMALE        10     43530       11617        3673.6

Hypothesis Test

    Null hypothesis:      Mean of (MALE - FEMALE) =  0
    Alternative:          Mean of (MALE - FEMALE) ^= 0

       t Statistic      Df      Prob > t
    -----------------------------------------
         2.910           9       0.0173


95% Confidence Interval for the Difference between Two Paired Means

          Lower Limit    Upper Limit
          -----------    -----------
             89.10          710.90
```

Figure 9.13
SAS analysis of salary differences

The 95% confidence interval for $\mu_d = (\mu_1 - \mu_2)$ for this small sample is

$$\bar{x}_d \pm t_{\alpha/2} \frac{s_d}{\sqrt{n_d}}$$

where $t_{\alpha/2} = t_{.025} = 2.262$ (obtained from Table II, Appendix B) is based on $n_d - 1 = 9$ degrees of freedom. Substituting the values of $\bar{x}_d$ and s_d shown on the printout, we obtain

$$\bar{x}_d \pm 2.262 \frac{s_d}{\sqrt{n_d}} = 400 \pm 2.262 \left(\frac{434.613}{\sqrt{10}} \right)$$
$$= 400 \pm 310.88 \approx 400 \pm 311 = (\$89, \$711)$$

[*Note:* This interval is also shown highlighted at the bottom of the SAS printout of Figure 9.13.] Our interpretation is that the true mean difference between the starting salaries of males and females falls between \$89 and \$711, with 95% confidence. Since the interval falls above 0, we infer that $\mu_1 - \mu_2 > 0$; that is, the mean salary for males exceeds the mean salary for females.

Look Back Remember that $\mu_d = \mu_1 - \mu_2$. So if $\mu_d > 0$, then $\mu_1 > \mu_2$. Alternatively, if $\mu_d < 0$, then $\mu_1 < \mu_2$.

■ **Now Work Exercise 9.43**

To measure the amount of information about $(\mu_1 - \mu_2)$ gained by using a paired difference experiment in Example 9.5 rather than an independent samples experiment, we can compare the relative widths of the confidence intervals obtained by the two methods. A 95% confidence interval for $(\mu_1 - \mu_2)$ obtained from a paired difference experiment is, from Example 9.5, (\$89, \$711). If we mistakenly analyzed the same data as though this were an independent samples experiment,* we would first obtain the descriptive statistics shown in the SAS printout of Figure 9.14. Then we substitute the sample means and standard deviations shown on the printout into the formula for a 95% confidence interval for $(\mu_1 - \mu_2)$ using independent samples. The result is

$$(\bar{x}_1 - \bar{x}_2) \pm t_{.025} \sqrt{ s_p^2 \left(\frac{1}{n_1} + \frac{1}{n_2} \right) }$$

where

$$s_p^2 = \frac{(n_1 - 1)s_1^2 + (n_2 - 1)s_2^2}{n_1 + n_2 - 2}$$

Group Statistics

	GENDER	N	Mean	Std. Deviation	Std. Error Mean
SALARY	M	10	43930.00	11665.148	3688.844
	F	10	43530.00	11616.946	3673.601

Independent Samples Test

		Levene's Test for Equality of Variances		t-test for Equality of Means					95% Confidence Interval of the Difference	
		F	Sig.	t	df	Sig. (2-tailed)	Mean Difference	Std. Error Difference	Lower	Upper
SALARY	Equal variances assumed	.000	.991	.077	18	.940	400.000	5206.046	-10537.496	11337.496
	Equal variances not assumed			.077	18.000	.940	400.000	5206.046	-10537.509	11337.509

Figure 9.14

SPSS analysis of salaries, assuming independent samples

*This is done only to provide a measure of the increase in the amount of information obtained by a paired design in comparison to an unpaired design. Actually, if an experiment were designed that used pairing, an unpaired analysis would be invalid because the assumption of independent samples would not be satisfied.

SPSS performed these calculations and obtained the interval ($-10,537.50$, $11,337.50$), highlighted in Figure 9.14.

Notice that the independent samples interval includes 0. Consequently, if we were to use this interval to make an inference about ($\mu_1 - \mu_2$), we would incorrectly conclude that the mean starting salaries of males and females do not differ! You can see that the confidence interval for the independent sampling experiment is about 35 times wider than for the corresponding paired difference confidence interval. Blocking out the variability due to differences in majors and grade point averages significantly increases the information about the difference in males' and females' mean starting salaries by providing a much more accurate (a smaller confidence interval for the same confidence coefficient) estimate of ($\mu_1 - \mu_2$).

You may wonder whether a paired difference experiment is always superior to an independent samples experiment. The answer is, most of the time, but not always. We sacrifice half the degrees of freedom in the t-statistic when a paired difference design is used instead of an independent samples design. This is a loss of information, and unless that loss is more than compensated for by the reduction in variability obtained by blocking (pairing), the paired difference experiment will result in a net loss of information about ($\mu_1 - \mu_2$). Thus, we should be convinced that the pairing will significantly reduce variability before performing a paired difference experiment. Most of the time, this will happen.

One final note: The pairing of the observations is determined *before* the experiment is performed (i.e., by the *design* of the experiment). A paired difference experiment is *never* obtained by pairing the sample observations *after* the measurements have been acquired.

Ethics IN Statistics

In a two-group analysis, intentionally pairing observations after the data have been collected in order to produce a desired result is considered *unethical statistical practice.*

What Do You Do When the Assumption of a Normal Distribution for the Population of Differences Is Not Satisfied?

Answer: Use the Wilcoxon signed rank test for the paired difference design (Chapter 14).

Exercises 9.31–9.54

Understanding the Principles

9.31 In a paired difference experiment, when should the observations be paired, before or after the data are collected?

9.32 What are the advantages of using a paired difference experiment over an independent samples design?

9.33 *True or False.* In a paired difference experiment, an independent sample t-test is appropriate.

9.34 Is a paired difference experiment always superior to an independent samples experiment?

Learning the Mechanics

9.35 A paired difference experiment yielded n_d pairs of observations. In each case, what is the rejection region for testing $H_0: \mu_d = 2$ against $H_a: \mu_d > 2$?
a. $n_d = 10, \alpha = .05$
b. $n_d = 20, \alpha = .10$
c. $n_d = 5, \alpha = .025$
d. $n_d = 9, \alpha = .01$

9.36 A paired difference experiment produced the following data:

$$n_d = 9 \quad \bar{x}_1 = 150 \quad \bar{x}_2 = 157 \quad \bar{x}_d = -7 \quad s_d^2 = 36$$

a. Determine the values of t for which the null hypothesis $\mu_1 - \mu_2 = 0$ would be rejected in favor of the alternative hypothesis $\mu_1 - \mu_2 < 0$. Use $\alpha = .10$.
b. Conduct the paired difference test described in part **a**. Draw the appropriate conclusions.

c. What assumptions are necessary so that the paired difference test will be valid?

d. Find a 90% confidence interval for the mean difference μ_d.

e. Which of the two inferential procedures, the confidence interval of part **d** or the test of hypothesis of part **b**, provides more information about the difference between the population means?

9.37 The data for a random sample of six paired observations are shown in the following table.

Pair	Sample from Population 1	Sample from Population 2
1	7	4
2	3	1
3	9	7
4	6	2
5	4	4
6	8	7

a. Calculate the difference between each pair of observations by subtracting observation 2 from observation 1. Use the differences to calculate $\bar{x}_d$ and s_d^2.

b. If μ_1 and μ_2 are the means of populations 1 and 2, respectively, express μ_d in terms of μ_1 and μ_2.

c. Form a 95% confidence interval for μ_d.

d. Test the null hypothesis $H_0: \mu_d = 0$ against the alternative hypothesis $H_a: \mu_d \neq 0$. Use $\alpha = .05$.

9.38 The data for a random sample of 10 paired observations are shown in the following table.

L09038

Pair	Population 1	Population 2
1	19	24
2	25	27
3	31	36
4	52	53
5	49	55
6	34	34
7	59	66
8	47	51
9	17	20
10	51	55

a. If you wish to test whether these data are sufficient to indicate that the mean for population 2 is larger than that for population 1, what are the appropriate null and alternative hypotheses? Define any symbols you use.

b. Conduct the test from part **a**, using $\alpha = .10$. What is your decision?

c. Find a 90% confidence interval for μ_d. Interpret this interval.

d. What assumptions are necessary to ensure the validity of the preceding analysis?

9.39 A paired difference experiment yielded the following results:

$$n_d = 49, \quad \bar{x}_d = 13, \quad s_d = 6.$$

a. Test $H_0: \mu_d = 10$ against $H_a: \mu_d \neq 10$, where $\mu_d = (\mu_1 - \mu_2)$. Use $\alpha = .05$.

b. Report the p-value for the test you conducted in part **a**. Interpret the p-value.

Applying the Concepts—Basic

9.40 Summer weight-loss camp. Camp Jump Start is an 8-week summer camp for overweight and obese adolescents. Counselors develop a weight-management program for each camper that centers on nutrition education and physical activity. In a study published in *Pediatrics* (Apr. 2010), the body mass index (BMI) was measured for each of 76 campers both at the start and end of camp. Summary statistics on BMI measurements are shown in the table.

	Mean	Standard Deviation
Starting BMI	34.9	6.9
Ending BMI	31.6	6.2
Paired Differences	3.3	1.5

Based on Huelsing, J., Kanafani, N., Mao, J., and White, N. H. "Camp Jump Start: Effects of a residential summer weight-loss camp for older children and adolescents." *Pediatrics*, Vol. 125, No. 4, Apr. 2010 (Table 3).

a. Give the null and alternative hypothesis for determining whether the mean BMI at the end of camp is less than the mean BMI at the start of camp.

b. How should the data be analyzed, as an independent-samples t-test or as a paired-difference t-test? Explain.

c. Calculate the test statistic using the formula for an independent-samples t-test. (*Note:* This is *not* how the test should be conducted.)

d. Calculate the test statistic using the formula for a paired-difference t-test.

e. Compare the test statistics, parts **c** and **d**. Which test statistic provides more evidence in support of the alternative hypothesis?

f. The p-value of the test, part **d**, was reported as $p < .0001$. Interpret this result assuming $\alpha = .01$.

g. Do the differences in BMI values need to be normally distributed in order for the inference, part **f**, to be valid? Explain.

h. Find a 99% confidence interval for the true mean change in BMI for Camp Jump Start campers. Interpret the result.

9.41 Packaging of a children's health food. Refer to the *Journal of Consumer Behaviour* (Vol. 10, 2011) study of packaging of a children's health food product, Exercise 8.42 (p. 419). Recall that a fictitious brand of a healthy food product—sliced apples—was packaged to appeal to children (a smiling cartoon apple on the front of the package). The researchers compared the appeal of this fictitious brand to a commercially available brand of sliced apples that was not packaged for children. Each of 408 schoolchildren rated both brands on a 5-point "willingness to eat" scale, with $1 = $ "not willing at all" and $5 = $ "very willing." The fictitious brand had a sample mean score of 3.69, while the commercially available brand had a sample mean score of 3.00. The researchers wanted to compare the population mean score for the fictitious brand, μ_F, to the population mean score for the commercially available brand, μ_C. They theorized that μ_F will be greater than μ_C.

a. Specify the null and alternative hypothesis for the test.

b. Explain how the researchers should analyze the data and why.

c. The researchers reported a test statistic value of 5.71. Interpret this result. Use $\alpha = .05$ to draw your conclusion.

d. Find the approximate p-value of the test.

e. Could the researchers have tested at $\alpha = .01$ and arrived at the same conclusion?

9.42 Twinned drill holes. A traditional method of verifying mineralization grades in mining is to drill twinned holes, i.e., the drilling of a new hole, or "twin," next to an earlier drillhole. The use of twinned drill holes was investigated in *Exploration and Mining Geology* (Vol. 18, 2009). Geologists use data collected at both holes to estimate the total amount of heavy minerals (THM) present at the drilling site. The data in the next table (based on information provided in the journal article) represent THM percentages for a sample of 15 twinned holes drilled at a diamond mine in Africa. The geologists want to know if there is any evidence of a difference in the true THM means of all original holes and their twin holes drilled at the mine.

a. Explain why the data should be analyzed as paired differences.

b. Compute the difference between the "1st hole" and "2nd hole" measurements for each drilling location.

c. Find the mean and standard deviation of the differences, part **b**.

DRILL2

d. Use the summary statistics, part **c**, to find a 90% confidence interval for the true mean difference ("1st hole" minus "2nd hole") in THM measurements.

e. Interpret the interval, part **d**. Can the geologists conclude that there is no evidence of a difference in the true THM means of all original holes and their twin holes drilled at the mine?

Location	1st Hole	2nd Hole
1	5.5	5.7
2	11.0	11.2
3	5.9	6.0
4	8.2	5.6
5	10.0	9.3
6	7.9	7.0
7	10.1	8.4
8	7.4	9.0
9	7.0	6.0
10	9.2	8.1
11	8.3	10.0
12	8.6	8.1
13	10.5	10.4
14	5.5	7.0
15	10.0	11.2

9.43 **Healing potential of handling museum objects.** Does handling a museum object have a positive impact on a sick patient's well-being? To answer this question, researchers at the University College London collected data from 32 sessions with hospital patients (*Museum & Society*, Nov. 2009). Each patient's health status (measured on a 100-point scale) was recorded both before and after handling museum objects such as archaeological artifacts and brass etchings. The data (simulated) are listed in the accompanying table.

Session	Before	After	Session	Before	After
1	52	59	17	65	65
2	42	54	18	52	63
3	46	55	19	39	50
4	42	51	20	59	69
5	43	42	21	49	61
6	30	43	22	59	66
7	63	79	23	57	61
8	56	59	24	56	58
9	46	53	25	47	55
10	55	57	26	61	62
11	43	49	27	65	61
12	73	83	28	36	53
13	63	72	29	50	61
14	40	49	30	40	52
15	50	49	31	65	70
16	50	64	32	59	72

a. Explain why the data should be analyzed as paired differences.

b. Compute the difference between the "before" and "after" measurements for each session.

c. Find the mean and standard deviation of the differences, part **b**.

d. Use the summary statistics, part **c**, to find a 90% confidence interval for the true mean difference ("before" minus "after") in health status scale measurements.

e. Interpret the interval, part **d**. Does handling a museum object have a positive impact on a sick patient's well-being?

9.44 **Laughter among deaf signers.** The *Journal of Deaf Studies and Deaf Education* (Fall 2006) published an article on vocalized laughter among deaf users of American Sign Language (ASL). In videotaped ASL conversations among deaf participants, 28 laughed at least once. The researchers wanted to know if they laughed more as speakers (while signing) or as audience members (while listening). For each of the 28 deaf participants, the number of laugh episodes as a speaker and the number of laugh episodes as an audience member were determined. One goal of the research was to compare the mean numbers of laugh episodes of speakers and audience members.

a. Explain why the data should be analyzed as a paired difference experiment.

b. Identify the study's target parameter.

c. The study yielded a sample mean of 3.4 laughter episodes for speakers and a sample mean of 1.3 laughter episodes for audience members. Is this sufficient evidence to conclude that the population means are different? Explain.

d. A paired difference t-test resulted in $t = 3.14$ and p-value $< .01$. Interpret the results in the words of the problem.

9.45 **The placebo effect and pain.** According to research published in *Science* (Feb. 20, 2004), the mere belief that you are receiving an effective treatment for pain can reduce the pain you actually feel. Researchers tested this placebo effect on 24 volunteers as follows: Each volunteer was put inside a magnetic resonance imaging (MRI) machine for two consecutive sessions. During the first session, electric shocks were applied to their arms and the blood oxygen level–dependent (BOLD) signal (a measure related to neural activity in the brain) was recorded during pain. The second session was identical to the first, except that, prior to applying the electric shocks, the researchers smeared a cream on the volunteer's arms. The volunteers were informed that the cream would block the pain when, in fact, it was just a regular skin lotion (i.e., a placebo). If the placebo is effective in reducing the pain experience, the BOLD measurements should be higher, on average, in the first MRI session than in the second.

a. Identify the target parameter for this study.

b. What type of design was used to collect the data?

c. Give the null and alternative hypotheses for testing the placebo effect theory.

d. The differences between the BOLD measurements in the first and second sessions were computed and summarized in the study as follows: $n_d = 24$, $\bar{x}_d = .21$, $s_d = .47$. Use this information to calculate the test statistic.

e. The p-value of the test was reported as p-value $= .02$. Make the appropriate conclusion at $\alpha = .05$.

Applying the Concepts—Intermediate

9.46 **Settlement of shallow foundations.** Structures built on a shallow foundation (e.g., a concrete slab-on-grade foundation) are susceptible to settlement. Consequently, accurate settlement prediction is essential in the design of the foundation. Several methods for predicting settlement of shallow foundations on cohesive soil were compared in *Environmental & Engineering Geoscience* (Nov. 2012). Settlement data for a sample of 13 structures built on a shallow foundation were collected. The actual settlement values (measured in millimeters) for each structure were compared to settlement predictions made using a formula that accounts for dimension, rigidity, and embedment depth of the foundation. The data are listed in the table.

Structure	Actual	Predicted
1	11	11
2	11	11
3	10	12
4	8	6
5	11	9
6	9	10
7	9	9
8	39	51
9	23	24
10	269	252
11	4	3
12	82	68
13	250	264

Source: Ozur, M. "Comparing methods for predicting immediate settlement of shallow foundations on cohesive soils based on hypothetical and real cases." *Environmental & Engineering Geoscience*, Vol. 18, No. 4, Nov. 2012 (from Table 4).

a. What type of design was employed to collect the data?

b. Use the information in the accompanying SAS print-out to construct a 99% confidence interval for the mean difference between actual and predicted settlement value. Give a practical interpretation of the interval.

c. Explain the meaning of "99% confidence" for this application.

The TTEST Procedure

Difference: ACTUAL - PREDICTED

N	Mean	Std Dev	Std Err	Minimum	Maximum
13	0.4615	8.3528	2.3166	-14.0000	17.0000

Mean	95% CL Mean		Std Dev	95% CL Std Dev	
0.4615	-4.5860	5.5091	8.3528	5.9897	13.7883

DF	t Value	Pr > |t|
12	0.20	0.8454

9.47 **Solar energy generation along highways.** The potential of using solar panels constructed above national highways to generate energy was explored in the *International Journal of Energy and Environmental Engineering* (Dec. 2013). Two-layer solar panels (with 1 meter separating the panels) were constructed above sections of both east-west and north-south highways in India. The amount of energy (kilowatt hours) supplied to the country's grid by the solar panels above the two types of highways was determined each month. The data for several randomly selected months are provided in the table. The researchers concluded that the "two-layer solar panel energy generation is more viable for

the north-south oriented highways as compared to east-west oriented roadways." Do you agree?

Month	East-West	North-South
February	8658	8921
April	7930	8317
July	5120	5274
September	6862	7148
October	8608	8936

Source: Sharma, P., and Harinarayana, T. "Solar energy generation potential along national highways." *International Journal of Energy and Environmental Engineering*, Vol. 49, No. 1, Dec. 2013 (Table 3).

9.48 **Estimating well scale deposits.** Scale deposits can cause a serious reduction in the flow performance of a well. A study published in the *Journal of Petroleum and Gas Engineering* (Apr. 2013) compared two methods of estimating the damage from scale deposits (called *skin factor*). One method of estimating the well skin factor uses a series of Excel spreadsheets, while the second method employs EPS computer software. Skin factor data was obtained from applying both methods to 10 randomly selected oil wells: 5 vertical wells and 5 horizontal wells. The results are supplied in the accompanying table.

a. Compare the mean skin factor values for the two estimation methods using all 10 sampled wells. Test at $\alpha = .05$. What do you conclude?

b. Repeat part **a**, but analyze the data for the 5 horizontal wells only.

c. Repeat part **a**, but analyze the data for the 5 vertical wells only.

Skin Factor Values

Well (Type)	Excel Spreadsheet	EPS Software
1 (Horizontal)	44.48	37.77
2 (Horizontal)	18.34	13.31
3 (Horizontal)	19.21	7.02
4 (Horizontal)	11.70	4.77
5 (Horizontal)	9.25	1.96
6 (Vertical)	317.40	281.74
7 (Vertical)	181.44	192.16
8 (Vertical)	154.65	140.84
9 (Vertical)	77.43	56.86
10 (Vertical)	49.37	45.01

Source: Rahuma, K. M., et al. "Comparison between spreadsheet and specialized programs in calculating the effect of scale deposition on the well flow performance." *Journal of Petroleum and Gas Engineering*, Vol. 4, No. 4, Apr. 2013 (Table 2).

9.49 **Acidity of mouthwash.** Acid has been found to be a primary cause of dental caries (cavities). It is theorized that oral mouthwashes contribute to the development of caries due to the antiseptic agent oxidizing into acid over time. This theory was tested in the *Journal of Dentistry, Oral Medicine and Dental Education* (Vol. 3, 2009). Three bottles of mouthwash, each of a different brand, were randomly selected from a drugstore. The pH level (where lower pH levels indicate higher acidity) of each bottle was measured on the date of purchase and after 30 days. The data are shown in the next table.

Conduct an analysis to determine if the mean initial pH level of mouthwash differs significantly from the mean pH level after 30 days. Use $\alpha = .05$ as your level of significance.

Mouthwash Brand	Initial pH	Final pH
LMW	4.56	4.27
SMW	6.71	6.51
RMW	5.65	5.58

Based on Chunhye, K. L., and Schmitz, B. C., "Determination of pH, total acid, and total ethanol in oral health products: Oxidation of ethanol and recommendations to mitigate its association with dental caries." *Journal of Dentistry, Oral Medicine and Dental Education*, Vol. 3, No. 1, 2009 (Table 1).

9.50 **Visual search and memory study.** In searching for an item (e.g., a roadside traffic sign, a lost earring, or a tumor in a mammogram), common sense dictates that you will not reexamine items previously rejected. However, researchers at Harvard Medical School found that a visual search has no memory (*Nature,* Aug. 6, 1998). In their experiment, nine subjects searched for the letter "T" mixed among several letters "L." Each subject conducted the search under two conditions: random and static. In the random condition, the locations of the letters were changed every 111 milliseconds; in the static condition, the locations of the letters remained unchanged. In each trial, the reaction time in milliseconds (i.e., the amount of time it took the subject to locate the target letter) was recorded.

a. One goal of the research was to compare the mean reaction times of subjects in the two experimental conditions. Explain why the data should be analyzed as a paired difference experiment.

b. If a visual search has no memory, then the main reaction times in the two conditions will not differ. Specify H_0 and H_a for testing the "no-memory" theory.

c. The test statistic was calculated as $t = 1.52$ with p-value $= .15$. Draw the appropriate conclusion.

9.51 **Linking dementia and leisure activities.** Does participation in leisure activities in your youth reduce the risk of Alzheimer's disease and other forms of dementia? To answer this question, a group of university researchers studied a sample of 107 same-sex Swedish pairs of twins (*Journal of Gerontology: Psychological Sciences and Social Sciences*, Sept. 2003). Each pair of twins was discordant for dementia; that is, one member of each pair was diagnosed with Alzheimer's disease while the other member (the control) was nondemented for at least five years after the sibling's onset of dementia. The level of overall leisure activity (measured on an 80-point scale, where higher values indicate higher levels of leisure activity) of each twin of each pair 20 years prior to the onset of dementia was obtained from the Swedish Twin Registry database. The leisure activity scores (simulated on the basis of summary information presented in the journal article) are saved in the **DEMENT** file. The first five and last five observations are shown in the following table.

Pair	Control	Demented
1	27	13
2	57	57
3	23	31
4	39	46
5	37	37
⋮	⋮	⋮
103	22	14
104	32	23
105	33	29
106	36	37
107	24	1

a. Explain why the data should be analyzed as a paired difference experiment.

b. Conduct the appropriate analysis, using $\alpha = .05$. Make an inference about which member of the pair, the demented or control (nondemented) twin, had the largest average level of leisure activity.

9.52 **Ethical sensitivity of teachers toward racial intolerance.** Many high schools have education programs that encourage teachers to embrace racial tolerance. To gauge the effectiveness of one such program that utilizes two videos of teachers engaging in racial stereotypes of their students, researchers recruited 238 high school professionals (including teachers and counselors) to participate in a study (*Journal of Moral Education*, Mar. 2010). Teachers watched the first video, then were given a pretest—the Quick-REST Survey—designed to measure ethical sensitivity toward racial intolerance. The teachers next participated in an all-day workshop on cultural competence. At the end of the workshop, the teachers watched the second video and again were given the Quick-REST Survey (the posttest). To determine whether the program was effective, the researchers compared the mean scores on the Quick-REST Survey using a paired-difference t-test. (*Note:* The higher the score on the Quick-REST Survey, the greater the level of racial tolerance.)

a. The researchers reported the sample means for the pretest and posttest as 75.85 and 80.35, respectively. Why is it dangerous to gauge the effectiveness of the program based only on these summary statistics?

b. The paired-difference t-test (posttest minus pretest) was reported as $t = 4.50$ with an associated observed significance level of p-value $< .001$. Interpret this result.

c. What assumptions, if any, are necessary for the validity of the inference, part **b**?

9.53 **Impact of red light cameras on car crashes.** To combat red-light-running crashes, many states are adopting photo-red enforcement programs. In these programs, red light cameras installed at dangerous intersections photograph the license plates of vehicles that run the red light. How effective are photo-red enforcement programs in reducing red-light-running crash incidents at intersections? The Virginia Department of Transportation (VDOT) conducted a comprehensive study of its newly adopted photo-red enforcement program and published the results in a June 2007 report. In one portion of the study, the VDOT provided crash data both before and after installation of red light cameras at several intersections. The data (measured as the number of crashes caused by red light running per intersection per year) for 13 intersections in Fairfax County, Virginia, are given in the table. Analyze the data for the VDOT. What do you conclude?

Intersection	Before Camera	After Camera
1	3.60	1.36
2	0.27	0
3	0.29	0
4	4.55	1.79
5	2.60	2.04
6	2.29	3.14
7	2.40	2.72
8	0.73	0.24
9	3.15	1.57
10	3.21	0.43
11	0.88	0.28
12	1.35	1.09
13	7.35	4.92

Based on Virginia Transportation Research Council, "Research report: The impact of red light cameras (photo-red enforcement) on crashes in Virginia." June 2007.

Applying the Concepts—Advanced

9.54 **Alcoholic fermentation in wines.** Determining alcoholic fermentation in wine is critical to the wine-making process. Must/wine density is a good indicator of the fermentation point, since the density value decreases as sugars are converted into alcohol. For decades, winemakers have measured must/wine density with a hydrometer. Although accurate, the hydrometer employs a manual process that is very time consuming. Consequently, large wineries are searching for more rapid measures of density measurement. An alternative method utilizes the hydrostatic balance instrument (similar to the hydrometer, but digital). A winery in Portugal collected must/wine density measurements on white wine samples randomly selected from the fermentation process for a recent harvest. For each sample, the density of the wine at 20°C was measured with both the hydrometer and the hydrostatic balance. The densities for 40 wine samples are saved in the **WINE40** file. The first five and last five observations are shown in the accompanying table. The winery will use the alternative method of mea-suring wine density only if it can be demonstrated that the mean difference between the density measurements of the two methods does not exceed .002. Perform the analysis for the winery. Provide the winery with a written report of your conclusions.

Sample	Hydrometer	Hydrostatic
1	1.08655	1.09103
2	1.00270	1.00272
3	1.01393	1.01274
4	1.09467	1.09634
5	1.10263	1.10518
⋮	⋮	⋮
36	1.08084	1.08097
37	1.09452	1.09431
38	0.99479	0.99498
39	1.00968	1.01063
40	1.00684	1.00526

Based on Cooperative Cellar of Borba (*Adega Cooperativ a de Borba*), Portugal.

9.4 Comparing Two Population Proportions: Independent Sampling

Suppose a presidential candidate wants to compare the preferences of registered voters in the northeastern United States with those in the southeastern United States. Such a comparison would help determine where to concentrate campaign efforts. The candidate hires a professional pollster to randomly choose 1,000 registered voters in the northeast and 1,000 in the southeast and interview each to learn her or his voting preference. The objective is to use this sample information to make an inference about the difference $(p_1 - p_2)$ between the proportion p_1 of *all* registered voters in the northeast and the proportion p_2 of *all* registered voters in the southeast who plan to vote for the presidential candidate.

The two samples represent independent binomial experiments. (See Section 4.4 for the characteristics of binomial experiments.) The binomial random variables are the numbers x_1 and x_2 of the 1,000 sampled voters in each area who indicate that they will vote for the candidate. The results are summarized in Table 9.6

We can now calculate the sample proportions $\hat{p}_1$ and $\hat{p}_2$ of the voters in favor of the candidate in the northeast and southeast, respectively:

Table 9.6	Results of Poll
Northeast	Southeast
$n_1 = 1,000$	$n_2 = 1,000$
$x_1 = 546$	$x_2 = 475$

$$\hat{p} = \frac{x_1}{n_1} = \frac{546}{1,000} = .546 \quad \hat{p}_2 = \frac{x_2}{n_2} = \frac{475}{1,000} = .475$$

The difference between the sample proportions $(\hat{p}_1 - \hat{p}_2)$ makes an intuitively appealing point estimator of the difference between the population $(p_1 - p_2)$. For this example, the estimate is

$$(\hat{p}_1 - \hat{p}_2) = .546 - .475 = .071$$

To judge the reliability of the estimator $(\hat{p}_1 - \hat{p}_2)$, we must observe its performance in repeated sampling from the two populations. That is, we need to know the sampling distribution of $(\hat{p}_1 - \hat{p}_2)$. The properties of the sampling distribution are given in the next box. Remember that $\hat{p}_1$ and $\hat{p}_2$ can be viewed as means of the number of successes per trial in the respective samples, so the Central Limit Theorem applies when the sample sizes are large.

Properties of the Sampling Distribution of $(\hat{p}_1 - \hat{p}_2)$

1. The mean of the sampling distribution of $(\hat{p}_1 - \hat{p}_2)$ is $(p_1 - p_2)$; that is,

$$E(\hat{p}_1 - \hat{p}_2) = p_1 - p_2$$

Thus, $(\hat{p}_1 - \hat{p}_2)$ is an unbiased estimator of $(p_1 - p_2)$.

2. The standard deviation of the sampling distribution of $(\hat{p}_1 - \hat{p}_2)$ is

$$\sigma_{(\hat{p}_1 - \hat{p}_2)} = \sqrt{\frac{p_1 q_1}{n_1} + \frac{p_2 q_2}{n_2}}$$

3. If the sample sizes n_1 and n_2 are large (see Section 7.4 for a guideline), the sampling distribution of $(\hat{p}_1 - \hat{p}_2)$ is approximately normal.

Since the distribution of $(\hat{p}_1 - \hat{p}_2)$ in repeated sampling is approximately normal, we can use the z-statistic to derive confidence intervals for $(p_1 - p_2)$ or to test a hypothesis about $(p_1 - p_2)$.

For the voter example, a 95% confidence interval for the difference $(p_1 - p_2)$ is

$$(\hat{p}_1 - \hat{p}_2) \pm 1.96\sigma_{(\hat{p}_1 - \hat{p}_2)}, \text{ or } (\hat{p}_1 - \hat{p}_2) \pm 1.96\sqrt{\frac{p_1 q_1}{n_1} + \frac{p_2 q_2}{n_2}}$$

The quantities $p_1 q_1$ and $p_2 q_2$ must be estimated in order to complete the calculation of the standard deviation $\sigma_{(\hat{p}_1 - \hat{p}_2)}$ and, hence, the calculation of the confidence interval. In Section 7.4, we showed that the value of pq is relatively insensitive to the value chosen to approximate p. Therefore, $\hat{p}_1\hat{q}_1$ and $\hat{p}_2\hat{q}_2$ will provide satisfactory approximations of $p_1 q_1$ and $p_2 q_2$, respectively. Then

$$\sqrt{\frac{p_1 q_1}{n_1} + \frac{p_2 q_2}{n_2}} \approx \sqrt{\frac{\hat{p}_1\hat{q}_1}{n_1} + \frac{\hat{p}_2\hat{q}_2}{n_2}}$$

and we will approximate the 95% confidence interval by

$$(\hat{p}_1 - \hat{p}_2) \pm 1.96\sqrt{\frac{\hat{p}_1\hat{q}_1}{n_1} + \frac{\hat{p}_2\hat{q}_2}{n_2}}$$

Substituting the sample quantities yields

$$(.546 - .475) \pm 1.96\sqrt{\frac{(.546)(.454)}{1,000} + \frac{(.475)(.525)}{1,000}}$$

or $.071 \pm .044$. Thus, we are 95% confident that the interval from .027 to .115 contains $(p_1 - p_2)$.

We infer that there are between 2.7% and 11.5% more registered voters in the northeast than in the southeast who plan to vote for the presidential candidate. It seems that the candidate should direct a greater campaign effort in the southeast than in the northeast.

Now Work Exercise 9.63

The general form of a confidence interval for the difference $(p_1 - p_2)$ between population proportions is given in the following box:

Large-Sample $100(1 - \alpha)\%$ Confidence Interval for $(p_1 - p_2)$: Normal (z) Statistic

$$(\hat{p}_1 - \hat{p}_2) \pm z_{\alpha/2}\sigma_{(\hat{p}_1 - \hat{p}_2)} = (\hat{p}_1 - \hat{p}_2) \pm z_{\alpha/2}\sqrt{\frac{p_1 q_1}{n_1} + \frac{p_2 q_2}{n_2}}$$

$$\approx (\hat{p}_1 - \hat{p}_2) \pm z_{\alpha/2}\sqrt{\frac{\hat{p}_1 \hat{q}_1}{n_1} + \frac{\hat{p}_2 \hat{q}_2}{n_2}}$$

The z-statistic,

$$z = \frac{(\hat{p}_1 - \hat{p}_2) - (p_1 - p_2)}{\sigma_{(\hat{p}_1 - \hat{p}_2)}}$$

is used to test the null hypothesis that $(p_1 - p_2)$ equals some specified difference, say, D_0. For the special case where $D_0 = 0$—that is, where we want to test the null hypothesis $H_0: (p_1 - p_2) = 0$ (or, equivalently, $H_0: p_1 = p_2$)—the best estimate of $p_1 = p_2 = p$ is obtained by dividing the total number of successes $(x_1 + x_2)$ for the two samples by the total number of observations $(n_1 + n_2)$, that is,

$$\hat{p} = \frac{x_1 + x_2}{n_1 + n_2}, \quad \text{or} \quad \hat{p} = \frac{n_1 \hat{p}_1 + n_2 \hat{p}_2}{n_1 + n_2}$$

The second equation shows that $\hat{p}$ is a weighted average of $\hat{p}_1$ and $\hat{p}_2$, with the larger sample receiving more weight. If the sample sizes are equal, then $\hat{p}$ is a simple average of the two sample proportions of successes.

We now substitute the weighted average $\hat{p}$ for both p_1 and p_2 in the formula for the standard deviation of $(\hat{p}_1 - \hat{p}_2)$:

$$\sigma_{(\hat{p} - \hat{p}_2)} = \sqrt{\frac{p_1 q_1}{n_1} + \frac{p_2 q_2}{n_2}} \approx \sqrt{\frac{\hat{p}\hat{q}}{n_1} + \frac{\hat{p}\hat{q}}{n_2}} = \sqrt{\hat{p}\hat{q}\left(\frac{1}{n_1} + \frac{1}{n_2}\right)}$$

The test is summarized in the following box:

Large-Sample Test of Hypothesis about $(p_1 - p_2)$: Normal (z) Statistic

Test statistic:
$$z_c = \frac{(\hat{p}_1 - \hat{p}_2)}{\sqrt{\frac{p_1 q_1}{n_1} + \frac{p_2 q_2}{n_2}}} \approx \frac{(\hat{p}_1 - \hat{p}_2)}{\sqrt{\hat{p}\hat{q}\left(\frac{1}{n_1} + \frac{1}{n_2}\right)}}$$

where $\hat{p} = (x_1 + x_2)/(n_1 + n_2)$

	One-Tailed Tests		Two-Tailed Test
	$H_0: p_1 - p_2 = 0$*	$H_0: p_1 - p_2 = 0$	$H_0: p_1 - p_2 = 0$
	$H_a: p_1 - p_2 < 0$	$H_a: p_1 - p_2 > 0$	$H_a: p_1 - p_2 \neq 0$
Rejection region:	$z_c < -z_\alpha$	$z_c > z_\alpha$	$|z_c| > z_{\alpha/2}$
p-value:	$P(z < z_c)$	$P(z > z_c)$	$2P(z > z_c)$ if z_c is positive
			$2P(t < z_c)$ if z_c is negative

Decision: Reject H_0 if $\alpha > p$-value or if test statistic (z_c) falls in rejection region where $P(z > z_\alpha) = \alpha, P(z > z_{\alpha/2}) = \alpha/2$, and $\alpha = P(\text{Type I error}) = P(\text{Reject } H_0 | H_0 \text{ true})$.

---

*The test can be adapted to test for a difference in proportions $D_0 \neq 0$. Because most applications require a test of equal proportions, $p_1 = p_2$ (i.e., $p_1 - p_2 = D_0 = 0$), we confine our attention to this case.

> **Conditions Required for Valid Large-Sample Inferences about $(p_1 - p_2)$**
> 1. The two samples are randomly selected in an independent manner from the two target populations.
> 2. The sample sizes, n_1 and n_2, are both large, so the sampling distribution of $(\hat{p}_1 - \hat{p}_2)$ will be approximately normal. (This condition will be satisfied if both $n_1\hat{p}_1 \geq 15, n_1\hat{q}_1 \geq 15$, and $n_2\hat{p}_2 \geq 15, n_2\hat{q}_2 \geq 15$.)

Example   9.6

A Large-Sample Test about $(p_1 - p_2)$ — Comparing Fractions of Smokers for Two Years

Problem In the past decade, intensive antismoking campaigns have been sponsored by both federal and private agencies. Suppose the American Cancer Society randomly sampled 1,500 adults in 2000 and then sampled 1,750 adults in 2010 to determine whether there was evidence that the percentage of smokers had decreased. The results of the two sample surveys are shown in Table 9.7, where x_1 and x_2 represent the numbers of smokers in the 2000 and 2010 samples, respectively. Do these data indicate that the fraction of smokers decreased over this 10-year period? Use $\alpha = .05$.

Solution If we define p_1 and p_2 as the true proportions of adult smokers in 2000 and 2010, respectively, then the elements of our test are

$$H_0: (p_1 - p_2) = 0$$
$$H_a: (p_1 - p_2) > 0$$

(The test is one tailed, since we are interested only in determining whether the proportion of smokers *decreased*.)

$$\text{Test statistic: } z = \frac{(\hat{p}_1 - \hat{p}_2) - 0}{\sigma_{(\hat{p}_1 - \hat{p}_2)}}$$

Rejection region using $\alpha = .05$: $z > z_\alpha = z_{.05} = 1.645$ (see Figure 9.15)

We now calculate the sample proportions of smokers:

$$\hat{p}_1 = \frac{555}{1,500} = .37 \qquad \hat{p}_2 = \frac{578}{1,750} = .33$$

Then

$$z = \frac{(\hat{p}_1 - \hat{p}_2) - 0}{\sigma_{(\hat{p}_1 - \hat{p}_2)}} \approx \frac{(\hat{p}_1 - \hat{p}_2)}{\sqrt{\hat{p}\hat{q}\left(\dfrac{1}{n_1} + \dfrac{1}{n_2}\right)}}$$

where

$$\hat{p} = \frac{x_1 + x_2}{n_1 + n_2} = \frac{555 + 578}{1,500 + 1,750} = .349$$

Note that $\hat{p}$ is a weighted average of $\hat{p}_1$ and $\hat{p}_2$, with more weight given to the larger (2010) sample. Thus, the computed value of the test statistic is

$$z = \frac{.37 - .33}{\sqrt{(.349)(.651)\left(\dfrac{1}{1,500} + \dfrac{1}{1,750}\right)}} = \frac{.040}{.0168} = 2.38$$

There is sufficient evidence at the $\alpha = .05$ level to conclude that the proportion of adults who smoke has decreased over the 10-year period.

Look Back We could place a confidence interval on $(p_1 - p_2)$ if we were interested in estimating the extent of the decrease.

■ **Now Work Exercise 9.68**

Table 9.7   Results of Smoking Survey

2000	2010
$n_1 = 1{,}500$	$n_2 = 1{,}750$
$x_1 = 555$	$x_2 = 578$

Figure 9.15
Rejection region for Example 9.6

Example	9.7

Finding The Observed Significance Level of a Test for $(p_1 - p_2)$

Problem Use a statistical software package to conduct the test presented in Example 9.6. Find and interpret the p-value of the test.

Solution We entered the sample sizes $(n_1$ and $n_2)$ and numbers of successes $(x_1$ and $x_2)$ into MINITAB and obtained the printout shown in Figure 9.16. The test statistic for this one-tailed test, $z = 2.37$, as well as the p-value of the test, are highlighted on the printout. Note that p-value $= .009$ is smaller than $\alpha = .05$. Consequently, we have strong evidence to reject H_0 and conclude that p_1 exceeds p_2.

Figure 9.16
MINITAB output for test of two proportions

```
Test and CI for Two Proportions

Sample    X      N    Sample p
1       555   1500   0.370000
2       578   1750   0.330286


Difference = p (1) - p (2)
Estimate for difference:  0.0397143
95% lower bound for difference:  0.0121024
Test for difference = 0 (vs > 0):   Z = 2.37   P-Value = 0.009

Fisher's exact test: P-Value = 0.010
```

As with a single population proportion, most studies designed to compare two population proportions employ large samples; consequently, the large-sample testing procedure based on the normal (z) statistic presented here will be appropriate for making inferences about $p_1 - p_2$. *However, in the case of small samples, tests for $p_1 - p_2$ based on the z-statistic may not be valid.* A test to compare proportions that can be applied to small samples utilizes *Fisher's exact test.* We discuss this method in an optional section in Chapter 13.

> **What Do You Do in the Small-Sample Case When Comparing Two Population Proportions?**
>
> *Answer:* Use Fisher's exact test (Chapter 13).

Statistics IN Action | Revisited | **Comparing Proportions**

In the first Statistics in Action Revisited in this chapter (p. 473), we demonstrated how the expert statistician used a comparison of two means to defend Visa in a libel case. Recall that ZixIt claims that a Visa executive's e-mail postings had a negative impact on ZixIt's attempt to develop a new online credit card processing system. Here, we demonstrate another way to analyze the data, one successfully presented in court by the statistician.

In addition to daily closing price and trading volume of ZixIt stock, the **ZIXITVISA** file also contains a qualitative variable that indicates whether the stock price increased or not (decreased or stayed the same) on the following day. This variable was created by the statistician to compare the proportion of days on which ZixIt stock went up for posting and nonposting days. Let p_1 represent the proportion of days where the ZixIt stock price increased following all nonposting days and p_2 represent the proportion of days where the

ZixIt stock price increased following posting days. Then, if the charges made by ZixIt are true (i.e., that postings will have a negative impact on ZixIt stock), p_1 will exceed p_2. Thus, a comparison of two population proportions is appropriate. Recall that during the 83-day period of interest, the executive posted e-mails on 43 days and had no postings on 40 days. Again, both sample sizes $(n_1 = 40$ and $n_2 = 43)$ are large, so we can apply the large-sample z-test or large-sample confidence interval procedure for independent samples. (Can you demonstrate this?) A MINITAB printout for this analysis is shown in Figure SIA9.3.

From the printout you can see that following the 40 nonposting days, the price increased on 20 days; following

(Continued)

Statistics IN Action
(continued)

Test and CI for Two Proportions: Up/Down, Posting

```
Event = UP

Posting   X   N   Sample p
NO       20  40   0.500000
POST     18  43   0.418605

Difference = p (NO) - p (POST)
Estimate for difference:  0.0813953
95% CI for difference:   (-0.132500, 0.295291)
Test for difference = 0 (vs not = 0):   Z = 0.75   P-Value = 0.456

Fisher's exact test: P-Value = 0.513
```

Figure SIA9.3
MINITAB comparison of two proportions analysis

the 43 posting days, the stock price increased on 18 days. Thus, the sample proportions are $p_1 = 20/40 = .5$ and $p_2 = 18/43 = .42$. Are these sample proportions different enough for us to conclude that the population proportions are different and that ZixIt's claim is true? Not according to the statistical analysis. Note that the 95% confidence interval for $(p_1 - p_2)$, $(-.133, .295)$, includes the value 0, and the p-value for the two-tailed test of $H_0: (p_1 - p_2) = 0$, p-value = .456,

exceeds, say, $\alpha = .05$. Both imply that the two population proportions are not significantly different. Also, neither sample proportion is significantly different from .5. (Can you demonstrate this?) Consequently, in courtroom testimony, the statistical expert used these results to conclude that "the direction of ZixIt's stock price movement following days with postings is random, just like days with no postings."

D *Data Set:* ZIXITVISA

Exercises 9.55–9.77

Understanding the Principles

9.55 What conditions are required for valid large-sample inferences about $p_1 - p_2$?

9.56 What is the problem with using the z-statistic to make inferences about $p_1 - p_2$ when the sample sizes are both small?

9.57 Consider making an inference about $p_1 - p_2$, where there are x_1 successes in n_1 binomial trials and x_2 successes in n_2 binomial trials.
  a. Describe the distributions of x_1 and x_2.
  b. For large samples, describe the sampling distribution of $(\hat{p}_1 - \hat{p}_2)$.

Learning the Mechanics

9.58 In each case, determine whether the sample sizes are large enough to conclude that the sampling distribution of $(\hat{p}_1 - \hat{p}_2)$ is approximately normal.
  a. $n_1 = 12, n_2 = 10, \hat{p}_1 = .50, \hat{p}_2 = .50$
  b. $n_1 = 12, n_2 = 10, \hat{p}_1 = .90, \hat{p}_2 = .02$
  c. $n_1 = n_2 = 40, \hat{p}_1 = .80, \hat{p}_2 = .70$
  d. $n_1 = 200, n_2 = 100, \hat{p}_1 = .50, \hat{p}_2 = .01$
  e. $n_1 = 200, n_2 = 200, \hat{p}_1 = .05, \hat{p}_2 = .09$

9.59 Construct a 95% confidence interval for $(p_1 - p_2)$ in each of the following situations:
  a. $n_1 = 300, \hat{p}_1 = .35; n_2 = 400, \hat{p}_2 = .42$
  b. $n_1 = 200, \hat{p}_1 = .69; n_2 = 230, \hat{p}_2 = .75$
  c. $n_1 = 120, \hat{p}_1 = .54; n_2 = 120, \hat{p}_2 = .39$

9.60 Independent random samples, each containing 800 observations, were selected from two binomial populations. The samples from populations 1 and 2 produced 400 and 480 successes, respectively.
  a. Test $H_0: (p_1 - p_2) = 0$ against $H_a: (p_1 - p_2) \neq 0$. Use $\alpha = .05$.
  b. Test $H_0: (p_1 - p_2) = 0$ against $H_a: (p_1 - p_2) \neq 0$. Use $\alpha = .01$.
  c. Test $H_0: (p_1 - p_2) = 0$ against $H_a: (p_1 - p_2) < 0$. Use $\alpha = .01$.
  d. Form a 90% confidence interval for $(p_1 - p_2)$.

9.61 Random samples of size $n_1 = 60$ and $n_2 = 50$ were drawn from populations 1 and 2, respectively. The samples yielded $\hat{p}_1 = .8$ and $\hat{p}_2 = .6$. Test $H_0: (p_1 - p_2) = .1$ against $H_a: (p_1 - p_2) > .1$, using $\alpha = .05$.

9.62 Sketch the sampling distribution of $(\hat{p}_1 - \hat{p}_2)$ based on independent random samples of $n_1 = 200$ and $n_2 = 100$ observations from two binomial populations with probabilities of success $p_1 = .5$ and $p_2 = .9$, respectively.

Applying the Concepts—Basic

9.63 **Bullying behavior study.** School bullying is a form of aggressive behavior that occurs when a student is exposed repeatedly to negative actions (e.g., name-calling, hitting, kicking, spreading slander) from another student. In order to study the effectiveness of an antibullying policy at Dutch elementary schools, a survey of over 2,000 elementary school children was conducted (*Health Education Research*, Feb. 2005). Each student was asked if he or she ever bullied another student. In a sample of 1,358 boys, 746 claimed they had never bullied another student. In a sample of 1,379 girls, 967 claimed they had never bullied another student.
  a. Estimate the true proportion of Dutch boys who have never bullied another student.
  b. Estimate the true proportion of Dutch girls who have never bullied another student.
  c. Estimate the difference in the proportions with a 90% confidence interval.

d. Make a statement about how likely the interval you used in part **c** contains the true difference in proportions.

e. Which group is more likely to bully another student, Dutch boys or Dutch girls?

9.64 Is steak your favorite barbeque food? July is National Grilling Month in the United States. A *Harris Poll* reported on a survey of Americans' grilling preferences. When asked about their favorite food prepared on a barbeque grill, 662 of 1,250 randomly sampled Democrats preferred steak, as compared to 586 of 930 randomly sampled Republicans.

a. Give a point estimate for the proportion of all Democrats who prefer steak as their favorite barbeque food.

b. Give a point estimate for the proportion of all Republicans who prefer steak as their favorite barbeque food.

c. Give a point estimate for the difference between the proportions of all Democrats and all Republicans who prefer steak as their favorite barbeque food.

d. Construct a 95% confidence interval for the difference between the proportions of all Democrats and all Republicans who prefer steak as their favorite barbeque food.

e. Give a practical interpretation of the interval, part **d**.

f. Explain the meaning of the phrase "95% confident" in your answer to part **e**.

9.65 Hospital administration of malaria patients. One of the most serious health problems in India is malaria. Consequently, Indian hospital administrators must have the resources to treat the high volume of malaria patients that are admitted. Research published in the *National Journal of Community Medicine* (Vol. 1, 2010) investigated whether the malaria admission rate is higher in some months than in others. In a sample of 192 hospital patients admitted in January, 32 were treated for malaria. In an independent sample of 403 patients admitted in May (five months later), 34 were treated for malaria.

a. Describe the two populations of interest in this study.

b. Give a point estimate of the difference in the malaria admission rates in January and May.

c. Find a 90% confidence interval for the difference in the malaria admission rates in January and May.

d. Based on the interval, part **c**, can you conclude that a difference exists in the true malaria admission rates in January and May? Explain.

9.66 Influencing performance in a serial addition task. A classic psychological test involves adding a set of numbers (e.g., $1,000 + 40 + 1,000 + 30 + 1,000 + 20 + 1,000 + 10$) to evaluate cognitive behavior. In a study published in *Advances in Cognitive Psychology* (Jan. 2013), undergraduate students were all given the serial addition task, with the numbers presented on a computer screen. One group ($n_1 = 60$ students) saw each of the numbers for 2 seconds on the screen. A second group ($n_2 = 60$ students) saw all the numbers on the screen at the same time, with the number 1,000 presented in bright red. The correct answer, of course, is 4,100. In the first group, 17 answered correctly; in the second group, 12 answered correctly.

a. Compute the proportion of students in Group 1 that answered correctly.

b. Compute the proportion of students in Group 2 that answered correctly.

c. Why is a statistical test of hypothesis (or confidence interval) required to compare the sample proportions, parts **a** and **b**?

d. Conduct a test (at $\alpha = .01$) to compare the proportions. What inference can the researchers make regarding the two serial addition tasks?

9.67 Web survey response rates. Response rates to Web surveys are typically low, partially due to users starting but not finishing the survey. The factors that influence response rates were investigated in *Survey Methodology* (Dec. 2013). In a designed study, Web users were directed to participate in one of several surveys with different formats. For example, one format utilized a welcome screen with a white background, and another format utilized a welcome screen with a red background. The "break-off rates," i.e., the proportion of sampled users who break off the survey before completing all questions, for the two formats are provided in the table.

	White Welcome Screen	Red Welcome Screen
Number of Web users	190	183
Number who break off survey	49	37
Break-off rate	.258	.202

Source: Haer, R., and Meidert, N. "Does the first impression count? Examining the effect of the welcome screen design on the response rate." *Survey Methodology,* Vol. 39, No. 2, Dec. 2013 (Table 4.1).

a. Verify the values of the break-off rates shown in the table.

b. The researchers theorize that the true break-off rate for Web users of the red welcome screen will be lower than the corresponding break-off rate for users of the white welcome screen. Give the null and alternative hypothesis for testing this theory.

c. Conduct the test, part **b**, at $\alpha = .10$. What do you conclude?

9.68 Planning-habits survey. *American Demographics* (Jan. 2002) reported the results of a survey on the planning habits of men and women. In response to the question "What is your preferred method of planning and keeping track of meetings, appointments, and deadlines?" 56% of the men and 46% of the women answered "I keep them in my head." A nationally representative sample of 1,000 adults participated in the survey; therefore, assume that 500 were men and 500 were women.

a. Set up the null and alternative hypotheses for testing whether the percentage of men who prefer keeping track of appointments in their head is larger than the corresponding percentage of women.

b. Compute the test statistic for the test.

c. Give the rejection region for the test, using $\alpha = .01$.

d. Find the *p*-value for the test.

e. Draw the appropriate conclusion.

Applying the Concepts—Intermediate

9.69 Salmonella in produce. Salmonella is the most common type of bacterial food-borne illness in the United States. How prevalent is salmonella in produce grown in the major agricultural region of Monterey, California? Researchers from the United States Department of

Agriculture (USDA) conducted tests for salmonella in produce grown in the region and published their results in *Applied and Environmental Microbiology* (Apr. 2011). In a sample of 252 cultures obtained from water used to irrigate the region, 18 tested positive for salmonella. In an independent sample of 476 cultures obtained from the region's wildlife (e.g., birds), 20 tested positive for salmonella. Is this sufficient evidence for the USDA to state that the prevalence of salmonella in the region's water differs from the prevalence of salmonella in the region's wildlife? Use $\alpha = .01$ to make your decision.

9.70 Study of armyworm pheromones. A study was conducted to determine the effectiveness of pheromones produced by two different strains of fall armyworms: the corn-strain and the rice-strain (*Journal of Chemical Ecology*, Mar. 2013). Both corn-strain and rice-strain male armyworms were released into a field containing a synthetic pheromone made from a corn-strain blend. A count of the number of males trapped by the pheromone was then determined. The experiment was conducted once in a corn field and then again in a grass field. The results are provided in the accompanying table.

a. Consider the corn field results. Construct a 90% confidence interval for the difference between the proportions of corn-strain and rice-strain males trapped by the pheromone.

b. Consider the grass field results. Construct a 90% confidence interval for the difference between the proportions of corn-strain and rice-strain males trapped by the pheromone.

c. Based on the confidence intervals, parts **a** and **b**, what can you conclude about the effectiveness of a corn-blend synthetic pheromone placed in a corn field? A grass field?

d. The researchers also want to compare the proportion of corn-strain males trapped in the corn field to the proportion of corn-strain males trapped in the grass field. Carry out this comparison using a hypothesis test (at $\alpha = .10$). What inference can you draw from the data?

e. Repeat part **d** for the proportions of rice-strain males trapped by the pheromone.

	Corn Field	Grass Field
Number of corn-strain males released	112	215
Number trapped	86	164
Number of rice-strain males released	150	669
Number trapped	92	375

9.71 Traffic sign maintenance. The Federal Highway Administration (FHWA) recently issued new guidelines for maintaining and replacing traffic signs. Civil engineers at North Carolina State University conducted a study of the effectiveness of various sign maintenance practices developed to adhere to the new guidelines and published the results in the *Journal of Transportation Engineering* (June 2013). One portion of the study focused on the proportion of traffic signs that fail the minimum FHWA retroreflectivity requirements. Of 1,000 signs maintained by the North Carolina Department of Transportation (NCDOT), 512 were deemed failures. Of 1,000 signs maintained by county-owned roads in North Carolina, 328 were deemed failures. Conduct a test of hypothesis to determine whether the true proportions of traffic signs that fail the minimum FHWA retroreflectivity requirements differ depending on whether the signs are maintained by the NCDOT or by the county. Test using $\alpha = .05$.

9.72 Angioplasty's benefits challenged. Each year, more than 1 million heart patients undergo an angioplasty. The benefits of an angioplasty were challenged in a recent study of 2,287 patients (2007 Annual Conference of the American College of Cardiology, New Orleans). All the patients had substantial blockage of the arteries, but were medically stable. All were treated with medication such as aspirin and beta-blockers. However, half the patients were randomly assigned to get an angioplasty and half were not. After five years, the researchers found that 211 of the 1,145 patients in the angioplasty group had subsequent heart attacks, compared with 202 of 1,142 patients in the medication-only group. Do you agree with the study's conclusion that "There was no significant difference in the rate of heart attacks for the two groups"? Support your answer with a 95% confidence interval.

9.73 "Tip-of-the-tongue" study. Trying to think of a word you know, but can't instantly retrieve, is called the "tip-of-the-tongue" phenomenon. *Psychology and Aging* (Sept. 2001) published a study of this phenomenon in senior citizens. The researchers compared 40 people between 60 and 72 years of age with 40 between 73 and 83 years of age. When primed with the initial syllable of a missing word (e.g., seeing the word *include* to help recall the word *incisor*), the younger seniors had a higher recall rate. Suppose 31 of the 40 seniors in the younger group could recall the word when primed with the initial syllable, while only 22 of the 40 seniors could recall the word. Compare the recall rates of the two groups, using $\alpha = .05$. Does one group of elderly people have a significantly higher recall rate than the other?

9.74 Vulnerability of relying party Web sites. When you sign on to your Facebook account, you are granted access to more than 1 million *relying party* (RP) Web sites. This single sign-on (SSO) scheme is enabled by OAuth 2.0, an open and standardized Web resource authorization protocol. Although the protocol claims to be secure, there is anecdotal evidence of critical vulnerabilities that allow an attacker to gain unauthorized access to the user's profile and allow the attacker to impersonate the victim on the RP Web site. Computer and systems engineers at the University of British Columbia investigated the vulnerability of relying party Web sites and presented their results at the *Proceedings of the 5th AMC Workshop on Computers & Communication Security* (Oct. 2012). RP Web sites were categorized as server-flow or client-flow Web sites. Of the 40 server-flow sites studied, 20 were found to be vulnerable to impersonation attacks. Of the 54 client-flow sites examined, 41 were found to be vulnerable to impersonation attacks. Do these results indicate that a client-flow Web site is more likely to be vulnerable to an impersonation attack than a client-flow Web site? Test using $\alpha = .01$.

9.75 Does sleep improve mental performance? Are creativity and problem solving linked to adequate sleep? This question was the subject of research conducted by German scientists at the University of Lübeck (*Nature,*

Jan. 22, 2004). One hundred volunteers were divided into two equal-sized groups. Each volunteer took a math test that involved transforming strings of eight digits into a new string that fit a set of given rules, as well as a third, hidden rule. Prior to taking the test, one group received eight hours of sleep, while the other group stayed awake all night. The scientists monitored the volunteers to determine whether and when they figured out the third rule. Of the volunteers who slept, 39 discovered the third rule; of the volunteers who stayed awake all night, 15 discovered the third rule. From the study results, what can you infer about the proportions of volunteers in the two groups who discover the third rule? Support your answer with a 90% confidence interval.

Applying the Concepts—Advanced

9.76 **Religious symbolism in TV commercials.** Gonzaga University professors conducted a study of television commercials and published their results in the *Journal of Sociology, Social Work, and Social Welfare* (Vol. 2, 2008). The key research question was: "Do television advertisers use religious symbolism to sell goods and services?" In a sample of 797 TV commercials collected in 1998, only 16 commercials used religious symbolism. Of the sample of 1,499 TV commercials examined in the more recent study, 51 commercials used religious symbolism. Conduct an analysis to determine if the percentage of TV commercials that use religious symbolism has changed since the 1998 study. If you detect a change, estimate the magnitude of the difference and attach a measure of reliability to the estimate.

9.77 **Teeth defects and stress in prehistoric Japan.** Linear enamel hypoplasia (LEH) defects are pits or grooves on the tooth surface that are typically caused by malnutrition, chronic infection, stress, and trauma. A study of LEH defects in prehistoric Japanese cultures was published in the *American Journal of Physical Anthropology* (May 2010). Three groups of Japanese people were studied: Yayoi farmers (early agriculturists), eastern Jomon foragers (broad-based economy), and western Jomon foragers (wet rice economy). LEH defect prevalence was determined from skulls of individuals obtained from each of the three cultures. Of the 182 Yayoi farmers in the study, 63.1% had at least one LEH defect; of the 164 Eastern Jomon foragers, 48.2% had at least one LEH defect; and, of the 122 Western Jomon foragers, 64.8% had at least one LEH defect. Two theories were tested. Theory 1 states that foragers with a broad-based economy will have a lower LEH defect prevalence than early agriculturists. Theory 2 states that foragers with a wet rice economy will not differ in LEH defect prevalence from early agriculturists. Use the results to test both theories, each at $\alpha = .01$.

Based on Temple, D. H. "Patterns of systemic stress during the agricultural transition in prehistoric Japan." *American Journal of Physical Anthropology*, Vol. 142, No. 1, May 2010.

9.5 Determining the Sample Size

You can find the appropriate sample size to estimate the difference between a pair of parameters with a specified sampling error (SE) and degree of reliability by using the method described in Section 7.5. That is, to estimate the difference between a pair of parameters correct to within SE units with confidence level $(1 - \alpha)$, let $z_{\alpha/2}$ standard deviations of the sampling distribution of the estimator equal SE. Then solve for the sample size. To do this, you have to solve the problem for a specific ratio between n_1 and n_2. Most often, you will want to have equal sample sizes—that is, $n_1 = n_2 = n$. We will illustrate the procedure with two examples.

Example **9.8** **Finding the Sample Sizes for Estimating** $(\mu_1 - \mu_2)$**—Comparing Mean Crop Yields** 	**Problem** New fertilizer compounds are often advertised with the promise of increased crop yields. Suppose we want to compare the mean yield μ_1 of wheat when a new fertilizer is used with the mean yield μ_2 from a fertilizer in common use. The estimate of the difference in mean yield per acre is to be correct to within .25 bushel with a confidence coefficient of .95. If the sample sizes are to be equal, find $n_1 = n_2 = n$, the number of 1-acre plots of wheat assigned to each fertilizer. **Solution** To solve the problem, you need to know something about the variation in the bushels of yield per acre. Suppose that, from past records, you know that the yields of wheat possess a range of approximately 10 bushels per acre. You could then approximate $\sigma_1 = \sigma_2 = \sigma$ by letting the range equal 4σ. Thus, $$4\sigma \approx 10 \text{ bushels}$$ $$\sigma \approx 2.5 \text{ bushels}$$

The next step is to solve the equation

$$z_{\alpha/2}\sigma_{(\bar{x}_1 - \bar{x}_2)} = SE, \text{ or } z_{\alpha/2}\sqrt{\frac{\sigma_1^2}{n_1} + \frac{\sigma_2^2}{n_2}} = SE$$

for n, where $n = n_1 = n_2$. Since we want our estimate to lie within SE = .25 of $(\mu_1 - \mu_2)$ with confidence coefficient equal to .95, we have $z_{\alpha/2} = z_{.025} = 1.96$. Then, letting $\sigma_1 = \sigma_2 = 2.5$ and solving for n, we get

$$1.96\sqrt{\frac{(2.5)^2}{n} + \frac{(2.5)^2}{n}} = .25$$

$$1.96\sqrt{\frac{2(2.5)^2}{n}} = .25$$

$$n = 768.32 \approx 769 \text{ (rounding up)}$$

Consequently, you will have to sample 769 acres of wheat for each fertilizer to estimate the difference in mean yield per acre to within .25 bushel.

Look Back Since $n = 769$ would necessitate extensive and costly experimentation, you might decide to allow a larger sampling error (say, SE = .50 or SE = 1) in order to reduce the sample size, or you might decrease the confidence coefficient. The point is that we can obtain an idea of the experimental effort necessary to achieve a specified precision in our final estimate by determining the approximate sample size *before* the experiment is begun.

■ **Now Work Exercise 9.82**

Example 9.9

Finding the Sample Sizes for Estimating $(p_1 - p_2)$ — Comparing Defect Rates of Two Machines

Problem A production supervisor suspects that a difference exists between the proportions p_1 and p_2 of defective items produced by two different machines. Experience has shown that the proportion defective for each of the two machines is in the neighborhood of .03. If the supervisor wants to estimate the difference in the proportions to within .005, using a 95% confidence interval, how many items must be randomly sampled from the output produced by each machine? (Assume that the supervisor wants $n_1 = n_2 = n$.)

Solution In this sampling problem, the sampling error SE = .005, and for the specified level of reliability, $z_{\alpha/2} = z_{.025} = 1.96$. Then, letting $p_1 = p_2 = .03$ and $n_1 = n_2 = n$, we find the required sample size per machine by solving the following equation for n:

$$z_{\alpha/2}\,\sigma_{(\hat{p}_1 - \hat{p}_2)} = SE$$

or

$$z_{\alpha/2}\sqrt{\frac{p_1 q_1}{n_1} + \frac{p_2 q_2}{n_2}} = SE$$

$$1.96\sqrt{\frac{(.03)(.97)}{n} + \frac{(.03)(.97)}{n}} = .005$$

$$1.96\sqrt{\frac{2(.03)(.97)}{n}} = .005$$

$$n = 8{,}943.2$$

Look Back This large n will likely result in a tedious sampling procedure. If the supervisor insists on estimating $(p_1 - p_2)$ correct to within .005 with 95% confidence, approximately 9,000 items will have to be inspected for each machine.

■ **Now Work Exercise 9.83a**

You can see from the calculations in Example 9.9 that $\sigma_{(\hat{p}_1 - \hat{p}_2)}$ (and hence the solution, $n_1 = n_2 = n$) depends on the actual (but unknown) values of p_1 and p_2. In fact, the required sample size $n_1 = n_2 = n$ is largest when $p_1 = p_2 = .5$. Therefore, if you have no prior information on the approximate values of p_1 and p_2, use $p_1 = p_2 = .5$ in the formula for $\sigma_{(\hat{p}_1 - \hat{p}_2)}$. If p_1 and p_2 are in fact close to .5, then the values of n_1 and n_2 that you have calculated will be correct. If p_1 and p_2 differ substantially from .5, then your solutions for n_1 and n_2 will be larger than needed. Consequently, using $p_1 = p_2 = .5$ when solving for n_1 and n_2 is a conservative procedure because the sample sizes n_1 and n_2 will be at least as large as (and probably larger than) needed.

The procedures for determining sample sizes necessary for estimating $(\mu_1 - \mu_2)$ or $(p_1 - p_2)$ for the case $n_1 = n_2$ and for μ_d in a paired difference experiment are given in the following boxes:

Determination of Sample Size for Estimating $(\mu_1 - \mu_2)$: Equal Sample Size Case

To estimate $(\mu_1 - \mu_2)$ to within a given sampling error SE and with confidence level $(1 - \alpha)$, use the following formula to solve for equal sample sizes that will achieve the desired reliability:

$$n_1 = n_2 = \frac{(z_{\alpha/2})^2(\sigma_1^2 + \sigma_2^2)}{(\text{SE})^2}$$

You will need to substitute estimates for the values of σ_1^2 and σ_2^2 before solving for the sample size. These estimates might be sample variances s_1^2 and s_2^2 from prior sampling (e.g., a pilot study) or from an educated (and conservatively large) guess based on the range—that is, $s \approx R/4$.

Determination of Sample Size for Estimating $p_1 - p_2$: Equal Sample Size Case

To estimate $(p_1 - p_2)$ to within a given sampling error SE and with confidence level $(1 - \alpha)$, use the following formula to solve for equal sample sizes that will achieve the desired reliability:

$$n_1 = n_2 = \frac{(z_{\alpha/2})^2(p_1 q_1 + p_2 q_2)}{(\text{SE})^2}$$

You will need to substitute estimates for the values of p_1 and p_2 before solving for the sample size. These estimates might be based on prior samples, obtained from educated guesses or, most conservatively, specified as $p_1 = p_2 = .5$.

Determination of Sample Size for Estimating μ_d

To estimate μ_d to within a given sampling error SE and with confidence level $(1 - \alpha)$, use the following formula to solve for the number of pairs, n_d, that will achieve the desired reliability:

$$n_d = (z_{\alpha/2})^2(\sigma_d^2)/(\text{SE})^2$$

You will need to substitute an estimate for the value of σ_d, the standard deviation of the paired differences, before solving for the sample size.

Note: When estimating either $(\mu_1 - \mu_2)$ or $(p_1 - p_2)$, you may desire one sample size to be a multiple of the other, e.g., $n_2 = a(n_1)$ where a is an integer. For example, you may want to sample twice as many experimental units in the second sample as in the first. Then $a = 2$ and $n_2 = 2(n_1)$. For this unequal sample size case, slight adjustments are made to the computing formulas. These formulas (proof omitted) are provided below for convenience.

> **Adjustment to Sample Size Formula for Estimating $(\mu_1 - \mu_2)$ When $n_2 = a(n_1)$**
>
> $$n_1 = \frac{(z_{\alpha/2})^2(a\sigma_1^2 + \sigma_2^2)}{a(\text{SE})^2} \qquad n_2 = a(n_1)$$

> **Adjustment to Sample Size Formula for Estimating $(p_1 - p_2)$ When $n_2 = a(n_1)$**
>
> $$n_1 = \frac{(z_{\alpha/2})^2(ap_1q_1 + p_2q_2)}{a(\text{SE})^2} \qquad n_2 = a(n_1)$$

Exercises 9.78–9.95

Understanding the Principles

9.78 In determining the sample sizes for estimating $\mu_1 - \mu_2$, how do you obtain estimates of the population variances $(\sigma_1)^2$ and $(\sigma_2)^2$ used in the calculations?

9.79 In determining the sample sizes for estimating $p_1 - p_2$, how do you obtain estimates of the binomial proportions p_1 and p_2 used in the calculations?

9.80 When will the sample size $n_1 = n_2 = n$ be the largest, while determining the sample size of estimate of $p_1 - p_2$?

Learning the Mechanics

9.81 Suppose you want to estimate the difference between two population means correct to within 3.2 with probability .95. If prior information suggests that the population variances are approximately equal to $\sigma_1^2 = \sigma_2^2 = 10$ and you want to select independent random samples of equal size from the populations, how large should the sample sizes, n_1 and n_2, be?

9.82 Find the appropriate values of n_1 and n_2 (assume that [NW] $n_1 = n_2$) needed to estimate $(\mu_1 - \mu_2)$ with
  a. A sampling error equal to 3.2 with 95% confidence. From prior experience, it is known that $\sigma_1 \approx 15$ and $\sigma_2 \approx 17$.
  b. A sampling error equal to 8 with 99% confidence. The range of each population is 60.
  c. A 90% confidence interval of width 1.0. Assume that $\sigma_1^2 \approx 5.8$ and $\sigma_2^2 \approx 7.5$.

9.83 Assuming that $n_1 = n_2$, find the sample sizes needed to [NW] estimate $(p_1 - p_2)$ for each of the following situations:
  a. SE = .01 with 99% confidence. Assume that $p_1 \approx .4$ and $p_2 \approx .7$.
  b. A 90% confidence interval of width .05. Assume there is no prior information available with which to obtain approximate values of p_1 and p_2.
  c. SE = .03 with 90% confidence. Assume that $p_1 \approx .2$ and $p_2 \approx .3$.

9.84 Enough money has been budgeted to collect independent random samples of size $n_1 = n_2 = 100$ from populations 1 and 2 in order to estimate $(p_1 - p_2)$. Prior information indicates that $p_1 = p_2 \approx .4$. Have sufficient funds been allocated to construct a 95% confidence interval for $(p_1 - p_2)$ of width .1 or less? Justify your answer.

Applying the Concepts—Basic

9.85 **Hygiene of handshakes, high fives, and fist bumps.** Refer to the *American Journal of Infection Control* (Aug. 2014) study of the hygiene of hand greetings, Exercise 9.24 (p. 479). The number of bacteria transferred from a gloved hand dipped into a culture of bacteria to a second gloved hand contacted by either a handshake, high five, or fist bump was recorded. Recall that the experiment was replicated only five times for each contact method and the data used to compare the mean percentage of bacteria transferred for any two contact methods. Therefore, for this independent-samples design, $n_1 = n_2 = 5$. Suppose you want to estimate the difference between the mean percentage of bacteria transferred for the handshake and fist bump greetings to within 10% using a 95% confidence interval.
  a. Define the parameter of interest in this study.
  b. Give the value of $z_{\alpha/2}$ for the confidence interval.
  c. What is the desired sampling error, SE?
  d. From the data provided in Exercise 9.24, find estimates of the variances, σ_1^2 and σ_2^2, for the two contact methods.
  e. Use the information in parts **a–d** to calculate the number of replicates for each contact method required to obtain the desired reliability. Assume an equal number of replicates.

9.86 **Influencing performance in a serial addition task.** Refer to the *Advances in Cognitive Psychology* (Jan. 2013) study of influencing performance in a classic psychological test that involved adding a set of numbers, Exercise 9.66 (p. 499). Recall that one group of students saw each of the numbers for 2 seconds on the screen while a second group saw all the numbers on the screen at the same time, with the number 1,000 presented in bright red. The researchers want to estimate the true difference in the proportions of students in the two groups that give the correct answer with a 95% confidence interval. They want to know how many students to sample in each group in order to obtain an estimate that is no more than .10 from the true difference.
  a. Identify the parameter of interest for this study.
  b. What is the desired confidence level?
  c. What is the desired sampling error?
  d. Find the sample size required to obtain the desired estimate. Assume $n_1 = n_2$.
  e. Repeat part **d**, but now assume that twice as many students will be sampled in the first group as in the second group, i.e., assume $n_1 = 2n_2$.

9.87 **Bulimia study.** Refer to the *American Statistician* (May 2001) study comparing the "fear of negative evaluation" (FNE) scores for bulimic and normal female students, presented in Exercise 9.19 (p. 477). Suppose you want to estimate $(\mu_B - \mu_N)$, the difference between the population means of the FNE scores for bulimic and normal female students, using a 95% confidence interval with a sampling error of two points. Find the sample sizes required to obtain such an estimate. Assume equal sample sizes of $\sigma_B^2 = \sigma_N^2 = 25$.

9.88 **Laughter among deaf signers.** Refer to the *Journal of Deaf Studies and Deaf Education* (Fall 2006) paired difference study on vocalized laughter among deaf users of sign language, presented in Exercise 9.44 (p. 490). Suppose you want to estimate $\mu_d = (\mu_S - \mu_A)$, the difference between the population mean number of laugh episodes of deaf speakers and deaf audience members, using a 90% confidence interval with a sampling error of .75. Find the number of pairs of deaf people required to obtain such an estimate, assuming that the variance of the paired differences is $\sigma_d^2 = 3$.

9.89 **Angioplasty's benefits challenged.** Refer to the study of patients with substantial blockage of the arteries presented at the 2007 Annual Conference of the American College of Cardiology, Exercise 9.72 (p. 500). Recall that half the patients were randomly assigned to get an angioplasty and half were not. The researchers compared the proportion of patients with subsequent heart attacks for the two groups and reported no significant difference between the two proportions. Although the study involved over 2,000 patients, the sample size may have been too small to detect a difference in heart attack rates.

  a. How many patients must be sampled in each group in order to estimate the difference in heart attack rates to within .015 with 95% confidence? (Use summary data from Exercise 9.64 in your calculation.)

  b. Comment on the practicality of carrying out the study with the sample sizes determined in part **a**.

  c. Comment on the practical significance of the difference detected in the confidence interval for the study, part **a**.

Applying the Concepts—Intermediate

9.90 **Last name and acquisition timing.** Refer to the *Journal of Consumer Research* (Aug. 2011) study of the *last name effect* in acquisition timing, Exercise 9.14 (p. 476). Recall that the mean response times (in minutes) to acquire basketball tickets were compared for two groups of MBA students: those students with last names beginning with one of the first 9 letters of the alphabet and those with last names beginning with one of the last 9 letters of the alphabet. How many MBA students from each group would need to be selected to estimate the difference in mean times to within 2 minutes of its true value with 95% confidence? (Assume equal sample sizes will be selected for each group and that the response time standard deviation for both groups is $\sigma \approx 9$ minutes.)

9.91 **Solar energy generation along highways.** Refer to the *International Journal of Energy and Environmental Engineering* (Dec. 2013) study of solar energy generation along highways, Exercise 9.47 (p. 491). Recall that the

researchers compared the mean monthly amount of solar energy generated by east-west and north-south oriented solar panels using a matched-pairs experiment. However, a small sample of only five months was used for the analysis. How many more months would need to be selected in order to estimate the difference in means to within 25 kilowatt hours with a 90% confidence interval? Use the information provided in the **SOLAR** file to find an estimate of the standard error required to carry out the calculation.

9.92 **Cable-TV home shoppers.** All cable television companies carry at least one home-shopping channel. Who uses these home-shopping services? Are the shoppers primarily men or women? Suppose you want to estimate the difference in the percentages of men and women who say they have used or expect to use televised home shopping. You want a 90% confidence interval of width .06 or less.

  a. Approximately how many people should be included in your samples?

  b. Suppose you want to obtain individual estimates for the two percentages of interest. Will the sample size found in part **a** be large enough to provide estimates of each percentage correct to within .02 with probability equal to .95? Justify your response.

9.93 **Do video game players have superior visual attention skills?** Refer to the *Journal of Articles in Support of the Null Hypothesis* (Vol. 6, 2009) study comparing the visual attention skill of video game and non-video game players, Exercise 9.22 (p. 478). Recall that there was no significant difference between the mean score on the *attentional blink* test of video game players and the corresponding mean for non–video game players. It is possible that selecting larger samples would yield a significant difference. How many video game and non–video game players would need to be selected in order to estimate the difference in mean score for the two groups to within 5 points with 95% confidence? (Assume equal sample sizes will be selected from the two groups and that the score standard deviation for both groups is $\sigma \approx 9$.)

9.94 **Rat damage in sugarcane.** Poisons are used to prevent rat damage in sugarcane fields. An agriculture department is investigating whether the rat poison should be located in the middle of the field or on the outer perimeter. One way to answer this question is to determine where the greater amount of damage occurs. If damage is measured by the proportion of cane stalks that have been damaged by rats, how many stalks from each section of the field should be sampled in order to estimate the true difference between proportions of stalks damaged in the two sections, to within .03 with 90% confidence?

9.95 **Scouting an NFL free agent.** In seeking a free-agent NFL running back, a general manager is looking for a player with high mean yards gained per carry and a small standard deviation. Suppose the GM wishes to compare the mean yards gained per carry for two free agents, on the basis of independent random samples of their yards gained per carry. Data from last year's pro football season indicate that $\sigma_1^2 = \sigma_2^2 \approx 10$ yards. If the GM wants to estimate the difference in means correct to within 4 yard with a confidence level of .95, how many runs would have to be observed for each player? (Assume equal sample sizes.)

9.6 Comparing Two Population Variances: Independent Sampling (Optional)

Many times, it is of practical interest to use the techniques developed in this chapter to compare the means or proportions of two populations. However, there are also important instances when we wish to compare two population variances. For example, when two devices are available for producing precision measurements (scales, calipers, thermometers, etc.), we might want to compare the variability of the measurements of the devices before deciding which one to purchase. Or when two standardized tests can be used to rate job applicants, the variability of the scores for both tests should be taken into consideration before deciding which test to use.

For problems like these, we need to develop a statistical procedure to compare population variances. The common statistical procedure for comparing population variances σ_1^2 and σ_2^2 makes an inference about the ratio σ_1^2/σ_2^2. In this section, we will show how to test the null hypothesis that the ratio σ_1^2/σ_2^2 equals 1 (the variances are equal) against the alternative hypothesis that the ratio differs from 1 (the variances differ):

$$H_0: \frac{\sigma_1^2}{\sigma_2^2} = 1 \quad (\sigma_1^2 = \sigma_2^2)$$

$$H_a: \frac{\sigma_1^2}{\sigma_2^2} \neq 1 \quad (\sigma_1^2 \neq \sigma_2^2)$$

To make an inference about the ratio σ_1^2/σ_2^2, it seems reasonable to collect sample data and use the ratio of the sample variances, s_1^2/s_2^2. We will use the test statistic

$$F = \frac{s_1^2}{s_2^2}$$

To establish a rejection region for the test statistic, we need to know the sampling distribution of s_1^2/s_2^2. As you will subsequently see, the sampling distribution of s_1^2/s_2^2 is based on two of the assumptions already required for the t-test:

1. The two sampled populations are normally distributed.

2. The samples are randomly and independently selected from their respective populations.

When these assumptions are satisfied and when the null hypothesis is true (i.e., when $\sigma_1^2 = \sigma_2^2$), the sampling distribution of $F = s_1^2/s_2^2$ is the **F-distribution** with $(n_1 - 1)$ numerator degrees of freedom and $(n_2 - 1)$ denominator degrees of freedom, respectively. The shape of the F-distribution depends on the number of degrees of freedom associated with s_1^2 and s_2^2—that is, on $(n_1 - 1)$ and $(n_2 - 1)$. An F-distribution with 7 and 9 df is shown in Figure 9.17. As you can see, the distribution is skewed to the right, since s_1^2/s_2^2 cannot be less than 0, but can increase without bound.

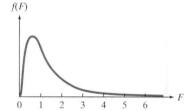

$f(F)$

Figure 9.17

An F-distribution with 7 numerator and 9 denominator degrees of freedom

BIOGRAPHY GEORGE W. SNEDECOR (1882–1974)

Snedecor's F-test

George W. Snedecor's education began at the University of Alabama, where he obtained his bachelor's degree in mathematics and physics. He went on to the University of Michigan for his master's degree in physics and finally earned his Ph.D. in mathematics at the University of Kentucky. Snedecor learned of an opening for an assistant professor of mathematics at the University of Iowa, packed his belongings in his car, and began driving to apply for the position. By mistake, he ended up in Ames, Iowa, home of Iowa State University—then an agricultural school that had no need for a mathematics teacher. Nevertheless, Snedecor stayed and founded a statistics laboratory, eventually teaching the first course in statistics at Iowa State in 1915. In 1933, Snedecor turned the statistics laboratory into the first-ever Department of Statistics in the United States. During his tenure as chair of the department, Snedecor published his landmark textbook, *Statistical Methods* (1937). The text contained the first published reference for a test of hypothesis to compare two variances. Although Snedecor named it the *F*-test in honor of statistician R. A. Fisher (who had developed the *F*-distribution a few years earlier) many researchers still refer to it as Snedecor's *F*-test. Now in its ninth edition, *Statistical Methods* (with William Cochran as coauthor) continues to be one of the most frequently cited texts in the statistics field. ■

We need to be able to find *F*-values corresponding to the tail areas of this distribution in order to establish the rejection region for our test of hypothesis because we expect the ratio *F* of the sample variances to be either very large or very small when the population variances are unequal. The upper-tail *F* values for $\alpha = .10, .05, .025$, and $.01$ can be found in Tables V, VI, VII, and VIII of Appendix B. Table VI is partially reproduced in Table 9.8. It gives *F* values that correspond to $\alpha = .05$ upper-tail areas for different degrees of freedom; the columns correspond to the number of degrees of freedom, v_1, for the numerator sample variance s_1^2, whereas the rows correspond to the number of degrees of freedom, v_2, for the denominator sample variance s_2^2. Thus, if the number of degrees of freedom denoted by the numerator is $v_1 = 7$ and the number of degrees of freedom denoted by the denominator is $v_2 = 9$, we look in the seventh column and ninth row to find $F_{.05} = 3.29$ (highlighted in Table 9.8). As shown in Figure 9.18, $\alpha = .05$

Table 9.8   Reproduction of Part of Table VI in Appendix B: Percentage Points of the *F*-distribution, $\alpha = .05$

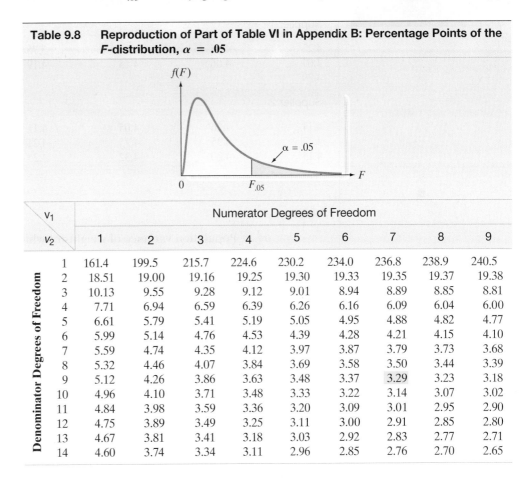

v_1	Numerator Degrees of Freedom								
v_2	1	2	3	4	5	6	7	8	9
1	161.4	199.5	215.7	224.6	230.2	234.0	236.8	238.9	240.5
2	18.51	19.00	19.16	19.25	19.30	19.33	19.35	19.37	19.38
3	10.13	9.55	9.28	9.12	9.01	8.94	8.89	8.85	8.81
4	7.71	6.94	6.59	6.39	6.26	6.16	6.09	6.04	6.00
5	6.61	5.79	5.41	5.19	5.05	4.95	4.88	4.82	4.77
6	5.99	5.14	4.76	4.53	4.39	4.28	4.21	4.15	4.10
7	5.59	4.74	4.35	4.12	3.97	3.87	3.79	3.73	3.68
8	5.32	4.46	4.07	3.84	3.69	3.58	3.50	3.44	3.39
9	5.12	4.26	3.86	3.63	3.48	3.37	3.29	3.23	3.18
10	4.96	4.10	3.71	3.48	3.33	3.22	3.14	3.07	3.02
11	4.84	3.98	3.59	3.36	3.20	3.09	3.01	2.95	2.90
12	4.75	3.89	3.49	3.25	3.11	3.00	2.91	2.85	2.80
13	4.67	3.81	3.41	3.18	3.03	2.92	2.83	2.77	2.71
14	4.60	3.74	3.34	3.11	2.96	2.85	2.76	2.70	2.65

(Denominator Degrees of Freedom — row labels down the left side)

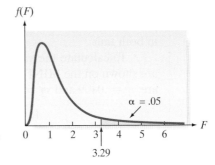

Figure 9.18
An *F*-distribution for $v_1 = 7$ and $v_2 = 9$ df; $\alpha = .05$

is the tail area to the right of 3.29 in the *F*-distribution with 7 and 9 df. That is, if $\sigma_1^2 = \sigma_2^2$, then the probability that the *F*-statistic will exceed 3.29 is $\alpha = .05$.

Now Work Exercise 9.100

Example	9.10	

MICEWTS

An *F*-Test Application—Comparing Weight Variations in Mice

Problem An experimenter wants to compare the metabolic rates of white mice subjected to different drugs. The weights of the mice may affect their metabolic rates; thus, the experimenter wishes to obtain mice that are relatively homogeneous with respect to weight. Five hundred mice will be needed to complete the study. Currently, 13 mice from supplier 1 and another 18 mice from supplier 2 are available for comparison. The experimenter weighs these mice and obtains the data shown in Table 9.9. Do these data provide sufficient evidence to indicate a difference in the variability of weights of mice obtained from the two suppliers? (Use $\alpha = .10$.) From the results of this analysis, what would you suggest to the experimenter?

Table 9.9   Weights (in ounces) of Experimental Mice

Supplier 1

4.23	4.35	4.05	3.75	4.41	4.37
4.01	4.06	4.15	4.19	4.52	4.21
4.29					

Supplier 2

4.14	4.26	4.05	4.11	4.31	4.12
4.17	4.35	4.25	4.21	4.05	4.28
4.15	4.20	4.32	4.25	4.02	4.14

 Data Set: MICEWTS

Solution Let

$$\sigma_1^2 = \text{Population variance of weights of white mice from supplier 1}$$
$$\sigma_2^2 = \text{Population variance of weights of white mice from supplier 2}$$

The hypotheses of interest are then

$$H_0: \frac{\sigma_1^2}{\sigma_2^2} = 1 \quad (\sigma_1^2 = \sigma_2^2)$$

$$H_a: \frac{\sigma_1^2}{\sigma_2^2} \neq 1 \quad (\sigma_1^2 \neq \sigma_2^2)$$

The nature of the *F*-tables given in Appendix B affects the form of the test statistic. To form the rejection region for a two-tailed *F*-test, we want to make certain that the upper tail is used because only the upper-tail values of *F* are shown in Appendix A. To accomplish that, *we will always place the larger sample variance in the numerator of the F-test statistic.* This has the effect of doubling the tabulated value for α, since we double the probability that the *F*-ratio will fall in the upper tail by always placing the larger sample variance in the numerator. That is, we establish a one-tailed rejection region by putting the larger variance in the numerator, rather than establishing rejection regions in both tails.

To calculate the value of the test statistic, we require the sample variances. These are shown on the MINITAB printout of Figure 9.19. The sample variances (highlighted) are $s_1^2 = .0409$ and $s_2^2 = .00964$. Therefore,

$$F = \frac{\text{Larger sample variance}}{\text{Smaller sample variance}} = \frac{s_1^2}{s_2^2} = \frac{(.0409)}{(.00964)} = 4.24$$

For our example, we have a numerator s_1^2 with df $= v_1 = n_1 - 1 = 12$ and a denominator s_2^2 with df $= v_2 = n_2 - 1 = 17$. Therefore, we will reject $H_0: \sigma_1^2 = \sigma_2^2$ for $\alpha = .10$ when the calculated value of *F* exceeds the tabulated value:

$$F_{\alpha/2} = F_{.05} = 2.38 \qquad \text{(see Figure 9.20)}$$

Descriptive Statistics: WEIGHT

Variable	SUPPLIER	N	Mean	StDev	Variance	Minimum	Maximum
WEIGHT	1	13	4.1992	0.2021	0.0409	3.7500	4.5200
	2	18	4.1878	0.0982	0.00964	4.0200	4.3500

Test for Equal Variances: WEIGHT versus SUPPLIER

95% Bonferroni confidence intervals for standard deviations

SUPPLIER	N	Lower	StDev	Upper
1	13	0.138579	0.202133	0.361467
2	18	0.070860	0.098193	0.156836

F-Test (normal distribution)
Test statistic = 4.24, p-value = 0.007

Levene's Test (any continuous distribution)
Test statistic = 4.31, p-value = 0.047

Figure 9.19
MINITAB summary statistics and *F*-test for mice weights

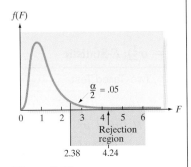

$f(F)$

$\frac{\alpha}{2} = .05$

Rejection region

2.38   4.24

Figure 9.20
Rejection region for Example 9.10

Note that $F = 4.24$ falls into the rejection region. Therefore, the data provide sufficient evidence to indicate that the population variances differ. It appears that the weights of mice obtained from supplier 2 tend to be more homogeneous (less variable) than the weights of mice obtained from supplier 1. On the basis of this evidence, we would advise the experimenter to purchase the mice from supplier 2.

Look Back What would you have concluded if the value of *F* calculated from the samples had not fallen into the rejection region? Would you conclude that the null hypothesis of equal variances is true? No, because then you risk the possibility of a Type II error (accepting H_0 if H_a is true) without knowing the value of β, the probability of accepting H_0: $\sigma_1^2 = \sigma_2^2$ if in fact it is false. Since we will not consider the calculation of β for specific alternatives in this text, when the *F*-statistic does not fall into the rejection region we simply conclude that the sample possesses insufficient evidence to refute the null hypothesis that $\sigma_1^2 = \sigma_2^2$.

■ **Now Work Exercise 9.105a**

Example **9.11** ⓓ
MICEWTS

The Observed Significance Level of an *F*-Test

Problem Find the *p*-value for the test in Example 9.10, using the *F*-tables in Appendix A. Compare this with the exact *p*-value obtained from a computer printout.

Solution Since the observed value of the *F*-statistic in Example 9.10 was found to be 4.24, the observed significance level of the test would equal the probability of observing a value of *F* at least as contradictory to H_0: $\sigma_1^2 = \sigma_2^2$ as $F = 4.24$, is, if in fact H_0 is true. Since we give the *F*-tables in Appendix B just for values of α equal to .10, .05, .025, and .01, we can only approximate the observed significance level. Checking Table VI, we find that $F_{.01} = 3.46$. Since the observed value of *F* is greater than $F_{.01}$, the observed significance level for the test is less than $2(.01) = .02$. (Note that we double the α value shown in Table VI because this is a two-tailed test.) The exact *p*-value, $p = .007$, is highlighted at the bottom of the MINITAB printout shown in Figure 9.19.

Look Back We double the α value in Table VI because this is a two-tailed test.

■ **Now Work Exercise 9.105b**

In addition to applying a hypothesis test for σ_1^2/σ_2^2, we can use the F-statistic to estimate the ratio with a confidence interval. The following boxes summarize the confidence interval and testing procedures:

A $(1 - \alpha) \times 100\%$ Confidence Interval for $(\sigma_1)^2/(\sigma_2)^2$

$$\left(\frac{s_1^2}{s_2^2}\right)\left(\frac{1}{F_{L,\alpha/2}}\right) < \left(\frac{\sigma_1^2}{\sigma_2^2}\right) < \left(\frac{s_1^2}{s_2^2}\right)F_{U,\alpha/2}$$

where $F_{L,\alpha/2}$ is the value of F that places an area $\alpha/2$ in the upper tail of an F-distribution with $v_1 = (n_1 - 1)$ numerator and $v_2 = (n_2 - 1)$ denominator degrees of freedom, and $F_{U,\alpha/2}$ is the value of F that places an area $\alpha/2$ in the upper tail of an F-distribution with $v_1 = (n_2 - 1)$ numerator and $v_2 = (n_1 - 1)$ denominator degrees of freedom.

Test of Hypothesis for Equal Population Variances*, $(\sigma_1^2 = \sigma_2^2)$: F-Statistic

	One-Tailed Tests		Two-Tailed Test
	$H_0: \sigma_1^2 = \sigma_2^2$	$H_0: \sigma_1^2 = \sigma_2^2$	$H_0: \sigma_1^2 = \sigma_2^2$
	$H_a: \sigma_1^2 < \sigma_2^2$	$H_a: \sigma_1^2 > \sigma_2^2$	$H_a: \sigma_1^2 \neq \sigma_2^2$
Test statistic:	$F_c = \dfrac{s_2^2}{s_1^2}$	$F_c = \dfrac{s_1^2}{s_2^2}$	$F_c = \dfrac{\text{Larger } s^2}{\text{Smaller } s^2}$
Rejection region:	$F_c > F_\alpha$	$F_c > F_\alpha$	$F_c > F_{\alpha/2}$
Numerator df:	$v_1 = n_2 - 1$	$v_1 = n_1 - 1$	$v_1 = n - 1$ for larger s^2
Denominator df:	$v_2 = n_1 - 1$	$v_2 = n_2 - 1$	$v_2 = n - 1$ for smaller s^2
p-value:	$P(F > F_c)$	$P(F > F_c)$	$P(F^* < 1/F_c) + P(F > F_c)$

Decision: Reject H_0 if $\alpha > p$-value or if test statistic (F_c) falls in rejection region where F is based on v_1 numerator df and v_2 denominator df, F^* is based on v_2 numerator df and v_1 denominator df, and $\alpha = P(\text{Type I error}) = P(\text{Reject } H_0 | H_0 \text{ true})$.

Conditions Required for Valid Inferences about $(\sigma_1)^2/(\sigma_2)^2$

1. The samples are random and independent.
2. Both populations are normally distributed.

To conclude this section, we consider the comparison of population variances as a check of the assumption $\sigma_1^2 = \sigma_2^2$ needed for the two-sample t-test. Rejection of the null hypothesis $\sigma_1^2 = \sigma_2^2$ would indicate that the assumption is invalid. [*Note:* Nonrejection of the null hypothesis does *not* imply that the assumption is valid.]

Example **9.12** Ⓓ READING

Checking the Assumption of Equal Variances

Problem In Example 9.4 (Section 9.2), we used the two-sample t-statistic to compare the mean reading scores of two groups of "slow learners" who had been taught to read by two different methods. The data are repeated in Table 9.10 for convenience. The use of the t-statistic was based on the assumption that the population variances of the test scores were equal for the two methods. Conduct a test of hypothesis to check this assumption at $\alpha = .10$.

---

*Although a test of a hypothesis of equality of variances is its most common application, the F-test can also be used to test a hypothesis that the ratio between the population variances is equal to some specified value $H_0: \sigma_1^2/\sigma_2^2 = k$. In that case, the test is conducted in exactly the same way as specified in the box, except that we use the test statistic

$$F = \left(\frac{s_1^2}{s_2^2}\right)\left(\frac{1}{k}\right)$$

Table 9.10 Reading Test Scores for "Slow Learners"

New Method				Standard Method			
80	80	79	81	79	62	70	68
76	66	71	76	73	76	86	73
70	85			72	68	75	66

Ⓓ *Data Set:* READING

Solution We want to test

$$H_0: \sigma_1^2/\sigma_2^2 = 1 \quad (\text{i.e., } \sigma_1^2 = \sigma_2^2)$$
$$H_a: \sigma_1^2/\sigma_2^2 \neq 1 \quad (\text{i.e., } \sigma_1^2 \neq \sigma_2^2)$$

The data were entered into SAS, and the SAS printout shown in Figure 9.21 was obtained. Both the test statistic, $F = .85$, and the two-tailed p-value, .8148, are highlighted on the printout. Since $\alpha = .10$ is less than the p-value, we do not reject the null hypothesis that the population variances of the reading test scores are equal. It is here that the temptation to misuse the F-test is strongest. *We cannot conclude that the data justify the use of the t-statistic.* Doing so would be equivalent to accepting H_0, and we have repeatedly warned against this conclusion because the probability of a Type II error, β, is unknown. The α level of .10 protects us only against rejecting H_0 if it is true. This use of the F-test may prevent us from abusing the t procedure when we obtain a value of F that leads to a rejection of the assumption that $\sigma_1^2 = \sigma_2^2$. But when the F-statistic does not fall into the rejection region, we know little more about the validity of the assumption than before we conducted the test.

```
                Two Sample Test for Variances of SCORE within METHOD

Sample Statistics

     METHOD
     Group        N      Mean     Std. Dev.    Variance
     ---------------------------------------------------------
     NEW          10     76.4       5.8348      34.04444
     STD          12     72.33333   6.3437      40.24242


Hypothesis Test

     Null hypothesis:        Variance 1 / Variance 2 =  1
     Alternative:            Variance 1 / Variance 2 ^= 1

                     - Degrees of Freedom -
           F          Numer.    Denom.              Pr > F
     ---------------------------------------------------------
         0.85           9         11               0.8148


95% Confidence Interval of the Ratio of Two Variances

          Lower Limit      Upper Limit
          -----------      -----------
            0.2358           3.3096
```

Figure 9.21

SAS F-test for testing assumption of equal variances

Look Back A 95% confidence interval for the ratio $(\sigma_1)^2/(\sigma_2)^2$ is shown at the bottom of the SAS printout of Figure 9.21. Note that the interval, (.2358, 3.3096), includes 1; hence, we cannot conclude that the ratio differs significantly from 1. Thus, the confidence interval leads to the same conclusion that the two-tailed test does: There is insufficient evidence of a difference between the population variances.

■ **Now Work Exercise 9.111**

What Do You Do If the Assumption of Normal Population Distributions Is Not Satisfied?

Answer: The F-test is much less robust (i.e., much more sensitive) to departures from normality than the t-test for comparing the population means (Section 9.2). If you have doubts about the normality of the population frequency distributions, use a *nonparametric method* (e.g., Levene's Test) for comparing the two population variances. A method can be found in the nonparametric statistics texts listed in the references for Chapter 14.

Exercises 9.96–9.117

Understanding the Principles

9.96 Describe the sampling distribution of $(s_1)^2/(s_2)^2$ for normal data and give the assumption about the population and sample.

9.97 Describe the confidence interval for $(\sigma_1)^2/(\sigma_2)^2$.

9.98 *True or false.* The F-statistic used for testing $H_0: \sigma_1^2 = \sigma_2^2$ against $H_a: \sigma_1^2 > \sigma_2^2$ is $F = (s_1)^2/(s_2)^2$.

9.99 *True or false.* $H_0: \sigma_1^2 = \sigma_2^2$ is equivalent to $H_0: \sigma_1^2/\sigma_2^2 = c$ where $c \neq 1$.

Learning the Mechanics

9.100 Use the F-tables in Appendix B or statistical software to
[NW] find each of the following F-values:
  a. $F_{.05}$, where $\nu_1 = 8$ and $\nu_2 = 5$
  b. $F_{.01}$, where $\nu_1 = 20$ and $\nu_2 = 14$
  c. $F_{.025}$, where $\nu_1 = 10$ and $\nu_2 = 5$
  d. $F_{.10}$, where $\nu_1 = 20$ and $\nu_2 = 5$

9.101 Given ν_1 and ν_2, find the following probabilities:
  a. $\nu_1 = 5, \nu_2 = 3, P(F \geq 14.88)$
  b. $\nu_1 = 6, \nu_2 = 12, P(F < 4.82)$
  c. $\nu_1 = 50, \nu_2 = 26, P(F < 2.357)$
  d. $\nu_1 = 9, \nu_2 = 50, P(F < 3.82)$

9.102 For each of the cases that follow, identify the rejection region that should be used to test $H_0: \sigma_1^2 = \sigma_2^2$ against $H_a: \sigma_1^2 > \sigma_2^2$. Assume that $\nu_1 = 30$ and $\nu_2 = 20$.
  a. $\alpha = .10$
  b. $\alpha = .05$
  c. $\alpha = .025$
  d. $\alpha = .01$

9.103 For each of the cases that follow, identify the rejection region that should be used to test $H_0: \sigma_1^2 = \sigma_2^2$ against $H_a: \sigma_1^2 \neq \sigma_2^2$. Assume that $\nu_1 = 9$ and $\nu_2 = 18$.
  a. $\alpha = .20$
  b. $\alpha = .10$
  c. $\alpha = .05$
  d. $\alpha = .02$

9.104 Specify the appropriate rejection region for testing against $H_0: \sigma_1^2 = \sigma_2^2$ in each of the following situations:
  a. $H_a: \sigma_1^2 > \sigma_2^2; \alpha = .05, n_1 = 20, n_2 = 25$
  b. $H_a: \sigma_1^2 < \sigma_2^2; \alpha = .05, n_1 = 15, n_2 = 10$
  c. $H_a: \sigma_1^2 \neq \sigma_2^2; \alpha = .10, n_1 = 31, n_2 = 21$
  d. $H_a: \sigma_1^2 < \sigma_2^2; \alpha = .01, n_1 = 41, n_2 = 31$
  e. $H_a: \sigma_1^2 \neq \sigma_2^2; \alpha = .05, n_1 = 16, n_2 = 7$

9.105 Independent random samples were selected from each of
[NW] two normally distributed populations, $n_1 = 16$ from population 1 and $n_2 = 25$ from population 2. The means and variances for the two samples are shown in the following table.

Sample 1	Sample 2
$n_1 = 16$	$n_2 = 25$
$\bar{x}_1 = 22.5$	$\bar{x}_2 = 28.2$
$s_1^2 = 2.87$	$s_2^2 = 9.85$

  a. Test the null hypothesis $H_0: \sigma_1^2 = \sigma_2^2$ against the alternative hypothesis $H_a: \sigma_1^2 \neq \sigma_2^2$. Use $\alpha = .05$.
  b. Find and interpret the p-value of the test.

9.106 Independent random samples were selected from each of
(D) two normally distributed populations, $n_1 = 6$ from popu-
[L09106] lation 1 and $n_2 = 4$ from population 2. The data are shown in the following table.

Sample 1	Sample 2
3.1	2.3
4.3	1.4
1.2	3.7
1.7	8.9
.6	
3.4	

  a. Test $H_0: \sigma_1^2 = \sigma_2^2$ against $H_a: \sigma_1^2 < \sigma_2^2$. Use $\alpha = .01$.
  b. Test $H_0: \sigma_1^2 = \sigma_2^2$ against $H_a: \sigma_1^2 \neq \sigma_2^2$. Use $\alpha = .10$.
  c. Test $H_0: \sigma_1^2 = \sigma_2^2$ against $H_a: \sigma_1^2 > \sigma_2^2$. Use $\alpha = .05$.

Applying the Concepts—Basic

9.107 Hippo grazing patterns in Kenya. In Kenya, human-induced land-use changes and excessive resource extraction have threatened the jungle ecosystem by reducing animal grazing areas and disrupting access to water sources. In *Landscape & Ecology Engineering* (Jan. 2013), researchers compared hippopotamus grazing patterns in two Kenyan areas: a national reserve and a community pastoral ranch. Each area was subdivided into plots of land. The plots were sampled (406 plots in the national reserve and 230 plots in the pastoral ranch), and the number of hippo trails from a water source was determined for each plot. Sample statistics are provided in the table. Suppose the researchers want to know if the variability in number of hippo trails from a water source in the National Reserve differs from the variability in number of hippo trails from a water source in the pastoral ranch.

	National Reserve	Pastoral Ranch
Sample size	406	230
Mean number of trails	.31	.13
Standard deviation	.40	.30

Source: Kanga, E. M., et al. "Hippopotamus and livestock grazing: Influences on riparian vegetation and facilitation of other herbivores in the Mara Region of Kenya." *Landscape & Ecology Engineering*, Vol. 9, No. 1, Jan. 2013.

  a. Find an interval estimate of σ_1^2/σ_2^2, the ratio of the variances associated with the two areas. Use a 90% confidence level.
  b. Can the researchers reliably conclude that the variability in number of hippo trails from a water source in the National Reserve differs from the variability in number of hippo trails from a water source in the pastoral ranch? Explain.
  c. Explain why a test of hypothesis at $\alpha = .10$ will result in the same inference, then carry out the test to verify your results.

9.108 Last name and acquisition timing. Refer to the *Journal of Consumer Research* (Aug. 2011) study of the *last name effect* in acquisition timing, Exercise 9.14 (p. 476). Recall that the mean response times (in minutes) to acquire basketball tickets were compared for two groups of MBA students: those students with last names beginning with one of the first 9 letters of the alphabet and those with last names beginning with one of the last 9 letters of the alphabet. Summary statistics for the two groups are repeated in the table. An analysis of the variances of the two groups is displayed in the accompanying MINITAB printout.

	First 9 Letters: A–I	Last 9 Letters: R–Z
Sample size	25	25
Mean response time (minutes)	25.08	19.38
Standard deviation (minutes)	10.41	7.12

Source: Carlson, K. A., & Conrad, J. M. "The last name effect: How last name influences acquisition timing." *Journal of Consumer Research*, Vol. 38, No. 2, Aug. 2011.

a. Locate a 95% confidence interval for the ratio of the true variances in response times for the two groups on the printout and interpret the result.

b. Based on the interval, can you conclude that one group has a larger variance in response time than the other? Explain.

c. How does your answer to part **b** affect the inference you made regarding the difference in mean response times in Exercise 9.14?

Test and CI for Two Variances

```
Method

Null hypothesis           Variance(1) / Variance(2) = 1
Alternative hypothesis    Variance(1) / Variance(2) not = 1
Significance level        Alpha = 0.05


Statistics

Sample   N    StDev   Variance
1       25   10.410   108.368
2       25    7.120    50.694

Ratio of standard deviations = 1.462
Ratio of variances = 2.138


95% Confidence Intervals

                                  CI for
Distribution    CI for StDev     Variance
of Data           Ratio           Ratio
Normal        (0.971, 2.202)   (0.942, 4.851)


Tests

                              Test
Method           DF1  DF2  Statistic  P-Value
F Test (normal)   24   24    2.14      0.069
```

9.109 Children's recall of TV ads. Refer to the *Journal of Advertising* (Spring 2006) study of children's recall of television commercials, Exercise 9.17 (p. 477). You used a small-sample t test to test the null hypothesis

$H_0: (\mu_1 - \mu_2) = 0$, where μ_1 = mean number of ads recalled by children in the video-only group and μ_2 = mean number of ads recalled by children in the A/V group. Summary statistics for the study are reproduced in the table. The validity of the inference derived from the test is based on the assumption of equal group variances, i.e., $\sigma_1^2 = \sigma_2^2$.

Video-Only Group	A/V Group
$n_1 = 20$	$n_2 = 20$
$\bar{x}_1 = 3.70$	$\bar{x}_2 = 3.30$
$s_1 = 1.98$	$s_2 = 2.13$

a. Set up the null and alternative hypothesis for testing this assumption.

b. Compute the test statistic.

c. Find the rejection region for the test using $\alpha = .10$.

d. Make the appropriate conclusion, in the words of the problem.

e. Comment on the validity of the inference derived about the difference in population means in Exercise 9.17.

9.110 Bulimia study. Refer to Exercise 9.19 (p. 477). The "fear of negative evaluation" (FNE) scores for the 11 bulimic females and 14 females with normal eating habits are saved in the **BULIMIA** file. The confidence interval you constructed in Exercise 9.19 requires that the variance of the FNE scores of bulimic females be equal to the variance of the FNE scores of normal females. Conduct a test (at $\alpha = .05$) to determine the validity of this assumption.

9.111 How do you choose to argue? Refer to the *Thinking and Reasoning* (Oct. 2006) study of the cognitive skills required for successful arguments, presented in Exercise 9.25 (p. 479). Recall that 52 psychology graduate students were equally divided into two groups. Group 1 was presented with arguments that always attempted to strengthen the favored position. Group 2 was presented with arguments that always attempted to weaken the nonfavored position. Summary statistics for the student ratings of the arguments are reproduced in the accompanying table. In Exercise 9.25, you compared the mean ratings for the two groups with a small-sample t-test, assuming equal variances. Determine the validity of this assumption at $\alpha = .05$.

	Group 1 (support favored position)	Group 2 (weaken opposing position)
Sample size	26	26
Mean	28.6	24.9
Standard deviation	12.5	12.2

Based on Kuhn, D., and Udell, W. "Coordinating own and other perspectives in argument." *Thinking and Reasoning*, Oct. 2006.

9.112 Hygiene of handshakes, high fives, and fist bumps. Refer to the *American Journal of Infection Control* (Aug. 2014) study of the hygiene of hand greetings, Exercise 9.24 (p. 479). The number of bacteria transferred from a gloved hand dipped into a culture of bacteria to a second gloved hand contacted by either a handshake, high five, or fist bump was recorded. Recall that the experiment was replicated five times for each contact method and the data used to compare the mean percentage of bacteria transferred for any two contact methods. The data are repeated in the table.

Handshake:	131	74	129	96	92
High five:	44	70	69	43	53
Fist bump:	15	14	21	29	21

a. What assumption about the variability of the data is required to make the mean comparisons for this small-sample study?
b. Conduct a test to determine if the assumption, part **a**, is reasonably satisfied when comparing the handshake to the fist bump. Test using $\alpha = .05$.
c. How does your answer, part **b**, affect the validity of the inferences made in Exercise 9.24?

Applying the Concepts—Intermediate

9.113 Cognitive impairment of schizophrenics. Refer to the *American Journal of Psychiatry* (Apr. 2010) study of the differences in cognitive function between normal individuals and patients diagnosed with schizophrenia, Exercise 9.16 (p. 476). Recall that the total time (in minutes) a subject spent on the Trail Making Test was used as a measure of cognitive function. Summary data for independent random samples of 41 schizophrenics and 49 normal individuals are reproduced below. Suppose the researchers theorize that schizophrenics will have a wider range in time on the Trail Making Test than will normal subjects. Is there evidence to support this theory? Test using $\alpha = .01$.

	Schizophrenia	Normal
Sample size	41	49
Mean time	104.23	62.24
Standard deviation	45.45	16.34

Based on Perez-Iglesias, R., et al. "White matter integrity and cognitive impairment in first-episode psychosis." *American Journal of Psychiatry*, Vol. 167, No. 4, Apr. 2010 (Table 1).

9.114 Oil content of fried sweet potato chips. Refer to the *Journal of Food Engineering* (Sept. 2013) study of the characteristics of fried sweet potato chips, Exercise 8.122 (p. 450). Recall that a sample of 6 sweet potato slices fried at 130° using a vacuum fryer yielded the following statistics on internal oil content (measured in gigagrams): $\bar{x}_1 = .178$ g/g and $s_1 = .011$ g/g. A second sample of 6 sweet potato slices was obtained, only these were subjected to a two-stage frying process (again, at 130°) in an attempt to improve texture and appearance. Summary statistics on internal oil content for this second sample follows: $\bar{x}_2 = .140$ g/g and $s_2 = .002$ g/g. The researchers want to compare the mean internal oil contents of sweet potato chips fried with the two methods using a *t*-test. Do you recommend the researchers carry out this analysis? Explain. (Recall your answer to Exercise 8.122.)

9.115 Is honey a cough remedy? Refer to the *Archives of Pediatrics and Adolescent Medicine* (Dec. 2007) study of honey as a children's cough remedy, Exercise 2.40 (p. 79). The data (cough improvement scores) for the 33 children in the DM dosage group and the 35 children in the honey dosage group are reproduced in the next table. The researchers

want to know if the variability in coughing improvement scores differs for the two groups. Conduct the appropriate analysis, using $\alpha = .10$. If the preferred treatment is the one with the smallest variation in improvement scores, which treatment is preferable?

Honey Dosage:	12	11	15	11	10	13	10	4	15	16	9	14
	10	6	10	8	11	12	12	8	9	5	12	
	12	9	11	15	10	15	9	13	8	12	10	8
DM Dosage:	4	6	9	4	7	7	7	9	12	10	11	6
	3	4	9	12	7	6	8	12	12	4	12	
	13	7	10	13	9	4	4	10	15	9		

Based on Paul, I. M., et al. "Effect of honey, dextromethorphan, and no treatment on nocturnal cough and sleep quality for coughing children and their parents." *Archives of Pediatrics and Adolescent Medicine*, Vol. 161, No. 12, Dec. 2007 (data simulated).

9.116 Shopping vehicle and judgment. Refer to the *Journal of Marketing Research* (Dec. 2011) study of shopping cart design, Exercise 9.21 (p. 478). Recall that design engineers want to know whether the mean choice of vice-over-virtue score is higher when a consumer's arm is flexed (as when carrying a shopping basket) than when the consumer's arm is extended (as when pushing a shopping cart). The average choice score for the $n_1 = 11$ consumers with a flexed arm was $\bar{x}_1 = 59$, while the average for the $n_2 = 11$ consumers with an extended arm was $\bar{x}_2 = 43$. In which scenario is the assumption required for a *t*-test to compare means more likely to be violated, $s_1 = 4$ and $s_2 = 2$ or $s_1 = 10$ and $s_2 = 15$? Explain.

9.117 Pentagon speeds up order-to-delivery times. Following the initial Persian Gulf War, the Pentagon changed its logistics processes to be more corporate-like. The extravagant "just-in-case" mentality was replaced with "just-in-time" systems. Emulating Federal Express and United Parcel Service, the Pentagon now expedites deliveries from factories to foxholes with the use of bar codes, laser cards, radio tags, and databases to track supplies. The following table contains order-to-delivery times (in days) for a sample of shipments from the United States to the Persian Gulf and a sample of shipments from the United States to Bosnia.

Persian Gulf	Bosnia
28.0	15.1
20.0	6.4
26.5	5.0
10.6	11.4
9.1	6.5
35.2	6.5
29.1	3.0
41.2	7.0
27.5	5.5

a. Determine whether the variances in order-to-delivery times for Persian Gulf and Bosnia shipments are equal. Use $\alpha = .05$.
b. Given your answer to part **a**, is it appropriate to construct a confidence interval for the difference between the mean order-to-delivery times? Explain.

CHAPTER NOTES

Key Terms

Note: Starred (*) terms are from the optional section in this chapter.

Blocking 484
F-distribution* 506
Nonparametric statistical tests 473
Paired difference experiment 484

Pooled sample estimator σ^2 468
Randomized block
  experiment 484
Standard error of the statistic 465

Key Symbols

$\mu_1 - \mu_2$	Difference between population means
μ_d	Paired difference in population means
$p_1 - p_2$	Difference between population proportions
σ_1^2/σ_2^2	Ratio of population variances
D_0	Hypothesized value of difference
$\bar{x}_1 - \bar{x}_2$	Difference between sample means
$\bar{x}_d$	Mean of sample differences
$\hat{p}_1 - \hat{p}_2$	Difference between sample proportions
s_1^2/s_2^2	*Ratio of sample variances
$\sigma_{(\bar{x}_1 - \bar{x}_2)}$	Standard error for $\bar{x}_1 - \bar{x}_2$
$\sigma_{\bar{d}}$	Standard error for $\bar{d}$
$\sigma_{(\hat{p}_1 - \hat{p}_2)}$	*Standard error for $\hat{p}_1 - \hat{p}_2$
F_α	*Critical value for F-distribution
v_1	*Numerator degrees of freedom for F-distribution
v_2	*Denominator degrees of freedom for F-distribution
SE	Sampling error in estimation

Key Ideas

Key Words for Identifying the Target Parameter

$\mu_1 - \mu_2$	Difference in Means or Averages
μ_d	Paired Difference in Means or Averages
$p_1 - p_2$	Difference in Proportions, Fractions, Percentages, Rates
σ_1^2/σ_2^2	*Ratio (or Difference) in Variances, Spreads

Determining the Sample Size

Estimating $\mu_1 - \mu_2$: $n_1 = n_2 = (z_{\alpha/2})^2(\sigma_1^2 + \sigma_2^2)/(\text{SE})^2$

Estimating $p_1 - p_2$: $n_1 = n_2 = (z_{\alpha/2})^2(p_1 q_1 + p_2 q_2)/(\text{SE})^2$

Estimating μ_d: $n = (z_{\alpha/2})^2 \sigma_d^2/(\text{SE})^2$

Conditions Required for Inferences about $\mu_1 - \mu_2$

Large samples:
1. Independent random samples
2. $n_1 \geq 30, n_2 \geq 30$

Small samples:
1. Independent random samples
2. Both populations normal
3. $\sigma_1^2 = \sigma_2^2$

*Conditions Required for Inferences about σ_1^2/σ_2^2

Large or small samples:
1. Independent random samples
2. Both populations normal

Conditions Required for Inferences about μ_d

Large samples:
1. Random sample of paired differences
2. $n_d \geq 30$

Small samples:
1. Random sample of paired differences
2. Population of differences is normal

Conditions Required for Inferences about $p_1 - p_2$

Large samples:
1. Independent random samples
2. $n_1 p_1 \geq 15, n_1 q_1 \geq 15$
3. $n_2 p_2 \geq 15, n_2 q_2 \geq 15$

Using a Confidence Interval for $(\mu_1 - \mu_2)$ or $(p_1 - p_2)$ to Determine whether a Difference Exists

1. If the confidence interval includes all *positive* numbers $(+, +): \rightarrow$ Infer $\mu_1 > \mu_2$ or $p_1 > p_2$
2. If the confidence interval includes all *negative* numbers $(-, -): \rightarrow$ Infer $\mu_1 < \mu_2$ or $p_1 < p_2$
3. If the confidence interval includes 0 $(-, +): \rightarrow$ Infer "no evidence of a difference."

Guide to Selecting a Two-Sample Hypothesis Test and Confidence Interval

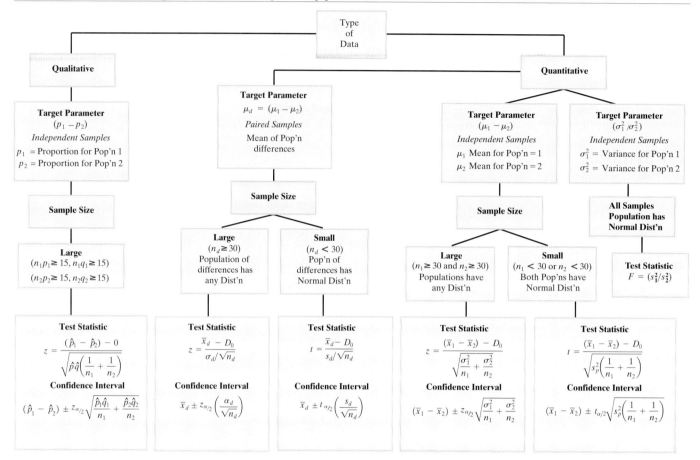

Supplementary Exercises 9.118–9.152

Note: Starred (*) exercises refer to the optional section in this chapter.

Understanding the Principles

9.118 List the assumptions necessary for each of the following inferential techniques:
  a. Large-sample inferences about the difference $(\mu_1 - \mu_2)$ between population means, using a two-sample z-statistic
  b. Small-sample inferences about $(\mu_1 - \mu_2)$, using an independent samples design and a two-sample t-statistic
  c. Small-sample inferences about $(\mu_1 - \mu_2)$, using a paired difference design and a single-sample t-statistic to analyze the differences
  d. Large-sample inferences about the differences $(p_1 - p_2)$ between binomial proportions, using a two-sample z-statistic
  *e. Inferences about the ratio σ_1^2/σ_2^2 of two population variances, using an F-test

9.119 For each of the following, identify the target parameter as $\mu_1 - \mu_2, p_1 - p_2$, or μ.
  a. Comparison of average marks in statistics of male and female graduates.
  b. Difference between mean waiting times at two metro rail ticket counters.
  c. Comparison of proportions of Democrats and Republicans who favor the legalization of marijuana.

*d. Comparison of variation in salaries of IPL players picked in the first round and the second round.
  e. Difference in dropout rates of school students in urban areas and rural areas.

Learning the Mechanics

9.120 Two independent random samples were selected from normally distributed populations with means and variances (μ_1, σ_1^2) and (μ_2, σ_2^2), respectively. The sample sizes, means, and variances are shown in the following table.

Sample 1	Sample 2
$n_1 = 15$	$n_2 = 20$
$\bar{x}_1 = 123$	$\bar{x}_2 = 116$
$s_1^2 = 15.65$	$s_2^2 = 60.05$

*a. Test $H_0: \sigma_1^2 = \sigma_2^2$ against $H_a: \sigma_1^2 \neq \sigma_2^2$. Use $\alpha = .05$.
  b. Would you be willing to use a t-test to test the null hypothesis $H_0: (\mu_1 - \mu_2) = 0$ against the alternative hypothesis $H_a: (\mu_1 - \mu_2) \neq 0$? Why?

9.121 Independent random samples were selected from two normally distributed populations with means μ_1 and μ_2, respectively. The sample sizes, means, and variances are shown in the following table.

Table for Exercise 9.121

Sample 1	Sample 2
$n_1 = 14$	$n_2 = 12$
$\bar{x}_1 = 19.8$	$\bar{x}_2 = 17.3$
$s_1^2 = 60.5$	$s_2^2 = 74.2$

a. Test $H_0: (\mu_1 - \mu_2) = 0$ against $H_a: (\mu_1 - \mu_2) > 0$. Use $\alpha = .05$.

b. Form a 99% confidence interval for $(\mu_1 - \mu_2)$.

c. How large must n_1 and n_2 be if you wish to estimate $(\mu_1 - \mu_2)$ to within two units with 99% confidence? Assume that $n_1 = n_2$.

9.122 Independent random samples were selected from two binomial populations. The size and number of observed successes for each sample are shown in the following table.

Sample 1	Sample 2
$n_1 = 200$	$n_2 = 200$
$x_1 = 70$	$x_2 = 90$

a. Test $H_0: (p_1 - p_2) = 0$ against $H_a: (p_1 - p_2) < 0$. Use $\alpha = .10$.

b. Form a 95% confidence interval for $(p_1 - p_2)$.

c. What sample sizes would be required if we wish to use a 95% confidence interval of width .01 to estimate $(p_1 - p_2)$?

9.123 Two independent random samples are taken from two populations. The results of these samples are summarized in the following table.

Sample 1	Sample 2
$n_1 = 148$	$n_2 = 135$
$\bar{x}_1 = 15.2$	$\bar{x}_2 = 11.3$
$s_1^2 = 3.0$	$s_2^2 = 2.1$

a. Form a 90% confidence interval for $(\mu_1 - \mu_2)$.

b. Test $H_0: (\mu_1 - \mu_2) = 0$ against $H_a: (\mu_1 - \mu_2) \neq 0$. Use $\alpha = .01$.

c. What sample size would be required if you wish to estimate $(\mu_1 - \mu_2)$ to within .2 with 90% confidence? Assume that $n_1 = n_2$.

9.124 A random sample of five pairs of observations was selected, one observation from a population with mean μ_1, the other from a population with mean μ_2. The data are shown in the following table.

Pair	Value from Population 1	Value from Population 2
1	28	22
2	31	27
3	24	20
4	30	27
5	22	20

a. Test the null hypothesis $H_0: \mu_d = 0$ against $H_a: \mu_d \neq 0$, where $\mu_d = \mu_1 - \mu_2$. Use $\alpha = .05$.

b. Form a 95% confidence interval for μ_d.

c. When are the procedures you used in parts **a** and **b** valid?

Applying the Concepts—Basic

9.125 Treating depression with St. John's wort. The *Journal of the American Medical Association* (Apr. 18, 2001) published a study of the effectiveness of using extracts of the herbal medicine St. John's wort in treating major depression. In an eight-week randomized, controlled trial, 200 patients diagnosed with major depression were divided into two groups, one of which ($n_1 = 98$) received St. John's wort extract while the other ($n_2 = 102$) received a placebo (no drug). At the end of the study period, 14 of the St. John's wort patients were in remission, compared with 5 of the placebo patients.

a. Compute the proportion of the St. John's wort patients who were in remission.

b. Compute the proportion of the placebo patients who were in remission.

c. If St. John's wort is effective in treating major depression, then the proportion of St. John's wort patients in remission will exceed the proportion of placebo patients in remission. At $\alpha = .01$, is St. John's wort effective in treating major depression?

d. Repeat part **c**, but use $\alpha = .10$.

e. Explain why the choice of α is critical for this study.

9.126 NHTSA new car crash tests. Refer to the National Highway Traffic Safety Administration (NHTSA) crash test data on new cars, saved in the **CRASH** file. Crash test dummies were placed in the driver's seat and front passenger's seat of a new car model, and the car was steered by remote control into a head-on collision with a fixed barrier while traveling at 35 miles per hour. Two of the variables measured for each of the 98 new cars in the data set are (1) the severity of the driver's chest injury and (2) the severity of the passenger's chest injury. (The more points assigned to the chest injury rating, the more severe the injury is.) Suppose the NHTSA wants to determine whether the true mean driver chest injury rating exceeds the true mean passenger chest injury rating and, if so, by how much.

a. State the parameter of interest to the NHTSA.

b. Explain why the data should be analyzed as matched pairs.

c. Find a 99% confidence interval for the true difference between the mean chest injury ratings of drivers and front-seat passengers.

d. Interpret the interval you found in part **c**. Does the true mean driver chest injury rating exceed the true mean passenger chest injury rating? If so, by how much?

e. What conditions are required for the analysis to be valid? Do these conditions hold for these data?

9.127 Oil spill impact on seabirds. Refer to the *Journal of Agricultural, Biological, and Environmental Statistics* (Sept. 2000) study of the impact of a tanker oil spill on the seabird population in Alaska, presented in Exercise 2.205 (p. 138). Recall that for each of 96 shoreline locations (called transects), the number of seabirds found, the length (in kilometers) of the transect, and whether the transect was in an oiled area were recorded. (The data are saved in the **EVOS** file.) *Observed seabird density* is defined as the observed count divided by the length of the transect. A comparison of the mean densities of oiled and unoiled transects is displayed in the MINITAB printout on the next page. Use this information to make an inference about the difference in the population mean seabird densities of oiled and unoiled transects.

MINITAB output for Exercise 9.127

Two-Sample T-Test and CI: Density, Oil

Two-sample T for Density

```
Oil   N  Mean  StDev  SE Mean
no   36  3.27   6.70     1.1
yes  60  3.50   5.97    0.77
```

Difference = mu (no) - mu (yes)
Estimate for difference: -0.221165
95% CI for difference: (-2.927767, 2.485436)
T-Test of difference = 0 (vs not =): T-Value = -0.16 P-Value = 0.871 DF = 67

9.128 Rating service at five-star hotels. A study published in the *Journal of American Academy of Business, Cambridge* (Mar. 2002) examined whether the perception of the quality of service at five-star hotels in Jamaica differed by gender. Hotel guests were randomly selected from the lobby and restaurant areas and asked to rate 10 service-related items (e.g., "the personal attention you received from our employees"). Each item was rated on a five-point scale (1 = "much worse than I expected," 5 = "much better than I expected"), and the sum of the items for each guest was determined. A summary of the guest scores is provided in the following table.

Gender	Sample Size	Mean Score	Standard Deviation
Males	127	39.08	6.73
Females	114	38.79	6.94

a. Construct a 90% confidence interval for the difference between the population mean service-rating scores given by male and female guests at Jamaican five-star hotels.

b. Use the interval you constructed in part **a** to make an inference about whether the perception of the quality of service at five-star hotels in Jamaica differs by gender.

*c. Is there evidence of a difference in the variation of guest scores for males and females? Test using $\alpha = .10$.

9.129 Effect of altitude on climbers. Dr. Philip Lieberman, a neuroscientist at Brown University, conducted a field experiment to gauge the effect of high altitude on a person's ability to think critically. The subjects of the experiment were five males who took part in an American expedition climbing Mount Everest. At the base camp, Lieberman read sentences to the climbers while they looked at simple pictures in a book. The length of time (in seconds) it took for each climber to match the picture with a sentence was recorded. Using a radio, Lieberman repeated the task when the climbers reached a camp 5 miles above sea level. At this altitude, he noted that the climbers took 50% longer to complete the task.

a. What is the variable measured in this experiment?

b. What are the experimental units?

c. Discuss how the data should be analyzed.

9.130 Executive workout dropouts. Refer to the *Journal of Sport Behavior* (2001) study of variety in exercise workouts, presented in Exercise 7.140 (p. 389). One group of 40 people varied their exercise routine in workouts, while a second group of 40 exercisers had no set schedule or regulations for their workouts. By the end of the study, 15 people had dropped out of the first exercise group and 23 had dropped out of the second group.

a. Find the dropout rates (i.e., the percentage of exercisers who had dropped out of the exercise group) for each of the two groups of exercisers.

b. Find a 90% confidence interval for the difference between the dropout rates of the two groups of exercisers.

c. Give a practical interpretation of the confidence interval you found in part **c**.

d. Suppose you want to reduce the sampling error in the 90% confidence interval to .1. Determine the number of exercisers to be sampled from each group in order to obtain such an estimate. Assume equal sample sizes, and assume that $p_1 \approx .4$ and $p_2 \approx .6$.

9.131 Racial profiling by the LAPD. *Racial profiling* is a term used to describe any police action that relies on ethnicity rather than behavior to target suspects engaged in criminal activities. Does the Los Angeles Police Department (LAPD) invoke racial profiling in stops and searches of Los Angeles drivers? This question was addressed in *Chance* (Spring 2006).

a. Data on stops and searches of both African-Americans and white drivers over a six-month period are summarized in the accompanying table. Conduct a test (at $\alpha = .05$) to determine whether there is a disparity in the proportions of African-American and white drivers who are searched by the LAPD after being stopped.

Race	Number Stopped	Number Searched	Number of "Hits"
African-American	61,688	12,016	5,134
White	106,892	5,312	3,006

Based on Khadjavi, L. S. "Driving while black in the City of Angels." *Chance*, Vol. 19, No. 2, Spring 2006 (Tables 1 and 2), pp. 43–46.

b. The LAPD defines a "hit rate" as the proportion of searches that result in a discovery of criminal activity. Use the data in the table to estimate the disparity in the hit rates for African-American and white drivers under a 95% confidence interval. Interpret the results.

9.132 Index of Biotic Integrity. The Ohio Environmental Protection Agency used the Index of Biotic Integrity (IBI) to measure the biological condition, or "health," of an aquatic region. The IBI is the sum of metrics that measure the presence, abundance, and health of fish in the region. (Higher values of the IBI correspond to healthier fish populations.) Researchers collected IBI measurements for sites located in different Ohio river basins (*Journal of Agricultural, Biological, and Environmental Sciences*, June 2005). Summary data for two river basins, Muskingum and Hocking, are given in the accompanying table.

River Basin	Sample Size	Mean	Standard Deviation
Muskingum	53	.035	1.046
Hocking	51	.340	.960

Based on Boone, E. L., Keying, Y., and Smith, E. P. "Evaluating the relationship between ecological and habitat conditions using hierarchical models." *Journal of Agricultural, Biological, and Environmental Sciences*, Vol. 10, No. 2, June 2005 (Table 01).

a. Use a 90% confidence interval to compare the mean IBI values of the two river basins. Interpret the interval.

b. Conduct a test of hypothesis (at $\alpha = .10$) to compare the mean IBI values of the two river basins. Explain why the result will agree with the inference you derived from the 90% confidence interval in part **a**.

9.133 The "winner's curse" in auction bidding. In auction bidding, the "winner's curse" is the phenomenon of the winning (or highest) bid price being above the expected value of the item being auctioned. *The Review of Economics and Statistics* (Aug. 2001) published a study on whether experience in bidding affects the likelihood of the winner's curse occurring. Two groups of bidders in a sealed-bid auction were compared: (1) superexperienced bidders and (2) less experienced bidders. In the superexperienced group, 29 of 189 winning bids were above the item's expected value; in the less experienced group, 32 of 149 winning bids were above the item's expected value.

a. Find an estimate of p_1, the true proportion of super-experienced bidders who fall prey to the winner's curse.

b. Find an estimate of p_2, the true proportion of less experienced bidders who fall prey to the winner's curse.

c. Construct a 90% confidence interval for $p_1 - p_2$.

d. Give a practical interpretation of the confidence interval you constructed in part **c**. Make a statement about whether experience in bidding affects the likelihood of the winner's curse occurring.

9.134 Environmental impact study. Some power plants are located near rivers or oceans so that the available water can be used to cool the condensers. Suppose that, as part of an environmental impact study, a power company wants to estimate the difference in mean water temperature between the discharge of its plant and the offshore waters. How many sample measurements must be taken at each site in order to estimate the true difference between means to within .3°C with 90% confidence? Assume that the range in readings will be about 4°C at each site and that the same number of readings will be taken at each site.

9.135 Animal-assisted therapy for heart patients. Refer to the *American Heart Association Conference* (Nov. 2005) study to gauge whether animal-assisted therapy can improve the physiological responses of heart failure patients, presented in Exercise 2.112 (p. 106). Recall that a sample of $n = 26$ heart patients was visited by a human volunteer accompanied by a trained dog; the anxiety level of each patient was measured (in points) both before and after the visits. The drop (before minus after) in anxiety level for patients is summarized as follows: $\bar{x}_d = 10.5$, $s_d = 7.6$. Does animal-assisted therapy significantly reduce the mean anxiety level of heart failure patients? Support your answer with a 95% confidence interval.

9.136 Life expectancy of Oscar winners. Does winning an Academy of Motion Picture Arts and Sciences award lead to long-term mortality for movie actors? In an article in the *Annals of Internal Medicine* (May 15, 2001), researchers sampled 762 Academy Award winners and matched each one with another actor of the same sex who was in the same winning film and was born in the same era. The life expectancies (ages) of the pairs of actors were compared.

a. Explain why the data should be analyzed as a paired difference experiment.

b. Set up the null hypothesis for a test to compare the mean life expectancies of Academy Award winners and nonwinners.

c. The sample mean life expectancies of Academy Award winners and nonwinners were reported as 79.7 years and 75.8 years, respectively. The *p*-value for comparing the two population means was reported as $p = .003$. Interpret this value in the context of the problem.

Applying the Concepts—Intermediate

9.137 Reading tongue twisters. According to *Webster's New World Dictionary*, a tongue twister is "a phrase that is hard to speak rapidly." Do tongue twisters have an effect on the length of time it takes to read silently? To answer this question, 42 undergraduate psychology students participated in a reading experiment (*Memory & Cognition*, Sept. 1997). Two lists, each composed of 600 words, were constructed. One list contained a series of tongue twisters, and the other list (called the *control*) did not contain any tongue twisters. Each student read both lists, and the length of time (in minutes) required to complete the lists was recorded. The researchers used a test of hypothesis to compare the mean reading response times for the tongue-twister and control lists.

a. Set up the null hypothesis for the test.

b. For each student, the researchers computed the difference between the reading response times for the tongue-twister and control lists. The mean difference was .25 minute with a standard deviation of .78 minute. Use the information to find the test statistic and *p*-value of the test.

c. Give the appropriate conclusion. Use $\alpha = .05$.

Based on Robinson, D. H., and Katayama, A. D. "At-lexical, articulatory interference in silent reading: The 'upstream' tongue-twister effect." Memory & Cognition, Vol. 25, No. 5, Sept. 1997, p. 663.

9.138 Patent infringement case. *Chance* (Fall 2002) described a lawsuit charging Intel Corp. with infringing on a patent for an invention used in the automatic manufacture of computer chips. In response, Intel accused the inventor of adding material to his patent notebook after the patent was witnessed and granted. The case rested on whether a patent witness's signature was written on top of or under key text in the notebook. Intel hired a physicist who used an X-ray beam to measure the relative concentrations of certain elements (e.g., nickel, zinc, potassium) at several spots on the notebook page. The zinc measurements for three notebook locations—on a text line, on a witness line, and on the intersection of the witness and text line—are provided in the following table.

Text line:	.335	.374	.440			
Witness line:	.210	.262	.188	.329	.439	.397
Intersection:	.393	.353	.285	.295	.319	

a. Use a test or a confidence interval (at $\alpha = .05$) to compare the mean zinc measurement for the text line with the mean for the intersection.

b. Use a test or a confidence interval (at $\alpha = .05$) to compare the mean zinc measurement for the witness line with the mean for the intersection.

c. From the results you obtained in parts a and b, what can you infer about the mean zinc measurements at the three notebook locations?

*d. Use a test (at $\alpha = .05$) to compare the variation in zinc measurements for the text line with the corresponding variation for the intersection.

*e. Use a test (at $\alpha = .05$) to compare the variation in zinc measurements for the witness line with the corresponding variation for the intersection.

*f. From your results in parts d and e, what can you infer about the variation in zinc measurements at the three notebook locations?

g. What assumptions are required for the inferences to be valid? Are they reasonably satisfied?

9.139 **Mongolian desert ants.** Refer to the *Journal of Biogeography* (Dec. 2003) study of ants in Mongolia (central Asia), presented in Exercise 2.68 (p. 91). Recall that botanists placed seed baits at 5 sites in the Dry Steppe region and 6 sites in the Gobi Desert and observed the number of ant species attracted to each site. These data are listed in the next table.

a. Is there evidence to conclude that a difference exists between the average number of ant species found at sites in the two regions of Mongolia? Draw the appropriate conclusion, using $\alpha = .05$.

b. Conduct a test of equal variances at the two regions, using $\alpha = .05$. Interpret the result.

Site	Region	Number of Ant Species
1	Dry Steppe	3
2	Dry Steppe	3
3	Dry Steppe	52
4	Dry Steppe	7
5	Dry Steppe	5
6	Gobi Desert	49
7	Gobi Desert	5
8	Gobi Desert	4
9	Gobi Desert	4
10	Gobi Desert	5
11	Gobi Desert	4

Based on Pfeiffer, M., et al. "Community organization and species richness of ants in Mongolia along an ecological gradient from steppe to Gobi desert." *Journal of Biogeography*, Vol. 30, No. 12, Dec. 2003.

9.140 **Identical twins reared apart.** Because they share an identical genotype, twins make ideal subjects for investigating the degree to which various environmental conditions affect personality. The classical method of studying this phenomenon, and the subject of a book by Susan Farber (*Identical Twins Reared Apart,* New York: Basic Books, 1981), is the study of identical twins separated early in life and reared apart. Much of Farber's discussion focuses on a comparison of IQ scores. The data for this analysis appear in the accompanying table. One member (A) of each of the $n = 32$ pairs of twins was reared by a natural parent; the other member (B) was reared by a relative or some other person. Is there a significant difference between the average IQ scores of identical twins when one member of the pair is reared by the natural parents and the other member of the pair is not? Use $\alpha = .05$ to draw your conclusion.

Pair ID	Twin A	Twin B	Pair ID	Twin A	Twin B
112	113	109	228	100	88
114	94	100	232	100	104
126	99	86	236	93	84
132	77	80	306	99	95
136	81	95	308	109	98
148	91	106	312	95	100
170	111	117	314	75	86
172	104	107	324	104	103
174	85	85	328	73	78
180	66	84	330	88	99
184	111	125	338	92	111
186	51	66	342	108	110
202	109	108	344	88	83
216	122	121	350	90	82
218	97	98	352	79	76
220	82	94	416	97	98

Based on Farber, S. L. *Identical Twins Reared Apart,* © 1981 by Basic Books, Inc.

9.141 **Killing insects with low oxygen.** Refer to the *Journal of Agricultural, Biological, and Environmental Statistics* (Sept. 2000) study of the mortality of rice weevils exposed to low oxygen, presented in Exercise 8.92 (p. 436). Recall that 31,386 of 31,421 rice weevils were found dead after exposure to nitrogen gas for 4 days. In a second experiment, 23,516 of 23,676 rice weevils were found dead after exposure to nitrogen gas for 3.5 days. Conduct a test of hypothesis to compare the mortality rates of adult rice weevils exposed to nitrogen at the two exposure times. Is there a significant difference (at $\alpha = .10$) in the mortality rates?

9.142 **Light-to-dark transition of genes.** *Synechocystis,* a type of bacterium that can grow and survive under a wide range of conditions, is used by scientists to model DNA behavior. In the *Journal of Bacteriology* (July 2002), scientists isolated genes of the bacterium responsible for photosynthesis and respiration and investigated the sensitivity of the genes to light. Each gene sample was grown in an incubator in "full light." The lights were then extinguished, and any growth of the sample was measured after 24 hours in the dark ("full dark"). The lights were then turned back on for 90 minutes ("transient light"), followed immediately by an additional 90 minutes in the dark ("transient dark"). Standardized growth measurements in each light–dark condition were obtained for 103 genes. The complete data set is saved in the **GENES** file. Data on the first 10 genes are shown in the following table.

Gene ID	Full-Dark	Tr-Light	Tr-Dark
SLR2067	−0.00562	1.40989	−1.28569
SLR1986	−0.68372	1.83097	−0.68723
SSR3383	−0.25468	−0.79794	−0.39719
SLL0928	−0.18712	−1.20901	−1.18618
SLR0335	−0.20620	1.71404	−0.73029
SLR1459	−0.53477	2.14156	−0.33174
SLL1326	−0.06291	1.03623	0.30392
SLR1329	−0.85178	−0.21490	0.44545
SLL1327	0.63588	1.42608	−0.13664
SLL1325	−0.69866	1.93104	−0.24820

Based on Gill, R. T., et al. "Genome-wide dynamic transcriptional profiling of the light to dark transition in *Synechocystis Sp.* PCC6803." *Journal of Bacteriology,* Vol. 184, No. 13, July 2002.

    a. Treat the data for the first 10 genes as a random sample collected from the population of 103 genes, and test the hypothesis that there is no difference between the mean standardized growth of genes in the full-dark condition and genes in the transient-light condition. Use $\alpha = .01$.

    b. Use a statistical software package to compute the mean difference in standardized growth of the 103 genes in the full-dark condition and the transient-light condition. Did the test you carried out in part **a** detect this difference?

    c. Repeat parts **a** and **b** for a comparison of the mean standardized growth of genes in the full-dark condition and genes in the transient-dark condition.

    d. Repeat parts **a** and **b** for a comparison of the mean standardized growth of genes in the transient-light condition and genes in the transient-dark condition.

9.143 **Effectiveness of drug tests of Olympic athletes.** Erythropoietin (EPO) is a banned drug used by athletes to increase the oxygen-carrying capacity of their blood. New tests for EPO were first introduced prior to the 2000 Olympic Games held in Sydney, Australia. *Chance* (Spring 2004) reported that of a sample of 830 world-class athletes, 159 did not compete in the World Championships (a year prior to the introduction of the new EPO test). Similarly, 133 of 825 potential athletes did not compete in the 2000 Olympic games. Was the new test effective in deterring an athlete from participating in the Olympics? Conduct the analysis (at $\alpha = .10$) and draw the proper conclusion.

9.144 **Swim maze study.** Merck Research Labs used the single-T swim maze to conduct an experiment to evaluate the effect of a new drug. Nineteen impregnated dam rats were allocated a dosage of 12.5 milligrams of the drug. One male and one female rat pup were randomly selected from each resulting litter to perform in the swim maze. Each pup was placed in the water at one end of the maze and allowed to swim until it escaped at the opposite end. If the pup failed to escape after a certain period of time, it was placed at the beginning of the maze and given another chance. The experiment was repeated until each pup accomplished three successful escapes. The accompanying table reports the number of swims required by each pup to perform three successful escapes. Is there sufficient

evidence of a difference between the mean number of swims required by male and female pups? Conduct the test (at $\alpha = .10$). Comment on the assumptions required for the test to be valid.

Litter	Male	Female	Litter	Male	Female
1	8	5	11	6	5
2	8	4	12	6	3
3	6	7	13	12	5
4	6	3	14	3	8
5	6	5	15	3	4
6	6	3	16	8	12
7	3	8	17	3	6
8	5	10	18	6	4
9	4	4	19	9	5
10	4	4			

Based on Thomas E. Bradstreet, "Favorite Data Sets from Early Phases of Drug Research - Part 2." *Proceedings of the Section on Statistical Education of the American Statistical Association.*

9.145 **Pig castration study.** Two methods of castrating male piglets were investigated in *Applied Animal Behaviour Science* (Nov. 1, 2000). Method 1 involved an incision in the spermatic cords, while Method 2 involved pulling and severing the cords. Forty-nine male piglets were randomly allocated to one of the two methods. During castration, the researchers measured the number of high-frequency vocal responses (squeals) per second over a 5-second period. The data are summarized in the accompanying table. Conduct a test of hypothesis to determine whether the population mean number of high-frequency vocal responses differs for piglets castrated by the two methods. Use $\alpha = .05$.

	Method 1	Method 2
Sample size	24	25
Mean number of squeals	.74	.70
Standard deviation	.09	.09

Based on Taylor, A. A., and Weary, D. M. "Vocal responses of piglets to castration: Identifying procedural sources of pain." *Applied Animal Behaviour Science,* Vol. 70, No. 1, Nov. 1, 2000.

9.146 **Identifying the target parameter.** For each of the following studies, give the parameter of interest and state any assumptions that are necessary for the inferences to be invalid.

    a. To investigate a possible link between jet lag and memory impairment, a University of Bristol (England) neurologist recruited 20 female flight attendants who worked flights across several time zones. Half of the attendants had only a short recovery time between flights, and half had a long recovery time between flights. The average size of the right temporal lobe of the brain for the short-recovery group was significantly smaller than the average size of the right temporal lobe of the brain for the long-recovery group.

    b. In a study presented at a meeting of the Association for the Advancement of Applied Sport Psychology, researchers revealed that the proportion of athletes who have a good self-image of their body is 20% higher than the corresponding proportion of nonathletes.

c. A University of Florida animal sciences professor has discovered that feeding chickens corn oil causes them to produce larger eggs. The weight of eggs produced by each of a sample of chickens on a regular feed diet was recorded. Then the same chickens were fed a diet supplemented by corn oil, and the weight of eggs produced by each was recorded. The mean weight of the eggs produced with corn oil was 3 grams heavier than the mean weight produced with the regular diet.

d. College students enrolled in music classes participated in a study to compare the teacher evaluations of music education majors and nonmusic majors (*Journal of Research in Music Education*). Independent random samples of 100 music majors and 100 nonmajors rated the overall performance of their teacher, using a six-point scale, where 1 was the lowest rating and 6 the highest. The nonmusic majors rated their teachers higher by an average of .33 point.

*9.147 Instrument precision. The quality control department of a paper company measures the brightness (a measure of reflectance) of finished paper on a periodic basis throughout the day. Two instruments that are available to measure the paper specimens are subject to error, but they can be adjusted so that the mean readings for a control paper specimen are the same for both instruments. Suppose you are concerned about the precision of the two instruments and want to compare the variability in the readings of instrument 1 with those of instrument 2. Five brightness measurements were made on a single paper specimen, using each of the two instruments. The data are shown in the following table.

Instrument 1	Instrument 2
29	26
28	34
30	30
28	32
30	28

a. Is the variance of the measurements obtained by instrument 1 significantly different from the variance of the measurements obtained by instrument 2?

b. What assumptions must be satisfied for the test in part **a** to be valid?

9.148 Testing electronic circuits. Japanese researchers have developed a compression–depression method of testing electronic circuits based on Huffman coding (*IEICE Transactions on Information & Systems*, Jan. 2005). The new method is designed to reduce the time required for input decompression and output compression—called the compression ratio. Experimental results were obtained by testing a sample of 11 benchmark circuits (all of different sizes) from a SUN Blade 1000 workstation. Each circuit was tested with the standard compression–depression method and the new Huffman-based coding method and the compression ratio recorded. The data are given in the next table.

a. Compare the two methods with a 95% confidence interval. Which method has the smaller mean compression ratio?

b. How many circuits need to be sampled in order to estimate the mean difference in compression ratio to within .03 with 95% confidence.

Data for Exercise 9.148

Circuit	Standard Method	Huffman Coding Method
1	.80	.78
2	.80	.80
3	.83	.86
4	.53	.53
5	.50	.51
6	.96	.68
7	.99	.82
8	.98	.72
9	.81	.45
10	.95	.79
11	.99	.77

Based on Ichihara, H., Shintani, M., and Inoue, T. "Huffman-based test response coding." *IEICE Transactions on Information & Systems*, Vol. E88-D, No. 1, Jan. 2005 (Table 3).

Applying the Concepts—Advanced

9.149 Gambling in public high schools. With the rapid growth in legalized gambling in the United States, there is concern that the involvement of youth in gambling activities is also increasing. University of Minnesota professor Randy Stinchfield compared the annual rates of gambling among Minnesota public school students (*Journal of Gambling Studies*, Winter 2001). Based on survey data, the following table shows the percentages of ninth-grade boys who gambled weekly or daily on any game (e.g., cards, sports betting, lotteries) for two years:

	Year 1	Year 2 (6 years later)
Number of ninth-grade boys in survey	21,484	23,199
Number who gambled weekly/daily	4,684	5,313

a. Are the percentages of ninth-grade boys who gambled weekly or daily on any game in the two years significantly different? (Use $\alpha = .01$.)

b. Professor Stinchfield states that "because of the large sample sizes, even small differences may achieve statistical significance, so interpretations of the differences should include a judgment regarding the magnitude of the difference and its public health significance." Do you agree with this statement? If not, why not? If so, obtain a measure of the magnitude of the difference between the two years and attach a measure of reliability to the difference.

9.150 MS and exercise study. A study published in *Clinical Kinesiology* (Spring 1995) was designed to examine the metabolic and cardiopulmonary responses during exercise of persons diagnosed with multiple sclerosis (MS). Leg-cycling and arm-cranking exercises were performed by 10 MS patients and 10 healthy (non-MS) control subjects. Each member of the control group was selected on the basis of gender, age, height, and weight to match (as closely as possible) with one member of the MS group. Consequently, the researchers compared the MS and non-MS groups by matched-pairs *t*-tests on such outcome variables as oxygen uptake, carbon dioxide output, and peak aerobic power. The data on the matching variables used in the experiment are shown in the table on the next page. Have the researchers successfully matched the MS and non-MS subjects?

Data for Exercise 9.150

Matched Pair	MS Subjects				Non-MS Subjects			
	Gender	Age (years)	Height (cm)	Weight (kg)	Gender	Age (years)	Height (cm)	Weight (kg)
1	M	48	171.0	80.8	M	45	173.0	76.3
2	F	34	158.5	75.0	F	34	158.0	75.6
3	F	34	167.6	55.5	F	34	164.5	57.7
4	M	38	167.0	71.3	M	34	161.3	70.0
5	M	45	182.5	90.9	M	39	179.0	96.0
6	F	42	166.0	72.4	F	42	167.0	77.8
7	M	32	172.0	70.5	M	34	165.8	74.7
8	F	35	166.5	55.3	F	43	165.1	71.4
9	F	33	166.5	57.9	F	31	170.1	60.4
10	F	46	175.0	79.9	F	43	175.0	77.9

From "Maximal aerobic exercise of individuals with multiple sclerosis using three modes of ergometry." *Clinical Kinesiology,* Vol. 49, No. 1, Spring 1995, p. 7. Reprinted with permission from W. Jeffrey Armstrong.

Critical Thinking Challenges

9.151 Self-managed work teams and family life. To improve quality, productivity, and timeliness, more and more American industries are utilizing self-managed work teams (SMWTs). A team typically consists of 5 to 15 workers who are collectively responsible for making decisions and performing all tasks related to a particular project. Researchers L. Stanley-Stevens (Tarleton State University), D. E. Yeatts, and R. R. Seward (both from the University of North Texas) investigated the connection between SMWTs, work characteristics, and workers' perceptions of positive spillover into family life (*Quality Management Journal*, Summer 1995). Survey data were collected from 114 AT&T employees who worked on 1 of 15 SMWTs at an AT&T technical division. The workers were divided into two groups: (1) those who reported a positive spillover of work skills to family life and (2) those who did not report any such positive work spillover. The two groups were compared on a variety of job and demographic characteristics, several of which are shown in the table (next column). All but the demographic characteristics were measured on a seven-point scale, ranging from 1 = "strongly disagree" to 7 = "strongly agree"; thus, the larger the number, the more the characteristic was indicated. The file named **SMWT** includes the values of the variables listed in the table for each of the 114 survey participants. The researchers' objectives were to compare the two groups of workers on each characteristic. In particular, they wanted to know which job-related characteristics are most highly associated with positive work spillover. Conduct a complete analysis of the data for the researchers.

Characteristic	Variable
Information Flow	Use of creative ideas (seven-point scale)
Information Flow	Utilization of information (seven-point scale)
Decision Making	Participation in decisions regarding personnel matters (seven-point scale)
Job	Good use of skills (seven-point scale)
Job	Task identity (seven-point scale)
Demographic	Age (years)
Demographic	Education (years)
Demographic	Gender (male or female)
Comparison	Group (positive spillover or no spillover)

9.152 Detection of rigged school milk prices. Each year, the state of Kentucky invites bids from dairies to supply half-pint containers of fluid milk products for its school districts. In several school districts in northern Kentucky (called the "tri-county" market), two suppliers—Meyer Dairy and Trauth Dairy—were accused of price fixing, that is, conspiring to allocate the districts so that the winning bidder was predetermined and the price per pint was set above the competitive price. These two dairies were the only two bidders on the milk contracts in the tri-county market for eight consecutive years. (In contrast, a large number of different dairies won the milk contracts for school districts in the remainder of the northern Kentucky market, called the "surrounding" market.) Did Meyer and Trauth conspire to rig their bids in the tri-county market? To answer this question, data on all bids received from the dairies competing for the milk contracts during the period in question are provided and saved in the **MILK** file. To help you in your analysis of the data, consider the following established economic theories that hold when collusive bidding exists: (1) The mean winning price in the rigged market will be higher than the mean winning price in the competitive surrounding market; (2) since dairies operating in a competitive market are likely not to share information about their bids, the variability in winning bid prices in the competitive market will exceed the corresponding variability in the rigged market; and (3) because dairies fixing prices usually control the market year after year, the rigged market *incumbency rate* (defined as the proportion of school districts that are won by the dairy that won the previous year) will exceed the incumbency rate of the competitive market and will likely exceed .90. [*Note*: Examining the data in the **MILK** file, you'll find that in 50 of the 51 potential vendor transitions for the tri-county market, the winning dairy from the previous year won the bid the next year; similarly, you'll find that in 91 of the 134 potential vendor transitions for the surrounding market, the same dairy won the bid the next year.] Several MINITAB printouts of the analysis of the data follow on the next page. Use this information to determine if there is evidence that Meyer and Trauth conspired to rig their bids in the tri-county market.

Minitab Printouts for Exercise 9.152

Two-Sample T-Test and CI: WWBID, Market

```
Two-sample T for WWBID

Market        N    Mean    StDev    SE Mean
SURROUND     254   0.1331  0.0158   0.00099
TRI-COUNTY   100   0.1431  0.0133   0.0013


Difference = mu (SURROUND) - mu (TRI-COUNTY)
Estimate for difference:  -0.009970
95% upper bound for difference:  -0.007232
T-Test of difference = 0 (vs <): T-Value = -6.02  P-Value = 0.000  DF = 213
```

Test and CI for Two Proportions

```
Sample   X    N    Sample p
1       91   134   0.679104
2       50    51   0.980392


Difference = p (1) - p (2)
Estimate for difference:  -0.301288
95% upper bound for difference:  -0.227669
Test for difference = 0 (vs < 0):  Z = -4.30  P-Value = 0.000
```

Test for Equal Variances for WWBID

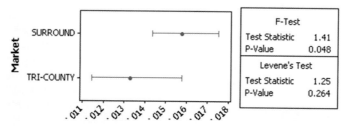

Activity   Paired vs. Unpaired Experiments

We have now discussed two methods of collecting data to compare two population means. In many experimental situations, a decision must be made either to collect two independent samples or to conduct a paired difference experiment. The importance of this decision cannot be overemphasized, since the amount of information obtained and the cost of the experiment are both directly related to the method of experimentation that is chosen.

Choose two populations (pertinent to your school major) that have unknown means and for which you could both collect two independent samples and collect paired observations. Before

conducting the experiment, state which method of sampling you think will provide more information (and why). Compare the two methods, first performing the independent sampling procedure by collecting 10 observations from each population (a total of 20 measurements) and then performing the paired difference experiment by collecting 10 pairs of observations.

Construct two 95% confidence intervals, one for each experiment you conduct. Which method provides the narrower confidence interval and hence more information on this performance of the experiment? Does your result agree with your preliminary expectations?

References

Freedman, D., Pisani, R., and Purves, R. *Statistics*. New York: W. W. Norton and Co., 1978.

Gibbons, J. D. *Nonparametric Statistical Inference*, 2nd ed. New York: McGraw-Hill, 1985.

Hollander, M., and Wolfe, D. A. *Nonparametric Statistical Methods*. New York: Wiley, 1973.

Mendenhall, W., Beaver, R. J., and Beaver, B. *Introduction to Probability and Statistics*, 13th ed. Belmont, CA: Brooks/Cole, 2009.

Satterthwaite, F. W. "An approximate distribution of estimates of variance components." *Biometrics Bulletin*, Vol. 2, 1946, pp. 110–114.

Snedecor, G. W., and Cochran, W. *Statistical Methods*, 7th ed. Ames, IA: Iowa State University Press, 1980.

Steel, R. G. D., and Torrie, J. H. *Principles and Procedures of Statistics*, 2nd ed. New York: McGraw-Hill, 1980.

USING TECHNOLOGY   MINITAB: Two-Sample Inferences

MINITAB can be used to make two-sample inferences about $\mu_1 - \mu_2$ or independent samples, μ_d for paired samples, $p_1 - p_2$ and σ_1^2/σ_2^2.

Comparing Means with Independent Samples

Step 1 Access the MINITAB worksheet that contains the sample data.

Step 2 Click on the "Stat" button on the MINITAB menu bar and then click on "Basic Statistics" and "2-Sample t," as shown in Figure 9.M.1. The resulting dialog box appears as shown in Figure 9.M.2.

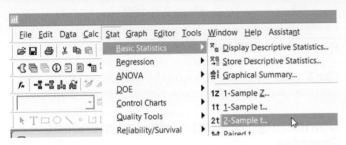

Figure 9.M.1 MINITAB menu options for comparing two means

Step 3a If the worksheet contains data for one quantitative variable (which the means will be computed on) and one qualitative variable (which represents the two groups or populations), select "Samples in one column" and then specify the quantitative variable in the "Samples" area and the qualitative variable in the "Subscripts" area. (See Figure 9.M.2.)

Figure 9.M.2 MINITAB 2-sample *t* dialog box

Step 3b If the worksheet contains the data for the first sample in one column and the data for the second sample in another column, select "Samples in different columns" and then specify the "First" and "Second" variables. Alternatively, if you have only summarized data (i.e., sample sizes, sample means, and sample standard deviations), select "Summarized data" and enter these summarized values in the appropriate boxes.

Step 4 Click the "Options" button on the MINITAB "2-Sample T" dialog box. Specify the confidence level for a confidence interval, the null-hypothesized value of the difference, $\mu_1 - \mu_2$, and the form of the alternative hypothesis (lower tailed, two tailed, or upper tailed) in the resulting dialog box, as shown in Figure 9.M.3.

Figure 9.M.3 MINITAB options dialog box

Step 5 Click "OK" to return to the "2-Sample T" dialog box and then click "OK" again to generate the MINITAB printout.

Important Note: The MINITAB two-sample *t*-procedure uses the *t*-statistic to conduct the test of hypothesis. When the sample sizes are small, check the "Assume equal variances" box in Figure 9.M.2. When the sample sizes are large, leave the "Assume equal variances" box unchecked; the *t*-value will be approximately equal to the large-sample *z*-value, and the resulting test will still be valid.

Comparing Means with Paired Samples

Step 1 Access the MINITAB worksheet that contains the sample data. The data file should contain two quantitative variables—one with the data values for the first group (or population) and one with the data values for the second group. (*Note:* The sample size should be the same for each group.)

Step 2 Click on the "Stat" button on the MINITAB menu bar and then click on "Basic Statistics" and "Paired t" (see Figure 9.M.1).

Step 3 On the resulting dialog box, select the "Samples in columns" option and specify the two quantitative variables of interest in the "First sample" and "Second sample" boxes, as shown in Figure 9.M.4. [Alternatively, if you have only summarized data of the paired differences, select the "Summarized data (differences)" option and enter the sample size, sample mean, and sample standard deviation in the appropriate boxes.]

Figure 9.M.4 MINITAB paired-samples *t* dialog box

Step 4 Click the "Options" button and specify the confidence level for a confidence interval, the null-hypothesized value of the difference μ_d, and the form of the alternative hypothesis (lower tailed, two tailed, or upper tailed) in the resulting dialog box. (See Figure 9.M.3.)

Step 5 Click "OK" to return to the "Paired *t*" dialog box and then click "OK" again to generate the MINITAB printout.

Comparing Proportions with Large Independent Samples

Step 1 Access the MINITAB worksheet that contains the sample data.

Step 2 Click on the "Stat" button on the MINITAB menu bar and then click on "Basic Statistics" and "2 Proportions," as shown in Figure 9.M.1.

Step 3 On the resulting dialog box (shown in Figure 9.M.5), select the data option ("Samples in different columns" or "Summarized data") and make the appropriate menu choices. (Figure 9.M.5 shows the menu options when you select "Summarized data.")

Figure 9.M.5 MINITAB 2 proportions dialog box

Step 4 Click the "Options" button and specify the confidence level for a confidence interval, the null-hypothesized value of the difference, and the form of the alternative hypothesis (lower tailed, two tailed, or upper tailed) in the resulting dialog box, as shown in Figure 9.M.6. (If you desire a pooled estimate of *p* for the test, be sure to check the appropriate box.)

Figure 9.M.6 MINITAB 2 proportions options

Step 5 Click "OK" to return to the "2 Proportions" dialog box and then click "OK" again to generate the MINITAB printout.

Comparing Variances with Independent Samples

Step 1 Access the MINITAB worksheet that contains the sample data.

Step 2 Click on the "Stat" buton on the MINITAB menu bar and then click on "Basic Statistics" and "2 Variances" (Figure 9.M.1).

Step 3 On the resulting dialog box (shown in Figure 9.M.7), the menu selections and options are similar to those for the two-sample *t*-test.

Figure 9.M.7 MINITAB 2 variances dialog box

Step 4 Click "OK" to produce the MINITAB *F*-test printout.

TI-83/TI-84 Plus Graphing Calculator: Two Sample Inferences

The TI-83/TI-84 plus graphing calculator can be used to conduct tests and form confidence intervals for the difference between two means with independent samples, the difference between two means with matched pairs, the difference between two proportions for large independent samples, and the ratio of two variances.

Confidence Interval for $\mu_1 - \mu_2$

Step 1 *Enter the data (Skip to Step 2 if you have summary statistics, not raw data)*

- Press **STAT** and select **1:Edit**

Note: If the lists already contain data, clear the old data. Use the up **ARROW** to highlight "**L1**."

- Press **CLEAR ENTER**
- Use the up **ARROW** to highlight "**L2**"
- Press **CLEAR ENTER**
- Use the **ARROW** and **ENTER** keys to enter the first data set into **L1**
- Use the **ARROW** and **ENTER** keys to enter the second data set into **L2**

Step 2 *Access the statistical tests menu*

```
EDIT CALC TESTS
4↑2-SampTTest…
5:1-PropZTest…
6:2-PropZTest…
7:ZInterval…
8:TInterval…
9:2-SampZInt…
0:2-SampTInt…
```

- Press **STAT**
- Arrow right to **TESTS**
- Arrow down to **2-SampTInt**
- Press **ENTER**

Step 3 *Choose "**Data**" or "**Stats**" ("Data" is selected when you have entered the raw data into the Lists. "Stats" is selected when you are given only the means, standard deviations, and sample sizes)*

- Press **ENTER**
- If you selected "Data," set **List1** to **L1** and **List2** to **L2**
- Set **Freq1** to **1** and set **Freq2** to **1**

```
2-SampTInt
 Inpt:Data Stats
 List1:L₁
 List2:L₂
 Freq1:1
 Freq2:1
 C-Level:.99
↓Pooled:No Yes
```

- Set **C-Level** to the confidence level
- If you are assuming that the two populations have equal variances, select **Yes** for **Pooled**
- If you are not assuming equal variances, select **No**
- Press **ENTER**
- Arrow down to "**Calculate**"
- Press **ENTER**
- If you selected "Stats," enter the means, standard deviations, and sample sizes
- Set **C-Level** to the confidence level
- If you are assuming that the two populations have equal variances, select **Yes** for **Pooled**
- If you are not assuming equal variances, select **No**
- Press **ENTER**
- Arrow down to "**Calculate**"
- Press **ENTER**

(The accompanying screen is set up for an example with a mean of 100, a standard deviation of 10, and a sample size of 15 for the first data set and a mean of 105, a standard deviation of 12, and a sample size of 18 for the second data set.)

The confidence interval will be displayed with the number of degrees of freedom, the sample statistics, and the pooled standard deviation (when appropriate).

```
2-SampTInt
↑x̄1:100
 Sx1:10
 n1:15
 x̄2:105
 Sx2:12
 n2:18
↓C-Level:.99
```

Hypothesis Test for $\mu_1 - \mu_2$

Step 1 *Enter the data (Skip to Step 2 if you have summary statistics, not raw data)*

- Press **STAT** and select **1:Edit**

Note: If the lists already contain data, clear the old data. Use the up **ARROW** to highlight "**L1**."

- Press **CLEAR ENTER**
- Use the up **ARROW** to highlight "**L2**"
- Press **CLEAR ENTER**
- Use the **ARROW** and **ENTER** keys to enter the first data set into **L1**
- Use the **ARROW** and **ENTER** keys to enter the second data set into **L2**

Step 2 *Access the statistical tests menu*

- Press **STAT**
- Arrow right to **TESTS**
- Arrow down to **2-SampTTest**
- Press **ENTER**

Step 3 *Choose "**Data**" or "**Stats**" ("Data" is selected when you have entered the raw data into the Lists. "Stats" is selected when you are given only the means, standard deviations, and sample sizes)*

- Press **ENTER**
- If you selected "Data," set **List1** to **L1** and **List2** to **L2**
- Set **Freq1** to **1** and set **Freq2** to **1**
- Use the **ARROW** to highlight the appropriate alternative hypothesis
- Press **ENTER**
- If you are assuming that the two populations have equal variances, select **Yes** for **Pooled**
- If you are not assuming equal variances, select **No**
- Press **ENTER**
- Arrow down to "**Calculate**"
- Press **ENTER**

```
2-SampTTest
 Inpt:Data Stats
 List1:L₁
 List2:L₂
 Freq1:1
 Freq2:1
 μ1:≠μ2 <μ2 >μ2
↓Pooled:No Yes
```

- If you selected "**Stats,**" enter the means, standard deviations, and sample sizes
- Use the **ARROW** to highlight the appropriate alternative hypothesis
- Press **ENTER**
- If you are assuming that the two populations have equal variances, select **Yes** for **Pooled**

- If you are not assuming equal variances, select **No**
- Press **ENTER**
- Arrow down to "**Calculate**"
- Press **ENTER**

(The screen that follows is set up for an example with a mean of 100, a standard deviation of 10, and a sample size of 15 for the first data set and a mean of 120, a standard deviation of 12, and a sample size of 18 for the second data set.)

```
2-SampTTest
↑x̄1:100
 Sx1:10
 n1:15
 x̄2:120
 Sx2:12
 n2:18■
↓μ1:≠μ2  <μ2  >μ2
```

The results of the hypothesis test will be displayed with the *p*-value, the number of degrees of freedom, the sample statistics, and the pooled standard deviation (when appropriate).

Confidence Interval for a Paired Difference Mean

Note: There is no paired difference option on the calculator. These instructions demonstrate how to calculate the differences and then use the 1-sample *t*-interval.

Step 1 *Enter the data and calculate the differences*

- Press **STAT** and select **1:Edit**

Note: If the lists already contain data, clear the old data. Use the up **ARROW** to highlight "**L1.**"

- Press **CLEAR ENTER**
- Use the up **ARROW** to highlight "**L2**"
- Press **CLEAR ENTER**
- Use the **ARROW** and **ENTER** keys to enter the first data set into **L1**
- Use the **ARROW** and **ENTER** keys to enter the second data set into **L2**
- The differences will be calculated in **L3**
- Use the up **ARROW** to highlight "**L3**"
- Press **CLEAR**—This will clear any old data, but **L3** will remain highlighted
- To enter the equation L3 = L1 − L2, use the following keystrokes:
- Press 2ND "**1**" (this will enter L1)
- Press the **MINUS** button
- Press 2ND "**2**" (this will enter L2)

(Notice the equation at the bottom of the screen.)

- Press **ENTER** (the differences should be calculated in L3)

```
L1      L2      L3    3
65      62      3
58      53      5
78      36      42
60      34      26
68      56      12
69      50      19
66      42      24
L3 =L1-L2■
```

Step 2 *Access the statistical tests menu*

- Press **STAT**
- Arrow right to **TESTS**
- Arrow down to **TInterval** (**even for large sample case**)
- Press **ENTER**

```
EDIT CALC TESTS
2↑T-Test…
3:2-SampZTest…
4:2-SampTTest…
5:1-PropZTest…
6:2-PropZTest…
7:ZInterval…
8↓TInterval…
```

Step 3 *Choose "Data"*

- Press **ENTER**
- Set **List** to **L3**
- Set **Freq** to **1**
- Set **C-Level** to the confidence level
- Arrow down to "**Calculate**"
- Press **ENTER**

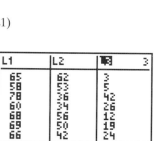
```
TInterval
Inpt:Data Stats
List:L3
Freq:1■
C-Level:.95
Calculate
```

The confidence interval will be displayed with the mean, standard deviation, and sample size of the differences.

Hypothesis Test for a Paired Difference Mean

Note: There is no paired difference option on the calculator. These instructions demonstrate how to calculate the differences and then use the 1-sample *t*-test.

Step 1 *Enter the data and calculate the differences*

- Press **STAT** and select **1:Edit**

Note: If the lists already contain data, clear the old data. Use the up **ARROW** to highlight "**L1.**"

- Press **CLEAR ENTER**
- Use the up **ARROW** to highlight "**L2**"
- Press **CLEAR ENTER**
- Use the **ARROW** and **ENTER** keys to enter the first data set into **L1**
- Use the **ARROW** and **ENTER** keys to enter the second data set into **L2**
- The differences will be calculated in **L3**
- Use the up **ARROW** to highlight "**L3**"
- Press **CLEAR**—This will clear any old data, but **L3** will remain highlighted
- To enter the equation L3 = L1 − L2, use the following keystrokes:
- Press 2ND "**1**" (this will enter L1)
- Press the **MINUS** button
- Press 2ND "**2**" (this will enter L2)

(Notice the equation at the bottom of the screen.)

- Press **ENTER** (the differences should be calculated in L3)

```
L1      L2      L3    3
65      62      3
58      53      5
78      36      42
60      34      26
68      56      12
69      50      19
66      42      24
L3 =L1-L2■
```

Step 2 *Access the statistical tests menu*

- Press **STAT**
- Arrow right to **TESTS**
- Arrow down to **T-Test** (**even for a large-sample case**)
- Press **ENTER**

```
EDIT CALC TESTS
1:Z-Test…
2:T-Test…
3:2-SampZTest…
4:2-SampTTest…
5:1-PropZTest…
6:2-PropZTest…
7↓ZInterval…
```

Step 3 *Choose "Data"*

- Press **ENTER**
- Enter the values for the hypothesis test, where $\mu_0 =$ the value for μ_d in the null hypothesis

- Set **List** to **L3**
- Set **Freq** to **1**
- Use the **ARROW** to highlight the appropriate alternative hypothesis
- Press **ENTER**
- Arrow down to "**Calculate**"
- Press **ENTER**

The test statistic and the *p*-value will be displayed, as will the sample mean, standard deviation, and sample size of the differences.

Confidence Interval for ($p_1 - p_2$)

Step 1 *Access the statistical tests menu*

- Press **STAT**
- Arrow right to **TESTS**
- Arrow down to **2-PropZInt**
- Press **ENTER**

```
EDIT CALC TESTS
6↑2-PropZTest…
7:ZInterval…
8:TInterval…
9:2-SampZInt…
0:2-SampTInt…
A:1-PropZInt…
B:2-PropZInt…
```

Step 2 *Enter the values from the sample information and **the confidence level***

where x_1 = number of successes in the first sample (e.g., 53)

n_1 = sample size for the first sample (e.g., 400)

x_2 = number of successes in the second sample (e.g., 78)

n_2 = sample size for the second sample (e.g., 500)

- Set **C-Level** to the confidence level
- Arrow down to "**Calculate**"
- Press **ENTER**

```
2-PropZInt
 (-.0695,.02249)
 p̂1=.1325
 p̂2=.156
 n1=400
 n2=500
```

Hypothesis Test for ($p_1 - p_2$)

Step 1 *Access the statistical tests menu*

- Press **STAT**
- Arrow right to **TESTS**
- Arrow down to **2-PropZTest**
- Press **ENTER**

```
EDIT CALC TESTS
1:Z-Test…
2:T-Test…
3:2-SampZTest…
4:2-SampTTest…
5:1-PropZTest…
6:2-PropZTest…
7↓ZInterval…
```

Step 2 *Enter the values from the sample information and **select the alternative hypothesis***

where x_1 = number of successes in the first sample (e.g., 53)

n_1 = sample size for the first sample (e.g., 400)

x_2 = number of successes in the second sample (e.g., 78)

n_2 = sample size for the second sample (e.g., 500)

- Use the **ARROW** to highlight the appropriate alternative hypothesis
- Press **ENTER**
- Arrow down to "**Calculate**"
- Press **ENTER**

```
2-PropZTest
 p1≠p2
 z=-.9933568641
 p=.3205360811
 p̂1=.1325
 p̂2=.156
↓p̂=.1455555556
```

Hypothesis Test for (σ_1^2/σ_2^2)

Step 1 *Enter the data (Skip to Step 2 if you have summary statistics, not raw data)*

- Press **STAT** and select **1:Edit**

Note: If the lists already contain data, clear the old data. Use the up **ARROW** to highlight "**L1.**"

- Press **CLEAR ENTER**
- Use the up **ARROW** to highlight "**L2**"
- Press **CLEAR ENTER**
- Use the **ARROW** and **ENTER** keys to enter the first data set into **L1**
- Use the **ARROW** and **ENTER** keys to enter the second data set into **L2**

Step 2 *Access the statistical tests menu*

- Press **STAT**
- Arrow right to **TESTS**
- Arrow down to **2-SampFTest**
- Press **ENTER**

```
EDIT CALC TESTS
0↑2-SampTInt…
A:1-PropZInt…
B:1-PropZInt…
C:X²-Test…
D:X²GOF-Test…
E:2-SampFTest…
F↓LinRegTTest…
```

Step 3 *Choose "**Data**" or "**Stats**" ("Data" is selected when you have entered the raw data into the Lists. "Stats" is selected when you are given only the means, standard deviations, and sample sizes)*

- Press **ENTER**
- If you selected "Data"
- Set **List1** to **L1** and **List2** to **L2**
- Set **Freq1** to **1** and set **Freq2** to **1**
- Use the **ARROW** to highlight the appropriate alternative hypothesis
- Press **ENTER**
- Arrow down to "**Calculate**"
- Press **ENTER**

```
2-SampFTest
 Inpt:Data Stats
 Sx1:10
 n1:23
 Sx2:12
 n2:33
 σ1:≠σ2 <σ2 >σ2
 Calculate Draw
```

- If you selected "Stats," enter the standard deviations and sample sizes
- Use the **ARROW** to highlight the appropriate alternative hypothesis
- Press **ENTER**
- Arrow down to "**Calculate**"
- Press **ENTER**

```
2-SampFTest
 Inpt:Data Stats
 List1:L1
 List2:L2
 Freq1:1
 Freq2:1
 σ1:≠σ2 <σ2 >σ2
 Calculate Draw
```

The results of the hypothesis test will be displayed with the *p*-value and the input data used.

10 Analysis of Variance

Comparing More than Two Means

CONTENTS

Where We've Been

- Presented methods for estimating and testing hypotheses about a single population mean
- Presented methods for comparing two population means

Where We're Going

- Discuss the critical elements in the *design* of a sampling experiment (10.1)
- Learn how to set up three of the more popular experimental designs for comparing more than two population means: *completely randomized, randomized block*, and *factorial designs* (10.2, 10.4–10.5)
- Show how to analyze data collected from a designed experiment using a technique called an *analysis of variance* (10.2, 10.4–10.5)
- Present a follow-up analysis to an ANOVA: ranking means (10.3)

The assumptions necessary to ensure the validity of the test are as follows:

1. The samples of 10 golf balls for each brand are selected randomly and independently.
2. The probability distributions of the distances for each brand are normal.
3. The variances of the distance probability distributions for each brand are equal.

b. The MINITAB printout for the data in Table 10.3 resulting from this completely randomized design is given in Figure 10.12. The values, MST = 931.5, MSE = 21.2, and $F = 43.99$, are highlighted on the printout. Since $F > 2.25$, we reject H_0. Also, the p-value of the test (.000) is highlighted on the printout. Since $\alpha = .10$ exceeds the p-value, we draw the same conclusion: Reject H_0. Therefore, at the .10 level of significance, we conclude that at least two of the brands differ with respect to mean distance traveled when struck by the driver.

Figure 10.12
MINITAB ANOVA for completely randomized design

One-way ANOVA: DISTANCE versus BRAND

Source	DF	SS	MS	F	P
BRAND	3	2794.4	931.5	43.99	0.000
Error	36	762.3	21.2		
Total	39	3556.7			

Look Ahead Now that we know that mean distances differ, a logical follow-up question is: "Which ball brand travels farther, on average, when hit with a driver?" In Section 10.3 we present a method for ranking treatment means in an ANOVA.

■ **Now Work Exercise 10.26**

The results of an **analysis of variance (ANOVA)** can be summarized in a simple tabular format similar to that obtained from the MINITAB program in Example 10.4. The general form of the table is shown in Table 10.4, where the symbols df, SS, and MS stand for degrees of freedom, sum of squares, and mean square, respectively. Note that the two sources of variation, Treatments and Error, add to the total sum of squares, SS(Total). The ANOVA summary table for Example 10.4 is given in Table 10.5, and the partitioning of the total sum of squares into its two components is illustrated in Figure 10.13.

Table 10.4   General ANOVA Summary Table for a Completely Randomized Design

Source	df	SS	MS	F
Treatments	$k - 1$	SST	$\text{MST} = \dfrac{\text{SST}}{k - 1}$	$\dfrac{\text{MST}}{\text{MSE}}$
Error	$n - k$	SSE	$\text{MSE} = \dfrac{\text{SSE}}{n - k}$	
Total	$n - 1$	SS(Total)		

Table 10.5   ANOVA Summary Table for Example 10.4

Source	df	SS	MS	F	p-Value
Brands	3	2,794.39	931.46	43.99	.000
Error	36	762.30	21.18		
Total	39	3,556.69			

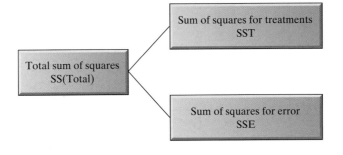

Figure 10.13
Partitioning of the total sum of squares for the completely randomized design

Example **10.5** Ⓓ

GOLFCRD

Checking the ANOVA Assumptions

Problem Refer to the completely randomized ANOVA design conducted in Example 10.4. Are the assumptions required for the test approximately satisfied?

Solution The assumptions for the test are repeated as follows:

1. The samples of golf balls for each brand are selected randomly and independently.
2. The probability distributions of the distances for each brand are normal.
3. The variances of the distance probability distributions for each brand are equal.

Since the sample consisted of 10 randomly selected balls of each brand and since the robotic golfer Iron Byron was used to drive all the balls, the first assumption of independent random samples is satisfied. To check the next two assumptions, we will employ two graphical methods presented in Chapter 2: histograms and box plots. A MINITAB histogram of driving distances for each brand of golf ball is shown in Figure 10.14, and SAS box plots are shown in Figure 10.15.

Figure 10.14

MINITAB histograms for golf ball driving distances

Figure 10.15

SAS box plots for golf ball distances

The normality assumption can be checked by examining the histograms in Figure 10.14. With only 10 sample measurements for each brand, however, the displays are not very informative. More data would need to be collected for each brand before we could assess whether the distances come from normal distributions. Fortunately, analysis of variance has been shown to be a very **robust method** when the assumption of normality is not satisfied exactly. That is, *moderate departures from normality do not have much effect on the significance level of the ANOVA F-test or on confidence coefficients*. Rather than spend the time, energy, or money to collect additional data for this experiment in order to verify the normality assumption, we will rely on the robustness of the ANOVA methodology.

Box plots are a convenient way to obtain a rough check on the assumption of equal variances. With the exception of a possible outlier for Brand D, the box plots in Figure 10.15 show that the spread of the distance measurements is about the same for each brand. Since the sample variances appear to be the same, the assumption of equal population variances for the brands is probably satisfied. Although robust with respect to the normality assumption, ANOVA is *not robust* with respect to the equal-variances assumption. Departures from the assumption of equal population variances can affect the associated measures of reliability (e.g., p-values and confidence levels). Fortunately, the effect is slight when the sample sizes are equal, as in this experiment.

■ **Now Work Exercise 10.36**

Although graphs can be used to check the ANOVA assumptions as in Example 10.5, no measures of reliability can be attached to these graphs. When you have a plot that is unclear as to whether an assumption is satisfied, you can use formal statistical tests that are beyond the scope of this text. Consult the references at the end of the chapter for information on these tests. When the validity of the ANOVA assumptions is in doubt, nonparametric statistical methods are useful.

What Do You Do When the Assumptions Are Not Satisfied for the Analysis of Variance for a Completely Randomized Design?

Answer: Use a nonparametric statistical method such as the Kruskal-Wallis H-Test of Section 14.5.

The procedure for conducting an analysis of variance for a completely randomized design is summarized in the next box. Remember that the hallmark of this design is independent random samples of experimental units associated with each treatment. We discuss a design with dependent samples in Section 10.4.

Steps for Conducting an ANOVA for a Completely Randomized Design

1. Make sure that the design is truly completely randomized, with independent random samples for each treatment.

2. Check the assumptions of normality and equal variances.

3. Create an ANOVA summary table that specifies the variabilities attributable to treatments and error, making sure that those variabilities lead to the calculation of the F-statistic for testing the null hypothesis that the treatment means are equal in the population. Use a statistical software program to obtain the numerical results. (If no such package is available, use the calculation formulas in Appendix C.)

(continued)

4. If the *F*-test leads to the conclusion that the means differ,
   a. Conduct a multiple-comparisons procedure for as many of the pairs of means as you wish to compare. (See Section 10.3.) Use the results to summarize the statistically significant differences among the treatment means.
   b. If desired, form confidence intervals for one or more individual treatment means.

5. If the *F*-test leads to the nonrejection of the null hypothesis that the treatment means are equal, consider the following possibilities:
   a. The treatment means are equal; that is, the null hypothesis is true.
   b. The treatment means really differ, but other important factors affecting the response are not accounted for by the completely randomized design. These factors inflate the sampling variability, as measured by MSE, resulting in smaller values of the *F*-statistic. Either increase the sample size for each treatment, or use a different experimental design (as in Section 10.4) that accounts for the other factors affecting the response.

   [*Note:* Be careful not to automatically conclude that the treatment means are equal since the possibility of a Type II error must be considered if you accept H_0.]

We conclude this section by making two important points about an analysis of variance. First, recall that we performed a hypothesis test for the difference between two means in Section 9.2 using a two-sample *t*-statistic for two independent samples. When two independent samples are being compared, the *t*- and *F*-tests are equivalent. To see this, apply the formula for *t* to the two samples of SAT scores in Table 10.2:

$$t = \frac{\bar{x}_1 - \bar{x}_2}{\sqrt{s_p^2\left(\frac{1}{n_1} + \frac{1}{n_2}\right)}} = \frac{590 - 550}{\sqrt{(62.5)\left(\frac{1}{5} + \frac{1}{5}\right)}} = \frac{40}{5} = 8$$

Here, we used the fact that $s_p^2 = \text{MSE}$, which you can verify by comparing the formulas. Recall that the calculated *F* for the two samples in Table 10.2 is $F = 64$. This value equals the square of the calculated *t* for the same samples ($t = 8$). Likewise, the critical *F*-value (5.32) equals the square of the critical *t*-value at the two-sided .05 level of significance ($t_{.025} = 2.306$ with 8 df). Since both the rejection region and the calculated values are related in the same way, the tests are equivalent. Moreover, the assumptions that must be met to ensure the validity of the *t*- and *F*-tests are the same:

1. The probability distributions of the populations of responses associated with each treatment must all be normal.

2. The probability distributions of the populations of responses associated with each treatment must have equal variances.

3. The samples of experimental units selected for the treatments must be random and independent.

In fact, the only real difference between the tests is that the *F*-test can be used to compare *more than two* treatment means, whereas the *t*-test is applicable to two samples only.

For our second point, refer to Example 10.4. Our conclusion that at least two of the brands of golf balls have different mean distances traveled when struck with a driver leads naturally to the following questions: Which of the brands differ? and How are the brands ranked with respect to mean distance?

One way to obtain this information is to construct a confidence interval for the difference between the means of any pair of treatments, using the method of Section 9.2. For example, if a 95% confidence interval for $\mu_A - \mu_C$ in Example 10.4 is found to be $(-24, -13)$, we are confident that the mean distance for Brand C exceeds the mean for Brand A (since all differences in the interval are negative). Constructing these confidence intervals for all possible pairs of brands allows you to rank the brand means. A method for conducting these *multiple comparisons*—one that controls for Type I errors—is presented in Section 10.3.

Statistics IN Action	Revisited

A One-Way Analysis of the Voice/Face Recognition Data in Experiment 1

Consider the *British Journal of Psychology* (Feb. 2014) study of associative priming using faces and voices of celebrity pairs (p. 531). The researchers tested the theory by conducting two designed experiments. Recall that in Experiment 1, 40 subjects were randomly assigned to either the within-modality condition (where the subjects were shown the *face* of the prime celebrity for 3 seconds and then the *face* of the target celebrity) or the cross-modality condition (where subjects listened to the *voice* of the prime celebrity for 3 seconds and then were shown the *face* of the target celebrity). Then within each condition, half of the subjects were randomly assigned to related celebrity pairs and half assigned to unrelated celebrity pairs. This design yielded four groups of subjects: (1) Within-modality/Related pairs, (2) Within-modality/Unrelated pairs, (3) Cross-modality/Related pairs, and (4) Cross-modality/Unrelated pairs (see Figure SIA10.1, p. 531).

All subjects were presented with 40 celebrity pairs and asked to identify the target celebrity in each pair. The researchers then recorded the accuracy rate (percentage of target celebrities correctly identified) and total response speed (in seconds) for each subject. Of interest is whether the means of these two variables differ depending on type of prime modality (within or cross) and relational pair (related or unrelated).

Clearly, an analysis of variance is required to compare the means. The quantitative variables, accuracy rate and response speed, represent two different dependent variables. Since the subjects were randomly assigned to one of four groups, a completely

randomized design was employed. The four groups represent the treatments in this experiment. Consequently, one way to analyze the data is to conduct a one-way ANOVA on the data saved in the **PRIME1** file. The null hypothesis of interest is:

$$H_0: \mu_{\text{Within/Related}} = \mu_{\text{Within/Unrelated}}$$
$$= \mu_{\text{Cross/Related}} = \mu_{\text{Cross/Unrelated}}$$

The results of an ANOVA for the first dependent variable, accuracy rate, are shown in the MINITAB printout, Figure SIA10.3. The ANOVA *F*-test *p*-value (highlighted) is .066. Therefore, at $\alpha = .05$, there is insufficient evidence to reject the null hypothesis; that is, there is insufficient evidence of a difference in the mean accuracy rates of the four groups of subjects. Consequently, neither prime modality nor relational pair appears to affect accuracy rate.

An ANOVA for the second dependent variable, response speed, is displayed in the MINITAB printout, Figure SIA10.4. The *p*-value = .000 (highlighted) is less than $\alpha = .05$; thus, there is sufficient evidence to reject the null hypothesis. From this result, we can infer that the mean response speeds of the four groups of subjects differ. To investigate this phenomenon further, examine the sample means (also highlighted on Figure SIA10.4). You can see that the mean response speeds for groups 3 and 4

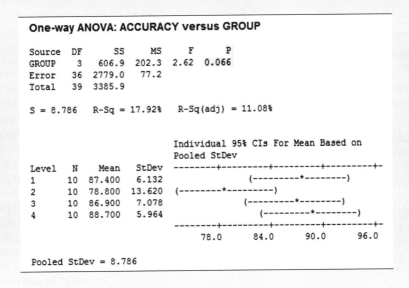

Figure SIA10.3
MINITAB One-Way ANOVA for Accuracy Rate

(Continued)

Statistics in Action
(continued)

Figure SIA10.4
MINITAB One-Way ANOVA
for Response Speed

One-way ANOVA: SPEED versus GROUP

Source	DF	SS	MS	F	P
GROUP	3	1443502	481167	26.52	0.000
Error	36	653061	18141		
Total	39	2096563			

S = 134.7 R-Sq = 68.85% R-Sq(adj) = 66.26%

Individual 95% CIs For Mean Based on
Pooled StDev

Level	N	Mean	StDev
1	10	763.3	100.7
2	10	895.0	129.9
3	10	1245.6	118.2
4	10	1131.5	177.7

```
------+---------+---------+---------+---
    (---*---)
        (----*---)
                            (---*----)
                       (----*---)
------+---------+---------+---------+---
    800       1000      1200      1400
```

Pooled StDev = 134.7

(i.e., the two cross-modality groups) are much larger than the corresponding means for groups 1 and 2 (i.e., the two within-modality groups). From this result, the researchers concluded that prime modality, but not relational pair, affects response speed.

In the next Statistics in Action Revisited section, we demonstrate how to rank the four group means, statistically, attaching a measure of reliability to our inference. Also, in the Statistics in Action Revisited section following Section 10.5, we present an alternative method of analyzing the data, one that will allow the researchers to determine if the impact of prime modality on response speed depends on whether the subjects are presented with related celebrity pairs or unrelated celebrity pairs.

Exercises 10.17–10.43

Understanding the Principles

10.17 What is a completely randomized design?

10.18 What is the main difference between a *t*-test and an *F*-test in ANOVA?

10.19 What conditions are required for a valid ANOVA *F*-test in a completely randomized design?

10.20 *True or False*. The ANOVA method is robust when the assumption of normality is not exactly satisfied in a completely randomized design.

Learning the Mechanics

10.21 Use Tables V, VI, VII, and VIII of Appendix B or statistical software to find each of the following *F* values:
    a. $F_{.05}, v_1 = 5, v_2 = 3$
    b. $F_{.01}, v_1 = 5, v_2 = 3$
    c. $F_{.10}, v_1 = 25, v_2 = 28$
    d. $F_{.025}, v_1 = 15, v_2 = 9$

10.22 Find the following probabilities:
    a. $P(F \le 3.58)$ for $v_1 = 6, v_2 = 8$
    b. $P(F > 2.98)$ for $v_1 = 25, v_2 = 18$
    c. $P(F > 2.28)$ for $v_1 = 20, v_2 = 16$
    d. $P(F \le 1.87)$ for $v_1 = 9, v_2 = 27$

10.23 Consider dot plots A and B (shown at the top of p. 551). Assume that the two samples represent independent random samples corresponding to two treatments in a completely randomized design.
    a. In which dot plot is the difference between the sample means small relative to the variability within the sample observations? Justify your answer.
    b. Calculate the treatment means (i.e., the means of samples 1 and 2) for both dot plots.
    c. Use the means to calculate the sum of squares for treatments (SST) for each dot plot.
    d. Calculate the sample variance for each sample and use these values to obtain the sum of squares for error (SSE) for each dot plot.
    e. Calculate the total sum of squares [SS(Total)] for the two dot plots by adding the sums of squares for treatment and error. What percentage of SS(Total) is accounted for by the treatments—that is, what percentage of the total sum of squares is the sum of squares for treatment—in each case?
    f. Convert the sums of squares for treatment and error to mean squares by dividing each by the appropriate number of degrees of freedom. Calculate the *F*-ratio of the mean square for treatment (MST) to the mean square for error (MSE) for each dot plot.

Dot Plots for Exercise 10.23

Plot A:

Plot B:

⊙ Treatment (Sample) 1
⊙ Treatment (Sample) 2

g. Use the F-ratios to test the null hypothesis that the two samples are drawn from populations with equal means. Take $\alpha = .05$.

h. What assumptions must be made about the probability distributions corresponding to the responses for each treatment in order to ensure the validity of the F-tests conducted in part **g**?

10.24 Refer to Exercise 10.23. Conduct a two-sample t-test (Section 9.2) of the null hypothesis that the two treatment means are equal for each dot plot. Use $\alpha = .05$ and two-tailed tests. In the course of the test, compare each of the following with the F-tests in Exercise 10.23:

a. The pooled variances and the MSEs

b. The t- and the F-test statistics

c. The tabled values of t and F that determine the rejection regions

d. The conclusions of the t- and F-tests

e. The assumptions that must be made in order to ensure the validity of the t- and F-tests

10.25 Refer to Exercises 10.23 and 10.24. Complete the following ANOVA table for each of the two dot plots:

Source	df	SS	MS	F
Treatments				
Error				
Total				

10.26 A partially completed ANOVA table for a completely randomized design is shown here:

Source	df	SS	MS	F
Treatments	6	18.4		
Error				
Total	41	45.2		

a. Complete the ANOVA table.

b. How many treatments are involved in the experiment?

c. Do the data provide sufficient evidence to indicate a difference among the population means? Test, using $\alpha = .10$.

d. Find the approximate observed significance level for the test in part **c**, and interpret it.

10.27 Suppose the total sum of squares for a completely randomized design with $p = 5$ treatments and $n = 30$ total measurements (6 per treatment) is equal to 600. In each of the following cases, conduct an F-test of the null hypothesis that the mean responses for the 5 treatments are the same. Use $\alpha = .05$.

a. Sum of squares for treatment (SST) is 30% of SS(Total)

b. SST is 40% of SS(Total)

c. SST is 70% of SS(Total)

d. What happens to the F-ratio as the percentage of the total sum of squares attributable to treatments is increased?

10.28 The data in the following table resulted from an experiment that utilized a completely randomized design:

Treatment 1	Treatment 2	Treatment 3
3.9	5.4	1.3
1.4	2.0	.7
4.1	4.8	2.2
5.5	3.8	
2.3	3.5	

a. Use statistical software (or the formulas in Appendix B) to complete the following ANOVA table:

Source	df	SS	MS	F
Treatments				
Error				
Total				

b. Test the null hypothesis that $\mu_1 = \mu_2 = \mu_3$, where μ_i represents the true mean for treatment i, against the alternative that at least two of the means differ. Use $\alpha = .01$.

Applying the Concepts—Basic

10.29 Treating cancer with yoga. According to a study funded by the National Institutes of Health, yoga classes can help cancer survivors sleep better. The study results were presented at the June 2010 American Society of Clinical Oncology's annual meeting. Researchers randomly assigned 410 cancer patients (who had finished cancer therapy) to receive either their usual follow-up care or attend a 75-minute yoga class twice per week. After four weeks, the researchers measured the level of fatigue and sleepiness experienced by each cancer survivor. Those who took yoga were less fatigued than those who did not.

a. Assume the patients are numbered 1 through 410. Use the random number generator of a statistical software package to randomly assign each patient to either receive the usual follow-up care or to attend yoga classes. Assign 205 patients to each treatment.

b. Consider the following treatment assignment scheme. The patients are ranked according to severity of cancer and the most severe patients are assigned to the yoga class while the others are assigned to receive their usual follow-up care. Comment on the validity of the results obtained from such an assignment.

10.30 Properties of cemented soils. The *Bulletin of Engineering Geology and the Environment* (Vol. 69, 2010) published a study of cemented sandy soils. The researchers applied one of three different sampling methods (rotary core, metal tube, or plastic tube) to randomly selected soil specimens, then measured the effective stress level (Newtons per meters-squared) of each specimen. Each method was applied to 10 soil specimens—a total of 30 measurements in all. For this completely randomized design, use a random number generator to assign the sampling methods to the soil specimens. List the soil specimens that are to be analyzed by each method.

10.31 How workers respond to wage cuts. A randomized field experiment was carried out to determine how workers respond to wage cuts and the results published in *Institute for the Study of Labor: Discussion Paper Series* (Mar. 2011). A company formed teams of two employees for a temporary promotion campaign. Both team members had identical individual tasks (sell promotional cards for entrance to a nightclub) and were initially paid the same hourly wage. After a short period of time on the job, teams were unknowingly randomly assigned to one of three treatments. In the "unilateral wage cut" treatment, one worker's pay was cut by 25%; in the "general wage cut" treatment, both workers' pay was cut by 25%; and in the "baseline" treatment, neither worker received a pay cut. The variable of interest was the decrease in the number of promotional cards sold after implementation of the pay cuts. The researchers wanted to know if the average decrease in cards sold would differ depending on whether one or more of the workers received a pay cut.

a. Identify the type of experimental design used in the study.

b. Identify the dependent variable of interest.

c. What is the factor in this experiment? The factor levels?

d. Specify the null hypothesis of interest to the researchers.

e. The ANOVA *F*-test was carried out and resulted in a *p*-value less than .001. Interpret these results using $\alpha = .01$.

10.32 Whales entangled in fishing gear. Entanglement of marine mammals (e.g., whales) in fishing gear is considered a significant threat to the species. A study published in *Marine Mammal Science* (Apr. 2010) investigated the type of net most likely to entangle a certain species of whale inhabiting the East Sea of Korea. A sample of 207 entanglements of whales in the area formed the data for the study. These entanglements were caused by one of three types of fishing gear: set nets, pots, and gill nets. One of the variables investigated was body length (in meters) of the entangled whale.

a. Set up the null and alternative hypotheses for determining whether the average body length of entangled whales differs for the three types of fishing gear.

b. An ANOVA *F*-test yielded the following results: $F = 34.81, p - \text{value} < .0001$. Interpret the results for $\alpha = .05$.

10.33 Does the media influence your attitude toward tanning? Dermatologists' primary recommendation to prevent skin cancer is minimal exposure to the sun. Yet models used in product advertisements are typically well-tanned. Do such advertisements influence a consumer's attitude toward tanning? University of California and California State University researchers designed an experiment to investigate this phenomenon and published their results in *Basic and Applied Social Psychology* (May 2010). College student participants were randomly assigned to one of three conditions: (1) view product advertisements featuring models with a tan, (2) view product advertisements featuring models without a tan, or (3) view products advertised with no models (control group). The objective was to determine whether the mean attitude toward tanning differs across the three conditions. A tanning attitude index (measured on a scale of 0 to 5 points) was recorded for each participant. The results are summarized in the accompanying table.

	Tanned Models	Models with No Tan	No Models
Sample size	56	56	56
Mean	2.40	2.11	2.50
Standard deviation	.85	.73	.82

Source: Mahler, H., et al. "Effects of media images on attitudes toward tanning." *Basic and Applied Social Psychology*, Vol. 32, No. 2, May 2010 (adapted from Table 1).

a. Identify the type of experimental design utilized by the researchers.

b. Identify the experimental units, dependent variable, and treatments for the design.

c. Set up the null hypothesis for a test to compare the treatment means.

d. The sample means shown in the table are obviously different. Explain why the researchers should not use these means alone to test the hypothesis, part **c**.

e. The researchers conducted an ANOVA on the data and reported the following results: $F = 3.60, p\text{-value} = .03$. Carry out the test, part **c**. Use $\alpha = .05$ to draw your conclusion.

f. What assumptions are required for the inferences derived from the test to be valid?

10.34 Evaluation of flexography printing plates. Flexography is a printing process used in the packaging industry. The process is popular because it is cost-effective and can be used to print on a variety of surfaces (e.g., paperboard, foil, and plastic). A study was conducted to determine if flexography exposure time has an impact on the quality of the printing (*Journal of Graphic Engineering and Design*, Vol. 3, 2012). Four different exposure times were studied: 8, 10, 12, and 14 minutes. A sample of 36 print images were collected at each exposure time level, for a total of 144 print images. The measure of print quality used was dot area (hundreds of dots per square millimeter). The data were subjected to an analysis of variance, with partial results shown in the accompanying table.

Source	df	SS	MS	F	p-value
Exposure	3	.010	—	—	< .001
Error	140	.029	—		
Total	143	.039			

a. Compute the missing entries in the ANOVA table.

b. Is there sufficient evidence to indicate that mean dot area differs depending on the exposure time? Use $\alpha = .05$.

10.35 Robots trained to behave like ants. Robotics researchers investigated whether robots could be trained to behave like ants in an ant colony (*Nature*, Aug. 2000). Robots were trained and randomly assigned to "colonies" (i.e., groups) consisting of 3, 6, 9, or 12 robots. The robots were assigned the tasks of foraging for "food" and recruiting another robot when they identified a resource-rich area. One goal of the experiment was to compare the mean energy expended (per robot) of the four different sizes of colonies.

a. What type of experimental design was employed?

b. Identify the treatments and the dependent variable.

c. Set up the null and alternative hypotheses of the test.

d. The following ANOVA results were reported: $F = 7.70$, numerator df $= 3$, denominator df $= 56$, p-value $< .001$. Conduct the test at a significance level of $\alpha = .05$ and interpret the result.

10.36 Soil scouring and overturned trees. Trees that grow in flood plains are susceptible to overturning. This is typically due to floodwaters exposing the tree roots (called soil *scouring*). Environmental engineers at Saitama University (Japan) investigated the impact of soil scouring on the characteristics of overturned and uprooted trees (*Landscape Ecology Engineering*, Jan. 2013). Tree pulling experiments were conducted in the floodplains of the Komagama river. Trees were randomly selected to be uprooted in each of three areas that had different scouring conditions: no scouring (NS), shallow scouring (SS), and deep scouring (DS). During the uprooting of the trees, the maximum resistive bending moment at the trunk base (kiloNewton-meters) was measured. Simulated data for five medium-sized trees selected at each area are shown in the table, followed by a MINITAB printout of the analysis of the data. Interpret the results. Does soil scouring affect the mean maximum resistive bending moment at the tree trunk base?

None	Shallow	Deep
23.68	11.13	4.27
8.88	29.19	2.36
7.52	13.66	8.48
25.89	20.47	12.09
22.58	23.24	3.46

One-way ANOVA: MOMENT versus SCOURING

```
Source    DF     SS     MS     F     P
SCOURING   2   528.5  264.3  5.40  0.021
Error     12   586.9   48.9
Total     14  1115.4

S = 6.993   R-Sq = 47.38%   R-Sq(adj) = 38.61%
```

Applying the Concepts — Intermediate

10.37 Study of recall of TV commercials. Do TV shows with violence and sex impair memory for commercials? To answer this question, researchers conducted a designed experiment in which 324 adults were randomly assigned to one of three viewer groups of 108 participants each (*Journal of Applied Psychology*, June 2002). One group watched a TV program with a violent content code (V) rating, the second group viewed a show with a sex content code (S) rating, and the last group watched a neutral TV program. Nine commercials were embedded into each TV show. After viewing the program, each participant was scored on his/her recall of the brand names in the commercial messages, with scores ranging from 0 (no brands recalled) to 9 (all brands recalled). The data (simulated from information provided in the article) are saved in the **TVADS** file. The researchers compared the mean recall scores of the three viewing groups with an analysis of variance for a completely randomized design.

a. Identify the experimental units in the study.

b. Identify the dependent (response) variable in the study.

c. Identify the factor and treatments in the study.

d. The sample mean recall scores for the three groups were $\bar{x}_v = 2.08$, $\bar{x}_s = 1.71$, and $\bar{x}_{\text{Neutral}} = 3.17$. Explain why one should not draw an inference about differences in the population mean recall scores on the basis of only these summary statistics.

e. An ANOVA on the data yielded the results shown in the accompanying MINITAB printout. Locate the test statistic and p-value on the printout.

f. Interpret the results from part **e**, using $\alpha = 0.01$. What can the researchers conclude about the three groups of TV ad viewers?

One-way ANOVA: VIOLENT, SEX, NEUTRAL

```
Source   DF      SS      MS      F      P
Factor    2   123.27   61.63  20.45  0.000
Error   321   967.35    3.01
Total   323  1090.62

S = 1.736   R-Sq = 11.30%   R-Sq(adj) = 10.75%
```

10.38 Improving driving performance while fatigued. Can a secondary task—such as a word association task—improve your performance when driving while fatigued? This was the question of interest in a *Human Factors* (May 2014) study. The researchers used a driving simulator to obtain their data. Each of 40 college students was assigned to drive a long distance in the simulator. However, the student-drivers were divided into four groups of 10 drivers each. Group 1 performed the verbal task continuously (*continuous verbal condition*); Group 2 performed the task only at the end of the drive (*late verbal condition*); Group 3 did not perform the task at all (*no verbal condition*); and Group 4 listened to a program on the car radio (*radio show condition*). At the end of the simulated drive, drivers were asked to recall billboards that they saw along the way. The percentage of billboards recalled by each student-driver is provided in the next table. Use the information in the accompanying SPSS printout to determine if the mean recall percentage differs for student-drivers in the four groups. Test using $\alpha = .01$.

Data for Exercise 10.38

Continuous Verbal	Late Verbal	No Verbal	Radio Show
14	57	64	37
63	64	83	45
10	66	54	87
29	18	59	62
37	95	60	14
60	52	39	46
43	58	56	59
4	92	73	45
36	85	78	45
47	47	73	50

SPSS Output for Exercise 10.38

ANOVA

RECALL

	Sum of Squares	df	Mean Square	F	Sig.
Between Groups	5921.700	3	1973.900	5.388	.004
Within Groups	13189.400	36	366.372		
Total	19111.100	39			

10.39 Effects of temperature on ethanol production. A low-cost and highly productive bio-fuel production method is high-temperature fermentation. However, heat stress can inhibit the amount of ethanol produced during the process. In *Engineering Life Sciences* (Mar. 2013), biochemical engineers carried out a series of experiments to assess the effect of temperature on ethanol production during fermentation. The maximum inhibitory concentration of ethanol (grams per liter) was measured at four different temperatures (30°, 35°, 40°, and 45° Celsius). The experiment was replicated 3 times, with the data shown in the table. (*Note:* The data are simulated based on information provided in the journal article.) Do the data indicate that high temperatures inhibit mean concentration of ethanol? Test using $\alpha = .10$.

Temperature			
30°	35°	40°	45°
103.3	101.7	97.2	55.0
103.4	102.0	96.9	56.4
101.0	101.1	96.2	54.9

10.40 Restoring self-control when intoxicated. Does coffee or some other form of stimulation really allow a person suffering from alcohol intoxication to "sober up"? Psychologists investigated the matter in *Experimental and Clinical Psychopharmacology* (Feb. 2005). A sample of 44 healthy male college students participated in the experiment. Each student was asked to memorize a list of 40 words (20 words on a green list and 20 words on a red list). The students were then randomly assigned to one of four different treatment groups (11 students in each group). Students in three of the groups were each given two alcoholic beverages to drink prior to performing a word completion task. Students in Group A received only the alcoholic drinks. Participants in Group AC had caffeine powder dissolved in their drinks. Group AR participants received a monetary award for correct responses on the word completion task. Students in Group P (the placebo group) were told that they would receive alcohol, but instead received two drinks containing a carbonated beverage (with a few drops of alcohol on the surface to provide an alcoholic scent). After consuming their drinks and resting for 25 minutes, the students performed the word completion task. Their scores (simulated on the basis of summary information from the article) are reported in the table. (*Note:* A task score represents the difference between the proportion of correct responses on the green list of words and the proportion of incorrect responses on the red list of words.)

AR	AC	A	P
.51	.50	.16	.58
.58	.30	.10	.12
.52	.47	.20	.62
.47	.36	.29	.43
.61	.39	−.14	.26
.00	.22	.18	.50
.32	.20	−.35	.44
.53	.21	.31	.20
.50	.15	.16	.42
.46	.10	.04	.43
.34	.02	−.25	.40

Based on Grattan-Miscio, K. E., and Vogel-Sprott, M. "Alcohol, intentional control, and inappropriate behavior: Regulation by caffeine or an incentive." *Experimental and Clinical Psychopharmacology*, Vol. 13, No. 1, Feb. 2005 (Table 1).

a. What type of experimental design is employed in this study?

b. Analyze the data for the researchers, using $\alpha = .05$. Are there differences among the mean task scores for the four groups?

c. What assumptions must be met in order to ensure the validity of the inference you made in part **b**?

10.41 Is honey a cough remedy? Pediatric researchers carried out a designed study to test whether a teaspoon of honey before bed calms a child's cough and published their results in *Archives of Pediatrics and Adolescent Medicine* (Dec. 2007). (This experiment was first described in Exercise 2.40, p. 79). A sample of 105 children who were ill with an upper respiratory tract infection and their parents participated in the study. On the first night, the parents rated their children's cough symptoms on a scale from 0 (no problems at all) to 6 (extremely severe) in five different areas. The total symptoms score (ranging from 0 to 30 points) was the variable of interest for the 105 patients. On the second night, the parents were instructed to give their sick child a dosage of liquid "medicine" prior to bedtime. Unknown to the parents, some were given a dosage of dextromethorphan (DM)—an over-the-counter cough medicine—while others were given a similar dose of honey. Also, a third group of parents (the control group) gave their sick children no dosage at all. Again, the parents rated their children's cough symptoms, and the improvement in total cough symptoms score was determined for each child. The data (improvement scores) for the study are shown in the next table. The goal of the researchers was to compare the mean improvement scores for the three treatment groups.

a. Identify the type of experimental design employed. What are the treatments?

b. Conduct an analysis of variance on the data and interpret the results.

Data for Exercise 10.41

Honey Dosage:	12	11	15	11	10	13	10	4	15	16	9	14	10
	6	10	8	11	12	12	8	12	9	11	15	10	15
	9	13	8	12	10	8	9	5	12				
DM Dosage:	4	6	9	4	7	7	7	9	12	10	11	6	3
	4	9	12	7	6	8	12	12	4	12	13	7	10
	13	9	4	4	10	15	9						
No Dosage (Control):	5	8	6	1	0	8	12	8	7	7	1	6	7
	7	12	7	9	7	9	5	11	9	5	6	8	8
	6	7	10	9	4	8	7	3	1	4	3		

Based on Paul, I. M., et al. "Effect of honey, dextromethorphan, and no treatment on nocturnal cough and sleep quality for coughing children and their parents," *Archives of Pediatrics and Adolescent Medicine*, Vol. 161, No. 12, Dec. 2007 (data simulated).

10.42 The "name game." Psychologists evaluated three methods of name retrieval in a controlled setting (*Journal of Experimental Psychology—Applied*, June 2000). A sample of 139 students was randomly divided into three groups, and each group of students used a different method to learn the names of the other students in the group. Group 1 used the "simple name game," in which a student states his/her name, and the names of all students who proceeded him/her. Group 2 used the "elaborate name game," a modification of the simple name game such that the students state not only their names, but also their favorite activity (e.g., sports). Group 3 used "pairwise introductions," according to which students are divided into pairs and each student must introduce the other member of the pair. One year later, all subjects were sent pictures of the students in their group and asked to state the full name of each. The researchers measured the percentage of names

Simple Name Game

24	43	38	65	35	15	44	44	18	27	0	38	50	31
7	46	33	31	0	29	0	0	52	0	29	42	39	26
51	0	42	20	37	51	0	30	43	30	99	39	35	19
24	34	3	60	0	29	40	40						

Elaborate Name Game

39	71	9	86	26	45	0	38	5	53	29	0	62	0
1	35	10	6	33	48	9	26	83	33	12	5	0	0
25	36	39	1	37	2	13	26	7	35	3	8	55	50

Pairwise Introductions

5	21	22	3	32	29	32	0	4	41	0	27	5	9
66	54	1	15	0	26	1	30	2	13	0	2	17	14
5	29	0	45	35	7	11	4	9	23	4	0	8	2
18	0	5	21	14									

Source: Morris, P. E., and Fritz, C. O. "The name game: Using retrieval practice to improve the learning of names," *Journal of Experimental Psychology—Applied*, Vol. 6, No. 2, June 2000 (data simulated from Figure 1).

recalled by each student respondent. The data (simulated on the basis of summary statistics provided in the research article) are shown in the table. Conduct an analysis of variance to determine whether the mean percentages of names recalled differ for the three name-retrieval methods. Use $\alpha = .05$.

Applying the Concepts—Advanced

10.43 Animal-assisted therapy for heart patients. Refer to the *American Heart Association Conference* (Nov. 2005) study to gauge whether animal-assisted therapy can improve the physiological responses of heart failure patients, presented in Exercise 2.112 (p. 106). Recall that 76 heart patients were randomly assigned to one of three groups. Each patient in group T was visited by a human volunteer accompanied by a trained dog, each patient in group V was visited by a volunteer only, and the patients in group C were not visited at all. The anxiety level of each patient was measured (in points) both before and after the visits. The accompanying table gives summary statistics for the drop in anxiety level for patients in the three groups. The mean drops in anxiety levels of the three groups of patients were compared with the use of an analysis of variance. Although the ANOVA table was not provided in the article, sufficient information is given to reconstruct it.

	Sample Size	Mean Drop	Std. Dev.
Group T: Volunteer + Trained Dog	26	10.5	7.6
Group V: Volunteer only	25	3.9	7.5
Group C: Control group (no visit)	25	1.4	7.5

Based on Cole, K., et al. "Animal assisted therapy decreases hemodynamics, plasma epinephrine and state anxiety in hospitalized heart failure patients." *American Journal of Critical Care*, 2007, 16: 575–585.

a. Compute SST for the ANOVA, using the formula (on p. 541)

$$\text{SST} = \sum_{i=1}^{3} n_i(\bar{x}_i - \bar{x})^2$$

where $\bar{x}$ is the overall mean drop in anxiety level of all 76 subjects. [*Hint:* $\bar{x} = (\Sigma_{i=1}^{3} n_i (\bar{x}_i))/76$.]

b. Recall that SSE for the ANOVA can be written as

$$\text{SSE} = (n_1 - 1)s_1^2 + (n_2 - 1)s_2^2 + (n_3 - 1)s_3^2$$

where s_1^2, s_2^2, and s_3^2 are the sample variances associated with the three treatments. Compute SSE for the ANOVA.

c. Use the results from parts **a** and **b** to construct the ANOVA table.

d. Is there sufficient evidence (at $\alpha = .01$) of differences among the mean drops in anxiety levels by the patients in the three groups?

e. Comment on the validity of the ANOVA assumptions. How might this affect the results of the study?

10.3 Multiple Comparisons of Means

Consider a completely randomized design with three treatments: A, B, and C. Suppose we determine, via the ANOVA F-test of Section 10.2, that the treatment means are statistically different. To complete the analysis, we want to rank the three treatment means. As mentioned in Section 10.2, we start by placing confidence intervals on the differences between various pairs of treatment means in the experiment. In the three-treatment experiment, for example, we would construct confidence intervals for the following differences: $\mu_A - \mu_B$, $\mu_A - \mu_C$, and $\mu_B - \mu_C$.

> **Determining the Number of Pairwise Comparisons of Treatment Means**
>
> In general, if there are k treatment means, there are
>
> $$c = k(k - 1)/2$$
>
> pairs of means that can be compared.

If we want to have $100(1 - \alpha)\%$ confidence that each of the c confidence intervals contains the true difference it is intended to estimate, we must use a smaller value of α for each individual confidence interval than we would use for a single interval. For example, suppose we want to rank the means of the three treatments A, B, and C, with 95% confidence that all three confidence intervals contain the true differences between the treatment means. Then, each individual confidence interval will need to be constructed using a level of significance smaller than $\alpha = .05$ in order to have 95% confidence that the three intervals collectively include the true differences.*

Now Work Exercise 10.43

To make **multiple comparisons of a set of treatment means**, we can use a number of procedures which, under various assumptions, ensure that the overall confidence level associated with all the comparisons remains at or above the specified $100(1 - \alpha)\%$ level. Three widely used techniques are the Bonferroni, Scheffé, and Tukey methods. For each of these procedures, the risk of making a Type I error applies to the comparisons of the treatment means in the experiment; thus, the value of α selected is called an **experimentwise error rate** (in contrast to a **comparisonwise error rate**).

> For a single comparison of two means in a designed experiment, the probability of making a Type I error (i.e., the probability of concluding that a difference in the means exists, given that the means are the same) is called a **comparisonwise error rate (CER)**.

> For multiple comparisons of means in a designed experiment, the probability of making at least one Type I error (i.e., the probability of concluding that at least one difference in means exists, given that the means are all the same) is called an **experimentwise error rate (EER)**.

*The reason each interval must be formed at a higher confidence level than that specified for the collection of intervals can be demonstrated as follows:

$$P\{\text{At least one of } c \text{ intervals fails to contain the true difference}\}$$
$$= 1 - P\{\text{All } c \text{ intervals contain the true differences}\}$$
$$= 1 - (1 - \alpha)^c \geq \alpha$$

Thus, to make this probability of at least one failure equal to α, we must specify the individual levels of significance to be less than α.

The choice of a multiple-comparison method in ANOVA will depend on the type of experimental design used and the comparisons that are of interest to the analyst. For example, **Tukey** (1949) developed his procedure specifically for pairwise comparisons when the sample sizes of the treatments are equal. The **Bonferroni** method (see Miller, 1981), like the Tukey procedure, can be applied when pairwise comparisons are of interest; however, Bonferroni's method does not require equal sample sizes. **Scheffé** (1953) developed a more general procedure for comparing all possible linear combinations of treatment means (called *contrasts*). Consequently, in making pairwise comparisons, the confidence intervals produced by Scheffé's method will generally be wider than the Tukey or Bonferroni confidence intervals.

BIOGRAPHY CARLO E. BONFERRONI (1892–1960)

Bonferroni Inequalities

During his childhood years in Turin, Italy, Carlo Bonferroni developed an aptitude for mathematics while studying music. He went on to obtain a degree in mathematics at the University of Turin. Bonferroni's first appointment as a professor of mathematics was at the University of Bari in 1923. Ten years later, he became chair of financial mathematics at the University of Florence, where he remained until his death. Bonferroni was a prolific writer, authoring over 65 research papers and books. His interest in statistics included various methods of calculating a mean and a correlation coefficient. Among statisticians, however, Bonferroni is most well known for developing his Bonferroni inequalities in probability theory in 1935. Later, other statisticians proposed using these inequalities to find simultaneous confidence intervals, which led to the development of the Bonferroni multiple-comparison method in ANOVA. Bonferroni balanced these scientific accomplishments with his music, becoming an excellent pianist and composer. ■

The formulas for constructing confidence intervals for differences between treatment means by the Tukey, Bonferroni, or Scheffé methods are provided in Appendix C. However, since these procedures (and many others) are available in the ANOVA programs of most statistical software packages, we will use the software to conduct the analyses. The programs generate a confidence interval for the difference between two treatment means for all possible pairs of treatments, based on the experimentwise error rate (α) selected by the analyst.

Example 10.6 Ⓓ GOLFCRD

Ranking Treatment Means—Golf Ball Experiment

Problem Refer to the completely randomized design of Example 10.4, in which we concluded that at least two of the four brands of golf balls are associated with different mean distances traveled when struck with a driver.

a. Use Tukey's multiple comparison procedure to rank the treatment means with an overall confidence level of 95%.

b. Estimate the mean distance traveled for balls manufactured by the brand with the highest rank.

Solution

a. To rank the treatment means with an overall confidence level of .95, we require the experimentwise error rate of $\alpha = .05$. The confidence intervals generated by Tukey's method appear at the bottom of the SAS ANOVA printout, shown in Figure 10.16. Note that for any pair of means μ_i and μ_j, SAS computes two confidence intervals— one for $(\mu_i - \mu_j)$ and one for $(\mu_j - \mu_i)$. Only one of these intervals is necessary to decide whether the means differ significantly.

In this example, we have $k = 4$ brand means to compare. Consequently, the number of relevant pairwise comparisons—that is, the number of nonredundant confidence intervals—is $c = 4(3)/2 = 6$. These six intervals, highlighted in Figure 10.16, are given in Table 10.6.

We are 95% confident that the intervals *collectively* contain all the differences between the true brand mean distances. Note that intervals containing 0, such as the (Brand A–Brand D) interval from −4.08 to 7.00, do not support a conclusion that the true brand mean distances differ. If both endpoints of the interval are positive, as with the (Brand B–Brand D) interval from 6.20 to 17.28, the implication is that the first

The ANOVA Procedure

Tukey's Studentized Range (HSD) Test for DISTANCE

NOTE: This test controls the Type I experimentwise error rate.

Alpha	0.05
Error Degrees of Freedom	36
Error Mean Square	21.17503
Critical Value of Studentized Range	3.80880
Minimum Significant Difference	5.5424

Comparisons significant at the 0.05 level are indicated by ***.

BRAND Comparison	Difference Between Means	Simultaneous 95% Confidence Limits	
C - B	8.890	3.348	14.432 ***
C - A	19.170	13.628	24.712 ***
C - D	20.630	15.088	26.172 ***
B - C	-8.890	-14.432	-3.348 ***
B - A	10.280	4.738	15.822 ***
B - D	11.740	6.198	17.282 ***
A - C	-19.170	-24.712	-13.628 ***
A - B	-10.280	-15.822	-4.738 ***
A - D	1.460	-4.082	7.002
D - C	-20.630	-26.172	-15.088 ***
D - B	-11.740	-17.282	-6.198 ***
D - A	-1.460	-7.002	4.082

Figure 10.16

SAS multiple–comparison printout for Example 10.6

Table 10.6   Pairwise Comparisons for Example 10.6

Brand Comparison	Confidence Interval
$(\mu_A - \mu_B)$	$(-15.82, -4.74)$
$(\mu_A - \mu_C)$	$(-24.71, -13.63)$
$(\mu_A - \mu_D)$	$(-4.08, 7.00)$
$(\mu_B - \mu_C)$	$(-14.43, -3.35)$
$(\mu_B - \mu_D)$	$(6.20, 17.28)$
$(\mu_C - \mu_D)$	$(15.09, 26.17)$

Mean:	249.3	250.8	261.1	270.0
Brand:	D	A	B	C

Figure 10.17

Summary of Tukey multiple comparisons

brand (B) mean distance exceeds the second (D). Conversely, if both endpoints of the interval are negative, as with the (Brand A–Brand C) interval from -24.71 to -13.63, the implication is that the second brand (C) mean distance exceeds the first brand (A) mean distance.

A convenient summary of the results of the Tukey multiple comparisons is a listing of the brand means from highest to lowest, with a solid line connecting those which are *not* significantly different. This summary is shown in Figure 10.17. The interpretation is that brand C's mean distance exceeds all others, brand B's mean exceeds that of brands A and D, and the means of brands A and D do not differ significantly. All these inferences are made with 95% confidence, the overall confidence level of the Tukey multiple comparisons.

b. Brand C is ranked highest; thus, we want a confidence interval for μ_C. Since the samples were selected independently in a completely randomized design, a confidence interval for an individual treatment mean is obtained with the one-sample t confidence interval of Section 7.3, using the standard deviation, $s = \sqrt{MSE}$, as the measure of sampling variability for the experiment. A 95% confidence interval on the mean distance traveled by brand C (apparently the "longest ball" of those tested) is

$$\bar{x}_C \pm t_{.025}\, s\sqrt{1/n}$$

where $n = 10$, $t_{.025} \approx 2$ (based on 36 degrees of freedom), and $s = \sqrt{MSE} = \sqrt{21.175} = 4.6$ (where MSE is obtained from Figure 10.16). Substituting, we obtain

$$270.0 \pm (2)(4.60)(\sqrt{.1})$$

$$270.0 \pm 2.9 \quad \text{or} \quad (267.1, 272.9)$$

Thus, we are 95% confident that the true mean distance traveled for brand C is between 267.1 and 272.9 yards when the ball is hit with a driver by Iron Byron.

Look Back The easiest way to create a summary table like Figure 10.17 is to first list the treatment means in rank order. Begin with the largest mean and compare it to (in order), the second largest mean, the third largest mean, etc., by examining the appropriate confidence intervals shown on the computer printout. If a confidence interval contains 0, then connect the two means with a line. (These two means are not significantly different.) Continue in this manner by comparing the second largest mean with the third largest, fourth largest, etc., until all possible $c = (k)(k-1)/2$ comparisons are made.

■ **Now Work Exercise 10.49**

Many of the available statistical software packages that have multiple-comparison routines will also produce the rankings shown in Figure 10.17. For example, the SAS printout in Figure 10.18 displays the ranking of the mean distances for the four golf ball brands, achieved with Tukey's method.

```
             Tukey's Studentized Range (HSD) Test for DISTANCE
NOTE: This test controls the Type I experimentwise error rate, but it generally has a higher Type
                     II error rate than REGWQ.

                  Alpha                                   0.05
                  Error Degrees of Freedom                  36
                  Error Mean Square                   21.17503
                  Critical Value of Studentized Range  3.80880
                  Minimum Significant Difference        5.5424


           Means with the same letter are not significantly different.

           Tukey Grouping          Mean      N    BRAND

                    A           269.950      10    C

                    B           261.060      10    B

                    C           250.780      10    A
                    C
                    C           249.320      10    D
```

Figure 10.18
Alternative SAS multiple-comparison output for Example 10.6

The experimentwise error rate (.05), MSE value (21.17503), and minimum significant difference (5.5424) used in the analysis are highlighted on the printout. The Tukey rankings of the means are displayed at the bottom of the figure. Instead of a solid line, note that SAS uses a "Tukey Grouping" letter to connect means that are not significantly different. You can see that the mean for brand C is ranked highest, followed by the mean for brand B. These two brand means are significantly different, since they are associated with a different "Tukey Grouping" letter. Brands A and D are ranked lowest, and their means are not significantly different (since they have the same "Tukey Grouping" letter).

Remember that the Tukey method—designed for comparing treatments with equal sample sizes—is just one of numerous multiple-comparison procedures available. Another technique may be more appropriate for the experimental design you employ. Consult the references for details on these other methods and when they should be applied. Guidelines for using the Tukey, Bonferroni, and Scheffé methods are given in the following box:

Guidelines for Selecting a Multiple-Comparison Method in ANOVA

Method	Treatment Sample Sizes	Types of Comparisons
Tukey	Equal	Pairwise
Bonferroni	Equal or unequal	Pairwise or general contrasts (number of contrasts known)
Scheffé	Equal or unequal	General contrasts

Note: For equal sample sizes and pairwise comparisons, Tukey's method will yield simultaneous confidence intervals with the smallest width, and the Bonferroni intervals will have smaller widths than the Scheffé intervals.

Ethics IN Statistics Running several multiple comparisons methods and reporting only the one that produces the desired outcome, without regard to the experimental design, is considered *unethical statistical practice*.

Ranking the Means of the Voice/Face Recognition Data in Experiment 1

Refer to the *British Journal of Psychology* (Feb. 2014) experiment designed to test the theory of associative priming using faces and voices of celebrity pairs. One goal was to compare the mean accuracy rates and mean response speeds for four groups of subjects: (1) Within-modality/Related pairs, (2) Within-modality/Unrelated pairs, (3) Cross-modality/Related pairs, and (4) Cross-modality/Unrelated pairs. In the previous Statistics in Action Revisited (p. 549), we applied a one-way analysis of variance to the **PRIME1** data. One ANOVA *F*-test found no statistical difference in the mean *accuracy rates* of the four groups. However, a second ANOVA *F*-test discovered a significant statistical difference in mean *response speeds* of the four groups.

Recall that the sample mean response speeds for groups 3 and 4 (i.e., the two cross-modality groups) are much larger than the corresponding sample means for groups 1 and 2 (i.e., the two within-modality groups). If this pattern holds for the corresponding population means, then the researchers can conclude that prime modality affects response speed.

We will rank the four population mean response speeds using a multiple comparison of means procedure, as outlined in this section. Since the sample sizes are equal ($n = 10$) for each group, we apply Tukey's procedure in order to attain the highest power (i.e., to attain the greatest chance of detecting a difference in means when differences actually exist). This method will also allow us to explicitly control the experimentwise error rate (i.e., the overall Type I error rate).

With four treatment means, there are $c = (4)(3)/2 = 6$ pairwise comparisons of interest. Using the symbol μ_j to represent the population mean response for group j, the six comparisons we desire are: $(\mu_1 - \mu_2)$, $(\mu_1 - \mu_3)$, $(\mu_1 - \mu_4)$, $(\mu_2 - \mu_3)$, $(\mu_2 - \mu_4)$, and $(\mu_3 - \mu_4)$. We used MINITAB to perform the multiple comparisons for the two dependent variables, both with an experimentwise error rate of .05. The results for accuracy rate and response speed are shown in Figures SIA10.5 and SIA10.6, respectively. Based on simultaneous

confidence intervals for each mean difference, MINITAB identifies the means that are significantly different with a different letter. Means with the same letter are not significantly different.

From Figure SIA10.5, we can infer that none of the accuracy rate means is significantly different, since all four groups have the same letter (A). Of course, this result is consistent with the ANOVA *F*-test conducted previously. Recall that the within-modality groups (1 and 2) were presented with the prime celebrity's face and the cross-modality groups (3 and 4) listened to the prime celebrity's voice. Now the theory of associative priming states that face recognition runs in parallel with voice recognition if the pairs are associated (e.g., related); that is, subjects observing related pairs should respond the same whether presented with the face first or the voice first but will respond differently if observing unrelated pairs. Since there are no differences in mean accuracy rates of the groups, the researchers inferred that associative priming does not occur with respect to recognition accuracy.

Examining Figure SIA10.6 for a comparison of the mean response speeds, we see that groups 3 and 4 are not significantly different (grouping letter A) and groups 1 and 2 are not significantly different (grouping letter B). However, groups 3 and 4 (prime voice conditions) have significantly larger means than groups 1 and 2 (prime face conditions). That is, subjects who listened to the prime celebrity's voice first (rather than seeing the face first) took, on average, longer to recognize the target celebrity's face. This led the researchers to a preliminary conclusion that, similar to accuracy, associative priming may not occur with respect to response speed. We discuss this inference in more detail in the Statistics in Action Revisited sections that follow.

```
Grouping Information Using Tukey Method

GROUP   N    Mean  Grouping
4      10  88.700  A
1      10  87.400  A
3      10  86.900  A
2      10  78.800  A

Means that do not share a letter are significantly different.
```

Figure SIA10.5
MINITAB Multiple Comparisons of Accuracy Rate Means

Statistics IN Action
(continued)

Figure SIA10.6
MINITAB Multiple Comparisons
of Response Speed Means

```
Grouping Information Using Tukey Method

GROUP   N    Mean  Grouping
3      10  1245.6  A
4      10  1131.5  A
2      10   895.0     B
1      10   763.3     B

Means that do not share a letter are significantly different.
```

Exercises 10.44–10.64

Understanding the Principles

10.44 Give three multiple comparisons procedures.

10.45 Define an experimentwise error rate.

10.46 Give a situation when it is most appropriate to apply Scheffe's multiple-comparison-of-means method.

10.47 For each of the following confidence intervals for the difference between two means, $(\mu_1 - \mu_2)$, which mean is significantly larger?
a. $(-12, 7)$
b. $(-12, -7)$
c. $(7, 12)$

Learning the Mechanics

10.48 Consider a completely randomized design with k treatments. Assume that all pairwise comparisons of treatment
[NW] means are to be made with the use of a multiple-comparison procedure. Determine the total number of pairwise comparisons for the following values of k:
a. $k = 3$ **b.** $k = 5$
c. $k = 4$ **d.** $k = 10$

10.49 A multiple-comparison procedure for comparing four
[NW] treatment means produced the confidence intervals shown here. Rank the means from smallest to largest. Which means are significantly different?

$(\mu_1 - \mu_2): (2, 15)$
$(\mu_1 - \mu_3): (4, 7)$
$(\mu_1 - \mu_4): (-10, 3)$
$(\mu_2 - \mu_3): (-5, 11)$
$(\mu_2 - \mu_4): (-12, -6)$
$(\mu_3 - \mu_4): (-8, -5)$

10.50 Consider a completely randomized design with five treatments: A, B, C, D, and E. The ANOVA *F*-test revealed significant differences among the means. A multiple-comparison procedure was used to compare all possible pairs of treatment means at $\alpha = .05$. The ranking of the five treatment means is summarized here. Identify which pairs of means are significantly different.
a. $\overline{A \quad C} \quad \overline{E \quad B \quad D}$ **b.** $\overline{A} \quad \overline{C \quad E \quad B \quad D}$
c. $\overline{A \quad C \quad E \quad B \quad D}$ **d.** $\overline{A \quad C \quad E} \quad \overline{B \quad D}$

Applying the Concepts—Basic

10.51 **How workers respond to wage cuts.** Refer to the *Institute for the Study of Labor: Discussion Paper Series* (Mar. 2011) study of how workers respond to wage cuts, Exercise 10.31

(p. 552). Recall that teams of workers were divided into three wage cut groups: "unilateral wage cut," "general wage cut," and "baseline" (no wage cut). The researchers determined that the average decrease in promotional cards sold differed depending on the wage cut treatment group. Now the researchers want to rank the treatment means by making all possible pairwise comparisons.
a. How many pairwise comparisons make up this phase of the study? List them.
b. Why is a multiple comparisons procedure like Tukey's recommended for the analysis?
c. The confidence interval for the difference between the means for "baseline" and "general wage cut," i.e., for $(\mu_{baseline} - \mu_{general})$, includes only positive values. Interpret this result.

10.52 **Whales entangled in fishing gear.** Refer to the *Marine Mammal Science* (April 2010) investigation of whales entangled by fishing gear, Exercise 10.32 (p. 552). The mean body lengths (meters) of whales entangled in each of the three types of fishing gear (set nets, pots, and gill nets) are reported below. Tukey's method was used to conduct multiple comparisons of the means with an experimentwise error rate of .01. Based on the results, which type of fishing gear will entangle the shortest whales, on average? The longest whales, on average?

Mean Length:	4.45	5.28	5.63
Fishing Gear:	Set nets	Gill nets	Pots

10.53 **Does the media influence your attitude toward tanning?** Refer to the *Basic and Applied Social Psychology* (May 2010) study of whether product advertisements influence consumers' attitudes toward tanning, Exercise 10.33 (p. 552). Recall that college students were randomly assigned to one of three conditions: view product advertisements featuring models with a tan, view product advertisements featuring models without a tan, or view products advertised with no models. An ANOVA *F*-test revealed that the mean attitude toward tanning differed across the conditions. The researchers followed up this analysis with a multiple comparisons of means using an experimentwise error rate of .05. These results are summarized below. Fully interpret the results. Does it appear that the type of product advertisement can influence a consumer's attitude toward tanning?

Mean:	2.11	2.40	2.50
Condition:	Models/No Tan	Tanned Models	No Models

10.54 Evaluation of flexography printing plates. Refer to the *Journal of Graphic Engineering and Design* (Vol. 3, 2012) study of the quality of flexography printing, Exercise 10.34 (p. 552). Recall that four different exposure times were studied—8, 10, 12, and 14 minutes—and that the measure of print quality used was dot area (hundreds of dots per square millimeter). Tukey's multiple comparisons procedure (at an experimentwise error rate of .05) was used to rank the mean dot areas of the four exposure times. The results are summarized below. Which exposure time yields the highest mean dot area? Lowest?

Mean Dot Area:	.571	.582	.588	.594
Exposure Time: (minutes)	12	10	14	8

10.55 Chemical properties of whole wheat breads. Whole wheat breads contain a high amount of phytic acid, which tends to lower the absorption of nutrient minerals. The *Journal of Agricultural and Food Chemistry* (Jan. 2005) published the results of a study to determine whether sourdough can increase the solubility of whole wheat bread. Four types of bread were prepared from whole meal flour: (1) yeast added, (2) sourdough added, (3) no yeast or sourdough added (control), and (4) lactic acid added. Data were collected on the soluble magnesium level (percent of total magnesium) during fermentation for samples of each type of bread and were analyzed with the use of a one-way ANOVA. The four mean soluble magnesium means were compared in pairwise fashion with Bonferroni's method. The results are summarized as follows:

Mean:	7%	12.5%	22%	27.5%
Type of Bread:	Control	Yeast	Lactate	Sourdough

a. How many pairwise comparisons are made in the Bonferroni analysis?
b. Which treatment(s) yielded the significantly highest mean soluble magnesium level? The lowest?
c. The experimentwise error rate for the analysis was .05. Interpret this value.

10.56 Robots trained to behave like ants. Refer to the *Nature* (Aug. 2000) study of robots trained to behave like ants, presented in Exercise 10.35 (p. 553). Multiple comparisons of mean energy expended for the four colony sizes were conducted with an experimentwise error rate of .05. The results are summarized as follows:

Sample mean:	.97	.95	.93	.80
Group size:	3	6	9	12

a. How many pairwise comparisons are conducted in this analysis?
b. Interpret the results shown.

Applying the Concepts—Intermediate

10.57 Guilt in decision making. The effect of guilt emotion on how a decision maker focuses on the problem was investigated in the *Journal of Behavioral Decision Making* (Jan. 2007). A sample of 77 volunteer students participated in one portion of the experiment, where each was randomly assigned to one of three emotional states (guilt, anger, or neutral) through a reading/writing task. Immediately after the task, the students were presented with a decision problem where the stated option has predominantly negative features (e.g., spending money on repairing a very old car). Prior to making the decision, the researchers asked each subject to list possible, more attractive, alternatives. The researchers then compared the mean number of alternatives listed across the three emotional states with an analysis of variance for a completely randomized design. A partial ANOVA summary table is shown below.

Source	df	F-value	p-value
Emotional State	2	22.68	0.001
Error	74		
Total	76		

a. What conclusion can you draw from the ANOVA results?
b. A multiple comparisons of means procedure was applied to the data using an experimentwise error rate of .05. Explain what the .05 represents.
c. The multiple comparisons yielded the following results. What conclusion can you draw?

Sample mean:	1.90	2.17	4.75
Emotional State:	*Angry*	*Neutral*	*Guilt*

10.58 Study of recall of TV commercials. Refer to the *Journal of Applied Psychology* (June 2002) completely randomized design study to compare the mean commercial recall scores of viewers of three TV programs, presented in Exercise 10.37 (p. 553). Recall that one program had a violent content code (V) rating, one had a sex content code (S) rating, and one was a neutral TV program. Using Tukey's method, the researchers conducted multiple comparisons of the three mean recall scores.

a. How many pairwise comparisons were made in this study?
b. The multiple-comparison procedure was applied to the **TVADS** data and the results are shown in the MINITAB printout on page 535. An experimentwise error rate of .05 was used. Locate the confidence interval for the comparison of the V and S groups. Interpret this result practically.
c. Repeat part **b** for the remaining comparisons. Which of the groups has the largest mean recall score?
d. In the journal article, the researchers concluded that "memory for [television] commercials is impaired after watching violent or sexual programming." Do you agree?

10.59 Baking properties of pizza cheese. Due to its unique stretchability, mozzarella is the most commonly used pizza cheese. How do the baking properties of mozzarella compare with those of other pizza cheese options? This was the question of interest in a study published in the *Journal of Food Science* (Aug. 2014). Seven different cheeses were compare: mozzarella, cheddar, colby, edam, emmental, gruyere, and provolone. A pizza was baked with each of the different cheeses, then images of the pizzas were taken with a digital video camera. From these images, the researchers calculated the color change index (CCI)—a

MINITAB output for Exercise 10.58

```
Grouping Information Using Tukey Method

             N   Mean  Grouping
NEUTRAL    108  3.167  A
VIOLENT    108  2.083     B
SEX        108  1.713     B

Means that do not share a letter are significantly different.


Tukey 95% Simultaneous Confidence Intervals
All Pairwise Comparisons

Individual confidence level = 98.01%


VIOLENT subtracted from:

          Lower  Center  Upper  -------+---------+---------+---------+--
SEX      -0.923  -0.370  0.183         (----*----)
NEUTRAL   0.530   1.083  1.636                     (----*----)
                                 -------+---------+---------+---------+--
                                     -1.2       0.0       1.2       2.4


SEX subtracted from:

          Lower  Center  Upper  -------+---------+---------+---------+--
NEUTRAL   0.901   1.454  2.007                       (---*----)
                                 -------+---------+---------+---------+--
                                     -1.2       0.0       1.2       2.4
```

measure of cheese color uniformity—for sampled digital pixels. The mean CCI values of the seven cheeses were compared using a completely randomized design ANOVA.
a. Identify the treatments for this experiment.
b. Identify the dependent variable for this experiment.
c. The ANOVA F-test resulted in p-value $< .05$. Practically interpret this result.
d. How many different pairwise comparisons of cheeses are possible?
e. A multiple comparisons procedure with an experiment-wise error rate of .05 was applied to the data, revealing the results at the bottom of the page. What conclusions can you draw from these results?

10.60 Writing styles in medical journals. When publishing research results, most medical journals advise authors to write using the "active" rather than the "passive" voice. The impact of the use of the passive voice in five top medical journals with differing writing style guidelines was investigated in *Applied Linguistics* (Dec. 2013). The five journals compared were the *Annals of Internal Medicine* (AIM), *British Medical Journal* (BMJ), *Journal of the American Medical Association* (JAMA), *Lancet*, and the *New England Journal of Medicine* (NEJM). Using a search engine, the researchers sampled (independently) a large number of declarative passive paragraphs from each journal and recorded the number of passives per sentence (NPPS) for each paragraph. A one-way analysis of

variance was conducted on the data. Interpret, fully, the results shown below.

$$\text{ANOVA } F = 35.5 \ (p\text{-value} < .001)$$

Mean NPPS:	.32	.38	.49	.50	.52
Journal:	BMJ	AIM	Lancet	JAMA	NEJM

10.61 Is honey a cough remedy? Refer to the *Archives of Pediatrics and Adolescent Medicine* (Dec. 2007) study of treatments for children's cough symptoms, Exercise 10.41 (p. 554). Do you agree with the statement (extracted from the article) "honey may be a preferable treatment for the cough and sleep difficulty associated with childhood upper respiratory tract infection"? Perform a multiple comparisons of means to answer the question.

10.62 Effect of scopolamine on memory. The drug scopolamine is often used as a sedative to induce sleep in patients. In *Behavioral Neuroscience* (Feb. 2004), medical researchers examined scopolamine's effects on memory with associated word pairs. A total of 28 human subjects, were given a list of related word pairs to memorize. For every word pair in the list (e.g., robber–jail), there was an associated word pair with the same first word, but a different second word (e.g., robber–police). The subjects were then randomly divided into three treatment groups. Group 1 subjects were administered an injection of scopolamine, group 2

Results for Exercise 10.59

Mean CCI:	.10	.15	.15	.18	.19	.30	3.62
Cheese:	Colby	Edam	Emmental	Cheddar	Gruyere	Provolone	Mozzarella

subjects were given an injection of glycopyrrolate (an active placebo), and group 3 subjects were not given any drug. Four hours later, subjects were shown 12 word pairs from the list and tested on how many they could recall. The data on number of pairs recalled (simulated on the basis of summary information provided in the research article) are listed below. Prior to the analysis, the researchers theorized that the mean number of word pairs recalled for the scopolamine subjects (group 1) would be less than the corresponding means for the other two groups.

Group 1 (Scopolamine):	5	8	8	6	6	6	6	8	6	4	5	6
Group 2 (Placebo):		8	10	12	10	9	7	9	10			
Group 3 (No drug):		8	9	11	12	11	10	12	12			

a. Explain why this is a completely randomized design.
b. Identify the treatments and response variable.
c. Find the sample means for the three groups. Is this sufficient information to support the researchers' theory? Explain.
d. Conduct an ANOVA F-test on the data. Is there sufficient evidence (at $\alpha = .05$) to conclude that the mean number of word pairs recalled differs among the three treatment groups?
e. Conduct multiple comparisons of the three means (using an experimentwise error rate of .05). Do the results support the researchers' theory? Explain.

Applying the Concepts—Advanced

10.63 Restoring self-control while intoxicated. Refer to the *Experimental and Clinical Psychopharmacology* (Feb. 2005) study of restoring self-control while intoxicated,

presented in Exercise 10.40 (p. 554). The researchers theorized that if caffeine can really restore self-control, then students in Group AC (the group that drank alcohol plus caffeine) will perform the same as students in Group P (the placebo group) on the word completion task. Similarly, if an incentive can restore self-control, then students in Group AR (the group that drank alcohol and got a reward for correct responses on the word completion task) will perform the same as students in Group P. Finally, the researchers theorized that students in Group A (the alcohol-only group) will perform worse on the word completion task than students in any of the other three groups. Access the data in the **DRINKS** file and conduct Tukey's multiple comparisons of the means, using an experimentwise error rate of .05. Are the researchers' theories supported?

10.64 Animal-assisted therapy for heart patients. Refer to the *American Heart Association Conference* (Nov. 2005) study to gauge whether animal-assisted therapy can improve the physiological responses of heart failure patients, presented in Exercise 10.43 (p. 555). You found evidence of a difference among the treatment means for the three treatments: Group T (volunteer plus trained dog), Group V (volunteer only), and Group C (control). Conduct a Bonferroni analysis to rank the three treatment means. Use an experimentwise error rate of $\alpha = .03$. Interpret the results for the researchers. [*Hint:* As shown in Appendix C, the Bonferroni formula for a confidence interval for the difference $(\mu_i - \mu_j)$ is

$$(\bar{x}_i - \bar{x}_j) \pm t_{\alpha*/2}(s)\sqrt{(1/n_i) + (1/n_j)}$$

where $\alpha* = 2\alpha/[(k)(k-1)]$ is the experimentwise error rate, and k is the total number of treatment means compared.]

10.4 The Randomized Block Design

If the completely randomized design results in nonrejection of the null hypothesis that the treatment means are equal because the sampling variability (as measured by MSE) is large, we may want to consider an experimental design that better controls the variability. In contrast to the selection of independent samples of experimental units specified by the completely randomized design, the *randomized block design* utilizes experimental units that are *matched sets*, assigning one from each set to each treatment. The matched sets of experimental units are called *blocks*. The theory behind the randomized block design is that the sampling variability of the experimental units in each block will be reduced, in turn reducing the measure of error, MSE.

> The **randomized block design** consists of a two-step procedure:
>
> 1. Matched sets of experimental units, called **blocks**, are formed, with each block consisting of k experimental units (where k is the number of treatments). The b blocks should consist of experimental units that are as similar as possible.
>
> 2. One experimental unit from each block is randomly assigned to each treatment, resulting in a total of $n = bk$ responses.

For example, if we wish to compare SAT scores of female and male high school seniors, we could select independent random samples of five females and five males, and analyze the results of the completely randomized design as outlined in Section 10.2. Or we could select matched pairs of females and males according to their scholastic records and analyze the SAT scores of the pairs. For instance, we could select pairs of students with approximately the same GPAs from the same high school. Five such pairs (blocks) are depicted in Table 10.7. Note that this is just a *paired difference experiment*, first discussed in Section 9.3.

Table 10.7 Randomized Block Design: SAT Score Comparison

Block	Female SAT Score	Male SAT Score	Block Mean
1 (School A, 2.75 GPA)	540	530	535
2 (School B, 3.00 GPA)	570	550	560
3 (School C, 3.25 GPA)	590	580	585
4 (School D, 3.50 GPA)	640	620	630
5 (School E, 3.75 GPA)	690	690	690
Treatment mean	606	594	

D *Data Set:* TAB10_7

As before, the variation between the treatment means is measured by squaring the distance between each treatment mean and the overall mean, multiplying each squared distance by the number of measurements for the treatment, and then summing over treatments:

$$\text{SST} = \sum_{i=1}^{k} b(\bar{x}_{T_i} - \bar{x})^2$$

$$= 5(606 - 600)^2 + 5(594 - 600)^2 = 360$$

Here, $\bar{x}_{T_i}$ represents the sample mean for the ith treatment, b (the number of blocks) is the number of measurements for each treatment, and k is the number of treatments.

The blocks also account for some of the variation among the different responses. That is, just as SST measures the variation between the female and male means, we can calculate a measure of variation among the five block means representing different schools and scholastic abilities. Analogous to the computation of SST, we sum the squares of the differences between each block mean and the overall mean, multiply each squared difference by the number of measurements for each block, and then sum over blocks to calculate the **sum of squares for blocks (SSB):**

$$\text{SSB} = \sum_{i=1}^{b} k(\bar{x}_{B_i} - \bar{x})^2$$

$$= 2(535 - 600)^2 + 2(560 - 600)^2 + 2(585 - 600)^2 +$$
$$2(630 - 600)^2 + 2(690 - 600)^2$$

$$= 30{,}100$$

Here, $\bar{x}_{B_i}$ represents the sample mean for the ith block and k (the number of treatments) is the number of measurements in each block. As we expect, the variation in SAT scores attributable to Schools and Levels of scholastic achievement is apparently large.

As before, we want to compare the variability attributed to treatments with that which is attributed to sampling. In a randomized block design, the sampling variability is measured by subtracting that portion attributed to treatments and blocks from the total sum of squares, SS(Total). The total variation is the sum of the squared differences of each measurement from the overall mean:

$$\text{SS(Total)} = \sum_{i=1}^{n} (x_i - \bar{x})^2$$

$$= (540 - 600)^2 + (530 - 600)^2 + (570 - 600)^2 + (550 - 600)^2 +$$
$$\cdots + (690 - 600)^2$$

$$= 30{,}600$$

The variation attributable to sampling error is then found by subtraction:

$$\text{SSE} = \text{SS(Total)} - \text{SST} - \text{SSB} = 30,600 - 360 - 30,100 = 140$$

In sum, the total sum of squares, 30,600, is divided into three components: 360 attributed to treatments (Gender), 30,100 attributed to blocks (Scholastic ability), and 140 attributed to sampling error.

The mean squares associated with each source of variability are obtained by dividing the sum of squares by the appropriate number of degrees of freedom. The partitioning of the total sum of squares and the total number of degrees of freedom for a randomized block experiment are summarized in Figure 10.19.

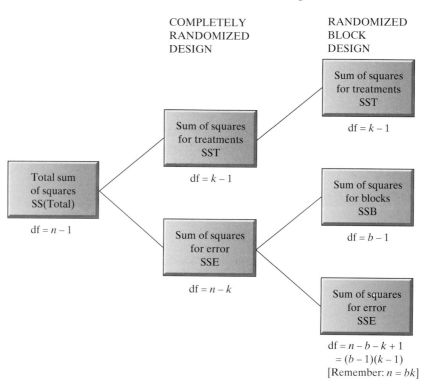

Figure 10.19

Partitioning of the total sum of squares for the randomized block design

To determine whether we can reject the null hypothesis that the treatment means are equal in favor of the alternative that at least two of them differ, we calculate

$$\text{MST} = \frac{\text{SST}}{k-1} = \frac{360}{2-1} = 360$$

$$\text{MSE} = \frac{\text{SSE}}{n-b-k+1} = \frac{140}{10-5-2+1} = 35$$

The *F*-ratio that is used to test the hypothesis is

$$F = \frac{360}{35} = 10.29$$

Comparing this ratio with the tabular *F*-value corresponding to $\alpha = .05$, $v_1 = (k-1) = 1$ degree of freedom in the numerator, and $v_2 = (n-b-k+1) = 4$ degrees of freedom in the denominator, we find that

$$F = 10.29 > F_{.05} = 7.71$$

which indicates that we should reject the null hypothesis and conclude that the mean SAT scores differ for females and males. All of these calculations, of course, can be obtained using statistical software. The output for the MINITAB analysis of the data in Table 10.7 is shown in Figure 10.20. The values of SST, SSE, MST, MSE, and *F* are highlighted on the printout.

Figure 10.20

MINITAB printout with ANOVA results for the data in Table 10.7

Two-way ANOVA: SAT versus GENDER, BLOCK

Source	DF	SS	MS	F	P
GENDER	1	360	360	10.29	0.033
BLOCK	4	30100	7525	215.00	0.000
Error	4	140	35		
Total	9	30600			

Comment: If you review Section 9.3, you will find that the analysis of a paired difference experiment results in a one-sample *t*-test on the differences between the treatment responses within each block. Applying the procedure to the differences between female and male scores in Table 10.7, we obtain

$$t = \frac{\bar{x}_d}{s_d/\sqrt{n_d}} = \frac{12}{\sqrt{70}/\sqrt{5}} = 3.207$$

At the .05 level of significance with $(n_d - 1) = 4$ degrees of freedom,

$$t = 3.207 > t_{.025} = 2.776$$

Since $t^2 = (3.207)^2 = 10.29$ and $t^2_{.025} = (2.776)^2 = 7.71$, we find that the paired difference *t*-test and the ANOVA *F*-test are equivalent, with both the calculated test statistics and the rejection region related by the formula $F = t^2$. The difference between the tests is that the paired difference *t*-test can be used to compare only two treatments in a randomized block design, whereas the *F*-test can be applied to *two or more* treatments in a randomized block design.

The *F*-test for a randomized block design is summarized in the following box:

ANOVA *F*-Test to Compare *k* Treatment Means: Randomized Block Design

$H_0: \mu_1 = \mu_2 = \cdots = \mu_k$

H_a: At least two treatment means differ

Test statistic: $F_c = \dfrac{\text{MST}}{\text{MSE}}$

Rejection region: $F_c > F_\alpha$

p-value: $P(F > F_c)$

where F_α is based on $(k - 1)$ numerator degrees of freedom and $(n - b - k + 1)$ denominator degrees of freedom.

Decision: Reject H_0 if $\alpha > p$-value or if test statistic (F_c) falls into the rejection region.

Conditions Required for a Valid ANOVA *F*-Test: Randomized Block Design

1. The *b* blocks are randomly selected, and all *k* treatments are applied (in random order) to each block.

2. The distributions of observations corresponding to all *bk* block–treatment combinations are approximately normal.

3. The *bk* block–treatment distributions have equal variances.

Note that the assumptions concern the probability distributions associated with each block–treatment combination. The experimental unit selected for each combination is assumed to have been randomly selected from all possible experimental units for that combination, and the response is assumed to be normally distributed with the same variance for each of the block–treatment combinations. For example, the *F*-test comparing female and male SAT score means requires the scores for each combination of gender and scholastic ability (e.g., females with 3.25 GPA) to be normally distributed with the same variance as the other combinations employed in the experiment.

For those who are interested, the calculation formulas for randomized block designs are given in Appendix B. Throughout this section, we will rely on statistical software packages to analyze randomized block designs and to obtain the necessary ingredients for testing the null hypothesis that the treatment means are equal.

Example **10.7**	**Problem** Refer to Examples 10.4–10.6. Suppose the USGA wants to compare the mean distances associated with the four brands of golf balls struck by a driver, but wishes to employ human golfers rather than the robot Iron Byron. Assume that 10 balls of each brand are to be utilized in the experiment.

Experimental Design Principles

a. Explain how a completely randomized design could be employed.

b. Explain how a randomized block design could be employed.

c. Which design is likely to provide more information about the differences among the brand mean distances?

Solution

a. Since the completely randomized design calls for independent samples, we can employ such a design by randomly selecting 40 golfers and then randomly assigning 10 golfers to each of the four brands. Finally, each golfer will strike the ball of the assigned brand, and the distance will be recorded. This design is illustrated in Figure 10.21.

a. Completely randomized design

Figure 10.21

Illustration of completely randomized design and randomized block design: comparison of four golf ball brands

BRAND

	A	B	C	D
1	Hit 3	Hit 1	Hit 4	Hit 2
2	Hit 2	Hit 4	Hit 3	Hit 1
⋮	⋮	⋮	⋮	⋮
10	Hit 4	Hit 3	Hit 1	Hit 2

GOLFERS

b. Randomized block design

b. The randomized block design employs blocks of relatively homogeneous experimental units. For example, we could randomly select 10 golfers and permit each golfer to hit four balls, one of each brand, in a random sequence. Then each golfer is a block, with each treatment (brand) assigned to each block (golfer). This design is summarized in Figure 10.21.

c. Because we expect much more variability among distances generated by "real" golfers than by Iron Byron, we would expect the randomized block design to control the variability better than the completely randomized design does. That is, with 40 different golfers, we would expect the sampling variability among the measured distances within each brand to be greater than that among the four distances generated by each of 10 golfers hitting one ball of each brand.

■ **Now Work Exercise 10.77 a,b**

Example	10.8	D

GOLFRBD

A Randomized Block Design ANOVA— Comparing Golf Ball Brands

Problem Refer to Example 10.7. Suppose the randomized block design of part **b** is employed, utilizing a random sample of 10 golfers, with each golfer using a driver to hit four balls, one of each brand, in a random sequence.

a. Set up a test of the research hypothesis that the brand mean distances differ. Use $\alpha = .05$.

b. The data for the experiment are given in Table 10.8. Use statistical software to analyze the data, and conduct the test set up in part **a.**

Table 10.8 Distance Data for Randomized Block Design

Golfer (Block)	Brand A	Brand B	Brand C	Brand D
1	202.4	203.2	223.7	203.6
2	242.0	248.7	259.8	240.7
3	220.4	227.3	240.0	207.4
4	230.0	243.1	247.7	226.9
5	191.6	211.4	218.7	200.1
6	247.7	253.0	268.1	244.0
7	214.8	214.8	233.9	195.8
8	245.4	243.6	257.8	227.9
9	224.0	231.5	238.2	215.7
10	252.2	255.2	265.4	245.2
Sample means	227.0	233.2	245.3	220.7

D *Data Set:* GOLFRBD

Solution

a. We want to test whether the data in Table 10.8 provide sufficient evidence to conclude that the brand mean distances differ. Denoting the population mean of the *i*th brand by μ_i, we test

$$H_0: \mu_1 = \mu_2 = \mu_3 = \mu_4$$
$$H_a: \text{The mean distances differ for at least two of the brands.}$$

The test statistic compares the variation among the four treatment (brand) means with the sampling variability within each of the treatments:

$$\text{Test statistic:} \quad F = \frac{\text{MST}}{\text{MSE}}$$

Rejection region: $F > F_\alpha = F_{.05}$, with $v_1 = (k - 1) = 3$ numerator degrees of freedom and $v_2 = (n - k - b + 1) = 27$ denominator degrees of freedom. From Table VI of Appendix B, we find that $F_{.05} = 2.96$. Thus, we will reject H_0 if $F > 2.96$.

The assumptions necessary to ensure the validity of the test are as follows: (1) The probability distributions of the distances for each brand–golfer combination are normal. (2) The variances of the distance probability distributions for each brand–golfer combination are equal.

b. SPSS was used to analyze the data in Table 10.8, and the result is shown in Figure 10.22. The values of MST and MSE (highlighted on the printout) are 1,099.552 and 20.245, respectively. The *F*-ratio for Brand (also highlighted on the printout) is

$F = 54.312$, which exceeds the tabled value of 2.96. We therefore reject the null hypothesis at the $\alpha = .05$ level of significance, concluding that at least two of the brands differ with respect to mean distance traveled when struck by the driver.

Tests of Between-Subjects Effects

Dependent Variable:DISTANCE

Source	Type III Sum of Squares	df	Mean Square	F	Sig.
Corrected Model	15372.539ᵃ	12	1281.045	63.276	.000
Intercept	2145032.910	1	2145032.910	105952.598	.000
BRAND	3298.657	3	1099.552	54.312	.000
GOLFER	12073.882	9	1341.542	66.265	.000
Error	546.621	27	20.245		
Total	2160952.070	40			
Corrected Total	15919.160	39			

a. R Squared = .966 (Adjusted R Squared = .950)

Figure 10.22

SPSS printout for randomized block design ANOVA of data in Table 10.8

Look Back The result of part **b** is confirmed by noting that the observed significance level of the test, highlighted on the printout, is $p \approx 0$.

■ **Now Work Exercise 10.77c**

The results of an ANOVA for a randomized block design can be summarized in a simple tabular format similar to that utilized for the completely randomized design in Section 10.2. The general form of the table is shown in Table 10.9, and that for Example 10.8 is given in Table 10.10. Note that the randomized block design is characterized by three sources of variation—treatments, blocks, and error—which sum to the total sum of squares. We hope that employing blocks of experimental units will reduce the error variability, thereby making the test for comparing treatment means more powerful.

When the F-test results in the rejection of the null hypothesis that the treatment means are equal, we will usually want to compare the various pairs of treatment means to determine which specific pairs differ. We can employ a multiple-comparison procedure as in Section 10.3. The number of pairs of means to be compared will again be $c = k(k - 1)/2$, where k is the number of treatment means. In Example 10.8, $c = 4(3)/2 = 6$; that is, there are six pairs of golf ball brand means to be compared.

Table 10.9   General ANOVA Summary Table for a Randomized Block Design

Source	df	SS	MS	F
Treatments	$k - 1$	SST	MST	MST/MSE
Blocks	$b - 1$	SSB	MSB	
Error	$n - k - b + 1$	SSE	MSE	
Total	$n - 1$	SS(Total)		

Table 10.10 ANOVA Table for Example 10.8

Source	df	SS	MS	F	p
Treatment (Brand)	3	3,298.66	1,099.55	54.31	.000
Block (Golfer)	9	12,073.88	1,341.54		
Error	27	546.62	20.25		
Total	39	15,919.16			

Example   10.9

Ranking Treatment Means in a Randomized Block Design—Comparing Golf Ball Brands

GOLFRBD

Figure 10.23

SPSS printout of Bonferroni confidence intervals for the randomized block design

Problem Bonferroni's procedure is used to compare the mean distances of the four golf ball brands in Example 10.8. The resulting confidence intervals, with an experimentwise error rate (EER) of $\alpha = .05$, are shown in the SPSS printout of Figure 10.23. Interpret the results.

Multiple Comparisons

DISTANCE
Bonferroni

(I) BRAND	(J) BRAND	Mean Difference (I-J)	Std. Error	Sig.	95% Confidence Interval Lower Bound	95% Confidence Interval Upper Bound
A	B	-6.1300*	2.01222	.031	-11.8586	-.4014
	C	-18.2800*	2.01222	.000	-24.0086	-12.5514
	D	6.3200*	2.01222	.024	.5914	12.0486
B	A	6.1300*	2.01222	.031	.4014	11.8586
	C	-12.1500*	2.01222	.000	-17.8786	-6.4214
	D	12.4500*	2.01222	.000	6.7214	18.1786
C	A	18.2800*	2.01222	.000	12.5514	24.0086
	B	12.1500*	2.01222	.000	6.4214	17.8786
	D	24.6000*	2.01222	.000	18.8714	30.3286
D	A	-6.3200*	2.01222	.024	-12.0486	-.5914
	B	-12.4500*	2.01222	.000	-18.1786	-6.7214
	C	-24.6000*	2.01222	.000	-30.3286	-18.8714

Based on observed means.
The error term is Mean Square(Error) = 20.245.

*. The mean difference is significant at the 0.05 level.

Solution Note that 12 confidence intervals are shown in Figure 10.23, rather than 6. SPSS (like SAS) computes intervals for both $\mu_i - \mu_j$ and $\mu_j - \mu_i, i \neq j$. Only half of these are necessary to conduct the analysis, and these are highlighted on the printout. The intervals (rounded) are summarized as follows:

$$(\mu_A - \mu_B): \quad (-11.9, -.4)$$
$$(\mu_A - \mu_C): \quad (-24.0, -12.6)$$
$$(\mu_A - \mu_D): \quad (.6, 12.0)$$
$$(\mu_B - \mu_C): \quad (-17.9, -6.4)$$
$$(\mu_B - \mu_D): \quad (6.7, 18.2)$$
$$(\mu_C - \mu_D): \quad (18.9, 30.3)$$

Note that we are 95% confident that all the brand means differ because none of the intervals contains 0. The listing of the brand means in Figure 10.24 has no lines connecting them because there are no nonsignificant differences at the .05 level.

Mean:	220.7	227.0	233.2	245.3
Brand:	D	A	B	C

Figure 10.24

Listing of brand means for randomized block design [*Note:* All differences are statistically significant.]

■ **Now Work Exercise 10.77d**

Unlike the completely randomized design, the randomized block design cannot, in general, be used to estimate individual treatment means. Whereas the completely randomized design employs a random sample for each treatment, the randomized block design does not necessarily do so: The experimental units within the blocks are assumed to be randomly selected, but the blocks themselves may not be randomly selected.

We can, however, test the hypothesis that the block means are significantly different. We simply compare the variability attributable to differences among the block means with that associated with sampling variability. The ratio of MSB to MSE is an

F-ratio similar to that formed in testing treatment means. The *F*-statistic is compared with a tabular value for a specific value of α, with $(b - 1)$ numerator degrees of freedom and $(n - k - b + 1)$ denominator degrees of freedom. The test is usually given on the same printout as the test for treatment means. Note in the SPSS printout in Figure 10.22 that the test statistic for comparing the block means is

$$F = \frac{\text{MSB}}{\text{MSE}} = \frac{\text{MS(Golfers)}}{\text{MS(Error)}} = \frac{1{,}341.54}{20.25} = 66.27$$

with a *p*-value of .000. Since $\alpha = .05$ exceeds this *p*-value, we conclude that the block means are different. The results of the test are summarized in Table 10.11.

Table 10.11 ANOVA Table for Randomized Block Design: Test for Blocks Included

Source	df	SS	MS	F	p
Treatments (Brands)	3	3,298.66	1,099.55	54.31	.000
Blocks (Golfers)	9	12,073.88	1,341.54	66.27	.000
Error	27	546.62	20.25		
Total	39	15,919.16			

In the golf example, the test for block means confirms our suspicion that the golfers vary significantly; therefore, the use of the block design was a good decision. However, be careful not to conclude that the block design was a mistake if the *F*-test for blocks does not result in rejection of the null hypothesis that the block means are the same. Remember that the possibility of a Type II error exists, and we are not controlling its probability as we are the probability α of a Type I error. If the experimenter believes that the experimental units are more homogeneous within blocks than between blocks, then he or she should use the randomized block design regardless of the results of a single test comparing the block means.

The procedure for conducting an analysis of variance for a randomized block design is summarized in the next box. Remember that the hallmark of this design is the utilization of blocks of homogeneous experimental units in which each treatment is represented.

Steps for Conducting an ANOVA for a Randomized Block Design

1. Be sure that the design consists of blocks (preferably, blocks of homogeneous experimental units) and that each treatment is randomly assigned to one experimental unit in each block.

2. If possible, check the assumptions of normality and equal variances for all block–treatment combinations. [*Note:* This may be difficult to do, since the design will likely have only one observation for each block–treatment combination.]

3. Create an ANOVA summary table that specifies the variability attributable to treatments, blocks, and error, and that leads to the calculation of the *F*-statistic to test the null hypothesis that the treatment means are equal in the population. Use a statistical software package or the calculation formulas in Appendix B to obtain the necessary numerical ingredients.

4. If the *F*-test leads to the conclusion that the means differ, employ the Bonferroni or Tukey procedure, or a similar procedure, to conduct multiple comparisons of as many of the pairs of means as you wish. Use the results to summarize the statistically significant differences among the treatment means. Remember that, in general, the randomized block design cannot be employed to form confidence intervals for individual treatment means.

5. If the *F*-test leads to the nonrejection of the null hypothesis that the treatment means are equal, several possibilities exist:

   a. The treatment means *are* equal: that is, the null hypothesis is true.

   b. The treatment means really differ, but other important factors affecting the response are not accounted for by the randomized block design. These factors inflate the sampling variability, as measured by MSE, resulting in smaller values of the *F*-statistic. Either increase the sample size for each treatment, or conduct an experiment that accounts for the other factors affecting the response (as is to be done in Section 10.5). Do not automatically reach the former conclusion, since the possibility of a Type II error must be considered if you accept H_0.

6. If desired, conduct the *F*-test of the null hypothesis that the block means are equal. Rejection of this hypothesis lends statistical support to the utilization of the randomized block design.

Note: It is often difficult to check whether the assumptions for a randomized block design are satisfied. When you feel that these assumptions are likely to be violated, a nonparametric procedure is advisable.

What Do You Do When the Assumptions Are Not Satisfied for the Analysis of Variance for a Completely Randomized Design?

Answer: Use a nonparametric statistical method such as the Friedman F_r test of Section 14.6.

Statistics IN Action Revisited

A Matched-Pairs Analysis of the Voice/Face Recognition Data in Experiment 2

Refer, once more, to the *British Journal of Psychology* (Feb. 2014) experiment designed to test the theory of associative priming using faces and voices of celebrity pairs. Recall that in Experiment 1 subjects were presented with a prime celebrity's face or voice for 3 seconds and then asked to identify the face of a second, target celebrity. In the cross-modality group of subjects (prime voice, target face), researchers were concerned that not enough time was allowed for listening to the prime voice. Consequently, a second experiment was carried out in which 20 (different) subjects in the cross-modality condition were given more time (8 seconds) to listen to the prime celebrity's voice. Unlike in the first experiment, these subjects were presented with *both* related celebrity pairs and unrelated celebrity pairs. Again, the two dependent variables measured were accuracy rate and response speed.

A diagram of Experiment 2 is shown in Figure SIA10.2 (p. 531) and the data (saved in the **PRIME2** file) shown in Table SIA10.1. Since measurements were recorded for both related celebrity pairs and unrelated celebrity pairs for all 20 subjects, this is considered a matched-pairs design. We learned in this section that a matched-pairs design is a randomized block design with just two treatments. Here, the two treatments are related and unrelated celebrity pairs and the blocks are the 20 subjects. With this design, the researchers'

objective is to determine if the mean responses for related and unrelated pairs differ in the cross-modality condition, after accounting (blocking) for subject-to-subject variation.

We used MINITAB to carry out the analysis of variance of the data saved in the **PRIME2** file. For both of the dependent variables, the null hypothesis of no treatment differences can be stated as: $H_0: \mu_{\text{Related}} = \mu_{\text{Unrelated}}$. The randomized block ANOVA results for accuracy rate and response speed are shown in Figures SIA10.7 and SIA10.8, respectively. The *p*-values for the ANOVA *F*-test for treatments (related/unrelated pairs) are highlighted on the printouts.

For comparing mean accuracy rates (Figure SIA10.7), the results are consistent with those from Experiment 1. At $\alpha = .05$, *p*-value = .358 implies no significant difference between the related pair and unrelated pair means. For comparing mean response speeds (Figure SIA10.8) using $\alpha = .05$, *p*-value = .000; thus, there is evidence of a difference between the related pair and unrelated pair population response speed means. The two sample response speed means (highlighted)

(Continued)

Table SIA10.1 Data for Experiment 2

| | Accuracy Rate (%) | | Response Speed (seconds) | |
Subject	Related	Unrelated	Related	Unrelated
1	95	92	1250	1346
2	93	93	1126	1244
3	94	87	1099	1171
4	91	89	1039	1195
5	87	85	1178	1200
6	94	92	1129	1189
7	91	89	1133	1266
8	87	87	1227	1189
9	88	86	1066	1201
10	89	96	1191	1202
11	91	95	1056	1275
12	92	90	917	1388
13	87	87	1119	1170
14	85	87	756	1200
15	86	85	1013	1070
16	98	94	1061	1302
17	85	87	1044	1405
18	92	90	1076	1134
19	94	97	1048	1192
20	91	89	1093	1121

Ⓓ PRIME2

Two-way ANOVA: ACCURACY versus PAIR, SUBJECT

```
Source   DF      SS       MS      F     P
PAIR      1    4.225   4.2250   0.89  0.358
SUBJECT  19  417.275  21.9618   4.62  0.001
Error    19   90.275   4.7513
Total    39  511.775


S = 2.180   R-Sq = 82.36%   R-Sq(adj) = 63.79%



                        Individual 95% CIs For Mean Based on Pooled StDev
PAIR        Mean    -+---------+---------+---------+--------
RELATED     90.50                  (--------------*-------------)
UNRELATED   89.85    (--------------*-------------)
                    -+---------+---------+---------+--------
                    88.90     89.60     90.30     91.00
```

Figure SIA10.7
MINITAB Randomized Block
ANOVA for Accuracy Rate

Two-way ANOVA: SPEED versus PAIR, SUBJECT

```
Source   DF     SS       MS      F      P
PAIR      1  201498   201498   20.29  0.000
SUBJECT  19  174322     9175    0.92  0.567
Error    19  188650     9929
Total    39  564471


S = 99.64   R-Sq = 66.58%   R-Sq(adj) = 31.40%



                    Individual 95% CIs For Mean Based on
                    Pooled StDev
PAIR        Mean    --------+---------+---------+---------+---------+-
RELATED   1081.05   (-------*-------)
UNRELATED 1223.00                               (-------*-------)
                    --------+---------+---------+---------+---------+-
                         1080      1140      1200      1260
```

Figure SIA10.8
MINITAB Randomized Block
ANOVA for Response Speed

indicate that subjects respond faster, on average, when presented with related celebrity pairs than with unrelated pairs.

Recall that in the previous experiment, no evidence of a difference in mean response speed was found in the cross-modality (prime voice, target face) conditions (i.e., groups 3 and 4). With this randomized block design, blocking on subjects, the difference is detected. And this difference supports the theory of associative priming since subjects presented with associated (related) pairs are faster at recognizing the target celebrity than subjects presented with unrelated pairs.

Exercises 10.65–10.84

Understanding the Principles

10.65 Between a completely randomized design and a randomized block design, which one is more efficient?

10.66 What do you do when the assumptions are not satisfied for the analysis of variance for a completely randomized design?

10.67 What conditions are required for valid inferences from a randomized block ANOVA design?

Learning the Mechanics

10.68 A randomized block design yielded the ANOVA table shown here:

Source	df	SS	MS	F
Treatments	5	501	375.9	9.112
Blocks	3	225	337.5	8.182
Error	8	110	41.25	
Total	18	836		

a. How many blocks and treatments were used in this experiment?

b. How many observations were collected in the experiment?

c. Specify the null and alternative hypotheses you would use to compare the treatment means.

d. What test statistic should be used to conduct the hypothesis test of part **c**?

e. Specify the rejection region for the test of parts **c** and **d**. Use $\alpha = .01$.

f. Conduct the test of parts **c–e**, and state the proper conclusion.

g. What assumptions are necessary to ensure the validity of the test you conducted in part **f**?

10.69 An experiment was conducted that used a randomized block design. The data from the experiment are displayed in the following table.

	Block		
Treatment	1	2	3
1	2	3	5
2	8	6	7
3	7	6	5

a. Fill in the missing entries in the following ANOVA table:

Source	df	SS	MS	F
Treatments		21.5555		
Blocks				
Error				
Total		30.2222		

b. Specify the null and alternative hypotheses you would use to investigate whether a difference exists among the treatment means.

c. What test statistic should be used in conducting the test of part **b**?

d. Describe the Type I and Type II errors associated with the hypothesis test of part **b**.

e. Conduct the hypothesis test of part **b**, using $\alpha = .05$.

10.70 Suppose an experiment utilizing a randomized block design has four treatments and nine blocks, for a total of $4 \times 9 = 36$ observations. Assume that the total sum of squares for the response is SS(Total) = 500. For each of the following partitions of SS(Total), test the null hypothesis that the treatment means are equal and the null hypothesis that the block means are equal (use $\alpha = .05$ for each test):

a. The sum of squares for treatments (SST) is 30% of SS(Total), and the sum of squares for blocks (SSB) is 20% of SS(Total).

b. SST is 20% of SS(Total), and SSB is 50% of SS(Total).

c. SST is 50% of SS(Total), and SSB is 20% of SS(Total).

d. SST is 40% of SS(Total), and SSB is 30% of SS(Total).

e. SST is 50% of SS(Total), and SSB is 20% of SS(Total).

10.71 A randomized block design was used to compare the mean responses for three treatments. Four blocks of three homogeneous experimental units were selected, and each treatment was randomly assigned to one experimental unit within each block. The data are shown in the following table, and a MINITAB ANOVA printout for this experiment is displayed below.

	Block			
Treatment	1	2	3	4
A	3.4	5.5	7.9	1.3
B	4.4	5.8	9.6	2.8
C	2.2	3.4	6.9	.3

MINITAB output for Exercise 10.71

```
Analysis of Variance for X

Source        DF     SS        MS      F       P
Treatment      2   12.032    6.016   50.96   0.000
Block          3   71.749   23.916  202.59   0.000
Error          6    0.708    0.118
Total         11   84.489


S = 0.343592   R-Sq = 99.16%   R-Sq(adj) = 98.46%


Grouping Information Using Tukey Method and 95.0% Confidence

Treatment  N  Mean  Grouping
B          4   5.7  A
A          4   4.5    B
C          4   3.2      C

Means that do not share a letter are significantly different.


Tukey 95.0% Simultaneous Confidence Intervals
Response Variable X
All Pairwise Comparisons among Levels of Treatment
Treatment = A  subtracted from:

Treatment   Lower   Center   Upper
B           0.379   1.125    1.8706
C          -2.071  -1.325   -0.5794
```

```
Treatment = B  subtracted from:

Treatment   Lower   Center   Upper
C          -3.196  -2.450   -1.704
```

a. Use the printout to fill in the entries in the following ANOVA table:

Source	df	SS	MS	F
Treatments				
Blocks				
Error				
Total				

b. Do the data provide sufficient evidence to indicate that the treatment means differ?

c. Do the data provide sufficient evidence to indicate that blocking was effective in reducing the experimental error? Use $\alpha = .05$.

d. Use the MINITAB printout above to rank the treatment means at $\alpha = .05$.

e. What assumptions are necessary to ensure the validity of the inferences made in parts **b**, **c**, and **d**?

Applying the Concepts—Basic

10.72 Corrosion prevention of buried steel structures. Refer to the *Materials Performance* (Mar. 2013) study to compare two tests for steel corrosion of underground piping, Exercise 1.17 (p. 48). The two tests, which engineers call "instant-off" and "instant-on" potential, were applied to buried piping at a petrochemical plant in Turkey. Recall

that both the "instant-off" and "instant-on" tests were used to make predictions of corrosion at each of 19 different randomly selected pipe locations. The researchers want to compare the mean accuracy of the corrosion predictions for the two tests in order to determine if one test is more desirable than the other when applied to buried steel piping.
a. What are the experimental units for this study?
b. What type of design is employed? Identify the features (e.g., treatments, blocks) of the design.
c. What is the dependent (response) variable of interest?
d. Illustrate the layout of this design using a graphic similar to Figure 10.21.
e. State H_0 and H_a in symbols.

10.73 Making high-stakes insurance decisions. The *Journal of Economic Psychology* (Sept. 2008) published a study on high-stakes insurance decisions. In part A of the experiment, 84 subjects were informed of the hazards (both fire and theft) of owning a valuable painting, but were not told the exact probabilities of the hazards occurring. The subjects then provided an amount they were willing to pay (WTP) for insuring the painting. In part B of the experiment, these same subjects were informed of the exact probabilities of the hazards (fire and theft) of owning a valuable sculpture. The subjects then provided a WTP amount for insuring the sculpture. The researchers were interested in comparing the mean WTP amounts for the painting and the sculpture.

a. Explain why the experimental design employed is a randomized block design.

b. Identify the dependent (response) variable, treatments, and blocks for the design.

c. Give the null and alternative hypotheses of interest to the researchers.

10.74 Solar energy generation along highways. The potential of solar panels on roofs built above national highways as a source of solar energy was investigated in the *International Journal of Energy and Environmental Engineering* (Dec. 2013). Computer simulation was used to estimate the monthly solar energy (kilowatt-hours) generated from solar panels installed across a 200-kilometer stretch of highway in India. Each month, the simulation was run under each of four conditions: single-layer solar panels, double-layer solar panels 1 meter apart, double-layer solar panels 2 meters apart, and double-layer solar panels 3 meters apart. The data for 12 months are shown in the table. In order to compare the mean solar energy values generated by the four panel configurations, a randomized block design ANOVA was conducted. A MINITAB printout of the analysis is provided below.

| Month | Single Layer | Double Layer | | |
		1-meter	2-meters	3-meters
January	7,308	8,917	9,875	10,196
February	6,984	8,658	9,862	9,765
March	7,874	9,227	11,092	11,861
April	7,328	7,930	9,287	10,343
May	7,089	7,605	8,422	9,110
June	5,730	6,350	7,069	7,536
July	4,531	5,120	5,783	6,179
August	4,587	5,171	5,933	6,422
September	5,985	6,862	8,208	8,925
October	7,051	8,608	10,008	10,239
November	6,724	8,264	9,238	9,334
December	6,883	8,297	9,144	9,808

Source: Sharma, P., and Harinarayana, T. "Solar energy generation potential along national highways." *International Journal of Energy and Environmental Engineering*, Vol. 49, No. 1, Dec. 2013 (Table 3).

Two-way ANOVA: Energy versus Condition, Month

```
Source      DF        SS        MS       F      P
Condition    3  49730750  16576917  115.54  0.000
Month       11  90618107   8238010   57.42  0.000
Error       33   4734730    143477
Total       47 145083587

S = 378.8   R-Sq = 96.74%   R-Sq(adj) = 95.35%
```

a. Identify the dependent variable, treatments, and blocks for this experiment.

b. What null hypothesis would you test in order to compare the mean solar energy values generated by the four panel configurations?

c. Carry out the test, part **b**. Give the *F*-value and associated *p*-value.

d. What conclusion can you draw from the analysis?

10.75 "Topsy-turvy" seasons in college football. Each week during the college football season the Associated Press (AP) ranks the top 25 Division I college football teams based on voting by sportswriters. Many recent upsets of top-rated teams have led to major changes in the weekly AP Poll over the past several seasons. Statisticians have created a formula for determining a weekly "topsy-turvy" (TT) index, designed to measure the degree to which the top 25 ranked teams changed from the previous week (*Chance*, Summer 2009). The greater the TT index, the greater the changes in the ranked teams. The statisticians calculated the TT index each week of the 15-week college football season for 6 recent seasons. In order to determine whether any of the 15 weeks in a season tends to be more or less topsy-turvy than others, they conducted an ANOVA on the data using a randomized block design. Here, weeks were considered the treatments and seasons were the blocks. The ANOVA summary table is shown below.

Source	df	SS	MS	F	p-Value
Weeks	14	3,562.3	254.4	2.57	.0044
Seasons	5	2,936.5	587.3	5.94	.0001
Error	70	6,914.4	98.8		
Total	89	13,413.1			

Source: Kaw, A. K., and Yalcin, A. "A metric to quantify college football's topsy-turvy season." *Chance*, Vol. 22, No. 3, Summer 2009 (Table 2). Reprinted with permission from *Chance*. Copyright 2009 by the American Statistical Association. All rights reserved.

a. Is there a significant difference (at $\alpha = .01$) in the mean TT index across the 15 weeks of the college football season?

b. Is there evidence (at $\alpha = .01$) that blocking on seasons was effective in removing an extraneous source of variation in the data?

c. Tukey's multiple comparisons method was applied to compare the mean TT index across the 15 weeks. How many pairwise comparisons are involved in this analysis?

d. Using an experimentwise error rate of .05, Tukey's analysis revealed that the last week of the regular season (week 14), when only conference championships are played, had a significantly smaller mean TT index than week 6. All other pairs of means were not significantly different. Use this result to make a comment about whether any of the 15 weeks in a college football season tend to be more or less topsy-turvy than others.

10.76 Treatment for tendon pain. Refer to the *British Journal of Sports Medicine* (Feb. 1, 2004) study of chronic Achilles tendon pain, presented in Exercise 10.12 (p. 538). Recall that each in a sample of 25 patients with chronic Achilles tendinosis was treated with heavy-load eccentric calf muscle training. Tendon thickness (in millimeters) was measured both before and following the treatment of each patient. These experimental data are shown in the next table.

a. What experimental design was employed in this study?

b. How many treatments are used in this study? How many blocks?

c. Give the null and alternative hypothesis for determining whether the mean tendon thickness before treatment differs from the mean after treatment.

d. Compute the test statistic for the test in part **c**, using the paired difference t formula of Section 9.3 .

e. Use statistical software (or the formulas in Appendix B) to compute the ANOVA F-statistic for the test in part **c**.

f. Compare the test statistics you computed in parts **d** and **e**. You should find that $F = t^2$.

g. Show that the two tests from parts **d** and **e** yield identical p-values.

h. Give the appropriate conclusion, using $\alpha = .05$.

Patient	Before Thickness (millimeters)	After Thickness (millimeters)
1	11.0	11.5
2	4.0	6.4
3	6.3	6.1
4	12.0	10.0
5	18.2	14.7
6	9.2	7.3
7	7.5	6.1
8	7.1	6.4
9	7.2	5.7
10	6.7	6.5
11	14.2	13.2
12	7.3	7.5
13	9.7	7.4
14	9.5	7.2
15	5.6	6.3
16	8.7	6.0
17	6.7	7.3
18	10.2	7.0
19	6.6	5.3
20	11.2	9.0
21	8.6	6.6
22	6.1	6.3
23	10.3	7.2
24	7.0	7.2
25	12.0	8.0

Based on Ohberg, L., et al. "Eccentric training in patients with chronic Achilles tendinosis: Normalized tendon structure and decreased thickness at follow up." *British Journal of Sports Medicine*, Vol. 38, No. 1, Feb. 1, 2004 (Table 2).

10.77 Interactive video games and physical fitness. Wii Fit is an
[NW] interactive video game marketed to consumers who want to increase their physical fitness level. The effectiveness of Wii Fit activities relative to other physical activities was investigated in the *Journal of Physical Activity and Health* (Vol. 7, 2010). A sample of 15 young adults (ages 21–38 years) participated in the study. Each adult completed a total of eight activities—rest, handheld gaming, Wii yoga, Wii muscle conditioning, Wii balance, Wii aerobics, brisk treadmill walking, and treadmill jogging. At the end of each session, the heart rate (beats per minute) of the participant was recorded. Since the goal is to compare mean heart rates across the eight activities and since each adult completed all activities, a randomized block design ANOVA was conducted.

a. Identify the treatments for this experiment.

b. Identify the blocks for this experiment.

c. The ANOVA F-test for treatments resulted in a p-value of .001. Interpret this result using $\alpha = .01$.

d. Multiple comparisons (with an experimentwise error rate of .05) of the mean heart rates for the eight activities revealed the following. (*Note:* Means with the same letter are not significantly different.) Provide a ranking of the treatment means.

	Mean Heart Rate
Rest	58.3 a
Handheld gaming	65.6 a
Wii balance	76.7 b
Wii yoga	77.6 b
Wii muscle conditioning	82.4 b
Wii aerobics	94.5 b
Brisk treadmill walking	108.3 c
Treadmill jogging	153.7 d

Applying the Concepts—Intermediate

10.78 Evaluating health care research reports. When evaluat-
[D] ing research reports in health care, a popular tool is the
RAMSTAR Assessment of Multiple Systematic Reviews (AMSTAR). AMSTAR, which incorporates 11 items (questions), has been widely accepted by professional health associations. A group of dental researchers have revised the assessment tool and named it R-AMSTAR (*The Open Dentistry Journal*, Vol. 4, 2010). The revised assessment tool was validated on five systematic reviews (named R1, R2, R3, R4, and R5) on rheumatoid arthritis. For each review, scores on the 11 R-AMSTAR items (all measured on a 4-point scale) were obtained. The data, are shown in the accompanying table.

Data for Exercise 10.78

Review	Item 1	Item 2	Item 3	Item 4	Item 5	Item 6	Item 7	Item 8	Item 9	Item 10	Item 11
R1	4.0	1.0	4.0	2.0	3.5	3.5	3.5	3.5	1.0	1.0	1.0
R2	3.5	2.5	4.0	4.0	3.5	4.0	3.5	2.5	3.5	1.5	1.0
R3	4.0	4.0	3.5	4.0	1.5	2.5	3.5	3.5	2.5	1.5	1.0
R4	3.5	2.0	4.0	4.0	2.0	4.0	3.5	3.0	3.5	1.0	1.0
R5	3.5	4.0	4.0	3.0	2.5	4.0	4.0	4.0	2.5	1.0	2.5

Source: Kung, J., Chiappelli, F., Cajulis, O., Avezona, R., Kossan, G., Chew, L., and Maida, C. A. "From systematic reviews to clinical recommendations to clinical-based health care: Validation of revised assessment of multiple-systematic reviews (R-AMSTAR) for grading of clinical relevance," *The Open Dentistry Journal*, Vol. 4, 2010, pp. 84–91.

SAS output for Exercise 10.78

The ANOVA Procedure

Dependent Variable: SCORE

Source	DF	Sum of Squares	Mean Square	F Value	Pr > F
Model	14	46.01818182	3.28701299	6.36	<.0001
Error	40	20.68181818	0.51704545		
Corrected Total	54	66.70000000			

R-Square	Coeff Var	Root MSE	SCORE Mean
0.689928	24.79513	0.719059	2.900000

Source	DF	Anova SS	Mean Square	F Value	Pr > F
REVIEW	4	2.51818182	0.62954545	1.22	0.3186
ITEM	10	43.50000000	4.35000000	8.41	<.0001

Means with the same letter are not significantly different.

Tukey Grouping		Mean	N	REVIEW
A		3.1818	11	R5
A				
A		3.0455	11	R2
A				
A		2.8636	11	R3
A				
A		2.8636	11	R4
A				
A		2.5455	11	R1

a. One goal of the study was to compare the mean item scores of the five reviews. Set up the null and alternative hypotheses for this test.

b. Examine the data in the table and explain why a randomized block ANOVA is appropriate to apply.

c. The SAS output for a randomized block ANOVA of the data (with Review as treatments and Item as blocks) appears above. Interpret the p-values of the tests shown.

d. The SAS printout also reports the results of a Tukey multiple-comparison analysis of the five Review means. Which pairs of means are significantly different? Do these results agree with your conclusion in part **c**?

e. The experimentwise error rate used in the analysis in part **d** is .05. Interpret this value.

10.79 Stress in cows prior to slaughter. What is the level of stress (if any) that cows undergo prior to being slaughtered? To answer this question, researchers designed an experiment involving cows bred in Normandy, France (*Applied Animal Behaviour Science,* June 2010). The heart rate (beats per minute) of a cow was measured at four different pre-slaughter phases—(1) first phase of visual contact with pen mates, (2) initial isolation from pen mates for prepping, (3) restoration of visual contact with pen mates, and (4) first contact with human prior to slaughter. Data for eight cows (simulated from information provided in the article) are shown in the next table. The researchers analyzed the data using an analysis of variance for a randomized block design. Their objective was to determine whether the mean heart rate of cows differed in the four pre-slaughter phases.

a. Identify the treatments and blocks for this experimental design.

b. Conduct the appropriate analysis using a statistical software package. Summarize the results in an ANOVA table.

c. Is there evidence of differences among the mean heart rates of cows in the four pre-slaughter phases? Test using $\alpha = .05$.

d. If warranted, conduct a multiple comparisons procedure to rank the four treatment means. Use an experiment wise error rate of $\alpha = .05$.

Cow	Phase			
	1	2	3	4
1	124	124	109	107
2	100	98	98	99
3	103	98	100	106
4	94	91	98	95
5	122	109	114	115
6	103	92	100	106
7	98	80	99	103
8	120	84	107	110

10.80 Taste-testing scales. When comparing the palatability of food items, numerical taste test scales are employed. For example, the Hedonic 9-point scale uses the following ratings: 1 = dislike extremely, 2 = dislike very much, 3 = dislike moderately, 4 = dislike slightly, 5 = neither like nor dislike, 6 = like slightly, 7 = like moderately, 8 = like very much, and 9 = like extremely. Another scale, called the general Labeled Magnitude Scale (gLMS), uses ratings that range from −100 (for strongest imaginable dislike) to +100 (for strongest imaginable like). Both of these scales were applied to food items in a study published in the *Journal of Food Science* (Feb. 2014). Students and staff at the University of Florida taste-tested five food/beverage

SAS Output for Exercise 10.80

Dependent Variable = Hedonic 9-point Scale

The GLM Procedure

Source	DF	Sum of Squares	Mean Square	F Value	Pr > F
Model	203	3823.677000	18.835847	9.66	<.0001
Error	796	1552.114000	1.949892		
Corrected Total	999	5375.791000			

R-Square	Coeff Var	Root MSE	HG Mean
0.711277	25.60765	1.396385	5.453000

Source	DF	Type I SS	Mean Square	F Value	Pr > F
PRODUCT	4	3386.686000	846.671500	434.21	<.0001
TASTER	199	436.991000	2.195935	1.13	0.1369

Source	DF	Type III SS	Mean Square	F Value	Pr > F
PRODUCT	4	3386.686000	846.671500	434.21	<.0001
TASTER	199	436.991000	2.195935	1.13	0.1369

items: black coffee (BC), cheesecake (CC), grapefruit juice (GF), orange juice (OJ), and pepperoni (PP). Half ($n = 200$) of the subjects used the 9-point scale, and half ($n = 200$) used gLMS to rate each of the food items. (These ratings, simulated from information provided in the article, are saved in the **TASTE** file.) One of the objectives of the research was to compare the average 9-point taste ratings of the five food/beverage items.

a. Explain why these data should be analyzed using an ANOVA for a randomized block design. Identify the treatments, blocks, and dependent variable for this design.

b. An SAS printout of the ANOVA is shown above, and a multiple comparisons of means summary is shown below. Interpret the results practically.

c. Refer to part **b**. Perform a similar analysis for the data on the gLMS rating scale. Again, interpret the results practically.

Mean Rating:	3.1	3.4	6.6	6.8	7.4
Food Item:	BC	GF	PP	OJ	CC

10.81 Irrelevant facial similarity effects on judgment. Facial similarity to previously encountered persons can be a potential source of bias in decision making. For example, in a job interview, the interviewer should judge the job candidate solely on his/her qualifications, not on whether the candidate looks similar to a previously rejected candidate. Research published in *Experimental Psychology* (Jan. 2014) investigated irrelevant facial similarity effects on judgment. Subjects (university students) were trained to evaluate employees' suitability for a job on a scale of

0 (not at all suitable) to 100 (very suitable). Then each subject rated each of three different job applicants. Each subject saw a picture of the applicant as well as a summary of the employee's application documents. Unknown to the subject, all the applicants were given average qualifications, but one applicant's picture was morphed to look similar to a candidate who had been given a low rating during training (low-performance morph), and another was morphed to look similar to a candidate who had been given a high rating during training (high-performance morph). The third candidate was given a neutral photo—one which had not been seen before by the subjects. Suitability data (simulated based on information provided in the article) for 10 subjects rating each of the 3 candidates are provided in the table. Are there differences in the mean ratings of the 3 candidates? If so, which candidate (low, neutral, or high) received the highest mean rating?

Subject	Low	Neutral	High
1	52	55	57
2	46	55	49
3	61	61	64
4	52	45	53
5	26	25	30
6	43	47	50
7	47	48	44
8	54	56	59
9	77	82	91
10	41	50	49

10.82 Effect of massage on boxers. Eight amateur boxers participated in an experiment to investigate the effect of massage on boxing performance (*British Journal of Sports Medicine*,

Apr. 2000). The punching power of each boxer (measured in newtons) was recorded in the round following each of four different interventions: (M1) in round 1, following a pre-bout sports massage; (R1) in round 1, following a pre-bout period of rest; (M5) in round 5, following a sports massage between rounds; and (R5) in round 5, following a period of rest between rounds. Based on information provided in the article, the data in the accompanying table were obtained. The main goal of the experiment was to compare the punching power means of the four interventions.

	Intervention			
Boxer	M1	R1	M5	R5
1	1,243	1,244	1,291	1,262
2	1,147	1,053	1,169	1,177
3	1,247	1,375	1,309	1,321
4	1,274	1,235	1,290	1,285
5	1,177	1,139	1,233	1,238
6	1,336	1,313	1,366	1,362
7	1,238	1,279	1,275	1,261
8	1,261	1,152	1,289	1,266

Based on Hemmings, B., Smith, M., Graydon, J., and Dyson, R. "Effects of massage on physiological restoration, perceived recovery, and repeated sports performance." *British Journal of Sports Medicine*, Vol. 34, No. 2, Apr. 2000 (adapted from Table 3).

a. Set up H_0 and H_a for this analysis.

b. Identify the treatments and blocks in the experiment.

c. Conduct the test set up in part **a**. What conclusions can you draw regarding the effect of massage on punching power?

10.83 Plants and stress reduction. Plant therapists believe that plants can reduce the stress levels of humans. A Kansas State University study was conducted to investigate this phenomenon. Two weeks prior to final exams, 10 undergraduate students took part in an experiment to determine what effect the presence of a live plant, a photo of a plant, or the absence of a plant has on the student's ability to relax while isolated in a dimly lit room. Each student participated in three sessions: one with a live plant, one with a photo of a plant, and one with no plant (the control).* During each session, finger temperature was measured at one-minute intervals for 20 minutes. Since increasing finger temperature indicates an increased level of relaxation, the maximum temperature (in degrees) was used as the response variable. For example, one student's finger measured 95.6° in the "Live Plant" condition, 92.6° in the "Plant Photo" condition, and 96.6° in the "No Plant" condition. The temperatures

under the three conditions for the other nine students follow: Student 2 (95.6°, 94.8°, 96.0°); Student 3 (96.0°, 97.2°, 96.2°); Student 4 (95.2°, 94.6°, 95.7°); Student 5 (96.7°, 95.5°, 94.8°); Student 6 (96.0°, 96.6°, 93.5°); Student 7 (93.7, 96.2°, 96.7°); Student 8 (97.0°, 95.8°, 95.4°); Student 9 (94.9°, 96.6°, 90.5°); Student 10 (91.4°, 93.5°, 96.6°). (These data are saved in the **PLANTS** file.) Conduct an ANOVA and make the proper inferences at $\alpha = .10$.

Based on data from Elizabeth Schreiber, Department of Statistics, Kansas State University, Manhattan, Kansas.

Applying the Concepts—Advanced

10.84 Listening ability of infants. *Science* (Jan. 1, 1999) reported on the ability of seven-month-old infants to learn an unfamiliar language. In one experiment, 16 infants were trained in an artificial language. Then each infant was presented with two 3-word sentences that consisted entirely of new words (e.g., "wo fe wo"). One sentence was consistent (i.e., constructed from the same grammar the infants got in the training session), and one sentence was inconsistent (i.e., constructed from grammar in which the infant was not trained). The variable measured in each trial was the time (in seconds) the infant spent listening to the speaker, with the goal being to compare the mean listening times of consistent and inconsistent sentences.

a. The data were analyzed as a randomized block design with the 16 infants representing the blocks and the two types of sentences (consistent and inconsistent) representing the treatments. Do you agree with this data analysis method? Explain.

b. Refer to part **a**. The test statistic for testing treatments was $F = 25.7$ with an associated observed significance level of $p < .001$. Interpret this result.

c. Explain why the data could also be analyzed as a paired difference experiment with a test statistic of $t = 5.07$.

d. The mean listening times and standard deviations for the two treatments are given here. Use this information to calculate the F-statistic for comparing the treatment means in a completely randomized ANOVA design. Explain why this test statistic provides weaker evidence of a difference between treatment means than the test in part **b** provides.

	Consistent Sentences	Inconsistent Sentences
Mean	6.3	9.0
Standard deviation	2.6	2.16

e. Explain why there is no need to control the experimentwise error rate in ranking the treatment means for this experiment.

*The experiment is simplified for this exercise. The actual experiment involved 30 students who participated in 12 sessions.

10.5   Factorial Experiments: Two Factors

All the experiments discussed in Sections 10.2 and 10.4 were **single-factor experiments**. The treatments were levels of a single factor, with the sampling of experimental units performed with either a completely randomized or a randomized block design. However, most responses are affected by more than one factor, and we will therefore often wish to design experiments involving more than one factor.

Consider an experiment in which the effects of two factors on the response are being investigated. Assume that factor A is to be investigated at a levels and factor B at b levels. Recalling that treatments are factor–level combinations, you can see that the experiment has, potentially, ab treatments that could be included in it. A *complete factorial experiment* is a factorial experiment in which all possible ab treatments are utilized.

> **A complete factorial experiment** is a factorial experiment in which every factor–level combination is utilized. That is, the number of treatments in the experiment equals the total number of factor–level combinations.

For example, suppose the USGA wants to determine the relationship not only between distance and brand of golf ball, but also between distance and the club used to hit the ball. If the USGA decides to use four brands and two clubs (say, driver and five-iron) in the experiment, then a complete factorial experiment would call for utilizing all $4 \times 2 = 8$ Brand–Club combinations. This experiment is referred to more specifically as a *complete 4×2 factorial*. A layout for a two-factor factorial experiment (henceforth, we use the term *factorial* to refer to a *complete factorial*) is given in Table 10.12. The factorial experiment is also referred to as a **two-way classification** because it can be arranged in the row-column format exhibited in Table 10.12.

Table 10.12 Schematic Layout of Two-Factor Factorial Experiment

	Level	Factor B at b Levels				
		1	2	3	...	b
Factor A	1	Trt. 1	Trt. 2	Trt. 3	...	Trt. b
at a levels	2	Trt. $b + 1$	Trt. $b + 2$	Trt. $b + 3$	...	Trt. $2b$
	3	Trt. $2b + 1$	Trt. $2b + 2$	Trt. $2b + 3$	...	Trt. $3b$
	⋮	⋮	⋮	⋮	...	⋮
	a	Trt. $(a-1)b + 1$	Trt. $(a-1)b + 2$	Trt. $(a-1)b + 3$	...	Trt. ab

In order to complete the specification of the experimental design, the treatments must be assigned to the experimental units. If the assignment of the ab treatments in the factorial experiment is random and independent, the design is completely randomized. For example, if the machine Iron Byron is used to hit 80 golf balls, 10 for each of the eight Brand–Club combinations, in a random sequence, the design would be completely randomized. By contrast, if the assignment is made within homogeneous blocks of experimental units, then the design is a randomized block design. For example, if 10 golfers are employed to hit each of the eight golf balls, and each golfer hits all eight Brand–Club combinations in a random sequence, then the design is a randomized block design, with the golfers serving as blocks. In the remainder of this section, we confine our attention to factorial experiments utilizing completely randomized designs.

If we utilize a completely randomized design to conduct a factorial experiment with ab treatments, we can proceed with the analysis in exactly the same way as we did in Section 10.2. That is, we calculate (or let the computer calculate) the measure of treatment mean variability (MST) and the measure of sampling variability (MSE) and use the F-ratio of these two quantities to test the null hypothesis that the treatment means are equal. However, if this hypothesis is rejected because we conclude that some differences exist among the treatment means, important questions remain: Are both factors affecting the response, or only one? If both, do they affect the response independently, or do they interact to affect the response?

For example, suppose the distance data indicate that at least two of the eight treatment (Brand–Club combinations) means differ in the golf experiment. Does the brand of ball (factor A) or the club utilized (factor B) affect mean distance, or do both affect it? Several possibilities are shown in Figure 10.25. In Figure 10.25a, the brand means are equal (only three are shown for the purpose of illustration), but the distances differ for the two levels of factor B (Club). Thus, there is no effect of Brand on distance, but a Club main effect is present. In Figure 10.25b, the Brand means differ, but the Club means are equal for each Brand. Here, a Brand main effect is present, but no effect of Club is present.

a. No A effect; B main effect

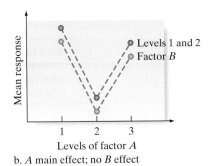

b. A main effect; no B effect

c. A and B main effects; no interaction

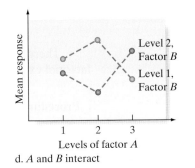

d. A and B interact

Figure 10.25

Illustration of possible treatment effects: factorial experiment

Figure 10.25c and Figure 10.25d illustrate cases in which both factors affect the response. In Figure 10.25c, the mean distance between clubs does not change for the three Brands, so the effect of Brand on distance is independent of Club. That is, the two factors Brand and Club *do not interact*. In contrast, Figure 10.25d shows that the difference between mean distances between clubs varies with Brand. Thus, the effect of Brand on distance depends on Club, and the two factors *do interact*.

In order to determine the nature of the treatment effect, if any, on the response in a factorial experiment, we need to break the treatment variability into three components: Interaction between factors A and B, Main effect of factor A, and Main effect of factor B. The **Factor interaction** component is used to test whether the factors combine to affect the response, while the **Factor main effect** components are used to determine whether the factors affect the response separately.

Now Work Exercise 10.99

The partitioning of the total sum of squares into its various components is illustrated in Figure 10.26. Notice that at stage 1 the components are identical to those in the one-factor, completely randomized designs of Section 10.2: The sums of squares for treatment and error sum to the total sum of squares. The degrees of freedom for treatments is equal to $(ab - 1)$, one less than the number of treatments. The degrees of freedom for error is equal to $(n - ab)$, the total sample size minus the number of treatments. Only at stage 2 of the partitioning does the factorial experiment differ from those previously discussed. Here, we divide the treatment sum of squares into its three components: interaction and the two main effects. These components can then be used to test the nature of the differences, if any, among the treatment means.

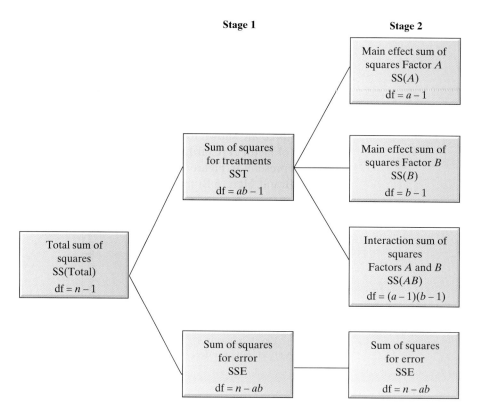

Figure 10.26

Partitioning the total sum of squares for a two-factor factorial

There are a number of ways to proceed in the testing and estimation of factors in a factorial experiment. We present one approach in the following box:

Procedure for Analysis of Two-Factor Factorial Experiment

1. Partition the total sum of squares into the treatment and error components (stage 1 of Figure 10.26). Use either a statistical software package or the calculation formulas in Appendix B to accomplish the partitioning.

2. Use the F-ratio of the mean square for treatments to the mean square for error to test the null hypothesis that the treatment means are equal.*
   a. If the test results in nonrejection of the null hypothesis, consider refining the experiment by increasing the number of replications or introducing other factors. Also, consider the possibility that the response is unrelated to the two factors.
   b. If the test results in rejection of the null hypothesis, then proceed to step 3.

3. Partition the treatment sum of squares into the main effect and the interaction sum of squares (stage 2 of Figure 10.26). Use either a statistical software package or the calculation formulas in Appendix B to accomplish the partitioning.

4. Test the null hypothesis that factors A and B do not interact to affect the response by computing the F-ratio of the mean square for interaction to the mean square for error.
   a. If the test results in nonrejection of the null hypothesis, proceed to step 5.
   b. If the test results in rejection of the null hypothesis, conclude that the two factors interact to affect the mean response. Then proceed to step 6a.

5. Conduct tests of two null hypotheses that the mean response is the same at each level of factor A and factor B. Compute two F-ratios by comparing the mean square for each factor main effect with the mean square for error.

---

*Some analysts prefer to proceed directly to test the interaction and main-effect components, skipping the test of treatment means. We begin with this test to be consistent with our approach in the one-factor completely randomized design.

a. If one or both tests result in rejection of the null hypothesis, conclude that the factor affects the mean response. Proceed to step 6b.

b. If both tests result in nonrejection, an apparent contradiction has occurred. Although the treatment means seemingly differ (step 2 test), the interaction (step 4) and main-effect (step 5) tests have not supported that result. Further experimentation is advised.

6. Compare the means:

a. If the test for interaction (step 4) is significant, use a multiple-comparison procedure to compare any or all pairs of the treatment means.

b. If the test for one or both main effects (step 5) is significant, use a multiple-comparison procedure to compare the pairs of means corresponding to the levels of the significant factor(s).

We assume that the completely randomized design is a **balanced design**, meaning that the same number of observations are made for each treatment. That is, we assume that r experimental units are randomly and independently selected for each treatment. The numerical value of r must exceed 1 in order to have any degrees of freedom with which to measure the sampling variability. [Note that if $r = 1$, then $n = ab$, and the number of degrees of freedom associated with error (Figure 10.26) is df $= n - ab = 0$.] The value of r is often referred to as the number of **replicates of the factorial experiment**, since we assume that all ab treatments are repeated, or replicated, r times. Whatever approach is adopted in the analysis of a factorial experiment, several tests of hypotheses are usually conducted. The tests are summarized in the following box.

Tests Conducted in Analyses of Factorial Experiments: Completely Randomized Design, r Replicates per Treatment

Test for Treatment Means

H_0: No difference among the ab treatment means

H_a: At least two treatment means differ

Test statistic: $F_c = \dfrac{\text{MST}}{\text{MSE}}$

Rejection region: $F_c \geq F_\alpha$; *p-value:* $P(F > F_c)$; where F is based on $(ab - 1)$ numerator and $(n - ab)$ denominator degrees of freedom [*Note: $n = abr$.*]

Test for Factor Interaction

H_0: Factors A and B do not interact to affect the response mean

H_a: Factors A and B do interact to affect the response mean

Test statistic: $F_c = \dfrac{\text{MS}(AB)}{\text{MSE}}$

Rejection region: $F_c \geq F_\alpha$; *p-value:* $P(F > F_c)$; where F is based on $(a - 1)(b - 1)$ numerator and $(n - ab)$ denominator degrees of freedom

Test for Main Effect of Factor A

H_0: No difference among the a mean levels of factor A

H_a: At least two factor A mean levels differ

Test statistic: $F_c = \dfrac{\text{MS}(A)}{\text{MSE}}$

Rejection region: $F_c \geq F_\alpha$; *p-value:* $P(F > F_c)$; where F is based on $(a - 1)$ numerator and $(n - ab)$ denominator degrees of freedom

(continued)

Test for Main Effect of Factor B

H_0: No difference among the b mean levels of factor B

H_a: At least two factor B mean levels differ

Test statistic: $F_c = \dfrac{MS(B)}{MSE}$

Rejection region: $F_c \geq F_\alpha$, p-value: $P(F > F_c)$; where F is based on $(b - 1)$ numerator and $(n - ab)$ denominator degrees of freedom

Conditions Required for Valid F-Tests in Factorial Experiments

1. The response distribution for each factor–level combination (treatment) is normal.
2. The response variance is constant for all treatments.
3. Random and independent samples of experimental units are associated with each treatment.

Example **10.10**
GOLFFAC1

Conducting a Factorial ANOVA— Golf Ball Driving Distances

Problem Suppose the USGA tests four different brands (A, B, C, D) of golf balls and two different clubs (driver, five-iron) in a 4×2 factorial design. Each of the eight Brand-Club combinations (treatments) is randomly and independently assigned to four experimental units, each consisting of a specific position in the sequence of hits by Iron Byron. The distance response is recorded for each of the 32 hits, and the results are shown in Table 10.13.

Table 10.13 Distance Data for 4 × 2 Factorial Golf Experiment

	Brand			
Club	A	B	C	D
Driver	226.4	238.3	240.5	219.8
	232.6	231.7	246.9	228.7
	234.0	227.7	240.3	232.9
	220.7	237.2	244.7	237.6
Five-iron	163.8	184.4	179.0	157.8
	179.4	180.6	168.0	161.8
	168.6	179.5	165.2	162.1
	173.4	186.2	156.5	160.3

Ⓓ Data Set: GOLFFAC1

a. Use statistical software to partition the total sum of squares into the components necessary to analyze this 4×2 factorial experiment.

b. Conduct the appropriate ANOVA tests and interpret the results of your analysis. Use $\alpha = .10$ for the tests you conduct.

c. If appropriate, conduct multiple comparisons of the treatment means. Use an experimentwise error rate of .10. Illustrate the comparisons with a graph.

Solution

a. The SAS printout that partitions the total sum of squares for this factorial experiment is given in Figure 10.27. The partitioning takes place in two stages. First, the total sum of squares is partitioned into the model (treatment) and error sums of squares at the top of the printout. Note that SST is 33,659.8 with 7 degrees of freedom and SSE is 822.2 with 24 degrees of freedom, adding to 34,482.0 and 31 degrees of freedom. In the second stage of partitioning, the treatment sum of squares is further divided into

Dependent Variable: DISTANCE

Source	DF	Sum of Squares	Mean Square	F Value	Pr > F
Model	7	33659.80875	4808.54411	140.35	<.0001
Error	24	822.24000	34.26000		
Corrected Total	31	34482.04875			

R-Square	Coeff Var	Root MSE	DISTANCE Mean
0.976155	2.896461	5.853204	202.0813

Source	DF	Type III SS	Mean Square	F Value	Pr > F
CLUB	1	32093.11125	32093.11125	936.75	<.0001
BRAND	3	800.73625	266.91208	7.79	0.0008
CLUB*BRAND	3	765.96125	255.32042	7.45	0.0011

Figure 10.27

SAS printout for factorial ANOVA of golf data

the main-effect and interaction sums of squares. From the highlighted values at the bottom of the printout, we see that SS(Club) is 32,093.1 with 1 degree of freedom, SS(Brand) is 800.7 with 3 degrees of freedom, and SS(Club $\times$ Brand) is 766.0 with 3 degrees of freedom, adding to 33,659.8 and 7 degrees of freedom.

b. Once partitioning is accomplished, our first test is

H_0: The eight treatment means are equal.

H_a: At least two of the eight means differ.

Test statistic: $F = \dfrac{MST}{MSE} = 140.35$ (highlighted, top of printout)

Observed significance level: $p < .0001$ (highlighted, top of printout)

Since $\alpha = .10$ exceeds p, we reject this null hypothesis and conclude that at least two of the Brand–Club combinations differ in mean distance.

After accepting the hypothesis that the treatment means differ, and therefore that the factors Brand and/or Club somehow affect the mean distance, we want to determine how the factors affect the mean response. We begin with a test of interaction between Brand and Club:

H_0: The factors Brand and Club do not interact to affect the mean response.

H_a: Brand and Club interact to affect mean response.

Test statistic: $F = \dfrac{MS(AB)}{MSE} = \dfrac{MS(\text{Brand} \times \text{Club})}{MSE}$

$= \dfrac{255.32}{34.26} = 7.45$ (highlighted, bottom of printout)

Observed significance level: $p = .0011$ (highlighted, bottom of printout)

Since $\alpha = .10$ exceeds the p-value, we conclude that the factors Brand and Club interact to affect mean distance.

Because the factors interact, we do not test the main effects for Brand and Club. Instead, we compare the treatment means in an attempt to learn the nature of the interaction.

c. Rather than compare all $8(7)/2 = 28$ pairs of treatment means, we test for differences only between pairs of brands within each club. That differences exist *between* clubs can be assumed. Therefore, only $4(3)/2 = 6$ pairs of means need to be compared for each club, or a total of 12 comparisons for the two clubs. The results of these comparisons, obtained from Tukey's method with an experimentwise error rate of $\alpha = .10$ for each club, are displayed in the SAS printout shown in Figure 10.28. For each club, the brand means are listed in descending order in the Figure 10.24, and those not significantly different are connected by the same letter in the "Tukey Grouping" column.

```
----------------------------------- CLUB=5IRON -----------------------------------

                            The ANOVA Procedure

                 Tukey's Studentized Range (HSD) Test for DISTANCE

     NOTE: This test controls the Type I experimentwise error rate, but it generally has a higher Type
                           II error rate than REGWQ.

                    Alpha                                     0.1
                    Error Degrees of Freedom                   12
                    Error Mean Square                    36.10792
                    Critical Value of Studentized Range   3.62071
                    Minimum Significant Difference         10.878

                 Means with the same letter are not significantly different.

              Tukey Grouping           Mean      N    BRAND

                           A        182.675      4    B

                           B        171.300      4    A
                           B
                           B        167.175      4    C
                           B
                           B        160.500      4    D
----------------------------------- CLUB=DRIVER ----------------------------------

                            The ANOVA Procedure

                 Tukey's Studentized Range (HSD) Test for DISTANCE

     NOTE: This test controls the Type I experimentwise error rate, but it generally has a higher Type
                           II error rate than REGWQ.

                    Alpha                                     0.1
                    Error Degrees of Freedom                   12
                    Error Mean Square                    32.41208
                    Critical Value of Studentized Range   3.62071
                    Minimum Significant Difference         10.307

                 Means with the same letter are not significantly different.

              Tukey Grouping           Mean      N    BRAND

                           A        243.100      4    C
                           A
                       B   A        233.725      4    B
                       B
                       B        229.750      4    D
                       B
                       B        228.425      4    A
```

Figure 10.28

SAS printout of Tukey rankings of golf ball brand means for each club

As shown in Figure 10.28, the picture is unclear with respect to Brand means. For the five-iron (top of figure), the brand B mean significantly exceeds all other brands. However, when the brand B golf balls are hit with a driver (bottom of figure), brand B's mean is not significantly different from any of the other brands. The Club × Brand interaction can be seen in the SPSS plot of means shown in Figure 10.29. Note that the difference between

Estimated Marginal Means of DISTANCE

Figure 10.29

SPSS means plot for factorial golf ball experiment

the mean distances of the two clubs (driver and five-iron) varies with the brand. The biggest difference appears for Brand C, while the smallest difference is for Brand B.

Look Back Note the nontransitive nature of the multiple comparisons. For example, for the driver, the brand C mean can be "the same" as the brand B mean, and the brand B mean can be "the same" as the brand D mean, yet the brand C mean can significantly exceed the brand D mean. The reason lies in the definition of "the same": We must be careful not to conclude that two means are equal simply because they are connected by a line or a letter. The line indicates only that *the connected means are not significantly different*. Thus, conclude (at the overall α level of significance) there are differences in those means that are not connected, while withholding judgement on those means that connected. The picture of which means differ and by how much will become clearer as we increase the number of replicates of the factorial experiment.

■ **Now Work Exercise 10.104**

As with completely randomized and randomized block designs, the results of a factorial ANOVA are typically presented in an ANOVA summary table. Table 10.14 gives the general form of the ANOVA table, while Table 10.15 gives the ANOVA table for the golf ball data analyzed in Example 10.10. A two-factor factorial is characterized by four sources of variation—factor A, factor B, $A \times B$ interaction, and error—that sum to the total sum of squares.

Table 10.14 General ANOVA Summary Table for a Two-Factor Factorial Experiment with r Replicates, Where Factor A Has a Levels and Factor B Has b Levels

Source	df	SS	MS	F
A	$a-1$	SSA	MSA	MSA/MSE
B	$b-1$	SSB	MSB	MSB/MSE
AB	$(a-1)(b-1)$	SSAB	MSAB	MSAB/MSE
Error	$ab(r-1)$	SSE	MSE	
Total	$n-1$	SS(Total)		

Table 10.15 ANOVA Summary Table for Example 10.10

Source	df	SS	MS	F
Brand	1	32,093.11	32,093.11	936.75
Club	3	800.74	266.91	7.79
Interaction	3	765.96	255.32	7.45
Error	24	822.24	34.26	
Total	31	34,482.05		

Example 10.11 Ⓓ
GOLFFAC2

More Practice on Conducting a Factorial ANOVA— Golf Ball Driving Distances

Problem Refer to Example 10.10. Suppose the same factorial experiment is performed on four other brands (E, F, G, and H), and the results are as shown in Table 10.16. Repeat the factorial analysis and interpret the results.

Solution An SPSS printout for the second factorial experiment is shown in Figure 10.30. We conduct several tests, as outlined in the box on page 590.

Test for Equality of Treatment Means

The F-ratio for Treatments is $F = 290.1$ (highlighted at the top of the printout), which exceeds the tabular value of $F_{.10} = 1.98$ for seven numerator and 24 denominator degrees of freedom. (Note that the same rejection regions will apply in this example as in Example 10.10, since the factors, treatments, and replicates are the same.) We conclude that at least two of the Brand-Club combinations have different mean distances.

Table 10.16 Distance Data for Second Factorial Golf Experiment

Club	Brand			
	E	F	G	H
Driver	238.6	261.4	264.7	235.4
	241.9	261.3	262.9	239.8
	236.6	254.0	253.5	236.2
	244.9	259.9	255.6	237.5
Five-iron	165.2	179.2	189.0	171.4
	156.9	171.0	191.2	159.3
	172.2	178.0	191.3	156.6
	163.2	182.7	180.5	157.4

Ⓓ *Data Set: GOLFFAC2*

Tests of Between-Subjects Effects

Dependent Variable:DISTANCE

Source	Type III Sum of Squares	df	Mean Square	F	Sig.
Corrected Model	49959.375ᵃ	7	7137.054	290.120	.000
Intercept	1423532.828	1	1423532.828	57866.453	.000
BRAND	3410.316	3	1136.772	46.210	.000
CLUB	46443.900	1	46443.900	1887.939	.000
BRAND * CLUB	105.158	3	35.053	1.425	.260
Error	590.407	24	24.600		
Total	1474082.610	32			
Corrected Total	50549.782	31			

a. R Squared = .988 (Adjusted R Squared = .985)

Figure 10.30

SPSS printout of second factorial ANOVA

Test for Interaction

We next test for interaction between Brand and Club. The F-value (highlighted on the SPSS printout) is

$$F = \frac{\text{MS}(\text{Brand} \times \text{Club})}{\text{MSE}} = 1.425$$

Since this F-ratio does *not* exceed the tabled value of $F_{.10} = 2.33$ with 3 and 24 df, we cannot conclude at the .10 level of significance that the factors interact. In fact, note that the observed significance level (highlighted on the SPSS printout) for the test of interaction is .260. Thus, at any level of significance lower than $\alpha = .26$, we could not conclude that the factors interact. We therefore test the main effects for Brand and Club.

Test for Brand Main Effect

We first test for the Brand main effect:

H_0: No difference exists among the true Brand mean distances.

H_a: At least two Brand mean distances differ.

Test statistic:

$$F = \frac{\text{MS}(\text{Brand})}{\text{MSE}} = \frac{1,136.772}{24.600} = 46.210 \quad (\text{highlighted on printout})$$

Observed significance level: $p = .000$ (highlighted on printout)

Since $\alpha = .10$ exceeds the p-value, we conclude that at least two of the brand means differ. We will subsequently determine which brand means differ by using Tukey's multiple-comparison procedure. But first, we want to test for the Club main effect:

Test for Club Main Effect

H_0: No differences exist between the Club mean distances.

H_a: The Club mean distances differ.

Test statistic:

$$F = \frac{\text{MS(Club)}}{\text{MSE}} = \frac{46{,}443.900}{24.600} = 1{,}887.94 \text{ (highlighted on printout)}$$

Observed significance level: $p = .000$ (highlighted on printout)

Since $\alpha = .10$ exceeds the *p*-value, we conclude that the two clubs are associated with different mean distances. Since only two levels of Club were utilized in the experiment, this *F*-test leads to the inference that the mean distance differs for the two clubs. It is no surprise (to golfers) that the mean distance for balls hit with the driver is significantly greater than the mean distance for those hit with the five-iron.

Ranking of Means

To determine which of the Brands' mean distances differ, we wish to compare the four Brand means by using Tukey's method at $\alpha = .10$. The results of these multiple comparisons are displayed in the SPSS printout shown in Figure 10.31. The Brand means are shown, grouped by subset, in the figure, with means that are not significantly different in the same subset. Brands G and F (subset 2) are associated with significantly greater mean distances than brands E and H (subset 1). However, since G and F are in the same Tukey subset, and since E and H are in the same subset, we cannot distinguish between brands G and F or between brands E and H by means of these data. The means that do not differ are illustated with lines in Figure 10.32.

Look Back Since the interaction between Brand and Club was not significant, we conclude that this difference among brands applies to both clubs. The sample means for all Club-Brand combinations are shown in the MINITAB graph of Figure 10.33 and appear to support the conclusions of the tests and comparisons. Note that the Brand means maintain their relative positions for each Club: Brands F and G dominate brands E and H for both the driver and the five-iron.

DISTANCE

Tukey HSD[a,b]

BRAND	N	Subset 1	Subset 2
H	8	199.2000	
E	8	202.4375	
F	8		218.4375
G	8		223.5875
Sig.		.568	.189

Means for groups in homogeneous subsets are displayed.
Based on observed means.
The error term is Mean Square(Error) = 24.600.

a. Uses Harmonic Mean Sample Size = 8.000.

b. Alpha = 0.05.

Figure 10.31
SPSS printout of Tukey rankings of brand means

Brand:	H	E	F	G
Mean:	199.2	202.4	218.4	223.6

Figure 10.32
Summary of Tukey Ranking of Brand Means

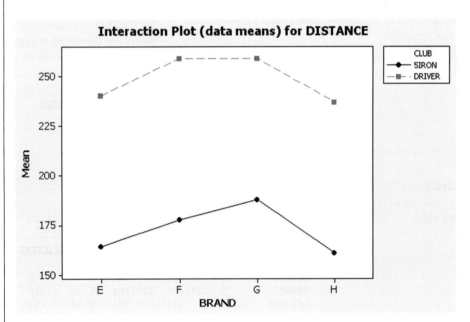

Figure 10.33
MINITAB means plot for second factorial golf ball experiment

■ Now Work Exercise 10.104e

Analysis-of-factorial experiments can become complex if the number of factors is increased. Even the two-factor experiment becomes more difficult to analyze if some factor combinations have different numbers of observations than others. We have provided an introduction to these important experiments by using two-factor factorials with equal numbers of observations for each treatment. Although similar principles apply to most factorial experiments, you should consult the references at the end of this chapter if you need to design and analyze more complex factorials.

Statistics IN Action **Revisited**

A Two-Way Analysis of the Voice/Face Recognition Data in Experiment 1

In Experiment 1 of the *British Journal of Psychology* (Feb. 2014) study of associative priming using faces and voices of celebrity pairs, 40 subjects were randomly assigned to one of four groups: (1) Within-modality/Related pairs, (2) Within-modality/Unrelated pairs, (3) Cross-modality/Related pairs, and (4) Cross-modality/Unrelated pairs. (See, again, Figure SIA10.1, p. 531) In a previous Statistics in Action Revisited, we treated the experiment as a completely randomized design with four treatments (groups) and conducted a one-way ANOVA.

The actual design employed by the researchers, however, is a factorial design. There are two factors—Prime Modality and Relational Pair—each at two levels. The two levels of modality are within-modality (prime face, target face) and cross-modality (prime voice, target face); the two levels of relational pair are related celebrities and unrelated celebrities. Thus, a 2 × 2 factorial design is utilized. A 2 × 2 factorial ANOVA will allow the researchers to answer the original three research questions, repeated as follows:

1. Does associative priming occur more often in the within-modality (face/face) condition than in the cross-modality (voice/face) condition?

2. Does associative priming occur more often with related pairs than with unrelated pairs?

3. Are the effects of modality condition and pair relation independent?

The first research question can be answered by conducting a main effect F-test for the Prime Modality factor, the second research question can be answered by conducting a main effect F-test for the Relational Pair factor, and the third research question can be answered by conducting an F-test for factor interaction because, by definition, if the impact of prime modality on the response variable depends on whether the subjects are presented with related celebrity pairs or unrelated celebrity pairs, then the factors *interact*. Of course, as outlined in this section, the test for interaction should be conducted first. If no interaction is detected, then we conduct the main effect tests for the two factors.

MINITAB printouts of the 2 × 2 factorial ANOVAs for the two dependent variables, accuracy rate and response speed, are displayed in Figures SIA10.9 and SIA10.10, respectively. For the dependent variable accuracy rate, the p-values (highlighted) for interaction as well as the two main effects all

---

Two-way ANOVA: ACCURACY versus PRIME, RELATED

Source	DF	SS	MS	F	P
PRIME	1	220.9	220.900	2.86	0.099
RELATED	1	115.6	115.600	1.50	0.229
Interaction	1	270.4	270.400	3.50	0.069
Error	36	2779.0	77.194		
Total	39	3385.9			

S = 8.786   R-Sq = 17.92%   R-Sq(adj) = 11.08%

Figure SIA10.9
MINITAB Factorial ANOVA for Accuracy Rate

Two-way ANOVA: SPEED versus PRIME, RELATED

Source	DF	SS	MS	F	P
PRIME	1	1291684	1291684	71.20	0.000
RELATED	1	774	774	0.04	0.837
Interaction	1	151044	151044	8.33	0.007
Error	36	653061	18141		
Total	39	2096563			

S = 134.7   R-Sq = 68.85%   R-Sq(adj) = 66.26%

Figure SIA10.10
MINITAB Factorial ANOVA for Response Speed

Statistics IN Action
(continued)

exceed $\alpha = .05$. Consequently, neither prime modality nor relational pair has an effect on mean accuracy rate. This result, consistent with the one-way analysis conducted previously, led the researchers to ultimately conclude that associative priming does not appear to occur with respect to recognition accuracy.

For the dependent variable response speed, the *p*-value (highlighted) for interaction is .007. Thus, at $\alpha = .05$, there is sufficient evidence to indicate that the effect of prime modality on the mean response speed depends on relational pair, and vice versa. Due to this significant interaction, the *p*-values for the main effect tests should be ignored regardless of their values.

A practical interpretation of this interaction can be obtained by examining the MINITAB plot of the $2 \times 2 = 4$ response speed means displayed in Figure SIA10.11. First,

note that the mean for cross-modality (prime voice) exceeds the mean for within-modality (prime face) but that the difference is larger for related pairs than for unrelated pairs. Second, in the cross-modality (prime voice) condition (i.e., the red line), mean response speed for related pairs exceeds the corresponding mean for unrelated pairs—although a *t*-test shows that this difference is not significantly different from 0. However, the opposite occurs in the within-modality (prime face) condition (i.e., the black line). There, the mean response speed for related pairs is smaller the corresponding mean for unrelated pairs—and this difference is statistically significant (shown via a *t*-test). Consequently, with respect to response speed, the researchers were able to infer the presence of associative priming in the within-modality (prime face) condition but not in the cross-modality (prime voice) condition.

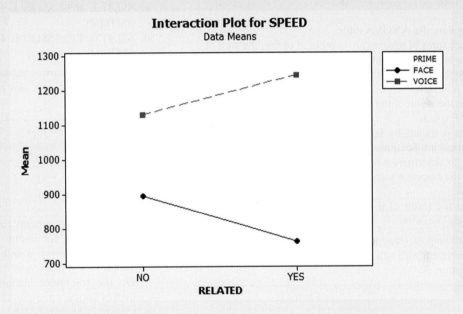

Figure SIA10.11
MINITAB Plot of Response Speed Means

Exercises 10.85–10.109

Understanding the Principles

10.85 Explain the difference between a single factor and a complete factorial experiment.

10.86 What is a balanced factorial design?

10.87 *True or False.* Completely randomized design is a balanced design.

10.88 *True or False.* Factor interactions are used to determine whether the factors affect the response separately.

10.89 Suppose you conduct a 4×3 factorial experiment.
a. How many factors are used in this experiment?
b. Can you determine the type(s) of factors—qualitative or quantitative—from the information given? Explain.
c. Can you determine the number of levels used for each factor? Explain.
d. Describe a treatment for the experiment, and determine the number of treatments used.

e. What problem is caused by using a single replicate of this experiment? How is the problem solved?

Learning the Mechanics

10.90 The partially complete ANOVA table given here is for a two-factor factorial experiment:

Source	df	SS	MS	F
A	1			.85
B	3	.97		
AB			.35	
Error				
Total	23	7.5		

a. Give the number of levels for each factor.

b. How many observations were collected for each factor–level combination?

c. Complete the ANOVA table.

d. Test to determine whether the treatment means differ. Use $\alpha = .10$.

e. Conduct the tests of factor interaction and mean effects, each at the $\alpha = .10$ level of significance. Which of the tests are warranted as part of the factorial analysis? Explain.

10.91 Following is a partially completed ANOVA table for a 3×4 factorial experiment with two replications:

Source	df	SS	MS	F
A		.8		
B		5.3		
AB		9.6		
Error				
Total		18.1		

a. Complete the ANOVA table.

b. Which sums of squares are combined to find the sum of squares for treatment? Do the data provide sufficient evidence to indicate that the treatment means differ? Use $\alpha = .05$.

c. Does the result of the test in part **b** warrant further testing? Explain.

d. What is meant by factor interaction, and what is the practical implication if it exists?

e. Test to determine whether these factors interact to affect the response mean. Use $\alpha = .05$, and interpret the result.

f. Does the result of the interaction test warrant further testing? Explain.

10.92 The following two-way table gives data for a 2×3 factorial experiment with two observations for each factor–level combination.

L10092

	Factor B		
Level	1	2	3
Factor A 1	3.1, 4.0	4.6, 4.2	6.4, 7.1
2	5.9, 5.3	2.9, 2.2	3.3, 2.5

a. Identify the treatments for this experiment. Calculate and plot the treatment means, using the response variable as the y-axis and the levels of factor B as the x-axis. Use the levels of factor A as plotting symbols. Do the treatment means appear to differ? Do the factors appear to interact?

b. The MINITAB ANOVA printout for this experiment is shown below. Test to determine whether the treatment means differ at the $\alpha = .05$ level of significance. Does the test support your visual interpretation from part **a**?

Two-way ANOVA: Response versus A, B

```
Source       DF    SS       MS       F       P
A             1   4.4408  4.44083  18.06   0.005
B             2   4.1267  2.06333   8.39   0.018
Interaction   2  18.0067  9.00333  36.62   0.000
Error         6   1.4750  0.24583
Total        11  28.0492

S = 0.4958   R-Sq = 94.74%   R-Sq(adj) = 90.36%
```

c. Does the result of the test in part **b** warrant a test for interaction between the two factors? If so, perform it, using $\alpha = .05$.

d. Do the results of the previous tests warrant tests of the two factor main effects? If so, perform them, using $\alpha = .05$.

e. Interpret the results of the tests. Do they support your visual interpretation from part **a**?

10.93 Suppose a 3×3 factorial experiment is conducted with three replications. Assume that SS(Total) = 1,000. For each of the following scenarios, form an ANOVA table, conduct the appropriate tests, and interpret the results:

a. The sum of squares of factor A main effect [SS(A)] is 20% of SS(Total), the sum of squares of factor B main effect [SS(B)] is 10% of SS(Total), and the sum of squares of interaction [SS(AB)] is 10% of SS(Total).

b. SS(A) is 10%, SS(B) is 10%, and SS(AB) is 50% of SS(Total).

c. SS(A) is 40%, SS(B) is 10%, and SS(AB) is 20% of SS(Total).

d. SS(A) is 40%, SS(B) is 40%, and SS(AB) is 10% of SS(Total).

10.94 The following two-way table gives data for a 2×2 factorial experiment with two observations per factor–level combination:

L10094

	Factor B	
Level	1	2
Factor A 1	29.6, 35.2	47.3, 42.1
2	12.9, 17.6	28.4, 22.7

a. Identify the treatments for this experiment. Calculate and plot the treatment means, using the response variable as the y-axis and the levels of factor B as the x-axis. Use the levels of factor A as plotting symbols. Do the treatment means appear to differ? Do the factors appear to interact?

b. Construct an ANOVA table for this experiment.

c. Test to determine whether the treatment means differ at the $\alpha = .05$ level of significance. Does the test support your visual interpretation from part **a**?

d. Does the result of the test in part **c** warrant a test for interaction between the two factors? If so, perform it, using

e. Do the results of the previous tests warrant tests of the two factor main effects? If so, perform them, using $\alpha = .05$.

f. Interpret the results of the tests. Do they support your visual interpretation from part **a**?

g. Given the results of your tests, which pairs of means, if any, should be compared?

Applying the Concepts—Basic

10.95 **Egg shell quality in laying hens.** In *Animal Feed Science and Technology* (June 2010) researchers investigated the effect of hen's age and limestone diet on eggshell quality. Two different diets were studied: fine limestone (FL) and coarse limestone (CL). Hens were classified as either younger hens (24–36 weeks old) or older hens (56–68 weeks old). The study used 120 younger hens and 120 older hens.

Within each age group, half the hens were fed a fine limestone diet and the other half a coarse limestone diet. Thus, there were 60 hens in each of the four combinations of age and diet. The characteristics of the eggs produced from the laying hens were recorded, including shell thickness.

a. Identify the type of experimental design employed by the researchers.

b. Identify the factors and the factor levels (treatments) for this design.

c. Identify the experimental unit.

d. Identify the dependent variable.

e. The researchers found no evidence of factor interaction. Interpret this result, practically.

f. The researchers found no evidence of a main effect for hen's age. Interpret this result, practically.

g. The researchers found statistical evidence of a main effect for limestone diet. Interpret this result, practically. (*Note:* The mean shell thickness for eggs produced by hens on a CL diet was larger than the corresponding mean for hens on an FL diet.)

10.96 Job satisfaction of STEM faculty. Are university faculty in science, technology, engineering, and math (STEM) disciplines more satisfied with their jobs than non-STEM faculty? And, if so, does this difference vary depending on gender? These were some of the questions of interest in a study published in the *Journal of Women and Minorities in Science and Engineering* (Vol. 18, 2012). A sample of 215 faculty members at a large public university participated in a survey. One question asked the degree to which the faculty members were satisfied with university policies and procedures. Responses were recorded on a 5-point numerical scale, with 1 = strongly disagree to 5 = strongly agree. Each participant was categorized according to gender (male or female) and discipline (STEM or non-STEM). Thus, a 2 × 2 factorial design was utilized.

a. Identify the treatments for this experiment.

b. For this study, what does it mean to say that discipline and gender interact?

c. A plot of the treatment means is shown below. Based on this graph only, would you say that discipline and gender interact?

d. Construct a partial ANOVA table for this study. (Give the sources of variation and degrees of freedom.)

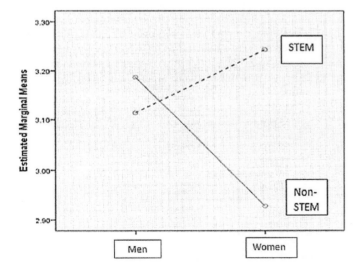

e. The journal article reported the *F*-test for interaction as $F = 4.10$ with *p*-value $= .04$. Interpret these results.

10.97 Temptation in consumer choice. Are you willing to pay more for a tempting vice option (e.g., eating a hamburger, surfing the Internet) than a virtuous option (e.g., eating broccoli, reading the newspaper)? The costs and benefits of consumer temptation choice were investigated in the *Journal of Marketing Research* (Feb. 2012). In one experiment, the researchers used a 2 × 2 factorial design involving a sample of 180 consumers. The first factor varied whether consumers were asked to imagine ordering a steak (*vice*) or organic pasta (*virtue*) at a restaurant. The second factor varied whether the restaurant described was *homogeneous* in its offerings (either "only tasty, less healthy dishes like steak" or "only healthy, less tasty dishes like organic pasta") or offered a *mixed menu* (both "tasty, less healthy dishes like steak" and "healthy, less tasty dishes like organic pasta"). Consumers were randomly assigned to one of the four order/menu conditions and then asked to give the maximum amount (in dollars) they would be willing to pay for their choice. A partial ANOVA table for the data is shown below.

Source	df	F-value	p-value
Order (vice/virtue)	----	----	----
Menu (homogeneous/mixed)	----	----	----
Order × Menu	----	11.25	< .001
Error	----		
Total	179		

a. Fill in the missing degrees of freedom values in the ANOVA table.

b. Give a practical interpretation of the test for interaction using $\alpha = .05$.

c. The *F*-values (and corresponding *p*-values) missing in the table were not provided in the journal article. Are these results required to complete the analysis of the data? Explain.

d. In a follow-up to the ANOVA, the researchers compared the mean willing to pay values for the homogeneous and mixed-menu conditions at each level of the order condition. In the virtue condition, the mean for the homogeneous menu ($11.08) was significantly lower than the mean for the mixed menu ($13.26). In the vice condition, the mean for the homogeneous menu ($17.11) was significantly higher than the mean for the mixed menu ($15.00). Demonstrate why these results support your answer to part **b**. Illustrate with a graph.

10.98 Insomnia and education. Refer to the *Journal of Abnormal Psychology* (Feb. 2005) study of whether insomnia is related to education status, presented in Exercise 1.30 (p. 50). A random-digit telephone dialing procedure was employed to collect data on 575 study participants. In addition to insomnia status (normal sleeper or chronic insomnia), the researchers classified each participant into one of four education categories (college graduate, some college, high school graduate, and high school dropout). One dependent variable of interest to the researchers was a quantitative measure of daytime functioning called the Fatigue

Severity Scale (FSS). The data were analyzed as a 2×4 factorial experiment, with Insomnia status and Education level as the two factors.

a. Determine the number of treatments for this study. List them.

b. The researchers reported that "the Insomnia $\times$ Education interaction was not statistically significant." Interpret this result practically. (Illustrate with a graph.)

c. The researchers discovered that the sample mean FSS for people with insomnia was greater than the sample mean FSS for normal sleepers and that this difference was statistically significant. Interpret this result practically.

d. The researchers reported that the main effect of Education was statistically significant. Interpret this result practically.

e. Refer to part **d**. In a follow-up analysis, the sample mean FSS values for the four Education levels were compared with the use of Tukey's method ($\alpha = .05$), with the results shown in the accompanying table. What do you conclude?

Mean:	3.3	3.6	3.7	4.2
Education:	College graduate	Some college	High school graduate	High school dropout

10.99 Purchase of fair-trade products. "Just-world" theory proposes that people receive the rewards and/or punishments that they deserve. Marketing researchers examined just-world theory in the context of fair trade (*Journal of Marketing*, Jan. 2012). In particular, the researchers wanted to know if manipulating the market conditions has an impact on whether consumers would purchase fair-trade products. A designed experiment with two manipulated market factors was employed. One factor was *justice reparation potential* (low or high); a second factor was *producer need* (moderate or high). A sample of business students were divided into four groups—34 students were randomly assigned to each of the $2 \times 2 = 4$ market condition treatments. After reading a news article and press release that manipulated their condition, students reported on their intention to purchase a fair-trade product. Intention was measured on a scale ranging from 0 to 6 points. The data for all 136 students (simulated based on information provided in the journal article) are saved in the **FTRADE** file. An ANOVA for the data is shown in the accompanying MINITAB printout.

```
Analysis of Variance for INT

Source    DF      SS      MS      F      P
JRP        1   1.360   1.360   1.25  0.266
NEED       1   0.622   0.622   0.57  0.451
JRP*NEED   1  22.405  22.405  20.55  0.000
Error    132 143.948   1.091
Total    135 168.335


S = 1.04428   R-Sq = 14.49%   R-Sq(adj) = 12.54%
```

a. For this designed experiment, explain (practically) what it means to have factor interaction.

b. Conduct the *F*-test for factor interaction using $\alpha = .01$. What do you conclude?

MINITAB Graph for Exercise 10.99

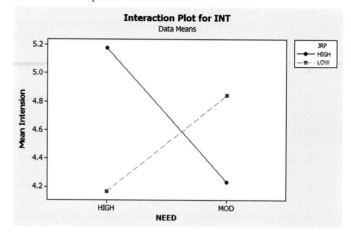

c. In the journal article, the researchers reported on the ANOVA *F*-tests for main effects. Is this necessary? Explain.

d. A plot of the sample means for the four treatments is shown below in the MINITAB graph. Explain why this graph supports your answer to part **b**.

e. The researchers hypothesized that when justice restoration potential is low, fair-trade purchase intentions will be lower when need is high rather than moderate. Conversely, when justice restoration potential is high, purchase intentions will be greater when need is high rather than moderate. Is there evidence to support this theory?

10.100 Do video game players have superior visual attention skills? Refer to the *Journal of Articles in Support of the Null Hypothesis* (Vol. 6, 2009) study of video game players, Exercise 9.22 (p. 478). In a second experiment, the two groups of male psychology students—32 video game players (VGP group) and 28 nonplayers (NVGP group)—were subjected to the *inattentional blindness* test. This test gauges students' ability to detect the appearance of unexpected or task-irrelevant objects in their visual field. The students were required to silently count the number of times moving block letters touch the sides of a display window while focusing on a small square in the center of the window. Within each group, half the students were assigned a task with 4 moving letters (low load) and half were assigned a task with 8 moving letters (high load). At the end of the task, the difference between the student's count of letter touches and the actual count (recorded as a percentage error) was determined. The data on percentage error were subjected to an ANOVA for a factorial design, with Group (VGP or NVGP) as one factor and Load (low or high) as the other factor. The results are summarized in the accompanying table.

Source	df	F	p-Value
Group	1	0.58	.449
Load	1	35.04	< .0005
Group × Load	1	1.37	.247
Error	56		
Total	59		

Based on Murphy, K., and Spencer, A. "Playing video games does not make for better visual attention skills." *Journal of Articles in Support of the Null Hypothesis*, Vol. 6, No. 1, 2009.

a. Interpret the *F*-test for Group × Load interaction. Use α = .10.

b. Interpret the *F*-test for Group main effect. Use α = .10.

c. Interpret the *F*-test for Load main effect. Use α = .10.

d. Based on your answers to parts **a–c**, do you believe video game players perform better on the inattentional blindness test than non-video game players?

10.101 Virtual-reality-based rehabilitation systems. In *Robotica* (Vol. 22, 2004), researchers described a study of the effectiveness of display devices for three virtual-reality (VR)-based hand rehabilitation systems. Display device A is a projector, device B is a desktop computer monitor, and device C is a head-mounted display. Twelve nondisabled right-handed male subjects were randomly assigned to the three VR display devices, four subjects in each group. In addition, within each group two subjects were randomly assigned to use an auxiliary lateral image and two subjects were not. Consequently, a 3 × 2 factorial design was employed, with the VR display device at three levels (A, B, or C) and an auxiliary lateral image at two levels (yes or no). Using the assigned VR system, each subject carried out a "pick-and-place" procedure and the collision frequency (number of collisions between moved objects) was measured.

a. Give the sources of variation and associated degrees of freedom in an ANOVA summary table for this design.

b. How many treatments are investigated in this experiment?

c. The factorial ANOVA resulted in the following *p*-values: Display main effect (.045), Auxiliary lateral image main effect (.003), and Interaction (.411). Interpret these results practically. Use α = .05 for each test you conduct.

Applying the Concepts—Intermediate

10.102 Baiting traps to maximize beetle catch. A field experiment was conducted to compare the effectiveness of different traps for catching beetles (*Journal of Chemical Ecology*, Vol. 94, 2011). Paraffin traps were baited either with or without linalool. Also, the color of the trap was varied as green or yellow. Seven traps for each combination of bait and color—a total of 28 traps—were set 1 meter from the ground in a random grid pattern. After 5 days, the total number of beetles captured by each trap was determined. The data (simulated from information provided in the journal article) are displayed in the following table. The researchers are investigating the effect of bait type and color on the mean number of beetles captured by the traps. Analyze the data for the researchers. What do you conclude?

	Yellow	Green
With linalool	17, 22, 13, 15, 14, 18, 11	4, 5, 0, 2, 3, 2, 0
Without linalool	29, 10, 6, 5, 12, 11, 13	1, 0, 2, 0, 0, 0, 1

10.103 Mowing effects on highway right-of-way. A vegetation height of greater than 30 centimeters on a highway right-of-way is generally considered a safety hazard to drivers. How often and at what height should the right-of-way be mowed in order to maintain a safe environment? This was the question of interest in an article published in the *Landscape Ecology Journal* (Jan. 2013). The researchers designed an experiment to estimate the effects of mowing frequency and mowing height on the mean height of vegetation in the highway right-of-way. Mowing frequency was set at three levels: once, twice, or three times per year. Mowing height of the equipment was also set at three levels: 5, 10, or 20 centimeters. A sample of 36 plots of land along a highway right-of-way were selected, and each was randomly assigned to receive one of the 3 × 3 = 9 mowing frequency/mowing height treatments. The design was balanced so that each treatment was applied to 4 plots of land. At the end of the year, the vegetation height (in centimeters) was measured for each plot. Simulated data are shown in the table. Conduct a complete analysis of variance of the data. What inferences can you make about the effects of mowing frequency and mowing height on vegetation height?

Mow Height	Mow Frequency	Vegetation Height (cm)
5	1	19.3, 17.3, 15.0, 16.7
10	1	16.0, 15.6, 16.9, 15.0
20	1	16.7, 17.9, 15.9, 13.7
5	2	22.4, 20.8, 24.5, 21.7
10	2	23.9, 23.6, 21.7, 23.8
20	2	24.7, 26.3, 27.2, 26.4
5	3	18.6, 17.9, 16.1, 19.4
10	3	22.2, 25.6, 21.8, 23.6
20	3	27.0, 25.3, 23.8, 28.0

10.104 Baker's versus brewer's yeast. The *Electronic Journal of Biotechnology* (Dec. 15, 2003) published an article on a comparison of two yeast extracts: baker's yeast and brewer's yeast. Brewer's yeast is a surplus by-product obtained from a brewery; hence, it is less expensive than primary-grown baker's yeast. Samples of both yeast extracts were prepared at four different temperatures (45, 48, 51, and 54°C); thus, a 2 × 4 factorial design with yeast extract at two levels and temperature at four levels was employed. The response variable was the autolysis yield (recorded as a percentage).

a. How many treatments are included in the experiment?

b. An ANOVA found sufficient evidence of factor interaction at α = .05. Interpret this result practically.

c. Give the null and alternative hypotheses for testing the main effects of yeast extract and temperature.

d. Explain why the tests referred to in part **c** should not be conducted.

e. Multiple comparisons of the four temperature means were conducted for each of the two yeast extracts. Interpret the results, shown as follows:

Baker's yeast:	Mean yield (%):	41.1	47.5	48.6	50.3
	Temperature (°C):	54	45	48	51

Brewer's yeast:	Mean yield (%):	39.4	47.3	49.2	49.6
	Temperature (°C):	54	51	48	45

10.105 Learning from picture book reading. *Developmental Psychology* (Nov. 2006) published an article that examined toddlers' ability to learn from reading picture books. The experiment involved 36 children at each of three different ages: 18, 24, and 30 months. The children were randomly assigned into one of three different reading book conditions: book with color photographs (Photos), book with colored pencil drawings (Drawings), and book with no photographs or drawing (Control). Thus, a 3×3 factorial experiment was employed (with age at three levels and reading book condition at three levels). After a book-reading session, the children were scored on their ability to reenact the target actions in the book. Scores ranged from 0 (low) to 3 (high). An ANOVA of the reenactment scores is summarized in the following table.

Source	df	F	p-value
Age	–	11.93	< .001
Book	–	23.64	< .001
Age × Book	–	2.99	< .05
Error	–		
Total	107		

a. Fill in the missing degrees-of-freedom (df) values in the table.
b. How many treatments are investigated in this experiment? List them.
c. Conduct a test for Age × Book interaction at $\alpha = .05$. Interpret the result practically.
d. On the basis of the test you conducted in part c, do you need to conduct tests for Age and Book main effects? Explain.
e. At each age level, the researchers performed multiple comparisons of the reading book condition means at $\alpha = .05$. The results are summarized in the accompanying table. What can you conclude from this analysis? Support your answer with a plot of the means.

Age = 18 months:	.40 Control	.75 Drawings	1.20 Photos
Age = 24 months:	.60 Control	1.61 Drawings	1.63 Photos
Age = 30 months:	.50 Control	2.20 Drawings	2.21 Photos

10.106 Violent lyrics and aggressiveness. In the *Journal of Personality and Social Psychology* (May 2003), psychologists investigated the potentially harmful effects of violent music lyrics. A total of 60 undergraduate college students participated in an experiment designed by the researchers. Half of the students were volunteers, and half were required to participate as part of their introductory psychology class. Each student listened to a song by the group "Tool," with half the students randomly assigned a song with violent lyrics and half assigned a song with nonviolent lyrics. Consequently, the experiment used a 2×2 factorial design with the factors Song

(violent, nonviolent) and Pool (volunteer, psychology class). After listening to the song, each student was given a list of word pairs and asked to rate the similarity of each word in the pair on a seven-point scale. One word in each pair was aggressive in meaning (e.g., *choke*) and the other was ambiguous (e.g., *night*). An aggressive cognition score was assigned on the basis of the average word-pair scores. (The higher the score, the more the subject associated an ambiguous word with a violent word.) The data (simulated) are shown in the accompanying table. Conduct a complete analysis of variance on the data.

	Volunteer	Psychology Class
Violent Song	4.1 3.5 3.4 4.1 3.7 2.8 3.4 4.0 2.5 3.0 3.4 3.5 3.2 3.1 3.6	3.4 3.9 4.2 3.2 4.3 3.3 3.1 3.2 3.8 3.1 3.8 4.1 3.3 3.8 4.5
Nonviolent Song	2.4 2.4 2.5 2.6 3.6 4.0 3.3 3.7 2.8 2.9 3.2 2.5 2.9 3.0 2.4	2.5 2.9 2.9 3.0 2.6 2.4 3.5 3.3 3.7 3.3 2.8 2.5 2.8 2.0 3.1

10.107 Eyewitnesses and mug shots. When an eyewitness to a crime examines a set of mug shots at a police station, the photos are usually presented in groups (e.g., 6 mug shots at a time). Criminologists investigated whether mug shot group size has an effect on the selections made by eyewitnesses (*Applied Psychology in Criminal Justice*, Apr. 2010). A sample of 90 college students was shown a video of a simulated theft. Shortly thereafter, each student was shown 180 mug shots and asked to select a photo which most closely resembled the thief. (Multiple photos could be selected.) The students were randomly assigned to view either 3, 6, or 12 mug shots at a time. Within each mug shot group size, the students were further randomly divided into three sets. In the first set, the researchers focused on the selections made in the first 60 photos shown; in the second set, the focus was on selections made in the middle 60 photos shown; and, in the third set, selections made in the last 60 photos were recorded. The dependent variable of interest was the number of mug shot selections. Simulated data for this 3×3 factorial ANOVA, with Mug Shot Group Size at 3 levels (3, 6, or 12 photos) and Photo Set at 3 levels (first 60, middle 60, and last 60), are saved in the **MUGSHOT** file. Fully analyze the data for the researchers. In particular, the researchers want to know if mug shot group size has an effect on the mean number of selections, and, if so, which group size leads to the most selections. Also, is a higher number of selections made in the first 60, middle 60, or last 60 photos viewed?

Applying the Concepts—Advanced

10.108 Testing a new pain-relief tablet. Refer to the *Tropical Journal of Pharmaceutical Research* (June 2003) study of the impact of binding agent, binding concentration, and relative density on the mean dissolution time of pain-relief tablets, presented in Exercise 10.16 (p. 538).

Recall that binding agent was set at two levels (khaya gum and PVP), binding concentration at two levels (.5% and 4.0%), and relative density at two levels (low and high); thus, a $2 \times 2 \times 2$ factorial design was employed. The sample mean dissolution times for the treatments associated with the factors binding agent and relative density when the other factor (binding concentration) is held fixed at .5% are $\bar{x}_{Gum/Low} = 4.70$, $\bar{x}_{Gum/High} = 7.95$, $\bar{x}_{PVP/Low} = 3.00$, and $\bar{x}_{PVP/High} = 4.10$. Do the results suggest that there is an interaction between binding agent and relative density? Explain.

10.109 Impact of flavor name on consumer choice. Do consumers react favorably to products with ambiguous colors or names? Marketing professors investigated this phenomenon in the *Journal of Consumer Research* (June 2005). As a "reward" for participating in an unrelated experiment, 100 consumers were told that they could have some jelly beans available in several cups on a table. Half the consumers were assigned to take jelly beans with common descriptive flavor names (e.g., watermelon green), while the other half were assigned to take jelly beans with ambiguous flavor names (e.g., monster green). Within each group, half of the consumers took the jelly beans and left (low cognitive load condition), while the other half were asked questions designed to distract them while they were taking their jelly beans (high cognitive load condition). Consequently, a 2×2 factorial experiment was employed—with Flavor name (common or ambiguous) and Cognitive load (low or high) as the two factors—with 25 consumers assigned to each of four treatments. The dependent variable of interest was the number of jelly beans taken by each consumer. The means and standard deviations of the four treatments are shown in the following table.

| | Ambiguous | | Common | |
	Mean	Std. Dev.	Mean	Std. Dev.
Low Load	18.0	15.0	7.8	9.5
High Load	6.1	9.5	6.3	10.0

Based on Miller, E. G., and Kahn, B. E. "Shades of meaning: The effect of color and flavor names on consumer choice." *Journal of Consumer Research*, Vol. 32, June 2005 (Table 1).

a. Calculate the total of the $n = 25$ measurements for each of the four categories in the 2×2 factorial experiment.
b. Calculate the correction for mean, CM. (See Appendix B for computational formulas.)
c. Use the results of parts **a** and **b** to calculate the sums of squares for Load, Name, and Load $\times$ Name interaction.
d. Calculate the sample variance for each treatment. Then calculate the sum of squares of deviations within each sample for the four treatments.
e. Calculate SSE. (*Hint:* SSE is the pooled sum of squares for the deviations calculated in part **d**.)
f. Now that you know SS(Load), SS(Name), SS(Load $\times$ Name), and SSE, find SS(Total).
g. Summarize the calculations in an ANOVA table.
h. The researchers reported the F-value for Load $\times$ Name interaction as $F = 5.34$. Do you agree?
i. Conduct a complete analysis of these data. Use $\alpha = .05$ for any inferential techniques you employ. Illustrate your conclusions graphically.
j. What assumptions are necessary to ensure the validity of the inferential techniques you utilized? State them in terms of this experiment.

CHAPTER NOTES

Key Terms

Guide to Selecting an Experimental Design

Guide to Conducting ANOVA *F*–Tests

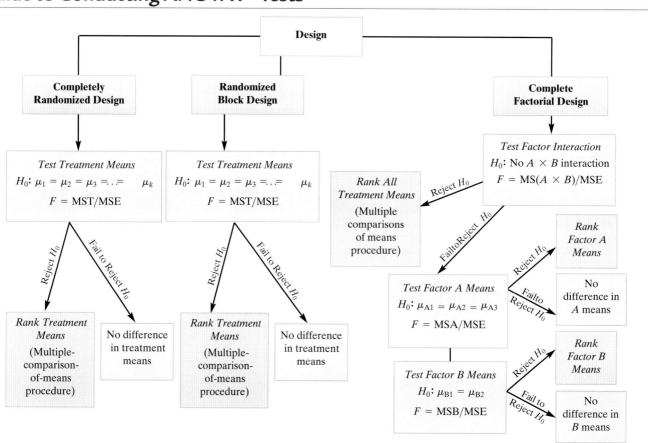

Key Symbols/Notation

ANOVA	Analysis of variance
SST	Sum of squares for treatments
MST	Mean square for treatments
SSB	Sum of squares for blocks
MSB	Mean square for blocks
SSE	Sum of squares for error
MSE	Mean square for error
$a \times b$ factorial	Factorial design with one factor at a levels and the other factor at b levels
SS(A)	Sum of squares for main-effect factor A
MS(A)	Mean square for main-effect factor A
SS(B)	Sum of squares for main-effect factor B
MS(B)	Mean square for main-effect factor B
SS(AB)	Sum of squares for factor $A \times B$ interaction
MS(AB)	Mean square for factor $A \times B$ interaction

Key Ideas

Key Elements of a Designed Experiment

1. *Response (dependent) variable*—quantitative
2. *Factors (independent variables)*—quantitative or qualitative
3. *Factor levels (values of factors)*—selected by the experimenter
4. *Treatments*—combinations of factor levels
5. *Experimental units*—assign treatments to experimental units and measure response for each

Balanced Design

Sample sizes for each treatment are equal.

Tests for Main Effects in a Factorial Design

appropriate only if the test for factor interaction is nonsignificant

Conditions Required for Valid F-Test in a Completely Randomized Design

1. All k treatment populations are approximately normal.
2. $\sigma_1^2 = \sigma_2^2 = \ldots = \sigma_k^2$

Conditions Required for Valid F-Tests in a Randomized Block Design

1. All treatment-block populations are approximately normal.
2. All treatment-block populations have the same variance.

Conditions Required for Valid F-Tests in a Complete Factorial Design

1. All treatment populations are approximately normal.
2. All treatment populations have the same variance.

Robust Method

Slight to moderate departures from normality have no impact on the validity on the ANOVA results.

Experimentwise Error Rate

Risk of making at least one Type I error when making multiple comparisons of means in ANOVA

Number of Pairwise Comparisons with k Treatment Means

$$c = k(k - 1)/2$$

Multiple-Comparison-of-Means Methods

Tukey
1. Balanced design
2. Pairwise comparisons of means

Bonferroni
1. Either balanced or unbalanced design
2. Pairwise comparisons of means

Scheffé
1. Either balanced or unbalanced design
2. General contrasts of means

Supplementary Exercises 10.110–10.141

Understanding the Principles

10.110 How do you determine the nature of the treatment effect on the response in factorial experiment?

10.111 Determine the number of pairwise comparisons of treatment means.

10.112 What are the treatments in a two-factor experiment with factor A at two levels and factor B at three levels?

10.113 What is the difference between a design study and an observational study?

Learning the Mechanics

10.114 A completely randomized design is utilized to compare four treatment means. The data are shown in the accompanying table.

a. Given that SST $= 36.95$ and SS(Total) $= 62.55$, complete an ANOVA table for this experiment.

b. Is there evidence that the treatment means differ? Use $\alpha = .10$.

Treatment 1	Treatment 2	Treatment 3	Treatment 4
8	6	9	12
10	9	10	13
9	8	8	10
10	8	11	11
11	7	12	11

10.115 An experiment utilizing a randomized block design was conducted to compare the mean responses for four treatments, A, B, C, and D. The treatments were randomly assigned to the four experimental units in each of five blocks. The data are shown in the next table.

Data for Exercise 10.115

Treatment	Block				
	1	2	3	4	5
A	8.6	7.5	8.7	9.8	7.4
B	7.3	6.3	7.3	8.4	6.3
C	9.1	8.3	9.0	9.9	8.2
D	9.3	8.2	9.2	10.0	8.4

a. Given that SS(Total) = 22.31, SS(Block) = 10.688, and SSE = .288, complete an ANOVA table for the experiment.
b. Do the data provide sufficient evidence to indicate a difference among treatment means? Test, using $\alpha = .05$
c. Does the result of the test in part **b** warrant further comparison of the treatment means? If so, how many pairwise comparisons need to be made?
d. Is there evidence that the block means differ? Use $\alpha = .05$.

10.116 The following table shows a partially completed ANOVA table for a two-factor factorial experiment.

Source	df	SS	MS	F
A	5	3.6		
B	3	10.2		
A × B			4.1	
Error		19.7		
Total	47			

a. Complete the ANOVA table.
b. How many levels were used for each factor? How many treatments were used? How many replications were performed?
c. Find the value of the sum of squares for treatments. Test to determine whether the data provide evidence that the treatment means differ. Use $\alpha = .05$.
d. Is further testing of the nature of factor effects warranted? If so, test to determine whether the factors interact. Use $\alpha = .05$. Interpret the result.

Applying the Concepts—Basic

10.117 **Accounting and Machiavellianism.** A study of Machiavellian traits in accountants was published in *Behavioral Research in Accounting* (Jan. 2008). *Machiavellian* describes negative character traits such as manipulation, cunning, duplicity, deception, and bad faith. A Machiavellian ("Mach") rating score was determined for each in a sample of accounting alumni of a large southwestern university. The accountants were then classified as having high, moderate, or low Mach rating scores. For one portion of the study, the researcher investigated the impact of both Mach score classification and gender on the average income of an accountant. For this experiment, identify each of the following:
a. experimental unit
b. response variable
c. factors
d. levels of each factor
e. treatments

10.118 **Extinct New Zealand birds.** Refer to the *Evolutionary Ecology Research* (July 2003) study of extinction in the New Zealand bird population, presented in Exercise 1.20 (p. 49). Recall that biologists measured the body mass (in grams) and type of habitat (aquatic, ground terrestrial, or aerial terrestrial) for each bird species. One objective is to compare the body mass means of birds with the three different types of habitat.
a. Identify the response variable of the study.
b. Identify the experimental units of the study.
c. Identify the factor(s) in the study.
d. Identify the treatments.

10.119 **Impact of paper color on exam scores.** A study published in *Teaching Psychology* (May 1998) examined how external clues influence student performance. Introductory psychology students were randomly assigned to one of four different midterm examinations. Form 1 was printed on blue paper and contained difficult questions, while form 2 was also printed on blue paper, but contained simple questions. Form 3 was printed on red paper with difficult questions; form 4 was printed on red paper with simple questions. The researchers were interested in the impact that Color (red or blue) and Question (simple or difficult) had on mean exam score.
a. What experimental design was employed in this study? Identify the factors and treatments.
b. The researchers conducted an ANOVA and found a significant interaction between Color and Question (p-value < .03). Interpret this result.
c. The sample mean scores (percentage correct) for the four exam forms are listed in the accompanying table. Plot the four means on a graph to illustrate the Color × Question interaction.

Form	Color	Question	Mean Score
1	Blue	Difficult	53.3
2	Blue	Simple	80.0
3	Red	Difficult	39.3
4	Red	Simple	73.6

10.120 **Rotary oil rigs.** An economist wants to compare the average monthly number of rotary oil rigs running in three states: California, Utah, and Alaska. In order to account for month-to-month variation, three months were randomly selected over a two-year period and the number of oil rigs running in each state in each month was obtained from data provided from *World Oil* (Jan. 2002) magazine. The data, shown in the accompanying table, were analyzed by means of a randomized block design. The MINITAB printout is provided on the next page.

Month	California	Utah	Alaska
1	27	17	11
2	34	20	14
3	36	15	14

MINITAB Output for Exercise 10.120

Two-way ANOVA: NumRigs versus State, Month/Year

```
Source       DF       SS       MS      F      P
State         2  617.556  308.778  38.07  0.002
Month/Year    2   30.889   15.444   1.90  0.262
Error         4   32.444    8.111
Total         8  680.889

S = 2.848   R-Sq = 95.23%   R-Sq(adj) = 90.47%


                  Individual 95% CIs For Mean Based on
                  Pooled StDev
State    Mean   ---------+---------+---------+---------+
AL    13.0000   (----*-----)
CAL   32.3333                         (----*-----)
UT    17.3333        (-----*----)
                  ---------+---------+---------+---------+
                      16.0      24.0      32.0      40.0
```

a. Why is a randomized block design preferred over a completely randomized design for comparing the mean number of oil rigs running monthly in California, Utah, and Alaska?

b. Identify the treatments for the experiment.

c. Identify the blocks for the experiment.

d. State the null hypothesis for the ANOVA F-test.

e. Locate the test statistic and p-value on the MINITAB printout above. Interpret the results.

f. A Tukey multiple comparison of means (at $\alpha = .05$) is summarized in the SPSS printout below. Which state(s) have the significantly largest mean number of oil rigs running monthly?

NUMRIGS

Tukey HSD[a,b]

STATE	N	Subset	
		1	2
AL	3	13.00	
UT	3	17.33	
CAL	3		32.33
Sig.		.262	1.000

Means for groups in homogeneous subsets are displayed.
Based on Type III Sum of Squares
The error term is Mean Square(Error) = 8.111.

a. Uses Harmonic Mean Sample Size = 3.000.

b. Alpha = .05.

10.121 College tennis recruiting with a team Web site. Most university athletic programs have a Web site with information on individual sports and a Prospective Student Athlete Form that allows high school athletes to submit information about their academic and sports achievements directly to the college coach. *The Sport Journal* (Winter 2004) published a study of how important team Web sites are to the recruitment of college tennis players. A survey was conducted of National Collegiate Athletic Association (NCAA) tennis coaches, of which 53 were from Division I schools, 20 were from Division II schools, and 53 were from Division III schools. Coaches were asked to respond to a series of statements, including "The Prospective Student Athlete Form on the Web site contributes very little to the recruiting process." Responses were measured on a seven-point scale (where

1 = strongly disagree and 7 = strongly agree). In order to compare the mean responses of tennis coaches from the three NCAA divisions, the data were analyzed with a completely randomized ANOVA design.

a. Identify the experimental unit, the dependent (response) variable, the factor, and the treatments in this study.

b. Give the null and alternative hypothesis for the ANOVA F-test.

c. The observed significance level of the test was found to be p-value $< .003$. What conclusion can you draw if you want to test at $\alpha = .05$?

d. The following results were obtained with the use of a multiple-comparison procedure with an experimentwise error rate of .05. Interpret the results practically.

Mean:	4.51	3.60	3.21
Division:	I	II	III

10.122 Alcohol-and-marriage study. An experiment was conducted to examine the effects of alcohol on the marital interactions of husbands and wives (*Journal of Abnormal Psychology*, Nov. 1998). A total of 135 couples participated in the experiment. The husband in each couple was classified as aggressive (60 husbands) or nonaggressive (75 husbands), on the basis of an interview and his response to a questionnaire. Before the marital interactions of the couples were observed, each husband was randomly assigned to one of three groups: receive no alcohol, receive several alcoholic mixed drinks, or receive placebos (nonalcoholic drinks disguised as mixed drinks). Consequently, a 2×3 factorial design was employed, with husband's aggressiveness at two levels (aggressive or nonaggressive) and husband's alcohol condition at three levels (no alcohol, alcohol, and placebo). The response variable observed during the marital interaction was severity of conflict (measured on a 100-point scale).

a. A partial ANOVA table is shown. Fill in the missing degrees of freedom.

Source	df	F
Aggressiveness (A)	–	16.43 ($p < .001$)
Alcohol Condition (C)	–	6.00 ($p < .01$)
A $\times$ C	–	–
Error	129	
Total	–	

b. Interpret the p-value of the F-test for Aggressiveness.

c. Interpret the p-value of the F-test for Alcohol Condition.

d. The F-test for interaction was omitted from the article. Discuss the dangers of making inferences based on the tests in parts **b** and **c** without knowing the result of the interaction test.

10.123 A new dental bonding agent. Refer to the *Trends in Biomaterials & Artificial Organs* (Jan. 2003) study of a bonding adhesive for teeth, presented in Exercise 8.154 (p. 455). Recall that the adhesive (called "Smartbond") has been developed to eliminate the necessity of a dry field. In one portion of the study, 30 extracted teeth were bonded with Smartbond and each was randomly assigned one of three different bonding times: 1 hour, 24 hours, or 48 hours. At the end of the bonding period, the breaking strength (in Mpa) of each tooth was determined. The data were analyzed with the use of analysis of variance in order to determine whether the true mean breaking strength of the new adhesive varies with the bonding time.

a. Identify the experimental units, treatments, and response variable for this completely randomized design.

b. Set up the null and alternative hypotheses for the ANOVA.

c. Find the rejection region using $\alpha = .01$.

d. The test results were $F = 61.62$ and p-value ≈ 0. Give the appropriate conclusion for the test.

e. What conditions are required for the test results to be valid?

f. The sample mean breaking strengths for the 3 bonding times were $\bar{x}_{1 \text{ hour}} = 3.32$ Mpa, $\bar{x}_{24 \text{ hours}} = 5.07$ Mpa, and $\bar{x}_{48 \text{ hours}} = 5.03$ Mpa. Using an experimentwise error rate of .05, Tukey's method detected no significant difference between the means at 24 and 48 hours; however, the mean at 1 hour was found to be significantly smaller than the other two means. Illustrate the results of the multiple comparisons of means by ordering the sample means and connecting means that are not significantly different. What practical conclusions can you draw from the analysis?

10.124 Contaminated fish in the Tennessee River. Refer to the U.S. Army Corps of Engineers data on contaminated fish, saved in the **DDT** file. The results of an ANOVA and Tukey multiple-comparison analysis to compare the three fish species (channel catfish, largemouth bass, and smallmouth buffalo) on the dependent variable length (in centimeters) are shown in the following MINITAB printout.

```
One-way ANOVA: LENGTH versus SPECIES

Source    DF    SS      MS      F      P
SPECIES    2  3533.1  1766.5  76.88  0.000
Error    141  3239.9    23.0
Total    143  6772.9

S = 4.794   R-Sq = 52.16%   R-Sq(adj) = 51.49%


Level            N    Mean   StDev
CHANNELCATFISH  96  44.729   4.581
LARGEMOUTHBASS  12  26.542   4.480
SMALLMOUTHBUFF  36  43.125   5.413

Pooled StDev = 4.794

Tukey 95% Simultaneous Confidence Intervals
All Pairwise Comparisons among Levels of SPECIES

Individual confidence level = 98.08%

SPECIES = CHANNELCATFISH subtracted from:

SPECIES          Lower   Center   Upper
LARGEMOUTHBASS  -21.664  -18.188  -14.711
SMALLMOUTHBUFF   -3.823   -1.604    0.615

SPECIES = LARGEMOUTHBASS subtracted from:

SPECIES          Lower   Center   Upper
SMALLMOUTHBUFF   12.798   16.583   20.368
```

a. Is there sufficient evidence (at $\alpha = .01$) to conclude that the mean lengths differ among the three fish species?

b. What is the experimentwise error rate for the Tukey multiple comparisons?

c. Locate the confidence interval for the difference $(\mu_{\text{bass}} - \mu_{\text{catfish}})$. Are the two means significantly different? If so, which mean is significantly larger?

d. Locate the confidence interval for the difference $(\mu_{\text{buffalo}} - \mu_{\text{catfish}})$. Are the two means significantly different? If so, which mean is significantly larger?

e. Locate the confidence interval for the difference $(\mu_{\text{buffalo}} - \mu_{\text{bass}})$. Are the two means significantly different? If so, which mean is significantly larger?

10.125 Are you lucky? Parapsychologists define "lucky" people as individuals who report that seemingly chance events consistently tend to work out in their favor. A team of British psychologists designed a study to examine the effects of luckiness and competition on performance in a guessing task (*The Journal of Parapsychology*, Mar. 1997). Each in a sample of 56 college students was classified as lucky, unlucky, or uncertain on the basis of their responses to a Luckiness Questionnaire. In addition, the participants were randomly assigned to either a competitive or a noncompetitive condition. All students were then asked to guess the outcomes of 50 flips of a coin. The response variable measured was percentage of coin flips correctly guessed.

a. A 2×3 factorial ANOVA design was conducted on the data. Identify the factors and their levels for this design.

b. The results of the ANOVA are summarized in the accompanying table. Interpret the results fully.

Source	df	F	p-Value
Luckiness (L)	2	1.39	.26
Competition (C)	1	2.84	.10
L × C	2	0.72	.72
Error	50		
Total	55		

Applying the Concepts—Intermediate

10.126 Income and road rage. The phenomenon of road rage has received much media attention in recent years. Is a driver's propensity to engage in road rage related to his/her income? Researchers at Mississippi State University attempted to answer this question by conducting a survey of a representative sample of over 1,000 U.S. adult drivers (*Accident Analysis and Prevention*, Vol. 34, 2002). Based on how often each driver engaged in certain road rage behaviors (e.g., making obscene gestures at, tailgating, and thinking about physically hurting another driver), a road rage score was assigned. (Higher scores indicate a greater pattern of road rage behavior.) The drivers were also grouped by annual income: under \$30,000, between \$30,000 and \$60,000, and over \$60,000. The data were subjected to an analysis of variance, with the results summarized in the next table.

Results for Exercise 10.126

Income Group	Sample Size	Mean Road Rage Score
Under $30,000	379	4.60
$30,000 to $60,000	392	5.08
Over $60,000	267	5.15
ANOVA results:	F-value $= 3.90$	p-value $< .01$

a. Is a driver's propensity to engage in road rage related to his/her income?

b. An experimentwise error rate of .01 was used to rank the three means. Give a practical interpretation of this error rate.

c. How many pairwise comparisons are necessary to compare the three means? List them.

d. A multiple-comparisons procedure revealed that the means for the two income groups Between $30 and $60 thousand and Over $60 thousand were not significantly different. All other pairs of means were found to be significantly different. Summarize these results in tabular form.

e. Which of the comparisons of part **c** will yield a confidence interval that does not contain 0?

10.127 Endangered dwarf shrubs. Rugel's pawpaw (yellow squirrel banana) is an endangered species of a dwarf shrub. Biologists conducted an experiment to determine the effects of fire on the shrub's growth (*Florida Scientist*, Spring 1997). Twelve experimental plots of land were selected in a pasture where the shrub is abundant. Within each plot, three pawpaws were randomly selected and treated as follows: One shrub was subjected to fire, another to clipping, and the third left unmanipulated (a control). After five months, the number of flowers produced by each of the 36 shrubs was determined. The objective of the study was to compare the mean number of flowers produced by pawpaws for the three treatments (fire, clipping, and control).

a. Identify the type of experimental design employed, including the treatments, response variable, and experimental units.

b. Illustrate the layout of the design, using a graphic similar to Figure 10.21.

c. The ANOVA of the data resulted in a test statistic of $F = 5.42$ for treatments with p-value $= .009$. Interpret this result.

d. The three treatment means were compared by Tukey's method at $\alpha = .05$. Interpret the results, shown as follows:

Mean number of flowers:	1.17	10.58	17.08
Treatment:	Control	Clipping	Burning

10.128 Exposure to low-frequency sound. *Infrasound* refers to sound frequencies below the audibility range of the human ear. A study of the physiological effects of infrasound was published in the *Journal of Low Frequency Noise, Vibration and Active Control* (Mar. 2004). In the experiment, one group of five university students (Group A) was exposed to infrasound at 4 hertz and 120 decibels for 1 hour, and a second group of five students (Group B) was exposed to infrasound at 2 hertz and 110 decibels, also for 1 hour. The heart rate (beats/minute) of each student was measured both before and after exposure. The experimental data are provided in the accompanying table. To determine the impact of infrasound, the researchers compared the mean heart rate before exposure to the mean heart rate after exposure.

Group A Students	Before Exposure	After Exposure	Group B Students	Before Exposure	After Exposure
A1	70	70	B1	73	79
A2	69	80	B2	68	60
A3	76	84	B3	61	69
A4	77	86	B4	72	77
A5	64	76	B5	61	66

Based on Qibai, C. Y. H., and Shi, H. "An investigation on the physiological and psychological effects of infrasound on persons." *Journal of Low Frequency Noise, Vibration and Active Control*, Vol. 23, No. 1, March 2004 (Tables I–IV).

a. Analyze the data on Group A students with an ANOVA for a randomized block design. Conduct the ANOVA test of interest with $\alpha = .05$.

b. Repeat part **a** for Group B students.

c. Now analyze the data via a paired difference t-test. Show that, for both groups of students, the results are equivalent to the randomized block ANOVA.

10.129 Commercial eggs produced from different housing systems. In the production of commercial eggs, four different types of housing systems for the chickens are used: cage, barn, free range, and organic. The characteristics of eggs produced from the four housing systems were investigated in *Food Chemistry* (Vol. 106, 2008). Twenty-eight commercial grade A eggs were randomly selected from supermarkets—10 of which were produced in cages, 6 in barns, 6 with free range, and 6 organically. A number of quantitative characteristics were measured for each egg, including shell thickness (millimeters), whipping capacity (percent overrun), and penetration strength (Newtons). The data (simulated from summary statistics provided in the journal article) are saved in the **EGGS2** file. For each characteristic, the researchers compared the means of the four housing systems. Conduct the ANOVA for each characteristic. Fully interpret the results. Identify the characteristics for which housing systems differ.

10.130 Mussel settlement patterns on algae. Mussel larvae are in great abundance in the drift material that washes up on Ninety Mile Beach in New Zealand. These larvae tend to settle on algae. Environmentalists at the University of Auckland investigated the impact of the type of algae on the abundance of mussel larvae in drift material (*Malacologia*, Feb. 8, 2002). Drift material from three different wash-up events on Ninety Mile Beach was collected; for each washup, the algae was separated into four strata: coarse-branching, medium-branching, fine-branching, and hydroid algae. Two samples were randomly selected for each of the $3 \times 4 = 12$ event–strata combinations, and the mussel density (percent per square centimeter) was measured for each. The data were analyzed as a complete 3×4 factorial design. The ANOVA summary table is as follows:

Source	df	F	p-Value
Event	2	.35	>.05
Strata	3	217.33	<.05
Interaction	6	1.91	>.05
Error	12		
Total	23		

a. Conduct the appropriate ANOVA F-tests (at $\alpha = .05$) and interpret the results.

b. Tukey multiple comparisons of the four algae strata means (at $\alpha = .05$) are summarized in the accompanying table. Which means are significantly different?

Mean $(\%/cm^2)$:	9	10	27	55
Algae stratum:	Coarse	Medium	Fine	Hydroid

10.131 Facial expression study. What do people infer from facial expressions of emotion? This was the research question of interest in an article published in the *Journal of Nonverbal Behavior* (Fall 1996). A sample of 36 introductory psychology students was randomly divided into six groups. Each group was assigned to view one of six slides showing a person making a facial expression.* The six expressions were (1) angry, (2) disgusted, (3) fearful, (4) happy, (5) sad, and (6) neutral. After viewing the slides, the students rated the degree of dominance they inferred from the facial expression (on a scale ranging from -15 to $+15$). The data (simulated from summary information provided in the article) are listed in the accompanying table.

a. Conduct an analysis of variance to determine whether the mean dominance ratings differ among the six facial expressions. Use $\alpha = .10$.

b. Use Tukey's method to rank the six dominance rating means. (Use $\alpha = .05$.)

Angry	Disgusted	Fearful	Happy	Sad	Neutral
2.10	.40	.82	1.71	.74	1.69
.64	.73	-2.93	$-.04$	-1.26	$-.60$
.47	$-.07$	$-.74$	1.04	-2.27	$-.55$
.37	$-.25$	.79	1.44	$-.39$	.27
1.62	.89	$-.77$	1.37	-2.65	$-.57$
$-.08$	1.93	-1.60	.59	$-.44$	-2.16

10.132 Effectiveness of geese decoys. Using decoys is a common method of hunting waterfowl. A study in the *Journal of Wildlife Management* (July 1995) compared the effectiveness of three different types of decoy—taxidermy-mounted decoys, plastic shell decoys, and full-bodied plastic decoys—in attracting Canada geese to sunken pit blinds. In order to account for an extraneous source of variation, three pit blinds were used as blocks in the experiment. Thus, a randomized block design with three treatments (types of decoy) and three blocks (pit blinds) was employed. The response variable was the percentage of a goose flock to approach within 46 meters of the pit blind on a given day. The data are given in the following table.[†]

Blind	Shell	Full Bodied	Taxidermy Mounted
1	7.3	13.6	17.8
2	12.6	10.4	17.0
3	16.4	23.4	13.6

Based on Harrey, W. F., Hindman, L. J., and Rhodes, W. E. "Vulnerability of Canada geese to taxidermy-mounted decoys." *Journal of Wildlife Management*, Vol. 59, No. 3, July 1995, p. 475 (Table 1).

a. Use statistical software (or the formulas in Appendix C) to construct an ANOVA table.

b. Interpret the F-statistic for comparing the response means of the three types of decoy.

c. What assumptions are necessary for the validity of the inference made in part **a**?

d. Why is it not necessary to conduct multiple comparisons of the response means for the three types of decoy?

10.133 Impact of vitamin-B supplement. In the *Journal of Nutrition* (July 1995), University of Georgia researchers examined the impact of a vitamin-B supplement (nicotinamide) on the kidney. The experimental "subjects" were 28 Zucker rats—a species that tends to develop kidney problems. Half of the rats were classified as obese and half as lean. Within each group, half were randomly assigned to receive a vitamin-B-supplemented diet and half were not. Thus, a 2×2 factorial experiment was conducted, with seven rats assigned to each of the four combinations of size (lean or obese) and diet (supplemental or not). One of the response variables measured was weight (in grams) of the kidney at the end of a 20-week feeding period. The data (simulated from summary information provided in the journal article) are shown in the accompanying table.

		Diet			
		Regular		Vitamin-B Supplement	
Rat Size	**Lean**	1.62	1.47	1.51	1.63
		1.80	1.37	1.65	1.35
		1.71	1.71	1.45	1.66
		1.81		1.44	
	Obese	2.35	2.84	2.93	2.63
		2.97	2.05	2.72	2.61
		2.54	2.82	2.99	2.64
		2.93		2.19	

a. Conduct an analysis of variance on the data. Summarize the results in an ANOVA table.

b. Conduct the appropriate ANOVA F-tests at $\alpha = .01$. Interpret the results.

*In the actual experiment, each group viewed all six facial expression slides and the design employed was a Latin Square (beyond the scope of this text).

[†]The actual design employed in the study was more complex than the randomized block design shown here. In the actual study, each number in the table represented the mean daily percentage of goose flocks attracted to the blind, averaged over 13–17 days.

10.134 Light-to-dark transition of genes. Refer to the *Journal of Bacteriology* (July 2002) study of the sensitivity of bacteria genes to light, presented in Exercise 9.142 (p. 520). Recall that scientists isolated 103 genes of the bacterium responsible for photosynthesis and respiration. Each gene was grown to midexponential phase in a growth incubator in "full light" and then was exposed to three alternative light/dark conditions: "full dark" (lights extinguished for 24 hours), "transient light" (lights kept on for another minutes), and "transient dark" (lights turned off for 90 minutes). At the end of each light/dark condition, the standardized growth measurement was determined for each of the 103 genes. The complete data set is saved in the **GENES** file. (Data for the first 10 genes are shown in the accompanying table.) Assume that the goal of the experiment is to compare the mean standardized growth measurements for the three light/dark conditions.

Gene ID	FULL-DARK	TR-LIGHT	TR-DARK
SLR2067	−0.00562	1.40989	−1.28569
SLR1986	−0.68372	1.83097	−0.68723
SSR3383	−0.25468	−0.79794	−0.39719
SLL0928	−0.18712	−1.20901	−0.18618
SLR0335	−0.20620	1.71404	−0.73029
SLR1459	−0.53477	2.14156	−0.33174
SLL1326	−0.06291	1.03623	−0.30392
SLR1329	−0.85178	−0.21490	0.44545
SLL1327	0.63588	1.42608	−0.13664
SLL1325	−0.69866	1.93104	−0.24820

Based on Gill, R. T., et al. "Genome-wide dynamic transcriptional profiling of the light to dark transition in *Synechocystis Sp*. PCC6803." *Journal of Bacteriology*, Vol. 184, No. 13, July 2002.

a. Explain why the data should be analyzed as a randomized block design.

b. Specify the null and alternative hypotheses for comparing the light/dark condition means.

c. Using a statistical software package, conduct the test you set up in part **b**. Interpret the results at $\alpha = .05$.

d. If necessary, employ a multiple-comparison procedure to rank the light/dark condition means. Use an experimentwise error rate of .05.

10.135 The thrill of a close game. Do women enjoy the thrill of a close basketball game as much as men do? To answer this question, male and female undergraduate students were recruited to participate in an experiment (*Journal of Sport & Social Issues*, Feb. 1997). The students watched one of eight live televised games of an NCAA basketball tournament. (None of the games involved a home team to which the students could be considered emotionally committed.) The "suspense" of each game was classified into one of four categories according to the closeness of scores at the game's conclusions: minimal (15-point or greater differential), moderate (10–14-point differential), substantial (5–9-point differential), and extreme (1–4-point differential). After the game, each student rated his/her enjoyment on an 11-point scale ranging from 0 (not at all) to 10 (extremely). The enjoyment rating data were analyzed as a 4×2 factorial design, with suspense (four levels) and gender (two levels) as the two factors. The $4 \times 2 = 8$ treatment means are shown in the following table.

	Gender	
Suspense	Male	Female
Minimal	1.77	2.73
Moderate	5.38	4.34
Substantial	7.16	7.52
Extreme	7.59	4.92

Based on Gan, Su-lin, et al. "The thrill of a close game: Who enjoys it and who doesn't?" *Journal of Sport & Social Issues*, Vol. 21, No. 1, Feb. 1997, pp. 59–60.

a. Plot the treatment means in a graph similar to Figure 10.29. Does the pattern of means suggest interaction between suspense and gender? Explain.

b. The ANOVA F-test for interaction yielded the following results: numerator df = 3, denominator df = 68, $F = 4.42$, p-value = .007. What can you infer from these results?

c. On the basis of the test carried out in part **b**, is the difference between the mean enjoyment levels of males and females the same, regardless of the suspense level of the game?

10.136 Ranging behavior of Spanish cattle. The cattle inhabiting the Biological Reserve of Doñana (Spain), live under free-range conditions, with virtually no human interference. The cattle population is organized into four herds (LGN, MTZ, PLC, and QMD). The *Journal of Zoology* (July 1995) investigated the ranging behavior of the four herds across the four seasons. Thus, a 4×4 factorial experiment was employed, with Herd and Season representing the two factors. Three animals from each herd during each season were sampled and the home range of each individual was measured (in square kilometers). The data were subjected to an ANOVA, with the results shown in the following table.

Source	df	F	p-Value
Herd (H)	3	17.2	$p < .001$
Season (S)	3	3.0	$p < .05$
H × S	9	1.2	$p > .05$
Error	32		
Total	47		

a. Conduct the appropriate ANOVA F-tests and interpret the results.

b. The researcher conducting the experiment ranked the four herd means independently of season. Do you agree with this strategy? Explain.

c. Refer to part **b** The Bonferroni rankings of the four herd means (at $\alpha = .05$) are shown in the next table. Interpret the results.

Mean (km)^2:	.75	1.0	2.7	3.8
Herd:	PLC	LGN	QMD	MTZ

10.137 Estimating the age of glacial drifts. Refer to the *American Journal of Science* (Jan. 2005) study of the chemical makeup of buried tills (glacial drifts) in Wisconsin, presented in Exercise 2.192 (p. 135). The ratio of the elements aluminum (Al) and beryllium (Be) in sediment is related to the duration of burial. Recall that the AI/Be ratios for a sample of 26 buried till specimens were determined. The till specimens were obtained from five different boreholes (labeled UMRB-1, UMRB-2, UMRB-3, SWRA, and SD). The data are shown here. Conduct an analysis of variance of the data. Is there sufficient evidence to indicate differences among the mean AI/Be ratios for the five boreholes? If so, which borehole has the highest mean ratio? Test, using $\alpha = .10$.

UMRB-1:	3.75	4.05	3.81	3.23	3.13	3.30	3.21
UMRB-2:	3.32	4.09	3.90	5.06	3.85	3.88	
UMRB-3:	4.06	4.56	3.60	3.27	4.09	3.38	3.37
SWRA:	2.73	2.95	2.25				
SD:	2.73	2.55	3.06				

Source: Adapted from *American Journal of Science*, Vol. 305, No. 1, Jan. 2005, p. 16 (Table 2).

Applying the Concepts—Advanced

10.138 Therapy for binge eaters. Do you experience episodes of excessive eating accompanied by being overweight? If so, you may suffer from binge eating disorder. Cognitive-behavioral therapy (CBT), in which patients are taught how to make changes in specific behavior patterns (e.g., exercise, eat only low-fat foods), can be effective in treating the disorder. A group of Stanford University researchers investigated the effectiveness of interpersonal therapy (IPT) as a second level of treatment for binge eaters (*Journal of Consulting and Clinical Psychology*, June 1995). The researchers employed a design that randomly assigned a sample of 41 overweight individuals diagnosed with binge eating disorder to either a treatment group (30 subjects) or a control group (11 subjects). Subjects in the treatment group received 12 weeks of CBT and then were subdivided into two groups. Those who responded successfully to CBT (17 subjects) were assigned to a weight-loss therapy (WLT) program for the next 12 weeks. Those CBT subjects who did not respond to treatment (13 subjects) received 12 weeks of IPT. The subjects in the control group received no therapy of any type. Thus, the study ultimately consisted of three groups of overweight binge eaters: the CBT-WLT group, the CBT-IPT group, and the control group. One outcome (response) variable measured for each subject was the number x of binge eating episodes per week. Summary statistics for each of the three groups at the end of the 24-week period are shown in the accompanying table. The data were analyzed as a completely randomized design with three treatments (CBT-WLT, CBT-IPT, and Control). Although the ANOVA tables were not provided in the article, sufficient information is given in the table to reconstruct them. [See Exercise 10.43 (p. 555).] Is CBT effective in reducing the mean number of binges experienced per week?

	CBT-WLT	CBT-IPT	Control
Sample size	17	13	11
Mean number of binges per week	0.2	1.9	2.9
Standard deviation	0.4	1.7	2.0

Based on Agras, W. S., et al. "Does interpersonal therapy help patients with binge eating disorder who fail to respond to cognitive-behavioral therapy?" *Journal of Consulting and Clinical Psychology*, Vol. 63, No. 3, June 1995, p. 358 (Table 1).

10.139 On the trail of the cockroach. Knowledge of how cockroaches forage for food is valuable for companies that develop and manufacture roach bait and traps. Many entomologists believe, however, that the navigational behavior of cockroaches scavenging for food is random. D. Miller of Virginia Tech University challenged the "random-walk" theory by designing an experiment to test a cockroach's ability to follow a trail of their fecal material (*Explore*, Research at the University of Florida, Fall 1998).

A methanol extract from roach feces—called a *pheromane*—was used to create a chemical trail. German cockroaches were released at the beginning of the trail, one at a time, and a video surveillance camera was used to monitor the roach's movements. In addition to the trail containing the fecal extract (the treatment), a trail using methanol only (the control) was created. To determine if trail-following ability differed among cockroaches of different age, sex, and reproductive status, four roach groups were used in the experiment: adult males, adult females, gravid (pregnant) females, and nymphs (immatures). Twenty roaches of each type were randomly assigned to the treatment trail, and 10 of each type were randomly assigned to the control trail. Thus, a total of 120 roaches were used in the experiment. The movement pattern of each cockroach was measured (in "pixels") as the average trail deviation. The data for the 120 cockroaches in the study are stored in the accompanying file. (The first 5 and last 5 observations in the data set are listed here.) Conduct a complete analysis of the data. Determine whether roaches can distinguish between the fecal extract and control trail and whether trail-following ability differs according to age, sex, and reproductive status.

Trail Deviation	Roach Group	Trail
3.1	Adult Male	Extract
42.0	Adult Male	Control
6.2	Adult Male	Extract
22.7	Adult Male	Control
34.0	Adult Male	Extract
⋮	⋮	⋮
23.8	Nymph	Extract
5.1	Nymph	Extract
3.8	Nymph	Extract
3.1	Nymph	Extract
2.8	Nymph	Extract

Critical Thinking Challenges

10.140 Exam performance study. *Teaching of Psychology* (Aug. 1998) published a study of whether a practice test helps students prepare for a final exam. Students in an introductory psychology class were grouped according to their class standing and whether they attended a review session or took a practice test prior to the final exam. The experimental design was a 3 × 2 factorial design, with Class Standing at 3 levels (low, medium, high) and Exam Preparation at 2 levels (practice exam, review session). There were 22 students in each of the 3 × 2 = 6 treatment groups. After completing the final exam, each student rated her or his exam preparation on an 11-point scale ranging from 0 (not helpful at all) to 10 (extremely helpful). The data for this experiment (simulated from summary statistics provided in the article) are saved in the **EXPREP** file. The first five and last five observations in the data set are listed in the accompanying table. Conduct a complete analysis of variance of the helpfulness ratings data, including (if warranted) multiple comparisons of means. Do your findings support the article's conclusion that "Students at all levels of academic ability benefit from a … practice exam"?

Exam Preparation	Class Standing	Helpfulness Rating
PRACTICE	LOW	6
PRACTICE	LOW	7
PRACTICE	LOW	7
PRACTICE	LOW	5
PRACTICE	LOW	3
⋮	⋮	⋮
REVIEW	HI	5
REVIEW	HI	2
REVIEW	HI	5
REVIEW	HI	4
REVIEW	HI	3

Based on Balch, W. R. "Practice versus review exams and final exam performance." *Teaching of Psychology*, Vol. 25, No. 3, Aug. 1998 (Table 1).

10.141 Anticorrosive behavior of steel coated with epoxy.

EPOXY Organic coatings that use epoxy resins are widely used to protect steel and metal against weathering and corrosion. Researchers at National Technical University in Athens, Greece, examined the steel anticorrosive behavior of different epoxy coatings formulated with zinc pigments in an attempt to find the epoxy coating with the best resistance to corrosion (*Pigment & Resin Technology*, Vol. 32, 2003). The experimental units were flat, rectangular panels cut from steel sheets. Each panel was coated with one of four different coating systems: S1, S2, S3, and S4. Three panels, labeled, S1-A, S1-B, S1-C, S2-A, S2-B, …, S4-C, were prepared for each coating system. The characteristics of the four coating systems are listed in the following table.

Coating System	First Layer	Second Layer
S1	Zinc dust	Epoxy paint, 100 micrometers thick
S2	Zinc phosphate	Epoxy paint, 100 micrometers thick
S2	Zinc phosphate with mica	Finish layer, 100 micrometers thick
S4	Zinc phosphate with mica	Finish layer, 200 micrometers thick

Each coated panel was immersed in deionized and deaerated water and then tested for corrosion. Since exposure time is likely to have a strong influence on anticorrosive behavior, the researchers attempted to remove this extraneous source of variation through the experimental design. Exposure times were fixed at 24 hours, 60 days, and 120 days. For each of the coating systems, one panel was exposed to water for 24 hours, one for 60 days, and one for 120 days, in random order. The design is illustrated in the following diagram.

Exposure Time	Coating System/Panel Exposed
24 Hours	S1-A, S2-C, S3-C, S4-B
60 Days	S1-C, S2-A, S3-B, S4-A
120 Days	S1-B, S2-B, S3-A, S4-C

Following exposure, the corrosion rate (in nanoamperes per square centimeter) was determined for each panel. The lower the corrosion rate, the greater the anticorrosion performance of the coating system. The data are shown in the next table. Are there differences among the epoxy treatment means? If so, which of the epoxy coating systems yields the lowest corrosion rate?

Exposure Time	System S1	System S2	System S3	System S4
24 Hours	6.7	7.5	8.2	6.1
60 Days	8.7	9.1	10.5	8.3
120 Days	11.8	12.6	14.5	11.8

Source: Kouloumbi, N., et al. "Anticorrosion performance of epoxy coatings on steel surface exposed to de-ionized water," *Pigment & Resin Technology*, Vol. 32, No. 2, 2003 (Table II).

Activity Comparing Supermarket Food Prices

Due to ever-increasing food costs, consumers are becoming more discerning in their choice of supermarkets. It is usually more convenient to shop at just one market as opposed to buying different items at different markets. Thus, it would be useful to compare the mean food expenditure for a market basket of food items from store to store. Since there is a great deal of variability in the prices of products sold at any supermarket, we will consider an experiment that blocks on products.

Choose three (or more) supermarkets in your area that you want to compare. Then choose approximately 10 (or more) food products you typically purchase. For each food item, record the price each store charges in the following manner:

Food Item 1	Food Item 2	⋯	Food Item 10
Price store 1	Price store 1	⋯	Price store 1
Price store 2	Price store 2	⋯	Price store 2
Price store 3	Price store 3	⋯	Price store 3

Use the data you obtain to test

H_0: Mean expenditures at the stores are the same.

H_a: Mean expenditures for at least two of the stores are different.

Also, test to determine whether blocking on food items is advisable in this kind of experiment. Interpret the results of your analysis fully.

References

Cochran, W. G., and Cox, G. M. *Experimental Designs*, 2nd ed. New York: Wiley, 1957.

Hsu, J. C. *Multiple Comparisons: Theory and Methods*. London: Chapman & Hall, 1996.

Kramer, C. Y. "Extension of multiple range tests to group means with unequal number of replications." *Biometrics*, Vol. 12, 1956, pp. 307–310.

Kutner, M., Nachtsheim, C., Neter, J., and Li, W. *Applied Linear Statistical Models*, 5th ed. New York: McGraw-Hill/Irwin, 2005.

Mason, R. L., Gunst, R. F., and Hess, J. L. *Statistical Design and Analysis of Experiments*. New York: Wiley, 1989.

Mendenhall, W. *Introduction to Linear Models and the Design and Analysis of Experiments*. Belmont, CA: Wadsworth, 1968.

Miller, R. G., Jr. *Simultaneous Statistical Inference*. New York: Springer-Verlag, 1981.

Rekab, K. & Shaikh, M. *Statistical Design of Experiments with Engineering Applications*, CRC publication, 2005.

Scheffé, H. "A method for judging all contrasts in the Analysis of Variance," *Biometrica*, Vol. 40, 1953, pp. 87–104.

Scheffé, H. *The Analysis of Variance*. New York: Wiley, 1959.

Snedecor, G. W., and Cochran, W. G. *Statistical Methods*, 7th ed. Ames, IA: Iowa State University Press, 1980.

Steele, R. G. D., and Torrie, J. H. *Principles and Procedures of Statistics: A Biometrical Approach*, 2nd ed. New York: McGraw-Hill, 1980.

Tukey, J. "Comparing individual means in the Analysis of Variance," *Biometrics*, Vol. 5, 1949, pp. 99–114.

Winer, B. J. *Statistical Principles in Experimental Design*, 2nd ed. New York: McGraw-Hill, 1971.

USING TECHNOLOGY | **MINITAB: Analysis of Variance**

MINITAB can conduct ANOVAs for all three types of experimental designs discussed in this chapter: completely randomized, randomized block, and factorial designs.

Completely Randomized Design

Step 1 Access the MINITAB worksheet file that contains the sample data. The data file should contain one quantitative variable (the response, or dependent, variable) and one factor variable with at least two levels.

Step 2 Click on the "Stat" button on the MINITAB menu bar and then click on "ANOVA" and "One-Way," as shown in Figure 10.M.1.

Figure 10.M.2 MINITAB one-way ANOVA dialog box

Figure 10.M.1 MINITAB menu options for one-way ANOVA

Step 3 On the resulting dialog screen Figure 10.M.2, specify the response variable in the "Response" box and the factor variable in the "Factor" box.

Step 4 Click the "Comparisons" button and select a multiple comparisons method and experimentwise error rate in the resulting dialog box (see Figure 10.M.3).

Step 5 Click "OK" to return to the "One-Way ANOVA" dialog screen and then click "OK" to generate the MINITAB printout.

Figure 10.M.3 MINITAB multiple comparisons dialog box

Randomized Block and Factorial Designs

Step 1 Access the MINITAB worksheet file that contains the sample data. The data file should contain one quantitative variable (the response, or dependent, variable) and two other variables that represent the factors and/or blocks.

Step 2 Click on the "Stat" button on the MINITAB menu bar, and then click on "ANOVA" and "Two-Way" (see Figure 10.M.1). The resulting dialog screen appears as shown in Figure 10.M.4.

Figure 10.M.4 MINITAB two-way ANOVA dialog box

Step 3 Specify the response variable in the "Response" box, the first factor variable in the "Row factor" box, and the second factor or block variable in the "Column factor" box. If the design is a randomized block, select the "Fit additive model" option, as shown in Figure 10.M.4. If the design is factorial, leave the "Fit additive model" option unselected.

Step 4 Click "OK" to generate the MINITAB printout.

Note: Multiple comparisons of treatment means are obtained by selecting "Stat," then "ANOVA," and then "General Linear Models." Specify the factors in the "Model" box and then select "Comparisons" and put the factor of interest in the "Terms" box. Press "OK" twice.

TI–83/TI–84 Plus Graphing Calculator: Analysis of Variance

The TI-83/TI-84 plus graphing calculator can be used to compute a one-way ANOVA for a completely randomized design but not a two-way ANOVA for either a randomized block or factorial design.

Completely Randomized Design

Step 1 *Enter each data set into its own list (i.e., sample 1 into L1, sample 2 into L2, sample 3 into L3, etc.)*

Step 2 *Access the statistical test menu*

- Press **STAT**
- Arrow right to **TESTS**
- Arrow down to **ANOVA**
- Press **ENTER**
- Type in each List name, separated by commas (*e.g., L1, L2, L3, L4*)
- Press **ENTER**

Step 3 *View display*

The calculator will display the F-test statistic, as well as the p-value, the factor degrees of freedom, sum of squares, mean square, and by arrowing down, the Error degrees of freedom, sum of squares, mean square, and the pooled standard deviation.

11 Simple Linear Regression

CONTENTS

Where We've Been

- Presented methods for estimating and testing population parameters (e.g., the mean, proportion, and variance) for a single sample
- Extended these methods to allow for a comparison of population parameters for multiple samples

Where We're Going

- Introduce the straight-line (*simple linear regression*) model as a means of relating one quantitative variable to another quantitative variable (11.1)
- Assess how well the simple linear regression model fits the sample data (11.2–11.4)
- Introduce the *correlation coefficient* as a means of relating one quantitative variable to another quantitative variable (11.5)
- Utilize the simple linear regression model to predict the value of one variable from a specified value of another variable (11.6, 11.7)

Statistics IN Action Can "Dowsers" Really Detect Water?

The act of searching for and finding underground supplies of water with the use of nothing more than a divining rod is commonly known as "dowsing." Although widely regarded among scientists as no more than a superstitious relic from medieval times, dowsing remains popular in folklore, and to this day, there are individuals who claim to have this mysterious skill. In fact, as recently as March 2014, a group of California farmers hired dowsers to find water during a drought.

Many dowsers claim that they respond to "earthrays" that emanate from the water source. Earthrays, say the dowsers, are a subtle form of radiation that is potentially hazardous to human health. As a result of these claims, in the mid-1980s the German government conducted a two-year experiment to investigate the possibility that dowsing is a genuine skill. If such a skill could be demonstrated, reasoned government officials, then dangerous levels of radiation in Germany could be detected, avoided, and disposed of.

A group of university physicists in Munich, Germany, was provided a grant of 400,000 marks (about $250,000) to conduct the study. Approximately 500 candidate dowsers were recruited to participate in preliminary tests of their skill. To avoid fraudulent claims, the 43 individuals who seemed to be the most successful in the preliminary tests were selected for the final, carefully controlled, experiment.

The researchers set up a 10-meter-long line on the ground floor of a vacant barn, along which a small wagon could be moved. Attached to the wagon was a short length of pipe, perpendicular to the test line, that was connected by hoses to a pump with running water. The location of the pipe along the line for each trial of the experiment was assigned by a computer-generated random number. On the upper floor of the barn, directly above the experimental line, a 10-meter test line was painted. In each trial, a dowser was admitted to this upper level and required, with his/her rod, stick, or other tool of choice, to ascertain where the pipe with running water on the ground floor was located.

Each dowser participated in at least one test series constituting a sequence of from 5 to 15 trials (typically, 10), with the pipe randomly repositioned after each trial. (Some dowsers undertook only 1 test series, whereas selected others underwent more than 10 test series.) Over the two-year experimental period, the 43 dowsers participated in a total of 843 tests. The experiment was "double blind" in that neither the observer (researcher) on the top floor nor the dowser knew the pipe's location, even after a guess was made. [*Note:* Before the experiment began, a professional magician inspected the entire arrangement for potential deception or cheating by the dowsers.]

For each trial, two variables were recorded: the actual location of the pipe (in decimeters from the beginning of the line) and the dowser's guess (also measured in decimeters). On the basis of an examination of these data, the German physicists concluded in their final report that although most dowsers did not do particularly well in the experiments, "some few dowsers, in particular tests, showed an extraordinarily high rate of success, which can scarcely if at all be explained as due to chance...a real core of dowser-phenomena can be regarded as empirically proven..." (Wagner, Betz, and König, 1990. Final Report 01 KB8602, Federal Ministry for Research and Technology).

This conclusion was critically assessed by Professor J. T. Enright of the University of California at San Diego (*Skeptical Inquirer*, Jan./Feb. 1999). In the Statistics in Action Revisited sections of this chapter, we demonstrate how Enright concluded the exact opposite of the German physicists. [*Note:* Enright's assessment of the German study led other researchers, notably Vogtand Hyman (2000), Carroll (2003), and Whitaker (2006), to conduct their own scientific studies debanking water dowsing.]

Statistics IN Action Revisited

- Estimating a Straight-Line Regression Model for the Dowsing Data (p. 623)
- Assessing How Well the Straight-Line Model Fits the Dowsing Data (p. 640)
- Using the Coefficients of Correlation and Determination to Assess the Dowsing Data (p. 651)
- Using the Straight-Line Model to Predict Pipe Location for the Dowsing Data (p. 659)

In Chapters 7–10, we described methods for making inferences about population means. The mean of a population has been treated as a *constant*, and we have shown how to use sample data to estimate or to test hypotheses about this constant mean. In many applications, the mean of a population is not viewed as a constant, but rather as a variable. For example, the mean sale price of residences in a large city might be treated as a variable that depends on the number of square feet of living space in the residence. The relationship might be

$$\text{Mean sale price} = \$30{,}000 + \$60\,(\text{Square feet})$$

This formula implies that the mean sale price of 1,000-square-foot homes is $90,000, the mean sale price of 2,000-square-foot homes is $150,000, and the mean sale price of 3,000-square-foot homes is $210,000.

In this chapter, we discuss situations in which the mean of the population is treated as a variable, dependent on the value of another variable. The dependence of the residential sale price on the number of square feet of living space is one illustration. Other examples include the dependence of the mean reaction time on the amount of a drug in the bloodstream, the dependence of the mean starting salary of a college graduate on the student's GPA, and the dependence of the mean number of years to which a criminal is sentenced on the number of previous convictions.

Here, we present the simplest of all models relating a populating mean to another variable: the *straight-line model*. We show how to use the sample data to estimate the straight-line relationship between the mean value of one quantitative variable, *y*, as it relates to a second quantitative variable, *x*. The methodology of estimating and using a straight-line relationship is referred to as *simple linear regression analysis*.

11.1 Probabilistic Models

An important consideration in taking a drug is how it may affect one's perception or general awareness. Suppose you want to model the length of time it takes to respond to a stimulus (a measure of awareness) as a function of the percentage of a certain drug in the bloodstream. The first question to be answered is this: "Do you think that an exact relationship exists between these two variables?" That is, do you think that it is possible to state the exact length of time it takes an individual (subject) to respond if the amount of the drug in the bloodstream is known? We think that you will agree with us that this is *not* possible, for several reasons: The reaction time depends on many variables other than the percentage of the drug in the bloodstream—for example, the time of day, the amount of sleep the subject had the night before, the subject's visual acuity, the subject's general reaction time without the drug, and the subject's age. Even if many variables are included in a model (the topic of Chapter 12), it is still unlikely that we would be able to predict the subject's reaction time *exactly*. There will almost certainly be some variation in response times due strictly to *random phenomena* that cannot be modeled or explained.

If we were to construct a model that hypothesized an exact relationship between variables, it would be called a **deterministic model**. For example, if we believe that *y*, the reaction time (in seconds), will be exactly one-and-one-half times *x*, the amount of drug in the blood, we write

$$y = 1.5x$$

This represents a **deterministic relationship** between the variables *y* and *x*. It implies that *y* can always be determined exactly when the value of *x* is known. *There is no allowance for error in this prediction.*

If, however, we believe that there will be unexplained variation in reaction times—perhaps caused by important, but unincluded, variables or by random phenomena—we discard the deterministic model and use a model that accounts for this **random error**. Our **probabilistic model** will include both a deterministic component and a random-error component. For example, if we hypothesize that the response time *y* is related to the percentage *x* of drug by

$$y = 1.5x + \text{Random error}$$

we are hypothesizing a **probabilistic relationship** between *y* and *x*. Note that the deterministic component of this probabilistic model is 1.5*x*.

Figure 11.1a shows the possible responses for five different values of *x*, the percentage of drug in the blood, when the model is deterministic. All the responses must fall exactly on the line because a deterministic model leaves no room for error.

Figure 11.1b shows a possible set of responses for the same values of *x* when we are using a probabilistic model. Note that the deterministic part of the model (the straight line itself) is the same. Now, however, the inclusion of a random-error component allows the response

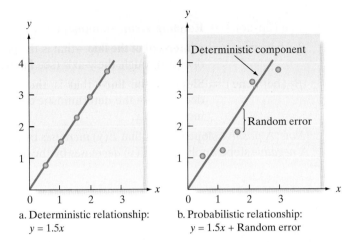

Figure 11.1

Possible reaction times y for five different drug percentages x

a. Deterministic relationship: $y = 1.5x$

b. Probabilistic relationship: $y = 1.5x + $ Random error

times to vary from this line. Since we know that the response time does vary randomly for a given value of x, the probabilistic model for y is more realistic than the deterministic model.

General Form of Probabilistic Models

$$y = \text{Deterministic component} + \text{Random error}$$

where y is the variable of interest. We always assume that the mean value of the random error equals 0. This is equivalent to assuming that the mean value of y, $E(y)$, equals the deterministic component of the model; that is,

$$E(y) = \text{Deterministic component}$$

BIOGRAPHY FRANCIS GALTON (1822–1911)

The Law of Universal Regression

Francis Galton was the youngest of seven children born to a middle-class English family of Quaker faith. A cousin of Charles Darwin, Galton attended Trinity College (Cambridge, England) to study medicine. Due to the death of his father, Galton was unable to obtain his degree. His competence in both medicine and mathematics, however, led Galton to pursue a career as a scientist. He made major contributions to the fields of genetics, psychology, meteorology, and anthropology. Some consider Galton to be the first social scientist for his applications of the novel statistical concepts of the time—in particular, regression and correlation. While studying natural inheritance in 1886, Galton collected data on heights of parents and adult children. He noticed the tendency for tall (or short) parents to have tall (or short) children, but that the children were not as tall (or short), on average as their parents. Galton called this phenomenon the "law of universal regression," for the average heights of adult children tended to "regress" to the mean of the population. With the help of his friend and disciple, Karl Pearson, Galton applied the straight-line model to the height data, and the term *regression model* was coined. ■

In this chapter, we present the simplest of probabilistic models—the **straight-line model**—which gets its name from the fact that the deterministic portion of the model graphs as a straight line. Fitting this model to a set of data is an example of **regression analysis,** or **regression modeling**. The elements of the straight-line model are summarized in the following box:

A First-Order (Straight-Line) Probabilistic Model

$$y = \beta_0 + \beta_1 x + \varepsilon$$

where

$y = $ **Dependent** *or* **response variable** (quantitative variable to be modeled or predicted)

$x = $ **Independent** *or* **predictor variable** (quantitative variable used as a predictor of y)*

$\beta_0 + \beta_1 x = E(y) = $ Deterministic component

(continued)

---

*The word *independent* should not be interpreted in a probabilistic sense as defined in Chapter 3. The phrase *independent variable* is used in regression analysis to refer to a predictor variable for the response y.

ε (epsilon) = **Random error** component

β_0 (beta zero) = **y-intercept of the line**—that is, the point at which the line intersects, or cuts through, the y-axis (see Figure 11.2)

β_1 (beta one) = **Slope of the line** that is, the change (amount of increase or decrease) in the deterministic component of y for every one-unit increase in x.

[*Note*: A *positive* **slope** implies that $E(y)$ *increases* by the amount β_1. (See Figure 11.2.) A *negative* slope implies that $E(y)$ *decreases* by the amount β_1.]

Figure 11.2
The straight-line model

In the probabilistic model, the deterministic component is referred to as the **line of means** because the mean of y, $E(y)$, is equal to the straight-line component of the model. That is,

$$E(y) = \beta_0 + \beta_1 x$$

Note that the Greek symbols β_0 and β_1 respectively represent the y-intercept and slope of the model. They are population parameters that will be known only if we have access to the entire population of (x, y) measurements. Together with a specific value of the independent variable x, they determine the mean value of y, which is just a specific point on the line of means (Figure 11.2).

| **Example** 11.1 | **Problem** Consider an experiment designed to estimate the linear relationship between the percentage of a certain drug in the bloodstream of a subject and the length of time it takes the subject to react to a stimulus. In particular, the researchers want to predict reaction time based on the amount of drug in the bloodstream. |

Modeling the Reaction Time to a Drug

a. In this study, identify the dependent and independent variables.

b. Explain why a probabilistic model is more appropriate than a deterministic model.

c. Write the equation of the straight-line, probabilistic model.

Solution

a. Since the researchers want to predict (or model) reaction time, then y = reaction is the dependent variable. And since the researchers want to use amount of drug in the bloodstream to make the prediction, x = amount of drug is the independent variable.

b. It would be unrealistic to expect all subjects with the same amount of drug in the bloodstream (x) to have the same reaction time (y). That is, we don't expect amount of drug to determine, exactly, a subject's reaction time due to subject-to-subject variation. Consequently, a probabilistic model is more appropriate than a deterministic model.

c. The probabilistic model takes the form $y = \beta_0 + \beta_1 x + \varepsilon$, where β_0 is the y-intercept of the line, β_1 is the slope of the line, and ε represents random error.

indicate that the mass of the spill tends to diminish as time increases? If so, how much will the mass diminish each minute?

11.36 Forecasting movie revenues with Twitter. Marketers are
Ⓓ keenly interested in how social media (e.g., Facebook,
TWEETS Twitter) may influence consumers who buy their products. Researchers at HP Labs (Palo Alto, CA) investigated whether the volume of chatter on Twitter.com could be used to forecast the box office revenues of movies (*IEEE International Conference on Web Intelligence and Intelligent Agent Technology*, 2010). Opening weekend box office revenue data (in millions of dollars) were collected for a sample of 24 recent movies. In addition, the researchers computed each movie's *tweet rate*, i.e., the average number of tweets (at Twitter.com) referring to the movie per hour one week prior to the movie's release. The data (simulated based on information provided in the study) are listed in the table. Assuming that movie revenue and tweet rate are linearly related, how much do you estimate a movie's opening weekend revenue to change as the tweet rate for the movie increases by an average of 100 tweets per hour?

Tweet Rate	Revenue (millions)
1365.8	142
1212.8	77
581.5	61
310.1	32
455	31
290	30
250	21
680.5	18
150	18
164.5	17
113.9	16
144.5	15
418	14
98	14
100.8	12
115.4	11
74.4	10
87.5	9
127.6	9
52.2	9
144.1	8
41.3	2
2.75	0.3

11.3 Model Assumptions

In Section 11.2, we assumed that the probabilistic model relating the drug reaction time y to the percentage x of drug in the bloodstream is

$$y = \beta_0 + \beta_1 x + \varepsilon$$

We also recall that the least squares estimate of the deterministic component of the model, $\beta_0 + \beta_1 x$, is

$$\hat{y} = \hat{\beta}_0 + \hat{\beta}_1 x = -.1 + .7x$$

Now we turn our attention to the random component ε of the probabilistic model and its relation to the errors in estimating β_0 and β_1. We will use a probability distribution to characterize the behavior of ε. We will see how the probability distribution of ε determines how well the model describes the relationship between the **dependent variable** y and the independent variable x.

Step 3 in a regression analysis requires us to specify the probability distribution of the random error ε. We will make four basic assumptions about the general form of this probability distribution:

Assumption 1 The mean of the probability distribution of ε is 0. That is, the average of the values of ε over an infinitely long series of experiments is 0 for each setting of the independent variable x. This assumption implies that the mean value of y, for a given value of x, is $E(y) = \beta_0 + \beta_1 x$.

Assumption 2 The variance of the probability distribution of ε is constant for all settings of the independent variable x. For our straight-line model, this assumption means that the variance of ε is equal to a constant—say, σ^2—for all values of x.

Assumption 3 The probability distribution of ε is normal.

Assumption 4 The values of ε associated with any two observed values of y are independent. That is, the value of ε associated with one value of y has no effect on any of the values of ε associated with any other y values.

The implications of the first three assumptions can be seen in Figure 11.9, which shows distributions of errors for three values of x, namely, 5, 10, and 15. Note that the relative frequency distributions of the errors are normal with a mean of 0 and a constant variance σ^2. (All of the distributions shown have the same amount of spread or variability.) The straight line shown in Figure 11.9 is the line of means; it indicates the mean value of y for a given value of x. We denote this mean value as $E(y)$. Then the line of means is given by the equation

$$E(y) = \beta_0 + \beta_1 x$$

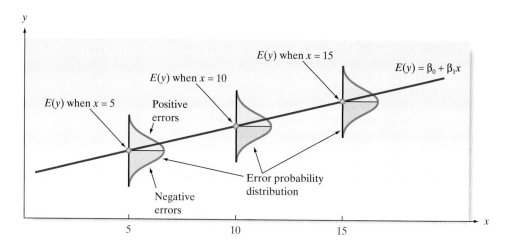

Figure 11.9
The probability distribution of ε

These assumptions make it possible for us to develop measures of reliability for the least squares estimators and to devise hypothesis tests for examining the usefulness of the least squares line. We have various techniques for checking the validity of these assumptions, and we have remedies to apply when they appear to be invalid. Several remedies are discussed in Chapter 12. Fortunately, the assumptions need not hold exactly in order for least squares estimators to be useful. The assumptions will be satisfied adequately for many applications encountered in practice.

It seems reasonable to assume that the greater the variability of the random error ε (which is measured by its variance σ^2), the greater will be the errors in the estimation of the model parameters β_0 and β_1 and in the error of prediction when $\hat{y}$ is used to predict y for some value of x. Consequently, you should not be surprised, as we proceed through this chapter, to find that σ^2 appears in the formulas for all confidence intervals and test statistics that we will be using.

Estimation of σ^2 for a (First-Order) Straight-Line Model

$$s^2 = \frac{\text{SSE}}{\text{Degrees of freedom for error}} = \frac{\text{SSE}}{n-2}$$

where $\text{SSE} = \sum (y_i - \hat{y}_i)^2 = \text{SS}_{yy} - \hat{\beta}_1 \text{SS}_{xy}$

in which

$$\text{SS}_{yy} = \sum (y_i - \bar{y})^2 = \sum y_i^2 - \frac{\left(\sum y_i\right)^2}{n}$$

To estimate the standard deviation σ of ε, we calculate

$$s = \sqrt{s^2} = \sqrt{\frac{\text{SSE}}{n-2}}$$

We will refer to s as the **estimated standard error of the regression model.**

In most practical situations, σ^2 is unknown and we must use our data to estimate its value. The best estimate of σ^2, denoted by s^2, is obtained by dividing the sum of the squares of the deviations of the y values from the prediction line, or

$$SSE = \sum (y_i - \hat{y}_i)^2$$

by the number of degrees of freedom associated with this quantity. We use 2 df to estimate the two parameters β_0 and β_1 in the straight-line model, leaving $(n - 2)$ df for the estimation of the error variance.

⚠ **CAUTION** When performing these calculations, you may be tempted to round the calculated values of SS_{yy}, $\hat{\beta}_1$, and SS_{xy}. Be certain to carry at least six significant figures for each of these quantities, to avoid substantial errors in calculating the SSE. ▲

Example **11.3**
STIMULUS

Estimating σ in Regression—Drug Reaction Data

Problem Refer to Example 11.2 and the simple linear regression of the drug reaction data in Table 11.1.

a. Compute an estimate of σ.

b. Give a practical interpretation of the estimate.

Solution

a. We previously calculated SSE = 1.10 for the least squares line $\hat{y} = -.1 + .7x$. Recalling that there were $n = 5$ data points, we have $n - 2 = 5 - 2 = 3$ df for estimating σ^2. Thus,

$$s^2 = \frac{SSE}{n-2} = \frac{1.10}{3} = .367$$

is the estimated variance, and

$$s = \sqrt{.367} = .61$$

is the standard error of the regression model.

b. You may be able to grasp s intuitively by recalling the interpretation of a standard deviation given in Chapter 2 and remembering that the least squares line estimates the mean value of y for a given value of x. Since s measures the spread of the distribution of y values about the least squares line and these errors of prediction are assumed to be normally distributed, we should not be surprised to find that most (about 95%) of the observations lie within $2s$, or $2(.61) = 1.22$, of the least squares line. For this simple example (only five data points), all five data points fall within $2s$ of the least squares line. In Section 11.6, we use s to evaluate the error of prediction when the least squares line is used to predict a value of y to be observed for a given value of x.

Dependent Variable: TIME_Y

Number of Observations Read	5
Number of Observations Used	5

Analysis of Variance

Source	DF	Sum of Squares	Mean Square	F Value	Pr > F
Model	1	4.90000	4.90000	13.36	0.0354
Error	3	1.10000	0.36667		
Corrected Total	4	6.00000			

Root MSE	0.60553	R-Square	0.8167
Dependent Mean	2.00000	Adj R-Sq	0.7556
Coeff Var	30.27650		

Parameter Estimates

Variable	DF	Parameter Estimate	Standard Error	t Value	Pr > |t|
Intercept	1	-0.10000	0.63509	-0.16	0.8849
DRUG_X	1	0.70000	0.19149	3.66	0.0354

Figure 11.10

SAS printout for the time–drug regression

Look Back The values of s^2 and s can also be obtained from a simple linear regression printout. The SAS printout for the drug reaction example is reproduced in Figure 11.10. The value of s^2 is highlighted on the printout (in the **Mean Square** column in the row labeled **Error**). The value $s^2 = .36667$, rounded to three decimal places, agrees with the one calculated by hand. The value of s is also highlighted in Figure 11.10 (next to the heading **Root MSE**). This value, $s = .60553$, agrees (except for rounding) with our hand-calculated value.

■ **Now Work Exercise 11.40a–b**

> **Interpretation of s, the Estimated Standard Deviation of ε**
>
> We expect most ($\approx 95\%$) of the observed y values to lie within $2s$ of their respective least squares predicted values, $\hat{y}$.

Exercises 11.37–11.52

Understanding the Principles

11.37 What are the four assumptions made about the probability distribution of ε in regression? Illustrate the assumptions with a graph.

11.38 Visually compare the scatterplots shown below. If a least squares line were determined for each data set, which do you think would have the smallest variance s^2? Explain.

a.

b.

c.
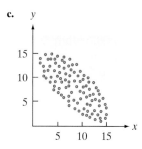

Learning the Mechanics

11.39 Calculate SSE and s^2 for each of the following cases:
 a. $n = 25, SS_{yy} = 110, SS_{xy} = 60, \hat{\beta}_1 = .80$

 b. $n = 50, \sum y^2 = 900, \sum y = 90,$
  $SS_{xy} = 2500, \hat{\beta}_1 = .25$

 c. $n = 25, \sum (y_i - \bar{y})^2 = 75,$
  $SS_{xy} = 100, SS_{xx} = 200$

11.40 Suppose you fit a least squares line to 12 data points and
[NW] the calculated value of SSE is .429.
 a. Find s^2, the estimator of σ^2 (the variance of the random error term ε).
 b. Find s, the estimate of σ.

 c. What is the largest deviation that you might expect between any one of the 12 points and the least squares line?

11.41 Refer to Exercises 11.18 and 11.21 (p. 625). Calculate SSE, s^2, and s for the least squares lines obtained in those exercises. Interpret the standard errors of the regression model for each.

Applying the Concepts—Basic

11.42 **Do nice guys really finish last?** Refer to the *Nature* (Mar. 20, 2008) study of whether "nice guys finish last," Exercise 11.22 (p. 625). Recall that Boston-area college students repeatedly played a version of the game "prisoner's dilemma," where competitors choose cooperation, defection, or costly punishment. At the conclusion of the games, the researchers recorded the average payoff and the number of times punishment was used for each player. Based on a scatterplot of the data, the simple linear regression relating average payoff (y) to punishment use (x) resulted in SSE = 1.04.
 a. Assuming a sample size of $n = 28$, compute the estimated standard deviation of the error distribution, s.
 b. Give a practical interpretation of s.

11.43 **Measuring the moon's orbit.** Refer to the *American Journal of Physics* (Apr. 2014) study of the moon's orbit, Exercise 11.23 (p. 626). Recall that the angular size (y) of the moon was modeled as a straight-line function of height above horizon (x). Find an estimate of σ, the standard deviation of the error of prediction, on the SAS printout (p. 626). Give a practical interpretation of this value.

11.44 **Game performance of water polo players.** Refer to the *Biology of Sport* (Vol. 31, 2014) study of the physiological performance of top-level water polo players, Exercise 11.24 (p. 626). Recall that data for eight Olympic male water polo players were used to model y = mean heart rate as a straight-line function of x = maximal oxygen uptake. Find an estimate of σ, the standard deviation of the error of prediction, on the MINITAB printout (p. 626). Give a practical interpretation of this value.

11.45 **New method for blood typing.** Refer to the *Analytical Chemistry* (May 2010) study in which medical researchers tested a new method of typing blood using low-cost paper, Exercise 11.25 (p. 627). The data were used to fit the straight-line model relating y = wicking length to x = antibody concentration. The SPSS printout follows.

Model Summary

Model	R	R Square	Adjusted R Square	Std. Error of the Estimate
1	.819[a]	.670	.587	2.35944

a. Predictors: (Constant), ABConc

ANOVA[b]

Model		Sum of Squares	df	Mean Square	F	Sig.
1	Regression	45.201	1	45.201	8.119	.046[a]
	Residual	22.268	4	5.567		
	Total	67.469	5			

a. Predictors: (Constant), ABConc

b. Dependent Variable: WickLength

a. Give the values of SSE, s^2, and s shown on the printout.

b. Give a practical interpretation of s. Recall that wicking length is measured in millimeters.

11.46 Quantitative models of music. *Chance* (Fall 2004) published a study on modeling a certain pitch of a musical composition. Data on 147 famous compositions were used to model the number of times (y) a certain pitch occurs—called entropy—as a straight-line function of year of birth (x) of the composer. On the basis of the scatterplot of the data, the standard deviation σ of the model is estimated to be $s = .1$. For a given year (x), about 95% of the actual entropy values (y) will fall within d units of their predicted values. Find the value of d.

11.47 Mongolian desert ants. Refer to the *Journal of Biogeography* (Dec. 2003) study of ant sites in Mongolia, presented in Exercise 11.26 (p. 627). The data were used to estimate the straight-line model relating annual rainfall (y) to maximum daily temperature (x).

a. Give the values of SSE, s^2, and s, shown on the MINITAB printout (p. 627).

b. Give a practical interpretation of the value of s.

Applying the Concepts—Intermediate

11.48 Detecting rapid visual targets. When two targets are presented close together in a rapid visual stream, the second target is often missed. Psychologists call this phenomenon the *attentional blink* (AB). A study published in *Advances in Cognitive Psychology* (July 2013) investigated whether simultaneous or preceding sounds could

reduce AB. Twenty subjects were presented a rapid visual stream of symbols and letters on a computer screen and asked to identify the first and second letters (the targets). After several trials, the subject's AB magnitude was measured as the difference between the percentages of first target and second target letters correctly identified. Each subject performed the task under each of three conditions. In the *Simultaneous* condition, a sound (tone) was presented simultaneously with the second target; in the *Alert* condition, a sound was presented prior to the coming of the second target; and in the *No-Tone* condition, no sound was presented with the second target. Scatterplots of AB magnitude for each possible pair of conditions are shown below as well as the least squares line for each.

a. Which pair of conditions produces the least squares line with the steepest estimated slope?

b. Which pair of conditions produces the least squares line with the largest SSE?

c. Which pair of conditions produces the least squares line with the smallest estimate of σ?

11.49 FCAT scores and poverty. Refer to the *Journal of Educational and Behavioral Statistics* (Spring 2004) study of scores on the Florida Comprehensive Assessment Test (FCAT), presented in Exercise 11.30 (p. 628).

a. Consider the simple linear regression relating math score (y) to percentage (x) of students below the poverty level. Find and interpret the value of s for this regression.

b. Consider the simple linear regression relating reading score (y) to percentage (x) of students below the poverty level. Find and interpret the value of s for this regression.

c. Which dependent variable, math score or reading score, can be more accurately predicted by percentage (x) of students below the poverty level? Explain.

11.50 Sweetness of orange juice. Refer to the study of the quality of orange juice produced at a juice manufacturing plant, Exercise 11.32 (p. 629). Recall that simple linear regression was used to predict the sweetness index (y) from the amount of pectin (x) in orange juice manufactured during a production run.

a. Give the values of SSE, s^2, and s for this regression.

b. Explain why it is difficult to give a practical interpretation to s^2.

c. Use the value of s to derive a range within which most (about 95%) of the errors of prediction of sweetness index fall.

AB Magnitude No-Tone

AB Magnitude No-Tone

AB Magnitude Alert

11.51 Ideal height of your mate. Refer to the *Chance* (Summer 2008) study of the height of the ideal mate, Exercise 11.33 (p. 629). The data were used to fit the simple linear regression model, $E(y) = \beta_0 + \beta_1 x$, where y = ideal partner's height (in inches) and x = student's height (in inches).

HEIGHT

   a. Fit the straight-line model to the data for the male students. Find an estimate for σ, the standard deviation of the error term, and interpret its value practically.

   b. Repeat part **a** for the female students.

   c. For which group, males or females, is student's height the more accurate predictor of ideal partner's height?

Applying the Concepts—Advanced

11.52 Life tests of cutting tools. To improve the quality of the output of any production process, it is necessary first to understand the capabilities of the process. (For example, see Gitlow, H., *Quality Management Systems: A Practical Guide*, 2000.) In a particular manufacturing process, the useful life of a cutting tool is linearly related to the speed at which the tool is operated. The data in the accompanying table were derived from life tests for the two different

TOOL

brands of cutting tools currently used in the production process. For which brand would you feel more confident using the least squares line to predict useful life for a given cutting speed? Explain.

Cutting Speed (meters per minute)	Useful Life (hours)	
	Brand A	Brand B
30	4.5	6.0
30	3.5	6.5
30	5.2	5.0
40	5.2	6.0
40	4.0	4.5
40	2.5	5.0
50	4.4	4.5
50	2.8	4.0
50	1.0	3.7
60	4.0	3.8
60	2.0	3.0
60	1.1	2.4
70	1.1	1.5
70	.5	2.0
70	3.0	1.0

11.4 Assessing the Utility of the Model: Making Inferences about the Slope β_1

Now that we have specified the probability distribution of ε and found an estimate of the variance σ^2, we are ready to make statistical inferences about the linear model's usefulness in predicting the response y. This is step 4 in our regression modeling procedure.

Refer again to the data of Table 11.1, and suppose the reaction times are *completely unrelated* to the percentage of drug in the bloodstream. What could then be said about the values of β_0 and β_1 in the hypothesized probabilistic model

$$y = \beta_0 + \beta_1 x + \varepsilon$$

if x contributes no information for the prediction of y? The implication is that the mean of y—that is, the deterministic part of the model $E(y) = \beta_0 + \beta_1 x$—does not change as x changes. In the straight-line model, this means that the true slope, β_1, is equal to 0. (See Figure 11.11.) Therefore, to test the null hypothesis that the linear model contributes no information for the prediction of y against the alternative hypothesis that the linear model is useful in predicting y, we test

$$H_0: \beta_1 = 0$$
$$H_a: \beta_1 \neq 0$$

If the data support the alternative hypothesis, we will conclude that x does contribute information for the prediction of y with the straight-line model [although the true

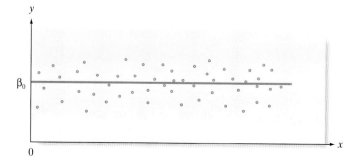

Figure 11.11

Graph of the straight-line model when the slope is zero, i.e., $y = \beta_0 + \varepsilon$

relationship between $E(y)$ and x could be more complicated than a straight line]. In effect, then, this is a test of the usefulness of the hypothesized model.

The appropriate test statistic is found by considering the sampling distribution of $\hat{\beta}_1$, the least squares estimator of the slope β_1, as shown in the following box:

Sampling Distribution of $\hat{\beta}_1$

If we make the four assumptions about ε (see Section 11.3), the sampling distribution of the least squares estimator $\hat{\beta}_1$ of the slope will be normal with mean β_1 (the true slope) and standard deviation

$$\sigma_{\hat{\beta}_1} = \frac{\sigma}{\sqrt{SS_{xx}}} \quad (\text{see Figure 11.12})$$

We estimate $\sigma_{\hat{\beta}_1}$ by $s_{\hat{\beta}_1} = \dfrac{s}{\sqrt{SS_{xx}}}$ and refer to $s_{\hat{\beta}_1}$ as the **estimated standard error of the least squares slope $\hat{\beta}_1$.**

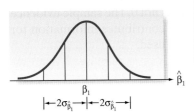

Figure 11.12
Sampling distribution of $\hat{\beta}_1$

Since σ is usually unknown, the appropriate test statistic is a t-statistic, formed as:

$$t = \frac{\hat{\beta}_1 - \text{Hypothesized value of } \beta_1}{s_{\hat{\beta}_1}} \quad \text{where} \quad s_{\hat{\beta}_1} = \frac{s}{\sqrt{SS_{xx}}}$$

Thus,

$$t = \frac{\hat{\beta}_1 - 0}{s/\sqrt{SS_{xx}}}$$

Note that we have substituted the estimator s for σ and then formed the estimated standard error $s_{\hat{\beta}_1}$ by dividing s by $\sqrt{SS_{xx}}$. The number of degrees of freedom associated with this t statistic is the same as the number of degrees of freedom associated with s. Recall that this number is $(n - 2)$ df when the hypothesized model is a straight line. (See Section 11.3.) The setup of our test of the usefulness of the straight-line model is summarized in the following two boxes.

A Test of Model Utility: Simple Linear Regression

Test statistic: $\quad t_c = \dfrac{\hat{\beta}_1}{s_{\hat{\beta}_1}} = \dfrac{\hat{\beta}_1}{(s/\sqrt{SS_{xx}})}$

	One-Tailed Tests		Two-Tailed Test
	$H_0: \beta_1 = 0$	$H_0: \beta_1 = 0$	$H_0: \beta_1 \neq 0$
	$H_a: \beta_1 < 0$	$H_a: \beta_1 > 0$	$H_a: \beta_1 \neq 0$
Rejection region:	$t_c < -t_\alpha$	$t_c > t_\alpha$	$\lvert t_c \rvert > t_{\alpha/2}$
p-value:	$P(t < t_c)$	$P(t > t_c)$	$2P(t > t_c)$ if t_c is positive
			$2P(t < t_c)$ if t_c is negative

Decision: Reject H_0 if $\alpha > p$-value or if test statistic (t_c) falls in rejection region where $P(t > t_\alpha) = \alpha$, $P(t > t_{\alpha/2}) = \alpha/2$, and t is based on $(n - 2)$ degrees of freedom

---

Conditions Required for a Valid Test: Simple Linear Regression

The four assumptions about ε listed in Section 11.3.

Example	11.4	D

STIMULUS

Testing the Regression Slope, β_1 — Drug Reaction Model

Problem Refer to the simple linear regression analysis of the drug reaction data performed in Examples 11.1–11.3. Conduct a test (at $\alpha = .05$) to determine whether the reaction time (y) is linearly related to the amount of drug (x).

Solution For the drug reaction example, $n = 5$. Thus, t will be based on $n - 2 = 3$ df, and the rejection region t (at $\alpha = .05$) will be

$$|t| > t_{.025} = 3.182$$

We previously calculated $\hat{\beta}_1 = .7$, $s = .61$, and $SS_{xx} = 10$. Thus,

$$t = \frac{\hat{\beta}_1}{s/\sqrt{SS_{xx}}} = \frac{.7}{.61/\sqrt{10}} = \frac{.7}{.19} = 3.7$$

Since this calculated t-value falls into the upper-tail rejection region (see Figure 11.13), we reject the null hypothesis and conclude that the slope β_1 is not 0. The sample evidence indicates that the percentage x of drug in the bloodstream contributes information for the prediction of the reaction time y when a linear model is used.

Figure 11.13

Rejection region and calculated t value for testing $H_0: \beta_1 = 0$ versus $H_a: \beta_1 \neq 0$

[*Note:* We can reach the same conclusion by using the observed significance level (*p*-value) of the test from a computer printout. The MINITAB printout for the drug reaction example is reproduced in Figure 11.14. The test statistic and the two-tailed *p*-value are highlighted on the printout. Since the *p*-value = .035 is smaller than $\alpha = .05$, we will reject H_0.]

```
Regression Analysis: TIME_Y versus DRUG_X

The regression equation is
TIME_Y = - 0.100 + 0.700 DRUG_X


Predictor      Coef   SE Coef      T      P
Constant    -0.1000    0.6351  -0.16  0.885
DRUG_X       0.7000    0.1915   3.66  0.035


S = 0.605530   R-Sq = 81.7%   R-Sq(adj) = 75.6%


Analysis of Variance

Source          DF      SS      MS      F      P
Regression       1  4.9000  4.9000  13.36  0.035
Residual Error   3  1.1000  0.3667
Total            4  6.0000
```

Figure 11.14

MINITAB printout for the time–drug regression

Look Back What conclusion can be drawn if the calculated t-value does not fall into the rejection region or if the observed significance level of the test exceeds α? We know from previous discussions of the philosophy of hypothesis testing that such a t-value does *not* lead us to accept the null hypothesis. That is, we do not conclude that $\beta_1 = 0$. Additional data might indicate that β_1 differs from 0, or a more complicated relationship may exist between x and y, requiring the fitting of a model other than the straight-line model. We discuss several such models in Chapter 12.

■ **Now Work Exercise 11.62c**

Interpreting p-Values for β Coefficients in Regression

Almost all statistical computer software packages report a *two-tailed* p-value for each of the β parameters in the regression model. For example, in simple linear regression, the p-value for the two-tailed test $H_0: \beta_1 = 0$ versus $H_a: \beta_1 \neq 0$ is given on the printout. If you want to conduct a *one-tailed* test of hypothesis, you will need to adjust the p-value reported on the printout as follows:

$$\textit{Upper-tailed test}(H_a: \beta_1 > 0): \quad p\text{-value} = \begin{cases} p/2 & \text{if } t > 0 \\ 1 - p/2 & \text{if } t < 0 \end{cases}$$

$$\textit{Lower-tailed test}(H_a: \beta_1 < 0): \quad p\text{-value} = \begin{cases} p/2 & \text{if } t < 0 \\ 1 - p/2 & \text{if } t > 0 \end{cases}$$

where p is the p-value reported on the printout and t is the value of the test statistic.

Another way to make inferences about the slope β_1 is to estimate it with a confidence interval, formed as shown in the following box:

A $100(1 - \alpha)\%$ Confidence Interval for the Simple Linear Regression Slope β_1

$$\hat{\beta}_1 \pm (t_{\alpha/2}) s_{\hat{\beta}_1}$$

where the estimated standard error of $\hat{\beta}_1$ is calculated by

$$s_{\hat{\beta}_1} = \frac{s}{\sqrt{SS_{xx}}}$$

and $t_{\alpha/2}$ is based on $(n - 2)$ degrees of freedom.

Conditions Required for a Valid Confidence Interval: Simple Linear Regression

The four assumptions about ε listed in Section 11.3.

For the simple linear regression for the drug reaction (Examples 11.1–11.3), $t_{\alpha/2}$ is based on $(n - 2) = 3$ degrees of freedom. Therefore, a 95% confidence interval for the slope β_1, the expected change in reaction time for a 1% increase in the amount of drug in the bloodstream, is

$$\hat{\beta}_1 \pm t_{.025} s_{\hat{\beta}_1} = .7 \pm 3.182 \left(\frac{s}{\sqrt{SS_{xx}}} \right) = .7 \pm 3.182 \left(\frac{.61}{\sqrt{10}} \right) = .7 \pm .61$$

Thus, the estimate of the interval for the slope parameter β_1 is from .09 to 1.31. [*Note:* This interval can also be obtained with statistical software and is highlighted on the SPSS printout shown in Figure 11.15.] In terms of this example, the implication is that we can be 95% confident that the *true* mean increase in reaction time per additional 1% of the drug is between .09 and 1.31 seconds. This inference is meaningful only over the sampled range of x—that is, from 1% to 5% of the drug in the bloodstream.

Figure 11.15

SPSS printout with 95% confidence intervals for the time–drug regression betas

Coefficients[a]

Model		Unstandardized Coefficients		Standardized Coefficients	t	Sig.	95% Confidence Interval for B	
		B	Std. Error	Beta			Lower Bound	Upper Bound
1	(Constant)	-.100	.635		-.157	.885	-2.121	1.921
	TIME_X	.700	.191	.904	3.656	.035	.091	1.309

a. Dependent Variable: DRUG_Y

Now Work Exercise 11.67

Since all the values in this interval are positive, it appears that β_1 is positive and that the mean of y, $E(y)$, increases as x increases. However, the rather large width of the confidence interval reflects the small number of data points (and, consequently, a lack of information) used in the experiment. We would expect a narrower interval if the sample size were increased.

We conclude this section with a comment on the other β-parameter in the straight-line model—the y-intercept, β_0. Why not conduct a test of hypothesis on β_0? For example, we could conduct the test $H_0: \beta_0 = 0$ against $H_a: \beta_0 \neq 0$. The p-value for this test appears on the printouts for SAS, SPSS, MINITAB, and most other statistical software packages. The answer lies in the interpretation of β_0. In the previous section, we learned that the y-intercept represents the mean value of y when $x = 0$. Thus, the test $H_0: \beta_0 = 0$ is equivalent to testing whether $E(y) = 0$ when $x = 0$. In the drug reaction simple linear regression, we would be testing whether the mean reaction time (y) is 0 seconds when the amount of drug in the blood (x) is 0%. The value $x = 0$ is typically not a meaningful value (as in the drug reaction example), or, $x = 0$ is typically outside the range of the sample data. In either of these cases, the test $H_0: \beta_0 = 0$ is not meaningful and should be avoided.

For those regression analyses where $x = 0$ is a meaningful value, one may desire to predict the value of y when $x = 0$. We discuss a confidence interval for such a prediction in Section 11.6.

⚠ CAUTION In simple linear regression, the test $H_0: \beta_0 = 0$ is only meaningful if the value $x = 0$ makes sense and is within the range of the sample data. ▲

Statistics IN Action | Revisited | **Assessing How Well the Straight-Line Model Fits the Dowsing Data**

In the previous Statistics in Action Revisited, we fit the straight-line model $E(y) = \beta_0 + \beta_1 x$, where x = dowser's guess (in meters) and y = pipe location (in meters) for each trial. The MINITAB regression printout is reproduced in Figure SIA11.3. The two-tailed p-value for testing the null hypothesis $H_0: \beta_1 = 0$ (highlighted on the printout) is p-value = .118. Even for an α-level as high as $\alpha = .10$, there is insufficient evidence to reject H_0. Consequently, the

dowsing data in Table SIA11.1 provide no statistical support for the German researchers' claim that the three best dowsers have an ability to find underground water with a divining rod.

This lack of support for the dowsing theory is made clearer with a confidence interval for the slope of the line. When $n = 26$, df $= (n - 2) = 24$ and $t_{.025} = 2.064$. Substituting the latter value and the relevant values shown on the MINITAB printout, we find that a 95% confidence interval for β_1 is

$$\hat{\beta}_1 \pm t_{.025}(s_{\hat{\beta}_1}) = .31 \pm (2.064)(.19)$$
$$= .31 \pm .39, \text{ or } (-.08, .70)$$

Thus, for every 1-meter increase in a dowser's guess, we estimate (with 95% confidence) that the change in the actual pipe location will range anywhere from a decrease of .08 meter to an increase of .70 meter. In other words, we're not sure whether the pipe location will increase or decrease along the 10-meter pipeline! Keep in mind also that the data in Table SIA11.1 represent the "best" performances of the three dowsers (i.e., the outcome of the dowsing experiment in its most favorable light). When the data for all trials are considered and plotted, there is not even a hint of a trend.

```
Regression Analysis: PIPE versus GUESS

The regression equation is
PIPE = 30.1 + 0.308 GUESS


Predictor    Coef   SE Coef     T      P
Constant    30.07    11.41    2.63  0.015
GUESS       0.3079   0.1900   1.62  0.118


S = 26.0298   R-Sq = 9.9%   R-Sq(adj) = 6.1%


Analysis of Variance

Source          DF      SS      MS     F      P
Regression       1   1778.9  1778.9  2.63  0.118
Residual Error  24  16261.2   677.6
Total           25  18040.2
```

Figure SIA11.3
MINITAB simple linear regression for dowsing data

Exercises 11.53–11.76

Understanding the Principles

11.53 In the equation $E(y) = \beta_0 + \beta_1 x$, what is the value of β_1 if x has no linear relationship to y?

11.54 What conditions are required for valid inferences about the β's in simple linear regression?

11.55 How do you adjust the p-value obtained from a computer printout when you perform an upper-tailed test of β_1 in simple linear regression when $t > 0$?

11.56 For each of the following 99% confidence intervals for β_1 in simple linear regression, decide whether there is evidence of a positive or negative linear relationship between y and x:
  a. $(8, 80)$ **b.** $(-51, -13)$ **c.** $(-37, 103)$

Learning the Mechanics

11.57 Construct both a 95% and a 90% confidence interval for β_1 for each of the following cases.
  a. $\hat{\beta}_1 = 32, s = 4, SS_{xx} = 32, n = 10$
  b. $\hat{\beta}_1 = 65, SSE = 1,860, SS_{xx} = 20, n = 14$
  c. $\hat{\beta}_1 = -8.4, SSE = 143, SS_{xx} = 49, n = 18$

11.58 Consider the following pairs of observations:

L11058

x	1	5	3	2	6	6	0
y	1	3	3	1	4	5	1

  a. Construct a scatterplot of the data.
  b. Use the method of least squares to fit a straight line to the seven data points in the table.
  c. Plot the least squares line on your scatterplot of part **a**.
  d. Specify the null and alternative hypotheses you would use to test whether the data provide sufficient evidence to indicate that x contributes information for the (linear) prediction of y.
  e. What is the test statistic that should be used in conducting the hypothesis test of part **d**? Specify the number of degrees of freedom associated with the test statistic.
  f. Conduct the hypothesis test of part **d**, using $\alpha = .05$.
  g. Construct a 95% confidence interval for β_1.

11.59 Consider the following pairs of observations:

L11059

y	4	2	5	3	2	4
x	1	4	5	3	2	4

  a. Construct a scatterplot of the data.
  b. Use the method of least squares to fit a straight line to the six data points.
  c. Graph the least squares line on the scatterplot of part **a**.
  d. Compute the test statistic for determining whether x and y are linearly related.
  e. Carry out the test you set up in part **d**, using $\alpha = .01$.
  f. Find a 99% confidence interval for β_1.

Applying the Concepts—Basic

11.60 Congress voting on women's issues. The *American Economic Review* (Mar. 2008) published research on how the gender mix of a U.S. legislator's children can influence the legislator's votes in Congress. Specifically, the researcher investigated how having daughters influences voting on women's issues. The American Association of University Women (AAUW) uses voting records of each member of Congress to compute an AAUW score, where higher scores indicate more favorable voting for women's rights. The researcher modeled AAUW score (y) as a function of the number of daughters (x) a legislator has. Data collected for the 434 members of the 107th Congress were used to fit the straight-line model, $E(y) = \beta_0 + \beta_1 x$.
  a. If it is true that having daughters influences voting on women's issues, will the sign of β_1 be positive or negative? Explain.
  b. The following statistics were reported in the article: $\hat{\beta}_1 = .27$ and $s_{\hat{\beta}_1} = .74$. Find a 95% confidence interval for β_1.
  c. Use the result, part **b**, to make an inference about the model.

11.61 Estimating repair and replacement costs of water pipes. Refer to the *IHS Journal of Hydraulic Engineering* (Sept. 2012) study of water pipes susceptible to breakage, Exercise 11.28 (p. 628). Recall that civil engineers used simple linear regression to model y = the ratio of repair to replacement cost of commercial pipe as a function of x = the diameter (in millimeters) of the pipe. The MINITAB printout of the analysis is reproduced below. Are the engineers able to conclude (at $\alpha = .05$) that the cost ratio increases linearly with pipe diameter? If so, provide a 95% confidence interval for the increase in cost ratio for every 1 millimeter increase in pipe diameter.

Regression Analysis: RATO versus DIAMETER

```
The regression equation is
RATO = 6.68 + 0.00479 DIAMETER


Predictor        Coef      SE Coef      T       P
Constant       6.6782      0.1208    55.29   0.000
DIAMETER    0.0047856   0.0003218    14.87   0.000


S = 0.221034    R-Sq = 95.3%    R-Sq(adj) = 94.8%


Analysis of Variance

Source          DF      SS        MS       F       P
Regression       1   10.807    10.807   221.20   0.000
Residual Error  11    0.537     0.049
Total           12   11.344
```

11.62 Generation Y's entitlement mentality. The current workforce is dominated by "Generation Y"—people born between 1982 and 1999. These workers have a reputation as having an entitlement mentality (e.g., they believe they have a right to a high-paying job without the work ethic). The reasons behind this phenomenon were investigated in *Proceedings of the Academy of Educational Leadership* (Vol. 16, 2011). A sample of 272 undergraduate business students was administered a questionnaire designed to capture the behaviors that lead to an entitlement mentality. The responses were used to measure the following two quantitative variables for each student: entitlement score (y)—where higher scores indicate a greater level of entitlement—and "helicopter parents" score (x)—where higher scores indicate that the student's parents had a higher level of involvement in their everyday experiences and problems.

a. Give the equation of a simple linear regression model relating y to x.

b. The researchers theorize that helicopter parents lead to an entitlement mentality. Based on this theory, would you expect β_0 to be positive or negative (or are you unsure)? Would you expect β_1 to be positive or negative (or are you unsure)? Explain.

[NW] c. The p-value for testing $H_0: \beta_1 = 0$ versus $H_0: \beta_1 > 0$ was reported as .002. Use this result to test the researchers' entitlement theory at $\alpha = .01$.

11.63 Lobster fishing study. Refer to the *Bulletin of Marine Science* (April 2010) study of teams of fishermen fishing for the red spiny lobster in Baja California Sur, Mexico, Exercise 9.20 (p. 478). Two variables measured for each of 8 teams from the Punta Abreojos (PA) fishing cooperative were y = total catch of lobsters (in kilograms) during the season and x = average percentage of traps allocated per day to exploring areas of unknown catch (called *search frequency*). These data are listed in the table.

Total Catch	Search Frequency
2,785	35
6,535	21
6,695	26
4,891	29
4,937	23
5,727	17
7,019	21
5,735	20

Source: From Shester, G. G. "Explaining catch variation among Baja California lobster fishers through spatial analysis of trap-placement decisions." *Bulletin of Marine Science*, Vol. 86, No. 2, Apr. 2010 (Table 1). Reprinted with permission from the University of Miami-*Bulletin of Marine Science.*

a. Graph the data in a scatterplot. What type of trend, if any, do you observe?

b. A simple linear regression analysis was conducted using SAS. Find the least squares prediction equation on the SAS printout at the bottom of the page. Interpret the slope of the least squares line.

c. Give the null and alternative hypotheses for testing whether total catch (y) is negatively linearly related to search frequency (x).

d. Find the p-value of the test, part **c**, on the SAS printout.

e. Give the appropriate conclusion of the test, part **c**, using $\alpha = .05$.

11.64 Beauty and electoral success. Are good looks an advantage when running for political office? This was the question of interest in an article published in the *Journal of Public Economics* (Feb. 2010). The researchers focused on a sample of 641 non-incumbent candidates for political office in Finland. Photos of each candidate were evaluated by non-Finnish subjects; each evaluator assigned a beauty rating—measured on a scale of 1 (lowest rating) to 5 (highest rating)—to each candidate. The beauty ratings for each candidate were averaged; then the average was divided by the standard deviation for all candidates to yield a beauty index for each candidate. (*Note:* A 1-unit increase in the index represents a 1-standard-deviation increase in the beauty rating.) The relative success (measured as a percentage of votes obtained) of each candidate was used as the dependent variable (y) in a regression analysis. One of the independent variables in the model was beauty index (x).

a. Write the equation of a simple linear regression relating y to x.

b. Does the y-intercept of the equation, part **a**, have a practical interpretation? Explain.

c. The article reported the estimated slope of the equation, part **a**, as 22.91. Give a practical interpretation of this value.

d. The standard error of the slope estimate was reported as 3.73. Use this information and the estimate from part **c** to conduct a test for a positive slope at $\alpha = .01$. Give the appropriate conclusion in the words of the problem.

SAS Output for Exercise 11.63

```
                        Dependent Variable: CATCH

                  Number of Observations Read        8
                  Number of Observations Used        8

                           Analysis of Variance

                                 Sum of        Mean
        Source          DF      Squares       Square    F Value   Pr > F

        Model            1      6888299      6888299       6.81   0.0402
        Error            6      6071019      1011837
        Corrected Total  7     12959318

               Root MSE          1005.90086    R-Square    0.5315
               Dependent Mean    5540.50000    Adj R-Sq    0.4535
               Coeff Var           18.15542

                          Parameter Estimates

                          Parameter     Standard
        Variable    DF     Estimate       Error     t Value   Pr > |t|

        Intercept    1    9658.24359   1617.76208      5.97    0.0010
        SEARCHFREQ   1    -171.57265     65.75780     -2.61    0.0402
```

Applying the Concepts—Intermediate

11.65 Ranking driving performance of professional golfers.
Ⓓ
PGA
Refer to *The Sport Journal* (Winter 2007) study of a new method for ranking the total driving performance of golfers on the Professional Golf Association (PGA) tour, presented in Exercise 11.29 (p. 628). You fit a straight-line model relating driving accuracy (y) to driving distance (x) to the data.
  a. Give the null and alternative hypotheses for testing whether driving accuracy (y) decreases linearly as driving distance (x) increases.
  b. Find the test statistic and p-value of the test you set up in part **a**.
  c. Make the appropriate conclusion at $\alpha = .01$.

11.66 FCAT scores and poverty. Refer to the *Journal of*
Ⓓ
FCAT
Educational and Behavioral Statistics (Spring 2004) study of scores on the Florida Comprehensive Assessment Test (FCAT), first presented in Exercise 11.30 (p. 628). Consider the simple linear regression relating math score (y) to percentage (x) of students below the poverty level.
  a. Test whether y is negatively related to x. Use $\alpha = .01$.
  b. Construct a 99% confidence interval for β_1. Interpret the result practically.

11.67 Sweetness of orange juice. Refer to Exercise 11.32
NW
Ⓓ
OJUICE
(p. 629) and the simple linear regression relating the sweetness index (y) of an orange juice sample to the amount of water-soluble pectin (x) in the juice. Find a 95% confidence interval for the true slope of the line. Interpret the result.

11.68 Ideal height of your mate. Refer to the *Chance* (Summer
Ⓓ
HEIGHT
2008) study of the height of the ideal mate, Exercise 11.33 (p. 629). You used the data to fit the simple linear regression model $E(y) = \beta_0 + \beta_1 x$, where y = ideal partner's height (in inches) and x = student's height (in inches), for both males and females.
  a. Find a 90% confidence interval for β_1 in the model for the male students. Give a practical interpretation of the result.
  b. Repeat part **a** for the female students.
  c. Which group, males or females, has the greater increase in ideal partner's height for every 1 inch increase in student's height?

11.69 Forest fragmentation study. Refer to the *Conservation*
Ⓓ
FRAG
Ecology (Dec. 2003) study on the causes of fragmentation of 54 South American forests, presented in Exercise 2.166 (p. 125). Recall that researchers developed two fragmentation indexes for each forest—one index for anthropogenic (human development activities) fragmentation and one for fragmentation from natural causes. Data on 5 of the 54 forests saved in the **FRAG** file are listed in the following table.

Ecoregion (forest)	Anthropogenic Index, y	Natural Origin Index, x
Araucaria moist forests	34.09	30.08
Atlantic Coast *restingas*	40.87	27.60
Bahia coastal forests	44.75	28.16
Bahia interior forests	37.58	27.44
Bolivian *Yungas*	12.40	16.75

Based on Wade, T. G., et al. "Distribution and causes of global forest fragmentation." *Conservation Ecology*, Vol. 72, No. 2, Dec. 2003 (Table 6).

  a. Ecologists theorize that a linear relationship exists between the two fragmentation indexes. Write the model relating y to x.
  b. Fit the model to the data in the **FRAG** file, using the method of least squares. Give the equation of the least squares prediction equation.
  c. Interpret the estimates of β_0 and β_1 in the context of the problem.
  d. Is there sufficient evidence to indicate that the natural origin index (x) and the anthropogenic index (y) are positively linearly related? Test, using $\alpha = .05$.
  e. Find and interpret a 95% confidence interval for the change in the anthropogenic index (y) for every 1-point increase in the natural origin index (x).

11.70 Effect of massage on boxers. Refer to the *British Journal*
Ⓓ
BOXING2
of Sports Medicine (Apr. 2000) study of the effect of massage on boxing performance, presented in Exercise 10.82 (p. 580). Two other variables measured on the boxers were blood lactate concentration (in mM) and the boxer's perceived recovery (on a 28-point scale). On the basis of information provided in the article, the data shown in the accompanying table were obtained for 16 five-round boxing performances in which a massage was given to the boxer between rounds. Conduct a test to determine whether blood lactate level (y) is linearly related to perceived recovery (x). Use $\alpha = .10$.

Blood Lactate Level	Perceived Recovery
3.8	7
4.2	7
4.8	11
4.1	12
5.0	12
5.3	12
4.2	13
2.4	17
3.7	17
5.3	17
5.8	18
6.0	18
5.9	21
6.3	21
5.5	20
6.5	24

Based on Hemmings, B., Smith, M., Graydon, J., and Dyson, R. "Effects of massage on physiological restoration, perceived recovery, and repeated sports performance." *British Journal of Sports Medicine*, Vol. 34, No. 2, Apr. 2000 (data adapted from Figure 3).

11.71 Eye anatomy of giraffes. Giraffes are believed to have excellent vision. The journal *African Zoology* (Oct. 2013) published a study of giraffe eye characteristics. Data were collected for a sample of 27 giraffes inhabiting southeastern Zimbabwe. Of interest to the researchers was how these eye characteristics relate to a giraffe's body mass. They fit a simple linear regression equation of the form $\ln(y) = \beta_0 + \beta_1 \ln(x) + \varepsilon$, where y represents an eye characteristic and x represents body mass (measured in kilograms). For this model, the slope β_1 represents the percentage change in y for every 1% increase in x.
  a. For the eye characteristic y = eye mass (grams), the regression equation yielded the following 95% confidence interval for β_1: $(.25, .30)$. Give a practical interpretation of this interval.

b. For the eye characteristic y = orbit axis angle (degrees), the regression equation yielded the following 95% confidence interval for β_1: $(-.14, -.05)$. Give a practical interpretation of this interval.

11.72 Pain empathy and brain activity. *Empathy* refers to being **D** able to understand and vicariously feel what others actually **EMPATHY** feel. Neuroscientists at University College of London investigated the relationship between brain activity and pain-related empathy in persons who watch others in pain (*Science*, Feb. 20, 2004). Sixteen couples participated in the experiment. The female partner watched while painful stimulation was applied to the finger of her male partner. Two variables were measured for each female: y = pain-related brain activity (measured on a scale ranging from -2 to 2) and x = score on the Empathic Concern Scale (0 to 25 points). The data are listed in the accompanying table. The research question of interest was "Do people scoring higher in empathy show higher pain-related brain activity?" Use simple linear regression analysis to answer this question.

Couple	Brain Activity (y)	Empathic Concern (x)
1	.05	12
2	−.03	13
3	.12	14
4	.20	16
5	.35	16
6	0	17
7	.26	17
8	.50	18
9	.20	18
10	.21	18
11	.45	19
12	.30	20
13	.20	21
14	.22	22
15	.76	23
16	.35	24

Based on Singer, T. et al. "Empathy for pain involves the affective but not sensory components of pain." *Science*, Vol. 303, Feb. 20, 2004. (Adapted from Figure 4.)

11.73 Characterizing bone with fractal geometry. In *Medical* **D** *Engineering & Physics* (May 2013), researchers used **RIBS** fractal geometry to characterize human cortical bone. A measure of the variation in the volume of cortical bone tissue—called fractal dimension—was determined for each in a sample of 10 human ribs. The researchers used fractal dimension scores to predict the bone tissue's stiffness index, called Young's Modulus (measured in gigapascals).

Young's Modulus (GPa)	Fractal Dimension
18.3	2.48
11.6	2.48
32.2	2.39
30.9	2.44
12.5	2.50
9.1	2.58
11.8	2.59
11.0	2.59
19.7	2.51
12.0	2.49

Source: Sanchez-Molina, D., et al. "Fractal dimension and mechanical properties of human cortical bone." *Medical Engineering & Physics,* Vol. 35, No. 5, May 2013 (Table 1).

The experimental data are shown in the first column. Consider the linear model $E(y) = \beta_0 + \beta_1 x$, where y = Young's Modulus and x = fractal dimension score. Find an interval estimate of the increase (or decrease) in Young's Modulus for every 1-point increase in a bone tissue's fractal dimension score. Use a confidence coefficient of .90.

11.74 The "name game." Refer to the *Journal of Experimental* **D** *Psychology—Applied* (June 2000) name-retrieval study, pre- **NAME2** sented in Exercise 11.34 (p. 630). Recall that the goal of the study was to investigate the linear trend between proportion of names recalled (y) and position (order) of the student (x) during the "name game." Is there sufficient evidence (at $\alpha = .01$) of a linear trend?

Applying the Concepts—Advanced

11.75 Does elevation affect hitting performance in baseball? **D** Refer to the *Chance* (Winter 2006) investigation of the ef- **PARKS** fects of elevation on slugging percentage in Major League Baseball, Exercise 2.160 (p. 124). Data were compiled on players' composite slugging percentages at each of 29 cities for the 2003 season, as well as on each city's elevation (feet above sea level). The data are saved in the **PARKS** file. (Selected observations are shown in the table below.) Consider a straight-line model relating slugging percentage (y) to elevation (x).

a. The model was fit to the data with the use of MINITAB, with the results shown in the printout below. Locate the estimates of the model parameters on the printout.

Regression Analysis: SLUGPCT versus ELEVATION

```
The regression equation is
SLUGPCT = 0.515 + 0.000021 ELEVATION

Predictor       Coef      SE Coef       T       P
Constant     0.515140    0.007954    64.76   0.000
ELEVATION   0.00002074  0.00000719    2.89   0.008

S = 0.0369803    R-Sq = 23.6%    R-Sq(adj) = 20.7%

Analysis of Variance

Source          DF      SS         MS       F       P
Regression       1   0.011390   0.011390   8.33   0.008
Residual Error  27   0.036924   0.001368
Total           28   0.048314
```

City	Slug Pct.	Elevation
Anaheim	.480	160
Arlington	.605	616
Atlanta	.530	1,050
Baltimore	.505	130
Boston	.505	20
⋮	⋮	⋮
Denver	.625	5,277
⋮	⋮	⋮
Seattle	.550	350
San Francisco	.510	63
St. Louis	.570	465
Tampa	.500	10
Toronto	.535	566

Based on Schaffer, J., and Heiny, E. L. "The effects of elevation on slugging percentage in Major League Baseball." *Chance*, Vol. 19, No. 1, Winter 2006 (adapted from Figure 2).

b. Is there sufficient evidence (at $\alpha = .01$) of a positive linear relationship between elevation (x) and slugging percentage (y)? Use the p-value shown on the printout to make the inference.

c. Construct a scatterplot of the data and draw the least squares line on the graph. Locate the data point for Denver on the graph. What do you observe?

d. Recall that the Colorado Rockies, who play their home games in Denver, are annually among the league leaders in slugging percentage. Baseball experts attribute this to the "thin air" of Denver—called the Mile High city due to its elevation. Remove the data point for Denver from the data set and refit the straight-line model to the remaining data. Repeat parts **a** and **b**. What conclusions can you draw about the "thin air" theory from this analysis?

11.76 Spreading rate of spilled liquid. Refer to the *Chemical Engineering Progress* (Jan. 2005) study of the rate at which a spilled volatile liquid will spread across a surface, Exercise 11.35 (p. 630). Is there sufficient evidence (at $\alpha = .05$) to indicate that the mass of the spill tends to diminish linearly as elapsed time increases? If so, give an interval estimate (with 95% confidence) of the decrease in spill mass for each minute of elapsed time.

11.5 The Coefficients of Correlation and Determination

In this section, we present two statistics that describe the adequacy of a model: the *coefficient of correlation* and the *coefficient of determination*.

Coefficient of Correlation

Recall (from optional Section 2.8) that a **bivariate relationship** describes a relationship— or correlation—between two variables x and y. Scatterplots are used to describe a bivariate relationship graphically. In this section, we will discuss the concept of **correlation** and how it can be used to measure the linear relationship between two variables x and y. A numerical descriptive measure of correlation is provided by the *coefficient of correlation*, r.

The **coefficient of correlation,*** r, is a measure of the strength of the *linear* relationship between two variables x and y. It is computed (for a sample of n measurements on x and y) as follows:

$$r = \frac{SS_{xy}}{\sqrt{SS_{xx} SS_{yy}}}$$

where

$$SS_{xy} = \sum (x - \bar{x})(y - \bar{y}) = \sum xy - \frac{(\sum x)(\sum y)}{n}$$

$$SS_{xx} = \sum (x - \bar{x})^2 = \sum x^2 - \frac{(\sum x)^2}{n}$$

$$SS_{yy} = \sum (y - \bar{y})^2 = \sum y^2 - \frac{(\sum y)^2}{n}$$

Note that the computational formula for the correlation coefficient r given above involves the same quantities that were used in computing the least squares prediction equation. In fact, since the numerators of the expressions for $\hat{\beta}_1$ and r are identical, it is clear that $r = 0$ when $\hat{\beta}_1 = 0$ (the case where x contributes no information for the prediction of y) and that r is positive when the slope is positive and negative when the slope is negative. Unlike $\hat{\beta}_1$, the correlation coefficient r is *scaleless* and assumes a value between -1 and $+1$, regardless of the units of x and y.

---

*The value of r is often called the *Pearson correlation coefficient* to honor its developer, Karl Pearson. (See Biography, p. 775.)

A value of r near or equal to 0 implies little or no linear relationship between y and x. In contrast, the closer r comes to 1 or -1, the stronger is the linear relationship between y and x. And if $r = 1$ or $r = -1$, all the sample points fall exactly on the least squares line. Positive values of r imply a positive linear relationship between y and x; that is, y increases as x increases. Negative values of r imply a negative linear relationship between y and x; that is, y decreases as x increases. Each of these situations is portrayed in Figure 11.16.

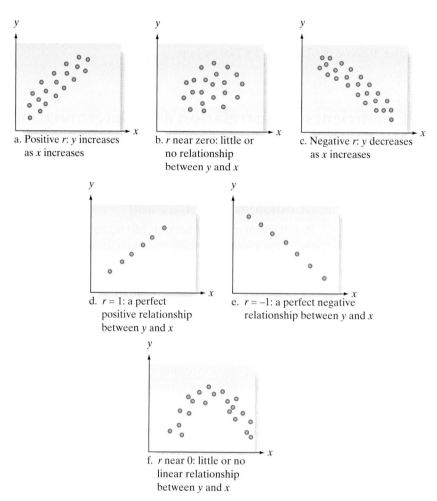

a. Positive r: y increases as x increases

b. r near zero: little or no relationship between y and x

c. Negative r: y decreases as x increases

d. $r = 1$: a perfect positive relationship between y and x

e. $r = -1$: a perfect negative relationship between y and x

f. r near 0: little or no linear relationship between y and x

Figure 11.16
Values of r and their implications

Now Work Exercise 11.79

We use the data in Table 11.1 for the drug reaction example to demonstrate how to calculate the coefficient of correlation, r. The quantities needed to calculate r are SS_{xy}, SS_{xx}, and SS_{yy}. The first two quantities have been calculated previously and are $SS_{xy} = 7$ and $SS_{xx} = 10$. The calculation for $SS_{yy} = \sum(y - \bar{y})^2$ is shown in the last column of the Excel spreadsheet, Figure 11.5 (p. 621). The result is $SS_{yy} = 6$.

We now find the coefficient of correlation:

$$r = \frac{SS_{xy}}{\sqrt{SS_{xx}\,SS_{yy}}} = \frac{7}{\sqrt{(10)(6)}} = \frac{7}{\sqrt{60}} = .904$$

The fact that r is positive and near 1 indicates that the reaction time tends to increase as the amount of drug in the bloodstream increases—*for the given sample of five subjects*. This is the same conclusion we reached when we found the calculated value of the least squares slope to be positive.

Using the Correlation Coefficient—Relating Crime Rate and Casino Employment

Problem Legalized gambling is available on several riverboat casinos operated by a city in Mississippi. The mayor of the city wants to know the correlation between the number of casino employees and the yearly crime rate. The records for the past 10 years are examined, and the results listed in Table 11.3 are obtained. Calculate the coefficient of correlation, r, for the data. Interpret the result.

Table 11.3   Data on Casino Employees and Crime Rate, Example 11.5

Year	Number x of Casino Employees (thousands)	Crime Rate y (number of crimes per 1,000 population)
2006	15	1.35
2007	18	1.63
2008	24	2.33
2009	22	2.41
2010	25	2.63
2011	29	2.93
2012	30	3.41
2013	32	3.26
2014	35	3.63
2015	38	4.15

D *Data Set:* CASINO

Solution Rather than use the computing formula given earlier, we resort to a statistical software package. The data of Table 11.3 were entered into a computer and MINITAB was used to compute r. The MINITAB printout is shown in Figure 11.17.

Correlations: EMPLOYEES, CRIMERAT

Pearson correlation of EMPLOYEES and CRIMERAT = 0.987
P-Value = 0.000

Figure 11.17
MINITAB correlation printout and scatterplot for Example 11.5

Ethics IN Statistics

Intentionally using the correlation coefficient only to make an inference about the relationship between two variables in situations where a nonlinear relationship may exist is considered *unethical statistical practice*.

The coefficient of correlation, highlighted at the top of the printout, is $r = .987$. Thus, the size of the casino workforce and crime rate in this city are very highly correlated—at least over the past 10 years. The implication is that a strong positive linear relationship exists between these variables. (See Figure 11.17.) We must be careful, however, not to jump to any unwarranted conclusions. For instance, the mayor may be tempted to conclude that hiring more casino workers next year will increase the crime rate—that is, that there is a *causal relationship* between the two variables. However, high correlation does not imply causality. The fact is, many things have probably contributed both to the increase in the casino workforce and to the increase in crime rate. The city's tourist trade has undoubtedly grown since riverboat casinos were legalized, and it is likely that the casinos have expanded both in services offered and in number. *We cannot infer a causal relationship on the basis of high sample correlation. When a high correlation is observed in the sample data, the only safe conclusion is that a linear trend may exist between x and y.*

Look Back Another variable, such as the increase in tourism, may be the underlying cause of the high correlation between x and y.

■ **Now Work Exercise 11.85a**

⚠ **CAUTION** Two caveats apply in using the sample correlation coefficient r to infer the nature of the relationship between x and y: (1) A *high correlation* does not necessarily imply that a causal relationship exists between x and y—only that a linear trend may exist; (2) a *low correlation* does not necessarily imply that x and y are unrelated—only that x and y are not strongly *linearly* related. ▲

Keep in mind that the correlation coefficient r measures the linear correlation between x values and y values in the sample, and a similar linear coefficient of correlation exists for the population from which the data points were selected. The **population correlation coefficient** is denoted by the symbol ρ (rho). As you might expect, ρ is estimated by the corresponding sample statistic r. Or, instead of estimating ρ, we might want to test the null hypothesis $H_0: \rho = 0$ against $H_a: \rho \neq 0$; that is, we can test the hypothesis that x contributes no information for the prediction of y by using the straight-line model against the alternative that the two variables are at least linearly related.

However, we already performed this *identical* test in Section 11.4 when we tested $H_0: \beta_1 = 0$ against $H_a: \beta_1 \neq 0$. That is, the null hypothesis $H_0: \rho = 0$ is equivalent to the hypothesis $H_0: \beta_1 = 0$.* When we tested the null hypothesis $H_0: \beta_1 = 0$ in connection with the drug reaction example, the data led to a rejection of the null hypothesis at the $\alpha = .05$ level. This rejection implies that the null hypothesis of a 0 linear correlation between the two variables (drug and reaction time) can also be rejected at the $\alpha = .05$ level. The only real difference between the least squares slope $\hat{\beta}_1$ and the coefficient of correlation, r, is the measurement scale. Therefore, the information they provide about the usefulness of the least squares model is to some extent redundant. For this reason, we will use the slope to make inferences about the existence of a positive or negative linear relationship between two variables.

For the sake of completeness, a summary of the test for linear correlation is provided in the following boxes.

A Test for Linear Correlation

$$\text{Test statistic: } t_c = \frac{r\sqrt{n-2}}{\sqrt{1-r^2}} = \frac{\hat{\beta}_1}{s_{\hat{\beta}_1}}$$

One-Tailed Tests

$H_0: \rho = 0$	$H_0: \rho = 0$
$H_a: \rho > 0$	$H_a: \rho < 0$

Two-Tailed Test

$H_0: \rho = 0$

$H_a: \rho \neq 0$

Rejection region: $t_c > t_\alpha \quad t_c < -t_\alpha$ \qquad $|t_c| > t_{\alpha/2}$

p-value: $P(t > t_c) \quad P(t < t_c)$ \qquad $2 \cdot P(t > t_c)$ if t_c is positive
$2 \cdot P(t < t_c)$ if t_c is negative

where the distribution of t depends on $(n-2)$ df.

Condition Required for a Valid Test of Correlation

The sample of (x, y) values is randomly selected from a normal population.

---

*The two tests are equivalent in simple linear regression only.

Coefficient of Determination

Another way to measure the usefulness of a linear model is to measure the contribution of x in predicting y. To accomplish this, we calculate how much the errors of prediction of y were reduced by using the information provided by x. To illustrate, consider the sample shown in the scatterplot of Figure 11.18a. If we assume that x contributes no information for the prediction of y, the best prediction for a value of y is the sample mean $\bar{y}$, which is shown as the horizontal line in Figure 11.18b. The vertical line segments in Figure 11.18b are the deviations of the points about the mean $\bar{y}$. Note that the sum of the squares of the deviations for the prediction equation $\hat{y} = \bar{y}$ is

$$SS_{yy} = \sum (y_i - \bar{y})^2$$

Now suppose you fit a least squares line to the same set of data and locate the deviations of the points about the line, as shown in Figure 11.18c. Compare the deviations about the prediction lines in Figures 11.18b and 11.18c You can see that

1. If x contributes little or no information for the prediction of y, the sums of the squares of the deviations for the two lines

$$SS_{yy} = \sum (y_i - \bar{y})^2 \quad \text{and} \quad SSE = \sum (y_i - \hat{y}_i)^2$$

   will be nearly equal.
2. If x does contribute information for the prediction of y, the SSE will be smaller than SS_{yy}. In fact, if all the points fall on the least squares line, then $SSE = 0$.

Consequently, the reduction in the sum of the squares of the deviations that can be attributed to x, expressed as a proportion of SS_{yy}, is

$$\frac{SS_{yy} - SSE}{SS_{yy}}$$

a. Scatterplot of data

b. Assumption: x contributes no information
for predicting y, $\hat{y} = \bar{y}$

c. Assumption: x contributes information
for predicting y, $\hat{y} = \hat{\beta}_0 + \hat{\beta}_1 x$

Figure 11.18

A comparison of the sum of squares of deviations for two models

Note that SS_{yy} is the "total sample variability" of the observations around the mean $\bar{y}$ and that SSE is the remaining "unexplained sample variability" after fitting the line $\hat{y}$. Thus, the difference $(SS_{yy} - SSE)$ is the "explained sample variability" attributable to the linear relationship with x. Thus, a verbal description of the proportion is

$$\frac{SS_{yy} - SSE}{SS_{yy}} = \frac{\text{Explained sample variability}}{\text{Total sample variability}}$$

$$= \text{Proportion of total sample variability explained by the linear relationship}$$

In simple linear regression, it can be shown that this proportion—called the *coefficient of determination*—is equal to the square of the simple linear coefficient of correlation, r.

The **coefficient of determination** is

$$r^2 = \frac{SS_{yy} - SSE}{SS_{yy}} = 1 - \frac{SSE}{SS_{yy}}$$

and represents the proportion of the total sample variability around $\bar{y}$ that is explained by the linear relationship between y and x. (In simple linear regression, it may also be computed as the square of the coefficient of correlation, r.)

Note that r^2 is always between 0 and 1 because r is between -1 and $+1$. Thus, an r^2 of .60 means that the sum of the squares of the deviations of the y values about their predicted values has been reduced 60% by the use of the least squares equation $\hat{y}$, instead of $\bar{y}$, to predict y.

Example 11.6 Ⓓ

Obtaining the Value of r^2—Drug Reaction Regression

Table 11.4

Percent x of Drug	Reaction Time y (seconds)
1	1
2	1
3	2
4	2
5	4

Ⓓ *Data Set:* STIMULUS

Problem Calculate the coefficient of determination for the drug reaction example. The data are repeated in Table 11.4 for convenience. Interpret the result.

Solution From previous calculations,

$$SS_{yy} = 6 \quad \text{and} \quad SSE = \sum (y - \hat{y})^2 = 1.10$$

Then, from our earlier definition, the coefficient of determination is

$$r^2 = \frac{SS_{yy} - SSE}{SS_{yy}} = \frac{6.0 - 1.1}{6.0} = \frac{4.9}{6.0} = .817$$

Another way to compute r^2 is to recall from earlier in this section that $r = .904$. Then we have $r^2 = (.904)^2 = .817$. A third way to obtain r^2 is from a computer printout. Its value is highlighted on the SPSS printout in Figure 11.19. Our interpretation is as follows: We know that using the percent x of drug in the blood to predict y with the least squares line

$$\hat{y} = -.1 + .7x$$

accounts for nearly 82% of the total sum of the squares of the deviations of the five sample y values about their mean. Or, stated another way, 82% of the sample variation in reaction time (y) can be "explained" by using the percent x of drug in a straight-line model.

Model Summary

Model	R	R Square	Adjusted R Square	Std. Error of the Estimate
1	.904ᵃ	.817	.756	.606

a. Predictors: (Constant), DRUG_X

Figure 11.19
Portion of SPSS printout for time-drug regression

■ **Now Work Exercise 11.87a**

> **Practical Interpretation of the Coefficient of Determination, r^2**
>
> $100(r^2)\%$ of the sample variation in y (measured by the total sum of the squares of the deviations of the sample y values about their mean $\bar{y}$) can be explained by (or attributed to) using x to predict y in the straight-line model.

Statistics IN Action ❘ Revised

Using the Coefficients of Correlation and Determination to Assess the Dowsing Data

In the previous Statistics in Action Revisited, we discovered that using a dowser's guess (x) in a straight-line model was not statistically useful in predicting actual pipe location (y). Both the coefficient of correlation and the coefficient of determination (highlighted on the MINITAB printouts in Figure SIA11.4) also support this conclusion. The value of the correlation coefficient, $r = .314$, indicates a fairly weak positive linear relationship between the variables. This value, however, is not statistically significant (p-value $= .118$). In other words, there is no evidence to indicate that the population correlation coefficient is different from 0. The coefficient of determination, $r^2 = .099$, implies that only about 10% of the sample variation in pipe location values can be explained by the simple linear model.

```
Regression Analysis: PIPE versus GUESS

The regression equation is
PIPE = 30.1 + 0.308 GUESS


Predictor    Coef   SE Coef     T      P
Constant    30.07     11.41   2.63  0.015
GUESS      0.3079    0.1900   1.62  0.118


S = 26.0298   R-Sq = 9.9%   R-Sq(adj) = 6.1%


Analysis of Variance

Source          DF      SS      MS      F      P
Regression       1  1778.9  1778.9   2.63  0.118
Residual Error  24 16261.2   677.6
Total           25 18040.2
```

Figure SIA11.4
MINITAB printouts with coefficients of correlation and determination for the dowsing data

```
Correlations: PIPE, GUESS

Pearson correlation of PIPE and GUESS = 0.314
P-Value = 0.118
```

Exercises 11.77–11.100

Understanding the Principles

11.77 *True or False.* The correlation coefficient is a measure of the strength of the linear relationship between x and y.

11.78 Describe the slope of the least squares regression line for the given information.
- **a.** $r = -.4$
- **b.** $r = .4$
- **c.** $r = 0$
- **d.** $r^2 = .81$

11.79 Explain what each of the following sample correlation coefficients tells you about the relationship between the x and y values in the sample:
[NW]
- **a.** $r = 1$
- **b.** $r = -1$
- **c.** $r = 0$
- **d.** $r = .90$
- **e.** $r = .10$
- **f.** $r = -.88$

11.80 *True or False.* A value of the correlation coefficient near 1 or near -1 implies a causal relationship between x and y.

Learning the Mechanics

11.81 Construct a scatterplot for each data set. Then calculate r and r^2 for each data set. Interpret their values.

a.

x	-2	-1	0	1	2
y	-1	2	3	6	9

b.

x	-2	-1	0	1	2
y	9	6	4	3	1

c.

x	2	3	3	4	4	4	5
y	3	2	4	2	3	4	3

d.

x	-2	-1	1	3	4
y	0	1	2	1	0

11.82 Calculate r^2 for the least squares line in Exercise 11.18 (p. 625).

11.83 Calculate r^2 for the least squares line in Exercise 11.21 (p. 625).

Ⓓ Applet Exercise 11.2

Use the applet entitled *Correlation by the Eye* to explore the relationship between the pattern of data in a scatterplot and the corresponding correlation coefficient.

   a. Run the applet several times. Each time, guess the value of the correlation coefficient. Then click *Show r* to see the actual correlation coefficient. How close is your value to the actual value of *r*? Click *New data* to reset the applet.

   b. Click the trash can to clear the graph. Use the mouse to place five points on the scatterplot that are approximately in a straight line. Then guess the value of the correlation coefficient. Click *Show r* to see the actual correlation coefficient. How close were you this time?

   c. Continue to clear the graph and plot sets of five points with different patterns among the points. Guess the value of *r*. How close do you come to the actual value of *r* each time?

   d. On the basis of your experiences with the applet, explain why we need to use more reliable methods of finding the correlation coefficient than just "eyeing" it.

Applying the Concepts—Basic

11.84 RateMyProfessors.com. A popular Web site among college students is RateMyProfessors.com (RMP). Established over 10 years ago, RMP allows students to post quantitative ratings of their instructors. In *Practical Assessment, Research & Evaluation* (May 2007), University of Maine researchers investigated whether instructor ratings posted on RMP are correlated with the formal in-class student evaluations of teaching (SET) that all universities are required to administer at the end of the semester. Data collected for $n = 426$ University of Maine instructors yielded a correlation between RMP and SET ratings of .68.

   a. Give the equation of a linear model relating SET rating (*y*) to RMP rating (*x*).

   b. Give a practical interpretation of the value $r = .68$.

   c. Is the estimated slope of the line, part **a**, positive or negative? Explain.

   d. A test of the null hypothesis $H_0: \rho = 0$ yielded a *p*-value of .001. Interpret this result.

   e. Compute the coefficient of determination, r^2, for the regression analysis. Interpret the result.

11.85 Last name and acquisition timing. Refer to the *Journal of Consumer Research* (Aug. 2011) study of the speed with which consumers decide to purchase a product, Exercise 9.14 (p. 476). Recall that the researchers theorized that consumers with last names that begin with letters later in the alphabet will tend to acquire items faster than those whose last names are earlier in the alphabet (i.e., the *last name effect*). Each in a sample of 50 MBA students was offered free tickets to attend a college basketball game for which there was a limited supply of tickets. The first letter of the last name of those who responded to an e-mail offer in time to receive the tickets was noted and given a numerical value (e.g., "A" = 1, "B" = 2, etc.). Each student's response time (measured in minutes) was also recorded.

NW   **a.** The researchers computed the correlation between the two variables as $r = -.271$. Interpret this result.

   b. The observed significance level for testing for a negative correlation in the population was reported as *p*-value = .018. Interpret this result for $\alpha = .05$.

   c. Does this analysis support the researchers' *last name effect* theory? Explain.

11.86 Taste-testing scales. Refer to the *Journal of Food Science* (Feb. 2014) taste-testing study, Exercise 10.80 (p. 579). Recall that the researchers evaluated the general Labeled Magnitude Scale (gLMS), used to rate the palatability of food items on a scale ranging from -100 (for strongest imaginable dislike) to $+100$ (for strongest imaginable like). The researchers called this rating the *perceived hedonic intensity*. A sample of 200 students and staff at the University of Florida used the scale to rate their most favorite and least favorite foods. In addition, each taster rated the sensory intensity of four different solutions: salt, sucrose, citric acid, and hydrochloride. The averages of these four ratings were used by the researchers to quantify individual variation in taste intensity—called *perceived sensory intensity*. These data are saved in the **TASTE** file. The accompanying MINITAB printout shows the correlation between perceived sensory intensity (PSI) and perceived hedonic intensity for both favorite (PHI-F) and least favorite (PHI-L) foods. According to the researchers, "the palatability of the favorite and least favorite foods varies depending on the perceived intensity of taste: Those who experience the greatest taste intensity (that is, supertasters) tend to experience more extreme food likes and dislikes." Do you agree? Explain.

Correlations: PSI, PHI-F

Pearson correlation of PSI and PHI-F = 0.401
P-Value = 0.000

Correlations: PSI, PHI-L

Pearson correlation of PSI and PHI-L = -0.375
P-Value = 0.000

11.87 Going for it on fourth down in the NFL. Each week coaches in the National Football League (NFL) face a decision during the game. On fourth down, should the team punt the ball or go for a first down? To aid in the decision-making process, statisticians at California State University, Northridge, developed a regression model for predicting the number of points scored (*y*) by a team that has a first down with a given number of yards (*x*) from the opposing goal line (*Chance*, Winter 2009). One of the models fit to data collected on five NFL teams from a recent season was the simple linear regression model, $E(y) = \beta_0 + \beta_1 x$. The regression yielded the following results: $\hat{y} = 4.42 - .048\,x$, $r^2 = .18$.

NW   **a.** Give a practical interpretation of the coefficient of determination, r^2.

   b. Compute the value of the coefficient of correlation, *r*, from the value of r^2. Is the value of *r* positive or negative? Why?

11.88 Lobster fishing study. Refer to the *Bulletin of Marine Science* (Apr. 2010) study of teams of fishermen fishing for the red spiny lobster in Baja California Sur, Mexico,

Exercise 11.63 (p. 642). Recall that simple linear regression was used to model y = total catch of lobsters (in kilograms) during the season as a function of x = average percentage of traps allocated per day to exploring areas of unknown catch (called *search frequency*).

a. Locate and interpret the coefficient of determination, r^2, on the SAS printout shown on p. 642.

b. Note that the coefficient of correlation, r, is not shown on the SAS printout. Is there information on the printout to determine whether total catch (y) is negatively linearly related to search frequency (x)? Explain.

11.89 Physical activity of obese young adults. Refer to the *International Journal of Obesity* (Jan. 2007) study of the physical activity of obese young adults, presented in Exercise 6.37 (p. 331). For two groups of young adults—13 obese and 15 of normal weight—researchers recorded the total number of registered movements (counts) of each young adult over a period of time. *Baseline* physical activity was then computed as the number of counts per minute (cpm). Four years later, physical activity measurements were taken again—called physical activity *at follow-up.*

a. For the 13 obese young adults, the researchers reported a correlation of $r = .50$ between baseline and follow-up physical activity, with an associated p-value of .07. Give a practical interpretation of this correlation coefficient and p-value.

b. Refer to part **a.** Construct a scatterplot of the 13 data points that would yield a value of $r = .50$.

c. For the 15 young adults of normal weight, the researchers reported a correlation of $r = -.12$ between baseline and follow-up physical activity, with an associated p-value of .66. Give a practical interpretation of this correlation coefficient and p-value.

d. Refer to part **c.** Construct a scatterplot of the 15 data points that would yield a value of $r = -12$.

11.90 Redshifts of quasi-stellar objects. Refer to the *Journal of Astrophysics & Astronomy* (Mar./Jun. 2003) study of redshifts in quasi-stellar objects presented in Exercise 11.27 (p. 627). Recall that simple linear regression was used to model the magnitude (y) of a quasi-stellar object as a function of the redshift level (x). In addition to the least squares line, $\hat{y} = 18.13 + 6.21x$, the coefficient of correlation was determined to be $r = .84$.

a. Interpret the value of r in the words of the problem.

b. What is the relationship between r and the estimated slope of the line?

c. Find and interpret the value of r^2.

Applying the Concepts—Intermediate

11.91 Evaluation of an imputation method for missing data. When analyzing large data sets with many variables, researchers often encounter the problem of missing data (e.g., non-response). Typically, an imputation method will be used to substitute in reasonable values (e.g., the mean of the variable) for the missing data. An imputation method that uses "nearest neighbors" as substitutes for the missing data was evaluated in *Data & Knowledge Engineering* (Mar. 2013). Two quantitative assessment measures of the imputation algorithm are normalized root mean square error (NRMSE) and classification bias. The researchers applied the imputation method to a sample of 3600 data sets with missing values and determined the NRMSE and

classification bias for each data set. The correlation coefficient between the two variables was reported as $r = .2838$.

a. Conduct a test to determine if the true population correlation coefficient relating NRMSE and bias is positive. Interpret this result practically.

b. A scatterplot for the data (extracted from the journal article) is shown below. Based on the graph, would you recommend using NRMSE as a linear predictor of bias? Explain why your answer does not contradict the result in part **a.**

11.92 Salary linked to height. Are short people shortchanged when it comes to salary? According to business professors T. A. Judge (University of Florida) and D. M. Cable (University of North Carolina), tall people tend to earn more money over their career than short people earn (*Journal of Applied Psychology*, June 2004). Using data collected from participants in the National Longitudinal Surveys, the researchers computed the correlation between average earnings (in dollars) and height (in inches) for several occupations. The results are given in the following table.

Occupation	Correlation, r	Sample Size, n
Sales	.41	117
Managers	.35	455
Blue Collar	.32	349
Service Workers	.31	265
Professional/Technical	.30	453
Clerical	.25	358
Crafts/Forepersons	.24	250

Source: Judge, T. A., and Cable, D. M. "The effect of physical height on workplace success and income: Preliminary test of a theoretical model." *Journal of Applied Psychology*, Vol. 89, No. 3, June 2004 (Table 5). Copyright © 2004 by the American Psychological Association. Reprinted with permission.

a. Interpret the value of r for people in sales occupations.

b. Compute r^2 for people in sales occupations. Interpret the result.

c. Give H_0 and H_a for testing whether average earnings and height are positively correlated.

d. Compute the test statistic for testing H_0 and H_a in part **c** for people in sales occupations.

e. Use the result you obtained in part **d** to conduct the test at $\alpha = .01$. State the appropriate conclusion.

f. Select another occupation and repeat parts **a–e.**

11.93 View of rotated objects. *Perception & Psychophysics* (July 1998) reported on a study of how people view three-dimensional objects projected onto a rotating two-dimensional image. Each in a sample of 25 university students viewed various depth-rotated objects (e.g., a hairbrush, a duck, and a shoe) until they recognized the object. The recognition exposure time—that is, the minimum time (in milliseconds) required for the subject to recognize the object—was recorded for each object. In addition, each subject rated the "goodness of view" of the object on a numerical scale, with lower scale values corresponding to better views. The following table gives the correlation coefficient r between recognition exposure time and goodness of view for several different rotated objects:

Object	r	t
Piano	.447	2.40
Bench	−.057	.27
Motorbike	.619	3.78
Armchair	.294	1.47
Teapot	.949	14.50

a. Interpret the value of r for each object.
b. Calculate and interpret the value of r^2 for each object.
c. The table also includes the t-value for testing the null hypothesis of no correlation (i.e., for testing $H_0: \beta_1 = 0$). Interpret these results using $x = .05$.

11.94 Eye anatomy of giraffes. Refer to the *African Zoology* (Oct. 2013) study of giraffe eye characteristics, Exercise 11.71 (p. 643). Recall that the researchers fit a simple linear regression equation of the form $\ln(y) = \beta_0 + \beta_1 \ln(x) + \varepsilon$, where y represents an eye characteristic and x represents body mass (measured in kilograms).
a. For the eye characteristic y = eye mass (grams), the regression equation yielded $r^2 = .948$. Give a practical interpretation of this result.
b. Refer to part **a** above and Exercise 11.71 part **a**. Find the value of the correlation coefficient, r, and interpret its value.
c. For the eye characteristic y = orbit axis angle (degrees), the regression equation yielded $r^2 = .375$. Give a practical interpretation of this result.
d. Refer to part **c** above and Exercise 11.71 part **b**. Find the value of the correlation coefficient, r, and interpret its value.

11.95 Do nice guys finish first or last? Refer to the *Nature* (Mar. 20, 2008) study of the use of punishment in cooperation games, Exercise 11.22 (p. 625). Recall that college students repeatedly played a version of the game "prisoner's dilemma" and the researchers recorded the average payoff (y) and the number of times punishment was used (x) for each player. A negative correlation was discovered between x and y.
a. Give the null and alternative hypotheses for testing whether average payoff and punishment use are negatively correlated.
b. The test, part **a**, yielded a p-value of .001. Interpret this result using $x = .05$.
c. Does the result, part **b**, imply that increasing punishment causes your payoff to decrease? Explain.

11.96 The "name game." Refer to the *Journal of Experimental Psychology—Applied* (June 2000) name-retrieval study, first presented in Exercise 11.34 (p. 630). Find and interpret the values of r and r^2 for the simple linear regression relating the proportion of names recalled (y) and the position (order) of the student (x) during the "name game."

11.97 Effect of massage on boxing. Refer to the *British Journal of Sports Medicine* (Apr. 2000) study of the effect of massage on boxing performance, presented in Exercise 11.70 (p. 643). Find and interpret the values of r and r^2 for the simple

linear regression relating the blood lactate concentration and the boxer's perceived recovery.

11.98 Child labor in diamond mines. The role of child laborers in Africa's colonial-era diamond mines was the subject of research published in the *Journal of Family History* (Vol. 35, 2010). One particular mining company lured children to the mines by offering incentives for adult male laborers to relocate their families close to the diamond mine. The success of the incentive program was examined by determining the annual *accompaniment rate*, i.e., the percentage of wives (or sons or daughters) who accompanied their husbands (or fathers) in relocating to the mine. The accompaniment rates over the years 1939–1947 are shown in the table below.
a. Find the correlation coefficient relating the accompaniment rates for wives and sons. Interpret this value.
b. Find the correlation coefficient relating the accompaniment rates for wives and daughters. Interpret this value.
c. Find the correlation coefficient relating the accompaniment rates for sons and daughters. Interpret this value.

Year	Wives	Sons	Daughters
1939	27.2	2.2	16.9
1940	40.1	1.5	15.7
1941	35.7	0.3	12.6
1942	37.8	3.5	22.2
1943	38.0	5.4	22.0
1944	38.4	11.0	24.3
1945	38.7	11.9	17.9
1946	29.8	8.6	17.7
1947	23.8	7.4	22.2

Source: Cleveland, T. "Minors in name only: Child laborers on the diamond mines of the *Companhia de Diamantes de Angola* (Diamang), 1917–1975." *Journal of Family History,* Vol. 35, No. 1, 2010 (Table 1).

11.99 Plants that grow on Swiss cliffs. Refer to the *Alpine Botany* (Nov. 2012) study of rare plants that grow on the limestone cliffs of the Northern Swiss Jura mountains, Exercise 2.165 (p. 125). Data on altitude above sea level (meters), plant population size (number of plants growing), and molecular variance (i.e., the variance in molecular weight of the plants) for a sample of 12 limestone cliffs are reproduced in the table. Recall that the researchers are interested in whether either altitude or population size is related to molecular variance.

Cliff Number	Altitude	Population Size	Molecular Variance
1	468	147	59.8
2	589	209	24.4
3	700	28	42.2
4	664	177	59.5
5	876	248	65.8
6	909	53	17.7
7	1032	33	12.5
8	952	114	27.6
9	832	217	35.9
10	1099	10	13.3
11	982	8	3.6
12	1053	15	3.2

Source: Rusterholz, H., Aydin, D., and Baur, B. "Population structure and genetic diversity of relict populations of *Alyssum montanum* on limestone cliffs in the Northern Swiss Jura mountains." *Alpine Botany,* Vol. 122, No. 2, Nov. 2012 (Tables 1 and 2).

a. Use simple linear regression to investigate the relationship between molecular variance (y) and altitude (x). Find and interpret the value of r^2.

b. Use simple linear regression to investigate the relationship between molecular variance (y) and population size (x). Find and interpret the value of r^2.

c. What are your recommendations to the researchers?

Applying the Concepts—Advanced

11.100 Pain tolerance study. A study published in *Psychosomatic Medicine* (Mar./Apr. 2001) explored the relationship between reported severity of pain and actual pain tolerance in 337 patients who suffer from chronic pain. Each patient reported his/her severity of chronic pain on a seven-point scale (1 = no pain, 7 = extreme pain). To obtain a pain tolerance level, a tourniquet was applied to the arm of each patient and twisted. The maximum pain level tolerated was measured on a quantitative scale.

a. According to the researchers, "Correlational analysis revealed a small but significant inverse relationship between [actual] pain tolerance and the reported severity of chronic pain." On the basis of this statement, is the value of r for the 337 patients positive or negative?

b. Suppose that the result reported in part **a** is significant at $\alpha = .05$. Find the approximate value of r for the sample of 337 patients.

Figure 11.20

Estimated mean value and predicted individual value of reaction time y for $x = 4$

11.6 Using the Model for Estimation and Prediction

If we are satisfied that a useful model has been found to describe the relationship between reaction time and percent of drug in the bloodstream, we are ready for step 5 in our regression modeling procedure: using the model for estimation and prediction.

The most common uses of a probabilistic model for making inferences can be divided into two categories. The first is the use of the model for estimating the mean value of y, E(y), for a specific value of x.

For our drug reaction example, we may want to estimate the mean response time for all people whose blood contains 4% of the drug.

The second use of the model entails predicting a new individual y value for a given x.

That is, we may want to predict the reaction time for a specific person who possesses 4% of the drug in the bloodstream.

In the first case, we are attempting to estimate the mean value of y for a very large number of experiments at the given x value. In the second case, we are trying to predict the outcome of a single experiment at the given x value. Which of these uses of the model—estimating the mean value of y or predicting an individual new value of y (for the same value of x)—can be accomplished with the greater accuracy?

Before answering this question, we first consider the problem of choosing an estimator (or predictor) of the mean (or a new individual) y value. We will use the least squares prediction equation

$$\hat{y} = \hat{\beta}_0 + \hat{\beta}_1 x$$

both to estimate the mean value of y and to predict a specific new value of y for a given value of x. For our example, we found that

$$\hat{y} = -.1 + .7x$$

so the estimated mean reaction time for all people when $x = 4$ (the drug is 4% of the blood content) is

$$\hat{y} = -.1 + .7(4) = 2.7 \text{ seconds}$$

The same value is used to predict a new y value when $x = 4$. That is, both the estimated mean and the predicted value of y are $\hat{y} = 2.7$ when $x = 4$, as shown in Figure 11.20.

The difference between these two uses of the model lies in the accuracies of the estimate and the prediction, best measured by the sampling errors of the least squares line when it is used as an estimator and as a predictor, respectively. These errors are reflected in the standard deviations given in the following box:

Sampling Errors for the Estimator of the Mean of y and the Predictor of an Individual New Value of y

1. The standard deviation of the sampling distribution of the estimator $\hat{y}$ of the mean value of y at a specific value of x, say x_p, is

$$\sigma_{\hat{y}} = \sigma \sqrt{\frac{1}{n} + \frac{(x_p - \bar{x})^2}{SS_{xx}}}$$

where σ is the standard deviation of the random error ε. We refer to $\sigma_{\hat{y}}$ as the **standard error of $\hat{y}$**.

2. The standard deviation of the prediction error for the predictor $\hat{y}$ of an individual new y value at a specific value of x is

$$\sigma_{(y - \hat{y})} = \sigma \sqrt{1 + \frac{1}{n} + \frac{(x_p - \bar{x})^2}{SS_{xx}}}$$

where σ is the standard deviation of the random error ε. We refer to $\sigma_{(y - \hat{y})}$ as the **standard error of prediction**.

The true value of σ is rarely known, so we estimate σ by s and calculate the estimation and prediction intervals as shown in the next two boxes:

A $100(1 - \alpha)\%$ Confidence Interval for the Mean Value of y at $x = x_p$

$$\hat{y} \pm t_{\alpha/2}(\text{Estimated standard error of } \hat{y})$$

or

$$\hat{y} \pm t_{\alpha/2} s \sqrt{\frac{1}{n} + \frac{(x_p - \bar{x})^2}{SS_{xx}}}$$

where $t_{\alpha/2}$ is based on $(n - 2)$ degrees of freedom.

A $100(1 - \alpha)\%$ Prediction Interval* for an Individual New Value of y at $x = x_p$

$$\hat{y} \pm t_{\alpha/2}(\text{Estimated standard error of prediction})$$

or

$$\hat{y} \pm t_{\alpha/2} s \sqrt{1 + \frac{1}{n} + \frac{(x_p - \bar{x})^2}{SS_{xx}}}$$

where $t_{\alpha/2}$ is based on $(n - 2)$ degrees of freedom.

Example 11.7
STIMULUS

Estimating the Mean of y—Drug Reaction Regression

Problem Refer to the simple linear regression on drug reaction. Find a 95% confidence interval for the mean reaction time when the concentration of the drug in the bloodstream is 4%.

Solution For a 4% concentration, $x = 4$ and the confidence interval for the mean value of y is

$$\hat{y} \pm t_{\alpha/2} s \sqrt{\frac{1}{n} + \frac{(x_p - \bar{x})^2}{SS_{xx}}} = \hat{y} \pm t_{.025} s \sqrt{\frac{1}{5} + \frac{(4 - \bar{x})^2}{SS_{xx}}}$$

---

*The term *prediction interval* is used when the interval formed is intended to enclose the value of a random variable. The term *confidence interval* is reserved for the estimation of population parameters (such as the mean).

where $t_{.025}$ is based on $n - 2 = 5 - 2 = 3$ degrees of freedom. Recall that $\hat{y} = 2.7$, $s = .61, \bar{x} = 3$, and $SS_{xx} = 10$. From Table III in Appendix A, $t_{.025} = 3.182$. Thus, we have

$$2.7 \pm (3.182)(.61)\sqrt{\frac{1}{5} + \frac{(4-3)^2}{10}} = 2.7 \pm (3.182)(.61)(.55)$$

$$= 2.7 \pm (3.182)(.34)$$

$$= 2.7 \pm 1.1$$

Therefore, when the percentage of drug in the bloodstream is 4%, we can be 95% confident that the mean reaction time for all possible subjects will range from 1.6 to 3.8 seconds.

Look Back Note that we used a small amount of data (a small sample size) for purposes of illustration in fitting the least squares line. The interval would probably be narrower if more information had been obtained from a larger sample.

━━━ ■ **Now Work Exercise 11.106a–d**

Example	**11.8**	Ⓓ

STIMULUS

Predicting an Individual Value of y — Drug Reaction Regression

Problem Refer again to the drug reaction regression. Predict the reaction time for the next performance of the experiment for a subject with a drug concentration of 4%. Use a 95% prediction interval.

Solution To predict the response time for an individual new subject for whom $x = 4$, we calculate the 95% prediction interval as

$$\hat{y} \pm t_{\alpha/2}s\sqrt{1 + \frac{1}{n} + \frac{(x_p - \bar{x})^2}{SS_{xx}}} = 2.7 \pm (3.182)(.61)\sqrt{1 + \frac{1}{5} + \frac{(4-3)^2}{10}}$$

$$= 2.7 \pm (3.182)(.61)(1.14)$$

$$= 2.7 \pm (3.182)(.70)$$

$$= 2.7 \pm 2.2$$

Therefore, when the drug concentration for an individual is 4%, we predict with 95% confidence that the reaction time for this new individual will fall into the interval from .5 to 4.9 seconds.

Look Back Like the confidence interval for the mean value of y, the prediction interval for y is quite large. This is because we have chosen a simple example (one with only five data points) to fit the least squares line. The width of the prediction interval could be reduced by using a larger number of data points.

━━━ ■ **Now Work Exercise 11.106e**

Both the confidence interval for $E(y)$ and the prediction interval for y can be obtained from a statistical software package. Figure 11.21 is a MINITAB printout showing the confidence interval and prediction interval, respectively, for the data in the drug example.

The 95% confidence interval for $E(y)$ when $x = 4$, highlighted under "95% CI" in Figure 11.21, is (1.645, 3.755). The 95% prediction interval for y when $x = 4$, highlighted in Figure 11.21 under "95% PI," is (.503, 4.897). These agree with the ones computed in Examples 11.7 and 11.8.

Figure 11.21

MINITAB printout giving 95% confidence interval for $E(y)$ and 95% prediction interval for y

```
Predicted Values for New Observations

New
Obs    Fit   SE Fit      95% CI          95% PI
  1  2.700   0.332  (1.645, 3.755)  (0.503, 4.897)


Values of Predictors for New Observations

New
Obs   DRUG_X
  1     4.00
```

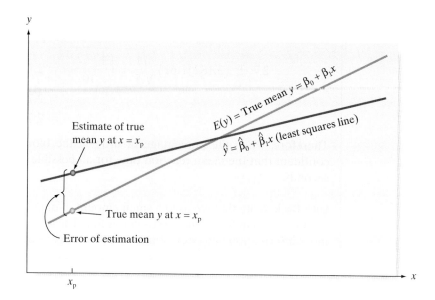

Figure 11.22

Error in estimating the mean value of y for a given value of x

Note that the prediction interval for an individual new value of y is *always* wider than the corresponding confidence interval for the mean value of y. Will this always be true? The answer is "Yes." The error in estimating the mean value of y, $E(y)$, for a given value of x, say, x_p, is the distance between the least squares line and the true line of means, $E(y) = \beta_0 + \beta_1 x$. This error, $[\hat{y} - E(y)]$, is shown in Figure 11.22. In contrast, *the error* $(y_p - \hat{y})$ *in predicting some future value of y is the sum of two errors*: the error in estimating the mean of y, $E(y)$, shown in Figure 11.22, plus the random error that is a component of the value of y that is to be predicted. (See Figure 11.23.) Consequently, the error in predicting a particular value of y will be larger than the error in estimating the mean value of y for a particular value of x. Note from their formulas that both the error of estimation and the error of prediction take their smallest values when $x_p = \bar{x}$. The farther x_p lies from $\bar{x}$, the larger will be the errors of estimation and prediction. You can see why this is true by noting the deviations for different values of x_p between the actual line of means $E(y) = \beta_0 + \beta_1 x$ and the predicted line of means $\hat{y} = \hat{\beta}_0 + \hat{\beta}_1 x$ shown in Figure 11.23. The deviation is larger at the extremes of the interval, where the largest and smallest values of x in the data set occur.

Both the confidence intervals for mean values and the prediction intervals for new values are depicted over the entire range of the regression line in Figure 11.24. You can see that the confidence interval is always narrower than the prediction interval and that they are both narrowest at the mean $\bar{x}$, increasing steadily as the distance $|x - \bar{x}|$

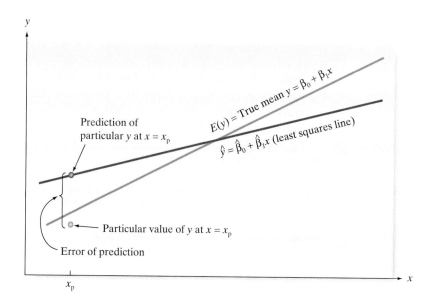

Figure 11.23

Error in predicting a future value of y for a given value of x

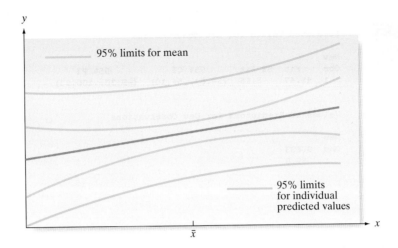

Figure 11.24

Confidence intervals for mean values and prediction intervals for new values

increases. In fact, when x is selected far enough away from $\bar{x}$ so that it falls outside the range of the sample data, it is dangerous to make any inferences about $E(y)$ or y. We call this the problem of extrapolation.

⚠ **CAUTION** Using the least squares prediction equation for extrapolation i.e., to estimate the mean value of y or to predict a particular value of y for values of x that fall outside the range of the values of x contained in your sample data, may lead to errors of estimation or prediction that are much larger than expected. Although the least squares model may provide a very good fit to the data over the range of x values contained in the sample, it could give a poor representation of the true model for values of x outside that region. ▲

The width of the confidence interval grows smaller as n is increased; thus, in theory, you can obtain as precise an estimate of the mean value of y as desired (at any given x) by selecting a large enough sample. The prediction interval for a new value of y also grows smaller as n increases, but there is a lower limit on its width. If you examine the formula for the prediction interval, you will see that the interval can get no smaller than $\hat{y} \pm z_{\alpha/2}\sigma$.* Thus, the only way to obtain more accurate predictions for new values of y is to reduce the standard deviation σ of the regression model. This can be accomplished only by improving the model, either by using a curvilinear (rather than linear) relationship with x or by adding new independent variables to the model (or both). Methods of improving the model are discussed in Chapter 12.

Now Work Exercise 11.106f

| Statistics IN Action | Revisited | **Using the Straight-Line Model to Predict Pipe Location for the Dowsing Data** |

The group of German physicists who conducted the dowsing experiments stated that the data for the three "best" dowsers empirically support the dowsing theory. If so, then the straight-line model relating a dowser's guess (x) to actual pipe location (y) should yield accurate predictions. The MINITAB printout shown in Figure SIA11.5 gives a 95% prediction interval for y when a dowser guesses $x = 50$ meters (the middle of the 100-meter-long waterpipe). The highlighted interval is $(-9.3, 100.23)$. Thus, we can be 95% confident that the actual pipe location will fall between -9.3 meters and

100.23 meters for this guess. Since the pipe is only 100 meters long, the interval in effect ranges from 0 to 100 meters—the entire length of the pipe! This result, of course, is due to the fact that the straight-line model is not a statistically useful predictor of pipe location, a fact we discovered in the previous Statistics in Action Revisited sections.

(Continued)

---

*The result follows from the facts that, for large n, $t_{\alpha/2} \approx z_{\alpha/2}$, $s \approx \sigma$, and the last two terms under the radical in the standard error of the predictor are approximately 0.

Statistics IN Action
(continued)

```
Predicted Values for New Observations

New
Obs    Fit  SE Fit      95% CI            95% PI
  1  45.47    5.15  (34.83, 56.10)   (-9.30, 100.23)


Values of Predictors for New Observations

New
Obs  GUESS
  1   50.0
```

Figure SIA11.5

MINITAB prediction interval
for dowsing data

Exercises 11.101–11.121

Understanding the Principles

11.101 Explain the difference between y and $E(y)$ for a given x.

11.102 *True or False.* For a given x, a confidence interval for $E(y)$ will always be wider than a prediction interval for y.

11.103 *True or False.* The greater the deviation between x and $\bar{x}$, the wider the prediction interval for y will be.

11.104 For each of the following, decide whether the proper inference is a prediction interval for y or a confidence interval for $E(y)$:
a. A jeweler wants to predict the selling price of a diamond stone on the basis of its size (number of carats).
b. A psychologist wants to estimate the average IQ of all patients who have a certain income level.

Learning the Mechanics

11.105 In fitting a least squares line to $n = 5$ data points, the following quantities were computed:
$$SS_{xx} = 31, \bar{x} = 2, SS_{yy} = 24, \bar{y} = 6, SS_{xy} = 26$$
a. Find the least squares line.
b. Graph the least squares line.
c. Calculate SSE.
d. Calculate s^2.
e. Find a 90% confidence interval for the mean value of y when $x_p = 4.1$.
f. Find a 90% prediction interval for y when $x_p = 2.7$.

11.106 Consider the following pairs of measurements.

x	1	2	3	4	5	6	7
y	3	5	4	6	7	7	10

a. Construct a scatterplot of these data.
b. Find the least squares line, and plot it on your scatterplot.
c. Find s^2.
d. Find a 90% confidence interval for the mean value of y when $x = 4$. Plot the upper and lower bounds of the confidence interval on your scatterplot.
e. Find a 90% prediction interval for a new value of y when $x = 4$. Plot the upper and lower bounds of the prediction interval on your scatterplot.

f. Compare the widths of the intervals you constructed in parts **d** and **e.** Which is wider and why?

11.107 Consider the following pairs of measurements.

x	4	6	0	5	2	3	2	6	2	1
y	3	5	−1	4	3	2	0	4	1	1

For these data, $SS_{xx} = 38.900$, $SS_{yy} = 33.600$, $SS_{xy} = 32.8$, and $\hat{y} = -.414 + .843x$.
a. Construct a scatterplot of the data.
b. Plot the least squares line on your scatterplot.
c. Use a 95% confidence interval to estimate the mean value of y when $x_p = 6$. Plot the upper and lower bounds of the interval on your scatterplot.
d. Repeat part **c** for $x_p = 3.2$ and $x_p = 0$.
e. Compare the widths of the three confidence intervals you constructed in parts **c** and **d**, and explain why they differ.

11.108 Refer to Exercise 11.107.
a. Using no information about x, estimate and calculate a 95% confidence interval for the mean value of y. [*Hint:* Use the one-sample t methodology of Section 7.3.]
b. Plot the estimated mean value and the confidence interval as horizontal lines on your scatterplot.
c. Compare the confidence intervals you calculated in parts **c** and **d** of Exercise 11.107 with the one you calculated in part **a** of this exercise. Does x appear to contribute information about the mean value of y?
d. Check the answer you gave in part **c** with a statistical test of the null hypothesis $H_0: \beta_1 = 0$ against $H_a: \beta_1 \neq 0$. Use $\alpha = .05$.

Applying the Concepts—Basic

11.109 **Do nice guys finish first or last?** Refer to the *Nature* (Mar. 20, 2008) study of the use of punishment in cooperation games, Exercise 11.22 (p. 625). Recall that simple linear regression was used to model a player's average payoff (y) as a straight-line function of the number of times punishment was used (x) by the player.
a. If the researchers want to predict average payoff for a single player who used punishment 10 times, how should they proceed?

b. If the researchers want to estimate the mean of the average payoffs for all players who used punishment 10 times, how should they proceed?

11.110 Measuring the moon's orbit. Refer to the *American Journal of Physics* (Apr. 2014) study of the moon's orbit, Exercise 11.23 (p. 626). Recall that the angular size (y) of the moon was modeled as a straight-line function of height above horizon (x). A MINITAB printout showing both a 95% prediction interval for y and a 95% confidence interval for E(y) when x = 50 degrees is displayed below.

a. Give a practical interpretation of the 95% prediction interval.

b. Give a practical interpretation of the 95% confidence interval.

c. A researcher wants to predict the angular size of the moon when the height above the horizon is 80 degrees. Do you recommend that the researcher use the least squares line shown in the printout to make the prediction? Explain.

```
The regression equation is
ANGLE = 321 + 0.0834 HEIGHT

Predictor      Coef    SE Coef       T       P
Constant    320.636      0.435  737.32   0.000
HEIGHT     0.083379   0.008868    9.40   0.000

S = 0.569262   R-Sq = 88.9%   R-Sq(adj) = 87.9%

Analysis of Variance

Source          DF      SS      MS      F      P
Regression       1  28.648  28.648  88.40  0.000
Residual Error  11   3.565   0.324
Total           12  32.212

Predicted Values for New Observations

New Obs    Fit  SE Fit        95% CI              95% PI
      1  324.805  0.162  (324.448, 325.163)  (323.502, 326.108)

Values of Predictors for New Observations

New Obs  HEIGHT
      1    50.0
```

11.111 Mongolian desert ants. Refer to the *Journal of Biogeography* (Dec. 2003) study of ant sites in Mongolia, presented in Exercise 11.26 (p. 627). You applied the method of least squares to estimate the straight-line model relating annual rainfall (y) and maximum daily temperature (x). A SAS printout giving 95% prediction intervals for the amount of rainfall at each of the 11 sites is shown at the bottom of the page. Select the interval associated with site (observation) 7 and interpret it practically.

11.112 Ranking driving performance of professional golfers. Refer to *The Sport Journal* (Winter 2007) study of a new method for ranking the total driving performance of golfers on the Professional Golf Association (PGA) tour, presented in Exercise 11.29 (p. 628). You fit a straight-line model relating driving accuracy (y) to driving distance (x). A MINITAB printout with prediction and confidence intervals for a driving distance of x = 300 yards is shown below.

```
Predicted Values for New Observations

New
Obs    Fit  SE Fit       95% CI              95% PI
  1  61.309  0.357  (60.586, 62.032)  (56.724, 65.894)


Values of Predictors for New Observations

New
Obs  DISTANCE
  1      300
```

a. Locate the 95% prediction interval for driving accuracy (y) on the printout, and give a practical interpretation of the result.

b. Locate the 95% prediction interval for mean driving accuracy (y) on the printout, and give a practical interpretation of the result.

c. If you are interested in knowing the average driving accuracy of all PGA golfers who have a driving distance of 300 yards, which of the intervals is relevant? Explain.

11.113 Sweetness of orange juice. Refer to the simple linear regression of sweetness index y and amount of pectin, x, for n = 24 orange juice samples, presented in Exercise 11.32 (p. 629). A 90% confidence interval for the mean sweetness index E(y) for each value of x is shown on the SPSS spreadsheet on the next page. Select an observation and interpret this interval.

SAS output for Exercise 11.111

Predictions

Obs	Region	Rain	Temp	Predicted Rain	Lower prediction limit of Rain	Upper prediction limit of Rain
1	DryStepp	196	5.7	201.977	179.525	224.430
2	DryStepp	196	5.7	201.977	179.525	224.430
3	DryStepp	179	7.0	180.704	163.694	197.714
4	DryStepp	197	8.0	164.340	150.594	178.085
5	DryStepp	149	8.5	156.157	143.513	168.802
6	GobiDese	112	10.7	120.156	106.038	134.274
7	GobiDese	125	11.4	108.701	92.298	125.104
8	GobiDese	99	10.9	116.883	102.172	131.595
9	GobiDese	125	11.4	108.701	92.298	125.104
10	GobiDese	84	11.4	108.701	92.298	125.104
11	GobiDese	115	11.4	108.701	92.298	125.104

	run	sweet	pectin	lower90m	upper90m
1	1	5.2	220	5.64898	5.83848
2	2	5.5	227	5.63898	5.81613
3	3	6.0	259	5.57819	5.72904
4	4	5.9	210	5.66194	5.87173
5	5	5.8	224	5.64337	5.82560
6	6	6.0	215	5.65564	5.85493
7	7	5.8	231	5.63284	5.80379
8	8	5.6	268	5.55553	5.71011
9	9	5.6	239	5.61947	5.78019
10	10	5.9	212	5.65946	5.86497
11	11	5.4	410	5.05526	5.55416
12	12	5.6	256	5.58517	5.73592
13	13	5.8	306	5.43785	5.65219
14	14	5.5	259	5.57819	5.72904
15	15	5.3	284	5.50957	5.68213
16	16	5.3	383	5.15725	5.57694
17	17	5.7	271	5.54743	5.70434
18	18	5.5	264	5.56591	5.71821
19	19	5.7	227	5.63898	5.81613
20	20	5.3	263	5.56843	5.72031
21	21	5.9	232	5.63125	5.80075
22	22	5.8	220	5.64898	5.83848
23	23	5.8	246	5.60640	5.76091
24	24	5.9	241	5.61587	5.77454

SPSS output for Exercise 11.113

11.114 Removing nitrogen from toxic wastewater. Highly toxic wastewater is produced during the manufacturing of dry-spun acrylic fiber. One way to lessen toxicity is to remove the nitrogen from the wastewater. A group of environmental engineers investigated a promising method—called anaerobic ammonium oxidation—for nitrogen removal and reported the results in the *Chemical Engineering Journal* (Apr. 2013). A sample of 120 specimens of toxic wastewater were collected and each treated with the nitrogen removal method. The amount of nitrogen removed (measured in milligrams per liter) was determined as well as the amount of ammonium (milligrams per liter) used in the removal process. These data (simulated from information provided in the journal article) are saved in the **NITRO** file. The data for the first 5 specimens are shown below. A simple linear regression analysis, where y = amount of nitrogen removed and x = amount of ammonium used, is also shown in the accompanying SAS printout.

a. Assess statistically the adequacy of the fit of the linear model. Do you recommend using the model for predicting nitrogen amount?

b. On the SAS printout, locate a 95% prediction interval for nitrogen amount when amount of ammonium used is 100 milligrams per liter. Practically interpret the result.

c. Will a 95% confidence interval for the mean nitrogen amount when amount of ammonium used is 100 milligrams per liter be wider or narrower than the interval, part **b**? Explain.

First 5 observations of 120

Nitrogen	Ammonium
18.87	67.40
17.01	12.49
23.88	61.96
10.45	15.63
36.03	83.66

SAS Output for Exercise 11.114

The SAS System

The REG Procedure
Model: MODEL1
Dependent Variable: NITRO_Y

Number of Observations Read	120
Number of Observations Used	120

Analysis of Variance

Source	DF	Sum of Squares	Mean Square	F Value	Pr > F
Model	1	88104	88104	1075.57	<.0001
Error	118	9665.80325	81.91359		
Corrected Total	119	97770			

Root MSE	9.05061	R-Square	0.9011
Dependent Mean	67.91083	Adj R-Sq	0.9003
Coeff Var	13.32720		

Parameter Estimates

Variable	DF	Parameter Estimate	Standard Error	t Value	Pr > |t|
Intercept	1	2.22255	2.16665	1.03	0.3071
AMMON_X	1	0.57637	0.01757	32.80	<.0001

Output Statistics

Obs	AMMON_X	Dependent Variable	Predicted Value	Std Error Mean Predict	95% CL Predict		Residual
1	100.0	.	59.8596	0.8619	41.8558	77.8634	.
2	67.4	18.8700	41.0699	1.1629	22.9999	59.1400	-22.1999
3	12.5	17.0100	9.4214	1.9655	-8.9190	27.7619	7.5886
4	62.0	23.8800	37.9345	1.2321	19.8465	56.0225	-14.0545
5	15.6	10.4500	11.2312	1.9156	-7.0885	29.5509	-0.7812
6	83.7	36.0300	50.4417	0.9830	32.4136	68.4698	-14.4117

Applying the Concepts—Intermediate

11.115 Game performance of water polo players. Refer to the *Biology of Sport* (Vol. 31, 2014) study of the physiological performance of top-level water polo players, Exercise 11.24 (p. 626). Recall that the in-game heart rate (*y*, expressed as a percentage of maximum heart rate) of a player was modeled as a straight-line function of the player's maximal oxygen uptake (*x*, VO₂max). A researcher desires an estimate of the average in-game heart rate of all top-level water polo players who have a maximal oxygen uptake of 150 VO₂max.

   a. Which interval is desired by the researcher, a 95% prediction interval for *y* or a 95% confidence interval for $E(y)$? Explain.

   b. Use statistical software to compute the desired interval.

   c. Give a practical interpretation of the interval.

11.116 Sound waves from a basketball. Refer to the *American Journal of Physics* (June 2010) study of sound waves in a spherical cavity, Exercise 11.31 (p. 629). You fit a straight-line model relating sound wave frequency (*y*) to number of resonances (*x*) for *n* = 24 echoes resulting from striking a basketball with a metal rod.

   a. Use the model to predict the sound wave frequency for the 10th resonance.

   b. Form a 90% confidence interval for the prediction, part **a**. Interpret the result.

   c. Suppose you want to predict the sound wave frequency for the 30th resonance. What are the dangers in making this prediction with the fitted model?

11.117 Ideal height of your mate. Refer to the *Chance* (Summer 2008) study of the height of the ideal mate, Exercise 11.33 (p. 629). The data were used to fit the simple linear regression model, $E(y) = \beta_0 + \beta_1 x$, where *y* = ideal partner's height (in inches) and *x* = student's height (in inches). One model was fitted for male students and one model was fitted for female students. Consider a student who is 66 inches tall.

   a. If the student is a female, use the model to predict the height of her ideal mate. Form a 95% confidence interval for the prediction and interpret the result.

   b. If the student is a male, use the model to predict the height of his ideal mate. Form a 95% confidence interval for the prediction and interpret the result.

   c. Which of the two inferences, parts **a** and **b**, may be invalid? Why?

11.118 The "name game." Refer to the *Journal of Experimental Psychology—Applied* (June 2000) name-retrieval study, presented in Exercise 11.34 (p. 630).

   a. Find a 99% confidence interval for the mean recall proportion for students in the fifth position during the "name game." Interpret the result.

   b. Find a 99% prediction interval for the recall proportion of a particular student in the fifth position during the "name game." Interpret the result.

   c. Compare the intervals you found in parts **a** and **b**. Which interval is wider? Will this always be the case? Explain.

11.119 Spreading rate of spilled liquid. Refer to the *Chemical Engineering Progress* (Jan. 2005) study of the rate at which a spilled volatile liquid will spread across a surface, presented in Exercise 11.35 (p. 630). Recall that simple linear regression was used to model *y* = mass of the spill as a function of *y* = elapsed time of the spill.

   a. Find a 99% confidence interval for the mean mass of all spills with an elapsed time of 15 minutes. Interpret the result.

   b. Find a 99% prediction interval for the mass of a single spill with an elapsed time of 15 minutes. Interpret the result.

   c. Compare the intervals you found in parts **a** and **b**. Which interval is wider? Will this always be the case? Explain.

11.120 Forecasting movie revenues with Twitter. Refer to the *IEEE International Conference on Web Intelligence and Intelligent Agent Technology* (2010) study of how social media (e.g., Twitter.com) may influence the products consumers buy, Exercise 11.36 (p. 631). Recall that opening weekend box office revenue (in millions of dollars) and *tweet rate* (average number of tweets referring to the movie per hour) were collected for a sample of 24 recent movies. The data are reproduced in the table. Use simple linear regression to find a 90% prediction interval for the revenue (*y*) of a movie with a tweet rate (*x*) of 150 tweets per hour. Give a practical interpretation of the interval.

Tweet Rate	Revenue (millions)
1365.8	142
1212.8	77
581.5	61
310.1	32
455	31
290	30
250	21
680.5	18
150	18
164.5	17
113.9	16
144.5	15
418	14
98	14
100.8	12
115.4	11
74.4	10
87.5	9
127.6	9
52.2	9
144.1	8
41.3	2
2.75	0.3

Applying the Concepts—Advanced

11.121 Life tests of cutting tools. Refer to the data on life tests of cutting tools, Exercise 11.52 (p. 636).

   a. Use a 90% confidence interval to estimate the mean useful life of a brand-A cutting tool when the cutting speed is 45 meters per minute. Repeat for brand B. Compare the widths of the two intervals and comment on the reasons for any difference.

b. Use a 90% prediction interval to predict the useful life of a brand-A cutting tool when the cutting speed is 45 meters per minute. Repeat for brand B. Compare the widths of the two intervals with each other and with the two intervals you calculated in part **a**. Comment on the reasons for any differences.

c. Note that the estimation and prediction you performed in parts **a** and **b** were for a value of x that was not included in the original sample. That is, the value $x = 45$ was not part of the sample. However, the value is within the range of x values in the sample, so that the regression model spans the x value for which the estimation

and prediction were made. In such situations, estimation and prediction represent **interpolations**.

Suppose you were asked to predict the useful life of a brand-A cutting tool for a cutting speed of $x = 100$ meters per minute. Since the given value of x is outside the range of the sample x values, the prediction is an example of extrapolation. Predict the useful life of a brand-A cutting tool that is operated at 100 meters per minute, and construct a 95% confidence interval for the actual useful life of the tool. What additional assumption do you have to make in order to ensure the validity of an extrapolation?

11.7 A Complete Example

In the previous sections, we presented the basic elements necessary to fit and use a straight-line regression model. In this section, we will assemble these elements by applying them in an example with the aid of computer software.

Suppose a fire insurance company wants to relate the amount of fire damage in major residential fires to the distance between the burning house and the nearest fire station. The study is to be conducted in a large suburb of a major city; a sample of 15 recent fires in this suburb is selected. The amount of damage, y, and the distance between the fire and the nearest fire station, x, are recorded for each fire. The results are given in Table 11.5 and saved in the **FIREDAM** file.

Table 11.5   Fire Damage Data

Distance from Fire Station, x (miles)	Fire Damage y (thousands of dollars)
3.4	26.2
1.8	17.8
4.6	31.3
2.3	23.1
3.1	27.5
5.5	36.0
.7	14.1
3.0	22.3
2.6	19.6
4.3	31.3
2.1	24.0
1.1	17.3
6.1	43.2
4.8	36.4
3.8	26.1

Ⓓ *Data Set:* FIREDAM

Step 1 First, we hypothesize a model to relate fire damage, y, to the distance from the nearest fire station, x. We hypothesize a straight-line probabilistic model:

$$y = \beta_0 + \beta_1 x + \varepsilon$$

Step 2 Next, we open the **FIREDAM** file and use statistical software to estimate the unknown parameters in the deterministic component of the hypothesized model. The SAS printout for the simple linear regression analysis is shown in Figure 11.25. The least squares estimate of the slope β_1 and intercept β_0, highlighted on the printout, are

$$\hat{\beta}_1 = 4.91933$$
$$\hat{\beta}_0 = 10.27793$$

and the least squares equation is (rounded)

$$\hat{y} = 10.28 + 4.92x$$

This prediction equation is graphed by MINITAB in Figure 11.26, along with a plot of the data points.

The least squares estimate of the slope, $\hat{\beta}_1 = 4.92$, implies that the estimated mean damage increases by $4,920 for each additional mile from the fire station. This interpretation is valid over the range of x, or from .7 to 6.1 miles from the station. The estimated y-intercept, $\hat{\beta}_0 = 10.28$, has the interpretation that a fire 0 miles from the fire station has an estimated mean damage of $10,280. Although this would seem to apply to the fire station itself, remember that the y-intercept is meaningfully interpretable only if $x = 0$ is within the sampled range of the independent variable. Since $x = 0$ is outside the range, β_0 has no practical interpretation.

Step 3 Now we specify the probability distribution of the random-error component ε. The assumptions about the distribution are identical to those listed in Section 11.3. Although we know that these assumptions are not completely satisfied (they rarely are for practical problems), we are willing to assume that they are approximately satisfied for this example. The estimate of the standard deviation σ of ε, highlighted on the SAS printout, is

$$s = 2.31635$$

Dependent Variable: DAMAGE

Analysis of Variance

Source	DF	Sum of Squares	Mean Square	F Value	Pr > F
Model	1	841.76636	841.76636	156.89	<.0001
Error	13	69.75098	5.36546		
Corrected Total	14	911.51733			

Root MSE	2.31635	R-Square	0.9235
Dependent Mean	26.41333	Adj R-Sq	0.9176
Coeff Var	8.76961		

Parameter Estimates

Variable	DF	Parameter Estimate	Standard Error	t Value	Pr > |t|	95% Confidence Limits	
Intercept	1	10.27793	1.42028	7.24	<.0001	7.20960	13.34625
DISTANCE	1	4.91933	0.39275	12.53	<.0001	4.07085	5.76781

Output Statistics

Obs	DISTANCE	Dep Var DAMAGE	Predicted Value	Std Error Mean Predict	95% CL Predict		Residual
1	3.4	26.2000	27.0037	0.5999	21.8344	32.1729	-0.8037
2	1.8	17.8000	19.1327	0.8340	13.8141	24.4514	-1.3327
3	4.6	31.3000	32.9068	0.7915	27.6186	38.1951	-1.6068
4	2.3	23.1000	21.5924	0.7112	16.3577	26.8271	1.5076
5	3.1	27.5000	25.5279	0.6022	20.3573	30.6984	1.9721
6	5.5	36.0000	37.3342	1.0573	31.8334	42.8351	-1.3342
7	0.7	14.1000	13.7215	1.1766	8.1087	19.3342	0.3785
8	3	22.3000	25.0359	0.6081	19.8622	30.2097	-2.7359
9	2.6	19.6000	23.0682	0.6550	17.8678	28.2686	-3.4682
10	4.3	31.3000	31.4311	0.7198	26.1908	36.6713	-0.1311
11	2.1	24.0000	20.6085	0.7566	15.3442	25.8729	3.3915
12	1.1	17.3000	15.6892	1.0444	10.1999	21.1785	1.6108
13	6.1	43.2000	40.2858	1.2587	34.5906	45.9811	2.9142
14	4.8	36.4000	33.8907	0.8450	28.5640	39.2175	2.5093
15	3.8	26.1000	28.9714	0.6320	23.7843	34.1585	-2.8714
16	3.5	.	27.4956	0.6043	22.3239	32.6672	.

Figure 11.25
SAS printout for fire damage regression

Figure 11.26
MINITAB scatterplot with least squares line for fire damage regression analysis

This implies that most of the observed fire damage (y) values will fall within approximately $2s = 4.64$ thousand dollars of their respective predicted values when the least squares line is used.

Step 4 We can now check the usefulness of the hypothesized model—in other words, whether x really contributes information for the prediction of y by the straight-line model. First, test the null hypothesis that the slope β_1 is 0—that is, that there is no linear relationship between fire damage and the distance from the nearest fire station—against the alternative hypothesis that fire damage increases as the distance increases. We test

$$H_0: \beta_1 = 0$$
$$H_a: \beta_1 > 0$$

The two-tailed observed significance level for testing $H_a: \beta_1 \neq 0$, highlighted on the SAS printout, is less than .0001. Thus, the p-value for our one-tailed test is less than half of this value (.00005). This small p-value leaves little doubt that mean fire damage and distance between the fire and the fire station are at least linearly related, with mean fire damage increasing as the distance increases.

We gain additional information about the relationship by forming a confidence interval for the slope β_1. A 95% confidence interval, highlighted on the SAS printout, is (4.071, 5.768). Thus, with 95% confidence, we estimate that the interval from $4,071 to $5,768 encloses the mean increase (β_1) in fire damage per additional mile in distance from the fire station.

Another measure of the utility of the model is the coefficient of determination, r^2. The value (also highlighted on the printout) is $r^2 = .9235$, which implies that about 92% of the sample variation in fire damage (y) is explained by the distance (x) between the fire and the fire station.

The coefficient of correlation, r, that measures the strength of the linear relationship between y and x is not shown on the SAS printout and must be calculated. Using the facts that $r = \sqrt{r^2}$ in simple linear regression and that r and $\hat{\beta}_1$ have the same sign, we calculate

$$r = +\sqrt{r^2} = \sqrt{.9235} = .96$$

The high correlation confirms our conclusion that β_1 is greater than 0; it appears that fire damage and distance from the fire station are positively correlated. All signs point to a strong linear relationship between y and x.

Step 5 We are now prepared to use the least squares model. Suppose the insurance company wants to predict the fire damage if a major residential fire were to occur 3.5 miles from the nearest fire station. The predicted value (highlighted at the bottom of the SAS printout) is $\hat{y} = 27.496$, while the 95% prediction interval (also highlighted) is (22.324, 32.667). Therefore, with 95% confidence, we predict fire damage in a major residential fire 3.5 miles from the nearest station to be between $22,324 and $32,667.

⚠ **CAUTION** We would not use this model to make predictions for homes less than .7 mile or more than 6.1 miles from the nearest fire station. A look at the data in Table 11.5 reveals that all the x-values fall between .7 and 6.1. It is dangerous to use the model to make predictions outside the region in which the sample data fall. A straight line might not provide a good model for the relationship between the mean value of y and the value of x when stretched over a wider range of x-values. ▲

Exercises 11.122–11.125

Applying the Concepts—Intermediate

11.122 Impact of dropping ping-pong balls. The impact of dropping hollow balls was investigated in the *American Journal of Physics* (Mar. 2014). Standard ping-pong balls were dropped vertically onto a force plate. Upon impact, two variables were measured: $y =$ coefficient of restitution, COR (measured as a ratio of the speed at impact and rebound speed) and $x =$ speed at impact (meters/second). Of the 19 balls dropped, 10 buckled at impact. The data (simulated from information provided in the article) are listed in the table.

a. Conduct a complete simple linear regression analysis of the relationship between coefficient of restitution (y) and impact speed (x). Write all your conclusions in the words of the problem.

b. The researcher believes that the rate of increase in the coefficient of restitution with impact speed differs depending on whether the ping-pong ball buckles. Do the data support this hypothesis? Explain.

Ball	COR y	Speed x	Buckle
1	.945	0.8	No
2	.950	1.0	No
3	.930	1.5	No
4	.920	1.8	No
5	.920	3.0	No
6	.930	3.4	No
7	.905	4.4	No
8	.915	5.0	No
9	.910	6.4	No
10	.900	4.4	Yes
11	.885	5.3	Yes
12	.870	5.4	Yes
13	.850	7.4	Yes
14	.795	7.2	Yes
15	.790	7.2	Yes
16	.800	8.0	Yes
17	.820	8.5	Yes
18	.810	9.4	Yes
19	.780	9.0	Yes

11.123 Thickness of dust on solar cells. The performance of a solar cell can deteriorate when atmospheric dust accumulates on the solar panel surface. In the *International Journal of Energy and Environmental Engineering* (Dec. 2012), researchers at the Renewable Energy Research Laboratory, University of Lucknow (India), estimated the relationship between the dust thickness and the efficiency of a solar cell. The thickness of dust (in millimeters) collected on a solar cell was measured three times per month over a year-long period. Each time the dust thickness was measured, the researchers also determined the percentage difference (before minus after dust collection) in efficiency of the solar panel. Data (monthly averages) for the 10 months when there was no rain are listed in the table.

a. Fit the linear model, $y = \beta_0 + \beta_1 x + \varepsilon$, to the data where y = efficiency and x = average dust thickness.

b. Evaluate the model statistically. Do you recommend using the model in practice?

Month	Efficiency (% change)	Average Dust Thickness (mm)
January	1.5666	0.00024
February	1.9574	0.00105
March	1.3707	0.00075
April	1.9563	0.00070
May	1.6332	0.00142
June	1.8172	0.00055
July	0.9202	0.00039
October	1.8790	0.00095
November	1.5544	0.00064
December	2.0198	0.00065

Source: Siddiqui, R., and Bajpai, U. "Correlation between thicknesses of dust collected on photovoltaic module and difference in efficiencies in composite climate," *International Journal of Energy and Environmental Engineering*, Vol. 4, No. 1, Dec. 2012 (Table 1).

11.124 An MBA's work-life balance. Many business schools offer courses that assist MBA students with developing good work-life balance habits and most large companies have developed work-life balance programs for their employees. The Graduate Management Admission Council (GMAC) conducted a survey of over 2,000 MBA alumni to explore the work-life balance issue. (For example, one question asked alumni to state their level of agreement with the statement, "My personal and work demands are overwhelming.") Based on these responses, the GMAC determined a work-life balance scale score for each MBA alumni. Scores ranged from 0 to 100, with lower scores indicating a higher imbalance between work and life. Many other variables, including average number of hours worked per week, were also measured. The data for the work-life balance study are saved in the **GMAC** file. (The first 15 observations are listed in the accompanying table.) Let x = average number of hours worked per week and y = work-life balance scale score for each MBA alumnus. Investigate the link between these two variables by conducting a complete simple linear regression analysis of the data. Summarize your findings in a professional report.

WLB Score	Hours
75.22	50
64.98	45
49.62	50
44.51	55
70.10	50
54.74	60
55.98	55
21.24	60
59.86	50
70.10	50
29.00	70
64.98	45
36.75	40
35.45	40
45.75	50

Based on "Work-life balance: An MBA alumni report." *Graduate Management Admission Council (GMAC) Research Report* (Oct. 13, 2005).

11.125 Legal advertising—does it pay? To gain a competitive edge, lawyers aggressively advertise their services. Does legal advertising really pay? To partially answer this question, consider the case of an actual law firm that specializes in personal injury (PI) cases. The firm spends thousands of dollars each month on advertising. The accompanying table shows the firm's new personal injury

Month	New PI Cases	6 Months Cumulative Adv. Exp.
7	11	$ 41,632.74
8	7	$ 38,227.39
9	13	$ 39,779.77
10	7	$ 37,490.22
11	9	$ 52,225.71
12	8	$ 56,249.15
13	18	$ 59,938.03
14	9	$ 65,250.59
15	25	$ 66,071.85
16	26	$ 81,765.94
17	27	$ 66,895.46
18	12	$ 71,426.16
19	14	$ 75,346.40
20	5	$ 81,589.97
21	22	$ 78,828.68
22	15	$ 78,415.73
23	12	$ 90,802.77
24	18	$ 95,689.44
25	20	$ 83,099.55
26	38	$ 82,703.75
27	13	$ 90,484.38
28	18	$102,084.54
29	21	$ 84,976.99
30	7	$ 95,314.29
31	16	$115,858.28
32	12	$108,557.00
33	15	$127,693.57
34	18	$122,761.67
35	30	$123,545.67
36	12	$119,388.26
37	30	$134,675.68
38	20	$133,812.93
39	19	$142,417.13
40	29	$149,956.61
41	58	$165,204.46

(Continued)

Month	New PI Cases	6 Months Cumulative Adv. Exp.
42	42	$156,725.72
43	24	$146,397.56
44	47	$197,792.64
45	24	$198,460.28
46	14	$206,662.87
47	31	$253,011.27
48	26	$249,496.28

Source: Info Tech, Inc., Gainesville, Florida.

cases each month over a 42-month period. Also shown is the total expenditure on advertising each month, and over the previous 6 months. Do these data provide support for the hypothesis that increased advertising expenditures are associated with more personal injury cases? Conduct a complete simple linear regression analysis of the data, letting y = new PI cases and x = 6-month cumulative advertising expenditure. Summarize your findings in a professional report.

CHAPTER NOTES

Key Terms

Bivariate relationship 645
Coefficient of correlation 645
Coefficient of determination 650
Confidence interval for
 mean of y 656
Correlation 645
Dependent (or response)
 variable 615
Deterministic model 614
Deterministic relationship 614
Errors of prediction 619
Estimated standard error of the
 least squares slope $\hat{\beta}_1$ 637
Estimated standard error of the
 regression model 632
Extrapolation 664
Independent (or predictor)
 variable 615
Interpolation 664
Least squares estimates 620
Least squares line (or regression
 line or least squares prediction
 equation) 619

Line of means 616
Method of least
 squares 619
Population correlation
 coefficient 648
Prediction interval for y 656
Probabilistic model 614
Probabilistic relationship 614
Random error 614
Regression analysis
 (modeling) 615
Regression residuals 619
Response variable 615
Sampling distribution
 of β_1 637
Scatterplot or scattergram 619
Slope 616
Standard error of
 prediction 656
Standard error of $\hat{y}$ 656
Straight-line (first-order)
 model 615
y-intercept 616

Key Symbols/Notation

y	Dependent variable (variable to be predicted)
x	Independent variable (variable used to predict)
$E(y)$	Expected (mean) of y
β_0	y-intercept of true line
β_1	slope of true line
$\hat{\beta}_0$	Least squared estimate of y-intercept
$\hat{\beta}_1$	Least squares estimate of slope
ε	Random error
$\hat{y}$	Predicted value of y for a given x-value
$(y - \hat{y})$	Estimated error of prediction
SSE	Sum of squared errors of prediction
r	Coefficient of correlation
r^2	Coefficient of determination
x_p	Value of x used to predict y

Key Ideas

Simple Linear Regression Variables

y = **Dependent** variable (quantitative)
x = **Independent** variable (quantitative)

Method of least squares properties

1. average error of prediction = 0
2. sum of squared errors is minimum

First-order (straight-line) model

$$E(y) = \beta_0 + \beta_1 x$$

where $E(y)$ = mean of y
β_0 = **y-intercept** of line (point where line intercepts y-axis)
β_1 = **slope** of line (change in y for every one-unit change in x)

Practical interpretation of y-intercept

Predicted y-value when $x = 0$
(no practical interpretation if $x = 0$ is either nonsensical or outside range of sample data)

Practical interpretation of slope

Increase (or decrease) in y for every one-unit increase in x

Coefficient of correlation, r

1. ranges between -1 and $+1$
2. measures strength of *linear relationship* between y and x

Coefficient of determination, r^2

1. ranges between 0 and 1
2. measures proportion of sample variation in y "explained" by the model.

Practical interpretation of model standard deviation s

Ninety-five percent of y-values fall within $2s$ of their respective predicted values

Comparing Intervals in Step 5

Width of *confidence interval for $E(y)$* will always be **narrower** than width of *prediction interval for y*.

Key Formulas

$$\hat{\beta}_1 = \frac{SS_{xy}}{SS_{xx}} \quad \text{where } SS_{xx} = \sum(x - \bar{x})^2 \text{ and}$$

$$SS_{xy} = \sum(x - \bar{x})(y - \bar{y})$$

$$\hat{\beta}_0 = \bar{y} - \hat{\beta}_1 \bar{x}$$

$$SSE = \sum(y - \hat{y})^2 = SS_{yy} - \hat{\beta}_1 SS_{xy}, \quad \text{where } SS_{yy} = \sum(y - \bar{y})^2$$

$$s = \sqrt{\frac{\text{SSE}}{n-2}}$$

$$t \,(\text{for testing } H_0: \beta_1 = 0) = \frac{\hat{\beta}_1}{s_{\hat{\beta}_1}} \quad \text{where } s_{\hat{\beta}_1} = \frac{s}{\sqrt{\text{SS}_{xx}}}$$

$$(1 - \alpha)\,100\% \text{ CI for } \beta_1 \colon \hat{\beta}_1 \pm t_{\alpha/2} s_{\hat{\beta}_1}$$

$$r = \frac{\text{SS}_{xy}}{\sqrt{\text{SS}_{xx}\text{SS}_{yy}}}$$

$$t \,(\text{for testing } H_0: \rho = 0) = \frac{r\sqrt{n-2}}{\sqrt{1-r^2}}$$

$$r^2 = \frac{\text{SS}_{yy} - \text{SSE}}{\text{SS}_{yy}}$$

$$(1 - \alpha)\,100\% \text{ CI for } E(y)\colon \quad \hat{y} \pm t_{\alpha/2} \sqrt{\frac{1}{n} + \frac{(x_p - \bar{x})^2}{\text{SS}_{xx}}}$$

$$(1 - \alpha)\,100\% \text{ PI for } y\colon \quad \hat{y} \pm t_{\alpha/2} \sqrt{1 + \frac{1}{n} + \frac{(x_p - \bar{x})^2}{\text{SS}_{xx}}}$$

Guide to Simple Linear Regression

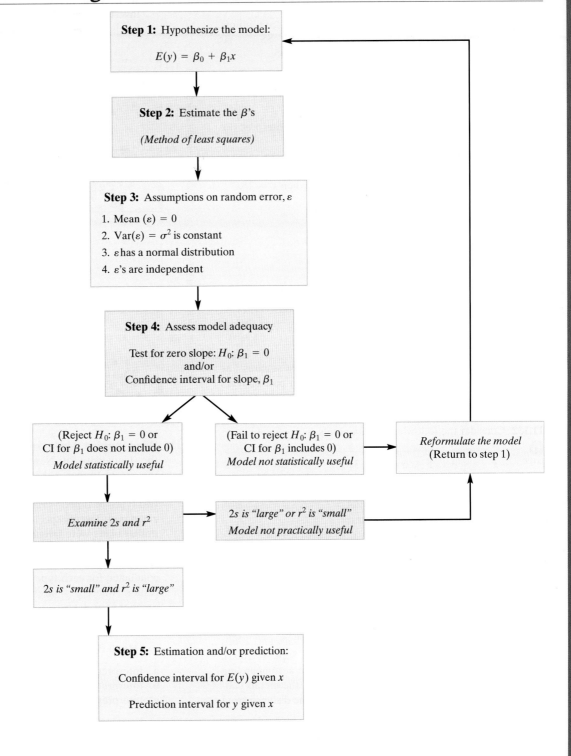

Supplementary Exercises 11.126–11.153

Understanding the Principles

11.126 Explain the difference between a probabilistic model and a deterministic model.

11.127 Give the general form of a straight-line model for $E(y)$.

11.128 Outline the five steps in a simple linear regression analysis.

11.129 *True or False.* In simple linear regression, about 95% of the y-values in the sample will fall within $2s$ of their respective predicted values.

Learning the Mechanics

11.130 In fitting a least squares line to $n = 20$ data points, the following quantities were computed: $SS_{xx} = 60$, $SS_{yy} = 225$, $SS_{xy} = -75$, $\bar{x} = 2.4$, and $\bar{y} = 45$.
  a. Find the least squares line.
  b. Graph the least squares line.
  c. Calculate SSE.
  d. Calculate s^2.
  e. Find a 90% confidence interval for β_1. Interpret this estimate.
  f. Find a 90% confidence interval for the mean value of y when $x = 15$.
  g. Find a 90% prediction interval for y when $x = 15$.

11.131 Consider the following sample data:

D
L11131

y	5	1	3
x	5	1	3

  a. Construct a scatterplot for the data.
  b. It is possible to find many lines for which $\Sigma(y - \hat{y}) = 0$. For this reason, the criterion $\Sigma(y - \hat{y}) = 0$ is not used to identify the "best-fitting" straight line. Find two lines that have $\Sigma(y - \hat{y}) = 0$.
  c. Find the least squares line.
  d. Compare the value of SSE for the least squares line with that of the two lines you found in part **b**. What principle of least squares is demonstrated by this comparison?

11.132 Consider the following 10 data points.

D
L11132

x	3	5	6	4	3	7	6	5	4	7
y	4	3	2	1	2	3	3	5	4	2

  a. Plot the data on a scatterplot.
  b. Calculate the values of r and r^2.
  c. Is there sufficient evidence to indicate that x and y are linearly correlated? Test at the $\alpha = .10$ level of significance.

Applying the Concepts—Basic

11.133 Arsenic in soil. In Denver, Colorado, environmentalists have discovered a link between high arsenic levels in soil and a crabgrass killer used in the 1950s and 1960s (*Environmental Science & Technology*, Sept. 1, 2000). The discovery was based, in part, on the accompanying scatterplots. The graphs plot the level of the metals cadmium and arsenic, respectively, against the distance from a former smelter plant for samples of soil taken from Denver residential properties.
  a. Normally, the metal level in soil decreases as distance from the source (e.g., a smelter plant) increases. Propose a straight-line model relating metal level y to distance x from the plant. On the basis of the theory, would you expect the slope of the line to be positive or negative?
  b. Examine the scatterplot for cadmium. Does the plot support the theory you set forth in part **a**?
  c. Examine the scatterplot for arsenic. Does the plot support the theory of part **a**? (*Note:* This finding led investigators to discover the link between high arsenic levels and the use of the crabgrass killer.)

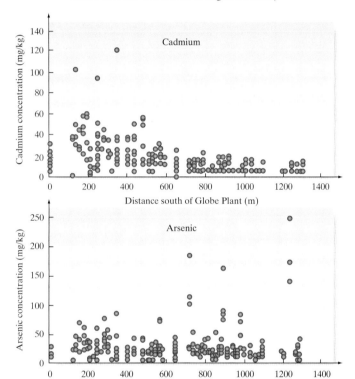

11.134 Wind turbine blade stress. Mechanical engineers at the University of Newcastle (Australia) investigated the use of timber in high-efficiency small wind turbine blades (*Wind Engineering*, Jan. 2004). The strengths of two types of timber—radiata pine and hoop pine—were compared. Twenty specimens (called "coupons") of each timber blade were fatigue tested by measuring the stress (in MPa) on the blade after various numbers of blade cycles. A simple linear regression analysis of the data, one conducted for each type of timber, yielded the following results (where $y = $ stress and $x = $ natural logarithm of number of cycles):

    Radiata Pine: $\hat{y} = 97.37 - 2.50x$, $r^2 = .84$
    Hoop Pine: $\hat{y} = 122.03 - 2.36x$, $r^2 = .90$

  a. Interpret the estimated slope of each line.
  b. Interpret the estimated y-intercept of each line.
  c. Interpret the value of r^2 for each line.

d. On the basis of these results, which type of timber blade appears to be stronger and more fatigue resistant? Explain.

11.135 **Sports news on local TV broadcasts.** *The Sports Journal* (Winter 2004) published the results of a study conducted to assess the factors that affect the time allotted to sports news on local television news broadcasts. Information on total time (in minutes) allotted to sports and on audience ratings of the TV news broadcast (measured on a 100-point scale) was obtained from a national sample of 163 news directors. A correlation analysis of the data yielded $r = .43$.

a. Interpret the value of the correlation coefficient r.

b. Find and interpret the value of the coefficient of determination r^2.

11.136 **Predicting sale prices of homes.** Real-estate investors, home buyers, and homeowners often use the appraised value of property as a basis for predicting the sale of that property. Data on sale prices and total appraised value of 78 residential properties sold recently in an upscale Tampa, Florida, neighborhood named Hunter's Green are saved in the **HOMES** file. Selected observations are listed in the table on p. 672.

MINITAB output for Exercise 11.136

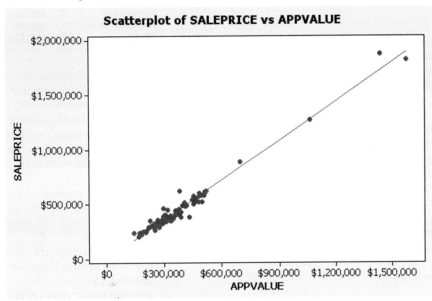

Correlations: SALEPRICE, APPVALUE

```
Pearson correlation of SALEPRICE and APPVALUE = 0.987
P-Value = 0.000
```

Regression Analysis: SALEPRICE versus APPVALUE

```
The regression equation is
SALEPRICE = 184 + 1.20 APPVALUE

Predictor      Coef  SE Coef       T      P
Constant        184     9834    0.02  0.985
APPVALUE    1.19956  0.02234   53.70  0.000

S = 44859.7   R-Sq = 97.4%   R-Sq(adj) = 97.4%

Analysis of Variance

Source         DF           SS           MS        F      P
Regression      1  5.80234E+12  5.80234E+12  2883.31  0.000
Residual Error 76  1.52942E+11   2012390103
Total          77  5.95528E+12

Predicted Values for New Observations

New
Obs     Fit  SE Fit         95% CI             95% PI
  1  480007    5105  (469839, 490175)  (390085, 569930)

Values of Predictors for New Observations

New
Obs  APPVALUE
  1    400000
```

Data for Exercise 11.136

Property	Sale Price	Appraised Value
1	$489,900	$418,601
2	1,825,000	1,577,919
3	890,000	687,836
4	250,00	191,620
5	1,275,000	1,063,901
⋮	⋮	⋮
74	325,000	292,702
75	516,000	407,449
76	309,300	272,275
77	370,000	347,320
78	580,000	511,359

Based on data from Hillsborough Country (Florida) Property Appraiser's Office.

a. Propose a straight-line model to relate the appraised property value (x) to the sale price (y) for residential properties in this neighborhood.

b. A MINITAB scatterplot of the data with the least squared line is shown at the top of the printout on p. 671. Does it appear that a straight-line model will be an appropriate fit to the data?

c. A MINITAB simple linear regression printout is also shown at the bottom of the printout on p. 671. Find the equation of the least squared line. Interpret the estimated slope and y-intercept in the words of the problem.

d. Locate the test statistic and p-value for testing $H_0: \beta_1 = 0$ against $H_a: \beta_1 > 0$. Is there sufficient evidence (at $\alpha = .01$) of a positive linear relationship between apprised property value (x) and sale price (y)?

e. Locate and interpret practically the values of r and r^2 on the printout.

f. Locate and interpret practically the 95% prediction interval for sale price (y) on the printout.

11.137 Baseball batting averages versus wins. Is the number of games won by a major league baseball team in a season related to the team's batting average? Consider data from the *Baseball Almanac* on the number of games won and the batting averages for the 14 teams in the American League for the 2013 Major League Baseball season. The data are listed in the next table.

Team	Games Won	Batting Avg. (average number of hits per 1,000 at bats)
New York	85	.242
Toronto	74	.252
Baltimore	85	.260
Boston	97	.277
Tampa Bay	92	.257
Cleveland	92	.255
Detroit	93	.283
Chicago	63	.249
Kansas City	86	.260
Minnesota	66	.242
Los Angeles	78	.264
Texas	91	.262
Seattle	71	.237
Oakland	96	.254

Based on data from *Baseball Almanac*, 2013; www.mlb.com.

Regression Analysis: Wins versus BatAve

```
The regression equation is
Wins = - 58.9 + 555 BatAve


Predictor    Coef   SE Coef      T      P
Constant   -58.86     50.76  -1.16  0.269
BatAve      554.5     197.5   2.81  0.016


S = 9.09171   R-Sq = 39.6%   R-Sq(adj) = 34.6%


Analysis of Variance

Source          DF       SS      MS     F      P
Regression       1   651.59  651.59  7.88  0.016
Residual Error  12   991.91   82.66
Total           13  1643.50
```

a. If you were to model the relationship between the mean (or expected) number of games won by a major league team and the team's batting average x, using a straight line, would you expect the slope of the line to be positive or negative? Explain.

b. Construct a scatterplot of the data. Does the pattern revealed by the scatterplot agree with your answer to part **a**?

c. A MINITAB printout of the simple linear regression is shown above. Find the estimates of the β's on the printout and write the equation of the least squares line.

d. Graph the least squares line on your scatterplot. Does your least squares line seem to fit the points on your scatterplot?

e. Interpret the estimates of β_0 and β_1 in the words of the problem.

f. Conduct a test (at $\alpha = .05$) to determine whether the mean (or expected) number of games won by a major league baseball team is positively linearly related to the team's batting average.

g. Find the coefficient of determination, r^2, and interpret its value.

h. Do you recommend using the model to predict the number of games won by a team during the 2013 season?

11.138 College protests of labor exploitation. Refer to the *Journal of World-Systems Research* (Winter 2004) study of student "sit-ins" for a "sweat-free campus" at universities, presented in Exercise 2.199 (p. 137). Data on the duration (in days) of each sit-in as well as the number of student arrests were measured. The data for 5 sit-ins in which there was at least one arrest are shown in the table. Let y = number of arrests and x = duration.

Sit-In	University	Duration (days)	Number of Arrests
12	Wisconsin	4	54
14	SUNY Albany	1	11
15	Oregon	3	14
17	Iowa	4	16
18	Kentucky	1	12

Based on Ross, R. J. S. "From antisweatshop to global justice to antiwar: How the new new left is the same and different from the old new left." *Journal of Word-Systems Research*, Vol. X, No. 1, Winter 2004 (Tables 1 and 3).

SPSS Output for Exercise 11.138

Model Summary

Model	R	R Square	Adjusted R Square	Std. Error of the Estimate
1	.601[a]	.361	.148	16.913

a. Predictors: (Constant), DURATION

ANOVA[b]

Model		Sum of Squares	df	Mean Square	F	Sig.
1	Regression	485.026	1	485.026	1.696	.284[a]
	Residual	858.174	3	286.058		
	Total	1343.200	4			

a. Predictors: (Constant), DURATION

b. Dependent Variable: ARRESTS

Coefficients[a]

Model		Unstandardized Coefficients		Standardized Coefficients	t	Sig.
		B	Std. Error	Beta		
1	(Constant)	2.522	16.352		.154	.887
	DURATION	7.261	5.576	.601	1.302	.284

a. Dependent Variable: ARRESTS

a. Give the equation of a straight-line model relating y to x.

b. SPSS was used to fit the model to the data for the 5 sit-ins. The printout is shown above. Give the least squares prediction equation.

c. Interpret the estimates of β_0 and β_1 in the context of the problem.

d. Find and interpret the value of s on the printout.

e. Find and interpret the value of r^2 on the printout.

f. Conduct a test to determine whether number of arrests is positively linearly related to duration. (Use $\alpha = .10$.)

11.139 Feeding habits of fish. Refer to the *Brain and Behavior Evolution* (Apr. 2000) study of the feeding behavior of black-bream fish, presented in Exercise 2.162 (p. 124). Recall that the zoologists recorded the number of aggressive strikes of two black-bream fish feeding at the bottom of an aquarium in the 10-minute period following the addition of food. The table listing the weekly number of strikes and the age of the fish (in days) is reproduced below.

Week	Number of Strikes	Age of Fish (days)
1	85	120
2	63	136
3	34	150
4	39	155
5	58	162
6	35	169
7	57	178
8	12	184
9	15	190

Based on Shand, J., et al. "Variability in the location of the retinal ganglion cell area centralis is correlated with ontogenetic changes in feeding behavior in the Blackbream, Acanthopagrus 'butcher'." *Brain and Behavior*, Vol. 55, No. 4, Apr. 2000 (Figure H).

a. Write the equation of a straight-line model relating number of strikes (y) to age of fish (x).

b. Fit the model to the data by the method of least squares and give the least squares prediction equation.

c. Give a practical interpretation of the value of $\hat{\beta}_0$ if possible.

d. Give a practical interpretation of the value of $\hat{\beta}_1$ if possible.

e. Test $H_0: \beta_1 = 0$ versus $H_a: \beta_1 < 0$, using $\alpha = .10$. Interpret the result.

11.140 English as a second language reading ability. What are the factors that allow a native Spanish-speaking person to understand and read English? A study published in the *Bilingual Research Journal* (Summer 2006) investigated the relationship of Spanish (first-language) grammatical knowledge to English (second-language) reading. The study involved a sample of $n = 55$ native Spanish-speaking adults who were students in an English as a second language college class. Each student took four standardized exams: Spanish grammar (SG), Spanish reading (SR), English grammar (EG), and English reading (ESLR). Simple linear regression was used to model the ESLR score (y) as a function of each of the other exam scores (x). The results are summarized in the next table.

a. At $\alpha = .05$, is there sufficient evidence to indicate that ESLR score is linearly related to SG score?

Independent variable (x)	r^2	p-value for testing $H_0: \beta_1 = 0$
SG score	.002	.739
SR score	.099	.012
EG score	.078	.022

b. At $\alpha = .05$, is there sufficient evidence to indicate that ESLR score is linearly related to SG score?

c. At $\alpha = .05$, is there sufficient evidence to indicate that ESLR score is linearly related to EG score?

d. Practically interpret the r^2 values.

11.141 Removing metal from water. In the *Electronic Journal of Biotechnology* (Apr. 15, 2004), Egyptian scientists studied a new method for removing heavy metals from water. Metal solutions were prepared in glass vessels, and then biosorption was used to remove the metal ions. Two variables were measured for each test vessel: y = metal uptake (milligrams of metal per gram of biosorbent) and x = final concentration of metal in the solution (milligrams per liter).

a. Write a simple linear regression model relating y to x.
b. For one metal, a simple linear regression analysis yielded $r^2 = .92$. Interpret this result.

Applying the Concepts—Intermediate

11.142 New method of estimating rainfall. Accurate measurements of rainfall are critical for many hydrological and meteorological projects. Two standard methods of monitoring rainfall use rain gauges and weather radar. Both, however, can be contaminated by human and environmental interference. In the *Journal of Data Science* (Apr. 2004), researchers employed artificial neural networks (i.e., computer-based mathematical models) to estimate rainfall at a meteorological station in Montreal. Rainfall estimates were made every 5 minutes over a 70-minute period by each of the three methods. The data (in millimeters) are listed in the table.

Time	Radar	Rain Gauge	Neural Network
8:00 a.m.	3.6	0	1.8
8:05	2.0	1.2	1.8
8:10	1.1	1.2	1.4
8:15	1.3	1.3	1.9
8:20	1.8	1.4	1.7
8:25	2.1	1.4	1.5
8:30	3.2	2.0	2.1
8:35	2.7	2.1	1.0
8:40	2.5	2.5	2.6
8:45	3.5	2.9	2.6
8:50	3.9	4.0	4.0
8:55	3.5	4.9	3.4
9:00 a.m.	6.5	6.2	6.2
9:05	7.3	6.6	7.5
9:10	6.4	7.8	7.2

Based on Hessami, M., et al. "Selection of an artificial neural network model for the post-calibration of weather radar rainfall estimation." *Journal of Data Science*, Vol. 2, No. 2, Apr. 2004. (Adapted from Figures 2 and 4.)

a. Propose a straight-line model relating rain gauge amount (y) to weather radar rain estimate (x).
b. Use the method of least squares to fit the model.
c. Graph the least squares line on a scatterplot of the data. Is there visual evidence of a relationship between the two variables? Is the relationship positive or negative?
d. Interpret the estimates of the y-intercept and slope in the words of the problem.
e. Find and interpret the value of s for this regression.
f. Test whether y is linearly related to x. Use $\alpha = .01$.
g. Construct a 99% confidence interval for β_1. Interpret the result practically.
h. Now consider a model relating rain gauge amount (y) to the artificial neural network rain estimate (x). Repeat parts **a–g** for this model.

11.143 Extending the life of an aluminum smelter pot. An investigation of the properties of bricks used to line aluminum smelter pots was published in *The American Ceramic Society Bulletin* (Feb. 2005). Six different commercial bricks were evaluated. The life span of a smelter pot depends on the porosity of the brick lining (the less porosity, the longer is the life); consequently, the researchers measured the apparent porosity of each brick specimen, as well as the mean pore diameter of each brick. The data are given in the next table.

Brick	Apparent Porosity (%)	Mean Pore Diameter (micrometers)
A	18.8	12.0
B	18.3	9.7
C	16.3	7.3
D	6.9	5.3
E	17.1	10.9
F	20.4	16.8

Based on Bonadia, P., et al. "Aluminosilicate refractories for aluminum cell linings." *The American Ceramic Society Bulletin*, Vol. 84, No. 2, Feb. 2005 (Table II).

a. Find the least squares line relating porosity (y) to mean pore diameter (x).
b. Interpret the y-intercept of the line.
c. Interpret the slope of the line.
d. Conduct a test of model adequacy. Use $\alpha = .10$.
e. Find r and r^2 and interpret these values.
f. Predict the apparent percentage of porosity for a brick with a mean pore diameter of 10 micrometers. Use a 90% prediction interval.

11.144 Relation of eye and head movements. How do eye and head movements relate to body movements when a person reacts to a visual stimulus? Scientists at the California Institute of Technology designed an experiment to answer this question and reported their results in *Nature* (Aug. 1998). Adult male rhesus monkeys were exposed to a visual stimulus (i.e., a panel of light-emitting diodes), and their eye, head, and body movements were electronically recorded. In one variation of the experiment, two variables were measured: active head movement (x, percent per degree) and body-plus-head rotation (y, percent per degree). The data for $n = 39$ trials were subjected to a simple linear regression analysis, with the following results: $\hat{\beta}_1 = .88$, $s_{\hat{\beta}_1} = .14$

a. Conduct a test to determine whether the two variables, active head movement x and body-plus-head rotation y, are positively linearly related. Use $\alpha = .05$.
b. Construct and interpret a 90% confidence interval for β_1.
c. The scientists want to know whether the true slope of the line differs significantly from 1. On the basis of your answer to part **b**, make the appropriate inference.

11.145 Mortality of predatory birds. Two species of predatory birds—collard flycatchers and tits—compete for nest holes during breeding season on the island of Gotland, Sweden. Frequently, dead flycatchers are found in nest boxes occupied by tits. A field study examined whether the risk of mortality to flycatchers is related to the degree of competition between the two bird species for nest sites (*The Condor*, May 1995). The table (p. 675) gives data on the number y of flycatchers killed at each of 14 discrete locations (plots) on the island, as well as on the nest box tit occupancy x (i.e., the percentage of nest boxes occupied by tits) at each plot. Consider the simple linear regression model $E(y) = \beta_0 + \beta_1 x$.

Plot	Number of Flycatchers Killed y	Nest Box Tit Occupancy x (%)
1	0	24
2	0	33
3	0	34
4	0	43
5	0	50
6	1	35
7	1	35
8	1	38
9	1	40
10	2	31
11	2	43
12	3	55
13	4	57
14	5	64

Based on Merila, J., and Wiggins, D. A. "Interspecific competition for nest holes causes adult mortality in the collard flycatcher." *The Condor*, Vol. 97, No. 2, May 1995, p. 449 (Figure 2), Cooper Ornithological Society.

a. Plot the data in a scatterplot. Does the frequency of flycatcher casualties per plot appear to increase linearly with increasing proportion of nest boxes occupied by tits?

b. Use the method of least squares to find the estimates of β_0 and β_1. Interpret their values.

c. Test the utility of the model, using $\alpha = .05$.

d. Find r and r^2 and interpret their values.

e. Find s and interpret the result.

f. Do you recommend using the model to predict the number of flycatchers killed? Explain.

11.146 Winning marathon times. In *Chance* (Winter 2000), statistician Howard Wainer and two students compared men's and women's winning times in the Boston Marathon. One of the graphs used to illustrate gender differences is reproduced below. The scatterplot graphs the winning times (in minutes) against the year in which the race was run. Men's times are represented by purple dots and women's times by red dots.

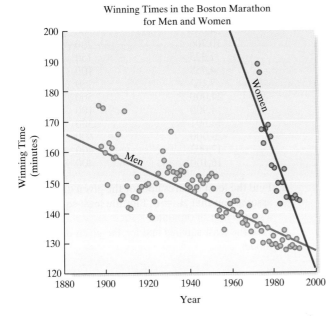

Winning Times in the Boston Marathon for Men and Women

a. Consider only the winning times for men. Is there evidence of a linear trend? If so, propose a straight-line model for predicting winning time (y) based on year (x). Would you expect the slope of this line to be positive or negative?

b. Repeat part **b** for women's times.

c. Which slope, the men's or the women's, will be greater in absolute value?

d. Would you recommend using the straight-line models to predict the winning time in the 2020 Boston Marathon? Why or why not?

e. Which model, the men's or the women's, is likely to have the smallest estimate of σ?

11.147 Quantum tunneling. At temperatures approaching absolute zero ($-273°C$), helium exhibits traits that seem to defy many laws of Newtonian physics. An experiment has been conducted with helium in solid form at various temperatures near absolute zero. The solid helium is placed in a dilution refrigerator along with a solid impure substance, and the fraction (in weight) of the impurity passing through the solid helium is recorded. (This phenomenon of solids passing directly through solids is known as *quantum tunneling*.) The data are given in the next table.

Temperature x (°C)	Proportion of Impurity
-262.0	.315
-265.0	.202
-256.0	.204
-267.0	.620
-270.0	.715
-272.0	.935
-272.4	.957
-272.7	.906
-272.8	.985
-272.9	.987

a. Find the least squares estimates of the intercept and slope. Interpret them.

b. Use a 95% confidence interval to estimate the slope β_1. Interpret the interval in terms of this application. Does the interval support the hypothesis that temperature contributes information about the proportion of impurity passing through helium?

c. Interpret the coefficient of determination for this model.

d. Find a 95% prediction interval for the percentage of impurity passing through solid helium at $-273°C$. Interpret the result.

e. Note that the value of x in part **d** is outside the experimental region. Why might this lead to an unreliable prediction?

11.148 Dance/movement therapy. In cotherapy, two or more therapists lead a group. An article in the *American Journal of Dance Therapy* (Spring/Summer 1995) examined the use of cotherapy in dance/movement therapy. Two of several variables measured on each of a sample of 136 professional dance/movement therapists were years x of formal training and reported success rate y (measured as a percentage) of coleading dance/movement therapy groups.

a. Propose a linear model relating y to x.

b. The researcher hypothesized that dance/movement therapists with more years in formal dance training will report higher perceived success rates in cotherapy relationships. State the hypothesis in terms of the parameter of the model you proposed in part **a**.

c. The correlation coefficient for the sample data was reported as $r = -.26$. Interpret this result.

d. Does the value of r in part **c** support the hypothesis in part **b**? Test, using $\alpha = .05$.

11.149 **Performance in online courses.** Florida State University information scientists assessed the impact of online courses on student performance (*Educational Technology & Society*, Jan. 2005). Each in a sample of 24 graduate students enrolled in an online advanced Web application course was asked, "How many courses per semester (on average) do you take online?" Each student's performance on weekly quizzes was also recorded. The information scientists found that the number of online courses and the weekly quiz grade were negatively correlated at $r = -.726$.

a. Give a practical interpretation of r.

b. The researchers concluded that there was a "significant negative correlation" between the number of online courses and the weekly quiz grade. Do you agree?

11.150 **Snow geese feeding trial.** Botanists at the University of Toronto conducted a series of experiments to investigate the feeding habits of baby snow geese (*Journal of Applied Ecology*, Vol. 32, 1995). Goslings were deprived of food until their guts were empty and then were allowed to feed for 6 hours on a diet of plants or Purina® Duck Chow®. For each feeding trial, the change in the weight of the gosling after 2.5 hours was recorded as a percentage of the bird's initial weight. Two other variables recorded were digestion efficiency (measured as a percentage) and amount of acid-detergent fiber in the digestive tract (also measured as a percentage). Data on 42 feeding trials are saved in the **SNOW** file. The first and last 5 observations are listed in the table below.

Feeding Trial	Diet	Weight Change (%)	Digestion Efficiency (%)	Acid-Detergent Fiber (%)
1	Plants	−6	0	28.5
2	Plants	−5	2.5	27.5
3	Plants	−4.5	5	27.5
4	Plants	0	0	32.5
5	Plants	2	0	32
⋮	⋮	⋮	⋮	⋮
38	Duck Chow	9	59	8.5
39	Duck Chow	12	52.5	8
40	Duck Chow	8.5	75	6
41	Duck Chow	10.5	72.5	6.5
42	Duck Chow	14	69	7

Based on Gadallah, F. L., and Jefferies, R. L. "Forage quality in brood rearing areas of the lesser snow goose and the growth of captive goslings." *Journal of Applied Biology*, Vol. 32, No. 2, 1995, pp. 281–282 (adapted from Figures 2 and 3).

a. The botanists were interested in the correlation between weight change (y) and digestion efficiency (x). Plot the data for these two variables in a scatterplot. Do you observe a trend?

b. Find the coefficient of correlation relating weight change y to digestion efficiency x. Interpret this value.

c. Conduct a test to determine whether weight change y is correlated with digestion efficiency x. Use $\alpha = .01$.

d. Repeat parts **b** and **c**, but exclude the data for trials that used duck chow. What do you conclude?

e. The botanists were also interested in the correlation between digestion efficiency y and acid-detergent fiber x. Repeat parts **a–d** for these two variables.

f. Fit the simple linear regression model to the data using $x =$ digestion efficiency.

g. Do you recommend using the model to predict weight change y? Explain.

h. Use the model to form a 95% confidence interval for the mean weight change of all baby snow geese with a digestion efficiency of $x = 15\%$. Interpret the interval.

Applying the Concepts—Advanced

11.151 **Regression through the origin.** Sometimes it is known from theoretical considerations that the straight-line relationship between two variables x and y passes through the origin of the xy-plane. Consider the relationship between the total weight y of a shipment of 50-pound bags of flour and the number x of bags in the shipment. Since a shipment containing $x = 0$ bags (i.e., no shipment at all) has a total weight of $y = 0$, a straight-line model of the relationship between x and y should pass through the point $x = 0, y = 0$. In such a case, you could assume that $\beta_0 = 0$ and characterize the relationship between x and y with the following model:

$$y = \beta_1 x + \varepsilon$$

The least squares estimate of β_1 for this model is

$$\hat{\beta}_1 = \frac{\Sigma x_i y_i}{\Sigma x_i^2}$$

From the records of past flour shipments, 15 shipments were randomly chosen and the data shown in the following table were recorded.

Weight of Shipment	Number of 50-Pound Bags in Shipment
5,050	100
10,249	205
20,000	450
7,420	150
24,685	500
10,206	200
7,325	150
4,958	100
7,162	150
24,000	500
4,900	100
14,501	300
28,000	600
17,002	400
16,100	400

a. Find the least squares line for the given data under the assumption that $\beta_0 = 0$. Plot the least squares line on a scatterplot of the data.

b. Find the least squares line for the given data, using the model

$$y = \beta_0 + \beta_1 x + \varepsilon$$

(i.e., do not restrict β_0 to equal 0). Plot this line on the same scatterplot you constructed in part **a**.

c. Refer to part **b**. Why might $\hat{\beta}_0$ be different from 0 even though the true value of β_0 is known to be 0?

d. The estimated standard error of $\hat{\beta}_0$ is equal to

$$s\sqrt{\frac{1}{n} + \frac{\bar{x}^2}{\text{SS}_{xx}}}$$

Use the t-statistic

$$t = \frac{\hat{\beta}_0 - 0}{s\sqrt{(1/n) + (\bar{x}^2/\text{SS}_{xx})}}$$

to test the null hypothesis $H_0: \beta_0 = 0$ against the alternative $H_a: \beta_0 \neq 0$. Take $\alpha = .10$. Should you include β_0 in your model?

11.152 Long-jump "takeoff error." The long jump is a track-and-field event in which a competitor attempts to jump a maximum distance into a sandpit after a running start. At the edge of the pit is a takeoff board. Jumpers usually try to plant their toes at the front edge of this board to maximize their jumping distance. The absolute distance between the front edge of the takeoff board and the spot where the toe actually lands on the board prior to jumping is called "takeoff error." Is takeoff error in the long jump linearly related to best jumping distance? To answer this question, kinesiology researchers videotaped the performances of 18 novice long jumpers at a high school track meet (*Journal of Applied Biomechanics*, May 1995). The average takeoff error x and the best jumping distance y (out of three jumps) for each jumper are recorded in the accompanying table. If a jumper can reduce his/her average takeoff error by .1 meter, how much would you estimate the jumper's best jumping distance to change? On the basis of your answer, comment on the usefulness of the model for predicting best jumping distance.

Jumper	Best Jumping Distance y (meters)	Average Takeoff Error x (meters)
1	5.30	.09
2	5.55	.17
3	5.47	.19
4	5.45	.24
5	5.07	.16
6	5.32	.22
7	6.15	.09
8	4.70	.12
9	5.22	.09
10	5.77	.09
11	5.12	.13
12	5.77	.16
13	6.22	.03
14	5.82	.50
15	5.15	.13
16	4.92	.04
17	5.20	.07
18	5.42	.04

Based on Berg, W. P., and Greer, N. L. "A kinematic profile of the approach run of novice long jumpers." *Journal of Applied Biomechanics*, Vol. 11, No. 2, May 1995, p. 147 (Table 1).

Critical Thinking Challenge

11.153 Spall damage in bricks. A recent civil suit revolved around a five-building brick apartment complex located in the Bronx, New York, which began to suffer *spalling* damage (i.e., a separation of some portion of the face of a brick from its body). The owner of the complex alleged that the bricks were manufactured defectively. The brick manufacturer countered that poor design and shoddy management led to the damage. To settle the suit, an estimate of the rate of damage per 1,000 bricks, called the spall rate, was required (*Chance*, Summer 1994). The owner estimated the spall rate by using several *scaffold-drop* surveys. (With this method, an engineer lowers a scaffold down at selected places on building walls and counts the number of visible spalls for every 1,000 bricks in the observation area.) The brick manufacturer conducted its own survey by dividing the walls of the complex into 83 wall segments and taking a photograph of each one. (The number of spalled bricks that could be made out from each photo was recorded, and the sum over all 83 wall segments was used as an estimate of total spall damage.) In this court case, the jury was faced with the following dilemma: On the one hand, the scaffold-drop survey provided the most accurate estimate of spall rates in a given wall segment. Unfortunately, however, the drop areas were not selected at random from the entire complex; rather, drops were made at areas with high spall concentrations, leading to an overestimate of the total damage. On the other hand, the photo survey was complete in that all 83 wall segments in the complex were checked for spall damage. But the spall rate estimated by the photos, at least in areas of high spall concentration, was biased low (spalling damage cannot always be seen from a photo), leading to an underestimate of the total damage.

The data in the table are the spall rates obtained from the two methods at 11 drop locations. Use the data, as did expert statisticians who testified in the case, to help the jury estimate the true spall rate at a given wall segment. Then explain how this information, coupled with the data (not given here) on all 83 wall segments, can provide a reasonable estimate of the total spall damage (i.e., total number of damaged bricks).

Drop Location	Drop Spall Rate (per 1,000 bricks)	Photo Spall Rate (per 1,000 bricks)
1	0	0
2	5.1	0
3	6.6	0
4	1.1	.8
5	1.8	1.0
6	3.9	1.0
7	11.5	1.9
8	22.1	7.7
9	39.3	14.9
10	39.9	13.9
11	43.0	11.8

Based on Fairley, W. B., et al. "Bricks, buildings, and the Bronx: Estimating masonry deterioration." *Chance*, Vol. 7. No. 3, Summer 1994, p. 36 (Figure 3).

[*Note:* The data points are estimated from the points shown on a scatterplot.]

Applying Simple Linear Regression to Your Favorite Data

Many dependent variables in all areas of research serve as the subjects of regression-modeling efforts. We list five such variables here.

1. Crime rate in various communities

2. Daily maximum temperature in your town

3. Grade point average of students who have completed one academic year at your college

4. Gross domestic product of the United States

5. Points scored by your favorite football team in a single game

   Choose one of these dependent variables, or choose some other dependent variable, for which you want to construct a prediction model. There may be a large number of independent variables that should be included in a prediction equation for the dependent variable you choose. List three potentially important independent variables, x_1, x_2, and x_3, that you think might be (individually) strongly related to your dependent variable. Next, obtain 10 data values, each of which consists of a measure of your dependent variable y and the corresponding values of x_1, x_2, and x_3.

a. Use the least squares formulas given in this chapter to fit three straight-line models—one for each independent variable—for predicting y.

b. Interpret the sign of the estimated slope coefficient $\hat{\beta}_1$ in each case, and test the utility of each model by testing $H_0: \beta_1 = 0$ against $H_a: \beta_1 \neq 0$. What assumptions must be satisfied to ensure the validity of these tests?

c. Calculate the coefficient of determination, r^2, for each model. Which of the independent variables predicts y best for the 10 sampled sets of data? Is this variable necessarily best in general (i.e., for the entire population)? Explain.

   Be sure to keep the data and the results of your calculations, since you will need them for the Activity section in Chapter 12.

References

Chatterjee, S., and Price, B. *Regression Analysis by Example*, 2nd ed. New York: Wiley, 1991.

Draper, N., and Smith, H. *Applied Regression Analysis*, 3rd ed. New York: Wiley, 1987.

Graybill, F. *Theory and Application of the Linear Model*. North Scituate, MA: Duxbury, 1976.

Kleinbaum, D., and Kupper, L. *Applied Regression Analysis and Other Multivariable Methods*, 2nd ed. North Scituate, MA: Duxbury, 1997.

Kutner, M., Nachtsheim, C., Neter, J., and Li, W. *Applied Linear Statistical Models*, 5th ed. New York: McGraw-Hill/Irwin, 2006.

Mendenhall, W. *Introduction to Linear Models and the Design and Analysis of Experiments*. Belmont, CA: Wadsworth, 1968.

Mendenhall, W., and Sincich, T. *A Second Course in Statistics: Regression Analysis*, 7th ed. Upper Saddle River, NJ: Prentice Hall, 2011.

Montgomery, D., Peck, E., and Vining, G. *Introduction to Linear Regression Analysis*, 3rd ed. New York: Wiley, 2001.

Mosteller, F., and Tukey, J. W. *Data Analysis and Regression: A Second Course in Statistics*. Reading, MA: Addison-Wesley, 1977.

Rousseeuw, P. J., and Leroy, A. M. *Robust Regression and Outlier Detection*. New York: Wiley, 1987.

Weisburg, S. *Applied Linear Regression*, 2nd ed. New York: Wiley, 1985.

MINITAB: Simple Linear Regression

Regression Analysis

Step 1 Access the MINITAB worksheet file that contains the two quantitative variables (dependent and independent variables).

Step 2 Click on the "Stat" button on the MINITAB menu bar, and then click on "Regression" and "Regression" again, as shown in Figure 11.M.1.

Step 3 On the resulting dialog box (see Figure 11.M.2), specify the dependent variable in the "Response" box and the independent variable in the "Predictors" box.

Figure 11.M.1 MINITAB menu options for regression

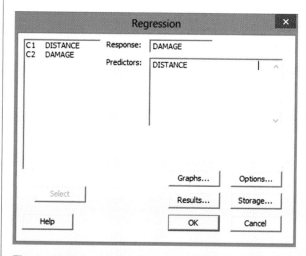

Figure 11.M.2 MINITAB regression dialog box

Step 4 To produce prediction intervals for y and confidence intervals for $E(y)$, click the "Options" button. The resulting dialog box is shown in Figure 11.M.3.

Figure 11.M.3 MINITAB regression options

Step 5 Check "Confidence limits" and/or "Prediction limits," specify the "Confidence level," and enter the value of x in the "Prediction intervals for new observations" box.

Step 6 Click "OK" to return to the main Regression dialog box and then click "OK" again to produce the MINITAB simple linear regression printout.

Correlation Analysis

Step 1 Click on the "Stat" button on the MINITAB main menu bar, then click on "Basic Statistics," and then click on "Correlation," as shown in Figure 11.M.4.

Figure 11.M.4 MINITAB menu options for correlation

Step 2 On the resulting dialog box (see Figure 11.M.5), enter the two variables of interest in the "Variables" box.

Figure 11.M.5 MINITAB correlation dialog box

Step 3 Click "OK" to obtain a printout of the correlation.

TI-83/TI-84 Plus Graphing Calculator: Simple Linear Regression

Finding the Least Squares Regression Equation

Step 1 *Enter the data*

• Press **STAT** and select **1:Edit**

Note: If a list already contains data, clear the old data.

• Use the up arrow to highlight the list name, "**L1**" or "**L2**"

• Press **CLEAR ENTER**

• Enter your x-data in **L1** and your y-data in **L2**

Step 2 *Find the equation*

• Press **STAT** and highlight **CALC**

• Press **4** for **LinReg($ax + b$)**

• Press **ENTER**

• The screen will show the values for a and b in the equation $y = ax + b$

Finding r and r^2

Use this procedure if r and r^2 do not already appear on the LinReg screen from part I:

Step 1 *Turn the diagnostics feature on*

• Press **2nd 0** for **CATALOG**

• Press the **ALPHA** key and x^{-1} for **D**

- Press the down **ARROW** until **DiagnosticsOn** is highlighted
- Press **ENTER** twice

Step 2 *Find the regression equation as shown in part I on p. 679* The values for r and r^2 will appear on the screen as well.

Graphing the Least Squares Line with the Scatterplot

Step 1 *Enter the data as shown in part I on p. 651*

Step 2 *Set up the data plot*

- Press **Y =** and **CLEAR** all functions from the **Y** register
- Press **2ndY =** for **STAT PLOT**
- Press **1** for **Plot1**
- Set the cursor so that **ON** is flashing and press **ENTER**
- For **Type**, use the **ARROW** and **ENTER** keys to highlight and select the scatterplot (first icon in the first row)

- For **Xlist**, choose the column containing the *x*-data
- For **Ylist**, choose the column containing the *y*-data

Step 3 *Find the regression equation and store the equation in Y1*

- Press **STAT** and highlight **CALC**
- Press **4** for **LinReg(ax + b)** (*Note*: Don't press **ENTER** here because you want to store the regression equation in Y1.)
- Press **VARS**
- Use the right arrow to highlight **Y-VARS**
- Press **ENTER** to select **1:Function**
- Press **ENTER** to select **1:Y1**
- Press **ENTER**

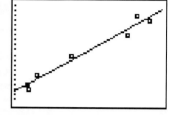

Step 4 *View the scatterplot and regression line*

- Press **ZOOM** and then press **9** to select **9:ZoomStat**

You should see the data graphed along with the regression line.

12 Multiple Regression and Model Building

CONTENTS

Where We've Been

- Introduced the straight-line model relating a dependent variable y to a single independent variable x
- Demonstrated how to estimate the parameters of the straight-line model by the method of least squares
- Showed how to statistically assess the adequacy of the model
- Showed how to use the model to estimate $E(y)$ and predict y for a given value of x

Where We're Going

- Introduce a *multiple-regression* model as a means of relating a dependent variable y to two or more independent variables. (12.1)
- Present several different multiple-regression models involving both quantitative and qualitative independent variables. (12.2, 12.5–12.8)
- Assess how well the multiple-regression model fits the sample data. (12.3)
- Demonstrate how to use the model for prediction (12.4)
- Present some model building techniques (12.9–12.10)
- Show how an analysis of a model's *residuals* can aid in detecting violations of the model's assumptions and in identifying modifications required by the model. (12.11)
- Alert the analyst to some regression pitfalls (12.12)

Statistics IN Action   Modeling Condominium Sales: What Factors Affect Auction Price?

This application involves an investigation of the factors that affect the sale price of oceanside condominium units. It represents an extension of an analysis of data collected by Herman Kelting. Although condo sale prices have increased dramatically since the time of the study, the relationship between these factors and sale prices remains about the same. Consequently, the data provide valuable insight into today's condominium sales market.

Sales data were obtained for a newly built oceanside condominium complex consisting of two adjacent, connected eight-floor buildings. The complex contains 209 units of equal size (approximately 500 square feet each). The locations of the buildings relative to the ocean, the swimming pool, the parking lot, etc., are shown in Figure SIA12.1.

Among the features of the complex that you should note are the following:

1. The units facing south, called *oceanview*, face the beach and ocean. In addition, units in building 1 have a good view of the pool. Units to the rear of the building, called *bayview*, face the parking lot and an area of land that, ultimately, borders a bay. The view from the upper floors of these units is primarily of wooded, sandy terrain. The bay is very distant and barely visible.

2. The only elevator in the complex is located at the east end of building 1, next to the office and the game room. People moving to or from the higher floor units in building 2 would probably use the elevator, then move through the passages to their units. Thus, units on the higher floors and at a greater distance from the elevator would be less convenient for their occupants, who would expend greater effort in moving baggage,

groceries, etc., and would be farther away from the game room, the office, and the swimming pool. These units also possess an advantage: Because traffic through the hallways in the area would be minimal, these units would be the most private.

3. Because lower-floor oceanside units open onto the beach, ocean, and pool, they are most suited to active people. They are within easy reach of the game room, and they are also easily reached from the parking area.

4. Checking Figure SIA12.1, you will see that the views in some of the units at the center of the complex—units ending in numbers 11 and 14—are partially blocked.

5. The condominium complex was completed at the time of the 1975 recession; sales were slow, and the developer was forced to sell about half of the units at auction approximately 18 months after the complex opened. Many unsold units were furnished by the developer and rented prior to the auction.

This condominium complex is particularly suited to our study. Because the single elevator is located at one end of the complex, it is the source of a remarkably high level of both inconvenience and privacy for the people occupying units on the top floors in building 2. Consequently, the data provide a

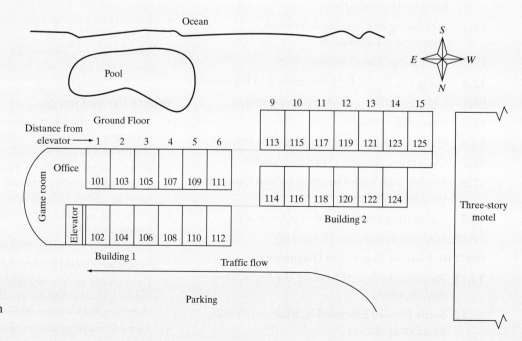

Figure SIA12.1
Layout of condominium complex

good opportunity to investigate the relationship that might exist between sale price, height of the unit (floor number), distance of the unit from the elevator, and presence or absence of an ocean view. In addition, the presence or absence of furniture in each of the units permits an investigation of the effect of the availability of furniture on the sale price. Finally, the units sold at auction are completely specified by the buyer and hence are consumer oriented, in contrast to most other real estate units, which are, to a high degree, seller oriented and specified by the broker.

The **CONDO** file contains data on each of the 209 units sold—106 at public auction and 103 at the developer's fixed price. The variables measured for each condominium unit are listed in Table SIA12.1. We want to build a model for the auction price and, ultimately, to use the model to predict the sale prices of future units.

In several Statistics in Action Revisited sections, we show how to analyze the data by means of a multiple-regression analysis.

Statistics IN Action   Revisited

- A First-Order Model for Condominium Sale Price (p. 704)
- Building a Model for Condominium Sale Price (p. 748)
- A Residual Analysis for the Condominium Sale Price Model (p. 773)

Table SIA12.1   Variables in the CONDO Data File

Variable Name	Type	Description
PRICE100	Quantitative	Sales price (hundreds of dollars)
FLOOR	Quantitative	Floor height $(1, 2, 3, \ldots, 8)$
DIST	Quantitative	Distance, in units, from the elevator $(1, 2, 3, \ldots, 15)$
VIEW	Qualitative	View $(1 = \text{ocean view}, 0 = \text{non–ocean view})$
END	Qualitative	Location of unit $(1 = \text{end of complex}, 0 = \text{not an end unit})$
FURNISH	Qualitative	Furniture status $(1 = \text{furnished}, 0 = \text{nonfurnished})$
AUCTION	Qualitative	Method of sale $(1 = \text{public auction}, 0 = \text{fixed price})$

D *Data Set:* CONDO

12.1   Multiple-Regression Models

Most practical applications of regression analysis utilize models that are more complex than the simple straight-line model. For example, a realistic probabilistic model for reaction time would include more than just the amount of a particular drug in the bloodstream. Factors such as age, a measure of visual perception, and sex of the subject are a few of the many variables that might be related to reaction time. Thus, we would want to incorporate these and other potentially important independent variables into the model in order to make accurate predictions.

Probabilistic models that include more than one independent variable are called **multiple-regression models**. The general form of these models is

$$y = \beta_0 + \beta_1 x_1 + \beta_2 x_2 + \cdots + \beta_k x_k + \varepsilon$$

The dependent variable y is now written as a function of k independent variables $x_1, x_2, \ldots, x_k$. The random-error term is added to make the model probabilistic rather than deterministic. The value of the coefficient β_i determines the contribution of the independent variable x_i, and β_0 is the y-intercept. The coefficients $\beta_0, \beta_1, \ldots, \beta_k$ are usually unknown because they represent population parameters.

At first glance, it might appear that the regression model just described would not allow for anything other than straight-line relationships between y and the independent variables, but this is not true. Actually, $x_1, x_2, \ldots, x_k$ can be functions of variables, as long as the functions do not contain unknown parameters. For example, the reaction time y of a subject to a visual stimulus could be a function of the independent variables

$$x_1 = \text{Age of the subject}$$
$$x_2 = (\text{Age})^2 = x_1^2$$
$$x_3 = 1 \text{ if male subject, } 0 \text{ if female subject}$$

The x_2-term is called a **higher order term**, since it is the value of a quantitative variable (x_1) squared (i.e., raised to the second power). The x_3-term is a (qualitative) **coded variable** representing a quality (gender). The **multiple-regression model** is quite versatile and can be made to model many different types of response variables.

> **The Multiple-Regression Model***
>
> $$y = \beta_0 + \beta_1 x_1 + \beta_2 x_2 + \cdots + \beta_k x_k + \varepsilon$$
>
> where
>
> > y is the dependent variable
> >
> > $x_1, x_2, \ldots, x_k$ are the independent variables
> >
> > $E(y) = \beta_0 + \beta_1 x_1 + \beta_2 x_2 + \cdots + \beta_k x_k$ is the deterministic portion of the model
> >
> > β_i determines the contribution of the independent variable x_i
>
> *Note:* The symbols $x_1, x_2, \ldots, x_k$ may represent higher order terms for quantitative predictors or terms that represent qualitative predictors.

As shown in the following box, the steps used to develop the multiple-regression model are similar to those used for the simple regression model:

> **Analyzing a Multiple-Regression Model**
>
> **Step 1** Hypothesize the deterministic component of the model. This component relates the mean $E(y)$ to the independent variables $x_1, x_2, \ldots, x_k$. Involved here is the choice of the independent variables to be included in the model (Sections 12.2, 12.5–12.10).
>
> **Step 2** Use the sample data to estimate the unknown parameters $\beta_0, \beta_1, \beta_2, \ldots, \beta_k$ in the model (Section 12.2).
>
> **Step 3** Specify the probability distribution of the random-error term ε, and estimate the standard deviation σ of this distribution (Section 12.3).
>
> **Step 4** Check that the assumptions about ε are satisfied, and make modifications to the model if necessary (Section 12.11).
>
> **Step 5** Statistically evaluate the usefulness of the model (Section 12.3).
>
> **Step 6** When you are satisfied that the model is useful, use it for prediction, estimation, and other purposes (Section 12.4).

The assumptions we make about the random error ε of the multiple-regression model are similar to those we make about the random error in a simple linear regression and are summarized as follows:

> **Assumptions about Random Error ε**
>
> For any given set of values of $x_1, x_2, \ldots, x_k$, the random error ε has a probability distribution with the following properties:
>
> 1. The mean is equal to 0.
> 2. The variance is equal to σ^2.
> 3. The probability distribution is a normal distribution.
> 4. Random errors are independent (in a probabilistic sense).

*Technically, this model is referred to as a multiple *linear* regression model, since the equation is a linear function of the β's.

This chapter is divided into three parts. In Part I, we consider the most basic multiple regression model involving only quantitative independent variables, called a *first-order model*. In Part II, we introduce several other different types of models that form the foundation of **model building** (i.e., useful model construction). Finally, in Part III, we discuss how to check the validity of the model assumptions on the random error term—called a *residual analysis*.

PART I: FIRST-ORDER MODELS WITH QUANTITATIVE INDEPENDENT VARIABLES

12.2 Estimating and Making Inferences about the β Parameters

A model that includes only terms denoting *quantitative* independent variables, called a **first-order model**, is described in the next box. Note that the first-order model does not include any higher order terms (such as x_1^2).

A First-Order Model in Five Quantitative Independent Variables*

$$E(y) = \beta_0 + \beta_1 x_1 + \beta_2 x_2 + \beta_3 x_3 + \beta_4 x_4 + \beta_5 x_5$$

where $x_1, x_2, \ldots, x_5$ are all quantitative variables that *are not* functions of other independent variables.

Note: β_i represents the slope of the line relating y to x_i when all the other x's are held fixed.

The method of fitting first-order models—and multiple-regression models in general—is identical to that of fitting the simple straight-line model: the *method of least squares*. That is, we choose the estimated model

$$\hat{y} = \hat{\beta}_0 + \hat{\beta}_1 x_1 + \cdots + \hat{\beta}_k x_k$$

that (1) has an average error of prediction of 0, i.e., $\Sigma(y - \hat{y}) = 0$, and (2) minimizes SSE $= \sum(y - \hat{y})^2$. As in the case of the simple linear model, the sample estimates $\hat{\beta}_0, \hat{\beta}_1, \ldots, \hat{\beta}_k$ are obtained as a solution of a set of simultaneous linear equations.[†]

The primary difference between fitting the simple- and multiple-regression models is computational difficulty. The $(k + 1)$ simultaneous linear equations that must be solved to find the $(k + 1)$ estimated coefficients $\hat{\beta}_0, \hat{\beta}_1, \ldots, \hat{\beta}_k$ are difficult (sometimes nearly impossible) to solve with a calculator. Consequently, we resort to the use of computers. Instead of presenting the tedious hand calculations required to fit the models, we present output from SAS, SPSS, and MINITAB.

BIOGRAPHY   GEORGE U. YULE (1871–1951)

Yule Processes

Born on a small farm in Scotland, George Yule received an extensive childhood education. After graduating from University College (London), where he studied civil engineering, Yule spent a year employed in engineering workshops. However, he made a career change in 1893, accepting a teaching position back at University College under the guidance of statistician Karl Pearson (see p. 803). Inspired by Pearson's work, Yule produced a series of important articles on the statistics of regression and correlation. Yule is considered the first to have applied the method of least squares in regression analysis and he developed the theory of multiple regression. He eventually was appointed a lecturer in statistics at Cambridge University and later became the president of the prestigious Royal Statistical Society. Yule made many other contributions to the field, including the invention of time-series analysis and the development of "Yule" processes and the "Yule" distribution. ■

---

*The terminology "first-order" is derived from the fact that each x in the model is raised to the first power.

[†]Students who are familiar with calculus should note that $\hat{\beta}_0, \hat{\beta}_1, \ldots, \hat{\beta}_k$ are the solutions of the set of equations $\partial \text{SSE}/\partial \hat{\beta}_0 = 0, \partial \text{SSE}/\partial \hat{\beta}_1 = 0, \ldots, \partial \text{SSE}/\partial \hat{\beta}_k = 0$. The solution is usually given in matrix form, but we do not present the details here. (See the references for details.)

12.4 Using the Model for Estimation and Prediction

In Section 11.6, we discussed the use of the least squares line in estimating the mean value of y, $E(y)$, for some particular value of x, say, $x = x_p$. We also showed how to use the same fitted model to predict, when $x = x_p$, some new value of y to be observed in the future. Recall that the least squares line yielded the same value for both the estimate of $E(y)$ and the prediction of some future value of y. That is, both are the result of substituting x_p into the prediction equation $\hat{y} = \hat{\beta}_0 + \hat{\beta}_1 x$ and calculating $\hat{y}_p$. There the equivalence ends, however: The confidence interval for the mean $E(y)$ is narrower than the prediction interval for y because of the additional uncertainty attributable to the random error ε that arises in predicting some future value of y.

These same concepts carry over to the multiple-regression model. Consider a first-order model relating sale price (y) of a residential property to land value (x_1), improvements (x_2), and home size (x_3). Suppose we want to estimate the mean sale price for a given property with $x_1 = \$15{,}000$, $x_2 = \$50{,}000$, and $x_3 = 1{,}800$ square feet. Assuming that the first-order model represents the true relationship between the sale price and the three independent variables, we want to estimate

$$E(y) = \beta_0 + \beta_1 x_1 + \beta_2 x_2 + \beta_3 x_3$$
$$= \beta_0 + \beta_1(15{,}000) + \beta_2(50{,}000) + \beta_3(1{,}800)$$

After obtaining the least squares estimates, $\beta_0, \beta_1, \beta_2$, and β_3, we find the estimate of $E(y)$ to be

$$\hat{y} = \hat{\beta}_0 + \hat{\beta}_1(15{,}000) + \hat{\beta}_2(50{,}000) + \hat{\beta}_3(1{,}800)$$

To form a confidence interval for the mean (or for an individual value of y), we need to know the standard deviation of the sampling distribution for the estimator $\hat{y}$. For multiple-regression models, the form of this standard deviation is rather complicated. However, the regression routines of statistical software packages allow us to obtain the confidence intervals for mean values or individual values of y for any given combination of values of the independent variables. We illustrate with an example.

Example 12.5 Ⓓ
GFCLOCK

Estimating $E(y)$ and Predicting y—Auction Price Model

Problem Refer to Examples 12.1–12.4 and the first-order model $E(y) = \beta_0 + \beta_1 x_1 + \beta_2 x_2$, where y = auction price of a grandfather clock, x_1 = age of the clock, and x_2 = number of bidders.

a. Estimate the average auction price for all 150-year-old clocks sold at an auction with 10 bidders. Use a 95% confidence interval. Interpret the result.

b. Predict the auction price for a single 150-year-old clock sold at an auction with 10 bidders. Use a 95% prediction interval. Interpret the result.

c. Suppose you want to predict the auction price for one clock that is 50 years old and has 2 bidders. How should you proceed?

Solution

a. Here, the key words *average* and *for all* imply that we want to estimate the mean of y, $E(y)$. We want a 95% confidence interval for $E(y)$ when $x_1 = 150$ years and $x_2 = 10$ bidders. A MINITAB printout for this analysis is shown in Figure 12.7. The confidence interval (highlighted under "**95% CI**") is (1381.4, 1481.9). Thus, we are 95% confident that the mean auction price for all 150-year-old clocks sold at an auction with 10 bidders lies between \$1,381.4 and \$1,481.9.

b. The key words *predict* and *for a single* imply that we want a 95% prediction interval for y when $x_1 = 150$ years and $x_2 = 10$ bidders. This interval (highlighted under "**95% PI**" on the MINITAB printout shown in Figure 12.7) is (1154.1, 1709.3). We say, with 95% confidence, that the auction price for a single 150-year-old clock sold at an auction with 10 bidders falls between \$1,154.1 and \$1,709.3.

Regression Analysis: PRICE versus AGE, NUMBIDS

```
The regression equation is
PRICE = - 1339 + 12.7 AGE + 86.0 NUMBIDS

Predictor       Coef   SE Coef       T       P
Constant     -1339.0     173.8   -7.70   0.000
AGE          12.7406    0.9047   14.08   0.000
NUMBIDS       85.953     8.729    9.85   0.000

S = 133.485   R-Sq = 89.2%   R-Sq(adj) = 88.5%

Analysis of Variance

Source           DF        SS        MS       F       P
Regression        2   4283063   2141531  120.19   0.000
Residual Error   29    516727     17818
Total            31   4799790

Predicted Values for New Observations

New
Obs      Fit   SE Fit        95% CI            95% PI
  1   1431.7     24.6  (1381.4, 1481.9)  (1154.1, 1709.3)

Values of Predictors for New Observations

New
Obs   AGE   NUMBIDS
  1   150      10.0
```

Figure 12.7

MINITAB printout with 95% confidence intervals for grandfather clock model

c. Now we want to predict the auction price y for a single (*one*) grandfather clock when $x_1 = 50$ years and $x_2 = 2$ bidders. Consequently, we desire a 95% prediction interval for y. However, before we form this prediction interval, we should check to make sure that the selected values of the independent variables, $x_1 = 50$ and $x_2 = 2$, are both reasonable and within their respective sample ranges. If you examine the sample data shown in Table 12.1 (p. 686), you will see that the range for age is $108 \le x_1 \le 194$ and the range for number of bidders is $5 \le x_2 \le 15$. Thus, both selected values fall well *outside* their respective ranges. Recall the *Caution* box in Section 11.6 (p. 659) warning about the dangers of using the model to predict y for a value of an independent variable that is not within the range of the sample data. Doing so may lead to an unreliable prediction.

Look Back If we want to make the prediction requested in part **c**, we would need to collect additional data on clocks with the requested characteristics (i.e., $x_1 = 50$ years and $x_2 = 2$ bidders) and then refit the model.

■ **Now Work Exercise 12.37**

Statistics IN Action | **Revisited** | # A First-Order Model for Condominium Sale Price

The developer of a Florida condominium complex wants to build a model for the sale price (y) of a condo unit (recorded in hundreds of dollars) and then use the model to predict the prices of future units sold. In addition to sale price, the **CONDO** file contains data on six potential predictor variables for a sample of 209 units sold. (See Table SIA12.1 on p. 683.) These independent variables are defined as follows:

$x_1 =$ floor location (i.e., floor height) of the unit (1, 2, 3, ..., or 8)
$x_2 =$ distance, in units, from the elevator (1, 2, 3 ..., or 15)
$x_3 = 1$ if ocean view, 0 if non–ocean view
$x_4 = 1$ if end unit, 0 if not an end unit

$x_5 = 1$ if furnished unit, 0 if unfurnished
$x_6 = 1$ if sold at public auction, 0 if sold at developer's price

MINITAB scatterplots (with the dependent variable, PRICE100, plotted against each of the potential predictors) for the data are shown in Figure SIA12.2. From the scatterplots, it appears that DISTANCE (x_2) and VIEW (x_3) may be the best predictors, since they both show stronger trends with sales price than the other independent variables. However, it would not be sound statistical practice to discard the other predictors based on graphs alone. Consequently, in this section, we will use all six independent variables in a multiple-regression model for sales price.

Consider the first-order multiple regression model

$$E(y) = \beta_0 + \beta_1 x_1 + \beta_2 x_2 + \beta_3 x_3 + \beta_4 x_4$$
$$+ \beta_5 x_5 + \beta_6 x_6$$

The MINITAB printout for the regression analysis is shown in Figure SIA12.3. The global *F*-statistic ($F = 49.99$) and associated *p*-value (.000) shown on the printout indicate that the overall model is statistically useful in predicting auction price. The value of the adjusted R^2, however, indicates that

the model explains only about 59% of the sample variation in price, and the standard deviation of the model ($s = 21.8$) implies that the model can predict price to within about $2s = 43.6$ hundred dollars (i.e., \$4,360). While the model appears to be "statistically" useful in predicting auction price, the moderate value of R_a^2 and relatively large $2s$ value indicate that the model may not yield accurate predictions.

[*Note:* Not all of the independent variables have statistically significant *t*-values. However, we caution against dropping the insignificant variables from the model at this stage. One reason (discussed in Section 12.3) is that performing a large number of *t*-tests will yield an inflated probability of at least one Type I error. In later sections of this chapter, we develop other reasons for why the multiple *t*-test approach is not a good strategy for determining which independent variables to keep in the model.]

Plots of PRICE100 vs FLOOR, DISTANCE, VIEW, END, FURNISH, AUCTION

Figure SIA12.2
MINITAB scatterplots for CONDO data

```
The regression equation is
PRICE100 = 187 - 1.16 FLOOR + 0.917 DISTANCE + 48.3 VIEW - 23.6 END
           + 6.95 FURNISH - 28.0 AUCTION


Predictor      Coef   SE Coef       T      P
Constant    186.990     4.693   39.84  0.000
FLOOR        -1.1599    0.7636   -1.52  0.130
DISTANCE      0.9168    0.3544    2.59  0.010
VIEW         48.250     3.281    14.70  0.000
END         -23.633     8.661    -2.73  0.007
FURNISH       6.945     3.243     2.14  0.033
AUCTION     -27.970     3.886    -7.20  0.000


S = 21.8163   R-Sq = 59.8%   R-Sq(adj) = 58.6%


Analysis of Variance

Source           DF       SS      MS      F      P
Regression        6   142750   23792  49.99  0.000
Residual Error  202    96142     476
Total           208   238893
```

Figure SIA12.3
MINITAB regression output for first-order model of condominium unit sale price

(Continued)

The MINITAB printout shown in Figure SIA12.4 gives a 95% prediction interval for auction price and a 95% confidence interval for the mean price for the following x-values:

$x_1 = 5$ (i.e., unit on the fifth floor)
$x_2 = 9$ (i.e., distance of nine units from the elevator)
$x_3 = 1$ (i.e., ocean view)
$x_4 = 0$ (i.e., not an end unit)
$x_5 = 0$ (i.e., an unfurnished unit)
$x_6 = 1$ (i.e., a unit sold at public auction)

The 95% confidence interval of (204.05, 215.39) implies that, for all condo units with these x-values, the mean auction price falls between 204.05 and 215.39 hundred dollars, with 95% confidence. The 95% prediction interval of (166.33, 253.11) implies that, for an individual condo unit with these x-values, the auction price falls between 166.33 and 253.11 hundred dollars, with 95% confidence. Note the wide range of the prediction interval. This is due to the

```
Predicted Values for New Observations

New Obs    Fit  SE Fit      95% CI             95% PI
     1  209.72    2.87  (204.05, 215.39)  (166.33, 253.11)


Values of Predictors for New Observations

New Obs  FLOOR  DISTANCE  VIEW      END    FURNISH  AUCTION
     1   5.00      9.00  1.00  0.000000  0.000000    1.00
```

Figure SIA12.4

MINITAB printout with 95% confidence and prediction intervals

large magnitude of the model's standard deviation $s = 21.8$ hundred dollars. Again, although the model is deemed statistically useful in predicting auction price, it may not be "practically" useful. To reduce the magnitude of s, we will need to improve the model's predictive ability. (We consider such a model in the next Statistics in Action Revisited section.)

 Data Set: CONDO

Exercises 12.31–12.41

Understanding the Principles

12.31 Explain why we use $\hat{y}$ as an estimate of $E(y)$ and to predict y.

12.32 Which interval will be narrower, a 95% confidence interval for $E(y)$ or a 95% prediction interval for y? (Assume that the values of the x's are the same for both intervals.)

Applying the Concepts—Basic

12.33 Characteristics of lead users. Refer to the *Creativity and Innovation Management* (Feb. 2008) study of lead users of children's computer games, Exercise 12.15 (p. 698). Recall that the researchers modeled lead-user rating (y, measured on a 5-point scale) as a function of gender ($x_1 = 1$ if female, 0 if male), age (x_2, years), degree of centrality (x_3, measured as the number of direct ties to other peers in the network), and betweenness centrality (x_4, measured as the number of shortest paths between peers). The least squares prediction equation was $\hat{y} = 3.58 + .01x_1 - .06x_2 - .01x_3 + .42x_4$.
a. Compute the predicted lead-user rating of a 10-year-old female child with 5 direct ties to other peers in her social network and with 2 shortest paths between peers.
b. Compute an estimate for the mean lead-user rating of all 8-year-old male children with 10 direct ties to other peers and with 4 shortest paths between peers.

12.34 Predicting runs scored in baseball. Refer to the *Chance* (Fall 2000) study of runs scored in Major League Baseball games, Exercise 12.19 (p. 698). Multiple regression was used to model total number of runs scored (y) of a team during the season as a function of number of walks (x_1), number of singles (x_2), number of doubles (x_3), number of triples (x_4), number of home runs (x_5), number of stolen bases (x_6), number of times caught stealing (x_7), number of strikeouts (x_8), and total number of outs (x_9). Using the β-estimates given in Exercise 12.19, predict the number of runs scored by your favorite Major League Baseball team last year. How close is the predicted value to the actual number of runs scored by your team? [*Note:* You can find data on your favorite team on the Internet at www.major-leaguebaseball.com.]

12.35 Reality TV and cosmetic surgery. Refer to the *Body Image: An International Journal of Research* (Mar. 2010) study of the impact of reality TV shows on one's desire to undergo cosmetic surgery, Exercise 12.25 (p. 700). Recall that psychologists used multiple regression to model desire to have cosmetic surgery (y) as a function of gender (x_1), self-esteem (x_2), body satisfaction (x_3), and impression of reality TV (x_4). The SAS printout below shows a confidence interval for $E(y)$ for each of the first five students in the study.

SAS Output for Exercise 12.35

Dependent Variable: DESIRE

Output Statistics

Obs	GENDER	SELFESTM	BODYSAT	IMPREAL	Dependent Variable	Predicted Value	Std Error Mean Predict	95% CL Mean	
1	0	24	3	4	11.0000	13.8655	0.2235	13.4243	14.3067
2	0	20	3	4	13.0000	14.0573	0.2519	13.5598	14.5547
3	0	25	4	5	11.0000	13.9883	0.2899	13.4159	14.5607
4	1	22	9	4	11.0000	9.8409	0.5318	8.7908	10.8910
5	0	8	1	6	18.0000	16.2634	0.5472	15.1830	17.3438

MIMITAB output for Exercise 12.37

```
Predicted Values for New Observations

New
Obs      Fit   SE Fit        95% CI              95% PI
  1  12632.5   237.3   (12157.9, 13107.1)   (11599.6, 13665.5)


Values of Predictors for New Observations

New
Obs    RPM   INLET-TEMP   EXH-TEMP   CPRATIO   AIRFLOW
  1   7500        1000        525      13.5      10.0
```

a. Interpret the confidence interval for $E(y)$ for student 1.
b. Interpret the confidence interval for $E(y)$ for student 4.

12.36 Usability Professionals, salary survey. The Usability Professionals' Association (UPA) supports people who research, design, and evaluate the user experience of products and services. The UPA conducted a salary survey of its members (*UPA Salary Survey*, Aug. 18, 2009). One of the report's authors, Jeff Sauro, investigated how much having a PhD affects salaries in this profession and discussed his analysis on the blog www.measuringusability.com. Sauro fit a first-order multiple regression model for salary (y, in dollars) as a function of years of experience (x_1), PhD status ($x_2 = 1$ if PhD, 0 if not), and manager status ($x_3 = 1$ if manager, 0 if not). The following prediction equation was obtained:

$$\hat{y} = 52{,}484 + 2{,}941x_1 + 16{,}880x_2 + 11{,}108x_3$$

a. Predict the salary of a UPA member with 10 years of experience who does not have a PhD but is a manager.
b. Predict the salary of a UPA member with 10 years of experience who does have a PhD but is not a manager.
c. Why is a 95% prediction interval preferred over the predicted values given in parts **a** and **b**?

12.37 Cooling method for gas turbines. Refer to the *Journal of Engineering for Gas Turbines and Power* (Jan. 2005) study of a high-pressure inlet fogging method for a gas turbine engine, presented in Exercise 12.27 (p. 701). Recall that you fitted a first-order model for heat rate (y) as a function

of speed (x_1), inlet temperature (x_2), exhaust temperature (x_3), cycle pressure ratio (x_4), and air mass flow rate (x_5). A MINITAB printout with both a 95% confidence interval for $E(y)$ and a prediction interval for y, for selected values of the x's, is shown above.
a. Interpret the 95% prediction interval for y in the words of the problem.
b. Interpret the 95% confidence interval for $E(y)$ in the words of the problem.
c. Will the confidence interval for $E(y)$ always be narrower than the prediction interval for y? Explain.

Applying the Concepts—Intermediate

12.38 California rain levels. An article published in *Geography* (July 1980) used multiple regression to predict annual rainfall levels in California. Data on the average annual precipitation (y), altitude (x_1), latitude (x_2), and distance from the Pacific coast (x_3) for 30 meteorological stations scattered throughout California are saved in the **CALRAIN** file. (Selected observations are listed in the table below.) Consider the first-order model $y = \beta_0 + \beta_1 x_1 + \beta_2 x_2 + \beta_3 x_3 + \varepsilon$.
a. Fit the model to the data and give the least squares prediction equation.
b. Is there evidence that the model is useful in predicting annual precipitation y? Test, using $\alpha = .05$.
c. Find a 95% prediction interval for y for the Giant Forest meteorological station (station 9). Interpret the interval.

Data for Exercise 12.38

Station	Avg. Annual Precipitation y (inches)	Altitude x_1 (feet)	Latitude x_2 (degrees)	Distance from Coast x_3 (miles)
1. Eureka	39.57	43	40.8	1
2. Red Bluff	23.27	341	40.2	97
3. Thermal	18.20	4152	33.8	70
4. Fort Bragg	37.48	74	39.4	1
5. Soda Springs	49.26	6752	39.3	150
⋮	⋮	⋮	⋮	⋮
26. San Diego	9.94	19	32.7	5
27. Daggett	4.25	2105	34.1	85
28. Death Valley	1.66	−178	36.5	194
29. Crescent City	74.87	35	41.7	1
30. Colusa	15.95	60	39.2	91

Source: Taylor, P. I. "A pedagogic application of multiple regression analysis." *Geography*, July 1980, Vol. 65, pp. 203–212. Reprinted with permission from the author.

12.39 Emotional intelligence and team performance. The *Engineering Project Organizational Journal* (Vol. 3, 2013) published the results of an exploratory study designed to gain a better understanding of how the emotional intelligence of individual team members relates directly to the performance of their team during an engineering project. Undergraduate students enrolled in the course *Introduction to the Building Industry* participated in the study. All students completed an emotional intelligence test and received an interpersonal score, stress management score, and mood score. Students were grouped into $n = 23$ teams and assigned a group project. However, each student received an individual project score. These scores were averaged to obtain the dependent variable in the analysis: mean project score (y). Three independent variables were determined for each team: range of interpersonal scores (x_1), range of stress management scores (x_2), and range of mood scores (x_3). Data (simulated from information provided in the article) are listed in the table.

Team	Intrapersonal (Range)	Stress (Range)	Mood (Range)	Project (Average)
1	14	12	17	88.0
2	21	13	45	86.0
3	26	18	6	83.5
4	30	20	36	85.5
5	28	23	22	90.0
6	27	24	28	90.5
7	21	24	38	94.0
8	20	30	30	85.5
9	14	32	16	88.0
10	18	32	17	91.0
11	10	33	13	91.5
12	28	43	28	91.5
13	19	19	21	86.0
14	26	31	26	83.0
15	25	31	11	85.0
16	40	35	24	84.0
17	27	12	14	85.5
18	30	13	29	85.0
19	31	24	28	84.5
20	25	26	16	83.5
21	23	28	12	85.0
22	20	32	10	92.5
23	35	35	17	89.0

a. Hypothesize a first-order model for project score (y) as a function of $x_1, x_2,$ and x_3.
b. Fit the model, part a, to the data using statistical software.
c. Is there sufficient evidence to indicate the overall model is statistically useful for predicting y? Test using $\alpha = .05$.
d. Evaluate the model statistics R_a^2 and $2s$.
e. Find and interpret a 95% prediction interval for y when $x_1 = 20, x_2 = 30,$ and $x_3 = 25$.

12.40 Boiler drum production. In a production facility, an accurate estimate of hours needed to complete a task is crucial to management in making such decisions as hiring the proper number of workers, quoting an accurate deadline for a client, or performing cost analyses regarding budgets. A manufacturer of boiler drums wants to use regression to predict the number of hours needed to erect the drums in future projects. To accomplish this task, data on 36 boilers were collected. In addition to hours (y), the variables measured were boiler capacity (x_1 = lb/hr), boiler design pressure (x_2 = pounds per square inch, or psi), boiler type (x_3 = 1 if industry field erected, 0 if utility field erected), and drum type (x_4 = 1 if steam, 0 if mud). The data are saved in the **BOILERS** file. (Selected observations are shown in the table below.)

Hours y	Boiler Capacity x_1	Design Pressure x_2	Boiler Type x_3	Drum Type x_4
3,137	120,000	375	1	1
3,590	65,000	750	1	1
4,526	150,000	500	1	1
10,825	1,073,877	2,170	0	1
4,023	150,000	325	1	1
⋮	⋮	⋮	⋮	⋮
4,206	441,000	410	1	0
4,006	441,000	410	1	0
3,728	627,000	1,525	0	0
3,211	610,000	1,500	0	0
1,200	30,000	325	1	0

Based on data provided by Dr. Kelly Uscategui, University of Connecticut.

a. Fit the model $E(y) = \beta_0 + \beta_1 x_1 + \beta_2 x_2 + \beta_3 x_3 + \beta_4 x_4$ to the data and give the prediction equation.
b. Conduct a test for the global utility of the model. Use $\alpha = .01$.
c. Find a 95% confidence interval for $E(y)$ when $x_1 = 150,000, x_2 = 500, x_3 = 1,$ and $x_4 = 0$. Interpret the result.
d. What type of interval would you use if you want to estimate the average number of hours required to erect all industrial mud boilers with a capacity of 150,000 lb/hr and a design pressure of 500 psi?

12.41 Arsenic in groundwater. Refer to the *Environmental Science & Technology* (Jan. 2005) study of the reliability of a commercial kit designed to test for arsenic in groundwater, presented in Exercise 12.23 (p. 700). You fit a first-order model for arsenic level (y) as a function of latitude, longitude, and depth. On the basis of the model statistics, the researchers concluded that the arsenic level is highest at a low latitude, high longitude, and low depth. Do you agree? If so, find a 95% prediction interval for arsenic level for the lowest latitude, highest longitude, and lowest depth that are within the range of the sample data. Interpret the result.

PART II: MODEL BUILDING IN MULTIPLE REGRESSION

12.5 Interaction Models

In Section 12.2, we demonstrated the relationship between $E(y)$ and the independent variables in a first-order model. When $E(y)$ is graphed against any one variable (say, x_1) for fixed values of the other variables, the result is a set of *parallel* straight lines. (See Figure 12.3, p. 688.) When this situation occurs (as it always does for a first-order model), we say that the relationship between $E(y)$ and any one independent variable *does not depend* on the values of the other independent variables in the model.

However, if the relationship between $E(y)$ and x_1 does, in fact, depend on the values of the remaining x's held fixed, then the first-order model is not appropriate for predicting y. In this case, we need another model that will take into account this dependence. Such a model includes the *cross products* of two or more x's.

For example, suppose that the mean value $E(y)$ of a response y is related to two quantitative independent variables x_1 and x_2 by the model

$$E(y) = 1 + 2x_1 - x_2 + x_1x_2$$

A graph of the relationship between $E(y)$ and x_1 for $x_2 = 0, 1,$ and 2 is displayed in the MINITAB graph shown in Figure 12.8.

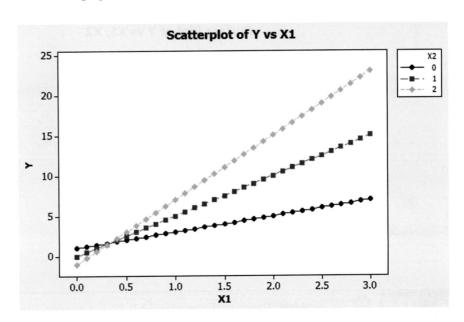

Figure 12.8

MINITAB graphs of $1 + 2x_1 - x_2 + 3x_1x_2$ for $x_2 = 0, 1, 2$

Note that the graph shows three nonparallel straight lines. You can verify that the slopes of the lines differ by substituting each of the values $x_2 = 0, 1,$ and 2 into the equation. For $x_2 = 0$,

$$E(y) = 1 + 2x_1 - (0) + x_1(0) = 1 + 2x_1 \quad (\text{slope} = 2)$$

For $x_2 = 1$,

$$E(y) = 1 + 2x_1 - (1) + x_1(1) = 3x_1 \quad (\text{slope} = 3)$$

For $x_2 = 2$,

$$E(y) = 1 + 2x_1 - (2) + x_1(2) = -1 + 4x_1 \quad (\text{slope} = 4)$$

Note that the slope of each line is represented by $\beta_1 + \beta_3x_2 = 2 + x_2$. Thus, the effect on $E(y)$ of a change in x_1 (i.e., the slope) now *depends* on the value of x_2. When this situation occurs, we say that x_1 and x_2 **interact**. The cross-product term, x_1x_2, is called an **interaction term**, and the model $E(y) = \beta_0 + \beta_1x_1 + \beta_2x_2 + \beta_3x_1x_2$ is called an **interaction model** with two quantitative variables.

An Interaction Model Relating $E(y)$ to Two Quantitative Independent Variables

$$E(y) = \beta_0 + \beta_1 x_1 + \beta_2 x_2 + \beta_3 x_1 x_2$$

where

$(\beta_1 + \beta_3 x_2)$ represents the change in $E(y)$ for every one-unit increase in x_1, holding x_2 fixed

$(\beta_2 + \beta_3 x_1)$ represents the change in $E(y)$ for every one-unit increase in x_2, holding x_1 fixed

A three-dimensional graph (generated by MINITAB) of an interaction model in two quantitative x's is shown in Figure 12.9. The interaction model traces a ruled surface (twisted plane) in three-dimensional space, unlike the flat planar surface displayed in Figure 12.4 (p. 689). If we slice the twisted plane at a fixed value of x_2, we obtain a straight line relating $E(y)$ to x_1; however, the slope of the line will change as we change the value of x_2. Consequently, *an interaction model is appropriate when the linear relationship between y and one independent variable depends on the value of the other independent variable.* The next example illustrates this idea.

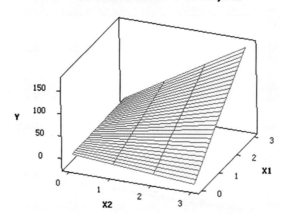

Figure 12.9

MINITAB 3-dimensional graph of $1 + 2x_1 - x_2 + 3x_1x_2$

Example   12.6   D

GFCLOCKS

Evaluating an Interaction Model—Clock Auction Prices

Problem Refer to Examples 12.1–12.4. Suppose the collector of grandfather clocks, having observed many auctions, believes that the *rate of increase* in the auction price with age will be driven upward by a large number of bidders. Thus, instead of a relationship like that shown in Figure 12.10a, in which the rate of increase in price with age is the same for any number of bidders, the collector believes that the relationship is like that shown in Figure 12.10b. Note that as the number of bidders increases from 5 to 15, the slope of the price-versus-age line increases.

Consequently, the following interaction model is proposed:

$$y = \beta_0 + \beta_1 x_1 + \beta_2 x_2 + \beta_3 x_1 x_2 + \varepsilon$$

The 32 data points listed in Table 12.1 were used to fit the model with interaction. A portion of the MINITAB printout is shown in Figure 12.11.

a. Use the global F-test at $\alpha = .05$ to test the overall utility of the model.

b. Test the hypothesis (at $\alpha = .05$) that the price–age slope increases as the number of bidders increases—that is, that age and number of bidders, x_2, interact positively.

c. Estimate the change in auction price y of a 150-year-old grandfather clock for each additional bidder.

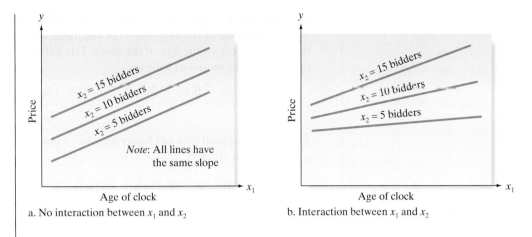

Figure 12.10

Examples of no-interaction and interaction models

a. No interaction between x_1 and x_2

b. Interaction between x_1 and x_2

Figure 12.11

MINITAB printout of interaction model of auction price

Regression Analysis: PRICE versus AGE, NUMBIDS, AGEBID

```
The regression equation is
PRICE = 320 + 0.88 AGE - 93.3 NUMBIDS + 1.30 AGEBID


Predictor    Coef   SE Coef      T      P
Constant    320.5     295.1   1.09  0.287
AGE         0.878     2.032   0.43  0.669
NUMBIDS    -93.26     29.89  -3.12  0.004
AGEBID      1.2978   0.2123   6.11  0.000


S = 88.9145    R-Sq = 95.4%    R-Sq(adj) = 94.9%


Analysis of Variance

Source          DF       SS       MS       F      P
Regression       3  4578427  1526142  193.04  0.000
Residual Error  28   221362     7906
Total           31  4799790
```

Solution

a. The global F-test is used to test the null hypothesis

$$H_0: \beta_1 = \beta_2 = \beta_3 = 0$$

The test statistic and p-value of the test (highlighted on the MINITAB printout) are $F = 193.04$ and $p = 0$, respectively. Since $\alpha = .05$ exceeds the p-value, there is sufficient evidence to conclude that the model fit is a statistically useful predictor of the auction price y.

b. The hypotheses of interest to the collector concern the interaction parameter β_3. Specifically,

$$H_0: \beta_3 = 0$$
$$H_a: \beta_3 > 0$$

Since we are testing an individual β parameter, a t-test is required. The test statistic and the two-tailed p-value (highlighted on the printout) are $t = 6.11$ and $p = 0$, respectively. The upper-tailed p-value, obtained by dividing the two-tailed p-value in half, is $0/2 = 0$. Since $\alpha = .05$ exceeds the p-value, the collector can reject H_0 and conclude that the rate of change of the mean price of the clocks with age increases as the number of bidders increases; that is, x_1 and x_2 interact positively. Thus, it appears that the interaction term should be included in the model.

c. To estimate the change in auction price y for every one-unit increase in number of bidders, x_2, we need to estimate the slope of the line relating y to x_2 when the age of the clock, x_1, is 150 years old. An analyst who is not careful may estimate this slope as $\hat{\beta}_2 = -93.26$. Although the coefficient of x_2 is negative, this does *not* imply that the auction price decreases as the number of bidders increases. Since interaction is present, the rate of change (slope) of the mean auction price with the number of bidders *depends* on x_1, the age of the clock. For a fixed value of age (x_1), we can rewrite the interaction model as follows:

$$E(y) = \beta_0 + \beta_1 x_1 + \beta_2 x_2 + \beta_3 x_1 x_2 = \underbrace{\{\beta_0 + \beta_1 x_1\}}_{y\text{-intercept}} + \underbrace{\{\beta_2 + \beta_3 x_1\}}_{\text{slope}} x_2$$

Thus, the estimated rate of change of y for a unit increase in x_2 (one new bidder) for a 150-year-old clock is

$$\text{Estimated } x_2 \text{ slope} = \hat{\beta}_2 + \hat{\beta}_3 x_1 = -93.26 + 1.30(150) = 101.74$$

In other words, we estimate that the auction price of a 150-year-old clock will *increase* by about \$101.74 for every additional bidder.

Look Back Although the rate of increase will vary as x_1 is changed, it will remain positive for the range of values of x_1 included in the sample. Extreme care is needed in interpreting the signs and sizes of coefficients in a multiple-regression model.

■ **Now Work Exercise 12.50**

Example 12.6 illustrates an important point about conducting t-tests on the β parameters in the interaction model. The "most important" β parameter in this model is the interaction β, β_3. [Note that this β is also the one associated with the highest-order term in the model, $x_1 x_2$.*] Consequently, we will want to test H_0: $\beta_3 = 0$ after we have determined that the overall model is useful in predicting y. Once interaction is detected (as in Example 12.6), however, tests on the first-order terms x_1 and x_2 should *not* be conducted, since they are meaningless tests; the presence of interaction implies that both x's are important.

⚠ **CAUTION** Once interaction has been deemed important in the model $E(y) = \beta_0 + \beta_1 x_1 + \beta_2 x_2 + \beta_3 x_1 x_2$, do not conduct t-tests on the β coefficients of the first-order terms x_1 and x_2. These terms should be kept in the model regardless of the magnitude of their associated p-values shown on the printout. ▲

We close this section with a comment: You will probably never know *a priori* whether interaction exists between two independent variables; consequently, you will need to fit and test the interaction term to determine its importance.

Exercises 12.42–12.59

Understanding the Principles

12.42 If two variables x_1 and x_2 do not interact, how would you describe their effect on the mean response $E(y)$?

12.43 Write an interaction model relating the mean value of y, $E(y)$, to
  a. two quantitative independent variables
  b. three quantitative independent variables [*Hint:* Include all possible two-way cross-product terms.]

Learning the Mechanics

12.44 Suppose the true relationship between $E(y)$ and the quantitative independent variables x_1 and x_2 is given by the model below.

$$E(y) = -5 + x_1 + 3x_2 - 3x_1 x_2$$

  a. Describe the corresponding three-dimensional response surface.
  b. Plot the linear relationship between y and x_2 for $x_1 = 0, 1, 2$, where $0 \le x_2 \le 5$. Choose the correct graph below.
  c. Explain why the lines you plotted in part **b** are not parallel.
  d. Use the lines you plotted in part **b** to explain how changes in the settings of x_1 and x_2 affect $E(y)$.
  e. Use your graph from part **b** to determine how much $E(y)$ changes when x_1 is changed from 1 to 2 and x_2 is simultaneously changed from 1 to 0.

12.45 Suppose you fit the interaction model

$$y = \beta_0 + \beta_1 x_1 + \beta_2 x_2 + \beta_3 x_1 x_2 + \varepsilon$$

*The order of a term is equal to the sum of the exponents of the quantitative variables included in the term. Thus, when x_1 and x_2 are both quantitative variables, the cross product, $x_1 x_2$, is a second-order term.

to $n = 32$ data points and obtain the results below:

$$SS_{yy} = 500 \qquad SSE = 21 \qquad \hat{\beta}_3 = 10 \qquad s_{\hat{\beta}_3} = 4$$

a. Find R^2 and interpret its value.

b. Is the model adequate for predicting y? Test at $\alpha = .05$.

c. Use a graph to explain the contribution of the x_1x_2 term to the model.

d. Is there evidence that x_1 and x_2 interact? Test at $\alpha = .05$.

12.46 MINITAB was used to fit the model

$$y = \beta_0 + \beta_1 x_1 + \beta_2 x_2 + \beta_3 x_1 x_2 + \varepsilon$$

to $n = 15$ data points. The resulting printout is shown below.

a. What is the prediction equation for the response surface?

b. Describe the geometric form of the response surface of part **a**.

c. Plot the prediction equation for the case when $x_2 = 1$. Do this twice more on the same graph for the cases when $x_2 = 3$ and $x_2 = 5$.

d. Explain what it means to say that x_1 and x_2 interact. Explain why the graph you plotted in part **c** suggests that x_1 and x_2 interact.

e. Specify the null and alternative hypotheses you would use to test whether x_1 and x_2 interact.

f. Conduct the hypothesis test of part **e**, using $\alpha = .01$.

```
The regression equation is
Y = -2.55 + 3.82 X1 + 2.63 X2 -1.29 X1X2


Predictor    Coef   SE Coef      T      P
Constant   -2.550    1.142   -2.23  0.043
X1          3.815    0.529    7.22  0.000
X2          2.630    0.344    7.64  0.000
X1X2       -1.285    0.159   -8.06  0.000


S = 0.713    R-Sq = 85.6%    R-Sq(adj) = 81.6%


Analysis of Variance

Source         DF      SS      MS      F      P
Regression      3  33.149  11.050  21.75  0.000
Residual Error 11   5.587   0.508
Total          14  38.736
```

Applying the Concepts—Basic

12.47 Tipping behavior in restaurants. By law, food servers at restaurants are not entitled to minimum wages because they are tipped by customers. Can food servers increase their tips by complimenting the customers they are waiting on? To answer this question, researchers collected data on the customer tipping behavior for a sample of 348 dining parties and reported their findings in the *Journal of Applied Social Psychology* (Vol. 40, 2010). Tip size (y, measured as a percentage of the total food bill) was modeled as a function of size of the dining party (x_1) and whether the server complimented the customers' choice of menu items (x_2). One theory states that the effect of size of the dining party on tip size is independent of whether the server compliments the customers' menu choices. A second theory hypothesizes that the effect of size of the dining party on tip size will be greater when the server compliments the

customers' menu choices as opposed to when the server refrains from complimenting menu choices.

a. Write a model for $E(y)$ as a function of x_1 and x_2 that corresponds to Theory 1.

b. Write a model for $E(y)$ as a function of x_1 and x_2 that corresponds to Theory 2.

c. The researchers summarized the results of their analysis with the following graph. Based on the graph, which of the two models would you expect to fit the data better? Explain.

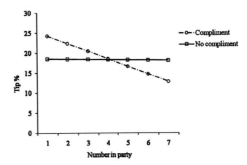

12.48 Study of high-arctic butterflies. The impact of global warming and subsequent snow melt on the flight season (migration dates) of butterflies inhabiting the High Arctic was investigated in *Current Zoology* (Apr. 2014). Annual data for a certain species of butterfly were collected for a 14-year period. The day of the onset of the flight season (y) was modeled as a function of the timing of the snowmelt (x_1, the first date when less than 10 cm of snow was measured) and average July temperature (x_2, in degrees Celsius). The interaction model, $E(y) = \beta_0 + \beta_1 x_1 + \beta_2 x_2 + \beta_3 x_1 x_2$, was fit to the data with the following results:

$$\hat{y} = -144.2 + 2.05x_1 + 35.4x_2 - .21x_1x_2, \quad R^2 = 73,$$
$$F\text{-test } p\text{-value} = .0032$$

a. Is the overall model statistically useful for predicting day of the onset of the flight season (y)? Test using $\alpha = .01$.

b. Give a practical interpretation of R^2.

c. A test of $H_0: \beta_3 = 0$ was rejected at $\alpha = .01$. Should the researchers conclude that timing of snowmelt and average July temperature interact to predict the day of the onset of the flight season?

d. Give an estimate of the change in the day of the onset of the flight season (y) for every 1-day increase in the timing of the snowmelt (x_1) for years with an average July temperature of $x_2 = 30$ degrees Celsius.

12.49 Whales entangled in fishing gear. Refer to the *Marine Mammal Science* (Apr. 2010) study of whales entangled in fishing gear, Exercise 12.17 (p. 698). Recall that the length (y) of an entangled whale (in meters) was modeled as a function of water depth of the entanglement (x_1, in meters) and distance of the entanglement from land (x_2, in miles).

a. Give the equation of an interaction model for length (y) as a function of the two independent variables.

b. The researchers theorize that the length of an entangled whale will increase linearly as the water depth increases. In terms of the parameters in the model, part **a**, write the slope of the line relating length (y) to water depth (x_1) for a distance of $x_2 = 10$ miles from land.

c. Repeat part **b** for a distance of $x_2 = 25$ miles from land.

12.50 Role of retailer interest in shopping behavior. Retail interest is defined by marketers as the level of interest a consumer has in a given retail store. Marketing professors at the University of Tennessee at Chattanooga and the University of Alabama investigated the role of retailer interest in consumers' shopping behavior (*Journal of Retailing*, Summer 2006). Using survey data collected on $n = 375$ consumers, the professors developed an interaction model for y = willingness of the consumer to shop at a retailer's store in the future (called "repatronage intentions") as a function of x_1 = consumer satisfaction and x_2 = retailer interest. The regression results are shown below.

Variable	$\hat{\beta}$	t-Value	p-Value
Satisfaction (x_1)	.426	7.33	<.01
Interest (x_2)	.044	0.85	>.10
Satisfaction × Interest (x_1x_2)	−.157	−3.09	<.01

$R^2 = .65, F = 226.35, p\text{-value} < .001$

a. Is the overall model statistically useful in predicting y? Test, using $\alpha = .05$.
b. Conduct a test for interaction at $\alpha = .05$.
c. Use the β-estimates to sketch the estimated relationship between repatronage intentions (y) and satisfaction (x_1) when retailer interest is $x_2 = 1$ (a low value).
d. Repeat part **c** for the case when retailer interest is $x_2 = 7$ (a high value).
e. Put the two lines you sketched in parts **c** and **d** on the same graph to illustrate the nature of the interaction.

12.51 Eye movement and spatial distortion. A study of how eye movement behavior can distort one's judgment of the location of an object was published in *Advances in Cognitive Psychology* (Vol. 6, 2010). The researchers had volunteers fixate their eyes on a cross in the middle of a computer screen. A probe was then spatially extended near the cross and each volunteer was asked to judge the location of the probe. Saccadic (i.e., fast voluntary) eye movement was monitored during each session. The researchers used spatial position of the probe (x_1, measured in degrees) and position of the cross (x_2, degrees) to predict the amplitude (y) of saccadic eye movement. The following model was fit to the data:

$$E(y) = \beta_0 + \beta_1x_1 + \beta_2x_2 + \beta_3x_1x_2$$

a. The model yielded $R^2 = .994$. Interpret this result.
b. In the words of the problem, what does it mean to say that "x_1 and x_2 interact"?
c. The least squares prediction equation was determined as: $\hat{y} = .91 + .70x_1 - .06x_2 - .03x_1x_2$. Illustrate interaction by graphing the relationship between predicted amplitude ($\hat{y}$) and cross position (x_2) for probe positions $x_1 = 3.5$ and $x_1 = 6.5$.

12.52 Psychology of waiting in line. While waiting in a long line for service (e.g., to use an ATM or at the post office), at some point you may decide to leave the line. The *Journal of Consumer Research* (Nov. 2003) published a study of consumer behavior while waiting in a line. College students (sample size $n = 148$) were asked to imagine that they were waiting in line at a post office to mail a package and that the estimated waiting time was 10 minutes or less. After a 10-minute wait, students were asked about their level of negative feelings (annoyed, anxious) on a scale of 1 (strongly disagree) to 9 (strongly agree). Before answering, however, the students were informed about how many people were ahead of them and behind them in the line. The researchers used regression to relate negative feelings score (y) to number ahead in line (x_1) and number behind in line (x_2).

a. The researchers fit an interaction model to the data. Write the hypothesized equation of this model.
b. In the words of the problem, explain what it means to say that "x_1 and x_2 interact to affect y."
c. A t-test for the interaction β resulted in a p-value greater than .25. Interpret this result.
d. From their analysis, the researchers concluded that "the greater the number of people ahead, the higher [is] the negative feeling score" and "the greater the number of people behind, the lower [is] the negative feeling score." Use this information to determine the signs of $\hat{\beta}_1$ and $\hat{\beta}_2$ in the model.

Applying the Concepts—Intermediate

12.53 Forecasting movie revenues with Twitter. Refer to the *IEEE International Conference on Web Intelligence and Intelligent Agent Technology* (2010) study on using the volume of chatter on Twitter.com to forecast movie box office revenue, Exercise 12.20 (p. 699). The researchers modeled a movie's opening weekend box office revenue (y) as a function of tweet rate (x_1) and ratio of positive to negative tweets (x_2) using a first-order model.

a. Write the equation of an interaction model for $E(y)$ as a function of x_1 and x_2.
b. In terms of the β's in the model, part **a**, what is the change in revenue (y) for every 1-tweet increase in the tweet rate (x_1), holding PN-ratio (x_2) constant at a value of 2.5?
c. In terms of the β's in the model, part **a**, what is the change in revenue (y) for every 1-tweet increase in the tweet rate (x_1), holding PN-ratio (x_2) constant at a value of 5.0?
d. In terms of the β's in the model, part **a**, what is the change in revenue (y) for every 1-unit increase in the PN-ratio (x_2), holding tweet rate (x_1) constant at a value of 100?
e. Give the null hypothesis for testing whether tweet rate (x_1) and PN-ratio (x_2) interact to affect revenue (y).

12.54 Reality TV and cosmetic surgery. Refer to the *Body Image: An International Journal of Research* (March 2010) study of the influence of reality TV shows on one's desire to undergo cosmetic surgery, Exercise 12.25 (p. 700). Recall that psychologists modeled desire to have cosmetic surgery (y) as a function of gender (x_1), self-esteem (x_2), body satisfaction (x_3), and impression of reality TV (x_4). For this exercise, consider only the independent variables gender (x_1) and impression of reality TV (x_4).

a. The research psychologists theorize that the impact of one's impression of reality TV on level of desire for cosmetic surgery will be greater for females than for males. Does this theory imply that the independent variables x_1 and x_4 interact, or that there is no interaction? Explain.
b. Fit the interaction model, $E(y) = \beta_0 + \beta_1x_1 + \beta_2x_4 + \beta_3x_1x_4$, to the data.
c. Use the results, part **b**, to carry out a test for interaction. Make your conclusion using $\alpha = .05$.

12.55 Factors that affect an auditor's judgment. A study was conducted to determine the effects of linguistic delivery style and client credibility on auditors' judgments (*Advances in Accounting and Behavioral Research*, 2003). Each of 200 auditors performed an analytical review of a client's financial statement. The researchers gave the auditors different information on the client's credibility and the linguistic delivery style of the client's explanation. Each auditor then provided an assessment of the likelihood that the client's explanation accounts for the fluctuation in the financial statement. The three variables of interest—credibility (x_1), linguistic delivery style (x_2), and likelihood (y)—were all measured on a numerical scale. Regression analysis was used to fit the interaction model $y = \beta_0 + \beta_1 x_1 + \beta_2 x_2 + \beta_3 x_1 x_2 + \varepsilon$. The results are summarized in the table at the bottom of the page.

a. Interpret the phrase "client credibility and linguistic delivery style interact" in the words of the problem.

b. Give the null and alternative hypotheses for testing the overall adequacy of the model.

c. Conduct the test suggested in part **b**, using the information in the table.

d. Give the null and alternative hypotheses for testing whether client credibility and linguistic delivery style interact.

e. Conduct the test suggested in part **d**, using the information in the table.

f. The researchers estimated the slope of the likelihood–linguistic delivery style line at a low level of client credibility $(x_1 = 22)$. Obtain this estimate and interpret it in the words of the problem.

g. The researchers also estimated the slope of the likelihood–linguistic delivery style line at a high level of client credibility $(x_1 = 46)$. Obtain this estimate and interpret it in the words of the problem.

12.56 Unconscious self-esteem study. Psychologists define *implicit* self-esteem as unconscious evaluations of one's worth or value. In contrast, *explicit* self-esteem refers to the extent to which a person consciously considers oneself as valuable and worthy. An article published in *Journal of Articles in Support of the Null Hypothesis* (Mar. 2006) investigated whether implicit self-esteem is really unconscious. A sample of 257 college undergraduate students completed a questionnaire designed to measure implicit self-esteem and explicit self-esteem. Thus, an implicit self-esteem score (x_1) and explicit self-esteem score (x_2) were obtained for each. (*Note*: Higher scores indicate higher levels of self-esteem.) Also, a second questionnaire was administered in order to obtain each subject's estimate of his/her level of implicit self-esteem. The score obtained from this questionnaire was called an estimated implicit self-esteem score (x_3). Finally, the researchers computed two measures of accuracy in estimating implicit self-esteem: $y_1 = (x_3 - x_1)$ and $y_2 = |x_3 - x_1|$.

a. The researchers fit the interaction model $E(y_1) = \beta_0 + \beta_1 x_1 + \beta_2 x_2 + \beta_3 x_1 x_2$. The *t*-test of the interaction term, β_3, was "nonsignificant," with a *p*-value $> .10$. However, both *t*-tests of β_1 and β_2 were statistically significant $(p\text{-value} < .001)$. Interpret these results practically.

b. The researchers also fit the interaction model $E(y_2) = \beta_0 + \beta_1 x_1 + \beta_2 x_2 + \beta_3 x_1 x_2$. The *t*-test on the interaction term, β_3, was "significant," with a *p*-value $< .001$. Interpret this result practically.

12.57 Cooling method for gas turbines. Refer to the *Journal of Engineering for Gas Turbines and Power* (Jan. 2005) study of a high-pressure inlet fogging method for a gas turbine engine, presented in Exercise 12.27 (p. 701). Recall that you fit a first-order model for heat rate (y) as a function of speed (x_1), inlet temperature (x_2), exhaust temperature (x_3), cycle pressure ratio (x_4), and air mass flow rate (x_5) to the data.

a. Researchers hypothesize that the linear relationship between heat rate (y) and temperature (both inlet and exhaust) depends on air mass flow rate. Write a model for heat rate that incorporates the researchers' theories.

b. Use statistical software to fit the interaction model you wrote in part **a**. Give the least squares prediction equation.

c. Conduct a test (at $\alpha = .05$) to determine whether inlet temperature and air mass flow rate interact to affect heat rate.

d. Conduct a test (at $\alpha = .05$) to determine whether exhaust temperature and air mass flow rate interact to affect heat rate.

e. Interpret practically the results of the tests you conducted in parts **c** and **d**.

12.58 Arsenic in groundwater. Refer to the *Environmental Science & Technology* (Jan. 2005) study of the reliability of a commercial kit to test for arsenic in groundwater, presented in Exercise 12.23 (p. 700). Recall that you fit a first-order model for arsenic level (y) as a function of latitude (x_1), longitude (x_2), and depth (x_3) to the data.

a. Write a model for arsenic level (y) that includes first-order terms for latitude, longitude, and depth, as well as terms for interaction between latitude and depth and interaction between longitude and depth.

b. Use statistical software to fit the interaction model you wrote in part **a**. Give the least squares prediction equation.

c. Conduct a test (at $\alpha = .05$) to determine whether latitude and depth interact to affect arsenic level.

Results for Exercise 12.55

	Beta Estimate	Std Error	t-statistic	p-value
Constant	15.865	10.980	1.445	0.150
Client credibility (x_1)	0.037	0.339	0.110	0.913
Linguistic delivery style (x_2)	−0.678	0.328	−2.064	0.040
Interaction $(x_1 x_2)$	0.036	0.009	4.008	<0.005

$F = 55.35 \ (p < 0.0005)$; Adjusted $R^2 = .450$

d. Conduct a test (at $\alpha = .05$) to determine whether longitude and depth interact to affect arsenic level.

e. Interpret practically the results of the tests you conducted in parts **c** and **d**.

12.59 Bubble behavior in subcooled flow boiling. Refer to the *Heat Transfer Engineering* (Vol. 34, 2013) study of bubble behavior in subcooled flow boiling, Exercise 12.28 (p. 701). Recall that bubble density (liters per meters squared) was modeled as a function of mass flux (kilograms per meters squared per second) and heat flux (megawatts per meters squared) using data saved in the **BUBBLE2** file.

a. Write an interaction model for bubble density (y) as a function of x_1 = mass flux and x_2 = heat flux.

b. Fit the interaction model, part **a**, to the data using statistical software. Give the least squares prediction equation.

c. Evaluate overall model adequacy by conducting a global F-test (at $\alpha = .05$) and interpreting the model statistics, R_a^2 and $2s$.

d. Conduct a test (at $\alpha = .05$) to determine whether mass flux and heat flux interact.

e. How much do you expect bubble density to decrease for every 1 kg/m^2-sec increase in mass flux when heat flux is set at .5 megawatts/m^2?

12.6 Quadratic and Other Higher Order Models

All of the models discussed in the previous sections proposed straight-line relationships between $E(y)$ and each of the independent variables in the model. In this section, we consider models that allow for curvature in the relationships. Each of these models is a **second-order model** because it includes an x^2-term.

First, we consider a model that includes only one independent variable x. The form of this model, called the **quadratic model**, is

$$y = \beta_0 + \beta_1 x + \beta_2 x^2 + \varepsilon$$

The term involving x^2, called a **quadratic term** (or **second-order term**), enables us to hypothesize curvature in the graph of the response model relating y to x. Graphs of the quadratic model for two different values of β_2 are shown in Figure 12.12. When the curve opens upward, the sign of β_2 is positive (see Figure 12.12a); when the curve opens downward, the sign of β_2 is negative (see Figure 12.12b).

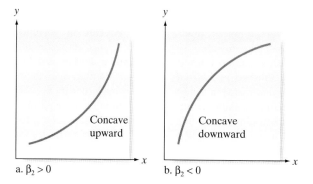

Figure 12.12
Graphs of two quadratic models

a. $\beta_2 > 0$ b. $\beta_2 < 0$

A Quadratic (Second-Order) Model in a Single Quantitative Independent Variable

$$E(y) = \beta_0 + \beta_1 x + \beta_2 x^2$$

where

β_0 is the y-intercept of the curve

β_1 is a shift parameter

β_2 is the rate of curvature

Example 12.7
ELECTRIC

Analyzing a Quadratic Model—Predicting Electrical Usage

Problem In all-electric homes, the amount of electricity expended is of interest to consumers, builders, and groups involved with energy conservation. Suppose we wish to investigate the monthly electrical usage y in all-electric homes and its relationship to the size x of the home. Moreover, suppose we think that monthly electrical usage in all-electric homes is related to the size of the home by the quadratic model

$$y = \beta_0 + \beta_1 x + \beta_2 x^2 + \varepsilon$$

To fit the model, the values of y and x are collected for 15 homes during a particular month. The data are shown in Table 12.2.

a. Construct a scatterplot of the data. Is there evidence to support the use of a quadratic model?

b. Use the method of least squares to estimate the unknown parameters β_0, β_1, and β_2 in the quadratic model.

c. Graph the prediction equation and assess how well the model fits the data, both visually and numerically.

d. Interpret the β estimates.

e. Is the overall model useful (at $\alpha = .01$) for predicting electrical usage y?

f. Is there sufficient evidence of concave-downward curvature in the electrical usage–home size relationship? Test, using $\alpha = .01$.

Solution

a. A scatterplot of the data of Table 12.2, produced with MINITAB, is shown in Figure 12.13. The figure illustrates that the electrical usage appears to increase in a curvilinear manner with the size of the home. This relationship provides some support for the inclusion of the quadratic term x^2 in the model.

b. We used SAS to fit the model to the data in Table 12.2. Part of the SAS regression output is displayed in Figure 12.14. The least squares estimates of the β parameters (highlighted) are $\hat{\beta}_0 = -806.7$, $\hat{\beta}_1 = 1.9616$, and $\hat{\beta}_2 = -.00034$. Therefore, the equation that minimizes the SSE for the data is

$$\hat{y} = -806.7 + 1.9616x - .00034x^2$$

c. Figure 12.15 is a MINITAB graph of the least squares prediction equation. Note that the graph provides a good fit to the data of Table 12.2. A numerical measure of fit is obtained with the adjusted coefficient of determination, R_a^2. From the SAS printout, $R_a^2 = .9735$. This implies that over 97% of the sample variation in

Table 12.2 Home Size–Electrical Usage Data

Size of Home x (sq. ft)	Monthly Usage, y (kilowatt-hours)
1,290	1,182
1,350	1,172
1,470	1,264
1,600	1,493
1,710	1,571
1,840	1,711
1,980	1,804
2,230	1,840
2,400	1,956
2,710	2,007
2,930	1,984
3,000	1,960
3,210	2,001
3,240	1,928
3,520	1,945

(D) *Data Set:* ELECTRIC

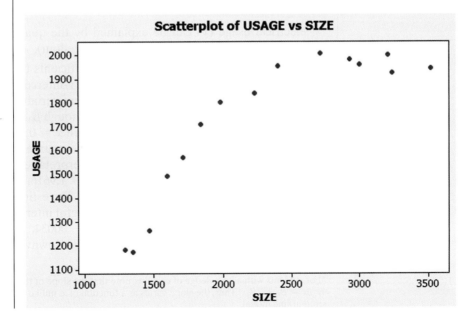

Figure 12.13
MINITAB scatterplot for electrical usage data

Figure 12.14

SAS regression output for electrical usage model

Dependent Variable: USAGE

Number of Observations Read	15	
Number of Observations Used	15	

Analysis of Variance

Source	DF	Sum of Squares	Mean Square	F Value	Pr > F
Model	2	1300900	650450	258.11	<.0001
Error	12	30240	2520.01514		
Corrected Total	14	1331140			

Root MSE	50.19975	R-Square	0.9773
Dependent Mean	1721.20000	Adj R-Sq	0.9735
Coeff Var	2.91656		

Parameter Estimates

Variable	DF	Parameter Estimate	Standard Error	t Value	Pr > |t|
Intercept	1	-806.71666	166.87209	-4.83	0.0004
SIZE	1	1.96162	0.15252	12.86	<.0001
SIZESQ	1	-0.00034044	0.00003212	-10.60	<.0001

Figure 12.15

MINITAB plot of least squares model for electrical usage

Fitted Line Plot
USAGE = - 806.7 + 1.962 SIZE
- 0.000340 SIZE**2

electrical usage (y) can be explained by the quadratic model (after adjusting for sample size and number of degrees of freedom).

d. The interpretation of the estimated coefficients in a quadratic model must be undertaken cautiously. First, the estimated y-intercept, $\hat{\beta}_0$, can be meaningfully interpreted only if the range of the independent variable includes zero—that is, if $x = 0$ is included in the sampled range of x. Although $\hat{\beta}_0 = -806.7$ seems to imply that the estimated electrical usage is negative when $x = 0$, this zero point is not in the range of the sample (the lowest value of x is 1,290 square feet), and the value is nonsensical (a home with 0 square feet); thus, the interpretation of $\hat{\beta}_0$ is not meaningful.

The estimated coefficient of x is $\hat{\beta}_1 = 1.9616$, but it no longer represents a slope in the presence of the quadratic term x^2.* The estimated coefficient of the first-order term x will not, in general, have a meaningful interpretation in the quadratic model.

The sign of the coefficient, $\hat{\beta}_2 = -.00034$, of the quadratic term, x^2, is the indicator of whether the curve is concave downward (mound shaped) or concave

---

*For students with a knowledge of calculus, note that the slope of the quadratic model is the first derivative $\partial y / \partial x = \beta_1 + 2\beta_2 x$. Thus, the slope varies as a function of x, unlike the constant slope associated with the straight-line model.

Figure 12.16

Potential misuse of quadratic model

upward (bowl shaped). A negative $\hat{\beta}_2$ implies downward concavity, as in this example (Figure 12.15), and a positive $\hat{\beta}_2$ implies upward concavity. Rather than interpreting the numerical value of $\hat{\beta}_2$ itself, we utilize a graphical representation, as in Figure 12.15, to describe the model.

Note that Figure 12.15 implies that the estimated electrical usage levels off as the home sizes increase beyond 2,500 square feet. In fact, the concavity of the model would lead to decreasing usage estimates if we were to display the model out to 4,000 square feet and beyond. (See Figure 12.16.) However, model interpretations are not meaningful outside the range of the independent variable, which has a maximum value of 3,520 square feet in this example. Thus, although the model appears to support the hypothesis that the *rate of increase* per square foot *decreases* for home sizes near the high end of the sampled values, the conclusion that usage will actually begin to decrease for very large homes would be a *misuse* of the model, since no homes of 3,600 square feet or more were included in the sample.

e. To test whether the quadratic model is statistically useful, we conduct the global *F*-test:

$$H_0: \beta_1 = \beta_2 = 0$$

H_a: At least one of the preceding coefficients is nonzero

From the SAS printout shown in Figure 12.14, the test statistic (highlighted) is $F = 258.11$, with an associated *p*-value $< .0001$. Thus, for any reasonable α, we reject H_0 and conclude that the overall model is a useful predictor of electrical usage *y*.

f. Figure 12.15 shows concave-downward curvature in the relationship between the size of a home and electrical usage in the sample of 15 data points. To determine whether this type of curvature exists in the population, we want to test

$$H_0: \beta_2 = 0 \text{ (no curvature exists in the response curve)}$$

$$H_a: \beta_2 < 0 \text{ (downward concavity exists in the response curve)}$$

The test statistic for testing β_2, highlighted on Figure 12.14, is $t = -10.60$, and the associated two-tailed *p*-value is less than .0001. Since this is a one-tailed test, the appropriate *p*-value is less than $.0001/2 = .00005$. Now, $\alpha = .01$ exceeds this *p*-value. Thus, there is very strong evidence of downward curvature in the population; that is, electrical usage increases more slowly per square foot for large homes than for small homes.

Look Back Note that the SAS printout in Figure 12.14 also provides the *t*-test statistic and corresponding two-tailed *p*-values for the tests of $H_0: \beta_0 = 0$ and $H_0: \beta_1 = 0$. Since the interpretation of these parameters is not meaningful for this model, the tests are not of interest.

■ **Now Work Exercise 12.67**

When two or more quantitative independent variables are included in a second-order model, we can incorporate squared terms for each x in the model, as well as the interaction between the two independent variables. A model that includes all possible second-order terms in two independent variables—called a **complete second-order model**—is given in the following box:

Complete Second-Order Model with Two Quantitative Independent Variables

$$E(y) = \beta_0 + \beta_1 x_1 + \beta_2 x_2 + \beta_3 x_1 x_2 + \beta_4 x_1^2 + \beta_5 x_2^2$$

Comments on the Parameters

β_0: y-intercept, the value of $E(y)$ when $x_1 = x_2 = 0$

β_1, β_2: Changing β_1 and β_2 causes the surface to shift along the x_1- and x_2-axes.

β_3: Controls the rotation of the surface

β_4, β_5: Signs and values of these parameters control the type of surface and the rates of curvature.

Three types of surfaces are produced by a second-order model:* a **paraboloid** that opens upward (Figure 12.17a), a paraboloid that opens downward (Figure 12.17b), and a **saddle-shaped surface** (Figure 12.17c).

A complete second-order model is the three-dimensional equivalent of a quadratic model in a single quantitative variable. Instead of tracing parabolas, however, it traces paraboloids and saddle-shaped surfaces. Since only a portion of the complete surface is used to fit the data, this model provides a very large variety of gently curving surfaces that can be used to fit data. It is a good choice for a model if you expect curvature in the response surface relating $E(y)$ to x_1 and x_2.

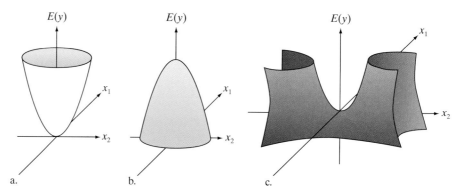

Figure 12.17

Graphs for three second-order surfaces

a.　　　　b.　　　　c.

Example **12.8**	**Problem** A social scientist would like to relate the number of hours worked per week (outside the home) by a married woman to the number of years of formal education she has completed and the number of children in her family.

A More Complex Second-Order Model—Predicting Hours Worked per Week

a. Identify the dependent variable and the independent variables.

b. Write the first-order model for this example.

c. Modify the model in part **b** so that it includes an interaction term.

d. Write a complete second-order model for $E(y)$.

Solution

a. The dependent variable is

$$y = \text{Number of hours worked per week by a married woman}$$

---

*The paraboloid opens upward (Figure 12.17a) when $\beta_4 + \beta_5 > 0$ and opens downward (Figure 12.17b) when $\beta_4 + \beta_5 < 0$; the saddle-shaped surface (Figure 12.17c) is produced when $\beta_3^2 > 4\beta_4\beta_5$.

The two independent variables, both quantitative in nature, are

x_1 = Number of years of formal education completed by the woman

x_2 = Number of children in the family

b. The first-order model is

$$E(y) = \beta_0 + \beta_1 x_1 + \beta_2 x_2$$

This model would probably not be appropriate in the current situation because x_1 and x_2 may interact or curvature terms corresponding to x_1^2 and x_2^2 may be needed to obtain a good model for $E(y)$.

c. Adding the interaction term, we obtain

$$E(y) = \beta_0 + \beta_1 x_1 + \beta_2 x_2 + \beta_3 x_1 x_2$$

This model should be better than the model in part **b**, since we have now allowed for interaction between x_1 and x_2.

d. The complete second-order model is

$$E(y) = \beta_0 + \beta_1 x_1 + \beta_2 x_2 + \beta_3 x_1 x_2 + \beta_4 x_1^2 + \beta_5 x_2^2$$

Since it would not be surprising to find curvature in the response surface, the complete second-order model would be preferred to the models in parts **b** and **c**. How can we tell whether the complete second-order model really does provide better predictions of hours worked than the models in parts **b** and **c**? The answers to these and similar questions are examined in Section 12.9.

■

Most relationships between $E(y)$ and two or more quantitative independent variables are second order and require the use of either the interactive or the complete second-order model to obtain a good fit to a data set. As in the case of a single quantitative independent variable, however, the curvature in the response surface may be very slight over the range of values of the variables in the data set. When this happens, a first-order model may provide a good fit to the data.

Exercises 12.60–12.80

Understanding the Principles

12.60 In the model $E(y) = \beta_0 + \beta_1 x + \beta_2 x^2$,
   a. Which β represents the y-intercept?
   b. Which β represents the shift?
   c. Which β represents the rate of curvature?

12.61 Write a second-order model relating the mean of y, $E(y)$, to
   a. one quantitative independent variable
   b. two quantitative independent variables
   c. three quantitative independent variables [*Hint:* Include all possible two-way cross-product terms and squared terms.]

Learning the Mechanics

12.62 Suppose you fit the quadratic model

$$E(y) = \beta_0 + \beta_1 x + \beta_2 x^2$$

to a set of $n = 20$ data points and found $R^2 = .85$, $SS_{yy} = 25.59$, and SSE = 3.82.
   a. Is there sufficient evidence to indicate that the model contributes information for predicting y? Test using $\alpha = .05$. Write the hypotheses for the test.

b. What null and alternative hypotheses would you test to determine whether upward curvature exists?
   c. What null and alternative hypotheses would you test to determine whether downward curvature exists?

12.63 Suppose you fit the second-order model

$$y = \beta_0 + \beta_1 x + \beta_2 x^2 + \varepsilon$$

to $n = 30$ data points. Your estimate of β_2 is $\hat{\beta}_2 = .52$, and the estimated standard error of the estimate is .18.
   a. Test $H_0: \beta_2 = 0$ against $H_a: \beta_2 \neq 0$. Use $\alpha = .05$.
   b. Suppose you want to determine only whether the quadratic curve opens upward; that is, as x increases, the slope of the curve would increase. Give the test statistic and the rejection region for the test for $\alpha = .05$. Do the data support the theory that the slope of the curve increases as x increases? Explain.

12.64 MINITAB was used to fit the complete second-order model

$$E(y) = \beta_0 + \beta_1 x_1 + \beta_2 x_2 + \beta_3 x_1 x_2 + \beta_4 x_1^2 + \beta_5 x_2^2$$

to $n = 39$ data points. (See the printout on p. 722)
   a. Is there sufficient evidence to indicate that at least one of the parameters β_1, β_2, β_3, β_4, and β_5 is nonzero? Test, using $\alpha = .05$.

The regression equation is
Y = -24.56 + 1.12 X1 + 27.99 X2 - 0.54 X1X2 - 0.004 X1SQ + 0.002 X2SQ

Predictor	Coef	SE Coef	T	P
Constant	-24.563	6.531	-3.76	0.001
X1	1.19848	0.1103	10.86	0.000
X2	27.988	79.489	0.35	0.727
X1X2	-0.5397	1.0338	-0.52	0.605
X1SQ	-0.0043	0.0004	-10.74	0.000
X2SQ	0.0020	0.0033	0.60	0.550

S = 2.762   R-Sq = 79.7%   R-Sq(adj) = 76.6%

Analysis of Variance

Source	DF	SS	MS	F	P
Regression	5	989.30	197.86	25.93	0.000
Residual Error	33	251.81	7.63		
Total	38	1241.11			

MINITAB output for
Exercise 12.64

b. Test $H_0: \beta_4 = 0$ against $H_a: \beta_4 \neq 0$. Use $\alpha = .01$.

c. Test $H_0: \beta_5 = 0$ against $H_a: \beta_5 \neq 0$. Use $\alpha = .01$.

d. Use graphs to explain the consequences of the tests in parts **b** and **c**.

12.65 Consider the following quadratic models:

$$(1)\ y = 1 + 4x + 2x^2$$
$$(2)\ y = 1 - 4x + 2x^2$$
$$(3)\ y = 1 + 2x^2$$
$$(4)\ y = 1 - 2x^2$$
$$(5)\ y = 1 + 5x^2$$

a. Graph each of these quadratic models, side by side, on the same sheet of graph paper.

b. What effect does the first-order term ($4x$) have on the graph of the curve?

c. What effect does the second-order term ($2x^2$) have on the graph of the curve?

Applying the Concepts—Basic

12.66 **Commercial refrigeration systems.** The role of maintenance in energy savings in commercial refrigeration was the topic of an article in the *Journal of Quality in Maintenance Engineering* (Vol. 18, 2012). The authors provided the following illustration of data relating the efficiency (relative performance) of a refrigeration system to the fraction of total charges for cooling the system required for optimal performance. Based on the data shown in the graph, hypothesize an appropriate model for relative performance (y) as a function of fraction of charge (x). What is the hypothesized sign (positive or negative) of the β_2 parameter in the model?

12.67 **Personality traits and job performance.** When attempting to predict job performance using personality traits, researchers typically assume that the relationship is linear. A study published in the *Journal of Applied Psychology* (Jan. 2011) investigated a curvilinear relationship between job task performance and a specific personality trait: conscientiousness. Using data collected for 602 employees of a large public organization, task performance was measured on a 30-point scale (where higher scores indicate better performance) and conscientiousness was measured on a scale of -3 to $+3$ (where higher scores indicate a higher level of conscientiousness).

a. The coefficient of correlation relating task performance score to conscientiousness score was reported as $r = .18$. Explain why the researchers should not use this statistic to investigate the curvilinear relationship between task performance and conscientiousness.

b. Give the equation of a curvilinear (quadratic) model relating task performance score (y) to conscientiousness score (x).

c. The researchers theorize that task performance will increase as level of conscientiousness increases but at a decreasing rate. Draw a sketch of this relationship.

d. If the theory in part **c** is supported, what is the expected sign of β_2 in the model, part **b**?

e. The researchers reported $\hat{\beta}_2 = -.32$ with an associated *p*-value of less than .05. Use this information to test the researchers' theory at $\alpha = .05$.

12.68 **Childhood obesity study.** The eating patterns of families of overweight preschool children were the subject of an article published in the *Journal of Education and Human Development* (Vol. 3, 2009). A sample of 10 overweight children living in a rural area of the United States was selected. A portion of the research focused on the body mass index of each child and his/her parent. (Body mass index—or BMI—is determined by dividing weight by height squared.) These data are provided in the table on p. 723. The researchers were interested in determining whether parent BMI could be used as a predictor of child BMI for overweight children.

a. For this study, identify the dependent and independent variables.

b. Construct a scatterplot for the data. What trend do you observe?

Data for Exercise 12.68

Child	Parent	Child	Parent
17.10	24.62	17.30	26.30
17.15	24.70	17.32	26.60
17.20	25.70	17.40	26.80
17.24	25.80	17.60	27.20
17.25	26.20	17.80	27.35

Based on Seal, N., and Seal, J. "Eating patterns of the rural families of overweight preschool children: A pilot study." *Journal of Education and Human Development*, Vol. 3, No. 1, 2009 (Figure 1).

c. A quadratic model was fit to the data, with the results shown in the SPSS printout below. Give the least squares prediction equation.

d. Is the overall model statistically useful? Test using $\alpha = .05$.

e. Is there evidence of upward curvature in the relationship between child BMI and parent BMI? Test using $\alpha = .05$.

f. Do you agree with the statement, "for obese children, child BMI increases at an increasing rate as parent BMI increases"?

Model Summary

Model	R	R Square	Adjusted R Square	Std. Error of the Estimate
1	.970[a]	.942	.925	.05887

a. Predictors: (Constant), ParBMIsq, ParentBMI

ANOVA[b]

Model		Sum of Squares	df	Mean Square	F	Sig.
1	Regression	.392	2	.196	56.520	.000[a]
	Residual	.024	7	.003		
	Total	.416	9			

a. Predictors: (Constant), ParBMIsq, ParentBMI
b. Dependent Variable: ChildBMI

Coefficients[a]

Model		Unstandardized Coefficients		Standardized Coefficients	t	Sig.
		B	Std. Error	Beta		
1	(Constant)	89.634	16.528		5.423	.001
	ParentBMI	-5.783	1.275	-25.290	-4.535	.003
	ParBMIsq	.115	.025	26.165	4.692	.002

a. Dependent Variable: ChildBMI

12.69 Going for it on fourth down in the NFL. Refer to the *Chance* (Winter 2009) study of fourth-down decisions by coaches in the National Football League (NFL), Exercise 11.87 (p. 652). Recall that statisticians at California State University, Northridge, fit a straight-line model for predicting the number of points scored (y) by a team that has a first down with a given number of yards (x) from the opposing goal line. A second model fit to data collected on five NFL teams from a recent season was the quadratic regression model, $E(y) = \beta_0 + \beta_1 x + \beta_2 x^2$. The regression yielded the following results: $\hat{y} = 6.13 + .141x - .0009x^2$, $R^2 = .226$.

a. If possible, give a practical interpretation of each of the β-estimates in the model.

b. Give a practical interpretation of the coefficient of determination, R^2.

c. In Exercise 11.87, the coefficient of correlation for the straight-line model was reported as $R^2 = .18$. Does this statistic alone indicate that the quadratic model is a better fit than the straight-line model? Explain.

d. What test of hypothesis would you conduct to determine if the quadratic model is a better fit than the straight-line model?

12.70 Assertiveness and leadership. Management professors at Columbia University examined the relationship between assertiveness and leadership (*Journal of Personality and Social Psychology*, Feb. 2007). The sample comprised 388 people enrolled in a full-time master's in business administration (MBA) program. On the basis of answers to a questionnaire, the researchers measured two variables for each subject: assertiveness score (x) and leadership ability score (y). A quadratic regression model was fit to the data, with the following results:

Independent Variable	β Estimate	t-Value	p-Value
x	.57	2.55	.01
x^2	-.88	-3.97	<.01

Model $R^2 = .12$

a. Conduct a test of overall model utility. Use $\alpha = .05$.

b. The researchers hypothesized that leadership ability will increase at a decreasing rate with assertiveness. Set up the null and alternative hypotheses to test this theory.

c. Use the reported results to conduct the test you set up in part **b**. Give your conclusion (at $\alpha = .05$) in the words of the problem.

12.71 Testing tires for wear. Underinflated or overinflated tires can increase tire wear. A new tire was tested for wear at different pressures, with the results shown in the following table.

Pressure x (pounds per square inch)	Mileage y (thousands)
30	29
31	32
32	36
33	38
34	37
35	33
36	26

a. Plot the data on a scatterplot.

b. If you were given only the information for $x = 30, 31, 32$, and 33, what kind of model would you suggest? For $x = 33, 34, 35$, and 36? For all the data?

12.72 Goal congruence in top management teams. Do chief executive officers (CEOs) and their top managers always agree on the goals of the company? Goal importance congruence between CEOs and vice presidents (VPs) was studied in the *Academy of Management Journal* (Feb. 2008). The researchers used regression to model a VP's attitude toward the goal of improving efficiency (y) as a function of the two quantitative independent variables, level of CEO leadership (x_1) and level of congruence between the CEO and the VP (x_2). A complete second-order model in x_1 and x_2 was fit to data collected for $n = 517$ top management team members at U.S. credit unions.

a. Write the complete second-order model for $E(y)$.

b. The coefficient of determination for the model, part **a**, was reported as $R^2 = .14$. Interpret this value.

c. The estimate of the β-value for the $(x_2)^2$ term in the model was found to be negative. Interpret this result, practically.

d. A t-test on the β-value for the interaction term in the model, x_1x_2, resulted in a p-value of .02. Practically interpret this result, using $\alpha = .05$.

12.73 Estimation of urban population by means of satellite images. Refer to the *Geographical Analysis* (Jan. 2007) study that demonstrated the use of satellite image maps for estimating urban population, presented in Exercise 12.18 (p. 698). A first-order model for census block population density (y) was fit as a function of the proportion of a block with low-density residential areas (x_1) and the proportion of a block with high-density residential areas (x_2). Now consider a complete second-order model for y.

a. Write the equation of the model.

b. Identify the terms in the model that allow for curvilinear relationships.

Applying the Concepts—Intermediate

12.74 Estimating repair and replacement costs of water pipes.
D
H2OPIPE Refer to the *IHS Journal of Hydraulic Engineering* (Sept. 2012) study of the repair and replacement of water pipes, Exercise 11.28 (p. 628). Recall that a team of civil engineers used regression analysis to model y = the ratio of repair to replacement cost of commercial pipe as a function of x = the diameter (in millimeters) of the pipe. Data for a sample of 13 different pipe sizes are reproduced in the accompanying table. In Exercise 11.28, you fit a straight-line model to the data. Now consider the quadratic model $E(y) = \beta_0 + \beta_1 x + \beta_2 x^2$. A MINITAB printout of the analysis follows in the next column.

Diameter	Ratio
80	6.58
100	6.97
125	7.39
150	7.61
200	7.78
250	7.92
300	8.20
350	8.42
400	8.60
450	8.97
500	9.31
600	9.47
700	9.72

Source: Suribabu, C. R., and Neelakantan, T. R. "Sizing of water distribution pipes based on performance measure and breakage-repair replacement economics." *IHS Journal of Hydraulic Engineering*, Vol. 18, No. 3, Sept. 2012 (Table 1).

a. Give the least squares prediction equation relating ratio of repair to replacement cost (y) to pipe diameter (x).

b. Conduct a global F-test for the model using $\alpha = .01$. What do you conclude about overall model adequacy?

c. Evaluate the adjusted coefficient of determination, R_a^2, for the model.

d. Give the null and alternative hypotheses for testing if the rate of increase of ratio (y) with diameter (x) is slower for larger pipe sizes.

e. Carry out the test, part **d**, using $\alpha = .01$.

f. Locate on the printout a 95% prediction interval for the ratio of repair to replacement cost for a pipe with a diameter of 250 millimeters. Interpret the result.

Regression Analysis: RATIO versus DIAMETER, DIAMSQ

```
The regression equation is
RATIO = 6.27 + 0.00791 DIAMETER - 0.000004 DIAMSQ


Predictor        Coef      SE Coef       T       P
Constant       6.2660       0.1554   40.31   0.000
DIAMETER     0.0079145    0.0009974    7.93   0.000
DIAMSQ      -0.00000426   0.00000132   -3.23   0.009


S = 0.162211   R-Sq = 97.7%   R-Sq(adj) = 97.2%


Analysis of Variance

Source            DF       SS       MS       F       P
Regression         2   11.0810   5.5405  210.56   0.000
Residual Error    10    0.2631   0.0263
Total             12   11.3441


Predicted Values for New Observations

New Obs    Fit  SE Fit        95% CI            95% PI
      1  7.9785  0.0580  (7.8494, 8.1077)  (7.5947, 8.3624)


Values of Predictors for New Observations

New Obs  DIAMETER  DIAMSQ
      1       250   62500
```

12.75 Estimating change-point dosage. A standard method for studying toxic substances and their effects on humans is to observe the responses of rodents exposed to various doses of the substance over time. In the *Journal of Agricultural, Biological, and Environmental Statistics* (June 2005), researchers used least squares regression to estimate the "change-point" dosage, defined as the largest dose level that has no adverse effects. Data were obtained from a dose-response study of rats exposed to the toxic substance aconiazide. A sample of 50 rats was evenly divided into five dosage groups: 0, 100, 200, 500, and 750 milligrams per kilogram of body weight. The dependent variable y measured was the weight change (in grams) after a 2-week exposure. The researchers fit the quadratic model $E(y) = \beta_0 + \beta_1 x + \beta_2 x^2$, where x = dosage level, with the following results: $\hat{y} = 10.25 + .0053x - .0000266x^2$.

a. Construct a rough sketch of the least squares prediction equation. Describe the nature of the curvature in the estimated model.

b. Estimate the weight change (y) for a rat given a dosage of 500 mg/kg of aconiazide.

c. Estimate the weight change (y) for a rat given a dosage of 0 mg/kg of aconiazide. (This dosage is called the "control" dosage level.)

d. Of the five groups in the study, find the largest dosage level x that yields an estimated weight change that is closest to, but below, the estimated weight change for the control group. This value is the change-point dosage.

12.76 Shopping on Black Friday. Refer to the *International Journal of Retail and Distribution Management* (Vol. 39, 2011) study of shopping on Black Friday (the day after Thanksgiving), Exercise 7.22 (p. 353). Recall that researchers conducted interviews with a sample of 38 women shopping on Black Friday to gauge their shopping habits. Two of the variables measured for each shopper were age (x) and number of years shopping on Black Friday (y). Data on these two variables for the 38 shoppers are listed in the accompanying table.

a. Fit the quadratic model, $E(y) = \beta_0 + \beta_1 x + \beta_2 x^2$, to the data using statistical software. Give the prediction equation.

b. Conduct a test of the overall adequacy of the model. Use $\alpha = .01$.

c. Conduct a test to determine if the relationship between age (x) and number of years shopping on Black Friday (y) is best represented by a linear or quadratic function. Use $\alpha = .01$.

Age	Years	Age	Years
32	5	21	5
27	3	52	10
40	12	40	18
62	35	38	5
47	20	56	8
53	30	60	5
24	8	35	15
27	2	50	25
47	24	56	10
40	25	20	2
45	11	20	4
22	11	21	4
25	5	22	5
60	35	50	10
22	3	30	6
50	15	28	16
70	22	25	7
50	10	30	6
21	6	49	30

Source: Thomas, J. B., and Peters, C. "An exploratory investigation of Black Friday consumption rituals." *International Journal of Retail and Distribution Management,* Vol. 39, No. 7, 2011 (Table I).

12.77 Satisfaction with membership in a new religious movement. How satisfied are people who have recently joined a new religious movement? To answer this question, German researchers collected data for a sample of 58 believers who had recently joined a new religious group (*Applied Psychology: An International Review*, Apr. 2010). The dependent variable of interest was *satisfaction level* (y), measured quantitatively on an 11-point scale (where $0 =$ totally dissatisfied and $10 =$ totally satisfied). Two independent variables were used to predict satisfaction level: *Needs* (x_1)—a measure of the level of needs one requires in a religion, and *Supplies* (x_2)—a measure of the level of supplies provided by the religion. In theory, if the level of needs matches the level of supplies, one will be highly satisfied with the religion.

a. The researchers fitted a complete second-order model for $E(y)$ as a function of x_1 and x_2. Write the equation of this model.

b. The regression results are reported in the table below. Interpret the value of R^2.

c. Use the R^2 statistic to conduct a test of overall model adequacy. Test using $\alpha = .10$.

d. Conduct a test to determine whether needs (x_1) is curvilinearly related to satisfaction (y). Test using $\alpha - .10$.

e. Conduct a test to determine whether supplies (x_2) is curvilinearly related to satisfaction (y). Test using $\alpha = .10$.

Independent Variable	Estimated Beta	Standard Error
Needs (x_1)	.952	.780
Supplies (x_2)	1.198	.766
Needs × Supplies $(x_1 x_2)$	.356	.429
Needs × Needs (x_1^2)	−.181	.302
Supplies × Supplies (x_2^2)	−.755	.413

$R^2 = .402$

12.78 Failure times of silicon wafer microchips. Researchers at National Semiconductor experimented with tin-lead solder bumps used to manufacture silicon wafer integrated circuit chips (International Wafer Level Packaging Conference, Nov. 3–4, 2005). The failure times of the microchips (in hours) were determined at different solder temperatures (degrees Celsius). The data for one experiment are given in the following table. The researchers want to predict failure time (y) based on solder temperature (x).

a. Construct a scatterplot for the data. What type of relationship, linear or curvilinear, appears to exist between failure time and solder temperature?

b. Fit the model, $E(y) = \beta_0 + \beta_1 x + \beta_2 x^2$, to the data. Give the least squares prediction equation.

c. Conduct a test to determine if there is upward curvature in the relationship between failure time and solder temperature. (Use $\alpha = .05$.)

Temperature (°C)	Time to Failure (hours)
165	200
162	200
164	1,200
158	500
158	600
159	750
156	1,200
157	1,500
152	500
147	500
149	1,100
149	1,150
142	3,500
142	3,600
143	3,650
133	4,200
132	4,800
132	5,000
134	5,200
134	5,400
125	8,300
123	9,700

Based on Gee, S., and Nguyen, L. "Mean time to failure in wafer level–CSP packages with SnPb and SnAgCu solder bmps." International Wafer Level Packaging Conference, San Jose, CA, Nov. 3–4, 2005 (adapted from Figure 7).

12.79 Public perceptions of health risks. In the *Journal of Experimental Psychology: Learning, Memory, and Cognition* (July 2005), University of Basel (Switzerland) psychologists tested the ability of people to judge the risk of an infectious disease. The researchers asked German college students to estimate the number of people who are infected with a certain disease in a typical year. The median estimates, as well as the actual incidence of the disease for each in a sample of 24 infections, are listed in the table. Consider the quadratic model $E(y) = \beta_0 + \beta_1 x + \beta_2 x^2$, where y = actual incidence rate and x = estimated rate.

Infection	Actual Incidence	Estimate
Polio	0.25	300
Diphtheria	1	1,000
Trachoma	1.75	691
Rabbit Fever	2	200
Cholera	3	17.5
Leprosy	5	0.8
Tetanus	9	1,000
Hemorrhagic Fever	10	150
Trichinosis	22	326.5
Undulant Fever	23	146.5
Well's Disease	39	370
Gas Gangrene	98	400
Parrot Fever	119	225
Typhoid	152	200
Q Fever	179	200
Malaria	936	400
Syphilis	1,514	1,500
Dysentery	1,627	1,000
Gonorrhea	2,926	6,000
Meningitis	4,019	5,000
Tuberculosis	12,619	1,500
Hepatitis	14,889	10,000
Gastroenteritis	203,864	37,000
Botulism	15	37,500

Based on Hertwig, R., Pachur, T., and Kurzenhauser, S. "Judgments of risk frequencies: Tests of possible cognitive mechanisms." *Journal of Experimental Psychology: Learning, Memory, and Cognition*, Vol. 31, No. 4, July 2005 (Table 1).

a. Fit the quadratic model to the data, and then conduct a test to determine whether the actual incidence is curvilinearly related to the estimated incidence. (Use $\alpha = .05$.)

b. Construct a scatterplot of the data. Locate the data point for botulism on the graph. What do you observe?

c. Repeat part **a**, but omit the data point for botulism from the analysis. Has the fit of the model improved? Explain.

Applying the Concepts—Advanced

12.80 Tree frog study. The optomotor responses of tree frogs were studied in the *Journal of Experimental Zoology* (Sept. 1993). Microspectrophotometry was used to measure the threshold quantal flux (the light intensity at which the optomotor response was first observed) of tree frogs tested at different spectral wavelengths. The data revealed the relationship between the logarithm of quantal flux (y) and wavelength (x), shown in the following graph.

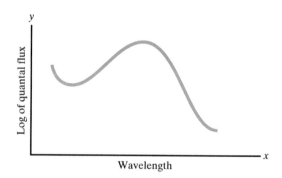

a. Explain why a first-order model would not be appropriate for modeling $E(y)$.

b. Explain why a second-order model would not be appropriate for modeling $E(y)$.

c. Demonstrate that the third-order model $E(y) = \beta_0 + \beta_1 x + \beta_2 x^2 + \beta_3 x^3$ may be the most appropriate model for $E(y)$.

12.7 Qualitative (Dummy) Variable Models

Multiple-regression models can also be written to include **qualitative** (or **categorical**) independent variables. **Qualitative variables**, unlike quantitative variables, cannot be measured on a numerical scale. Therefore, we must code the values of the qualitative variable (called **levels**) as numbers before we can fit the model. These coded qualitative variables are called **dummy** (or **indicator**) **variables**, since the numbers assigned to the various levels are arbitrarily selected.

To illustrate, suppose a female executive at a certain company claims that male executives earn higher salaries, on average, than female executives with the same education, experience, and responsibilities. To support her claim, she wants to model the salary y of an executive, using a qualitative independent variable representing the gender of the executive (male or female).

A convenient method of coding the values of a qualitative variable at two levels involves assigning a value of 1 to one of the levels and a value of 0 to the other. For example, the **dummy variable** used to describe gender could be coded as follows:

$$x = \begin{cases} 1 & \text{if male} \\ 0 & \text{if female} \end{cases}$$

The choice of which level is assigned to 1 and which is assigned to 0 is arbitrary. The model then takes the following form:

$$E(y) = \beta_0 + \beta_1 x$$

The advantage of using a 0–1 coding scheme is that the β coefficients are easily interpreted. The foregoing model allows us to compare the mean executive salary $E(y)$ for males with the corresponding mean for females:

$$\text{Males } (x = 1): \quad E(y) = \beta_0 + \beta_1(1) = \beta_0 + \beta_1$$
$$\text{Females } (x = 0): \quad E(y) = \beta_0 + \beta_1(0) = \beta_0$$

These two means are illustrated in the bar graph in Figure 12.18.

First note that β_0 represents the mean salary for females (say, μ_F). When a 0–1 coding convention is used, β_0 will always represent the mean response associated with the level of the qualitative variable assigned the value 0 (called the **base level**). The difference between the mean salary for males and the mean salary for females, $\mu_M - \mu_F$, is represented by β_1; that is,

$$\mu_M - \mu_F = (\beta_0 + \beta_1) - (\beta_0) = \beta_1$$

This difference is shown in Figure 12.18.* With a 0–1 coding convention, β_1 will always represent the difference between the mean response for the level assigned the value 1 and the mean for the base level. Thus, for the executive salary model, we have

$$\beta_0 = \mu_F$$
$$\beta_1 = \mu_M - \mu_F$$

The model relating a mean response $E(y)$ to a qualitative independent variable at two levels is summarized in the following box.

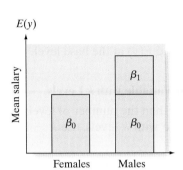

Figure 12.18

Bar chart comparing $E(y)$ for males and females

A Model Relating $E(y)$ to a Qualitative Independent Variable with Two Levels

$$E(y) = \beta_0 + \beta_1 x$$

where

$$x = \begin{cases} 1 & \text{if level A} \\ 0 & \text{if level B} \quad \text{(base level)} \end{cases}$$

Interpretation of β's:

$\beta_0 = \mu_B$ (mean for base level)

$\beta_1 = \mu_A - \mu_B$ (mean for level assigned "1" minus mean for base level)

Note: μ_i represents the value of $E(y)$ for level i.

Now carefully examine the model with a single qualitative independent variable at two levels because we will use exactly the same pattern for any number of levels. Moreover, the interpretation of the parameters will always be the same.

One level (say, level A) is selected as the **base level**. Then, for the 0–1 coding[†] for the dummy variables,

$$\mu_A = \beta_0$$

---

*Note that β_1 could be negative. If β_1 were negative, the height of the bar corresponding to males would be reduced (rather than increased) from the height of the bar for females by the amount β_1. Figure 12.18 is constructed under the assumption that β_1 is a positive quantity.

[†]You do not have to use a 0–1 system of coding for the dummy variables. *Any* two-value system will work, but the interpretation given to the model parameters will depend on the code. Using the 0–1 system makes the model parameters easy to interpret.

The coding for all dummy variables is as follows: To represent the mean value of y for a particular level, let that dummy variable equal 1; otherwise, the dummy variable is set equal to 0. Using this system of coding, we obtain

$$\mu_B = \beta_0 + \beta_1$$
$$\mu_C = \beta_0 + \beta_2$$

and so on. Because $\mu_A = \beta_0$, any other model parameter will represent the difference between means for that level and the base level; that is,

$$\beta_1 = \mu_B - \mu_A$$
$$\beta_2 = \mu_C - \mu_A$$

and so on. Consequently, each β multiplied by a dummy variable represents the difference between $E(y)$ at one level of the qualitative variable and $E(y)$ at the base level.

A Model Relating $E(y)$ to One Qualitative Independent Variable with k Levels

Always use a number of dummy variables that is one less than the number of levels of the qualitative variable. Thus, for a qualitative variable with k levels, use $k - 1$ dummy variables:

$$E(y) = \beta_0 + \beta_1 x_1 + \beta_2 x_2 + \cdots + \beta_{k-1} x_{k-1}$$

where x_i is the dummy variable for level $i + 1$ and

$$x_i = \begin{cases} 1 & \text{if } y \text{ is observed at level } i + 1 \\ 0 & \text{otherwise} \end{cases}$$

Then, for this system of coding,

$$\mu_1 = \beta_0$$
$$\mu_2 = \beta_0 + \beta_1$$
$$\mu_3 = \beta_0 + \beta_2$$
$$\mu_4 = \beta_0 + \beta_3$$
$$\vdots \quad \vdots$$
$$\mu_k = \beta_0 + \beta_{k-1}$$

and

$$\beta_1 = \mu_2 - \mu_1$$
$$\beta_2 = \mu_3 - \mu_1$$
$$\beta_3 = \mu_4 - \mu_1$$
$$\vdots \quad \vdots$$
$$\beta_{k-1} = \mu_k - \mu_1$$

Note: μ_i represents the value of $E(y)$ for level i.

Example 12.9 Ⓓ
GOLFCRD

A Model with One Qualitative Independent Variable: Golf Ball Driving Distances

Problem Refer to Example 10.4 (p. 544). Recall that the USGA wants to compare the mean driving distances of four different golf ball brands (A, B, C, and D). Iron Byron, the USGA's robotic golfer, is used to hit a sample of 10 balls of each brand. The distance data are reproduced in Table 12.3.

a. Hypothesize a regression model for driving distance y, using Brand as an independent variable.

b. Interpret the β's in the model.

c. Fit the model to the data and give the least squares prediction equation. Show that the β-estimates can also be obtained from the sample means.

d. Use the model to determine whether the mean driving distances for the four brands are significantly different at $\alpha = .05$.

Solution

a. Note that golf ball brand (A, B, C, and D) is a qualitative variable (measured on a nominal scale). According to the previous box, for a four-level qualitative variable we require three dummy variables in the regression model. The model relating $E(y)$, where y is the distance the ball is driven by Iron Byron, to this single qualitative variable, golf ball Brand, is

$$E(y) = \beta_0 + \beta_1 x_1 + \beta_2 x_2 + \beta_3 x_3$$

Table 12.3	Driving Distances (in feet) for Four Golf Ball Brands		
Brand A	Brand B	Brand C	Brand D
251.2	263.2	269.7	251.6
245.1	262.9	263.2	248.6
248.0	265.0	277.5	249.4
251.1	254.5	267.4	242.0
260.5	264.3	270.5	246.5
250.0	257.0	265.5	251.3
253.9	262.8	270.7	261.8
244.6	264.4	272.9	249.0
254.6	260.6	275.6	247.1
248.8	255.9	266.5	245.9

D *Data Set:* GOLFCRD

where

$$x_1 = \begin{cases} 1 & \text{if Brand B} \\ 0 & \text{if not} \end{cases} \quad x_2 = \begin{cases} 1 & \text{if Brand C} \\ 0 & \text{if not} \end{cases} \quad x_3 = \begin{cases} 1 & \text{if Brand D} \\ 0 & \text{if not} \end{cases}$$

We arbitrarily choose Brand A to be the base level.

b. Since Brand A is the base level, β_0 represents the mean driving distance for Brand A (i.e., $\beta_0 = \mu_A$). The other β's are differences in means, namely,

$$\beta_1 = \mu_B - \mu_A$$
$$\beta_2 = \mu_C - \mu_A$$
$$\beta_3 = \mu_D - \mu_A$$

where μ_A, μ_B, μ_C, and μ_D are the mean distances for Brands A, B, C, and D, respectively.

c. The MINITAB printout of the regression analysis is shown in Figure 12.19. The least squares estimates of the β's are highlighted on the printout, yielding the following least squares prediction equation:

$$\hat{y} = 250.78 + 10.28x_1 + 19.17x_2 - 1.46x_3$$

Regression Analysis: DISTANCE versus X1, X2, X3

```
The regression equation is
DISTANCE = 251 + 10.3 X1 + 19.2 X2 - 1.46 X3

Predictor     Coef  SE Coef        T      P
Constant   250.780    1.455   172.34  0.000
X1          10.280    2.058     5.00  0.000
X2          19.170    2.058     9.32  0.000
X3          -1.460    2.058    -0.71  0.483

S = 4.60163   R-Sq = 78.6%   R-Sq(adj) = 76.8%

Analysis of Variance

Source          DF       SS      MS      F      P
Regression       3  2794.39  931.46  43.99  0.000
Residual Error  36   762.30   21.18
Total           39  3556.69
```

Descriptive Statistics: DISTANCE

```
Variable  BRAND   N    Mean  StDev  Minimum  Maximum
DISTANCE    A    10  250.78   4.74   244.60   260.50
            B    10  261.06   3.87   254.50   265.00
            C    10  269.95   4.50   263.20   277.50
            D    10  249.32   5.20   242.00   261.80
```

Figure 12.19

MINITAB output for model with dummy variables

The interpretations of the β's in part **b** allow us to obtain the β estimates from the sample means associated with the different levels of the qualitative variable.*

Since $\beta_0 = \mu_A$, then the estimate of β_0 is the estimated mean driving distance for Brand A (base level). This sample mean, highlighted at the bottom of the MINITAB printout, is 250.78; thus, $\hat{\beta}_0 = 250.78$.

Now $\beta_1 = \mu_B - \mu_A$; therefore, the estimate of β_1 is the difference between the sample mean driving distances for Brands B and A. Based on the sample means highlighted at the bottom of the MINITAB printout, we have $\hat{\beta}_1 = 261.06 - 250.78 = 10.28$.

Similarly, the estimate of $\beta_2 = \mu_C - \mu_A$ is the difference between the sample mean distances for Brands C and A. From the sample means highlighted at the bottom of the MINITAB printout, we have $\hat{\beta}_2 = 269.95 - 250.78 = 19.17$.

Finally, the estimate of $\beta_3 = \mu_D - \mu_A$ is the difference between the sample mean distances for Brands D and A. Using the sample means highlighted at the bottom of the MINITAB printout, we have $\hat{\beta}_3 = 249.32 - 250.78 = -1.46$.

d. Testing the null hypothesis that the means for the four brands are equal (i.e., $\mu_A = \mu_B = \mu_C = \mu_D$) is equivalent to testing

$$H_0: \beta_1 = \beta_2 = \beta_3 = 0$$

You can see this by observing that if $\beta_1 = \mu_B - \mu_A = 0$, then $\mu_A = \mu_B$. Similarly, $\beta_2 = \mu_C - \mu_A = 0$ implies that $\mu_A = \mu_C$, and $\beta_3 = \mu_D - \mu_A = 0$ implies that $\mu_A = \mu_D$. The alternative hypothesis is

$$H_a: \text{At least one of the parameters } \beta_1, \beta_2, \text{ and } \beta_3 \text{ differs from 0}$$

which implies that at least two of the four means ($\mu_A, \mu_B, \mu_C,$ and μ_D) differ.

To test this hypothesis, we conduct the global F-test on the model. The value of the F-statistic for testing the adequacy of the model, $F = 43.99$, and the observed significance level of the test, $p = .000$, are both highlighted in Figure 12.19. Since $\alpha = .05$ exceeds the p-value, we reject H_0 and conclude that at least one of the parameters differs from 0. Or, equivalently, we conclude that the data provide sufficient evidence to indicate that the mean driving distance does vary from one golf ball brand to another.

Look Back This global F-test is equivalent to the analysis-of-variance F-test in Chapter 10 for a completely randomized design.

■ **Now Work Exercise 12.84**

⚠ **CAUTION** A common mistake by regression analysts is the use of a single dummy variable x for a qualitative variable at k levels, where $x = 1, 2, 3, \ldots, k$. Such a regression model will have unestimable β's and β's that are difficult to interpret. Remember, in modeling $E(y)$ with a single qualitative independent variable, the number of 0–1 dummy variables to include in the model will always be one less than the number of levels of the qualitative variable. ▲

Exercises 12.81–12.99

Understanding the Principles

12.81 Write a regression model relating the mean value of y to a qualitative independent variable that can assume two levels. Interpret all the terms in the model.

12.82 Write a regression model relating $E(y)$ to a qualitative independent variable that can assume three levels. Interpret all the terms in the model.

Learning the Mechanics

12.83 The model $E(y) = \beta_0 + \beta_1 x_1 + \beta_2 x_2 + \beta_3 x_3$, where

$$x_1 = \begin{cases} 1 & \text{if level 2} \\ 0 & \text{if not} \end{cases}$$

$$x_2 = \begin{cases} 1 & \text{if level 3} \\ 0 & \text{if not} \end{cases}$$

$$x_3 = \begin{cases} 1 & \text{if level 4} \\ 0 & \text{if not} \end{cases}$$

---

*The least squares method and the sample means method will yield equivalent β estimates when the sample sizes associated with the different levels of the qualitative variable are equal.

was used to relate $E(y)$ to a single qualitative variable with four levels. This model was fit to $n = 45$ data points, and the result was and the following result was obtained:

$$\hat{y} = 16.3 - 7x_1 + 15x_2 + 3x_3$$

a. Use the least squares prediction equation to find the estimate of $E(y)$ for each level of the qualitative independent variable.

b. Specify the null and alternative hypotheses you would use to test whether $E(y)$ is the same for all four levels of the independent variable.

12.84 MINITAB was used to fit the model

NW

$$y = \beta_0 + \beta_1 x_1 + \beta_2 x_2 + \varepsilon$$

where

$$x_1 = \begin{cases} 1 & \text{if level 2} \\ 0 & \text{if not} \end{cases}$$

$$x_2 = \begin{cases} 1 & \text{if level 3} \\ 0 & \text{if not} \end{cases}$$

to $n = 15$ data points. The results are shown in the accompanying MINITAB printout.

```
The regression equation is
Y = 80.0 + 16.8 X1 + 40.4 X2


Predictor    Coef   SE Coef     T      P
Constant   80.000     4.082  19.60  0.000
X1         16.800     5.774   2.91  0.013
X2         40.400     5.774   7.00  0.000


S = 9.129    R-Sq = 80.5%   R-Sq(adj) = 77.2%

Analysis of Variance

Source          DF      SS      MS      F      P
Regression       2  4118.9  2059.5  24.72  0.000
Residual Error  12  1000.0    83.3
Total           14  5118.9
```

a. Report the least squares prediction equation.

b. Interpret the values of β_1 and β_2.

c. Interpret the following hypotheses in terms of $\mu_1, \mu_2,$ and μ_3:

H_0: $\beta_1 = \beta_2 = 0$

H_a: At least one of the parameters β_1 and β_2 differs from 0

d. Conduct the hypothesis test of part c.

Applying the Concepts—Basic

12.85 Modeling number of children. As part of a study on how children affect parents' depression in old age, researchers fit a model for total number of children in a family (*Social Science & Medicine*, Jan. 2014). One of the key independent variables used in the model is whether the first birth was a single birth or a multiple (twin, triplet) birth.

a. Create a dummy variable for type of first birth.

b. Write a model for total number of children (y) as a function of type of first birth.

c. Give a practical interpretation of each of the β-coefficients in the model.

12.86 Production technologies, terroir, and quality of Bordeaux wine. The production of quality wine is strongly influenced by the natural endowments of the grape-growing region—called the "terroir." *The Economic Journal* (May 2008) published an empirical study of the factors that yield a quality Bordeaux winc. A quantitative measure of wine quality (y) was modeled as a function of several qualitative independent variables, including grape-picking method (manual or automated), soil type (clay, gravel, or sand), and slope orientation (east, south, west, southeast, or southwest).

a. Create the appropriate dummy variables for each of the qualitative independent variables.

b. Write a model for wine quality (y) as a function of grape-picking method. Interpret the β's in the model.

c. Write a model for wine quality (y) as a function of soil type. Interpret the β's in the model.

d. Write a model for wine quality (y) as a function of slope orientation. Interpret the β's in the model.

12.87 Whales entangled in fishing gear. Refer to the *Marine Mammal Science* (Apr. 2010) study of whales entangled in fishing gear, Exercise 12.17 (p. 698). These entanglements involved one of three types of fishing gear: set nets, pots, and gill nets. Consequently, the researchers used gear type as a predictor of the body length (y, in meters) of the entangled whale. Consider the regression model $E(y) = \beta_0 + \beta_1 x_1 + \beta_2 x_2$, where $x_1 = \{1$ if set net, 0 if not$\}$ *and* $x_2 = \{1$ if pots, 0 if not$\}$. [*Note*: Gill nets is the "base" level of gear type.]

a. The researchers want to know the mean body length of whales entangled in gill nets. Give an expression for this value in terms of the β's in the model.

b. Practically interpret the value of β_1 in the model.

c. In terms of the β's in the model, how would you test to determine if the mean body lengths of entangled whales differ for the three types of fishing gear?

12.88 Do blondes raise more funds? During fundraising, does the physical appearance of the solicitor affect the level of capital raised? An economist at the University of Nevada–Reno designed an experiment to answer this question and published the results in *Economic Letters* (Vol. 100, 2008). Each in a sample of 955 households was contacted by a female solicitor and asked to contribute to the Center for Natural Hazards Mitigation Research. The level of contribution (in dollars) was recorded as well as the hair color of the solicitor (blonde Caucasian, brunette Caucasian, or minority female).

a. Consider a model for the mean level of contribution, $E(y)$, that allows for different means depending on the hair color of the solicitor. Create the appropriate number of dummy variables for hair color. (Use minority female as the base level.)

b. Write the equation of the model, part a, incorporating the dummy variables.

c. In terms of the β's in the model, what is the mean level of contribution for households contacted by a blonde Caucasian solicitor?

d. In terms of the β's in the model, what is the difference between the mean level of contribution for households contacted by a blonde solicitor and those contacted by a minority female?

e. One theory posits that blonde solicitors will achieve the highest mean contribution level but there will be no difference between the mean contribution levels attained by brunette Caucasian and minority females. If this theory is true, give the expected signs of the β's in the model.

f. The researcher found the β estimate for the dummy variable for blonde Caucasian to be positive and significantly different from 0 (p-value $< .01$). The β estimate for the dummy variable for brunette Caucasian was also positive but not significantly different from 0 (p-value $> .10$). Do these results support the theory, part **e**?

12.89 Impact of race on football card values. University of Colorado sociologists investigated the impact of race on the value of professional football players' "rookie" cards (*Electronic Journal of Sociology*, 2007). The sample consisted of 148 rookie cards of National Football League (NFL) players who were inducted into the Football Hall of Fame. The price of a card (in dollars) was modeled as a function of several qualitative independent variables: race of player (black or white), availability of the card (high or low), and position of the player (quarterback, running back, wide receiver, tight end, defensive lineman, linebacker, defensive back, or offensive lineman).

a. Create the appropriate dummy variables for each of the qualitative independent variables.

b. Write a model for price (y) as a function of race. Interpret the β's in the model.

c. Write a model for price (y) as a function of the availability of the card. Interpret the β's in the model.

d. Write a model for price (y) as a function of the player's position. Interpret the β's in the model.

12.90 Chemical composition of rainwater. Researchers at the University of Aberdeen (Scotland) developed a statistical model for estimating the chemical composition of water (*Journal of Agricultural, Biological, and Environmental Statistics*, Mar. 2005). For one application, the nitrate concentration y (milligrams per liter) in a water sample collected after a heavy rainfall was modeled as a function of water source (groundwater, subsurface flow, or overground flow).

a. Write a model for $E(y)$ as a function of the qualitative independent variable.

b. Give an interpretation of each of the β parameters in the model you wrote in part **a**.

12.91 Detecting quantitative traits in genes. In gene therapy, it is important to know the location of a gene for a disease on the genome (genetic map). Although many genes yield a specific trait (e.g., disease or not), others cannot be categorized, since they are quantitative in nature (e.g., extent of disease). Researchers at the University of North Carolina at Wilmington developed statistical models that link quantitative genetic traits to locations on the genome (*Chance*, Summer 2006). The extent of a certain disease is determined by the absence (A) or presence (B) of a gene marker at each of two locations, L1 and L2, on the genome. For example, AA represents absence of the marker at both locations, while AB represents absence at location L1, but presence at location L2.

a. How many different gene marker combinations are possible at the two locations?

b. Using dummy variables, write a model for extent of the disease, y, as a function of gene marker combination.

c. Interpret the β-values in the model you wrote in part **b.**

d. Give the null hypothesis for testing whether the overall model from part **b** is statistically useful for predicting extent of the disease, y.

Applying the Concepts—Intermediate

12.92 Ascorbic acid reduces goat stress. Refer to the *Animal Science Journal* (May 2014) study on the use of ascorbic acid (AA) to reduce the stress in goats during transportation, Exercise 10.7 (p. 537). Recall that 24 healthy goats were randomly divided into four groups (A, B, C, and D) of six animals each. Goats in group A were administered a dosage of AA 30 minutes prior to transportation; goats in group B were administered a dosage of AA 30 minutes following transportation; group C goats were not given any AA prior to or following transportation; and goats in group D were not given any AA and were not transported. Weight was measured before and after transportation and the weight loss (in kilograms) determined for each goat.

a. Write a model for mean weight loss, $E(y)$, as a function of AA dosage group (A, B, C, or D). Use group D as the base level.

b. Interpret the β's in the model, part **a.**

c. Recall that the researchers discovered that mean weight loss is reduced in goats administered AA compared with goats not given any AA. Based on this result, determine the sign (positive or negative) of as many of the β's in the model, part **a**, as possible.

12.93 Major depression and personality disorders. A team of physicians, psychiatrists, and psychologists investigated whether depressed patients exhibit more or fewer personality disorder symptoms than nondepressed patients in the *American Journal of Psychiatry* (May 2010). A study group of over 400 psychiatric patients was monitored over a six-year period. At the start of the study, each was diagnosed as having (1) major depression only, (2) personality disorder only, or (3) both major depression and personality disorder. Of interest to the researchers was the number of personality disorder criteria met at the end of the study. Consider a regression model for the number of personality disorders (y).

a. Write a model for $E(y)$ as a function of the qualitative variable, patient diagnosis group.

b. If there are no differences among the mean number of personality disorders for the three patient groups, what are the values of the β's in the model, part **a**?

c. How could you test to determine if the mean number of personality disorders for the major depression–only patients is less than the corresponding mean for the patients with both major depression and personality disorder?

12.94 Study of recall of TV commercials. Refer to the *Journal of Applied Psychology* (June 2002) study of recall of television commercials, presented in Exercise 10.37 (p. 553). Participants were assigned to watch one of three types of TV programs, with nine commercials embedded in each show. Group V watched a TV program with a violent-content code rating, Group S viewed a show with a sex-content code rating, and Group N watched a neutral TV program with neither a V nor an S rating. The dependent variable measured for each participant was the score (y) on his/her recall of the brand names mentioned in the commercial messages, with scores ranging from 0 (no brands recalled) to 9 (all brands recalled).

a. Write a model for $E(y)$ as a function of viewer group.

b. Fit the model you wrote in part **a** to the data. Give the least squares prediction equation.

c. Conduct a test of overall model utility at $\alpha = .01$. Interpret the results. Show that the results agree with the analysis performed in Exercise 10.37.

d. The sample mean recall scores for the three groups were $\bar{y}_V = 2.08$, $\bar{y}_S = 1.71$, and $\bar{y}_N = 3.17$. Show how to find these sample means by using only the β-estimates obtained in part **b**.

12.95 Expert testimony in homicide trials of battered women. An article published in the *Duke Journal of Gender Law & Policy* (Summer 2003) examined the impact of expert testimony on the outcome of homicide trials that involve battered woman syndrome. On the basis of data collected on individual juror votes from past trials, the article reported that "when expert testimony was present, women jurors were more likely than men to change a verdict from not guilty to guilty after deliberations." Assume that when no expert testimony was present, male jurors were more likely than women to change a verdict from not guilty to guilty after deliberations. These results were obtained from a multiple-regression model for likelihood of changing a verdict from not guilty to guilty after deliberations, y, as a function of juror gender (male or female) and expert testimony (yes or no). Give the model for $E(y)$ that hypothesizes the relationships reported in the article. Illustrate the model with a sketch.

12.96 Homework assistance for college students. Do college professors who provide their students with assistance on homework help improve student grades? This was the research question of interest in the *Journal of Accounting Education* (Vol. 25, 2007). A sample of 175 accounting students took a pretest on a topic not covered in class, then each was given a homework problem to solve on the same topic. The students were assigned to one of three homework assistance groups. Some students received the completed solution, some were given check figures at various steps of the solution, and some received no help at all. After finishing the homework, the students were all given a posttest on the subject. The dependent variable of interest was the knowledge gain (or test score improvement). These data are saved in the **ACCHW** file.

a. Propose a model for the knowledge gain (y) as a function of the qualitative variable, homework assistance group.

b. In terms of the β's in the model, give an expression for the difference between the mean knowledge gains of students in the "completed solution" and "no help" groups.

c. Fit the model to the data and give the least squares prediction equation.

d. Conduct the global F-test for model utility using $\alpha = .05$. Interpret the results practically.

12.97 Extinct New Zealand birds. Refer to the *Evolutionary Ecology Research* (July 2003) study of the patterns of extinction in the New Zealand bird population, presented in Exercise 2.24 (p. 70). Recall that the **NZBIRDS** file contains qualitative data on flight capability (volant or flightless), habitat (aquatic, ground terrestrial, or aerial terrestrial), nesting site (ground, cavity within ground, tree, or cavity above ground), nest density (high or low), diet (fish, vertebrates, vegetables, or invertebrates), and extinct status (extinct, absent from island, present), and quantitative data on body mass (grams) and egg length (millimeters) for 132 bird species at the time of the Maori colonization of New Zealand.

a. Write a model for mean body mass as a function of flight capability.

b. Write a model for mean body mass as a function of diet.

c. Write a model for mean egg length as a function of nesting site.

d. Fit the model you wrote in part **a** to the data and interpret the estimates of the β's.

e. Conduct a test to determine whether the model from part **a** is statistically useful (at $\alpha = .01$) for estimating mean body mass.

f. Fit the model you wrote in part **b** to the data and interpret the estimates of the β's.

g. Conduct a test to determine whether the model from part **b** is statistically useful (at $\alpha = .01$) for estimating mean body mass.

h. Fit the model you wrote in part **c** to the data and interpret the estimates of the β's.

i. Conduct a test to determine whether the model from part **c** is statistically useful (at $\alpha = .01$) for estimating mean egg length.

12.98 Permeability of sandstone during weathering. Refer to the *Geographical Analysis* (Vol. 42, 2010) study of the permeability of sandstone exposed to the weather, Exercise 2.69 (p. 91). Recall that blocks of sandstone were cut into 300 equal-sized slices and the slices randomly divided into three groups of 100 slices each. Slices in group A were not exposed to any type of weathering; slices in group B were repeatedly sprayed with a 10% salt solution (to simulate wetting by driven rain) under temperate conditions; and slices in group C were soaked in a 10% salt solution and then dried (to simulate blocks of sandstone exposed during a wet winter and dried during a hot summer). All sandstone slices were then tested for permeability, a measure of pressure decay in milliDarcies (mD). The data for the study (simulated) are saved in the **SAND** file.

a. Write a model for mean permeability, $E(y)$, as a function of sandstone group (A, B, or C).

b. Measures of central tendency for the permeability measurements of each sandstone group are displayed in the MINITAB printout below. Use this information to estimate the β parameters of the model, part **a**.

MINITAB Output for Exercise 12.98

Descriptive Statistics: PermA, PermB, PermC

Variable	N	Mean	Median	Mode	N for Mode
PermA	100	73.62	70.45	59.9, 60, 60.1, 60.4	2
PermB	100	128.54	139.30	146.4, 146.6, 147.9, 148.3	3
PermC	100	83.07	78.65	70.9	3

The data contain at least five mode values. Only the smallest four are shown.

c. Fit the model, part **a**, to the data in the **SAND** file. Use the output to verify your β estimates in part **b**.

Applying the Concepts—Advanced

12.99 Community responses to a violent crime. How communities respond to a disaster or a violent crime was the subject of research published in the *American Journal of Community Psychology* (Vol. 44, 2009). Psychologists at the University of California tracked monthly violent crime incidents in two Texas cities, Jasper and Center, before and after the murder of a Jasper citizen that had racial overtones and heavy media coverage. (Center, Texas, was selected as comparison city since it had roughly the same population and racial makeup as Jasper.) Using monthly data on violent crimes, the researchers fit the regression model:

$$E(y) = \beta_0 + \beta_1 x_1 + \beta_2 x_2 + \beta_3 x_1 x_2,$$

where y = violent crime rate (number of crimes per 1,000 population), $x_1 = \{1$ if Jasper, 0 if Center$\}$, and $x_2 = \{1$ if after the murder, 0 if before the murder$\}$.

a. In terms of the β's in the model, what is the mean violent crime rate in Center, Texas, for months following the murder?

b. In terms of the β's in the model, what is the mean violent crime rate in Jasper, Texas, for months following the murder?

c. For months following the murder, find the difference between the mean violent crime rate for Jasper and Center. (Use your answers to parts **a** and **b**.)

d. Repeat part **c** for months before the murder.

e. Note that the differences, parts **c** and **d**, are not the same. Explain why this illustrates the notion of interaction between x_1 and x_2.

f. A test for $H_0: \beta_3 = 0$ yielded a p-value $< .001$. Using $\alpha = .01$, interpret this result.

g. The regression resulted in the following β-estimates: $\hat{\beta}_1 = -429$, $\hat{\beta}_2 = -169$, $\hat{\beta}_3 = 255$. Use these estimates to illustrate that average monthly violent crime decreased in Center after the murder but increased in Jasper.

12.8   Models with Both Quantitative and Qualitative Variables (Optional)

Suppose you want to relate the mean monthly sales $E(y)$ of a company to the monthly advertising expenditure x for three different advertising media (say, newspaper, radio, and television) and you wish to use first-order (straight-line) models to model the responses for all three media. Graphs of these three relationships might appear as shown in Figure 12.20.

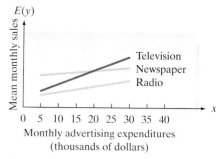

Since the lines in Figure 12.20 are hypothetical, a number of practical questions arise. Is one advertising medium as effective as any other? That is, do the three mean sales lines differ for the three advertising media? Do the increases in mean sales per dollar input in advertising differ for the three advertising media? That is, do the *slopes* of the three lines differ? Note that the two practical questions have been rephrased into questions about the parameters that define the three lines of Figure 12.20. To answer these questions, we must write a single regression model that will characterize the three lines of the figure and that, by testing hypotheses about the lines, will answer the questions.

Figure 12.20
Graphs of the relationship between mean sales $E(y)$ and advertising expenditure x

The response described previously, monthly sales, is a function of *two* independent variables, one quantitative (advertising expenditure x_1) and one qualitative (type of medium). We will proceed in stages to build a model relating $E(y)$ to these variables and will show graphically the interpretation we would give to the model at each stage. This approach will help you see the contributions of the various terms in the model.

1. The straight-line relationship between mean sales $E(y)$ and advertising expenditure is the same for all three media; that is, a single line will describe the relationship between $E(y)$ and advertising expenditure x_1 for all the media. (See Figure 12.21.) Thus,

$$E(y) = \beta_0 + \beta_1 x_1,$$

where

$$x_1 = \text{Advertising expenditure}$$

Figure 12.21
The relationship between $E(y)$ and x_1 is the same for all media

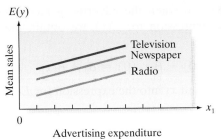

Figure 12.22
Parallel response lines for the three media

2. The straight lines relating mean sales $E(y)$ to advertising expenditure x_1 differ from one medium to another, but the rate of increase in mean sales per increase in dollar advertising expenditure x_1 is the same for all media; that is, the lines are parallel but possess different y-intercepts. (See Figure 12.22.) Hence,

$$E(y) = \beta_0 + \beta_1 x_1 + \beta_2 x_2 + \beta_3 x_3$$

where

$$x_1 = \text{Advertising expenditure}$$

$$x_2 = \begin{cases} 1 & \text{if radio medium} \\ 0 & \text{if not} \end{cases}$$

$$x_3 = \begin{cases} 1 & \text{if television medium} \\ 0 & \text{if not} \end{cases}$$

Notice that this model is essentially a combination of a first-order model with a single quantitative variable and a model with a single qualitative variable. That is,

First-order model with a single
quantitative variable: $\quad\quad E(y) = \beta_0 + \beta_1 x_1$

Model with single qualitative
variable at three levels: $\quad\quad E(y) = \beta_0 + \beta_2 x_2 + \beta_3 x_3$

where $x_1, x_2,$ and x_3 are as just defined. The model described here implies no interaction between the two independent variables, which are advertising expenditure x_1 and the qualitative variable (type of advertising medium). The change in $E(y)$ for a one-unit increase in x_1 is identical (the slopes of the lines are equal) for all three advertising media. The terms corresponding to each of the independent variables are called **main-effect terms** because they imply no interaction.

3. The straight lines relating mean sales $E(y)$ to advertising expenditure x_1 differ for the three advertising media; that is, both the line intercepts and the slopes differ. (See Figure 12.23.) This interaction model is obtained by adding terms involving the cross-product terms, one each from each of the two independent variables:

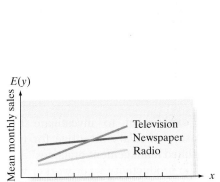

Figure 12.23
Different response lines for the three media

$$\underset{\substack{\text{Main effect,}\\\text{advertising}\\\text{expenditure}}}{} \quad \underset{\substack{\text{Main effect,}\\\text{type of}\\\text{medium}}}{} \quad \underset{\text{Interaction}}{}$$

$$E(y) = \beta_0 + \overbrace{\beta_1 x_1} + \overbrace{\beta_2 x_2 + \beta_3 x_3} + \overbrace{\beta_4 x_1 x_2 + \beta_5 x_1 x_3}$$

Note that each of the preceding models is obtained by adding terms to Model 1, the single first-order model used to model the responses for all three media. Model 2 is obtained by adding the main-effect terms for type of medium, the qualitative variable. Model 3 is obtained by adding the interaction terms to Model 2.

Example   12.10

Interpreting the β's in a Model with Mixed Variables

Problem Refer to the mean monthly sales models. Substitute the appropriate values of the dummy variables in Model 3 above to obtain the equations of the three response lines in Figure 12.23.

Solution The complete model that characterizes the three lines in Figure 12.23 is

$$E(y) = \beta_0 + \beta_1 x_1 + \beta_2 x_2 + \beta_3 x_3 + \beta_4 x_1 x_2 + \beta_5 x_1 x_3$$

where

$$x_1 = \text{Advertising expenditure}$$

$$x_2 = \begin{cases} 1 & \text{if radio medium} \\ 0 & \text{if not} \end{cases}$$

$$x_3 = \begin{cases} 1 & \text{if television medium} \\ 0 & \text{if not} \end{cases}$$

Examining the coding, you can see that $x_2 = x_3 = 0$ when the advertising medium is newspaper. Substituting these values into the expression for $E(y)$, we obtain the newspaper medium line:

$$E(y) = \beta_0 + \beta_1 x_1 + \beta_2(0) + \beta_3(0) + \beta_4 x_1(0) + \beta_5 x_1(0) = \beta_0 + \beta_1 x_1$$

Similarly, we substitute the appropriate values of x_2 and x_3 into the expression for $E(y)$ to obtain the radio medium line $(x_2 = 1, x_3 = 0)$,

$$E(y) = \beta_0 + \beta_1 x_1 + \beta_2(1) + \beta_3(0) + \beta_4 x_1(1) + \beta_5 x_1(0)$$

$$= \underbrace{(\beta_0 + \beta_2)}_{\text{y-intercept}} + \underbrace{(\beta_1 + \beta_4)x_1}_{\text{Slope}}$$

and the television medium line $(x_2 = 0, x_3 = 1)$,

$$E(y) = \beta_0 + \beta_1 x_1 + \beta_2(0) + \beta_3(1) + \beta_4 x_1(0) + \beta_5 x_1(1)$$

$$= \underbrace{(\beta_0 + \beta_3)}_{\text{y-intercept}} + \underbrace{(\beta_1 + \beta_5)x_1}_{\text{Slope}}$$

Look Back If you were to fit Model 3, obtain estimates of $\beta_0, \beta_1, \beta_2, \ldots, \beta_5$, and substitute them into the equations for the three media lines, you would obtain exactly the same prediction equations as you would if you were to fit three separate straight lines, one to each of the three sets of media data. You may ask why we would not fit the three lines separately. Why bother fitting a model that combines all three lines (Model 3) into the same equation? The answer is that you need to use this procedure if you wish to use statistical tests to compare the three media lines. We need to be able to express a practical question about the lines in terms of a hypothesis that a set of parameters in the model equals 0. (We demonstrate this procedure in the next section.) You could not do that if you were to perform three separate regression analyses and fit a line to each set of media data.

◼ **Now Work Exercise 12.103**

Example 12.11

Testing for Two Different Slopes—Worker Productivity Data

Problem An industrial psychologist conducted an experiment to investigate the relationship between worker productivity and a measure of salary incentive for two manufacturing plants; one plant operates under "disciplined management practices" and the other plant uses a traditional management style. The productivity y per worker was measured by recording the number of machined castings that a worker could produce in a four-week period of 40 hours per week. The incentive was the amount x_1 of bonus (in cents per casting) paid for all castings produced in excess of 1,000 per worker for the four-week period. Nine workers were selected from each plant, and three from each group of nine were assigned to receive a 20¢ bonus per casting, three a 30¢ bonus, and three a 40¢ bonus. The productivity data for the 18 workers, three for each type of plant and incentive combination, are shown in Table 12.4.

Table 12.4   Productivity Data (Number of Castings) for Example 12.11

Management Style	Incentive								
	20¢/casting			30¢/casting			40¢/casting		
Traditional	1,435	1,512	1,491	1,583	1,529	1,610	1,601	1,574	1,636
Disciplined	1,575	1,512	1,488	1,635	1,589	1,661	1,645	1,616	1,689

Ⓓ *Data Set:* CASTING

a. Write a model for mean productivity $E(y)$, assuming that the relationship between $E(y)$ and incentive x_1 is first order.

b. Fit the model, and graph the prediction equations for the traditional and disciplined plants.

c. Do the data provide sufficient evidence to indicate that the rate of increase in worker productivity is different for disciplined and traditional plants? Test at $\alpha = .10$.

Solution

a. If we assume that a first-order model* is adequate to detect a change in mean productivity as a function of incentive x_1, then the model that produces two straight lines, one for each management style, is

$$E(y) = \beta_0 + \beta_1 x_1 + \beta_2 x_2 + \beta_3 x_1 x_2$$

where

$$x_1 = \text{Incentive} \qquad x_2 = \begin{cases} 1 & \text{if disciplined management style} \\ 0 & \text{if traditional management style} \end{cases}$$

b. A MINITAB printout for the regression analysis is shown in Figure 12.24. Reading the parameter estimates highlighted on the printout, you can see that

$$\hat{y} = 1{,}365.83 + 6.217x_1 + 47.78x_2 + .033x_1 x_2$$

The prediction equation for the plant using traditional management style can be obtained (see the coding) by substituting $x_2 = 0$ into the general prediction equation. Then

$$\hat{y} = \hat{\beta}_0 + \hat{\beta}_1 x_1 + \hat{\beta}_2(0) + \hat{\beta}_3 x_1(0) = \hat{\beta}_0 + \hat{\beta}_1 x_1$$
$$= 1{,}365.83 + 6.217x_1$$

Similarly, the prediction equation for the plant with a disciplined management style can be obtained by substituting $x_2 = 1$ into the general prediction equation. Then

$$\hat{y} = \hat{\beta}_0 + \hat{\beta}_1 x_1 + \hat{\beta}_2 x_2 + \hat{\beta}_3 x_1 x_2$$
$$= \hat{\beta}_0 + \hat{\beta}_1 x_1 + \hat{\beta}_2(1) + \hat{\beta}_3 x_1(1)$$
$$= \underbrace{(\hat{\beta}_0 + \hat{\beta}_2)}_{y\text{-intercept}} + \underbrace{(\hat{\beta}_1 + \hat{\beta}_3)}_{\text{Slope}} x_1$$
$$= (1{,}365.83 + 47.78) + (6.217 + .033)x_1$$
$$= 1{,}413.61 + 6.250x_1$$

Figure 12.24

MINITAB printout of the complete model for the casting data

```
The regression equation is
CASTINGS = 1366 + 6.22 INCENTIVE + 47.8 SDUMMY + 0.03 INC_SDUM


Predictor      Coef   SE Coef      T      P
Constant    1365.83     51.84  26.35  0.000
INCENTIVE     6.217     1.667   3.73  0.002
SDUMMY       47.78      73.31   0.65  0.525
INC_SDUM      0.033     2.358   0.01  0.989


S = 40.8387    R-Sq = 71.1%    R-Sq(adj) = 64.9%


Analysis of Variance

Source          DF     SS     MS      F      P
Regression       3  57332  19111  11.46  0.000
Residual Error  14  23349   1668
Total           17  80682
```

---

*Although the model contains a term involving $x_1 x_2$, it is first-order (graphs as a straight line) in the quantitative variable x_1. The variable x_2 is a dummy variable that introduces or deletes terms in the model. The order of a model is determined only by the quantitative variables that appear in the model.

A MINITAB graph of these prediction equations is shown in Figure 12.25. Note that the slopes of the two lines are nearly identical (6.217 for traditional and 6.250 for disciplined).

Figure 12.25

MINITAB plot of prediction equations for two plants

c. If the rate of increase in productivity with incentive (i.e., the slope) for the disciplined management style plant is different from the corresponding slope for the traditional plant, then the interaction β (i.e., β_3) will differ from 0. Consequently, we want to test

$$H_0: \beta_3 = 0$$
$$H_a: \beta_3 \neq 0$$

This test is conducted with the use of the t-test of Section 12.2. The test statistic and the corresponding p-value are highlighted on the MINITAB printout:

$$t = .014 \qquad p\text{-value} = .989$$

Since $\alpha = .10$ is less than the p-value, we fail to reject H_0, meaning that there is insufficient evidence to conclude that the traditional and disciplined shapes differ. Thus, the test supports our observation of two nearly identical slopes in part **b**.

Look Back Since interaction is not significant, we will drop the $x_1 x_2$ term from the model and use the simpler model $E(y) = \beta_0 + \beta_1 x_1 + \beta_2 x_2$ to predict productivity.

■ **Now Work Exercise 12.106**

Models with both quantitative and qualitative x's may also include higher order (e.g., second-order) terms. In the problem of relating mean monthly sales $E(y)$ of a company to monthly advertising expenditure x_1 and type of medium, suppose we think that the relationship between $E(y)$ and x_1 is curvilinear. We will construct the model stage by stage to enable you to compare the procedure with the stage-by-stage construction of the first-order model at the beginning of this section. The graphical interpretations will help you understand the contributions of the model terms.

1. The mean sales curves are identical for all three advertising media; that is, a single second-order curve will suffice to describe the relationship between $E(y)$ and x_1 for all the media. (See Figure 12.26.) Thus,

$$E(y) = \beta_0 + \beta_1 x_1 + \beta_2 x_1^2$$

where x_1 = Advertising expenditure

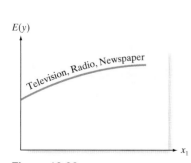

Figure 12.26

The relationship between $E(y)$ and x_1 is the same for all media

2. The response curves possess the same shapes but different y-intercepts. (See Figure 12.27.) Hence,

$$E(y) = \beta_0 + \beta_1 x_1 + \beta_2 x_1^2 + \beta_3 x_2 + \beta_4 x_3$$

where

$$x_1 = \text{Advertising expenditure}$$

$$x_2 = \begin{cases} 1 & \text{if radio medium} \\ 0 & \text{if not} \end{cases}$$

$$x_3 = \begin{cases} 1 & \text{if television medium} \\ 0 & \text{if not} \end{cases}$$

3. The response curves for the three advertising media are different (i.e., Advertising expenditure and Type of medium interact), as shown in Figure 12.28. Then

$$E(y) = \beta_0 + \beta_1 x_1 + \beta_2 x_1^2 + \beta_3 x_2 + \beta_4 x_3 + \beta_5 x_1 x_2 + \beta_6 x_1 x_3 + \beta_7 x_1^2 x_2 + \beta_8 x_1^2 x_3$$

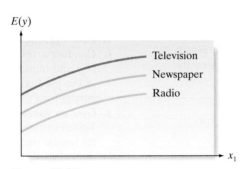

Figure 12.27
The response curves have the same shapes, but different y-intercepts

Figure 12.28
The response curves for the three media differ

Now that you know how to write a model with two independent variables—one qualitative and one quantitative—we ask a question: Why do it? Why not write a separate second-order model for each type of medium where $E(y)$ is a function of only advertising expenditure? As stated earlier, one reason we wrote the single model representing all three response curves is so that we can test to determine whether the curves are different. We illustrate this procedure in optional Section 12.9. A second reason for writing a single model is that we obtain a pooled estimate of σ^2, the variance of the random-error component ε. If the variance of ε is truly the same for each type of medium, the pooled estimate is superior to three separate estimates calculated by fitting a separate model for each type of medium.

Exercises 12.100–12.116

Understanding the Principles

12.100 Consider a multiple-regression model for a response y with one quantitative independent variable x_1 and one qualitative variable at three levels.
  a. Write a first-order model that relates the mean response $E(y)$ to the quantitative independent variable.
  b. Add the main-effect terms for the qualitative independent variable to the model of part **a**. Specify the coding scheme you use.
  c. Add terms to the model of part **b** to allow for interaction between the quantitative and qualitative independent variables.
  d. Under what circumstances will the response lines of the model in part **c** be parallel?
  e. Under what circumstances will the model in part **c** have only one response line?

12.101 Refer to Exercise 12.100.
  a. Write a complete second-order model that relates $E(y)$ to the quantitative variable.
  b. Add the main-effect terms for the qualitative variable (at three levels) to the model of part **a**.
  c. Add terms to the model of part **b** to allow for interaction between the quantitative and qualitative independent variables.
  d. Under what circumstances will the response curves of the model have the same shape, but different y-intercepts?
  e. Under what circumstances will the response curves of the model be parallel lines?
  f. Under what circumstances will the response curves of the model be identical?

12.102 Write a model that relates $E(y)$ to two independent variables, one quantitative and one qualitative (at four levels). Construct a model that allows the associated response curves to be second order but does not allow for interaction between the two independent variables.

Learning the Mechanics

12.103 Consider the model

NW

$$y = \beta_0 + \beta_1 x_1 + \beta_2 x_2 + \beta_3 x_3 + \varepsilon$$

where x_1 is a quantitative variable and x_2 and x_3 are dummy variables describing a qualitative variable at three levels, using the coding scheme

$$x_2 = \begin{cases} 1 & \text{if level 2} \\ 0 & \text{otherwise} \end{cases} \quad x_3 = \begin{cases} 1 & \text{if level 3} \\ 0 & \text{otherwise} \end{cases}$$

The resulting least squares prediction equation is

$$\hat{y} = 44.8 + 2.2x_1 + 9.4x_2 + 15.6x_3$$

a. What is the response line (equation) for $E(y)$ when $x_2 = x_3 = 0$? When $x_2 = 1$ and $x_3 = 0$? When $x_2 = 0$ and $x_3 = 1$?

b. What is the least squares prediction equation associated with level 1? Level 2? Level 3? Plot these on the same graph.

12.104 Consider the model

$$y = \beta_0 + \beta_1 x_1 + \beta_2 x_1^2 + \beta_3 x_2 + \beta_4 x_3 + \beta_5 x_1 x_2$$
$$+ \beta_6 x_1 x_3 + \beta_7 x_1^2 x_2 + \beta_8 x_1^2 x_3 + \varepsilon$$

where x_1 is a quantitative variable and

$$x_2 = \begin{cases} 1 & \text{if level 2} \\ 0 & \text{otherwise} \end{cases} \quad x_3 = \begin{cases} 1 & \text{if level 3} \\ 0 & \text{otherwise} \end{cases}$$

The resulting least squares prediction equation is

$$\hat{y} = 45.6 - 4.7x_1 + .07x_1^2 - 3.7x_2 - 5.9x_3 + 3.1x_1 x_2$$
$$+ 2.5x_1 x_3 - .09x_1^2 x_2 - .03x_1^2 x_3$$

a. What is the response curve for $E(y)$ when $x_2 = 0$ and $x_3 = 0$? When $x_2 = 1$ and $x_3 = 0$? When $x_2 = 0$ and $x_3 = 1$?

b. On the same graph, plot the least squares prediction equation associated with level 1, with level 2, and with level 3. Choose the correct graph below.

Applying the Concepts—Basic

12.105 Reality TV and cosmetic surgery. Refer to the *Body Image: An International Journal of Research* (Mar. 2010) study of the impact of reality TV shows on a college student's decision to undergo cosmetic surgery, Exercise 12.25 (p. 700). Consider the interaction model $E(y) = \beta_0 + \beta_1 x_1 + \beta_2 x_4 + \beta_3 x_1 x_4$, where y = desire to have cosmetic surgery (25-point scale), $x_1 = \{1 \text{ if male, 0 if female}\}$, and x_4 = impression of reality TV (7-point scale). The model was fit to the data and the resulting SAS printout appears at the bottom of the page.

D
IMAGE

a. Give the least squares prediction equation.

b. Find the predicted level of desire (y) for a male college student with an impression-of-reality-TV-scale score of 5.

c. Conduct a test of overall model adequacy. Use $\alpha = .10$.

d. Give a practical interpretation of R_a^2.

e. Give a practical interpretation of s.

f. Conduct a test (at $\alpha = .10$) to determine if gender (x_1) and impression of reality TV show (x_4) interact in the prediction of level of desire for cosmetic surgery (y).

g. Give an estimate of the change in desire (y) for every 1-point increase in impression of reality TV show (x_4) for female students.

h. Repeat part **g** for male students.

12.106 Impact of dropping ping-pong balls. Refer to the *American Journal of Physics* (Mar. 2014) study of the impact of dropping ping-pong balls, Exercise 11.122 (p. 666). Recall that 19 standard ping-pong balls were dropped vertically onto a force plate. The data on y = coefficient of restitution (COR, measured as a ratio of the speed at impact and rebound speed), x_1 = speed at impact (meters/second), and $x_2 = \{1 \text{ if ball buckled, 0 if not}\}$ are reproduced in the table on p. 741. Consider the interaction model $E(y) = \beta_0 + \beta_1 x_1 + \beta_2 x_2 + \beta_3 x_1 x_2$. A MINITAB printout of the regression analysis is also shown on p. 741.

NW

D
PONG

Dependent Variable: DESIRE

Number of Observations Read	170
Number of Observations Used	170

Analysis of Variance

Source	DF	Sum of Squares	Mean Square	F Value	Pr > F
Model	3	747.00146	249.00049	45.09	<.0001
Error	166	916.78678	5.52281		
Corrected Total	169	1663.78824			

Root MSE	2.35007	R-Square	0.4490
Dependent Mean	12.56471	Adj R-Sq	0.4390
Coeff Var	18.70371		

Parameter Estimates

Variable	DF	Parameter Estimate	Standard Error	t Value	Pr > |t|
Intercept	1	11.77891	0.67363	17.49	<.0001
GENDER	1	-1.97223	1.17922	-1.67	0.0963
IMPREAL	1	0.58462	0.16163	3.62	0.0004
IMPREAL_GENDER	1	-0.55330	0.27611	-2.00	0.0467

SAS output for Exercise 12.105

Data for Exercise 12.106

Ball	COR y	SPEED x_1	BUCKLE x_2
1	.945	0.8	0 (No)
2	.950	1.0	0 (No)
3	.930	1.5	0 (No)
4	.920	1.8	0 (No)
5	.920	3.0	0 (No)
6	.930	3.4	0 (No)
7	.905	4.4	0 (No)
8	.915	5.0	0 (No)
9	.910	6.4	0 (No)
10	.900	4.4	1 (Yes)
11	.885	5.3	1 (Yes)
12	.870	5.4	1 (Yes)
13	.850	7.4	1 (Yes)
14	.795	7.2	1 (Yes)
15	.790	7.2	1 (Yes)
16	.800	8.0	1 (Yes)
17	.820	8.5	1 (Yes)
18	.810	9.4	1 (Yes)
19	.780	9.0	1 (Yes)

a. Give the equation of the hypothesized line relating COR (y) to impact speed (x_1) for ping-pong balls that did not buckle. What is the estimated slope of this line?

b. Repeat part **a** for ping-pong balls that buckled.

c. The researcher believes that the rate of increase in COR with impact speed differs depending on whether the ping-pong ball buckles. Do the data support this hypothesis? Conduct the appropriate test using $\alpha = .05$.

Regression Analysis: COR versus SPEED, X2, SPEED_X2

```
The regression equation is
COR = 0.944 - 0.00637 SPEED + 0.0398 X2 - 0.0151 SPEED_X2


Predictor    Coef    SE Coef     T      P
Constant    0.94433  0.01256   75.17  0.000
SPEED      -0.006374 0.003545  -1.80  0.092
X2          0.03977  0.03121    1.27  0.222
SPEED_X2   -0.015089 0.005260  -2.87  0.012


S = 0.0194826   R-Sq = 90.7%   R-Sq(adj) = 88.8%


Analysis of Variance

Source          DF      SS       MS       F      P
Regression       3   0.055556  0.018519  48.79  0.000
Residual Error  15   0.005694  0.000380
Total           18   0.061250
```

12.107 Do blondes raise more funds? Refer to the *Economic Letters* (Vol. 100, 2008) study of whether the color of a female solicitor's hair affects the level of capital raised, Exercise 12.88 (p. 731). Recall that 955 households were contacted by a female solicitor to raise funds for hazard mitigation research. In addition to the household's level of contribution (in dollars) and the hair color of the solicitor (blonde Caucasian, brunette Caucasian, or minority female), the researcher also recorded the beauty rating of the solicitor (measured quantitatively on a 10-point scale).

a. Write a first-order model (with no interaction) for mean contribution level, $E(y)$, as a function of a solicitor's hair color and her beauty rating.

b. Refer to the model, part **a**. For each hair color, express the change in contribution level for each 1-point increase in solicitor's beauty rating in terms of the model parameters.

c. Write an interaction model for mean contribution level, $E(y)$, as a function of a solicitor's hair color and her beauty rating.

d. Refer to the model, part **c**. For each hair color, express the change in contribution level for each 1-point increase in solicitor's beauty rating in terms of the model parameters.

e. Refer to the model, part **c**. Illustrate the interaction with a graph.

12.108 Impact of race on football card values. Refer to the *Electronic Journal of Sociology* (2007) study of the impact of race on the value of professional football players' "rookie" cards, presented in Exercise 12.89 (p. 732). Recall that the sample consisted of 148 rookie cards of National Football League (NFL) players who were inducted into the Football Hall of Fame. The researchers modeled the natural logarithm of card price (y) as a function of the following independent variables:

Race: $x_1 = 1$ if black, 0 if white
Card availability: $x_2 = 1$ if high, 0 if low
Card vintage: $x_3 =$ year card printed
Finalist: $x_4 =$ natural logarithm of number of times player
          was on final Hall of Fame ballot
Position-QB: $x_5 = 1$ if quarterback, 0 if not
Position-RB: $x_7 = 1$ if running back, 0 if not
Position-WR: $x_8 = 1$ if wide receiver, 0 if not
Position-TE: $x_9 = 1$ if tight end, 0 if not
Position-DL: $x_{10} = 1$ if defensive lineman, 0 if not
Position-LB: $x_{11} = 1$ if linebacker, 0 if not
Position-DB: $x_{12} = 1$ if defensive back, 0 if not
[*Note*: For Position, offensive lineman is the base level]

---

a. The model $E(y) = \beta_0 + \beta_1 x_1 + \beta_2 x_2 + \beta_3 x_3 + \beta_4 x_4 + \beta_5 x_5 + \beta_6 x_6 + \beta_7 x_7 + \beta_8 x_8 + \beta_9 x_9 + \beta_{10} x_{10} + \beta_{11} x_{11} + \beta_{11} x_{11} + \beta_{12} x_{12}$ was fit to the data, with the following results: $R^2 = .705$, adj. $R^2 = .681$, $F = 26.9$. Interpret the results practically. Make an inference about the overall adequacy of the model.

b. Refer to part **a**. Statistics for the race variable were reported as follows: $\hat{\beta}_1 = -147$, $s_{\hat{\beta}_1} = .145$, $t = -1.014$, p-value $= .312$. Use this information to make an inference about the impact of race on the value of professional football players' rookie cards.

c. Refer to part **a**. Statistics for the card vintage variable were reported as follows: $\hat{\beta}_3 = -.074$, $s_{\hat{\beta}_2} = .007$, $t = -10.92$, p-value $= .000$. Use this information to make an inference about the impact of card vintage on the value of professional football players' rookie cards.

d. Write a first-order model for $E(y)$ as a function of card vintage (x_4) and position (x_5–x_{12}) that allows for the relationship between price and vintage to vary with position.

12.109 Whales entangled in fishing gear. Refer to the *Marine Mammal Science* (Apr. 2010) study of whales entangled in fishing gear, Exercise 12.87, (p. 731). Now consider a model for the length (y) of an entangled whale (in meters) that is a function of water depth of the entanglement (in meters) and gear type (set nets, pots, or gill nets).

a. Write a main-effects-only model for $E(y)$.

b. Sketch the relationships hypothesized by the model, part **a**. (*Hint:* Plot length on the vertical axis and water depth on the horizontal axis.)

c. Add terms to the model, part **a**, that include interaction between water depth and gear type. (*Hint:* Be sure to interact each dummy variable for gear type with water depth.)

d. Sketch the relationships hypothesized by the model, part **c**.

e. In terms of the β's in the model of part **c**, give the rate of change of whale length with water depth for set nets.

f. Repeat part **e** for pots.

g. Repeat part **e** for gill nets.

h. In terms of the β's in the model of part **c**, how would you test to determine if the rate of change of whale length with water depth is the same for all three types of fishing gear?

12.110 Winning marathon times. Refer to the *Chance* (Winter 2000) study of men's and women's winning times in the Boston Marathon, presented in Exercise 11.146 (p. 675). Suppose the researchers want to build a model for predicting winning time (y) of the marathon as a function of year (x_1) in which race is run and gender of winning runner (x_2).

a. Set up the appropriate dummy variables (if necessary) for x_1 and x_2.

b. Write the equation of a model that proposes parallel straight-line relationships between winning time (y) and year (x_1), one line for each gender.

c. Write the equation of a model that proposes nonparallel straight-line relationships between winning time (y) and year (x_1), one line for each gender.

d. Which of the models do you think will provide the best predictions of winning time (y)? Base your answer on the graph displayed in Exercise 11.146.

Applying the Concepts—Intermediate

12.111 Personality traits and job performance. Refer to the *Journal of Applied Psychology* (Jan. 2011) study of the relationship between task performance and conscientiousness, Exercise 12.67 (p. 722). Recall that the researchers used a quadratic model to relate y = task performance score (measured on a 30-point scale) to x_1 = conscientiousness score (measured on a scale of -3 to $+3$). In addition, the researchers included job complexity in the model, where $x_2 = \{1 \text{ if highly complex job}, 0 \text{ if not}\}$. The complete model took the form

$$E(y) = \beta_0 + \beta_1 x_1 + \beta_2 (x_1)^2 + \beta_3 x_2 + \beta_4 x_1 x_2 + \beta_5 (x_1)^2 x_2$$

a. For jobs that are not highly complex, write the equation of the model for $E(y)$ as a function of x_1. (Substitute $x_2 = 0$ into the equation.)

b. Refer to part **a**. What do each of the β's represent in the model?

c. For highly complex jobs, write the equation of the model for $E(y)$ as a function of x_1. (Substitute $x_2 = 1$ into the equation.)

d. Refer to part **c**. What do each of the β's represent in the model?

e. Does the model support the researchers' theory that the curvilinear relationship between task performance score (y) and conscientiousness score (x_1) depends on job complexity (x_2)? Explain.

12.112 Agreeableness, gender, and wages. Do agreeable individuals get paid less, on average, than those who are less agreeable on the job? And is this gap greater for males than for females? These questions were addressed in the *Journal of Personality and Social Psychology* (Feb. 2012). Several variables were measured for each in a sample of individuals enrolled in the National Survey of Midlife Development in the United States. Three of these variables are: (1) level of agreeableness score (where higher scores indicate a greater level of agreeableness), (2) gender (male or female), and (3) annual income (dollars). The researchers modeled mean income, $E(y)$, as a function of both agreeableness score (x_1) and a dummy variable for gender $(x_2 = 1$ if male, 0 if female). Data for a sample of 100 individuals (simulated, based on information provided in the study) are saved in the **WAGAP** file. The first 10 observations are listed in the accompanying table.

Data for First 10 Individuals in Study

Income	Agree Score	Gender
44,770	3.0	1
51,480	2.9	1
39,600	3.3	1
24,370	3.3	0
15,460	3.6	0
43,730	3.8	1
48,330	3.2	1
25,970	2.5	0
17,120	3.5	0
20,140	3.2	0

a. Consider the model $E(y) = \beta_0 + \beta_1 x_1 + \beta_2 x_2$. The researchers theorize that for either gender, income will decrease as agreeableness score increases. If this theory is true, what is the expected sign of β_1 in the model?

b. The researchers also theorize that the rate of decrease of income with agreeableness score will be steeper for males than for females (i.e., the income gap between males and females will be greater the less agreeable the individuals are). Can this theory be tested using the model, part **a**? Explain.

c. Consider the interaction model $E(y) = \beta_0 + \beta_1 x_1 + \beta_2 x_2 + \beta_3 x_1 x_2$. If the theory, part **b**, is true, give the expected sign of β_1 and the expected sign of β_3.

d. Fit the model, part **c**, to the sample data. Check the signs of the estimated β coefficients. How do they compare to the expected values, part **c**?

e. Refer to the interaction model, part **c**. Give the null and alternative hypotheses for testing whether the rate of decrease of income with agreeableness score is steeper for males than for females.

f. Conduct the test, part **e**. Use $\alpha = .05$. Is the researchers' theory supported?

12.113 Workplace bullying and intention to leave. Workplace bullying is defined as harassment, persistent criticism, withholding of key information, spreading of rumors, or intimidation on the job. In *Human Resource Management Journal* (Oct. 2008), researchers employed multiple regression to examine whether perceived organizational support would moderate the relationship between workplace bullying and victims' intention to leave the firm. The dependent variable in the analysis, intention to leave (y), was measured on a quantitative scale. The two key independent variables in the study were bullying (x_1, measured on a quantitative scale) and perceived organizational support (measured qualitatively as "low," "neutral," or "high").

a. Set up the dummy variables required to represent perceived organizational support (POS) in the regression model.

b. Write a model for $E(y)$ as a function of bullying and POS that hypothesizes three parallel straight lines, one for each level of POS.

c. Write a model for $E(y)$ as a function of bullying and POS that hypothesizes three nonparallel straight lines, one for each level of POS.

d. The researchers discovered that the effect of bullying on intention to leave was greater at the low level of POS than at the high level of POS. Which of the two models, parts **b** and **c**, supports these findings?

12.114 Lead levels in mountain moss. A study of the atmospheric pollution on the slopes of the Blue Ridge Mountains (in Tennessee) was conducted. The file **MOSS** contains the levels of lead found in 70 fern moss specimens (in micrograms of lead per gram of moss tissue) collected from the mountain slopes, as well as the elevation of the moss specimen (in feet) and the direction (1 if east, 0 if west) of the slope face. The first five and last five observations of the data set are listed in the following table.

Specimen	Lead Level	Elevation	Slope Face
1	3.475	2,000	0
2	3.359	2,000	0
3	3.877	2,000	0
4	4.000	2,500	0
5	3.618	2,500	0
⋮	⋮	⋮	⋮
66	5.413	2,500	1
67	7.181	2,500	1
68	6.589	2,500	1
69	6.182	2,000	1
70	3.706	2,000	1

Based on Schilling, J. "Bioindication of atmospheric heavy metal deposition in the Blue Ridge using the moss, *Thuidium delicatulum*." master-of-science thesis, Spring 2000.

a. Write the equation of a first-order model relating mean lead level $E(y)$ to elevation (x_1) and slope face (x_2). Include interaction between elevation and slope face in the model.

b. Graph the relationship between mean lead level and elevation for the different slope faces that is hypothesized by the model you wrote in part **a**.

c. In terms of the β's of the model from part **a**, give the change in lead level for every 1-foot increase in elevation for moss specimens on the east slope.

d. Fit the model from part **a** to the data, using an available statistical software package. Is the overall model statistically useful in predicting lead level? Test, using $\alpha = .10$.

12.115 Chemical composition of rainwater. Refer to the *Journal of Agricultural, Biological, and Environmental Statistics* (Mar. 2005) study of the chemical composition of rainwater, presented in Exercise 12.90 (p. 732). Recall that the nitrate concentration y (milligrams per liter) in a sample of rainwater was modeled as a function of water source (groundwater, subsurface flow, or overground flow). Now consider adding a second independent variable, silica concentration (milligrams per liter), to the model.

a. Write a first-order model for $E(y)$ as a function of the independent variables. Assume that the rate of increase of nitrate concentration with silica concentration is the same for all three water sources. Sketch the relationships hypothesized by the model on a graph.

b. Write a first-order model for $E(y)$ as a function of the independent variables, but now assume that the rate of increase of nitrate concentration with silica concentration differs for the three water sources. Sketch the relationships hypothesized by the model on a graph.

Applying the Concepts—Advanced

12.116 Iron supplement for anemia. Many women suffer from anemia. A female physician who is also an avid jogger wanted to know if women who exercise regularly have a different mean red blood cell count than women who do not. She also wanted to know if the amount of a particular iron supplement a woman takes has any effect and whether the effect is the same for both groups. Write a model that will reflect the relationship between red blood cell count and the two independent variables just described, assuming that

a. the effect of the iron supplement on mean blood cell count is the same regardless of whether a woman exercises regularly.

b. the effect of the iron supplement on mean blood cell count depends on whether a woman exercises regularly.

12.9 Comparing Nested Models (Optional)

To be successful model builders, we require a statistical method that will allow us to determine (with a high degree of confidence) which one among a set of candidate models best fits the data. In this section, we present such a technique for *nested models*.

> Two models are **nested** if one model contains all the terms of the second model and at least one additional term. The more complex of the two models is called the **complete** model, and the simpler of the two is called the **reduced** model.

To illustrate the concept of nested models, consider the straight-line interaction model for the mean auction price $E(y)$ of a grandfather clock as a function of two quantitative variables: age of the clock (x_1) and number of bidders (x_2). The interaction model fit in Example 12.6 is

$$E(y) = \beta_0 + \beta_1 x_1 + \beta_2 x_2 + \beta_3 x_1 x_2$$

If we assume that the relationship between auction price (y), age (x_1), and bidders (x_2) is curvilinear, then the complete second-order model is more appropriate:

$$E(y) = \overbrace{\beta_0 + \beta_1 x_1 + \beta_2 x_2 + \beta_3 x_1 x_2}^{\text{Terms in interaction model}} + \overbrace{\beta_4 x_1^2 + \beta_5 x_2^2}^{\text{Quadratic terms}}$$

Note that the curvilinear model contains quadratic terms for x_1 and x_2, as well as the terms in the interaction model. Therefore, the models are nested models. In this case, the interaction model is nested within the more inclusive curvilinear model. Thus, the curvilinear model is the *complete* model and the interaction model is the *reduced* model.

Suppose we want to know whether the curvilinear model contributes more information relevant to the prediction of y than the straight-line interaction model does. This is equivalent to determining whether the quadratic betas β_4 and β_5 should be retained in the model. To test whether these terms should be retained, we set up the null and alternative hypotheses as follows:

$H_0: \beta_4 = \beta_5 = 0$ (i.e., quadratic terms are not important in predicting y).
H_a: At least one of the parameters β_4 and β_5 is nonzero (i.e., at least one of the quadratic terms is useful in predicting y).

Note that the terms being tested are those additional terms in the complete (curvilinear) model that are not in the reduced (straight-line interaction) model.

We presented the t-test for a single β coefficient (Section 12.2) and the global F-test for *all* the β parameters (except β_0) in the model (Section 12.3). We now need a test for a *subset* of the β parameters in the complete model. The test procedure is intuitive. First, we use the method of least squares to fit the reduced model and calculate the corresponding sum of squares for error, SSE_R (the sum of squares of the deviations between the observed and the predicted y-values). Next, we fit the complete model and calculate its sum of squares for error, SSE_C. Then we compare SSE_R with SSE_C by calculating the difference, $\text{SSE}_R - \text{SSE}_C$. If the additional terms in the complete model are significant, then SSE_C should be much smaller than SSE_R, and the difference $\text{SSE}_R - \text{SSE}_C$ will be large.

Since SSE will always decrease when new terms are added to the model, the question is whether the difference $\text{SSE}_R - \text{SSE}_C$ is large enough to conclude that it is due to more than just an increase in the number of model terms and to chance. The formal statistical test utilizes an F-statistic, as shown in the following box:

F-Test for Comparing Nested Models

Reduced model: $E(y) = \beta_0 + \beta_1 x_1 + \cdots + \beta_g x_g$

Complete model: $E(y) = \beta_0 + \beta_1 x_1 + \cdots + \beta_g x_g + \beta_{g+1} x_{g+1} + \cdots + \beta_k x_k$

H_0: $\beta_{g+1} = \beta_{g+2} = \cdots = \beta_k = 0$

H_a: At least one of the β parameters specified in H_0 is nonzero.

Test statistic:
$$F_c = \frac{(SSE_R - SSE_C)/(k - g)}{SSE_C/[n - (k + 1)]}$$
$$= \frac{(SSE_R - SSE_C)/\#\ \beta\text{'s tested in } H_0}{MSE_C}$$

where

SSE_R = Sum of squared errors for the reduced model

SSE_C = Sum of squared errors for the complete model

MSE_C = Mean square error (s^2) for the complete model

$k - g$ = Number of β parameters specified in H_0 (i.e., number of β's tested)

$k + 1$ = Number of β parameters in the complete model (including β_0)

n = Total sample size

Rejection region: $F_c > F_\alpha$

p-value: $P(F > F_c)$

where F is based on $\nu_1 = k - g$ numerator degrees of freedom and $\nu_2 = n - (k + 1)$ denominator degrees of freedom

---

When the assumptions listed in Section 12.1 about the random-error term are satisfied, this F-statistic has an F-distribution with ν_1 and ν_2 df. Note that ν_1 is the number of β parameters being tested and ν_2 is the number of degrees of freedom associated with s^2 in the complete model.

Example 12.12

FLOWER

Analyzing a Complete Second-Order Model—Carnation Growth Data

Problem A botanist conducted an experiment to study the growth of carnations as a function of the temperature x_1 (°F) in a greenhouse and the amount of fertilizer x_2 [kilograms (kg) per plot] applied to the soil. Twenty-seven plots of equal size were treated with fertilizer in amounts varying between 50 and 60 kg per plot and were mechanically kept at constant temperatures between 80 and 100°F. Small carnation plants [approximately 15 centimeters (cm) in height] were planted in each plot, and their height y (cm) was measured after a six-week growing period. The resulting data are shown in Table 12.5.

Table 12.5 **Temperature (x_1), Amount of Fertilizer (x_2), and Height (y) of Carnations**

x_1	x_2	y	x_1	x_2	y	x_1	x_2	y
80	50	50.8	90	50	63.4	100	50	46.6
80	50	50.7	90	50	61.6	100	50	49.1
80	50	49.4	90	50	63.4	100	50	46.4
80	55	93.7	90	55	93.8	100	55	69.8
80	55	90.9	90	55	92.1	100	55	72.5
80	55	90.9	90	55	97.4	100	55	73.2
80	60	74.5	90	60	70.9	100	60	38.7
80	60	73.0	90	60	68.8	100	60	42.5
80	60	71.2	90	60	71.3	100	60	41.4

Data Set: FLOWER

a. Fit a complete second-order model to the data.

b. Sketch the fitted model in three dimensions.

c. Do the data provide sufficient evidence to indicate that the second-order terms β_3, β_4, and β_5 contribute information relevant to the prediction of y?

Solution

a. The complete second-order model is

$$E(y) = \beta_0 + \beta_1 x_1 + \beta_2 x_2 + \beta_3 x_1 x_2 + \beta_4 x_1^2 + \beta_5 x_2^2$$

The data in Table 12.5 were used to fit this model, and a portion of the SAS output is shown in Figure 12.29.

```
                        Dependent Variable: HEIGHT

                  Number of Observations Read        27
                  Number of Observations Used        27

                            Analysis of Variance

                                  Sum of        Mean
  Source                   DF    Squares       Square    F Value    Pr > F

  Model                     5   8402.26454   1680.45291   596.32    <.0001
  Error                    21     59.17843      2.81802
  Corrected Total          26   8461.44296


          Root MSE             1.67870    R-Square    0.9930
          Dependent Mean      66.96296    Adj R-Sq    0.9913
          Coeff Var            2.50690

                          Parameter Estimates

                        Parameter      Standard
  Variable     DF        Estimate         Error    t Value    Pr > |t|

  Intercept     1      -5127.89907     110.29601     -46.49    <.0001
  TEMP          1         31.09639       1.34441      23.13    <.0001
  FERT          1        139.74722       3.14005      44.50    <.0001
  TEM_FERT      1         -0.14550       0.00969     -15.01    <.0001
  TEMPSQ        1         -0.13339       0.00685     -19.46    <.0001
  FERTSQ        1         -1.14422       0.02741     -41.74    <.0001


        Test HIGHORD Results for Dependent Variable HEIGHT

                                     Mean
  Source              DF            Square    F Value    Pr > F

  Numerator            3        2204.11003     782.15    <.0001
  Denominator         21           2.81802
```

Figure 12.29

SAS printout of complete second-order model for height.

The least squares prediction equation (rounded) is

$$\hat{y} = -5{,}127.90 + 31.10 x_1 + 139.75 x_2 - .146 x_1 x_2 - .133 x_1^2 - 1.14 x_2^2$$

b. A three-dimensional graph of this prediction model, called a **response surface**, is shown in Figure 12.30. Note that the height seems to be greatest for temperatures of

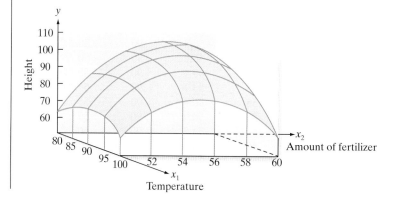

Figure 12.30

Plot of second-order least squares model for Example 12.12

about 85–90°F and for applications of about 55–57 kg of fertilizer per plot.* Further experimentation in these ranges might lead to a more precise determination of the optimal temperature–fertilizer combination.

c. To determine whether the data provide sufficient information to indicate that the second-order terms contribute information for the prediction of y, we wish to test

$$H_0: \beta_3 = \beta_4 = \beta_5 = 0$$

against the alternative hypothesis

H_a: At least one of the parameters β_3, β_4, and β_5 differs from 0.

The first step in conducting the test is to drop the second-order terms out of the complete (second-order) model and fit the reduced model

$$E(y) = \beta_0 + \beta_1 x_1 + \beta_2 x_2$$

to the data. The SAS printout for this model is shown in Figure 12.31.

The sums of squares for error, highlighted in Figures 12.29 and 12.31 for the complete and reduced models, respectively, are

$$\text{SSE}_C = 59.17843$$
$$\text{SSE}_R = 6{,}671.50852$$

and s^2 for the complete model (highlighted on Figure 12.29) is

$$s^2 = \text{MSE}_C = 2.81802$$

Recall that $n = 27, k = 5$, and $g = 2$. Therefore, the calculated value of the F-statistic, based on $\nu_1 = (k - g) = 3$ numerator df and $\nu_2 = [n - (k + 1)] = 21$ denominator df, is

$$F = \frac{(\text{SSE}_R - \text{SSE}_C)/(k - g)}{\text{SSE}_C/[n - (k + 1)]} = \frac{(\text{SSE}_R - \text{SSE}_C)/(k - g)}{\text{MSE}_C}$$

where $\nu_1 = (k - g)$ is equal to the number of parameters involved in H_0. Therefore,

$$\textit{Test statistic: } F = \frac{(6{,}671.50852 - 59.17843)/3}{2.81802} = 782.15$$

The final step in the test is to compare this computed value of F with the tabulated value based on $\nu_1 = 3$ and $\nu_2 = 21$ df. If we choose $\alpha = .05$, then $F_{.05} = 3.07$ and the rejection region is

$$\textit{Rejection region: } F > 3.07.$$

Figure 12.31

SAS printout of first-order model for height

```
                        Dependent Variable: HEIGHT

                   Number of Observations Read        27
                   Number of Observations Used        27

                             Analysis of Variance

                                  Sum of        Mean
       Source            DF      Squares       Square    F Value   Pr > F

       Model              2    1789.93444    894.96722     3.22    0.0577
       Error             24    6671.50852    277.97952
       Corrected Total   26    8461.44296

                   Root MSE          16.67272    R-Square     0.2115
                   Dependent Mean    66.96296    Adj R-Sq     0.1458
                   Coeff Var         24.89842

                             Parameter Estimates

                          Parameter      Standard
       Variable    DF      Estimate         Error    t Value   Pr > |t|

       Intercept    1     106.08519      55.94500       1.90     0.0700
       TEMP         1      -0.91611       0.39298      -2.33     0.0285
       FERT         1       0.78778       0.78596       1.00     0.3262
```

*Students with a knowledge of calculus should note that we can solve for the exact temperature and amount of fertilizer that maximize height in the least squares model by solving $\partial\hat{y}/\partial x_1 = 0$ and $\partial\hat{y}/\partial x_2 = 0$ for x_1 and x_2. Sample estimates of these estimated optimal values are $x_1 = 86.25°F$ and $x_2 = 55.58$ kg per plot.

Since the computed value of F falls in the rejection region (i.e., it exceeds $F_{.05} = 3.07$), we reject H_0 and conclude that at least one of the second-order terms contributes information relevant to the prediction of y. Thus, the second-order model appears to provide better predictions of y than does a first-order model.

Look Back Using special commands, you can get SAS to perform the desired nested-model F-test. The test statistic and p-value for the preceding test are highlighted at the bottom of the SAS printout in Figure 12.29.

■ **Now Work Exercise 12.120**

The nested-model F-test can be used to determine whether *any* subset of terms should be included in a complete model by testing the null hypothesis that a particular set of β parameters simultaneously equals 0. For example, we may want to test whether a set of interaction terms for quantitative variables or a set of main-effect terms for a qualitative variable should be included in a model. If we reject H_0, the complete model is the better of the two nested models.

Suppose the F-test in Example 12.12 yielded a test statistic that did not fall into the rejection region. Although we must be cautious about accepting H_0, most practitioners of regression analysis adopt the principle of *parsimony*. That is, in situations where two competing models are found to have essentially the same predictive power (as in this case), the model with the lesser number of β's (i.e., the more parsimonious model) is selected. On the basis of this principle, we would drop the three second-order terms and select the straight-line (reduced) model over the second-order (complete) model.

> A **parsimonious model** is a general linear model with a small number of β parameters. In situations where two competing models have essentially the same predictive power (as determined by an F-test), choose the more parsimonious of the two.

> **Guidelines for Selecting Preferred Model in a Nested Model F-Test**
>
> | Conclusion | Preferred Model |
> |---|---|
> | Reject H_0 | → Complete Model |
> | Fail to reject H_0 | → Reduced Model |

When the candidate models in model building are nested models, the F-test developed in this section is the appropriate procedure to apply to compare the models. However, if the models are not nested, this F-test is not applicable. In such a situation, the analyst must base the choice of the best model on statistics such as R_a^2 and s. It is important to remember that decisions based on these and other numerical descriptive measures of the adequacy of a model cannot be supported with a measure of reliability and are often highly subjective in nature.

Statistics IN Action | Revisited | Building a Model for Condominium Sale Price

In the previous Statistics in Action Revisited section (p. 704), we used the six independent variables listed in Table SIA12.1 to fit a first-order model for the auction price (y) of a condominium unit. Although the model was deemed statistically useful in predicting y, the standard deviation of the model ($s = 21.8$ hundred dollars) was probably too large for the model to be "practically" useful. A more complicated model—one involving higher order terms (interactions and squared terms)—needs to be considered. We start with a second-order model involving only the two quantitative independent variables FLOOR (x_1) and DISTANCE (x_2). The model assumes curvilinear (rather than straight-line) relationships and is given by the equation

$$E(y) = \beta_0 + \beta_1 x_1 + \beta_2 x_2 + \beta_3 x_1 x_2 + \beta_4 (x_1)^2 + \beta_5 (x_2)^2$$

The SAS printout for this model is shown in Figure SIA12.5. Note that the global F-test for the model is statistically significant (p-value $<.0001$).

Are the higher (second)-order terms in the model, namely, $\beta_3 x_1 x_2$, $\beta_4 (x_1)^2$,

```
                      Dependent Variable: PRICE100

                 Number of Observations Read        209
                 Number of Observations Used        209

                        Analysis of Variance

                              Sum of         Mean
    Source            DF      Squares        Square    F Value   Pr > F

    Model              5       60858         12172      13.88    <.0001
    Error            203      178035      877.01813
    Corrected Total  208      238893

            Root MSE            29.61449    R-Square     0.2548
            Dependent Mean     201.28708    Adj R-Sq     0.2364
            Coeff Var           14.71256

                        Parameter Estimates

                       Parameter      Standard
    Variable     DF     Estimate        Error     t Value   Pr > |t|

    Intercept     1    229.80506      13.17675     17.44     <.0001
    FLOOR         1     -8.76315       4.49100     -1.95     0.0524
    DISTANCE      1     -7.33316       2.31456     -3.17     0.0018
    FLR_DIST      1     -0.17739       0.20153     -0.88     0.3798
    FLOORSQ       1      0.76065       0.44691      1.70     0.0903
    DISTSQ        1      0.66948       0.13221      5.06     <.0001


    Test HIORDER Results for Dependent Variable PRICE100

                              Mean
    Source            DF      Square     F Value    Pr > F

    Numerator          3    8500.32594     9.69     <.0001
    Denominator      203     877.01813
```

Figure SIA12.5

SAS printout of the second-order model for condo sale price — quantitative variables only

and $\beta_5(x_2)^2$, necessary? If not, we can simplify the model by dropping these curvature terms. The hypothesis of interest is $H_0: \beta_3 = \beta_4 = \beta_5 = 0$. To test this subset of β's, we compare the second-order model with a model that lacks the interaction and curvilinear terms. The reduced model takes the form

$$E(y) = \beta_0 + \beta_1 x_1 + \beta_2 x_2$$

The results of this nested model (or partial) F-test are shown at the bottom of the SAS printout in Figure SIA12.5. The p-value of the test (highlighted) is less than .0001. Since this p-value is smaller than $\alpha = .01$, there is sufficient evidence to reject H_0. That is, there is evidence to indicate that at least one of the three higher order terms is a useful predictor of auction price.

To improve the model, we now add terms for the qualitative variables VIEW (x_3), END (x_4), FURNISH (x_5), and AUCTION (x_6). The developer theorizes that the impact of floor height and distance from the elevator on price will vary with the unit's view. Consequently, we also add interaction between floor and view and between distance and view. The complete model takes the form

$$E(y) = \beta_0 + \beta_1 x_1 + \beta_2 x_2 + \beta_3 x_1 x_2 + \beta_4(x_1)^2$$
$$+ \beta_5(x_2)^2 + \beta_6 x_3 + \beta_7 x_3 x_1 + \beta_8 x_3 x_2$$
$$+ \beta_9 x_3 x_1 x_2 + \beta_{10} x_3(x_1)^2 + \beta_{11} x_3(x_2)^2$$
$$+ \beta_{12} x_4 + \beta_{13} x_5 + \beta_{14} x_6$$

The SAS printout for this complete model is shown in Figure SIA12.6. The overall model is statistically useful (p-value <.0001 for global F-test), explaining about 68% (adjusted $R^2 = .6815$) of the sample variation in auction prices. The model standard deviation, $s = 19$, implies that we can predict price to within about 38 hundred dollars. Both the adjusted R^2 and $2s$ values are improvements over the corresponding values for the first-order model of the previous Statistics in Action Revisited (p. 704).

To test the developer's theory of how the view affects the sales price relationship, we conduct a nested-model F-test of all the VIEW (x_3) interaction terms. The null hypothesis of interest is $H_0: \beta_7 = \beta_8 = \beta_9 = \beta_{10} = \beta_{11} = 0$, and the reduced model takes the form

$$E(y) = \beta_0 + \beta_1 x_1 + \beta_2 x_2 + \beta_3 x_1 x_2$$
$$+ \beta_4(x_1)^2 + \beta_5(x_2)^2 + \beta_6 x_3 + \beta_{12} x_4$$
$$+ \beta_{13} x_5 + \beta_{14} x_6$$

The p-value of the test (highlighted at the bottom of the SAS printout in Figure SIA12.6) is less than .0001. Since this p-value is smaller than $\alpha = .01$, there is sufficient evidence to conclude that at least one of the view interaction terms is useful in predicting the auction price. This implies, as theorized by the developer, that the price–floor and price–distance relationships depend on the unit's view (ocean view or not).

(continued)

Statistics IN Action
(continued)

Dependent Variable: PRICE100

Number of Observations Read   209
Number of Observations Used   209

Analysis of Variance

Source	DF	Sum of Squares	Mean Square	F Value	Pr > F
Model	14	167924	11995	32.79	<.0001
Error	194	70968	365.81593		
Corrected Total	208	238893			

Root MSE	19.12632	R-Square	0.7029
Dependent Mean	201.28708	Adj R-Sq	0.6815
Coeff Var	9.50201		

Parameter Estimates

Variable	DF	Parameter Estimate	Standard Error	t Value	Pr > |t|
Intercept	1	188.72646	13.15224	14.35	<.0001
FLOOR	1	-4.61416	4.49901	-1.03	0.3064
DISTANCE	1	-2.35297	2.43410	-0.97	0.3349
FLR_DIST	1	-0.33458	0.20177	-1.66	0.0989
FLOORSQ	1	1.01858	0.42699	2.39	0.0180
DISTSQ	1	0.29095	0.14111	2.06	0.0406
VIEW	1	74.33636	17.73275	4.19	<.0001
VU_FLOOR	1	-4.77034	5.90393	-0.81	0.4201
VU_DIST	1	-1.62826	3.17204	-0.51	0.6083
VU_FLR_DIST	1	0.07496	0.27422	0.27	0.7849
VU_FLRSQ	1	-0.26670	0.57992	-0.46	0.6461
VU_DISTSQ	1	0.13340	0.18392	0.73	0.4691
END	1	-16.27750	7.90038	-2.06	0.0407
FURNISH	1	7.96051	2.99043	2.66	0.0084
AUCTION	1	-25.59314	3.66294	-6.99	<.0001

Figure SIA12.6

SAS regression printout for the complete second-order model of condo sale price—qualitative variables added

Test VUINT Results for Dependent Variable PRICE100

Source	DF	Mean Square	F Value	Pr > F
Numerator	5	2565.21064	7.01	<.0001
Denominator	194	365.81593		

Exercises 12.117–12.134

Understanding the Principles

12.117 Determine which pairs of models that follow are nested models. For each pair of nested models, identify the complete and reduced model.
  a. $E(y) = \beta_0 + \beta_1 x_1 + \beta_2 x_2$
  b. $E(y) = \beta_0 + \beta_1 x_1$
  c. $E(y) = \beta_0 + \beta_1 x_1 + \beta_2 x_1^2$
  d. $E(y) = \beta_0 + \beta_1 x_1 + \beta_2 x_2 + \beta_3 x_1 x_2$
  e. $E(y) = \beta_0 + \beta_1 x_1 + \beta_2 x_2 + \beta_3 x_1 x_2 + \beta_4 x_1^2 + \beta_5 x_2^2$

12.118 Explain why the *F*-test used to compare complete and reduced models is a one-tailed, upper-tailed test.

12.119 What is a parsimonious model?

Learning the Mechanics

12.120 Suppose you fit the regression model
  NW
$$y = \beta_0 + \beta_1 x_1 + \beta_2 x_2 + \beta_3 x_1 x_2 + \beta_4 x_1^2 + \beta_5 x_2^2 + \varepsilon$$
to $n = 30$ data points and you wish to test
$$H_0: \beta_3 = \beta_4 = \beta_5 = 0$$
  a. State the alternative hypothesis H_a.
  b. Give the reduced model appropriate for conducting the test.

c. What are the numerator and denominator degrees of freedom associated with the *F*-statistic?
  d. Suppose the SSE's for the reduced and complete models are $SSE_R = 1,250.2$ and $SSE_C = 1,125.2$, respectively. Conduct the hypothesis test and interpret the results. Use $\alpha = .05$.

12.121 The complete model
$$y = \beta_0 + \beta_1 x_1 + \beta_2 x_2 + \beta_3 x_3 + \beta_4 x_4 + \varepsilon$$
was fit to $n = 22$ data points, with SSE = 146.99. The reduced model $y = \beta_0 + \beta_1 x_1 + \beta_2 x_2 + \varepsilon$ was also fit, with SSE = 165.37.
  a. How many β parameters are in the complete model? The reduced model?
  b. Specify the null and alternative hypotheses you would use to investigate whether the complete model contributes more information for the prediction of y than the reduced model.
  c. Conduct the hypothesis test of part **b.** Use $\alpha = .10$. Determine the test statistic.

Applying the Concepts—Basic

12.122 Willingness to donate organs. How willing are you to donate your organs upon death? This was the question of interest in research published in *Death Studies*

(Vol. 38, 2014). A sample of $n = 307$ Australians participated in the study. Willingness to donate organs upon death was measured for each participant on a 7-point scale (1 = very unlikely, 7 = very likely). Multiple regression was used to model *willingness* (y) as a function of general *attitude* (x_1) toward donating organs (28-point scale), *subjective norm* (x_2)—assessment of whether others would approve of organ donation (14-point scale), *prototype favorability* (x_3)—opinion of those who do donate organs (14-point scale), *prototype similarity* (x_4)—similarity to organ donors (14-point scale), *age* (x_5) in years, and *gender* (x_6).

a. Write a first-order, main-effects model for $E(y)$ as a function of the six independent variables.

b. The model, part **a**, resulted in $R^2 = .39$ and global F p-value < .001. Interpret these results.

c. What term would you add to the model, part **a**, to allow for the effect of prototype favorability on willingness to depend on prototype similarity?

d. The model, part **c**, resulted in $R^2 = .41$ and global F p-value < .001. Interpret these results.

e. Are the models, parts **a** and **c**, nested models?

f. A nested model F-test comparing the models resulted in a test statistic of $F = 10.94$ with a p-value of .001. Interpret these results.

12.123 Shared leadership in airplane crews. Refer to the *Human Factors* (Mar. 2014) study of shared leadership by the cockpit and cabin crews of a commercial airplane, Exercise 9.13 (p. 476). Recall that simulated flights were taken by 84 six-person crews, where each crew consisted of a two-person cockpit (captain and first officer) and a four-person cabin team (three flight attendants and a purser). During the simulation, smoke appeared in the cabin and the reactions of the crew were monitored for teamwork. One key variable in the study was team goal attainment score, measured on a 0-to-60-point scale. Multiple regression analysis was used to model team goal attainment (y) as a function of the independent variables job experience of purser (x_1), job experience of head flight attendant (x_2), gender of purser (x_3), gender of head flight attendant (x_4), leadership score of purser (x_5), and leadership score of head flight attendant (x_6).

a. Write a complete, first-order model for $E(y)$ as a function of the six independent variables.

b. Consider a test of whether the leadership score of either the purser or head flight attendant (or both) is statistically useful for predicting team goal attainment. Give the null and alternative hypotheses as well as the reduced model for this test.

c. The two models were fit to the data for the $n = 60$ successful cabin crews with the following results: $R^2 = .02$ for reduced model, $R^2 = .25$ for complete model. Based on this information only, give your opinion regarding the null hypothesis for successful cabin crews.

d. The p-value of the subset F-test for comparing the two models for successful cabin crews was reported in the article as $p < .05$. Formally test the null hypothesis using $\alpha = .05$. What do you conclude?

e. The two models were also fit to the data for the $n = 24$ unsuccessful cabin crews with the following results: $R^2 = .14$ for reduced model, $R^2 = .15$ for complete model. Based on this information only, give your opinion regarding the null hypothesis for unsuccessful cabin crews.

f. The p-value of the subset F-test for comparing the two models for unsuccessful cabin crews was reported in the article as $p > .10$. Formally test the null hypothesis using $\alpha = .05$. What do you conclude?

12.124 Mental health of a community. An article in the *Community Mental Health Journal* (Aug. 2000) used multiple-regression analysis to model the level of community adjustment of clients of the Department of Mental Health and Addiction Services in Connecticut. The dependent variable, community adjustment (y), was measured quantitatively on the basis of staff ratings of the clients. (Lower scores indicate better adjustment.) The complete model was a first-order model with 21 independent variables. The independent variables were categorized as demographic (four variables), diagnostic (seven variables), treatment (four variables), and community (six variables).

a. Write the equation of $E(y)$ for the complete model.

b. Give the null hypothesis for testing whether the seven diagnostic variables contribute information relevant to the prediction of y.

c. Give the equation of the reduced model appropriate for the test suggested in part **b**.

d. The test in part **b** was carried out and resulted in a test statistic of $F = 59.3$ and p-value <.0001. Interpret this result in the words of the problem.

12.125 Workplace bullying and intention to leave. Refer to the *Human Resource Management Journal* (Oct. 2008) study of workplace bullying, Exercise 12.113 (p. 743). Recall that multiple regression was used to model an employee's intention to leave (y) as a function of bullying (x_1, measured on a quantitative scale) and perceived organizational support (measured qualitatively as "low POS," "neutral POS," or "high POS"). In Exercise 12.113**b**, you wrote a model for $E(y)$ as a function of bullying and POS that hypothesizes three parallel straight lines, one for each level of POS. In Exercise 12.113**c**, you wrote a model for $E(y)$ as a function of bullying and POS that hypothesizes three nonparallel straight lines, one for each level of POS.

a. Explain why the two models are nested. Which is the complete model? Which is the reduced model?

b. Give the null hypothesis for comparing the two models.

c. If you reject H_0 in part **b**, which model do you prefer? Why?

d. If you fail to reject H_0 in part **b**, which model do you prefer? Why?

12.126 Cooling method for gas turbines. Refer to the *Journal of Engineering for Gas Turbines and Power* (Jan. 2005) study of a high-pressure inlet fogging method for a gas turbine engine, presented in Exercise 12.27 (p. 701). Consider a model for the heat rate (kilojoules per kilowatt per hour) produced by a gas turbine as a function of cycle speed (revolutions per minute) and cycle pressure ratio.

a. Write a complete second-order model for heat rate (y).

b. Give the null and alternative hypotheses for determining whether the curvature terms in the complete second-order model are statistically useful in predicting the heat rate (y).

MINITAB output for Exercise 12.126

Complete Model

The regression equation is
HEATRATE = 15583 + 0.078 RPM - 523 CPRATIO + 0.00445 RPM_CPR - 0.000000 RPMSQ
      | 0.04 CPRSQ

S = 563.513   R-Sq = 88.5%   R-Sq(adj) = 87.5%

Analysis of Variance

Source	DF	SS	MS	F	P
Regression	5	148526859	29705372	93.55	0.000
Residual Error	61	19370350	317547		
Total	66	167897208			

Reduced Model

The regression equation is
HEATRATE = 12065 + 0.170 RPM - 146 CPRATIO - 0.00242 RPM_CPR

S = 633.842   R-Sq = 84.9%   R-Sq(adj) = 84.2%

Analysis of Variance

Source	DF	SS	MS	F	P
Regression	3	142586570	47528857	118.30	0.000
Residual Error	63	25310639	401756		
Total	66	167897208			

c. For the test in part **b**, identify the complete and reduced models.

d. Portions of the MINITAB printouts for the two models are shown above. Find the values of SSE_R, SSE_C, and MSE_C on the printouts.

e. Compute the value of the test statistics for the test of part **b**.

f. Find the rejection region for the test of part **b**. Use $\alpha = .10$.

g. State the conclusion of the test in the words of the problem.

12.127 **Study of supervisor-targeted aggression.** "Moonlighters" are workers who hold two jobs at the same time. What are the factors that affect the likelihood of a moonlighting worker becoming aggressive toward his or her supervisor? This was the research question of interest in the *Journal of Applied Psychology* (July 2005). Completed questionnaires were obtained from $n = 105$ moonlighters, and the data were used to fit several multiple-regression models for supervisor-targeted aggression score (y). Two of the models (with R^2 values in parentheses) are shown in the table below.

a. Interpret the R^2 values for the models.

b. Give the null and alternative hypotheses for comparing the fits of Models 1 and 2.

c. Are the two models nested? Explain.

d. The nested F-test for comparing the two models resulted in $F = 42.13$ and p-value $< .001$. What can you conclude from these results?

e. A third model was fit, one that hypothesizes all possible pairs of interactions between self-esteem, history of aggression, interactional injustice at primary job, and abusive supervisor at primary job. Give the equation of this model (Model 3).

f. A nested F-test to compare Models 2 and 3 resulted in p-value $> .10$. What can you conclude from this result?

Applying the Concepts—Intermediate

12.128 **Reality TV and cosmetic surgery.** Refer to the *Body Image: An International Journal of Research* (Mar. 2010) study of the influence of reality TV shows on one's desire to undergo cosmetic surgery, Exercise 12.25 (p. 700). Recall that psychologists modeled desire to have cosmetic surgery (y) as a function of gender (x_1), self-esteem (x_2), body satisfaction (x_3), and impression of reality TV (x_4). The psychologists theorize that one's impression of reality TV will "moderate" the impact that each of the first three independent variables has on one's desire to have cosmetic surgery. If so, then x_4 will interact with each of the other independent variables.

Models for Exercise 12.127

Model 1: $E(y) = \beta_0 + \beta_1(\text{Age}) + \beta_2(\text{Gender})$
         $+ \beta_3(\text{Interaction injustice at secondary job}) + \beta_4(\text{Abusive supervisor at secondary job})$
($R^2 = .101$)

Model 2: $E(y) = \beta_0 + \beta_1(\text{Age}) + \beta_2(\text{Gender})$
         $+ \beta_3(\text{Interactional injustice at secondary job}) + \beta_4(\text{Abusive supervisor at secondary job})$
         $+ \beta_5(\text{Self-esteem}) + \beta_6(\text{History of aggression}) + \beta_7(\text{Interactional injustice at primary job})$
         $+ \beta_8(\text{Abusive supervisor at primary job})$
($R^2 = .555$)

a. Give the equation of the model for $E(y)$ that matches the theory.

b. Fit the model, part **a**, to the data saved in the **IMAGE** file. Evaluate the overall utility of the model.

c. Give the null hypothesis for testing the psychologists' theory.

d. Conduct a nested model F-test to test the theory. What do you conclude?

12.129 Personality traits and job performance. Refer to the *Journal of Applied Psychology* (Jan. 2011) study of the relationship between task performance and conscientiousness, Exercise 12.111 (p. 742). Recall that y = task performance score (measured on a 30-point scale) was modeled as a function of x_1 = conscientiousness score (measured on a scale of -3 to $+3$) and x_2 = {1 if highly complex job, 0 if not} using the complete model

$$E(y) = \beta_0 + \beta_1 x_1 + \beta_2 (x_1)^2 + \beta_3 x_2 + \beta_4 x_1 x_2 + \beta_5 (x_1)^2 x_2$$

a. Specify the null hypothesis for testing the overall adequacy of the model.

b. Specify the null hypothesis for testing whether task performance score (y) and conscientiousness score (x_1) are curvilinearly related.

c. Specify the null hypothesis for testing whether the curvilinear relationship between task performance score (y) and conscientiousness score (x_1) depends on job complexity (x_2).

d. Explain how each of the tests, parts **a–c**, should be conducted (i.e., give form of the test statistic and the reduced model).

12.130 Glass as a waste encapsulant. The encapsulation of waste in glass is considered to be a promising solution to the problem of low-level nuclear waste. A study was undertaken jointly by the Department of Materials Science and Engineering at the University of Florida and the U.S. Department of Energy to assess the utility of glass as a waste encapsulant.* Corrosive chemical solutions (called corrosion baths) were prepared and applied directly to glass samples containing one of three types of waste (TDS-3A, FE, and AL); the chemical reactions were observed over time. A few of the key variables measured were

y = Amount of silicon (in parts per million) found in solution at end of experiment (This is both a measure of the degree of breakdown in the glass and a proxy for the amount of radioactive species released into the environment.)

x_1 = Temperature (°C) of the corrosion bath

x_2 = 1 if waste type TDS-3A, 0 if not

x_3 = 1 if waste type FE, 0 if not

(Waste type AL is the base level.) Suppose we want to model amount y of silicon as a function of temperature (x_1) and type of waste (x_2, x_3).

a. Write a model that proposes parallel straight-line relationships between amount of silicon and temperature, one line for each of the three types of waste.

b. Add terms for the interaction between temperature and waste type to the model from part **a**.

c. Refer to the model from part **b**. For each type of waste, give the slope of the line relating amount of silicon to temperature.

d. Explain how you could test for the presence of temperature–type-of-waste interaction.

12.131 Whales entangled in fishing gear. Refer to the *Marine Mammal Science* (Apr. 2010) study of whales entangled in fishing gear, Exercise 12.109 (p. 741). A first-order model for the length (y) of an entangled whale that is a function of water depth of the entanglement (x_1) and gear type (set nets, pots, or gill nets) is written as follows: $E(y) = \beta_0 + \beta_1 x_1 + \beta_2 x_2 + \beta_3 x_3 + \beta_4 x_1 x_2 + \beta_5 x_1 x_3$, where x_2 = {1 if set net, 0 if not} and x_2 = {1 if pot, 0 if not}. Consider this model the complete model in a nested model F-test.

a. Suppose you want to determine if there are any differences in the mean lengths of entangled whales for the three gear types. Give the appropriate null hypothesis to test.

b. Refer to part **a**. Give the reduced model for the test.

c. Refer to parts **a** and **b**. If you reject the null hypothesis, what would you conclude?

d. Suppose you want to determine if the rate of change of whale length (y) with water depth (x_1) is the same for all three types of fishing gear. Give the appropriate null hypothesis to test.

e. Refer to part **d**. Give the reduced model for the test.

f. Refer to parts **d** and **e**. If you fail to reject the null hypothesis, what would you conclude?

12.132 Agreeableness, gender, and wages. Refer to the *Journal of Personality and Social Psychology* (Feb. 2012) study of on-the-job agreeableness and wages, Exercise 12.112 (p. 742). The researchers modeled mean income, $E(y)$, as a function of both agreeableness score (x_1) and a dummy variable for gender (x_2 = 1 if male, 0 if female). Suppose the researchers theorize that for either gender, income will decrease at a decreasing rate as agreeableness score increases. Consequently, they want to fit a second-order model.

a. Consider the model $E(y) = \beta_0 + \beta_1 x_1 + \beta_2 (x_1)^2 + \beta_3 x_2$. If the researchers' belief is true, what is the expected sign of β_2 in the model?

b. Draw a sketch of the model, part **a**, showing how gender affects the income–agreeableness score relationship.

c. Write a complete second-order model for $E(y)$ as a function of x_1 and x_2.

d. Draw a sketch of the model, part **c**, showing how gender affects the income–agreeableness score relationship.

e. What null hypothesis would you test in order to compare the two models, parts **a** and **c**?

f. Fit the models to the sample data saved in the **WAGAP** file and carry out the test, part **e**. What do you conclude? (Test, using $\alpha = .10$.)

12.133 Predicting thrust force of a metal drill. In *Frontiers in Automobile and Mechanical Engineering* (Nov. 2010), a model was developed to predict the thrust force when

*The background information for this exercise was provided by Dr. David Clark, Department of Materials Science and Engineering, University of Florida.

drilling into a hybrid metal composite. Three variables related to thrust force are spindle speed (revolutions per minute), feed rate (millimeters per minute), and fraction weight of silicon carbide in composite (percentage). Experimental data were collected by varying these three variables at two levels each: speed (1000 and 3000 rpm), rate (50 and 150 mm/min), and weight (5 and 15 percent). For each combination of these values, thrust force (Newtons) of the drill was measured. The data (adapted from information in the article) are listed below.

Experiment	SPEED	RATE	PCTWT	FORCE
1	1000	50	5	510
2	3000	50	5	540
3	1000	150	5	710
4	3000	150	5	745
5	1000	50	15	615
6	3000	50	15	635
7	1000	150	15	810
8	3000	150	15	850
9	1000	50	5	500
10	3000	50	5	545
11	1000	150	5	720
12	3000	150	5	730
13	1000	50	15	600
14	3000	50	15	615
15	1000	150	15	825
16	3000	150	15	840

a. Write the equation of an interaction model for thrust force as a function of spindle speed, feed rate, and fraction weight. Include all possible two-variable interaction terms in the model.
b. Give a function of the model parameters (i.e., the slope) that represent the change in force for every 1%

increase in weight when rate is fixed at 50 mm/minute and speed is fixed at 1000 rpm.
c. Give a function of the model parameters (i.e., the slope) that represent the change in force for every 1% increase in weight when rate is fixed at 150 mm/minute and speed is fixed at 1000 rpm.
d. Fit the interaction model, part **a**, to the data using statistical software. Give the least squares prediction equation.
e. Conduct a test to determine whether any of the interaction terms are statistically useful for predicting thrust force. Use $\alpha = .05$.

Applying the Concepts—Advanced

12.134 Emotional distress in firefighters. The *Journal of Human Stress* (Summer 1987) reported on a study of the "psychological response of firefighters to chemical fire." It is thought that the complete second-order model

$$E(y) = \beta_0 + \beta_1 x_1 + \beta_2 x_1^2 + \beta_3 x_2 + \beta_4 x_1 x_2 + \beta_5 x_1^2 x_2$$

where

$$y = \text{Emotional distress}$$
$$x_1 = \text{Experience (years)}$$
$$x_2 = 1 \text{ if exposed to chemical fire, } 0 \text{ if not}$$

will be adequate to describe the relationship between emotional distress and years of experience for two groups of firefighters: those exposed to a chemical fire and those not exposed.

a. How would you determine whether the *rate* of increase of emotional distress with experience is different for the two groups of firefighters?
b. How would you determine whether there are differences in mean emotional distress *levels* that are attributable to exposure group?

12.10 Stepwise Regression (Optional)

Consider the problem of predicting the salary y of an executive. Perhaps the biggest problem in building a model to describe executive salaries is choosing the important independent variables to be included. The list of potentially important independent variables is extremely long (e.g., age, experience, tenure, education level, etc.), and we need some objective method of screening out those that are not important.

The problem of deciding which of a large set of independent variables to include in a model is a common one. Trying to determine which variables influence the profit of a firm, affect the blood pressure of humans, or are related to a student's performance in college are only a few examples.

A systematic approach to building a model with a large number of independent variables is difficult because the interpretation of multivariable interactions and higher order terms is tedious. We therefore turn to a screening procedure, available in most statistical software packages, known as *stepwise regression*.

The most commonly used **stepwise regression** procedure works as follows: The user first identifies the response y and the set of potentially important independent variables $x_1, x_2, \ldots, x_k$, where k is generally large. [*Note:* This set of variables could include both first-order and higher order terms. However, we often include only the main effects of both quantitative variables (first-order terms) and qualitative variables (dummy variables), since the inclusion of second-order terms greatly increases the number of independent variables.] The response and independent variables are then entered into the computer software, and the stepwise procedure begins.

Step 1 The software program fits all possible one-variable models of the form

$$E(y) = \beta_0 + \beta_1 x_i$$

to the data, where x_i is the ith independent variable, $i = 1, 2, \ldots, k$. For each model, the t-test (or the equivalent F-test) for a single β parameter is conducted to test the null hypothesis

$$H_0: \beta_1 = 0$$

against the alternative hypothesis

$$H_a: \beta_1 \neq 0$$

The independent variable that produces the largest (absolute) t-value is then declared the best one-variable predictor of y.* Call this independent variable x_1.

Step 2 The stepwise program now begins to search through the remaining $(k - 1)$ independent variables for the best two-variable model of the form

$$E(y) = \beta_0 + \beta_1 x_1 + \beta_2 x_i$$

This is done by fitting all two-variable models containing x_1 and each of the other $(k - 1)$ options for the second variable x_i. The t-values for the test $H_0: \beta_2 = 0$ are computed for each of the $(k - 1)$ models (corresponding to the remaining independent variables x_i, $i = 2, 3, \ldots, k$), and the variable having the largest t is retained. Call this variable x_2. At this point, some software packages diverge in methodology. The better packages now go back and check the t-value of $\hat{\beta}_1$ after $\hat{\beta}_2 x_2$ has been added to the model. If the t-value has become nonsignificant at some specified α level (say, $\alpha = .05$), the variable x_1 is removed and a search is made for the independent variable with a β parameter that will yield the most significant t-value in the presence of $\hat{\beta}_2 x_2$. Other packages do not recheck the significance of $\hat{\beta}_1$, but proceed directly to step 3.†

The reason the t-value for x_1 may change from step 1 to step 2 is that the meaning of the coefficient $\hat{\beta}_1$ changes. In step 2, we are approximating a complex response surface in two variables by a plane. The best-fitting plane may yield a different value for $\hat{\beta}_1$ than that obtained in step 1. Thus, both the value of $\hat{\beta}_1$ and its significance usually change from step 1 to step 2. For this reason, the software packages that recheck the t-values at each step are preferred.

Step 3 The stepwise procedure now checks for a third independent variable to include in the model with x_1 and x_2. That is, we seek the best model of the form

$$E(y) = \beta_0 + \beta_1 x_1 + \beta_2 x_2 + \beta_3 x_i$$

To do this, we fit all the $(k - 2)$ models using x_1, x_2, and each of the $(k - 2)$ remaining variables x_i as a possible x_3. The criterion is again to include the independent variable with the largest t-value. Call this best third variable x_3. The better programs now recheck the t-values corresponding to the x_1- and x_2-coefficients, replacing the variables that yield nonsignificant t-values. This procedure is continued until no further independent variables can be found that yield significant t-values (at the specified α level) in the presence of the variables already in the model.

The result of the stepwise procedure is a model containing only those terms with t-values that are significant at the specified α level. Thus, in most practical situations, only several of the large number of independent variables remain. However, it is very important *not* to jump to the conclusion that all the independent variables that are important in predicting y have been identified or that the unimportant independent variables have

---

*Note that the variable with the largest t-value is also the one with the largest (absolute) Pearson product moment correlation r (Section 11.5) with y.

†*Forward selection* is the name given to stepwise routines that *do not* recheck the significance of each previously entered independent variable. This is in contrast to *stepwise selection* routines that recheck the significance of each entered term. A third approach is to use *backward selection*, where initially all terms are entered and then eliminated one by one.

been eliminated. Remember, the stepwise procedure is using only *sample estimates* of the true model coefficients (β's) to select the important variables. An extremely large number of single β parameter *t*-tests have been conducted, and the probability is very high that one or more errors have been made in including or excluding variables. That is, we have very likely included some unimportant independent variables in the model (Type I errors) and eliminated some important ones (Type II errors).

There is a second reason we might not have arrived at a good model. When we choose the variables to be included in the stepwise regression, we often omit high-order terms (to keep the number of variables manageable). Consequently, we may have initially omitted several important terms from the model. Thus, we should recognize step-wise regression for what it is: an **objective variable-screening procedure**.

Successful model builders will now consider second-order terms (for quantitative variables) and other interactions among variables screened by the stepwise procedure. Indeed, it would be best to develop this response surface model with a second set of data independent of those used for the screening, so that the results of the stepwise procedure can be partially verified with new data. This is not always possible, however, because in many modeling situations only a small amount of data is available.

Do not be deceived by the impressive-looking *t*-values that result from the step-wise procedure: it has retained only the independent variables with the largest *t*-values. Also, be certain to consider second-order terms in systematically developing the pre-diction model. Finally, if you have used a first-order model for your stepwise procedure, remember that it may be greatly improved by the addition of higher order terms.

⚠ **CAUTION** Be wary of using the results of stepwise regression to make inferences about the relationship between $E(y)$ and the independent variables in the resulting first-order model. First, an extremely large number of *t*-tests have been conducted, lead-ing to a high probability of making one or more Type I or Type II errors. Second, the stepwise model does not include any higher order or interaction terms. Stepwise regres-sion should be used only when necessary—that is, when you want to determine which of a large number of potentially important independent variables should be used in the model-building process. ▲

Example 12.13 Ⓓ
EXECSAL

Stepwise Regression— Modeling Executive Salary

Problem An international management consulting company develops multiple-regression models for executive salaries of its client firms. The consulting company has found that models that use the natural logarithm of salary as the dependent variable have better predictive power than those using salary as the dependent variable.* A preliminary step in the construction of these models is the determination of the most important independent variables. For one firm, 10 potential independent variables (7 quantitative and 3 qualitative) were measured in a sample of 100 executives. The data, described in Table 12.6, are saved in the **EXECSAL** file. Since it would be very difficult to construct a complete second-order model with all of the 10 independent variables, use stepwise regression to decide which of the 10 variables should be included in the building of the final model for the logarithm of executive salaries.

Solution We will use stepwise regression with the main effects of the 10 independent variables to identify the most important variables. The dependent variable y is the natural logarithm of the executive salaries. The MINITAB stepwise regression printout is shown in Figure 12.32.

Note that the first variable included in the model is x_1, years of experience. At the second step, x_3, a dummy variable for the qualitative variable bonus eligibility, is entered into the model. In steps 3, 4, and 5, the variables x_4 (number of employees supervised), x_2 (years of education), and x_5 (corporate assets), respectively, are selected for inclusion in the model. MINITAB stops after five steps because no other independent variables met the criterion for admission into the model. As a default, MINITAB uses $\alpha = .15$ in

*This is probably because salaries tend to be incremented in *percentages* rather than dollar values. When a response variable undergoes percentage changes as the independent variables are varied, the logarithm of the response variable will be more suitable as a dependent variable.

Table 12.6	Independent Variables in the Executive Salary Example
Independent Variable	Description
x_1	Experience (years)—quantitative
x_2	Education (years)—quantitative
x_3	Bonus eligibility (1 if yes, 0 if no)—qualitative
x_4	Number of employees supervised—quantitative
x_5	Corporate assets (millions of dollars)—quantitative
x_6	Board member (1 if yes, 0 if no)—qualitative
x_7	Age (years)—quantitative
x_8	Company profits (past 12 months, millions of dollars)—quantitative
x_9	Has international responsibility (1 if yes, 0 if no)—qualitative
x_{10}	Company's total sales (past 12 months, millions of dollars)—quantitative

Ⓓ *Data Set:* EXECSAL

Figure 12.32
MINITAB stepwise regression printout for executive salary data

```
Alpha-to-Enter: 0.15   Alpha-to-Remove: 0.15


Response is Y on 10 predictors, with N = 100


Step              1       2        3        4        5
Constant     11.091  10.968   10.783   10.278    9.962

X1           0.0278  0.0273   0.0273   0.0273   0.0273
T-Value       12.62   15.13    18.80    24.68    26.50
P-Value       0.000   0.000    0.000    0.000    0.000

X3                    0.197    0.233    0.232    0.225
T-Value                7.10    10.17    13.30    13.74
P-Value               0.000    0.000    0.000    0.000

X4                           0.00048  0.00055  0.00052
T-Value                         7.32    10.92    11.06
P-Value                        0.000    0.000    0.000

X2                                     0.0300   0.0291
T-Value                                  8.38     8.72
P-Value                                 0.000    0.000

X5                                              0.00196
T-Value                                            3.95
P-Value                                           0.000

S             0.161   0.131    0.106   0.0807   0.0751
R-Sq          61.90   74.92    83.91    90.75    92.06
R-Sq(adj)     61.51   74.40    83.41    90.36    91.64
Mallows C-p   343.9   195.5     93.8     16.8      3.6
```

the *t*-tests conducted. In other words, if the *p*-value associated with a β coefficient exceeds $\alpha = .15$, the variable is *not* included in the model.

The results of the stepwise regression suggest that we should concentrate on the preceding five independent variables. Models with second-order terms and interactions should be proposed and evaluated to determine the best model for predicting executive salaries.

■ **Now Work Exercise 12.137**

We conclude this section with some advice on the use of stepwise regression.

⚠ **RECOMMENDATION** Do *not* use the stepwise regression model as the final model for predicting *y*. Recall that the stepwise procedure tends to perform a large number of *t*-tests, inflating the overall probability of a Type I error, and does not automatically include higher order terms (e.g., interactions and squared terms) in the final model. Use stepwise regression as a variable-screening tool when there exists a large number of potentially important independent variables. Then begin building models for *y*, using the variables identified by stepwise regression. ▲

Exercises 12.135–12.145

Understanding the Principles

12.135 Explain the difference between a stepwise model and a standard regression model.

12.136 Give two caveats associated with using the stepwise regression results as the final model for predicting *y*.

Learning the Mechanics

12.137 Suppose there are six independent variables $x_1, x_2, x_3,$ $x_4, x_5,$ and x_6 that might be useful in predicting a response *y*. A total of $n = 50$ observations is available, and it is decided to employ stepwise regression to help in selecting the independent variables that appear to be useful. The computer fits all possible one-variable models of the form

$$E(y) = \beta_0 + \beta_1 x_i$$

where x_i is the *i*th independent variable, $i = 1, 2, \ldots, 6$. The information in the following table is provided from the computer printout:

Independent Variable	$\hat{\beta}_i$	$s_{\hat{\beta}_i}$
x_1	1.6	.42
x_2	−.9	.01
x_3	3.4	1.14
x_4	2.5	2.06
x_5	−4.4	.73
x_6	.3	.35

a. Which independent variable is declared the best one-variable predictor of *y*? Explain.
b. Would this variable be included in the model at this stage? Explain.
c. Describe the next phase that a stepwise procedure would execute.

Applying the Concepts—Basic

12.138 **Distribution of yellowhammer birds.** An article in the *Journal of Animal Ecology* (July 2006) discussed some of the caveats of using stepwise regression. An example presented used stepwise regression to find the best predictors of the number of yellowhammers inhabiting lowland farms in the United Kingdom. Nine potential independent variables were considered: (1) hedge presence, (2) tree-line presence, (3) ditch presence, (4) adjacent road, (5) width of margin, (6) adjacent pasture, (7) adjacent silage ley, (8) winter rape, and (9) adjacent beans.
a. How many one-variable models are fit in step 1 of the stepwise regression?
b. After selecting ditch presence in step 1, how many two-variable models are fit in step 2 of the stepwise regression?

c. After selecting adjacent pasture in step 2, how many three-variable models are fit in step 3 of the stepwise regression?
d. Through the first three steps of the stepwise regression, determine the total number of *t*-tests performed. Assuming each test uses an $\alpha = .05$ level of significance, give an estimate of the probability of at least one Type I error in the stepwise regression.

12.139 **Study of sex offenders.** A study of sex offenders in the Canadian Federal Prison System was published in the *British Journal of Criminology* (May 2014). The following data were collected for each of 59 male sex offenders: race, type of crime (violent or nonviolent), age at first conviction, and total number of criminal convictions. Suppose you want to use the data to model the total number of convictions (*y*) as a function of race, type of crime, and age at first conviction. Do you recommend using stepwise regression to find the "best" model for predicting *y*? Explain. If not, outline a strategy for finding the best model.

12.140 **Accuracy of software effort estimates.** Periodically, software engineers must provide estimates of their effort in developing new software. In the *Journal of Empirical Software Engineering* (Vol. 9, 2004), multiple regression was used to predict the accuracy of these effort estimates. The dependent variable, defined as the relative error in estimating effort,

$$y = (\text{Actual effort} - \text{Estimated effort})/(\text{Actual effort})$$

was determined for each in a sample of $n = 49$ software development tasks. Eight independent variables were evaluated as potential predictors of relative error using stepwise regression. Each of these was formulated as a dummy variable, as shown in the table.

Company role of estimator: $x_1 = 1$ if developer, 0 if project leader
Task complexity: $x_2 = 1$ if low, 0 if medium/high
Contract type: $x_3 = 1$ if fixed price, 0 if hourly rate
Customer importance: $x_4 = 1$ if high, 0 if low/medium
Customer priority: $x_5 = 1$ if time of delivery, 0 if cost or quality
Level of knowledge: $x_6 = 1$ if high, 0 if low/medium
Participation: $x_7 = 1$ if estimator participates in work, 0 if not
Previous accuracy: $x_8 = 1$ if more than 20% accurate, 0 if less than 20% accurate

---

a. In step 1 of the stepwise regression, how many different one-variable models are fitted to the data?
b. In step 1, the variable x_1 is selected as the "best" one-variable predictor. How is this determined?

c. In step 2 of the stepwise regression, how many different two-variable models (where x_1 is one of the variables) are fitted to the data?

d. The only two variables selected for entry into the stepwise regression model were x_1 and x_8. The stepwise regression yielded the following prediction equation:

$$\hat{y} = .12 - .28x_1 + .27x_8$$

Give a practical interpretation of the β estimates multiplied by x_1 and x_8.

e. Why should a researcher be wary of using the model, part **d**, as the final model for predicting effort (y)?

12.141 An analysis of footprints in sand. Fossilized human footprints provide a direct source of information on the gait dynamics of extinct species. To gain insight into this phenomenon, a group of scientists used human subjects (16 young adults) to generate footprints in sand (*American Journal of Physical Anthropology*, Apr. 2010). One dependent variable of interest was *heel depth* (y) of the footprint (in millimeters). The scientists wanted to find the best predictors of depth from among six possible independent variables. Three variables were related to the human subject (*foot mass, leg length*, and *foot type*) and three variables were related to walking in sand (*velocity, pressure*, and *impulse*). A stepwise regression run on these six variables yielded the following results:

Selected independent variables: *pressure* and *leg length*
$R^2 = .771$, Global F-test p-value $< .001$

---

a. Write the hypothesized equation of the final stepwise regression model.

b. Interpret the value of R^2 for the model.

c. Conduct a test of the overall utility of the final stepwise model.

d. At minimum, how many t-tests on individual β's were conducted to arrive at the final stepwise model?

e. Based on your answer to part **d**, comment on the probability of making at least one Type I error during the stepwise analysis.

12.142 Yield strength of steel alloy. Industrial engineers at the University of Florida used regression modeling as a tool to reduce the time and cost associated with developing new metallic alloys (*Modelling and Simulation in Materials Science and Engineering*, Vol. 13, 2005). To illustrate, the engineers build a regression model for the tensile yield strength (y) of a new steel alloy. The potential important predictors of yield strength are listed in the following table.

---

x_1 = Carbon amount (% weight)
x_2 = Manganese amount (% weight)
x_3 = Chromium amount (% weight)
x_4 = Nickel amount (% weight)
x_5 = Molybdenum amount (% weight)
x_6 = Copper amount (% weight)
x_7 = Nitrogen amount (% weight)
x_8 = Vanadium amount (% weight)
x_9 = Plate thickness (millimeters)
x_{10} = Solution treating (millimeters)
x_{11} = Ageing temperature (degrees Celsius)

---

a. The engineers used stepwise regression to search for a parsimonious set of predictor variables. Do you agree with this decision? Explain.

b. The stepwise regression selected the following independent variables: x_1 = Carbon, x_2 = Manganese, x_3 = Chromium, x_5 = Molybdenum, x_6 = Copper, x_8 = Vanadium, x_9 = Plate thickness, x_{10} = Solution treating, and x_{11} = Ageing temperature. On the basis of this information, determine the total number of first-order models that were fit in the stepwise routine.

c. Refer to part **b**. All the variables listed there were statistically significant in the stepwise model, with $R^2 = .94$. Consequently, the engineers used the estimated stepwise model to predict yield strength. Do you agree with this decision? Explain.

Applying the Concepts—Intermediate

12.143 Bus rapid-transit study. The Center for Urban Transportation Research (CUTR) at the University of South Florida conducted a survey of Bus rapid transit (BRT) customers in Miami (*Transportation Research Board* Annual Meeting, Jan. 2003). Data on the following variables (all measured on a five-point scale, where 1 = "very unsatisfied" and 5 = "very satisfied") were collected for a sample of over 500 bus riders: overall satisfaction with BRT (y), safety on bus (x_1), seat availability (x_2), dependability (x_3), travel time (x_4), cost (x_5), information/maps (x_6), convenience of routes (x_7), traffic signals (x_8), safety at bus stops (x_9), hours of service (x_{10}), and frequency of service (x_{11}). CUTR analysts used stepwise regression to model overall satisfaction (y).

a. How many models are fitted at step 1 of the stepwise regression?

b. How many models are fitted at step 2 of the stepwise regression?

c. How many models are fitted at step 11 of the stepwise regression?

d. The stepwise regression selected the following eight variables to include in the model (in order of selection): $x_{11}, x_4, x_2, x_7, x_{10}, x_1, x_9$, and x_3. Write the equation for $E(y)$ that results.

e. The model in part **d** was tested and resulted in $R^2 = .677$. Interpret this value.

f. Explain why the CUTR analysts should be cautious in concluding that the "best" model for $E(y)$ has been found.

12.144 Modeling species abundance. A marine biologist was hired by the EPA to determine whether the hot-water runoff from a particular power plant located near a large gulf is having an adverse effect on the marine life in the area. The biologist's goal is to acquire a prediction equation for the number of marine animals located at certain designated areas, or stations, in the gulf. On the basis of past experience, the EPA considered the following environmental factors as predictors for the number of animals at a particular station:

x_1 = Temperature of water (TEMP)
x_2 = Salinity of water (SAL)
x_3 = Dissolved oxygen content of water (DO)
x_4 = Turbidity index, a measure of the turbidity of the water (TI)
x_5 = Depth of the water at the station (ST_DEPTH)
x_6 = Total weight of sea grasses in sampled area (TGRSWT)

SPSS output for Exercise 12.144

Variables Entered/Removed[a]

Model	Variables Entered	Variables Removed	Method
1	ST_DEPTH	·	Stepwise (Criteria: Probability-of-F-to-enter <= .050, Probability-of-F-to-remove >= .100).
2	TGRSWT	·	Stepwise (Criteria: Probability-of-F-to-enter <= .050, Probability-of-F-to-remove >= .100).
3	TI	·	Stepwise (Criteria: Probability-of-F-to-enter <= .050, Probability-of-F-to-remove >= .100).

a. Dependent Variable: LOGNUM

Model Summary

Model	R	R Square	Adjusted R Square	Std. Error of the Estimate
1	.329[a]	.122	.121	.7615773
2	.427[b]	.182	.180	.7348470
3	.432[c]	.187	.184	.7348469

a. Predictors: (Constant), ST_DEPTH

b. Predictors: (Constant), ST_DEPTH, TGRSWT

c. Predictors: (Constant), ST_DEPTH, TGRSWT, TI

As a preliminary step in the construction of this model, the biologist used a stepwise regression procedure to identify the most important of these six variables. A total of 716 samples was taken at different stations in the gulf, producing the SPSS printout shown above. (The response measured was y, the logarithm of the number of marine animals found in the sampled area.)

a. According to the printout, which of the independent variables should be used in the model?

b. Are we able to assume that the marine biologist has identified all the important independent variables for the prediction of y? Why?

c. Using the variables identified in part **a**, write the first-order model with interaction that may be used to predict y.

d. How would the marine biologist determine whether the model specified in part **c** is better than the first-order model?

e. Note the small value of R^2. What action might the biologist take to improve the model?

12.145 Using corn in a duck diet. Corn is high in starch content; consequently, it is considered excellent feed for domestic chickens. Does corn possess the same potential in feeding ducks bred for broiling? This was the subject of research published in *Animal Feed Science and Technology* (Apr. 2010). The objective of the study was to establish a prediction model for the true metabolizable energy (TME) of corn regurgitated from ducks. The researchers considered 11 potential predictors of TME: dry matter (DM), crude protein (CP), ether extract (EE), ash (ASH), crude fiber (CF), neutral detergent fiber (NDF), acid detergent fiber (ADF), gross energy (GE), amylose (AM), amylopectin (AP), and amylopectin/amylose (AMAP). Stepwise regression was used to find the best subset of predictors. The final stepwise model yielded the following results:

$$\widehat{TME} = 7.70 + 2.14(AMAP) + .16(NDF),$$
$$R^2 = .988, s = .07, \text{global } F \text{ } p\text{-value} = .001$$

a. Determine the number of t-tests performed in step 1 of the stepwise regression.

b. Determine the number of t-tests performed in step 2 of the stepwise regression.

c. Give a full interpretation of the final stepwise model regression results.

d. Explain why it is dangerous to use the final stepwise model as the "best" model for predicting TME.

e. Using the independent variables selected by the stepwise routine, write a complete second-order model for TME.

f. Refer to part **e**. How would you determine if the terms in the model that allow for curvature are statistically useful for predicting TME?

PART III: MULTIPLE REGRESSION DIAGNOSTICS

12.11 Residual Analysis: Checking the Regression Assumptions

When we apply regression analysis to a set of data, we never know for certain whether the assumptions of Section 12.1 are satisfied. How far can we deviate from the assumptions and still expect regression analysis to yield results that will have the reliability stated in this chapter? How can we detect departures (if they exist) from the assumptions, and what can we do about them? We provide some answers to these questions in this section.

Recall from Section 12.1 that, for any given set of values of $x_1, x_2, \ldots, x_k$, we assume that the random-error term ε has the following properties:

1. mean equal to 0
2. constant variance (σ^2)
3. normal probability distribution
4. probabilistically independent

It is unlikely that these assumptions are ever satisfied exactly in a practical application of regression analysis. Fortunately, experience has shown that least squares regression analysis produces reliable statistical tests, confidence intervals, and prediction intervals, as long as the departures from the assumptions are not too great. In this section, we present some methods for determining whether the data indicate significant departures from the assumptions.

Because the assumptions all concern the random-error component ε of the model, the first step is to estimate the random error. Since the actual random error associated with a particular value of y is the difference between the actual y value and its unknown mean, we estimate the error by the difference between the actual y value and the *estimated* mean. This estimated error is called the *regression residual*, or simply the **residual**, and is denoted by $\hat{\varepsilon}$. The actual error ε and residual $\hat{\varepsilon}$ are shown in Figure 12.33.

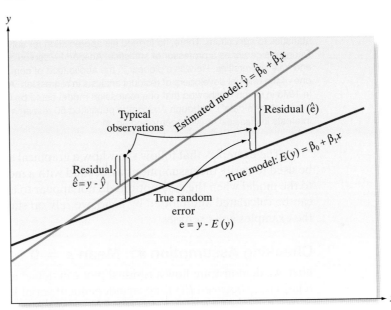

Figure 12.33

Actual random error ε and regression residual $\hat{\varepsilon}$

A **regression residual** $\hat{\varepsilon}$ is defined as the difference between an observed y-value and its corresponding predicted value:

$$\hat{\varepsilon} = (y - \hat{y}) = y - (\hat{\beta}_0 + \hat{\beta}_1 x_1 + \hat{\beta}_2 x_2 + \cdots + \hat{\beta}_k x_k)$$

Since the true mean of y (i.e., the true regression model) is not known, the actual random error cannot be calculated. However, because the residual is based on the estimated mean (the least squares regression model), it can be calculated and used to estimate the random error and to check the regression assumptions. Such checks are generally referred to as **residual analyses**. Two useful properties of residuals are given in the next box.

Properties of Regression Residuals

1. *The mean of the residuals is equal to 0.* This property follows from the fact that the sum of the differences between the observed y values and their least squares predicted $\hat{y}$ values is equal to 0:

$$\sum (\text{Residuals}) = \sum (y - \hat{y}) = 0$$

(continued)

2. *The standard deviation of the residuals is equal to the standard deviations of the fitted regression model.* This property follows from the fact that the sum of the squared residuals is equal to SSE, which, when divided by the error degrees of freedom, is equal to the variance of the fitted regression model. The square root of the variance is both the standard deviation of the residuals and the standard deviation of the regression model:

$$\sum (\text{Residuals})^2 = \sum (y - \hat{y})^2 = \text{SSE}$$

$$s = \sqrt{\frac{\sum (\text{Residuals})^2}{n - (k + 1)}} = \sqrt{\frac{\text{SSE}}{n - (k + 1)}} = \sqrt{\text{MSE}}$$

BIOGRAPHY   FRANCIS J. ANSCOMBE (1918–2001)

Anscombe's Data

British citizen Frank Anscombe grew up in a small town near the English Channel. He attended Trinity College in Cambridge, England, on a merit scholarship, graduating with first-class honors in mathematics in 1939. He earned his master's degree in 1943. During World War II, Anscombe worked for the British Ministry of Supply, developing a mathematical solution for aiming antiaircraft rockets at German bombers and buzz bombs. Following the war, Anscombe worked at the Rothamsted Experimental Station, applying statistics to agriculture. There, he formed his appreciation for solving problems with social relevance. During his career as a professor of statistics, Anscombe served on the faculty of Cambridge, Princeton, and Yale Universities. He was a pioneer in the application of computers to statistical analysis and was one of the original developers of residual analysis in regression. Anscombe is famous for a paper he wrote in 1973 in which he showed that one regression model could be fitted by four very different data sets ("Anscombe's data"). Although Anscombe published 50 research articles on statistics, he also had serious interests in classical music, poetry, and art. ■

The examples that follow show how a graphical analysis of regression residuals can be used to verify the assumptions associated with a model and to support improvements to the model when the assumptions do not appear to be satisfied. Although the residuals can be calculated and plotted by hand, we rely on statistical software for these tasks in the examples and exercises.

Checking Assumption #1: Mean $\varepsilon = 0$

First, we demonstrate how a residual plot can detect a model in which the hypothesized relationship between $E(y)$ and an independent variable x is misspecified. The assumption of mean error of 0 is violated in these types of models.*

Example   12.14

Analyzing Residuals: Electrical Usage Model

Problem Refer to the problem of modeling the relationship between home size (x) and electrical usage (y) in Example 12.7 (p. 717). The data for $n = 15$ homes are repeated in Table 12.7. MINITAB printouts for a straight-line model and a quadratic model fitted to the data are shown in Figures 12.34a and 12.34b, respectively. The residuals from these models are highlighted in the printouts. The residuals are then plotted on the vertical axis against the variable x, size of home, on the horizontal axis in Figures 12.35a and 12.35b.

a. Verify that each residual is equal to the difference between the observed y-value and the estimated mean value $\hat{y}$.

b. Analyze the plots of the residuals.

Solution

a. Consider the first usage value in the data set, $y = 1,182$. For the straight-line model, the predicted value for this observation is $\hat{y} = 1,362.2$. Consequently, the residual is

$$\hat{\varepsilon} = (y - \hat{y}) = 1,182 - 1,362.2 = -180.2 \qquad \text{(highlighted in Figure 12.34a)}$$

---

*For a misspecified model, the hypothesized mean of y, denoted by $E_h(y)$, will not equal the true mean of y, $E(y)$. Since $y = E_h(y) + \varepsilon$, then $\varepsilon = y - E_h(y)$ and $E(\varepsilon) = E\{y - E_h(y)\} = E(y) - E_h(y) \neq 0$.

Table 12.7 Home Size–Electrical Usage Data	
Size of Home, x (sq. ft)	Monthly Usage, y (kilowatt-hours)
1,290	1,182
1,350	1,172
1,470	1,264
1,600	1,493
1,710	1,571
1,840	1,711
1,980	1,804
2,230	1,840
2,400	1,956
2,710	2,007
2,930	1,984
3,000	1,960
3,210	2,001
3,240	1,928
3,520	1,945

Ⓓ Data Set: ELECTRIC

Similarly, the residual for the first y-value from the quadratic model (Figure 12.34b) is

$$\hat{\varepsilon} = 1{,}182 - 1{,}157.2 = 24.8 \qquad \text{(highlighted in Figure 12.34b)}$$

The two residuals agree (after rounding) with the first values given in the column labeled "**Residual**" in Figures 12.34a and 12.34b, respectively. Although the residuals both correspond to the same observed y value, 1,182, they differ because the predicted mean value changes, depending on whether the straight-line model or quadratic model is used. Similar calculations produce the remaining residuals.

b. The MINITAB plot of the residuals for the straight-line model (Figure 12.35a) reveals a nonrandom pattern. The residuals exhibit a curved shape, with the residuals for the small values of x below the horizontal 0 (mean of the residuals) line, the residuals corresponding to the middle values of x above the 0 line, and the residuals for the largest values of x again below the 0 line. The indication is that the mean value of the random error *within* each of these ranges of x (small, medium, large) may not be equal to 0. Such a pattern usually indicates that curvature needs to be added to the model.

When the second-order term is added to the model, the nonrandom pattern disappears. In Figure 12.35b, the residuals appear to be randomly distributed around the 0 line, as expected. Also, the variation of the residuals around the 0 line appears to be much smaller for the quadratic model. In fact, $s \approx 50$ for the quadratic model is much smaller than $s \approx 155$ for the straight-line model. The implication is that the quadratic model provides a considerably better model for predicting electrical usage.

Regression Analysis: USAGE versus SIZE

```
The regression equation is
USAGE = 903 + 0.356 SIZE

Predictor     Coef   SE Coef      T      P
Constant     903.0     132.1   6.83  0.000
SIZE       0.35594   0.05477   6.50  0.000

S = 155.251   R-Sq = 76.5%   R-Sq(adj) = 74.7%

Analysis of Variance

Source           DF        SS       MS      F      P
Regression        1   1017803  1017803  42.23  0.000
Residual Error   13    313338    24103
Total            14   1331140

Obs  SIZE   USAGE     Fit  SE Fit  Residual  St Resid
  1  1290  1182.0  1362.2    68.3    -180.2     -1.29
  2  1350  1172.0  1383.5    65.6    -211.5     -1.50
  3  1470  1264.0  1426.2    60.6    -162.2     -1.13
  4  1600  1493.0  1472.5    55.4      20.5      0.14
  5  1710  1571.0  1511.7    51.4      59.3      0.41
  6  1840  1711.0  1557.9    47.3     153.1      1.04
  7  1980  1804.0  1607.8    43.7     196.2      1.32
  8  2230  1840.0  1696.8    40.3     143.2      0.96
  9  2400  1956.0  1757.3    40.5     198.7      1.33
 10  2710  2007.0  1867.6    46.0     139.4      0.94
 11  2930  1984.0  1945.9    52.9      38.1      0.26
 12  3000  1960.0  1970.8    55.5     -10.8     -0.07
 13  3210  2001.0  2045.6    64.0     -44.6     -0.32
 14  3240  1928.0  2056.3    65.3    -128.3     -0.91
 15  3520  1945.0  2155.9    78.0    -210.9     -1.57
```

Figure 12.34a
MINITAB printout for straight-line model of electrical usage

Regression Analysis: USAGE versus SIZE, SIZESQ

```
The regression equation is
USAGE = - 807 + 1.96 SIZE - 0.000340 SIZESQ

Predictor        Coef     SE Coef      T      P
Constant       -806.7       166.9  -4.83  0.000
SIZE           1.9616      0.1525  12.86  0.000
SIZESQ    -0.00034044  0.00003212 -10.60  0.000

S = 50.1998   R-Sq = 97.7%   R-Sq(adj) = 97.3%

Analysis of Variance

Source           DF        SS      MS      F      P
Regression        2   1300900  650450  258.11  0.000
Residual Error   12     30240    2520
Total            14   1331140

Obs  SIZE   USAGE     Fit  SE Fit  Residual  St Resid
  1  1290  1182.0  1157.2    29.3      24.8      0.61
  2  1350  1172.0  1221.0    26.2     -49.0     -1.14
  3  1470  1264.0  1341.2    21.2     -77.2     -1.70
  4  1600  1493.0  1460.3    18.0      32.7      0.70
  5  1710  1571.0  1552.2    17.1      18.8      0.40
  6  1840  1711.0  1650.1    17.6      60.9      1.30
  7  1980  1804.0  1742.6    19.0      61.4      1.32
  8  2230  1840.0  1874.7    21.2     -34.7     -0.76
  9  2400  1956.0  1940.2    21.7      15.8      0.35
 10  2710  2007.0  2009.0    20.0      -2.0     -0.04
 11  2930  1984.0  2018.2    18.4     -34.2     -0.73
 12  3000  1960.0  2014.1    18.4     -54.1     -1.16
 13  3210  2001.0  1982.1    21.5      18.9      0.42
 14  3240  1928.0  1975.1    22.5     -47.1     -1.05
 15  3520  1945.0  1880.0    36.2      65.0      1.87
```

Figure 12.34b
MINITAB printout for quadratic model of electrical usage

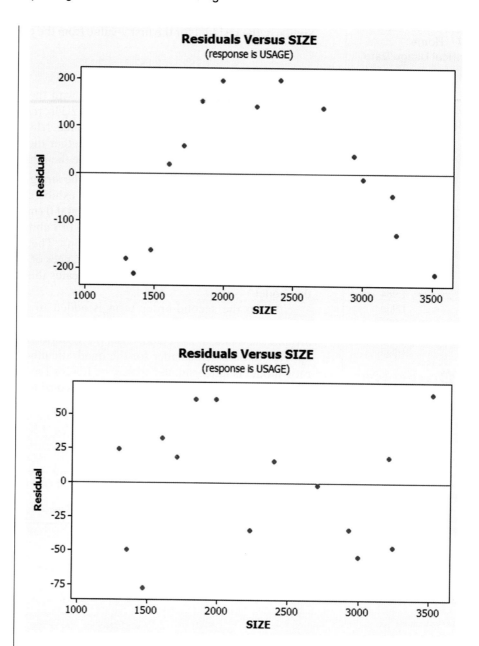

Figure 12.35a
MINITAB residual plot
for straight-line model of
electrical usage

Figure 12.35b
MINITAB residual plot
for quadratic model of
electrical usage

Look Back The residual analysis verifies our conclusions from Example 12.7, where we found the quadratic term, $\beta_2 x^2$, to be statistically significant.

■ **Now Work Exercise 12.154a**

Checking Assumption #2: Constant Error Variance

Residual plots can also be used to detect violations of the assumption of constant error variance. For example, a plot of the residuals versus the predicted value $\hat{y}$ may display one of the patterns shown in Figure 12.36. In these figures, the range in values of the residuals increases (or decreases) as $\hat{y}$ increases, thus indicating that the variance of the random error, ε, becomes larger (or smaller) as the estimate of $E(y)$ increases in value. Because $E(y)$ depends on the x values in the model, this implies that the variance of ε is not constant for all settings of the x's.

In the next example, we demonstrate how to use this plot to detect a nonconstant variance and suggest a useful remedy.

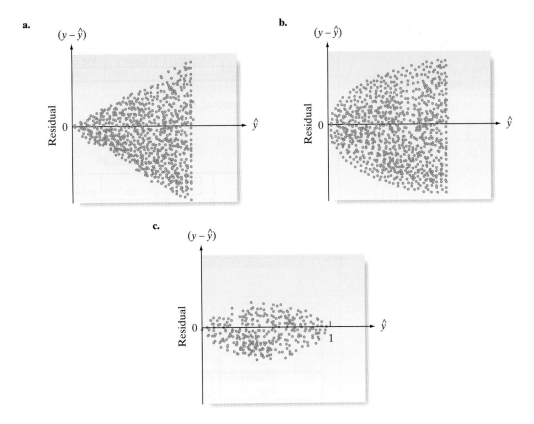

Figure 12.36
Residual plots showing changes
in the variance of ε

Example 12.15 Ⓓ
SOCWORK

**Using Residuals
to Check Equal
Variances**

Problem The data in Table 12.8 are the salaries, y, and years of experience, x, for a sample of 50 social workers. The first-order model $E(y) = \beta_0 + \beta_1 x$ was fitted to the data using by SPSS. The SPSS printout is shown in Figure 12.37, followed by a plot of the residuals versus $\hat{y}$ in Figure 12.38. Interpret the results. Make model modifications, if necessary.

Solution The SPSS printout, Figure 12.37, suggests that the first-order model provides an adequate fit to the data. The R^2-value indicates that the model explains about 78.7% of the sample variation in salaries. The t-value for testing β_1, 13.31, is highly significant (p-value ≈ 0) and indicates that the model contributes information for the prediction of y. However, an examination of the residuals plotted against $\hat{y}$ (Figure 12.38) reveals a potential problem. Note the "cone" shape of the residual variability; the size of the residuals increases as the estimated mean salary increases, implying that the constant variance assumption is violated.

One way to stabilize the variance of ε is to refit the model using a transformation on the dependent variable y. With economic data (e.g., salaries), a useful **variance-stabilizing transformation** is the natural logarithm of y, denoted $\ln(y)$.* We fit the model

$$\ln(y) = \beta_0 + \beta_1 x + \varepsilon$$

to the data of Table 12.8. Figure 12.39 shows the SPSS regression analysis printout for the $n = 50$ measurements, while Figure 12.40 shows a plot of the residuals from the log model.

You can see that the logarithmic transformation has stabilized the error variances. Note that the cone shape is gone; there is no apparent tendency of the residual variance

---

*Other variance-stabilizing transformations that are used successfully in practice are $\sqrt{y}$ and $\sin^{-1}\sqrt{y}$. Consult the references for more details on these transformations.

Table 12.8 Salary Data for Example 12.15

Years of Experience, x	Salary, y
7	$26,075
28	79,370
23	65,726
18	41,983
19	62,308
15	41,154
24	53,610
13	33,697
2	22,444
8	32,562
20	43,076
21	56,000
18	58,667
7	22,210
2	20,521
18	49,727
11	33,233
21	43,628
4	16,105
24	65,644
20	63,022
20	47,780
15	38,853
25	66,537
25	67,447
28	64,785
26	61,581
27	70,678
20	51,301
18	39,346
1	24,833
26	65,929
20	41,721
26	82,641
28	99,139
23	52,624
17	50,594
25	53,272
26	65,343
19	46,216
16	54,288
3	20,844
12	32,586
23	71,235
20	36,530
19	52,745
27	67,282
25	80,931
12	32,303
11	38,371

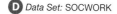 *Data Set:* SOCWORK

Model Summary[b]

Model	R	R Square	Adjusted R Square	Std. Error of the Estimate
1	.887[a]	.787	.782	8642.441

a. Predictors: (Constant), ESP

b. Dependent Variable: SALARY

ANOVA[b]

Model		Sum of Squares	df	Mean Square	F	Sig.
1	Regression	1.3E+10	1	1.324E+10	177.257	.000[a]
	Residual	3.6E+09	48	74691793.28		
	Total	1.7E+10	49			

a. Predictors: (Constant), ESP

b. Dependent Variable: SALARY

Coefficients[a]

Model		Unstandardized Coefficients		Standardized Coefficients	t	Sig.
		B	Std. Error	Beta		
1	(Constant)	11368.72	3160.317		3.597	.001
	ESP	2141.381	160.839	.887	13.314	.000

a. Dependent Variable: SALARY

Figure 12.37
SPSS regression printout for first-order model of salary

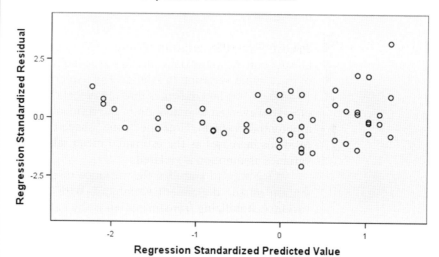

Figure 12.38
SPSS residual plot for first-order model of salary

to increase as mean salary increases. We therefore are confident that inferences using the $\ln(y)$ model are more reliable than those using the untransformed model.

Look Back With transformed models, the analyst should be wary when interpreting model statistics such as $\hat{\beta}_1$ and s. These interpretations must take into account that the dependent variable is not some function of y. For example, the antilogarithm of $\hat{\beta}_1$ in the

Model Summary[b]

Model	R	R Square	Adjusted R Square	Std. Error of the Estimate
1	.929[a]	.864	.861	.1541127

a. Predictors: (Constant), ESP

b. Dependent Variable: LNSALARY

ANOVA[b]

Model		Sum of Squares	df	Mean Square	F	Sig.
1	Regression	7.212	1	7.212	303.660	.000[a]
	Residual	1.140	48	.024		
	Total	8.352	49			

a. Predictors: (Constant), ESP

b. Dependent Variable: LNSALARY

Coefficients[a]

Model		Unstandardized Coefficients		Standardized Coefficients	t	Sig.
		B	Std. Error	Beta		
1	(Constant)	9.841	.056		174.631	.000
	ESP	.050	.003	.929	17.426	.000

a. Dependent Variable: LNSALARY

Figure 12.39

SPSS regression printout for model of log salary

Figure 12.40

SPSS residual plot for model of ln(salary)

ln(y) model for salary (y) represents the percentage change in salary for every 1 year increase in experience (x).

■ **Now Work Exercise 12.154b**

Checking Assumption #3: Errors Normally Distributed

Several graphical methods are available for assessing whether the random error ε has an approximate normal distribution. Recall (Section 5.4) that stem-and-leaf displays, histograms, and normal probability plots are useful for checking whether the data are normally distributed. We illustrate these techniques in an example. But first, we discuss a related problem—using residuals to check for *outliers*.

If the assumption of normally distributed errors is satisfied, then we expect approximately 95% of the residuals to fall within 2 standard deviations of the mean of 0, and almost all of the residuals to lie within 3 standard deviations of the mean of 0. Residuals that are extremely far from 0 and disconnected from the bulk of the other residuals are called **regression outliers** and should receive special attention from the analyst.

> A **regression outlier** is a residual that is larger than $3s$ (in absolute value).

Example   12.16   Ⓓ

GFCLOCK2

Identifying Outliers—Grandfather Clock Price Model

Problem Refer to Example 12.6 (p. 710), in which we modeled the auction price y of a grandfather clock as a function of age x_1 and number of bidders, x_2. The data for this example are repeated in Table 12.9, with one important difference: The auction price of the clock at the top of the sixth column has been changed from $2,131 to $1,131 (highlighted in the table). The interaction model

$$E(y) = \beta_0 + \beta_1 x_1 + \beta_2 x_2 + \beta_3 x_1 x_2$$

is again fit to these (modified) data, with the MINITAB printout shown in Figure 12.41. The residuals are shown highlighted in the printout and then are plotted against the number of bidders, x_2, in Figure 12.42. Analyze the residual plot.

Table 12.9   Altered Auction Price Data

Age x_1	Number of Bidders, x_2	Auction Price y	Age x_1	Number of Bidders, x_2	Auction Price y
127	13	$1,235	170	14	$1,131
115	12	1,080	182	8	1,550
127	7	845	162	11	1,884
150	9	1,522	184	10	2,041
156	6	1,047	143	6	845
182	11	1,979	159	9	1,483
156	12	1,822	108	14	1,055
132	10	1,253	175	8	1,545
137	9	1,297	108	6	729
113	9	946	179	9	1,792
137	15	1,713	111	15	1,175
117	11	1,024	187	8	1,593
137	8	1,147	111	7	785
153	6	1,092	115	7	744
117	13	1,152	194	5	1,356
126	10	1,336	168	7	1,262

Ⓓ *Data Set:* GFCLOCK2

Solution The residual plot dramatically reveals the one altered measurement. Note that one of the two residuals at $x_2 = 14$ bidders falls more than three standard deviations below 0. (This observation is highlighted on Figure 12.41.) Note also that no other residual falls more than two standard deviations from 0.

What do we do with outliers once we identify them? First, we try to determine the cause. Were the data entered into the computer incorrectly? Was the observation recorded incorrectly when the data were collected? If so, we correct the observation and rerun the analysis. Another possibility is that the observation is not representative of the conditions we are trying to model. For example, in this case the low price may be attributable to extreme damage to the clock or to a clock of inferior quality compared with the others. In these cases, we probably would exclude the observation from the analysis. In many cases, you may not be able to determine the cause of the outlier. Even so, you may want to rerun the regression analysis excluding the outlier, in order to assess the effect of that observation on the results of the analysis.

```
The regression equation is
PRICE = - 513 + 8.17 AGE + 19.9 NUMBIDS + 0.320 AGE_BIDS


Predictor    Coef   SE Coef      T      P
Constant    -512.8    665.9   -0.77  0.448
AGE          8.165    4.585    1.78  0.086
NUMBIDS      19.89    67.44    0.29  0.770
AGE_BIDS    0.3196   0.4790    0.67  0.510


S = 200.598   R-Sq = 72.9%   R-Sq(adj) = 70.0%


Analysis of Variance

Source            DF       SS       MS      F      P
Regression         3  3033587  1011196  25.13  0.000
Residual Error    28  1126703    40239
Total             31  4160290


Obs  AGE   PRICE      Fit  SE Fit  Residual  St Resid
  1  127  1235.0   1310.4    59.3     -75.4     -0.39
  2  115  1080.0   1105.9    62.1     -25.9     -0.14
  3  127   845.0    947.5    61.1    -102.5     -0.54
  4  150  1522.0   1322.5    37.1     199.5      1.01
  5  156  1047.0   1179.5    60.3    -132.5     -0.69
  6  182  1979.0   1831.9    82.9     147.1      0.81
  7  156  1822.0   1598.0    61.9     224.0      1.17
  8  132  1253.0   1185.8    39.7      67.2      0.34
  9  137  1297.0   1178.9    39.0     118.1      0.60
 10  113   946.0    913.9    58.6      32.1      0.17
 11  137  1713.0   1561.0    78.4     152.0      0.82
 12  117  1024.0   1072.6    53.1     -48.6     -0.25
 13  137  1147.0   1115.2    44.3      31.8      0.16
 14  153  1092.0   1149.2    59.0     -57.2     -0.30
 15  117  1152.0   1187.2    69.7     -35.2     -0.19
 16  126  1336.0   1117.6    43.4     218.4      1.12
 17  170  1131.0   1914.4   116.7    -783.4     -4.80R
 18  182  1550.0   1597.7    62.8     -47.7     -0.25
 19  162  1884.0   1598.3    57.0     285.7      1.49
 20  184  2041.0   1776.6    70.7     264.4      1.41
 21  143   845.0   1048.4    58.9    -203.4     -1.06
 22  159  1483.0   1421.8    40.6      61.2      0.31
 23  108  1055.0   1130.7    97.9     -75.7     -0.43
 24  175  1545.0   1522.7    55.4      22.3      0.12
 25  108   729.0    695.5    99.6      33.5      0.19
 26  179  1792.0   1642.7    57.6     149.3      0.78
 27  111  1175.0   1224.0   107.2     -49.0     -0.29
 28  187  1593.0   1651.3    68.6     -58.3     -0.31
 29  111   785.0    781.1    80.9       3.9      0.02
 30  115   744.0    822.7    75.5     -78.7     -0.42
 31  194  1356.0   1480.7   133.6    -124.7     -0.83 X
 32  168  1262.0   1374.0    57.7    -112.0     -0.58

R denotes an observation with a large standardized residual.
X denotes an observation whose X value gives it large influence.
```

Figure 12.41
MINITAB regression printout for altered grandfather clock data

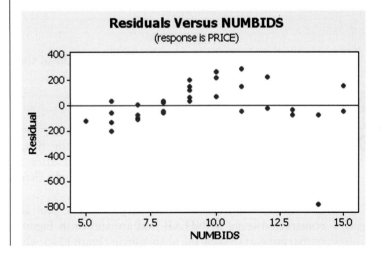

Figure 12.42
MINITAB residual plot of altered grandfather clock data

```
The regression equation is
PRICE = 474 - 0.46 AGE - 114 NUMBIDS + 1.48 AGE_BIDS


Predictor      Coef   SE Coef       T       P
Constant      474.0     298.2    1.59   0.124
AGE          -0.465     2.107   -0.22   0.827
NUMBIDS     -114.12     31.23   -3.65   0.001
AGE_BIDS     1.4781    0.2295    6.44   0.000


S = 85.8286    R-Sq = 95.2%    R-Sq(adj) = 94.7%


Analysis of Variance

Source          DF        SS        MS        F       P
Regression       3   3933417   1311139   177.99   0.000
Residual Error  27    198897      7367
Total           30   4132314
```

Figure 12.43

MINITAB regression printout when outlier is deleted

Figure 12.43 shows the printout when the outlier observation is excluded from the grandfather clock analysis, and Figure 12.44 shows the new plot of the residuals against the number of bidders. Now only one of the residuals lies beyond two standard deviations from 0, and none lies beyond three standard deviations. Also, the statistics indicate a much better model without the outlier. Most notably, the standard deviation (s) has decreased from 200.6 to 85.83, indicating a model that will provide more precise estimates and predictions (narrower confidence and prediction intervals) for clocks that are similar to those in the reduced sample.

Figure 12.44

MINITAB residual plot when outlier is deleted

Look Back Remember that if the outlier is removed from the analysis when in fact it belongs to the same population as the rest of the sample, the resulting model may provide misleading estimates and predictions.

■ **Now Work Exercise 12.156**

The next example checks the assumption of the normality of the random error component.

Example 12.17
GFCLOCK2

Using Residuals to Check for Normal Errors

Problem Refer to Example 12.16. Analyze the distribution of the residuals in the grandfather clock example, both before and after the outlier residual is removed. Determine whether the assumption of a normally distributed error term is reasonable.

Solution A histogram and normal probability plot for the two sets of residuals are constructed using MINITAB and are shown in Figures 12.45 and 12.46. Note that the outlier appears to skew the histogram in Figure 12.45, whereas the histogram in Figure 12.46

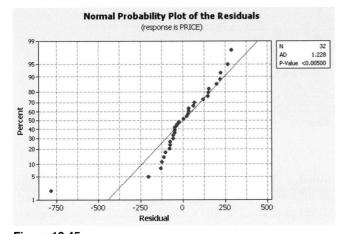

Figure 12.45
MINITAB graphs of regression residuals for grandfather clock model (outlier included)

Figure 12.46
MINITAB graphs of regression residuals for grandfather clock model (outlier deleted)

appears to be more mound shaped. Similarly, the pattern of residuals in the normal probability plot in Figure 12.46 (outlier deleted) is more nearly a straight line than the pattern in Figure 12.45 (outlier included). Thus, the normality assumption appears to be more plausible after the outlier is removed.

Look Back Although graphs do not provide formal statistical tests of normality, they do provide a descriptive display. Consult the references for methods to conduct statistical tests of normality using the residuals.

■ **Now Work Exercise 12.154d**

Of the four assumptions, the assumption that the random error is normally distributed is the least restrictive when we apply regression analysis in practice—that is, moderate departures from a normal distribution have very little effect on the validity of the statistical tests, confidence intervals, and prediction intervals presented in this chapter. In this case, we say that regression analysis is **robust** with respect to nonnormal errors. However, great departures from normality cast doubt on any inferences derived from the regression analysis.

Checking Assumption #4: Errors Independent

The assumption of independent errors is violated when successive errors are correlated. This typically occurs when the data for both the dependent and independent variables are observed sequentially over a period of time—called **time-series data**. Time-series data have a unique characteristic; the experimental unit represents a unit of time (e.g., a year, a month, a quarter). There are both graphical and formal statistical tests available for checking the assumption of independent regression errors. For example, a simple graph is to plot the residuals against time. If the residuals tend to group alternately into positive and negative clusters (as shown in Figure 12.47), then it is likely that the errors are correlated and the assumption is violated. If **correlated errors** are detected, one solution is to construct a **time-series model** for $E(y)$. These methods are beyond the scope of this text. (Consult the references for examples of time-series models.)

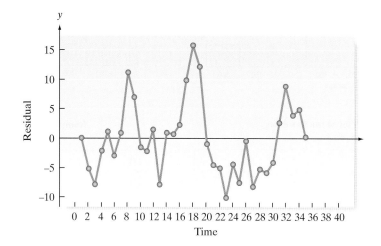

Figure 12.47

Hypothetical residual plot for time-series data

Summary

Residual analysis is a useful tool for the regression analyst, not only to check the assumptions but also to provide information about how the model can be improved. A summary of the residual analyses presented in this section to check the assumption that the random error ε is normally distributed with mean 0 and constant variance is presented in the next box.

Steps in a Residual Analysis

1. Check for a misspecified model by plotting the residuals against each of the quantitative independent variables. Analyze each plot, looking for a curvilinear trend. This shape signals the need for a quadratic term in the model. Try a second-order term in the variable against which the residuals are plotted.

2. Examine the residual plots for outliers. Draw lines on the plots at two and three standard deviations below and above the 0 line. Examine residuals outside the three-standard-deviation lines as potential outliers, and check to see that approximately 5% of the residuals exceed the two-standard-deviation lines. Determine whether each outlier can be explained as an error in data collection or transcription, whether it corresponds to a member of a population different from that of the remainder of the sample, or whether it simply represents an unusual observation. If the observation is determined to be an error, fix it or remove it. Even if you can't determine the cause, you may want to rerun the regression analysis without the observation to determine its effect on the analysis.

3. Check for nonnormal errors by plotting a frequency distribution of the residuals, using a stem-and-leaf display, histogram, or normal probability plot. Check to see if obvious departures from normality exist. Extreme skewness of the frequency distribution may be due to outliers or could indicate the need for a transformation of the dependent variable. (Normalizing transformations are beyond the scope of this book, but you can find discussions of the procedure in the references.)

---

4. Check for unequal error variances by plotting the residuals against the predicted values $\hat{y}$. If you detect a cone-shaped pattern or some other pattern indicating that the variance of ε is not constant, refit the model, using an appropriate variance-stabilizing transformation on y, such as $\ln(y)$. (Consult the references for other useful variance-stabilizing transformations.)

Statistics IN Action | **Revisited** | ## A Residual Analysis for the Condominium Sale Price Model

In the previous Statistics in Action Revisited section (p. 748), we found a second-order model to be both a statistically and practically useful model for predicting the auction price (y) of a condominium unit. Before using the model in practice, we need to examine the residuals to be sure that the standard regression assumptions are reasonably satisfied.

Figures SIA12.7 and SIA12.8 are MINITAB graphs of the residuals from the model. Except for a few outliers, the histogram shown in Figure SIA12.7 appears to be approximately normally distributed; consequently, the assumption of normal errors is reasonably satisfied. The scatterplot of the residuals against $\hat{y}$ shown in Figure SIA12.8 reveals no distinct pattern;

thus, the assumption of a constant error variance is reasonably satisfied.

To find the outliers, we used MINITAB to compute the residuals of the model. A list of the unusual residual values is shown in the MINITAB printout in Figure SIA12.9. Four condo units (numbers 111, 184, 185, and 188; the values of the standardized residuals are highlighted on the printout) have residual values that are greater than 3 in absolute

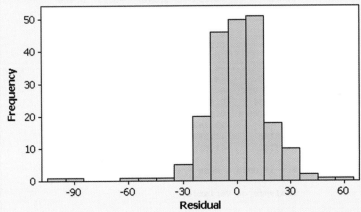

Figure SIA12.7

MINITAB histogram of residuals from second-order model of condo sale price

Figure SIA12.8

MINITAB plot of residuals versus predicted values from second-order model of condo sale price

(Continued)

Unusual Observations

Obs	FLOOR	PRICE100	Fit	SE Fit	Residual	St Resid
11	8.00	279.00	240.71	7.62	38.29	2.18R
26	3.00	269.00	338.18	1.75	10.52	2.19R
37	6.00	294.00	246.22	5.34	47.78	2.60R
111	5.00	265.00	202.98	3.69	62.02	3.30R
183	3.00	150.00	197.54	4.36	-47.54	-2.55R
184	5.00	140.00	238.04	6.02	-98.04	-5.40R
185	5.00	130.00	192.19	4.37	-62.19	-3.34R
186	6.00	182.00	224.70	5.17	-42.70	-2.32R
188	5.00	164.00	252.16	4.66	-88.16	-4.75R

R denotes an observation with a large standardized residual.

Figure SIA12.9
MINITAB list of outliers for the second-order model of condo sale price

value. (Research reveals that all of these units are located on the fifth floor.) If we delete these observations from the data set and refit the model, we obtain the results shown in the MINITAB printout in Figure SIA12.10. You can see that the adjusted-R^2 value is now .793, an increase of about 9% from the previous value. Also, the standard deviation of the model is now $s = 15.1$, a reduction of about four hundred dollars from the previous value. Obviously, removing the outliers from the analysis yields a "better" prediction equation. If we can justify removing these four outliers (except for number 111, they were all undersold units), then we can use the model shown in Figure SIA12.10 to obtain more accurate predictions of future condo sale prices.

The regression equation is
PRICE100 = 185 - 2.59 FLOOR - 1.98 DISTANCE - 0.387 FLR_DST + 0.912 FLOORSQ
           + 0.264 DISTSQ + 83.9 VIEW - 4.01 VU_FLR - 4.61 VU_DIST
           + 0.154 VU_FLR_DST - 0.426 VU_FLRSQ + 0.304 VU_DISTSQ - 13.1 END
           + 11.5 FURNISH - 31.8 AUCTION

Predictor	Coef	SE Coef	T	P
Constant	184.93	10.39	17.81	0.000
FLOOR	-2.591	3.570	-0.73	0.469
DISTANCE	-1.977	1.923	-1.03	0.305
FLR_DST	-0.3871	0.1593	-2.43	0.016
FLOORSQ	0.9120	0.3384	2.70	0.008
DISTSQ	0.2644	0.1115	2.37	0.019
VIEW	83.88	14.03	5.98	0.000
VU_FLR	-4.007	4.705	-0.85	0.396
VU_DIST	-4.608	2.529	-1.82	0.070
VU_FLR_DST	0.1535	0.2166	0.71	0.479
VU_FLRSQ	-0.4261	0.4624	-0.92	0.358
VU_DISTSQ	0.3041	0.1465	2.08	0.039
END	-13.142	6.244	-2.10	0.037
FURNISH	11.549	2.405	4.80	0.000
AUCTION	-31.816	2.954	-10.77	0.000

S = 15.0925   R-Sq = 80.7%   R-Sq(adj) = 79.3%

Analysis of Variance

Source	DF	SS	MS	F	P
Regression	14	181271	12948	56.84	0.000
Residual Error	190	43279	228		
Total	204	224550			

Figure SIA12.10
MINITAB printout of second-order model of condo sale price with outliers deleted

12.12 Some Pitfalls: Estimability, Multicollinearity, and Extrapolation

You should be aware of several potential problems when you construct a prediction model for some response y. A few of the most important ones are discussed in this final section.

Problem 1: Parameter Estimability

Suppose you want to fit a model relating annual crop yield y to the total expenditure for fertilizer, x. We propose the first-order model

$$E(y) = \beta_0 + \beta_1 x$$

Now suppose we have three years of data and $1,000 is spent on fertilizer each year. The data are shown in Figure 12.48. You can see the problem: The parameters of the model cannot be estimated when all the data are concentrated at a single x value. Recall that it takes two points (x-values) to fit a straight line. Thus, the parameters are not estimable when only one x is observed.

A similar problem would occur if we attempted to fit the quadratic model

$$E(y) = \beta_0 + \beta_1 x + \beta_2 x^2$$

to a set of data for which only one or two different x-values were observed. (See Figure 12.49.) At least three different x-values must be observed before a quadratic model can be fitted to a set of data (i.e., before all three parameters are estimable).

In general, the number of levels of observed x-values must be one more than the order of the polynomial in x that you want to fit.

For controlled experiments, the researcher can select one of the experimental designs in Chapter 10 that will permit estimation of the model parameters. Even when the values of the independent variables cannot be controlled by the researcher, the independent variables are almost always observed at a sufficient number of levels to permit estimation of the model parameters. When the statistical software you use suddenly refuses to fit a model, however, the problem is probably inestimable parameters.

Figure 12.48
Yield and fertilizer expenditure data: three years

Figure 12.49
Only two x values observed: quadratic model is not estimable

Problem 2: Multicollinearity

Often, two or more of the independent variables used in a regression model contribute redundant information. That is, the independent variables are correlated with each other. For example, suppose we want to construct a model to predict the gas mileage rating of a truck as a function of its load x_1 (in tons) and the horsepower x_2 (in foot-pounds per second) of its engine. We would expect heavy loads to require greater horsepower and to result in lower mileage ratings. Thus, although both x_1 and x_2 contribute information for the prediction of mileage rating y, some of the information is overlapping because x_1 and x_2 are correlated.

When the independent variables are correlated, we say *multicollinearity* exists. In practice, it is not uncommon to observe correlations among the independent variables. However, a few problems arise when serious multicollinearity is present in the regression variables.

> **Multicollinearity** exists when two or more of the independent variables used in a regression are correlated.

First, high correlations among the independent variables increase the likelihood of rounding errors in the calculations of the β estimates, standard errors, and so forth. Second, and more important, the regression results may be confusing and misleading. Consider the model for gasoline mileage rating (y) of a truck

$$E(y) = \beta_0 + \beta_1 x_1 + \beta_2 x_2$$

where x_1 = load and x_2 = horsepower. Fitting the model to a sample data set, we might find that the t-tests for testing β_1 and β_2 are both nonsignificant at the α = .05 level, while the F-test for H_0: $\beta_1 = \beta_2 = 0$ is highly significant ($p = .001$). The tests may seem to be contradictory, but really they are not. On the one hand, the t-tests indicate that the contribution of one variable, say, $x_1 = $ load, is not significant after the effect of $x_2 = $ horsepower has been accounted for (because x_2 is also in the model). On the other hand, the significant F-test tells us that at least one of the two variables is making a contribution to the prediction of y (i.e., either β_1 or β_2 or both differ from 0). In fact, both are probably contributing, but the contribution of one overlaps with that of the other.

Multicollinearity can also have an effect on the signs of the parameter estimates. More specifically, a value of $\hat{\beta}_i$ may have the opposite sign from what is expected. In the truck gasoline mileage example, we expect heavy loads to result in lower mileage ratings and we expect higher horsepowers to result in lower mileage ratings; consequently, we expect the signs of both the parameter estimates to be negative. Yet, we may actually see a positive value of $\hat{\beta}_1$ and be tempted to claim that heavy loads result in *higher* mileage ratings. This is the danger of interpreting a β coefficient when the independent variables are correlated. Because the variables contribute redundant information, the effect of $x_1 = $ load on $y = $ mileage rating is measured only partially by β_1.

How can you avoid the problems of multicollinearity in regression analysis? One way is to conduct a designed experiment (Chapter 10), so that the levels of the x variables are uncorrelated. Unfortunately, time and cost constraints may prevent you from collecting data in this manner. Consequently, most data are collected observationally. Since observational data frequently consist of correlated independent variables, you will need to recognize when multicollinearity is present and, if necessary, make modifications in the regression analysis.

Several methods are available for detecting multicollinearity in regression. A simple technique is to calculate the coefficient of correlation, r, between each pair of independent variables in the model and use the procedure outlined in Section 11.5 to test for significantly correlated variables. If one or more of the r values are statistically different from 0, the variables in question are correlated and a multicollinearity problem may exist.* The degree of multicollinearity will depend on the magnitude of the value of r, as shown in the box.

Using the Correlation Coefficient r to Detect Multicollinearity

Extreme Multicollinearity: $|r| \geq .8$

Moderate Multicollinearity: $.2 \leq |r| < .8$

Low Multicollinearity: $|r| < .2$

Other indications of the presence of multicollinearity include those just mentioned, namely, nonsignificant t-tests for the individual parameter estimates when the F-test for overall model adequacy is significant, and estimates with opposite signs from what is expected.[†]

Detecting Multicollinearity in the Regression Model

Multicollinearity may exist if there are

1. Significant correlations between pairs of independent variables

2. Nonsignificant t-tests for all (or nearly all) of the individual β parameters when the F-test for overall model adequacy is significant

3. Signs opposite from what is expected in the estimated β parameters

---

*Remember that r measures only the pairwise correlation between x values. Three variables, $x_1, x_2,$ and $x_3,$ may be highly correlated as a group, but may not exhibit large pairwise correlations. Thus, multicollinearity may be present even when all pairwise correlations are not significantly different from 0.

[†]More formal methods for detecting multicollinearity, such as variance-inflation factors (VIFs), are available. Independent variables with a VIF of 10 or above are usually considered to be highly correlated with one or more of the other independent variables in the model. Calculation of VIFs is beyond the scope of this introductory text. (Consult the chapter references for a discussion of VIFs and other formal methods of detecting multicollinearity.)

Example 12.18 Ⓓ
FTC

Detecting Multicollinearity— Modeling Carbon Monoxide Content

Problem The Federal Trade Commission (FTC) annually ranks varieties of domestic cigarettes according to their tar, nicotine, and carbon monoxide content. The U.S. surgeon general considers each of these three substances hazardous to a smoker's health. Past studies have shown that increases in the tar and nicotine content of a cigarette are accompanied by an increase in the carbon monoxide emitted from the cigarette smoke. Table 12.10 presents data on tar, nicotine, and carbon monoxide content (in milligrams) and weight (in grams) for a sample of 25 (filter) brands tested in a recent year. Suppose we want to model carbon monoxide content y as a function of tar content x_1, nicotine content x_2, and weight x_3, using the model

$$E(y) = \beta_0 + \beta_1 x_1 + \beta_2 x_2 + \beta_3 x_3$$

The model is fitted to the 25 data points in Table 12.10, and a portion of the SAS printout is shown in Figure 12.50. Examine the printout. Do you detect any signs of multicollinearity?

Table 12.10 FTC Cigarette Data for Example 12.18

Tar (x_1)	Nicotine (x_2)	Weight (x_3)	Carbon Monoxide (y)
14.1	.86	.9853	13.6
16.0	1.06	1.0938	16.6
29.8	2.03	1.1650	23.5
8.0	.67	.9280	10.2
4.1	.40	.9462	5.4
15.0	1.04	.8885	15.0
8.8	.76	1.0267	9.0
12.4	.95	.9225	12.3
16.6	1.12	.9372	16.3
14.9	1.02	.8858	15.4
13.7	1.01	.9643	13.0
15.1	.90	.9316	14.4
7.8	.57	.9705	10.0
11.4	.78	1.1240	10.2
9.0	.74	.8517	9.5
1.0	.13	.7851	1.5
17.0	1.26	.9186	18.5
12.8	1.08	1.0395	12.6
15.8	.96	.9573	17.5
4.5	.42	.9106	4.9
14.5	1.01	1.0070	15.9
7.3	.61	.9806	8.5
8.6	.69	.9693	10.6
15.2	1.02	.9496	13.9
12.0	.82	1.1184	14.9

Source: Federal Trade Commission.   Ⓓ *Data Set*: FTC

Solution First, note that the F-test for overall model utility is highly significant. The test statistic ($F = 78.98$) and the observed significance level (p-value $< .0001$) are highlighted on the SAS printout shown in Figure 12.50. Therefore, we can conclude at, say, $\alpha = .01$, that at least one of the parameters β_1, β_2, and β_3 in the model is nonzero. The t-tests for two of three individual β's, however, are nonsignificant. (The p-values for these tests are highlighted on the printout.) Unless tar (x_1) is the only one of the three variables that is useful in predicting carbon monoxide content, these results are the first indication of a potential multicollinearity problem.

The negative values for $\hat{\beta}_2$ and $\hat{\beta}_3$ (highlighted on the printout) are a second clue to the presence of multicollinearity. From past studies, the FTC expects carbon monoxide content (y) to increase when either nicotine content (x_2) or weight (x_3) increases; that is, the FTC expects *positive* relationships between y and x_2 and between y and x_3, not negative ones.

All signs indicate that a serious multicollinearity problem exists.*

*Note also that the variance-inflation factors (VIFs) for both tar and nicotine, given on the SAS printout, Figure 12.50, exceed 10.

```
                          Dependent Variable: CO

                   Number of Observations Read        25
                   Number of Observations Used        25

                           Analysis of Variance

                               Sum of        Mean
      Source            DF     Squares       Square    F Value    Pr > F

      Model              3   495.25781    165.08594      78.98    <.0001
      Error             21    43.89259      2.09012
      Corrected Total   24   539.15040


                  Root MSE              1.44573    R-Square    0.9186
                  Dependent Mean       12.52800    Adj R-Sq    0.9070
                  Coeff Var            11.53996


                           Parameter Estimates

                   Parameter      Standard                        Variance
   Variable   DF    Estimate         Error    t Value   Pr > |t|  Inflation

   Intercept   1     3.20219       3.46175       0.93     0.3655         0
   TAR         1     0.96257       0.24224       3.97     0.0007  21.63071
   NICOTINE    1    -2.63166       3.90056      -0.67     0.5072  21.89992
   WEIGHT      1    -0.13048       3.88534      -0.03     0.9735   1.33386


               Pearson Correlation Coefficients, N = 25
                   Prob > |r| under H0: Rho=0

                         TAR        NICOTINE       WEIGHT

      TAR            1.00000        0.97661       0.49077
                                    <.0001         0.0127

      NICOTINE       0.97661        1.00000       0.50018
                      <.0001                       0.0109

      WEIGHT         0.49077        0.50018       1.00000
                      0.0127         0.0109
```

Figure 12.50

SAS printout for model of CO content, Example 12.18

Look Back To confirm our suspicions, we had SAS produce the coefficient of correlation, r, for each of the three pairs of independent variables in the model. The resulting output is shown (highlighted) at the bottom of Figure 12.50. You can see that tar (x_1) and nicotine (x_2) are highly correlated $(r = .9766)$, while weight (x_3) is moderately correlated with the other two x's $(r \approx .5)$. All three correlations have p-values less than .05; consequently, all three are significantly different from 0 at $\alpha = .05$.

■ **Now Work Exercise 12.157**

Once you have detected multicollinearity, you can choose from among several alternative measures available for solving the problem. Several of these are outlined in the next box. The appropriate measure to take depends on the severity of the multicollinearity and the ultimate goal of the regression analysis. Some researchers, when confronted with highly correlated independent variables, choose to include only one of the correlated variables in the final model. If you are interested only in using the model for estimation and prediction (step 6), you may decide not to drop any of the independent variables from the model. We have seen that it is dangerous to interpret the individual β parameters in the presence of multicollinearity. However, confidence intervals for $E(y)$ and prediction intervals for y generally remain unaffected *as long as the values of the x's used to predict y follow the same pattern of multicollinearity exhibited in the sample data.* That is, you must take strict care to ensure that the values of the x-variables fall within the range of the sample data.

Solutions to Some Problems Created by Multicollinearity in Regression*

1. Drop one or more of the correlated independent variables from the model. One way to decide which variables to keep in the model is to employ stepwise regression (Section 12.10).

---

*Several other solutions are available. For example, in the case where higher order regression models are fitted, the analyst may want to code the independent variables so that higher order terms (e.g., x^2) for a particular x-variable are not highly correlated with x. One transformation that works is $z = (x - \bar{x})/s$. Other, more sophisticated procedures for addressing multicollinearity (such as *ridge regression*) are beyond the scope of this text. (Consult the references at the end of the chapter.)

b. Using the variables specified in the model presented, describe how you would conduct the multiple-regression analysis. (Propose a more complicated model and describe the appropriate model tests, including a residual analysis.)

12.205 FLAG study of bid collusion. Road construction contracts in the state of Florida are awarded on the basis of competitive, sealed bids; the contractor who bids the lowest price wins the contract. During the 1980s, the Office of the Florida Attorney General (FLAG) suspected numerous contractors of practicing bid collusion (i.e., setting the winning bid price above the fair, or competitive, price in order to increase their own profit margin). By comparing the prices bid (and other important bid variables) of the fixed (i.e., rigged) contracts with the competitively bid contracts, FLAG was able to establish invaluable benchmarks for detecting future bid rigging. FLAG collected data on 279 road construction contracts. For each contract, the following variables were measured (the data are saved in the **FLAG** file):

1. Price of contract ($) bid by lowest bidder
2. Department of Transportation (DOT) engineer's estimate of fair contract price ($)
3. Ratio of low (winning) bid price to DOT engineer's estimate of fair price
4. Status of contract (1 if fixed, 0 if competitive)
5. District (1, 2, 3, 4, or 5) in which construction project is located
6. Number of bidders on contract
7. Estimated number of days required to complete work
8. Length of road project (miles)
9. Percentage of costs allocated to liquid asphalt
10. Percentage of costs allocated to base material
11. Percentage of costs allocated to excavation
12. Percentage of costs allocated to mobilization
13. Percentage of costs allocated to structures
14. Percentage of costs allocated to traffic control
15. Subcontractor utilization (1 if yes, 0 if no)

Use the methodology of this chapter to build a model for low-bid contract price (y). Comment on how the status of the bid affects the price.

Activity Collecting Data and Fitting a Multiple-Regression Model

Note: The use of statistical software is required for this project.

This is a continuation of the Activity section in Chapter 11, in which you selected three independent variables as predictors of a dependent variable of your choice and obtained at least 10 data values. Now, by means of an available software package, fit the multiple-regression model

$$y = \beta_0 + \beta_1 x_1 + \beta_2 x_2 + \beta_3 x_3 + \varepsilon$$

where

y = Dependent variable you chose

x_1 = First independent variable you chose

x_2 = Second independent variable you chose

x_3 = Third independent variable you chose

a. Compare the coefficients $\hat{\beta}_1$, $\hat{\beta}_2$, and $\hat{\beta}_3$ with their corresponding slope coefficients in the Activity of Chapter 11, where you fit three separate straight-line models. How do you account for the differences?

b. Calculate the coefficient of determination, R^2, and conduct the F-test of the null hypothesis H_0: $\beta_1 = \beta_2 = \beta_3 = 0$. What is your conclusion?

c. Check the data for multicollinearity. If multicollinearity exists, how should you proceed?

d. Now increase your list of 3 variables to include approximately 10 that you think would be useful in predicting the dependent variable. With the aid of statistical software, employ a stepwise regression program to choose the important variables among those you have listed. To test your intuition, list the variables in the order you think they will be selected before you conduct the analysis. How does your list compare with the stepwise regression results?

e. After the group of 10 variables has been narrowed to a smaller group of variables by the stepwise analysis, try to improve the model by including interactions and quadratic terms. Be sure to consider the meaning of each interaction or quadratic term before adding it to the model. (A quick sketch can be very helpful.) See if you can systematically construct a useful model for prediction. If you have a large data set, you might want to hold out the last observations to test the predictive ability of your model after it is constructed. (As noted in Section 12.10, using the same data to construct *and* to evaluate predictive ability can lead to invalid statistical tests and a false sense of security.)

References

Barnett, V., and Lewis, T. *Outliers in Statistical Data*. New York: Wiley, 1978.

Belsley, D. A., Kuh, E., and Welsch, R. E. *Regression Diagnostics: Identifying Influential Data and Sources of Collinearity*. New York: Wiley, 1980.

Chatterjee, S., and Price, B. *Regression Analysis by Example*, 2nd ed. New York: Wiley, 1991.

Draper, N., and Smith, H. *Applied Regression Analysis*, 2nd ed. New York: Wiley, 1981.

Graybill, F. *Theory and Application of the Linear Model*. North Scituate, MA: Duxbury, 1976.

Kelting, H. "Investigation of condominium sale prices in three market scenarios: Utility of stepwise, interactive, multiple regression analysis and implications for design and appraisal methodology." Unpublished paper, University of Florida, Gainesville, FL, 1979.

Kutner, M., Nachtsheim, C., Neter, J., and Li, W. *Applied Linear Statistical Models*, 5th ed. New York: McGraw-Hill/Irwin, 2005.

Mendenhall, W. *Introduction to Linear Models and the Design and Analysis of Experiments*, Belmont, CA: Wadsworth, 1968.

Mendenhall, W., and Sincich, T. *A Second Course in Statistics: Regression Analysis*, 7th ed. Upper Saddle River, NJ: Prentice Hall, 2011.

Mosteller, F., and Tukey, J. W. *Data Analysis and Regression: A Second Course in Statistics*. Reading, MA: Addison-Wesley, 1977.

Rousseeuw, P. J., and Leroy, A. M. *Robust Regression and Outlier Detection*. New York: Wiley, 1987.

Weisberg, S. *Applied Linear Regression*, 2nd ed. New York: Wiley, 1985.

USING TECHNOLOGY   MINITAB: Multiple Regression

Multiple Regression

Step 1 Access the MINITAB worksheet file that contains the dependent and independent variables.

Step 2 Click on the "Stat" button on the MINITAB menu bar, and then click on "Regression" and "Regression" again, as shown in Figure 12.M.1.

Figure 12.M.1 MINITAB menu options for regression

Step 3 The resulting dialog box appears as shown in Figure 12.M.2 Specify the dependent variable in the "Response" box and the independent variables in the "Predictors" box. [*Note:* If your model includes interaction and/or squared terms, you must create and add these higher-order variables to the MINITAB worksheet *prior* to running a regression analysis. You can do this by clicking the "Calc" button on the MINITAB main menu and selecting the "Calculator" option.]

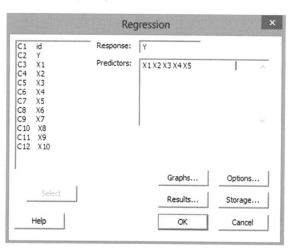

Figure 12.M.2 MINITAB regression dialog box

Step 4 To produce prediction intervals for y and confidence intervals for $E(y)$, click the "Options" button and select the appropriate menu items in the resulting menu list. (See Figure 12.M.3.)

Figure 12.M.3 MINITAB regression options

Step 5 Residual plots are obtained by clicking the "Graphs" button and making the appropriate selections on the resulting menu. (See Figure 12.M.4.)

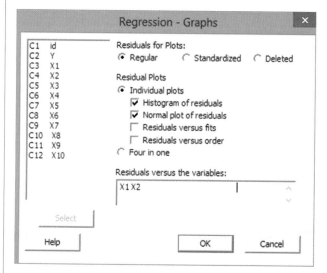

Figure 12.M.4 MINITAB regression graphs options

Step 6 To return to the main Regression dialog box from any of these optional screens, click "OK."

Step 7 When you have made all your selections, click "OK" on the main Regression dialog box to produce the MINITAB multiple-regression printout.

Stepwise Regression

Step 1 Click on the "Stat" button on the main menu bar; then click on "Regression," and click on "Stepwise." (See Figure 12.M.1.) The resulting dialog box appears like the one in Figure 12.M.2

Step 2 Specify the dependent variable in the "Response" box and the independent variables in the stepwise model in the "Predictors" box.

Step 3 As an option, you can select the value of α to use in the analysis by clicking on the "Methods" button and specifying the value. (The default is $\alpha = .15$.)

Step 4 Click "OK" to view the stepwise regression results.

TI-83/TI-84 Plus Graphing Calculator: Multiple Regression

Note: Only simple linear and quadratic regression models can be fit using the TI-83/TI-84 Plus graphing calculator.

Quadratic Regression
I. Finding the Quadratic Regression Equation

Step 1 *Enter the data*

- Press **STAT** and select **1:Edit**

  [*Note:* If the list already contains data, clear the old data. Use the up arrow to highlight "**L1**" or "**L2**."]

- Press **CLEAR ENTER**
- Use the **ARROW** and **ENTER** keys to enter the data set into **L1** and **L2**

Step 2 *Find the quadratic regression equation*

- Press **STAT** and highlight **CALC**
- Press **5** for **QuadReg**
- Press **ENTER**
- The screen will show the values for a, b, and c in the equation
- If the diagnostics are on, the screen will also give the value for r^2
- To turn the diagnostics feature on:
- Press **2nd 0** for **CATALOG**
- Press the **ALPHA** key and x^{-1} for **D**
- Press the down **ARROW** until **DiagnosticsOn** is highlighted
- Press **ENTER** twice

II. Graphing the Quadratic Curve with the Scatterplot

Step 1 *Enter the data as shown in part I in the first column*

Step 2 *Set up the data plot*

- Press **Y =** and **CLEAR** all functions from the Y registers
- Press **2nd Y =** for **STAT PLOT**
- Press **1** for **Plot**
- Set the cursor so that **ON** is flashing, and press **ENTER**
- For **Type**, use the **ARROW** and **ENTER** keys to highlight and select the scatterplot (first icon in the first row)
- For **Xlist**, choose the column containing the *x*-data
- For **Ylist**, choose the column containing the *y*-data

Step 3 *Find the regression equation and store the equation in Y1*

- Press **STAT** and highlight **CALC**
- Press **5** for **QuadReg**

  (*Note:* Don't press ENTER here because you want to store the regression equation in Y1.)

- Press **VARS**
- Use the right arrow to highlight **Y-VARS**
- Press **ENTER** to select **1:Function**
- Press **ENTER** to select **1:Y1**
- Press **ENTER**

Step 4 *View the scatterplot and regression line*

- Press **ZOOM** and then press **9** to select **9:ZoomStat**

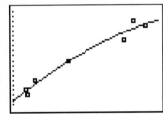

III. Plotting Residuals

When computing a regression equation on the graphing calculator, the residuals are automatically computed and saved to a list called **RESID. RESID** can be found under the **LIST menu (2nd STAT)**.

Step 1 *Enter the data*

- Press **STAT** and select **1:Edit**

  [*Note:* If the list already contains data, clear the old data. Use the up arrow to highlight "**L1**" or "**L2**."]

- Press **CLEAR ENTER**
- Use the **ARROW** and **ENTER** keys to enter the data set into **L1** and **L2**

Step 2 *Compute the regression equation*

- Press **STAT** and highlight **CALC**
- Press **4** for **LinReg(ax + b)**
- Press **ENTER**

Step 3 *Set up the data plot*

- Press **Y =** and **CLEAR** all functions from the Y registers

- Press **2nd Y =** for **STATPLOT**

- Press **1** for **Plot1**

- Set the cursor so that **ON** is flashing, and press **ENTER**

- For **Type**, use the **ARROW** and **ENTER** keys to highlight and select the scatterplot (first icon in the first row)

- Move the cursor to **Xlist** and choose the column containing the *x*-data

- Move the cursor to **Ylist**

- Press **2nd STAT** for **LIST**

- Use the down arrow to highlight the listname **RESID** and press **ENTER**

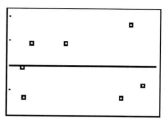

Step 4 *View the scatterplot of the residuals*

- Press **ZOOM 9** for **ZoomStat**

13 Categorical Data Analysis

CONTENTS

Where We've Been

- Presented methods for making inferences about the population proportion associated with a two-level qualitative variable (i.e., a binomial variable)
- Presented methods for making inferences about the difference between two binomial proportions

Where We're Going

- Discuss qualitative (i.e., categorical) data with more than two outcomes (13.1)
- Present a *chi-square* hypothesis test for comparing the category proportions associated with a single qualitative variable—called a *one-way analysis* (13.2)
- Present a chi-square hypothesis test for relating two qualitative variables—called a *two-way analysis* (13.3)
- Caution about the misuse of chi-square tests (13.4)

Statistics IN Action The Case of the Ghoulish Transplant Tissue

In the United States, tissue implants are routinely performed to aid patients in various types of surgery, including joint replacements, spinal surgery, sports-related surgeries (tendons and ligaments), and others. The process of obtaining a tissue transplant involves several parties. First, of course, is the donor, who has agreed to have tissue removed upon death and whose family has approved the donation. The tissue is then "harvested" by an approved tissue bank. Next, the harvested tissue is sent to a processor that sterilizes the tissue. Finally, the processor sends it either directly to the hospital/surgeon doing the implant or to a distributor that inventories the tissue and ultimately sends it on to the hospital/surgeon. The entire process is highly regulated by the Federal Trade Commission (FTC), particularly the harvesting and processing aspects.

Given this background, we consider an actual case that began in the early 2000s when the owner of a tissue bank — Biomedical Tissue Services (BTS) — became a ringleader of a group of funeral home directors that harvested tissue illegally and without permission of donors or their families. In some cases, the cadavers were cancerous or infected with HIV or hepatitis, all of which would, of course, disqualify them as donors. BTS then sent the tissue to processors without divulging that it had obtained the tissue illegally. (Note: The owner is currently serving 18 to 24 years in a New York prison.) The unsuspecting processors sterilized the tissue and sent it on for use as surgical implants. When the news story broke about how the tissue had been obtained, the processors and their distributors were required to send recall notices using an FTC recall letter to the hospital/surgeons who had received the tissue. Some of the BTS tissue was recovered; however, much of the tissue had already been

implanted, and hospitals and surgeons were required to inform patients receiving implants of the potentially infectious tissue. Although few patients subsequently became infected, a number filed suit against the distributors and processors (and BTS), asking for monetary damages.

After the bulk of the lawsuits had been either tried or settled, a dispute arose between a processor and one of its distributors regarding ultimate responsibility for payment of damages to litigating patients. In particular, the processor claimed that the distributor should be held more responsible for the damages because in its recall package, the distributor had of its own volition included some salacious, inflammatory newspaper articles describing in graphic detail the "ghoulish" acts that had been committed. None of the patients who received implants that had been sterilized by this processor ever became infected, but many still filed suit.

To establish its case against the distributor, the processor collected data on the patients who had received implants of BTS tissue it had processed and the number of those patients who subsequently filed suit: The data revealed that of a total of 7,914 patients, 708 filed suit. A consulting statistician subdivided this information according to whether the recall notice had been sent to the patient's surgeon by the processor or one of its distributors that had sent only the notice or by the distributor that had included the newspaper articles. The breakdown was as shown in Table SIA13.1.

Table SIA13.1 Data for the Tainted Tissue Case*

Recall Notice Sender	Number of Patients	Number of Lawsuits
Processor/Other Distributor	1,751	51
Distributor in Question	6,163	657
Totals	7,914	708

Do these data provide evidence of a difference in the probability that a patient would file a lawsuit depending on which party sent the recall notice? If so, and if the probability is significantly higher for the distributor in question, then the processor can argue in court that the distributor that sent the inflammatory newspaper articles is more responsible for the damages.

We apply the statistical methodology presented in this chapter to solve the case of the ghoulish transplant tissue in the following Statistics in Action Revisited example.

Statistics IN Action Revisited

- Testing whether Likelihood of a Lawsuit Is Related to Recall Notice Sender (p. 817)

*For confidentiality purposes, the parties in the case cannot be identified. Permission to use the data in this Statistics in Action has been granted by the consulting statistician.

13.1 Categorical Data and the Multinomial Experiment

Recall from Section 1.4 (p. 38) that observations on a qualitative variable can only be categorized. For example, consider the highest level of education attained by a professional hockey player. Level of education is a qualitative variable with several categories, including some high school, high school diploma, some college, college undergraduate degree, and graduate degree. If we were to record education level for all professional hockey players, the result of the categorization would be a count of the numbers of players falling into the respective categories.

When the qualitative variable of interest results in one of two responses (e.g., yes or no, success or failure, favor or do not favor), the data—called *counts*—can be analyzed with the binomial probability distribution discussed in Section 4.4. However, qualitative variables, such as level of education, that allow for more than two categories for a response are much more common, and these must be analyzed by a different method.

Qualitative data with more than two levels often result from a **multinomial experiment**. The characteristics for a multinomial experiment with k outcomes are described in the next box. You can see that the binomial experiment of Chapter 4 is a multinomial experiment with $k = 2$.

Properties of the Multinomial Experiment

1. The experiment consists of n identical trials.
2. There are k possible outcomes to each trial. These outcomes are sometimes called **classes**, **categories**, or **cells**.
3. The probabilities of the k outcomes, denoted by $p_1, p_2, \ldots, p_k$, where $p_1 + p_2 + \cdots + p_k = 1$, remain the same from trial to trial.
4. The trials are independent.
5. The random variables of interest are the **cell counts** $n_1, n_2, \ldots, n_k$ of the number of observations that fall into each of the k categories.

Example   13.1

Identifying a Multinomial Experiment

Problem Consider the problem of determining the highest level of education attained by each of a sample of $n = 40$ National Hockey League (NHL) players. Suppose we categorize level of education into one of five categories—some high school, high school diploma, some college, college undergraduate degree, and graduate degree—and count the number of the 40 players that fall into each category. Is this a multinomial experiment, to a reasonable degree of approximation?

Solution Checking the five properties of a multinomial experiment shown in the box, we have the following:

1. The experiment consists of $n = 40$ identical trials, each of which is undertaken to determine the education level of an NHL player.
2. There are $k = 5$ possible outcomes to each trial, corresponding to the five education-level responses.
3. The probabilities of the $k = 5$ outcomes $p_1, p_2, p_3, p_4,$ and p_5, where p_i represents the true probability that an NHL player attains level-of-education category i, remain the same from trial to trial (to a reasonable degree of approximation).
4. The trials are independent; that is, the education level attained by one NHL player does not affect the level attained by any other player.
5. We are interested in the count of the number of hockey players who fall into each of the five education-level categories. These five cell counts are denoted $n_1, n_2, n_3, n_4,$ and n_5.

Thus, the properties of a multinomial experiment are satisfied.

In this chapter, we are concerned with the analysis of categorical data—specifically, the data that represent the counts for each category of a multinomial experiment. In Section 13.2, we learn how to make inferences about the probabilities of categories for data classified according to a single qualitative (or categorical) variable. Then, in Sections 13.3 and 13.4 we consider inferences about categorical probabilities for data classified according to two qualitative variables. The statistic used for these inferences is one that possesses, approximately, the familiar chi-square distribution.

13.2 Testing Categorical Probabilities: One-Way Table

In this section, we consider a multinomial experiment with k outcomes that correspond to the categories of a *single* qualitative variable. The results of such an experiment are summarized in a **one-way table**. The term *one-way* is used because only one variable is classified. Typically, we want to make inferences about the true percentages that occur in the k categories on the basis of the sample information in the one-way table.

To illustrate, suppose three political candidates are running for the same elective position. Prior to the election, we conduct a survey to determine the voting preferences of a random sample of 150 eligible voters. The qualitative variable of interest is *preferred candidate*, which has three possible outcomes: candidate 1, candidate 2, and candidate 3. Suppose the number of voters preferring each candidate is tabulated and the resulting count data appear as in Table 13.1.

Table 13.1	Results of Voter Preference Survey	
	Candidate	
1	2	3
61 votes	53 votes	36 votes

Note that our voter preference survey satisfies the properties of a multinomial experiment for the qualitative variable, preferred candidate. The experiment consists of randomly sampling $n = 150$ voters from a large population of voters containing an unknown proportion p_1 that favors candidate 1, a proportion p_2 that favors candidate 2, and a proportion p_3 that favors candidate 3. Each voter sampled represents a single trial that can result in one of three outcomes: The voter will favor candidate 1, 2, or 3 with probabilities p_1, p_2, and p_3, respectively. (Assume that all voters will have a preference.) The voting preference of any single voter in the sample does not affect the preference of any other; consequently, the trials are independent. Finally, you can see that the recorded data are the numbers of voters in each of the three preference categories. Thus, the voter preference survey satisfies the five properties of a multinomial experiment.

In this survey, and in most practical applications of the multinomial experiment, the k outcome probabilities $p_1, p_2, \ldots, p_k$ are unknown and we want to use the survey data to make inferences about their values. The unknown probabilities in the voter preference survey are

$$p_1 = \text{Proportion of all voters who favor candidate 1}$$
$$p_2 = \text{Proportion of all voters who favor candidate 2}$$
$$p_3 = \text{Proportion of all voters who favor candidate 3}$$

To decide whether the voters, in total, have a preference for any one of the candidates, we will test the null hypothesis that the candidates are equally preferred (i.e., $p_1 = p_2 = p_3 = \frac{1}{3}$) against the alternative hypothesis that one candidate is preferred (i.e., at least one of the probabilities p_1, p_2, and p_3 exceeds $\frac{1}{3}$). Thus, we want to test

$H_0: p_1 = p_2 = p_3 = \frac{1}{3}$ (no preference)

$H_a:$ At least one of the proportions exceeds $\frac{1}{3}$ (a preference exists)

If the null hypothesis is true and $p_1 = p_2 = p_3 = \frac{1}{3}$, then the expected value (mean value) of the number of voters who prefer candidate 1 is given by

$$E_1 = np_1 = (n)\frac{1}{3} = (150)\frac{1}{3} = 50$$

Similarly, $E_2 = E_3 - 50$ if the null hypothesis is true and no preference exists.

The **chi-square test** measures the degree of disagreement between the data and the null hypothesis:

$$\chi^2 = \frac{[n_1 - E_1]^2}{E_1} + \frac{[n_2 - E_2]^2}{E_2} + \frac{[n_3 - E_3]^2}{E_3}$$
$$= \frac{(n_1 - 50)^2}{50} + \frac{(n_2 - 50)^2}{50} + \frac{(n_3 - 50)^2}{50}$$

Note that the farther the observed numbers $n_1, n_2,$ and n_3 are from their expected value (50), the larger χ^2 will become. That is, large values of χ^2 imply that the null hypothesis is false.

We have to know the distribution of χ^2 in repeated sampling before we can decide whether the data indicate that a preference exists. When H_0 is true, χ^2 can be shown to have (approximately) the familiar chi-square distribution of Section 8.8. For this one-way classification, the χ^2 distribution has $(k - 1)$ degrees of freedom.* The rejection region for the voter preference survey for $\alpha = .05$ and $k - 1 = 3 - 1 = 2$ df is

$$Rejection\ region:\ \chi^2 > \chi^2_{.05}$$

This value of $\chi^2_{.05}$ (found in Table IV of Appendix B) is 5.99147. (See Figure 13.1.) The computed value of the test statistic is

$$\chi^2 = \frac{(n_1 - 50)^2}{50} + \frac{(n_2 - 50)^2}{50} + \frac{(n_3 - 50)^2}{50}$$
$$= \frac{(61 - 50)^2}{50} + \frac{(53 - 50)^2}{50} + \frac{(36 - 50)^2}{50} = 6.52$$

Since the computed $\chi^2 = 6.52$ exceeds the critical value of 5.99147, we conclude at the $\alpha = .05$ level of significance that there does exist a voter preference for one or more of the candidates.

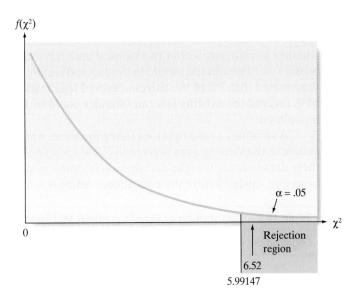

Figure 13.1

Rejection region for voter preference survey

---

*The derivation of the number of degrees of freedom for χ^2 involves the number of linear restrictions imposed on the count data. In the present case, the only constraint is that $\Sigma n_i = n$, where n (the sample size) is fixed in advance. Therefore, df $= k - 1$. For other cases, we will give the number of degrees of freedom for each usage of χ^2 and refer the interested reader to the references for more detail.

Now that we have evidence to indicate that the proportions $p_1, p_2,$ and p_3 are unequal, we can use the methods of Section 7.4 to make inferences concerning their individual values. [*Note:* We cannot use the methods of Section 9.4 to compare two proportions because the cell counts are dependent random variables.] The general form for a test of hypothesis concerning multinomial probabilities is shown in the following box:

A Test of a Hypothesis about Multinomial Probabilities: One-Way Table

$H_0: p_1 = p_{1,0}, \ p_2 = p_{2,0}, \dots, p_k = p_{k,0}$

where $p_{1,0}, p_{2,0}, \dots, p_{k,0}$ represent the hypothesized values of the multinomial probabilities

H_a: At least one of the probabilities does not equal its hypothesized value

Test statistic: $\chi_c^2 = \sum \dfrac{[n_i - E_i]^2}{E_i}$

where $E_i = np_{i,0}$ is the **expected cell count**—that is, the expected number of outcomes of type i, assuming that H_0 is true. The total sample size is n.

Rejection region: $\chi_c^2 > \chi_\alpha^2$, where χ_α^2 has $(k - 1)$ df

p-value: $P(\chi^2 > \chi_c^2)$

Conditions Required for a Valid χ^2 Test: One-Way Table

1. A multinomial experiment has been conducted. This is generally satisfied by taking a random sample from the population of interest.

2. The sample size n will be large enough so that, for every cell, the expected cell count E_i will be equal to 5 or more.*

Example   13.2   Ⓓ

A One-Way χ^2 Test—Effectiveness of a TV Program on Marijuana

Problem Suppose an educational television station in a state that has not legalized marijuana has broadcast a series of programs on the physiological and psychological effects of smoking marijuana. Now that the series is finished, the station wants to see whether the citizens within the viewing area have changed their minds about how the possession of marijuana should be considered legally. Before the series was shown, it was determined that 7% of the citizens favored legalization, 18% favored decriminalization, 65% favored the existing law (an offender could be fined or imprisoned), and 10% had no opinion.

A summary of the opinions (after the series was shown) of a random sample of 500 people in the viewing area is given in Table 13.2. Test at the $\alpha = .01$ level to see whether these data indicate that the distribution of opinions differs significantly from the proportions that existed before the educational series was aired.

Table 13.2   Distribution of Opinions about Marijuana Possession

Legalization	Decriminalization	Existing Laws	No Opinion
39	99	336	26

Ⓓ *Data Set:* MARIJ

---

*The assumption that all expected cell counts are at least 5 is necessary in order to ensure that the χ^2 approximation is appropriate. Exact methods for conducting the test of hypothesis exist and may be used for small expected cell counts, but these methods are beyond the scope of this text.

Solution Define the proportions after the airing to be

$$p_1 = \text{Proportion of citizens favoring legalization}$$
$$p_2 = \text{Proportion of citizens favoring decriminalization}$$
$$p_3 = \text{Proportion of citizens favoring existing laws}$$
$$p_4 = \text{Proportion of citizens with no opinion}$$

Then the null hypothesis representing no change in the distribution of percentages is

$$H_0: p_1 = .07, p_2 = .18, p_3 = .65, p_4 = .10$$

and the alternative is

H_a: At least one of the proportions differs from its null hypothesized value

Thus, we have

$$\text{Test statistic: } \chi^2 = \sum \frac{[n_i - E_i]^2}{E_i}$$

where

$$E_1 = np_{1,0} = 500(.07) = 35$$
$$E_2 = np_{2,0} = 500(.18) = 90$$
$$E_3 = np_{3,0} = 500(.65) = 325$$
$$E_4 = np_{4,0} = 500(.10) = 50$$

Since all these values are larger than 5, the χ^2 approximation is appropriate. Also, if the citizens in the sample were randomly selected, then the properties of the multinomial probability distribution are satisfied.

Rejection region: For $\alpha = .01$ and df $= k - 1 = 3$, reject H_0 if $\chi^2 > \chi^2_{.01}$, where (from Table IV in Appendix B) $\chi^2_{.01} = 11.3449$.

We now calculate the test statistic:

$$\chi^2 = \frac{(39 - 35)^2}{35} + \frac{(99 - 90)^2}{90} + \frac{(336 - 325)^2}{325} + \frac{(26 - 50)^2}{50} = 13.249$$

Since this value exceeds the table value of χ^2 (11.3449), the data provide sufficient evidence ($\alpha = .01$) that the opinions on the legalization of marijuana have changed since the series was aired.

The χ^2 test can also be conducted with the use of an available statistical software package. Figure 13.2 is an SPSS printout of the analysis of the data in Table 13.2. The test statistic and p-value of the test are highlighted on the printout. Since $\alpha = .01$ exceeds $p = .004$, there is sufficient evidence to reject H_0.

OPINION

	Observed N	Expected N	Residual
LEGAL	39	35.0	4.0
DECRIM	99	90.0	9.0
EXISTLAW	336	325.0	11.0
NONE	26	50.0	-24.0
Total	500		

Test Statistics

	OPINION
Chi-Square[a]	13.249
df	3
Asymp. Sig.	.004

a. 0 cells (.0%) have expected frequencies less than 5. The minimum expected cell frequency is 35.0.

Figure 13.2

SPSS analysis of data in Table 13.2

Look Back If the conclusion for the χ^2 test is "fail to reject H_0," then there is insufficient evidence to conclude that the distribution of opinions differs from the proportions stated in H_0. Be careful not to "accept H_0" and conclude that $p_1 = .07$, $p_2 = .18$, $p_3 = .65$, and $p_4 = .10$. The probability (β) of a Type II error is unknown.

■ **Now Work Exercise 13.9**

If we focus on one particular outcome of a multinomial experiment, we can use the methods developed in Section 7.4 for a binomial proportion to establish a confidence interval for any one of the multinomial probabilities.* For example, if we want a 95% confidence interval for the proportion of citizens in the viewing area who have no opinion about the issue, we calculate

$$\hat{p}_4 \pm 1.96\sigma_{\hat{p}_4}$$

where

$$\hat{p}_4 = \frac{n_4}{n} = \frac{26}{500} = .052 \quad \text{and} \quad \sigma_{\hat{p}_4} \approx \sqrt{\frac{\hat{p}_4(1 - \hat{p}_4)}{n}}$$

Thus, we get

$$.052 \pm 1.96\sqrt{\frac{(.052)(.948)}{500}} = .052 \pm .019$$

or (.033, .071). Consequently, we estimate that between 3.3% and 7.1% of the citizens now have no opinion on the issue of the legalization of marijuana. The series of programs may have helped citizens who formerly had no opinion on the issue to form an opinion, since it appears that the proportion of "no opinions" is now less than 10%.

*Note that focusing on one outcome has the effect of lumping the other $(k - 1)$ outcomes into a single group. Thus, we obtain, in effect, two outcomes—or a binomial experiment.

Exercises 13.1–13.21

Understanding the Principles

13.1 What are the characteristics of a multinomial experiment? Compare the characteristics with those of a binomial experiment.

13.2 What conditions must n satisfy to make the χ^2 test for a one-way table valid?

Learning the Mechanics

13.3 Use Table IV of Appendix B, to find each of the following χ^2 values:
 a. $\chi^2_{.995}$ for df = 28
 b. $\chi^2_{.950}$ for df = 2
 c. $\chi^2_{.050}$ for df = 23
 d. $\chi^2_{.025}$ for df = 13

13.4 Use Table IV of Appendix B to find the following probabilities:
 a. $P(\chi^2 \leq 20.5992)$ for df = 30
 b. $P(\chi^2 > 36.7807)$ for df = 22
 c. $P(\chi^2 \geq 53.5400)$ for df = 80
 d. $P(\chi^2 < 14.6837)$ for df = 9

13.5 Find the rejection region for a one-dimensional χ^2-test of a null hypothesis concerning $p_1, p_2, \ldots, p_k$ if
 a. $k = 11$; $\alpha = .01$
 b. $k = 9$; $\alpha = .01$
 c. $k = 6$; $\alpha = .05$

13.6 A multinomial experiment with $k = 3$ cells and $n = 400$ produced the data shown in the accompanying table. Do these data provide sufficient evidence to contradict the null hypothesis that $p_1 = .50$, $p_2 = .25$, and $p_3 = .25$? Test, using $\alpha = .05$.

		Cell	
	1	2	3
n_i	228	80	92

13.7 A multinomial experiment with $k = 4$ cells and $n = 191$ produced the data shown in the following table.

		Cell		
	1	2	3	4
n_i	51	47	56	37

 a. Do these data provide sufficient evidence to conclude that the multinomial probabilities differ? Test, using $\alpha = .05$.
 b. What are the Type I and Type II errors associated with the test of part **a**?
 c. Construct a 95% confidence interval for the multinomial probability associated with cell 2.

13.8 A multinomial experiment with $k = 4$ cells and $n = 400$ produced the data shown in the accompanying table. Do these data provide sufficient evidence to contradict the null hypothesis that $p_1 = .2, p_2 = .4, p_3 = .1$, and $p_4 = .3$? Test, using $\alpha = .05$.

	Cell			
	1	2	3	4
n_i	64	156	41	139

Applying the Concepts—Basic

13.9 **Jaw dysfunction study.** A report on dental patients with
[NW] temporomandibular (jaw) joint dysfunction (TMD) was published in *General Dentistry* (Jan/Feb. 2004). A random sample of 60 patients was selected for an experimental treatment of TMD. Prior to treatment, the patients filled out a survey on two nonfunctional jaw habits—bruxism (teeth grinding) and teeth clenching—that have been linked to TMD. Of the 60 patients, 3 admitted to bruxism, 11 admitted to teeth clenching, 30 admitted to both habits, and 16 claimed they had neither habit.
   a. Describe the qualitative variable of interest in the study. Give the levels (categories) associated with the variable.
   b. Construct a one-way table for the sample data.
   c. Give the null and alternative hypotheses for testing whether the percentages associated with the admitted habits are the same.
   d. Calculate the expected numbers for each cell of the one-way table.
   e. Calculate the appropriate test statistic.
   f. Give the rejection region for the test at $\alpha = .05$
   g. Give the appropriate conclusion in the words of the problem.
   h. Find and interpret a 95% confidence interval for the true proportion of dental patients who admit to both habits.

13.10 **Beetles and slime molds.** Myxomycetes are mushroom-
[D] like slime molds that are a food source for insects. The
SLIME *Journal of Natural History* (May 2010) published the results of a study that investigated which of six species of slime molds are most attractive to beetles inhabiting an Atlantic rain forest. A sample of 19 beetles feeding on slime mold was obtained and the species of slime mold was determined for each beetle. The numbers of beetles captured on each of the six species are given in the accompanying table. The researchers want to know if the relative frequency of occurrence of beetles differs for the six slime mold species.

Slime mold species:	LE	TM	AC	AD	HC	HS
Number of beetles:	3	2	7	3	1	3

   a. Identify the categorical variable (and its levels) of interest in this study.
   b. Set up the null and alternative hypotheses of interest to the researchers.
   c. Find the test statistic and corresponding *p*-value.

   d. The researchers found "no significant differences in the relative frequencies of occurrence" using $\alpha = .05$. Do you agree?
   e. Comment on the validity of the inference, part **d**. (Determine the expected cell counts.)

13.11 **Do social robots walk or roll?** Refer to the *International Conference on Social Robotics* (Vol. 6414, 2010) study of how engineers design social robots, Exercise 2.7 (p. 66). Recall that a social (or service) robot is designed to entertain, educate, and care for human users. In a random sample of 106 social robots obtained through a Web search, the researchers found that 63 were built with legs only, 20 with wheels only, 8 with both legs and wheels, and 15 with neither legs nor wheels. Prior to obtaining these sample results, a robot design engineer stated that 50% of all social robots produced have legs only, 30% have wheels only, 10% have both legs and wheels, and 10% have neither legs nor wheels.
   a. Explain why the data collected for each sampled social robot are categorical in nature.
   b. Specify the null and alternative hypotheses for testing the design engineer's claim.
   c. Assuming the claim is true, determine the number of social robots in the sample you expect to fall into each design category.
   d. Use the results to compute the chi-square test statistic.
   e. Make the appropriate conclusion using $\alpha = .05$.

13.12 **Museum management.** Refer to the *Museum Management*
[D] *and Curatorship* (June 2010) worldwide survey of 30 lead-
MMC ing museums of contemporary art, Exercise 2.22 (p. 69). Recall that each museum manager was asked to provide the performance measure used most often for internal evaluation. A summary of the results is provided in the table. The data were analyzed using a chi-square test for a multinomial experiment. The results are shown in the MINITAB printout on the next page.
   a. Is there evidence to indicate that one performance measure is used more often than any of the others? Test using $\alpha = .10$.
   b. Find a 90% confidence interval for the proportion of museums worldwide that use total visitors as their performance measure. Interpret the result.

Performance Measure	Number of Museums
Total visitors	8
Paying visitors	5
Big shows	6
Funds raised	7
Members	4

13.13 **Gender in two-child families.** Refer to the *Human Biology*
[D] (Feb. 2009) study on the gender of children in two-child
BOYGRL families, Exercise 4.29 (p. 222). The article reported on the results of the National Health Interview Survey (NHIS) of 42,888 two-child families. The table below gives the number of families with each gender configuration.

Gender Configuration	Number of Families
Girl-girl (GG)	9,523
Boy-girl (BG)	11,118
Girl-boy (GB)	10,913
Boy-boy (BB)	11,334

MINITAB Output for Exercise 13.12

Chi-Square Goodness-of-Fit Test for Observed Counts in Variable: NUMBER

Using category names in PERFORM

Category	Observed	Test Proportion	Expected	Contribution to Chi-Sq
Total visitors	8	0.2	6	0.666667
Paying visitors	5	0.2	6	0.166667
Big shows	6	0.2	6	0.000000
Funds raised	7	0.2	6	0.166667
Members	4	0.2	6	0.666667

N	DF	Chi-Sq	P-Value
30	4	1.66667	0.797

a. If it is just as likely to have a boy as a girl, find the probability of each of the gender configurations for a two-child family.

b. Use the probabilities, part **a**, to determine the expected number of families for each gender configuration.

c. Compute the chi-square test statistic for testing the hypothesis that it is just as likely to have a boy as a girl.

d. Interpret the result, part **c**, if you conduct the test using $\alpha = .10$.

e. Recent research indicates that the ratio of boys to girls in the world population is not 1 to 1, but instead higher (e.g., 1.06 to 1). Using a ratio of 1.06 to 1, the researchers showed that the probabilities of the different gender configurations are: GG−.23795, BG−.24985, GB−.24985, and BB−.26235. Repeat parts **b–d** using these probabilities.

Applying the Concepts—Intermediate

13.14 Mobile device typing strategies. Researchers estimate that in a typical month about 75 billion text messages are sent in the United States. Text messaging on mobile devices (e.g., cell phones, smartphones) often requires typing in awkward positions that may lead to health issues. A group of Temple University public health professors investigated this phenomenon and published their results in *Applied Ergonomics* (Mar. 2012). One portion of the study focused on the typing styles of mobile device users. Typing style was categorized as (1) device held with both hands/both thumbs typing, (2) device held with right hand/right thumb typing, (3) device held with left hand/left thumb typing, (4) device held with both hands/right thumb typing, (5) device held with left hand/right index finger typing, or (6) other. In a sample of 859 college students observed typing on their mobile devices, the professors observed 396, 311, 70, 39, 18, and 25, respectively, in the six categories. Is this sufficient evidence to conclude that the proportions of mobile device users in the six texting style categories differ? Use $\alpha = .10$ to answer the question.

13.15 Curbing street gang gun violence. Refer to the *Journal of Quantitative Criminology* (Mar. 2014) study of street gun violence in Boston, Exercise 2.18 (p. 69). Recall that over a 5-year period (2006–2010), 80 shootings involving a Boston street gang were sampled and analyzed. Of these, 37 occurred in 2006, 30 in 2007, 4 in 2008, 5 in 2009, and 4 in 2010. A program designed to reduce street gang violence was implemented at the end of 2007. If the program is effective, then researchers believe that the percentage of all gang shootings over the 5-year period will break down as follows: 40% in 2006, 40% in 2007, 10% in 2008, 5% in 2009, and 5% in 2010. Give your opinion on the effectiveness of the program.

13.16 Characteristics of ice-melt ponds. Refer to the National Snow and Ice Data Center (NSIDC) collection of data on 504 ice-melt ponds in the Canadian Arctic, presented in Exercise 12.200 (p. 793). One variable of interest to environmental engineers studying the ponds is the type of ice observed in each. Ice type is classified as first-year ice, multiyear ice, or landfast ice. The SAS summary table for the types of ice of the 504 ice-melt ponds is reproduced below.

a. Use a 90% confidence interval to estimate the proportion of ice-melt ponds in the Canadian Arctic that have first-year ice.

b. Suppose environmental engineers hypothesize that 15% of Canadian Arctic ice-melt ponds have first-year ice, 40% have landfast ice, and 45% have multiyear ice. Test the engineers' theory, using $\alpha = .01$.

The FREQ Procedure

ICETYPE	Frequency	Percent	Cumulative Frequency	Cumulative Percent
First-year	88	17.46	88	17.46
Landfast	196	38.89	284	56.35
Multi-year	220	43.65	504	100.00

13.17 Detecting Alzheimer's disease at an early age. Geneticists at Australian National University are studying whether the cognitive effects of Alzheimer's disease can be detected at an early age (*Neuropsychology,* Jan. 2007). One portion of the study focused on a particular strand of DNA extracted from each in a sample of 2,097 young adults between the ages of 20 and 24. The DNA strand was classified into one of three genotypes: $E4^+/E4^+$, $E4^+/E4^-$, and $E4^-/E4^-$. The number of young adults with each genotype is shown in the table. Suppose that in adults who are not afflicted with Alzheimer's disease, the distribution of genotypes for this strand of DNA is 2% with $E4^+/E4^+$, 25% with $E4^+/E4^-$,

Genotype:	$E4^+/E4^+$	$E4^+/E4^-$	$E4^-/E4^-$
Number of young adults:	56	517	1,524

and 73% with E4$^-$/E4$^-$. If differences in this distribution are detected, then this strand of DNA could lead researchers to an early test for the onset of Alzheimer's. Conduct a test (at $\alpha = .05$) to determine if the distribution of E4/E4 genotypes for the population of young adults differs from the norm.

13.18 **Traffic sign maintenance.** Refer to the *Journal of Transportation Engineering* (June 2013) study of traffic sign maintenance, Exercise 9.71 (p. 500). Recall that civil engineers estimated the proportion of traffic signs maintained by the North Carolina Department of Transportation (NCDOT) that fail minimum retroreflectivity requirements. The researchers were also interested in the proportions of NCDOT signs with background colors white (regulatory signs), yellow (warning/caution), red (stop/yield/wrong way), and green (guide/information). In a random sample of 1,000 road signs maintained by the NCDOT, 373 were white, 447 were yellow, 88 were green, and 92 were red. Suppose that NCDOT stores new signs in a warehouse for use as replacement signs; of these, 35% are white, 45% are yellow, 10% are green, and 10% are red. Does the distribution of background colors for all road signs maintained by NCDOT match the color distribution of signs in the warehouse? Test, using $\alpha = .05$.

13.19 **Interactions in a children's museum.** Refer to the *Early Childhood Education Journal* (Mar. 2014) study of interactions in a children's museum, Exercise 2.19 (p. 69). Recall that interactions by visitors to the museum were classified as (1) show-and-tell, (2) learning/teaching, (3) refocusing, (4) participatory play, or (5) advocating/disciplining. Over a 3-month period, the researchers observed 170 meaningful interactions, of which 81 were led by children and 89 were led by adult caregivers. The number of interactions observed in each category is provided in the accompanying table.

Type of Interaction	Child-Led	Adult-Led
Show-and-tell	26	0
Learning/Teaching	21	64
Refocusing	21	10
Participatory Play	12	9
Advocating/Disciplining	1	6
Totals	81	89

Source: McMunn-Dooley, C. and Welch, M. M. "Nature of interactions among young children and adult caregivers in a children's museum." *Early Childhood Education Journal,* Vol. 42, No. 2, Mar. 2014 (adapted from Figure 2).

a. For child-led interactions, is there evidence of differences in the proportions associated with the types of interactions? Test, using $\alpha = .01$.

b. Repeat part **a** for adult-led interactions.

Applying the Concepts—Advanced

13.20 **Political representation of religious groups.** Do those elected to the U.S. House of Representatives really "represent" their constituents demographically? This was a question of interest in *Chance* (Summer 2002). One of several demographics studied was religious affiliation. The table in the next column gives the proportion of the U.S. population for several religions, as well as the number of the 435 seats in the House of Representatives affiliated with that religion. Give your opinion on whether the members of the House of Representatives are statistically representative of the religious affiliation of their constituents in the United States.

Religion	Proportion of U.S. Population	Number of Seats in House
Catholic	.28	117
Methodist	.04	61
Jewish	.02	30
Other	.66	227
Totals	1.00	435

13.21 **Analysis of a Scrabble game.** In the board game Scrabble™, a player initially draws a "hand" of seven tiles at random from 100 tiles. Each tile has a letter of the alphabet, and the player attempts to form a word from the letters in his or her hand. A few years ago, a handheld electronic version of the game, called ScrabbleExpress™, was developed and sold. However, pure Scrabble players complained that the handheld game produced too few vowels in the 7-letter draws. For each of the 26 letters (and "blank" for any letter), the accompanying table gives the true relative frequency of the letter in the board game. To investigate the validity of the complaint, assume that we record the frequency of occurrence of the letter in a sample of 350 tiles (i.e., 50 "hands") randomly drawn in the electronic game. These results are also shown in the table.

a. Do the data support the claim that ScrabbleExpress™ produces letters at a different rate than the Scrabble™ board game? Test, using $\alpha = .05$.

b. Use a 95% confidence interval to estimate the true proportion of letters drawn in the electronic game that are vowels. Compare the results with the true relative frequency of a vowel in the board game.

Letter	Relative Frequency in Board Game	Frequency in Electronic Game
A	.09	20
B	.02	8
C	.02	15
D	.04	14
E	.12	16
F	.02	10
G	.03	16
H	.02	11
I	.09	12
J	.01	9
K	.01	13
L	.04	10
M	.02	15
N	.06	17
O	.08	10
P	.02	16
Q	.01	6
R	.06	13
S	.04	14
T	.06	12
U	.04	11
V	.02	17
W	.02	15
X	.01	8
Y	.02	16
Z	.01	7
# (blank)	.02	19
Total		350

13.3 Testing Categorical Probabilities: Two-Way (Contingency) Table

In Section 13.2, we introduced the multinomial probability distribution and considered data classified according to a single criterion. We now consider multinomial experiments in which the data are classified according to two criteria—that is, *classification with respect to two qualitative factors*.

Consider a study similar to one in the *Journal of Marketing* on the impact of using celebrities in television advertisements. The researchers investigated the relationship between the gender of a viewer and the viewer's brand awareness. Three hundred TV viewers were randomly selected and each asked to identify products advertised by male celebrity spokespersons. The data are summarized in the **two-way table** shown in Table 13.3. This table, called a **contingency table**, presents multinomial count data classified on two scales, or **dimensions, of classification**: gender of viewer and brand awareness.

Table 13.3   Contingency Table for Marketing Example

Brand Awareness		Male	Female	Totals
		Gender		
Brand Awareness	**Could Identify Product**	95	41	136
	Could Not Identify Product	50	114	164
	Totals	145	155	300

Ⓓ *Data Set:* CELEB

The symbols representing the cell counts for the multinomial experiment in Table 13.3 are shown in Table 13.4a, and the corresponding cell, row, and column probabilities are shown in Table 13.4b. Thus, n_{11} represents the number of viewers who are male and could identify the brand, and p_{11} represents the corresponding cell probability. Note the symbols for the row and column totals and also the symbols for the probability totals. The latter are called **marginal probabilities** for each row and column. The marginal probability p_{r1} is the probability that a TV viewer identifies the product; the marginal probability p_{c1} is the probability that a TV viewer is male. Thus,

$$p_{r1} = p_{11} + p_{12} \text{ and } p_{c1} = p_{11} + p_{21}$$

Table 13.4a Observed Counts for Contingency Table 13.3

		Gender		
		Male	Female	Totals
Brand Awareness	**Could Identify Product**	n_{11}	n_{12}	R_1
	Could Not Identify Product	n_{21}	n_{22}	R_2
	Totals	C_1	C_2	n

Table 13.4b Probabilities for Contingency Table 13.3

		Gender		
		Male	Female	Totals
Brand Awareness	**Could Identify Product**	p_{11}	p_{12}	p_{r1}
	Could Not Identify Product	p_{21}	p_{22}	p_{r2}
	Totals	p_{c1}	p_{c2}	1

We can see, then, that this really is a multinomial experiment with a total of 300 trials, $(2)(2) = 4$ cells or possible outcomes, and probabilities for each cell as shown in Table 13.4b. Since the 300 TV viewers are randomly chosen, the trials are considered independent and the probabilities are viewed as remaining constant from trial to trial.

Suppose we want to know whether the two classifications of gender and brand awareness are dependent. That is, if we know the gender of the TV viewer, does that information give us a clue about the viewer's brand awareness? In a probabilistic sense, we know (Chapter 3) that the independence of events A and B implies that $P(AB) = P(A)P(B)$. Similarly, in the contingency table analysis, if the **two classifications are independent**, the probability that an item is classified into any particular cell of the table is the product of the corresponding marginal probabilities. Thus, under the hypothesis of independence, in Table 13.4b we must have

$$p_{11} = p_{r1}p_{c1} \quad p_{12} = p_{r1}p_{c2}$$
$$p_{21} = p_{r2}p_{c1} \quad p_{22} = p_{r2}p_{c2}$$

To test the hypothesis of independence, we use the same reasoning employed in the one-dimensional tests of Section 13.2. First, we calculate the *expected*, or *mean, count in each cell*, assuming that the null hypothesis of independence is true. We do this by noting that the expected count in a cell of the table is just the total number of multinomial trials, n, times the cell probability. Recall that n_{ij} represents the **observed count** in the cell located in the ith row and jth column. Then the expected cell count for the upper left-hand cell (first row, first column) is

$$E_{11} = np_{11}$$

or, when the null hypothesis (the classifications are independent) is true,

$$E_{11} = np_{r1}p_{c1}$$

Since these true probabilities are not known, we estimate p_{r1} and p_{c1} by the same proportions $\hat{p}_{r1} = R_1/n$ and $\hat{p}_{c1} = C_1/n$. Thus, the estimate of the expected value E_{11} is

$$\hat{E}_{11} = n\left(\frac{R_1}{n}\right)\left(\frac{C_1}{n}\right) = \frac{R_1 C_1}{n}$$

Similarly, for each i, j,

$$\hat{E}_{ij} = \frac{(\text{Row total})(\text{Column total})}{\text{Total sample size}}$$

Hence,

$$\hat{E}_{12} = \frac{R_1 C_2}{n}$$

$$\hat{E}_{21} = \frac{R_2 C_1}{n}$$

$$\hat{E}_{22} = \frac{R_2 C_2}{n}$$

Finding Expected Cell Counts for a Two-Way Contingency Table

The estimate of the expected number of observations falling into the cell in row i and column j is given by

$$\hat{E}_{ij} = \frac{R_i C_j}{n}$$

where R_i = total for row i, C_j = total for column j, and n = sample size.

Tabulated statistics: AWARE, GENDER

Using frequencies in NUMBER

Rows: AWARE   Columns: GENDER

	Male	Female	All
ID-Product	95	41	136
	65.7	70.3	136.0
No-ID	50	114	164
	79.3	84.7	164.0
All	145	155	300
	145.0	155.0	300.0

Cell Contents: Count
                         Expected count

Pearson Chi-Square = 46.135, DF = 1, P-Value = 0.000
Likelihood Ratio Chi-Square = 47.362, DF = 1, P-Value = 0.000

Figure 13.3

MINITAB contingency table analysis of data in Table 13.3

Using the data in Table 13.3, we find that

$$\hat{E}_{11} = \frac{R_1 C_1}{n} = \frac{(136)(145)}{300} = 65.73$$

$$\hat{E}_{12} = \frac{R_1 C_2}{n} = \frac{(136)(155)}{300} = 70.27$$

$$\hat{E}_{21} = \frac{R_2 C_1}{n} = \frac{(164)(145)}{300} = 79.27$$

$$\hat{E}_{22} = \frac{R_2 C_2}{n} = \frac{(164)(155)}{300} = 84.73$$

These estimated expected values are more easily obtained using computer software. Figure 13.3 is a MINITAB printout of the analysis, with the expected values highlighted.

We now use the χ^2 statistic to compare the observed and expected (estimated) counts in each cell of the contingency table:

$$\chi^2 = \frac{[n_{11} - \hat{E}_{11}]^2}{\hat{E}_{11}} + \frac{[n_{12} - \hat{E}_{12}]^2}{\hat{E}_{12}} + \frac{[n_{21} - \hat{E}_{21}]^2}{\hat{E}_{21}} + \frac{[n_{22} - \hat{E}_{22}]^2}{\hat{E}_{22}}$$

$$= \sum \frac{[n_{ij} - \hat{E}_{ij}]^2}{\hat{E}_{ij}}$$

(*Note:* The use of $\sum$ in the context of a contingency table analysis refers to a sum over all cells in the table.)

Substituting the data of Table 13.3 and the expected values into this expression, we get

$$\chi^2 = \frac{(95 - 65.73)^2}{65.73} + \frac{(41 - 70.27)^2}{70.27} + \frac{(50 - 79.27)^2}{79.27} + \frac{(114 - 84.73)^2}{84.73} = 46.14$$

Note that this value is also shown (highlighted) in Figure 13.3.

Large values of χ^2 imply that the observed counts do not closely agree and hence that the hypothesis of independence is false. To determine how large χ^2 must be before it is too large to be attributed to chance, we make use of the fact that the sampling distribution of χ^2 is approximately a χ^2 probability distribution when the classifications are independent.

When testing the null hypothesis of independence in a two-way contingency table, the appropriate degrees of freedom will be $(r - 1)(c - 1)$, where r is the number of rows and c is the number of columns in the table. For the brand awareness example, the number of degrees of freedom for χ^2 is $(r - 1)(c - 1) = (2 - 1)(2 - 1) = 1$. Then, for $\alpha = .05$, we reject the hypothesis of independence when

$$\chi^2 > \chi^2_{.05} = 3.84146$$

Since the computed $\chi^2 = 46.14$ exceeds the value 3.84146, we conclude that viewer gender and brand awareness are dependent events. This result may also be obtained by noting that the p-value of the test (highlighted on Figure 13.3) is approximately 0.

The pattern of **dependence** can be seen more clearly by expressing the data as percentages. We first select one of the two classifications to be used as the base variable. In the preceding example, suppose we select gender of the TV viewer as the classificatory variable to be the base. Next, we represent the responses for each level of the second categorical variable (brand awareness here) as a percentage of the subtotal for the base variable. For example, from Table 13.3, we convert the response for males who identify the brand (95) to a percentage of the total number of male viewers (145). That is,

$$\left(^{95}/_{145}\right)100\% = 65.5\%$$

All of the entries in Table 13.3 are similarly converted, and the values are shown in Table 13.5. The value shown at the right of each row is the row's total, expressed as a percentage of the total number of responses in the entire table. Thus, the percentage of TV viewers who identify the product is $\left(\frac{136}{300}\right)100\% = 45.3\%$ (rounded to the nearest percent).

Table 13.5	Percentage of TV Viewers Who Identify Brand, by Gender			
		Gender		
		Male	Female	Totals
Brand Awareness	**Could Identify Product**	65.5	26.5	45.3
	Could Not Identify Product	34.5	73.5	54.7
	Totals	100	100	100

If the gender and brand awareness variables are independent, then the percentages in the cells of the table are expected to be approximately equal to the corresponding row percentages. Thus, we would expect the percentage of viewers who identify the brand for each gender to be approximately 45% if the two variables are independent. The extent to which each gender's percentage departs from this value determines the dependence of the two classifications, with greater variability of the row percentages meaning a greater degree of dependence. A plot of the percentages helps summarize the observed pattern. In the SPSS bar graph in Figure 13.4, we show the gender of the viewer (the base variable) on the horizontal axis and the percentage of TV viewers who identify the brand (green bars) on the vertical axis. The "expected" percentage under the assumption of independence is shown as a horizontal line.

Figure 13.4 clearly indicates the reason that the test resulted in the conclusion that the two classifications in the contingency table are dependent. The percentage of male TV viewers who identify the brand promoted by a male celebrity is more than twice as high as the percentage of female TV viewers who identify the brand. Statistical measures of the degree of dependence and procedures for making comparisons of pairs of levels for classifications are beyond the scope of this text, but can be found in the references. We will utilize descriptive summaries such as Figure 13.4 to examine the degree of dependence exhibited by the sample data.

Figure 13.4
SPSS bar graph showing percent of viewers who identify TV product

The general form of a two-way contingency table containing r rows and c columns (called an $r \times c$ contingency table) is shown in Table 13.6. Note that the observed count in the ijth cell is denoted by n_{ij}, the ith row total is r_i, the jth column total is c_j, and the total sample size is n. Using this notation, we give the general form of the contingency table test for independent classifications in the box.

Table 13.6   General $r \times c$ Contingency Table

		Column				Row Totals
		1	2	. . .	c	
Row	**1**	n_{11}	n_{12}	. . .	n_{1c}	R_1
	2	n_{21}	n_{22}	. . .	n_{2c}	R_2
	⋮	⋮	⋮		⋮	⋮
	r	n_{r1}	n_{r2}	. . .	n_{rc}	R_r
Column Totals		C_1	C_2	. . .	C_c	n

General Form of a Two-Way (Contingency) Table Analysis: A Test for Independence

H_0: The two classifications are independent

H_a: The two classifications are dependent

Test statistic: $\chi_c^2 = \sum \dfrac{[n_{ij} - \hat{E}_{ij}]^2}{\hat{E}_{ij}}$

where $\hat{E}_{ij} = \dfrac{R_i C_j}{n}$

Rejection region: $\chi_c^2 > \chi_\alpha^2$, where χ_α^2 has $(r - 1)(c - 1)$ df
p-value: $P(\chi^2 > \chi_c^2)$

> **Conditions Required for a Valid χ^2 Test: Contingency Tables**
>
> 1. The n observed counts are a random sample from the population of interest. We may then consider this to be a multinomial experiment with $r \times c$ possible outcomes.
> 2. The sample size n will be large enough so that, for every cell, the expected count $\hat{E}(n_{ij})$ will be equal to 5 or more.

Example 13.3 D

Conducting a Two-Way Analysis—Marital Status and Religion

Problem　A social scientist wants to determine whether the marital status (divorced or not divorced) of U.S. men is independent of their religious affiliation (or lack thereof). A sample of 500 U.S. men is surveyed, and the results are tabulated as shown in Table 13.7.

Table 13.7　Survey Results (Observed Counts), Example 13.3

		Religious Affiliation					
		A	B	C	D	None	Totals
Marital Status	**Divorced**	39	19	12	28	18	116
	Married, never divorced	172	61	44	70	37	384
	Totals	211	80	56	98	55	500

D *Data Set:* MARREL

a. Test to see whether there is sufficient evidence to indicate that the marital status of men who have been or are currently married is dependent on religious affiliation. Take $\alpha = .01$.

b. Graph the data and describe the patterns revealed. Is the result of the test supported by the graph?

Solution

a. The first step is to calculate estimated expected cell frequencies under the assumption that the classifications are independent. Rather than compute these values by hand, we resort to a computer. The SAS printout of the analysis of Table 13.7 is displayed in Figure 13.5, each cell of which contains the observed (top) and expected (bottom)

```
                        The FREQ Procedure

                    Table of MARITAL by RELIGION

    MARITAL      RELIGION

    Frequency|
    Expected |A      |B      |C      |D      |NONE   | Total

    DIVORCED |    39 |    19 |    12 |    28 |    18 |   116
             |48.952| 18.56 |12.992|22.736| 12.76 |

    NEVER    |   172 |    61 |    44 |    70 |    37 |   384
             |162.05| 61.44 |43.008|75.264| 42.24 |

    Total        211     80      56      98      55     500


         Statistics for Table of MARITAL by RELIGION

         Statistic                      DF     Value      Prob

         Chi-Square                      4     7.1355    0.1289
         Likelihood Ratio Chi-Square     4     6.9854    0.1367
         Mantel-Haenszel Chi-Square      1     6.4943    0.0108
         Phi Coefficient                       0.1195
         Contingency Coefficient               0.1186
         Cramer's V                            0.1195


                    Fisher's Exact Test

            Table Probability (P)      6.936E-06
            Pr <= P                       0.1251

              Sample Size = 500
```

Figure 13.5

SAS contingency table printout for Example 13.3

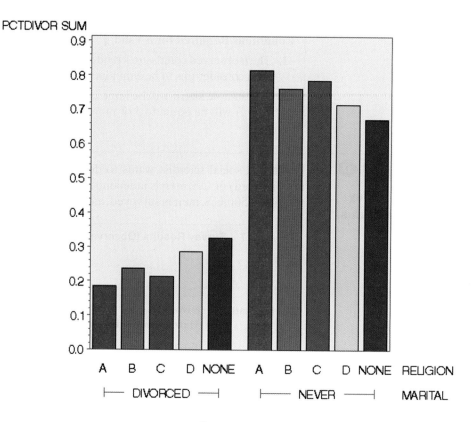

Figure 13.6

SAS side-by-side bar graphs showing percentage of divorced and never divorced males by religion

frequency in that cell. Note that $\hat{E}_{11}$, the estimated expected count for the Divorced, A cell, is 48.952. Similarly, the estimated expected count for the Divorced, B cell, is $\hat{E}_{12} = 18.56$. Since all the estimated expected cell frequencies are greater than 5, the χ^2 approximation for the test statistic is appropriate. Assuming that the men chosen were randomly selected from all married or previously married American men, the characteristics of the multinomial probability distribution are satisfied.

The null and alternative hypotheses we want to test are

H_0: The marital status of U.S. men and their religious affiliation are independent

H_a: The marital status of U.S. men and their religious affiliation are dependent

The test statistic, $\chi^2 = 7.135$, is highlighted at the bottom of the printout, as is the observed significance level (p-value) of the test. Since $\alpha = .01$ is less than $p = .129$, we fail to reject H_0; that is, we cannot conclude that the marital status of U.S. men depends on their religious affiliation. (Note that we could not reject H_0 even with $\alpha = .10$.)

b. The marital status frequencies can be expressed as percentages of the number of men in each religious affiliation category. The expected percentage of divorced men under the assumption of independence is $\left(^{116}/_{500}\right)100\% = 23\%$. A SAS graph of the percentages is shown in Figure 13.6. Note that the percentages of divorced men (see the bars in the "DIVORCED" block of the SAS graph) deviate only slightly from that expected under the assumption of independence, supporting the result of the test in part **a**. That is, neither the descriptive bar graph nor the statistical test provides evidence that the male divorce rate depends on (varies with) religious affiliation.

■ **Now Work Exercise 13.29**

Contingency Tables with Fixed Marginals

In the *Journal of Marketing* study on celebrities in TV ads, a single random sample was selected from the target population of all TV viewers and the outcomes—values of gender and brand awareness—were recorded for each viewer. For this type of study, the researchers had no *a priori* knowledge of how many observations would fall into the categories of the qualititative variables. In other words, prior to obtaining the sample, the researchers did not know how many males or how many brand identifiers would make up the sample. Oftentimes, it is advantageous to select a random sample from each of the levels of one of the qualitative variables.

For example, in the *Journal of Marketing* study, the researchers may want to be sure of an equivalent number of males and females in their sample. Consequently, they will select independent random samples of 150 males and 150 females. (In fact, this was the sampling plan for the actual study.) Summary data for this type of study yield a **contingency table with fixed marginals** since the column totals for one qualitative variable (e.g., gender) are known in advance.* The goal of the analysis does not change—determine whether the two qualitative variables (e.g., gender and brand awareness) are dependent.

The procedure for conducting a chi-square analysis for a contingency table with fixed marginals is identical to the one outlined above, since it can be shown (proof omitted) that the χ^2 test statistic for this type of sampling also has an approximate chi-square distribution with $(r-1)(c-1)$ degrees of freedom. One reason why you might choose this alternative sampling plan is to obtain sufficient observations in each cell of the contingency table to ensure that the chi-square approximation is valid. Remember, this will usually occur when the expected cell counts are all greater than or equal to 5. By selecting a large sample (150 observations) for each gender in the *Journal of Marketing* study, the researchers improved the odds of obtaining large expected cell counts in the contingency table.

Testing whether Likelihood of a Lawsuit Is Related to Recall Notice Sender

We return to the case involving tainted transplant tissue (see p. 800). Recall that a processor of the tainted tissue filed a lawsuit against a tissue distributor, claiming that the distributor was more responsible for paying damages to litigating transplant patients. Why? Because the distributor in question had sent recall notices (as required by the FTC) to hospitals and surgeons with unsolicited newspaper articles describing in graphic detail the "ghoulish" acts that had been committed. According to the processor, by including the articles in the recall package, this distributor inflamed the tissue recipients, increasing the likelihood that patients would file a lawsuit.

To prove its case in court, the processor needed to establish a statistical link between the likelihood of a lawsuit and the sender of the recall notice. More specifically, can the processor show that the probability of a lawsuit is higher for those patients of surgeons who received the recall notice with the inflammatory articles than for those patients of surgeons who received only the recall notice?

A statistician, serving as an expert consultant for the processor, reviewed data for the 7,914 patients who received recall notices (of which 708 filed suit). These data are saved in the **GHOUL1** file. For each patient, the file contains information on the SENDER of the recall notice (Processor or Distributor) and whether a LAWSUIT was filed (Yes or No).

Since both of these variables are qualitative and we want to know whether the probability of a LAWSUIT depends on the SENDER of the recall notice, a contingency table analysis is appropriate.

Figure SIA13.1 shows the MINITAB contingency table analysis. The null and alternative hypotheses for the test are

H_0: Lawsuit and Sender are independent

H_a: Lawsuit and Sender are dependent

Both the chi-square test statistic (100.5) and *p*-value of the test (.000) are highlighted on the printout. If we conduct the test at $\alpha = .01$, there is sufficient evidence to reject H_0. That is, the data provide evidence to indicate that the likelihood of a tainted transplant patient filing a lawsuit is associated with the sender of the recall notice.

To determine which sender had the higher percentage of patients to file a lawsuit, examine the row percentages (highlighted) in the contingency table of Figure SIA13.1 You can see that of the 1,751 patients sent recall notices by the

(Continued)

---

*Data from this type of study are also known as *product binomial data*.

Tabulated statistics: SENDER, LAWSUIT

Rows: SENDER   Columns: LAWSUIT

```
                   No      Yes      All

Distributor      5506      657     6163
                89.34    10.66   100.00
                 5612      551     6163
                1.989   20.244        *

Processor        1700       51     1751
                97.09     2.91   100.00
                 1594      157     1751
                7.001   71.252        *

All              7206      708     7914
                91.05     8.95   100.00
                 7206      708     7914
                    *        *        *
```

Cell Contents: Count
                    % of Row
                    Expected count
                    Contribution to Chi-square

Figure SIA13.1

MINITAB Contingency Table Analysis—Likelihood of Lawsuit vs. Recall Notice Sender

Pearson Chi-Square = 100.485, DF = 1, P-Value = 0.000
Likelihood Ratio Chi-Square = 124.748, DF = 1, P-Value = 0.000

processor, 51 (or 2.91%) filed lawsuits. In contrast, of the 6,163 patients sent recall notices by the distributor in question, 657 (or 10.66%) filed lawsuits. Thus, the probability of a patient filing a lawsuit is almost five times higher for the distributor's patients than for the processor's patients.

Before testifying on these results in court, the statistician decided to do one additional analysis: He eliminated from the sample data any patients whose surgeon had been sent notices by both parties. Why? Since these patients' surgeons received both recall notices, the underlying reason for filing a lawsuit would be unclear. Did the patient file simply because he or she received tainted transplant tissue, or was the filing motivated by the inflammatory articles that accompanied the recall notice? After eliminating these patients, the data looked like those shown in Table SIA13.2. A MINITAB contingency table analysis on this reduced data set (saved in the **GHOUL2** file) is shown in Figure SIA13.2.

Like in the previous analysis, the chi-square test statistic (110.2) and *p*-value of the test (.000)—both highlighted on the printout—imply that the likelihood of a tainted transplant patient filing a lawsuit is associated with the sender of the recall notice, at $\alpha = .01$. Also, the percentage of patients filing lawsuits when sent a recall notice by the distributor (10.62%)

is again five times higher than the percentage of patients filing lawsuits when sent a recall notice by the processor (2.04%).

The results of both analyses were used to successfully support the processor's claim in court. Nonetheless, we need to point out one caveat to the contingency table analyses. Be careful not to conclude that the data are proof that the inclusion of the inflammatory articles *caused* the probability of litigation to increase. Without controlling all possible variables that may be related to filing a lawsuit (e.g., a patient's socioeconomic status, whether a patient has filed a lawsuit in the past), we can only say that the two qualitative variables, lawsuit status and recall notice sender, are statistically associated. However, the fact that the likelihood of a lawsuit is almost five times higher when the notice is sent by the distributor shifts the burden of proof to the distributor to explain why this occurred and to convince the court that it should not be held accountable for paying the majority of the damages.

Alternative analysis: As mentioned in Section 10.3, a 2 × 2 contingency table analysis is equivalent to a comparison of two population proportions. In the tainted tissue case, we want to compare p_1, the proportion of lawsuits filed by patients who were sent recall notices by the processor, to p_2, the proportion

Table SIA13.2 Data for the Tainted Tissue Case, Dual Recall Notices Eliminated

Recall Notice Sender	Number of Patients	Number of Lawsuits
Processor/Other Distributor	1,522	31
Distributor in Question	5,705	606
Totals:	7,227	637

Source: Info Tech, Inc. Gainesville, FL

Statistics in Action
(continued)

```
Tabulated statistics: SENDER, LAWSUIT

Rows: SENDER    Columns: LAWSUIT

                No      Yes      All

Distributor    5099      606     5705
               89.38    10.62   100.00
               5202      503     5705
               2.045   21.160        *

Processor      1491       31     1522
               97.96     2.04   100.00
               1388      134     1522
               7.667   79.315        *

All            6590      637     7227
               91.19     8.81   100.00
               6590      637     7227
                  *        *        *

Cell Contents:       Count
                     % of Row
                     Expected count
                     Contribution to Chi-square
```

```
Pearson Chi-Square = 110.187, DF = 1, P-Value = 0.000
Likelihood Ratio Chi-Square = 144.862, DF = 1, P-Value = 0.000
```

Figure SIA13.2

MINITAB Contingency Table Analysis, with Dual Recall Notices Eliminated

of lawsuits filed by patients who were sent recall notices by the distributor that included the inflammatory articles. Both a test of the null hypothesis, $H_0: (p_1 - p_2) = 0$, and a 95% confidence interval for the difference, $(p_1 - p_2)$, using the reduced sample data are shown (highlighted) on the MINITAB printout, Figure SIA13.3.

The *p*-value for the test (.000) indicates that the two proportions are significantly different at $\alpha = .05$. The 95%

confidence interval, $(-.097, -.075)$, shows that the proportion of lawsuits associated with patients who were sent recall notices from the distributor ranges between .075 and .097 higher than the corresponding proportion for the processor. Both results support the processor's case, namely, that the patients who were sent recall notices with the inflammatory news articles were more likely to file a lawsuit than those who were sent only recall notices.

```
Test and CI for Two Proportions

Sample    X      N    Sample p
1        31   1522   0.020368
2       606   5705   0.106223


Difference = p (1) - p (2)
Estimate for difference:  -0.0858547
95% CI for difference:  (-0.0965452, -0.0751641)
Test for difference = 0 (vs not = 0):  Z = -10.50  P-Value = 0.000

Fisher's exact test: P-Value = 0.000
```

Figure SIA13.3

Exercises 13.22–13.46

Understanding the Principles

13.22 What is a two-way (contingency) table?

13.23 What is a contingency table with fixed marginals?

13.24 *True or False.* One goal of a contingency table analysis is to determine whether the two classifications are independent or dependent.

13.25 What conditions are required for a valid chi-square test of data from a contingency table?

Learning the Mechanics

13.26 Find the rejection region for a test of independence of two classifications where the contingency table contains *r* rows and *c* columns.
   a. $r = 5$, $c = 6$, $\alpha = .05$
   b. $r = 5$, $c = 3$, $\alpha = .10$
   c. $r = 2$, $c = 5$, $\alpha = .01$

13.27 Consider the following 2 × 3 (i.e., $r = 2$ and $c = 3$) contingency table.

		Column		
		1	2	3
Row	**1**	9	34	53
	2	16	30	25

a. Specify the null and alternative hypotheses that should be used in testing the independence of the row and column classifications.
b. Specify the test statistic and the rejection region that should be used in conducting the hypothesis test of part **a**. Use $\alpha = .01$.
c. Assuming the row classification and the column classification are independent, find estimates for the expected cell counts.
d. Conduct the hypothesis test of part **a**. Interpret your result.

13.28 Refer to Exercise 13.27.
a. Convert the frequency responses to percentages by calculating the percentage of each column total falling in each row. Also, convert the row totals to percentages of the total number of responses. Display the percentages in a table.
b. Create a bar graph with row 1 percentage on the vertical axis and column number on the horizontal axis. Show the row 1 total percentage as a horizontal line on the graph.
c. What pattern do you expect to see if the rows and columns are independent? Does the plot support the result of the test of independence in Exercise 13.27?

13.29 Test the null hypothesis of independence of the two classifications A and B of the 3 × 3 contingency table shown here. Use $\alpha = .05$.

		B		
		B_1	B_2	B_3
	A_1	40	72	42
A	A_2	63	53	70
	A_3	31	38	30

13.30 Refer to Exercise 13.29. Convert the responses to percentages by calculating the percentage of each B class total falling into each A classification. Also, calculate the percentage of the total number of responses that constitute each of the A classification totals.
a. Create a bar graph with row A_1 percentage on the vertical axis and B classification on the horizontal axis. Does the graph support the result of the test of hypothesis in Exercise 13.29? Explain.
b. Repeat part **a** for the row A_2 percentages.
c. Repeat part **a** for the row A_3 percentages.

Applying the Concepts—Basic

13.31 Children's perceptions of their neighborhood. In *Health Education Research* (Feb. 2005), nutrition scientists at Deakin University (Australia) investigated children's perceptions of their environments. Each in a sample of 147 ten-year-old children drew maps of their home and neighborhood environment. The researchers examined the maps for certain themes (e.g., presence of a pet, television in the bedroom, opportunities for physical activity). The results, broken down by gender, for two themes (presence of a dog and TV in the bedroom) are shown in the next two tables.
a. Find the sample proportion of boys who drew a dog on their maps.
b. Find the sample proportion of girls who drew a dog on their maps.
c. Compare the proportions you found in parts **a** and **b**. Does it appear that the likelihood of drawing a dog on the neighborhood map depends on gender?
d. Give the null hypothesis for testing whether the likelihood of a drawing a dog on the neighborhood map depends on gender.
e. Use the MINITAB printout below to conduct the test mentioned in part **d** at $\alpha = .05$.
f. Conduct a test to determine whether the likelihood of drawing a TV in the bedroom is different for boys and girls. Use $\alpha = .05$.

Presence of a Dog	Number of Boys	Number of Girls
Yes	6	11
No	71	59
Total	77	70

Presence of TV in Bedroom	Number of Boys	Number of Girls
Yes	11	9
No	66	61
Total	77	70

Based on Hume, C., Salmon, J., and Ball, K. "Children's perceptions of their home and neighborhood environments, and their association with objectively measured physical activity: A qualitative and quantitative study." *Health Education Research*, Vol. 20, No. 1, Feb. 2005 (Table III).

```
Tabulated statistics: DOG, GENDER

Using frequencies in NUMBER


Rows: DOG    Columns: GENDER

         Boy    Girl     All

No        71      59      130
        68.10   61.90   130.00

Yes        6      11       17
         8.90    8.10    17.00

All       77      70      147
        77.00   70.00   147.00

Cell Contents:       Count
                     Expected count


Pearson Chi-Square = 2.250, DF = 1, P-Value = 0.134
Likelihood Ratio Chi-Square = 2.268, DF = 1, P-Value = 0.132
```

13.32 Eyewitnesses and mug shots. Refer to the *Applied Psychology in Criminal Justice* (Apr. 2010) study of mug shot choices by eyewitnesses to a crime, Exercise 10.107 (p. 598). Recall that a sample of 96 college students was shown a video of a simulated theft, then asked to select the mug shot that most closely resembled the thief. The students were randomly assigned to view either 3, 6, or 12 mug shots at a time, with 32 students in each group. The number of students in the 3-, 6-, or 12-photos-per-page groups who selected the target mugshot were 19, 19, and 15, respectively.

a. For each photo group, compute the proportion of students who selected the target mug shot. Which group yielded the lowest proportion?

b. Create a contingency table for these data, with photo group in the rows and whether or not the target mug shot was selected in the columns.

c. Refer to part **b**. Are there differences in the proportions who selected the target mug shot among the three photo groups? Test, using $\alpha = .10$.

13.33 Stereotyping deceptive and authentic news stories. Major newspapers lose their credibility (and subscribers) when they are found to have published deceptive or misleading news stories. In *Journalism and Mass Communication Quarterly* (Summer 2007), University of Texas researchers investigated whether certain stereotypes (e.g., negative references to certain nationalities) occur more often in deceptive news stories than in authentic news stories. The researchers analyzed 183 news stories that were proven to be deceptive in nature and 128 news stories that were considered authentic. Specifically, the researchers determined whether each story was negative, neutral, or positive in tone. The accompanying table gives the number of news stories found in each tone category.

	Authentic News Stories	Deceptive News Stories
Negative Tone	59	111
Neutral Tone	49	61
Positive Tone	20	11
Total	128	183

Based on Lasorsa, D., and Dai, J. "When news reporters deceive: The production of stereotypes." *Journalism and Mass Communication Quarterly*, Vol. 84, No. 2, Summer 2007 (Table 2).

a. Find the sample proportion of negative tone news stories that is deceptive.

b. Find the sample proportion of neutral news stories that is deceptive.

c. Find the sample proportion of positive news stories that is deceptive.

d. Compare the sample proportions, parts **a–c**. Does it appear that the proportion of news stories that is deceptive depends on story tone?

e. Give the null hypothesis for testing whether the authenticity of a news story depends on tone.

f. Use the SPSS printout in the next column to conduct the test, part **e**. Test at $\alpha = .05$.

TONE * STORY Crosstabulation

			STORY Authentic	STORY Deceptive	Total
TONE	Negative	Count	59	111	170
		Expected Count	70.0	100.0	170.0
	Neutral	Count	49	61	110
		Expected Count	45.3	64.7	110.0
	Positive	Count	20	11	31
		Expected Count	12.8	18.2	31.0
Total		Count	128	183	311
		Expected Count	128.0	183.0	311.0

Chi-Square Tests

	Value	df	Asymp. Sig. (2-sided)
Pearson Chi-Square	10.427[a]	2	.005
Likelihood Ratio	10.348	2	.006
N of Valid Cases	311		

a. 0 cells (.0%) have expected count less than 5. The minimum expected count is 12.76.

13.34 Healing heart patients with music, imagery, touch, and prayer. "Frontier medicine" is a term used to describe medical therapies (e.g., energy healing, therapeutic prayer, spiritual healing) for which there is no plausible explanation. *The Lancet* (July 16, 2005) published the results of a study designed to test the effectiveness of two types of frontier medicine—music, imagery, and touch (MIT) therapy and therapeutic prayer—in healing cardiac care patients. Patients were randomly assigned to receive one of four types of treatment: (1) prayer, (2) MIT, (3) prayer and MIT, and (4) standard care (no prayer and no MIT). Six months after therapy, the patients were evaluated for a major adverse cardiovascular event (e.g., a heart attack). The results of the study are summarized in the accompanying table.

Therapy	Number of Patients with Major Cardiovascular Events	Number of Patients with No Events	Total
Prayer	43	139	182
MIT	47	138	185
Prayer and MIT	39	150	189
Standard	50	142	192

Based on Krucoff, M. W., et al. "Music, imagery, touch, and prayer as adjuncts to interventional cardiac care: The Monitoring and Actualization of Noetic Trainings (MANTRA) II randomized study." *The Lancet*, Vol. 366, July 16, 2005 (Table 4).

a. Identify the two qualitative variables (and associated levels) measured in the study.

b. State H_0 and H_a for testing whether a major adverse cardiovascular event depends on type of therapy.

c. Use the MINITAB printout on p. 822 to conduct the test mentioned in part **b** at $\alpha = .10$. On the basis of this test, what can the researchers infer about the effectiveness of music, imagery, and touch therapy and the effectiveness of healing prayer in heart patients?

MINITAB Output for Exercise 13.34

Tabulated statistics: THERAPY, EVENT

Using frequencies in NUMBER

```
Rows: THERAPY   Columns: EVENT

               No     Yes     All

MIT           138     47      185
              140.7   44.3    185.0

Prayer        139     43      182
              138.4   43.6    182.0

Prayer&MIT    150     39      189
              143.8   45.2    189.0

Standard      142     50      192
              146.1   45.9    192.0

All           569     179     748
              569.0   179.0   748.0

Cell Contents:        Count
                      Expected count
```

Pearson Chi-Square = 1.828, DF = 3, P-Value = 0.609
Likelihood Ratio Chi-Square = 1.855, DF = 3, P-Value = 0.603

13.35 Package design influences taste. Can the package design of a food product influence how the consumer will rate the taste of the product? A team of experimental psychologists reported on a study that examined how rounded or angular package shapes and high- or low-pitched sounds can convey information about the taste (sweetness and sourness) of a product (*Food Quality and Preference*, June 2014). Study participants were presented with one of two types of packaging displayed on a computer screen monitor: rounded shape with a low-pitched sound or angular shape with a high-pitched sound. Assume that half of the participants viewed the rounded packaging and half viewed the angular packaging. After viewing the product, each participant rated whether the packaging was more appropriate for either a sweet- or a sour-tasting food product. A summary of the results (numbers of participants) for a sample of 80 participants is shown in the following contingency table. (These data are simulated based on the results reported in the article.)

		Package Design/Pitch	
		Angular/High	Rounded/Low
Taste	Sweet	35	7
Choice	Sour	5	33

a. Specify the null and alternative hypotheses for testing whether the package design and sound pitch combination influences the consumer's opinion on the product taste.

b. Assuming the null hypothesis is true, find the expected number in each cell of the table.

c. Use the expected numbers and observed counts in the table to compute the chi-square test statistic.

d. The observed significance level (*p*-value) of the test is approximately 0. (You can verify this result using statistical software.) For any reasonably chosen value of α, give the appropriate conclusion.

Applying the Concepts—Intermediate

13.36 "Cry wolf" effect in air traffic controlling. Researchers at Alion Science Corporation and New Mexico State University collaborated on a study of how air traffic controllers respond to false alarms (*Human Factors*, Aug. 2009). The researchers theorize that the high rate of false alarms regarding midair collisions leads to the "cry wolf" effect, i.e., the tendency for air traffic controllers to ignore true alerts in the future. The investigation examined data on a random sample of 437 conflict alerts. Each alert was first classified as a "true" or "false" alert. Then, each was classified according to whether there was a human controller response to the alert. The number of the 437 alerts that fall into each of the combined categories is given as follows: True alert/No response–3; True alert/Response–231; False alert/No response–37; False alert/Response–166. This summary information is saved in the **ATC** file. Do the data indicate that the response rate of air traffic controllers to midair collision alarms differs for true and false alerts? Test using $\alpha = .05$. What inference can you make concerning the "cry wolf" effect?

Based on Wickens, C. D., Rice, S., Keller, D., Hutchins, S., Hughes, J., and Clayton, K., "False alerts in air traffic control conflict alerting system: Is there a 'cry wolf' effect?" *Human Factors*, Vol. 51, Issue 4, Aug. 2009 (Table 2).

13.37 Influencing performance in a serial addition task. Refer to the *Advances in Cognitive Psychology* (Jan. 2013) study of performance in a classic psychological test involving adding a set of numbers, Exercise 9.66 (p. 499). Recall that 300 undergraduate students were given a serial addition task, with the numbers (1,000 + 40 + 1,000 + 30 + 1,000 + 20 + 1,000 + 10) presented on a computer screen. However, the students were divided into five groups of 60 students each, and the presentation of the numbers (total display time and color of the number 1,000) was varied in each group: Group 1—2 seconds/black, Group 2—2 seconds/bright red, Group 3—15 seconds/bright red, Group 4—1 second/black, and Group 5—15 seconds/black. The number of students who gave the correct response of 4,100 was determined as well as the number who gave an incorrect response. The results are summarized (number of responses) in the accompanying table. Does presentation of the numbers to be added influence performance (correct response rate) on the serial addition task? Test, using $\alpha = .01$.

Presentation	Responses of 4,100	Responses of 5,000	Other Incorrect Responses
Group 1 (2 sec/black)	17	33	10
Group 2 (2 sec/red)	8	42	10
Group 3 (15 sec/red)	12	36	12
Group 4 (1 sec/black)	13	42	5
Group 5 (15 sec/black)	20	31	9

Source: Giannouli, V. "Number perseveration in healthy subjects: Does prolonged stimulus exposure influence performance on a serial addition task?" *Advances in Cognitive Psychology*, Vol. 9, No. 1, Jan. 2013 (adapted from Table 3).

13.38 Job satisfaction of women in construction. The hiring of women in construction and construction-related jobs has steadily increased over the years. A study was conducted to provide employers with information designed to reduce the potential for turnover of female employees (*Journal of Professional Issues in Engineering Education & Practice*, Apr. 2013). A survey questionnaire was e-mailed to members of the National Association of Women in Construction (NAWIC). A total of 447 women responded to survey questions on job challenge and satisfaction with life as an employee. The results (number of females responding in the different categories) are summarized in the accompanying table. What conclusions can you draw from the data regarding the association between an NAWIC member's satisfaction with life as an employee and her satisfaction with job challenge?

		Life as an Employee	
		Satisfied	Dissatisfied
Job	Satisfied	364	33
Challenge	Dissatisfied	24	26

Source: Malone, E. K., and Issa, R. A. "Work-life balance and organizational commitment of women in the U.S. construction industry." *Journal of Professional Issues in Engineering Education & Practice,* Vol. 139, No. 2, Apr. 2013 (Table 11).

13.39 Detecting Alzheimer's disease at an early age. Refer to the *Neuropsychology* (Jan. 2007) study of whether the cognitive effects of Alzheimer's disease can be detected at an early age, Exercise 13.17 (p. 808). Recall that a particular strand of DNA was classified into one of three genotypes: $E4^+/E4^+$, $E4^+/E4^-$, and $E4^-/E4^-$. In addition to a sample of 2,097 young adults (20–24 years), two other age groups were studied: a sample of 2,182 middle-aged adults (40–44 years) and a sample of 2,281 elderly adults (60–64 years). The accompanying table gives a breakdown of the number of adults with the three genotypes in each age category for the total sample of 6,560 adults. The researchers concluded that "there were no significant genotype differences across the three age groups" using $\alpha = .05$. Do you agree?

Age Group	$E4^+/E4^+$ Genotype	$E4^+/E4^-$ Genotype	$E4^-/E4^-$ Genotype	Sample Size
20–24	56	517	1,524	2,097
40–44	45	566	1,571	2,182
60–64	48	564	1,669	2,281

Source: Jorm, A. F., et al. "APOE genotype and cognitive functioning in a large age-stratified population sample." *Neuropsychology,* Vol. 21, No. 1, Jan. 2007 (Table 1). Copyright © 2007 by the American Psychological Association. Reprinted with permission.

13.40 Mobile device typing strategies. Refer to the *Applied Ergonomics* (Mar. 2012) study of mobile device typing strategies, Exercise 13.14 (p. 808). Recall that typing style of mobile device users was categorized as (1) device held with both hands/both thumbs typing, (2) device held with right hand/right thumb typing, (3) device held with left hand/left thumb typing, (4) device held with both hands/right thumb typing, (5) device held with left hand/right index finger typing, or (6) other. The researchers' main objective was to determine if there are gender differences

in typing strategies. Typing strategy and gender were observed for each in a sample of 859 college students observed typing on their mobile devices. The data are summarized in the accompanying table. Is this sufficient evidence to conclude that the proportions of mobile device users in the six texting style categories depend on whether a male or a female is texting? Use $\alpha = .10$ to answer the question.

Typing Strategy	Number of Males	Number of Females
Both hands hold/both thumbs type	161	235
Right hand hold/right thumb type	118	193
Left hand hold/left thumb type	29	41
Both hands hold/right thumb type	10	29
Left hand hold/right index type	6	12
Other	11	14

Source: Gold, J. E., et al. "Postures, typing strategies, and gender differences in mobile device usage: An observational study." *Applied Ergonomics,* Vol. 43, No. 2, Mar. 2012 (Table 2).

13.41 Classifying air threats with heuristics. The *Journal of Behavioral Decision Making* (Jan. 2007) published a study on the use of heuristics to classify the threat level of approaching aircraft. Of special interest was the use of a fast and frugal heuristic—a computationally simple procedure for making judgments with limited information—named "Take-the-Best-for-Classification" (TTB-C). The subjects were 48 men and women, some from a Canadian Forces reserve unit, others university students. Each subject was presented with a radar screen on which simulated approaching aircraft were identified with asterisks. By using the computer mouse to click on the asterisk, one could receive further information about the aircraft. The goal was to identify the aircraft as "friend" or "foe" as fast as possible. Half the subjects were given cue-based instructions for determining the type of aircraft, while the other half were given pattern-based instructions. The researcher also classified the heuristic strategy used by the subject as TTB-C, Guess, or Other. Data on the two variables Instruction type and Strategy, measured for each of the 48 subjects, are saved in the **TTBC** file. (Data on the first and last five subjects are shown in the accompanying table below). Do the data provide sufficient evidence at $\alpha = .05$ to indicate that choice of heuristic strategy depends on type of instruction provided? How about at $\alpha = .01$?

Instruction	Strategy
Pattern	Other
Pattern	Other
Pattern	Other
Cue	TTBC
Cue	TTBC
⋮	⋮
Pattern	TTBC
Cue	Guess
Cue	TTBC
Cue	Guess
Pattern	Guess

Based on Bryant, D. J. "Classifying simulated air threats with fast and frugal heuristics." *Journal of Behavioral Decision Making.* Vol. 20, Jan. 2007 (Appendix C).

13.42 Reading comprehension of Texas students. An analysis of reading test scores of students at a rural Texas school district was carried out in *Current Issues in Education* (Jan. 2014). Students were classified as attending elementary, middle, or high school and whether they passed a reading comprehension test. The data for the sample of 1,012 students are summarized in the accompanying table. Does the passing rate on the reading comprehension test in Texas differ for elementary, middle, and high school students? Use $\alpha = .10$.

	Elementary School	Middle School	High School
Number Passing	372	418	143
Number Failing	44	25	10
Totals	416	443	153

Source: Bigham, G. D., and Riney, M. R. "Trend analysis techniques to assist school leaders in making critical curriculum and instruction decisions." *Current Issues in Education*, Vol. 17, No. 1, Jan. 2014 (Table 6).

13.43 Versatility with resistor-capacitor circuits. Research published in the *International Journal of Electrical Engineering Education* (Oct. 2012) investigated the versatility of engineering students' knowledge of circuits with one resistor and one capacitor connected in series. Students were shown four different configurations of a resistor-capacitor circuit and then given two tasks. First, each student was asked to state the voltage at the nodes on the circuit and, second, each student was asked to graph the dynamic behavior of the circuit. Suppose that in a sample of 160 engineering students, 40 were randomly assigned to analyze Circuit 1, 40 assigned to Circuit 2, 40 assigned to Circuit 3, and 40 assigned to Circuit 4. The researchers categorized task grades as follows: correct voltages and graph, incorrect voltages but correct graph, incorrect graph but correct voltages, incorrect voltages and incorrect graph. A summary of the results (based on information provided in the journal article) is shown in the table. Does any one circuit appear to be more difficult to analyze than any other circuit? Support your answer with a statistical test of hypothesis.

		Circuit 1	Circuit 2	Circuit 3	Circuit 4
	Both Correct	31	10	5	4
	Incorrect Voltage	0	3	11	12
Answer	Incorrect Graph	5	17	16	14
	Both Incorrect	4	10	8	10
	Total Number of Students	40	40	40	40

13.44 Interactions in a children's museum. Refer to the *Early Childhood Education Journal* (Mar. 2014) study of interactions in a children's museum, Exercise 13.19 (p. 809). Summary information for the 170 meaningful interactions sampled is reproduced in the following table. Do the proportions associated with the different types of interactions depend on whether the interaction was child-led or adult-led? Test, using $\alpha = .01$.

Type of Interaction	Child-Led	Adult-Led
Show-and-tell	26	0
Learning/Teaching	21	64
Refocusing	21	10
Participatory Play	12	9
Advocating/Disciplining	1	6
Totals	81	89

Source: McMunn-Dooley, C., and Welch, M. M. "Nature of interactions among young children and adult caregivers in a children's museum." *Early Childhood Education Journal*, Vol. 42, No. 2, Mar. 2014 (adapted from Figure 2).

Applying the Concepts—Advanced

13.45 Efficacy of an HIV vaccine. New, effective AIDS vaccines have been developed through the process of "sieving"—that is, sifting out infections with some strains of HIV. Consider a vaccine designed to eliminate a particular strain of the virus. To test the efficacy of the vaccine, a clinical trial was conducted. The trial consisted of 7 AIDS patients vaccinated with the new drug and 31 AIDS patients who were treated with a placebo (no vaccination). Of the 7 vaccinated patients, 2 tested positive and 5 tested negative at the end of a follow-up period. Of the 31 unvaccinated patients, 22 tested positive and 9 tested negative. (These data are saved in the **HIVVAC1** file.)

a. Construct a contingency table for the data. Then, conduct a test to determine whether the vaccine is effective in treating this strain of HIV. Use $\alpha = .05$.

b. Are the assumptions for the test you carried out in part **a** satisfied? What are the consequences if the assumptions are violated?

c. In the case of a 2×2 contingency table, R. A. Fisher (1935) developed a procedure for computing the exact *p*-value for the test (called *Fisher's exact test*). The method utilizes the hypergeometric probability distribution of Chapter 4 (p. 248). Consider the hypergeometric probability

$$\frac{\binom{7}{2}\binom{31}{22}}{\binom{38}{24}}$$

which represents the probability that 2 out of 7 vaccinated AIDS patients test positive and 22 out of 31 unvaccinated patients test positive—that is, the probability of the result of the clinical trial, given that the null hypothesis of independence is true. Compute this probability (called the *probability of the contingency table*).

d. Refer to part **c**. Now consider two other possible results from the clinical trial that are different from the original results. *Result 2*: Only 1 of the 7 vaccinated patients tests positive; 23 of the 31 unvaccinated patients test positive. *Result 3*: None of the 7 vaccinated patients tests positive; 24 of the 31 unvaccinated patients test positive. Create two contingency tables, one for each of these results. (These data are saved in the **HIVVAC2** and **HIVVAC3** files respectively.) Note that these two contingency tables have the same marginal totals as the original table in part **a**. Explain why each of these tables provides more evidence to reject the null hypothesis of independence than does the original table. Then, compute the probability of each table, using the hypergeometric formula.

e. The *p*-value of Fisher's exact test is the probability of observing a result at least as unsupportive of the null hypothesis as is the observed contingency table, given the same marginal totals. Sum the probabilities of parts **c** and **d** to obtain the *p*-value of Fisher's exact test. (To verify your calculations, check the *p*-value labeled **Left-sided Pr < = F** at the bottom of the SAS printout shown below.) Interpret this value in the context of the vaccine trial.

```
                The FREQ Procedure

         Table of GROUP by MNSTRAIN

    GROUP      MNSTRAIN

    Frequency
    Expected   NEG      POS        Total

    UN              9        22        31
              11.421   19.579

    V               5         2         7
              2.5789    4.4211

    Total          14        24        38


   Statistics for Table of GROUP by MNSTRAIN

  Statistic                   DF     Value      Prob

  Chi-Square                   1    4.4112    0.0357
  Likelihood Ratio Chi-Square  1    4.2893    0.0384
  Continuity Adj. Chi-Square   1    2.7773    0.0956
  Mantel-Haenszel Chi-Square   1    4.2952    0.0382
  Phi Coefficient                  -0.3407
  Contingency Coefficient           0.3225
  Cramer's V                       -0.3407

  WARNING: 50% of the cells have expected counts less
           than 5. Chi-Square may not be a valid test.


             Fisher's Exact Test

  Cell (1,1) Frequency (F)              9
  Left-sided Pr <= F              0.0498
  Right-sided Pr >= F             0.9940

  Table Probability (P)           0.0438
  Two-sided Pr <= P               0.0772

            Sample Size = 38
```

13.46 Examining the "Monty Hall Dilemma." In Exercise 3.195 (p. 209) you solved the game show problem of whether to switch your choice of three doors, one of which hides a prize, after the host reveals what is behind a door that is not chosen. (Despite the natural inclination of many to keep one's first choice, the correct answer is that you should switch your choice of doors.) This problem is sometimes called the "Monty Hall Dilemma," named for Monty Hall, the host of the popular TV game show *Let's Make a Deal*. In *Thinking & Reasoning* (July, 2007), Wichita State University professors set up an experiment designed to influence subjects to switch their original choice of doors. Each subject participated in 23 trials. In trial 1, three (boxes) representing doors were presented on a computer screen; only one box hid a prize. In each subsequent trial, an additional box was presented, so that in trial 23, twenty-five boxes were presented. In each trial, after a box was selected, all of the remaining boxes except for one either (1) were shown to be empty (*Empty* condition), (2) disappeared (*Vanish* condition), (3) disappeared and the chosen box was enlarged (*Steroids* condition), or (4) disappeared and the remaining box not chosen was enlarged (*Steroids2* condition). Twenty-seven subjects were assigned to each condition. The number of subjects who ultimately switched boxes is tallied, by condition, in the following table for both the first trial and the last trial.

	First Trial (1)		Last Trial (23)	
Condition	Switch Boxes	No Switch	Switch Boxes	No Switch
Empty	10	17	23	4
Vanish	3	24	12	15
Steroids	5	22	21	6
Steroids2	8	19	19	8

Based on Howard, J. N., Lambdin, C. G., and Datteri, D. L. "Let's make a deal: Quality and availability of second-stage information as a catalyst for change." *Thinking & Reasoning*, Vol. 13, No. 3, July 2007 (Table 2).

a. For a selected trial, does the likelihood of switching boxes depend on condition?
b. For a given condition, does the likelihood of switching boxes depend on trial number?
c. On the basis of the results you obtained in parts **a** and **b**, what factors influence a subject to switch choices?

13.4 A Word of Caution about Chi-Square Tests

Because the χ^2 statistic for testing hypotheses about multinomial probabilities is one of the most widely applied statistical tools, it is also one of the most abused statistical procedures. Consequently, the user should always be certain that the experiment satisfies the assumptions underlying each procedure. Furthermore, the user should be certain that the sample is drawn from the correct population—that is, from the population about which the inference is to be made.

The use of the χ^2 probability distribution as an approximation to the sampling distribution for χ^2 should be avoided when the expected counts are very small. The approximation can become very poor when these expected counts are small; thus, the true α level may be quite different from the tabular value. As a rule of thumb, an expected cell count of at least 5 means that the χ^2 probability distribution can be used to determine an approximate critical value.

If the χ^2 value does not exceed the established critical value of χ^2, *do not accept the hypothesis of independence.* You would be risking a Type II error (accepting H_0 when it is false), and the probability β of committing such an error is unknown. The usual alternative hypothesis is that the classifications are dependent. Because the number of ways in which two classifications can be dependent is virtually infinite, it is difficult to calculate one or even several values of β to represent such a broad alternative hypothesis. Therefore, we avoid concluding that two classifications are independent, even when χ^2 is small.

Finally, if a contingency table χ^2 value does exceed the critical value, we must be careful to avoid inferring that a *causal* relationship exists between the classifications. Our alternative hypothesis states that the two classifications are statistically dependent—and a statistical dependence does not imply causality. Therefore, *the existence of a causal relationship cannot be established by a contingency table analysis.*

CHAPTER NOTES

Key Terms

Categories 801
Cell counts 801
Cells 801
Chi-square test 803
Classes 801
Contingency table 810
Contingency table with fixed
   marginals 817
Dependence 813
Dimensions of
   classification 810

Expected cell count 804
Independence of two
   classifications 811
Marginal probabilities 810
Multinomial
   experiment 801
Observed cell count 811
One-way table 802
Two-way table 810

Key Symbols/Notation

$p_{i,0}$	Value of multinomial probability p_1 hypothesized in H_0
χ^2	Chi-square test statistic used in analysis of categorical data
n_i	Number of observed outcomes in cell i of a one-way table
E_i	Expected number of outcomes in cell i of a one-way table
p_{ij}	Probability of an outcome in row i and column j of a two-way table
n_{ij}	Number of observed outcomes in row i and column j of a two-way table
E_{ij}	Expected number of outcomes in row i and column j of a two-way table
R_i	Total number of outcomes in row i of a two-way table
C_j	Total number of outcomes in column j of a two-way table

Key Ideas

Multinomial Data

Qualitative data that fall into more than two categories (or classes)

Properties of a Multinomial Experiment

1. n identical trials
2. k possible outcomes to each trial
3. probabilities of the k outcomes $(p_1, p_2, \ldots, p_k)$, where $p_1 + p_2 + \cdots + p_k = 1$, remain the same from trial to trial
4. trials are independent
5. variables of interest: *cell counts* (i.e., number of observations falling into each outcome category), denoted $n_1, n_2, \ldots, n_k$

One-Way Table

Summary table for a *single* qualitative variable

Two-Way (Contingency) Table

Summary table for *two* qualitative variables

Chi-Square (χ^2) Statistic

used to test category probabilities in one-way and two-way tables

Chi-square tests for independence

should **not** be used to *infer a causal relationship between two QLs*

Conditions Required for Valid χ^2 Tests

1. multinomial experiment
2. sample size n is large (expected cell counts are all greater than or equal to 5)

Categorical Data Analysis Guide

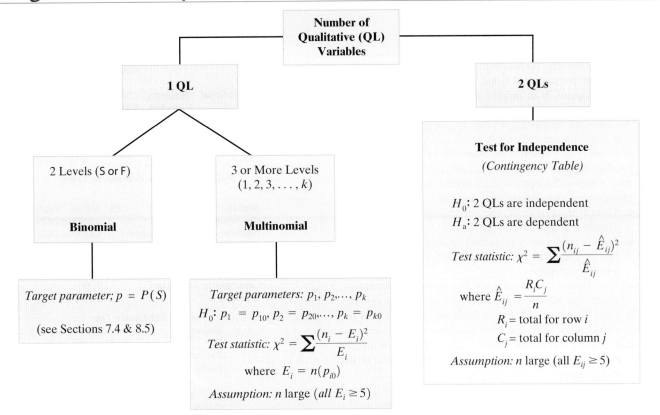

Supplementary Exercises 13.47–13.73

Understanding the Principles

13.47 *True or False.* Rejecting the null hypothesis in a chi-square test for independence implies that a causal relationship exists between the two categorical variables.

13.48 What is the difference between a one-way chi-square analysis and a two-way chi-square analysis?

Learning the Mechanics

13.49 A random sample of 250 observations was classified according to the row and column categories shown in the following table:

		Column		
		1	2	3
	1	30	30	10
Row	**2**	10	20	60
	3	20	60	10

a. Do the data provide sufficient evidence to conclude that the rows and columns are dependent? Test, using $\alpha = .05$.

b. Would the analysis change if the row totals were fixed before the data were collected?

c. Do the assumptions required for the analysis to be valid differ according to whether the row (or column) totals are fixed? Explain.

d. Convert the table entries to percentages by using each column total as a base and calculating each row response as a percentage of the corresponding column total. In addition, calculate the row totals and convert them to percentages of all 250 observations.

e. Create a bar graph with the percentages from row 1 on the vertical axis and the column number on the horizontal axis. Draw a horizontal line corresponding to the total percentage for row 1. Does the graph support the result of the test conducted in part **a**?

f. Repeat part **e** for the percentages from row 2.

g. Repeat part **e** for the percentages from row 3.

13.50 A random sample of 200 observations was classified into the categories shown in the following table:

	Category				
	1	2	3	4	5
n_i	38	36	48	37	41

a. Do the data provide sufficient evidence that the categories are not equally likely? Use $\alpha = .10$.

b. Form a 90% confidence interval for p_2, the probability that an observation will fall into category 2.

Applying the Concepts—Basic

13.51 Safety of hybrid cars. According to the Highway Loss Data
Institute (HLDI), "hybrid [automobiles] have a safety edge over their conventional twins when it comes to shielding their occupants from injuries in crashes" (*HLDI Bulletin*, Sept. 2011). Consider data collected by the HLDI on Honda Accords over the past eight years. In a sample of 50,132 collision claims for conventional Accords, 5,364 involved injuries; in a sample of 1,505 collision claims for hybrid Accords, 137 involved injuries. You want to use this information to determine whether the injury rate for hybrid Accords is less than the injury rate for conventional Accords.

a. Identify the two qualitative variables measured for each Honda Accord collision claim.

b. Form a contingency table for this data, giving the number of claims in each combination of the qualitative variable categories.

c. Give H_0 and H_a for testing whether injury rate for collision claims depends on Accord model (hybrid or conventional).

d. Find the expected number of claims in each cell of the contingency table, assuming that H_0 is true.

e. Compute the χ^2 test statistic and compare your answer to the test statistic shown on the SAS printout in the next column.

f. Find the rejection region for the test using $\alpha = .05$ and compare your answer to the critical value shown on the accompanying SAS printout.

g. Make the appropriate conclusion using both the rejection region method and the p-value (shown on the SAS printout).

h. Find a 95% confidence interval for the difference between the injury rates of conventional and hybrid Honda Accords. (See Section 9.4.) Use the interval to infer whether the injury rate for hybrid Accords is less than the injury rate for conventional Accords.

13.52 Excavating ancient pottery. Refer to the *Chance* (Fall 2000) study of ancient Greek pottery, presented in Exercise 2.186 (p. 134). Recall that 837 pottery pieces were uncovered at the excavation site. The table describing the types of pottery found is reproduced here.

Pot Category	Number Found
Burnished	133
Monochrome	460
Painted	183
Other	61
Total	837

Based on Berg, I., and Bliedon, S. "The pots of Phyiakopi: Applying statistical techniques to archaeology." *Chance*, Vol. 13, No. 4, Fall 2000.

a. Describe the qualitative variable of interest in the study. Give the levels (categories) associated with the variable.

b. Assume that the four types of pottery occur with equal probability at the excavation site. What are the values of p_1, p_2, p_3, and p_4, the probabilities associated with the four pottery types?

SAS Output for Exercise 13.51

The FREQ Procedure

Table of Model by Claim

Model		Claim		
		Injury	No Injury	Total
Conventional	Frequency	5364	44768	50132
	Expected	5340.7	44791	
Hybrid	Frequency	137	1368	1505
	Expected	160.33	1344.7	
Total	Frequency	5501	46136	51637

Statistics for Table of Model by Claim

Statistic	DF	Value	Prob
Chi-Square	1	3.9139	0.0479
Likelihood Ratio Chi-Square	1	4.0893	0.0432
Continuity Adj. Chi-Square	1	3.7480	0.0529
Mantel-Haenszel Chi-Square	1	3.9138	0.0479
Phi Coefficient		0.0087	
Contingency Coefficient		0.0087	
Cramer's V		0.0087	

Pearson Chi-Square Test	
Chi-Square	3.9139
DF	1
Asymptotic Pr > ChiSq	0.0479
Exact Pr >= ChiSq	0.0510

Fisher's Exact Test	
Cell (1,1) Frequency (F)	5364
Left-sided Pr <= F	0.9801
Right-sided Pr >= F	0.0246
Table Probability (P)	0.0047
Two-sided Pr <= P	0.0510

Sample Size = 51637

c. Give the null and alternative hypotheses for testing whether one type of pottery is more likely to occur at the site than any of the other types.

d. Find the test statistic for testing the hypotheses stated in part **c**.

e. Find and interpret the p-value of the test. State the conclusion in the words of the problem if you use $\alpha = .10$.

13.53 Risk factor for lumbar disease. One of the most common musculoskeletal disorders is lumbar disk disease (LDD). Medical researchers reported finding a common genetic risk factor for LDD (*Journal of the American Medical Association*, Apr. 11, 2001). The study included 171 Finnish patients diagnosed with LDD (the patient group) and 321 without LDD (the control group). Of the 171 LDD patients, 21 were discovered to have the genetic trait. Of the 321 people in the control group, 15 had the genetic trait.

a. Consider the two categorical variables group and presence/absence of genetic trait. Form a 2 × 2 contingency table for these variables.

b. Conduct a test to determine whether the genetic trait occurs at a higher rate in LDD patients than in the controls. Use $\alpha = .01$.

c. Construct a bar graph that will visually support your conclusion in part **b**.

13.54 Late-emerging reading disabilities. Studies of children with reading disabilities typically focus on "early-emerging" difficulties identified prior to the fourth grade. Psychologists at Haskins Laboratories recently studied children with "late-emerging" reading difficulties (i.e., children who appeared to undergo a fourth-grade "slump" in reading achievement) and published their findings in the *Journal of Educational Psychology* (June 2003). A sample of 161 children was selected from fourth and fifth graders at elementary schools in Philadelphia. In addition to recording the grade level, the researchers determined whether each child had a previously undetected reading disability. Sixty-six children were diagnosed with a reading disability. Of these children, 32 were fourth graders and 34 were fifth graders. Similarly, of the 95 children with normal reading achievement, 55 were fourth graders and 40 were fifth graders.

a. Identify the two qualitative variables (and corresponding levels) measured in the study.

b. From the information provided, form a contingency table.

c. Assuming that the two variables are independent, calculate the expected cell counts.

d. Find the test statistic for determining whether the proportions of fourth and fifth graders with reading disabilities differs from the proportions of fourth and fifth graders with normal reading skills.

e. Find the rejection region for the test if $\alpha = .10$.

f. Is there a link between reading disability and grade level? Give the appropriate conclusion of the test.

13.55 Politics and religion. University of Maryland professor Ted R. Gurr examined the political strategies used by ethnic groups worldwide in their fight for minority rights (*Political Science & Politics*, June 2000). Each in a sample of 275 ethnic groups was classified according to world region and highest level of political action reported. The data are summarized in the contingency table below. Conduct a test at $\alpha = .10$ to determine whether political strategy of ethnic groups depends on world region. Support your answer with a graph.

13.56 Sociology fieldwork methods. Refer to the *Teaching Sociology* (July 2006) study of the fieldwork methods used by qualitative sociologists, presented in Exercise 2.195 (p. 136). Recall that fieldwork methods can be categorized as follows: Interview, Observation plus Participation, Observation Only, and Grounded Theory. The table in the next column shows the number of papers published over the past seven years in each category. Suppose a sociologist claims that 70%, 15%, 10%, and 5% of the fieldwork methods involve interview, observation plus participation, observation only, and grounded theory, respectively. Do the data support or refute the claim? Explain.

Fieldwork Method	Number of Papers
Interview	5,079
Observation + Participation	1,042
Observation Only	848
Grounded Theory	537

Based on Hood, J. C. "Teaching against the text: The case of qualitative methods." *Teaching Sociology*, Vol. 34, July 2006 (Exhibit 2).

13.57 The *Family and Consumer Sciences Research Journal* (Mar. 2005) published a study of college students and drinking. One of the researchers' main objectives was to segment college students according to their rates of alcohol consumption. A segmentation was developed based on survey responses to the questions on frequency of drinking and average number of drinks per occasion for a sample of 657 students. Four types, or groups, of college drinkers emerged: non/seldom drinkers, social drinkers, typical binge drinkers, and heavy binge drinkers. Frequencies are shown in the accompanying SPSS printout. Is this sufficient evidence to indicate that the true proportions in the population of college students are different?

Chi-Square Test

Frequencies

TYPE

	Observed N	Expected N	Residual
Non/Seldom	118	164.3	-46.3
Social	282	164.3	117.8
Typical Binge	163	164.3	-1.3
Heavy Binge	94	164.3	-70.3
Total	657		

Test Statistics

	TYPE
Chi-Square[a]	127.493
df	3
Asymp. Sig.	.000

a. 0 cells (.0%) have expected frequencies less than 5. The minimum expected cell frequency is 164.3.

13.58 Refer to the *Family and Consumer Sciences Research Journal* (Mar. 2005) study of college students and drinking, Exercise 13.57. A second objective of the researchers was to establish a statistical link between the frequency

Table for Exercise 13.55

		Political Strategy		
		No Political Action	Mobilization, Mass Action	Terrorism, Rebellion, Civil War
World Region	Latin American	24	31	7
	Post-Communist	32	23	4
	South, Southeast, East Asia	11	22	26
	Africa/Middle East	39	36	20

Table 1 from "Nonviolence ethnopolitics: Strategies for the attainment of group rights and autonomy" by Ted Robert Gurr. *Political Science & Politics*, Vol. 33, No. 2, June 2000. Copyright © 2000 The American Political Science Association. Reprinted with the permission of Cambridge University Press.

SPSS output for Exercise 13.58

FREQNCY * AMOUNT Crosstabulation

			AMOUNT						Total
			None	1 drink	2-3 drinks	4-6 drinks	7-9 drinks	More	
FREQNCY	None	Count	118	0	0	0	0	0	118
		Expected Count	21.2	11.0	51.2	17.8	11.5	5.4	118.0
		% within FREQNCY	100.0%	.0%	.0%	.0%	.0%	.0%	100.0%
	OnceMonth	Count	0	30	43	20	0	0	93
		Expected Count	16.7	8.6	40.3	14.0	9.1	4.2	93.0
		% within FREQNCY	.0%	32.3%	46.2%	21.5%	.0%	.0%	100.0%
	TwiceWeek	Count	0	26	158	46	35	11	276
		Expected Count	49.6	25.6	119.7	41.6	26.9	12.6	276.0
		% within FREQNCY	.0%	9.4%	57.2%	16.7%	12.7%	4.0%	100.0%
	More	Count	0	5	84	33	29	19	170
		Expected Count	30.5	15.8	73.7	25.6	16.6	7.8	170.0
		% within FREQNCY	.0%	2.9%	49.4%	19.4%	17.1%	11.2%	100.0%
Total		Count	118	61	285	99	64	30	657
		Expected Count	118.0	61.0	285.0	99.0	64.0	30.0	657.0
		% within FREQNCY	18.0%	9.3%	43.4%	15.1%	9.7%	4.6%	100.0%

Chi-Square Tests

	Value	df	Asymp. Sig. (2-sided)
Pearson Chi-Square	756.606[a]	15	.000
Likelihood Ratio	706.412	15	.000
Linear-by-Linear Association	322.813	1	.000
N of Valid Cases	657		

a. 1 cells (4.2%) have expected count less than 5. The minimum expected count is 4.25.

of drinking (none, once per month, twice per week, or more) and the amount of alcohol consumed (none, 1 drink, 2–3 drinks, 4–6 drinks, 7–9 drinks, or more). That is, the researchers sought a link between frequency of drinking alcohol over the previous one-month period and average number of drinks consumed per occasion. An SPSS contingency table analysis relating frequency of drinking to average amount of alcohol consumed is shown above. These results led the researchers to report that "The frequent drinkers were more likely to consume more [alcohol] on each occasion, a tendency that clearly makes them heavy drinkers." Do you agree? Explain.

13.59 Purchasing souvenirs. A major tourist activity is shopping. Travel researchers estimate that nearly one-third of total travel expenditures are used on shopping for souvenirs (*Journal of Travel Research,* May 2011). To investigate the impact of gender on souvenir shopping, a survey of 3,200 tourists was conducted. One question asked how often the tourist purchases photographs, postcards, or paintings of the region visited. Responses were recorded as "always," "often," "occasionally," or "rarely or never." The table shows the percentages of tourists responding in each category by gender.

Photos, Postcards, Paintings	Male Tourist	Female Tourist
Always	16%	28%
Often	27	31
Occasionally	35	29
Rarely or Never	22	12
Totals	100%	100%

Source: Wilkins, H. "Souvenirs: What and why we buy." *Journal of Travel Research,* Vol. 50, No. 3, May 2011 (adapted from Table 2).

a. Based on the percentages shown in the table, do you think male and female tourists differ in their responses to purchasing photographs, postcards, or paintings? Why are these percentages alone insufficient to draw a conclusion about the true response category proportions?

b. Assume that 1,500 males and 1,700 females participated in the survey. Use these sample sizes and the percentages in the table to compute the counts of tourists in each of the Response/Gender categories. This represents the contingency table for the study.

c. Specify the null and alternative hypotheses for testing whether male and female tourists differ in their responses to purchasing photographs, postcards, or paintings.

d. An SPSS printout of the contingency table analyses is shown below. Locate the test statistic and p-value on the printout.

e. Make the appropriate conclusion using $\alpha = .01$.

RESPONSE * GENDER Crosstabulation

			GENDER		Total
			Female	Male	
RESPONSE	1:Always	Count	476	240	716
		Expected Count	380.4	335.6	716.0
	2:Often	Count	527	405	932
		Expected Count	495.1	436.9	932.0
	3:Occasionally	Count	493	525	1018
		Expected Count	540.8	477.2	1018.0
	4:Rarely/Never	Count	204	330	534
		Expected Count	283.7	250.3	534.0
Total		Count	1700	1500	3200
		Expected Count	1700.0	1500.0	3200.0

Chi-Square Tests

	Value	df	Asymp. Sig. (2-sided)
Pearson Chi-Square	112.433[a]	3	.000
Likelihood Ratio	113.788	3	.000
N of Valid Cases	3200		

a. 0 cells (0.0%) have expected count less than 5. The minimum expected count is 250.31.

Applying the Concepts—Intermediate

13.60 Masculinity and crime. Refer to the *Journal of Sociology* (July 2003) study on the link between the level of masculinity and criminal behavior in men, presented in Exercise 9.27 (p. 480). The researcher identified events that a sample of newly incarcerated men were involved in and classified each event as "violent" (involving the use of a weapon, the throwing of objects, punching, choking, or kicking) or "avoided-violent" (involving pushing, shoving, grabbing, or threats of violence that did not escalate into a violent event). Each man (and corresponding event) was also classified as possessing "high-risk masculinity" (scored high on the Masculinity–Femininity Scale test and low on the Traditional Outlets of Masculinity Scale test) or "low-risk masculinity." The data on 1,507 events are summarized in the following table.

	Violent Events	Avoided-Violent Events	Totals
High-Risk Masculinity	236	143	379
Low-Risk Masculinity	801	327	1,128
Totals	1,037	470	1,507

Based on Krienert, J. L. "Masculinity and crime: A quantitative exploration of Messerschmidt's hypothesis." *Journal of Sociology,* Vol. 7, No. 2, July 2003 (Table 4).

a. Identify the two categorical variables measured (and their levels) in the study.
b. Identify the experimental units.
c. If the type of event (violent or avoided-violent) is independent of high-/low-risk masculinity, how many of the 1,507 events would you expect to be violent and involve a high-risk-masculine man?
d. Repeat part **c** for the other combinations of event type and high-/low-risk masculinity.
e. Calculate the χ^2 statistic for testing whether event type depends on high- low-risk masculinity.
f. Give the appropriate conclusion of the test mentioned in part **e,** using $\alpha = .05$.

13.61 Multiple sclerosis drug. Interferons are proteins produced naturally by the human body that help fight infections and regulate the immune system. A drug developed from interferons, called Avonex, is now available for treating patients with multiple sclerosis (MS). In a clinical study, 85 MS patients received weekly injections of Avonex over a two-year period. The number of exacerbations (i.e., flare-ups of symptoms) was recorded for each patient and is summarized in the accompanying table. For MS patients who take a placebo (no drug) over a similar two-year period, it is known from previous studies that 26% will experience no exacerbations, 30% one exacerbation, 11% two exacerbations, 14% three exacerbations, and 19% four or more exacerbations.

Number of Exacerbations	Number of Patients
0	32
1	26
2	15
3	6
4 or more	6

Based on data from Biogen, Inc.

a. Conduct a test to determine whether the exacerbation distribution of MS patients who take Avonex differs from the percentages reported for placebo patients. Use $\alpha = .10$.
b. Find a 90% confidence interval for the true percentage of Avonex MS patients who remain free of exacerbations during a two-year period.
c. Refer to part **b**. Is there evidence that Avonex patients are more likely to have no exacerbations than placebo patients? Explain.

13.62 Flight response of geese to helicopter traffic. Offshore oil drilling near an Alaskan estuary has led to increased air traffic—mostly large helicopters—in the area. The U.S. Fish and Wildlife Service commissioned a study to investigate the impact these helicopters have on the flocks of Pacific brant geese that inhabit the estuary in the fall before migrating (*Statistical Case Studies: A Collaboration between Academe and Industry*, 1998). Two large helicopters were flown repeatedly over the estuary at different altitudes and lateral distances from the flock. The flight responses of the geese (recorded as "low" or "high"), the altitude (in hundreds of meters), and the lateral distance (also in hundreds of meters) for each of 464 helicopter overflights were recorded and are saved in the **GEESE** file. The data for the first 10 overflights are shown in the following table.

Overflight	Altitude	Lateral Distance	Flight Response
1	0.91	4.99	HIGH
2	0.91	8.21	HIGH
3	0.91	3.38	HIGH
4	9.14	21.08	LOW
5	1.52	6.60	HIGH
6	0.91	3.38	HIGH
7	3.05	0.16	HIGH
8	6.10	3.38	HIGH
9	3.05	6.60	HIGH
10	12.19	6.60	HIGH

Source: From Erickson, W., Nick, T., and Ward, D. "Investigating Flight Response of Pacific Brant to Helicopters at Izembek Lagoon, Alaska by Using Logistic Regression." *Statistical Case Studies: A Collaboration between Academe and Industry*, ASA-SIAM Series on Statistics and Applied Probability, 1998. Copyright © 1998 Society for Industrial and Applied Mathematics. Reprinted with permission. All rights reserved.

a. The researchers categorized altitude as follows: less than 300 meters, 300–600 meters, and 600 or more meters. Summarize the data by creating a contingency table for altitude category and flight response.
b. Conduct a test to determine whether flight response of the geese depends on altitude of the helicopter. Test, using $\alpha = .01$.
c. The researchers categorized lateral distance as follows: less than 1,000 meters, 1,000–2,000 meters, 2,000–3,000 meters, and 3,000 or more meters. Summarize the data in the **GEESE** file by creating a contingency table for lateral distance category and flight response.
d. Conduct a test to determine whether flight response of the geese depends on lateral distance of helicopter from the flock. Test, using $\alpha = .01$.
e. The current Federal Aviation Authority (FAA) minimum altitude standard for flying over the estuary is 2,000 feet (approximately 610 meters). On the basis of the results obtained in parts **a–d,** what changes to the

FAA regulations do you recommend in order to minimize the effects to Pacific brant geese?

13.63 Defensible landscapes in archaeology. The defensibility of a given landscape in the Northwest United States was studied in the *Journal of Anthropological Archaeology* (May 2014). Archaeologists typically define "defensive" locations as those that are hidden with high elevation and escape routes. A question arose regarding the defensibility of trench embankments. Each in a sample of 1,914 archaeological sites was classified as "highly defensible" or not. In addition, whether the site was located in a trench embankment was determined. The results of the categorizations (number of sites in each category) are summarized below.

	Highly Defensible	Not Highly Defensible	Total
Trench Embankment	9	14	23
Non-trench Embankment	311	1,580	1,891

a. Is this sufficient evidence to conclude that the proportion of trench embankment sites that are "highly defensible" differs from the corresponding proportion of non-trench embankment sites? That is, is the defensibility status of a site dependent on site type (trench embankment or not)?

b. Fisher's exact test (see Exercise 13.45) resulted in a *p*-value of *p* = .008. Interpret this result, and then explain why this test is preferred over the χ^2 test in part **a**.

13.64 Gangs and homemade weapons. The National Gang Crime Research Center (NGCRC) has developed a six-level gang classification system for both adults and juveniles. The six categories are shown in the accompanying table. The classification system was developed as a potential predictor of a gang member's propensity for violence in prison, jail, or a correctional facility. To test the system, the NGCRC collected data on approximately 10,000 confined offenders and assigned each a score from the gang classification system. (*Journal of Gang Research*, Winter 1997). One of several other variables measured by the NGCRC was whether the offender had ever carried a homemade weapon (e.g., knife) while in custody. The data on gang score and homemade weapon are summarized in the table below. Conduct a test to determine whether carrying a homemade weapon in custody depends on gang classification score. (Use α = .01.) Support your conclusion with a graph.

Gang Classification Score	Homemade Weapon Carried	
	Yes	No
0 (Never joined a gang, no close friends in a gang)	255	2,551
1 (Never joined a gang, 1–4 close friends in a gang)	110	560
2 (Never joined a gang, 5 or more friends in a gang)	151	636
3 (Inactive gang member)	271	959
4 (Active gang member, no position of rank)	175	513
5 (Active gang member, holds position of rank)	476	831

Source: From Knox, G. W., et al. "A gang classification system for corrections." *Journal of Gang Research*, Vol. 4, No. 2, Winter 1997, p. 54 (Table 4). Reprinted with permission from National Gang Crime Research Center.

13.65 IQ and mental deficiency. A person is diagnosed with a mental deficiency if, before the age of 18, his/her score on a standard IQ test is no higher than 70 (two standard deviations below the mean of 100). Researchers at Cornell and West Virginia Universities examined the impact of rising IQ scores on diagnoses of mental deficiency (MD) (*American Psychologist*, Oct. 2003). IQ data were collected from different school districts across the United States, and the students were tested with either the Wechsler Intelligence Scale for Children—Revised (WISC-R) or the Wechsler Intelligence Scale for Children—Third Revision (WISC-III) IQ tests. The researchers focused on those students with IQs just above the mental deficiency cutoff (between 70 and 85), based on the original IQ test. These "borderline" MD students were then retested one year later with one of the IQ tests. The accompanying table gives the number of students diagnosed with mental deficiency on the basis of the retest. Conduct a chi-square test for independence to determine whether the proportion of students diagnosed with MD depends on the IQ test/retest method. Use α = .01.

Test/Retest	Diagnosed with MD	Above MD Cutoff IQ	Total
WISC-R / WISC-R	25	167	192
WISC-R / WISC-III	54	103	157
WISC-III / WISC-III	36	141	177

Source: Kanaya, T., Scullin, M. H., and Ceci, S. J. "The Flynn effect and U.S. Policies." *American Psychologist*, Vol. 58, No. 10, Oct. 2003 (Figure 1). Copyright © 2003 by the American Psychological Association. Reprinted with permission.

13.66 Orientation clue experiment. *Human Factors* (Dec. 1988) published a study of color brightness as a body orientation clue. Ninety college students reclining on their backs in the dark were disoriented when positioned on a rotating platform under a slowly rotating disk that blocked their field of vision. The subjects were asked to say "Stop" when they felt as if they were right-side up. The position of the brightness pattern on the disk in relation to each student's body orientation was then recorded. Subjects selected only three disk brightness patterns as subjective vertical clues: (1) brighter side up, (2) darker side up, and (3) brighter and darker sides aligned on either side of the subjects' heads. The frequency counts for the experiment are given in the accompanying table. Conduct a test to compare the proportions of subjects who fall into the three disk-orientation categories. Assume that you want to determine whether the three proportions differ. Use α = .05.

Disk Orientation		
Brighter Side Up	Darker Side Up	Bright and Dark Sides Aligned
58	15	17

13.67 Coupon usage study. A hot topic in marketing research is the exploration of a technology-based self-service (TBSS) encounter, in which various technologies (e.g., ATMs, online banking, self-scanning at retail stores) allow the customer to perform all or part of the service. Marketing professor Dan Ladik of the University of Suffolk investigated whether there were differences in customer characteristics and customer satisfaction between users of discount coupons distributed through the mail (nontechnology users) and users of coupons distributed via the Internet (TBSS users). A questionnaire measured several qualitative variables (defined in the accompanying table) for each of 440 coupon users. The data are saved in the **TBSS** file.

Variable Name	Levels (Possible Values)
Coupon User Type	Mail, Internet, or Both
Gender	Male or Female
Education	High School, Vo-Tech/College, 4-year College Degree, or Graduate School
Work Status	Full Time, Part Time, Not Working, Retired
Coupon Satisfaction	Satisfied, Unsatisfied, Indifferent

a. Consider the variable Coupon User Type. Conduct a test (at $\alpha = .05$) to determine whether the proportions of mail-only users, Internet-only users, and users of both media are statistically different. Illustrate the results with a graph.

b. The researcher wants to know whether there are differences in customer characteristics (i.e., Gender, Education, Work Status, and Coupon Satisfaction) among the three types of coupon users. For each characteristic, conduct a contingency table analysis (at $\alpha = .05$) to determine whether Coupon User Type is related to that characteristic. Illustrate your results with graphs.

13.68 Subarctic plant study. The traits of seed-bearing plants indigenous to subarctic Finland were studied in *Arctic, Antarctic, and Alpine Research* (May 2004). Plants were categorized according to *type* (dwarf shrub, herb, or grass), *abundance of seedlings* (no seedlings, rare seedlings, or abundant seedlings), *regenerative group* (no vegetative reproduction, vegetative reproduction possible, vegetative reproduction ineffective, or vegetative reproduction effective), *seed weight class* (0–.1, .1–.5, .5–1.0, 1.0–5.0, and >5.0 milligrams), and *diaspore morphology* (no structure, pappus, wings, fleshy fruits, or awns/hooks). The data on a sample of 73 plants are saved in the **SEED** file.

a. A contingency table for plant type and seedling abundance, produced by MINITAB, is shown in the next column. (*Note:* NS = no seedlings, SA = seedlings abundant, and SR = seedlings rare.) Suppose you want to perform a chi-square test of independence to

Tabulated statistics: Abundance, Type

```
Rows: Abundance   Columns: Type

            DwarfShrub   Grasses   Herbs   All

NS              3           1         1      5
SA              5          14        32     51
SR              5           2        10     17
All            13          17        43     73

Cell Contents:        Count
```

determine whether seedling abundance depends on plant type. Find the expected cell counts for the contingency table. Are the assumptions required for the test satisfied?

b. Reformulate the contingency table by combining the NS and SR categories of seedling abundance. Find the expected cell counts for this new contingency table. Are the assumptions required for the test satisfied?

c. Reformulate the contingency table of part **b** by combining the dwarf shrub and grasses categories of plant type. Find the expected cell counts for this contingency table. Are the assumptions required for the test satisfied?

d. Carry out the chi-square test for independence on the contingency table you came up with in part **c**, using $\alpha = .10$. What do you conclude?

13.69 Susceptibility to hypnosis. A standardized procedure for determining a person's susceptibility to hypnosis is the Stanford Hypnotic Susceptibility Scale, Form C (SHSS:C). Recently, a new method called the Computer-Assisted Hypnosis Scale (CAHS), which uses a computer as a facilitator of hypnosis, has been developed. Each scale classifies a person's hypnotic susceptibility as low, medium, high, or very high. Researchers at the University of Tennessee compared the two scales by administering both tests to each of 130 undergraduate volunteers (*Psychological Assessment*, Mar. 1995). The hypnotic classifications are summarized in the table at the bottom of the page. A contingency table analysis will be performed to determine whether CAHS level and SHSS level are independent.

a. Check to see if the assumption of expected cell counts of 5 or more is satisfied. Should you proceed with the analysis? Explain.

Table for Exercise 13.69

		CAHS Level				
		Low	Medium	High	Very High	Totals
SHSS: C Level	**Low**	32	14	2	0	48
	Medium	11	14	6	0	31
	High	6	14	19	3	42
	Very High	0	2	4	3	9
	Totals	49	44	31	6	130

Copyright © 1995 by the American Psychological Association. Adapted with permission. The official citation that should be used in referencing this material is "The Computer-Assisted Hypnosis Scale: Standardization and Norming of a Computer-administered Measure of Hypnotic Ability" by Carolyn D. Grant and Michael R. Nash. *Psychological Assessment*, Vol. 7, No. 1, Mar. 1995, p. 53. The use of APA information does not imply endorsement by APA.

b. One way to satisfy the assumption of part **a** is to combine the data for two or more categories (e.g., high and very high) in the contingency table. Form a new contingency table by combining the data for the high and very high categories in both the rows and the columns.

c. Calculate the expected cell counts in the new contingency table you formed in part **c**. Is the assumption now satisfied?

d. Perform the chi-square test on the new contingency table. Use $\alpha = .05$. Interpret the results.

13.70 Guilt in decision making. The effect of guilt emotion on how a decision maker focuses on the problem was investigated in the Jan. 2007 issue of the *Journal of Behavioral Decision Making*. A total of 171 volunteer students participated in the experiment, where each was randomly assigned to one of three emotional states (guilt, anger, or neutral) through a reading/writing task. Immediately after the task, the students were presented with a decision problem where the stated option has predominantly negative features (e.g., spending money on repairing a very old car). The results (number responding in each category) are summarized in the accompanying table. Is there sufficient evidence (at $\alpha = .10$) to claim that the option choice depends on emotional state?

Emotional State	Choose Stated Option	Do Not Choose Stated Option	Totals
Guilt	45	12	57
Anger	8	50	58
Neutral	7	49	56
Totals	60	111	171

Based on Gangemi, A., and Mancini, F. "Guilt and focusing in decision-making." *Journal of Behavioral Decision Making*, Vol. 20, Jan. 2007 (Table 2).

Applying the Concepts—Advanced

13.71 Goodness-of-fit test. A statistical analysis is to be done on a set of data consisting of 1,000 monthly salaries. The analysis requires the assumption that the sample was drawn from a normal distribution. A preliminary test, called the χ^2 *goodness-of-fit test*, can be used to help determine whether it is reasonable to assume that the sample is from a normal distribution. Suppose the mean and standard deviation of the 1,000 salaries are hypothesized to be $1,200 and $200, respectively. Using the standard normal table, we can approximate the probability of a salary being in the intervals listed in the table in the next column. The third column represents the expected number of the 1,000 salaries to be found in each interval if the sample was drawn from a normal distribution with $\mu = \$1,200$ and $\sigma = \$200$. Suppose the last column contains the actual observed frequencies in the sample. Large differences between the observed and expected frequencies cast doubt on the normality assumption.

Interval	Probability	Expected Frequency	Observed Frequency
Less than $800	.025	25	28
$800 < $1,000	.150	150	155
$1,000 < $1,200	.325	325	315
$1,200 < $1,400	.325	325	312
$1,400 < $1,600	.150	150	148
$1,600 or above	.025	25	42

a. Compute the χ^2 statistic on the basis of the observed and expected frequencies.

b. Find the tabulated χ^2 value when $\alpha = .05$ and there are five degrees of freedom. (There are $k - 1 = 5$ df associated with this χ^2 statistic.)

c. On the basis of the χ^2 statistic and the tabulated χ^2 value, is there evidence that the salary distribution is non-normal?

d. Find the approximate observed significance level for the test in part **c**.

13.72 Testing normality. Suppose a random variable is hypothesized to be normally distributed with a mean of 0 and a standard deviation of 1. A random sample of 200 observations of the variable yields frequencies in the intervals listed in the table shown at the bottom of the page. Do the data provide sufficient evidence to contradict the hypothesis that x is normally distributed with $\mu = 0$ and $\sigma = 1$? Use the technique developed in Exercise 13.71.

Critical Thinking Challenge

13.73 A "rigged" election? *Chance* (Spring 2004) presented data from a recent election held to determine the board of directors of a local community. There were 27 candidates for the board, and each of 5,553 voters was allowed to choose 6 candidates. The claim was that "a fixed vote with fixed percentages [was] assigned to each and every candidate, making it impossible to participate in an honest election." Votes were tallied in six time slots: after 600 total votes were in, after 1,200, after 2,444, after 3,444, after 4,444, and, finally, after 5,553 votes. The data on three of the candidates (Smith, Coppin, and Montes) are shown in the accompanying table. A residential organization believes that "there was nothing random about the count and tallies for each time slot, and specific unnatural or rigged percentages were being assigned to each and every candidate." Give your opinion. Is the probability of a candidate receiving votes independent of the time slot, and if so, does this imply a rigged election?

Time Slot	1	2	3	4	5	6
Votes for Smith	208	208	451	392	351	410
Votes for Coppin	55	51	109	98	88	104
Votes for Montes	133	117	255	211	186	227
Total Votes	600	600	1,244	1,000	1,000	1,109

Based on Gelman, A. "55,000 residents desperately need your help!" *Chance*, Vol. 17, No. 2, Spring 2004 (Figures 1 and 5).

Table for Exercise 13.72

Interval	$x < -2$	$-2 \le x < -1$	$-1 \le x < 0$	$0 \le x < 1$	$1 \le x < 2$	$x \ge 2$
Frequency	10	22	55	68	29	16

Activity Binomial versus Multinomial Experiments

In this activity, you will study the difference between binomial and multinomial experiments.

1. A television station has hired an independent research group to determine whether television viewers in the area prefer its local news program to the news programs of two other stations in the same city. Explain why a multinomial experiment would be appropriate, and design a poll that satisfies the five properties of a multinomial experiment. State the null and alternative hypotheses for the corresponding χ^2 test.

2. Suppose the television station believes that a majority of local viewers prefers its news program to those of its two competitors. Explain why a binomial experiment would be appropriate to support this claim, and design a poll that satisfies the five properties of a binomial experiment. State the null and alternative hypotheses for the corresponding test.

3. Generalize the situations in Exercises 1 and 2 in order to describe conditions under which a multinomial experiment can be rephrased as a binomial experiment. Is there any advantage in doing so? Explain.

References

Agresti, A. *Categorical Data Analysis*. New York: Wiley, 1990.

Cochran, W. G. "The χ^2 test of goodness of fit." *Annals of Mathematical Statistics*, 1952, 23.

Conover, W. J. *Practical Nonparametric Statistics*, 2nd ed. New York: Wiley, 1980.

Fisher, R. A. "The logic of inductive inference (with discussion)." *Journal of the Royal Statistical Society*, Vol. 98, 1935, pp. 39–82.

Hollander, M., and Wolfe, D. A. *Nonparametric Statistical Methods*. New York: Wiley, 1973.

Savage, I. R. "Bibliography of nonparametric statistics and related topics." *Journal of the American Statistical Association*, 1953, 48.

USING TECHNOLOGY MINITAB: Chi-Square Analyses

MINITAB can conduct chi-square tests for both one-way and two-way (contingency) tables.

One-Way Table

Step 1 Access the MINITAB worksheet file that contains the sample data for the qualitative variable of interest. [*Note*: The data file can have actual values (levels) of the variable for each observation, or, alternatively, two columns—one listing the levels of the qualitative variable and the other column with the observed counts for each level.]

Step 2 Click on the "Stat" button on the MINITAB menu bar, and then click on "Tables" and "Chi-Square Goodness-of-Fit Test (One Variable)," as shown in Figure 13.M.1. The resulting dialog box appears as shown in Figure 13.M.2.

Figure 13.M.2 MINITAB one-way chi-square dialog box

Step 3 If your data have one column of values for your qualitative variable, select "Categorical data" and specify the variable name (or column) in the box. If your data have summary information in two columns (see the first column), select "Observed counts" and specify the column with the counts and the column with the variable names in the respective boxes.

Step 4 Select "Equal proportions" for a test of equal proportions, or select "Specific proportions" and enter the hypothesized proportion next to each level in the resulting box.

Step 5 Click "OK" to generate the MINITAB printout.

Two-Way Table

Step 1 Access the MINITAB worksheet file that contains the sample data. The data file should contain two qualitative variables, with category values for each of the n observations in the

Figure 13.M.1 MINITAB menu options for a one-way chi-square analysis

data set. Alternatively, the worksheet can contain the cell counts for each of the categories of the two qualitative variables.

Step 2 Click on the "Stat" button on the MINITAB menu bar and then click on "Tables" and "Cross Tabulation and Chi-Square" (see Figure 13.M.1). The resulting dialog box appears as shown in Figure 13.M.3.

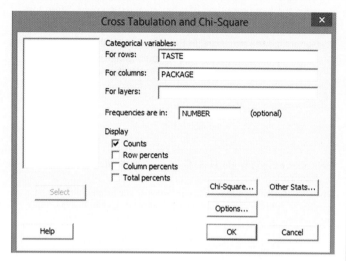

Figure 13.M.3 MINITAB cross tabulation dialog box

Step 3 Specify one qualitative variable in the "For rows" box and the other qualitative variable in the "For columns" box. [*Note:* If your worksheet contains cell counts for the categories, enter the variable with the cell counts in the "Frequencies are in" box.]

Step 4 Select the summary statistics (e.g., counts, percentages) you want to display in the contingency table.

Step 5 Click the "Chi-Square" button. The resulting dialog box is shown in Figure 13.M.4.

Figure 13.M.4 MINITAB chi-square dialog box

Step 6 Select "Chi-Square analysis" and "Expected cell counts" and click "OK."

Step 7 When you return to the "Cross Tabulation" menu screen, click "OK" to generate the MINITAB printout.
Note: If your MINITAB worksheet contains only the cell counts for the contingency table in columns, click the "Chi-Square Test (Two-Way Table in Worksheet)" menu option (see Figure 13.M.1) and specify the columns in the "Columns containing the table" box. Click "OK" to produce the MINITAB printout.

TI-83/TI-84 Plus Graphing Calculator: Chi-Square Analyses

The TI-83/TI-84 Plus graphing calculator can be used to conduct a chi-square test for a two-way (contingency) table but cannot conduct a chi-square test for a one-way table.

Two-Way (Contingency) Table

Step 1 *Access the matrix menu to enter the observed values*

- Press **2nd x⁻¹** for **MATRX**
- Arrow right to **EDIT**
- Press **ENTER**
- Use the **ARROW** key to enter the row and column dimensions of your observed Matrix
- Use the **ARROW** key to enter your observed values into Matrix [A]

Step 2 *Access the matrix menu to enter the expected values*

- Press **2nd x⁻¹** for **MATRX**
- Arrow right to **EDIT**
- Arrow down to **2:[B]**
- Press **ENTER**
- Use the **ARROW** key to enter the row and column dimensions of your expected matrix (The dimensions will be the same as in Matrix A)
- Use the **ARROW** key to enter your expected values into Matrix [B]

Step 3 *Access the statistical tests menu and perform the chi-square test*

- Press **STAT**
- Arrow right to **TESTS**
- Arrow down to χ^2 **Test**
- Press **ENTER**
- Arrow down to **Calculate**
- Press **ENTER**

Appendix A: Summation Notation

Denote the measurements of a quantitative data set as follows: $x_1, x_2, x_3, \ldots, x_n$, where x_1 is the first measurement in the data set, x_2 is the second measurement in the data set, x_3 is the third measurement in the data set, . . . , and x_n is the nth (and last) measurement in the data set. Thus, if we have five measurements in a set of data, we will write x_1, x_2, x_3, x_4, x_5 to represent the measurements. If the actual numbers are 5, 3, 8, 5, and 4, we have $x_1 = 5, x_2 = 3, x_3 = 8, x_4 = 5$, and $x_5 = 4$.

Most of the formulas used in this text require a summation of numbers. For example, one sum is the sum of all the measurements in the data set, or $x_1 + x_2 + x_3 + \cdots + x_n$. To shorten the notation, we use the symbol Σ for the summation—that is, $x_1 + x_2 + x_3 + \cdots + x_n = \sum_{i=1}^{n} x_i$. Verbally translate $\sum_{i=1}^{n} x_i$ as follows:

"The sum of the measurements, whose typical member is x_i, beginning with the member x_1 and ending with the member x_n."

Suppose, as in our earlier example, that $x_1 = 5, x_2 = 3, x_3 = 8, x_4 = 5$, and $x_5 = 4$. Then the sum of the five measurements, denoted $\sum_{i=1}^{n} x_i$, is obtained as follows:

$$\sum_{i=1}^{5} x_i = x_1 + x_2 + x_3 + x_4 + x_5$$

$$= 5 + 3 + 8 + 5 + 4 = 25$$

Another important calculation requires that we square each measurement and then sum the squares. The notation for this sum is $\sum_{i=1}^{n} x_i^2$. For the preceding five measurements, we have

$$\sum_{i=1}^{5} x_i^2 = x_1^2 + x_2^2 + x_3^2 + x_4^2 + x_5^2$$

$$= 5^2 + 3^2 + 8^2 + 5^2 + 4^2$$

$$= 25 + 9 + 64 + 25 + 16 = 139$$

In general, the symbol following the summation sign represents the variable (or function of the variable) that is to be summed.

The Meaning of Summation Notation $\sum_{i=1}^{n} x_i$

Sum the measurements on the variable that appears to the right of the summation symbol, beginning with the 1st measurement and ending with the nth measurement.

Example	A.1

Finding a Sum

Problem A sample data set contains the following six measurements: 5, 1, 3, 0, 2, 1. Find

$$\sum_{i=1}^{n} x_i$$

Solution For these data, the sample size $n = 6$. The measurements are denoted $x_1 = 5$, $x_2 = 1, x_3 = 3, x_4 = 0, x_5 = 2$, and $x_6 = 1$. Then

$$\sum_{i=1}^{n} x_i = x_1 + x_2 + x_3 + x_4 + x_5 + x_6 = 5 + 1 + 3 + 0 + 2 + 1 = 12$$

Example	A.2

Finding a Sum of Squares

Problem Refer to Example A.1. Find

$$\sum_{i=1}^{n} x_i^2$$

Solution Now we desire the sum of squares of the six measurements.

$$\sum_{i=1}^{n} x_i^2 = x_1^2 + x_2^2 + x_3^2 + x_4^2 + x_5^2 + x_6^2 = 5^2 + 1^2 + 3^2 + 0^2 + 2^2 + 1^2$$

$$= 25 + 1 + 9 + 0 + 4 + 1 = 40$$

Example	A.3

Finding a Sum of Differences

Problem Refer to Example A.1. Find

$$\sum_{i=1}^{n} (x_i - 2)$$

Solution Here, we subtract 2 from each measurement, then sum the six differences.

$$\sum_{i=1}^{n} (x_i - 2) = (x_1 - 2) + (x_2 - 2) + (x_3 - 2) + (x_4 - 2) + (x_5 - 2) + (x_6 - 2)$$

$$= (5 - 2) + (1 - 2) + (3 - 2) + (0 - 2) + (2 - 2) + (1 - 2)$$

$$= 3 + (-1) + 1 + (-2) + 0 + (-1) = 0$$

Appendix B: Tables

Table I   Binomial Probabilities

Tabulated values are $\sum\limits_{x=0}^{k} p(x)$. *(Computations are rounded at the third decimal place.)*

a. $n = 5$

k \ p	.01	.05	.10	.20	.30	.40	.50	.60	.70	.80	.90	.95	.99
0	.951	.774	.590	.328	.168	.078	.031	.010	.002	.000	.000	.000	.000
1	.999	.977	.919	.737	.528	.337	.188	.087	.031	.007	.000	.000	.000
2	1.000	.999	.991	.942	.837	.683	.500	.317	.163	.058	.009	.001	.000
3	1.000	1.000	1.000	.993	.969	.913	.812	.663	.472	.263	.081	.023	.001
4	1.000	1.000	1.000	1.000	.998	.990	.969	.922	.832	.672	.410	.226	.049

b. $n = 6$

k \ p	.01	.05	.10	.20	.30	.40	.50	.60	.70	.80	.90	.95	.99
0	.941	.735	.531	.262	.118	.047	.016	.004	.001	.000	.000	.000	.000
1	.999	.967	.886	.655	.420	.233	.109	.041	.011	.002	.000	.000	.000
2	1.000	.998	.984	.901	.744	.544	.344	.179	.070	.017	.001	.000	.000
3	1.000	1.000	.999	.983	.930	.821	.656	.456	.256	.099	.016	.002	.000
4	1.000	1.000	1.000	.998	.989	.959	.891	.767	.580	.345	.114	.033	.001
5	1.000	1.000	1.000	1.000	.999	.996	.984	.953	.882	.738	.469	.265	.059

c. $n = 7$

k \ p	.01	.05	.10	.20	.30	.40	.50	.60	.70	.80	.90	.95	.99
0	.932	.698	.478	.210	.082	.028	.008	.002	.000	.000	.000	.000	.000
1	.998	.956	.850	.577	.329	.159	.063	.019	.004	.000	.000	.000	.000
2	1.000	.996	.974	.852	.647	.420	.227	.096	.029	.005	.000	.000	.000
3	1.000	1.000	.997	.967	.874	.710	.500	.290	.126	.033	.003	.000	.000
4	1.000	1.000	1.000	.995	.971	.904	.773	.580	.353	.148	.026	.004	.000
5	1.000	1.000	1.000	1.000	.996	.981	.937	.841	.671	.423	.150	.044	.002
6	1.000	1.000	1.000	1.000	1.000	.998	.992	.972	.918	.790	.522	.302	.068

(continued)

Table VIII (continued)

ν_1										
ν_2	10	12	15	20	24	30	40	60	120	∞
1	6,056	6,106	6,157	6,209	6,235	6,261	6,287	6,313	6,339	6,366
2	99.40	99.42	99.43	99.45	99.46	99.47	99.47	99.48	99.49	99.50
3	27.23	27.05	26.87	26.69	26.60	26.50	26.41	26.32	26.22	26.13
4	14.55	14.37	14.20	14.02	13.93	13.84	13.75	13.65	13.56	13.46
5	10.05	9.89	9.72	9.55	9.47	9.38	9.29	9.20	9.11	9.02
6	7.87	7.72	7.56	7.40	7.31	7.23	7.14	7.06	6.97	6.88
7	6.62	6.47	6.31	6.16	6.07	5.99	5.91	5.82	5.74	5.65
8	5.81	5.67	5.52	5.36	5.28	5.20	5.12	5.03	4.95	4.86
9	5.26	5.11	4.96	4.81	4.73	4.65	4.57	4.48	4.40	4.31
10	4.85	4.71	4.56	4.41	4.33	4.25	4.17	4.08	4.00	3.91
11	4.54	4.40	4.25	4.10	4.02	3.94	3.86	3.78	3.69	3.60
12	4.30	4.16	4.01	3.86	3.78	3.70	3.62	3.54	3.45	3.36
13	4.10	3.96	3.82	3.66	3.59	3.51	3.43	3.34	3.25	3.17
14	3.94	3.80	3.66	3.51	3.43	3.35	3.27	3.18	3.09	3.00
15	3.80	3.67	3.52	3.37	3.29	3.21	3.13	3.05	2.96	2.87
16	3.69	3.55	3.41	3.26	3.18	3.10	3.02	2.93	2.84	2.75
17	3.59	3.46	3.31	3.16	3.08	3.00	2.92	2.83	2.75	2.65
18	3.51	3.37	3.23	3.08	3.00	2.92	2.84	2.75	2.66	2.57
19	3.43	3.30	3.15	3.00	2.92	2.84	2.76	2.67	2.58	2.49
20	3.37	3.23	3.09	2.94	2.86	2.78	2.69	2.61	2.52	2.42
21	3.31	3.17	3.03	2.88	2.80	2.72	2.64	2.55	2.46	2.36
22	3.26	3.12	2.98	2.83	2.75	2.67	2.58	2.50	2.40	2.31
23	3.21	3.07	2.93	2.78	2.70	2.62	2.54	2.45	2.35	2.26
24	3.17	3.03	2.89	2.74	2.66	2.58	2.49	2.40	2.31	2.21
25	3.13	2.99	2.85	2.70	2.62	2.54	2.45	2.36	2.27	2.17
26	3.09	2.96	2.81	2.66	2.58	2.50	2.42	2.33	2.23	2.13
27	3.06	2.93	2.78	2.63	2.55	2.47	2.38	2.29	2.20	2.10
28	3.03	2.90	2.75	2.60	2.52	2.44	2.35	2.26	2.17	2.06
29	3.00	2.87	2.73	2.57	2.49	2.41	2.33	2.23	2.14	2.03
30	2.98	2.84	2.70	2.55	2.47	2.39	2.30	2.21	2.11	2.01
40	2.80	2.66	2.52	2.37	2.29	2.20	2.11	2.02	1.92	1.80
60	2.63	2.50	2.35	2.20	2.12	2.03	1.94	1.84	1.73	1.60
120	2.47	2.34	2.19	2.03	1.95	1.86	1.76	1.66	1.53	1.38
∞	2.32	2.18	2.04	1.88	1.79	1.70	1.59	1.47	1.32	1.00

Numerator Degrees of Freedom

Denominator Degrees of Freedom

Table IX Critical Values of T_L and T_U for the Wilcoxon Rank Sum Test: Independent Samples

Test statistic is the rank sum associated with the smaller sample (if equal sample sizes, either rank sum can be used).

a. $\alpha = .025$ one-tailed; $\alpha = .05$ two-tailed

n_1 / n_2	3		4		5		6		7		8		9		10	
	T_L	T_U	T_L	T_U	T_L	T_U	T_L	T_U	T_L	T_U	T_L	T_U	T_L	T_U	T_L	T_U
3	5	16	6	18	6	21	7	23	7	26	8	28	8	31	9	33
4	6	18	11	25	12	28	12	32	13	35	14	38	15	41	16	44
5	6	21	12	28	18	37	19	41	20	45	21	49	22	53	24	56
6	7	23	12	32	19	41	26	52	28	56	29	61	31	65	32	70
7	7	26	13	35	20	45	28	56	37	68	39	73	41	78	43	83
8	8	28	14	38	21	49	29	61	39	73	49	87	51	93	54	98
9	8	31	15	41	22	53	31	65	41	78	51	93	63	108	66	114
10	9	33	16	44	24	56	32	70	43	83	54	98	66	114	79	131

b. $\alpha = .05$ one-tailed; $\alpha = .10$ two-tailed

n_1 / n_2	3		4		5		6		7		8		9		10	
	T_L	T_U	T_L	T_U	T_L	T_U	T_L	T_U	T_L	T_U	T_L	T_U	T_L	T_U	T_L	T_U
3	6	15	7	17	7	20	8	22	9	24	9	27	10	29	11	31
4	7	17	12	24	13	27	14	30	15	33	16	36	17	39	18	42
5	7	20	13	27	19	36	20	40	22	43	24	46	25	50	26	54
6	8	22	14	30	20	40	28	50	30	54	32	58	33	63	35	67
7	9	24	15	33	22	43	30	54	39	66	41	71	43	76	46	80
8	9	27	16	36	24	46	32	58	41	71	52	84	54	90	57	95
9	10	29	17	39	25	50	33	63	43	76	54	90	66	105	69	111
10	11	31	18	42	26	54	35	67	46	80	57	95	69	111	83	127

Source: From F. Wilcoxon and R. A. Wilcox, "Some Rapid Approximate Statistical Procedures," 1964, 20–23.

Table X Critical Values of T_0 in the Wilcoxon Paired Difference Signed Rank Test

One-Tailed	Two-Tailed	n = 5	n = 6	n = 7	n = 8	n = 9	n = 10
$\alpha = .05$	$\alpha = .10$	1	2	4	6	8	11
$\alpha = .025$	$\alpha = .05$		1	2	4	6	8
$\alpha = .01$	$\alpha = .02$			0	2	3	5
$\alpha = .005$	$\alpha = .01$				0	2	3

		n = 11	n = 12	n = 13	n = 14	n = 15	n = 16
$\alpha = .05$	$\alpha = .10$	14	17	21	26	30	36
$\alpha = .025$	$\alpha = .05$	11	14	17	21	25	30
$\alpha = .01$	$\alpha = .02$	7	10	13	16	20	24
$\alpha = .005$	$\alpha = .01$	5	7	10	13	16	19

		n = 17	n = 18	n = 19	n = 20	n = 21	n = 22
$\alpha = .05$	$\alpha = .10$	41	47	54	60	68	75
$\alpha = .025$	$\alpha = .05$	35	40	46	52	59	66
$\alpha = .01$	$\alpha = .02$	28	33	38	43	49	56
$\alpha = .005$	$\alpha = .01$	23	28	32	37	43	49

		n = 23	n = 24	n = 25	n = 26	n = 27	n = 28
$\alpha = .05$	$\alpha = .10$	83	92	101	110	120	130
$\alpha = .025$	$\alpha = .05$	73	81	90	98	107	117
$\alpha = .01$	$\alpha = .02$	62	69	77	85	93	102
$\alpha = .005$	$\alpha = .01$	55	61	68	76	84	92

		n = 29	n = 30	n = 31	n = 32	n = 33	n = 34
$\alpha = .05$	$\alpha = .10$	141	152	163	175	188	201
$\alpha = .025$	$\alpha = .05$	127	137	148	159	171	183
$\alpha = .01$	$\alpha = .02$	111	120	130	141	151	162
$\alpha = .005$	$\alpha = .01$	100	109	118	128	138	149

		n = 35	n = 36	n = 37	n = 38	n = 39	
$\alpha = .05$	$\alpha = .10$	214	228	242	256	271	
$\alpha = .025$	$\alpha = .05$	195	208	222	235	250	
$\alpha = .01$	$\alpha = .02$	174	186	198	211	224	
$\alpha = .005$	$\alpha = .01$	160	171	183	195	208	

		n = 40	n = 41	n = 42	n = 43	n = 44	n = 45
$\alpha = .05$	$\alpha = .10$	287	303	319	336	353	371
$\alpha = .025$	$\alpha = .05$	264	279	295	311	327	344
$\alpha = .01$	$\alpha = .02$	238	252	267	281	297	313
$\alpha = .005$	$\alpha = .01$	221	234	248	262	277	292

		n = 46	n = 47	n = 48	n = 49	n = 50	
$\alpha = .05$	$\alpha = .10$	389	408	427	446	466	
$\alpha = .025$	$\alpha = .05$	361	379	397	415	434	
$\alpha = .01$	$\alpha = .02$	329	345	362	380	398	
$\alpha = .005$	$\alpha = .01$	307	323	339	356	373	

Source: From F. Wilcoxon and R. A. Wilcox, "Some Rapid Approximate Statistical Procedures," 1964, p. 28

Table XI   Critical Values of Spearman's Rank Correlation Coefficient

The α values correspond to a one-tailed test of H_0: $\rho = 0$. The value should be doubled for two-tailed tests.

n	α = .05	α = .025	α − .01	α − .005	n	α = .05	α = .025	α − .01	α − .005
5	.900	–	–	–	18	.399	.476	.564	.625
6	.829	.886	.943	–	19	.388	.462	.549	.608
7	.714	.786	.893	–	20	.377	.450	.534	.591
8	.643	.738	.833	.881	21	.368	.438	.521	.576
9	.600	.683	.783	.833	22	.359	.428	.508	.562
10	.564	.648	.745	.794	23	.351	.418	.496	.549
11	.523	.623	.736	.818	24	.343	.409	.485	.537
12	.497	.591	.703	.780	25	.336	.400	.475	.526
13	.475	.566	.673	.745	26	.329	.392	.465	.515
14	.457	.545	.646	.716	27	.323	.385	.456	.505
15	.441	.525	.623	.689	28	.317	.377	.448	.496
16	.425	.507	.601	.666	29	.311	.370	.440	.487
17	.412	.490	.582	.645	30	.305	.364	.432	.478

Table XII Critical Values of the Studentized Range, $\alpha = .05$

v \\ k	2	3	4	5	6	7	8	9	10	11	12	13	14	15	16	17	18	19	20
1	17.97	26.98	32.82	37.08	40.41	43.12	45.40	47.36	49.07	50.59	51.96	53.20	54.33	55.36	56.32	57.22	58.04	58.83	59.56
2	6.08	8.33	9.80	10.88	11.74	12.44	13.03	13.54	13.99	14.39	14.75	15.08	15.38	15.65	15.91	16.14	16.37	16.57	16.77
3	4.50	5.91	6.82	7.50	8.04	8.48	8.85	9.18	9.46	9.72	9.95	10.15	10.35	10.52	10.69	10.84	10.98	11.11	11.24
4	3.93	5.04	5.76	6.29	6.71	7.05	7.35	7.60	7.83	8.03	8.21	8.37	8.52	8.66	8.79	8.91	9.03	9.13	9.23
5	3.64	4.60	5.22	5.67	6.03	6.33	6.58	6.80	6.99	7.17	7.32	7.47	7.60	7.72	7.83	7.93	8.03	8.12	8.21
6	3.46	4.34	4.90	5.30	5.63	5.90	6.12	6.32	6.49	6.65	6.79	6.92	7.03	7.14	7.24	7.34	7.43	7.51	7.59
7	3.34	4.16	4.68	5.06	5.36	5.61	5.82	6.00	6.16	6.30	6.43	6.55	6.66	6.76	6.85	6.94	7.02	7.10	7.17
8	3.26	4.04	4.53	4.89	5.17	5.40	5.60	5.77	5.92	6.05	6.18	6.29	6.39	6.48	6.57	6.65	6.73	6.80	6.87
9	3.20	3.95	4.41	4.76	5.02	5.24	5.43	5.59	5.74	5.87	5.98	6.09	6.19	6.28	6.36	6.44	6.51	6.58	6.64
10	3.15	3.88	4.33	4.65	4.91	5.12	5.30	5.46	5.60	5.72	5.83	5.93	6.03	6.11	6.19	6.27	6.34	6.40	6.47
11	3.11	3.82	4.26	4.57	4.82	5.03	5.20	5.35	5.49	5.61	5.71	5.81	5.90	5.98	6.06	6.13	6.20	6.27	6.33
12	3.08	3.77	4.20	4.51	4.75	4.95	5.12	5.27	5.39	5.51	5.61	5.71	5.80	5.88	5.95	6.02	6.09	6.15	6.21
13	3.06	3.73	4.15	4.45	4.69	4.88	5.05	5.19	5.32	5.43	5.53	5.63	5.71	5.79	5.86	5.93	5.99	6.05	6.11
14	3.03	3.70	4.11	4.41	4.64	4.83	4.99	5.13	5.25	5.36	5.46	5.55	5.64	5.71	5.79	5.85	5.91	5.97	6.03
15	3.01	3.67	4.08	4.37	4.60	4.78	4.94	5.08	5.20	5.31	5.40	5.49	5.57	5.65	5.72	5.78	5.85	5.90	5.96
16	3.00	3.65	4.05	4.33	4.56	4.74	4.90	5.03	5.15	5.26	5.35	5.44	5.52	5.59	5.66	5.73	5.79	5.84	5.90
17	2.98	3.63	4.02	4.30	4.52	4.70	4.86	4.99	5.11	5.21	5.31	5.39	5.47	5.54	5.61	5.67	5.73	5.79	5.84
18	2.97	3.61	4.00	4.28	4.49	4.67	4.82	4.96	5.07	5.17	5.27	5.35	5.43	5.50	5.57	5.63	5.69	5.74	5.79
19	2.96	3.59	3.98	4.25	4.47	4.65	4.79	4.92	5.04	5.14	5.23	5.31	5.39	5.46	5.53	5.59	5.65	5.70	5.75
20	2.95	3.58	3.96	4.23	4.45	4.62	4.77	4.90	5.01	5.11	5.20	5.28	5.36	5.43	5.49	5.55	5.61	5.66	5.71
24	2.92	3.53	3.90	4.17	4.37	4.54	4.68	4.81	4.92	5.01	5.10	5.18	5.25	5.32	5.38	5.44	5.49	5.55	5.59
30	2.89	3.49	3.85	4.10	4.30	4.46	4.60	4.72	4.82	4.92	5.00	5.08	5.15	5.21	5.27	5.33	5.38	5.43	5.47
40	2.86	3.44	3.79	4.04	4.23	4.39	4.52	4.63	4.73	4.82	4.90	4.98	5.04	5.11	5.16	5.22	5.27	5.31	5.36
60	2.83	3.40	3.74	3.98	4.16	4.31	4.44	4.55	4.65	4.73	4.81	4.88	4.94	5.00	5.06	5.11	5.15	5.20	5.24
120	2.80	3.36	3.68	3.92	4.10	4.24	4.36	4.47	4.56	4.64	4.71	4.78	4.84	4.90	4.95	5.00	5.04	5.09	5.13
∞	2.77	3.31	3.63	3.86	4.03	4.17	4.29	4.39	4.47	4.55	4.62	4.68	4.74	4.80	4.85	4.89	4.93	4.97	5.01

Table XIII Critical Values of the Studentized Range, $\alpha = .01$

ν \ k	2	3	4	5	6	7	8	9	10	11	12	13	14	15	16	17	18	19	20
1	90.03	135.0	164.3	185.6	202.2	215.8	227.2	237.0	245.6	253.2	260.0	266.2	271.8	277.0	281.8	286.3	290.0	294.3	298.0
2	14.04	19.02	22.29	24.72	26.63	28.20	29.53	30.68	31.69	32.59	33.40	34.13	34.81	35.43	36.00	36.53	37.03	37.50	37.95
3	8.26	10.62	12.17	13.33	14.24	15.00	15.64	16.20	16.69	17.13	17.53	17.89	18.22	18.52	18.81	19.07	19.32	19.55	19.77
4	6.51	8.12	9.17	9.96	10.58	11.10	11.55	11.93	12.27	12.57	12.84	13.09	13.32	13.53	13.73	13.91	14.08	14.24	14.40
5	5.70	6.98	7.80	8.42	8.91	9.32	9.67	9.97	10.24	10.48	10.70	10.89	11.08	11.24	11.40	11.55	11.68	11.81	11.93
6	5.24	6.33	7.03	7.56	7.97	8.32	8.61	8.87	9.10	9.30	9.48	9.65	9.81	9.95	10.08	10.21	10.32	10.43	10.54
7	4.95	5.92	6.54	7.01	7.37	7.68	7.94	8.17	8.37	8.55	8.71	8.86	9.00	9.12	9.24	9.35	9.46	9.55	9.65
8	4.75	5.64	6.20	6.62	6.96	7.24	7.47	7.68	7.86	8.03	8.18	8.31	8.44	8.55	8.66	8.76	8.85	8.94	9.03
9	4.60	5.43	5.96	6.35	6.66	6.91	7.13	7.33	7.49	7.65	7.78	7.91	8.03	8.13	8.23	8.33	8.41	8.49	8.57
10	4.48	5.27	5.77	6.14	6.43	6.67	6.87	7.05	7.21	7.36	7.49	7.60	7.71	7.81	7.91	7.99	8.08	8.15	8.23
11	4.39	5.15	5.62	5.97	6.25	6.48	6.67	6.84	6.99	7.13	7.25	7.36	7.46	7.56	7.65	7.73	7.81	7.88	7.95
12	4.32	5.05	5.50	5.84	6.10	6.32	6.51	6.67	6.81	6.94	7.06	7.17	7.26	7.36	7.44	7.52	7.59	7.66	7.73
13	4.26	4.96	5.40	5.73	5.98	6.19	6.37	6.53	6.67	6.79	6.90	7.01	7.10	7.19	7.27	7.35	7.42	7.48	7.55
14	4.21	4.89	5.32	5.63	5.88	6.08	6.26	6.41	6.54	6.66	6.77	6.87	6.96	7.05	7.13	7.20	7.27	7.33	7.39
15	4.17	4.84	5.25	5.56	5.80	5.99	6.16	6.31	6.44	6.55	6.66	6.76	6.84	6.93	7.00	7.07	7.14	7.20	7.26
16	4.13	4.79	5.19	5.49	5.72	5.92	6.08	6.22	6.35	6.46	6.56	6.66	6.74	6.82	6.90	6.97	7.03	7.09	7.15
17	4.10	4.74	5.14	5.43	5.66	5.85	6.01	6.15	6.27	6.38	6.48	6.57	6.66	6.73	6.81	6.87	6.94	7.00	7.05
18	4.07	4.70	5.09	5.38	5.60	5.79	5.94	6.08	6.20	6.31	6.41	6.50	6.58	6.65	6.72	6.79	6.85	6.91	6.97
19	4.05	4.67	5.05	5.33	5.55	5.73	5.89	6.02	6.14	6.25	6.34	6.43	6.51	6.58	6.65	6.72	6.78	6.84	6.89
20	4.02	4.64	5.02	5.29	5.51	5.69	5.84	5.97	6.09	6.19	6.28	6.37	6.45	6.52	6.59	6.65	6.71	6.77	6.82
24	3.96	4.55	4.91	5.17	5.37	5.54	5.69	5.81	5.92	6.02	6.11	6.19	6.26	6.33	6.39	6.45	6.51	6.56	6.61
30	3.89	4.45	4.80	5.05	5.24	5.40	5.54	5.65	5.76	5.85	5.93	6.01	6.08	6.14	6.20	6.26	6.31	6.36	6.41
40	3.82	4.37	4.70	4.93	5.11	5.26	5.39	5.50	5.60	5.69	5.76	5.83	5.90	5.96	6.02	6.07	6.12	6.16	6.21
60	3.76	4.28	4.59	4.82	4.99	5.13	5.25	5.36	5.45	5.53	5.60	5.67	5.73	5.78	5.84	5.89	5.93	5.97	6.01
120	3.70	4.20	4.50	4.71	4.87	5.01	5.12	5.21	5.30	5.37	5.44	5.50	5.56	5.61	5.66	5.71	5.75	5.79	5.83
∞	3.64	4.12	4.40	4.60	4.76	4.88	4.99	5.08	5.16	5.23	5.29	5.35	5.40	5.45	5.49	5.54	5.57	5.61	5.65

Appendix C: Calculation Formulas for Analysis of Variance

C.1 Completely Randomized Design

$$CM = \text{Correction for mean}$$

$$= \frac{(\text{Total of all observations})^2}{\text{Total number of observations}} = \frac{(\Sigma y_i)^2}{n}$$

$$SS(\text{Total}) = \text{Total sum of squares}$$

$$= (\text{Sum of squares of all observations}) - CM = \Sigma y_i^2 - CM$$

$$SST = \text{Sum of squares for treatments}$$

$$= \left(\begin{array}{c} \text{Sum of squares of treatment totals with} \\ \text{each square divided by the number of} \\ \text{observations for that treatment} \end{array} \right) - CM$$

$$= \frac{T_1^2}{n_1} + \frac{T_2^2}{n_2} + \cdots + \frac{T_k^2}{n_k} - CM$$

$$SSE = \text{Sum of squares for error} = SS(\text{Total}) - SST$$

$$MST = \text{Mean square for treatments} = \frac{SST}{k-1}$$

$$MSE = \text{Mean square for error} = \frac{SSE}{n-k}$$

$$F = \text{Test statistic} = \frac{MST}{MSE}$$

where

$$n = \text{Total number of observations}$$

$$k = \text{Number of treatments}$$

$$T_i = \text{Total for treatment } i \ (i = 1, 2, \ldots, k)$$

C.2 Randomized Block Design

$$CM = \text{Correction for mean}$$

$$= \frac{(\text{Total of all observations})^2}{\text{Total number of observations}} = \frac{(\Sigma y_i)^2}{n}$$

$$SS(\text{Total}) = \text{Total sum of squares}$$

$$= (\text{Sum of squares of all observations}) - CM = \Sigma y_i^2 - CM$$

$$SST = \text{Sum of squares for treatments}$$

$$= \left(\begin{array}{c} \text{Sum of squares of treatment totals with} \\ \text{each square divided by } b, \text{ the number of} \\ \text{observations for that treatment} \end{array} \right) - CM$$

$$= \frac{T_1^2}{b} + \frac{T_2^2}{b} + \cdots + \frac{T_k^2}{b} - CM$$

$$SST = \text{Sum of squares for blocks}$$

$$= \left(\begin{array}{c} \text{Sum of squares of block totals with} \\ \text{each square divided by } k, \text{ the number} \\ \text{of observations in that block} \end{array} \right) - CM$$

$$= \frac{B_1^2}{k} + \frac{B_2^2}{k} + \cdots + \frac{B_b^2}{k} - CM$$

$$SSE = \text{Sum of squares for error} = SS(\text{Total}) - SST - SSB$$

$$MST = \text{Mean square for treatments} = \frac{SST}{k - 1}$$

$$MSB = \text{Mean square for blocks} = \frac{SSB}{b - 1}$$

$$MSE = \text{Mean square for error} = \frac{SSE}{n - k - b + 1}$$

$$F = \text{Test statistic} = \frac{MST}{MSE}$$

where

$$n = \text{Total number of observations}$$
$$b = \text{Number of blocks}$$
$$k = \text{Number of treatments}$$
$$T_i = \text{Total for treatment } i \ (i = 1, 2, \ldots, k)$$
$$B_i = \text{Total for block } i \ (i = 1, 2, \ldots, b)$$

C.3 Two-Factor Factorial Experiment

$$CM = \text{Correction for mean}$$

$$= \frac{(\text{Total of all } n \text{ measurements})^2}{n} = \frac{\left(\sum_{i=1}^{n} y_i \right)^2}{n}$$

$$SS(\text{Total}) = \text{Total sum of squares}$$

$$= (\text{Sum of squares of all } n \text{ measurements}) - CM = \sum_{i=1}^{n} y_i^2 - CM$$

$\text{SS}(A) = $ Sum of squares for main effects, factor A

$$= \left(\begin{array}{c} \text{Sum of squares of the totals } A_1, A_2, \dots, A_a \\ \text{divided by the number of measurements} \\ \text{in a single total, namely } br \end{array}\right) - \text{CM}$$

$$= \frac{\displaystyle\sum_{i=1}^{a} A_i^2}{br} - \text{CM}$$

$\text{SS}(B) = $ Sum of squares for main effects, factor B

$$= \left(\begin{array}{c} \text{Sum of squares of the totals } B_1, B_2, \dots, B_b \\ \text{divided by the number of measurements} \\ \text{in a single total, namely } ar \end{array}\right) - \text{CM}$$

$$= \frac{\displaystyle\sum_{i=1}^{b} B_i^2}{ar} - \text{CM}$$

$\text{SS}(AB) = $ Sum of squares for AB interaction

$$= \left(\begin{array}{c} \text{Sum of squares of the cell totals} \\ AB_{11}, AB_{12}, \dots, AB_{ab} \text{ divided by} \\ \text{the number of measurements in} \\ \text{a single total, namely } r \end{array}\right) - \text{SS}(A) - \text{SS}(B) - \text{CM}$$

$$= \frac{\displaystyle\sum_{j=1}^{b}\sum_{i=1}^{a} AB_{ij}^2}{r} - \text{SS}(A) - \text{SS}(B) - \text{CM}$$

$\text{SSE} = $ Sum of squares for error $= \text{SS(total)} - \text{SS}(A) - \text{SS}(B) - \text{SS}(AB)$

$\text{MS}(A) = \dfrac{\text{SS}(A)}{a-1} = $ Mean square for A

$\text{MS}(B) = \dfrac{\text{SS}(B)}{b-1} = $ Mean square for B

$\text{MS}(AB) = \dfrac{\text{SS}(AB)}{(a-1)(b-1)} = $ Mean square for AB

$\text{MSE} = \dfrac{\text{SSE}}{ab(r-1)} = $ Mean square for error

$F(A) = $ Test statistic for $A = \dfrac{\text{MS}(A)}{\text{MSE}}$

$F(B) = $ Test statistic for $B = \dfrac{\text{MS}(B)}{\text{MSE}}$

$F(AB) = $ Test statistic for $AB = \dfrac{\text{MS}(AB)}{\text{MSE}}$

where

$a = $ Number of levels of factor A

$b = $ Number of levels of factor B

$r = $ Number of replicates (observations per treatment)

$A_i = $ Total for level i of factor A ($i = 1, 2, \dots, a$)

$B_i = $ Total for level i of factor B ($i = 1, 2, \dots, b$)

$AB_{ij} = $ Total for treatment (ij), i.e., for ith level of factor A and ith level of factor B

C.4   Tukey's Multiple Comparisons Procedure (Equal Sample Sizes)

Step 1 Select the desired experimentwise error rate, α

Step 2 Calculate

$$\omega = q_\alpha(k, v)\frac{s}{\sqrt{n_t}}$$

where

k = Number of sample means (i.e., number of treatments)

$s = \sqrt{\text{MSE}}$

v = Number of degrees of freedom associated with MSE

n_t = Number of observations in each of the k samples (i.e., number of observations per treatment)

$q_\alpha(k, v)$ = Critical value of the Studentized range (Tables XII and XIII of Appendix B)

Step 3 Calculate and rank the k sample means.

Step 4 Place a bar over those pairs of treatment means that differ by less than ω. A pair of treatments not connected by an overbar (i.e., differing by more than ω) implies a difference in the corresponding population means.

Note: The confidence level associated with all inferences drawn from the analysis is $(1 - \alpha)$.

C.5   Bonferroni Multiple Comparisons Procedure (Pairwise Comparisons)

Step 1 Calculate for each treatment pair (i, j)

$$B_{ij} = t_{\alpha/(2c)}s\sqrt{\frac{1}{n_i} + \frac{1}{n_j}}$$

where

k = Number of sample (treatment) means in the experiment

c = Number of pairwise comparisons

[*Note:* If all pairwise comparisons are to be made, then $c = k(k - 1)/2$]

$s = \sqrt{\text{MSE}}$

v = Number of degrees of freedom associated with MSE

n_i = Number of observations in sample for treatment i

n_j = Number of observations in sample for treatment j

$t_{\alpha/(2c)}$ = Critical value of t distribution with v df and tail area $\alpha/(2c)$ (Table III in Appendix B)

Step 2 Rank the sample means and place a bar over any treatment pair (i, j) whose sample means differ by less than B_{ij}. Any pair of means not connected by an overbar implies a difference in the corresponding population means.

Note: The level of confidence associated with all inferences drawn from the analysis is at least $(1 - \alpha)$.

C.6 Scheffé's Multiple Comparisons Procedure (Pairwise Comparisons)

Step 1 Calculate Scheffé's critical difference for each pair of treatments (i, j):

$$S_{ij} = \sqrt{(k-1)(F_\alpha)(\text{MSE})\left(\frac{1}{n_1} + \frac{1}{n_j}\right)}$$

where

k = Number of sample (treatment) means

MSE = Mean squared error

n_i = Number of observations in sample for treatment i

n_j = Number of observations in sample for treatment j

F_α = Critical value of F distribution with $k-1$ numerator df and v denominator df (Tables V, VI, VII, and VIII of Appendix B)

v = Number of degrees of freedom associated with MSE

Step 2 Rank the k sample means and place a bar over any treatment pair (i, j) that differs by less than S_{ij}. Any pair of sample means not connected by an overbar implies a difference in the corresponding population means.

Short Answers to Selected Odd Exercises

Chapter 1

1.3 The selection process of collecting data from a larger set is called sampling. **1.11** quantitative; qualitative **1.13 a.** all young women who recently participated in a STEM program **b.** 159 women recruited to complete an on-line survey **c.** 27% of all young women who recently participated in a STEM program felt program increased their interest in science. **1.15 a.** earthquake sites **b.** sample **c.** ground motion (qualitative); magnitude (quantitative); ground acceleration (quantitative) **1.17 a.** pipe locations **b.** 19 different randomly selected pipe locations **c.** set of all possible pipe locations **d.** inferential statistics **1.19 a.** qualitative **b.** quantitative **c.** quantitative **1.21** Town, Type of water supply, and Presence of hydrogen sulphide are qualitative; all others are quantitative **1.23** descriptive **1.25 a.** qualitative **b.** quantitative **c.** quantitative **d.** quantitative **e.** quantitative **f.** qualitative **g.** qualitative **1.27 a.** designed experiment **b.** smokers **c.** quantitative **d.** population: all smokers in the U.S.; sample: 50,000 smokers in trial **e.** the difference in mean age at which each of the scanning methods first detects a tumor. **1.29 a.** designed experiment **b.** amateur boxers **c.** heart rate (quantitative); blood lactate level (quantitative) **d.** no difference between the two groups of boxers **e.** no **1.31 a.** population: all students; sample: 155 volunteer students **b.** designed experiment **c.** higher proportion of students in guilty-state group chose not to repair the car than those in neutral-state and anger-state **d.** representativeness of sample **1.33 a.** quantitative: age and dating experience; qualitative: gender and willingness to tell **b.** possible non-representative sample **1.35 a.** survey **b.** qualitative **c.** non-representative sample

Chapter 2

2.5 a. $X - 8, Y - 9, Z - 3$ **b.** $X - .40, Y - .45, Z - .15$ **2.7 a.** bar graph **b.** type of robotic limbs. **c.** legs only **d.** none: .142; both: .075; legs only: .594; wheels only: .189 **2.9 a.** qualitative **b.** Unknown−5, Unworn−2, Slight−4, Light/Mod−2, Mod−3, Mod/Heavy−1, Heavy−1 **c.** Unknown−.278, Unworn−.111, Slight−.222, Light/Mod−.111, Mod−.167, Mod/Heavy−.056, Heavy−.056 **f.** Unknown **2.11** relative frequencies: Black−.175, White−.705, Sumatran−.0035, Javan−.0014, OneHorned−.115 **c.** .88; .12 **2.13 a.** Explorer **b.** remote code execution **2.15 a.** .389 **b.** yes **c.** multi-year ice is most common **2.17** LEO–most government owned (43.7%); GEO—most commercially owned (69.1%) **2.19** children have many types of interactions; adults have mainly one type—teaching **2.21 a.** response 1: .316; response 2: .462; response 3: .206; response 4: .215 **2.23 b.** public—40% contaminated; private—21.4% contaminated **c.** bedrock—31.3% contaminated; unconsolidated—31.8% contaminated $z = 1.57$ **2.29** about 6 classes **2.31 a.** 23 **2.33** frequencies: 50, 75, 125, 100, 25, 50, 50, 25 **2.35 a.** frequency histogram **b.** fup/fumic ratio **c.** .034 **d.** .695 **2.37** 2 **2.39 c.** 40−49 **2.41 b.** .941 **2.43** 67% of frequencies exceed 3,500 hertz **2.45** most of the PMI's range from 3 to 7.5 **2.47** histogram looks similar to graph; "inside job" not likely **2.49** mean, median, mode **2.51** mode **2.53** mean **2.55** mean = 2.72; median = 2.65 **2.57 a.** 8.5 **b.** 25 **c.** .78 **d.** 13.44 **2.59 a.** 2.5; 3; 3 **b.** 3.08; 3; 3 **c.** 49.6; 49; 50 **2.61** mean = 9.72; median = 10.94 **2.63 a.** 67.76; accurate **b.** 68; accurate **c.** 64; not accurate **d.** little skewness **2.65 a.** 16.5; increase **b.** 16.16; no change **c.** no mode **2.67 a.** skewed right **b.** skewed left **c.** skewed right **d.** symmetric **e.** skewed left **f.** skewed left **2.69 a.** average permeability value is 73.62 mD; half of the permeability values are less than 70.45 mD **b.** average permeability value is 128.54 mD; half of the permeability values are less than 139.30 mD **c.** average permeability value is 83.07 mD; half of the permeability values are less than 78.65 mD **d.** the permeability value that occurs most often (3 times) is 70.9 mD **e.** type B **2.71 a.** mean = −1.091; median = −.655 **b.** −8.11 **c.** mean = −.519; median = −.520; mean **2.73** largest value minus smallest value **2.77** data set B **2.79 a.** 5, 3.7, 1.92 **b.** 99, 1949.25, 44.15 **c.** 98, 1307.84, 36.16 **2.81** data set 1: −5, −4, −3, −2, −1, 1, 2, 3, 4, 5; data set 2: −10, −9, −8, −7, −6, 6, 7, 8, 9, 10 **2.83 a.** 3, 1.3, 1.14 **b.** 3, 1.3, 1.14 **c.** 3, 1.3, 1.14 **2.85 a.** 51.26 **b.** 128.57 **c.** 11.34 **d.** σ^2 and σ **2.87 a.** 43.41; points squared **b.** 6.59; points **2.89 a.** 2.86 **b.** 3.26 **c.** 2.94 **d.** DM; honey **2.91 a.** 6.45; increase **b.** 3.883; increase **c.** 1.97; increase **2.93** mean = median = mode **2.95 a.** dollars; quantitative **b.** at least 3/4; at least 8/9; nothing; nothing **2.97 a.** ≈68% **b.** ≈95% **c.** ≈all **2.99** range/6 = 104.17, range/4 = 156.25; no **2.101 a.** $\bar{x} = 94.4$, $s = 5.34$ **b.** (89.1, 99.8); (83.8, 105.1); (78.4, 110.4) **c.** 76.3%, 94.1%, 95.7; Chebyshev **2.103 a.** unknown **b.** ≈84% **2.105** 68% (≈675 senior managers) **2.107 a.** (0, 300.5) **b.** (0, 546) **c.** handrubbing appears to be more effective **2.109 a.** at least 8/9 of the velocities will fall within 936 ± 30 **b.** No **2.111 a.** yes **b.** no **2.113** do not purchase **2.115 a.** 25%; 75% **b.** 50%; 50% **c.** 80%; 20% **d.** 16%; 84% **2.117 a.** 2 **b.** .5 **c.** 0 **d.** −2.5 **e.** sample: a, b, d; population: c **f.** above: a, b; below: d **2.119** $\mu = 60$, $\sigma = 10$ **2.121** 31st percentile **2.123 a.** $z = -3.83$ **b.** $z = -.46$ **2.125 a.** 23 **b.** $z = 3.28$ **2.127 a.** $z = -1.57$ **b.** $z = -3.36$ **2.129 a.** $z = 5.08$ **b.** $z = -.81$ **c.** yes **2.131 a.** $z = 2.0$: 3.7; $z = -1.0$: 2.2; $z = .5$: 2.95; $z = -2.5$: 1.45 **b.** 1.9 **c.** $z = 1.0$ and 2.0; GPA = 3.2 and 3.7; mound-shaped; symmetric **2.133** upper and lower quartiles **2.135** unusually large or small observation **2.137 a.** 4 **b.** $Q_U \approx 6$, $Q_L \approx 3$ **c.** 3 **d.** skewed right **e.** 50%; 75% **f.** 12, 13, and 16 **2.139** two suspect outliers: 85 and 100 **2.141** CC: 3 outliers; CT: 2 outliers; TT: none **2.143 a.** −1.26 **b.** No **2.145 a.** no, $z = 1.57$ **b.** yes, $z = -3.36$ **2.147 a.** 69, 73, 74, 76, 78, 81, 81, 83, 84, 84 and 85 **b.** 69, 73, 73, 74, 76 and 78 **c.** no; highly skewed data **2.149 b.** $\sigma_{2011} = 121.5$; $\sigma_{2014} = 134$ **c.** no **2.151 a.** 3 outliers: 117.3, 118.5, 122.4 **b.** 1 outlier: 50.4 **c.** no outliers **d.** Group A: mean and standard deviation will decrease; Group B: mean will increase, standard deviation will decrease **2.153** yes; $z = -1.125$ for syntactic complexity score and $z = -.96$ for lexical diversity score **2.155** association between two underlying variables. **2.157** weak positive linear trend **2.159** positive linear trend **2.161** negative linear trend **2.163** negative trend, nonlinear **2.165 a.** weak negative linear trend **b.** weak positive linear trend **2.167 a.** negative trend **b.** positive trend with both plant coverage and diversity **2.169** Yes; accuracy decreases as driving distance increases **2.171 b.** information for last year incomplete **c.** complete data for 2006 and following years **2.173 a.** mean **b.** median **2.175** skewed data sets **2.177** small data sets **2.179 a.** −1, 1, 2 **b.** −2, 2, 4 **c.** 1, 3, 4 **d.** .1, .3, .4

2.183 a. $\bar{x} = 5.67, s^2 = 1.07, s = 1.03$ **b.** $\bar{x} = -\$1.5, s^2 = 11.5, s = \$3.39$ **c.** $\bar{x} = .413\%, s^2 = .088, s = .297\%$ **d.** 3; \$10; .7375\%
2.185 yes, positive **2.187 a.** $39/266 = .147$ **b.** level 2−.286, level 3−.188, level 4−.327, level 5−.041, level 6−.011 **e.** level 4
2.189 a. histogram **b.** 15 **c.** .197 **d.** .539 **2.191 b.** $z = -1.06$ **2.193 b.** probably none **2.195** about twice as many interviews as all
other types combined **2.197 a.** 19 ± 195 **b.** 7 ± 147 **c.** SAT-Math **2.199 a.** no **b.** Yes **c.** not reliable (only a few observations
plotted) **2.201** Over half of the whistle types were "Type a" **2.203 a.** no outliers **2.205 a.** Seabirds—quantitative; length—
quantitative; oil—qualitative **b.** transect **c.** oiled: 38%; unoiled: 62% **d.** distributions are similar **d.** 3.27 ± 13.4 **f.** 3.50 ± 11.94
g. unoiled **2.207 b.** yes **c.** A1775A: $xQ = 19,462.2, s = 532.29$; A1775B: $xQ = 22,838.5, s = 560.95$ **d.** cluster A1775A
2.209 yes; $z = -2.5$ **2.211 a.** median **b.** mean **2.213** results not reliable; sample not representative

Chapter 3

3.1 observing a single outcome **3.3** collection of all sample points **3.5** (1) probability falls between 0 and 1; (2)
sum of probabilities of sample points equals 1 **3.7** sum probabilities of sample points in A **3.9 a.** .4 **b.** .1 **c.** .6
3.11 $P(A) = .55; P(B) = .50; P(C) = .70$ **3.13 a.** 10 **b.** 20 **c.** 15,504 **3.15 a.** $(R_1R_2), (R_1R_3), (R_2R_3), (R_1B_1), (R_1B_2), (R_2B_1), (R_2B_2),$
$(R_3B_1), (R_3B_2), (B_1B_2)$ **b.** 1/10 for each sample point **c.** $P(A) = 1/10, P(B) = 6/10, P(C) = 3/10$ **3.17 a.** legs only, wheels only,
both legs and wheels, neither legs nor wheels **b.** $P(\text{Legs only}) = .594, P(\text{Wheels only}) = .189, P(\text{Both}) = .075, P(\text{Neither}) = .142$
c. .264 **d.** .669 **3.19 a.** Blue, Orange, Green, Yellow, Brown, and Red **b.** $P(\text{Blue}) = .24, P(\text{Orange}) = .20, P(\text{Green}) = .16,$
$P(\text{Yellow}) = .14, P(\text{Brown}) = .13, P(\text{Red}) = .13$ **c.** .13 **d.** .43 **e.** .76 **3.21 a.** .01 **b.** yes **3.23** .615; .943 **3.25** .389 **3.27 a.** .091
b. .273 **3.29 a.** 6 **b.** .282, .065, .339, .032, .008, .274 **c.** .686 **3.31 a.** 28 **b.** 1/28 **3.33** $693/1,686,366 = .000411$ **3.35 a.** 15 **b.** 20 **c.** 15 **d.** 6
e. 64 **3.37** no, probabilities are equal **3.39** if either A or B or both occur **3.41** all sample points that are not in A
3.43 $P(A \cup B) = P(A) + P(B)$ **3.45 b.** $P(A) = 7/8, P(B) = 1/2, P(A \cup B) = 7/8, P(A^c) = 1/8, P(A \cap B) = 1/2$ **c.** 7/8 **d.**
no **3.47 a.** $^3/_4$ **b.** 13/20 **c.** 1 **d.** 2/5 **e.** $^1/_4$ **f.** 7/20 **g.** 1 **h.** 1/4 **3.49 a.** A: $\{(2, 6), (3, 5), (4, 4), (5, 3), (6, 2)\}$; B: $\{(1, 6), (2, 6), (3, 6),$
$(4, 6), (5, 6), (6, 1), (6, 2), (6, 3), (6, 4), (6, 5), (6, 6)\}$; $A \cap B = \{(2, 6), (6, 2)\}$; $A \cup B = \{(1, 6), (2, 6), (3, 5), (3, 6), (4, 4),$
$(4, 6), (5, 3), (5, 6), (6, 1), (6, 2), (6, 3), (6, 4), (6, 5), (6,6)\}$; A^c: $\{(1, 1), (1, 2), (1, 3), (1, 4), (1, 5), (1, 6), (2, 1), (2, 2), (2, 3),$
$(2, 4), (2, 5), (3, 1), (3, 2), (3, 3), (3, 4), (3, 6), (4, 4), (4, 2), (4, 3), (4, 5), (4, 6), (5, 1), (5, 2), (5, 4), (5, 5), (5, 6), (6, 1), (6, 3),$
$(6, 4), (6, 5), (6, 6)\}$ **b.** $P(A) = 5/36, P(B) = 11/36, P(A \cap B) = 1/18, P(A \cup B) = 7/18, P(A^c) = 31/36$
c. 7/18 **d.** no **3.51 a.** .65 **b.** .72 **c.** .25 **d.** .08 **e.** .35 **f.** .72 **g.** .65 **h.** A and C, B and C, C and D **3.53** .858 **3.55 b.** .43 **c.** .57
3.57 a. $\{\text{Non-REM/Non-REM}\}, \{\text{Non-REM/REM}\}, \{\text{Non-REM/Wake}\}, \{\text{REM/Non-REM}\}, \{\text{REM/REM}\},$
$\{\text{REM/Wake}\}, \{\text{Wake/Non-REM}\}, \{\text{Wake/REM}\}, \{\text{Wake/Wake}\}$ **b.** .6385, .0032, .0347, .0069, .1524, .0035, .0318, .0072, .1218
c. .1628 **d.** .16 **e.** .0032 **3.59 a.** .333 **b.** .351 **c.** .263 **d.** .421 **3.61 a.** .156 **b.** .617 **c.** .210 **d.** .449 **e.** yes **3.63 a.** .19 **b.** .405 **c.** .275
3.65 $7/45 = .156$ **3.69 a.** $P(A \cap B) = P(A)P(B)$ **b.** $P(A \cap B) = P(B|A)P(A) = P(A|B)P(B)$ **3.71 a.** false **b.** true
c. false **3.73 a.** .5 **b.** .25 **c.** no **3.75 a.** .12 **b.** .2 **c.** .68 **3.77 a.** $P(A) = .4; P(B) = .8; P(A \cap B) = .3$ **b.** $P(E_1|A) = .25,$
$P(E_2|A) = .25, P(E_3|A) = .5$ **c.** .75 **3.79** no **3.81 a.** .75 **b.** .30 **3.83** .3195 **3.85 a.** .26 **b.** .013 **3.87** .40
3.89 .013; independent events **3.91** .35 **3.93 a.** .23 **b.** .729 **3.95 a.** $38/132 = .2879$ **b.** $29/123 = .236$ **3.97 a.** .406 **b.** .294
3.99 $P(A|B) = 0$ **3.101 a.** .1 **b.** .522 **c.** The person is guessing **3.103** $.5\beta + \alpha\beta - \alpha\beta^2$ **3.107** combinations rule does not take
order into account; the permutation rule does **3.109 a.** 256 **b.** 1024 **c.** 4096 **d.** 4^n **3.111 a.** 70 **b.** 10 **c.** 561 **d.** 36 **e.** $\binom{t}{n} = \dfrac{t!}{n!(t-n)!}$

3.113 a. 10 **b.** 6 **c.** 36 **3.115 a.** 56 **b.** 1,680 **c.** 6,720 **3.117** 12; yes **3.119 a.** 120 **b.** 72 **3.121** 1,500,625 **3.123 a.** 42
b. 1008 **3.125 a.** 1/6 **b.** 1/3 **c.** 1/3 **3.127 a.** 18 **b.** 4/18 **3.129 a.** 1/42 **b.** 1/42 **c.** 5/21 **3.131 a.** 2,598,960 **b.** .002 **c.** .00394
d. .0000154 **3.135 a.** .031 **b.** .345 **c.** .376 **d.** .082 **e.** .918 **3.137** .55 **3.139 a.** .5 **b.** .99 **c.** .847 **3.141** 1/159
3.143 $P(\text{dolomite}|\text{reading} > 60) = .1082$ **3.145** .966 **3.147 a.** $P(T^c|E) > P(T|E)$ **3.149 a.** $A \cup B$ **b.** B^c
c. $A \cap B$ **d.** $A^c|B$ **3.151 a.** 0 **b.** no **3.153** .33 **3.155 c.** $P(A) = 1/3; P(B) = 2/3$ **e.**
$P(A^c) = 2/3; P(B^c) = 1/3; P(A \cap B) = 1/3; P(A \cup B) = 2/3; P(A|B) = 1/2; P(B|A) = 1$ **f.** no, no **3.157 c.**
$P(A) = 1/3; P(B) = 2/3$ **e.** $P(A^c) = 2/3; P(B^c) = 1/3; P(A \cap B) = 1/3; P(A \cup B) = 2/3; P(A|B) = 1/2; P(B|A) = 1$
f. no, no **3.159 a.** (None), (1 or 2), (3–5), (6–9), (10 or more) **b.** $P(\text{None}) = .25, P(1 \text{ or } 2) = .31, P(3–5) = .25,$
$P(6–9) = .05, P(10 \text{ or more}) = .14$ **c.** .44 **3.161** .05 **3.163 a.** false **b.** true **c.** true **d.** false **3.165 a.** .261 **b.** Trunk–.85,
Leaves–.10, Branch–.05 **3.167 a.** $\{\text{Single, shore parallel; Other; Planar}\}$ **b.** .1/3, 1/3, 1/3 **c.** 2/3 **d.** $\{\text{No dunes/flat};$
Bluff/scarp; Single dune; Not observed$\}$ **e.** 1/3, 1/6, 1/3, 1/6 **f.** 2/3 **3.169 a.** School laboratory, In Transit, Chemical plant,
Non-chemical plant, Other **b.** .06, .26, .21, .35, .12 **c.** .06 **d.** .56 **e.** .74 **f.** .000216 **g.** no **3.171 a.** $\{11, 13, 15, 17, 29, 31, 33, 35\}$
b. $\{2, 4, 6, 8, 10, 11, 13, 15, 17, 20, 22, 24, 26, 28, 29, 31, 33, 35, 1, 3, 5, 7, 9, 19, 21, 23, 25, 27\}$ **c.** $P(A) = 9/19, P(B) = 9/19,$
$P(A \cap B) = 4/19, P(A \cup B) = 14/19, P(C) = 9/19$ **d.** $\{11, 13, 15, 17\}$ **e.** 14/19, no **f.** 2/19 **g.** $\{1, 2, 3, \ldots, 29, 31, 33, 35\}$
h. 16/19 **3.173 a.** .158 **b.** .316 **c.** .526 **d.** #3 **3.175 a.** 6 **b.** 1/3 **3.177 a.** .5625, .375, .0625 **b.** .6, .3375, .0625 **c.** dependent **3.179 a.** .116
b. .728 **3.181 a.** .72036 **b.** .27964 **c.** .8704 **d.** .4608 **e.** .1296 **f.** at least 5 **3.183** .2362; .1942; .5696 **3.185** .748 **3.187 a.** 1 to 4 **b.** .4 **c.**
.375 **3.189 a.** .20 **b.** .0016 **c.** .4093 **d.** .001, .000000000001, .9960 **3.191** .9975 **3.193** 4.4739×10^{-28} **3.195** Marilyn

Chapter 4

4.3 a. discrete **b.** continuous **c.** discrete **d.** continuous **e.** continuous **f.** continuous **4.5 a.** continuous **b.** discrete **c.** discrete
d. discrete **e.** discrete **f.** continuous **4.7** 0, 1, 2, 3, …; discrete **4.9** continuous; discrete **4.11** number of people receiving protein supplement **4.13** number of patients in a day **4.17** table, graph, formula **4.19 a.** .25 **b.** .35 **c.** .8 **4.21 a.** .7 **b.** .3 **c.** 1 **d.** .1 **e.** .8
4.23 b. $P(0) = 1/8, P(1) = 3/8, P(2) = 3/8, P(3) = 1/8$ **d.** 1/2 **4.25 a.** 0, 2, 3, 4 **b.** $P(0) = .330, P(2) = .547, P(3) = .047,$
$P(4) = .075$ **c.** 1.54 **4.27 b.** .34 **4.29 a.** BB, BG, GB, GG **b.** $^1/_4, ^1/_4, ^1/_4, ^1/_4,$ **c.** $P(0) = .25, P(1) = .5, P(2) = .25$ **d.** $P(0) = .222,$
$P(1) = .513, P(2) = .265$ **4.31 a.** $P(\text{any letter}) = 1/26$ **b.** no **4.33 a.** .1875 **b.** .1055 **c.** .75 **4.35 a.** $P(6) = .282, P(7) = .065,$
$P(8) = .339, P(9) = .032, P(10) = .008, P(11) = .274$ **b.** .274 **4.37 a.** voters 1, 2, 3, 4, 7, 8, 10: 2; voter 5: 0; voter 6: 1, voter 9: 3
b. $P(0) = .10, P(1) = .10, P(2) = .70, P(3) = .10$ **c.** .10 **d.** no **4.39 a.** 0, 1, 2 **b.** $P(0) = .625, P(1) = .250, P(2) = .125$

4.41 yes   **4.43 a.** 1.6 **b.** 3.84 **c.** 1.96 **d.** No **e.** Yes   **4.45 a.** $\mu_x = 4$ **b.** distribution of x **c.** $\mu_x = 4$, $\sigma^2 = .6$; $\mu_y = 4$, $\sigma^2 = .2$
4.47 3.93   **4.49** 1.7   **4.51** 12.83   **4.53 a.** MC, MS, MB, MO, ML, CS, CB, CO, CL, SB, SO, SL, BO, BL, OL **b.** equally likely, with $p = 1/15$ **d.** $P(0) = 6/15$, $P(1) = 8/15$, $P(2) = 1/15$ **e.** .667   **4.55** $-\$0.263$ **4.59 a.** 7 **b.** .4   **4.61 a.** 10 **b.** 15 **c.** 1 **d.** 1 **e.** 10
4.63 a. .4096 **b.** .3456 **c.** .027 **d.** .0081 **e.** .3456 **f.** .027   **4.65 a.** .00058 **b.** .1509 **c.** 58   **4.67 a.** $p = .5$ **b.** $p < .5$ **c.** $p > .5$
4.69 a. believe finding/picking up penny is good luck **c.** .33 **d.** .9134 **e.** .0000022 **f.** .6917   **4.71 b.** 40 **c.** 24 **d.** (30.2, 49.8)
4.73 a. 750 **b.** 13.69 **c.** (722.62, 777.38)   **4.75 b.** .455 **c.** .999023   **4.77 a.** .375 **a.** .5   **4.79 a.** .21279 **b.** .10 **c.** rare event if $p = .1$
4.81 a. .1 **b.** .7 **c.** .4783 **d.** .0078 **e.** yes   **4.83 b.** $\mu = 2.4$, $\sigma = 1.47$ **c.** $p = .9$, $q = .10$, $n = 24$, $\mu = 21.6$, $\sigma = 1.47$
4.87 5   **4.89 b.** $\mu = 5$, $\sigma = 2.2361$ **4.91 a.** .6767 **b.** .5940 **c.** .2707 **d.** .1353 **e.** .1353 **f.** .857   **4.93 b.** $\mu = .5$, $\sigma = .707107$
c. .9098   **4.95 a.** .368 **b.** .264 **c.** .920   **4.97 a.** .122456 **b.** .687754   **4.99 a.** .0111 **b.** .05 **c.** $\mu = 4.25$, $\sigma = 2.12$ **4.101 a.** .017352
b. 6 **c.** no; $p(0) = .002479$ **4.103 a.** .25 **b.** $(-.75, 1.25)$; .9735 **c.** very unlikely   **4.105** yes, .96   **4.111 a.** .4 **b.** .0714 **c.** 1 **d.** .3
4.113 a. .1143 **b.** .3810 **c.** .9286 **d.** .4522 **e.** .1191 **f.** 0   **4.115 a.** Hypergeometric **b.** Binomial   **4.117 a.** hypergeometric **b.**
$r = 20$, $N = 100$, and $n = 3$ **c.** .508 **d.** .391 **e.** .094 **f.** .007   **4.119 a.** .383 **b.** .0002   **4.121** .1614

4.123 .1848   **4.125** hypergeometric; $p(x) = \dfrac{\binom{r}{x}\binom{N-r}{n-x}}{\binom{N}{n}} = \dfrac{\binom{4}{x}\binom{6}{3-x}}{\binom{10}{3}}$ **4.127 a.** $\mu = 114.168$, $\sigma = 4.261977$ **b.** $z = 6.530303$

4.129 a. Poisson **b.** binomial **c.** binomial   **4.131 a.** .1344108 **b.** .668102 **c.** .331898 **d.** .2704 **e.** .71256 **f.** 17.6; 3.52; 1.8762 **g.** .9364767
4.133 a. .4096 **b.** .0259 **c.** .0625 **d.** .311 **e.** .20 **f.** .2311   **4.135 a.** .168 **b.** .1804 **c.** .251   **4.137 b.** 4.65   **4.139 b.** $n = 20$, $p = .8$ **c.** .174 **d.** .804
e. 16   **4.141** binomial   **4.143 a.** 1 **b.** .24 **c.** .38 **d.** 1.86 **e.** 1.0099 **f.** .93   **4.145 a.** .00006 **b.** .0726 **c.** .9994   **4.147 a.** .0883 **b.** .1585
4.149 .642   **4.151 a.** 2.236 **b.** no, $p = .014$ **4.153** .059   **4.155 a.** $P(\text{any outcome}) = .05$ **b.** 52.5 **c.** $(-5.16, 110.16)$ **d.** $P(0) = .525$,
$P(10) = .0025$, $P(15) = .005$, $P(20) = .0075$, $P(25) = .01$, $P(30) = .0125$, $P(35) = .015$, $P(40) = .0175$,
$P(45) = .02$, $P(50) = .0225$, $P(55) = .025$, $P(60) = .0275$, $P(65) = .03$, $P(70) = .0325$, $P(75) = .035$, $P(80) = .0375$,
$P(85) = .04$, $P(90) = .0425$, $P(95) = .045$, $P(100) = /0.475$ **e.** balanced (fair) wheel **f.** $\mu = 33.25$, $\sigma = 38.36$ **g.** .525
h. $P(0) = .2$, $P(20) = P(25) = P(30) = P(35) = P(40) = P(45) = P(50) = P(55) = P(60) = .0275$,
$P(65) = P(70) = P(75) = P(80) = P(85) = P(90) = P(95) = P(100) = .05$ **i.** .20 **j.** .65   **4.157** ≈ 292 **4.159** .137

Chapter 5
5.1 False **5.3 b.** $\mu = 30$, $\sigma = 5.774$ **c.** $(18.452, 41.548)$ **5.5 b.** $\mu = 5$, $\sigma = .577$ **c.** .577 **d.** .61 **e.** .65 **f.** 0   **5.7 a.** 0 **b.** 1 **c.** 1   **5.9 a.** 2 **b.** .25
c. .375   **5.11 a.** .6 **b.** .3   **5.13 a.** .1333, .5714 **b.** .2667, 0   **5.15 a.** .333 **b.** .133 **c.** 113.5   **5.17 a.** continuous **c.** $\mu = 8$, $\sigma = .2887$ **d.** .5
e. 0 **f.** .5 **g.** .625   **5.19** .4444   **5.21** true   **5.23** standard normal   **5.25 a.** 0 **b.** .8413 **c.** .8413 **d.** .1587 **e.** .6826 **f.** .9544 **g.** .6934
h. .6378   **5.27 a.** 2 **b.** -2 **c.** 0 **d.** 2.5 **e.** 3   **5.29 a.** 1.282. **b.** 1.1881. **c.** -2.326 **d.** .842. **e.** 2.054   **5.31 a.** .3830 **b.** .3023 **c.** .1525 **d.** .7333
e. .1314 **f.** .9545   **5.33 a.** 27.18 **b.** 48.42 **c.** 63.26   **5.35** 175.6   **5.37 a.** $-.25$ **b.** .44 **c.** .0122   **5.39 a.** .8185 **b.** .9082 **c.** 2.55
5.41 a. .4207 **b.** .000337   **5.43** .0548   **5.45 a.** .9414 **b.** .9996   **5.47 a.** no; $P(x \le 9) = .8413$ **b.** no; $P(x \le 2) = .0139$
5.49 .7019   **5.51 a.** $z_L = -.675$, $z_U = .675$ **b.** -2.68, 2.68 **c.** -4.69, 4.69 **d.** .0074, 0   **5.53 f.** $(62.26, 65.74)$; $(61.01, 66.99)$;
$(59.72, 68.28)$; $(58.90, 69.10)$; $(57.30, 70.70)$ **5.57 a.** .383. **b.** .987. **c.** 789.   **5.59** plot c   **5.61 a.** not normal (skewed right)
b. $Q_L = 37.5$, $Q_U = 152.8$, $s = 95.8$ **c.** IQR/$s = 1.204$ **5.63** no   **5.65** no; IQR/$s = .82$ **5.67 a.** histogram too peaked
b. more than 95%   **5.69** both   **5.71** both distributions approx. normal   **5.73** no   **5.75** TT approx. normal   **5.79 a.** yes
b. $\mu = 10$, $\sigma^2 = 6$ **c.** .726 **d.** .7291   **5.81 a.** .327; .411 **b.** .327; .411 **c.** .483; .498   **5.83 a.** .179 **b.** .536 **c.** .676
5.85 a. 450 **b.** 315 **c.** .704 **d.** no   **5.87 a.** 17.5 **b.** 3.62 **c.** 2.07 **d.** .0192   **5.89 a.** 750 **b.** 13.693 **c.** 4.381 **d.** 1.0   **5.91** ≈ 0 **5.93** no;
$P = .023$ **5.95** event not likely to occur; probability less than .9   **5.99 a.** .367879 **b.** .082085 **c.** .000533 **d.** .000533   **5.101 a.** .5507
b. .3679 **c.** .3297 **d.** .5488   **5.103** $\mu = 3$, $\sigma = 3$, .9502   **5.105 a.** 3012. **b.** .9093.   **5.107 a.** .283 **b.** yes   **5.109 a.** 25 **b.** 625
c. $\approx .95$ **d.** .9502 **e.** .8187   **5.111 a.** 17 **b.** .5862   **5.113 a.** $P(x > 500) = P(x > 300)P(x > 200) = .135335$
5.115 a. exponential **b.** uniform **c.** random   **5.117 a.** .9893 **b.** .0107 **c.** .9394 **d.** .3207 **e.** .9125 **f.** .0735   **5.119 a.** .5987 **b.** 0228 **c.** 533
d. .4013 **e.** .9938 **f.** 0   **5.121 a.** .3821 **b.** .5398 **c.** .0 **d.** .1395 **e.** .0045 **f.** .4602   **5.123 a.** .221 **b.** .779 **c.** 0 **d.** .777 **e.** .525   **5.125 a.** no **b.** 2.353
c. $Q_L = 3.625$, $Q_U = 7.375$, IQR/$s = 1.59$ **d.** approx. normal   **5.127** 83.15%   **5.129 a.** .9406 **b.** .9406 **c.** .1140   **5.131 a.** 3000; 2550
b. .0239 **c.** no   **5.133 a.** .8413 **b.** .7528   **5.135** IQR/$s = 1.56$; approx. normal   **5.137 a.** .0735 **b.** .3651 **c.** 7.29   **5.139** .334   **5.141** .0154;
very unlikely   **5.143 a.** 0 **b.** .366 **c.** unlikely, $P(x < 30) \approx 0$ **5.145** no   **5.147 a.** $d = 56.24$ cm **b.** .434598   **5.149 a.** (i) .384, (ii) .49,
(iii) .212, (iv) .84 **b.** (i) .3849, (ii) .4938, (iii) .2119, (iv) .8413 **c.** (i) .0009, (ii) .0038, (iii) .0001, (iv) .0013   **5.151** 58 minutes.   **5.153** 5.068
5.155 0; unlikely   **5.157 a.** .05, .20, .50, .20, .05 **b.** z-scores **c.** identical

Chapter 6
6.3 c. 1/16   **6.5 c.** .04 **d.** no
6.7 a.

M	1	1.5	2	2.5	3	3.5	4	4.5	5
$P(M)$	.04	.12	.17	.20	.20	.14	.08	.04	.01

6.13 unbiased, minimum variance   **6.15 a.** 5 **b.** $E(\bar{x}) = 5$ **c.** $E(M) = 4.778$ **d.** $\bar{x}$ **6.17 a.** Yes **b.** Median   **6.19 b.** 1.24 **c.** $E(s^2) = 1.3$
e. $E(s) = .9257$ **6.21** mean and standard deviation of sampling distribution of $\bar{x}$ **6.23** smaller   **6.27 a.** $\mu = 225$, $\sigma = 5$
b. $\mu = 225$, $\sigma = 2.143$ **c.** $\mu = 225$, $\sigma = 1$ **d.** $\mu = 225$, $\sigma = .671$ **e.** $\mu = 225$, $\sigma = .474$ **f.** $\mu = 225$, $\sigma = .3$
6.29 a. $\mu = 5.2$, $\sigma^2 = 3.36$, $\sigma = 1.833$ **c.** $\mu_{\bar{x}} = 5.2$, $\sigma_{\bar{x}} = 1.833$ **6.31 a.** $\mu_{\bar{x}} = 31$, $\sigma_{\bar{x}} = 2.5$ **b.** approx. normal **c.** .8849 **d.** .0465

e. .1314 **f.** .9452   **6.33** As n increases, variance decreases.   **6.35 a.** 353 **b.** 20 **c.** approx. normal **d.** ≈ 0 **6.37 a.** $\mu_x = 320$, $\sigma_x = 10$, approx. normal **b.** .1359 **c.** 0   **6.39 a.** 1800 **b.** 14,400 **c.** approx. normal **d.** .7888 **e.** .0018   **6.41** No, $P(\bar{x} < 103) = .0107$
6.43 a. 10; .0002 **b.** Central Limit Theorem **c.** .0170   **6.45 a.** .0034 **b.** $\mu > 6$ **6.47** handrubbing: $P(\bar{x} < 30) = .2743$; handwashing: $P(\bar{x} < 30) = .0047$; sample used handrubbing   **6.49 a.** $\mu_{\hat{p}} = .14$, $\sigma_{\hat{p}} = .0055$ **b.** $\mu_{\hat{p}} = .14$, $\sigma_{\hat{p}} = .0110$
c. $\mu_{\hat{p}} = .14$, $\sigma_{\hat{p}} = .0155$ **6.51 a.** $\mu_{\hat{p}} = .3$, $\sigma_{\hat{p}} = .0548$ **b.** approx. normal **c.** .55 **d.** .2912   **6.53** .221, .279   **6.55**
a. $\mu_{\hat{p}} = .4$, $\sigma_{\hat{p}} = .0476$ **b.** approx. normal **c.** .00003 **d.** yes   **6.57 a.** ≈ 0 **b.** .0094   **6.59 a.** .1562 **b.** 0 **c.** 0   **6.61** .1335
6.63 false   **6.65** true   **6.67** true   **6.69 a.** $E(A) = \alpha$ **b.** choose estimator with smallest variance   **6.71 a.** .5 **b.** .0767 **c.** .0021
d. .7651   **6.73 a.** .9953 **b.** .9846 **c.** .6102   **6.79 a.** 106 **b.** 2.73 **c.** approx. normal **d.** -2.20 **e.** .0136   **6.81 a.** $\mu_{\hat{p}} = .9$, $\sigma_{\hat{p}} = .019$
b. (.862, .938) **c.** .9544   **6.83 a.** .7486 **b.** .2486 **c.** .7486   **6.85** .9525   **6.87 a.** No, prob. ≈ 0 **b.** likely that $\mu > 4.59$
6.89 a. $\mu_{\bar{x}} = .53$, $\sigma_{\bar{x}} = .0273$, approx. normal **b.** .0336 **c.** after   **6.91** .8980

Chapter 7

7.5 yes   **7.7 a.** 1.645 **b.** 2.58 **c.** 1.96 **d.** 1.28   **7.9 a.** 28 ± 4.53 **b.** 102 ± 2.77 **c.** 15 ± 3.92 **d.** 4.05 ± 3.92 **e.** no   **7.11 a.** 76.7 ± 1.3
c. 76.7 ± 1.55 **d.** increases **e.** yes   **7.13 a.** .95 **b.** 95% confident that the mean HRV for all police officers diagnosed with hypertension is between 4.1 and 124.5; 95% confident that the mean HRV for all police officers who are not hypertensive is between 148 and 192.6
d. smaller   **7.15 a.** 19.3 **b.** 19.3 ± 3.44 **d.** random sample, large n **7.17** 39 ± 1.55 **7.19 a.** $.36 \pm .0099$ **c.** first-year: $.303 \pm .0305$;
landfast: $.362 \pm .0177$; multi-year: $.381 \pm .0101$ **7.21 a.** (1.92, 1.9996) **b.** no   **7.23 a.** μ **b.** no; apply Central Limit Theorem
c. (0.4786, 0.8167) **d.** yes   **7.25 a.** $1.13 \pm .67$ **b.** yes   **7.27 a.** males: 16.79 ± 2.35; females: 10.79 ± 1.67 **b.** .0975 **c.** males
7.29 Central Limit Theorem no longer applies; σ unknown   **7.31 a.** large: normal; small: t-distribution **b.** large: normal; small: unknown
7.33 a. 2.228 **b.** 2.552 **c.** -3.499 **d.** -1.761 **7.35 a.** 5 ± 1.88 **b.** 5 ± 2.39 **c.** 5 ± 3.75 **d.** (a) $5 \pm .78$, (b) $5 \pm .94$, (c) 5 ± 1.28; width
decreases   **7.37 a.** 45.63 **b.** 6.59 **c.** (40.12, 51.14) **d.** normal **e.** .95   **7.39** (2889, 4397)   **7.41 a.** μ = population mean trap spacing **b.** 89.86
c. population of trap spacings has unknown distribution **d.** 89.86 ± 10.77 **f.** random sample; population of trap spacings is normal
7.43 a. (652.76, 817.40) **d.** yes   **7.45** $.11 \pm .08$ **7.47 a.** (.243, .281) **b.** no   **7.49 a.** both untreated: 20.9 ± 1.06; male treated: 20.3 ± 1.25;
female treated: 22.9 ± 1.79; both treated: $18.6 \pm .79$ **b.** female treated   **7.51 a.** 37.3 ± 7.70 **c.** One hour before: $\mu > 25.5$
7.53 mean of sampling distribution of $\hat{p}$ is equal to p **7.55 a.** yes **b.** $.80 \pm .05$ **7.57 a.** yes **b.** no **c.** no **d.** no   **7.59 a.** .504
b. (.478, .530)   **7.61 a.** all American adults **b.** 1,000 adults surveyed **c.** p = proportion of adults who say Starbucks coffee is overpriced
d. $.73 \pm .03$ **7.63 a.** set of all gun ownership status (yes/no) values for all U.S. adults **b.** true percentage of all adults who own a gun
c. .26 **d.** $.26 \pm .02$ **7.65 a.** $.63 \pm .03$ **b.** yes   **7.67 a.** (.471, .717) **b.** likely not valid   **7.69** $.64 \pm .05$ **7.71 a.** no **b.** $.009 \pm .010$
7.73 no, both intervals contain .12   **7.75** true   **7.77** 174   **7.79 a.** 752 **b.** 334   **7.81** 46   **7.83 a.** $n = 36; W = .65; n = 64; W = .49$;
$n = 81; W = .44; n = 225; W = .261; n = 900; W = .131$ **7.85 a.** no **b.** yes **c.** 60   **7.87** 21   **7.89** 286   **7.91** 40   **7.93** 229
7.95 1,041   **7.97** 385   **7.99** chi-square   **7.101** $n - 1$ **7.103 a.** (5.38, 10.53) **b.** (.00093, .00353) **c.** (657.12, 1765.59) **d.** (1.08, 14.41)
7.105 (4.27, 65.90)   **7.107 a.** (8.6, 45.7) **b.** (2.94, 6.76) **c.** random sample from a normal population   **7.109 a.** (.1304, .1936) **b.** approx.
normal   **7.111** (.0069, .0270)   **7.113** (1.4785, 2.9533)   **7.115** (43.76, 1200.99)   **7.117 c.** 8,348   **7.119 a.** μ **b.** μ **c.** p **d.** p **e.** p **f.** σ^2
7.121 a. 19.0228, 2.70039 **b.** 32.8523, 8.90655 **c.** 78.2307, 27.2493   **7.123 a.** $.56 \pm .06$ **b.** 2,372   **7.125** (1) p; (2) μ; (3) μ; (4) μ
7.127 $.219 \pm .024$ **7.129 a.** small sample **b.** 36   **7.131 a.** (71.67, 79.13) **b.** 21   **7.133 a.** .009 **b.** 2.306 **c.** $.009 \pm .0037$ **f.** 97
7.135 a. $1.07 \pm .24$ **c.** (.145, .568) **d.** normal **e.** 129   **7.137 a.** $.660 \pm .029$ **b.** $.301 \pm .035$ **7.139 a.** $.025 \pm .122$ **b.** no evidence of
inbreeding   **7.141** 171   **7.143 a.** no **b.** 593   **7.145 a.** .094 **b.** yes. $.094 \pm .037$ **7.147** 95% confident that proportion of health care
workers with latex allergy who suspect he/she has allergy is between .327 and .541   **7.149 a.** 781   **7.151 a.** yes **b.** missing measure of
reliability **c.** 95% CI for μ: $.932 \pm .037$

Chapter 8

8.1 the null hypothesis is the status quo; the alternative hypothesis is the research hypothesis   **8.3** maximum probability of type I
error   **8.5** answers will vary   **8.7** not true   **8.9 a.** .025, **b.** .05, **c.** $\approx .005$, **d.** $\approx .10$, **e.** .10, **f.** $\approx .01$ **8.11** H_0: $p = .45$, H_a: p $\neq .45$
8.13 a. H_0: $\mu = 2400$, H_a: $\mu > 2400$ **b.** the probability of concluding that the mean gain in fees is greater than \$2,400 when, in fact, it is
equal to \$2,400 is .05 **c.** $z > 1.645$ **8.15** H_0: $p = .05$, H_a: p $< .05$ **8.17 a.** p **b.** H_0: $p = .70$ **c.** conclude the true proportion of times
the avatar can detect deceitful activity correctly is not .70, when, in fact, the true proportion is .70. **d.** conclude the true proportion of
times the avatar can detect deceitful activity correctly is .70, when, in fact, the true proportion is not .70.
8.19 a. H_0: $\mu = 150$, H_a: $\mu < 150$ **b.** conclude true average melting point is less than 150°C when, in fact, the true average is 150°C.
c. conclude true average melting point is 150°C when, in fact, the true average is less than 150°C.   **8.21 a.** H_0: No intrusion occurs **b.**
H_a: Intrusion occurs **c.** $\alpha = .001$, $\beta = .5$ **8.23** d and e   **8.25** .0401; reject H_0 **8.27** .099; fail to reject H_0 **8.29 a.** rejec H_0
b. reject H_0 **c.** reject H_0 **d.** fail to reject H_0 **8.31** Central Limit Theorem   **8.33 a.** $z = 2.2$; reject H_0 **b.** $z = 2.2$; reject H_0
8.35 a. $z > 2.65$ **b.** .004   **8.37 a.** H_0: $\mu = 75$, H_a: $\mu \neq 75$ **c.** 0 **d.** reject H_0 **f.** none   **8.39 a.** H_0: $\mu = 3$, H_a: $\mu > 3$ **b.** $z = 3.07$ **c.** .0011 **d.**
reject H_0 **e.** $\alpha > .0011$ **8.41** H_0: $\mu = 2.2$, H_a: $\mu \neq 2.2$; $z = -11.87$; p-value ≈ 0; reject H_0 **8.43** $z = 4.17$, reject H_0
8.45 a. $z = 4.03$, reject H_0 **8.47** $z = .42$, fail to reject H_0 **8.49 a.** $z = -3.72$, reject H_0 **8.51** when population variance is not
known   **8.55 a.** $|t| < 2.160$ **b.** $t > 2.500$ **c.** $t > 1.397$ **d.** $t < -2.718$ **e.** $|t| > 1.729$ **f.** $t < -2.353$ **8.57 a.** $t = -2.064$; fail to reject H_0
b. $t = -2.064$; fail to reject H_0 **c.** (a) $.05 < p$-value $< .1$ (b) $.10 < p$-value $< .20$ **8.59 a.** H_0: $\mu = 95$, H_a: $\mu \neq 95$
b. variation in sample data not taken into account **c.** $t = -1.163$ **d.** p-value $= .289$ **e.** $\alpha = .10$; probability of concluding mean trap
spacing differs from 95 when, in fact, it is equal to 95 is .10 **f.** fail to reject H_0 **g.** random sample from a normal population of trap
spacing measurements **h.** yes   **8.61** $t = -.32$; fail to reject H_0 **8.63** yes; $z = 3.23$, p-value $= .015$, reject H_0: $\mu = 15$
8.65 yes; $z = 12.91$, p-value $= .003$, reject H_0: $\mu = 300$ **8.67** $t = -1.80$, p-value $= .115$, fail to reject H_0: $\mu = .5$ **8.69** yes,
$t = -2.53$ **8.71 a.** $t = .21$, p-value $= .84$, fail to reject H_0: $\mu = 0$ **b.** average of positive and negative scores will tend to cancel
8.73 binomial   **8.75 a.** yes **b.** no **c.** no **d.** no **e.** no   **8.77 a.** $z = -1.5$ **c.** fail to reject H_0 **d.** $p = .0668$ **8.79 a.** $z = 1.13$, fail to reject H_0
b. .1292   **8.81 a.** H_0: $p = .5$, H_a: p $> .5$, **b.** .84 **c.** $z = 4.81$ **d.** ≈ 0 **e.** reject H_0 **8.83** support; p-value $= .704$

8.85 yes; $z = 1.71$ **8.87** yes; $z = 3.14$, reject H_0: $p = .70$ **8.89** $z = 4.89$, reject H_0: $p = .70$ **8.91** no, $z = 1.52$ **8.93** 10 **8.95** power $= 1 - \beta \Rightarrow \beta = .15$ **8.97 b.** 2,032.9 **d.** .2578 **e.** .7422 **8.99 a.** approx. normal, $\mu_{\bar{x}} = 50$, $\sigma_{\bar{x}} = 2.5$ **b.** approx. normal, $\mu_{\bar{x}} = 45$, $\sigma_{\bar{x}} = 2.5$ **c.** .2358 **d.** .7642 **8.101 a.** approx. normal, $\mu_{\hat{p}} = .7$, $\sigma_{\hat{p}} = .0458$ **b.** approx. normal, $\mu_{\hat{p}} = .65$, $\sigma_{\hat{p}} = .0477$ **c.** .2266 **d.** .1685 **8.103** power increases **8.105** .1814 **8.107** .7764 **8.109** normality of underlying observations **8.111** true **8.113 a.** $\chi^2 < 4.60$ or $\chi^2 > 32.80$ **b.** $\chi^2 > 41.64$ **c.** $\chi^2 > 23.68$ **d.** $\chi^2 < 5.23$ **e.** $\chi^2 > 14.68$ **f.** $\chi^2 < 10.86$ **8.115 a.** $\chi^2 = 176.64$ **8.117 a.** H_0: $\sigma^2 = 100$, H_a: $\sigma^2 < 100$ **b.** $\chi^2 < 167.361$ **d.** $\chi^2 = 154.81$, p-value $= .009$ **e.** reject H_0 **f.** population of rates of return is normally distributed **8.119 a.** σ^2, the variance of the population of rock bounces **b.** H_0: $\sigma^2 = 10$, H_a: $\sigma^2 \neq 10$ **c.** $\chi^2 = 20.12$ **d.** $\chi^2 > 21.0261$ or $\chi^2 < 5.22603$ **e.** .130 **f.** do not reject H_0 **g.** population of rock bounces is normally distributed **8.121 a.** $\chi^2 < 20.7065$ or $\chi^2 > 66.7659$ **b.** $\chi^2 = 63.72$ **c.** do not reject H_0 **8.123** $\chi^2 = 10.96$, do not reject H_0 **8.125** $\chi^2 = 19.65$ **8.127** $\chi^2 = 12.61$, fail to reject H_0 **8.129** observed significance level **8.131** H_0, H_a, α **8.133** alternative **8.135 a.** $z < -1.645$, reject H_0 **b.** $z < -1.96$, reject H_0 **c.** .39 $\pm$ .074 **d.** .39 $\pm$ .056 **e.** 549 **8.137 a.** $\chi^2 = 119.025$, reject H_0 **b.** $\chi^2 = 119.025$, reject H_0 **8.139 a.** H_0: $p = .45$ **b.** H_0: $\mu = 2.5$ **8.141 a.** $t = -1.93$, reject H_0 **b.** (65.46,80.54) **8.143 a.** H_0: $\mu = 85$, H_a: $\mu \neq 85$ **b.** conclude that the true mean Mach rating score differs from 85 when, in fact, it is equal to 85 **c.** probability of concluding that true mean Mach rating score differs from 85 when, in fact, it is equals to 85 is .10 **d.** $|z| > 1.645$ **e.** $z = 12.80$ **f.** reject H_0 **g.** no **8.145 a.** H_0: $p = .5$, H_a: $p \neq .5$ **b.** $z = 5.99$ **c.** yes **d.** reject H_0 **8.147** yes, $z = 12.36$ **8.149 a.** Type I: conclude true mean response for all N.Y. public school children is not 3 when mean equals 3; Type II: conclude true mean response equals 3 when mean is not equal to 3 **b.** $z = -85.52$, reject H_0 **c.** $z = -85.52$, reject H_0 **8.151 a.** H_0: $\mu = 16$, H_a: $\mu < 16$ **b.** $z = -4.31$, reject H_0 **8.153 a.** $z = .70$, fail to reject H_0 **b.** yes **8.155** yes $z = 1.74$, **8.157 a.** no, $z = 1.41$ **b.** small **8.159 a.** no **b.** $\beta = .3897$, power $= .6103$ **c.** increases **8.161** $z = 15.46$, reject H_0: $p = .167$ **8.163** $t = 3.725$, p-value $= .0058$, reject H_0 **8.165 b.** .37, .24 **8.167 a.** yes, $z = -2.33$ **b.** .1075 **8.169 a.** no at $\alpha = .01$, $z = 1.85$ **b.** yes at $\alpha = .01$, $z = 3.09$

Chapter 9

9.1 $\sqrt{\dfrac{\sigma_1^2}{n_1} + \dfrac{\sigma_2^2}{n_2}}$. **9.3** d **9.5** b. **9.7 a.** 16,.5 **b.** 12, .4 **c.** 4, .64 **d.** yes **9.9 a.** .5989 **b.** $t = -2.39$, reject H_0 **c.** $-1.24 \pm .98$ **d.** confidence interval **9.11 a.** fail to reject H_0 **b.** $p = .0625$, **9.13 a.** reject H_0; sufficient evidence of a difference **b.** fail to reject H_0; insufficient evidence of a difference **9.15 a.** $(\mu_1 - \mu_2)$ **b.** H_0: $(\mu_1 - \mu_2) = 0$, H_a: $(\mu_1 - \mu_2) > 0$ **c.** yes; fail to reject H_0 **9.17 a.** H_0: $(\mu_1 - \mu_2) = 0$, H_a: $(\mu_1 - \mu_2) \neq 0$ **b.** $t = .62$ **c.** $|t| > 1.684$ **d.** fail to reject H_0; supports theory **9.19 a.** $(-.60, 7.95)$ **b.** independent random samples from normal populations with equal variances **9.21 a.** $t = 11.87$, reject H_0 **b.** $t = 2.94$, reject H_0 **9.23** $t = .57$, p-value $= .573$, fail to reject H_0: $(\mu_1 - \mu_2) = 0$ **9.25** $t = 1.08$, fail to reject H_0 **9.27 a.** $\mu_1 - \mu_2$ **b.** no **c.** .2262 **d.** fail to reject H_0 **9.29 a.** no standard deviations reported **b.** $s_1 = s_2 = 5$ **c.** $s_1 = s_2 = 6$ **9.31** before **9.33** False **9.35 a.** $t > 1.833$ **b.** $t > 1.328$ **c.** $t > 2.776$ **d.** $t > 2.896$ **9.37 a.** $\bar{x}_d = 2$, $s_d^2 = 2$ **b.** $\mu_d = \mu_1 - \mu_2$ **c.** 2 ± 1.484 **d.** $t = 3.46$, reject H_0 **9.39 a.** $z = 3.5$ **b.** $p \approx .0006$ **9.41 a.** H_0: $\mu_d = 0$, H_a: $\mu_d > 0$ **b.** paired difference **c.** reject H_0 **d.** ≈ 0 **e.** yes **9.43 a.** two measurements for each patient—before and after **c.** $\bar{x}_d = -7.63$, $s_d = 5.27$ **d.** -7.63 ± 1.53 **e.** yes; evidence that $\mu_{\text{before}} < \mu_{\text{after}}$ **9.45 a.** μ_d **b.** paired difference **c.** H_0: $\mu_d = 0$, H_a: $\mu_d > 0$ **d.** $t = 2.19$ **e.** reject H_0 **9.47** yes, $t = -7.34$ **9.49** $t = 2.92$, fail to reject H_0; no evidence of a difference **9.51 a.** two scores for each twin pair **b.** 95% CI for μ_d: 1.95 $\pm$ 1.91; control group has larger mean **9.53** $t = 3.00$, reject H_0; after camera mean is smaller than before camera mean **9.55** large, independent samples **9.57 a.** binomial **b.** normal **9.59 a.** $-.07 \pm .073$ **b.** $-.06 \pm .088$ **c.** $.15 \pm .125$ **9.61** $z = 1.15$, **9.63 a.** .55 **b.** .70 **c.** $-.15 \pm .03$ **d.** 90% confidence **e.** Dutch boys **9.65 a.** (1) all Indian hospital patients admitted in January, (2) all Indian hospital patients admitted in May **b.** .083 **c.** (.033, .133) **d.** yes **9.67 b.** H_0: $(p_1 - p_2) = 0$, H_a: $(p_1 - p_2) > 0$ **c.** $z = 1.27$, fail to reject H_0 **9.69** no, $z = 1.70$ **9.71** $z = 8.34$, reject H_0 **9.73** yes; $z = 2.13$, reject H_0 at $\alpha = .05$ **9.75** .48 $\pm$ .144; proportion greater for those who slept **9.77** Theory 1: $z = 2.79$, reject H_0, support theory; Theory 2: $z = -.30$, fail to reject H_0, support theory **9.79** if no prior information, use $p_1 = p_2 = .5$ **9.81** $n_1 = n_2 = 75$ **9.83 a.** $n_1 = n_2 = 29,954$ **b.** $n_1 = n_2 = 2,165$ **c.** $n_1 = n_2 = 1,113$ **9.85 a.** $\mu_1 - \mu_2$ **b.** 1.96 **c.** 10 **d.** handshakes: 615.3; fist bumps: 36 **e.** 27 **9.87** $n_1 = n_2 = 49$ **9.89 a.** $n_1 = n_2 = 5051$ **b.** impractical to obtain such a large sample **c.** difference almost meaningless **9.91** 28 **9.93** $n_1 = n_2 = 25$ **9.95** $n_1 = n_2 = 48$

9.97 $\left(\dfrac{s_1^2}{s_2^2}\right)\left(\dfrac{1}{F_{L,\alpha/2}}\right) < \left(\dfrac{\sigma_1^2}{\sigma_2^2}\right) < \left(\dfrac{s_1^2}{s_2^2}\right)\dfrac{1}{F_{U,\alpha/2}}$ **9.99** false **9.101 a.** .025 **b.** .99 **c.** .01 **d.** .999 **9.103 a.** $F > 2.004$ **b.** $F > 2.456$

c. $F > 2.929$ **d.** $F > 3.597$ **9.105 a.** $F = 3.43$, reject H_0 **b.** p-value $< .02$, reject H_0 **9.107 a.** (1.46, 2.15) **b.** yes **c.** $F = 1.78$, reject H_0 **9.109 a.** H_0: $\sigma_1^2 = \sigma_2^2$ H_a: $\sigma_1^2 \neq \sigma_2^2$ **b.** $F = 1.16$ **c.** $F > 2.17$ **d.** fail to reject H_0 **e.** valid **9.111** $F = 1.05$, fail to reject H_0; assumption is valid **9.113** yes; $F = 7.74$, reject H_0 **9.115** $F = 1.30$, fail to reject H_0; no evidence of a difference in variation **9.117 a.** $F = 8.29$, reject H_0 **b.** no **9.119 a.** $\mu_1 - \mu_2$ **b.** $\mu_1 - \mu_2$ **c.** $p_1 - p_2$ **d.** σ_1^2 / σ_2^2 **e.** $p_1 - p_2$ **9.121 a.** $t = .78$, fail to reject H_0 **b.** 2.5 $\pm$ 9 **c.** $n_1 = n_2 = 225$ **9.123 a.** 3.9 $\pm$ .3 **b.** $z = 20.60$, reject H_0 **c.** $n_1 = n_2 = 346$ **9.125 a.** .143 **b.** .049 **c.** $z = 2.27$, fail to reject H_0 **d.** reject H_0 **9.127 a.** p-value $= .871$, fail to reject H_0: $\mu_{\text{no}} - \mu_{\text{yes}} = 0$ **9.129 a.** time needed **b.** climbers **c.** paired experiment **9.131 a.** $z = 94.35$, reject H_0 **b.** $-.139 \pm .016$ **9.133 a.** .153 **b.** .215 **c.** $-.062 \pm .070$ **9.135** yes; (7.43, 13.57) **9.137 a.** H_0: $\mu_d = 0$, H_a: $\mu_d \neq 0$ **b.** $z = 2.08$, p-value $= .0376$ **c.** reject H_0 **9.139 a.** no, $t = .18$ **b.** $F = 1.37$, fail to reject H_0 **9.141** yes, $z = 11.05$ **9.143** $z = 1.62$, fail to reject H_0 **9.145** $z = 1.56$, fail to reject H_0 **9.147 a.** yes, $F = 10$ **b.** both populations normal **9.149 a.** yes; $z = -2.79$, reject H_0 **b.** yes **9.151** use of creative ideas ($z = 8.85$); good use of job skills ($z = 4.76$)

Chapter 10

10.1 A, B, C   **10.3** designed study—values of independent variables controlled   **10.5 a.** observational **b.** designed **c.** designed **d.** observational **e.** observational **f.** observational   **10.7 a.** goat **b.** weight loss **c.** transport drug group **d.** A, B, C, and D   **10.9 a.** patient **b.** HAM-D score **c.** drug combination group **d.** 1, 2, 3, and 4   **10.11 a.** healthy adult **b.** postural index **c.** gender and strength knowledge **d.** gender: male, female; strength knowledge: yes, no **d.** male/yes, male/no, female/yes, female/no   **10.13 a.** cockatiel **b.** yes **c.** experimental group **d.** 1, 2, 3 **e.** 3 **f.** total consumption   **10.15 a.** Temperature (45, 48, 51, and 54°C); Type of yeast (baker's, brewer's) **b.** autolysis yield **c.** 8   **10.17** consists of the independent random selection of experimental units representing each treatment   **10.19** normal treatment populations, with equal variances   **10.21 a.** 9.01 **b.** 28.24 **c.** 1.65 **d.** 3.77   **10.23 a.** plot b **b.** 9; 14 **c.** 75; 75 **d.** 20; 144 **e.** 95 (78.95%); 219 (34.25%) **f.** MST = 75, MSE = 2, F = 37.5; MST = 75, MSE = 14.4, F = 5.21 **g.** reject H_0; reject H_0 **h.** both populations normal with equal variances   **10.25** plot a: df(T) = 1, df(E) = 10, df(Total) = 11, SST = 75, SSE = 20, SS(Total) = 95, MST = 75, MSE = 2, F = 37.5; plot b: df(T) = 1, df(E) = 10, df(Total) = 11, SST = 75, SSE = 144, SS(Total) = 219, MST = 75, MSE = 14.4, F = 5.21   **10.27 a.** F = 2.68; do not reject H_0 **b.** F = 4.17; reject H_0 **c.** F = 14.58; **d.** increases   **10.29 b.** not valid   **10.31 a.** completely randomized **b.** decrease in cards sold **c.** type of pay cut: unilateral, general, baseline **d.** $H_0: \mu_1 = \mu_2 = \mu_3$, H_a: At least 2 μ's differ **e.** reject H_0 **10.33 a.** completely randomized **b.** college students; tanning attitude; tanned models, models with no tan, no model **c.** $H_0: \mu_1 = \mu_2 = \mu_3$ **d.** variation not accounted for **e.** reject H_0 **f.** independent random samples; normal populations; equal variances   **10.35 a.** completely randomized **b.** treatments: 3, 6, 9, 12 robots; dep. variable.: energy expanded **c.** $H_0: \mu_3 = \mu_6 = \mu_9 = \mu_{12}$, H_a: At least 2 μ's differ **d.** reject H_0 **10.37 a.** TV viewers **b.** recall score **c.** program rating; V, S, neutral **d.** variances not taken into account **e.** F = 20.45, p-value = .000 **f.** reject H_0; mean recall scores differ among program groups   **10.39** yes; F = 2026.3   **10.41 a.** completely randomized; honey dosage, DM dosage, no dosage **b.** F = 17.51, p-value = .000, reject H_0: $\mu_{\text{Honey}} = \mu_{\text{DM}} = \mu_{\text{Control}}$ **10.43 a.** 1132.2 **b.** 4144 **c.**

Source	df	SS	MS	F
Treatment	2	1,132.2	566.10	9.97
Error	73	4,144.0	56.77	
Total	75			

d. yes, F = 9.97 **e.** valid   **10.45** probability of at least one Type I error   **10.47 a.** no significant difference **b.** μ_2 **c.** μ_1 **10.49** $\mu_1 > \mu_2, \mu_1 > \mu_3, \mu_4 > \mu_2, \mu_4 > \mu_3$ **10.51 a.** 3; unilateral/general, unilateral/baseline, general/baseline **b.** $\mu_{\text{baseline}} > \mu_{\text{general}}$ **10.53** yes, $\mu_{\text{NoTan}} < \mu_{\text{Tanned}}$ **10.55 a.** 6 **b.** sourdough; control & yeast   **10.57 a.** reject H_0: $\mu_{\text{Angry}} = \mu_{\text{Guilt}} = \mu_{\text{Neutral}}$ **b.** P(at least two means differ | none are different) = .05 **c.** $\mu_{\text{Guilt}} > \mu_{\text{Angry}}, \mu_{\text{Guilt}} > \mu_{\text{Neutral}}$ **10.59 a.** 7 different cheeses **b.** CCI **c.** reject H_0 **d.** 21 **e.** $\mu_{\text{Mozzarella}}$ is largest   **10.61** yes; μ_{DM} is largest   **10.63** yes; F = 10.29, p-value = .000, reject H_0: $\mu_A = \mu_{AR} = \mu_{AC} = \mu_P$; $\mu_P < (\mu_A, \mu_{AR}, \mu_{AC})$ **10.65** randomized block design   **10.67** all block-treatment combinations have normal populations with equal variances   **10.69 a.** df(T) = 2, df(B) = 2, df(E) = 4, df(Total) = 8, SSB = .8889, SSE = 7.7778, MST = 10.7778, MSB = .4444, MSE = 1.9444, F(T) = 5.54, F(B) = .23 **b.** $H_0: \mu_1 = \mu_2 = \mu_3$ **c.** F = 5.54 **d.** Type I error = conclude means differ when the means are equal; Type II error = conclude means are equal when the means differ **e.** do not reject H_0 **10.71 a.** df(T) = 2, df(B) = 3, df(E) = 6, df(Total) = 11, SST = 12.03, SSB = 71.75, SSE = .71, SS(Total) = 84.49, MST = 6.02, MSB = 23.92, MSE = .12, F(T) = 50.96, F(B) = 202.59 **b.** yes, p-value = .000 **c.** yes, p-value = .000 **d.** $\mu_C < \mu_A < \mu_B$ **10.73 b.** response: WTP; treatments: 2 conditions; blocks: subjects **c.** $H_0: \mu_P = \mu_S$ **10.75 a.** yes; F = 2.579, p-value = .0044 **b.** yes; F = 5.94, p-value = .0001 **c.** 105 **d.** only weeks 6 and 14 are more topsy-turvy than other weeks   **10.77 a.** 8 activities **b.** 15 adults **c.** reject H_0 **d.** $\mu_{\text{Treadmill}}$ is largest   **10.79 a.** 4 pre-slaughter phases **b.** df(T) = 3, df(B) = 7, df(E) = 21, df(Total) = 31, SST = 521, SSB = 1923, SSE = 1005, SS(Total) = 3449, MST = 173.7, MSB = 274.7, MSE = 47.94, F(T) = 3.63, F(B) = 5.74 **c.** yes; F = 3.63, p-value = .030 **d.** $\mu_{\text{Phase-1}} > \mu_{\text{Phase-2}}$ **10.81 a.** yes, F = 5.04 **b.** $\mu_{\text{High}} > \mu_{\text{Low}}$ **10.83** do not reject H_0, F = .02   **10.85** treatments are levels of a single factor; number of treatments = total number of factor–level combinations.   **10.87** true   **10.89 a.** 2 **b.** no **c.** yes; 4 and 3 **d.** 12 **e.** df(Error) = 0; replication   **10.91 a.** df(A) = 2, df(B) = 3, df(AB) = 6, df(E) = 12, df(Total) = 23, SSE = 2.4, MS(A) = .40, MS(B) = 1.77, MS(AB) = 1.60, MSE = .20, F(A) = 2.00, F(B) = 8.83, F(AB) = 8.00 **b.** $SSA, SSB, SSAB$; yes F = 7.14 **c.** yes **d.** effects of one factor on the dependent variable are not the same at different levels of the second factor **e.** F = 8.00, reject H_0 **f.** no   **10.93 a.** F(AB) = .75; F(A) = 3; F(B) = 1.5 **b.** F(AB) = 7.5; F(A) = 3; F(B) = 3 **c.** F(AB) = 3; F(A) = 12; F(B) = 3 **d.** F(AB) = 4.5; F(A) = 36; F(B) = 36   **10.95 a.** complete factorial design **b.** Age (young, old); Diet (fine limestone, coarse limestone) **c.** hen **d.** shell thickness **e.** effect of diet on thickness is the same for each age **f.** mean thickness is not different for young and old hens **g.** shell thickness is affected by diet   **10.97 a.** df(Order) = 1, df(Menu) = 1, df(Order × Menu) = 1, df(Error) = 176; **b.** sufficient evidence to say effect of Order on mean pay depends on Menu **c.** no, since interaction is significant   **10.99 a.** effect of JRP on mean intention depends on Need **b.** sufficient evidence of interaction **c.** no   **10.101 a.** df(Display) = 2, df(Lateral) = 1, df(Display × Lateral) = 2, df(Error) = 6, df(Total) = 11; **b.** 6 **c.** at α = .05, no evidence of interaction; evidence of both Dispay and Lateral main effects   **10.103** evidence of Height-Frequency interaction; F = 10.18   **10.105 a.** 2; 2; 4; 99 **b.** 9 **c.** evidence of interaction **d.** no **e.** 18 months: $\mu_{\text{control}} < \mu_{\text{photos}}$; 24 months:

$\mu_{\text{control}} < \mu_{\text{drawings}}$, $\mu_{\text{control}} < \mu_{\text{photos}}$; 30 months: $\mu_{\text{control}} < \mu_{\text{drawings}}$, $\mu_{\text{control}} < \mu_{\text{photos}}$ **10.107** interaction non-significant ($F = 1.77$, p-value $= .142$); Group main effect significant ($F = 7.59$, p-value $= .001$); Set main effect significant ($F = 31.11$, p-value $= 000$); mean for group size 3 is largest; mean for first photo set is largest **10.109 a.** Low/Ambig: 450; Low/Common: 195; High/Ambig: 152.5; High/Common: 157.5 **b.** 9,120.25 **c.** SS(Load) $= 1,122.25$, SS(Name) $= 625$, SS(Load $\times$ Name) $= 676$ **d.** Low/Ambig: 225, 5400; Low/Common: 90.25, 2166; High/Ambig: 90.25, 2166; High/Common: 100, 2400 **e.** 12,132 **f.** 14,555.25
g.

Source	df	SS	MS	F
LOAD	1	1,122.25	1,122.25	8.88
NAME	1	625	625	4.95
LOAD $\times$ NAME	1	676	676	5.35
Error	96	12,132	126.375	
Total	99	14,555.25		

h. yes **i.** significant interaction
10.115 a.

Source	df	SS	MS	F
Treatment	3	11.332	3.777	157.38
Block	4	10.688	2.672	111.33
Error	12	.288	.024	
Total	19	22.308		

b. yes, reject H_0 **c.** yes; 6 **d.** yes, reject H_0 **10.117 a.** accountant **b.** income **c.** Mach rating and Gender **d.** Mach rating (high, moderate, low); Gender (male, female) **e.** high/male, moderate/male, low/male, high/female, moderate/female, low/female **10.119 a.** 2×2 factorial; Color (blue, red), question (difficult, simple) **b.** evidence of interaction ($\alpha = .05$) **10.121 a.** exp. units: coaches; dep. variable: 7-point rating; factor: division; treatments: I, II, III **b.** H_0: $\mu_I = \mu_{II} = \mu_{III}$ **c.** reject H_0 **d.** μ_I is largest **10.123 a.** exp. units: teeth; treatments: 3 bonding times; response: breaking strength **b.** H_0: $\mu_1 = \mu_{24} = \mu_{48}$ **c.** $F > 4.68$ **d.** reject H_0 **e.** normal populations with equal variances **10.125 a.** Luckiness (lucky, unlucky, uncertain); Competition (competitive, noncompetitive)

Source	df	SS	MS	F
A	1	658.845	658.845	46.65
B	1	255.380	255.380	18.08
AB	1	2.000	2.000	0.14
Error	4	56.490	14.122	
Total	7	972.715		

b. no significant effects at $\alpha = .05$ **c.** yes; $F(AB) = .14$, do not reject H_0 **d.** yes; $F(A) = .46.45$, reject H_0; $F(B) = .18.08$, reject H_0 **e.** yes **f.** compare two A means; compare two B means **10.127 a.** randomized block design **c.** reject H_0 at $\alpha > .009$
d. $\mu_{\text{control}} < \mu_{\text{burning}}$ **10.129** Thickness: $F = 11.74$, p-value $= .000$, reject H_0: $\mu_{\text{Barn}} = \mu_{\text{Cage}} = \mu_{\text{Free}} = \mu_{\text{Organic}}$; Overrun: $F = 31.36$, p-value $= 000$, reject H_0: $\mu_{\text{Barn}} = \mu_{\text{Cage}} = \mu_{\text{Free}} = \mu_{\text{Organic}}$; Strength: $F = 1.70$, p-value $= .193$, do not reject H_0: $\mu_{\text{Barn}} = \mu_{\text{Cage}} = \mu_{\text{Free}} = \mu_{\text{Organic}}$; thickness and overrun **10.131 a.** $F = 3.96$; reject H_0 **b.** $\mu_{\text{Sad}} < \mu_{\text{Happy}}$, $\mu_{\text{Sad}} < \mu_{\text{Angry}}$
10.133 a.

Source	df	SS	MS	F	P
Diet (D)	1	0.0124	0.0124	0.22	.645
Size (S)	1	8.0679	8.0679	141.18	.000
D $\times$ S	1	0.0364	0.0364	0.64	0.432
Error	24	1.3715	0.0571		
Total	27	9.4883			

b. D $\times$ S interaction: fail to reject H_0; main effect Diet: fail to reject H_0; main effect Size: reject H_0 **10.135 b.** evidence of interaction ($\alpha = .01$) **c.** no **10.137** yes, $F = 7.25$ **10.139** no evidence of Trail-Group interaction ($F = 1.02$); evidence of Trail main effect ($F = 63.34$), $\mu_{\text{Control}} > \mu_{\text{Extract}}$; evidence of Group main effect ($F = 5.91$), $\mu_{\text{Gravid}} > (\mu_{\text{Females}}, \mu_{\text{Males}})$ **10.141** yes, $F = 34.12$; System 1 or System 4

Chapter 11

11.7 $\beta_0 = 68/13$, $\beta_1 = 11/13$, $y = 68/13 + (11/13)x$ **11.11 a.** dependent: size; independent: distance **c.** $y = \beta_0 + \beta_1 x + \varepsilon$
11.13 a. dependent: ratio; independent: diameter **c.** $y = \beta_0 + \beta_1 x + \varepsilon$ **11.15** difference between the observed and predicted
11.17 true **11.19 b.** $\hat{y} = 6.60 + .24x$ **11.21 c.** $\hat{\beta}_1 = .918$, $\hat{\beta}_0 = .020$ **e.** -1 to 7 **11.23 a.** yes; positive **d.** $\hat{y} = 320.64 + .084x$
e. 3.565 **11.25 a.** $y = \beta_0 + \beta_1 x + \varepsilon$ **b.** $\hat{y} = 19.393 - 8.036x$ **c.** y-intercept: when concentration $= 0$, predicted wicking

length $= 19.393$ mm; slope: for every 1-unit increase in concentration, wicking length decreases 8.036 mm   **11.27 a.** positive
11.29 a. $y = \beta_0 + \beta_1 x + \varepsilon$ **b.** $\hat{y} = 250.14 - .627x$ **e.** slope   **11.31 a.** $y = \beta_0 + \beta_1 x + \varepsilon$ **b.** positive **c.** $\hat{\beta}_1 = 210.8$: for every additional
resonance, frequency increases 210.8; $\hat{\beta}_0 = .020$: no practical interpretation   **11.33 a.** $\hat{y} = 86.0 - .260x$ **b.** yes **c.** positive trend for
females; positive trend for males **d.** $\hat{y} = 39.3 + .493x$; for every 1-inch increase in height for females, ideal partner's height increases
.493 inch **e.** $\hat{y} = 23.3 + .596x$; for every 1-inch increase in height for males, ideal partner's height increases .596 inch **f.** yes
11.35 yes, $\hat{y} = 5.22 - .114x$; decrease by .114 pound   **11.37** (1) mean error $- 0$, (2) error variance is constant for all x,
(3) errors are normally distributed, (4) errors are independent   **11.39 a.** 62; 2.69565 **b.** 113; 2.3542 **c.** 25; 1.087
11.41 11.18: SSE $= 34.30$, $s^2 = 6.86$, $s = 2.62$; 11.21: SSE $= 5.135$, $s^2 = 1.03$, $s = 1.01$ **11.43** $s = .57$; $\approx 95\%$ of angular sizes
fall within 1.14 pixels of their respective predicted values   **11.45 a.** SSE $= 22.268$, $s^2 = 5.567$, $s = 2.3594$ **b.** $\approx 95\%$ of wicking
length values fall within 4.72 mm of their respective predicted values   **11.47 a.** SSE $= 2760$, $s^2 = 306.6$, and s $= 17.51$
11.49 a. 5.37 **b.** 3.42 **c.** reading score   **11.51 a.** $\hat{y} = 23.3 + .596x$; $s = 2.06$ **b.** $\hat{y} = 39.3 + .493x$; $s = 2.32$ **c.** males   **11.53** 0
11.55 divide the value in half   **11.57 a.** 95%: 32 $\pm$ 1.63; 90%: 32 $\pm$ 1.32 **b.** 95%: 65 $\pm$ 6.07; 90%: 65 $\pm$ 4.96
c. 95%: $-8.4 \pm .91$; 90%: $-8.4 \pm .75$ **11.59 b.** $\hat{y} = 2.554 + .246x$ **d.** $t = .627$ **e.** fail to reject H_0 **f.** $.246 \pm 1.81$
11.61 yes, $t = 14.87$; (.0041, .0055)   **11.63 a.** negative linear trend **b.** $\hat{y} = 9,658.24 - 171.573x$; for each 1-unit increase in search
frequency, the total catch is estimated to decrease by 171.573 kg **c.** $H_0: \beta_1 = 0$, $H_a: \beta_1 < 0$ **d.** $.0402/2 = .0201$ **e.** reject H_0
11.65 a. $H_0: \beta_1 = 0$, $H_a: \beta_1 < 0$ **b.** $t = -13.23$, p-value $= 0$ **c.** reject H_0 **11.67** $-.0023 \pm .0019$; 95% confident that change in
sweetness index for each 1-unit change in pectin is between $-.0042$ and $-.0004$ **11.69 a.** $y = \beta_0 + \beta_1 x + \varepsilon$ **b.** $\hat{y} = -8.524 + 1.665x$
d. yes, $t = 7.25$ **e.** $1.67 \pm .46$ **11.71 a.** 95% confident that for each 1% increase in body mass, the percentage change in eye mass
increases between 0.25% and 0.30% **b.** 95% confident that for each 1% increase in body mass, the percentage change in orbit axis
angle decreases between 0.05% and 0.14%.   **11.73** $(-7.83, 3.71)$ **11.75 a.** $\hat{\beta}_0 = .515$, $\hat{\beta}_1 = .000021$ **b.** yes **c.** very influential
d. $\hat{\beta}_0 = .515$, $\hat{\beta}_1 = .000020$, p-value $= .332$, fail to reject H_0 **11.77** true   **11.79 a.** perfect positive linear **b.** perfect negative linear
c. no linear **d.** strong positive linear **e.** weak positive linear **f.** strong negative linear   **11.81 a.** $r = .9897$, $r^2 = .9796$
b. $r = -.9851$, $r^2 = .9704$ **c.** $r = 0$, $r^2 = 0$ **d.** $r = 0$, $r^2 = 0$ **11.83** .877   **11.85 a.** weak, negative linear relationship between the first
letter of the last name and response time **b.** reject $H_0: \rho = 0$ **c.** yes   **11.87 a.** 18% of sample variation in points scored can be
explained by the linear model **b.** $-.424$ **11.89 a.** moderate positive linear relationship; not significantly different from 0 at
$\alpha = .05$ **c.** weak negative linear relationship; not significantly different from 0 at $\alpha = .10$ **11.91 a.** $t = 17.75$, reject H_0 **b.** no
11.93 b. piano: $r^2 = .1998$; bench: $r^2 = .0032$; motorbike: $r^2 = .3832$, armchair: $r^2 = .0864$; teapot: $r^2 = .9006$ **c.** Reject H_0 for all
objects except bench and armchair   **11.95 a.** $H_0: \beta_1 = 0$, $H_a: \beta_1 < 0$ **b.** reject H_0 **c.** no; be careful not to infer a causal relationship
11.97 $r = .570$, $r^2 = .325$ **11.99 a.** $r^2 = 0.4756$; 47.56% of the sample variation in molecular variance can be explained by
altitude in the linear model **b.** $r^2 = 0.5236$; 52.36% of the sample variation in molecular variance can be explained by
population size in the linear model   **11.101** $E(y)$ represents mean of y for all experimental units with same x-value
11.103 true   **11.105 a.** $\hat{y} = -1.028 + .806x$ **b.** 1.5 **c.** 4.626 **d.** 1.542 **e.** 2.2766 ± 1.3785 **f.** 1.1482 ± 3.3917 **11.107**
c. 4.65 ± 1.12 **d.** $2.28 \pm .63$; $-.414 \pm 1.717$ **11.109 a.** Find a prediction interval for y when $x = 10$ **b.** Find a confidence
interval for $E(y)$ when $x = 10$ **11.111** (92.298, 125.104)   **11.113** run 1: 90% confident that for all runs with a pectin
value of 220, mean sweetness index will fall between 5.65 and 5.84   **11.115 a.** 95% confidence interval for $E(y)$ **b.** (48.4, 64.6)
c. 95% confident that the mean in-game heart rate of all top-level water polo players who have a maximal oxygen uptake of 150
VO$_2$max is between 48.4% and 64.6%.   **11.117 a.** (67.07, 76.41); 95% confident that ideal partner's height is between 67.07 and
76.41 in. when a female's height is 66 in. **b.** (58.39, 66.83); 95% confident that ideal partner's height is between 58.39 and 66.83 in.
when a male's height is 66 in. **c.** males; 66 in. is outside range of male heights in sample   **11.119 a.** (2.955, 4.066) **b.** (1.020, 6.000)
c. prediction interval; yes   **11.121 a.** Brand A: 3.35 $\pm$ .59; Brand B: 4.46 $\pm$ .30 **b.** Brand A: 3.35 $\pm$ 2.22; Brand B: 4.46 $\pm$ 1.12 **c.**
$-.65 \pm 3.61$ **11.123 a.** $\hat{y} = 1.411 + 349x$ **b.** $t = 1.05$, fail to reject $H_0: \beta_1 = 0$; $r^2 = 0.1219$; no   **11.125** yes;
$\hat{y} = 7.77 + .000113x$, $t = 4.04$, reject $H_0: \beta_1 = 0$ **11.127** $E(y) = \beta_0 + \beta_1 x$ **11.129** true   **11.131 b.** $\hat{y} = x$; $\hat{y} = 3$
c. $\hat{y} = x$ **d.** least squares line has the smallest SSE   **11.133 a.** $y = \beta_0 + \beta_1 x + \varepsilon$; negative **b.** yes **c.** no   **11.135 b.** .185
11.137 a. positive **b.** yes **c.** $\hat{y} = -58.86 + 554.5x$ **e.** slope: for each 01-point increase in batting average, estimated number of games
won will increase by 5.54 **f.** $t = 2.81$, reject $H_0: \beta_1 = 0$ **g.** .396; $\approx 40\%$ of sample variation in games won is explained by the linear
model **h.** yes   **11.139 a.** $y = \beta_0 + \beta_1 x + \varepsilon$ **b.** $\hat{y} = 175.70 - .8195x$ **e.** $t = -3.43$, reject H_0 **11.141 a.** $y = \beta_0 + \beta_1 x + \varepsilon$ **b.** 92% of
sample variation in metal uptake is explained by the linear model   **11.143 a.** $\hat{y} = 6.31 + 0.967x$ **b.** no practical interpretation
c. For every 1 micrometer increase in mean pore diameter, porosity is estimated to increase by 0.967% **d.** $t = 2.68$, reject H_0
e. $r = .802$, $r^2 = .643$ **f.** (8.56, 23.40)   **11.145 a.** yes **b.** $\hat{\beta}_0 = -3.05$, $\hat{\beta}_1 = .108$ **c.** $.t = 4.00$, reject H_0 **d.** $r = .756$, $r^2 = .572$ **e.** 1.09
f. yes   **11.147 a.** $\hat{\beta}_0 = -13.49$, $\hat{\beta}_1 = -.0528$ **b.** $-.0528 \pm .0178$; yes **c.** $r^2 = .854$ **d.** (.5987, 1.2653)   **11.149 a.** moderately strong
negative linear relationship between the number of online courses taken and weekly quiz grade **b.** yes, $t = -4.95$
11.151 a. $\hat{y} = 46.4x$ **b.** $\hat{y} = 478.44 + 45.15x$ **c.** no, $t = .91$ **11.153** $\hat{y} = 2.55 + 2.76x$; $t = 12.66$ reject H_0

Chapter 12
12.1 a. $E(y) = \beta_0 + \beta_1 x_1 + \beta_2 x_2$ **b.** $E(y) = \beta_0 + \beta_1 x_1 + \beta_2 x_2 + \beta_3 x_3 + \beta_4 x_4$ **c.** $E(y) = \beta_0 + \beta_1 x_1 + \beta_2 x_2 + \beta_3 x_3 + \beta_4 x_4 + \beta_5 x_5$
12.5 test the null hypothesis that all the beta parameters (except β_0) are equal to 0   **12.7 a.** $t = 1.79$, do not reject H_0 **b.** $t = 3.59$,
reject H_0 **12.9** df $= n - (k + 1)$ **12.11 a.** yes **b.** yes, $F = 48.42$ **12.13 a.** $F = 31.98$, reject H_0 **b.** yes, $t = -5.61$
12.15 a. sum of errors $= 0$; SSE is minimized **b.** For each unit increase in "betweenness centrality," lead-user rating is estimated
to increase by .42, holding all other x's constant **c.** reject H_0 **12.17 a.** $E(y) = \beta_0 + \beta_1 x_1 + \beta_2 x_2$ **b.** test $H_0: \beta_2 = 0$ vs. $H_a: \beta_2 > 0$
c. reject H_0 **12.19 a.** $\hat{y} = 39.7 + .25x_1 + .41x_2 + .84x_3 + 1.63x_4 + 1.09x_5 + .05x_6 - .04x_7 - .09x_8 - .12x_9$
b. $t = -1.00$, do not reject H_0 **c.** $t = -.09$, do not reject H_0 **d.** 1.09, .41   **12.21 a.** $E(y) = \beta_0 + \beta_1 x_1 + \beta_2 x_2 + \beta_3 x_3$ **c.** reject
$H_0: \beta_1 = \beta_2 = \beta_3 = 0$ **d.** reject H_0 **e.** fail to reject H_0 **f.** fail to reject H_0 **12.23 a.** $E(y) = \beta_0 + \beta_1 x_1 + \beta_2 x_2 + \beta_3 x_3$
b. $\hat{y} = -86,868 - 2,218.8x_1 + 1,542.2x_2 - .3496x_3$ **d.** 103.3 **e.** $R^2 = .128$, $R_a^2 = .120$ **f.** $F = 15.80$, reject H_0 **g.** yes

12.25 a. $\hat{y} = 14.0107 - 2.1865$ GENDER $- .04794$ SELFESTM $- .3223$ BODYSAT $+ .4931$ IMPREAL **c.** yes, reject H_0 (p-value ≈ 0) **d.** $R_a^2 = .485$; 48.5% of sample variation in desire level can be explained by the model (after accounting for sample size and size of the model) **e.** reject $H_0: \beta_3 = 0$ in favor of $H_0: \beta_3 < 0$ (p-value $= .013$) **f.** $(.24, .74)$

12.27 a. $E(y) = \beta_0 + \beta_1 x_1 + \beta_2 x_2 + \beta_3 x_3 + \beta_4 x_4 + \beta_5 x_5$ **b.** $\hat{y} = 13,614.5 + .089 x_1 - 9.20 x_2 + 14.39 x_3 + .35 x_4 - .85 x_5$ **d.** 458.83 **e.** .917 **f.** yes, $F = 147.30$ **12.29 a.** $\hat{y} = 1.81231 + .10875 x_1 + .00017 x_2$ **c.** $(.046, .172)$ **d.** $(.00011, .00023)$ **e.** $\hat{y} = 1.20785 + .06343 x_1 + .00056 x_2$; $(.016, .111)$, $(.00025, .00087)$ **12.33 a.** 3.76 **b.** 4.68   **12.35 a.** 95% confident that true mean desire falls between 13.42 and 14.31 for all females with a self-esteem score $= 24$, a body satisfaction score $= 3$, and a reality TV impression score $= 4$ **b.** 95% confident that true mean desire falls between 8.79 and 10.89 for all males with a self-esteem score $= 22$, a body satisfaction score $= 9$, and a reality TV impression score $= 4$ **12.37 a.** 95% confident that true heat rate falls between 11,599.6 and 13,665.5 kj/kw/hr for an engine with a speed $= 7,500$ rpm, inlet temperature $= 1,000°$C, exhaust temperature $= 525°$C, cycle pressure ratio $= 13.5$, and air mass flow rate $= 10$ kg/s **b.** 95% confident that true mean heat rate falls between 12,157.9 and 13,107.1 kj/kw/hr for all engines with a speed $= 7,500$ rpm, inlet temperature $= 1,000°$C, exhaust temperature $= 525°$C, cycle pressure ratio $= 13.5$, and air mass flow rate $= 10$ kg/s **c.** yes   **12.39 a.** $E(y) \beta_0 + \beta_1 x_1 + \beta_2 x_2 + \beta_3 x_3$ **b.** $\hat{y} = 86.902 - 0.20986 x_1 + 0.15154 x_2 + 0.07330 x_3$ **c.** no, $F = 2.66$ **d.** 18.5% of the sample variation in mean project score is explained by the model; about 95% of the observed mean project score values fall within 5.93 points of their predicted values **e.** $(82.6, 95.6)$; 95% confident that the actual mean project score is between 82.60 and 95.57 points when the intrapersonal range is 20, the stress range is 30, and the mood range is 25.   **12.41** $(24.03, 440.64)$ **12.43 a.** $E(y) = \beta_0 + \beta_1 x_1 + \beta_2 x_2 + \beta_3 x_1 x_2$ **b.** $E(y) = \beta_0 + \beta_1 x_1 + \beta_2 x_2 + \beta_3 x_3 + \beta_4 x_1 x_2 + \beta_5 x_1 x_3 + \beta_6 x_2 x_3$ **12.45 a.** .958 **b.** yes, $F = 212.9$ **c.** $F = 212.9$ **d.** yes; $t = 2.5$ **12.47 a.** $E(y) = \beta_0 + \beta_1 x_1 + \beta_2 x_2$ **b.** $E(y) = \beta_0 + \beta_1 x_1 + \beta_2 x_2 + \beta_3 x_1 x_2$ **c.** Theory 2   **12.49 a.** $E(y) = \beta_0 + \beta_1 x_1 + \beta_2 x_2 + \beta_3 x_1 x_2$ **b.** $\beta_1 + 10\beta_3$ **c.** $\beta_1 + 25\beta_3$ **12.51 a.** 99.4% of sample variation in amplitude (y) can be explained by the model **b.** the relationship between amplitude (y) and cross position (x_2) depends on probe position (x_1) **c.** slope of line for $x_1 = 3.5$: $-.165$; slope of line for $x_1 = 6.5$: $-.255$;   **12.53 a.** $E(y) = \beta_0 + \beta_1 x_1 + \beta_2 x_2 + \beta_3 x_1 x_2$ **b.** $(\beta_1 + 2.5\beta_3)$ **c.** $(\beta_1 + 5\beta_3)$ **d.** $(\beta_2 + 100\beta_3)$ **e.** $H_0: \beta_3 = 0$ **12.55 a.** affect of client credibility on likelihood depends on the level of linguistic delivery style **b.** $H_0: \beta_1 = \beta_2 = \beta_3 = 0$ **c.** $F = 55.35$, reject H_0 **d.** $H_0: \beta_3 = 0$ **e.** $t = 4.01$, reject H_0 **f.** .114 **g.** .978   **12.57 a.** $E(y) = \beta_0 + \beta_1 x_1 + \beta_2 x_2 + \beta_3 x_3 + \beta_4 x_4 + \beta_5 x_5 + \beta_6 x_2 x_5 + \beta_7 x_3 x_5$ **b.** $\hat{y} = 13,646 + .046 x_1 - 12.68 x_2 + 23.00 x_3 - 3.02 x_4 + 1.29 x_5 + .016 x_2 x_5 - .04 x_3 x_5$ **c.** $t = 4.40$, reject H_0 **d.** $t = -3.77$, reject H_0 **12.59 a.** $E(y) = \beta_0 + \beta_1 x_1 + \beta_2 x_2 + \beta_3 x_1 x_2$ **b.** $\hat{y} = -63,238 + 18.77 x_1 + 445,486 x_2 - 139.83 x_1 x_2$ **c.** $F = 110.44$, reject H_0; $R^2 = 0.951$; $2s = 48,720.6$ **d.** $t = -2.47$, reject H_0 **e.** -51.15 **12.61 a.** $E(y) = \beta_0 + \beta_1 x + \beta_2 x^2$ **b.** $E(y) = \beta_0 + \beta_1 x_1 + \beta_2 x_2 + \beta_3 x_1 x_2 + \beta_4 (x_1)^2 + \beta_5 (x_2)^2$ **c.** $E(y) = \beta_0 + \beta_1 x_1 + \beta_2 x_2 + \beta_3 x_3 + \beta_4 x_1 x_2 + \beta_5 x_1 x_3 + \beta_6 x_2 x_3 + \beta_7 (x_1)^2 + \beta_8 (x_2)^2 + \beta_9 (x_3)^2$ **12.63 a.** $t = 2.889$, reject H_0 **b.** $t = 2.889$, $t > 2.074$, reject H_0 **12.65 b.** moves graph to right or left **c.** controls whether graph opens upward or downward.   **12.67 a.** correlation limited to measuring strength of linear relationship **b.** $E(y) = \beta_0 + \beta_1 x + \beta_2 x^2$ **d.** negative **e.** reject H_0, theory supported   **12.69 b.** 22.6% of sample variation in points scored (y) can be explained by the model **c.** no **d.** $H_0: \beta_2 = 0$ **12.71 b.** 1st-order model, $\beta_1 > 0$; 1st-order model, $\beta_1 < 0$; 2nd-order model   **12.73 a.** $E(y) = \beta_0 + \beta_1 x_1 + \beta_2 x_2 + \beta_3 x_1 x_2 + \beta_4 (x_1)^2 + \beta_5 (x_2)^2$ **b.** β_4 and β_5 **12.75 b.** 6.25 **c.** 10.25 **d.** 200   **12.77 a.** $E(y) = \beta_0 + \beta_1 x_1 + \beta_2 x_2 + \beta_3 x_1 x_2 + \beta_4 (x_1)^2 + \beta_5 (x_2)^2$ **b.** 40.2% of sample variation in satisfaction level (y) can be explained by the model **c.** $F = 6.99$, reject $H_0: \beta_1 = \beta_2 = \beta_3 = \beta_4 = \beta_5 = 0$ **d.** $t = -.60$, do not reject $H_0: \beta_4 = 0$ **e.** $t = -1.83$, reject $H_0: \beta_5 = 0$ **12.79 a.** $\hat{y} = -288 + 1.395 x + .0000351 x^2$, $t = .36$, fail to reject H_0 **b.** outlier **c.** yes; $R_{adj}^2 = .996$, model statistically useful (p-value $= .000$), evidence of curvature (p-value $= .000$)   **12.81** $E(y) = \beta_0 + \beta_1 x$; $x = 1$ if level 2, 0 if level 1   **12.83 a.** 16.3, 9.3, 31.3, 19.3 **b.** $H_0: \beta_1 = \beta_2 = \beta_3 = 0$ **12.85 a.** $x = \{1$ if multiple first birth, 0 if single first birth$\}$ **b.** $E(y) = \beta_0 + \beta_1 x$ **c.** $\beta_0 = $ mean number of children for family with a single first birth, $\beta_1 = $ difference between mean number of children for family with a multiple first birth and a single first birth   **12.87 a.** β_0 **b.** $\mu_{SetNet} - \mu_{GillNet}$ **c.** test $H_0: \beta_1 = \beta_2 = 0$ **12.89 a.** Race: $x_1 = \{1$ if black, 0 if white$\}$; Availability: $x_2 = \{1$ if high, 0 if low$\}$; Position: $x_3 = \{1$ if QB, 0 if not$\}$, $x_4 = \{1$ if RB, 0 if not$\}$, $x_5 = \{1$ if WR, 0 if not$\}$, $x_6 = \{1$ if TE, 0 if not$\}$, $x_7 = \{1$ if DL, 0 if not$\}$, $x_8 = \{1$ if LB, 0 if not$\}$, $x_9 = \{1$ if DB, 0 if not$\}$ **b.** $E(y) = \beta_0 + \beta_1 x_1$ **c.** $E(y) = \beta_0 + \beta_1 x_2$ **d.** $E(y) = \beta_0 + \beta_1 x_3 + \beta_2 x_4 + \beta_3 x_5 + \beta_4 x_6 + \beta_5 x_7 + \beta_6 x_8 + \beta_7 x_9$ **12.91 a.** 4; AA, AB, BA, BB **b.** $E(y) = \beta_0 + \beta_1 x_1 + \beta_2 x_2 + \beta_3 x_3$, where $x_1 = \{1$ if AA, 0 if not$\}$, $x_2 = \{1$ if AB, 0 if not$\}$, $x_3 = \{1$ if BA, 0 if not$\}$ **d.** $H_0: \beta_1 = \beta_2 = \beta_3 = 0$ **12.93 a.** $E(y) = \beta_0 + \beta_1 x_1 + \beta_2 x_2$, where $x_1 = \{1$ if major depression only, 0 if not$\}$, $x_2 = \{1$ if personality disorder only, 0 if not$\}$ **b.** $\beta_1 = \beta_2 = 0$ **c.** test $H_0: \beta_1 = 0$ vs. $H_a: \beta_1 < 0$ **12.95** $E(y) = \beta_0 + \beta_1 x_1 + \beta_2 x_2 + \beta_3 x_1 x_2$ **12.97 a.** $E(y) = \beta_0 + \beta_1 x$, where $x = \{1$ if flightless, 0 otherwise$\}$ **b.** $E(y) = \beta_0 + \beta_1 x_1 + \beta_2 x_2 + \beta_3 x_3$, where $x_1 = \{1$ if vertebrates, 0 if not$\}$, $x_2 = \{1$ if vegetables, 0 if not$\}$, $x_3 = \{1$ if invertebrates, 0 if not$\}$ **c.** $E(y) = \beta_0 + \beta_1 x_1 + \beta_2 x_2 + \beta_3 x_3$, where $x_1 = \{1$ if cavity within ground, 0 if not$\}$, $x_2 = \{1$ if trees, 0 if not$\}$, $x_3 = \{1$ if cavity above ground, 0 if not$\}$ **d.** $\hat{y} = 641 + 30,647 x$ **e.** $F = 33.05$, reject H_0 **f.** $\hat{y} = 903 + 2,997 x_1 + 26,206 x_2 - 660 x_3$ **g.** $F = 8.43$, reject H_0 **h.** $\hat{y} = 73.732 - 9.132 x_1 - 45.01 x_2 - 39.51 x_3$ **i.** $F = 8.07$, reject H_0 **12.99 a.** $\beta_0 + \beta_2$ **b.** $\beta_0 + \beta_1 + \beta_2 + \beta_3$ **c.** $\beta_1 + \beta_3$ **d.** β_0; $\beta_0 + \beta_1$; β_1 **f.** evidence of interaction **g.** change in crime rate after murder in Jasper: $\hat{\beta}_2 + \hat{\beta}_3 = -169 + 255 = 86$; change in crime rate after murder in Center: $\hat{\beta}_2 = -169$ **12.101 a.** $E(y) = \beta_0 + \beta_1 x_1 + \beta_2 (x_1)^2$ **b.** $E(y) = \beta_0 + \beta_1 x_1 + \beta_2 (x_1)^2 + \beta_3 x_2 + \beta_4 x_3$ **c.** $E(y) = \beta_0 + \beta_1 x_1 + \beta_2 (x_1)^2 + \beta_3 x_2 + \beta_4 x_3 + \beta_5 x_1 x_2 + \beta_6 (x_1)^2 x_2 + \beta_7 x_1 x_3 + \beta_8 (x_1)^2 x_3$ **d.** $\beta_5 = \beta_6 = \beta_7 = \beta_8 = 0$ **e.** $\beta_2 = \beta_5 = \beta_6 = \beta_7 = \beta_8 = 0$ **f.** $\beta_3 = \beta_4 = \beta_5 = \beta_6 = \beta_7 = \beta_8 = 0$ **12.103 a.** $E(y) = \beta_0 + \beta_1 x_1$; $E(y) = (\beta_0 + \beta_2) + \beta_1 x_1$; $E(y) = (\beta_0 + \beta_3) + \beta_1 x_1$ **b.** $\hat{y} = 44.8 + 2.2 x_1$; $\hat{y} = 54.2 + 2.2 x_1$; $\hat{y} = 60.4 + 2.2 x_1$ **12.105 a.** $\hat{y} = 11.779 - 1.972 x_1 + .585 x_4 - .553 x_1 x_4$ **b.** 9.97 **c.** $F = 45.09$ (p-value $< .0001$), reject H_0 **d.** 43.9% of sample variation in level of desire (y) can be explained by the model **e.** $\approx 95\%$ of sampled desire levels fall within 4.70 points of their respective predicted values **f.** $t = -2.00$ (p-value $= .0467$), reject H_0 **g.** .585 **h.** .032   **12.107 a.** $E(y) = \beta_0 + \beta_1 x_1 + \beta_2 x_2 + \beta_3 x_3$, where $x_1 = \{1$ if blonde Caucasian, 0 if not$\}$, $x_2 = \{1$ if brunette Caucasian, 0 if not$\}$, $x_3 = $ beauty rating **b.** for all three types: β_3 **c.** $E(y) = \beta_0 + \beta_1 x_1 + \beta_2 x_2 + \beta_3 x_3 + \beta_4 x_1 x_3 + \beta_5 x_2 x_3$ **d.** minority: β_3; blonde: $\beta_3 + \beta_4$; brunette: $\beta_3 + \beta_5$

12.109 a. $E(y) = \beta_0 + \beta_1 x_1 + \beta_2 x_2 + \beta_3 x_3$, where $x_1 = $ water depth, $x_2 = \{1$ if set net, 0 if not$\}$, $x_3 = \{1$ if pots, 0 if not$\}$
c. $E(y) = \beta_0 + \beta_1 x_1 + \beta_2 x_2 + \beta_3 x_3 + \beta_4 x_1 x_2 + \beta_5 x_1 x_3$ **e.** $\beta_1 + \beta_4$ **f.** $\beta_1 + \beta_5$ **g.** β_1 **h.** test $H_0: \beta_4 = \beta_5 = 0$
12.111 a. $E(y) = \beta_0 + \beta_1 x_1 + \beta_2(x_1)^2$ **b.** β_0: y-intercept; β_1: shift; β_2: rate of curvature **c.** $(\beta_0 + \beta_3) + (\beta_1 + \beta_4)x_1 + (\beta_2 + \beta_5)(x_1)^2$
d. $\beta_0 + \beta_3$: y-intercept; $\beta_1 + \beta_4$: shift; $\beta_2 + \beta_5$: rate of curvature **e.** yes **12.113 a.** $x_2 = \{1$ if low, 0 if not$\}$,
$x_3 = \{1$ if neutral, 0 if not$\}$, $\{$base level $=$ high$\}$ **b.** $E(y) = \beta_0 + \beta_1 x_1 + \beta_2 x_2 + \beta_3 x_3$ **c.** $E(y) = \beta_0 + \beta_1 x_1 + \beta_2 x_2 + \beta_3 x_3 + \beta_4 x_1 x_2$
$+ \beta_5 x_1 x_3$ **d.** part c **12.115 a.** $E(y) = \beta_0 + \beta_1 x_1 + \beta_2 x_2 + \beta_3 x_3$ **b.** $E(y) = \beta_0 + \beta_1 x_1 + \beta_2 x_2 + \beta_3 x_3 + \beta_4 x_1 x_3 + \beta_5 x_2 x_3$
12.117 (a and b); (a and d); (a and e); (b and c); (b and d); (b and e); (c and e); (d and e) **12.119** model with a small number of β parameters
12.121 a. 5; 3 **b.** $H_0: \beta_3 = \beta_4 = 0$ **c.** $F = 1.06$, do not reject H_0 **12.123 a.** $E(y) = \beta_0 + \beta_1 x_1 + \beta_2 x_2 + \beta_3 x_3 + \beta_4 x_4 + \beta_5 x_5 + \beta_6 x_6$
b. $H_0: \beta_5 = \beta_6 = 0$; $E(y) = \beta_0 + \beta_1 x_1 + \beta_2 x_2 + \beta_3 x_3 + \beta_4 x_4$ **c.** likely to reject H_0 **d.** reject H_0 **e.** not likely to reject H_0
f. fail to reject H_0 **12.125 a.** complete: 12.103c; reduced: 12.103b **b.** $H_0: \beta_4 = \beta_5 = 0$ **c.** model 12.103c **e.** model 12.103b
12.127 b. $H_0: \beta_3 = \beta_4 = \beta_6 = \beta_7 = \beta_8 = 0$ **c.** yes **d.** reject H_0; complete model better **e.** $E(y) = \beta_0 + \beta_1 x_1 + \beta_2 x_2 + \beta_3 x_3$
$+ \beta_4 x_4 + \beta_5 x_5 + \beta_6 x_6 + \beta_7 x_7 + \beta_8 x_8 + \beta_9 x_5 x_6 + \beta_{10} x_5 x_7 + \beta_{11} x_5 x_8 + \beta_{12} x_6 x_7 + \beta_{13} x_6 x_8 + \beta_{14} x_7 x_8$ **f.** do not reject H_0: no evidence
of interaction **12.129 a.** $H_0: \beta_1 = \beta_2 = \beta_3 = \beta_4 = \beta_5 = 0$ **b.** $H_0: \beta_2 = \beta_5 = 0$ **c.** $H_0: \beta_5 = 0$ **12.131 a.** $H_0: \beta_2 = \beta_3 = \beta_4 = \beta_5 = 0$
b. $E(y) = \beta_0 + \beta_1 x_1$ **c.** mean lengths of entangled whales for 3 gear types differ **d.** $H_0: \beta_4 = \beta_5 = 0$ **e.** $E(y) = \beta_0 + \beta_1 x_1 + \beta_2 x_2 + \beta_3 x_3$
f. rate of change of whale length with water depth is same for all 3 gear types **12.133 a.** $E(y) = \beta_0 + \beta_1 x_1 + \beta_2 x_2 + \beta_3 x_3 +$
$\beta_4 x_1 x_2 + \beta_5 x_1 x_3 + \beta_6 x_2 x_3$, where $x_1 = $ speed, $x_2 = $ rate, and $x_3 = $ weight **b.** $\beta_3 + 1000\beta_5 + 50\beta_6$ **c.** $\beta_3 + 1000\beta_5 + 150\beta_6$
d. $\hat{y} = 343 + 0.0181 x_1 + 1.99 x_2 + 9.37 x_3 - 0.000013 x_1 x_2 - 0.000375 x_1 x_3 + 0.0125 x_2 x_3$ **e.** $F = .77$, fail to reject H_0 **12.137 a.** x_2
b. yes **c.** fit all models of the form $E(y) = \beta_0 + \beta_1 x_2 + \beta_2 x_j$ **12.139** No; find best predictors using stepwise, then begin model
building **12.141 a.** $E(y) = \beta_0 + \beta_1 x_1 + \beta_2 x_2$, where $x_1 = $ pressure and $x_2 = $ leg length **b.** 77.1% of sample variation in heel depth
(y) can be explained by the model **c.** p-value $< .001$, reject H_0 **d.** 15 **e.** very high **12.143 a.** 11 **b.** 10 **c.** 1 **d.** $E(y) = \beta_0 + \beta_1 x_{11} +$
$\beta_2 x_4 + \beta_3 x_2 + \beta_4 x_7 + \beta_5 x_{10} + \beta_6 x_1 + \beta_7 x_9 + \beta_8 x_3$ **12.145 a.** 11 **b.** 10 **c.** model is statistically useful (p-value $= .001$) **d.** large
number of t-tests performed; no higher order terms (e.g., interactions) in model **e.** $E(y) = \beta_0 + \beta_1 x_1 + \beta_2 x_2 +$
$\beta_3 x_1 x_2 + \beta_4(x_1)^2 + \beta_5(x_2)^2$ **f.** test $H_0: \beta_3 = \beta_4 = \beta_5 = 0$ **12.147** residual that lies more than 3 standard deviations from 0
12.149 false **12.151** (1) Global F test signficant, but all t-tests insignificant; (2) β estimates with opposite signs from expected;
high pairwise correlations among x's **12.153** yes, x_1 is highly correlated with both x_2 and x_3 **12.155** no **12.157** yes; drop 1
or more of the moderately/highly correlated variables **12.159 a.** no **b.** yes **12.161** multicollinearity **12.163 a.** reasonably satisfied
b. likely violated **c.** 8 outliers **d.** likely violated **e.** no multicollinearity **12.165** no **12.167** no **12.169** $E(y) = \beta_0 + \beta_1 x_1 +$
$\beta_2 x_2 + \beta_3 x_3$; $x_1 = \{1$ if level 2, 0 if not$\}$; $x_2 = \{1$ if level 3, 0 if not$\}$; $x_3 = \{1$ if level 4, 0 if not$\}$ **12.171 a.** $E(y) = \beta_0 + \beta_1 x_1 +$
$\beta_2 x_2 + \beta_3 x_3$, where $x_1 = $ quantitative, $x_2 = \{1$ if level 2, 0 if not$\}$, $x_3 = \{1$ if level 3, 0 if not$\}$ **b.** $E(y) = \beta_0 + \beta_1 x_1 + \beta_2 x_1^2 + \beta_3 x_2 +$
$\beta_4 x_3 + \beta_5 x_1 x_2 + \beta_6 x_1 x_3 + \beta_7 x_1^2 x_2 + \beta_8 x_1^2 x_3$ **12.173 a.** yes, $F = 28.33$ **b.** $t = -2.03$, reject H_0 **c.** $t = .17$, do not reject H_0 **d.** $t = 2.17$,
reject H_0 **12.175** yes; $x_1 = 80$, $x_2 = .9$, $x_3 = 1,850$ are outside the range of sample data. **12.177** df(Error) $= 0$ **12.179 a.** 32% of
sample variation in ln (CO_2 emissions) can be explained by the model **b.** $F = 3.72$, reject H_0 **c.** no **d.** $H_0: \beta_1 = 0$ **e.** $t = 2.52$, reject H_0
f. x_4 is highly correlated with x_5 **12.181 a.** $E(y) = \beta_0 + \beta_1 x_1 + \beta_2 x_2 + \beta_3 x_3 + \beta_4 x_4 + \beta_5 x_5$ **b.** model is statistically useful
c. $E(y) = \beta_0 + \beta_1 x_1 + \beta_2 x_2 + \beta_3 x_3 + \beta_4 x_4 + \beta_5 x_5 + \beta_6 x_6 + \beta_7 x_7$ **12.183 a.** $E(y) = \beta_0 + \beta_1 x_1 + \beta_2 x_2 + \beta_3 x_1 x_2$ **c.** $\beta_3 < 0$
12.185 a. 76% of sample variation in SAT-Math scores is explained by the model **b.** (13.12, 24.88); 95% confident that difference in
mean SAT-Math scores between coached and non-coached students is between 13.12 and 24.88 points, holding PSAT scores constant
c. effective **d.** $E(y) = \beta_0 + \beta_1 x_1 + \beta_2 x_2 + \beta_3 x_3 + \beta_4 x_4 + \beta_5 x_5 + \beta_6 x_6 + \beta_7 x_7 + \beta_8 x_8 + \beta_9 x_9 + \beta_{10} x_{10}$ **e.** H_0:
$\beta_3 = \beta_4 = \cdots = \beta_{10} = 0$ **f.** at least one of the additional variables is important **h.** (8.12, 19.88) **i.** yes **j.** $E(y) = \beta_0 + \beta_1 x_1 + \beta_2 x_2 +$
$\beta_3 x_3 + \beta_4 x_4 + \beta_5 x_5 + \beta_6 x_6 + \beta_7 x_7 + \beta_8 x_8 + \beta_9 x_9 + \beta_{10} x_{10} + \beta_{11} x_1 x_2 + \beta_{12} x_2 x_3 + \beta_{13} x_2 x_4 + \beta_{14} x_2 x_5 + \beta_{15} x_2 x_6 + \beta_{16} x_2 x_7 +$
$\beta_{17} x_2 x_8 + \beta_{18} x_2 x_9 + \beta_{19} x_2 x_{10}$ **k.** $H_0: \beta_{11} = \beta_{12} = \cdots = \beta_{19} = 0$; nested model F-test **12.187 a.** $E(y) = \beta_0 + \beta_1 x_1 + \beta_2 x_2$
b. for every 1 degree increase in temperature, size is estimated to decrease .002 millimeters, holding date fixed **c.** $t = -.72$, fail to
reject H_0 **12.189 a.** $\hat{y} = 12.180 - .0265 x_1 - .4578 x_2$ **c.** $t = -.50$, do not reject H_0 **d.** $-.4578 \pm .3469$ **e.** $R^2 = .529$, $R_a^2 = .505$; R_{adj}^2
f. yes, $F = 21.88$ **g.** $E(y) = \beta_0 + \beta_1 x_1 + \beta_2 x_2 + \beta_3 x_1 x_2$, where $x_2 = \{1$ if plant, 0 if duck chow$\}$ **h.** $\hat{y} = 8.14 - .016 x_1 - 10.4 x_2 +$
$.095 x_1 x_2$ **i.** .079 **j.** $-.016$ **k.** $t = .67$, fail to reject H_0 **12.191 a.** no **b.** no **c.** yes **d.** possibly **12.193 a.** negative **b.** no, $F = 1.60$
c. no, $F = 1.61$ **12.195 a.** $E(y) = \beta_0 + \beta_1 x$ **b.** $E(y) = \beta_0 + \beta_1 x + \beta_2(x)^2$ **c.** 2nd-order **d.** $t = 3.55$, yes **e.** 2nd-order
12.197 a. $E(y) = \beta_0 + \beta_1 x_1 + \beta_2 x_6 + \beta_3 x_7$, where $x_6 = \{1$ if good, 0 if not$\}$, $x_7 = \{1$ if fair, 0 if not$\}$ **c.** excellent:
$\hat{y} = 188,875 + 15,617 x_1$; good: $\hat{y} = 85,829 + 15,617 x_1$; fair: $\hat{y} = 36,388 + 15,617 x_1$ **e.** yes, $F = 5.82$ **f.** x_1, x_3 and x_5 are highly correlated **g.** assumptions are satisfied **12.199 a.** 7 **b.** 6 **c.** 5 **12.201 a.** $E(y) = \beta_0 + \beta_1 x_1 + \beta_2 x_2 + \beta_3 x_3 + \beta_4 x_4 + \beta_5 x_5 + \beta_6 x_1 x_2$
b. $H_0: \beta_4 = 0$ **c.** reject H_0 **d.** yes **12.203 a.** number of trees-QN (x_1); transect location-QL: $x_2 = \{1$ if SPF, 0 if not$\}$ and
$x_3 = \{1$ if SAF, 0 if not$\}$; land use-QL: $x_4 = \{1$ if pasture, 0 if arable$\}$ **b.** $E(y) = \beta_0 + \beta_1 x_1$ **c.** $E(y) = \beta_0 + \beta_1 x_1 + \beta_2 x_2 + \beta_3 x_3$
d. $E(y) = \beta_0 + \beta_1 x_1 + \beta_2 x_2 + \beta_3 x_3 + \beta_4 x_4$; β_1 **e.** $E(y) = \beta_0 + \beta_1 x_1 + \beta_2 x_2 + \beta_3 x_3 + \beta_4 x_4 + \beta_5 x_2 x_4 + \beta_6 x_3 x_4$; no
f. $E(y) = \beta_0 + \beta_1 x_1 + \beta_2 x_2 + \beta_3 x_3 + \beta_4 x_4 + \beta_5 x_2 x_4 + \beta_6 x_3 x_4 + \beta_7 x_1 x_2 + \beta_8 x_1 x_3 + \beta_9 x_1 x_4 + \beta_{10} x_1 x_2 x_4 + \beta_{11} x_1 x_3 x_4$;
LAF/Arable: β_1; LAF/Pasture: $\beta_1 + \beta_9$; SPF/Arable: $\beta_1 + \beta_7$; SPF/Pasture: $\beta_1 + \beta_7 + \beta_{10}$; SAF/Arable: $\beta_1 + \beta_8$; SAF/Pasture:
$\beta_1 + \beta_8 + \beta_{11}$ **12.205** model includes $x_1 = $ DOT estimate, either $x_2 = $ low bid ratio or $x_3 = \{1$ if fixed, 0 if competitive$\}$,
and $x_5 = $ estimated days to complete

Chapter 13

13.1 (1) n identical trials, (2) k possible outcomes to each trial, (3) probabilities of outcomes sum to 1, (4) probabilities remain the
same from trial to trial, (5) trials are independent **13.3 a.** 12.4613 **b.** .102587 **c.** 35.1725 **d.** 24.7356 **13.5 a.** $\chi^2 > 23.2093$
b. $\chi^2 > 20.0902$ **c.** $\chi^2 > 11.0705$ **13.7 a.** no, $\chi^2 = 4.079$ **13.9 a.** jaw habits; bruxism, clenching, bruxism and clenching, neither
c. $H_0: p_1 = p_2 = p_3 = p_4 = .25$, H_a: At least one p_i differs from .25 **d.** 15 **e.** $\chi^2 = 25.73$ **f.** $\chi^2 > 7.81473$ **g.** reject H_0 **h.** (.37, .63)

13.11 b. H_0: $p_1 = .5, p_2 = .3, p_3 = .1, p_4 = .1$ **c.** 53; 31.8; 10.6; 10.6 **d.** $\chi^2 = 8.73$ **e.** reject H_0 **13.13 a.** $P(GG) = P(BG) =$ $P(GB) = P(BB) = 1/4$ **b.** 10,722 **c.** 187.04 **d.** reject H_0 **e.** $E_{GG} = 10,205.2$, $E_{BG} = 10,715.6$, $E_{GB} = 10,715.6$, $E_{BB} = 11,251.7$; $\chi^2 = 64.95$; reject H_0 **13.15** $\chi^2 = 3.16$, do not reject H_0; no evidence to refute percentages associated with effective program **13.17** $\chi^2 = 4.84$, do not reject H_0 **13.19 a.** yes, $\chi^2 = 24.12$ **b.** yes, $\chi^2 = 153.3$ **13.21 a.** yes, $\chi^2 = 190.9$ **b.** $.197 \pm .042$ **13.23** column totals for one qualitative variable are known in advance   **13.25** expected cell counts are all at least 5 **13.27 a.** H_0: Row & Column are independent, H_a: Row & Column are dependent **b.** $\chi^2 = 9.201034$
c.

	Column 1	Column 2	Column 3
Row 1	15.43	44	40.57
Row 2	11.57	33	30.43

d. $\chi^2 = 12.04$, reject H_0 **13.29** $\chi^2 = 11.92$, reject H_0 **13.31 a.** .078 **b.** .157 **c.** possibly **d.** H_0: Presence of Dog & Gender are independent **e.** $\chi^2 = 2.25$, fail to reject H_0 **f.** $\chi^2 = .102$, fail to reject H_0 **13.33 a.** .653 **b.** .555 **c.** .367 **d.** yes **e.** H_0: Authenticity of news story & Tone are independent **f.** $\chi^2 = 10.427$ (p-value $= .005$), reject H_0 **13.35 a.** H_0: Package design/sound pitch & Consumer opinion are independent **b.** $E_{11} = 21$, $E_{12} = 21$, $E_{21} = 19$, $E_{22} = 19$ **c.** $\chi^2 = 39.3$ **d.** reject H_0 **13.37** $\chi^2 = 11.85$, fail to reject H_0 **13.39** yes; $\chi^2 = 3.29$, fail to reject H_0 **13.41** yes, $\chi^2 = 7.38$, p − value $= .025$; no   **13.43** yes, $\chi^2 = 64.24$ **13.45 a.** $\chi^2 = 4.407$, reject H_0 **b.** no **c.** .0438 **d.** .0057; .0003 **e.** p-value $= .0498$, reject H_0 **13.47** false   **13.49 a.** yes, $\chi^2 = 89.89$ **b.** no **c.** yes
d.

	Col. 1	Col. 2	Col. 3	Totals
Row 1	.500	.273	.125	.280
Row 2	.167	.182	.750	.360
Row 3	.333	.555	.125	.360

13.51 a. Accord model and Collision status
b.

	Injury Status		
Automobile Type	Injuries	No Injuries	Total
Accord	5,364	44,768	50,132
Accord Hybrid	137	1,368	1,505
Total	5,501	46,136	51,637

c. H_0: Accord model & Collision status are independent **d.** $E_{11} = 5340.7$, $E_{12} = 44,791.3$, $E_{21} = 160.3$, $E_{22} = 1344.7$ **e.** $\chi^2 = 3.91$ **f.** $\chi^2 > 3.84146$ **g.** reject H_0 **h.** (.001, .031)
13.53 a.

LLD			
Genetic Trait	Yes	No	Total
Yes	21	15	36
No	150	306	456
Total	171	321	492

b. $\chi^2 = 9.52$, reject H_0 **13.55** $\chi^2 = 35.41$, reject H_0 **13.57** yes, $\chi^2 = 127.5$ **13.59 a.** yes; no measure of reliability
b.

Tourist			
Photos, postcards, paintings	Male	Female	Totals
Always	240	476	716
Often	405	527	932
Occasionally	525	493	1018
Rarely or Never	330	204	534
Totals	1500	1700	3200

c. H_0: Gender & Purchase choice are independent **d.** $\chi^2 = 112.43$, p-value $= 0$ **d.** reject H_0 **13.61 a.** $\chi^2 = 17.16$, reject H_0 **b.** $.376 \pm .086$ **c.** yes   **13.63 a.** yes, $\chi^2 = 8.40$ **b.** reject H_0; expected numbers do not need to exceed 5 to apply test **13.65** $\chi^2 = 23.46$, reject H_0 **13.67 a.** $\chi^2 = 164.90$, reject H_0 **b.** Gender: $\chi^2 = 6.80$, reject H_0; Education: $\chi^2 = 6.59$, fail to reject H_0; Work: $\chi^2 = 11.69$, fail to reject H_0; Satisfaction: $\chi^2 = 30.42$, reject H_0 **13.69 a.** no **b.**

	CAHS		
SHSS:C	Low	Medium	High
Low	32	14	2
Medium	11	14	6
High	6	16	29

c. yes **d.** $\chi^2 = 46.70$, reject H_0 **13.71 a.** $\chi^2 = 12.942$ **b.** 11.0705 **c.** yes **d.** $.025 < p\text{-value} < .01$ **13.73** $\chi^2 = 2.28$; insufficient evidence to reject the null hypothesis of independence

Chapter 14 (available online)

14.1 population not normal   **14.3 a.** .179 **b.** .009 **c.** .010 **d.** .006 **e.** .0329   **14.5** $S = 16$, $p\text{-value} = .115$, do not reject H_0
14.7 a. $H_0: \eta = 15$, $H_a: \eta \neq 15$ **b.** data not normal **c.** $S = 14$ **d.** .0309; reject H_0 **14.9 a.** $H_0: \eta = 2.8$, $H_a: \eta > 2.8$ **b.** $S = 21$, $z = 2.01$
c. .222 **d.** do not reject H_0 **14.11** yes, $S = 180$, $p\text{-value} = 0$ **14.13 a.** invalid inference **b.** sign test of $H_0: \eta = 95$ vs. $H_a: \eta \neq 95$
c. $S = 5$ **d.** .454 **e.** do not reject H_0 **14.15** no; $S = 6$, $p\text{-value} = .145$ **14.17** $S = 8$, $p\text{-value} = .820$, do not reject H_0
14.19 true   **14.21 a.** $T_1 \leq 41$, $T_1 \geq 71$ **b.** $T_1 \geq 50$ **c.** $T_1 \leq 43$ **d.** $|z| > 1.96$ **14.23** yes, $z = -2.47$ **14.25 a.** $T_1 = 62.5$, reject H_0
b. yes, $T_1 = 62.5$ **14.27 a.** $H_0: D_{\text{Text}} = D_{\text{NoText}}$, $H_a: D_{\text{Text}} > D_{\text{NoText}}$ **b.** reject H_0; texting group   **14.29 a.** rank sum test
b. $H_0: D_{\text{Low}} = D_{\text{High}}$, $H_a: D_{\text{Low}} > D_{\text{High}}$ **c.** $T_2 \leq 41$ **d.** do not reject H_0 **14.31 a.** $H_0: D_{\text{Bulimic}} = D_{\text{Normal}}$, $H_a: D_{\text{Bulimic}} > D_{\text{Normal}}$
c. 174.5 **d.** 150.5 **e.** $z > 1.28$ **f.** $z = 1.72$, reject H_0 **14.33** Private vs. Public: $p\text{-value} = .1109$, no significant difference;
Unconsolidated vs. Bedrock: $p\text{-value} = .0337$, significant difference at $\alpha = 05$ **14.35** yes, $z = 2.86$ **14.37 a.** $p\text{-value} = .0368$,
reject H_0 **b.** $p\text{-value} = .0005$, reject H_0 **14.39 a.** data not normal **b.** $z = -6.47$, reject H_0 **c.** $z = -.39$, do not reject H_0
d. $z = -1.35$, do not reject H_0 **14.41** data must be ranked   **14.43 a.** $H_a: D_A > D_B$ **b.** $T_- = 3.5$, reject H_0 **14.45 a.** $H_a: D_A > D_B$
b. $z = 3.0$, reject H_0 **c.** .0013   **14.47 a.** before and after measurements not independent **b.** scores not normal **c.** reject H_0, ichthyotherapy
is effective   **14.49 a.** $z = -4.638$ **b.** yes, $p\text{-value} \approx 0$ **14.51 a.** $H_0: D_{\text{good}} = D_{\text{average}}$, $H_a: D_{\text{good}} \neq D_{\text{average}}$ **d.** $T = 3$ **e.** $T \leq 4$
f. reject H_0 **14.53** $T_+ = 3.5$, reject H_0; program was effective   **14.55** $p - \text{value} = .003$, reject H_0; inventory method.
14.57 yes, $T_- = 50.5$ **14.59** samples size for each distribution is more than 5   **14.61 a.** completely randomized design
b. H_0: Three probability distributions are identical **c.** $H > 9.21034$ **d.** $H = 14.53$, reject H_0 **14.63 a.** group variances unequal
b. Kruskal-Wallis; H_0: Three probability distributions are identical **c.** $R_{\text{ground}} = 108.5$, $R_{\text{office}} = 97$, $R_{\text{air}} = 25.5$ **d.** $H = 12.78$
e. $H > 4.60517$ **f.** reject H_0 **14.65 a.** $H_0: D_{1\text{dog}} = D_{2\text{dogs}} = D_{3\text{dogs}}$ **b.** $H > 5.99147$ **c.** do not reject H_0 **14.67 b.** 84 **c.** 145
d. 177 **e.** $H = 18.4$ **f.** reject H_0 **g.** $z = -3.36$, reject H_0 **14.69 a.** $H = 5.67$, do not reject H_0 **b.** $z = -2.36$, reject H_0
14.71 $H = 19.03$, reject H_0 **14.73 a.** normality assumption violated **b.** $H = 13.66$, reject H_0 **14.75** ranking b   **14.77 a.** 8 **b.** H_0:
The probability distributions for the four treatments are identical **c.** $F_r = 12.1$, reject H_0 **d.** $p\text{-value} < .010$ **14.79** $F_r = 13$,
reject H_0 **14.81 a.** $R_1 = 25.5$, $R_2 = 11.0$, $R_3 = 18.5$, $R_4 = 25.0$ **b.** 10.39, yes **c.** .016 **d.** reject H_0 **14.83 a.** a possibly nonnormal data
b. $H = 14.37$, $p\text{-value} = .002$ **c.** reject H_0 **14.85 a.** $F_r > 9.21034$ **b.** reject H_0 **c.** do not reject H_0 **d.** reject H_0 **14.87 a.** invalid
b. H_0: The probability distributions for the 15 weeks are identical **c.** $F_r > 29.1413$ **d.** $p - \text{value} > .01$ **14.89** $F_r = 11.10$,
reject H_0 at $\alpha = .05$ **14.91** $-1; 1$ **14.93 a.** .005 **b.** .025 **c.** .975 **d.** .02   **14.95 a.** .4 **b.** $-.9$ **c.** $-.2$ **d.** .2   **14.97 c.** $r_s = .95$ **d.** reject H_0
14.99 a. $r_s = .934$ **c.** yes, **d.** $p\text{-value} = .00035$ **14.101 c.** .713 **d.** $|r_s| > .425$ **e.** reject H_0 **14.103** Private: do not reject
H_0, $p\text{-value} = .103$; Public: reject H_0, $p\text{-value} = .002$ **14.105** yes   **14.107** $r_s = .341$ reject H_0, $p - \text{value} = .0145$
14.109 a. invalid inference **b.** .507 **d.** yes   **14.111 a.** rank sum test **b.** sign test **c.** Kruskal-Wallis test **d.** Spearman's test **e.** signed rank test
f. Friedman's test   **14.113 a.** no, $r_s = .40$ **b.** yes, $T_- = 1.5$ **14.115** yes, $F_r = 14.9$ **14.117** $S = 8$, $p\text{-value} = .0391$, reject H_0
14.119 $T = 27.5$, do not reject H_0 **14.121 a.** $r_s = -.733$ **b.** do not reject H_0 **b.** $z = 6.07$, $p\text{-value} = 0$ **c.** reject H_0
14.123 a. $H_0: D_{\text{Driver}} = D_{\text{Passenger}}$ **b.** $T_- = 23$ **c.** $T_- \leq 19$ **d.** do not reject H_0 **14.125 a.** zinc measurements non-normal
b. $T_1 = 18$, do not reject H_0 **c.** $T_2 = 32$, do not reject H_0 **d.** $H = 2.26$, do not reject H_0 **e.** no significant difference
14.127 yes, $F_r = 6.35$ **14.129** $H = 16.26$, reject H_0 **14.131 a.** .714 **b.** reject H_0 **14.133 a.** data not normal **b.** Kruskal-Wallis
c. $H = 11.20$ **d.** reject H_0 **14.135** no, $S = 10$, $p\text{-value} = .3145$ **14.137** .592; .189

Index

Additive rule of probability, 165–166
Adjusted multiple coefficient of
  determination, 694
Alternative hypothesis
  explanation of, 398, 402
  formulation of, 403–406
  steps to select, 403
Analysis of variance. *See* ANOVA
  (analysis of variance)
ANOVA (analysis of variance)
  checking assumptions for, 546–547
  explanation of, 547–548
  factorial, 589
  on graphing calculator, 611
  on MINITAB, 549, 610–611
  randomized block design table, 572
  steps to conduct, 547–548
ANOVA *F*-tests
  to compare treatment means,
    543, 567
  conditions required for, 544, 600
  example of, 544–545
  explanation of, 543–544, 694, 695
  for global usefulness, 695
Anscombe, Francis J., 762
Arithmetic means. *See* Means

Balanced design, 539, 585
Bar graphs, 60, 61
Base level, 727–730
Bayes, Thomas, 198
Bayesian statistical methods,
  198–199
Bayes's rule, 198–199
Bernoulli, Jacob, 230
Bias
  nonresponse, 42, 43, 46
  sampling, 44–46
  selection, 46
Biased estimators, 320
Biased sample, 45, 46
Binomial distribution
  application of, 234–235
  approximating with normal
    distribution, 290–293
  characteristics of, 231
  derivation of, 231–233
  explanation of, 233
Binomial experiments, 229–233
Binomial probabilities, 291–293
Binomial random variables
  applications of, 234–239, 493
  assessment of, 230–231
  characteristics of, 229, 290–291
  explanation of, 229
  formula for, 233
  mean, variance, and standard
    deviation for, 235
Binomial tables
  explanation of, 235–236
  use of, 236–239, 290

Bivariate relationships
  explanation of, 121, 645
  methods to graph, 121–123
Blocking, 484
Blocks, 564–567
Bonferroni, Carlo E., 557
Bonferroni multiple comparisons
  procedure, 557, 559, 571
Bound, 36
Box, 113
Box plots
  comparison of, 115–116
  computer-generated, 114–115
  elements of, 113
  explanation of, 112
  interpretation of, 114
  use of, 112–113, 117

Categorical data. *See also* Qualitative
    data
  graphing and summarizing, 61–63
  methods to describe, 59–61
  multinomial experiment
    and, 801
Categorical probabilities
  chi-square test and, 803, 825–826
  one-way table and, 802–806
  two-way table and, 810–817
Categories, 801
Cell counts, 801, 810, 811
Cells, 801, 811
Census, 33
Central limit theorem
  application of, 327–328
  explanation of, 325, 413
  to find probability, 326–327
  sample size and, 494
  sampling distribution of sample
    mean and, 345
Central tendency. *See also* Measures of
    central tendency
  explanation of, 82, 224–225
Chebyshev, P. L., 99
Chebyshev's rule
  for discrete random variable, 226
  explanation of, 99
  use of, 99
Chi-square distribution, 379–381, 445
Chi-square tests, 825–826
  explanation of, 803
  on graphing calculator, 836
  on MINITAB, 835–836
Class, 59
Class frequency, 59
Class intervals, 72–74
Class percentage, 60
Class relative frequency, 59
Classes, 801
Cluster sampling, 41, 230
Clusters, 41
Coded variable, 684

Coefficient of correlation. *See*
    Correlation coefficient
Coefficient of determination. *See*
    Determination, coefficient of
Combinations rule, 155–156, 191, 193
Combinatorial mathematics
  to count number of sample points,
    232–233
  explanation of, 155, 156
Comparisonwise error rate (CER), 556
Complementary events, 163–165
Complements, rule of
  explanation of, 163–164
  use of, 292
Complete model, 744
Complete second-order model, 720
Completely randomized design
  ANOVA assumptions and, 546–547
  ANOVA *F*-test and, 544–547
  explanation of, 539, 600
  objective of, 540
  variability and, 540–543
Compound events, 160
Conditional probabilities
  example of, 230
  explanation of, 172
  formula for, 172–175
  in two-way table, 173–175
Confidence coefficients, 347–348, 350
Confidence intervals
  to compare two population means,
    469–470
  for differences between treatment
    means, 556, 557
  estimation of slope β_1 and, 636–640
  explanation of, 37, 344
  general form of, 495
  on graphing calculator, 394–395
  guide to forming, 386
  large-sample, 345–347, 365–369,
    463, 465
  on MINITAB, 392–393, 657
  paired difference, 485
  for population mean: normal (z)
    statistic, 345–350
  for population mean: student's
    t-statistic, 355–360
  for population proportion,
    365–369, 495
  for population variance, 379–382
  selection of, 516
  small-sample, 358–360, 469–473
Confidence level, 347
Constant error variance, 764
Contingency table, 810. *See also*
    Two-way table
Continuity, correction for, 290
Continuous probability
    distributions
  explanation of, 261
  method to find, 262

878

Credits

Chapter 1

p. 32 Source: Human Factors, Vol. 56, No. 31, May 2014; **p. 32** Source: Academy of Management Journal, Oct. 2007; **p. 49** James T. McClave, Terry Sincich; **pp. 53–56** Screenshots from Minitab Corporation. Courtesy of Minitab Corporation.

Chapter 2

p. 59 Based on Li. E.C, Williams. S.F and Volpe. R.D, "The Effects of Topic and Listener Familiarity of Discourse Variables in Procedural and Narrative Discource Tasks", The Journal of Communication Disorders, Vol: 28, No: 01, Mar. 1995, Page: 44 (Table 1). **p. 60** Summary Table from SPSS. Reprint Courtesy of International Business Machines Corporation, © International Business Machines Corporation; **p. 67** Based on Journal of Moral Education, March 2010; **p. 67** Source: International Rhino Foundation, 2014; **pp. 67–68** Based on Sangkuk Lee, Journal of Family History (Jan 2010); **p. 68** Screenshot from SAS. © 1999 SAS In stitute Inc. Reprinted with Permission; **p. 68** Source: Union of Concerned Scientists, Nov 2012; **p. 68** Source: Wu. J, et al. Track Allocation Optimization at a Railway Station: Mean-Variation Model and Case Study, Journal of Transportation Engineering, Vol: 39, No: 5, May 2013 (Extracted from Table 4); **p. 69** Source: Braga. A.A, Hureau. D.M & Papachristos. A.V, Deterring Gang-Involved Gun Violence: Measuring the Impact of Boston's Operation Ceasefire on Street Gang Behaviour, Journal of Quantitative Criminology, Vol: 30, No: 01, March 2014 (adapted from Figure 3); **p. 69** Source: McMunn-Dooley. C & Welch. M.M, Nature of Interactions Among Young Children and Adult Caregivers in a Children's Museum, Early Childhood Education Journal, Vol: 42, No: 02, March 2014 (adapted from Figure 2); **p. 69** Source: Roskes. M. et al., The Right Side? Under Time Pressure, Approach Motivation Leads to Right Oriented Bias, Psychological Science, Vol: 22, No: 11, Nov 2011 (adapted from Figure 2); **p. 70** Based on Ayotte. J.D, Argue. D.M and McGarry. F.J, Methyl Tert-Butyl Ether Occurrence and Related Factors in Public and Private Wells in Southeast New Hampshire, Environmental Science and Technology, Vol: 39, No: 1, Jan 2005, Page: 9–16; **p. 78** Based on Boyer. D.M, Evans. A.R and Jernvall. J, Evidence of Dietary Differentiation Among Late Paleocene-Early Eocene Plesiadapids (Mammalia, Primates), American Journal of Physical Anthropology, Vol: 142, © 2010 (Table A3); **p. 78** Source: Harell. K, The Fallen and Their Swords: A New Explanation for the Rise of the Shaft Graves, American Journal of Archaeology, Vol: 118, No: 1, January 2014 (Figure 1); **p. 78** Source: Patson. T, Teaching Stage Fright? Implications for Music Educators, British Journal of Music Education, Vol: 31, No: 1, March 2014 (Adapted from Figure 1); **p. 79** Based on Paul. I.M, et al. Effect of Honey, Dextromethorphan, and no Treatment on Nocturnal Cough and Sleep Quality for Coughing and their Parents, Archives of Pediatrics and Adolescent Medicine, Vol: 161, No: 12, Dec. 2007 (Data Simulated); **p. 79** Based on National Center for Environmental Health, Centers for Disease Control and Prevention, Aug 5, 2013; **pp. 79–80** Based on Thery. M, et al. Specific Color Sensitivities of Prey and Predator Explain Camouflage in Different Visual Systems, Behavioral Ecology, Vol: 16, No: 1, Jan 2005 (Table 1); **p. 80** Based on Russell. D.A, Basketballs as Spherical Acoustic Cavities, American Journal of Physics, Vol: 78, No: 6, June 2010 (Table 1); **p. 80** Based on Hayes. T. and Lewis. D.A, Anatomical Specialization of the Anterior Motor Speech Area: Hemisphere Differences in Magnopyramidal Neurons, Brain and Language, Vol: 49, No: 3, June 1995, Page: 292 (Table 1); **p. 80** Based on Randles. R.H, On Neutral Responses (Zeros) in the Sign Test and Ties in the Wilcoxon-Mann-Whitney Test, The American Statistician, Vol: 55, No: 2, May 2001 (Figure 3); **pp. 80–81** Based on College Entrance Examination Board, 2014; **p. 89** Based on Paronuzzi. P, Rockfall-Induced Block Propagation on a Soil Slope, Northern Italy, Environmental Geology, Vol: 58, 2009 (Table 2); **p. 89** Source: Harell. K, The Fallen and Their Swords: A New Explanation for the Rise of the Shaft Graves, American Journal of Archaeology, Vol: 118, No: 1, January 2014 (Figure 1); **p. 89** Source: Patson. T, Teaching Stage Fright? Implications for Music Educators, British Journal of Music Education, Vol: 31, No: 1, March 2014 (Adapted from Figure 1); **p. 90** Based on Paul. I.M, et al. Effect of Honey, Dextromethorphan, and no Treatment on Nocturnal Cough and Sleep Quality for Coughing and their Parents, Archives of Pediatrics and Adolescent Medicine, Vol: 161, No: 12, Dec. 2007 (Data Simulated); **p. 90** Based on Boyer. D.M, Evans. A.R and Jernvall. J, Evidence of Dietary Differentiation Among Late Paleocene-Early Eocene Plesiadapids (Mammalia, Primates), American Journal of Physical Anthropology, Vol: 142, © 2010 (Table A3); **pp. 90–91** Based on Wiseman. F, et al. A New Method for Ranking Total Driving Perfrmance on the PGA Tour, Sports Journal, Vol: 10, No: 1, Winter 2007 (Table 2); **p. 91** Based on Pfeiffer. M. et al. Community Organization and Species Richness of Ants in Mongolia Along an Ecological Gradient from Steppe to Gobi Desert, Journal of Biography, Vol: 30, No: 12, Dec 2003 (Table 1 and 2); **p. 92** Based on Statistical Abstract of the United States, 2012 (Table 942), U.S Energy Information Administration, Electric Power Annual; **p. 97** Source: Harell. K, The Fallen and Their Swords: A New Explanation for the Rise of the Shaft Graves, American Journal of Archaeology, Vol: 118, No: 1, January 2014 (Figure 1); **p. 97** Source: Patson. T, Teaching Stage Fright? Implications for Music Educators, British Journal of Music Education, Vol: 31, No: 1, March 2014 (Adapted from Figure 1); **pp. 97–98** Based on Paronuzzi. P, Rockfall-Induced Block Propagation on a Soil Slope, Northern Italy, Environmental Geology, Vol: 58, 2009 (Table 2); **p. 98** Based on Paul. I.M, et al. Effect of Honey, Dextromethorphan, and no Treatment on Nocturnal Cough and Sleep Quality for Coughing and their Parents, Archives of Pediatrics and Adolescent Medicine, Vol: 161, No: 12, Dec. 2007 (Data Simulated); **p. 98** Based on Boyer. D.M, Evans. A.R and Jernvall. J, Evidence of Dietary Differentiation Among Late Paleocene-Early Eocene Plesiadapids (Mammalia, Primates), American Journal of Physical Anthropology, Vol: 142, 2010 (Table A3); **pp. 104–105** Source: Chew. S, et al. Do Social Robots Walk or Roll?, International Conference on Social Robotics, Vol: 6414, 2010 (adapted from Figure 2); **pp. 106–107** Based on Cole. K, et al. Animal Assisted Therapy Decreases Hemodynamics, Plasma Epinephrine and State Anxiety in Hospitalized Heart Failure Patients, American Journal of Critical Care, 16: 575–585; **p. 120** Based on Grassberger.

M and Hoch. W, Ichthyotherapy as Alternative Treatment for Patients with Psoriasis: A Pilot Study, Evidence-Based Research in Complementary and Alternative Medicine, Vol: 3, No: 4, Dec 2004 (Figure 3), National Center for Complementary and Alternative Medicine; **p. 120** Based on Paronuzzi. P, Rockfall-Induced Block Propagation on a Soil Slope, Northern Italy, Environmental Geology, Vol: 58, 2009 (Table 2); **p. 122** Based on Bayonet Point Hospital, Coronary Care Unit; **p. 124** Based on Schaffer. J and Heiny. E.L, The Effects of Elevation on Slugging Percentage in Major League Baseball, Chance, Vol: 19, No: 1, Winter 2006 (Figure 2); **p. 124** Source: Shester. G.G, Explaining Catch Variation Among Baja California Lobster Fishers Therough Spatial Analysis of Trap-Placement Decisions, Bulletin of Marine Science, Vol: 86, No: 2, April 2010 (Table 1); **p. 124** Based on Shand. J, et al. Variability in the Location of the Retinal Ganglion Cell Area Centralis is Correlated with Ontogenic Changes in Feeding Behaviour in the Blackbream, Acanthopagrus Butcher, Brain and Behaviour, Vol: 55, No: 4, April 2000 (Figure 10H); **p. 124** Based on Khan. M.S, Paper Diagnostic for Instant Blood Typing, Analytical Chemistry, Vol: 82, No: 10, May 2010 (Figure 4b); **pp. 124–125** Based on Russell. D.A, Basketballs as Spherical Acoustic Cavities, American Journal of Physics, Vol: 78, No: 6, June 2010 (Table 1); **p. 125** Based on Russell. D.A, Basketballs as Spherical Acoustic Cavities, American Journal of Physics, Vol: 78, No: 6, June 2010 (Table 1); **p. 125** Source: Rusterholz. H, Aydin. D and Baur. B, Population Structure and Genetic Diversity of Relict Populations of Alyssum Montanum on Limestone Cliffs in the Northern Swiss Jura Mountains, Alpine Botany, Vol: 122, No: 2, November 2012, (Tables 1 and 2); **p. 125** Based on Wade. T.G, et al. Distribution and Causes of Global Forest Fragmentation, Conservation Ecology, Vol: 72, No: 2, Dec 2003 (Table 6); **pp. 125–126** Based on Barry. J, Estimating Rates of Spreading and Evaporation of Volatile Liquids, Chemical Engineering Progress, Vol: 101, No: 1, Jan 2005; **p. 126** Based on The Dock and Harbour Authority; **p. 134** Table 2: Books Categorized by Level from Extensive Reading in Japanese by Claire Ikumi Hitosugi and Richard R. Day from Reading In a Foreign Language, Vol: 16, No: 01. Copyright © 2004. Used by permission of National Foreign Language Resource Center; **p. 134** Based on Berg. I and Bliedon. S, The Pots of Phylakopi: Applying Statistical Techniques to Archaeology, Chance, Vol: 13, No: 04, Fall 2000; **p. 134** Table 4: Number of Books Read in the Ten-Week Period from Extensive Reading in Japanese by Claire Ikumi Hitosugi and Richard R. Day from Reading In a Foreign Language, Vol: 16, No: 01. Copyright © 2004. Used by permission of National Foreign Language Resource Center; **pp. 135–136** Based on American Journal of Science, Vol: 305, No: 1, Jan 2005, Page: 16 (Table 2); **p. 136** Based on Hood. J.C, Teaching Against the Text: The Case of Qualitative Methods, Teaching Sociology, Vol: 34, Issue: 3, Page: 207 (Exhibit 2); **p. 136** Based on Lichen Radionuclide Baseline Research Project, 2003, Page: 25, Orion, University of Alaska-Fairbanks; **p. 137** High Dispersion Clusters of Galaxies from Observations of High Dispersion Clusters of Galaxies: Observations of High Dispersion Clusters of Galaxies: Constraints on Cold Dark Matter, The Astronomical Journal, Vol: 110, No: 1, Page: 37 by John M Hill, William R Oegerle, M.J. Fitchett from The Astronomical Journal. Copyright © 1995. IN ALL CASES for works published under any of the Creative Commons licenses the words "© AAS. Reproduced with permission"; **p. 137** Based on Ross. R.J.S, From Antisweatshop to Global Justice to Antiwar: How the New Left is the Same and Different from the Old New Left, Journal of World Systems Research, Vol: X, No: 1, Winter 2004 (Table 1 and 3). **pp. 137–138** Based on McCowan. B and Reiss. D, Quantitative Comparison of Whistle Repertoires from Captive Adult Bottlenose Dolphins (Delphiniae Tursiops Truncates): A Re-Evaluation of the Signature Whistle Hypothesis, Ethology, Vol: 100, No: 3, July 1995, Page: 200 (Table 2); **p. 138** Based on Arriaga. R.L, et al. Scores on the MacArthur Communicative Development Inventory of Children from Low-Income and Middle Income Families, Applied Psycholinguistics, Vol: 19, No: 2, June 1998, Page: 217 (Table 7); **p. 138** Based on Yang. W.H, A Freckle Criterion for the Solidification of Super alloys with a Tilted Solidification Front, JOM: Journal of the Minerals, Metals and Materials Society, Vol: 56, No: 9, Sept 2032; **p. 139** Based on McDonald.T.L, Erickson. W.P and McDonald. L.L, Analysis of Count Data from Before-After Control-Impact Studies, Journal of Agricultural, Biological Statistics, Vol: 5, No: 3, Sept 2028, Page: 277–8 (Table A.1); **p. 139** Based on Hill. T.P, The First Digit Phenomenon, American Scientist, Vol: 86, No: 4, July-Aug 1998, Page: 363 (Figure 5); **p. 139** Based on Oegerle. W.R, Hill. J.M and Fitchett. M.J, Observations of High Dispersion Clusters of Galaxies: Constraints on Cold Dark Matter, The Astronomical Journal, Vol: 110, No: 1, July 1995 (Table 1 and Figure 1); **pp. 142–14** Screenshots from Minitab Corporation. Courtesy of Minitab Corporation; **pp. 152–153** Screenshots from Texas Instruments. Courtesy of Texas Instruments.

Chapter 3

p. 159 Based on Hamilton. R. W. "Why do People Suggest What They do not Want? Using Context Effects to Influence Others' Choices. Journal of Consumer Research, Vol: 29, March 2003, Table 1; **p. 160** Source: Parade Magazine, Feb. 27, 1994; **p. 169** Based on Caffo. B. S, et al. "An Overview of Observational Sleep Research With Application to Sleep Stage Transitioning", Chance, Vol: 22, No: 01, Winter 2009 (Table 2); **pp. 169,184** Based on Caffo. B. S, et al. "An Overview of Observational Sleep Research With Application to Sleep Stage Transitioning", Chance, Vol: 22, No: 01, Winter 2009 (Table 2); **p. 170** Based on Henshaw. S. K, and Finer. L. B. "The Accessibility of Abortion Services in the United States, 2001", Perspectives on Sexual and Reproductive Health, Vol: 35, No: 01, Jan/Feb 2003 (Table 4); **pp. 170, 183** Based on Gangemi. A and Mancini. F. "Guilt and Focussing in Decision-Making", Journal of Behavioral Decision Making, Vol: 20, Jan 2007 (Table 2); **pp. 171, 185** Based on Bartos. L. et al. "Estimation of the Probability of Fighting in Fallow Deer (Dama Dama) During the Rut", Aggressive Behavior, Vol: 33, Jan/Feb. 2007, pp. 35–13; **p. 200** Source: Tangen, J.M, et al. "Identifying Fingerprint Expertise", Psychological Science, Vol: 22, No: 08, August 2011 (Figure 1); **p. 204** Based on "Cyberchondriac" Poll, The Harris Poll, July 29, 2008 (Table 3); **p. 205** Based on Chen. J. R, et al. "Emergency Response of Toxic Chemicals in Taiwan: The System and Case Studies", Process Safety Progress, Vol: 23, No: 03, Sept. 2004 (Figure 5a); **p. 205** Based on Identification and Characterization of Erosional Hotspots, William & Mary Virginia Institute of Marine Science, U.S Army Corps of Engineers Project Report, Mar. 18, 2002; **p. 206** Based on Berg. K, Hines. M and Allen. S, "Wheelchair Users at Home: Few Home Modifications and Many Injurious Falls", American Journal of Public Health, Vol: 92, No: 01, Jan. 2002 (Table 1); **p. 206** Based on Shapiro. J.A, Jacobs. E.J and Thun. M.J, "Cigar Smoking in Men and Risk of Death From Tobacco-Related Cancers", Journal of the National Cancer Institute, Vol: 92, No: 04, Feb. 16, 2000 (Table 2); **p. 208** Based on Hood. J. C, "Teaching Against the Text: The Case of Qualitative Methods." Teaching Sociology, Vol: 34, Issue: 03 (Exhibit 2). **p. 209** Based on Rodgers. J.L and Doughty. D, "Does Having Boys or Girls Run in the Family?", Chance, Vol: 14, No: 04, Fall 2001 (Table 3). **p. 210** Based on CNN.com. **p. 211** Screenshots from Texas Instruments. Courtesy of Texas Instruments.

Chapter 4

p. 223 Based on Hamilton, R.W "Why do people suggest what they do not want? Using context effects to influence others' choices." Journal of consumer Research, Vol.29, Mar 2003(Table 1); **p. 228** Source: Sosa, A. .J., et al. "Life history of Megamelus scutellaris with description of immature stages," Annals of the Entomological Society of America, Vol. 98, No. 1, Jan. 2005 (adapted from Table 1); **pp. 228–229** Based on Identification and Characterization of Erosional Hotspots, William& Mary Virginia Institute of Marine Science, U.S.Army Corps of Engineers Project Report, March 18,2002; **p. 212** Vladimir Wrangel/Shutterstock; **p. 213** JohnKwan/Shutterstock; **p. 224** Irina Tischenko/Shutterstock; **p. 225** MistikaS/iStockphoto; **p. 237** Ibsky/Shutterstock; **pp. 257–259** Screenshots from Minitab Corporation. Courtesy of Minitab Corporation.

Chapter 5

p. 284 Probability Plot from SPSS. Reprint Courtesy of International Business Machines Corporation, © International Business Machines Corporation; **p. 287** Figure 3. Frequency Distributions of the Difference in Elevation Estimates taken from Uncertainty in Digital Elevation Models of Axel Heiberg Island, Artic Canada Vol. 36, No. 2, 2004, p. 281. Copyright ©2004. Used by permission of Regents of the University of Colorado; **p. 287** Based on Pooyan Asadollahi and Fulvio Tonon, "Constitutive model for rock fractures: Revisiting Barton's empirical model." Engineering Geology, Vol 113, no. 1. pp. 39–32; **p. 287** Based on Borman, P.J., et al., "Design and analysis of method equivalence studies" Analytical Chemistry, Vol. 81, No 24, December 15, 2009 (Table 3); **p. 303** Figure 1. Histograms of Observed Velocities for A2029 taken from Observations of High Dispersion Clusters of Galaxies, Author: John.M.Hill. Constraints on Cold Dark Matter, Astronomical Journal Vol.110, p. 60 Copyright©1995. IN ALL CASES for works published under any of the Creative Commons licenses the words "© AAS. Reproduced with permission". **p. 304** Based on Cozzolino, J. M., and Mikola. F. J. "Applications of the Piecewise Constant Pareto Distribution." Research Review. Summer 1998, pp. 67–59; **p. 305** Based on Irony, T. Z., Lauretto, M., Pereira, C., and Stern, .1. M. "A Wethull Wearout Test: Full Bayesian Approach," paper presented at Mathematical Sciences Colloquium, Binghamton University, Binghamton, UK, December 2001; **pp. 307–309** Screenshots from Minitab Corporation. Courtesy of Minitab Corporation.

Chapter 6

p. 333 Histogram from SPSS. Reprint Courtesy of International Business Machines Corporation, © International Business Machines Corporation; **p. 341** Screenshots from Minitab Corporation. Courtesy of Minitab Corporation.

Chapter 7

p. 346 Screenshot from SAS. © 1999 SAS Institute Inc. Reprinted with Permission; **p. 353** Source: Thomas, J.B. and Peters, C. "An Exploratory Investigation of Black Friday Consumption Rituals", International Journal of Retail and Distribution Management, Vol. 39, No. 7, 2011 (Table I); **p. 354** Based on Gray.M. & Saucrheek .J.A. "Speed training program for high school football players" The Sport Journal, Vol 7. No. I, Winter 2004 (Table 2); **p. 354** Based on Leeman.R F Fenton M. &VolpiceIli.J "Impaired Control and Undergraduate Problem Drinking" Alcohol & Alcoholism, Vol 42, No 1.Jan/Feb 2007 (Table 1); **p. 357** Reprint Courtesy of International Business Machines Corporation, © International Business Machines Corporation; **p. 362** Source: Patston, T. "Teaching stage fright? — Implications for music educators", British Journal of Music Education, Vol. 31, No. 1, March 2014 (adapted from Figure 1); **p. 363** Based on Shestcr. G G "Explaining catch variation among Baja Californialobster fishers through spatial analysis of trap-placement decisions" Bulletin of Marine, Vol 86. No. 2.April 2010 (table 1). pp. 507–498; **p. 363** Based on Astronomical Applications Dept. U.S. Naval Observatory. Washington.DC; **p. 363** Source: Chew, S., et al. "Do social robots walk or roll?", International Conference on Social Robotics, Vol. 6414, 2010 (adapted from Figure 2); **p. 364** Based on Gotoh.T. Noda. H. & lto. S. "Cardinium symbionts cause cytoplasmic incompatibility in spider mites." Heredity. Vol. 98. No.1, Jan 2007 (Table 1); **p. 364** Based on Tujek, J. 6; Pacola.E "Algorithms for skidding distance modeling on a raster Digital Terrain Model." Journal of forest Engineering Vol.10, No.1, July 1999 (Table 1); **p. 364** Source: Harell, K, "The Fallen and Their Swords: A New Explanation for the Rise of the Shaft Graves", American Journal of Archaeology, Vol. 118, No. 1, January 2014 (Figure 1); **p. 383** Based on Paronuzzi.P. "Rockfall Induced Block Propogation on a Soil Slope, Northern Italy." Environmental Geology. Vol.58.2009 (Table 2); **p. 384** Based on Boyer. D.M., Evans.A.R. and Jernvatt.J. "Evidence of Dietary Differenciation Among Late Paleocene-Early Eocene Plesiadapids (Mammalia, PRimates)." American Journal of Physical Anthropology. Vol. 142.2010(Table A3); **p. 384** Based on Shester.G.G, "Explaining Catch Variation Among Baja California Lobster Fishers Through Spatial Analysis of Trap-Placement Decisions." Bulletin of Marine Science, Vol.86, No.2, April 2010(Table 1).pp. 507–498; **p. 385** Based on Paul I.M., et al., "Effect of Honey, Dextromethorphan. and No Treatment on Nocturnal Cough and Sleep Quality For Coughing Children and Their Parents," Archives of Pediatrries and Adolescent Medicine, Vol. 161.No. 12. Dec. 2007 (Data Simulated); **p. 388** Based on Abt.E., et al. "Relative Contribution of Outdoor and Indoor Particle sources to indoor Concentrations. "Environmental Science & Technology, Vol. 34, No. 17.Sept. 1. 2000 (Table 3); **p. 389** Based on Chiapelli.E., et al., "Evidence based research in complementary and alternative medicine: Treatment of patients with Alzheimer's disease," eCAM, Vol.3, No.4, Nov 2006 (Table 1); **p. 389** Based on Hayes.T.L., and Lewis. D.A. "Anatomical Specialization of the Anterior Motor Speech Area: Hemispheric Differences in Magnopyramidal Neurons." Brain and Language. Vol.49.No. 3.p. 320 (Table 1); **p. 391** Based on Barnett.A. "Misapplications Review: Jail Terms." Interfaces. Vol.25. No.2.Mar. Apr.1995. p. 48; **pp. 392–395** Screenshots from Minitab Corporation. Courtesy of Minitab Corporation.

Chapter 8

p. 418 Screenshots from SPSS. Reprint Courtesy of International Business Machines Corporation, © International Business Machines Corporation; **p. 421** Source: The Differential Effects of Explicit and Implicit Instruction on EFL Learners' Use of Epistemic Stance, Applied Linguistics, Vol. 35, No.1, February 2014 (Table 6); **p. 425** Based on Shester.G.G Explaining catch Variation Among Bajaj

California Lobster Fishers Through Spatial Analysis of Trap Palcement Decision. Bulletin of Marine Science, Vol. 86, No. 2, April 2010 (Table 1). pp. 507–498; **p. 425** Source: Patston, T. Teaching Stage Fright?–Implication for Educators, British Journal of Music Education, Vol. 31, No. 1, March 2014; **p. 426** Based on Doug M. Boyer, Alistair R. Evans and Jukka Jernvall, Evidence of dietary differentiation among late Paleocene–early Eocene plesiadapids (Mammalia, primates) (pages 194–210), American Journal of Physical Anthropology Volume 142, Issue 2, 2010; **p. 427** Based on Tujek. J. and Pacola. E. Algorithms for Skidding Distance Modeling on a Ratser Digital Terrain Model Journal Forest Engineering. Vol. 10. No. 1. July 1999; **p. 427** Source: Senkova, O. & Otani, H. Category Clustering Calculator for Free Recall, Advences in Cognitive Psychology, Vol. 8, No.4, October 2012; **p. 428** Based on Jan L. Souman, Ilja Frissen, Manish N. Sreenivasa and Marc O. Ernst, Walking Staright Into Circles, Current Biology. Vol. 19. No. 18. Sept. 29, 2009; **p. 428** Based on Langman. O. C. et al. Control of Dissolved Oxygen in Northern Temperate Lakes Over Scales Ranging from Minutes To Days, Aquatic Biology, Vol. 9 pp. 221–202, May 2010 (Table 1); **p. 435** Based on R. Kyle Bocinsky Extrinsic Site Defensibility and Landscape-Based Archaeological Inference: An Example from the Northwest Coast, Journal of Anthropological Archaeology, Vol. 35, 2014; **p. 435** Screenshot from SAS. © 1999 SAS Institute Inc. Reprinted with Permission; **p. 449** Based on Paronuzzi, P. Rockfall- Induced Block Propogation on a Soil Slope, Northern Italy, Environmental Geology Vol. 58, 2009; **p. 450** Source: Harrell, K. The Fallen and The Sowrds: A New Explaination for the Rise of the Shaft Graves, American Journal of Archaeology, Vol. 118, No. 1, January 2014; **p. 450** Source: Phil J. Borman, Marion J. Chatfield, Ivana Damjanov and Patrick Jackson, Design and Analysis of Method Equivalence Studies, Analytical Chemistry, Vol. 81, No. 24 December 15, 2009; **p. 456** Based on Abhijit Basu and David S. McKay Lunar Soil Evolution Processes and Apollo 16 Core 60013/60014, Meteorities, Vol. 30 No.2 Mar.1995, p. 194; **pp. 458–460** Screenshots from Minitab Corporation. Courtesy of Minitab Corporation.

Chapter 9

p. 470 Screenshot from SAS. © 1999 SAS Institute Inc. Reprinted with Permission; **p. 476** Source: Bienefeld, N. & Grote, G. Shared Leadership in Multiteam Systems: How Cockpit and Cabin Crews Lead Each Other to Safety, Human Factors, Vol.65, No.2, March 2014; **p. 476** Source: Carlson, K.A. & Conard, J.M. The Last Name Effect: How Last Name Influences Acquisition Timing, Journal of Consumer Research, Vol. 38, No. 2, August 2011; **p. 477** Source: Perez Iglesis. R, et al. White Matter Integrity and Cognitive Impairment in First Episode Psychosis, American Journal of Psychiatry, Vol. 167, No.4 April 2010; **p. 477** Based on Maher, J.K., Hu, M.Y., and Koble, R.H. Childern's Recall of Television ad elements. Journal of Advertising, Vol. 35, No. 1, Spring 2006; **p. 477** Source: Gutierrez – Salomon, Gambro, A & Angulo, O. Influence of Sample Presentation Protocol on the Result of Consumer Test, Journal of Sensory Studies, Vol. 29 No. 3 June 2014; **p. 478** Based on Shester.G.G Explaining catch Variation Among Bajaj California Lobster Fishers Through Spatial Analysis of Trap Palcement Decision. Bulletin of Marine Science, Vol. 86, No. 2, April 2010 (Table 1). pp. 507–498; **p. 479** Source: Phil J. Borman, Marion J. Chatfield, Ivana Damjanov and Patrick Jackson, Design and Analysis of Method Equivalence Studies, Analytical Chemistry, Vol. 81, No. 24 December 15, 2009; **p. 479** Based on Kuhn, D., and Udell. W Coordinating Own and Other Perspective in argument, Thinking and Reasoning, October 2006; **p. 489** Based on Huelsing, J., Kanafani, N., Mao, J., and White, N.H, Camp Jump Start: Effect of a Residential Summer Weight-Loss Camp for Older Children and Adolescents, Pediatrics, Vol. 125, No. 4. April 2010; **p. 491** Source: Ozur, M. Comparing Methods for Predicting Immediate Settlement of Shallow Foundation on Cohesive Soils based on Hypothetical and real Cases, Environmental & Engineering Geoscience, Vol. 18, No. 4, November 2012; **p. 491** Source: Sharma, P. and Harinarayana, T. Solar Energy Generation Potential Along National Highways, International Journal of Energy and Environmental Engineering, Vol 49, No 1, December 2013; **p. 491** Source: Rahuma, K.M., et al. Comparison Between Spreadsheet and Specialized Programs in Calculating the Effect of Scale Deposition on the Well Flow Performance, Journal of Petroleum and Gas Engineering, Vol 4, No 4, April 2013; **p. 492** Based on Chunhye, K.L,. and Schmitz, M.C,. Determination of pH, Total Acid and Total Ethanol in Oral Health Products: Oxidation of Ethanol and Recommendation to Mitigate its Association with Dental Caries, Journal of Dentistry, Oral Medicine and Dental Education, Vol.3 No.1 2009; **p. 493** Based on Virginia Transportation Research Council, Research Report: The Impact of Red Light Cameras (Photo Red Enforcement) on Crashes in Virginia, June 2007; **p. 493** Based on Cooperative Cellar of Borba (Adega Cooperative a de Borba). Portugal; **p. 499** Source: Haer, R and Meidert, N. Does the First Impression Count? Examining the Effect of the Welcome Screen Design on the Response Rates, Survey Methodology, Vol 39, No 2, December 2013; **p. 501** Based on Temple, D. H. Patterns of Systemic Stress During the Agricultural Transition in Prehistoric Japan, American Journal of Physical Anthrapology, Vol 142, No 1, May 2010; **p. 512** Source: Kanga, E. M. et al. Hippopotamus and Livestock Grazing: Influences on Riparian Vegetation and Facilitation of Other Herbivores in Mara Region of Kenya, Landscape & Ecology Engineering, Vol 9, No 1, January 2013; **p. 513** Source: Carlson, K.A. and Conrad, J.M. The Last Name Effect: How Last Name Infuences Acquisition Timing, Journal of Consumer Research, Vol 38, No2, August 2011; **p. 513** Based on Kuhn, D. and Udell, W. Coordinating Own and Other Perspectives in Argument, Thinking and Reasoning, October 2006; **p. 514** Based on Perez-Iglosias, R, et al. While Matter Intergrity and Cognitive Impairment in First -Episode Psychosis, American Journal of Psychiatry, Vol 167, No 4, April 2010; **p. 514** Based on Paul, I.M., et al. Effect of Honey Dextromethorphan and No Treatment on Nocturnal Cough and Sleep Quality for Coughing Children and Their Parents, Archives of Pediatrics and Adolescent Medicine, Vol 161, No 12, December 2007; **p. 518** Based on Khadjavi, L.S. Driving While Black in the City Angels Chance Vol 19, No 2, Spring 2006, pp. 71–46; **p. 519** Based on Boone, E.I., Keying, Y., and Smith, E.P. Evaluating the Relationship Between Ecologocal and Habotat Condition Using Hierarchical Models, Journal of Agricultural, Biological and Environmental Sciences, Vol. 10. No. 2. June 2005; **p. 519** Based on Robinson, D. H. and Katayama, A. D., At Lexical, Articulatory Interference in Silent Reading: The Upstream Tongue Twister Effect, Memory & Cognition, Vol 25, No 5, September 1997, p. 691; **p. 520** Based on Farber, S. L. Identical Twins Reared Apart © 1981 by Basic Books, Inc; **p. 520** Based on Pfeiffer, M., et al. Community Organization and Species Richness of Ants in Mongolia Along an Ecological Gradient From Steppe to Gopi Desert, Journal of Biogeography, Vol. 30 No. 12, Dec 2003; **p. 521** Based on Taylor, A.A., and Weary, D.M., Vocal Responses of Piglets to Castration: Identifying Procedural Source of Pain, Applied Animal Behaviour Science, Vol. 70, No.1 November 1, 2000; **p. 521** Based on Gill R. T., et al. Genome wide Dynamic Transcriptional Proliling of the Light to Dark Transition in Synechocystis Sp. PCC6803, Journal of Bacteriology, Vol 184, No 13, July 2002; **p. 521** Based on Thomas E. Bradstreet, Favorite Data

Sets From Early Phases of Drug Research, Part 2, Proceedings of the Suction on Statistical Education of the American Statistical Association; **p. 522** Based on Lehibara, H., Shintani, M. and Inoue, T. Hulfman Based Test Response Coding, IEICE Transactions on Information & Systems, Vol E88-D, No 1, January 2005; **p. 523** From Maximal Acrobic Exercise of Individuals with Multiple Sclerosis Using Three Modes of Ergometry, Clinical Kinesiology, Vol 49, No 1, Spring 1995, p. 35. Copyright © 1995 by Dr. W. Jeffrey Armstrong. Used by permission of Dr. W. Jeffrey Armstrong; **pp. 525–529** Screenshots from Minitab Corporation. Courtesy of Minitab Corporation.

Chapter 10

p. 546 Screenshot from SAS. © 1999 SAS Institute Inc. Reprinted with Permission; **p. 552** Source: Mahler, H., et al., Effects of Media Imges on Attitudes Towards Tanning, Basic and Applied Social Psychology, Vol. 32, No. 2, May 2010; **p. 554** Based on Grattan Miscio K.E, and Vogel Sprott M. Alcohol International Control and Inappropriate Behaviour: Regulation by Caffeine or an incentive. Experimental and Clinical Psychopharmacology, Vol. 13, No.1 February 2005. **p. 555** Based on Paul, I.M., et al., Effect of Honey Dextromethorphan and no Treatment on Nocturnal Cough and Sleep Quantity for Coughing Children and their Parents, Archives of Pediatrics and Adolescent Medicine, Vol. 161 No.12 Dec 2007; **p. 555** Source: Moris P.E., and Fritz C.O The Name Game: Using retrieval Practice to Improve the Learning of Names, Journal of Experimental Psychology–Applied. Vol. 6 No. 2 june 2000; **p. 555** Based on Cole, K. et al., Animal Assisted Therapy Decreases Hemodynamics Plasma Epinephrine and State Anxiety in Hospitalised Heart Failure Patients, American Journal of Critical Care, 2007. 16:575–585; **p. 570** Screenshots from SPSS. Reprint Courtesy of International Business Machines Corporation, © International Business Machines Corporation; **pp. 573–574, 592–593, 610–611** Screenshots from Minitab Corporation. Courtesy of Minitab Corporation; **p. 577** Source: Sharma, P. and Harinarayana, T. Solar Energy Generation Potential Along National Highways, International Journal of Energy and Environmental Engineering, Vol 49, No 1, December 2013; **p. 578** Based on Ohberg, L., et al. Eccentric Training in Patients with Chronic Achilles Tendinosis: Normalized Tendon Structure and Decreased Thickness at Follow Up. British Journal of Sport Medicine, Vol. 38, No.1, Feb. 1 2004; **p. 578** Source: Jason Kung, Francesco Chiappelli, Olivia O Cajulis, Raisa Avezova, George Kossan, Laura Chew., From Systematic Reviews to Clinical Recommendation to Clinical Based Health Care: Validation of Revised Assessment of Multiple Systematic Reviews Clinical Relevance. The Open Dentistry Journal. Vol. 4 2010. pp. 84–91; **p. 581** Based on Hemmings B., Smith, M., Graydon, J. and Dyson, R. Effects of massage on Psychological Restoration, Perceived Recovery and Repeated Sports Performance, British Journal of Sports Medicine, Vol. 34, No.2, Apr 2000; **p. 581** Based on Data from Elizabeth Schreiber. Department of Statistics Kansas State University, Manhattan, Kansas; **p. 596** Based on Murphy. K .and Spencer.A "Playing Video games Does Not Make for Better Visual Attention Skills" Journal of Articles in Support of the Null Hypothesis, Vol.6. No. 1. 2009; **p. 599** Based on Miller.E.G., and Kahn. B.E, "Shades of meaning: The effect of color and flavor names on Consumer Choice," Journal of Consumer Research. Vol.32. June 2005 (Table 1); **p. 605** Based on Qibai.C.Y.H. and Shi.H. "An investigation on the physiological and psychological effects of infrasound on persons." Journal of Low Frequency Noiae, Vibration and Active Control. Vol.23. No.1 .March 2004 (Tables i–iv); **p. 606** Based on Harrey. W. F. Hindman. L .J., and Rhodes. W. e. "Vulnerability of Canada geese to taxidermy-mounted decoys". Journal of Wildlife Management. Vol.59. No. 3. July 1995. p. 503(Table 1); **p. 608** Based on Agras.W.S., et al., "Does interpersonal therapy help patients with binge eating disorder who fail to respond to cognitive behavioural therapy?" Journal of Consulting and Clininal Psychology, Vol.63.No .3, June 1995. p. 386(Table 1); **p. 607** Based on Gill.R.T., et al., "Genome Wide Dynamic Transcriptional Profiling of the Light to Dark Transition on Synechocystis Sp.PCC6803." Journal of Bacteriology. Vol.184. No. 13.July 2002; **p. 607** Based on Gan.Su-lin .et al. "The Thrill of a Close Game: Who Enjoys it and Who Doedn't?" Journal of Sport & Social Issues.Vol. 21. No.1, Feb. 1997, pp. 87–60; **p. 608** Source Adapted from American Journal of Science, Vol. 305, No.1 Jan 2005 p. 44; **p. 609** Source: Kouloumbi.N., et al. "Antocorrosion Performance of Epoxy Coatings on Steel Surface Exposed to De-Ionised Water," Pigment & Resin Technology, Vol.32, No.2,2003(Table 2); **p. 609** Based on Balch.W.r. "Practise Versus Review Exams and Final Exam Performance," Teaching of Psychology.Vol. 25, No. 3, Aug 1998 (Table 1).

Chapter 11

p. 621 Screenshots from Microsoft ® Excel ®. Used by permission of Microsoft Corporation; **p. 621** Screenshots from Microsoft ® Excel ®. Used by permission of Microsoft Corporation; **p. 622** Screenshot from SAS. © 1999 SAS Institute Inc. Reprinted with Permission; **p. 622** Screenshots from SPSS. Reprint Courtesy of International Business Machines Corporation, © International Business Machines Corporation; **p. 624** Based on Enright. J. T. "Testing dowsing: The failure of the Munich experiments." Skeptical Inquirer Jan./Feb. I999, p. 73 (Figure 6a); **p. 627** Based on Khan, M.S., et al., "Paper Diagnostic for Instant Blood Typing," Analytical Chemistry, Vol. 82, No. 10, May 2010; **p. 627** Based on Pfeiffer.M,. "Community Organization and Species Richness of Ants in Mongolia Along an Ecological Gradient from Steppe to Gobi Desert." Journal of Biogeography, Vol.30, No. 12, Dec. 2003 (Tables 1 and 2); **p. 628** Source: Suribabu, C.R. 8- Neelakantan, T.R. "Sizing of water distribution pipes based on perfo rmance measure and breakage-repair replacement economics", IHS Journal of Hydraulic Engineering, Vol. 18, No. 3, September 2012 (Table 1); **p. 628** Based on Frederick Wiseman, Ph. D., Mohammed Hahibullah, Ph.D., and Mustafa Yilmaz, Ph. D, Sports Journal, Vol. 10, No. 1; **p. 628–629** Based on Tekwe, C. D., et al., "An Empirical Comparison of Statistical Models for Value- Added Assessment of School Performance." Journal of Educational and Behavioural Statistics, Vol. 29, No.1. Spring 2004 (Table 2); **p. 629** Based on Russell, D.A. "Basketballs as Spherical Acoustic Cavities," American Journal of Physics, Vol. 48, No. 6, June 2010 (Table 1); **p. 629–630** Based on Lee.G. Velleman. P. and Wainer.H. "Giving the Finger to Dating Services" Chance, Vol. 21. No. 3. Summer 2008 (Adapted from Figure 3); **p. 630** Based on Morris, P.E., and Fritz, C.O. "The Name Game: Using Retrieval Practise To Improve The Learning Of Names." Journal of Experimental Psychology-Applied. Vol. 6, No. 2, June 2000 (Data Simulated from Figure 2); **p. 630** Based on Barry.L. "Estimating Rates of Spreading and Evapouration of Volatile Liquids." Chemical Engineering Progress, Vol. 101, No. 1, Jan. 2005; **p. 642** Problem 11.63, Lobster fishing Study taken from "Explaining Catch Variation Among Baja California Lobster Fishers Through Spatial Analysis of Trap-Placement Decisions." Bulletin of Marine Science, Vol. 86, No.2, Copyright©2010. Used by permission of University of Miami; **p. 643** Based on Hemmings, B., Smith, M., Graydon, J., and Dyson, R, "Effects of Massage on Physiological Restoration, Perceived Recovery, and Repeated Sports Performance." British Journal of Sports Medicine. Vol. 34, No. 2, Apr. 2000 (data adapted from Figure 3); **p. 643** Based on Wade, T.G., et al., "Distribution and Causes of Global Forest Fragmentation." Conservation Ecology, Vol. 72, No 2, Dec. 2003 (Table 6); **p. 644** Based on Singer, T.et al. "Empathy for Pain Involves the Affective but not Sensory Components of Pain" Science, Vol. 303, Feb. 20 2004. (Adapted from Figure 4); **p. 644** Based on Schaffer, J., and Heiny, E.L., "The Effects of Elevation on Slugging Percentage in Major League Baseball" Chance, Vol. 19, No. 1, Winter 2006 (adapted from Figure 2); **p. 644** Source: Sanchez-Molina, D., et al. "Fractal dimension and mechanical properties of human corticalbone", Medical Engineering & Physics, Vol. 35, No. 5, May 2013 (Table 1); **p. 644** Problem 11.75, Salary Linked to Height taken from "The Effect of Physical Height on Work Place Success and Income: Preliminary Test of a Theoretical Model." Journal of Applied Psychology, Vol. 89, No. 3. Copyright ©2004. Used by permission of American Psychological Association; **p. 654** Source: Cleveland, T., "Minors in name only: Child laborers on the diamond mines of the Companhia de Diamantes de Angola (Diamang), 1917–1975". Journal of Family History, Vol. 35, No. 1, 2010 (Table 1); **p. 654** Source: Rusterholz, H., Aydin, D. & Baur, B. "Population structure and genetic diversity of relict populations of Alyssum montanum on limestone cliffs in the Northern Swiss Jura mountains", Alpine Botany, Vol. 122, No. 2, November 2012 (Tables 1 and 2); **p. 667** Based on "Work-Life Balance: An MBA Alumni Report." Graduate Management Admission Council (GMAC), Research Report (Oct. 13, 2005); **p. 667** Source: Siddiqui, R. & Bajpai, U. "Correlation between thicknesses of dust collected on photovoltaic module and difference in efficiencies in composite climate", International Journal of Energy and Environmental Engineering, Vol. 4, No. 1, December 2012 (Table 1); **pp. 667–668** Source: Infotech, Inc, Gainesville, Florida; **pp. 671–672** Based on data from Hillsbough Country (florida) Property Appraiser's Officer; **p. 672** Based on data from Baseball Almanac,2013; www.mlb.com; **p. 672** Based on Ross, R.J.S. "From Antisweatshop to Global Justice to Antiwar: How the New New Left is the Same and Different from Old New Left." Journal of Word-Systems Research, Vol.X, No. 1, Winter 2004 (Tables 1 and 3); **p. 673** Based on Shand, J., et al. "Variability in the Location of the Retinal Ganglion Cell Area Centralis is Correlated with Ontogenetic Changes in Feeding Behaviour in the Blackbream. Acanthopagrus' butcher', Vol. 55. No. 4, Apr.2000 (Figure H); **p. 674** Based on Hessami, M. et al. "Selection of an Artifical Neural Network Model for the Post-Calibration of Weather Radar Rainfall Estimation." Journal of Data Science, Vol. 2, Apr. 2004. (Adapted from Figures 2 and 4); **pp. 674–675** Based on Merila, J, and Wiggins, D.A. "Interspecific Competiton for Nest Holes Causes Adult Mortality in the Collard Flycatcher," The Condor, Vol.97, No. 2, May 1995, pp-449 (Figure 2), Cooper Ornithologival Society; **p. 676** Based on Gadallah, F.L and Jefferies, R.L "Forage Quality in Broad Rearing Areas of the Lesser Snow Goose and the Growth of Capacitive Goslings," Journal of Applied Biology, Vol. 32, No. 2,1995, pp. 309–282 (adapted from Figures 2 and 3); **p. 677** Based on Berg, W.P., and Greer, N.L., "An Kinematic Approach run of Novice long Jumpers." Journal of Applied Biomechanics, vol. 11, No. 2, May 1995, pp-147 (Table 1); **p. 677** Based on Fairley, W. B., et al." Bricks, buildings, and the Bronx: Estimating Masonry Deterioration." Chance Vol. 7. No. 3, Summer 1994. p. 64 (Figure 3); **pp. 678–680** Screenshots from Minitab Corporation. Courtesy of Minitab Corporation.

Chapter 12

p. 693 Screenshot from SAS. © 1999 SAS Institute Inc. Reprinted with Permission; **pp. 698–699** James T. McClave; **p. 699** Based on Petrick J F, "A examination of golf vacationers' novelty", Annals of Tourism Research, Vol.29, No.2, April 2002; **p. 700** Source: Mohapatra S, Nando G B, "Chemical Modification of Natural Rubber in the Latex Stage by Grafting Cardanol, A Waste from the Cashew Industry and a Renewable Resource", Industrial & Engineering Chemistry Research, Vol.52, No.17, May 2013,(Table 2 and Table 3); **p. 701** Based on Bhargava R and Meher-Homji, C.B, "Parametric Analysis for Gas Turbines with Inlet Evaporative and Overspray Fogging", "Journal of Engineering for Gas Turbines and Power", Vol. 127, No. 1, Jan 2005.Table from pp 184–157; **p. 701** Source: Puli U, Rajvanshi A K & Das S K, "Investigation of Bubble Behaviour in a Subcooled Flow Boiling of Water in a Horizontal Annulus Using High-Speed Flow Visualisation", Heat Transfer Engineering, Vol.34, No.10, 2003(Table 8); **p. 702** James T. McClave; **p. 707** From A Pedagogic Application of Multiple Regression Analysis by Taylor P.J in Geography, Vol. 65, pp 203–212. Copyright © 1980 by Taylor P.J. Used by permission of Peter J Taylor; **p. 708** Based on data provided by Dr. Kelly Uscategui, University of Connecticut; **pp. 722–723** Based on Seal N and Seal J, "Eating Patterns of the Rural Families of Overweight Preschool Children: A Pilot Study", Journal of Education and Human Development, Vol. 1, No. 3, 2009 (Figure 1); **p. 724** Source: Suribabu C R and Neelakantan T R, "Sizing of Water Distribution Pipes Based on Performance Measure and Breakage-Repair Replacement Economics", IHS Journal of Hydraulic Engineering, Vol.18, No.3, September 2012, Table 1; **p. 725** Source: Thomas J B and Peters C, "An Exploratory Investigation of Black Friday Consumption Rituals", International Journal of Retail and Distribution Management, Vol.39, No. 7, 2011, Table1; **p. 725** Based on Gee. 5 and Nguyen. L. "Mean time to failure in wafer level CSP packages with SnPh and SnAgCu solder hmps." International Wafer Level Packaging Conference, San Jose, CA. Nov. 3-1, 2005(adapted from Figure 7); **p. 726** Based on l-lert\vig, R. Pachur, T. and Kurzenhauser. S. "Judgments of risk frequencies: Tests of possible cognitive mechanisms." JournaI of Experimental Psycology: Learning, Memory and Cognition, Vol. 31.No. 4, July 2005; **p. 743** Based on Schilling, J "Bioindication of Atmospheric Heavy Metal Deposition in the Blue Ridge Using the Moss. Thuidium Delicatulum", master-ol-science thesis, spring 2000; **p. 746** Screenshots from SPSS. Reprint Courtesy of International Business Machines Corporation, © International Business Machines Corporation; **p. 777** Source: Federal Trade Commission; **p. 781** Based on Gee, S., and Nguyen, L. "Mean time to failure in wafer level CSP packages with SnPb and SnAgCu solder bmps." International Wafer Level Packaging Conference, San Jose. CA, Nov 3-4, 2005 (adapted from Figure 7); **p. 785** Based on Grimes. P. and Kentor J. "Exporting the greenhouse: Foreign capital penetration and CO2 emissions 1980-1996. Journal of World Systems Research, Vol lX, No. 2 Summer 2003 (Table I); **p. 785** Based on Grimes. P. and Kentor J. "Exporting the greenhouse: Foreign capital penetration and CO2 emissions 1980-1996. Journal of World Systems Research, Vol lX, No. 2 Summer 2003 (Appendix B); **p. 786** Based on Jovanovic J and King. S. S. "Boys and girls in the performance-based science classroom: Who's doing the performing." American Educational Research Journal. Vol. 35, No. 3, Fall I998. p. -489 (Table 8); **pp. 787–788** Based on Berry, K. A. et al., "Interpreting what is rural and urban for western U. S.

counties". Professional Geographer, Vol. 52, No. I, Feb 2002(Table 2); **pp. 788–789** Based on Gadallab F. L and Jefferies R.L. "Forage Quality in Brood Rearing Areas of the Lesser Snow Goose and the Growth of Captive Goslings. "Journal of Applied Biology, Vol. 32, No. 2,1995, pp. 309–282 (adapted from figures 2 and 3); **p. 789** Based on Takizawa .K et al. "Characteristics of C3 Radicals in High Density C4F8 Plasmas Studied by Laser-induced Fluorescence Spectroscopy". Journal of Applied Physics. Vol. 88, No. 11, Dec 1,2000 (Figure 7); **p. 790** Based on "Comparing mosquito repellents" from "Buzz off: Insect repellents: Which keep bugs at bay" Consumer Reports. June 2000; **p. 791** Based on Internet Movie Database(www.imdb.com); **pp. 790–791** Based on Wielgus.R. B and Bunnell F. L. "Tests of hypotheses for sexual segregation in grizzly bear. "Journal of Wild life Management, Vol 59, No. 3, July 1995, p 555 (Table 1); **p. 791–792** Based on Baek K ll. and Skinner. D. Z. "Quantitative real-time PCR method to detect changes in speci?c transcript and total RNA amounts." Electronic Journal of Biotechnology, Vol 7. No. 1.April 15.2004 (adapted from Figure 2); **pp. 796–798** Screenshots from Minitab Corporation. Courtesy of Minitab Corporation.

Chapter 13

p. 805 Screenshots from SPSS. Reprint Courtesy of International Business Machines Corporation, © International Business Machines Corporation; **p. 809** James T. McClave; **p. 809** Source: McMunn-Dooley, C. & Welch, M.M. "Nature of Interactions Among Young Children and Adult Caregivers in a Children's Museum", Early Childhood Education Journal, Vol. 42, No. 2, March 2014 (adapted from Figure 2); **p. 810** Screenshot from SAS. © 1999 SAS Institute Inc. Reprinted with Permission; **p. 818** Source: Info Tech, Inc., Gainesville, FL; **p. 820** Based on Hume. C, Salmon. J and Ball. K. "Children's perceptions of their home and neighborhood environments, and their association with objectively measured physical activity: A qualitative and quantitative study". Health Education Research, Vol. 21, No. 1, February 2005 (Table III); **p. 821** Based on Lasorsa. D, and Dai J. "When news reporters deceive The production of stereotypes". Journalism and Mass Communication Quarterly, Vol 84, No. 2. Summer 2007 (Table 2); **p. 821** Based on Krucoff, M. W et al. "Music, imagery, touch and prayer as adjuncts to interventional cardiac care: The Monitoring and Actualization of Noetic Trainings (MANTRA) II randomized study." The Lancet, Vol. 366.July 16,2005 (Table 4); **p. 822** Based on Wickens C. D, Rice, S., Keller, D., Hutchins, S., Hughes J, and Clayton, K., "False alerts in air traffic control conflict alerting system: Is there a 'cry wolf' effect?" Human Factors, Vol. 51. Issue 4, August 2009 (Table 2); **p. 822** Source: Giannouli, V. "Number perseveration in healthy subjects: Does prolonged stimulus exposure influence performance on a serial addition task?", Advances in Cognitive Psychology, Vol. 9, No. 1, Jan. 2013 (adapted from Table 3); **p. 823** Source: Malone, E.K. 8. Issa, R.A. "Work-Life Balance and Organizational Commitment of Women in the U.S. Construction Industry", Journal of Professional Issues in Engineering Education & Practice, Vol. 139, No. 2, April 2013 (Table 11); **p. 823** APOE genotype and cognitive functioning in a large age-stratified population sample by Jorm A. F et al. from Neuropsycology, Vol. 21, No. 1.January 2007 (Table 1). Copyright © 2007 by American PsychologicalAssociation. Used by permission of American Psycological Association; **p. 823** Based on Bryant, D. J. "Classifying simulated air threats with fast and frugal heuristics." Journal of Behavioural Decision Making, Vol. 20, January 2007 (Appendix C); **p. 823** Source: Gold, J.E., et al. "Postures, typing strategies, and gender differences in mobile device usage: An observational study", Applied Ergonomics, Vol. 43, No. 2, March 2012 (Table 2); **p. 824** Source: Bigham, G.D. & Riney, M.R. "Trend Analysis Techniques to Assist School Leaders in Making Critical Curriculum and Instruction Decisions", Current Issues in Education, Vol. 17, No. 1, January 2014 (Table 6); **p. 824** Source: McMunn-Dooley, C. & Welch, M.M. "Nature of Interactions Among Young Children and Adult Caregivers in a Children's Museum", Early Childhood Education Journal, Vol. 42, No. 2, March 2014 (adapted from Figure 2); **p. 825** Based on Howard J.N, Lambdin C. G and Datteri, D. L. "Let's make a deal: Quality and availability of second-stage information as a catalyst for change", Thinking & Reasoning, Vol 13. No. 3, July 2007 (Table 2); **p. 828** Based on Berg .I and Bliedon. S. "The pots of Phyiakopi: Applying statistical techniques to archaeology". Chance, Vol 13, No. 4. Fall 2000; **p. 829** Based on Hood J. C. "Teaching, against the text: The case of qualitative methods". Teaching Sociology, Vol. 34, July 2006 (Exhibit 2); **p. 829** Reprinted with the permission of Cambridge University Press; **p. 830** Source: Wilkins, H. "Souvenirs: What and why we buy", Journal of Travel Research, Vol. 50, No. 3, May 2011 (adapted from Table 2); **p. 831** Investigating Flight Response of Pacific Brant to Helicopters at Izembek Lagoon, Alaska by Using Logistic Regression by Erickson W, Nick T, Ward D from Statistical Case Studies: A Collaboration Between Academe and Industry, ASA-SIAM Series on Statistics and Applied Probability, I998. Copyright © I998 by Society for Industrial and Applied Mathematics. Used by permission of Society for Industrial and Applied Mathematics; **p. 831** Based on Krienert J.l, "Masculinity and Crime: A Quantitative Exploration of Messerschmidt's Hypothesis", Journal of Sociology, Vol. 7, No.2, July 2003 (Table 4); **p. 832** The Flynn Effect and U.S Policies by Kanaya .T, Scullin M.H, Ceci S.J from American Psychologist, Vol.58, No. 10, Oct. 2003(Figure 1). Copyright © 2003 by American Psycological Association. Used by permission of American Psycological Association; **p. 832** A gang classification system for corrections by Knox, G. W., et al. from Journal of Gang Research, Vol. 4, No. 2, Winter I997. p. 82 (Table 4). Copyright © 1997 by National Gang Crime Research Center. Used by permission of National Gang Crime Research Center; **p. 833** The Computer-Assisted Hypnosis Scale: Standardization and Norming of a Computer-administered Measure of Hypnotic Ability by Carolyn D Grant and Michael R. Nash from Psycological Assessment Vol.7. No. I March I995, p 53 .Copyright © 1995 by American Psychological Association. Used by permission of American Psycological Association; **p. 834** Based on Gelman. "A 55,000 residents desperately need your help!", Chance, Vol 17. No 2, Spring 2004 (Figures 1 and 5); **p. 834** Based on Gangemi.A., and Mancini. F. "Guilt and focusing in decision-making" Journal of Behavioural Decision Making, Vol. 20, Jan 2007 (Table 2); **pp. 835–836** Screenshots from Minitab Corporation. Courtesy of Minitab Corporation.

Chapter 14 (available online)

p. 14-7 Source: Patston, T, 'Teaching stage fright?–Implications for music educators British Journal ofMusic Education, Vol. 31, No. 1, March 2014 (adapted from Figure 1); **p. 14-8** Based on Boyer D.M, Evans A.R and Jernvall J, "Evidence of Dietary Differentiation Among Late Paleocene–Early Eocene Plesiadapids (Mammalia, Primates)", American Journal of Anthropology, Vol. 142, © 2010, Table A3; **p. 14-9,** Based on Thery.M, et al., "Specific Color Sensitivities of Prey and Predator Explain Camouflage in Different Visual